Encyclopedia of Database Systems

Ling Liu • M. Tamer Özsu
Editors

Encyclopedia of Database Systems

Second Edition

Volume 1

A–C

With 1374 Figures and 143 Tables

 Springer

Editors
Ling Liu
Georgia Institute of Technology College
of Computing
Atlanta, GA, USA

M. Tamer Özsu
University of Waterloo School of Computer Science
Waterloo, ON, Canada

ISBN 978-1-4614-8266-6   ISBN 978-1-4614-8265-9 (eBook)
ISBN 978-1-4614-8264-2 (print and electronic bundle)
https://doi.org/10.1007/978-1-4614-8265-9

Library of Congress Control Number: 2018938558

Printed on acid-free paper

This Springer imprint is published by the registered company Springer Science+Business Media, LLC part of Springer Nature.
The registered company address is: 233 Spring Street, New York, NY 10013, U.S.A.

Preface to the Second Edition

Since the release of the first volume of this Encyclopedia, big data has emerged as a central feature of information technology innovation in many business, science, and engineering fields. Databases are one of the fundamental core technologies for big data systems and big data analytics. In order to extract features and derive values from big data, it must be stored, processed, and analyzed in a timely manner. Not surprisingly, big data not only fuels the development and deployment of database systems and database technologies, it also opens doors to new opportunities and new challenges in the field of databases. As data grows in volume, velocity, variety, and with the attendant veracity issues, there is a growing demand for volume-scalable databases, velocity-adaptive databases, and variety-capable databases that can handle data quality issues properly. As machine learning and artificial intelligence renew their momentum with the power of big data, there is an increasing demand for new generation of database systems that are built for extracting features from databases as efficient and effective as conventional database systems are capable of for querying databases.

The first edition of the *Encyclopedia of Database Systems* is a comprehensive, multivolume collection of over 1,250 in-depth entries (3,067 including synonyms), covering important concepts on all aspects of database systems, including areas of current interest and research results of historical significance. This second edition of *Encyclopedia of Database Systems* expands the first edition by enriching the content of existing entries, expanding existing topic areas with new entries, adding a set of cutting-edge topic areas, including cloud data management, crowdsourcing, data analytics, data provenance management, graph data management, social networks, and uncertain data management to name a few. The new entries and the new topic areas were determined through discussions and consultations with the Advisory Board of the *Encyclopedia of Database Systems*. Each of the new topic areas was managed by a new Area Editor who, together with the editor-in-chief, further developed the content for each area, soliciting experts in the field as contributors to write the entries, and performed the necessary technical editing. We also reviewed the entries from the first edition and revised them as needed to bring them up-to-date.

We would like to thank the members of the Advisory Board, the Editorial Board, and all of the authors for their contributions to this second edition. We would also like to thank Springer's editors and staff, including Susan Lagerstrom-Fife, Michael Hermann, and Sonja Peterson for their assistance and support throughout the project, and Annalea Manalili for her involvement in the early period of this project.

In closing, we trust the Encyclopedia can serve as a valuable source for students, researchers, and practitioners who need a quick and authoritative reference to the subject on database systems. Suggestions and feedbacks to further improve the Encyclopedia are welcome from readers and from the community.

Preface to the First Edition

We are in an information era where generating and storing large amounts of data are commonplace. A growing number of organizations routinely handle terabytes and exabytes of data, and individual digital data collections easily reach multiple gigabytes. Along with the increases in volume, the modality of digitized data that requires efficient management and the access modes to these data have become more varied. It is increasingly common for business and personal data collections to include images, video, voice, and unstructured text; the retrieval of these data comprises various forms, including structured queries, keyword search, and visual access. Data have become a highly valued asset for governments, industries and individuals, and the management of these data collections remains a critical technical challenge.

Database technology has matured over the past four decades and is now quite ubiquitous in many applications that deal with more traditional business data. The challenges of expanding data management to include other data modalities while maintaining the fundamental tenets of database management (data independence, data integrity, data consistency, etc.) are issues that the community continues to work on. The lines between database management and other fields such as information retrieval, multimedia retrieval, and data visualization are increasingly blurred.

This multi-volume *Encyclopedia of Database Systems* provides easy access to important concepts on all aspects of database systems, including areas of current interest and research results of historical significance. It is a comprehensive collection of over 1,250 in-depth entries (3,067 including synonyms) that present coverage of the important concepts, issues, emerging technology and future trends in the field of database technologies, systems, and applications. The content of the *Encyclopedia* was determined through wide consultations. We were assisted by an Advisory Board in coming up with the overall structure and content. Each of these areas were put under the control of Area Editors (70 in total) who further developed the content for each area, soliciting experts in the field as contributors to write the entries, and performed the necessary technical editing. Some of them even wrote entries themselves. Nearly 1,000 authors were involved in writing entries.

The intended audience for the *Encyclopedia* is technically broad and diverse. It includes anyone concerned with database system technology and its applications. Specifically, the *Encyclopedia* can serve as a valuable and authoritative reference for students, researchers and practitioners who need a quick and authoritative reference to the subject of databases, data management, and database systems. We anticipate that many people will benefit from this reference work, including database specialists, software developers, scientists and engineers who need to deal with (structured, semi-structured or unstructured) large datasets. In addition, database and data mining researchers and scholars in the many areas that apply database technologies, such as artificial intelligence, software engineering, robotics and computer vision, machine learning, finance and marketing are expected to benefit from the *Encyclopedia*.

We would like to thank the members of the Advisory Board, the Editorial Board, and the individual contributors for their help in creating this *Encyclopedia*. The success of the *Encyclopedia* could not have been achieved without the expertise and the effort of the many contributors. Our sincere thanks also go to Springer's editors and staff, including Jennifer Carlson, Susan Lagerstrom-Fife, Oona Schmid, and Susan Bednarczyk for their support throughout the project.

Finally, we would very much like to hear from readers for any suggestions regarding the *Encyclopedia's* content. With a project of this size and scope, it is quite possible that we may have missed some concepts. It is also possible that some entries may benefit from revisions and clarifications. We are committed to issuing periodic updates and we look forward to the feedback from the community to improve the *Encyclopedia*.

Ling Liu

M. Tamer Özsu

List of Topics

Association Rule Mining

Section Editor: *Jian Pei*

Workflow Management

Section Editor: *Barbara Pernici*

Stream Mining

Section Editor: *Divesh Srivastava*

Distributed Database Systems

Section Editor: *Kian-Lee Tan*

About the Editors

Ling Liu Georgia Institute of Technology College of Computing, Atlanta, GA, USA

Ling Liu is Professor of Computer Science in the College of Computing at Georgia Institute of Technology. She holds a Ph.D. (1992) in Computer and Information Science from Tilburg University, The Netherlands. Dr. Liu directs the research programs in the Distributed Data Intensive Systems Lab (DiSL), examining various aspects of data intensive systems, ranging from big data systems, cloud computing, databases, Internet and mobile systems and services, machine learning, to social and crowd computing, with the focus on performance, availability, security, privacy, and trust. Prof. Liu is an elected IEEE Fellow and a recipient of IEEE Computer Society Technical Achievement Award (2012). She has published over 300 international journal and conference articles and is a recipient of the best paper award from numerous top venues, including ICDCS, WWW, IEEE Cloud, IEEE ICWS, and ACM/IEEE CCGrid. In addition to serving as general chair and PC chairs of numerous IEEE and ACM conferences in big data, distributed computing, cloud computing, data engineering, and very large databases fields, Prof. Liu served as the Editor-in-Chief of *IEEE Transactions on Service Computing* (2013–2016) and also served on editorial boards of over a dozen international journals. Her current research is sponsored primarily by NSF and IBM.

M. Tamer Özsu University of Waterloo School of Computer Science, Waterloo, ON, Canada

M. Tamer Özsu is Professor of Computer Science at the David R. Cheriton School of Computer Science and the Associate Dean of Research of the Faculty of Mathematics at the University of Waterloo. He was the Director of the Cheriton School of Computer Science from January 2007 to June 2010.

His research is in data management focusing on large-scale data distribution and management of nontraditional data, currently focusing on graph and RDF data. His publications include the book *Principles of Distributed Database Systems* (with Patrick Valduriez), which is now in its third edition. He was the Founding Series Editor of *Synthesis Lectures on Data Management* (Morgan & Claypool) and is now the Editor-in-Chief of *ACM Books*. He serves on the editorial boards of three journals and two book series.

He is a Fellow of the Royal Society of Canada, American Association for the Advancement of Science (AAAS), Association for Computing Machinery (ACM), and the Institute of Electrical and Electronics Engineers (IEEE). He is an elected member of the Science Academy, Turkey, and a member of Sigma Xi. He was awarded the ACM SIGMOD Test-of-Time Award in 2015, the ACM SIGMOD Contributions Award in 2006, and the Ohio State University College of Engineering Distinguished Alumnus Award in 2008.

Advisory Board

Ramesh Jain Department of Computer Science, School of Information and Computer Sciences, University of California Irvine, Irvine, CA, USA

Peter MG Apers Centre for Telematics and Information Technology, University of Twente, Enschede, The Netherlands

Timos Sellis Data Science Research Institute, Swinburne University of Technology, Hawthorn, VIC, Australia

Matthias Jarke Informatik 5 Information Systems, RWTH-Aachen University, Aachen, Germany

Ricardo Baeza-Yate Department of Information and Communication Technologies, University of Pompeu Fabra, Barcelona, Spain

Jai Menon Cloudistics, Reston, VA, USA

Beng Chin Ooi School of Computing, National University of Singapore, Singapore, Singapore

Elisa Bertino Department of Computer Science, Purdue University, West Lafeyette, IN, USA

Erhard Rahm Fakultät für Mathematik und Informatik, Institut für Informatik, Universität Leipzig, Leipzig, Germany

Gerhard Weikum Department 5: Databases and Information Systems, Max-Planck-Institut für Informatik, Saarbrücken, Germany

Stefano Ceri Department of Electronics, Information and Bioengineering, Politecnico di Milano, Milano, Italy

Asuman Dogac SRDC Software Research and Development and Consultancy Ltd., Cankaya/Ankara, Turkey

Hans-Joerg Schek Department of Computer Science, ETH Zürich, Zürich, Switzerland

Alon Halevy Recruit Institute of Technology, Mountain View, CA, USA

Jennifer Widom Frederick Emmons Terman School of Engineering, Stanford University, Stanford, CA, USA

John Mylopoulos Department of Computer Science, University of Toronto, Toronto, ON, Canada

Jiawei Han Department of Computer Science, University of Illinois at Urbana-Champaign, Urbana, IL, USA

Lizhu Zhou Department of Computer Science and Technology, Tsinghua University, Beijing, China

Theo Härder Department of Computer Science, University of Kaiserslautern, Kaiserslautern, Germany

Serge Abiteboul INRIA and ENS, Paris, France

Frank Tompa David R. Cheriton School of Computer Science, University of Waterloo, Waterloo, ON, Canada

Patrick Valduriez INRIA and LIRMM, Montpellier, France

Gustavo Alonso Department of Computer Science, ETH Zürich, Zürich, Switzerland

Krithi Ramamritham Department of Computer Science and Engineering, Indian Institute of Technology Bombay, Mumbai, India

Area Editors

Peer-to-Peer Data Management

Karl Aberer Department of Computer Science, École Polytechnique Fédérale de Lausanne (EPFL), Lausanne, Switzerland

Database Management System Architectures

Anastasia Ailamaki Department of Computer Science, Ecole Polytechnique Fédérale de Lausanne, Lausanne, Switzerland

Information Retrieval Models

Giambattista Amati Fondazione Ugo Bordoni, Rome, Italy

XML Data Management

Sihem Amer-Yahia CNRS, University Grenoble Alpes, Saint Martin D'Hères, France

Database Middleware

Cristiana Amza Electrical and Computer Engineering, University of Toronto, Toronto, ON, Canada

Database Management Utilities

Philippe Bonnet Department of Computer Science, IT University of Copenhagen, Copenhagen, Denmark

Visual Interfaces

Tiziana Catarci Department of Computer Engineering, Automation and Management, Sapienza – Università di Roma, Rome, Italy

Stream Data Management

Ugur Cetintemel Department of Computer Science, Brown University, Providence, RI, USA

Querying over Data Integration Systems

Kevin Chang Department of Computer Science, University of Illinois at
Urbana-Champaign, Urbana-Champaign, IL, USA

Self Management

Surajit Chaudhuri Microsoft Corporation, Redmond, CA, USA

Text Mining

Zheng Chen Microsoft Corporation, Beijing, China

Extended Transaction Models

Panos K. Chrysanthis Department of Computer Science, School of Computing and Information, University of Pittsburgh, Pittsburgh, PA, USA

Privacy-Preserving Data Mining

Chris Clifton Department of Computer Science, Purdue University, West Lafayette, IN, USA

Digital Libraries

Amr El Abbadi Department of Computer Science, UC Santa Barbara, Santa Barbara, CA, USA

Data Models

David W. Embley Department of Computer Science, Brigham Young University, Provo, UT, USA

Complex Event Processing

Opher Etzion Department of Information Systems, Yezreel Valley College, Jezreel Valley, Israel

Database Security and Privacy

Elena Ferrari Department of Computer Science, Università degli Studi dell'Insubria, Varese, Italy

Semantic Web and Ontologies

Avigdor Gal Industrial Engineering and Management, Technion – Israel Institute of Technology, Haifa, Israel

Data Cleaning

Venkatesh Ganti Alation, Redwood City, CA, USA

Web Data Extraction

Georg Gottlob Computing Lab, Oxford University, Oxford, UK

Sensor Networks

Le Gruenwald School of Computer Science, University of Oklahoma, Norman, OK, USA

Data Clustering

Dimitrios Gunopulos Department of Informatics and Telecommunications, National and Kapodistrian University of Athens, Athens, Greece

Scientific Databases

Amarnath Gupta San Diego Supercomputer Center, University of California San Diego, La Jolla, CA, USA

Geographical Information Systems

Ralf Hartmut Güting Department of Computer Science, FernUniversität in Hagen, Hagen, Germany

Data Visualization

Hans Hinterberger Department of Computer Science, ETH Zurich, Zurich, Switzerland

Web Services and Service Oriented Architectures

Hans-Arno Jacobsen Department of Electrical and Computer Engineering, University of Toronto, Toronto, ON, Canada

Metadata Management

Manfred Jeusfeld IIT, University of Skövde, Skövde, Sweden

Health Informatics Databases

Vipul Kashyap CIGNA Healthcare, Bloomfield, CT, USA

Visual Data Mining

Daniel A. Keim Computer Science Department, University of Konstanz, Konstanz, Germany

Data Replication

Bettina Kemme School of Computer Science, McGill University, Montreal, QC, Canada

Storage Structures and Systems

Masaru Kitsuregawa Institute of Industrial Science, University of Tokyo, Tokyo, Japan

Views and View Management

Yannis Kotidis Department of Informatics, Athens University of Economics and Business, Athens, Greece

Structured Text Retrieval

Jaap Kamps Faculty of Humanities, University of Amsterdam, Amsterdam, The Netherlands

Information Quality

Yang W. Lee School of Business, Northeastern University, Boston, MA, USA

Relational Theory

Leonid Libkin School of Informatics, University of Edinburgh, Edinburgh, UK

Information Retrieval Evaluation Measures

Weiyi Meng Department of Computer Science, State University of New York at Binghamton, Binghamton, NY, USA

Logical Data Integration

Renée J. Miller Department of Computer Science, University of Toronto, Toronto, ON, Canada

Database Design

Alexander Borgida Department of Computer Science, Rutgers University, New Brunswick, NJ, USA

Text Indexing Techniques

Mario A. Nascimento Department of Computing Science, University of Alberta, Edmonton, AB, Canada

Data Quality

Felix Naumann Hasso Plattner Institute, University of Potsdam, Potsdam, Germany

Web Search and Crawl

Cong Yu Google Research, New York, NY, USA

Multimedia Databases

Vincent Oria Department of Computer Science, New Jersey Institute of Technology, Newark, NJ, USA

Shin'ichi Satoh Digital Content and Media Sciences ReseaMultimedia Information Research Division, National Institute of Informatics, Tokyo, Japan

Active Databases

M. Tamer Özsu Cheriton School of Computer Science, University of Waterloo, Waterloo, ON, Canada

Spatial, Spatiotemporal, and Multidimensional Databases

Dimitris Papadias Department of Computer Science and Engineering, Hong Kong University of Science and Technology, Kowloon, China

Data Warehouse

Torben Bach Pedersen Department of Computer Science, Aalborg University, Aalborg, Denmark

Stefano Rizzi DISI – University of Bologna, Bologna, Italy

Association Rule Mining

Jian Pei School of Computing Science, Simon Fraser University, Burnaby, BC, Canada

Workflow Management

Barbara Pernici Department di Elettronica e Informazione, Politecnico di Milano, Milan, Italy

Query Processing and Optimization

Evaggelia Pitoura Department of Computer Science and Engineering, University of Ioannina, Ioannina, Greece

Data Management for the Life Sciences

Louiqa Raschid Robert H. Smith School of Business, University of Maryland, College Park, MD, USA

Information Retrieval Operations

Edie Rasmussen Library, Archival and Information Studies, The University of British Columbia, VC, Canada

Query Languages

Tore Risch Department of Information Technology, Uppsala University, Uppsala, Sweden

Database Tuning and Performance

Dennis Shasha Department of Computer Science, New York University, New York, NY, USA

Classification and Decision Trees

Kyuseok Shim School of Electrical Engineering and Computer Science, Seoul National University, Seoul, Republic of Korea

Temporal Databases

Christian S. Jensen Department of Computer Science, Aalborg University, Aalborg, Denmark

Richard T. Snodgrass Department of Computer Science, University of Arizona, Tucson, AZ, USA

Stream Mining

Divesh Srivastava AT&T Labs-Research, Bedminster, NJ, USA

Distributed Database Systems

Kian-Lee Tan Department of Computer Science, National University of Singapore, Singapore, Singapore

Logics and Databases

Val Tannen Department of Computer and Information Science, University of Pennsylvania, Philadelphia, PA, USA

Structured and Semi-structured Document Databases

Frank Tompa David R. Cheriton School of Computer Science, University of Waterloo, Waterloo, ON, Canada

Indexing

Vassilis J. Tsotras Department of Computer Science and Engineering, University of California-Riverside, Riverside, CA, USA

Parallel Database Systems

Patrick Valduriez INRIA and LIRMM, Montpellier, France

Advanced Storage Systems

Kaladhar Voruganti Equinix, San Francisco, CA, USA

Transaction Management

Gottfried Vossen Department of Information Systems, Westfälische Wilhelms-Universität, Münster, Germany

Mobile and Ubiquitous Data Management

Ouri Wolfson Department of Computer Science, University of Illinois at Chicago, Chicago, IL, USA

Multimedia Information Retrieval

Jeffrey Xu Yu Department of Systems Engineering and Engineering Management, The Chinese University of Hong Kong, Hong Kong, China

Approximation and Data Reduction Techniques

Xiaofang Zhou School of Information Technology and Electrical Engineering, University of Queensland, Brisbane, Australia

Social Networks

Nick Koudas Department of Computer Science, University of Toronto, Toronto, ON, Canada

Cloud Data Management

Amr El Abbadi Department of Computer Science, UC Santa Barbara, Santa Barbara, CA, USA

Data Analytics

Fatma Özcan IBM Research – Almaden, San Jose, CA, USA

Data Management Fundamentals

Ramez Elmasri Department of Computer Science and Engineering, The University of Texas at Arlington, Arlington, TX, USA

NoSQL Databases

M. Tamer Özsu Cheriton School of Computer Science, University of Waterloo, Waterloo, ON, Canada

Ling Liu College of Computing, Georgia Institute of Technology, Atlanta, GA, USA

Graph Data Management

Lei Chen Department of Computer Science and Engineering, The Hong Kong University of Science and Technology, Hong Kong, China

Data Provenance Management

Juliana Freire Computer Science and Engineering, New York University, New York, NY, USA

Ranking Queries

Ihab F. Ilyas Cheriton School of Computer Science, University of Waterloo, Waterloo, ON, Canada

Uncertain Data Management

Minos Garofalakis Technical University of Crete, Chania, Greece

Crowd Sourcing

Reynold Cheng Computer Science, The University of Hong Kong, Hong Kong, China

List of Contributors

Daniel Abadi Yale University, New Haven, CT, USA

Sofiane Abbar Qatar Computing Research Institute, Doha, Qatar

Alberto Abelló Polytechnic University of Catalonia, Barcelona, Spain

Serge Abiteboul Inria, Paris, France

Maribel Acosta Institute AIFB, Karlsruhe Institute of Technology, Karlsruhe, Germany

Ioannis Aekaterinidis University of Patras, Rio, Patras, Greece

Nitin Agarwal University of Arkansas, Little Rock, AR, USA

Charu C. Aggarwal IBM T. J. Watson Research Center, Yorktown Heights, NY, USA

Lalitha Agnihotri McGraw-Hill Education, New York, NY, USA

Marcos K. Aguilera VMware Research, Palo Alto, CA, USA

Yanif Ahmad Department of Computer Science, Brown University, Providence, RI, USA

Gail-Joon Ahn Arizona State University, Tempe, AZ, USA

Anastasia Ailamaki Informatique et Communications, Ecole Polytechnique Fédérale de Lausanne, Lausanne, Switzerland

Ablimit Aji Analytics Lab, Hewlett Packard, Palo Alto, CA, USA

Alexander Alexandrov Database and Information Management (DIMA), Institute of Software Engineering and Theoretical Computer Science, Berlin, Germany

Yousef J. Al-Houmaily Institute of Public Administration, Riyadh, Saudi Arabia

Mohammed Eunus Ali Department of Computer Science and Engineering, Bangladesh University of Engineering and Technology (BUET), Dhaka, Bangladesh

Robert B. Allen Drexel University, Philadelphia, PA, USA

Gustavo Alonso ETH Zürich, Zurich, Switzerland

Omar Alonso Microsoft Silicon Valley, Mountain View, CA, USA

Bernd Amann Pierre & Marie Curie University (UPMC), Paris, France

Giambattista Amati Fondazione Ugo Bordoni, Rome, Italy

Sihem Amer-Yahia CNRS, Univ. Grenoble Alps, Grenoble, France

Laboratoire d'Informatique de Grenoble, CNRS-LIG, Saint Martin-d'Hères, Grenoble, France

Rainer von Ammon Center for Information Technology Transfer GmbH (CITT), Regensburg, Germany

Robert A. Amsler CSC, Falls Church, VA, USA

Yael Amsterdamer Department of Computer Science, Bar Ilan University, Ramat Gan, Israel

Cristiana Amza Department of Electrical and Computer Engineering, University of Toronto, Toronto, ON, Canada

George Anadiotis VU University Amsterdam, Amsterdam, The Netherlands

Mihael Ankerst Ludwig-Maximilians-Universität München, Munich, Germany

Sameer Antani National Institutes of Health, Bethesda, MD, USA

Grigoris Antoniou Foundation for Research and Technology-Hellas (FORTH), Heraklion, Greece

Arvind Arasu Microsoft Research, Redmond, WA, USA

Danilo Ardagna Politechnico di Milano University, Milan, Italy

Walid G. Aref Purdue University, West Lafayette, IN, USA

Marcelo Arenas Pontifical Catholic University of Chile, Santiago, Chile

Nikos Armenatzoglou Department of Computer Science and Engineering, Hong Kong University of Science and Technology, Kowloon, Hong Kong, Hong Kong

Samuel Aronson Harvard Medical School – Partners Healthcare Center for Genetics and Genomics, Boston, MA, USA

Paavo Arvola University of Tampere, Tampere, Finland

Colin Atkinson Software Engineering, University of Mannheim, Mannheim, Germany

Noboru Babaguchi Osaka University, Osaka, Japan

Shivnath Babu Duke University, Durham, NC, USA

Nathan Backman Computer Science, Buena Vista University, Storm Lake, IA, USA

Kenneth Paul Baclawski Northeastern University, Boston, MA, USA

Ricardo Baeza-Yates NTENT, USA - Univ. Pompeu Fabra, Spain - Univ. de Chile, Chile

James Bailey University of Melbourne, Melbourne, VIC, Australia

Peter Bailis Department of Computer Science, Stanford University, Palo Alto, CA, USA

Sumeet Bajaj Stony Brook University, Stony Brook, NY, USA

Peter Bak IBM Watson Health, Foundational Innovation, Haifa, Israel

Magdalena Balazinska University of Washington, Seattle, WA, USA

Krisztian Balog University of Stavanger, Stavanger, Norway

Farnoush Banaei-Kashani Computer Science and Engineering, University of Colorado Denver, Denver, CO, USA

Jie Bao Data Management, Analytics and Services (DMAS) and Ubiquitous Computing Group (Ubicomp), Microsoft Research Asia, Beijing, China

Stefano Baraldi University of Florence, Florence, Italy

Mauro Barbieri Phillips Research Europe, Eindhoven, The Netherlands

Denilson Barbosa University of Alberta, Edmonton, AL, Canada

Pablo Barceló University of Chile, Santiago, Chile

Luciano Baresi Dipartimento di Elettronica, Informazione e Bioingegneria – Politecnico di Milano, Milano, Italy

Ilaria Bartolini Department of Computer Science and Engineering (DISI), University of Bologna, Bologna, Italy

Saleh Basalamah Computer Science, Umm Al-Qura University, Mecca, Makkah Province, Saudi Arabia

Sugato Basu Google Inc, Mountain View, CA, USA

Carlo Batini University of Milano-Bicocca, Milan, Italy

Michal Batko Masaryk University, Brno, Czech Republic

Peter Baumann Jacobs University, Bremen, Germany

Robert Baumgartner Vienna University of Technology, Vienna, Austria

Sean Bechhofer University of Manchester, Manchester, UK

Steven M. Beitzel Telcordia Technologies, Piscataway, NJ, USA

Ladjel Bellatreche LIAS/ISAE-ENSMA, Poitiers University, Futuroscope, France

Omar Benjelloun Google Inc., New York, NY, USA

Véronique Benzaken University Paris 11, Orsay Cedex, France

Rafael Berlanga Department of Computer Languages and Systems, Universitat Jaume I, Castellón, Spain

Mikael Berndtsson University of Skövde, The Informatics Research Centre, Skövde, Sweden

University of Skövde, School of Informatics, Skövde, Sweden

Philip A. Bernstein Microsoft Corporation, Redmond, WA, USA

Damon Andrew Berry University of Massachusetts, Lowell, MA, USA

Leopoldo Bertossi Carleton University, Ottawa, ON, Canada

Claudio Bettini Dipartimento di Informatica, Università degli Studi di Milano, Milan, Italy

Nigel Bevan Professional Usability Services, London, UK

Bharat Bhargava Purdue University, West Lafayette, IN, USA

Arnab Bhattacharya Indian Institute of Technology, Kanpur, India

Ernst Biersack Eurecom, Sophia Antipolis, France

Alberto Del Bimbo University of Florence, Florence, Italy

Carsten Binnig Computer Science-Database Systems, Brown University, Providence, RI, USA

Christian Bizer Web-based Systems Group, University of Mannheim, Mannheim, Germany

Alan F. Blackwell University of Cambridge, Cambridge, UK

Carlos Blanco GSyA and ISTR Research Groups, Department of Computer Science and Electronics, Faculty of Sciences, University of Cantabria, Santander, Spain

Marina Blanton University of Notre Dame, Notre Dame, IN, USA

Toine Bogers Department of Communication and Psychology, Aalborg University Copenhagen, Copenhagen, Denmark

Philip Bohannon Yahoo! Research, Santa Clara, CA, USA

Michael H. Böhlen Free University of Bozen-Bolzano, Bozen-Bolzano, Italy

University of Zurich, Zürich, Switzerland

Christian Böhm University of Munich, Munich, Germany

Peter Boncz CWI, Amsterdam, The Netherlands

Philippe Bonnet Department of Computer Science, IT University of Copenhagen, Copenhagen, Denmark

Alexander Borgida Rutgers University, New Brunswick, NJ, USA

Vineyak Borkar CTO and VP of Engineering, X15 Software, San Francisco, CA, USA

Chavdar Botev Yahoo Research!, Cornell University, Ithaca, NY, USA

Sara Bouchenak University of Grenoble I – INRIA, Grenoble, France

Luc Bouganim INRIA Saclay and UVSQ, Le Chesnay, France

Nozha Boujemaa INRIA Paris-Rocquencourt, Le Chesnay, France

Shawn Bowers University of California-Davis, Davis, CA, USA

Stéphane Bressan National University of Singapore, School of Computing, Department of Computer Science, Singapore, Singapore

Martin Breunig University of Osnabrueck, Osnabrueck, Germany

Scott A. Bridwell University of Utah, Salt Lake City, UT, USA

Thomas Brinkhoff Institute for Applied Photogrammetry and Geoinformatics (IAPG), Oldenburg, Germany

Nieves R. Brisaboa Database Laboratory, Department of Computer Science, University of A Coruña, A Coruña, Spain

Andrei Broder Yahoo! Research, Santa Clara, CA, USA

Nicolas Bruno Microsoft Corporation, Redmond, WA, USA

François Bry University of Munich, Munich, Germany

Yingyi Bu Chinese University of Hong Kong, Hong Kong, China

Alejandro Buchmann Darmstadt University of Technology, Darmstadt, Germany

Thilina Buddhika Colorado State University, Fort Collins, CO, USA

Chiranjeeb Buragohain Amazon.com, Seattle, WA, USA

Thorsten Büring Ludwig-Maximilians-University Munich, Munich, Germany

Benjamin Bustos Department of Computer Science, University of Chile, Santiago, Chile

David J. Buttler Lawrence Livermore National Laboratory, Livermore, CA, USA

Yanli Cai Shanghai Jiao Tong University, Shanghai, China

Diego Calvanese Research Centre for Knowledge and Data (KRDB), Free University of Bozen-Bolzano, Bolzano, Italy

Guadalupe Canahuate The Ohio State University, Columbus, OH, USA

K. Selcuk Candan Arizona State University, Tempe, AZ, USA

Turkmen Canli University of Illinois at Chicago, Chicago, IL, USA

Alan Cannon Napier University, Edinburgh, UK

Cornelia Caragea Computer Science and Engineering, University of North Texas, Denton, TX, USA

Barbara Carminati Department of Theoretical and Applied Science, University of Insubria, Varese, Italy

Sheelagh Carpendale University of Calgary, Calgary, AB, Canada

Michael W. Carroll Villanova University School of Law, Villanova, PA, USA

Ben Carterette University of Massachusetts Amherst, Amherst, MA, USA

Marco A. Casanova Pontifical Catholic University of Rio de Janeiro, Rio de Janeiro, Brazil

Giuseppe Castagna C.N.R.S. and University Paris 7, Paris, France

Tiziana Catarci Dipartimento di Ingegneria Informatica, Automatica e Gestionale "A.Ruberti", Sapienza Università di Roma, Rome, Italy

James Caverlee Department of Computer Science, Texas A&M University, College Station, TX, USA

Emmanuel Cecchet EPFL, Lausanne, Switzerland

Wojciech Cellary Department of Information Technology, Poznan University of Economics, Poznan, Poland

Ana Cerdeira-Pena Database Laboratory, Department of Computer Science, University of A Coruña, A Coruña, Spain

Michal Ceresna Lixto Software GmbH, Vienna, Austria

Ugur Cetintemel Department of Computer Science, Brown University, Providence, RI, USA

Soumen Chakrabarti Indian Institute of Technology of Bombay, Mumbai, India

Don Chamberlin IBM Almaden Research Center, San Jose, CA, USA

Allen Chan IBM Toronto Software Lab, Markham, ON, Canada

Chee-Yong Chan National University of Singapore, Singapore, Singapore

K. Mani Chandy California Institute of Technology, Pasadena, CA, USA

Edward Y. Chang Google Research, Mountain View, CA, USA

Kevin Chang Department of Computer Science, University of Illinois at Urbana-Champaign, Urbana, IL, USA

Adriane Chapman University of Southampton, Southampton, UK

Surajit Chaudhuri Microsoft Research, Microsoft Corporation, Redmond, WA, USA

Elizabeth S. Chen Partners HealthCare System, Boston, MA, USA

James L. Chen University of Chicago, Chicago, IL, USA

Jin Chen Computer Engineering Research Group, University of Toronto, Toronto, ON, Canada

Jinjun Chen Swinburne University of Technology, Melbourne, VIC, Australia

Jinchuan Chen Key Laboratory of Data Engineering and Knowledge Engineering, Ministry of Education, Renmin University of China, Beijing

Lei Chen Hong Kong University of Science and Technology, Hong Kong, China

Peter P. Chen Louisiana State University, Baton Rouge, LA, USA

James Cheney University of Edinburgh, Edinburgh, UK

Hong Cheng Department of Systems Engineering and Engineering Management, The Chinese University of Hong Kong, Hong Kong, China

Reynold Cheng Computer Science, The University of Hong Kong, Hong Kong, China

Vivying S. Y. Cheng Hong Kong University of Science and Technology, Hong Kong, China

InduShobha N. Chengalur-Smith University at Albany – SUNY, Albany, NY, USA

Mitch Cherniack Brandeis University, Wattham, MA, USA

Yun Chi NEC Laboratories America, Cupertino, CA, USA

Fernando Chirigati NYU Tandon School of Engineering, Brooklyn, NY, USA

Rada Chirkova North Carolina State University, Raleigh, NC, USA

Laura Chiticariu Scalable Natural Language Processing, IBM Research – Almaden, San Jose, CA, USA

Jan Chomicki Department of Computer Science and Engineering, State University of New York at Buffalo, Buffalo, NY, USA

Fred Chong Computer Science, University of Chicago, Chicago, IL, USA

Stephanie Chow University of Ontario Institute of Technology, Oshawa, ON, Canada

Peter Christen Research School of Computer Science, The Australian National University, Canberra, Australia

Vassilis Christophides INRIA Paris-Roquencourt, Paris, France

Panos K. Chrysanthis Department of Computer Science, University of Pittsburgh, Pittsburgh, PA, USA

Paolo Ciaccia Computer Science and Engineering, University of Bologna, Bologna, Italy

John Cieslewicz Google Inc., Mountain View, CA, USA

Gianluigi Ciocca University of Milano-Bicocca, Milan, Italy

Eugene Clark Harvard Medical School – Partners Healthcare Center for Genetics and Genomics, Boston, MA, USA

Charles L. A. Clarke University of Waterloo, Waterloo, ON, Canada

William R. Claycomb CERT Insider Threat Center, Software Engineering Institute, Carnegie Mellon University, Pittsburgh, PA, USA

Eliseo Clementini University of L'Aquila, L'Aquila, Italy

Chris Clifton Department of Computer Science, Purdue University, West Lafayette, IN, USA

Edith Cohen AT&T Labs-Research, Florham Park, NJ, USA

Sara Cohen The Rachel and Selim Benin School of Computer Science and Engineering, The Hebrew University of Jerusalem, Jerusalem, Israel

Sarah Cohen-Boulakia University Paris-Sud, Orsay Cedex, France

Carlo Combi Department of Computer Science, University of Verona, Verona, VR, Italy

Mariano P. Consens University of Toronto, Toronto, ON, Canada

Dianne Cook Iowa State University, Ames, IA, USA

Graham Cormode Computer Science, University of Warwick, Warwick, UK

Antonio Corral University of Almeria, Almeria, Spain

Maria Francesca Costabile Department of Computer Science, University of Bari, Bari, Italy

Nick Craswell Microsoft Research Cambridge, Cambridge, UK

Fabio Crestani University of Lugano, Lugano, Switzerland

Marco Antonio Cristo FUCAPI, Manaus, Brazil

Maxime Crochemore King's College London, London, UK

Université Paris-Est, Paris, France

Andrew Crotty Database Group, Brown University, Providence, RI, USA

Matthew G. Crowson University of Chicago, Chicago, IL, USA

Michel Crucianu Conservatoire National des Arts et Métiers, Paris, France

Philippe Cudré-Mauroux Massachusetts Institute of Technology, Cambridge, MA, USA

Sonia Leila Da Silva Cerveteri, Italy

Peter Dadam University of Ulm, Ulm, Germany

Mehmet M. Dalkiliç Indiana University, Bloomington, IN, USA

Nilesh Dalvi Airbnb, San Francisco, CA, USA

Marina Danilevsky IBM Almaden Research Center, San Jose, CA, USA

Minh Dao-Tran Institute of Information Systems, Vienna University of Technology, Vienna, Austria

Gautam Das Department of Computer Science and Engineering, University of Texas at Arlington, Arlington, TX, USA

Mahashweta Das Visa Research, Palo Alto, CA, USA

Sudipto Das Microsoft Research, Redmond, WA, USA

Manoranjan Dash Nanyang Technological University, Singapore, Singapore

Anupam Datta Computer Science Department and Electrical and Computer Engineering Department, Carnegie Mellon University, Pittsburgh, PA, USA

Anwitaman Datta Nanyang Technological University, Singapore, Singapore

Ian Davidson University of California-Davis, Davis, CA, USA

Susan B. Davidson Department of Computer and Information Science, University of Pennsylvania, Philadelphia, PA, USA

Todd Davis Department of Computer Science and Software Engineering, Concordia University, Montreal, QC, Canada

Maria De Marsico Sapienza University of Rome, Rome, Italy

Edleno Silva De Moura Federal University of Amazonas, Manaus, Brazil

Antonios Deligiannakis University of Athens, Athens, Greece

Alex Delis University of Athens, Athens, Greece

Alan Demers Cornell University, Ithaca, NY, USA

Jennifer Dempsey University of Arizona, Tucson, AZ, USA

Raytheon Missile Systems, Tucson, AZ, USA

Ke Deng University of Queensland, Brisbane, QLD, Australia

Amol Deshpande University of Maryland, College Park, MD, USA

Zoran Despotovic NTT DoCoMo Communications Laboratories Europe, Munich, Germany

Alin Deutsch University of California-San Diego, La Jolla, CA, USA

Yanlei Diao University of Massachusetts Amherst, Amherst, MA, USA

Suzanne W. Dietrich Arizona State University, Phoenix, AZ, USA

Nevenka Dimitrova Philips Research, Briarcliff Manor, New York, USA

Bolin Ding University of Illinois at Urbana-Champaign, Urbana, IL, USA

Chris Ding University of Texas at Arlington, Arlington, TX, USA

Alan Dix Lancaster University, Lancaster, UK

Belayadi Djahida National High School for Computer Science (ESI), Algiers, Algeria

Hong-Hai Do SAP AG, Dresden, Germany

Gillian Dobbie University of Auckland, Auckland, New Zealand

Alin Dobra University of Florida, Gainesville, FL, USA

Vlastislav Dohnal Masaryk University, Brno, Czech Republic

Mario Döller University of Applied Science Kufstein, Kufstein, Austria

Carlotta Domeniconi George Mason University, Fairfax, VA, USA

Josep Domingo-Ferrer Universitat Rovira i Virgili, Tarragona, Catalonia, Spain

Guozhu Dong Wright State University, Dayton, OH, USA

Xin Luna Dong Amazon, Seattle, WA, USA

Chitra Dorai IBM T. J. Watson Research Center, Hawthorne, NY, USA

Zhicheng Dou Nankai University, Tianjin, China

Ahlame Douzal CNRS, Univ. Grenoble Alps, Grenoble, France

Yang Du Northeastern University, Boston, MA, USA

Susan Dumais Microsoft Research, Redmond, WA, USA

Marlon Dumas University of Tartu, Tartu, Estonia

Schahram Dustdar Technical University of Vienna, Vienna, Austria

Curtis E. Dyreson Utah State University, Logan, UT, USA

Johann Eder Department of Informatics-Systems, Alpen-Adria-Universität Klagenfurt, Klagenfurt, Austria

Milad Eftekhar University of Toronto, Toronto, ON, Canada

Thomas Eiter Institute of Information Systems, Vienna University of Technology, Vienna, Austria

Ibrahim Abu El-Khair Information Science Department, School of Social Sciences, Umm Al-Qura University, Mecca, Saudi Arabia

Ahmed K. Elmagarmid Purdue University, West Lafayette, IN, USA

Qatar Computing Research Institute, HBKU, Doha, Qatar

Ramez Elmasri Computer Science, The University of Texas at Arlington, Arlington, TX, USA

Aaron J. Elmore Department of Computer Science, University of Chicago, Chicago, IL, USA

Sameh Elnikety Microsoft Research, Redmond, WA, USA

David W. Embley Brigham Young University, Provo, UT, USA

Vincent Englebert University of Namur, Namur, Belgium

AnnMarie Ericsson University of Skövde, Skövde, Sweden

Martin Ester Simon Fraser University, Burnaby, BC, Canada

Opher Etzion IBM Software Group, IBM Haifa Labs, Haifa University Campus, Haifa, Israel

Patrick Eugster Purdue University, West Lafayette, IN, USA

Ronald Fagin IBM Almaden Research Center, San Jose, CA, USA

Ju Fan DEKE Lab and School of Information, Renmin University of China, Beijing, China

Wei Fan IBM T.J. Watson Research, Hawthorne, NY, USA

Wenfei Fan University of Edinburgh, Edinburgh, UK

Beihang University, Beijing, China

Hui Fang University of Delaware, Newark, DE, USA

Alan Fekete University of Sydney, Sydney, NSW, Australia

Jean-Daniel Fekete INRIA, LRI University Paris Sud, Orsay Cedex, France

Pascal Felber University of Neuchatel, Neuchatel, Switzerland

Paolino Di Felice University of L'Aguila, L'Aguila, Italy

Hakan Ferhatosmanoglu The Ohio State University, Columbus, OH, USA

Eduardo B. Fernandez Florida Atlantic University, Boca Raton, FL, USA

Eduardo Fernández-Medina GSyA Research Group, Department of Information Technologies and Systems, Institute of Information Technologies and Systems, Escuela Superior de Informática, University of Castilla-La Mancha, Ciudad Real, Spain

Paolo Ferragina Department of Computer Science, University of Pisa, Pisa, Italy

Elena Ferrari DiSTA, University of Insubria, Varese, Italy

Dennis Fetterly Google, Inc., Mountain View, CA, USA

Stephen E. Fienberg Carnegie Mellon University, Pittsburgh, PA, USA

Michael Fink Institute of Information Systems, Vienna University of Technology, Vienna, Austria

Peter M. Fischer Computer Science Department, University of Freiburg, Freiburg, Germany

Simone Fischer-Hübner Karlstad University, Karlstad, Sweden

Fabian Flöck GESIS – Leibniz Institute for the Social Sciences, Köln, Germany

Avrilia Floratou Microsoft, Sunnyvale, CA, USA

Leila De Floriani University of Genova, Genoa, Italy

Christian Fluhr CEA LIST, Fontenay-aux, Roses, France

Greg Flurry IBM SOA Advanced Technology, Armonk, NY, USA

Edward A. Fox Virginia Tech, Blacksburg, VA, USA

Chiara Francalanci Politecnico di Milano University, Milan, Italy

Andrew U. Frank Vienna University of Technology, Vienna, Austria

Michael J. Franklin University of California-Berkeley, Berkeley, CA, USA

Keir Fraser University of Cambridge, Cambridge, UK

Juliana Freire NYU Tandon School of Engineering, Brooklyn, NY, USA

NYU Center for Data Science, New York, NY, USA

New York University, New York, NY, USA

Elias Frentzos University of Piraeus, Piraeus, Greece

Johann-Christoph Freytag Humboldt University of Berlin, Berlin, Germany

Ophir Frieder Georgetown University, Washington, DC, USA

Oliver Frölich Lixto Software GmbH, Vienna, Austria

Ada Wai-Chee Fu Chinese University of Hong Kong, Hong Kong, China

Xiang Fu University of Southern California, Los Angeles, CA, USA

Kazuhisa Fujimoto Hitachi Ltd., Tokyo, Japan

Tim Furche University of Munich, Munich, Germany

Ariel Fuxman Microsoft Research, Mountain View, CA, USA

Silvia Gabrielli Bruno Kessler Foundation, Trento, Italy

Isabella Gagliardi National Research Council (CNR), Milan, Italy

Avigdor Gal Faculty of Industrial Engineering and Management, Technion–Israel Institute of Technology, Haifa, Israel

Alex Galakatos Database Group, Brown University, Providence, RI, USA

Department of Computer Science, Brown University, Providence, RI, USA

Wojciech Galuba EPFL, Lausanne, Switzerland

Johann Gamper Free University of Bozen-Bolzano, Bolzano, Italy

Weihao Gan University of Southern California, Los Angeles, CA, USA

Vijay Gandhi University of Minnesota, Minneapolis, MN, USA

Venkatesh Ganti Microsoft Research, Microsoft Corporation, Redmond, WA, USA

Dengfeng Gao IBM Silicon Valley Lab, San Jose, CA, USA

Like Gao Teradata Corporation, San Diego, CA, USA

Wei Gao Qatar Computing Research Institute, Doha, Qatar

Minos Garofalakis Technical University of Crete, Chania, Greece

Wolfgang Gatterbauer University of Washington, Seattle, WA, USA

Buğra Gedik Department of Computer Engineering, Bilkent University, Ankara, Turkey

IBM T.J. Watson Research Center, Hawthorne, NY, USA

Floris Geerts University of Antwerp, Antwerp, Belgium

Johannes Gehrke Cornell University, Ithaca, NY, USA

Betsy George Oracle (America), Nashua, NH, USA

Lawrence Gerstley PSMI Consulting, San Francisco, CA, USA

Michael Gertz Heidelberg University, Heidelberg, Germany

Giorgio Ghelli Dipartimento di Informatica, Università di Pisa, Pisa, Italy

Gabriel Ghinita National University of Singapore, Singapore, Singapore

Giuseppe De Giacomo Dip. di Ingegneria Informatica Automatica e Gestionale Antonio Ruberti, Sapienza Università di Roma, Rome, Italy

Phillip B. Gibbons Computer Science Department and the Electrical and Computer Engineering Department, Carnegie Mellon University, Pittsburgh, PA, USA

Sarunas Girdzijauskas EPFL, Lausanne, Switzerland

Fausto Giunchiglia University of Trento, Trento, Italy

Kazuo Goda The University of Tokyo, Tokyo, Japan

Max Goebel Vienna University of Technology, Vienna, Austria

Bart Goethals University of Antwerp, Antwerp, Belgium

Martin Gogolla University of Bremen, Bremen, Germany

Aniruddha Gokhale Vanderbilt University, Nashville, TN, USA

Lukasz Golab University of Waterloo, Waterloo, ON, Canada

Matteo Golfarelli DISI – University of Bologna, Bologna, Italy

Arturo González-Ferrer Innovation Unit, Instituto de Investigación Sanitaria del Hospital Clínico San Carlos (IdISSC), Madrid, Spain

Michael F. Goodchild University of California-Santa Barbara, Santa Barbara, CA, USA

Georg Gottlob Computing Laboratory, Oxford University, Oxford, UK

Valerie Gouet-Brunet CNAM Paris, Paris, France

Ramesh Govindan University of Southern California, Los Angeles, CA, USA

Tyrone Gradison Proficiency Labs, Ashland, OR, USA

Goetz Graefe Google, Inc., Mountain View, CA, USA

Gösta Grahne Concordia University, Montreal, QC, Canada

Fabio Grandi Alma Mater Studiorum Università di Bologna, Bologna, Italy

Tyrone Grandison Proficiency Labs, Ashland, OR, USA

Peter M. D. Gray University of Aberdeen, Aberdeen, UK

Todd J. Green University of Pennsylvania, Philadelphia, PA, USA

Georges Grinstein University of Massachusetts, Lowell, MA, USA

Tom Gruber RealTravel, Emerald Hills, CA, USA

Le Gruenwald School of Computer Science, University of Oklahoma, Norman, OK, USA

Torsten Grust University of Tübingen, Tübingen, Germany

Dirk Van Gucht Indiana University, Bloomington, IN, USA

Carlos Guestrin Carnegie Mellon University, Pittsburgh, PA, USA

Dimitrios Gunopulos Department of Computer Science and Engineering, The University of California at Riverside, Bourns College of Engineering, Riverside, CA, USA

Amarnath Gupta San Diego Supercomputer Center, University of California San Diego, La Jolla, CA, USA

Himanshu Gupta Stony Brook University, Stony Brook, NY, USA

Cathal Gurrin Dublin City University, Dublin, Ireland

Ralf Hartmut Güting Fakultät für Mathematik und Informatik, Fernuniversität Hagen, Hagen, Germany

Computer Science, University of Hagen, Hagen, Germany

Marc Gyssens Hasselt University, Hasselt, Belgium

Peter J. Haas IBM Almaden Research Center, San Jose, CA, USA

Karl Hahn BMW AG, Munich, Germany

Jean-Luc Hainaut University of Namur, Namur, Belgium

Alon Halevy The Recruit Institute of Technology, Mountain View, CA, USA

Google Inc., Mountain View, CA, USA

Maria Halkidi University of Piraeus, Piraeus, Greece

Terry Halpin Neumont University, South Jordan, UT, USA

Jiawei Han University of Illinois at Urbana-Champaign, Urbana, IL, USA

Alan Hanjalic Delft University of Technology, Delft, The Netherlands

David Hansen The Australian e-Health Research Centre, Brisbane, QLD, Australia

Jörgen Hansson University of Skövde, Skövde, Sweden

Nikos Hardavellas Carnegie Mellon University, Pittsburgh, PA, USA

Theo Härder University of Kaiserslautern, Kaiserslautern, Germany

David Harel The Weizmann Institute of Science, Rehovot, Israel

Jayant R. Haritsa Indian Institute of Science, Bangalore, India

Stavros Harizopoulos HP Labs, Palo Alto, CA, USA

Per F. V. Hasle Royal School of Library and Information Science, University of Copenhagen, Copenhagen S, Denmark

Jordan T. Hastings Department of Geography, University of California-Santa Barbara, Santa Barbara, CA, USA

Alexander Hauptmann Carnegie Mellon University, Pittsburgh, PA, USA

Helwig Hauser University of Bergen, Bergen, Norway

Manfred Hauswirth Open Distributed Systems, Technical University of Berlin, Berlin, Germany

Fraunhofer FOKUS, Galway, Germany

Ben He University of Glasgow, Glasgow, UK

Thomas Heinis Imperial College London, London, UK

Pat Helland Microsoft Corporation, Redmond, WA, USA

Joseph M. Hellerstein University of California-Berkeley, Berkeley, CA, USA

Jean Henrard University of Namur, Namur, Belgium

John Herring Oracle USA Inc, Nashua, NH, USA

Nicolas Hervé INRIA Paris-Rocquencourt, Le Chesnay, France

Marcus Herzog Vienna University of Technology, Vienna, Austria

Jean-Marc Hick University of Namur, Namur, Belgium

Jan Hidders University of Antwerp, Antwerpen, Belgium

Djoerd Hiemstra University of Twente, Enschede, The Netherlands

Linda L. Hill University of California-Santa Barbara, Santa Barbara, CA, USA

Alexander Hinneburg Institute of Computer Science, Martin-Luther-University Halle-Wittenberg, Halle/Saale, Germany

Hans Hinterberger Department of Computer Science, ETH Zurich, Zurich, Switzerland

Howard Ho IBM Almaden Research Center, San Jose, CA, USA

Erik Hoel Environmental Systems Research Institute, Redlands, CA, USA

Vasant Honavar Iowa State University, Ames, IA, USA

Mingsheng Hong Cornell University, Ithaca, NY, USA

Katja Hose Department of Computer Science, Aalborg University, Aalborg, Denmark

Haruo Hosoya The University of Tokyo, Tokyo, Japan

Vagelis Hristidis Department of Computer Science and Engineering, University of California, Riverside, Riverside, CA, USA

Wynne Hsu National University of Singapore, Singapore, Singapore

Yu-Ling Hsueh Computer Science and Information Engineering Department, National Chung Cheng University, Taiwan, Republic of China

Jian Hu Microsoft Research Asia, Haidian, China

Kien A. Hua University of Central Florida, Orlando, FL, USA

Xian-Sheng Hua Microsoft Research Asia, Beijing, China

Jun Huan University of Kansas, Lawrence, KS, USA

Haoda Huang Microsoft Research Asia, Beijing, China

Michael Huggett University of British Columbia, Vancouver, BC, Canada

Patrick C. K. Hung University of Ontario Institute of Technology, Oshawa, ON, Canada

Jeong-Hyon Hwang Department of Computer Science, University at Albany – State University of New York, Albany, NY, USA

Noha Ibrahim Grenoble Informatics Laboratory (LIG), Grenoble, France

Ichiro Ide Graduate School of Informatics, Nagoya University, Nagoya, Aichi, Japan

Sergio Ilarri University of Zaragoza, Zaragoza, Spain

Ihab F. Ilyas Cheriton School of Computer Science, University of Waterloo, Waterloo, ON, Canada

Alfred Inselberg Tel Aviv University, Tel Aviv, Israel

Yannis Ioannidis University of Athens, Athens, Greece

Ekaterini Ioannou Faculty of Pure and Applied Sciences, Open University of Cyprus, Nicosia, Cyprus

Panagiotis G. Ipeirotis New York University, New York, NY, USA

Zachary G. Ives Computer and Information Science Department, University of Pennsylvania, Philadelphia, PA, USA

Hans-Arno Jacobsen Department of Electrical and Computer Engineering, University of Toronto, Toronto, ON, Canada

H. V. Jagadish University of Michigan, Ann Arbor, MI, USA

Alejandro Jaimes Telefonica R&D, Madrid, Spain

Ramesh Jain University of California, Irvine, CA, USA

Sushil Jajodia George Mason University, Fairfax, VA, USA

Greg Janée University of California-Santa Barbara, Santa Barbara, CA, USA

Kalervo Järvelin University of Tampere, Tampere, Finland

Christian S. Jensen Department of Computer Science, Aalborg University, Aalborg, Denmark

Eric C. Jensen Twitter, Inc., San Francisco, CA, USA

Manfred Jeusfeld IIT, University of Skövde, Skövde, Sweden

Aura Frames, New York City, NY, USA

Heng Ji New York University, New York, NY, USA

Zhe Jiang University of Alabama, Tuscaloosa, AL, USA

Ricardo Jiménez-Peris Distributed Systems Lab, Universidad Politecnica de Madrid, Madrid, Spain

Hai Jin Service Computing Technology and System Lab, Cluster and Grid Computing Lab, School of Computer Science and Technology, Huazhong University of Science and Technology, Wuhan, China

Jiashun Jin Carnegie Mellon University, Pittsburgh, PA, USA

Ruoming Jin Department of Computer Science, Kent State University, Kent, OH, USA

Ryan Johnson Carnegie Mellon University, Pittsburg, PA, USA

Theodore Johnson AT&T Labs – Research, Florham Park, NJ, USA

Christopher B. Jones Cardiff University, Cardiff, UK

Rosie Jones Yahoo! Research, Burbank, CA, USA

James B. D. Joshi University of Pittsburgh, Pittsburgh, PA, USA

Vanja Josifovski Uppsala University, Uppsala, Sweden

Marko Junkkari University of Tampere, Tampere, Finland

Jan Jurjens The Open University, Buckinghamshire, UK

Mouna Kacimi Max-Planck Institute for Informatics, Saarbrücken, Germany

Tamer Kahveci University of Florida, Gainesville, FL, USA

Panos Kalnis National University of Singapore, Singapore, Singapore

Jaap Kamps University of Amsterdam, Amsterdam, The Netherlands

James Kang University of Minnesota, Minneapolis, MN, USA

Carl-Christian Kanne University of Mannheim, Mannheim, Germany

Aman Kansal Microsoft Research, Redmond, WA, USA

Murat Kantarcıoğlu University of Texas at Dallas, Richardson, TX, USA

Ben Kao Department of Computer Science, The University of Hong Kong, Hong Kong, China

George Karabatis University of Maryland, Baltimore Country (UMBC), Baltimore, MD, USA

Grigoris Karvounarakis LogicBlox, Atlanta, GA, USA

George Karypis University of Minnesota, Minneapolis, MN, USA

Vipul Kashyap Clinical Programs, CIGNA Healthcare, Bloomfield, CT, USA

Yannis Katsis University of California-San Diego, La Jolla, CA, USA

Raghav Kaushik Microsoft Research, Redmond, WA, USA

Gabriella Kazai Microsoft Research Cambridge, Cambridge, UK

Daniel A. Keim Computer Science Department, University of Konstanz, Konstanz, Germany

Jaana Kekäläinen University of Tampere, Tampere, Finland

Anastasios Kementsietsidis IBM T.J. Watson Research Center, Hawthorne, NY, USA

Bettina Kemme School of Computer Science, McGill University, Montreal, QC, Canada

Jessie Kennedy Napier University, Edinburgh, UK

Vijay Khatri Operations and Decision Technologies Department, Kelley School of Business, Indiana University, Bloomington, IN, USA

Ashfaq Khokhar University of Illinois at Chicago, Chicago, IL, USA

Daniel Kifer Yahoo! Research, Santa Clara, CA, USA

Stephen Kimani Director ICSIT, Jomo Kenyatta University of Agriculture and Technology (JKUAT), Juja, Kenya

Sofia Kleisarchaki CNRS, Univ. Grenoble Alps, Grenoble, France

Craig A. Knoblock University of Southern California, Marina del Rey, Los Angeles, CA, USA

Christoph Koch Cornell University, Ithaca, New York, NY, USA

EPFL, Lausanne, Switzerland

Solmaz Kolahi University of British Columbia, Vancouver, BC, Canada

George Kollios Boston University, Boston, MA, USA

Christian Koncilia Institute of Informatics-Systems, University of Klagenfurt, Klagenfurt, Austria

Roberto Konow Department of Computer Science, University of Chile, Santiago, Chile

Marijn Koolen Research and Development, Huygens ING, Royal Netherlands Academy of Arts and Sciences, Amsterdam, The Netherlands

David Koop University of Massachusetts Dartmouth, Dartmouth, MA, USA

Poon Wei Koot Nanyang Technological University, Singapore, Singapore

Julius Köpke Department of Informatics-Systems, Alpen-Adria-Universität Klagenfurt, Klagenfurt, Austria

Flip R. Korn AT&T Labs–Research, Florham Park, NJ, USA

Harald Kosch University of Passau, Passau, Germany

Cartik R. Kothari Biomedical Informatics, Ohio State University, College of Medicine, Columbus, OH, USA

Yannis Kotidis Department of Informatics, Athens University of Economics and Business, Athens, Greece

Spyros Kotoulas VU University Amsterdam, Amsterdam, The Netherlands

Manolis Koubarakis University of Athens, Athens, Greece

Konstantinos Koutroumbas Institute for Space Applications and Remote Sensing, Athens, Greece

Bernd J. Krämer University of Hagen, Hagen, Germany

Tim Kraska Department of Computer Science, Brown University, Providence, RI, USA

Werner Kriechbaum IBM Development Lab, Böblingen, Germany

Hans-Peter Kriegel Ludwig-Maximilians-University, Munich, Germany

Chandra Krintz Department of Computer Science, University of California, Santa Barbara, CA, USA

Rajasekar Krishnamurthy IBM Almaden Research Center, San Jose, CA, USA

Peer Kröger Ludwig-Maximilians-Universität München, Munich, Germany

Thomas Kühne School of Engineering and Computer Science, Victoria University of Wellington, Wellington, New Zealand

Krishna Kulkarni Independent Consultant, San Jose, CA, USA

Ravi Kumar Yahoo Research, Santa Clara, CA, USA

Nicholas Kushmerick VMWare, Seattle, WA, USA

Alan G. Labouseur School of Computer Science and Mathematics, Marist College, Poughkeepsie, NY, USA

Alexandros Labrinidis Department of Computer Science, University of Pittsburgh, Pittsburgh, PA, USA

Zoé Lacroix Arizona State University, Tempe, AZ, USA

Alberto H. F. Laender Federal University of Minas Gerais, Belo Horizonte, Brazil

Bibudh Lahiri Iowa State University, Ames, IA, USA

Laks V. S. Lakshmanan University of British Columbia, Vancouver, BC, Canada

Mounia Lalmas Yahoo! Inc., London, UK

Lea Landucci University of Florence, Florence, Italy

Birger Larsen Royal School of Library and Information Science, Copenhagen, Denmark

Mary Lynette Larsgaard University of California-Santa Barbara, Santa Barbara, CA, USA

Per-Åke Larson Microsoft Corporation, Redmond, WA, USA

Robert Laurini INSA-Lyon, University of Lyon, Lyon, France

LIRIS, INSA-Lyon, Lyon, France

Georg Lausen University of Freiburg, Freiburg, Germany

Jens Lechtenbörger University of Münster, Münster, Germany

Thierry Lecroq Université de Rouen, Rouen, France

Dongwon Lee The Pennsylvania State University, Park, PA, USA

Victor E. Lee John Carroll University, University Heights, OH, USA

Yang W. Lee College of Business Administration, Northeastern University, Boston, MA, USA

Pieter De Leenheer Vrije Universiteit Brussel, Collibra NV, Brussels, Belgium

Wolfgang Lehner Dresden University of Technology, Dresden, Germany

Domenico Lembo Dip. di Ingegneria Informatica Automatica e Gestionale Antonio Ruberti, Sapienza Università di Roma, Rome, Italy

Ronny Lempel Yahoo! Research, Haifa, Israel

Maurizio Lenzerini Dip. di Ingegneria Informatica Automatica e Gestionale Antonio Ruberti, Sapienza Università di Roma, Rome, Italy

Kristina Lerman University of Southern California, Marina del Rey, Los Angeles, CA, USA

Ulf Leser Humboldt University of Berlin, Berlin, Germany

Carson Kai-Sang Leung Department of Computer Science, University of Manitoba, Winnipeg, MB, Canada

Mariano Leva Dipartimento di Ingegneria Informatica, Automatica e Gestionale "A.Ruberti", Sapienza – Università di Roma, Roma, Italy

Stefano Levialdi Sapienza University of Rome, Rome, Italy

Brian Levine University of Massachusetts, Amherst, MA, USA

Changqing Li Duke University, Durham, NC, USA

Chen Li University of California – Irvine, School of Information and Computer Sciences, Irvine, CA, USA

Chengkai Li University of Texas at Arlington, Arlington, TX, USA

Hua Li Microsoft Research Asia, Beijing, China

Jinyan Li Nanyang Technological University, Singapore, Singapore

Ninghui Li Purdue University, West Lafayette, IN, USA

Ping Li Cornell University, Ithaca, NY, USA

Qing Li City University of Hong Kong, Hong Kong, China

Xue Li The University of Queensland, Brisbane, QLD, Australia

Yunyao Li IBM Almaden Research Center, San Jose, CA, USA

Ying Li Cognitive People Solutions, IBM Human Resources, Armonk, NY, USA

Xiang Lian Department of Computer Science, Kent State University, Kent, OH, USA

Leonid Libkin School of Informatics, University of Edinburgh, Edinburgh, Scotland, UK

Sam S. Lightstone IBM Canada Ltd, Markham, ON, Canada

Jimmy Lin University of Maryland, College Park, MD, USA

Tsau Young Lin Department of Computer Science, San Jose State University, San Jose, CA, USA

Xuemin Lin University of New South Wales, Sydney, NSW, Australia

Tok Wang Ling National University of Singapore, Singapore, Singapore

Bing Liu University of Illinois at Chicago, Chicago, IL, USA

Danzhou Liu University of Central Florida, Orlando, FL, USA

Guimei Liu Institute for Infocomm Research, Singapore, Singapore

Huan Liu Data Mining and Machine Learning Lab, School of Computing, Informatics, and Decision Systems Engineering, Arizona State University, Tempe, AZ, USA

Jinze Liu University of Kentucky, Lexington, KY, USA

Lin Liu Department of Computer Science, Kent State University, Kent, OH, USA

Ning Liu Microsoft Research Asia, Beijing, China

Qing Liu CSIRO, Hobart, TAS, Australia

Xiangyu Liu Xiamen University, Xiamen, China

Vebjorn Ljosa Broad Institute of MIT and Harvard, Cambridge, MA, USA

David Lomet Microsoft Research, Redmond, WA, USA

Cheng Long School of Electronics, Electrical Engineering and Computer Science, Queen's University Belfast, Kowloon, Hong Kong

Boon Thau Loo ETH Zurich, Zurich, Switzerland

Phillip Lord Newcastle University, Newcastle-Upon-Tyne, UK

Nikos A. Lorentzos Informatics Laboratory, Department of Agricultural Economics and Rural Development, Agricultural University of Athens, Athens, Greece

Lie Lu Microsoft Research Asia, Beijing, China

Bertram Ludäscher University of California-Davis, Davis, CA, USA

Yan Luo University of Illinois at Chicago, Chicago, IL, USA

Yves A. Lussier University of Chicago, Chicago, IL, USA

Ioanna Lykourentzou CRP Henri Tudor, Esch-sur-Alzette, Luxembourg

Craig MacDonald University of Glasgow, Glasgow, UK

Ashwin Machanavajjhala Cornell University, Ithaca, NY, USA

Samuel Madden Massachusetts Institute of Technology, Cambridge, MA, USA

Paola Magillo University of Genova, Genoa, Italy

Ahmed R. Mahmood Computer Science, Purdue University, West Lafayette, IN, USA

David Maier Portland State University, Portland, OR, USA

Ratul kr. Majumdar Department of Computer Science and Engineering, Indian Institute of Technology Bombay, Mumbai, India

Jan Małuszyński Linköping University, Linköping, Sweden

Nikos Mamoulis University of Hong Kong, Hong Kong, China

Stefan Manegold CWI, Amsterdam, The Netherlands

Murali Mani Worcester Polytechnic, Worcester, MA, USA

Serge Mankovski CA Labs, CA Inc., Thornhill, ON, Canada

Ioana Manolescu INRIA Saclay–Île de France, Orsay, France

Yannis Manolopoulos Aristotle University of Thessaloniki, Thessaloniki, Greece

Florian Mansmann University of Konstanz, Konstanz, Germany

Svetlana Mansmann University of Konstanz, Konstanz, Germany

Shahar Maoz The Weizmann Institute of Science, Rehovot, Israel

Patrick Marcel Département Informatique, Laboratoire d'Informatique, Université François Rabelais Tours, Blois, France

Amélie Marian Computer Science Department, Rutgers University, New Brunswick, NJ, USA

Volker Markl IBM Almaden Research Center, San Jose, CA, USA

David Martin Nuance Communications, Sunnyvale, CA, USA

Maria Vanina Martinez University of Maryland, College Park, MD, USA

Maristella Matera Politecnico di Milano, Milan, Italy

Michael Mathioudakis Université de Lyon, CNRS, INSA-Lyon, LIRIS, UMR5205, F-69621, France

Marta Mattoso Federal University of Rio de Janeiro, Rio de Janeiro, Brazil

Andrea Maurino University of Milano-Bicocca, Milan, Italy

Jose-Norberto Mazón University of Alicante, Alicante, Spain

John McCloud CERT Insider Threat Center, Software Engineering Institute, Carnegie Mellon University, Pittsburgh, PA, USA

Kevin S. McCurley Google Research, Mountain View, CA, USA

Andrew McGregor Microsoft Research, Silicon Valley, Mountain View, CA, USA

Timothy McPhillips University of California-Davis, Davis, CA, USA

Massimo Mecella Dipartimento di Ingegneria Informatica, Automatica e Gestionale "A.Ruberti", Sapienza – Università di Roma, Roma, Italy

Brahim Medjahed The University of Michigan–Dearborn, Dearborn, MI, USA

Carlo Meghini The Italian National Research Council, Pisa, Italy

Tao Mei Microsoft Research Asia, Beijing, China

Jonas Mellin University of Skövde, The Informatics Research Centre, Skövde, Sweden

University of Skövde, School of Informatics, Skövde, Sweden

Massimo Melucci University of Padua, Padua, Italy

Niccolò Meneghetti Computer Science and Engineering Department, University at Buffalo, Buffalo, NY, USA

Weiyi Meng Department of Computer Science, State University of New York at Binghamton, Binghamton, NY, USA

Ahmed Metwally LinkedIn Corp., Mountain View, CA, USA

Jan Michels Oracle Corporation, Redwood Shores, CA, USA

Gerome Miklau University of Massachusetts, Amherst, MA, USA

Alessandra Mileo Insight Centre for Data Analytics, Dublin City University, Dublin, Ireland

Harvey J. Miller University of Utah, Salt Lake City, UT, USA

Renée J. Miller Department of Computer Science, University of Toronto, Toronto, ON, Canada

Tova Milo School of Computer Science, Tel Aviv University, Tel Aviv, Israel

Umar Farooq Minhas Microsoft Research, Redmond, WA, USA

Paolo Missier School of Computing Science, Newcastle University, Newcastle upon Tyne, UK

Prasenjit Mitra The Pennsylvania State University, University Park, PA, USA

Michael Mitzenmacher Harvard University, Boston, MA, USA

Mukesh Mohania IBM Research, Melbourne, VIC, Australia

Mohamed F. Mokbel Department of Computer Science and Engineering, University of Minnesota-Twin Cities, Minneapolis, MN, USA

Angelo Montanari University of Udine, Udine, Italy

Reagan W. Moore School of Information and Library Science, University of North Carolina at Chapel Hill, Chapel Hill, NC, USA

Konstantinos Morfonios Oracle, Redwood City, CA, USA

Peter Mork The MITRE Corporation, McLean, VA, USA

Mirella M. Moro Departamento de Ciencia da Computaçao, Universidade Federal de Minas Gerais – UFMG, Belo Horizonte, MG, Brazil

Kyriakos Mouratidis Singapore Management University, Singapore, Singapore

Kamesh Munagala Duke University, Durham, NC, USA

Ethan V. Munson Department of EECS, University of Wisconsin-Milwaukee, Milwaukee, WI, USA

Shawn Murphy Massachusetts General Hospital, Boston, MA, USA

John Mylopoulos Department of Computer Science, University of Toronto, Toronto, ON, Canada

Marta Patiño-Martínez Distributed Systems Lab, Universidad Politecnica de Madrid, Madrid, Spain

ETSI Informáticos, Universidad Politécnica de Madrid (UPM), Madrid, Spain

Frank Nack University of Amsterdam, Amsterdam, The Netherlands

Marc Najork Google, Inc., Mountain View, CA, USA

Ullas Nambiar Zensar Technologies Ltd, Pune, India

Alexandros Nanopoulos Aristotle University, Thessaloniki, Greece

Vivek Narasayya Microsoft Corporation, Redmond, WA, USA

Mario A. Nascimento Department of Computing Science, University of Alberta, Edmonton, AB, Canada

Alan Nash Aleph One LLC, La Jolla, CA, USA

Harald Naumann Vienna University of Technology, Vienna, Austria

Gonzalo Navarro Department of Computer Science, University of Chile, Santiago, Chile

Wolfgang Nejdl L3S Research Center, University of Hannover, Hannover, Germany

Thomas Neumann Max-Planck Institute for Informatics, Saarbrücken, Germany

Bernd Neumayr Department for Business Informatics – Data and Knowledge Engineering, Johannes Kepler University Linz, Linz, Austria

Frank Neven Hasselt University and Transnational University of Limburg, Diepenbeek, Belgium

Chong-Wah Ngo City University of Hong Kong, Hong Kong, China

Peter Niblett IBM United Kingdom Limited, Winchester, UK

Naoko Nitta Osaka University, Osaka, Japan

Igor Nitto Department of Computer Science, University of Pisa, Pisa, Italy

Cheng Niu Microsoft Research Asia, Beijing, China

Vilém Novák Institute for Research and Applications of Fuzzy Modeling, University of Ostrava, Ostrava, Czech Republic

Chimezie Ogbuji Cleveland Clinic Foundation, Cleveland, OH, USA

Peter Øhrstrøm Aalborg University, Aalborg, Denmark

Christine M. O'Keefe CSIRO Preventative Health National Research Flagship, Acton, ACT, Australia

Paul W. Olsen Department of Computer Science, The College of Saint Rose, Albany, NY, USA

Dan Olteanu Department of Computer Science, University of Oxford, Oxford, UK

Behrooz Omidvar-Tehrani Interactive Data Systems Group, Ohio State University, Columbus, OH, USA

Patrick O'Neil University of Massachusetts, Boston, MA, USA

Beng Chin Ooi School of Computing, National University of Singapore, Singapore, Singapore

Iadh Ounis University of Glasgow, Glasgow, UK

Mourad Ouzzani Qatar Computing Research Institute, HBKU, Doha, Qatar

Fatma Özcan IBM Research – Almaden, San Jose, CA, USA

M. Tamer Özsu Cheriton School of Computer Science, University of Waterloo, Waterloo, ON, Canada

Esther Pacitti INRIA and LINA, University of Nantes, Nantes, France

Chris D. Paice Lancaster University, Lancaster, UK

Noël de Palma INPG – INRIA, Grenoble, France

Nathaniel Palmer Workflow Management Coalition, Hingham, MA, USA

Themis Palpanas Paris Descartes University, Paris, France

Biswanath Panda Cornell University, Ithaca, NY, USA

Ippokratis Pandis Carnegie Mellon University, Pittsburgh, PA, USA

Amazon Web Services, Seattle, WA, USA

Dimitris Papadias Department of Computer Science and Engineering, Hong Kong University of Science and Technology, Kowloon, Hong Kong, Hong Kong

Spiros Papadimitriou IBM T.J. Watson Research Center, Hawthorne, NY, USA

Apostolos N. Papadopoulos Aristotle University of Thessaloniki, Thessaloniki, Greece

Yannis Papakonstantinou University of California-San Diego, La Jolla, CA, USA

Jan Paredaens University of Antwerp, Antwerpen, Belgium

Christine Parent University of Lausanne, Lausanne, Switzerland

Josiane Xavier Parreira Siemens AG, Galway, Austria

Gabriella Pasi Department of Informatics, Systems and Communication, University of Milano-Bicocca, Milan, Italy

Chintan Patel Columbia University, New York, NY, USA

Jignesh M. Patel University of Wisconsin-Madison, Madison, WI, USA

Norman W. Paton University of Manchester, Manchester, UK

Cesare Pautasso University of Lugano, Lugano, Switzerland

Torben Bach Pedersen Department of Computer Science, Aalborg University, Aalborg, Denmark

Fernando Pedone Università della Svizzera Italiana (USI), Lugano, Switzerland

Jovan Pehcevski INRIA Paris-Rocquencourt, Le Chesnay Cedex, France

Jian Pei School of Computing Science, Simon Fraser University, Burnaby, BC, Canada

Ronald Peikert ETH Zurich, Zurich, Switzerland

Mor Peleg Department of Information Systems, University of Haifa, Haifa, Israel

Fuchun Peng Yahoo! Inc., Sunnyvale, CA, USA

Peng Peng Alibaba, Yu Hang District, Hangzhou, China

Liam Peyton University of Ottawa, Ottawa, ON, Canada

Dieter Pfoser Department of Geography and Geoinformation Science, George Mason University, Fairfax, VA, USA

Danh Le Phuoc Open Distributed Systems, Technical University of Berlin, Berlin, Germany

Mario Piattini University of Castilla-La Mancha, Ciudad Real, Spain

Benjamin C. Pierce University of Pennsylvania, Philadelphia, PA, USA

Karen Pinel-Sauvagnat IRIT laboratory, University of Toulouse, Toulouse, France

Leo L. Pipino University of Massachusetts, Lowell, MA, USA

Peter Pirolli Palo Alto Research Center, Palo Alto, CA, USA

Evaggelia Pitoura Department of Computer Science and Engineering, University of Ioannina, Ioannina, Greece

Benjamin Piwowarski University of Glasgow, Glasgow, UK

Vassilis Plachouras Yahoo! Research, Barcelona, Spain

Catherine Plaisant University of Maryland, College Park, MD, USA

Claudia Plant University of Vienna, Vienna, Austria

Christian Platzer Technical University of Vienna, Vienna, Austria

Dimitris Plexousakis Foundation for Research and Technology-Hellas (FORTH), Heraklion, Greece

Neoklis Polyzotis University of California Santa Cruz, Santa Cruz, CA, USA

Raymond K. Pon University of California, Los Angeles, CA, USA

Lucian Popa IBM Almaden Research Center, San Jose, CA, USA

Alexandra Poulovassilis University of London, London, UK

Sunil Prabhakar Purdue University, West Lafayette, IN, USA

Cecilia M. Procopiuc AT&T Labs, Florham Park, NJ, USA

Enrico Puppo Department of Informatics, Bioengineering, Robotics and Systems Engineering, University of Genova, Genoa, Italy

Ross S. Purves University of Zurich, Zurich, Switzerland

Vivien Quéma CNRS, INRIA, Saint-Ismier Cedex, France

Christoph Quix RWTH Aachen University, Aachen, Germany

Sriram Raghavan IBM Almaden Research Center, San Jose, CA, USA

Erhard Rahm University of Leipzig, Leipzig, Germany

Habibur Rahman Department of Computer Science and Engineering, University of Texas at Arlington, Arlington, TX, USA

Krithi Ramamritham Department of Computer Science and Engineering, Indian Institute of Technology Bombay, Mumbai, India

Maya Ramanath Max-Planck Institute for Informatics, Saarbrücken, Germany

Georgina Ramírez Yahoo! Research Barcelona, Barcelona, Spain

Edie Rasmussen Library, Archival and Information Studies, The University of British Columbia, Vancouver, BC, Canada

Indrakshi Ray Colorado State University, Fort Collins, CO, USA

Colin R. Reeves Coventry University, Coventry, UK

Payam Refaeilzadeh Google Inc., Los Angeles, CA, USA

D. R. Reforgiato University of Maryland, College Park, MD, USA

Bernd Reiner Technical University of Munich, Munich, Germany

Frederick Reiss IBM Almaden Research Center, San Jose, CA, USA

Harald Reiterer University of Konstanz, Constance, Germany

Matthias Renz Ludwig-Maximilians-Universität München, Munich, Germany

Andreas Reuter Heidelberg Laureate Forum Foundation, Schloss-Wolfsbrunnenweg 33, Heidelberg, Germany

Peter Revesz University of Nebraska-Lincoln, Lincoln, NE, USA

Mirek Riedewald Cornell University, Ithaca, NY, USA

Rami Rifaieh University of California-San Diego, San Diego, CA, USA

Stefanie Rinderle-Ma University of Vienna, Vienna, Austria

Tore Risch Department of Information Technology, Uppsala University, Uppsala, Sweden

Thomas Rist University of Applied Sciences, Augsburg, Germany

Stefano Rizzi DISI, University of Bologna, Bologna, Italy

Stephen Robertson Microsoft Research Cambridge, Cambridge, UK

Roberto A. Rocha Partners eCare, Partners HealthCare System, Wellesley, MA, USA

John F. Roddick Flinders University, Adelaide, SA, Australia

Thomas Roelleke Queen Mary University of London, London, UK

Didier Roland University of Namur, Namur, Belgium

Oscar Romero Polytechnic University of Catalonia, Barcelona, Spain

Rafael Romero University of Alicante, Alicante, Spain

Riccardo Rosati Dip. di Ingegneria Informatica Automatica e Gestionale Antonio Ruberti, Sapienza Università di Roma, Rome, Italy

Timothy Roscoe ETH Zurich, Zurich, Switzerland

Kenneth A. Ross Columbia University, New York, NY, USA

Prasan Roy Sclera, Inc., Walnut, CA, USA

Senjuti Basu Roy Department of Computer Science, New Jersey Institute of Technology, Tacoma, WA, USA

Sudeepa Roy Department of Computer Science, Duke University, Durham, NC, USA

Yong Rui Microsoft China R&D Group, Redmond, WA, USA

Dan Russler Oracle Health Sciences, Redwood Shores, CA, USA

Georgia Tech Research Institute, Atlanta, Georgia, USA

Michael Rys Microsoft Corporation, Sammamish, WA, USA

Giovanni Maria Sacco Dipartimento di Informatica, Università di Torino, Torino, Italy

Tetsuya Sakai Waseda University, Tokyo, Japan

Kenneth Salem University of Waterloo, Waterloo, ON, Canada

Simonas Šaltenis Aalborg University, Aalborg, Denmark

George Samaras University of Cyprus, Nicosia, Cyprus

Giuseppe Santucci University of Rome, Rome, Italy

Maria Luisa Sapino University of Turin, Turin, Italy

Sunita Sarawagi IIT Bombay, Mumbai, India

Anatol Sargin University of Augsburg, Augsburg, Germany

Mohamed Sarwat School of Computing, Informatics, and Decision Systems Engineering, Arizona State University, Tempe, AZ, USA

Kai-Uwe Sattler Technische Universität Ilmenau, Ilmenau, Germany

Monica Scannapieco University of Rome, Rome, Italy

Matthias Schäfer University of Konstanz, Konstanz, Germany

Sebastian Schaffert Salzburg Research, Salzburg, Austria

Ralf Schenkel Campus II Department IV – Computer Science, Professorship for databases and information systems, University of Trier, Trier, Germany

Raimondo Schettini University of Milano-Bicocca, Milan, Italy

Peter Scheuermann Department of ECpE, Iowa State University, Ames, IA, USA

Ulrich Schiel Federal University of Campina Grande, Campina Grande, Brazil

Markus Schneider University of Florida, Gainesville, FL, USA

Marc H. Scholl University of Konstanz, Konstanz, Germany

Michel Scholl Cedric-CNAM, Paris, France

Tobias Schreck Department of Computer Science and Biomedical Engineering, Institute of Computer Graphics and Knowledge Visualization, Graz University of Technology, Graz, Austria

Michael Schrefl University of Linz, Linz, Austria

Erich Schubert Heidelberg University, Heidelberg, Germany

Matthias Schubert Ludwig-Maximilians-University, Munich, Germany

Christoph G. Schuetz Department for Business Informatics – Data and Knowledge Engineering, Johannes Kepler University Linz, Linz, Austria

Heiko Schuldt Department of Mathematics and Computer Science, Databases and Information Systems Research Group, University of Basel, Basel, Switzerland

Heidrun Schumann University of Rostock, Rostock, Germany

Felix Schwagereit University of Koblenz-Landau, Koblenz, Germany

Nicole Schweikardt Johann Wolfgang Goethe-University, Frankfurt am Main, Frankfurt, Germany

Fabrizio Sebastiani Qatar Computing Research Institute, Doha, Qatar

Nicu Sebe University of Amsterdam, Amsterdam, Netherlands

Monica Sebillo University of Salerno, Salerno, Italy

Thomas Seidl RWTH Aachen University, Aachen, Germany

Manuel Serrano University of Alicante, Alicante, Spain

Amnon Shabo (Shvo) University of Haifa, Haifa, Israel

Mehul A. Shah Amazon Web Services (AWS), Seattle, WA, USA

Nigam Shah Stanford University, Stanford, CA, USA

Cyrus Shahabi University of Southern California, Los Angeles, CA, USA

Jayavel Shanmugasundaram Yahoo Research!, Santa Clara, NY, USA

Marc Shapiro Inria Paris, Paris, France

Sorbonne-Universités-UPMC-LIP6, Paris, France

Mohamed Sharaf Electrical and Computer Engineering, University of Toronto, Toronto, ON, Canada

Mehdi Sharifzadeh Google, Santa Monica, CA, USA

Jayant Sharma Oracle USA Inc, Nashua, NH, USA

Guy Sharon IBM Research Labs-Haifa, Haifa, Israel

Dennis Shasha Department of Computer Science, New York University, New York, NY, USA

Shashi Shekhar Department of Computer Science, University of Minnesota, Minneapolis, MN, USA

Jialie Shen Singapore Management University, Singapore, Singapore

Xuehua Shen Google, Inc., Mountain View, CA, USA

Dou Shen Microsoft Corporation, Redmond, WA, USA

Baidu, Inc., Beijing City, China

Heng Tao Shen School of Information Technology and Electrical Engineering, The University of Queensland, Brisbane, QLD, Australia

University of Electronic Science and Technology of China, Chengdu, Sichuan Sheng, China

Rao Shen Yahoo!, Sunnyvale, CA, USA

Frank Y. Shih New Jersey Institute of Technology, Newark, NJ, USA

Arie Shoshani Lawrence Berkeley National Laboratory, Berkeley, CA, USA

Pavel Shvaiko University of Trento, Trento, Italy

Wolf Siberski L3S Research Center, University of Hannover, Hannover, Germany

Ronny Siebes VU University Amsterdam, Amsterdam, The Netherlands

Laurynas Šikšnys Department of Computer Science, Aalborg University, Aalborg, Denmark

Adam Silberstein Yahoo! Research Silicon Valley, Santa Clara, CA, USA

Fabrizio Silvestri Yahoo Inc, London, UK

Alkis Simitsis HP Labs, Palo Alto, CA, USA

Simeon J. Simoff University of Western Sydney, Sydney, NSW, Australia

Elena Simperl Electronics and Computer Science, University of Southampton, Southampton, UK

Radu Sion Stony Brook University, Stony Brook, NY, USA

Mike Sips Stanford University, Stanford, CA, USA

Cristina Sirangelo IRIF, Paris Diderot University, Paris, France

Yannis Sismanis IBM Almaden Research Center, Almaden, CA, USA

Hala Skaf-Molli Computer Science, University of Nantes, Nantes, France

Spiros Skiadopoulos University of Peloponnese, Tripoli, Greece

Richard T. Snodgrass Department of Computer Science, University of Arizona, Tucson, AZ, USA

Dataware Ventures, Tucson, AZ, USA

Cees Snoek University of Amsterdam, Amsterdam, The Netherlands

Mohamed A. Soliman Datometry Inc., San Francisco, CA, USA

Il-Yeol Song College of Computing and Informatics, Drexel University, Philadelphia, PA, USA

Ruihua Song Microsoft Research Asia, Beijing, China

Jingkuan Song Columbia University, New York, NY, USA

Stefano Spaccapietra EPFL, Lausanne, Switzerland

Greg Speegle Department of Computer Science, Baylor University, Waco, TX, USA

Padmini Srinivasan The University of Iowa, Iowa City, IA, USA

Venkat Srinivasan Virginia Tech, Blacksburg, VA, USA

Divesh Srivastava AT&T Labs – Research, AT&T, Bedminster, NJ, USA

Steffen Staab Institute for Web Science and Technologies – WeST, University of Koblenz-Landau, Koblenz, Germany

Constantine Stephanidis Foundation for Research and Technology-Hellas (FORTH), Heraklion, Greece

University of Crete, Heraklion, Greece

Robert Stevens University of Manchester, Manchester, UK

Andreas Stoffel University of Konstanz, Konstanz, Germany

Michael Stonebraker Massachusetts Institute of Technology, Cambridge, MA, USA

Umberto Straccia The Italian National Research Council, Pisa, Italy

Martin J. Strauss University of Michigan, Ann Arbor, MI, USA

Diane M. Strong Worcester Polytechnic Institute, Worcester, MA, USA

Jianwen Su University of California-Santa Barbara, Santa Barbara, CA, USA

Kazimierz Subieta Polish-Japanese Institute of Information Technology, Warsaw, Poland

V. S. Subrahmanian University of Maryland, College Park, MD, USA

Dan Suciu University of Washington, Seattle, WA, USA

S. Sudarshan Indian Institute of Technology, Bombay, India

Torsten Suel Yahoo! Research, Sunnyvale, CA, USA

Jian-Tao Sun Microsoft Research Asia, Beijing, China

Subhash Suri University of California-Santa Barbara, Santa Barbara, CA, USA

Jaroslaw Szlichta University of Ontario Institute of Technology, Oshawa, ON, Canada

Stefan Tai University of Karlsruhe, Karlsruhe, Germany

Kian-Lee Tan Department of Computer Science, National University of Singapore, Singapore, Singapore

Pang-Ning Tan Michigan State University, East Lansing, MI, USA

Wang-Chiew Tan University of California-Santa Cruz, Santa Cruz, CA, USA

Letizia Tanca Computer Science, Politecnico di Milano, Milan, Italy

Lei Tang Chief Data Scientist, Clari Inc., Sunnyvale, CA, USA

Wei Tang Teradata Corporation, El Segundo, CA, USA

Egemen Tanin Computing and Information Systems, University of Melbourne, Melbourne, VIC, Australia

Val Tannen Department of Computer and Information Science, University of Pennsylvania, Philadelphia, PA, USA

Abdullah Uz Tansel Baruch College, CUNY, New York, NY, USA

Yufei Tao Chinese University of Hong Kong, Hong Kong, China

Sandeep Tata IBM Almaden Research Center, San Jose, CA, USA

Nesime Tatbul Intel Labs and MIT, Cambridge, MA, USA

Christophe Taton INPG – INRIA, Grenoble, France

Behrooz Omidvar Tehrani Laboratoire d'Informatique de Grenoble, Saint-Martin d'Hères, France

Paolo Terenziani Dipartimento di Scienze e Innovazione Tecnologica (DiSIT), Università del Piemonte Orientale "Amedeo Avogadro", Alessandria, Italy

Alexandre Termier LIG (Laboratoire d'Informatique de Grenoble), HADAS team, Université Joseph Fourier, Saint Martin d'Hères, France

Evimaria Terzi Computer Science Department, Boston University, Boston, MA, USA

IBM Almaden Research Center, San Jose, CA, USA

Bernhard Thalheim Christian-Albrechts University, Kiel, Germany

Martin Theobald Institute of Databases and Information Systems (DBIS), Ulm University, Ulm, Germany

Stanford University, Stanford, CA, USA

Sergios Theodoridis University of Athens, Athens, Greece

Yannis Theodoridis University of Piraeus, Piraeus, Greece

Saravanan Thirumuruganathan Department of Computer Science and Engineering, University of Texas at Arlington, Arlington, TX, USA

Qatar Computing Research Institute, Hamad Bin Khalifa University, Doha, Qatar

Stephen W. Thomas Dataware Ventures, Kingston, ON, Canada

Alexander Thomasian Thomasian and Associates, Pleasantville, NY, USA

Christian Thomsen Department of Computer Science, Aalborg University, Aalborg, Denmark

Bhavani Thuraisingham The University of Texas at Dallas, Richardson, TX, USA

Srikanta Tirthapura Iowa State University, Ames, IA, USA

Wee Hyong Tok National University of Singapore, Singapore, Singapore

David Toman University of Waterloo, Waterloo, ON, Canada

Frank Tompa David R. Cheriton School of Computer Science, University of Waterloo, Waterloo, ON, Canada

Alejandro Z. Tomsic Sorbonne-Universités-UPMC-LIP6, Paris, France

Inria Paris, Paris, France

Rodney Topor Griffith University, Nathan, Australia

Riccardo Torlone University of Rome, Rome, Italy

Kristian Torp Aalborg University, Aalborg, Denmark

Nicola Torpei University of Florence, Florence, Italy

Nerius Tradišauskas Aalborg University, Aalborg, Denmark

Goce Trajcevski Department of ECpE, Iowa State University, Ames, IA, USA

Peter Triantafillou University of Patras, Rio, Patras, Greece

Silke Trißl Humboldt University of Berlin, Berlin, Germany

Andrew Trotman University of Otago, Dunedin, New Zealand

Juan Trujillo Lucentia Research Group, Department of Information Languages and Systems, Facultad de Informática, University of Alicante, Alicante, Spain

Beth Trushkowsky Department of Computer Science, Harvey Mudd College, Claremont, CA, USA

Panayiotis Tsaparas Department of Computer Science and Engineering, University of Ioannina, Ioannina, Greece

Theodora Tsikrika Center for Mathematics and Computer Science, Amsterdam, The Netherlands

Vassilis J. Tsotras University of California-Riverside, Riverside, CA, USA

Mikalai Tsytsarau University of Trento, Povo, Italy

Peter A. Tucker Whitworth University, Spokane, WA, USA

Anthony K. H. Tung National University of Singapore, Singapore, Singapore

Deepak Turaga IBM Research, San Francisco, CA, USA

Theodoros Tzouramanis University of the Aegean, Samos, Greece

Antti Ukkonen Helsinki University of Technology, Helsinki, Finland

Mollie Ullman-Cullere Harvard Medical School – Partners Healthcare Center for Genetics and Genomics, Boston, MA, USA

Ali Ünlü University of Augsburg, Augsburg, Germany

Antony Unwin Augsburg University, Augsburg, Germany

Susan D. Urban Arizona State University, Phoenix, AZ, USA

Jaideep Vaidya Rutgers University, Newark, NJ, USA

Alejandro A. Vaisman Instituto Tecnológico de Buenos Aires, Buenos Aires, Argentina

Shivakumar Vaithyanathan IBM Almaden Research Center, San Jose, CA, USA

Athena Vakali Aristotle University, Thessaloniki, Greece

Patrick Valduriez INRIA, LINA, Nantes, France

Maarten van Steen VU University, Amsterdam, The Netherlands

W. M. P. van der Aalst Eindhoven University of Technology, Eindhoven, The Netherlands

Christelle Vangenot EPFL, Lausanne, Switzerland

Stijn Vansummeren Hasselt University and Transnational University of Limburg, Diepenbeek, Belgium

Vasilis Vassalos Athens University of Economics and Business, Athens, Greece

Michael Vassilakopoulos University of Thessaly, Volos, Greece

Panos Vassiliadis University of Ioannina, Ioannina, Greece

Michalis Vazirgiannis Athens University of Economics and Business, Athens, Greece

Olga Vechtomova University of Waterloo, Waterloo, ON, Canada

Erik Vee Yahoo! Research, Silicon Valley, CA, USA

Jari Veijalainen University of Jyvaskyla, Jyvaskyla, Finland

Yannis Velegrakis Department of Information Engineering and Computer Science, University of Trento, Trento, Italy

Suresh Venkatasubramanian University of Utah, Salt Lake City, UT, USA

Rossano Venturini Department of Computer Science, University of Pisa, Pisa, Italy

Victor Vianu University of California-San Diego, La Jolla, CA, USA

Maria-Esther Vidal Computer Science, Universidad Simon Bolivar, Caracas, Venezuela

Millist Vincent University of South Australia, Adelaide, SA, Australia

Giuliana Vitiello University of Salerno, Salerno, Italy

Michail Vlachos IBM T.J. Watson Research Center, Hawthorne, NY, USA

Akrivi Vlachou Athena Research and Innovation Center, Institute for the Management of Information Systems, Athens, Greece

Hoang Vo Computer Science, Stony Brook University, Stony Brook, NY, USA

Hoang Tam Vo IBM Research, Melbourne, VIC, Australia

Agnès Voisard Fraunhofer Institute for Software and Systems Engineering (ISST), Berlin, Germany

Kaladhar Voruganti Advanced Development Group, Network Appliance, Sunnyvale, CA, USA

Gottfried Vossen Department of Information Systems, Westfälische Wilhelms-Universität, Münster, Germany

Daisy Zhe Wang Computer and Information Science and Engineering (CISE), University of Florida, Gainesville, FL, USA

Feng Wang City University of Hong Kong, Hong Kong, China

Fusheng Wang Stony Brook University, Stony Brook, NY, USA

Jianyong Wang Tsinghua University, Beijing, China

Jun Wang Queen Mary University of London, London, UK

Meng Wang Microsoft Research Asia, Beijing, China

X. Sean Wang School of Computer Science, Fudan University, Shanghai, China

Xin-Jing Wang Microsoft Research Asia, Beijing, China

Micros Facebook, CA, USA

Zhengkui Wang InfoComm Technology, Singapore Institute of Technology, Singapore, Singapore

Matthew O. Ward Worcester Polytechnic Institute, Worcester, MA, USA

Segev Wasserkrug IBM Research Labs-Haifa, Haifa, Israel

Hans Weda Phillips Research Europe, Eindhoven, The Netherlands

Gerhard Weikum Department 5: Databases and Information Systems, Max-Planck-Institut für Informatik, Saarbrücken, Germany

Michael Weiner Regenstrief Institute, Inc., Indiana University School of Medicine, Indianapolis, IN, USA

Michael Weiss Carleton University, Ottawa, ON, Canada

Ji-Rong Wen Microsoft Research Asia, Beijing, China

Chunhua Weng Columbia University, New York, NY, USA

Mathias Weske University of Potsdam, Potsdam, Germany

Thijs Westerveld Teezir Search Solutions, Ede, Netherlands

Till Westmann Oracle Labs, Redwood City, CA, USA

Karl Wiggisser Institute of Informatics-Systems, University of Klagenfurt, Klagenfurt, Austria

Jef Wijsen University of Mons, Mons, Belgium

Mark D. Wilkinson University of British Columbia, Vancouver, BC, Canada

Graham Wills SPSS Inc., Chicago, IL, USA

Ian H. Witten University of Waikato, Hamilton, New Zealand

Kent Wittenburg Mitsubishi Electric Research Laboratories, Inc., Cambridge, MA, USA

Eric Wohlstadter University of British Columbia, Vancouver, BC, Canada

Dietmar Wolfram University of Wisconsin-Milwaukee, Milwaukee, WI, USA

Ouri Wolfson Mobile Information Systems Center (MOBIS), The University of Illinois at Chicago, Chicago, IL, USA

Department of CS, University of Illinois at Chicago, Chicago, IL, USA

Janette Wong IBM Canada Ltd, Markham, ON, Canada

Raymond Chi-Wing Wong Department of Computer Science and Engineering, The Hong Kong University of Science and Technology, Clear Water Bay, Kowloon, Hong Kong

Peter T. Wood Birkbeck, University of London, London, UK

David Woodruff IBM Almaden Research Center, San Jose, CA, USA

Marcel Worring University of Amsterdam, Amsterdam, The Netherlands

Adam Wright Partners HealthCare, Boston, MA, USA

Sai Wu Zhejiang University, Hangzhou, Zhejiang, People's Republic of China

Yuqing Wu Indiana University, Bloomington, IN, USA

Alex Wun University of Toronto, Toronto, ON, Canada

Ming Xiong Bell Labs, Murray Hill, NJ, USA

Google, Inc., New York, NY, USA

Guandong Xu University of Technology Sydney, Sydney, Australia

Hua Xu Columbia University, New York, NY, USA

Jun Yan Microsoft Research Asia, Haidian, China

Xifeng Yan IBM T. J. Watson Research Center, Hawthorne, NY, USA

Jun Yang Duke University, Durham, NC, USA

Li Yang Western Michigan University, Kalamazoo, MI, USA

Ming-Hsuan Yang University of California at Merced, Merced, CA, USA

Seungwon Yang Virginia Tech, Blacksburg, VA, USA

Yang Yang Center for Future Media and School of Computer Science and Engineering, University of Electronic Science and Technology of China, Chengdu, Sichuan, China

Yun Yang Swinburne University of Technology, Melbourne, VIC, Australia

Yu Yang City University of Hong Kong, Hong Kong, China

Yong Yao Cornell University, Ithaca, NY, USA

Mikalai Yatskevich University of Trento, Trento, Italy

Xun Yi Computer Science and Info Tech, RMIT University, Melbourne, VIC, Australia

Hiroshi Yoshida VLSI Design and Education Center, University of Tokyo, Tokyo, Japan

Fujitsu Limited, Yokohama, Japan

Masatoshi Yoshikawa University of Kyoto, Kyoto, Japan

Matthew Young-Lai Sybase iAnywhere, Waterloo, ON, Canada

Google, Inc., Mountain View, CA, USA

Hwanjo Yu University of Iowa, Iowa City, IA, USA

Ting Yu North Carolina State University, Raleigh, NC, USA

Cong Yu Google Research, New York, NY, USA

Philip S. Yu Computer Science Department, University of Illinois at Chicago, Chicago, IL, USA

Jeffrey Xu Yu Department of Systems Engineering and Engineering Management, The Chinese University of Hong Kong, Hong Kong, China

Pingpeng Yuan Service Computing Technology and System Lab, Cluster and Grid Computing Lab, School of Computer Science and Technology, Huazhong University of Science and Technology, Wuhan, China

Vladimir Zadorozhny University of Pittsburgh, Pittsburgh, PA, USA

Matei Zaharia Douglas T. Ross Career Development Professor of Software Technology, MIT CSAIL, Cambridge, MA, USA

Ilya Zaihrayeu University of Trento, Trento, Italy

Mohammed J. Zaki Rensselaer Polytechnic Institute, Troy, NY, USA

Carlo Zaniolo University of California-Los Angeles, Los Angeles, CA, USA

Hugo Zaragoza Yahoo! Research, Barcelona, Spain

Stan Zdonik Brown University, Providence, RI, USA

Demetrios Zeinalipour-Yazti Department of Computer Science, Nicosia, Cyprus

Hans Zeller Hewlett-Packard Laboratories, Palo Alto, CA, USA

Pavel Zezula Masaryk University, Brno, Czech Republic

Cheng Xiang Zhai University of Illinois at Urbana-Champaign, Urbana, IL, USA

Aidong Zhang State University of New York, Buffalo, NY, USA

Benyu Zhang Microsoft Research Asia, Beijing, China

Donghui Zhang Paradigm4, Inc., Waltham, MA, USA

Dongxiang Zhang School of Computer Science and Engineering, University of Electronic Science and Technology of China, Sichuan, China

Ethan Zhang University of California, Santa Cruz, CA, USA

Jin Zhang University of Wisconsin Milwaukee, Milwaukee, WI, USA

Kun Zhang Xavier University of Louisiana, New Orleans, LA, USA

Lei Zhang Microsoft Research Asia, Beijing, China

Lei Zhang Microsoft Research, Redmond, WA, USA

Li Zhang Peking University, Beijing, China

Meihui Zhang Information Systems Technology and Design, Singapore University of Technology and Design, Singapore, Singapore

Qing Zhang The Australian e-health Research Center, Brisbane, Australia

Rui Zhang University of Melbourne, Melbourne, VIC, Australia

Dataware Ventures, Tucson, AZ, USA

Dataware Ventures, Redondo Beach, CA, USA

Yanchun Zhang Victoria University, Melbourne, VIC, Australia

Yi Zhang Yahoo! Inc,, Santa Clara, CA, USA

Yue Zhang University of Pittsburgh, Pittsburgh, PA, USA

Zhen Zhang University of Illinois at Urbana-Champaign, Urbana, IL, USA

Feng Zhao Microsoft Research, Redmond, WA, USA

Ying Zhao Tsinghua University, Beijing, China

Baihua Zheng Singapore Management University, Singapore, Singapore

Yi Zheng University of Ontario Institute of Technology, Oshawa, ON, Canada

Yu Zheng Data Management, Analytics and Services (DMAS) and Ubiquitous Computing Group (Ubicomp), Microsoft Research Asia, Beijing, China

Zhi-Hua Zhou National Key Lab for Novel Software Technology, Nanjing University, Nanjing, China

Jingren Zhou Alibaba Group, Hangzhou, China

Li Zhou Partners HealthCare System Inc., Boston, MA, USA

Xiaofang Zhou School of Information Technology and Electrical Engineering, University of Queensland, Brisbane, QLD, Australia

Huaiyu Zhu IBM Almaden Research Center, San Jose, CA, USA

Xiaofeng Zhu Guangxi Normal University, Guilin, Guangxi, People's Republic of China

Xingquan Zhu Florida Atlantic University, Boca Raton, FL, USA

Cai-Nicolas Ziegler Siemens AG, Munich, Germany

Hartmut Ziegler University of Konstanz, Konstanz, Germany

Esteban Zimányi CoDE, Université Libre de Bruxelles, Brussels, Belgium

Arthur Zimek Ludwig-Maximilians-Universität München, Munich, Germany

Department of Mathematics and Computer Science, University of Southern Denmark, Odense, Denmark

Roger Zimmermann Department of Computer Science, School of Computing, National University of Singapore, Singapore, Republic of Singapore

Lei Zou Institute of Computer Science and Technology, Peking University, Beijing, China

A

Absolute Time

Christian S. Jensen[1] and Richard T. Snodgrass[2,3]
[1]Department of Computer Science, Aalborg
University, Aalborg, Denmark
[2]Department of Computer Science, University of
Arizona, Tucson, AZ, USA
[3]Dataware Ventures, Tucson, AZ, USA

Definition

A temporal database contains time-referenced, or timestamped, facts. A time reference in such a database is *absolute* if its value is independent of the context, including the current time, *now*.

Key Points

An example is "Mary's salary was raised on March 30, 2007." The fact here is that Mary's salary was raised. The absolute time reference is March 30, 2007, which is a time instant at the granularity of day.

Another example is "Mary's monthly salary was $15,000 from January 1, 2006 to November 30, 2007." In this example, the absolute time reference is the time period [January 1, 2006–November 30, 2007].

Absolute time can be contrasted with *relative time*.

Cross-References

▶ Now in Temporal Databases
▶ Relative Time
▶ Time Instant
▶ Time Period
▶ Temporal Database
▶ Temporal Granularity

Recommended Reading

1. Bettini C, Dyreson CE, Evans WS, Snodgrass RT, Wang XS. The glossary of time granularity concepts. In: Jajodia S, Etzion O, Sripada S, editors. Temporal databases: research and practice. LNCS, vol. 1399. Berlin: Springer; 1998, p. 406–413.
2. Jensen CS, Dyreson CE, editors. The consensus glossary of temporal database concepts – February 1998 version. In: Etzion O, Jajodia S, Sripada S, editors. Temporal databases: research and practice. LNCS, vol. 1399. Berlin: Springer; 1998. p. 367–405.

Abstract Versus Concrete Temporal Query Languages

Jan Chomicki[1] and David Toman[2]
[1]Department of Computer Science and
Engineering, State University of New York at
Buffalo, Buffalo, NY, USA
[2]University of Waterloo, Waterloo, ON, Canada

Synonyms

Historical query languages

Definition

Temporal query languages are a family of
query languages designed to query (and access
in general) time-dependent information stored
in temporal databases. The languages are
commonly defined as extensions of standard
query languages for non-temporal databases
with *temporal features*. The additional features
reflect the way dependencies of data on time are
captured by and represented in the underlying
temporal data model.

Historical Background

Most databases store time-varying information.
On the other hand, SQL is often the language of
choice for developing applications that utilize the
information in these databases. Plain SQL, however, does not seem to provide adequate support
for temporal applications.

Example To represent the *employment histories*
of persons, a common relational design would use
a schema

   Employment (From Date, To Date, EID,
   Company)

with the intended meaning that a person identified
by *EID* worked for *Company* continuously from
FromDate to *ToDate*. Note that while the above
schema is a standard relational schema, the additional assumption that the values of the attributes
FromDate and *ToDate* represent *continuous periods* of time is itself *not* a part of the relational
model.

Formulating even simple queries over such a
schema is non-trivial. For example, the query
GAPS: "*List all persons with gaps in their employment history, together with the gaps*" leads to
a rather complex formulation in, e.g., SQL over
the above schema (this is left as a challenge to
readers who consider themselves SQL experts;
for a list of appealing, but incorrect solutions,
including the reasons why, see [9]) The difficulty
arises because a single tuple in the relation is
conceptually a *compact representation of a set of
tuples*, each tuple stating that an employment fact
was true on a particular day.

The tension between the conceptual abstract
temporal data model (in the example, the property
that employment facts are associated with individual *time instants*) and the need for an efficient
and compact representation of temporal data (in
the example, the representation of continuous
periods by their start and end instants) has been
reflected in the development of numerous temporal data models and temporal query languages [3].

Foundations

Temporal query languages are commonly defined using *temporal extensions* of existing non-
temporal query languages, such as relational calculus, relational algebra, or SQL. The temporal
extensions can be categorized in two, mostly
orthogonal, ways:

* *The choice of the actual temporal values manipulated by the language*. This choice is primarily determined by the underlying temporal
  data model. The model also determines the
  associated operations on these values. The
  meaning of temporal queries is then defined
  in terms of temporal values and operations on

them, and their interactions with *data* (non-temporal) values in a temporal database.

- *The choice of syntactic constructs to manipulate temporal values in the language.* This distinction determines whether the temporal values in the language are accessed and manipulated *explicitly*, in a way similar to other values stored in the database, or whether the access is *implicit*, based primarily on *temporally extending* the meaning of constructs that already exist in the underlying non-temporal language (while still using the operations defined by the temporal data model).

Additional design considerations relate to *compatibility* with existing query languages, e.g., the notion of temporal upward compatibility.

However, as illustrated above, an additional hurdle stems from the fact that many (early) temporal query languages allowed the users to manipulate a *finite underlying representation* of temporal databases rather than the actual temporal values/objects in the associated temporal data model. A typical example of this situation would be an approach in which the temporal data model is based on time instants, while the query language introduces interval-valued attributes. Such a discrepancy often leads to a complex and unintuitive semantics of queries.

In order to clarify this issue, Chomicki has introduced the notions of *abstract* and *concrete* temporal databases and query languages [2]. Intuitively, *abstract temporal query languages* are defined at the conceptual level of the temporal data model, while their *concrete* counterparts operate directly on an actual *compact encoding* of temporal databases. The relationship between abstract and concrete temporal query languages is also implicitly present in the notion of snapshot equivalence [7]. Moreover, Bettini et al. [1] proposed to distinguish between *explicit* and *implicit* information in a temporal database. The explicit information is stored in the database and used to derive the implicit information through *semantic assumptions*. Semantic assumptions related to fact persistence play a role similar

to mappings between concrete and abstract databases, while other assumptions are used to address time-granularity issues.

Abstract Temporal Query Languages

Most temporal query languages derived by temporally extending the relational calculus can be classified as abstract temporal query languages. Their semantics are defined in terms of abstract temporal databases which, in turn, are typically defined within the point-stamped temporal data model, in particular *without* any additional hidden assumptions about the meaning of tuples in instances of temporal relations.

Example The *employment histories* in an abstract temporal data model would most likely be captured by a simpler schema "*Employment(Date, EID, Company)*", with the intended meaning that a person identified by *EID* was working for *Company* on a particular *Date*. While instances of such a schema can potentially be very large (especially when a fine granularity of time is used), formulating queries is now much more natural.

Choosing abstract temporal query languages over concrete ones resolves the first design issue: the temporal values used by the former languages are time instants equipped with an appropriate temporal ordering (which is typically a linear order over the instants), and possibly other predicates such as temporal distance. The second design issue – access to temporal values – may be resolved in two different ways, as exemplified by two different query languages. They are as follows:

- Temporal Relational Calculus (TRC): a two-sorted first-order logic with variables and quantifiers explicitly ranging over the time and data domains.
- First-order Temporal Logic (FOTL): a language with an implicit access to timestamps using temporal connectives.

> **Example**
> The GAPS query is formulated as follows:TRC: $\exists t_1, t_3.t_1 < t_2 < t_3 \wedge \exists c.$ Employment $(t_1, x, c) \wedge (\neg \exists c.$ Employment $(t_2, x, c)) \wedge \exists c.$ Employment (t_3, x, c) FOTL: $\blacklozenge \exists c.$ Employment $(x, c) \wedge (\neg \exists c.$ Employment $(x, c)) \wedge \blacklozenge \exists c.$ Employment (x, c)

Here, the explicit access to temporal values (in TRC) using the variables t_1, t_2, and t_3 can be contrasted with the implicit access (in FOTL) using the temporal operators $\blacklozenge$ (read "sometime in the past") and $\lozenge$ (read "sometime in the future"). The conjunction in the FOTL query represents an implicit temporal join. The formulation in TRC leads immediately to an equivalent way of expressing the query in SQL/TP [9], an extension of SQL based on TRC.

Example The above query can be formulated in SQL/TP as follows:

```
SELECT t.Date, e1.EID
FROMEmployment e1, Time t,
                    Employment e2
WHERE e1.EID = e2.EID
AND e1.Date < e2.Date
  AND NOT EXISTS (SELECT *
    FROM Employment e3
    WHERE e1.EID = e3.EID
            AND t.Date = e3.Date
      AND e1.Date < e3.Date
            AND e3.Date < e2.Date)
```

The unary constant relation *Time* contains all time instants in the time domain (in our case, all *Date*s) and is only needed to fulfill syntactic SQL-style requirements on attribute ranges. However, despite the fact that the instance of this relation is not finite, the query can be efficiently evaluated [9].

Note also that in all of the above cases, the formulation is *exactly the same* as if the underlying temporal database used the *plain* relational model (allowing for attributes ranging over time instants).

The two languages, FOTL and TRC, are the counterparts of the snapshot and timestamp

models (cf. the entry ▶ Point-Stamped Temporal Models) and are the roots of many other temporal query languages, ranging from the more TRC-like temporal extensions of SQL to more FOTL-like temporal relational algebras (e.g., the conjunction in temporal logic directly corresponds to a temporal join in a temporal relational algebra, as both of them induce an *implicit equality* on the associated time attributes).

Temporal integrity constraints over point-stamped temporal databases can also be conveniently expressed in TRC or FOTL.

Multiple Temporal Dimensions and Complex Values

While the abstract temporal query languages are typically defined in terms of the point-based temporal data model, they can similarly be defined with respect to complex temporal values, e.g., pairs (or tuples) of time instants or even sets of time instants. In these cases, particularly in the case of set-valued attributes, it is important to remember that the set values are treated as *indivisible objects*, and hence truth (i.e., query semantics) is associated with the entire objects, but not necessarily with their components/subparts.

Concrete Temporal Query Languages

Although abstract temporal query languages provide a convenient and clean way of specifying queries, they are not immediately amenable to implementation. The main problem is that, in practice, the facts in temporal databases persist over periods of time. Storing all true facts individually *for every time instant* during a period would be prohibitively expensive or, in the case of infinite time domains such as *dense time*, even impossible.

Concrete temporal query languages avoid these problems by operating directly on the compact encodings of temporal databases. The most commonly used encoding is the one that uses *intervals*. However, in this setting, a tuple that associates a fact with such an interval is a compact representation of the association between the same fact and *all the time instants that belong to this interval*. This observation leads to the design choices that are commonly present in such languages:

- Coalescing is used, explicitly or implicitly, to consolidate representations of (sets of) time instants associated *with the same fact*. In the case of interval-based encodings, this leads to coalescing adjoining or overlapping intervals into a single interval. Note that coalescing only changes the *concrete representation* of a temporal relation, not its meaning (i.e., the abstract temporal relation); hence it has no counterpart in abstract temporal query languages.
- Implicit *set operations* on time values are used in relational operations. For example, conjunction (join) typically uses set intersection to generate a compact representation of the time instants attached to the facts in the result of such an operation.

Example For the running example, a concrete schema for the employment histories would typically be defined as "*Employment(VT, EID, Company)*" where *VT* is a valid time attribute ranging over periods (intervals). The GAPS query can be formulated in a calculus-style language corresponding to TSQL2 (see the entry on ▸ TSQL2) along the following lines:

$$\exists I_1, \ I_2. \left[\exists c \cdot \text{Employment}\,(I_1, x, c)\right] \wedge \\ \left[\exists c \cdot \text{Employment}\,(I_2, x, c)\right] \wedge I_1 \text{ precedes} \\ I_2 \wedge I = \left[\text{end}\,(I_1) + 1, \text{begin}\,(I_2) - 1\right].$$

In particular, the variables I_1 and I_2 range over periods and the *precedes* relationship is one of Allen's interval relationships. The final conjunct,

$$I = \left[\text{end}\,(I_1) + 1, \text{begin}\,(I_2) - 1\right],$$

creates a new period corresponding to the time instants related to a person's *gap in employment*; this interval value is explicitly constructed from the end and start points of I_1 and I_2, respectively. For the query to be correct, however, the results of evaluating the bracketed subexpressions, e.g., "$[\exists c.\text{Employment}(I_1, x, c)]$," have to

be *coalesced*. Without the insertion of the explicit coalescing operators, the query is *incorrect*. To see that, consider a situation in which a person p_0 is first employed by a company c_1, then by c_2, and finally by c_3, without any gaps in employment. Then without coalescing of the bracketed subexpressions of the above query, p_0 will be returned as a part of the result of the query, which is incorrect. Note also that it is not enough for the underlying (concrete) database to be coalesced.

The need for an explicit use of coalescing often makes the formulation of queries in some concrete SQL-based temporal query languages cumbersome and error-prone.

An orthogonal issue is the difference between explicit and implicit access to temporal values. This distinction also carries over to the concrete temporal languages. Typically, the various temporal extensions of SQL are based on the assumption of an explicit access to temporal values (often employing a built-in *valid time* attribute ranging over intervals or temporal elements), while many temporal relational algebras have chosen to use the implicit access based on temporally extending standard relational operators such as temporal join or temporal projection.

Compilation and Query Evaluation

An alternative to allowing users direct access to the encodings of temporal databases is to develop techniques that allow the evaluation of *abstract temporal queries* over these encodings. The main approaches are based on *query compilation* techniques that map abstract queries to concrete queries, while preserving query answers. More formally:

$$Q\left(\|E\|\right) = \|\text{eval}(Q)(E)\|,$$

where Q an abstract query, $eval(Q)$ the corresponding concrete query, E is a concrete temporal database, and $\|.\|$ a mapping that associates encodings (concrete temporal databases) with their abstract counterparts (cf. Fig. 1). Note that a single abstract temporal database, D, can be encoded using several *different* instances of the

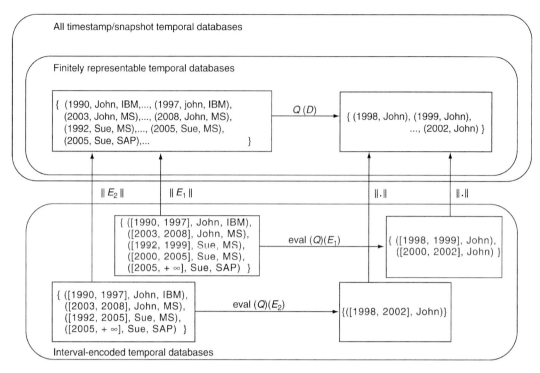

Abstract Versus Concrete Temporal Query Languages, Fig. 1 Query evaluation over interval encodings of point-stamped temporal databases

corresponding concrete database, e.g., E_1 and E_2 in Fig. 1.

Most of the practical temporal data models adopt a common approach to physical representation of temporal databases: with every fact (usually represented as a tuple), a *concise encoding* of the set of time points at which the fact holds is associated. The encoding is commonly realized by *intervals* [6, 7] or temporal elements (finite unions of intervals). For such an encoding it has been shown that both First-Order Temporal Logic [4] and Temporal Relational Calculus [8] queries can be *compiled* to first-order queries over a natural relational representation of the interval encoding of the database. Evaluating the resulting queries yields the interval encodings of the answers to the original queries, as if the queries were directly evaluated on the point-stamped temporal database. Similar results can be obtained for more complex encodings, e.g., periodic sets, and for abstract temporal query languages that adopt the duplicate semantics matching the SQL standard, such as SQL/TP [9].

Key Applications

Temporal query languages are primarily used for querying temporal databases. However, because of their generality they can be applied in other contexts as well, e.g., as an underlying conceptual foundation for querying sequences and data streams [5].

Cross-References

► Allen's Relations
► Bitemporal Relation
► Constraint Databases
► Key
► Nested Transaction Models
► Point-Stamped Temporal Models

Recommended Reading

1. Bettini C, Wang XS, Jajodia S. Temporal semantic assumptions and their use in databases. Knowl Data Eng. 1998;10(2):277–96.
2. Chomicki J. Temporal query languages: a survey. In: Proceedings of the 1st International Conference on Temporal Logic; 1994. p. 506–34.
3. Chomicki J, Toman D. Temporal databases. In: Fischer M, Gabbay D, Villa L, editors. Handbook of temporal reasoning in artificial intelligence. Elsevier Foundations of Artificial Intelligence; 2005. p. 429–67.
4. Chomicki J, Toman D, Böhlen MH. Querying ATSQL databases with temporal logic. ACM Trans Database Syst. 2001;26(2):145–78.
5. Law Y-N, Wang H, Zaniolo C. Query languages and data models for database sequences and data streams. In: Proceedings of the 30th International Conference on Very Large Data Bases; 2004. p. 492–503.
6. Navathe SB, Ahmed R. Temporal extensions to the relational model and SQL. In: Tansel A, Clifford J, Gadia S, Jajodia S, Segev A, Snodgrass RT, editors. Temporal databases: theory, design, and implementation. Menlo Park: Benjamin/Cummings; 1993. p. 92–109.
7. Snodgrass RT. The temporal query language TQuel. ACM Trans Database Syst. 1987;12(2):247–98.
8. Toman D. Point vs. interval-based query languages for temporal databases. In: Proceedings of the 15th ACM SIGACT-SIGMOD-SIGART Symposium on Principles of Database Systems; 1996. p. 58–67.
9. Toman D. Point-based temporal extensions of SQL. In: Proceedings of the 5th International Conference on Deductive and Object Oriented Databases; 1997. p. 103–21.

Abstraction

Bernhard Thalheim
Christian-Albrechts University, Kiel, Germany

Synonyms

Aggregation; Association; Classification; Component abstraction; Composition; Generalization; Grouping; Implementation abstraction; Specialization

Definition

Abstraction allows developers to concentrate on the essential, relevant, or important parts of an application. It uses a mapping to a model from things in reality or from virtual things. The model has the truncation property, i.e., it lacks some of the details in the original, and a pragmatic property, i.e., the model use is only justified for particular model users, tools of investigation, and periods of time. Database engineering uses construction abstraction, context abstraction, and refinement abstraction. Construction abstraction is based on the principles of hierarchical structuring, constructor composition, and generalization. Context abstraction assumes that the surroundings of a concept are commonly understood by a community or within a culture and focuses on the concept, turning away attention from its surroundings such as the environment and setting. Refinement abstraction uses the principle of modularization and information hiding. Developers typically use conceptual models or languages for representing and conceptualizing abstractions. The enhanced entity-relationship model schema are typically depicted by an EER diagram.

Key Points

Database engineering distinguishes three kinds of abstraction: construction abstraction, context abstraction, and refinement abstraction.

Constructor composition depends on the constructors as originally introduced by J. M. Smith and D.C.W. Smith. Composition constructors must be well founded and their semantics must be derivable by inductive construction. There are three main methods for construction: development of ordered structures on the basis of hierarchies, construction by combination or association, and construction by classification into groups or collections. The set constructors $\subset$ (subset), $\times$ (product), and P (powerset) for subset, product and nesting are complete for the construction of sets.

Subset constructors support hierarchies of object sets in which one set of objects is a subset of some other set of objects. Subset hierarchies are usually a rooted tree. Product constructors support associations between object sets. The schema is decomposed into object sets related to each other by association or relationship types. Power set constructors support a classification of object sets into clusters or groups of sets – typically according to their properties.

Context abstraction allows developers to commonly concentrate on those parts of an application that are essential for some perspectives during development and deployment of systems. Typical types of context abstraction are component abstraction, separation of concern, interaction abstraction, summarization, scoping, and focusing on typical application cases.

Component abstraction factors out repeating, shared or local patterns of components or functions from individual concepts. It allows developers to concentrate on structural or behavioral aspects of similar elements of components. Separation of concern allows developers to concentrate on those concepts under development and to neglect all other concepts that are stable or not under consideration. Interaction abstraction allows developers to concentrate on parts of the model that are essential for interaction with other systems or users. Summarisation maps the conceptualizations within the scope to more abstract concepts. Scoping is typically used to select those concepts that are necessary for current development and removes those concepts which that do not have an impact on the necessary concepts.

Database models may cover a large variety of different application cases. Some of them reflect exceptional, abnormal, infrequent and untypical application situations. Focusing on typical application cases explicitly separates models intended for the normal or typical application case from those that are atypical. Atypical application cases are not neglected but can be folded into the model whenever atypical situations are considered.

The context abstraction concept is the main concept behind federated databases. Context of databases can be characterized by schemata, version, time, and security requirements. Sub-schemata, types of the schemata or views on the schemata, are associated with explicit import/export bindings based on a name space. Parametrization lets developers consider collections of objects. Objects are identifiable under certain assumptions and completely identifiable after instantiation of all parameters.

Interaction abstraction allows developers to display the same set of objects in different forms. The view concept supports this visibility concept. Data is abstracted and displayed in various levels of granularity. Summarization abstraction allows developers to abstract from details that are irrelevant at a certain point. Scope abstraction allows developers to concentrate on a number of aspects. Names or aliases can be multiply used with varying structure, functionality and semantics.

Refinement abstraction mainly concerns implementation and modularisation. It allows developers to selectively retain information about structures. Refinement abstraction is defined on the basis of the development cycle (refinement of implementations). It refines, summarizes and views conceptualizations, hides or encapsulates details, or manages collections of versions. Each refinement step transforms a schema to a schema of finer granularity. Refinement abstraction may be modeled by refinement theory and infomorphisms.

Encapsulation removes internal aspects and concentrates on interface components. Blackbox or graybox approaches hide all aspects of the objects being considered. Partial visibility may be supported by modularization concepts. Hiding supports differentiation of concepts into public, private (with the possibility to be visible as "friends") and protected (with visibility to subconcepts). It is possible to define a number of visibility conceptualizations based on inflection. Inflection is used for the injection of combinable views into the given view, for tailoring, ordering and restructuring of views, and for enhancement of views by database functionality. Behavioral transparency is supported by the glassbox approach. Security views are based on hiding. Versioning allows developers to manage a number of concepts which can be considered to be versions of each other.

Cross-References

▶ Entity Relationship Model
▶ Extended Entity-Relationship Model
▶ Language Models
▶ Object Data Models
▶ Object-Role Modeling
▶ Specialization and Generalization

Recommended Reading

1. Börger E. The ASM refinement method. Form Asp Comput. 2003;15(2–3):237–57.
2. Smith JM, Smith DCW. Data base abstractions: aggregation and generalization. ACM Trans Database Syst. 1977;2(2):105–33.
3. Thalheim B. Entity-relationship modeling – foundations of database technology. Springer; 2000.

Access Control

Elena Ferrari
DiSTA, University of Insubria, Varese, Italy

Synonyms

Authorization verification

Definition

Access control deals with preventing unauthorized operations on the managed data. Access control is usually performed against a set of *authorizations* stated by Security Administrators (SAs) or users according to the *access control policies* of the organization. Authorizations are then processed by the *access control mechanism* (or *reference monitor*) to decide whether each access request can be authorized or should be denied.

Historical Background

Access control models for DBMSs have been greatly influenced by the models developed for the protection of operating system resources (see, for instance, the model proposed by Lampson [1], also known as the *access matrix* model, since authorizations are represented as a matrix). However, much of the early work on database protection was on inference control in statistical databases.

Then, in the 1970s, as research in relational databases began, attention was directed towards access control issues. As part of the research on System R at IBM Almaden Research Center, there was much work on access control for relational database systems [2, 3], which strongly influenced access control models and mechanisms of current commercial relational DBMSs. Around the same time, some early work on multilevel secure database management systems (MLS/DBMSs) was reported. However, it was only after the Air Force Summer Study in 1982 [4] that developments on MLS/DBMSs began. For instance, the early prototypes based on the integrity lock mechanisms developed at the MITRE Corporation. Later, in the mid-1980s, pioneering research was carried out at SRI International and Honeywell Inc. on systems such as SeaView and LOCK Data Views [5]. Some of the technologies developed by these research efforts were transferred to commercial products by corporations such as Oracle, Sybase, and Informix. In the 1990s, numerous other developments were made to meet the access control requirements of new applications and environments, such as the World Wide Web, data warehouses, data mining systems, multimedia systems, sensor systems, workflow management systems, and collaborative systems. This resulted in several extensions to the basic access control models previously developed, by including the support for temporal constraints, derivation rules, positive and negative authorizations, strong and weak authorizations, and content and context-dependent authorizations [6]. Role-based access control has been proposed [7] to simplify authorization management within companies and organizations. In the 2000s, there have been numerous developments in access control. Some of them have been driven by developments in web data management. For example, standards such as XML (eXtensible Markup Language) and RDF (Resource Description Framework) require proper access control mechanisms. Also, web services and the social web have become extremely popular and therefore research has been carried out to address the related access control issues. Access control has also being examined for new application areas, such as knowledge management [8], data outsourcing, GIS and location-based services [9], peer-to-peer computing, and stream data management [10], and social networks [11]. Today, we are in the era of Big Data [12], Cloud Computing [13], and NoSQL databases which have opened new opportunities for research in the access control field.

Foundations

The basic building block on which access control relies is a set of *authorizations* [6] which state, who can access which resource, and under which mode. Authorizations are specified according to a set of *access control policies*, which define the high-level rules according to which access control must occur. In its basic form, an authorization is, in general, specified on the basis of three components (s, o, p), and specifies that subject s is authorized to exercise privilege p on object o. The three main components of an authorization have the following meaning:

- *Authorization subjects*: they are the "active" entities in the system to which authorizations are granted. Subjects can be further classified into the following, not mutually exclusive, categories: *users*, that is, single individuals connecting to the system; *groups*, that is, sets of users; *roles*, that is, named collection of privileges needed to perform specific activities within the system; and *processes*, executing programs on behalf of users.

- *Authorization objects*: they are the "passive" components (i.e., resources) of the system to which protection from unauthorized accesses should be given. The set of objects to be protected clearly depends on the considered environment. For instance, files and directories are examples of objects of an operating system environment, whereas in a relational DBMS, examples of resources to be protected are relations, views, and attributes. Authorizations can be specified at different granularity levels, that is, on a whole object or only on some of its components. This is a useful feature when an object (e.g., a relation) contains information (e.g., tuples) of different sensitivity levels and therefore requires a differentiated protection.

- *Authorization privileges*: they state the types of operations (or access modes) that a subject can exercise on the objects in the system. As for objects, the set of privileges also depends on the resources to be protected. For instance, read, write, and execute privileges are typical of an operating system environment, whereas in a relational DBMS privileges refer to SQL commands (e.g., select, insert, update, delete). Moreover, new environments such as social networks are characterized by new access modes, for instance, share and post access rights.

Depending on the considered domain and the way in which access control is enforced, objects, subjects, and/or privileges can be hierarchically organized. The hierarchy can be exploited to propagate authorizations and therefore to simplify authorization management by limiting the set of authorizations that must be explicitly specified. For instance, when objects are hierarchically organized, the hierarchy usually represents

Access Control, Fig. 1
Access control: main
components

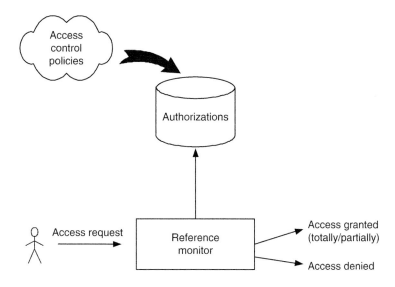

a "part-of" relation, that is, the hierarchy reflects the way objects are organized in terms of other objects. In contrast, the privilege hierarchy usually represents a subsumption relation among privileges. Privileges towards the bottom of the hierarchy are subsumed by privileges towards the top (for instance, the write privilege is at a higher level in the hierarchy with respect to the read privilege, since write subsumes read operations). Also roles and groups can be hierarchically organized. The group hierarchy usually reflects the membership of a group to another group. In contrast, the role hierarchy usually reflects the relative position of roles within an organization. The higher the level of a role in the hierarchy, the higher its position in the organization.

Authorizations are stored into the system and are then used to verify whether an access request can be authorized or not. How to represent and store authorizations depends on the protected resources. For instance, in a relational DBMS, authorizations are modeled as tuples stored into system catalogs. In contrast, when resources to be protected are XML documents, authorizations are usually encoded using XML itself. Finally, the last key component of the access control infrastructure is the *access control mechanism* (or *reference monitor*), which is a trusted software module in charge of enforcing access control. It intercepts each access request submitted to the system (for instance, SQL statements in case of relational DBMSs) and, on the basis of the

specified authorizations, it determines whether the access can be partially or totally authorized or should be denied. The reference monitor should be *non-bypassable*. Additionally, the hardware and software architecture should ensure that the reference monitor is *tamper proof*, that is, it cannot be maliciously modified (or at least that any improper modification can be detected). The main components of access control are illustrated in Fig. 1.

A basic distinction when dealing with access control is between *discretionary* and *mandatory* access control [6]. Discretionary access control (DAC) governs the access of subjects to objects on the basis of subjects' identity and a set of explicitly specified authorizations that state, for each subject, the set of objects that he/she can access in the system and the allowed access modes. When an access request is submitted to the system, the access control mechanism verifies whether or not the access can be authorized according to the specified authorizations. The system is discretionary in the sense that a subject, by proper configuring the set of authorizations, is both able to enforce various access control requirements and to dynamically change them when needed (simply by updating the authorization state). In contrast, mandatory access control (MAC) specifies the accesses that subjects can exercise on the objects in the system, on the basis of subjects and objects security classification. Security classes usually form a partially ordered set. This type of security has also been referred to as *multilevel security*, and database systems that enforce multilevel access control are called *Multilevel Secure Database Management Systems* (MLS/DBMSs). When mandatory access control is enforced, authorizations are implicitly specified, by assigning subjects and objects proper security classes. The decision on whether or not to grant an access depends on the access mode and the relation existing between the classification of the subject requesting the access and that of the requested object. In addition to DAC and MAC, role-based access control (RBAC) has been more recently proposed [7]. RBAC is an alternative to discretionary and mandatory access control, mainly conceived for regulating accesses within companies and organizations. In RBAC, permissions are associated with roles, instead of with users, and users acquire permissions through their membership to roles. The set of authorizations can be inferred by the sets of user-role and role-permission assignments. RBAC models have been shown to be policy-neutral since, by appropriately configuring the set of roles,one can support both mandatory and discretionary policies.

Key Applications

Access control techniques are applied in almost all environments that need to grant a controlled access to their resources, including, but not limited, to the following: DBMSs, Data Stream Management Systems, Operating Systems, Workflow Management Systems, Digital Libraries, GIS, Multimedia DBMSs, E-commerce services, Publish-subscribe systems, Data warehouses, Social Networks.

Future Directions

Altough access control is a mature area with consolidated results, the evolution of DBMSs and the requirements of new applications and environments pose new challenges to the research community. Some of the most recent research issues in the field are discussed below.

Social networks. On-line social networks (OSNs) represent one of the biggest revolution in the Computer Science field. Social Networks, as many other Web 2.0 technologies, have rapidly transformed the Web from a simple tool for publishing textual data into a complex collaborative knowledge management system to be used both for personal purposes and for business activities. Despite the clear advantages of OSNs in terms of information diffusion, they raised the need for giving content owners more control on the distribution of their resources, which may be accessed by a community far wider than they expected. So far, this issue has been mainly addressed by commercial OSNs and research proposals through Relationship-based Access Control (ReBAC) [11]. ReBAC takes into account the existence of a particular relationship or a particular sequence of relationships between users and/or resources and expresses access control policies in terms of such user-to-user (U2U), user-to-resource (U2R), or resource-to-resource (R2R) relationships. Despite the fact that the ReBAC model and its requirements are nowadays rather clearly specified, this paradigm has been applied incompletely or only partially by most of the available commercial OSNs, where the user is provided only a limited number of options to protect his/her personal data. Additionally, with difference to traditional contexts, OSN users create joint content; for instance, Alice uploads a photo, Bob tags it to say that Dave appears in it and Ann comments on Bob's tag. This calls for new and efficient ways to protect digital artifacts with multiple stakeholders within an OSN. Therefore, we need alternative paradigms to perform access control, wrt traditional ones, where additional information, such as for instance trust relationships and risks, are considered in the access control decision process.

Big Data. Big Data platforms are now considered the new frontier for innovation, competition, and productivity, and the integrated analysis of large volumes of data they make possible is becoming a strategic asset for many companies and organizations. Using innovative distributed computational paradigms and simple but effective data models, Big Data storage and analysis services feature high levels of scalability, performance, and availability. However, the analyzed data sources can contain data of any category, including personal, sensitive, and identifiable information. Privacy and confidentiality of the managed data is therefore among the most challenging aspects to be addressed within Big Data platforms [12], but today Big Data platforms only provide poor privacy enforcement mechanisms. The variety of existing Big Data platforms, along with their data models and query languages, make the definition of privacy-enhanced solutions a very challenging task. This is mainly due to the strict performance requirements, the heterogeneity of the data, the speed at which data are generated and must be analyzed, and the distributed nature of these systems.

Cloud computing. In the era of cloud computing, resources and applications are provided as a service over the Internet. Main benefits of the cloud computing paradigm are well known and range from cost reduction, scalability, better quality of service, and more effective allocation of internal resources. In this scenario, an important role is played by data management services, where a new emerging option is represented by the Database as a Service (DbaaS) paradigm. DbaaS is regulated by the same principles as Software as a Service (SaaS): data owners do not have to install and maintain the data management system on their own. In contrast, this is done by the service provider, whereas data owners only pay according to the system usage. Data outsourcing enacted by DBaaS poses challenging access control issues. The challenge is how to ensure data confidentiality and integrity, even if data are not directly managed by the owner but by a third party. The solutions that can be adopted can be classified into two main categories, depending on the trust residing in the service provider. In the case of trusted providers, that is, providers that correctly enforce the access control policies of the data owner, what is mainly required is the extension of traditional access control models to fulfill the needs of the cloud computing environment (see [13] for more details). In contrast, under the untrusted provider model, data protection should be ensured both wrt the users querying the data and the providers themself, since no assumption is made on their trustworthiness. Both these data protection requirements have been achieved so far by mainly exploiting cryptographic-based techniques. The idea is that data are encrypted by the owner before their delivering to the provider. Since the provider does not receive any decryption key, it is not able to read the data it manages, whereas users receive keys only for the data portions they are allowed to access according to the owner access control policies. The most challenging issues in this scenario are related to the efficiency of key management and the development of query processing techniques for encrypted data.

Cross-References

► Access Control Policy Languages
► Discretionary Access Control
► Mandatory Access Control
► Multilevel Secure Database Management System
► Role-Based Access Control
► Secure Data Outsourcing
► Storage Security

Recommended Reading

1. Lampson BW. Protection. Fifth Princeton symposium on information science and systems (Reprinted in). ACM Operat Syst Rev. 1974;8(1):18–24.
2. Fagin R. On an authorization mechanism. ACM Trans Database Syst. 1978;3(3):310–9.
3. Griffiths PP, Wade BW. An authorization mechanism for a relational database system. ACM Trans Database Syst. 1976;1(3):242–55.
4. Air Force Studies Board, Committee on Multilevel Data Management Security. Multilevel data management security. National Research Council; 1983.
5. Castano S, Fugini MG, Martella G, Samarati P. Database security. Addison-Wesley & ACM Press; 1995.
6. Ferrari E. Access control in data management systems. Synthesis lectures on data management. Morgan & Claypool Publishers; 2010.

7. Ferraiolo DF, Sandhu RS, Gavrila SI, Kuhn DR, Chandramouli R. Proposed NIST standard for role-based access control. ACM Trans Inf Syst Secur. 2001;4(3):224–74.

8. Bertino E, Khan LR, Sandhu RS, Thuraisingham BM. Secure knowledge management: confidentiality, trust, and privacy. IEEE Trans Syst Man Cybern A. 2006;36(3):429–38.

9. Bertino E, Kirkpatrick MS. Location-based access control systems for mobile users: concepts and research directions. In: Proceedings of the 4th ACM IGSPATIAL International Workshop on Security and Privacy in GIS and LBS; 2011.

10. Carminati B, Ferrari E, Tan KL. A framework to enforce access control over data streams. ACM Trans Inf Syst Secur. 2011;8(3):337–52.

11. Carminati B, Ferrari E, Viviani M. Security and trust in online social networks, synthesis lectures on information security, privacy and trust. Morgan & Claypool; 2013.

12. Kuner C, Cate F, Millard C, Svantesson D. The challenge of big data for data protection. Int Data Priv Law. 2012;2(2).

13. Takabi H, Joshi James BD, Gail-Joon A. Security and privacy challenges in cloud computing environments. IEEE Secur Priv. 2010;8(6):24–31.

14. Ferrari E, Thuraisingham BM. Security and privacy for web databases and services. In: Advances in Database Technology, Proceedings of the 9th International Conference on Extending Database Technology; 2004. p. 17–28.

Access Control Administration Policies

Elena Ferrari
DiSTA, University of Insubria, Varese, Italy

Synonyms

Authorization administration policies; Authorization administration privileges

Definition

Administration policies regulate who can modify the authorization state, that is, who has the right to grant and revoke authorizations.

Historical Background

Authorization management is a an important issue when dealing with access control and, as such, research on this topic is strongly related to the developments in access control. A milestone in the field is represented by the research carried out in the 1970s at IBM in the framework of the System R project. In particular, the work by Griffiths and Wade [9] defines a semantics for authorization revocation, which had greatly influenced the way in which authorization revocation has been implemented in commercial Relational DBMSs. Administrative policies for Object-oriented DBMSs have been studied in [8]. Later on, some extensions to the System R access control administration model, have been defined [3], with the aim of making it more flexible and adaptable to a variety of access control requirements. Additionally, as the research on extending the System R access control model with enhanced functionalities progresses, authorization administration has been studied for these extensions, such as temporal authorizations [2], strong and weak and positive and negative authorizations [5]. Also, administrative policies for new environments and data models such as WFMSs [1] and XML data [12] have been investigated. Back in the 1990s, when research on role-based access control began, administration policies for RBAC were investigated [6, 10, 11, 13]. Some of the ideas developed as part of this research were adopted by the SQL standard [7].

Foundations

Access control administration deals with granting and revoking of authorizations. This function is usually regulated by proper *administration policies*. Usually, if mandatory access control is enforced, the adopted administration policies are very simple, so that the Security Administrator (SA) is the only one authorized to change the classification level of subjects and objects. In contrast, discretionary and role-based access control are characterized by more articulated administration policies, which can be classified according to the following categories [3]:

- *SA administration.* According to this policy, only the SA can grant and revoke authorizations. Although the SA administration policy has the advantage of being very simple and easily implemented, it has the disadvantage of being highly centralized (even though different SAs can manage different portions of the database) and is seldom used in current DBMSs, apart from very simple systems.
- *Object owner administration.* This is the policy commonly adopted by DBMSs and operating systems. Under this policy, whoever creates an object become its owner and he/she is the only one authorized to grant and revoke authorizations on the object.
- *Joint administration.* Under this policy, particularly suited for collaborative environments, several subjects are jointly responsible for administering specific authorizations. For instance, under the joint administration policy it can be a requirement that the authorization to write a certain document is given by two different users, such as two different job functions within an organization. Authorizations for a subject to access a data object requires that all the administrators of the object issue a grant request.

The object owner administration policy can be further combined with *administration delegation*, according to which the administrator of an object can grant other subjects the right to grant and revoke authorizations on the object. Delegation can be specified for selected privileges, for example only for read operations. Most current DBMSs support the owner administration policy with delegation. For instance, the Grant command provided by the SQL standard [7] supports a Grant Option optional clause. If a privilege p is granted with the grant option on an object o, the subject receiving it is not only authorized to exercise p on object o but he/she is also authorized to grant other subjects authorizations for p on object o with or without the grant option. Moreover, SQL provides an optional Admin Option clause, which has the same meaning as the Grant option clause but it applies to roles instead of to standard authorizations. If a subject is granted the authorization to play a role with the admin option he/she not only receives all the authorizations associated with the role, but he/she can also authorize other subjects to play that role.

If administration delegation is supported, different administrators can grant the same authorization to the same subject. A subject can therefore receive an authorization for the same privilege on the same object by different sources. An important issue is therefore related to the management of revoke operations, that is, what happens when a subject revokes some of the authorizations he/she previously granted. For instance, consider three users: Ann, Tom, and Alice. Suppose that Ann grants Tom the privilege to select tuples from the Employee relation with the grant option and that, by having this authorization, Tom grants Alice the same privilege on the Employee relation. What happens to the authorization of Alice when Ann revokes Tom the privilege to select tuples from the Employee relation? The System R authorization model [9] adopts the most conscious approach with respect to security by enforcing *recursive revocation*: whenever a subject revokes an authorization on a relation from another subject, all the authorizations that the revokee had granted because of the revoked authorization are recursively removed from the system. The revocation is iteratively applied to all the subjects that received an authorization from the revokee. In the example above, Alice will lose the privilege to select tuples from the Employee relation when Ann revokes this privilege to Tom.

Implementing recursive revocation requires keeping track of the *grantor* of each authorization, that is, the subject who specifies the authorization, since the same authorization can be granted by different subjects, as well as of its *timestamp*, that is, the time when it was specified. To understand why the timestamp is important in correctly implementing recursive revocation, consider the graph in Fig. 1a, which represents the authorization state for a specific privilege p on a specific object o. Nodes represent subjects,

Access Control Administration Policies, Fig. 1 Recursive revocation

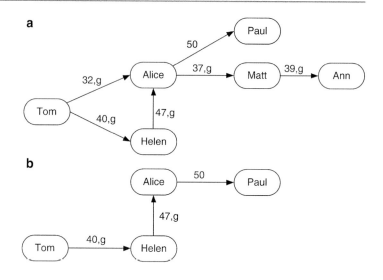

and an edge from node n_1 to node n_2 means that n_1 has granted privilege p on object o to n_2. The edge is labeled with the timestamp of the granted privilege and, optionally, with symbol "g," if the privilege has been granted with the grant option. Suppose that Tom revokes the authorization to Alice. As a result, the authorizations also held by Matt and Ann are recursively revoked because they could not have been granted if Alice did not receive authorization from Tom at time 32. In contrast, the authorization held by Paul is not revoked since it could have been granted even without the authorization granted by Tom to Alice at time 32, because of the privilege Alice had received by Helen at time 47. The authorization state resulting from the revoke operation is illustrated in Fig. 1b. Although recursive revocation has the advantage of being the most conservative solution with regard to security, it has the drawback of in some cases the unnecessarily revoking of too many authorizations. For instance, in an organization, the authorizations a user possesses are usually related to his/her job functions within the organization, rather than to his/her identity. If a user changes his/her tasks (for instance, because of a promotion), it is desirable to remove only the authorizations of the user, without revoking all the authorizations granted by the user before changing his/her job function. For this reason, research has been carried out to devise alternative

semantics for the revoke operation with regard to recursive revocation. Bertino et al. [4] have proposed an alternative type of revoke operation, called *noncascading revocation*. According to this, no recursive revocation is performed upon the execution of a revoke operation. Whenever a subject revokes a privilege on an object from another subject, all authorizations which the subject may have granted using the privilege received by the revoker are not removed. Instead, they are restated as if they had been granted by the revoker.

SQL [7] adopts the object owner administration policy with delegation. A revoke request can either be issued to revoke an authorization from a subject for a particular privilege on a given object, or to revoke the authorization to play a given role. SQL supports two different options for the revoke operation. If the revoke operation is requested with the Restrict clause, then the revocation is not allowed if it causes the revocation of other privileges and/or the deletion of some objects from the database schema. In contrast, if the Cascade option is specified, then the system implements a revoke operation similar to the recursive revocation of the System R, but without taking into account authorization timestamps. Therefore, an authorization is recursively revoked only if the grantor no longer holds the grant/admin option for that, because of the requested revoke operation. Otherwise, the authorization is not deleted, regardless of the time the grantor had received the grant/admin option for that authorization. To illustrate the differences with regard to recursive revocation, consider once again Fig. 1a, and suppose that Tom revokes privilege p on object o to Alice with the Cascade option. With difference to the System R access control model, this revoke operation does not cause any other changes to the authorization state. The authorization granted by Alice to Matt is not deleted, because Alice still holds the grant option for that access (received by Helen).

Key Applications

Access control administration policies are fundamental in every environment where access control services are provided.

Cross-References

► Access Control
► Discretionary Access Control
► Role-Based Access Control

Recommended Reading

1. Atluri V, Bertino E, Ferrari E, Mazzoleni P. Supporting delegation in secure workflow management systems. In: Proceedings of the 17th IFIP WG 11.3 Conference on Data and Application Security; 2003. p. 190–202.
2. Bertino E, Bettini C, Ferrari E, Samarati P. Decentralized administration for a temporal access control model. Inf Syst. 1997;22(4):223–48.
3. Bertino E, Ferrari E. Administration policies in a multipolicy authorization system. In: Proceedings of the 11th IFIP WG 11.3 Conference on Database Security; 1997. p. 341–55.
4. Bertino E, Samarati P, Jajodia S. An extended authorization model. IEEE Trans Knowl Data Eng. 1997;9(1):85–101.
5. Bertino E, Jajodia S, Samarati P. A flexible authorization mechanism for relational data management systems. ACM Trans Inf Syst. 1999;17(2):101–40.
6. Crampton J, Loizou G. Administrative scope: a foundation for role-based administrative models. ACM Trans Inf Syst Secur. 2003;6(2):201–31.
7. Database languages – SQL,ISO/IEC 9075–*; 2003.
8. Fernandez EB, Gudes E, Song H. A model for evaluation and administration of security in object-oriented databases. IEEE Trans Knowl Data Eng. 1994;6(2):275–92.
9. Griffiths PP, Wade BW. An authorization mechanism for a relational database system. ACM Trans Database Syst. 1976;1(3):242–55.
10. Oh S, Sandhu RS, Zhang X. An effective role administration model using organization structure. ACM Trans Inf Syst Secur. 2006;9(2):113–37.
11. Sandhu RS, Bhamidipati V, Munawer Q. The ARBAC97 model for role-based administration of roles. ACM Trans Inf Syst Secur. 1999;2(1):105–35.
12. Seitz L, Rissanen E, Sandholm T, Sadighi Firozabadi B, Mulmo O. Policy administration control and delegation using XACML and delegent. In: Proceedings of the 6th IEEE/ACM International Workshop on Grid Computing; 2005. p. 49–54.
13. Zhang L, Ahn G, Chu B. A rule-based framework for role-based delegation and revocation. ACM Trans Inf Syst Secur. 2003;6(3):404–41.

Access Control Policy Languages

Athena Vakali
Aristotle University, Thessaloniki, Greece

Synonyms

Authorization policy languages

Definition

An access control policy language is a particular set of grammar, syntax rules (logical and mathematical), and operators which provides an abstraction-layer for access control policy specifications. Such languages combine individual rules into a single policy set, which is the basis for (user/subject) authorization decisions on accessing content (object) stored in various information resources. The operators of an access control policy language are used on attributes of the subject, resource (object), and their underlying application framework to facilitate identifying the policy that (most appropriately) applies to a given action.

Historical Background

The evolution of access control policy languages is inline with the evolving large-scale highly distributed information systems and the Internet, which turned the tasks of authorizing and controlling of accessing on a global enterprise (or on Internet) framework increasingly challenging and difficult. Obtaining a solid and accurate view of the policy in effect across its many and diverse systems and devices has guided the development of access control policy languages accordingly.

Access control policy languages followed the Digital Rights Management (DRM) standardization efforts, which had focused in introducing DRM technology into commercial and mainstream products. Originally, access control was practiced in the most popular RDBMSs by policy languages that were SQL based. Certainly, the access control policy languages evolution was highly influenced by the wide adoption of XML (late 1990s) mainly in the enterprise world and its suitability for supporting access control acts. XML's popularity resulted in an increasing need to support more flexible provisional access decisions than the initial simplistic authorization acts which were limited in an accept/deny decision. In this context, proposals of various access control policy languages were very active starting around the year 2000. This trend seemed to stabilize around 2005.

The historical pathway of such languages should highlight the following popular and general-scope access control policy languages:

- 1998: the Digital Property Rights Language (DPRL, Digital Property Rights Language, http://xml.coverpages.org/dprl.html) mostly addressed to commercial and enterprise communities was specified for describing rights, conditions, and fees to support commerce acts
- 2000: XML Access Control Language (XACL, XML Access Control Language, http://xml.coverpages.org/xacl.html) was the first XML-based access control language for the provisional authorization model
- 2001: two languages were publicized:

  – The eXtensible rights Markup Language (XrML, The Digital Rights Language for Trusted Content and Services, http://www.xrml.org/) promoted as the digital rights language for trusted content and services
  – The Open Digital Rights Language (ODRL, Open Digital Rights Language, http://odrl.net/) for developing and promoting an open standard for rights expressions for transparent use of digital content in all sectors and communities
- 2002: the eXtensible Media Commerce Language (XMCL, eXtensible Media Commerce Language, http://www.w3.org/TR/xmcl/) to communicate usage rules in an implementation independent manner for interchange between business systems and DRM implementations
- 2003: the eXtensible Access Control Markup Language (XACML, eXtensible Access Control Markup Language, http://www.oasis-open.org/committees/xacml/) was accepted as a new OASIS, Organization for the Advancement of Structured Information Standards, http://www.oasis-open.org/, Open Standard language, designed as an XML specification with emphasis on expressing policies for information access over the Internet.
- 2005: Latest version XACML 2.0 appeared and policy languages which are mostly suited for Web services appear. These include WS-SecurityPolicy, http://www-128.ibm.com/developerworks/library/specification/ws-secpol/, which defines general security policy assertions to be applied into Web services security frameworks.

Foundations

Since Internet and networks in general are currently the core media for data and knowledge exchange, a primary issue is to assure authorized access to (protected) resources located in such infrastructures. To support access control policies and mechanisms, the use of an appropriate and suitable language is the core requirement in order to express all of the various components of access control policies, such as subjects, objects, constraints, etc. Initial attempts for expressing access control policies (consisting of authorizations) involved primary "participants" in a policy, namely the *subject* (client requesting access), the *object* (protected resource), and the *action* (right or type of access).

To understand the access control policy languages the context in which they are applied must be explained. Hence, the following notions which appear under varying terminology must be noted:

- *Content/objects*: Any physical or digital content which may be of different formats, may be divided into subparts and must be uniquely identified. Objects may also be encrypted to enable secure distribution of content.
- *Permissions/rights/actions*: Any task that will enforce permissions for accessing, using and acting over a particular content/object. They may contain constraints (limits), requirements (obligations), and conditions (such as exceptions, negotiations).
- *Subjects/users/parties*: Can be humans (end users), organizations, and defined roles which aim in consuming (accessing) content.

Under these three core entities, the policies are formed under a particular language to express offers and agreements. Therefore, the initial format of such languages authorization was (subject, object, and action) defining which subject can conduct what type of action over what object. However, with the advent of databases, networking, and distributed computing, users have witnessed (as presented in the section "Historical Background") a phenomenal increase in the automation of organizational tasks covering several physical locations, as well as the computerization of information related services [1, 2]. Therefore, new ideas have been added into modern access control models, like time, tasks, origin, etc. This was evident in the evolution of languages which initially supported an original syntax for policies limited in a three-tuple (subject, Subject primitive allows user IDs, groups, and/or role names. ob-

ject, Object primitive allows granularity as fine as a single element within an XML document, and action, Action primitive consists of four kinds of actions: read, write, create, and delete.) which then was found quite simplistic and limited and it was extended to include non-XML documents, to allow roles and collections as subjects and to support more actions (such as approve, execute, etc.).

Table 1 summarizes the most important characteristics of the popular general scope access control policy languages. It is evident that these languages differentiate on the subjects/users types, on the protected object/content type (which is considered as trusted when it is addressed to trusted audience/users) and on the capabilities of access control acts, which are presented under various terms and formats (rights, permissions, privileges, etc.). Moreover, this table highlights the level at which the access control may be in effect for each language, i.e., the broad categorization into fine- and coarse-grained protection granularity, respectively, refers to either partitions/detailed or full document/object protection capability. Moreover, the extensibility of languages which support Web-based objects and content is noted.

The need for moreover flexible policy languages is evident by cases such as the *OPL* which supports a wide range of access control principles in XML directly, by providing dedicated language constructs for each supported principle [3]. OPL is based on a module concept, and it can easily cope with the language complexity that usually comes with a growing expressiveness and it is suitable for enterprise frameworks.

To expand on the above, specific-scope languages have also emerged mainly to support research-oriented applications and tools. The most representative of such languages include:

- X-Sec [4]: To support the specification of subject credentials and security policies in Author-X and Decentral Author-X [5]. X-Sec adopts the idea of credentials which is similar to roles in that one user can be characterized by more than one credentials.

Access Control Policy Languages, Table 1 Summary of most popular access control policy languages

Language/technology	Subject types	Object types	Protection granularity	Accessing core formats	Focus
DPRL/XML DTDs	Registered users	Digital XML data sources, stored on repositories	Fine-grained	Digital licenses assigned for a time-limited period	
XACL/XML syntax	Group or organization members	Particular XML documents	Fine-grained	Set of particular specified privileges	
XrML/XML schema	Registered users and/or parties	digital XML data sources	Fine-grained	Granted rights under specified conditions	
ODRL/open-source schema-valid XML syntax	Any user	Trusted or untrusted content	Coarse-grained	Digital or physical rights	
XMCL/XML namespaces	Registered users	Trusted multimedia content	Coarse-grained	Specified keyword-based licenses	Particular business models
XACML/XML schema	Any users organized in categories	Domain-specific input	Fine grained	Rule-based permissions	
WS-Security policy/XML, SOAP	Any Web users/Web services	Digital data sources	Fine-grained	Protection acts at SOAP Web services messages level	Web services security

- XAS Syntax: Designed to support the ACP (Access Control Processor) tool [6]. It is a simplified XML-based syntax for expressing authorizations.
- RBXAC: A specification XML-based language supporting the role-based access control model [7].
- XACL: Which was originally based on a provisional authorization model and it has been designed to support ProvAuth (Provisional Authorizations) tool. Its main function is to specify security policies to be enforced upon accesses to XML documents.
- Cred-XACL [8]: A recent access control policy language focusing on credentials support on distributed systems and the Internet.

The core characteristics of these specific-scope languages are given in Table 2, which summarizes them with respect to their approach for objects and subjects management, their policies practicing and their subscription and ownership mechanisms. Such a summary is important in order to understand the "nature" of each such language in terms of objects and subjects

identification, protection (sources) granularity and (subject) hierarchies, policies expression and accessing modes under prioritization, and conflict resolution constraints. Finally, it should be noted that these highlighted characteristics are important in implementing security service tasks which support several security requirements from both the system and the sources perspective.

Key Applications

Access control policy languages are involved in the transparent and innovative use of digital resources which are accessed in applications related to key nowadays areas such as publishing, distributing and consuming of electronic publications, digital images, audio and movies, learning objects, computer software and other creations in digital form.

Relationship-Based Access Control (ReBAC) has moreover characterized tracking of interpersonal relationships among social networks users, and the expression of access control policies in terms of these relationships has been facilitated

Access Control Policy Languages, Table 2 Specific-scope access control languages characteristics

	X-Sec	XACL	RBXAC	XAS syntax
Objects				
Protected resources	XML documents and DTDs	XML documents and DTDs	XML documents	XML documents and DTDs
Identification	XPath	XPath	XPath	XPath
Protection granularity	Content, attribute	Element	Content, attribute	Element
Subjects				
Identification	XML-expressed credentials	Roles, UIDs, groups	Roles	User ID, location
Grouping of subjects	No	Yes	No	Yes
Subjects hierarchy	No	Yes	Role trees	Yes
Support public subject	No	Yes	No	Yes
Policies				
Expressed in	Policy base	XACL policy file	Access control files	XAS
Closed/open	Closed	Both	Closed	Closed
Permissions/denials	Both	Both	Permissions	Both
Access modes	Authoring, browsing	Read, write, create, delete	RI, WI, RC, WC	Read
Propagation	No-prop, first-level, cascade	No/up/down	According to role tree	Local, recursive
Priority	Implicit rules	ntp, ptp, dtd	–	Hard, soft
Conflict resolution	Yes	According to priorities and implicit rules	–	Implicitly, explicitly
Other issues				
Subscription-based	Yes	Yes	Yes	Yes
Ownership	No	No	Yes	No

by devising a policy language [9], based on modal logic, for composing access control policies that support delegation of trust.

Access control languages have also ben utilized in several other ways such as in the case of securing RDF graphs via an underlying query language which on the basis of a redaction mechanism provides fine grained RDF access control [10]. Such access control languages require critical features support (such as policy resolution, cascading policies etc.).

Future Directions

From the evolution of access control policy languages, it appears that, in the future, emphasis will be given on languages that are mostly suited for Web-accessed repositories, databases, and in-formation sources. This trend is now apparent from the increasing interest on languages that control accessing on Web services and Web data sources. At the same time, it manages the challenges posed by acknowledging and identifying users/subjects on the Web, especially in emerging and evolving domains such as in social networks.

URL to Code

Code, examples, and application scenarios may be found for: ODRL application scenarios at http://www.w3.org/TR/odrl/#46354 and http://odrl.net/, XrML at http://www.xrml.org/, XMCL at http://www.w3.org/TR/xmcl/, XACML at http://www.oasis-open.org/committees/xacml/, http://xml.coverpages.org/xacml.html and WS-SecurityPolicy at http://www-128.ibm.com/developerworks/library/specification/ws-secpol/.

Cross-References

► Access Control
► Database Security
► Role-Based Access Control
► Secure Database Development

Recommended Reading

1. Stoupa K, Vakali A. Policies for web security services, chapter III. In: Ferrari E, Thuraisingham B, editors. Web and information security. Hershey: Idea-Group Publishing; 2006.
2. Vuong NN, Smith GS, Deng Y. Managing security policies in a distributed environment using eXtensible markup language (XML). In: Proceedings of the 16th ACM Symposium on Applied Computing; 2001. p. 405–11.
3. Alm C, Wolf R, Posegga J. The OPL access control policy language. In: Fischer-Hübner S, et al., editors. TrustBus 2009, LNCS 5695. Berlin/Heidelberg: Springer; 2009. p. 138–48.
4. Bertino E, Castano S, Ferrari E. On specifying security policies for web documents with an XML-based language. In: Proceedings of the 6th ACM Symposium on Access Control Models and Technologies; 2001. p. 57–65.
5. Bertino E, Castano S, Ferrari E. Securing XML documents with author-X. IEEE Internet Comput. 2001;5(3):21–31.
6. Damiani E, De Capitani di Vimercati S, Paraboschi S, Samarati P. Design and implementation of an access control processor for XML documents. In: Proceedings of the 9th International World Wide Web Conference; 2000. p. 59–75.
7. He H, Wong RK. A role-based access control model for XML repositories. In: Proceedings of the 1st International Conference on Web Information Systems Engineering; 2000. p. 138–45.
8. Stoupa K. Access control techniques in distributed systems and the Internet. Ph.D. Thesis, Aristotle University, Department of Informatics; 2007.
9. Fong PWL. Relationship-based access control: protection model and policy language. In: Proceedings of the 1st ACM Conference on Data and Application Security and Privacy; 2011. p. 191–202.
10. Rachapalli J, Khadilkar V, Kantarcioglu M, Thuraisingham B. Redaction based RDF access control language. In: Proceedings of the 19th ACM Symposium on Access Control Models and Technologies; 2014. p. 177–80.
11. Qi N, Kud M. Access control policy languages in XML, applications and trends. Springer Science+Business Media, LLC; 2008. p. 55–71.

Access Path

Evaggelia Pitoura
Department of Computer Science and Engineering, University of Ioannina, Ioannina, Greece

Synonyms

Access methods; Access path

Definition

An access path specifies the path chosen by a database management system to retrieve the requested tuples from a relation. An access path may be either (i) a sequential scan of the data file or (ii) an index scan with a matching selection condition when there are indexes that match the selection conditions in the query. In general, an index matches a selection condition, if the index can be used to retrieve all tuples that satisfy the condition.

Key Points

Access paths are the alternative ways for retrieving specific tuples from a relation. Typically, there is more than one way to retrieve tuples because of the availability of indexes and the potential presence of conditions specified in the query for selecting the tuples. Typical access methods include sequential access of unordered data files (heaps) as well as various kinds of indexes. All commercial database systems implement heaps and B+ tree indexes. Most of them also support hash indexes for equality conditions.

To choose an access path, the optimizer first determines which matching access paths are available by examining the conditions specified by the query. Then, it estimates the selectivity of each access path using any available statistics for the index and data file. The *selectivity of an access path* is the number of pages (both index and data pages) accessed when the specific access path is used to retrieve the requested tuples. The access path having the smallest selectivity is called the most *selective access path*. Clearly, using the most selective access path minimizes the cost of data retrieval. Additional information can be found in [1].

Cross-References

▸ Index Structures for Biological Sequences
▸ Query Optimization
▸ Selectivity Estimation

Recommended Reading

1. Selinger PG, Astrahan MM, Chamberlin DD, Lorie RA, Price TG. Access path selection in a relational database management system. In: Proceedings of the ACM SIGMOD International Conference on Management of Data; 1979. p. 23–34.

ACID Properties

Gottfried Vossen
Department of Information Systems,
Westfälische Wilhelms-Universität, Münster,
Germany

Synonyms

ACID properties; Atomicity; Consistency preservation; Durability; Isolation; Persistence

Definition

The conceptual *ACID properties* (short for atomicity, isolation, consistency preservation, and durability) of a transaction together provide the key abstraction which allows application developers to disregard irregular or even malicious effects from concurrency or failures of transaction executions, as the transactional server in charge guarantees the consistency of the underlying data and ultimately the correctness of the application [1–3]. For example, in a banking context where debit/credit transactions are executed, this means that no money is ever lost in electronic funds transfers and customers can rely on electronic receipts and balance statements. These cornerstones for building highly dependable information systems can be successfully applied outside the scope of online transaction processing and classical database applications as well.

Key Points

The *ACID properties* are what a database server guarantees for transaction executions, in particular in the presence of multiple concurrently running transactions and in the face of failure situations; they comprise the following four properties (whose initial letters form the word "ACID"):

Atomicity. From the perspective of a client and an application program, a transaction is executed completely or not at all, i.e., in an all-or-nothing fashion. So the effects of a program under execution on the underlying data server(s) will only become visible to the outside world or to other program executions if and when the transaction reaches its "commit" operation. This case implies that the transaction could be processed completely and no errors whatsoever were discovered while it was processed. On the other hand, if the program is abnormally terminated before reaching its commit operation, the data in the underlying data servers will be left in or automatically brought back to the state in which it was before the transaction started, i.e., the data appears as if the transaction had never been invoked at all.

Consistency preservation: Consistency constraints that are defined on the underlying data servers (e.g., keys, foreign keys) are preserved by a transaction; so a transaction leads from one consistent state to another. Upon the commit of a transaction, all integrity constraints defined for the underlying database(s) must be satisfied; however, between the beginning and the end of a transaction, inconsistent intermediate states are tolerated and may even be unavoidable. This property generally cannot be ensured in a completely automatic manner. Rather, it is necessary that the application is programmed such that the code between the beginning and the commit of a transaction will eventually reach a consistent state.

Isolation: A transaction is isolated from other transactions, i.e., each transaction behaves as if it was operating alone with all resources to itself. In particular, each transaction will "see" only consistent data in the underlying data sources. More specifically, it will see only data modifications that result from committed transactions, and it will see them only in their entirety and never any effects of an incomplete transaction. This is the decisive property that allows to hide the fallacies and pitfalls of concurrency from the application developers. A sufficient condition for isolation is that concurrent executions are equivalent to sequential ones, so that all transactions appear as if they were executed one after the other rather than in an interleaved manner; this condition is made precise through serializability.

Durability: When the application program from which a transaction derives is notified that the transaction has been successfully completed (i.e., when the commit point of the transaction has been reached), all updates the transaction has made in the underlying data servers are guaranteed to survive subsequent software or hardware failures. Thus, updates of committed transactions are durable (until another transaction later modifies the same data items) in that they persist even across failures of the affected data server(s).

Therefore, a transaction is a set of operations executed on one or more data servers which are issued by an application program and are guaranteed to have the ACID properties by the runtime system of the involved servers. The "ACID contract" between the application program and the data servers requires the program to demarcate the boundaries of the transaction as well as the

desired outcome – successful or abnormal termination – of the transaction, both in a dynamic manner. There are two ways a transaction can finish: it can commit, or it can abort. If it commits, all its changes to the database are installed, and they will remain in the database until some other application makes further changes. Furthermore, the changes will seem to other programs to take place together. If the transaction aborts, none of its changes will take effect, and the DBMS will roll back by restoring previous values to all the data that was updated by the application program. A programming interface of a transactional system consequently needs to offer three types of calls: (i) "begin transaction" to specify the beginning of a transaction, (ii) "commit transaction" to specify the successful end of a transaction, and (iii) "rollback transaction" to specify the unsuccessful end of a transaction with the request to abort the transaction.

The core requirement for a transactional server is to provide the ACID guarantees for sets of operations that belong to the same transaction issued by an application program requires that the server. This requires a *concurrency control* component to guarantee the isolation properties of transactions, for both committed and aborted transactions, and a *recovery* component to guarantee the atomicity and durability of transactions. The server may or may not provide explicit support for consistency preservation. In addition to the ACID contract, a transactional server should meet a number of technical requirements: a transactional data server (which most often will be a database system) must provide *good performance* with a given hardware/software configuration or, more generally, a good cost/performance ratio when the configuration is not yet fixed. Performance typically refers to the two metrics of *high throughput*, which is defined as the number of successfully processed transactions per time unit, and of *short response times*, where the response time of a transaction is defined as the time span between issuing the transaction and its successful completion as perceived by the client.

While the ACID properties are crucial for many applications in which the transaction concept arises, some of them are too restrictive when the transaction model is extended beyond the read/write context. For example, business processes can be cast into various forms of *business transactions*, i.e., long-running transactions for which atomicity and isolation are generally too strict. In these situations, additional or alternative guarantees need to be employed.

Cross-References

▶ Atomicity
▶ Eventual Consistency
▶ Extended Transaction Models and the ACTA Framework
▶ Multilevel Recovery and the ARIES Algorithm
▶ Serializability
▶ Snapshot Isolation
▶ SQL Isolation Levels

Recommended Reading

1. Bernstein PA, Hadzilacos V, Goodman N. Concurrency control and recovery in database systems. Reading: Addison-Wesley; 1987.
2. Bernstein PA, Newcomer E. Principles of transaction processing for the systems professional. San Francisco: Morgan Kaufmann; 1997.
3. Gray J, Reuter A. Transaction processing: concepts and techniques. San Francisco: Morgan Kaufmann; 1993.

Active Database Coupling Modes

Mikael Berndtsson and Jonas Mellin
University of Skövde, The Informatics Research Centre, Skövde, Sweden
University of Skövde, School of Informatics, Skövde, Sweden

Definition

Coupling modes specify execution points for ECA rule conditions and ECA rule actions with respect to the triggering event and the transaction model.

Historical Background

Coupling modes for ECA rules were first suggested in the HiPAC project [2, 3].

Foundations

Coupling modes are specified for event-condition couplings and for condition-action couplings. In detail, the event-condition coupling specifies when the condition should be evaluated with respect to the triggering event, and the condition-action coupling specifies when the rule action should be executed with respect to the evaluated rule condition (if condition is evaluated to true).

The three most common coupling modes are immediate, deferred, and decoupled. The immediate coupling mode preempts the execution of the transaction and immediately initiates condition evaluation and action execution. In the deferred coupling mode, condition evaluation and action execution is deferred to the end of the transaction (before transaction commit). Finally, in decoupled (also referred to as detached) coupling mode, condition evaluation and action execution is performed in separate transactions.

Specifying event-condition couplings and condition-action couplings in total isolation from each other is not a good idea. What first might seem to be one valid coupling mode for event-condition and one valid coupling mode for condition-action can be an invalid coupling mode when used together. Thus, when combining event-condition couplings and condition-action couplings, not all combinations of coupling modes are valid. The HiPAC project [2, 3] proposed seven valid coupling modes; see Table 1.

- *Immediate, immediate*: the rule condition is evaluated immediately after the event, and the rule action is executed immediately after the rule condition.
- *Immediate, deferred*: the rule condition is evaluated immediately after the event, and the execution of the rule action is deferred to the end of the transaction.

Active Database Coupling Modes, Table 1 Coupling modes

| | Condition-action | | |
Event-condition	Immediate	Deferred	Decoupled
Immediate	Condition evaluated and action executed after event	Condition evaluated after event, action executed at end of transaction	Condition evaluated after event, action executed in a separate transaction
Deferred	Not valid	Condition evaluated and action executed at end of transaction	Condition evaluated at end of transaction, action executed in a separate transaction
Decoupled	In a separate transaction: condition evaluated and action executed after event	Not valid	Condition evaluated in one separate transaction, action executed in another separate transaction

- *Immediate, decoupled*: the rule condition is evaluated immediately after the event, and the rule action is decoupled in a totally separate and parallel transaction.
- *Deferred, deferred*: both the evaluation of the rule condition and the execution of the rule action are deferred to the end of the transaction.
- *Deferred, decoupled*: the evaluation of the rule condition is deferred to the end of the transaction, and the rule action is decoupled in a totally separate and parallel transaction.
- *Decoupled, immediate*: the rule condition is decoupled in a totally separate and parallel transaction, and the rule action is executed (in the same parallel transaction) immediately after the rule condition.
- *Decoupled, decoupled*: the rule condition is decoupled in a totally separate and parallel transaction, and the rule action is decoupled in another totally separate and parallel transaction.

The two invalid coupling modes are:

- *Deferred, immediate*: this combination violates the semantics of ECA rules, that is, rule conditions must be evaluated before rule actions are executed. One cannot preempt the execution of the transaction immediately after the event and execute the rule action and at the same time postpone the condition evaluation to the end of the transaction.
- *Decoupled, deferred*: this combination violates transaction boundaries, that is, one can-

not decouple the condition evaluation in a separate and parallel transaction and at the same time postpone the execution of the rule action to the end of the original transaction, since one cannot know when the condition evaluation will take place. Thus, there is a risk that the action execution in the original transaction will run before the condition has been evaluated in the parallel transaction.

Rule actions executed in decoupled transactions can either be dependent upon or independent of the transaction in which the event took place.

The research project REACH (REal-time ACtive Heterogeneous System) [1] introduced two additional coupling modes for supporting side effects of rule actions that are irreversible. The new coupling modes are variants of the detached casually dependent coupling mode: sequential casually dependent and exclusive casually dependent. In sequential casually dependent, a rule is executed in a separate transaction. However, the rule execution can only begin once the triggering transaction has committed. In exclusive casually dependent, a rule is executed in a detached parallel transaction, and it can commit only if the triggering transaction failed.

Cross-References

- ▶ Active Database Execution Model
- ▶ ECA Rules

Recommended Reading

1. Branding H, Buchmann A, Kudrass T, Zimmermann J. Rules in an open system: the REACH rule system. In: Proceedings of the 1st International Workshop on Rules in Database Systems, Workshops in Computing; 1994. p. 111–26.
2. Dayal U, Blaustein BA, Buchmann SC, et al. The HiPAC project: combining active databases and timing constraints. ACM SIGMOD Rec. 1988a;17(1):51–70.
3. Dayal U, Blaustein B, Buchmann A, Chakravarthy S, et al. HiPAC: a research project in active, time-constrained database management. Technical report. CCA-88-02. Cambridge: Xerox Advanced Information Technology; 1988b.

Active Database Execution Model

Mikael Berndtsson and Jonas Mellin
University of Skövde, The Informatics Research Centre, Skövde, Sweden
University of Skövde, School of Informatics, Skövde, Sweden

Definition

The execution model of an active database describes how a set of ECA rules behave at run time.

Key Points

The execution model describes how a set of ECA rules (i.e., active database rule base) behave at run time [2, 4]. Any execution model of an active database must have support for (i) detecting event occurrences, (ii) evaluating conditions, and (iii) executing actions.

If an active database supports composite event detection, it needs a policy that describes how a composite event is computed. A typical approach is to use the event consumption modes as described in Snoop [1]: recent, chronicle, continuous, and cumulative. In the recent event context, only the most recent constituent events will be used to form composite events. In the chronicle event context, events are consumed in chronicle order. The earliest unused initiator/terminator pair are used to form the composite event. In the continuous event context, each initiator starts the detection of a new composite event, and a terminator may terminate one or more composite event occurrences. The difference between continuous and chronicle event contexts is that in the continuous event context, one terminator can detect more than one occurrence of the composite event. In the cumulative event context, all events contributing to a composite event are accumulated until the composite event is detected. When the composite event is detected, all contributing events are consumed. Another approach to these event consumption modes is to specify a finer semantics for each event by using logical events as suggested in [3].

Once an event has been detected, there are several execution policies related to rule conditions and rule actions that must be in place in the execution model. Thus an execution model for an active database should provide answers to the following questions [2, 4, 5]:

- When should the condition be evaluated and when should the action should be executed with respect to the triggering event and the transaction model? This is usually specified by coupling modes.
- What happens if an event triggers several rules?
  - Are all rules evaluated, a subset, or only one rule?
  - Are rules executed in parallel, according to rule priority, or non-deterministically?
- What happens if one's rules trigger another set of rules?
  - What happens if the rule action of one rule negates the rule condition of an already triggered rule?
  - Can cycles appear? For example, can a rule trigger itself?

The answers to the above questions are important to know, as they dictate how a ECA rule system will behave at run time. If the answers to the above questions are not known, then the behavior of the ECA rule application becomes unpredictable.

Cross-References

► Active Database Coupling Modes
► Active Database Rulebase
► Composite Event
► Database Trigger
► ECA Rules

Recommended Reading

1. Chakravarthy S, Krishnaprasad V, Anwar E, Kim SK. Composite events for active databases: semantics contexts and detection. In: Proceedings of the 20th International Conference on Very Large Data Bases; 1994. p. 606–17.
2. Dayal U, Blaustein B, Buchmann A, Chakravarthy S. et al. HiPAC: a research project in active, time-constrained database management. Technical report CCA-88-02. Cambridge: Xerox Advanced Information Technology; 1988.
3. Gehani N, Jagadish HV, Smueli O. Event specification in an active object-oriented database. In: Proceedings of the ACM SIGMOD International Conference on Management of Data; 1992. p. 81–90.
4. Paton NW, Diaz O. Active Database Systems. ACM Comput. Surv. 1999;31(1):63–103.
5. Widom J, Finkelstein S. Set-oriented production rules in relational database systems. In: Proceedings of the ACM SIGMOD International Conference on Management of Data; 1990. p. 259–70.

Active Database Knowledge Model

Mikael Berndtsson and Jonas Mellin
University of Skövde, The Informatics Research Centre, Skövde, Sweden
University of Skövde, School of Informatics, Skövde, Sweden

Definition

The knowledge model of an active database describes what can be said about the ECA rules, that is, what types of events are supported, what types of conditions are supported, and what types of actions are supported?

Key Points

The knowledge model describes what types of events, conditions, and actions are supported in an active database. Another way to look at the knowledge model is to imagine what types of features are available in an ECA rule definition language.

A framework of dimensions for the knowledge model is presented in [3]. Briefly, each part of an ECA rule is associated with dimensions that describe supported features. Thus, an event can be described as either a primitive event or a composite event, how it was generated (source), whether the event is generated for all instances in a given set or only for a subset (event granularity), and what types (if event is a composite event) of operators and event consumption modes are used in the detection of the composite event.

Conditions are evaluated against a database state. There are three different database states that a rule condition can be associated with [3]: (i) the database state at the start of the transaction, (ii) the database state when the event was detected, and (iii) the database state when the condition is evaluated.

There are four different database states that a rule action can be associated with [3]: (i) the database state at the start of the transaction, (ii) the database state when the event was detected and (iii) the database state when the condition is evaluated, and (iv) the database state just before action execution. The type of rule actions range from internal database updates (e.g., update a table) to external programs (e.g., send email).

Within the context of the knowledge model, it is also useful to consider how ECA rules are represented, for example, inside classes, as data members, or first-class objects. Representing ECA rules as first-class objects [1, 2] is a popular choice, since rules can be treated as any other object in the database and traditional database operations can be used to manipulate the ECA rules. Thus, representing ECA rules as first-class objects implies that ECA rules are not dependent upon the existence of other objects.

The knowledge model of an active database should also describe whether the active database supports passing of parameters between the ECA rule parts, for example, passing of parameters from the event part to the condition part.

Related to the knowledge model is the execution model that describes how ECA rules behave at run time.

Cross-References

▶ Active Database Execution Model
▶ ECA Rules

Recommended Reading

1. Dayal U, Blaustein B, Buchmann A. et al. S.C. HiPAC: a research project in active, time-constrained database management. Technical report CCA-88-02, Xerox Advanced Information Technology, Cambridge; 1988a.
2. Dayal U, Buchmann A, McCarthy D. Rules are objects too: a knowledge model for an active, object-oriented database system. In: Proceedings of the 2nd International Workshop on Object-Oriented Database Systems; 1988b. p. 129–43.
3. Paton NW, Diaz O. Active database systems. ACM Comput Surv. 1999;31(1):63–103.

Active Database Management System Architecture

Jonas Mellin and Mikael Berndtsson
University of Skövde, The Informatics Research Centre, Skövde, Sweden
University of Skövde, School of Informatics, Skövde, Sweden

Synonyms

ADBMS; ADBMS framework; ADBMS infrastructure

Definition

The active database management system (ADBMS) architecture is the software organization of a DBMS with active capabilities. That is, the architecture defines support for active capabilities expressed in terms of services, significant components providing the services, as well as critical interaction among these services.

Historical Background

Several architectures have been proposed: HiPAC [1, 2], REACH [3], ODE [4], SAMOS [5], SMILE [6], and DeeDS [1]. Each of these architectures emphasizes particular issues concerning the actual DBMS that they are based on as well as the type of support for active capabilities. Paton and Diaz [7] provide an excellent survey on this topic. Essentially, these architectures propose that the *active capabilities* of an ADBMS require the services specified in Table 1. It is assumed that queries to the database are encompassed in transactions, and hence transactions imply queries as well as database manipulation operations such as insertion, updates, and deletion of tuples.

The services in Table 1 interact as depicted in Fig. 1. Briefly, transactions are submitted to the scheduling service that updates the dispatch table read by the transaction processing service. When these transactions are processed by the transaction processing service, events are generated. These events are signaled to the event monitoring service that analyzes them. Events that are associated with rules (subscribed events) are signaled to the rule evaluation service that evaluates the conditions of *triggered rules* (i.e., rules associated with signaled events). The actions of the rules whose conditions are true are submitted for scheduling and are executed as dictated by the scheduling policy. These actions execute as part of some transaction according to the coupling mode and can, in turn, generate events. This is a general description of the service interaction, and it can be optimized by refining it for a specific purpose, for example, in immediate

Active Database Management System Architecture, Table 1   Services in active database management systems

Service	Responsibility
Event monitoring	The event monitoring service is responsible for collecting events, analyzing events, and disseminating results of the analysis (in terms of events) to subscribers, in particular, ECA rules
Rule evaluation	The rule evaluation service is responsible for invoking condition evaluation of triggered ECA rules and submitting actions for execution to the scheduler
Scheduling service	The scheduling service is responsible for readying and ordering schedulable activities such as ECA rule actions, transactions, etc., for execution

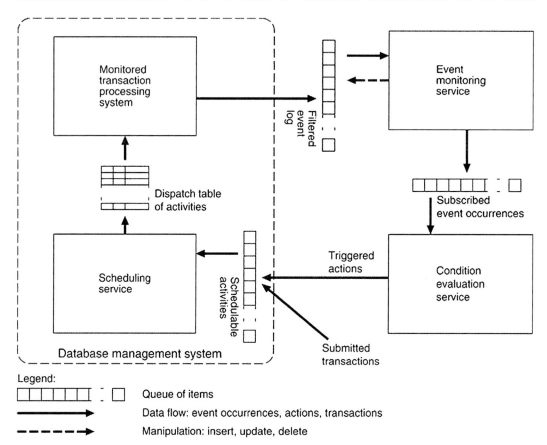

Active Database Management System Architecture, Fig. 1   Service interaction view of architecture (Based on architecture by Paton and Diaz [7])

coupling mode no queues between the services are actually needed.

In more detail, transactions are submitted to the scheduling service via a queue of schedulable activities; this queue of schedulable activities is processed, and a dispatch table of schedulable activities is updated. This scheduling service encompasses scheduling of transactions as well as ECA rule actions in addition to other necessary schedulable activities. It is desirable for the scheduling service to encompass all these types of schedulable activities, because they impact each other, since they compete for the same resources. The next step in the processing chain is the monitored transaction processing service, which includes the transaction management, lock management, and log management [8, Chap. 5], as well as a database query engine (cf. query processor [9, Chap. 1]), but not the scheduling service. Another way to view the transaction processing service is as a passive database management system without the transaction scheduling service. The transaction processing service is denoted "monitored," since it generates events that are handled by the active capabilities. The monitored transaction processing service executes transactions and ECA rule actions according to the dispatch table. When transactions execute, event occurrences are signaled to the event monitoring service via a filtered event log. When event monitoring executes, it updates the filtered event log and submits subscribed events to the rule evaluation service. An example of event log filtering is that if a composite event occurrence is detected, then for optimization reasons (cf. dynamic programming) this event occurrence is stored in the filtered event log. Another example is that when events are no longer needed, then they are pruned; for example, when a transaction is aborted, then all event occurrences can be pruned unless intertransaction events are allowed (implying that dirty reads may occur). The rule evaluation service reads the queue of subscribed events, finds the triggered rules, and evaluates their conditions. These conditions may be queries, logical expressions, or arbitrary code depending on the active database system [2]. The rule evaluation results in a set of actions that is submitted to the scheduling service for execution.

The general view of active capabilities (in Fig. 1) can be refined and implemented in different ways. As mentioned, it is possible to optimize an implementation by removing the queues between the services if only immediate coupling mode is considered; this results in less overhead but restricts the expressibility of ECA rules significantly. A service can be implemented via one or more servers. These servers can be replicated to different physical nodes for performance or dependability reasons (e.g., availability, reliability).

In active databases, a set of issues have a major impact on refinement and implementation of the general service-oriented view depicted in Fig. 1. These issues are (i) coupling modes; (ii) interaction with typical database management services such as transaction management, lock management, and recovery management (both

precrash such as logging and checkpointing and postcrash such as the actually recovery) (cf., e.g., transaction processing by Gray and Reuter [8, Chap. 4]); (iii) when and how to invoke services; and (iv) active capabilities in distributed active databases.

The coupling modes control how rule evaluation is invoked in response to events and how the ECA rule actions are submitted, scheduled, dispatched, and executed for rules whose conditions are true (see entry "Coupling modes" for more detailed description). There are different alternatives to interaction with a database system. One alternative is to place active database services on top of existing database management systems. However, this is problematic if the database management system is not extended with active capabilities [3]. For example, the deferred coupling mode requires that when a transaction is requested to commit, then queued actions should be evaluated. This requires the transaction management to interact with the rule evaluation and scheduling services during commit processing (e.g., by using back hooks in the database management system). Further, to be useful, the detached coupling mode has a set of significant varieties [3] that require the possibility to express constraints between transactions.

The nested transaction model [10] is a sound basis for active capabilities. For example, deferred actions can be executed as subtransactions that can be committed or aborted independently of the parent transaction. Nested transactions still require that existing services are modified. Alternatively rule evaluation can be performed as subtransactions.

To achieve implicit events the database schema translation process needs to automatically instrument the monitored systems. An inferior solution is to extend an existing schema with instrumented entities, for example, each class in an object-oriented database can be inherited to an instrumented class. In this example, there is no way to enforce that the instrumented classes are actually used. The problem is to modify the database schema translation process, since this is typically an intrinsic part in commercial DBMSs.

Concerning issue (iii), the services must be allocated to existing resources and scheduled together with the transactions. Typically, the services are implemented as a set of server processes, and transactions are performed by transaction programs running as processes (cf., [11]). These processes are typically scheduled, dispatched, and executed as a response to the requests from outside the database management system or as a direct or indirect response to a time-out. Each service is either invoked when something is stored in the queue or table or explicitly invoked, for example, when the system clock is updated to reflect the new time. The issues concerning scheduling are paramount in any database management system for real-time system purposes [8].

Event monitoring can either be (i) implicitly invoked whenever an event occurs or it can be (ii) explicitly invoked. This is similar to coupling modes, but it is between the event sources (e.g., transaction processing service and application) and the event monitoring service rather than in between the services of the active capabilities. Case (i) is prevalent in most active database research, but it has a negative impact in terms of determinism of the result of event monitoring. For example, the problem addressed in the event specification and event detection entry concerning the unintuitive semantics of the disjunctive event operator is a result of implicit invocation. In distributed and real-time systems, explicit invocation is preferable in case (ii), since it provides the operating system with the control when something should be evaluated. Explicit invocation solves the problem of disjunction operator (see event specification and event detection entries), since the event expressions defining composite event types can be explicitly evaluated when all events have been delivered to event monitoring rather than implicitly evaluated whenever an event is delivered.

In explicit invocation of event monitoring, the different event contexts can be treated in different ways. For example, in recent event context, only the most recent result is of interest in implicit invocation. However, in terms of explicit invocation, all possible most recent event occurrences may be of interest, not only the last one. For example, it may be desirable to keep the most recent event occurrence per time slot rather than per terminating event.

Issue (iv) has been addressed in, for example, DeeDS [1], COBEA [6], Hermes [12], and X2TS [13]. Further, it has been addressed in event-based systems for mobile networks by Mühl et al. [14]. Essentially, it is necessary to perform event detection in a moving time window, where the end of the time window is the current time. All events that are older than the beginning of the time window can be removed and ignored. Further, the heterogeneity must be addressed, and there are XML-based solutions (e.g., Common Base Events [15]).

Another issue that is significant in distributed active databases is the time and order of events. For example, in Snoop [16] it is suggested to separate global and local event detection, because of the difference in the time granularity of the local view of time and the global (distributed) view of time.

Foundations

For a particular application domain, common significant requirements and properties as well as prerequisites of available resources need to be considered to refine the general architecture. Depending on the requirements, properties, and prerequisites, different compromises are reached. One example is the use of composite event detection in active real-time databases. In REACH [3], composite event detection is disallowed for real-time transactions. The reason for this is that during composite event detection, contributing events are locked and this locking affects other transaction in a harmful way with respect to meeting deadlines. A different approach is proposed in DeeDS [1], where events are stored in the database and cached in a special filtered event log; during event composition, events are not locked thus enabling the use of composite event detection for transaction with critical deadlines. The cost is that isolation of transactions can be violated unless it is handled by the active capabilities.

Availability is an example of a property that significantly affects the software architecture. For example, availability is often considered significant in distributed systems, that is, even though physical nodes may fail, communications links may be down, or the other physical nodes may be overloaded; one should get, at least, some defined level of service from the system. An example of availability requirements is that emergency calls in phone switches should be prioritized over nonemergency calls, a fact that entails that existing phone call connections can be disconnected to let an emergency call through. Another example to improve availability is pursued in DeeDS [1], where eventual consistency is investigated as a mean to improve availability of data. The cost is that data can temporarily be inconsistent.

As addressed in the aforementioned examples, different settings affect the architecture. Essentially, there are two approaches that can be mixed: (i) refine or invent new method, tools, and techniques to solve a problem, and these method, tools, and techniques can stem from different but relevant research areas and (ii) refine the requirements or prerequisites to solve the problem (e.g., weaken the ACID properties of transactions).

Key Applications

The architecture of ADBMSs is of special interest to developers of database management systems and their applications. In particular, software engineering issues are of major interest. Researchers performing experiments can make use of this architecture to enable valid experiments, study effects of optimizations, etc.

Concerning real examples of applications, only simple things such as using rules for implementing alerters has been tested, for example, when an integrity constraint is violated. SQL Triggers implement simple ECA rules in immediate coupling mode between event monitoring and rule evaluation as well as between rule evaluation and action execution.

Researchers have aimed for various application domains such as:

- Stock market
- Inventory control
- Bank applications

Essentially, any application domain in which there is an interest to move functionality from the applications to the database schema in order to reduce the interdependence between applications and databases is a key application.

Future Directions

There are no silver bullets in computer science or software engineering, and each refinement of the architecture (in Fig. 1) is a compromise providing or enabling certain features and properties. For example, by allowing only detached coupling mode, it is easier to achieve timeliness, an important property of real-time systems; however, the trade-off is that it is difficult to specify integrity rules in terms of ECA rules, since the integrity checks are performed in a different transaction. The consequence is that dirty transactions as well as compensating transactions that perform recovery from violated integrity rules must be allowed.

It is desirable to study architectures addressing how to meet specific requirement of the application area (e.g., accounting information in mobile ad hoc networks) and the specific environment in which the active database are used (e.g., distributed systems, real-time systems, mobile ad hoc networks, limited resource equipment). The major criteria for a successful architecture (e.g., by refining an existing architecture) are if anyone can gain something from using it. For example, Borr [17] reported that by refining their architecture by employing transaction processing, they improved productivity, reliability, as well as average throughput in their heterogenous distributed reliable applications.

An area that has received little attention in active database is optimization of processing. For example, how can queries to the database be optimized with condition evaluation if conditions are expressed as arbitrary queries? Another question is how to group actions to optimize performance? So far, the emphasis has been on expressibility as well as techniques on how to enable active support in different settings. Another area that has received little attention is recovery processing, both precrash and postcrash recovery. For example, how should recovery with respect to detached but dependent transactions be managed?

Intertransaction events and rules have been proposed by, for example, Buchmann et al. [3]. How should this be managed with respect to the isolation levels proposed by Gray and Reuter [8, Chap. 7]?

There are several other areas with which active database technology can be combined. Historical examples include real-time databases, temporal databases, main-memory databases, and geographical information systems. One area that has received little attention is how to enable reuse of database schemas.

Cross-References

► Active Database Coupling Modes
► Active Database Execution Model
► Active Database Knowledge Model
► Event Detection
► Event Specification

Recommended Reading

1. Andler S, Hansson J, Eriksson J, Mellin J, Berndtsson M, Eftring B. DeeDS towards a distributed active and real-time database system. ACM SIGMOD Rec. 1996;25(1): 38.
2. Eriksson J. Real-time and active databases: a survey. In: Proceedings of the 2nd International Workshop on Active, Real-Time, and Temporal Database Systems; 1997. p. 1–23.
3. Buchmann AP, Zimmermann J, Blakeley JA, Wells DL. Building an integrated active OODBMS: requirements, architecture, and design decisions. In: Proceedings of the 11th International Conference on Data Engineering; 1995. p. 117–28.
4. Lieuwen DF, Gehani N, Arlein R. The ODE active database: trigger semantics and implementation. In: Proceedings of the 12th International Conference on Data Engineering; 1996. p. 412–20.
5. Gatziu S. Events in an active object-oriented database system. PhD thesis, University of Zurich, Switzerland; 1994.
6. Ma C, Bacon J. COBEA: A CORBA-based event architecture. In: Proceedings of the 4th USENIX Conference on Object-Oriented Technologies and System; 1998. p. 117–32.
7. Paton N, Diaz O. Active database systems. ACM Comput Surv. 1999;31(1):63–103.
8. Berndtsson M, Hansson J. Issues in active real-time databases. In: Proceedings of the 1st International Workshop on Active and Real-Time Database System; 1995. p. 142–50.
9. Ullman JD. Principles of database systems. Rockville: Computer Science; 1982.
10. Moss JEB. Nested transactions: an approach to reliable distributed computing. Cambridge: MIT; 1985.
11. Gray J, Reuter A. Transaction processing: concepts and techniques. Los Altos: Morgan Kaufmann; 1994.
12. Pietzuch P, Bacon JH. A distributed event-based middleware architecture. In: Proceedings of the 22nd International Conference on Distributed Computing Systems Workshop; 2002. p. 611–8.
13. Chakravarthy S, Blaustein B, Buchmann AP, Carey M, Dayal U, Goldhirsch D, Hsu M, Jauhuri R, Ladin R, Livny M, McCarthy D, McKee R, Rosenthal A. HiPAC: a research project in active time-constrained database management. Technical report XAIT-89-02, Xerox Advanced Information Technology; 1989.
14. Mühl G, Fiege L, Pietzuch PR. Distributed event-based systems. Berlin: Springer; 2006.
15. Common Base Events. http://www.ibm.com/developerworks/library/specification/ws-cbe/
16. Dayal U, Blaustein B, Buchmann A, Chakravarthy S, Hsu M, Ladin R, McCarty D, Rosenthal A, Sarin S, Carey MJ, Livny M, Jauharu R. The HiPAC project: combining active databases and timing constraints. ACM SIGMOD Rec. 1988;17(1)
17. Borr AJ. Robustness to crash in a distributed database: a non shared-memory multi-processor approach. In: Proceedings of the 10th International Conference on Very Large Data Bases; 1984. p. 445–53.
18. Chakravarthy S, Krishnaprasad V, Anwar E, Kim SK. Composite events for active database: semantics, contexts, and detection. In: Proceedings of the 20th International Conference on Very Large Data Bases; 1994. p. 606–17.
19. Jaeger U. Event detection in active databases. PhD thesis, University of Berlin; 1997.

20. Liebig CM, Malva AB. Integrating notifications and transactions: concepts and X2TS prototype. In: Proceedings of the 2nd International Workshop on Engineering Distributed Objects; 2000. p. 194–214.

Active Database Rulebase

AnnMarie Ericsson[1], Mikael Berndtsson[2,3], and Jonas Mellin[2,3]
[1]University of Skövde, Skövde, Sweden
[2]University of Skövde, The Informatics Research Centre, Skövde, Sweden
[3]University of Skövde, School of Informatics, Skövde, Sweden

Definition

An active database rulebase is a set of ECA rules that can be manipulated by an active database.

Key Points

An active database rulebase is a set of ECA rules that can be manipulated by an active database. Thus, an ADB rulebase is not static, but it evolves over time. Typically, ECA rules can be added, deleted, modified, enabled, and disabled. Each update of the ADB rulebase can potentially lead to different behaviors of the ECA rules at run time, in particular with respect to termination and confluence.

Termination concerns whether a set of rules is guaranteed to terminate. A set of rules may have a nonterminating behavior if rules are triggering each other in a circular order, for example, if the execution of rule R1 triggers rule R2 and the execution of rule R2 triggers rule R1. A set of rules is confluent if the outcome of simultaneously triggered rules is unique and independent of execution order.

Active Database, Active Database (Management) System

Mikael Berndtsson and Jonas Mellin
University of Skövde, The Informatics Research Centre, Skövde, Sweden
University of Skövde, School of Informatics, Skövde, Sweden

Definition

An active database (aDB) or active database (management) system (aDBS/aDBMS) is a database (management) system that supports reactive behavior through ECA rules.

Historical Background

The term active database was first used in the early 1980s [12]. Some related active database work was also done within the area of expert database systems in the mid 1980s, but it was not until the mid/late 1980s that the research on supporting ECA rules in database systems took off, for example [10, 12, 18]. During the 1990s, the area was extensively explored through more than 20 suggested active database prototypes and a large body of publications:

- Seven workshops were held between 1993 and 1997: RIDS [12, 16, 17], RIDE-ADS [20], Dagstuhl Seminar [5], and ARTDB [3, 4].
- Two special issues of journals [8, 9] and one special issue of ACM SIGMOD Record [1].
- Two textbooks [13, 19] and one ACM Computing Survey paper [15].

Cross-References

► ECA Rules

In addition, the groups within the ACT-NET consortium (A European research network of Excellence on active databases 1993–1996) reached a consensus on what constitutes an active database management system with the publication of the Active Database System Manifesto [2].

Most of the active databases are monolithic and assume a centralized environment; consequently, the majority of the prototype implementations do not consider distributed issues. Initial work on how active databases are affected by distributed issues is reported in [7].

Foundations

An active database can automatically react to events such as database transitions, time events, and external signals in a timely and efficient manner. This is in contrast to traditional database systems, which are passive in their behaviors, so that they only execute queries and transactions when they are explicitly requested to do so. Previous approaches to support reactive behavior can broadly be classified into:

• Periodically polling the database
• Embedding or encoding event detection and related action execution in the application code

The first approach implies that the queries must be run exactly when the event occurs. The frequency of polling can be increased in order to detect such an event, but if the polling is too frequent, then the database is overloaded with queries and will most often fail. On the other hand, if the frequency is too low, the event will be missed.

The second approach implies that every application which updates the database needs to be augmented with condition checks in order to detect events. For example, an application may be extended with code to detect whether the quantity of certain items has fallen below a given level. From a software engineering point of view, this approach is inappropriate, since a change in a

condition specification implies that every application that uses the modified condition needs to be updated.

Neither of the two previous approaches can satisfactorily support reactive behavior in a database context [10]. An active database system avoids the previous disadvantages by moving the support for reactive behavior inside the database (management) system. Reactive behavior in an active database is supported by ECA rules that have the following semantics: when an event is detected, evaluate a condition, and if the condition is true, execute an action.

Similar to describing an object by its static features and dynamic features, an active database can be described by its knowledge model (static features) and execution model (dynamic features). Thus, by investigating the knowledge model and execution model of an active database, one can identify what type of ECA rules that can be defined and how the active database behave at run-time.

Key Applications

An aDB or aDBS/aDBMS is useful for any non-mission critical application that requires reactive behavior.

Future Directions

Looking back, the RIDS workshop marks the end of the active database period, since there are very few active database publications after 1997. However, the concept of ECA rules has resurfaced and has been picked up by other research communities such as Complex Event Processing and Semantic Web. In contrast to typical active database approaches that assume a centralized environment, the current research on ECA rules within Complex Event Processing and Semantic Web assumes that the environment is distributed and heterogeneous. Thus, as suggested within the REWERSE project [3], one cannot assume that the event, condition, and action parts of an ECA rule are defined in one single ECA rule language.

For example, the event part of an ECA rule can be defined in one language (e.g., Snoop), whereas the condition part and action part are defined in a completely different rule language.

The popularity of using XML for manipulating data has also led to proposals of ECA rule markup languages. These ECA rule markup languages are used for storing information about ECA rules and facilitate exchange of ECA rules between different rule engines and applications.

One research question that remains from the active database period is how to model and develop applications that use ECA rules. Some research on modeling ECA rules has been carried out, but there is no widely agreed approach for modeling ECA rules explicitly in UML, or how to derive ECA rules from existing UML diagrams.

Cross-References

▶ Active Database Execution Model
▶ Active Database Knowledge Model
▶ Complex Event Processing
▶ ECA Rules

Recommended Reading

1. ACM SIGMOD (Special Interest Group on Management of Data) Record. Special issue on rule management and processing in expert databases; 1989.
2. ACT-NET consortium the active database management system manifesto: a rulebase of ADBMS features. ACM SIGMOD Rec. 1996; 25(3):40–9.
3. Alferes JJ, Amador R, May W. A general language for evolution and reactivity in the semantic web. In: Proceedings of the 3rd Workshop on Principles and Practice of Semantic Web Reasoning; 2005. p. 101–15.
4. Andler SF, Hansson J, editors. In: Proceedings of the 2nd International Workshop on Active, Real-Time, and Temporal Database Systems; 1998.
5. Berndtsson M, Hansson J. Workshop report: the first international workshop on active and real-time database systems. SIGMOD Rec. 1996;25(1):64–6.
6. Buchmann A, Chakravarthy S, Dittrich K. Active databases. Dagstuhl seminar No. 9412, Report No. 86; 1994.
7. Bültzingsloewen G, Koschel A, Lockemann PC, Walter HD. ECA functionality in a distributed environment, chap. 8. In: Monographs in computer science. New York: Springer; 1999. p. 147–75.
8. Chakravarthy S, editor. IEEE Data Engineering Bulletin. Special issue on active databases, vol. 15. 1992.
9. Chakravarthy S, Widom J, editors. Special issue on the active database systems. J Intell Inf Syst. 1996;7(2):109–10.
10. Dayal U, Blaustein B, Buchmann A, et al. S.C. HiPAC: a research project in active, Time-constrained database management. Technical report CCA-88-02, Xerox advanced information technology, Cambridge; 1988.
11. Dittrich KR, Kotz AM, Mulle JA. An event/trigger mechanism to enforce complex consistency constraints in design databases. ACM SIGMOD Rec. 1986;15(3):22–36.
12. Geppert A, Berndtsson M, editors. In: Proceedings of the 3rd International Workshop on Rules in Database Systems; 1997.
13. Morgenstern M. Active databases as a paradigm for enhanced computing environments. In: Proceedings of the 9th International Conference on Very Data Bases; 1983. p. 34–42.
14. Paton NW, editor. Active rules in database systems. Monographs in computer science. Springer; 1999.
15. Paton NW, Diaz O. Active database systems. ACM Comput Surv. 1999;31(1):63–103.
16. Paton NW, Williams MW, editors. In: Proceedings of the 1st International Workshop on Rules in Database Systems; 1994.
17. Sellis T, editor. In: Proceedings of the 2nd International Workshop on Rules in Database Systems; 1995.
18. Stonebraker M, Hearst M, Potamianos S. Commentary on the POSTGRES rules system. SIGMOD Rec. 1989;18(3):5–11.
19. Widom J, Ceri S, editors. Active database systems: triggers and rules for advanced database processing. Morgan kaufmann; 1996.
20. Widom J, Chakravarthy S, editors. In: Proceedings of the 4th International Workshop on Research Issues in Data Engineering – Active Database Systems; 1994.

Active Storage

Kazuo Goda
The University of Tokyo, Tokyo, Japan

Synonyms

Active Disks; Intelligent Disks

Definition

Active Storage is a computer system architecture which utilizes processing power in disk drives to execute application code. Active Storage was introduced in separate academic papers [1–3] in 1998. The term Active Storage is sometimes identified merely with the computer systems proposed in these papers. Two synonyms, Active Disk and Intelligent Disk, are also used to refer to Active Storage. The basic idea behind Active Storage is to offload computation and data traffic from host computers to the disk drives themselves such that the system can achieve significant performance improvements for data intensive applications such as decision support systems and multimedia applications.

Key Points

A research group at Carnegie Mellon University proposed, in [3], a storage device called Active Disk, which has the capability of downloading application-level code and running it on a processor embedded on the device. Active Disk has a performance advantage for I/O bound scans, since processor-per-disk processing can potentially reduce data traffic on interconnects to host computers and yield great parallelism of scans. E. Riedel et al. carefully studied the potential benefits of using Active Disks for four types of data intensive applications, and introduced analytical performance models for comparing traditional server systems and Active Disks. They also prototyped ten Active Disks, each having a DEC Alpha processor and two Seagate disk drives, and demonstrated almost linear scalability in the experiments.

A research group at University of California at Berkeley discussed a vision of Intelligent Disks (IDISKs) in [2]. The approach of Intelligent Disk is similar to that of Active Disk. Keeton et al. carefully studied the weaknesses of shared-nothing clusters of workstations and then explored the possibility of replacing the cluster nodes with Intelligent Disks for large-scale decision support applications. Intelligent Disks assumed higher complexity of applications and

hardware resources in comparison with CMU's Active Disks.

Another Active Disk was presented by a research group at the University of California at Santa Barbara and University of Maryland in [1]. Acharya et al. carefully studied programming models to exploit disk-embedded processors efficiently and safely and proposed algorithms for typical data intensive operations such as selection and external sorting, which were validated by simulation experiments.

These three works are often recognized as opening the gate for new researches of Intelligent Storage Systems in the post-"database machines" era.

Cross-References

▶ Database Machine
▶ Intelligent Storage Systems

Recommended Reading

1. Acharya A, Mustafa U, Saltz JH. Active disks: programming model, algorithms and evaluation. In: Proceedings of the 8th International Conference Architectural Support for Programming Languages and Operating System; 1998. p. 81–91.
2. Keeton K, Patterson DA, Hellerstein JM. A case for intelligent disks (IDISKs). SIGMOD Rec. 1998;27(3):42–52.
3. Riedel E, Gibson GA, Faloutsos C. Active storage for large-scale data mining and multimedia. In: Proceedings of the 24th International Conference on Very Large Data Bases; 1998. p. 62–73.

Active XML

Serge Abiteboul[1], Omar Benjelloun[2], and Tova Milo[3]
[1]Inria, Paris, France
[2]Google Inc., New York, NY, USA
[3]School of Computer Science, Tel Aviv University, Tel Aviv, Israel

Synonyms

Active document; AXML

Definition

Active XML documents (AXML documents, for short) are XML documents [12] that may include embedded calls to Web services [13]. Hence, AXML documents are a combination of regular "extensional" XML data with data that is defined "Intentionally," i.e., as a description that enables obtaining data dynamically (by calling the corresponding service).

AXML documents evolve in time when calls to their embedded services are triggered. The calls may bring data once (when invoked) or continually (e.g., if the called service is a continuous one, such as a subscription to an RSS feed). They may even update existing parts of the document (e.g., by refreshing previously fetched data).

Historical Background

The AXML language was originally proposed at INRIA around 2002. Work around AXML has been going there in the following years. A survey of the research on AXML is given in [13]. The software, primarily under the form of an AXML system, is available as open source software. Resources on Active XML may be found on the project's website [11].

The notion of embedding function calls into data is old. Embedded functions are already present in relational systems as stored procedures. Of course, method calls form a key component of object databases. For the Web, scripting languages such as PHP or JSP have made popular the integration of processing inside HTML or XML documents. Combined with standard database interfaces such as JDBC and ODBC, functions are used to integrate results of (SQL) queries. This idea can also be found in commercial software products, for instance, in Microsoft Office XP, SmartTags inside Office documents can be linked to Microsoft's.NET platform for Web services.

The originality of the AXML approach is that it proposed to *exchange* such documents, building on the fact that Web services may be invoked from anywhere. In that sense, this is truly a language for distributed data management.

Another particularity is that the logic (the AXML language) is a subset of the AXML algebra.

Looking at the services in AXML as queries, the approach can be viewed as closely related to recent works based on XQuery [14] where the query language is used to describe query plans. For instance, the DXQ project [7] developed at ATT and UCSD emphasizes the distributed evaluation of XQuery queries. Since one can describe documents in an XQquery syntax, such approaches encompass in some sense AXML documents where the service calls are XQuery queries.

The connection with deductive databases is used in [1] to study the diagnosis problems in distributed networks. A similar approach is followed in [8] for declarative network routing.

It should be observed that the AXML approach touches upon most database areas. In particular, the presence of intentional data leads to views, deductive databases, and data integration. The activation of calls contained in a document essentially leads to active databases. AXML services may be activated by external servers, which relates to subscription queries and stream databases. Finally, the evolution of AXML documents and their inherent changing nature lead to an approach of workflows and service choreography in the style of business artifacts [10].

The management of AXML document raises a number of issues. For instance, the evaluation of queries over active documents is studied in [2]. The "casting" of a document to a desired type is studied in [9]. The distribution of documents between several peers and their replication is the topic of [4].

Foundations

An AXML document is an (syntactically valid) XML document, where service calls are denoted by special XML elements labeled *call*. An example AXML document is given in Fig. 1. The figure shows first the XML serialized syntax, then a more abstract view of the same document as a labeled tree. The document in the figure describes a (simplified) newspaper homepage consisting of (i) some extensional information (the name of the

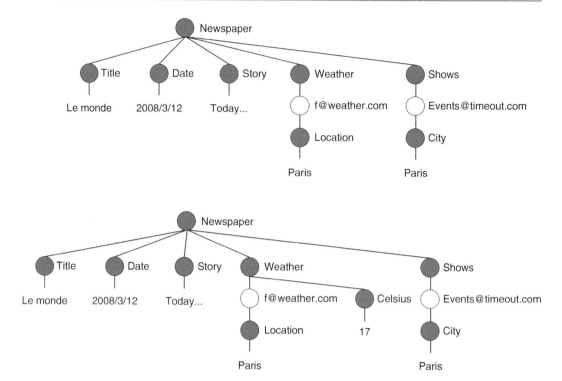

Active XML, Fig. 1   An AXML document

newspaper, the current date, and a news story) and (ii) some intentional information (service calls for the weather forecast and for the current exhibits). When the services are called, the tree evolves. For example, the tree at the bottom is what results from a call to the service *f* at weather. com to obtain the temperature in Paris.

AXML documents fit nicely in a peer-to-peer architecture, where each peer is a persistent store of AXML documents, and may act both as a client, by invoking the service calls embedded in its AXML documents and as a server by providing Web services over these documents.

Two fundamental issues arise when dealing with AXML documents. The first one is related to the exchange of AXML documents between peers, and the second one is related to query evaluation over such data.

Documents Exchange: When exchanged between two applications/peers, AXML documents have a crucial property: since Web services can be called from anywhere on the Web, data can either be materialized before sending or sent in its intentional form and left to the receiver to materialize if and when needed. Just like *XML Schemas*

do for standard XML, *AXML schemas* let the user specify the desired format of the exchanged data, including which parts should remain intentional and which should be materialized. Novel algorithms allow the sender to determine (statically or dynamically) which service invocations are required to "cast" the document to the required data exchange format [9].

Query evaluation: Answering a query on an AXML document may require triggering some of the service calls it contains. These services may, in turn, query other AXML documents and trigger some other services, and so on. This recursion, based on the management of intentional data, leads to a framework in the style of *deductive databases*. Query evaluation on AXML data can therefore benefit from techniques developed in deductive databases such as Magic Sets [6]. Indeed, corresponding AXML query optimization techniques were proposed in [1, 2].

Efficient query processing is, in general, a critical issue for Web data management. AXML, when properly extended, becomes an algebraic language that enables query processors installed on different peers to collaborate by exchanging streams of (A)XML data [14]. The crux of the approach is (i) the introduction of generic services (i.e., services that can be provided by several peers, such as query processing) and (ii) some explicit control of distribution (e.g., to allow delegating part of some work to another peer).

Key Applications

AXML and the AXML algebra target all distributed applications that involve the management of distributed data. AXML is particularly suited for data integration (from databases and other data resources exported as Web services) and for managing (active) views on top of data sources. In particular, AXML can serve as a formal foundation for mash-up systems. Also, the language is useful for (business) applications based on evolving documents in the style of business artifacts and on the exchange of such information. The fact that the exchange is based on flows of XML messages makes it also well adapted to the management of distributed streams of information.

Cross-References

▶ Business Process Execution Language
▶ Web Services
▶ XML
▶ XML Information Integration
▶ XML Programming
▶ XML Publish/Subscribe
▶ XML Types

Recommended Reading

1. Abiteboul S, Abrams Z, Milo T. Diagnosis of asynchronous discrete event systems – datalog to the rescue! In: Proceedings of the 24th ACM SIGACT-SIGMOD-SIGART Symposium on Principles of Database Systems; 2005. p. 358–67.
2. Abiteboul S, Benjelloun O, Cautis B, Manolescu I, Milo T, Preda N. Lazy query evaluation for active XML. In: Proceedings of the ACM SIGMOD International Conference on Management of Data; 2004. p. 227–38.
3. Abiteboul S, Benjelloun O, Milo T. The active XML project, an overview. VLDB J. 2008;17(5):1019–40.
4. Abiteboul S, Bonifati A, Cobena G, Manolescu I, Milo T. Dynamic XML documents with distribution and replication. In: Proceedings of the ACM SIGMOD International Conference on Management of Data; 2003. p. 527–38.
5. Abiteboul S, Manolescu I, Taropa E. A framework for distributed XML data management. In: Advances in Database Technology, Proceedings of the 10th International Conference on Extending Database Technology; 2006.
6. Bancilhon F, Maier D, Sagiv Y, Ullman JD. Magic sets and other strange ways to implement logic programs. In: Proceedings of the 5th ACM SIGACT-SIGMOD Symposium on Principles of Database Systems; 1986. p. 1–15.
7. DXQ. Managing distributed system resources with distributed XQuery. http://db.ucsd.edu/dxq/
8. Loo BT, Condie T, Garofalakis M, Gay DE, Hellerstein JM, Maniatis P, Ramakrishnan R, Roscoe T, Stoica I. In: Proceedings of the ACM SIGMOD International Conference on Management of Data; 2006. p. 97–108.
9. Milo T, Abiteboul S, Amann B, Benjelloun O, Ngoc FD. Exchanging intentional XML data. In: Proceedings of the ACM SIGMOD International Conference on Management of Data; 2003. p. 289–300.
10. Nigam A, Caswell NS. Business artifacts: an approach to operational specification. IBM Syst J. 2003;42(3):428–45.
11. The Active XML homepage. http://www.activexml.net/
12. The Extensible Markup Language (XML) 1.0 (2nd edn). http://www.w3.org/TR/REC-xml
13. The W3C Web Services Activity. http://www.w3.org/2002/ws
14. The XQuery language. http://www.w3.org/TR/xquery

Active, Real-Time, and Intellective Data Warehousing

Mukesh Mohania[1], Ullas Nambiar[2], Hoang Tam Vo[1], Michael Schrefl[3], and Millist Vincent[4]
[1]IBM Research, Melbourne, VIC, Australia
[2]Zensar Technologies Ltd, Pune, India
[3]University of Linz, Linz, Austria
[4]University of South Australia, Adelaide, SA, Australia

Synonyms

Right-time data warehousing

Definition

Active data warehousing is the technical ability to capture transactions when they change and integrate them into the warehouse, along with maintaining batch or scheduled cycle refreshes. An active data warehouse offers the possibility of automating routine tasks and decisions. The active data warehouse exports decisions automatically to the online transaction processing (OLTP) systems.

Real-time data warehousing describes a system that reflects the state of the source systems in real time. If a query is run against the real-time data warehouse to understand a particular facet about the business or entity described by the warehouse, the answer reflects the fully up-to-date state of that entity. Most data warehouses have data that are highly latent and thus reflect the business at a point in the past. In contrast, a real-time data warehouse has low latency data and provides current (or real-time) data.

Simply put, a real-time data warehouse can be built using an active data warehouse with a very low latency constraint added to it. An alternate view is to consider active data warehousing as being a design methodology suited to tactical decision-making based on very current data, while (near) real-time data warehousing is a collection of technologies that refresh a data warehouse frequently. In particular, a (near) real-time data warehouse is one that acquires, cleanses, transforms, stores, and disseminates information in minutes or seconds after the data has arrived at the source systems. An active data warehouse, on the contrary, operates in a non-real-time response mode with one or more OLTP systems.

Intellective data warehousing represents the next generation of systems that incorporate three key cognitive aspects, namely, understanding, contextual awareness, and continuous learning to provide complete data-to-knowledge pipeline without notable human intervention [20]. Intellective data warehouse systems leverage recent advances in machine learning and information retrieval to simplify the ingestion and characterization of data and to collect and manage the critical metadata needed for integration, analytical query processing, result summarization, provenance, explanation, and facet analysis.

Historical Background

A *data warehouse* is a decision support database that is periodically updated by extracting, transforming, and loading operational data from several OLTP databases. In the data warehouse, OLTP data is arranged using the (multi)dimensional data modeling approach (see [1] for a basic approach and [2] for details on translating an OLTP data model into a dimensional model), which classifies data into *measures* and *dimensions*. Several multidimensional data models have been proposed [3–6], whereas an in-depth comparison is provided by Pedersen and Jensen in [5]. The basic unit of interest in a data warehouse is a *measure* or *fact* (e.g., sales), which represents countable, semi-summable, or summable information concerning a business process. An instance of a measure is called *measure value*. A measure can be analyzed from different perspectives, which are called the *dimensions* (e.g., location, product, time) of the data warehouse [7]. A dimension consists of a set of *dimension levels* (e.g., time: Day, Week, Month, Quarter, Season, Year, ALLTimes), which are organized in multiple hierarchies or *dimension paths* [6] (e.g., Time[Day] → Time[Month] → Time[Quarter] → Time[Year] → Time[ALLTimes]; Time[Day] → Time[Week] → Time[Season] → Time[ALLTimes]). The hierarchies of a dimension form a lattice having at least one top dimension level and one bottom dimension level. The measures that can be analyzed by the same set of dimensions are described by a *base cube* or *fact table*. A base cube uses level instances of the lowest dimension levels of each of its dimensions to identify a measure value. The relationship between a set of measure values and the set of identifying level instances is called *cell*. Loading data into the data warehouse means that new *cells* will be added to *base cubes* and new level instances will be added to *dimension levels*. If a dimension D is related to a measure *m* by means of a base cube, then the hierarchies of *D* can be used to aggregate the *measure values* of *m* using operators like SUM, COUNT, or AVG. Aggregating measure values along the hierarchies of different dimensions (i.e., *rollup*) creates a multidimensional view on data, which is known as *data cube* or *cube*. Deaggregating the measures of a cube to a lower dimension level (i.e., *drilldown*) creates a more detailed cube. Selecting the subset of a cube's cells that satisfy a certain selection condition (i.e., *slicing*) also creates a more detailed cube.

The data warehouses are used by analysts to find solutions for decision tasks by using OLAP (online analytical processing) systems [7]. The decision tasks can be split into three, viz., *nonroutine, semi-routine*, and *routine*. Nonroutine tasks occur infrequently and/or do not have a generally accepted decision criteria. For example, strategic business decisions such as introducing a new brand or changing an existing business policy are nonroutine tasks. Routine tasks, on the other hand, are well-structured problems for which generally accepted procedures exist, and they occur frequently and at predictive intervals. Examples can be found in the areas of product assortment (change price, withdraw product, etc.), in customer relationship management (grant loyalty discounts, etc.), and in many administrative areas (accept/reject paper based on review scores). Semi-routine tasks are tasks that require a nonroutine solution, e.g., a paper rated contradictorily must be discussed by program committee. Since most tasks are likely to be routine, it is logical to automate processing of such tasks to reduce the delay in decision-making.

Active data warehouses [8] were designed to enable data warehouses to support automatic decision-making when faced with routine decision tasks and routinizable elements of semi-routine decision tasks. The active data warehouse design extends the technology behind active database systems. Active database technology transforms passive database systems into reactive systems that respond to database and external events through the use of rule processing features [9, 10]. Limited versions of active rules exist in commercial database products [11, 12].

Real-time data warehousing captures business activity data as it occurs. As soon as the business activity is complete and there is data about it, the completed activity data flows into the data warehouse and becomes available instantly. In other words, real-time data warehousing is a framework for deriving information from data as the data becomes available. Traditionally, data warehouses were regarded as an environment for analyzing historical data, either to understand what has happened or simply to log the changes as they happened. However, of late, businesses want to use them to predict the future, e.g., to predict customers likely to churn, and thereby seek better control of the business. Nevertheless, until recently, it was not practical to have *zero-latency* data warehouses – the process of extracting data had too much of an impact on the source systems concerned, and the various steps needed to cleanse and transform the data required multiple temporary tables and took several hours to run. However, the increased visibility of (the value of) warehouse data, and the take-up by a wider audience within the organization, has led to several product developments by IBM [13], Oracle [14], and other vendors that make real-time data warehousing now possible.

Right-time data warehousing is a more sophisticated approach that makes new data quickly available for data warehouses while retaining the insert speeds of bulk loading [24]. The essence of this approach is using a main-memory-based catalyst to provide intermediate storage ("memory tables") for data warehouse tables, which are eventually loaded to its final target (the physical data warehouse tables) at the right time. This approach allows for fast insertions of new data from the source systems. A policy can define when to materialize the new data in the main-memory catalyst into the data warehouse; however, the data can be queried on demand while being held by memory tables as well for end users to immediately access changes in the data sources.

Foundations

Enabling Active Data Warehousing

The two example scenarios below describe typical situations in which active rules can be used to automate decision-making:

Scenario 1: Reducing the price of an article. Twenty days after a soft drink has been launched on a market, analysts compare the quantities sold during this period with a standardized indicator. This indicator requires the total quantities sold during the 20-day period do not drop below a threshold of 10,000 sold items. If the analyzed sales figures are below this threshold, the price of the newly launched soft drink will be reduced by 15%.

Scenario 2: Withdrawing articles from a market. At the end of every quarter, high-priced soft drinks which are sold in Upper Austrian stores will be analyzed. If the sales figures of a high-priced soft drink have continuously dropped, the article will be withdrawn from the Upper Austrian market. Analysts inspect sales figures at different granularities of the time dimension and at different granularities of the location dimension. Trend, average, and variance measures are used as indicators in decision-making.

Rules that mimic the analytical work of a business analyst are called *analysis rules* [8]. The components of analysis rules constitute the *knowledge model* of an active data warehouse (and also a real-time data warehouse). The knowledge model determines *what* an analyst must consider when he specifies an active rule to automate a routine decision task.

An analysis rule consists of (i) the *primary dimension level* and (ii) the *primary condition*, which identify the objects for which decision-making is necessary; (iii) the *event*, which triggers rule processing; (iv) the *analysis graph*, which specifies the cubes for analysis; (v) the *decision steps*, which represent the conditions under which a decision can be made; and (vi) the *action*, which represents the rule's decision task. Below is a brief description of the components of an analysis rule. A detailed discussion is given in [8].

Event: Events are used to specify the time points at which analysis rules should be carried out. Active data warehouses provide three kinds of events: (i) OLTP method events, (ii) relative temporal events, and (iii) calendar events. OLTP method events describe basic events in the data warehouse's sources. Relative temporal events are used to define a temporal distance between such a basic event and carrying out an analysis rule. Calendar events represent fixed points in time at which an analysis rule may be carried out. Structurally, every event instance is characterized by an occurrence time and by an event identifier. In its event part, an analysis rule refers to a calendar event or to a relative temporal event.

An *OLTP method event* describes an event in the data warehouse's source systems that is of interest to analysis rules in the active data warehouse. Besides occurrence time and event identifier, the attributes of an OLTP method event are a reference to the dimension level for which the OLTP method event occurred and the parameters of the method invocation. To make OLTP method events available in data warehouses, a data warehouse designer has to define the schema of OLTP method events and extend the data warehouse's extract/transform/load mechanism. Since instances of OLTP method events are loaded some time after their occurrence, analysis rules cannot be triggered directly by OLTP method events.

Temporal events determine the time points at which decision-making has to be initiated. Scenario 1 uses the relative temporal event "20 days after launch" while scenario 2 uses the periodic temporal event "end of quarter." The *conditions* for decision-making are based on indicators, which have been established in manual decision-making. Each condition refers to a multidimensional cube, and therefore "analyzing" means to evaluate the condition on this cube. Scenario 1 uses a quantity-based indicator, whereas scenario 2 uses value-based indicators for decision-making. The decision whether to carry out the rule's *action* depends on the result of evaluating the conditions. The action of scenario 1 is to reduce the price of an article, whereas the action of scenario 2 is to withdraw an article from a market.

Primary condition: Several analysis rules may share the same OLTP method as their action. These rules may be carried out at different time points and may utilize different multidimensional analyses. Thus, a certain analysis rule usually analyzes only a subset of the level instances that belong to the rule's primary dimension level. The primary condition is used to determine for a level instance of the primary dimension level whether multidimensional analysis should be carried out by the analysis rule. The primary condition is specified as a Boolean expression, which refers to the describing attributes of the primary dimension level. If omitted, the primary condition evaluates to TRUE.

Action: The purpose of an analysis rule is to automate decision-making for objects that are available in OLTP systems and in the data warehouse. A *decision* means to invoke (or not to invoke) a method on a certain object in an OLTP system. In its action part, an analysis rule may refer to a single OLTP method of the primary dimension level, which represents a transaction in an OLTP system. These methods represent the *decision space* of an active data warehouse. To make the transactional behavior of an OLTP object type available in the active data warehouse, the data warehouse designer must provide (i) the specifications of the OLTP object type's methods together with required parameters, (ii) the preconditions that must be satisfied before the OLTP method can be invoked in the OLTP system, and (iii) a conflict resolution mechanism, which solves contradictory decisions of different analysis rules. Since different analysis rules can make a decision for the same level instance of the rules' primary dimension level during the same active data warehouse cycle, a *decision conflict* may occur. Such conflicts are considered as inter-rule conflicts. To detect inter-rule conflicts, a *conflict table* covering the OLTP methods of the decision space is used. The tuples of the conflict table have the form $<m1, m2, m3>$, where $m1$ and $m2$ identify two conflicting methods and $m3$ specifies the conflict resolution method that will be finally executed in OLTP systems. If a conflict cannot be solved automatically, it has to be reported to analysts for manual conflict resolution.

Analysis graph: When an analyst queries the data warehouse to make a decision, he or she follows an incremental top-down approach in creating and analyzing cubes. Analysis rules follow the same approach. To automate decision-making, an analysis rule must "know" the cubes that are needed for multidimensional analysis. These cubes constitute the *analysis graph*, which is specified once by the analyst. The n dimensions of each cube of the analysis graph are classified into one *primary dimension*, which represents the level instances of the primary dimension level, and $n - 1$ *analysis dimensions*, which represent the multidimensional space for analysis. Since a level instance of the primary dimension level is described by one or more cells of a cube, multidimensional analysis means to compare, aggregate, transform, etc., the measured values of these cells. Two kinds of multidimensional analysis are carried out at each cube of the analysis graph: (i) select the level instances of the primary dimension level whose cells comply with the decision-making condition (e.g., withdraw an article if the sales total of the last quarter is below USD 10,000) and (ii) select the level instances of the primary dimension level whose cells comply with the condition under which more detailed analyses (at finer grained cubes) are necessary (e.g., continue analysis if the sales total of the last quarter is below USD 500,000). The multidimensional analysis that is carried out on the cubes of the analysis graph are called decision steps. Each decision step analyzes the data of exactly one cube of the analysis graph. Hence, analysis graph and decision steps represent the knowledge for multidimensional analysis and decision-making of an analysis rule.

Enabling Real-Time Data Warehousing

As mentioned earlier, real-time data warehouses are active data warehouses that are loaded with data having (near) zero latency. Data warehouse vendors have used multiple approaches such as *hand-coded scripting* and data extraction, transformation, and loading (ETL) [15] solutions to serve the data acquisition needs of a data warehouse. However, as users move toward real-time data warehousing, there is a limited choice of technologies that facilitate real-time data delivery. The challenge is to determine the right technology approach or combination of solutions that best meets the data delivery needs. Selection criteria should include considerations for frequency of data, acceptable latency, data volumes, data integrity, transformation requirements, and processing overhead. To solve the real-time challenge, businesses are turning to technologies such as enterprise application integration (EAI) [16] and transactional data management (TDM) [17], which offer high-performance, low impact movement of data, even at large volumes with sub-second speed. EAI has a greater implementation complexity and cost of maintenance and handles smaller volumes of data. TDM provides the ability to capture transactions from OLTP systems and to apply mapping, filtering, and basic transformations and delivers to the data warehouse directly. A more detailed study of the challenges and possible solutions involved in implementing a real-time data warehouse is given in [18], while best practices for real-time data warehousing have recently been described in [14].

Near real-time ETL. There exist applications that do not have high demand for real-time data. In this case, true real-time data warehousing is not strictly required, and it is sufficient to simply increase the frequency of data loading, e.g., from daily to twice a day.

Direct trickle feed. This approach enables true real-time data by continuously moving new data from source systems and updating the fact tables in the data warehouse. However, constant updates on fact tables also lead to degrading performance of the data warehouse due to the contention with other queries by reporting and OLAP tools accessing these fact tables simultaneously.

Trickle and flip. This approach reduces the impact of trickle feed on query performance of the data warehouse by storing updates to the fact tables in staging tables of the same format. These staging tables are then duplicated and swapped with the fact tables on a periodic basis, ranging from hourly to minutes for bringing the data warehouse up to date.

External real-time data cache. All the approaches discussed above require the data warehouse to take additional load handling the incoming real-time data. A real-time data cache external to the data warehouse can be used instead to load real-time data from source systems, hence resolving the query performance and scalability problem by routing any query accessing real-time data to the cache. This cache can be implemented as a main-memory catalyst as proposed in [24].

Log shipping. This technique is originally designed for database replication and recovery purpose. Updates to the primary database are tracked in a transaction log file which are periodically transferred to a secondary database for restoring at this replica. As this technique is efficient in detecting changes to data (i.e., inspecting the log rather than scanning the entire database), it is usually used in the above real-time data warehousing approaches for moving changed data from source systems to or near the data warehouse.

Data stream. The moving of changed data from source systems to the data warehouse can be considered as real-time data event streams. Therefore, stream analysis and complex event processing techniques can be used for analyzing data in real-time, thus eliminating the reliance on batched or offline updating of the data warehouse [25].

Integrating OLTP and OLAP. In the recent years, there has been a growing trend in the integrating both OLTP and OLAP into a single system. For example, a hybrid cloud storage for supporting both transactional and analytical workloads was proposed in [26], whereas main-memory database technologies such as Hyper [27] and SAP HANA [28] have also evolved to serve both OLTP and OLAP within the same database engine. In these systems, OLTP operations access the latest version of the data, whereas the OLAP data analysis tasks execute on a recent consistent snapshot of the database.

Key Applications

Active and real-time data warehouses enable businesses across all industry verticals to gain competitive advantage by allowing them to run *analytics* solutions over the most recent data of interest that is captured in the warehouse. This will provide them with the ability to make intelligent business decisions and better understand and predict customer and business trends based on accurate, up-to-the-second data. By introducing real-time flows of information to data warehouses, companies can increase supply chain visibility, gain a complete view of business performance, and increase service levels, ultimately increasing customer retention and brand value.

The following are some additional business benefits of active and real-time data warehousing:

- *Real-time analytics:* Real-time analytics is the ability to use all available data to improve performance and quality of service at the moment they are required. It consists of dynamic analysis and reporting, right at the moment (or very soon after) the resource (or information) entered the system. In a practical sense, real time is defined by the need of the consumer (business) and can vary from a few seconds to few minutes. In other words, more frequent than daily can be considered real time, because it crosses the overnight-update barrier. With increasing availability of active and real-time data warehouses, the technology for capturing and analyzing real-time data is increasingly becoming available. Learning how to apply it effectively becomes the differentiator. Early detection of fraudulent activity in financial transactions is a potential environment for applying real-time analytics. For example, credit card companies monitor transactions and activate counter measures when a customer's credit transactions fall outside the range of expected patterns. Nevertheless, being able to correctly identify fraud while not offending a well-intentioned valuable customer is a critical necessity that adds complexity to the potential solution.
- *Maximize ERP investments:* With a real-time data warehouse in place, companies can maximize their enterprise resource planning (ERP)

technology investment by turning integrated data into business intelligence. ETL solutions act as an integral bridge between ERP systems that collect high volumes of transactions and business analytics to create data reports.

- *Increase supply chain visibility:* Real-time data warehousing helps streamline supply chains through highly effective business-to-business communications and identifies any weak links or bottlenecks, enabling companies to enhance service levels and gain a competitive cdge.
- *Live 360° view of customers:* The active database solutions enable companies to capture, transform, and flow all types of customer data into a data warehouse, creating one seamless database that provides a 360° view of the customer. By tracking and analyzing all modes of interaction with a customer, companies can tailor new product offerings, enhance service levels, and ensure customer loyalty and retention.

Future Directions

Data warehousing has greatly matured as a technology discipline; however, enterprises that undertake data warehousing initiatives continue to face fresh challenges that evolve with the changing business and technology environment. Most future needs and challenges will come in the areas of active and real-time data warehousing solutions. Listed below are some future challenges:

- *Integration of OLTP and OLAP systems into a single main-memory database system:* Main-memory technologies such as SAP HANA and other in-memory products from Oracle, IBM, Microsoft, Teradata, and Pivotal have started to gain maturity and major adoptions in industry. The capability to work in both analytical and transactional workload environments make these systems highly relevant in scenarios in which real-time in-memory data marts are desirable [29]. These high-speed data marts are complementing existing large-scale data warehouses in today's enterprises and providing near real-time analytics and new business insights.

- *Real-time and on-demand integration with heterogeneous data sources:* The number of enterprise data sources is growing rapidly, with new types of sources emerging every year. Enterprises want to integrate the unstructured data generated from customer emails, chat and voice call transcripts, feedbacks, and surveys with other internal data in order to get a complete picture of their customers and integrate internal processes. Other sources for valuable data include ERP programs, operational data stores, packaged and homegrown analytic applications, and existing data marts. The process of real-time and on-demand integration of these sources into a data warehouse can be complicated and is made even more difficult when an enterprise merges with or acquires another enterprise.

- *Integrating with CRM tools:* Customer relationship management (CRM) is one of the most popular business initiatives in enterprises today. CRM helps enterprises attract new customers and develop loyalty among existing customers with the end result of increasing sales and improving profitability. Increasingly, enterprises want to use the holistic view of the customer to deliver value-added services to the customer based on her overall value to the enterprise. This would include automatically identifying when an important life event is happening and sending out emails with necessary information and/or relevant products, gauging the mood of the customer based on recent interactions, and alerting the enterprise before it is too late to retain the customer and most important of all identifying customers who are likely to accept suggestions about upgrades of existing products/services or be interested in newer versions. The data warehouse is essential in this integration process, as it collects data from all channels and customer touch points and presents a unified view of the customer to sales, marketing, and customer-care employees. Going forward, data warehouses will have to provide support for analytics tools that are embedded into the warehouse, analyze the various customer interactions continuously, and then use the insights to trigger *actions* that enable delivery of the abovementioned value-added services. Clearly, this requires an active data warehouse to be tightly integrated with the CRM systems. If the enterprise has low latency for insight detection and value-added service delivery, then a real-time data warehouse would be required.

- *Built-in data mining and analytics tools:* Users are also demanding more sophisticated business intelligence tools. For example, if a telecom customer calls to cancel his call-waiting feature, real-time analytic software can detect this and trigger a special offer of a lower price in order to retain the customer. The need is to develop a new generation of data mining algorithms that work over data warehouses that integrate heterogeneous data and have self-learning features. These new algorithms must automate data mining and make it more accessible to mainstream data warehouse users by providing explanations with results, indicating when results are not reliable and automatically adapting to changes in underlying predictive models.

- *Intellective data warehousing:* Over the last several decades, data warehousing technologies have played an important role in assisting business decision-makers to derive valuable insights from various data sources. Nevertheless, faced with an ever-increasing quantity and diversity of data in their enterprises, from internal and external sources, today's CDOs (chief data officers) are looking for next-generation systems that can provide complete data-to-knowledge pipeline without notable human intervention at each step including data acquisition; selection, cleaning, and transformation; extraction and integration; mining, OLAP, and analytics; and result summarization, provenance, and explanation [21]. Furthermore, in contrast to traditional data warehouse systems which are mainly query-based and rely much on users thinking hard to find smart questions for querying relevant information from the system, the next-generation *intellective data warehouse systems* are changing the nature of data discovery for their capability to understand the content of data and find related information with respect to interesting facets or potentially useful associations [22]. In addition, these systems can present both the known information and facet analysis to the users in novel ways such as augmented reality visualization.

The need of an intellective data warehouse is further illustrated in the following example. A near up-to-date quantity of a product available across multiple stores can be queried from a centralized data warehouse where data from the stores' databases are periodically pushed into. Now, given a new order from a customer, the system may improve customer satisfaction if it is able to understand every line order and provide real-time discount for specific products that are still available in large quantity in the stores. Another exemplary need of cognition in this application is automatic fulfillment of customer orders using data mining and association. Consider a scenario where the customer's order cannot be fulfilled by the local store (i.e., the query result for local inventory is null). In this case, instead of rejecting the order immediately, which creates customer dissatisfaction, it would be desirable if the system is able to perform reasoning and find other stores that can post the purchased items to the customer's address within the requested time frame. In addition, the ETL process in a data warehouse can also benefit from the capability of an intellective system that can understand input data and reason the relevance and quality of the incoming data for automated data curation.

Enabling intellective data warehousing: To date, the database, information retrieval, machine learning, and artificial intelligence communities have been working in silos when creating intelligent information systems. There is a tremendous opportunity to marry recent advances in these areas and bring cognitive capability including understanding, contextual awareness, and continuous learning into data warehouse systems in order to make them intelligent and powerful. These systems recognize semantic relationships between data items and continuously build and update its knowledge base for answering user's query with rich information beyond the known data. They evaluate multiple features on the fly to provide facet analysis and useful contextual-associated information pertaining to the user query's results. Recently, various search engines [23] have started to support similar capability, but limited to keyword-based entity-seeking queries.

Cross-References

▶ Cube Implementations
▶ Data Warehousing Systems: Foundations and Architectures
▶ Extraction, Transformation, and Loading
▶ Interoperability in Data Warehouses
▶ Multidimensional Modeling
▶ Online Analytical Processing
▶ Query Processing in Data Warehouses

Recommended Reading

1. Kimball R, Strethlo K. Why decision support fails and how to fix it. ACM SIGMOD Rec. 1995;24(3):91–7.
2. Golfarelli M, Maio D, Rizzi S. Conceptual design of data warehouses from E/R schemes. In: Proceedings of the 31st Annual Hawaii International Conference on System Sciences; 1998. p. 334–43.
3. Lehner W. Modeling large scale OLAP scenarios. In: Advances in database technology. In: Proceedings of the 6th International Conference on Extending Database Technology; 1998. p. 153–67.
4. Samtani S, Mohania M, Kumar V, Kambayashi Y. Recent advances and research problems in data warehousing. In: Proceedings of the Workshops on Data Warehousing and Data Mining: Advances in Database Technologies; 1998. p. 81–92.
5. Pedersen TB, Jensen CS. Multidimensional data modeling for complex data. In: Proceedings of the 15th International Conference on Data Engineering; 1999. p. 336–45.
6. Vassiliadis P. Modeling multidimensional databases, cubes and cube operations. In: Proceedings of the 10th International Conference on Scientific and Statistical Database Management; 1998. p. 53–62.
7. Mohania M, Samtani S, Roddick J, Kambayashi Y. Advances and research directions in data-warehousing technology. Australas J Inf Syst. 1999;7(1):2.

8. Thalhammer T, Schrefl M, Mohania M. Active data warehouses: complementing OLAP with analysis rules. Data Knowl Eng. 2001;39(3):241–69.

9. ACT-NET Consortium. The active database management system manifesto: a rulebase of ADBMS features. ACM SIGMOD Rec. 1996;25(3).

10. Simon E, Dittrich A. Promises and realities of active database systems. In: Proceedings of the 21th International Conference on Very Large Data Bases; 1995. p. 642–53.

11. Brobst S. Active data warehousing: a new breed of decision support. In: Proceedings of the 13th International Workshop on Data and Expert System Applications; 2002. p. 769–72.

12. Borbst S, Rarey J. The five stages of an active data warehouse evolution. Teradata Mag. 2001;3(1):38–44.

13. IBM Data Warehousing. https://www.ibm.com/analytics/us/en/data-management/data-warehouse.

14. Best practices for Real-time Data Warehousing. An oracle white paper. 2014. http://www.oracle.com/us/products/middleware/data-integration/realtime-data-warehousing-bp-2167237.pdf

15. Kimball R, Caserta J. The data warehouse ETL toolkit: practical techniques for extracting, cleaning, conforming, and delivering data. Wiley; 2004.

16. Linthicum RS. Enterprise application integration. Addison-Wesley; 1999.

17. Improving SOA with Goldengate TDM Technology. GoldenGate White Paper; 2007.

18. Langseth J. Real-time data warehousing: challenges and solutions. DSSResources.COM; 2004.

19. Paton NW, Diaz O. Active database systems. ACM Comput Surv. 1999; 31(1):63–103.

20. High R. The era of cognitive systems: an inside look at IBM Watson and how it works. IBM Corporation Redbooks; 2012.

21. Abadi D, Agrawal R, Ailamaki A, Balazinska M, Bernstein P, Carey M, Chaudhuri S, Dean J, Doan A, Franklin M, Gehrke J, Haas L, Halevy A, Hellerstein J, Ioannidis Y, Jagadish H, Kossmann D, Madden S, Mehrotra S, Milo T, Naughton J, Ramakrishnan R, Markl V, Olston C, Ooi BC, Re C, Suciu D, Stonebraker M, Walter T, Widom J. The Beckman report on database research. Commun ACM. 2016;59(2):92–9.

22. Jonas J, Sokol L. Data finds data, Chapter 7. In: Segaran T, Hammerbacher J, editors. Beautiful data: the stories behind elegant data solutions. O'Reilly Media; 2009.

23. Chirigati F, Liu J, Korn F, Wu YW, Yu C, Zhang H. Knowledge exploration using tables on the web. PVLDB. 2016;10(3):193–204.

24. Thomsen C, Pedersen TB, Lehner W. RiTE: providing on-demand data for right-time data warehousing. In: Proceedings of the IEEE 24th International Conference on Data Engineering; 2008. p. 456–65.

25. Agrawal D. The reality of real-time business intelligence. In: Castellanos M, Dayal U, Sellis T, editors. Proceedings of the 2nd International Workshop on Business Intelligence for the Real-Time Enterprise; 2009. p. 75–88.

26. Cao Y., Chen C., Guo F., Jiang D., Lin Y., Ooi B. C., Vo H. T., Wu S., Xu Q. ES2: a cloud data storage system for supporting both OLTP and OLAP. In: Proceedings of the 27th International Conference on Data Engineering; 2011. p. 291–302.

27. Kemper A, Neumann T. HyPer: a hybrid OLTP and OLAP main memory database system based on virtual memory snapshots. In: Proceedings of the 27th International Conference on Data Engineering; 2011. p. 195–206.

28. Sikka V, Färber F, Lehner W, Cha SK, Peh T, Bornhövd C. Efficient transaction processing in SAP HANA database: the end of a column store myth. In: Proceedings of the ACM SIGMOD International Conference on Management of Data; 2012. p. 731–42.

29. SAP HANA data warehousing. 2015. http://www.computerweekly.com/feature/SAP-Hana-as-a-data-warehouse

Activity

Nathaniel Palmer
Workflow Management Coalition, Hingham, MA, USA

Synonyms

Node; Step; Task; Work element

Definition

A description of a piece of work that forms one logical step within a process. An activity may be a manual activity, which does not support computer automation, or a workflow (automated) activity. A workflow activity requires human and/or machine resources to support process execution; where human resource is required an activity is allocated to a workflow participant.

Key Points

A process definition generally consists of many process activities which are logically related in terms of their contribution to the overall realization of the business process.

An activity is typically the smallest unit of work which is scheduled by a workflow engine during process enactment (e.g., using transition and pre/post- conditions), although one activity may result in several work items being assigned (to a workflow participant).

Wholly manual activities may form part of a business process and be included within its associated process definition, but do not form part of the automated workflow resulting from the computer supported execution of the process.

An activity may therefore be categorized as "manual," or "automated." Within this document, which is written principally in the context of workflow management, the term is normally used to refer to an automated activity.

Cross-References

▶ Activity Diagrams
▶ Actors/Agents/Roles
▶ Workflow Model

Activity Diagrams

Luciano Baresi
Dipartimento di Elettronica, Informazione e Bioingegneria – Politecnico di Milano, Milano, Italy

Synonyms

Control flow diagrams; Data flow diagrams; Flowcharts; Object flow diagrams

Definition

Activity diagrams, also known as control flow and object flow diagrams, are one of the UML (Unified Modeling Language [11]) behavioral diagrams. They provide a graphical notation to define the sequential, conditional, and parallel composition of lower-level behaviors. These diagrams are suitable for business process modeling and can easily be used to capture the logic of a single use case, the usage of a scenario, or the detailed logic of a business rule. They model the workflow behavior of an entity (system) in a way similar to *state diagrams* where the different activities are seen as the states of doing something. Although they could also model the internal logic of a complex operation, this is not their primary use, and tangled operations should always be decomposed into simpler ones [1, 2].

An activity [3] represents a behavior that is composed of individual elements called *actions*. Actions have incoming and outgoing edges that specify control and data flow from and to other nodes. Activities may form invocation hierarchies by invoking other activities, ultimately resolving to individual actions.

The execution of an activity implies that each contained action be executed zero, one, or more times depending on the execution conditions and the structure of the activity. The execution of an action is initiated by the termination of other actions, the availability of particular objects and data, or the occurrence of external events. The execution is based on token flow (like Petri nets). A token contains an object, datum, or locus of control and is present in the activity diagram at a particular node. When an action begins execution, tokens are accepted from some or all of its input edges, and a token is placed on the node. When an action completes execution, a token is removed from the node, and tokens are moved to some or all of its output edges.

Historical Background

OMG (Object Management Group, [10]) proposed and standardized *activity diagrams* by borrowing concepts from flow-based notations and some formal methods. As for the first class, these diagrams mimic *flowcharts* [6] in their idea of step-by-step representation of algorithms and processes, but they also resemble data and control flow diagrams [4]. The former provide a hierarchical and graphical representation of the "flow" of data through a system inspired by the idea of *data flow graph*. They show the flow of data from external entities into the system, how these data are moved from one computation to another, and how they are logically stored. Similarly, *object flow diagrams* show the relationships among input objects, methods, and output objects in object-based models. Control flow diagrams represent the paths that can be traversed while executing a program. Each node in the graph represents a basic block, be it a single line or an entire function, and edges render how the execution moves among them.

Moving to the second group, activity diagrams are similar to state diagrams [8], where the evolution of a system is rendered by the identification of the states, which characterize the element's life cycle, and of the transitions between them. A state transition can be constrained by the occurrence of an event and by an additional condition; its firing can cause the execution of an associated action. Mealy, Moore, and Harel propose different variations: Mealy assumes that actions be only associated with transitions, Moore only considers actions associated with states, and Harel's *state charts* [7] merge the two approaches with actions on both states and transitions and enhance their flat model with nested and concurrent states.

The dynamic semantics of activity diagrams is clearly inspired by Petri nets [9], which are a simple graphical formalism to specify the behavior of concurrent and parallel systems. The nodes are partitioned into places and transitions, with arcs that can only connect nodes of different type. Places may contain any number of tokens, and a distribution of tokens over the places of a net is called marking. A transition can only fire when there is at least a token in all its input places (i.e., those places connected to the transition by means of incoming edges), and its firing removes a token for all these places and produces a new one in each output place (i.e., a place connected to the transition through an outgoing edge). P/T nets only consider tokens as placeholders, while colored nets augment them

with typed data and thus with firing conditions that become more articulated and can predicate on the tokens' values in the input places.

Activity diagrams also borrow from SDL (Specification and Description Language, [5]). This is a specification language for the unambiguous description of the behavior of reactive and distributed systems. Originally, the notation was conceived for the specification of telecommunication systems, but currently its application is wider and includes process control and real-time applications in general. A system is specified as a set of interconnected abstract machines, which are extensions of finite state machines.

Scientific Fundamentals

Figure 1 addresses the well-known problem of order management and proposes a first *activity diagram* whose aim is twofold: it presents a possible formalization of the process, and it also introduces many of the concepts supplied by these diagrams.

Each atomic step is called **action**, with an **initial node** and **activity final** nodes to delimit their ordering as sequences, parallel threads, or conditional flows. A **fork** splits a single execution thread into a set of parallel ones, while a **join**, along with an optional **join specification** to constrain the unification, is used to re-synchronize the different threads into a single execution. Similarly, a **decision** creates alternative paths, and a **merge** reunifies them. To avoid misunderstandings, each path must be decorated with the condition, in brackets, that must be verified to make the execution take that path.

The diagram of Fig. 1 also exemplifies the use of **connectors** to render **flows/edges** that might tangle the representation. This is nothing but an example, but the solution is interesting to avoid drawing flows that cross other elements or move all around the diagram. Another key feature is the use of a rake to indicate that action fill order is actually an **activity invocation** and hides a hierarchical decomposition of actions into activities.

Besides the control flow, activity diagrams can also show the data/object flow among the actions.

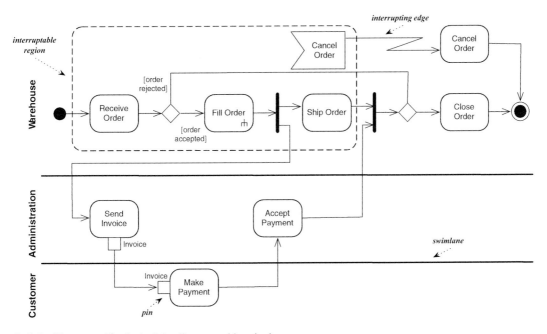

Activity Diagrams, Fig. 1 Example activity diagram

Activity Diagrams, Fig. 2 Activity diagrams with swim lanes

The use of **object nodes** allows users to state the artifacts exchanged between two actions, even if they are not directly connected by an edge. In many cases, control and object flows coincide, but this is not mandatory.

Activities can also comprise **input** and **output parameters** to render the idea that the activity's execution initiates when the inputs are

available, and produces some outputs. For example, activity fill order of Fig. 2, which can be seen as a refinement of the invocation in Fig. 1, requires that at least one request be present, but then it considers the parameter as a **stream** and produces shipment information and rejected items. While the first outcome is the "normal" one, the second object is produced only in case

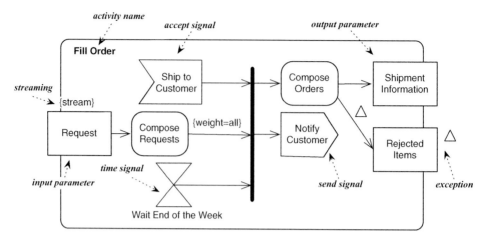

Activity Diagrams, Fig. 3 Example signals

of **exceptions** (rendered with a small triangle on both the object and the flow that produces it). In a stream, the flow is annotated from action compose requests to the join with its **weight** to mean that the subsequent processing must consider *all* the requests received when the composition starts.

The execution can also consider signals as enablers or outcomes of special-purpose actions. For example, Fig. 2 shows the use of an **accept signal**, to force that the composition of orders (compose orders) be initiated by an external event, a **time signal**, to make the execution wait for a given timeframe (be it absolute or relative), and a **send signal**, to produce a notification to the customer as soon as the action starts.

Basic diagrams can also be enriched with **swimlanes** to partition the different actions with respect to their responsibilities. Figure 3 shows a simple example: The primitive actions are the same as those of Fig. 1, but now they are associated with the three players in charge of activating the behaviors in the activity. The standard also supports hierarchical and multidimensional partitioning, that is, hierarchies of responsible actors or matrix-based partitions.

The warehouse can also receive cancel order notifications to asynchronously interrupt the execution as soon as the external event arrivers. This is obtained by declaring an **interruptable region**, which contains the accept signal node and

generates the interrupt that stops the computation in that region and moves the execution directly to action cancel order by means of an **interrupting edge**. More generally, this is a way to enrich diagrams with specialized *exception handlers* similarly to many modern programming and workflow languages. The figure also introduces **pins** as a compact way to render the objects exchanged between actions: empty boxes correspond to discrete elements, while filled ones refer to streams.

The discussion so far considers the case in which the outcome of an action triggers a single execution of another action, but in some cases, conditions may exist in which the "token" is structured and a single result triggers multiple executions of the same action. For example, if the example of Fig. 1 were slightly modified and after receiving an order, the user wants to check the items in it, a single execution of action receive order would trigger multiple executions of validate item. This situation is depicted in the left-hand side of Fig. 4, where the star * renders the information described so far.

The same problem can be addressed in a more complete way (right-hand side of figure) with an **expansion region**. The two arrays are supposed to store the input and output elements. In some cases, the number of input and output tokens is the same, but it might also be the case that the behavior in the region filters the incoming elements.

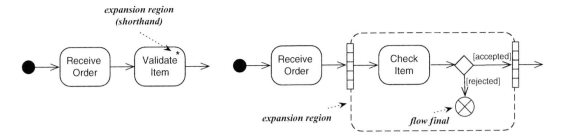

Activity Diagrams, Fig. 4 Expansion region

In the left-hand side of Fig. 4, it is assumed that some items are accepted and fill the output array, while others are rejected and thus their execution flow ends there. This situation requires that a **flow final** be used to state that only the flow is ended and not the whole activity. Flow final nodes are a means to interrupt particular flows in this kind of regions but also in loops or other similar cases.

The execution leaves an expansion region as soon as all the output tokens are available, that is, as soon as all the executions of the behavior embedded in the region are over. Notice that these executions can be carried out both concurrently (by annotating the rectangle with stereotype **concurrent**) or iteratively (with stereotype **iterative**). The next action considers the whole set of tokens as a single entity.

Further details about exceptions and other advanced elements like pre- and post-conditions associated with single actions or whole activities, central buffers, and data stores are not discussed here, but the reader is referred to [11] for a thorough presentation.

Key Applications

Activity diagrams are usually employed to describe complex behaviors. This means that they are useful to model tangled processes, describe the actions that need to take place and when they should occur in use cases, render complicated algorithms, and model applications with parallel and alternative flows. Nowadays, these necessities belong to ICT specialists, like software engineering, requirements experts, and information

systems architects, but also to experts in other fields (e.g., business analysts or production engineers) that need this kind of graphical notations to describe their solutions.

Activity diagrams can be used in isolation, when the user needs a pure control (data) flow notation, but they can also be adopted in conjunction with other modeling techniques such as interaction diagrams, state diagrams, or other UML diagrams. However, activity diagrams should not take the place of other diagrams. For example, even if the border between activity and state diagrams is sometimes blurred, activity diagrams provide a procedural decomposition of the problem under analysis, while state diagrams mostly concentrate on how studied elements behave. Moreover, activity diagrams do not give details about how objects behave or how they collaborate.

Cross-References

▶ Web Services
▶ Workflow Model

Recommended Reading

1. Arlow J, Neustadt I. UML 2 and the unified process: practical object-oriented analysis and design. 3rd ed. Boston: Addison-Wesley Professional; 2005.
2. Booch G, Rumbaugh J, Jacobson I. The unified modeling language user guide. 2nd ed. Boston: Addison-Wesley Professional; 2005.
3. Fowler M. UML distilled: a brief guide to the standard object modeling language. 3rd ed. Boston: Addison-Wesley Professional; 2003.
4. Gane C, Sarson T. Structured system analysis. Englewood Cliffs: Prentice-Hall; 1979.

5. Gaudin E, Najm E, Reed R. Proceedings of the SDL 2007: Design for Dependable Systems, 13th International SDL Forum, Paris, Sept 2007, vol. 4745. Lecture Notes in Computer Science; 2007.
6. Goldstine H. The computer from Pascal to Von Neumann. Princeton: Princeton University Press; 1972. p. 266–67.
7. Harel D, Naamad A. The STATEMATE semantics of statecharts. ACM Trans Softw Eng Methodol. 1996;5(4):293–333.
8. Hopcroft J, Ullman J. Introduction to automata theory, languages, and computation. Harlow: Addison-Wesley; 2002.
9. Murata T. Petri nets: properties, analysis, and applications. Proc IEEE. 1989;77(4):541–80.
10. Object Management Group. http://www.omg.org/
11. OMG. Unified modeling language. http://www.uml.org/

Actors/Agents/Roles

Nathaniel Palmer
Workflow Management Coalition, Hingham, MA, USA

Synonyms

End user; Player; Work performer; Workflow participant

Definition

A resource that performs the work represented by a workflow activity instance.

This work is normally manifested as one or more work items assigned to the workflow participant via the worklist.

Key Points

These terms are normally applied to a human resource, but it could conceptually include machine-based resources such as an intelligent agent.

Where an activity requires no human resource and is handled automatically by a computer application, the normal terminology for the machine-based resource is invoked application.

An actor, agent, or role may be identified directly within the business process definition or (more normally) is identified by reference within the process definition to a role, which can then be filled by one or more of the resources available to the workflow system to operate in that role during process enactment.

Cross-References

▶ Activity
▶ Workflow Model

Adaptive Interfaces

Maristella Matera
Politecnico di Milano, Milan, Italy

Synonyms

Context-aware interfaces; Personalized interfaces

Definition

A specific class of user interfaces that are able to change in some way in response to different characteristics of the user, of the usage environment, or of the task the user is supposed to accomplish. The aim is to improve the user's experience, by providing both interaction mechanisms and contents that best suit the specific situation of use.

Key Points

There are a number of ways in which interface adaptivity can be exploited to support user interaction. The interaction dimensions that are adapted vary among functionality (e.g., error correction or active help), presentation (user presentation of input to the system, system presentation of information to the user), and user tasks (e.g., task simplification based on the user's capabilities). Adaptivity along such dimensions is achieved by capturing and representing into some models a number of characteristics: the user's characteristics (preferences, experience, etc.), the tasks that the user accomplishes through the system, and the characteristics of the information with which the user must be provided.

Due to current advances in communication and network technologies, adaptivity is now gaining momentum. Different types of mobile devices indeed offer support to access – at any time, from anywhere, and with any media – services and contents customized to the users' preferences and usage environments. In this new context, *content personalization*, based on user profile, has demonstrated its benefits for both users and content providers and has been commonly recognized as fundamental factor for augmenting the overall effectiveness of applications. Going one step further, the new challenge in adaptive interfaces is now *context awareness*. It can be interpreted as a natural evolution of personalization, addressing not only the user's identity and preferences but also the environment that hosts users, applications, and their interaction, i.e., the *context*. Context awareness, hence, aims at enhancing the application usefulness by taking into account a wide range of properties of the context of use.

Cross-References

▶ Visual Interaction

Adaptive Middleware for Message Queuing Systems

Christophe Taton[1], Noël de Palma[1], and Sara Bouchenak[2]
[1]INPG – INRIA, Grenoble, France
[2]University of Grenoble I – INRIA, Grenoble, France

Synonyms

Adaptive message-oriented middleware; Autonomous message-oriented middleware; Autonomous message queuing systems

Definition

Distributed database systems are usually built on top of middleware solutions, such as message queuing systems. Adaptive message queuing systems are able to improve the performance of such a middleware through load balancing and queue provisioning.

Historical Background

The use of message oriented middlewares (MOMs) in the context of the Internet has evidenced a need for highly scalable and highly available MOM. A very promising approach to the above issue is to implement performance management as an autonomic software. The main advantages of this approach are: (i) Providing a high-level support for deploying and configuring applications reduces errors and administrator's efforts. (ii) Autonomic management allows the required reconfigurations to be performed without human intervention, thus improving the system reactivity and saving administrator's time. (iii) Autonomic management is a means to save hardware resources, as resources can be allocated only when required (dynamically upon failure or load peak) instead of pre-allocated.

Several parameters may impact the performance of MOMs. Self-optimization makes use of these parameters to improve the performance of the MOM. The proposed self-optimization approach is based on a queue clustering solution: a *clustered queue* is a set of queues each running on different servers and sharing clients. Self-optimization takes place in two parts: (i) the optimization of the clustered queue load-balancing and (ii) the dynamic provisioning of a queue in the clustered queue. The first part allows the overall improvement of the clustered queue performance while the second part optimizes the resource usage inside the clustered queue. Thus the idea is to create an autonomic system that fairly distributes client connections among the queues belonging to the clustered queue and dynamically adds and removes queues in the clustered queue depending on the load. This would allow to use the adequate number of queues at any time.

Foundations

Clustered Queues

A queue is a staging area that contains messages which have been sent by message producers and are waiting to be read by message consumers. A message is removed from the queue once it has been read. For scalability purpose, a queue can be replicated forming a clustered queue. The clustered queue feature provides a load balancing mechanism. A clustered queue is a cluster of queues (a given number of queue destinations knowing each other) that are able to exchange messages depending on their load. Each queue of a cluster periodically reevaluates its load factor and sends the result to the other queues of the cluster. When a queue hosts more messages than it is authorized to do, and according to the load factors of the cluster, it distributes the extra messages to the other queues. When a queue is requested to deliver messages but is empty, it requests messages from the other queues of the cluster. This mechanism guarantees that no queue is hyper-active while some others are lazy, and tends to distribute the work load among the servers involved in the cluster.

Clustered Queue Performance

Clustered queues are standard queues that share a common pool of message producers and consumers, and that can exchange message to balance the load. All the queues of a clustered queue are supposed to be directly connected to each other. This allows message exchanges between the queues of a cluster in order to empty flooded queues and to fill draining queues.

The clustered queue Q_c is connected to N_c message producers and to M_c message consumers. Q_c is composed of standard queues Q_i ($i \in [1..k]$). Each queue Q_i is in charge of a subset of N_i message producers and of a subset of M_i message consumers:

$$\begin{cases} N_c = \sum_i N_i \\ M_c = \sum_i M_i \end{cases}$$

The distribution of the clients between the queues Q_i is described as follows: x_i (resp. y_i) is the fraction of message producers (resp. consumers) that are directed to Q_i.

$$\begin{cases} N_i = x_i \cdot N_c \\ M_i = y_i \cdot M_c \end{cases} , \quad \begin{cases} \sum_i x_i = 1 \\ \sum_i y_i = 1 \end{cases}$$

The standard queue Q_i to which a consumer or producer is directed to cannot be changed after the client connection to the clustered queue. This way, the only action that may affect the client distribution among the queues is the selection of an adequate queue when the client connection is opened.

The clustered queue Q_c is characterized by its aggregate message production rate p_c and its aggregate message consumption rate c_c. The clustered queue Q_c also has a virtual clustered queue length l_c that aggregates the length of all contained standard queues:

$$l_c = \sum_i l_i = p_c - c_c, \quad \begin{cases} p_c = \sum i\ p_i \\ c_c = \sum i\ c_i \end{cases}$$

The clustered queue length l_c obeys to the same law as a standard queue:

1. Q_c is globally stable when $\Delta l_c = 0$. This configuration ensures that the clustered queue is globally stable. However Q_c may observe local unstabilities if one of its queues is draining or is flooded.

2. If $\Delta l_c > 0$, the clustered queue will grow and eventually saturate; then message producers will have to wait.

3. If $\Delta l_c < 0$, the clustered queue will shrink until it is empty; then message consumers will also have to wait.

Now, considering that the clustered queue is globally stable, several scenarios that illustrate

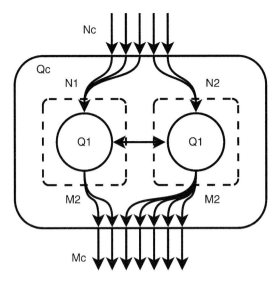

Adaptive Middleware for Message Queuing Systems, Fig. 1 Clustered queue Q_c

the impact of client distribution on performance are given below.

Optimal client distribution of the clustered queue Q_c is achieved when clients are fairly distributed among the k queues Q_i. Assuming that all queues and hosts have equivalent processing capabilities and that all producers (*resp.* consumers) have equivalent message production (*resp.* consumption) rates (and that all produced messages are equivalent: message cost is uniformly distributed), this means that:

$$\begin{cases} x_i = 1/k \\ y_i = 1/k \end{cases}, \quad \begin{cases} N_i = \frac{N_c}{k}, \\ M_i = \frac{M_c}{k} \end{cases}$$

In these conditions, all queues Q_i are stable and the queue cluster is balanced. As a consequence, there are no internal queue-to-queue message exchanges, and performance is optimal. Queue clustering then provides a quasi-linear speedup.

The worst clients distribution appears when one queue only has message producers or only has message consumers. In the example depicted in Fig. 1, this is realized when:

$$\begin{cases} x_1=1 \\ y_1=0 \end{cases}, \begin{cases} x_2 = 0 \\ y_2 = 1 \end{cases}, \begin{cases} N_1 = N_c \\ M_1 = 0 \end{cases}, \begin{cases} N_2 = 0 \\ M_2 = M_c \end{cases}$$

Indeed, this configuration implies that the whole message production is directed to queue Q_1. Q_1 then forwards all messages to Q_2 that in turn delivers messages to the message consumers.

Local instability is observed when some queues Q_i of Q_c are unbalanced. This is characterized by a mismatch between the fraction of producers and the fraction of consumers directed to Q_i:

$$x_i \neq y_i$$

In the example showed in Fig. 1, Q_c is composed of two standard queues Q_1 and Q_2. A scenario of local instability can be envisioned with the following clients distribution:

$$\begin{cases} x_1 = 2/3 \\ y_1 = 1/3 \end{cases}, \begin{cases} x_2 = 1/3 \\ y_2 = 2/3 \end{cases}$$

This distribution implies that Q_1 is flooding and will have to enqueue messages, while Q_2 is draining and will see its consumer clients wait. However the queue cluster Q_c ensures the global stability of the system thanks to internal message exchanges from Q_1 to Q_2.

A stable and unfair distribution can be observed when the clustered queue is globally and locally stable, but the load is unfairly balanced within the queues. This happens when the client distribution is non-uniform.

In the example presented in Fig. 1, this can be realized by directing more clients to Q_1 than Q_2:

$$\begin{cases} x_1 = 2/3 \\ y_1 = 2/3 \end{cases}, \begin{cases} x_2 = 1/3 \\ y_2 = 1/3 \end{cases}$$

In this scenario, queue Q_1 processes two third of the load, while queue Q_2 only processes one third. Suc situation can lead to bad performance since Q_1 may saturates while Q_2 is lazy.

It is worthwhile to indicate that these scenarios may all happen since clients join and leave the system in an uncontrolled way. Indeed, the global stability of a (clustered) queue is under responsability of the application developper. For instance, the queue can be flooded for a period; it is assumed that it will get inverted and draining after, thus providing global stability over time.

Provisioning

The previous scenario of stable and non-optimal distribution raises the question of the capacity of a queue. The capacity C_i of standard queue Q_i is expressed as an optimal number of clients. The queue load L_i is then expressed as the ratio between its current number of clients and its capacity:

$$L_i = \frac{N_i + M_i}{C_i}$$

1. $L_i < 1$: queue Q_i is underloaded and thus lazy; the message throughput delivered by the queue can be improved and resources are wasted.
2. $L_i > 1$: queue Q_i is overloaded and may saturate; this induces a decreased message throughput and eventually leads to thrashing.
3. $L_i = 1$: queue Q_i is fairly loaded and delivers its optimal message throughput.

These parameters and indicators are transposed to queue clusters. The clustered queue Q_c is characterized by its aggregated capacity C_c and its global load L_c:

$$C_c = \sum_i C_i, \quad L_c = \frac{N_c + M_c}{C_c} = \frac{\sum_i L_i \cdot C_i}{\sum_i C_i}$$

The load of a clustered queue obeys to the same law as the load of a standard queue.

However a clustered queue allows to control k, the number of inside standard queues, and thus to control its aggregated capacity $C_c = \sum_{i=1}^{k} C_i$. This control is indeed operated with a re-evaluation of the clustered queue provisioning.

1. When $L_c < 1$, the clustered queue is underloaded: if the clients distribution is optimal, then all the standard queues inside the cluster will be underloaded. However, as the client distribution may be non-optimal, some of the single queues may be overloaded, even if the cluster is globally lazy. If the load is too low, then some queues may be removed from the cluster.
2. When $L_c > 1$, the clustered queue is overloaded: even if the distribution of clients over the queues is optimal, there will exist at least one standard queue that will be overloaded. One way to handle this case is to re-provision the clustered queue by inserting one or more queues into the cluster.

Control Rules for a Self-Optimizing Clustered Queue

The global clients distribution D of the clustered queue Q_c is captured by the fractions of message producers x_i and consumers y_i. The optimal clients distribution D_{opt} is realized when all queues are stable ($\forall i \; x_i = y_i$) and when the load is fairly balanced over all queues ($\forall i, j \, x_i = x_j, y_i = y_j$). This implies that the optimal distribution is reached when $x_i = y_i = 1/k$.

$$D = \begin{bmatrix} x_1 & y_1 \\ \vdots & \vdots \\ x_k & y_k \end{bmatrix}, \quad D_{opt} = \begin{bmatrix} 1/k & 1/k \\ \vdots & \vdots \\ 1/k & 1/k \end{bmatrix}$$

Local instabilities are characterized by a mismatch between the fraction of message producers x_i and consumers y_i on a standard queue. The purpose of this rule is the stability of all standard queues so as to minimize internal queue-to-queue message transfer.

1. $[(R_1)]$ $x_i > y_i$: Q_i is flooding with more message production than consumption and should then seek more consumers and/or fewer producers.
2. $[(R_2)]$ $x_i < y_i$: Q_i is draining with more message consumption than production and should then seek more producers and/or fewer consumers.

Load balancing rules control the load applied to a single standard queue. The goal is then to enforce a fair load balancing over all queues.

1. [(R_3)] $L_i > 1$: Q_i is overloaded and should avoid accepting new clients as it may degrade its performance.
2. [(R_4)] $L_i < 1$: Q_i is underloaded and should request more clients so as to optimize resource usage.

Global provisioning rules control the load applied to the whole clustered queue. These rules target the optimal size of the clustered queue while the load applied to the system evolves.

1. [(R_5)] $L_c > 1$: the queue cluster is overloaded and requires an increased capacity to handle all its clients in an optimal way.
2. [(R_6)] $L_c < 1$: the queue cluster is underloaded and could accept a decrease in capacity.

Key Applications

Adaptive middleware for message queuing systems helps building autonomous distributed systems to improve their performance while minimizing their resource usage, such as distributed Internet services and distributed information systems.

Cross-References

▶ Distributed Database Systems
▶ Distributed DBMS
▶ Message Queuing Systems

Recommended Reading

1. Aron M, Druschel P, Zwaenepoel W. Cluster reserves: a mechanism for resource management in cluster-based network servers. In: Proceedings of the 2000 ACM SIGMETRICS International Conference on Measurement and Modeling of Computer Systems; 2000. p. 90–101.
2. Menth M, Henjes R. Analysis of the message waiting time for the fioranoMQ JMS server. In: Proceedings of the 23rd International Conference on Distributed Computing Systems; 2006. p. 1.
3. Shen K, Tang H, Yang T, Chu L. Integrated resource management for cluster-based internet services. In: Proceedings of the 5th USENIX Symposium on Operating System Design and Implementation; 2002.
4. Urgaonkar B, Shenoy P. Sharc: managing CPU and network bandwidth in shared clusters. IEEE Trans Parall Distrib Syst. 2004;15(1):2–17.
5. Zhu H, Ti H, Yang Y. Demand-driven service differentiation in cluster-based network servers. In: Proceedings of the 20th Annual Joint Conference of the IEEE Computer and Communications Societies; 2001. p. 679–88.

Adaptive Query Processing

Evaggelia Pitoura
Department of Computer Science and Engineering, University of Ioannina, Ioannina, Greece

Synonyms

Adaptive query optimization; Autonomic query processing; Eddies;

Definition

While in traditional query processing, a query is first optimized and then executed, *adaptive query processing* techniques use runtime feedback to modify query processing in a way that provides better response time, more efficient CPU utilization or more useful incremental results. Adaptive query processing makes query processing more robust to optimizer mistakes, unknown statistics, and dynamically changing data, runtime and workload characteristics. The spectrum of adaptive query processing techniques is quite broad: they may span the executions of multiple queries or adapt within the execution of a single query; they may affect the query plan being executed or just the scheduling of operations within the plan.

Key Points

Conventional query processing follows an optimize-then-execute strategy: after generating alternative query plans, the query optimizer selects the most cost-efficient among them and passes it to the execution engine that directly executes it, typically with little or no runtime decision-making. As queries become more complex, this strategy faces many limitations such as missing statistics, unexpected correlations, and dynamically changing data, runtime, and workload characteristics. These problems are aggregated in the case of long-running queries over data streams as well as in the case of queries over multiple potentially heterogeneous data sources across wide-area networks. Adaptive query processing tries to address these shortcomings by using feedback during query execution to tune query processing. The goal is to increase throughput, improve response time or provide more useful incremental results.

To implement adaptivity, regular query execution is supplemented with a control system for monitoring and analyzing at run-time various parameters that affect query execution. Based on this analysis, certain decisions are made about how the system behavior should be changed. Clearly, this may introduce considerable overheads.

The complete space of adaptive query processing techniques is quite broad and varied. Adaptability may be applied to query execution of multiple queries or just a single one. It may also affect the whole query plan being executed or just the scheduling of operations within the plan. Adaptability techniques also differ on how much they interleave plan generation and execution. Some techniques interleave planning and execution just a few times, by just having the plan re-optimized at specific points, whereas other techniques interleave planning and execution to the point where they are not even clearly distinguishable.

A number of fundamental adaptability techniques include:

- *Horizontal partitioning*, where different plans are used on different portions of the data. Partitioning may be explicit or implicit in the functioning of the operator.
- *Query execution by tuple routing*, where query execution is treated as the process of routing tuples through operators and adaptability is achieved by changing the order in which tuples are routed.
- *Plan partitioning*, where execution progresses in stages, by interleaving optimization and execution steps at a number of well-defined points during query execution.
- *Runtime binding decisions*, where certain plan choices are deferred until runtime, allowing the execution engine to select among several alternative plans by potentially re-invoking the optimizer.
- *In-operator adaptive logic,* where scheduling and other decisions are made part of the individual query operators, rather than the optimizer.

Many adaptability techniques rely on a symmetric hash join operator that offers a non-blocking variant of join by building hash tables on both the input relations. When an input tuple is read, it is stored in the appropriate hash table and probed against the opposite table, thus producing incremental output. The symmetric hash join operator can process data from either input, depending on availability. It also enables additional adaptivity, since it has frequent moments of symmetry, that is, points at which the join order can be changed without compromising correctness or losing work.

The eddy operator provides an example of fine-grained run-time control by tuple routing through operators. An eddy is used as a tuple router; it monitors execution, and makes routing decisions for the tuples. Eddies achieve adaptability by simply changing the order in which the tuples are routed through the operators. The degree of adaptability achieved depends on the type of the operators. Pipelined operators, such as the symmetric hash join, offer the most freedom, whereas, blocking operators, such as the sort-merge join, are less suitable since they do not produce output before consuming the input relations in their entirety.

Cross-References

▶ Adaptive Stream Processing
▶ Cost Estimation

- ▶ Multi-query Optimization
- ▶ Query Optimization
- ▶ Query Processing

Recommended Reading

1. Avnur R, Hellerstein JM. Eddies: continuously adaptive query processing. In: Proceedings of the ACM SIGMOD International Conference on Management of Data; 2000. p. 261–72.
2. Babu S, Bizarro P. Adaptive query processing in the looking glass. In: Proceedings of the 2nd Biennial Conference on Innovative Data Systems Research; 2005. p. 238–49.
3. Deshpande A, Ives ZG, Raman V. Adaptive query processing. Found. Trends Databases. 2007;1(1):1–140.

Adaptive Stream Processing

Zachary G. Ives
Computer and Information Science Department, University of Pennsylvania, Philadelphia, PA, USA

Synonyms

Adaptive query processing

Definition

When querying long-lived data streams, the characteristics of the data may change over time or data may arrive in bursts – hence, the traditional model of optimizing a query prior to executing it is insufficient. As a result, most data stream management systems employ feedback-driven *adaptive stream processing*, which continuously re-optimizes the query execution plan based on data and stream properties, in order to meet certain performance or resource consumption goals. Adaptive stream processing is a special case of the more general problem of *adaptive query processing*, with the special prop-erty that intermediate results are bounded in size (by stream windows), but where query processing may have quality-of-service constraints.

Historical Background

The field of adaptive stream processing emerged in the early 2000s, as two separate developments converged. *Adaptive* techniques for database query processing had become an area of increasing interest as Web and integration applications exceeded the capabilities of conventional static query processing [10]. Simultaneously, a number of data stream management systems [1, 6, 8, 12] were emerging, and each of these needed capabilities for query optimization. This led to a common approach of developing feedback-based re-optimization strategies for stream query computation. In contrast to Web-based adaptive query processing techniques, the focus in adaptive stream processing has especially been on maintaining quality of service under overload conditions.

Foundations

Data stream management systems (DSMSs) typically face two challenges in query processing. First, the data to be processed comes from remote feeds that may be subject to significant variations in distribution or arrival rates over the lifetime of the query, meaning that no single query evaluation strategy may be appropriate over the entirety of execution. Second, DSMSs may be *under-provisioned* in terms of their ability to handle bursty input at its maximum rate and yet may still need to meet certain *quality-of-service* or resource constraints (e.g., they may need to ensure data is processed within some latency bound). These two challenges have led to two classes of adaptive stream processing techniques: those that attempt to *minimize the cost* of computing query results from the input data (the problem traditionally faced by query optimization) and those that attempt to manage query processing, possibly at reduced accuracy, in the presence of *limited resources.* This article provides an overview of significant work in each area.

Minimizing Computation Cost

The problem of adaptive query processing to minimize computation cost has been well studied in a variety of settings [10]. What makes the adaptive stream processing setting unique (and unusually tractable) is the fact that joins are performed over *sliding windows* with size bounds: as the data stream exceeds the window size, old data values are expired. This means intermediate state within a query plan operator has constant maximum size, as opposed to being bounded by the size of the input data. Thus a windowed join operator can be modeled as a pair of filter operators, each of which joins its input with the bounded intermediate state produced from the other input. Optimization of joins in data stream management systems becomes a minor variation on the problem of optimizing selection or filtering operators; hence certain theoretical optimality guarantees can actually be made.

Eddies

Eddies [2, 11, 14] are composite dataflow operators that model select-project-join expressions. An eddy consists of a *tuple router*, plus a set of primitive *query operators* that run concurrently, and each has input queues. Eddies come in several variations; the one proposed for distributed stream management uses *state modules* (SteMs) [14, 11]. Figure 1 shows an example of such an eddy for a simplified stream SQL query, which joins three streams and applies a selection predicate over them.

Eddy creation. The eddy is created prior to execution by an optimizer: every selection operation (σ_P in the example) is converted to a corresponding operator; additionally, each base relation to be joined is given a *state module*, keyed on the join attribute, to hold the intermediate state for each base relation [14] ($\bowtie_R$, $\bowtie_S$, $\bowtie_T$). If a base relation appears with multiple different join attributes, then it may require multiple SteMs. In general, the state module can be thought of as one of the hash tables within a symmetric or pipelined hash join. The optimizer also determines whether the semantics of the query force certain operators to execute before others. Such constraints are expressed in an internal routing table, illustrated on the right side of the figure. As a tuple is processed, it is annotated with a *tuple signature* specifying what input streams' data it contains and what operator may have last modified it. The routing table is a map from the tuple signature to a set of *valid routing destinations*, those operators that can successfully process a tuple with that particular signature.

Query execution/tuple routing. Initially, a tuple from an input data stream (R, S, or T) flows into the eddy router. The eddy (i) adds the data to the associated SteM or SteMs and (ii) consults the routing table to determine the set of possible destination operators. It then chooses a destination (using a *policy* to be described later) and sends the tuple to the operator. The operator then either *filters* the tuple or *produces* one or more output tuples, as a result of applying selection conditions

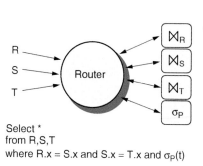

Tuple signature		
Contains	From	Valid routing destinations
{R}	{}	{$\bowtie$S, $\bowtie$T, σ_P}
{S}	{}	{$\bowtie$R, $\bowtie$T, σ_P}
{T}	{}	{$\bowtie$R, $\bowtie$S, σ_P}
{S,T}	{$\bowtie$S,$\bowtie$T}	{$\bowtie$R, σ_P}
{R}	{σ_P}	{$\bowtie$S, $\bowtie$T}
...	...	

Select *
from R,S,T
where R.x = S.x and S.x = T.x and σ_P(t)

Adaptive Stream Processing, Fig. 1 Illustration of eddy with SteMs or joining with the data within a SteM. Output tuples are marked as having been processed by the operator that produced them. If they have been processed by all operators, they will be sent to the query output, and if not, they will be sent back to the eddy's router and to one of the remaining operators.

Routing policies. The problem of choosing among alternate routing destinations has been addressed with a variety of strategies.

Tickets and lottery scheduling [2]. In this scheme, each operator receives a *ticket* for each tuple it receives from the router, and it returns the ticket each time it outputs a tuple to the router. Over time, each operator is expected to have a number of tickets proportional to $(1 - p)$ where p is the operator's selectivity. The router holds a *lottery* among valid routing destinations, where each operator's chance of winning is proportional to its number of tickets. Additionally, as a flow control mechanism, each operator has an input queue, and if this queue fills, then the operator may not participate in the lottery.

Deterministic with batching [9]. A later scheme was developed to reduce the per-tuple overhead of eddies by choosing destinations for batches of tuples. Here, each operator's selectivity is explicitly monitored, and each predicate is assumed to be independent. Periodically, a *rank ordering* algorithm is used to choose a destination for a *batch* of tuples: the rank ordering algorithm sorts predicates in decreasing order of $c_i/(1 - p_i)$, where c_i is the cost of the applying predicate σ_i and p_i is its selectivity.

Content-based routing [7]. CBR attempts to learn correlations between attribute values and selectivities. Using sampling, the system determines for each operator the attribute most strongly correlated with its selectivity – this is termed the *classifier attribute*. CBR then builds a table characterizing all operators' selectivities for different values of each classifier attribute. Under this policy, when the eddy needs to route a tuple, it first looks up the tuple's classifier attribute values in the table and determines the destination operators' selectivities. It routes the tuple probabilistically, choosing a next operator

with probability inversely proportional to its selectivity.

Other optimization strategies. An alternative strategy that does not use the eddy framework is the *adaptive greedy* [5] (A-greedy) algorithm. A-greedy continuously monitors the selectivities of query predicates using a *sliding window profile*, a table with one Boolean attribute for each predicate in the query, and sampling. As a tuple is processed by the query, it may be chosen for sampling into the sliding window profile – if so, it is tested against every query predicate. The vector of Boolean results is added as a row to the sliding window profile. Then the sliding window profile is then used to create a *matrix view* $V[i, j]$ containing, for each predicate σ_i, the number of tuples in the profile that satisfy $\sigma_1 \ldots \sigma_{i-1}$ but not σ_j. From this matrix view, the re-optimizer seeks to maintain the constraint that the ith operation over an input tuple must have the lowest cost/selectivity ratio $c_i/(1 - p(S_i|S_1,\ldots, S_{i-1}))$. The overall strategy has one of the few performance guarantees in the adaptive query processing space: if data properties were to converge, then performance would be within a factor of 4 of optimal [5].

Managing Resource Consumption

A common challenge in data stream management systems is limiting the use of resources – or accommodating limited resources while maintaining quality of service – in the case of bursty data. We discuss three different problems that have been studied: load shedding to ensure input data is processed by the CPU as fast as it arrives, minimizing buffering and memory consumption during data bursts, and minimizing network communication with remote streaming sites.

Load Shedding

Allows the system to selectively drop data items to ensure it can process data as it arrives. Both the Aurora and STREAM DSMSs focused heavily on adaptive load shedding.

Aurora. In the Aurora DSMS [15], load shedding for a variety of query types is supported: the main requirement is that the user has a *utility function* describing the value of output data relative to how much of it has been dropped. The system seeks to place load shedding operators in the query plan in a way that maximizes the user's utility function, while the system achieves sufficient throughput. Aurora precomputes conditional load shedding plans, in the form of a *load shedding road map* (LRSM), containing a sequence of plans that shed progressively more load; this enables the runtime system to rapidly move to strategies that shed more or less load.

LRSMs are created using the following heuristics: first, load shedding points are only inserted at data input points or at points in which data is split to two or more operators. Second, for each load shedding point, a *loss/gain ratio* is computed: this is the reduction in output utility divided by the gain in cycles, $R(p \cdot L - D)$, where R is the input rate into the drop point, p is the ratio of tuples to be dropped, L is the amount of system load flowing from the drop point, and D is the cost of the drop operator. Drop operators are injected at load shedding points in decreasing order of loss/gain ratio. Two different types of drops are considered using the same framework: *random drop*, in which an operator is placed in the query plan to randomly drop some fraction p of tuples, and *semantic drop*, which drops the p tuples of lowest utility. Aurora assumes for the latter case that there exists a utility function describing the relative worth of different attribute values.

Stanford STREAM. The Stanford STREAM system [4] focuses on aggregate (particularly SUM) queries. Again the goal is to process data at the rate it arrives while minimizing the inaccuracy in query answers: specifically, the goal is to minimize the *maximum relative error across all queries*, where the relative error of a query is the difference between actual and approximate value, divided by the actual value.

A statistics manager monitors computation and provides estimates of each operators selectivity and its running time, as well as the mean value and standard deviation of each query q_i's aggregate operator. For each q_i, STREAM computes an error threshold C_i, based on the mean, standard deviation, and number of values. (The results are highly technical, so the reader is referred to [4]

for more details.) A sampling rate P_i is chosen for query q_i that satisfies $P_i \geq C_i/\in_i$, where $\in_i$ is the allowable relative error for the query.

As in Aurora's load shedding scheme, STREAM only inserts load shedding operators at the inputs or at the start of shared segments. Moreover, if a node has a set of children who all need to shed load, then a portion of the load shedding can be "pulled up" to the parent node, and all other nodes can be set to shed some amount of additional load *relative* to this. Based on this observation, STREAM creates a query dataflow graph in which each path from source to sink initially traverses through a load shedding operator whose sampling rate is determined by the desired error rate, followed by additional load shedding operators whose sampling rate is expressed relative to that first operator. STREAM iterates over each path, determines a sampling rate for the initial load shedding operator to satisfy the load constraint, and then computes the maximum relative error for any query. From this, it can set the load shedding rates for individual operators.

Memory Minimization

STREAM also addresses the problem of minimizing the amount of space required to buffer data in the presence of burstiness [3]. The Chain algorithm begins by defining a *progress chart* for each operator in the query plan: this chart plots the relative size of the operator output versus the time it takes to compute. A point is plotted at time 0 with the full size of the input, representing the start of the query; then each operator is given a point according to its cost and relative output size. Now a *lower envelope* is plotted on the progress chart: starting with the initial point at time 0, the *steepest* line is plotted to any operator to the right of this point; from the point at the end of the first line, the next steepest line is plotted to a successor operator; etc. Each line segment (and the operators whose points are plotted beside it) represents a chain, and operators within a chain are scheduled together. During query processing, at each time "tick," the scheduler considers all tuples that have been output by any chain. The tuple that lies on the segment with the *steepest* _slope_ is the one that is scheduled next; as a tiebreaker, the earliest of such tuple is scheduled. This Chain algorithm is proven to be near optimal (differing by at most one unit of memory per operator path for queries where selectivity is at most one).

Minimizing Communication

In some cases, the constrained resource is the network rather than CPU or memory. Olston et al. [13] develop a scheme for reducing network I/O for AVERAGE queries, by using accuracy bounds. Each remote object O is given a _bound width_ w_O: the remote site will only notify the central query processor if O's value V falls outside this bound. Meanwhile, the central site maintains a _bound cache_ with the last value and the bound width for every object.

If given a _precision constraint_ δ_j for each query Q_j, then if the query processor is to provide query answers within δ_j, the sum of the bound widths for the data objects of Q_j must not exceed δ_j times the number of objects. The challenge lies in the selection of widths for the objects.

Periodically, the system tries to tighten all bounds, in case values have become more stable; objects whose values fall outside the new bounds get reported back to the central site. Now some of those objects' bounds must be loosened in a way that maintains the precision constraints over all queries. Each object O is given a _burden score_ equal to $c_O/(p_O\, w_O)$, where c_O is the cost of sending the object, w_O is its bound width, and p_O is the frequency of updates since the previous width adjustment. Using an approximation method based on an iterative linear equation solver, Olston et al. compute a _burden target_ for each query, i.e., the lowest overall burden score required to always meet the query's precision constraint. Next, each object is assigned a _deviation_, which is the maximum difference between the object's burden score and any query's burden target. Finally, a queried objects' bounds are adjusted in decreasing order of deviation, and each object's bound is increased by the largest amount that still conforms to the precision constraint for every query.

Key Applications

Data stream management systems have seen significant adoption in areas such as sensor monitoring and processing of financial information. When there are associated quality-of-service constraints that might require load shedding, or when the properties of the data are subject to significant change, adaptive stream processing becomes vitally important.

Future Directions

One of the most promising directions of future study is how to best use a combination of offline modeling, selective probing (in parallel with normal query execution), and feedback from query execution to find optimal strategies quickly. Algorithms with certain optimality guarantees are being explored in the online learning and theory communities (e.g., the k-armed bandit problem), and such work may lead to new improvements in adaptive stream processing.

Cross-References

▶ Distributed Data Streams
▶ Query Processor
▶ Stream Processing

Recommended Reading

1. Abadi DJ, Carney D, Cetintemel U, Cherniack M, Convey C, Lee S, Stonebraker M, Tatbul N, Zdonik S. Aurora: a new model and architecture for data stream management. VLDB J. 2003;12(2):120–39.
2. Avnur R, Hellerstein JM. Eddies: continuously adaptive query processing. In: Proceedings of the ACM SIGMOD International Conference on Management of Data; 2000. p. 261–72.
3. Babcock B, Babu S, Datar M, Motwani R. Chain: operator scheduling for memory minimization in data stream systems. In: Proceedings of the ACM SIGMOD International Conference on Management of Data; 2003. p. 253–64.
4. Babcock B, Datar M, Motwani R. Load shedding for aggregation queries over data streams. In: Proceed-

ings of the 20th International Conference on Data Engineering; 2004. p. 350.

5. Babu S, Motwani R, Munagala K, Nishizawa I, Widom J. Adaptive ordering of pipelined stream filters. In: Proceedings of the ACM SIGMOD International Conference on Management of Data; 2004. p. 407–18.

6. Balazinska M, BalaKrishnan H, Stonebraker M. Demonstration: load management and high availability in the Medusa distributed stream processing system. In: Proceedings of the ACM SIGMOD International Conference on Management of Data; 2004. p. 929–30.

7. Bizarro P, Babu S, De Witt DJ, Widom J. Content-based routing: different plans for different data In: Proceedings of the 31st International Conference on Very Large Data; 2005. p. 757–68.

8. Chandrasekaran S, Cooper O, Deshpande A, Franklin MJ, Hellerstein JM, Hong W, Krishnamurthy S, Madden S, Raman V, Reiss F, Shah MA. TelegraphCQ: continuous dataflow processing for an uncertain world. In: Proceedings of the 1st Biennial Conference on Innovative Data Systems Research; 2003.

9. Deshpande A. An initial study of overheads of eddies. ACM SIGMOD Rec. 2004;33(1): 44–9.

10. Deshpande A, Ives Z, Raman V. Adaptive query processing. Found. Trends Databases. 2007;1(1): 1–140.

11. Madden S, Shah MA, Hellerstein JM, Raman V. Continuously adaptive continuous queries over streams. In: Proceedings of the ACM SIGMOD International Conference on Management of Data; 2002. p. 49–60.

12. Motwani R, Widom J, Arasu A, Babcock B, Babu S, Datar M, Manku G, Olston C, Rosenstein J, Varma R. Query processing, resource management, and approximation in a data stream management system. In: Proceedings of the 1st Biennial Conference on Innovative Data Systems Research; 2003.

13. Olston C, Jiang J, Widom J. Adaptive filters for continuous queries over distributed data streams. In: Proceedings of the ACM SIGMOD International Conference on Management of Data; 2003. p. 563–74.

14. Raman V, Deshpande A, Hellerstein JM. Using state modules for adaptive query processing. In: Proceedings of the 19th International Conference on Data Engineering; 2003. p. 353–66.

15. Tatbul N, Cetintemel U, Zdonik SB, Cherniack M, Stonebraker M. Load shedding in a data stream manager. In: Proceedings of the 29th International Conference on Very Large Data Bases; 2003. p. 309–20.

Administration Model for RBAC

Yue Zhang and James B. D. Joshi
University of Pittsburgh, Pittsburgh, PA, USA

Synonyms

ARBAC97; SARBAC

Definition

The central ideal of administration model for RBAC is to use the role itself to manage roles. There are two well-known families of administration RBAC models.

Administrative RBAC

The Administrative RBAC family of models known as ARBAC97 [3] introduces administrative roles that are used to manage the regular roles. These roles can form a role hierarchy and may have constraints. ARBAC97 consists of three administrative models, the user-role assignment (URA97) model, the permission-role assignment (PRA97) model, and the role-role administration (RRA97) model. URA97 defines which administrative roles can assign which users to which regular roles by means of the relation: *can_assign*. Similarly, PRA97 defines which administrative roles can assign which permissions to which regular roles by means of the relation: *can_assignp*. Each of these relations also has a counterpart for revoking the assignment (e.g., *can_revoke*). RRA97 defines which administrative roles can change the structure (add roles, delete roles, add edges, etc.) of which range of the regular roles using the notion of *encapsulated range* and the relation: *can_modify*.

Scoped Administrative RBAC

The SARBAC model uses the notion of *administrative scope* to ensure that any operations executed by a role *r* will not affect other roles due to the hierarchical relations among them

[1]. There are no special administrative roles in SARBAC, and each regular role has a *scope* of other regular roles called *administrative scope* that can be managed by it. Each role can only be managed by its administrators. For example, a senior-most role should be able to manage all its junior roles.

Key Points

ARBAC model is the first known role-based administration model and uses the notion of *range* and *encapsulated range*. Role *range* is essentially a set of regular roles. To avoid undesirable side effects, RRA97 requires that all role ranges in the *can_modify* relation be encapsulated, which means the range should have exactly one senior-most role and one junior-most role. Sandhu et al. later extended the ARBAC97 model into AR-BAC99 model where the notion of mobile and immobile user/permission was introduced [4]. Oh et al. later extended ARBAC99 to ARBAC02 by adding the notion of organizational structure to redefine the user-role assignment and the role-permission assignment [2]. Recently, Zhang et al. have proposed an ARBAC07 model that extends the family of ARBAC models to deal with an RBAC model that allows hybrid hierarchies to co-exit [6].

SARBAC

The most important notion in SARBAC is that of the *administrative scope*, which is similar to the notion of *encapsulated range* in ARBAC97. A role *r* is said to be within the administrative scope of another role *a* if every path upwards from *r* goes through *a*; and *a* is said to be the administrator of *r*. SARBAC also consists of three models: SARBAC-RHA, SARBAC-URA, and SARBAC-PRA. In SARBAC-RHA, each role can only administer the roles that are within its own administrative scope. The operations include adding roles, deleting roles, adding permissions, and deleting permissions. The semantics for SARBAC-URA and SARBAC-PRA is similar to URA97 and PRA97. The administra-

tive scope can change dynamically. Zhang et al. have extended SARBAC to also deal with hybrid hierarchy [5].

Cross-References

▶ Role-Based Access Control

Recommended Reading

1. Crampton J, Loizou G. Administrative scope: a foundation for role-based administrative models. ACM Trans Inf Syst Secur. 2003;6(2):201–31.
2. Oh S, Sandhu R. A model for role administration using organization structure. In: Proceedings of the 7th ACM Symposium on Access Control Models and Technologies; 2002. p. 155–62.
3. Sandhu R, Bhamidipati V, Munawer Q. The ARBAC97 model for role-based administration of roles. ACM Trans Inf Syst Secur. 1999;2(1):105–35.
4. Sandhu R, Munawer Q. The ARBAC99 model for administration of roles (1999). In: Proceedings of the 15th Computer Security Applications Conference; 1999. p. 229.
5. Zhang Y, James B, Joshi D. SARBAC07: scoped administration model for RBAC with hybrid hierarchy. In: Proceedings of the 3rd International Symposium on Information Assurance and Security; 2007, p. 149–54.
6. Zhang Y, Joshi JBD. ARBAC07: a role based administration model for RBAC with hybrid hierarchy. In: Proceedings of the IEEE International Conference Information Reuse and Integration; 2007, p. 196–202.

Administration Wizards

Philippe Bonnet[1] and Dennis Shasha[2]
[1]Department of Computer Science, IT University of Copenhagen, Copenhagen, Denmark
[2]Department of Computer Science, New York University, New York, NY, USA

Definition

Modern database systems provide a collection of utilities and programs to assist a database administrator with tasks such as database installation and configuration, import/export, indexing (index wizards are covered in the self-management entry), and backup/restore.

Historical Background

Database Administrators have been skeptical of any form of automation as long as they could control the performance and security of a relatively straightforward installation. The advent of enterprise data management towards the end of the 1990s, where few administrators became responsible for many, possibly diverse database servers, has led to the use of graphical automation tools. In the mid-1990s, third party vendors introduced such tools. With SQL Server 6.5, Microsoft was the first constructor to provide an administration wizard.

Foundations

Installation and Configuration
Database servers are configured using hundreds of parameters that control everything buffer size, file layout, concurrency control options and so on. They are either set statically in a configuration file before the server is started, or dynamically while the server is running. Out-of-the-box database servers are equipped with a limited set of typical configurations.

The installation/configuration wizard is a graphical user interface that guides the administrator through the initial server configuration. The interface provides high-level choices (e.g., OLTP vs. OLAP workload), or simple questions (e.g., number of concurrent users) that are mapped onto database configuration values (log buffer size and thread pool size respectively).

Data Import/Export
Import/export wizards are graphical tools that help database administrators map a database schema with an external data format (e.g., XML, CSV, PDF), or generate scripts that automate the transfer of data between a database and an external data source (possibly another database server).

Back-Up/Restore
Back-up/restore wizards automate the back-up procedure given a few input arguments: complete/incremental backup, scope of the back-up/restore operations (file, tablespace, database), target directory.

Key Applications

Automation of the central database administration tasks.

Recommended Reading

1. Bersinic D, Gile S. Portable DBA: SQL Server. New York: McGraw Hill; 2004.
2. Schumacher R. DBA tools today. DBMS Magazine, January 1997.

Advanced Information Retrieval Measures

Tetsuya Sakai
Waseda University, Tokyo, Japan

Definition

Advanced information retrieval measures are effectiveness measures for various types of information access tasks that go beyond traditional document retrieval. Traditional document retrieval measures are suitable for *set retrieval* (measured by precision, recall, F-measure, etc.) or *ad hoc ranked retrieval*, the task of ranking documents by relevance (measured by average precision, etc.). Whereas, advanced information retrieval measures may work for *diversified search* (the task of retrieving relevant and diverse documents), *aggregated search* (the task of retrieving from multiple sources/media and merging the results), *one-click access* (the task of returning a textual multidocument summary instead of a list of URLs in response to a query), and *multiquery sessions* (information-seeking activities that involve query reformulations), among other tasks. Some advanced measures are based on user models that arguably better reflect real user behaviors than standard measures do.

Historical Background

More Graded Relevance Measures

Historically, information retrieval evaluation mainly dealt with binary relevance: relevant or nonrelevant. But after Järvelin and Kekäläinen proposed *normalized discounted cumulative gain* (nDCG) in 2002 [5], it quickly became the most widely used evaluation measure that utilizes graded relevance. For example, it is known that search engine companies routinely use (variants of) nDCG for internal evaluations. Interestingly, nDCG happens to be a variant of the *normalized sliding ratio* measure proposed by Pollock in 1968 [8]. The advent of web search, where some web pages can be much more important than others (e.g., in home page finding), is probably one factor that made nDCG so popular.

nDCG, however, is not the only graded-relevance measure available. *Q-measure*, a simple variant of *average precision*, was proposed in 2004 [15]. (It is called Q-measure because it was originally designed for factoid question answering with equivalence classes of correct answers.) Q-measure has been used in several tasks of NTCIR (NII Testbeds and Community for Information access Research http://research.nii.ac.jp/ntcir/): cross-lingual information retrieval, information retrieval for question answering, geotemporal information retrieval, and community question answering. *Expected Reciprocal Rank* (ERR), proposed in 2009 [2], has an intuitive user behavior model particularly suitable for *navigational* search where only a few relevant documents are

required [3]. ERR is now a popular evaluation measure.

Other graded-relevance measures not covered here include *Rank-Biased Precision* (RBP) proposed in 2008 [7] and *Graded Average Precision* (GAP) proposed in 2010 [9].

Diversified Search Measures

If the user's query is *ambiguous* or *underspecified*, it is difficult for a search engine to guess the *intent* of the user. To make the same first search result page an acceptable entry point for different users with different intents but sharing the same query, *search result diversification* has recently become a popular research topic. In 2003, Zai, Cohen, and Lafferty [17] defined *subtopic retrieval* as well as an evaluation measure called *subtopic recall*: given a query, how many of its possible intents does the set of retrieved documents cover? Subtopic recall is also known as *intent recall* [13]. Diversified search effectiveness measures, α-nDCG [4] and *ERR-IA* [3], were proposed originally in 2008 and in 2009, respectively, and TREC launched the web track diversity task in 2009, using these measures for evaluating participating systems. As a third alternative to evaluating diversified search, *D-measure* [13] (an instance of which is D-nDCG) was proposed in 2011: unlike α-nDCG and ERR-IA, this new family of diversity measures is free from the *NP-complete* problem of computing the *ideal ranked list*. D-nDCG and its variant D♯-nDCG (a linear combination of D-nDCG and intent recall) have been used in the NTCIR INTENT task since 2011.

Beyond Document Retrieval: Beyond Single Queries

For some types of queries and search environments, returning a concise, direct answer may be more effective than returning the so-called ten blue links. Hence information retrieval evaluation should probably go beyond document retrieval and beyond evaluation based on document-level relevance. The task of returning a direct answer in response to a query is similar to question answering and query-focussed text summarization; for evaluating these related tasks, the evaluation unit is often *nuggets* or *semantic content units* which are more fine grained than documents or passages. *S-measure* [14] proposed in 2011 is an evaluation measure that in a way bridges the gap between textual output evaluation and information retrieval evaluation: while S-measure is based on nugget-level relevance, it also has a discounting mechanism similar to that of nDCG. While nDCG and other information retrieval measures discount the value of a relevant document based on its rank, S-measure discounts the value of a relevant piece of text based on its position within the textual output.

U-measure [12, 15], proposed in 2013, generalizes the idea of S-measure: it can evaluate question answering, query-focussed summarization, ad hoc and diversified document retrieval, and even multiquery sessions. The key concept behind U-measure is the *trail text*, which represents all pieces of texts read by the user during an information-seeking activity. U-measure discounts the value of each relevant text fragment based on how much text the user has read so far. Unlike common evaluation measures, U-measure does not rely on the *linear traversal* assumption, which says that the user always scans the ranked list from top to bottom. For example, given some click data, U-measure can quantify the difference between a linear traversal and a nonlinear traversal with the same set of clicks. Measures related to U-measure include *Time-Biased Gain* (TBG) by Smucker and Clarke [16] and session-based measures described in Kanoulas et al. [6], but the measures as described in these papers assume linear traversal. Another advantage of U-measure over rank-based effectiveness measures is that it can (just like TBG) consider realistic user behaviors such as reading snippets and reading documents of various lengths. Just like ERR, U-measure possesses the diminishing return property.

Evaluating Evaluation Measures

So which evaluation measures should researchers choose? The most important selection criterion is whether the measures are measuring what we want to measure: this may be tested to some extent through user studies. Another important aspect is the statistical stability of the measures: do the measures give us reliable conclusions? One moderately popular method for comparing measures from this viewpoint is the *discriminative power* proposed in 2006 [15]: a significance test is conducted for every pair of systems, and the sorted p-values (i.e., the probability of obtaining the observed difference or something more extreme under the null hypothesis) are plotted against the system pairs. Another way to compare the statistical stability of measures is to utilize the *sample size design* technique [11]: to achieve a given level of *statistical power*, how many topics does each evaluation measure require? The latter method enables us to compare measures in terms of *practical significance*, as the number of topics is basically proportional to the total relevance assessment cost. From the statistical viewpoint, it is known that measures like nDCG and Q-measure are much more reliable than others such as ERR; as for diversity measures, D-nDCG is much more reliable than α-nDCG and ERR-IA [11, 15]. While the diminishing return property of ERR is intuitive and important, this very property makes the measure rely on a small number of data points (i.e., retrieved relevant documents) and thereby hurts statistical stability.

Scientific Fundamentals

In an ad hoc information retrieval task (i.e., ranking documents by relevance to a given query), it is assumed that we have a test collection consisting of a set of topics (i.e., search requests), a set of (graded) relevance assessments for each topic, and a target document corpus. The topic set is assumed to be a sample from the population, and systems are often compared in terms of *mean Q-measure*, *mean* ERR, and so on. In the case of diversified search, it is assumed that each topic has a known set of intents and that a set of (graded) relevance assessments is available *for each intent*. In addition, the intent probability given the query is assumed to be known for each intent. Systems are compared in terms of mean α-nDCG, mean ERR-IA, mean D-nDCG, etc. Confidence intervals, p-values, and effect sizes should be reported with these evaluation results [10].

In addition to the information available from a test collection, U-measure (just like TBG) requires the document lengths of retrieved documents or their estimates. U-measure may also be used with time-stamped clicks instead of relevance assessments whereby nonlinear traversals can be quantified unlike common effectiveness measures.

Key Applications

Advanced information retrieval measures are very important for designing effective search engines. The effective search engines should accommodate not only desktop PC users but also mobile and wearable device users.

Future Directions

The future of information retrieval evaluation has been discussed recently in a SIGIR Forum paper [1]; it is possible that information retrieval evaluation will move more toward *information* retrieval rather than document retrieval or "ten blue links" and toward mobile and ubiquitous information access rather than desktop.

Data Sets

Test collections for evaluating various information retrieval and access tasks are available from evaluation forums such as TREC (http://trec.nist. gov), NTCIR (http://research.nii.ac.jp/ntcir/), and CLEF (http://www.clef-initiative.eu/).

Tools for evaluating advanced information retrieval measures include `trec_eval` (http://trec.nist.gov/trec_eval/), `ndeval` (https://github.com/trec-web/trec-web-2013/tree/master/src/eval), and `NTCIREVAL` (http://research.nii.ac.jp/ntcir/tools/ntcireval-en.html). A few tools for evaluating evaluation measures in terms of

discriminative power are available from http://www.f.waseda.jp/tetsuya/tools.html.

Cross-References

- ▶ α-nDCG
- ▶ Average Precision
- ▶ D-Measure
- ▶ ERR-IA
- ▶ Expected Reciprocal Rank
- ▶ F-Measure
- ▶ Precision
- ▶ Q-Measure
- ▶ Recall
- ▶ Standard Effectiveness Measures
- ▶ U-Measure

Recommended Reading

1. Allan J, Croft B, Moffat A, Sanderson M, editors. Frontiers, challenges and opportunities for information retrieval: report from SWIRL 2012. SIGIR Forum. 2012;46(1):2–32.
2. Chapelle O, Metzler D, Zhang Y, Grinspan P. Expected reciprocal rank for graded relevance. In: Proceedings of the 18th ACM International Conference on Information and Knowledge Management; 2009. p. 621–30.
3. Chapelle O, Ji S, Liao C, Velipasaoglu E, Lai L, Wu SL. Intent-based diversification of web search results: metrics and algorithms. Inf Retr. 2011;14(6):572–92.
4. Clarke CLA, Craswell N, Soboroff I, Ashkan A. A comparative analysis of cascade measures for novelty and diversity. In: Proceedings of the 4th ACM International Conference on Web Search and Data Mining; 2011. p. 75–84.
5. Järvelin K, Kekäläinen J. Cumulated gain-based evaluation of IR techniques. ACM Trans Information Syst. 2002;20(4):422–46.
6. Kanoulas E, Carterette B, Clough PD, Sanderson M. Evaluating multi-query sessions. In: Proceedings of the 34th Annual International ACM SIGIR Conference on Research and Development in Information Retrieval; 2011. p. 1026–53.
7. Moffat A, Zobel J. Rank-biased Precision for measurement of retrieval effectiveness. ACM Trans Information Syst. 2008;27(1):2:1–2:27.
8. Pollock SM. Measures for the comparison of information retrieval systems. Am Doc. 1968;19(4):387–97.
9. Robertson SE, Kanoulas E, Yilmaz E. Extending average Precision to graded relevance judgments. In:

Proceedings of the 33rd Annual International ACM SIGIR Conference on Research and Development in Information Retrieval; 2010. p. 603–10.

10. Sakai T. Statistical reform in information retrieval? SIGIR Forum. 2014;48(1):3–12.

11. Sakai, T. Inf Retrieval J (2016) 19(3):256. https://doi.org/10.1007/s10791-015-9273-z

12. Sakai T, Dou Z. Summaries, ranked retrieval and sessions: a unified framework for information access evaluation. In: Proceedings of the 36th Annual International ACM SIGIR Conference on Research and Development in Information Retrieval; 2013. p. 473–82.

13. Sakai T, Song R. Evaluating diversified search results using per-intent graded relevance. In: Proceedings of the 34th Annual International ACM SIGIR Conference on Research and Development in Information Retrieval; 2011. p. 1043–52.

14. Sakai T, Kato MP, Song YI. Click the search button and be happy: evaluating direct and immediate information access. In: Proceedings of the 20th ACM International Conference on Information and Knowledge Management; 2011. p. 621–30.

15. Sakai T. Metrics, statistics, tests. In: PROMISE winter school 2013: bridging between information retrieval and databases, Bressanone. LNCS, vol 8173. 2014.

16. Smucker MD, Clarke CLA. Time-based calibration of effectiveness measures. In: Proceedings of the 35th Annual International ACM SIGIR Conference on Research and Development in Information Retrieval; 2012. p. 95–104.

17. Zhai C, Cohen WW, Lafferty J. Beyond independent relevance: methods and evaluation metrics for subtopic retrieval. In: Proceedings of the 26th Annual International ACM SIGIR Conference on Research and Development in Information Retrieval; 2003. p. 10–7.

Aggregation: Expressiveness and Containment

Sara Cohen
The Rachel and Selim Benin School of Computer Science and Engineering, The Hebrew University of Jerusalem, Jerusalem, Israel

Definition

An *aggregate function* is a function that receives as input a multiset of values and returns a single value. For example, the aggregate function *count* returns the number of input values. An *aggregate query* is simply a query that mentions an aggregate function, usually as part of its output. Aggregate queries are commonly used to retrieve concise information from a database, since they can cover many data items while returning few. Aggregation is allowed in SQL, and the addition of aggregation to other query languages, such as relational algebra and Datalog, has been studied.

The problem of determining *query expressiveness* is to characterize the types of queries that can be expressed in a given query language. The study of query expressiveness for languages with aggregation is often focused on determining how aggregation increases the ability to formulate queries. It has been shown that relational algebra with aggregation (which models SQL) has a *locality property*.

Query containment is the problem of determining, for any two given queries q and q', whether $q(D) \subseteq q'(D)$, for all databases D, where $q(D)$ is the result of applying q to D. Similarly, the *query equivalence problem* is to determine whether $q(D) = q'(D)$ for all databases D. For aggregate queries, it seems that characterizing query equivalence may be easier than characterizing query containment. In particular, almost all known results on query containment for aggregate queries are derived by a reduction from query equivalence.

Historical Background

The SQL standard defines five aggregate functions, namely, *count*, *sum*, *min*, *max*, and *avg* (average). Over time, it has become apparent that users would like to aggregate data in additional ways. Therefore, major database systems have added new built-in aggregate functions to meet this need. In addition, many database systems now allow the user to extend the set of available aggregate functions by defining his own aggregate functions.

Aggregate queries are typically used to summarize detailed information. For example, consider a database with the relations *Dept(*deptId, *deptName)* and *Emp(*empId, *deptId, salary)*. The following SQL query returns the number of employees, and the total department expenditure on salaries, for each department which has an average salary above $10,000.

```
(Q1) SELECT deptID, count(empID),
     sum(salary) FROM Dept, Emp
     WHERE Dept.deptID = Emp.deptID
     GROUP BY Dept.deptID
     HAVING avg(salary) > 10,000
```

Typically, aggregate queries have three special components. First, the *GROUP BY* clause is used to state how intermediate tuples should be grouped before applying aggregation. In this example, tuples are grouped by their value of *deptID*, i.e., all tuples with the same value for this attribute form a single group. Second, a *HAVING* clause can be used to determine which groups are of interest, e.g., those with average salary above $10,000. Finally, the outputted aggregate functions are specified in the *SELECT* clause, e.g., the number of employees and the sum of salaries.

The inclusion of aggregation in SQL has motivated the study of aggregation in relational algebra, as an abstract modeling of SQL. One of the earliest studies of aggregation was by Klug [11], who extended relational algebra and relational calculus to allow aggregate functions and showed the equivalence of these two languages. Aggregation has also been added to Datalog. This has proved challenging since it is not obvious what semantics should be adopted in the presence of recursion [15].

Foundations

Expressiveness
The study of query expressiveness deals with determining what can be expressed in a given query language. The expressiveness of query languages with aggregation has been studied both for the language of relational algebra and for Datalog, which may have recursion.

Various papers have studied the expressive power of nonrecursive languages, extended with aggregation, e.g., [7, 9, 13]. The focus here will be on [12], which has the cleanest, general proofs

for the expressive power of languages modeling SQL.

In [12], the expressiveness of variants of relational algebra, extended with aggregation, was studied. First, [12] observes that the addition of aggregation to relational algebra strictly increases its expressiveness. This is witnessed by the query $Q2$:

```
(Q2) SELECT 1
     FROM R1
     WHERE (SELECT COUNT(*) FROM R) >
     (SELECT COUNT(*) FROM S)
```

Observe that $Q2$ returns 1 if R contains more tuples than S and otherwise an empty answer. It is known that first-order logic cannot compare cardinalities, and hence neither can relational algebra. Therefore, SQL with aggregation is strictly more expressive than SQL without aggregation.

The language $\mathrm{ALG}_{\mathrm{aggr}}$ is presented in [12]. Basically, $\mathrm{ALG}_{\mathrm{aggr}}$ is relational algebra, extended by arbitrary aggregation and arithmetic functions. In $\mathrm{ALG}_{\mathrm{aggr}}$, nonnumerical selection predicates are restricted to using only the equality relation (and not order comparisons). A *purely relational query* is one which is applied only to nonnumerical data. It is shown that all purely relational queries in $\mathrm{ALG}_{\mathrm{aggr}}$ are *local*. Intuitively, the answers to local queries are determined by looking at small portions of the input.

The formal definition of local queries follows. Let D be a database. The *Gaifman graph* $G(D)$ of D is the undirected graph on the values appearing in D, with $(a,b) \in G(D)$ if a and b belong to the same tuple of some relation in D. Let $\vec{a} = (a_1,...,a_k)$ be a tuple of values, each of which appears in D. Let r be an integer, and let $S_r^D\left(\vec{a}\right)$ be the set of values b such that $dist\,(a_i,b) \leq r$ in $G(D)$, for some i. The *r-neighborhood* $N_r^D\left(\vec{a}\right)$ of $\vec{a}$ is a new database in which the relations of D are restricted to contain only the values in $S_r^D\left(\vec{a}\right)$. Then, $\vec{a}$ and $\vec{b}$ are *(D,r)-equivalent* if there is an isomorphism $h : N_r^D\left(\vec{a}\right) \to N_r^D\left(\vec{b}\right)$ such that $h\left(\vec{a}\right) = \vec{b}$. Finally, a q is *local* if there exists a number r such that for all

D, if $\left(\overrightarrow{a}\right)$ and $\left(\overrightarrow{b}\right)$ are (D, r)-equivalent, then $\overrightarrow{a} \in q(D)$ if and only if $\overrightarrow{b} \in q(D)$.

There are natural queries that are not local. For example, transitive closure (also called reachability) is not local. Since all queries in $\mathrm{ALG_{aggr}}$ are local, this implies that transitive closure cannot be expressed in $\mathrm{ALG_{aggr}}$.

In addition to $\mathrm{ALG_{aggr}}$, [12] introduces the languages $\mathrm{ALG}_{\mathrm{aggr}}^{\leq,\mathrm{N}}$ and $\mathrm{ALG}_{\mathrm{aggr}}^{\leq,\mathrm{Q}}$. $\mathrm{ALG}_{\mathrm{aggr}}^{\leq,\mathrm{N}}$ and $\mathrm{ALG}_{\mathrm{aggr}}^{\leq,\mathrm{Q}}$ are the extensions of $\mathrm{ALG_{aggr}}$ which allow order comparisons in the selection predicates and allow natural numbers and rational numbers, respectively, in the database. It is not known whether transitive closure can be expressed in $\mathrm{ALG}_{\mathrm{aggr}}^{\leq,\mathrm{N}}$. More precisely, [12] shows that if transitive closure is not expressible in $\mathrm{ALG}_{\mathrm{aggr}}^{\leq,\mathrm{N}}$, then the complexity class Uniform $\mathrm{TC^0}$ is properly contained in the complexity class NLOGSPACE. Since the latter problem (i.e., determining strict containment of $\mathrm{TC^0}$ in NLOGSPACE) is believed to be very difficult to prove, so is the former. Moreover, this result holds even if the arithmetic functions are restricted to $\{+, \cdot, <, 0, 1\}$ and the aggregate functions are restricted to $\{sum\}$. On the other hand, $\mathrm{ALG}_{\mathrm{aggr}}^{\leq,\mathrm{Q}}$ extended by arbitrary aggregation and arithmetic functions can express all computable queries.

The languages $\mathrm{ALG_{aggr}}$, $\mathrm{ALG}_{\mathrm{aggr}}^{\leq,\mathrm{N}}$, and $\mathrm{ALG}_{\mathrm{aggr}}^{\leq,\mathrm{Q}}$ are based on relational algebra and, therefore, do not allow recursion. The Datalog language allows queries to be defined as programs, containing recursion. The meaning of an aggregate function within a recursive program is not always well defined. One solution is to restrict the program to have only *stratified aggregation*. Stratification means that if a derived predicate p is defined by applying an aggregate function on a derived predicate q, then the definition of q does not depend, syntactically, upon the definition of p. For example, consider the following Datalog program, P_1.

$$p\left(X, \mathrm{sum}(Y)\right) \leftarrow q\left(X, Y\right)$$

$$q\left(X, Y\right) \leftarrow a\left(X, Y\right)$$

$$q\left(X, Z\right) \leftarrow q\left(X, Y\right), q\left(Y, Z\right)$$

The program P_1 is stratified. Replacing the final rule in P_1 with

$$q\left(X, Z\right) \leftarrow q\left(X, Y\right), p\left(Y, Z\right)$$

would yield a program with nonstratified aggregation.

The expressiveness of stratified aggregation was studied in [14]. Only the aggregate functions *sum*, *avg.*, *min*, *max*, and *count* were allowed. It is shown that stratified aggregation cannot express *summarized explosion* (i.e., the number of instances of a part needed to construct a bigger part). On the other hand, if the language is extended to allow the function $+$, as well as the constants 0 and 1, then all computable queries on the integer domain can be expressed. This is correct even if the only aggregate function allowed is *max*. Additional results of this type, i.e., expressibility of other fragments of stratified Datalog, also appear in [14].

Query Containment

The equivalence and containment problems for aggregate queries have been studied for nonrecursive Datalog programs. A survey of the containment and equivalence problems for aggregate queries, containing references to most works on this topic, appears in [2].

Deriving general characterizations of containment (or equivalence) for aggregate queries is difficult, since each aggregate function tends to have its own idiosyncrasies. For example, *count* is sensitive to the number of occurrences of each value, but not to the values themselves, whereas *max* ignores repeated values but is sensitive to the exact values appearing. As another example, *sum* ignores the value 0, whereas *prod* ignores 1. In addition, *prod* always returns 0 if it is applied to a bag containing 0.

Due to aggregate function quirks, it is often the case that equivalent queries are no longer so, if the aggregate function appearing in their head changes. To demonstrate, consider the two pairs of queries q_1, q'_1 and q_2, q'_2.

$$q_1\left(X, \mathrm{count}\right) \leftarrow a\left(X, Y\right)$$

$$q'_1(X, \text{count}) \leftarrow a(X, Y), a(X, Z)$$

$$q_2(X, \max(Y)) \leftarrow a(X, Y)$$

$$q'_2(X, \max(Y)) \leftarrow a(X, Y), a(X, Z)$$

The queries q_1 and q_2 (and similarly q'_1 and q'_2) have the same conditions in their body and differ only on the output aggregate function. One may show that q_1 is not equivalent to q'_1 (nor is there containment in either direction), as witnessed by the database

$$D_1 = \{a(\text{c}, 0), a(\text{c}, 1), a(\text{d}, 0)\}$$

over which $q_1(D_1) = \{(c, 2), (d, 1)\}$ and $q'_1(D_1) = \{(c, 4), (d, 1)\}$. On the other hand, $q_2 \equiv q'_2$ does hold.

The different oddities of aggregate functions make finding a general solution for the equivalence and containment problems very difficult. Thus, characterizations for equivalence of aggregate queries often are defined separately for each aggregate function. Most known characterizations for equivalence are based on checking for the existence of special types of mappings between the queries. For example, conjunctive queries (i.e., Datalog programs consisting of a single rule and no negation) with the aggregate function *count* are equivalent if and only if they are isomorphic [1, 4].

For other types of aggregate functions, as well as for *count* queries with comparisons or disjunctions, isomorphism is not a necessary condition for equivalence. To demonstrate, each pair of queries q_i, q'_i below is equivalent, yet not isomorphic:

$$q_3(\text{count}) \leftarrow b(X), b(Y), b(Z), X < Y, X < Z$$
$$q'_3(\text{count}) \leftarrow b(X), b(Y), b(Z), X < Z, Y < Z$$

$$q_4(\text{sum}(Y)) \leftarrow b(Y), b(Z), Y > 0, Z > 0$$
$$q'_4(\text{sum}(Y)) \leftarrow b(Y), b(Z), Y \geq 0, Z > 0$$

$$q_5(\text{avg}(Y)) \leftarrow b(Y)$$
$$q'_5(\text{avg}(Y)) \leftarrow b(Y), b(Z)$$

$$q_6(\max(Y)) \leftarrow b(Y), b(Z_1), b(Z_2), Z_1 < Z_2$$
$$q'_6(\max(Y)) \leftarrow b(Y), b(Z), Z < Y$$

Characterizations for equivalence are known for queries of the above types. Specifically, characterizations have been presented for equivalence of conjunctive queries with the aggregate functions *count*, *sum*, *max*, and *count-distinct* [4], and these were extended in [5] to queries with disjunctive bodies. Equivalence of conjunctive queries with *avg* and with *percent* was characterized in [8].

It is sometimes possible to define classes of aggregate functions and then present general characterizations for equivalence of queries with any aggregate function within the class of functions. Such characterizations are often quite intricate since they must deal with many different aggregate functions. A characterization of this type was given in [6] to decide equivalence of aggregate queries with *decomposable* aggregate functions, even if the queries contain negation. Intuitively, an aggregate function is decomposable if partially computed values can easily be combined together to return the result of aggregating an entire multiset of values, e.g., as is the case for *count*, *sum*, and *max*.

Interestingly, when dealing with aggregate queries, it seems that the containment problem is more elusive than the equivalence problem. In fact, for aggregate queries, containment is decided by reducing to the equivalence problem. A reduction of containment to equivalence is presented for queries with *expandable* aggregate functions in [3]. Intuitively, for expandable aggregate functions, changing the number of occurrences of values in bags B and B' does not affect the correctness of the formula $\alpha(B) = \alpha(B')$, as long as the proportion of each value in each bag remains the same, e.g., as is the case for *count*, *sum*, *max*, *count-distinct*, and *avg*.

The study of aggregate queries using the *count* function is closely related to the study of nonaggregate queries evaluated under *bag-set semantics*. Most past research on query containment and equivalence for nonaggregate queries assumed that queries are evaluated under *set semantics*. In set semantics, the output of a query does not contain duplicated tuples. (This corresponds to SQL queries with the *DISTINCT* operator.) Under *bag-set semantics*, the result of a query is a multiset of values, i.e., the same value may appear many times. A related semantics is *bag semantics* in which both the database and the query results may contain duplication.

To demonstrate the different semantics, recall the database D_1 defined above. Consider evaluating, over D_1, the following variation of q_1:

$$q_1''(X) \leftarrow a\,(X, Y)$$

Under set semantics $q''_1(D_1) = \{(c), (d)\}$, and under bag-set semantics $q''_1(D_1) = \{\{(c), (c), (d)\}\}$. Note the correspondence between bag-set semantics and using the *count* function, as in q_1, where *count* returns exactly the number of duplicates of each value. Due to this correspondence, solutions for the query containment problem for queries with the *count* function immediately give rise to solutions for the query containment problem for nonaggregate queries evaluated under bag-set semantics and vice versa.

The first paper to directly study containment and equivalence for nonaggregate queries under bag-set semantics was [1], which characterized equivalence for conjunctive queries. This was extended in [4] to queries with comparisons, in [5] to queries with disjunctions, and in [6] to queries with negation.

Key Applications

Query Optimization
The ability to decide query containment and equivalence is believed to be a key component in query optimization. When optimizing a query, the database can use equivalence characterizations to remove redundant portions of the query or to find an equivalent, yet cheaper, alternative query.

Query Rewriting
Given a user query q, and previously computed queries $v_1, \ldots, v_n$, the query rewriting problem is to find a query r that (i) is equivalent to q and (ii) uses the queries $v_1, ..., v_n$ instead of accessing the base relations. (Other variants of the query rewriting problem have also been studied.) Due to Condition (i), equivalence characterizations are needed to solve the query rewriting problem. Query rewriting is useful as an optimization technique, since it can be cheaper to use past results, instead of evaluating a query from scratch. Integrating information sources is another problem that can be reduced to the query rewriting problem.

Future Directions

Previous work on query containment does not consider queries with *HAVING* clauses. Another open problem is containment for queries evaluated under bag-set semantics. In this problem, one wishes to determine if the bag returned by q is always sub-bag of that returned by q'. (Note that this is different from the corresponding problem of determining containment of queries with *count*, which has been solved). It has shown [10] that bag-set containment is undecidable for conjunctive queries containing inequalities. However, for conjunctive queries without any order comparisons, the general problem of determining bag-set containment is still an open problem.

Cross-References

▶ Answering Queries Using Views
▶ Bag Semantics
▶ Data Aggregation in Sensor Networks
▶ Expressive Power of Query Languages
▶ Locality
▶ Query Containment
▶ Query Optimization (in Relational Databases)
▶ Rewriting Queries Using Views

Recommended Reading

1. Chaudhuri S, Vardi MY. Optimization of *real* conjunctive queries. In: Proceedings of the 12th ACM SIGACT-SIGMOD-SIGART Symposium on Principles of Database Systems; 1993. p. 59–70.

2. Cohen S. Containment of aggregate queries. ACM SIGMOD Rec. 2005;34(1):77–85.

3. Cohen S, Nutt W, Sagiv Y. Containment of aggregate queries. In: Proceedings of the 9th International Conference on Database Theory; 2003. p. 111–25.

4. Cohen S, Nutt W, Sagiv Y. Deciding equivalences among conjunctive aggregate queries. J ACM. 2007;54(2)

5. Cohen S, Nutt W, Serebrenik A. Rewriting aggregate queries using views. In: Proceedings of the 18th ACM SIGACT-SIGMOD-SIGART Symposium on Principles of Database Systems; 1999. p. 155–66.

6. Cohen S, Sagiv Y, Nutt W. Equivalences among aggregate queries with negation. ACM Trans Comput Log. 2005;6(2):328–60.

7. Consens MP, Mendelzon AO. Low complexity aggregation in graphlog and datalog. Theor Comput Sci. 1993;116(1–2):95–116.

8. Grumbach S, Rafanelli M, Tininini L. On the equivalence and rewriting of aggregate queries. Acta Inf. 2004;40(8):529–84.

9. Hella L, Libkin L, Nurmonen J, Wong L. Logics with aggregate operators. J ACM. 2001;48(4):880–907.

10. Jayram TS, Kolaitis PG, Vee E. The containment problem for real conjunctive queries with inequalities. In: Proceedings of the 25th ACM SIGACT-SIGMOD-SIGART Symposium on Principles of Database Systems; 2006. p. 80–9.

11. Klug AC. Equivalence of relational algebra and relational calculus query languages having aggregate functions. J ACM. 1982;29(3):699–717.

12. Libkin L. Expressive power of SQL. Theor Comput Sci. 2003;3(296):379–404.

13. Libkin L, Wong L. Query languages for bags and aggregate functions. J Comput Syst Sci. 1997;55(2):241–72.

14. Mumick IS, Shmueli O. How expressive is stratified aggregation? Ann Math Artif Intell. 1995;15(3–4):407–34.

15. Ross KA, Sagiv Y. Monotonic aggregation in deductive database. J Comput Syst Sci. 1997;54(1):79–97.

Aggregation-Based Structured Text Retrieval

Theodora Tsikrika
Center for Mathematics and Computer Science, Amsterdam, The Netherlands

Definition

Text retrieval is concerned with the retrieval of documents in response to user queries. This is achieved by (i) representing documents and queries with indexing features that provide a characterisation of their information content, and (ii) defining a function that uses these representations to perform retrieval. *Structured text retrieval* introduces a finer-grained retrieval paradigm that supports the representation and subsequent retrieval of the individual document components defined by the document's *logical structure*. *Aggregation-based structured text retrieval* defines (i) the representation of each document component as the aggregation of the representation of its own information content and the representations of information content of its structurally related components, and (ii) retrieval of document components based on these (aggregated) representations.

The aim of aggregation-based approaches is to improve retrieval effectiveness by capturing and exploiting the interrelations among the components of structured text documents. The representation of each component's own information content is generated at indexing time. The recursive aggregation of these representations, which takes place at the level of their indexing features, leads to the generation, either at indexing or at query time, of the representations of those components that are structurally related with other components.

Aggregation can be defined in numerous ways; it is typically defined so that it enables retrieval to focus on those document components more specific to the query or to each document's best entry points, i.e., document components that contain relevant information and from which users can browse to further relevant components.

Historical Background

A well-established Information Retrieval (IR) technique for improving the effectiveness of *text retrieval* (i.e., retrieval at the document level) has been the generation and subsequent combination of multiple representations for each document [3]. To apply this useful technique to the *text retrieval* of structured text documents, the typical approach has been to exploit their *logical structure* and consider that the individual representations of their components can act as the different representations to be combined [11]. This definition of the representation of a structured text document as the combination of the representations of its components was also based on the intuitive idea that the information content of each document consists of the information content of its sub-parts [2, 6].

As the above description suggests, these combination-based approaches, despite restricting retrieval only at the document level, assign representations not only to documents, but also to individual document components. To generate these representations, structured text documents can simply be viewed as series of non-overlapping components (Fig. 1a), such as title, author, abstract, body, etc. [13]. The proliferation of *SGML* and *XML* documents, however, has led to the consideration of hierarchical components (Fig. 1b), and their interrelated representations [1]. For these (disjoint or nested) document components, the combination of their representations can take place (i) directly at the level of their indexing features, which typically correspond to terms and their statistics (e.g., [13]), or (ii) at the level of retrieval scores computed independently for each component (e.g., [15]). Overall, these combination-based approaches have proven effective for the

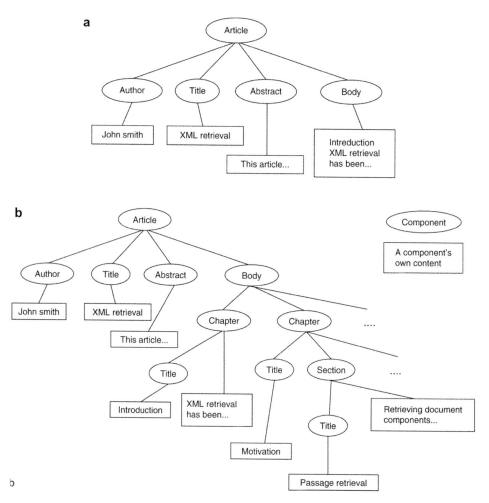

Aggregation-Based Structured Text Retrieval, Fig. 1 Two views on the logical structure of a structured text document

text *retrieval* of structured text documents [11, 13, 15].

Following the recent shift towards the *structured text retrieval* paradigm [2], which supports the retrieval of document components (including whole documents), it was only natural to try to adapt these combination-based approaches to this new requirement for retrieval at the sub-document level. Here, the focus is on each document component: its representation corresponds to the combination of its own representation with the representations of its structurally related components, and its retrieval is based on this combined representation. Similarly to the case of combination-based approaches for *text retrieval*, two strands of research can be identified: (i) approaches that operate at the level of the components' indexing features (e.g., [12]), referred to as *aggregation-based structured text retrieval* (described in this entry), and (ii) approaches that operate at the level of retrieval scores computed independently for each component (e.g., [14]), referred to as *propagation*-based structured text retrieval.

Figure 2b illustrates the premise of aggregation- and propagated-based approaches for the simple structured text document depicted in Fig. 2a. Since these approaches share some of their underlying motivations and assumptions, there has been a cross-fertilisation of ideas between the two. This also implies that this entry is closely related to the entry on *propagation*-based structured text retrieval.

Foundations

Structured text retrieval supports, in principle, the representation and subsequent retrieval of document components of any granularity; in practice, however, it is desirable to take into account only document components that users would find informative in response to their queries [1, 2, 4, 6]. Such document components are referred to as *indexing units* and are usually chosen (manually or automatically) with respect to the requirements of each application. Once the *indexing units* have been determined, each can be assigned a repre-

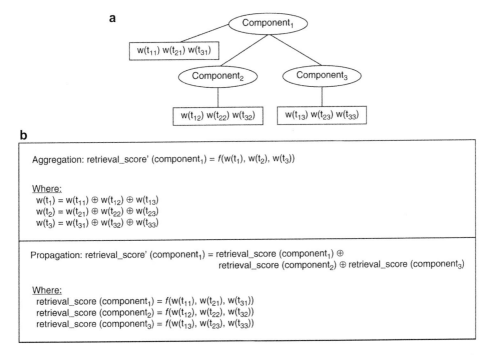

Aggregation-Based Structured Text Retrieval, Fig. 2 Simple example illustrating the differences between aggregation- and propagation-based approaches

sentation of its information content, and, hence, become individually retrievable.

Aggregation-based structured text retrieval approaches distinguish two types of *indexing units*: *atomic* and *composite*. Atomic components correspond to *indexing units* that cannot be further decomposed, i.e., the leaf components in Fig. 1b. The representation of an atomic component is generated by considering only its own information content. Composite components, on the other hand, i.e., the non-leaf nodes in Fig. 1b, correspond to *indexing units* which are related to other components, e.g., consist of sub-components. In addition to its own information content, a composite component is also dependent on the information content of its structurally related components. Therefore, its representation can be derived via the *aggregation* of the representation of its own information content with the representations of the information content of its structurally related components; this aggregation takes place at the level of their indexing features. Given the representations of atomic components and of composite components' own information content, aggregation-based approaches recursively generate the aggregated representations of composite components and, based on them, perform retrieval of document components of varying granularity.

In summary, each aggregation-based approach needs to define the following: (i) the representation of each component's own information content, (ii) the aggregated representations of composite components, and (iii) the retrieval function that uses these representations. Although these three steps are clearly interdependent, the major issues addressed in each step need to be outlined first, before proceeding with the description of the key aggregation-based approaches in the field of *structured text retrieval*.

1. *Representing each component's own information content*: In the field of *text retrieval*, the issue of representing documents with indexing features that provide a characterisation of their information content has been extensively studied in the context of several *IR retrieval mod-*

els (e.g., Boolean, vector space, probabilistic, language models, etc.). For text documents, these indexing features typically correspond to term statistics. Retrieval functions produce a ranking in response to a user's query, by taking into account the statistics of query terms together with each document's length. The term statistics most commonly used correspond to the *term frequency* $tf(t, d)$ of term t in document d and to the *document frequency df* (t, C) of term t in the document collection C, leading to standard $tf \times idf$ weighting schemes.

Structured text retrieval approaches need to generate representations for all components corresponding to *indexing units*. Since these components are nested, it is not straightforward to adapt these *term statistics* (particularly *document frequency*) at the component level [10]. Aggregation-based approaches, on the other hand, directly generate representations only for components that have their own information content, while the representations of the remaining components are obtained via the aggregation process. Therefore, the first step is to generate the representations of atomic components and of the composite components' own information content, i.e., the content not contained in any of their structurally related components. This simplifies the process, since only *disjoint* units need to be represented [6], as illustrated in Fig. 3 where the dashed boxes enclose the components to be represented (cf. [5]).

Text retrieval approaches usually consider that the information content of a document corresponds only to its *textual content*, and possibly its metadata (also referred to as *attributes*). In addition to that, structured *text retrieval* approaches also aim at representing the information encoded in the logical structure of documents. Representing this *structural information*, i.e., the interrelations among the documents and their components, enables retrieval in response to both *content-only queries* and *content-and-structure queries*.

Aggregation-based approaches that only represent the textual content typically adapt standard

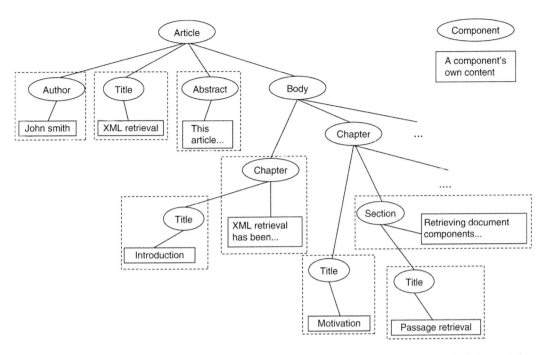

Aggregation-Based Structured Text Retrieval, Fig. 3 Representing the components that contain their own information

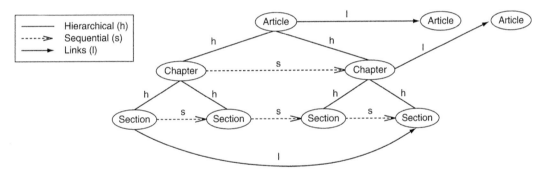

Aggregation-Based Structured Text Retrieval, Fig. 4 Different types of structural relationships between the components of a structured text document

representation formalisms widely employed in *text retrieval* approaches to their requirements for representation at the component level (e.g., [9, 11]). Those that consider richer representations of information content apply more expressive formalisms (e.g., various logics [2, 4]).

2. *Aggregating the representations*: The concept underlying aggregation-based approaches is that of *augmentation* [4]: the information content of a document component can be aug-

mented with that of its structurally related components. Given the already generated representations (i.e., the representations of atomic components and of composite components' own information content), the augmentation of composite components is performed by the aggregation process.

The first step in the aggregation process is the identification of the structurally related components of each composite component. Three basic types of structural relationships (Fig. 4) can be distinguished: hierarchical (*h*), sequential (*s*), and links (*l*). Hierarchical connections express the *composition* relationship among components, and induce the tree representing the *logical structure* of a structured document. Sequential connections capture the order imposed by the document's author(s), whereas links to components of the same or different documents reference (internal or external) sources that offer similar information. In principle, all these types of structural relationships between components can be taken into account by the aggregation process (and some aggregation-based approaches are generic enough to accommodate them, e.g., [7]). In practice, however, the hierarchical structural relations are the only ones usually considered. This leads to the aggregated representations of composite components being recursively generated in an ascending manner.

The next step is to define the *aggregation operator* (or *aggregation function*). Since the aggregation of the textual content of related components is defined at the level of the indexing features of their representations, the aggregation function is highly dependent on the model (formalism) chosen to represent each component's own content. This aggregation results in an (aggregated) representation modeled in the same formalism, and can be seen as being performed at two stages (although these are usually combined into one step): the aggregation of *index expressions* [2] (e.g., terms, conjunctions of terms, etc.), and of the *uncertainty* assigned to them (derived mainly by their statistics).

An aggregation function could also take into account: (i) *augmentation factors* [6], which capture the fact that the textual content of the structurally related components of a composite component is not included in that components own content and has to be "propagated" in order to become part of it, (ii) *accessibility factors* [4], which specify how the representation of a component is influenced by its connected components (a measure of the contribution of, say, a section to its embedding chapter [2]), and (iii) the overall importance of a component in a document's structure [7] (e.g., it can be assumed that

a title contains more informative content than a small sub-section [13]). Finally, the issue of the possible aggregation of the attributes assigned to related components needs to be addressed [2].

The above aggregation process can take place either at indexing time (*global aggregation*) or at query time (*local aggregation*). Global aggregation is performed for all composite *indexing units* and considers all indexing features involved. Since this strategy does not scale well and can quickly become highly inefficient, local aggregation strategies are primarily used. These restrict the aggregation only to indexing features present in the query (i.e., query terms), and, starting from components retrieved in terms of their own information content, perform the aggregation only for these components' ancestors.

3. *Retrieval*: The retrieval function operates both on the representations of atomic components and on the aggregated representations of composite components. Its definition is highly dependent on the formalism employed in modeling these representations. In conjunction with the definition of the aggregation function, the retrieval function operationalizes the notion of *relevance* for a *structured text retrieval* system. It can, therefore, determine whether retrieval focuses on those document components more specific to the query [2], or whether the aim is to support the users' browsing activities by identifying each documents best entry points [7] (i.e., document components that contain relevant information which users can browse to further relevant components).

Aggregation-Based Approaches

One of the most influential aggregation-based approaches has been developed by Chiaramella et al. [2] in the context of the FERMI project (http://www.dcs.gla.ac.uk/fermi/). Aiming at supporting the integration of IR, hypermedia, and database systems, the FERMI model introduced some of the founding principles of *structured text retrieval* (including the notion of retrieval focussed to the most specific components). It follows the logical view on IR, i.e., it models the retrieval process as inference, and it employs predicate logic as its underlying formalism. The model defines a generic representation of content, attributes, and structural information associated with the *indexing units*. This allows for rich querying capabilities, including support for both *content-only queries* and *content-and-structured queries*. The indexing features of structured text documents can be defined in various ways, e.g., as sets of terms or as logical expressions of terms, while the semantics of the aggregation function depend on this definition. Retrieval can then be performed by a function of the *specificity* of each component with respect to the query.

The major limitation of the FERMI model is that it does not incorporate the uncertainty inherent to the representations of content and structure. To address this issue, Lalmas [8] adapted the FERMI model by using propositional logic as its basis, and extended it by modeling the uncertain representation of the textual content of components (estimated by a $tf \times idf$ weighting scheme) using Dempster-Shafer's theory of evidence. The structural information is not explicitly captured by the formalism; therefore, the model does not provide support for *content-and-structured queries*. The aggregation is performed by Dempster's combination rule, while retrieval is based on the belief values of the query terms.

Fuhr et al. [4] also extended the FERMI model using a combination of (a restricted form of) predicate logic with probabilistic inference. Their model captures the uncertainty in the representations of content, structure, and attributes. Aggregation of index expressions is based on a four-valued logic, allowing for the handling of incomplete information and of inconsistencies arising by the aggregation (e.g., when two components containing contradictory information are aggregated). Aggregation of term weights is performed according to the rules of probability theory, typically by adopting term independence assumptions. This approach introduced the notion of *accessibility factor* being taken into account. Document components are retrieved based on the computed probabilities of query terms occurring in their (aggregated) representations.

Following its initial development in [4], Fuhr and his colleagues investigated further this logic-based probabilistic aggregation model in [5, 6]. They experimented with modeling aggregation by different Boolean operators; for instance, they noted that, given terms propagating in the document tree in a bottom-up fashion, a probabilistic-OR function would always result in higher weights for components further up the hierarchy. As this would lead (in contrast to the objectives of *specificity*-oriented retrieval) to the more general components being always retrieved, they introduced the notion of *augmentation factors*. These could be used to "downweight" the weights of terms (estimated by a $tf \times idf$ scheme) that are aggregated in an ascending manner. The effectiveness of their approach has been assessed in the context of the *Initiative for the Evaluation of XML retrieval (INEX)* [6].

Myaeng et al. [11] also developed an aggregation-based approach based on probabilistic inference. They employ Bayesian networks as the underlying formalism for explicitly modeling the (hierarchical) structural relations between components. The document components are represented as nodes in the network and their relations as (directed) edges. They also capture the uncertainty associated with both textual content (again estimated by $tf \times idf$ term statistics) and structure. Aggregation is performed by probabilistic inference, and retrieval is based on the computed beliefs. Although this model allows for document component scoring, in its original publication [11] it is evaluated in the context of *text retrieval* at the document level.

Following the recent widespread application of statistical language models in the field of *text retrieval*, Ogilvie and Callan [8] adapted them to the requirements of *structured text retrieval*. To this end, each document component is modeled by a language model; a unigram language model estimates the probability of a term given some text. For atomic components, the language model is estimated by their own text by employing a maximum likelihood estimate (MLE). For instance, the probability of term t given the language model θ_T of text T in a component can be estimated by: $P(t|\theta_T) = (1 - \omega)P_{MLE}$ $(t|\theta_T) + \omega P_{MLE} (t|\theta_{collection})$, where ω is a parameter controlling the amount of smoothing of the background collection model. For composite components $comp_i$, the aggregation of language models is modeled as a linear interpolation: $P\left(t|\theta'_{comp_i}\right) = \lambda^{c'}_{comp_i} P\left(t|\theta_{comp_i}\right) + \sum_{j\in children(comp_i)} \lambda^c_j P\left(t|\theta_j\right)$, where $\lambda^{c'}_{comp_i} + \sum_{j\in children(comp_i)} \lambda^c_j = 1$. These λs model the contribution of each language model (i.e., document component) in the aggregation, while their estimation is a non-trivial issue. Ranking is typically produced by estimating the probability that each component generated the query string (assuming an underlying multinomial model). The major advantage of the language modeling approach is that it provides guidance in performing the aggregation and in estimating the term weights.

A more recent research study has attempted to apply BM25 (one of the most successful *text retrieval* term weighting schemes) to *structured text retrieval*. Robertson et al. [13] initially adapted BM25 to structured text documents with non-hierarchical components (see Fig. 1a), while investigating the effectiveness of retrieval at the document level. Next, they [9] adapted BM25 to deal with nested components (see Fig. 1b), and evaluated it in the context of the *INitiative for the Evaluation of XML retrieval (INEX)*.

A final note on these aggregation-based approaches is that most aim at focusing retrieval on those document components more specific to the query. However, there are approaches that aim at modeling the criteria determining what constitutes a best entry point. For instance, Kazai et al. [7] model aggregation as a fuzzy formalisation of linguistic quantifiers. This means that an indexing feature (term) is considered in an aggregated representation of a composite component, if it represents LQ of its structurally related components, where LQ a linguistic quantifier, such as "at least one," "all," "most," etc. By using these aggregated representations, the retrieval function determines that a component is relevant to a query if LQ of its structurally related components are relevant, in essence implementing different

criteria of what can be regarded as a best entry point.

Key Applications

Aggregation-based approaches can be used in any application requiring retrieval according to the structured text retrieval paradigm. In addition, such approaches are also well suited to the retrieval of multimedia documents. These documents can be viewed as consisting of (disjoint or nested) components each containing one or more media. Aggregation can be performed by considering atomic components to only contain a single medium, leading to retrieval of components of varying granularity. This was recognized early in the field of structured text retrieval and some of the initial aggregation-based approaches, e.g., [2, 4], were developed for multimedia environments.

Experimental Results

For most of the presented approaches, particularly for research conducted in the context of the *INitiative for the Evaluation of XML retrieval (INEX)*, there is an accompanying experimental evaluation in the corresponding reference.

Data Sets

A testbed for the evaluation of structured text retrieval approaches has been developed as part of the efforts of the *INitiative for the Evaluation of XML retrieval (INEX)* (http://inex.is.informatik. uni-duisburg.de/).

URL to Code

The aggregation-based approach developed in [8] has been implemented as part of the open source *Lemur* toolkit (for language modeling and IR), available at: http://www.lemurproject.org/.

Cross-References

▶ Content-and-Structure Query
▶ Content-Only Query
▶ Indexing Units of Structured Text Retrieval
▶ Information Retrieval Models
▶ INitiative for the Evaluation of XML Retrieval
▶ Logical Document Structure
▶ Propagation-Based Structured Text Retrieval
▶ Relevance
▶ Specificity
▶ Structured Document Retrieval
▶ Text Indexing and Retrieval

Recommended Reading

1. Chiaramella Y. Information retrieval and structured documents. In Lectures on information retrieval, Third European Summer-School, Revised Lectures, LNCS, Agosti M, Crestani F, and Pasi G (eds.). Vol. 1980. Springer; 2001, p. 286–309.
2. Chiaramella Y, Mulhem P, and Fourel F. A model for multimedia information retrieval. Technical Report FERMI, ESPRIT BRA 8134, University of Glasgow, Scotland; 1996.
3. Croft WB. Combining approaches to information retrieval. In Advances in information retrieval: Recent research from the center for intelligent information retrieval, Croft WB (ed.). The Information retrieval series, Vol. 7. Kluwer Academic, Dordrecht; 2000, p. 1–36.
4. Fuhr N, Gövert N, and Rölleke T. DOLORES: A system for logic-based retrieval of multimedia objects. In: Proceedings of the 21st Annual International ACM SIGIR Conference on Research and Development in Information Retrieval; 1998. p. 257–65.
5. Fuhr N and Großjohann K. XIRQL: A query language for information retrieval in XML documents. In: Proceedings of the 24th Annual International ACM SIGIR Conference on Research and Development in Information Retrieval; 2001. p. 172–80.
6. Gövert N, Abolhassani M, Fuhr N, and Großjohann K. Content-oriented XML retrieval with HyREX. In: Proceedings of the 1st International Workshop of the Initiative for the Evaluation of XML Retrieval; 2002. p. 26–32.
7. Kazai G, Lalmas M, and Rölleke T A model for the representation and focussed retrieval of structured documents based on fuzzy aggregation. In: Proceedings of the 8th International Symposium on String Processing and Information Retrieval; 2001. p. 123–35.
8. Lalmas M. Dempster-Shafer's theory of evidence applied to structured documents: Modelling uncertainty.

In: Proceedings of the 20th Annual International ACM SIGIR Conference on Research and Development in Information Retrieval; 1997. p. 110–18.
9. Lu W, Robertson SE, and MacFarlane A. Field-weighted XML retrieval based on BM25. In: Proceedings of the 4th International Workshop of the Initiative for the Evaluation of XML Retrieval; 2006. p. 161–71.
10. Mass Y and Mandelbrod M. Retrieving the most relevant XML components. In: Proceedings of the 2nd International Workshop of the Initiative for the Evaluation of XML Retrieval; 2003. p. 53–58.
11. Myaeng SH, Jang DH, Kim MS, and Zhoo ZC. A flexible model for retrieval of SGML documents. In: Proceedings of the 21st Annual International ACM SIGIR Conference on Research and Development in Information Retrieval; 1998. p. 138–45.
12. Ogilvie P and Callan J. Hierarchical language models for retrieval of XML components. In Advances in XML Information Retrieval and Evaluation. In: Proceedings of the 3rd International Workshop of the Initiative for the Evaluation of XML Retrieval; 2004. p. 224–37.
13. Robertson SE, Zaragoza H, and Taylor M. Simple BM25 extension to multiple weighted fields. In: Proceedings of the 13th ACM International Conference on Information and Knowledge Management; 2004. p. 42–9.
14. Sauvagnat K, Boughanem M, and Chrisment C. Searching XML documents using relevance propagation. In: Proceedings of the 11th International Symposium on String Processing and Information Retrieval; 2004. p. 242–54.
15. Wilkinson R. Effective retrieval of structured documents. In: Proceedings of the 17th Annual International ACM SIGIR Conference on Research and Development in Information Retrieval; 1994. p. 311–17.

Air Indexes for Spatial Databases

Baihua Zheng
Singapore Management University, Singapore, Singapore

Definition

Air indexes refer to indexes employed in wireless broadcast environments to address scalability issue and to facilitate power saving on mobile devices [4]. To retrieve a data object in wireless broadcast systems, a mobile client has to continuously monitor the broadcast channel until the data arrives. This will consume a lot of energy since the client has to remain active during its waiting time. The basic idea of air indexes is that by including index information about the arrival times of data items on the broadcast channel, mobile clients are able to predict the arrivals of their desired data. Thus, they can stay in power saving mode during waiting time and switch to active mode only when the data of their interests arrives.

Historical Background

In spatial databases, clients are assumed to be interested in data objects having spatial features (e.g., hotels, ATM, gas stations). "Find me the nearest restaurant" and "locate all the ATMs that are within 100 miles of my current location" are two examples. A central server is allocated to keep all the data, based on which the queries issued by the clients are answered. There are basically two approaches to disseminating spatial data to clients: (i) *on-demand access*: a mobile client submits a request, which consists of a query and the query's issuing location, to the server. The server returns the result to the mobile client via a dedicated point-to-point channel. (ii) *periodic broadcast*: data are periodically broadcast on a wireless channel open to the public. After a mobile client receives a query from its user, it tunes into the broadcast channel to receive the data of interest based on the query and its current location.

On-demand access is particularly suitable for light-loaded systems when contention for wireless channels and server processing is not severe.

However, as the number of users increases, the system performance deteriorates rapidly. Compared with on-demand access, broadcast is a more scalable approach since it allows simultaneous access by an arbitrary number of mobile clients. Meanwhile, clients can access spatial data without reporting to the server their current location and hence the private location information is not disclosed.

In the literature, two performance metrics, namely *access latency* and *tuning time*, are used to measure access efficiency and energy conservation, respectively [4]. The former means the time elapsed between the moment when a query is issued and the moment when it is satisfied, and the latter represents the time a mobile client stays active to receive the requested data. As energy conservation is very critical due to the limited battery capacity on mobile clients, a mobile device typically supports two operation modes: *active mode* and *doze mode*. The device normally operates in active mode; it can switch to doze mode to save energy when the system becomes idle.

With data broadcast, clients listen to a broadcast channel to retrieve data based on their queries and hence are responsible for query processing. Without any index information, a client has to download all data objects to process spatial search, which will consume a lot of energy since the client needs to remain active during a whole broadcast cycle. A broadcast cycle means the minimal duration within which all the data objects are broadcast at least once. A solution to this problem is *air indexes* [4]. The basic idea is to broadcast an index before data objects (see Fig. 1 for an example). Thus, query processing can be performed over the index instead of actual data objects. As the index is much smaller than the data objects and is selectively accessed

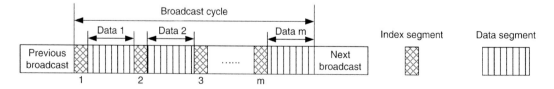

Air Indexes for Spatial Databases, Fig. 1 Air indexes in wireless broadcast environments

to perform a query, the client is expected to download less data (hence incurring less tuning time and energy consumption) to find the answers. The disadvantage of air indexing, however, is that the broadcast cycle is lengthened (to broadcast additional index information). As a result, the access latency would be worsen. It is obvious that the larger the index size, the higher the overhead in access latency.

An important issue in air indexes is how to multiplex data and index on the sequential-access broadcast channel. Figure 1 shows the well-known $(1, m)$ scheme [4], where the index is broadcast in front of every $1/m$ fraction of the dataset. To facilitate the access of index, each data page includes an offset to the beginning of the next index. The general access protocol for processing spatial search involves following three steps: (i) *initial probe*: the client tunes into the broadcast channel and determines when the next index is broadcast; (ii) *index search*: The client tunes into the broadcast channel again when the index is broadcast. It selectively accesses a number of index pages to find out the spatial data object and when to download it; and (iii) *data retrieval*: when the packet containing the qualified object arrives, the client downloads it and retrieves the object.

To disseminate spatial data on wireless channels, well-known spatial indexes (e.g., R-trees) are candidates for air indexes. However, unique characteristics of wireless data broadcast make the adoption of existing spatial indexes inefficient (if not impossible). Specifically, traditional spatial indexes are designed to cluster data objects with spatial locality. They usually assume a resident storage (such as disk and memory) and adopt search strategies that minimize I/O cost. This is achieved by *backtracking* index nodes during search. However, the broadcast order (and thus the access order) of index nodes is extremely important in wireless broadcast systems because data and index are only available to the client when they are broadcast on air. Clients cannot randomly access a specific data object or index node but have to wait until the next time it is broadcast. As a result, each backtracking operation extends the access latency by one more

cycle and hence becomes a constraint in wireless broadcast scenarios.

Figure 2 depicts an example of spatial query. Assume that an algorithm based on R-tree first visits root node, then the node R_2, and finally R_1, while the server broadcasts nodes in the order of root, R_1, and R_2. If a client wants to backtrack to node R_1 after it retrieves R_2, it will have to wait until the next cycle because R_1 has already been broadcast. This significantly extends the access latency and it occurs every time a navigation order is different from the broadcast order. As a result, new air indexes which consider both the constraints of the broadcast systems and features of spatial queries are desired.

Foundations

Several air indexes have been recently proposed to support broadcast of spatial data. These studies can be classified into two categories, according to the nature of the queries supported. The first category focuses on retrieving data associated with some specified geographical range, such as "Starbucks Coffee in New York City's Times Square" and "Gas stations along Highway 515." A representative is the index structure designed for *DAYS* project [1]. It proposes a location hierarchy and associates data with locations. The index structure is designed to support query on various types of data with different location granularity. The authors intelligently exploit an important property of the locations, i.e., containment relationship among the objects, to determine the relative location of an object with respect to its parent that contains the object. The containment relationship limits the search range of available data and thus facilitates efficient processing of the supported queries. In brief, a broadcast cycle consists of several sub-cycles, with each containing data belonging to the same type. A major index (one type of index buckets) is placed at the beginning of each sub-cycle. It provides information related to the types of data broadcasted, and enables clients to quickly jump into the right sub-cycle which contains her interested data. Inside a sub-cycle, minor indexes (another type of index buckets) are

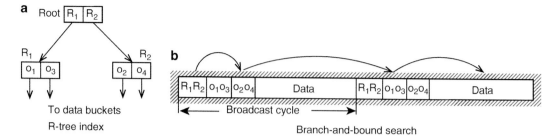

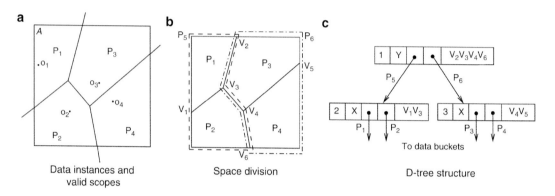

Air Indexes for Spatial Databases, Fig. 2 Linear access on wireless broadcast channel

Air Indexes for Spatial Databases, Fig. 3 Index construction using the D-tree

interleaved with data buckets. Each minor index contains multiple pointers pointing to the data buckets with different locations. Consequently, a search for a data object involves accessing a major index and several minor indexes.

The second category focuses on retrieving data according to specified distance metric, based on client's current location. An example is nearest neighbor (NN) search based on Euclidian distance. According to the index structure, indexes of this category can be further clustered into two groups, i.e., *central tree-based* structure and *distributed* structure. In the following, we review some of the representative indexes of both groups.

D-tree is a paged binary search tree to index a given solution space in support of planar point queries [6]. It assumes a data type has multiple data instances, and each instance has a certain *valid scope* within which this instance is the only correct answer. For example, restaurant is a data type, and each individual restaurant represents an instance. Take NN search as an example, Fig. 3a

illustrates four restaurants, namely o_1, o_2, o_3, and o_4, and their corresponding valid scopes p_1, p_2, p_3, and p_4. Given any query location q in, say, p_3, o_3 is the restaurant to which q is nearest. D-tree assumes the valid scopes of different data instances are known and it focuses only on planar point queries which locate the query point into a valid scope and return the client the corresponding data instance.

The D-tree is a binary tree built based on the divisions between data regions (e.g., valid scopes). A space consisting of a set of data regions is recursively partitioned into two complementary subspaces containing about the same number of regions until each subspace has one region only. The partition between two subspaces is represented by one or more polylines. The overall orientation of the partition can be either x-dimensional or y-dimensional, which is obtained, respectively, by sorting the data regions based on their lowest/uppermost y-coordinates, or leftmost/rightmost x-coordinates. Figure 3b shows the partitions for the running example. The polyline $pl(v_2, v_3, v_4, v_6)$ partitions the original space into p_5 and p_6, and polylines $pl(v_1, v_3)$ and $pl(v_4, v_5)$ further partition p_5 into p_1 and p_2, and p_6 into p_3 and p_4, respectively. The first polyline is y-dimensional and the remaining two are x-dimensional. Given a query point q, the search algorithm works as follows. It starts from the root and recursively follows either the left subtree or the right subtree that bounds the query point until a leaf node is reached. The associated data instance is then returned as the final answer.

Grid-partition index is specialized for NN problem [8]. It is motivated by the observation that an object is the NN only to the query points located inside its Voronoi Cell. Let $O = \{o_1, o_2, \ldots, o_n\}$ be a set of points. $V(o_i)$, the *Voronoi cell* (VC) for o_i, is defined as the set of points q in the space such that $dist(q, o_i) < dist(q, o_j)$, $\forall j \neq i$. That is, $V(o_i)$ consists of the set of points for which o_i is the NN. As illustrated in Fig. 3a, p_1, p_2, p_3, and p_4 denote the VCs for four objects, o_1, o_2, o_3, and o_4, respectively. Grid-partition index tries to reduce the search space for a query at the very beginning by partitioning the space into disjoint grid cells. For each grid cell, all the objects that could be NNs of at least one query point inside the grid cell are indexed, i.e., those objects whose VCs overlap with the grid cell are associated with that grid cell.

Figure 4a shows a possible grid partition for the running example, and the index structure is depicted in Fig. 4b. The whole space is divided into four grid cells; i.e., G_1, G_2, G_3, and G_4. Grid cell G_1 is associated with objects o_1 and o_2, since their VCs, p_1 and p_2, overlap with G_1; likewise, grid cell G_2 is associated with objects o_1, o_2, o_3, and so on. If a given query point is in grid cell G_1, the NN can be found among the objects associated with G_1 (i.e., o_1 and o_2), instead of among the whole set of objects. Efficient search algorithms and partition approaches have been proposed to speed up the performance.

Conventional spatial index R-tree has also been adapted to support kNN search in broadcast environments [2]. For R-tree index, the kNN search algorithm would visit index nodes and objects sequentially as backtracking is not feasible on the broadcast. This certainly results in

a considerably long tuning time especially when the result objects are located in later part of the broadcast. However, if clients know that there are at least k objects in the later part of the broadcast that are closer to the query point than the currently found ones, they can safely skip the downloading of the intermediate objects currently located. This observation motivates the design of the enhanced kNN search algorithm which caters for the constraints of wireless broadcast. It requires each index node to carry a count of the underlying objects (object count) referenced by the current node. Thus, clients do not blindly download intermediate objects.

Hilbert Curve Index (HCI) is designed to support general spatial queries, including window queries, kNN queries, and continuous nearest-neighbor (CNN) queries in wireless broadcast environments. Motivated by the linear streaming property of the wireless data broadcast channel and the optimal spatial locality of the Hilbert Curve (HC), HCI organizes data according to Hilbert Curve order [7, 9], and adopts B^+-tree as the index structure. Figure 5 depicts a 8×8 grid, with solid dots representing data objects. The numbers next to the data points, namely *index value*, represent the visiting orders of different points at Hilbert Curve. For instance, data point with (1,1) as the coordinates has the index value of 2, and it will be visited before data point with (2,2) as the coordinates because of the smaller index value.

The *filtering and refining* strategy is adopted to answer all the queries. For window query, the basic idea is to decide a candidate set of points along the Hilbert curve which includes all the points within the query window and later to filter out those outside the window. Suppose the rectangle shown in Fig. 5 is a query window. Among all the points within the search range, the first point is point a and the last is b, sorted according to their occurring orders on the Hilbert curve, and both of them are lying on the boundary of the search range. Therefore, all the points inside this query window should lie on the Hilbert curve segmented by points a and b. In other words, data points with index values between 18 and 29, but not the others, are the candidates. During the

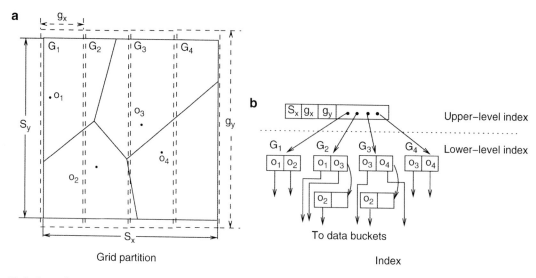

Air Indexes for Spatial Databases, Fig. 4 Index construction using the grid-partition

Air Indexes for Spatial Databases, Fig. 5 Hilbert curve index

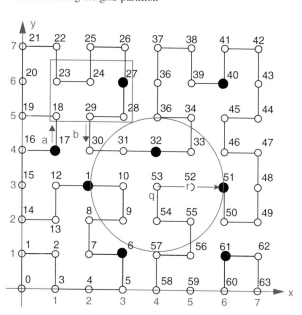

access, the client can derive the coordinates of data points based on the index values and then retrieve those within the query window.

For *k*NN query, the client first retrieves those *k* nearest objects to the query point along the Hilbert curve and then derives a range which for sure bounds at least *k* objects. In the filtering phase, a window query which bounds the search range is issued to filter out those unqualified. Later in the refinement phase, *k* nearest objects are identified according to their distance to the query point. Suppose an NN query at point *q* (i.e.,

index value 53) is issued. First, the client finds its nearest neighbor (i.e., point with index value 51) along the curve and derives a circle centered at *q* with *r* as the radius (i.e., the green circle depicted in Fig. 5). Since the circle bounds point 51, it is certain to contain the nearest neighbor to point *q*. Second, a window query is issued to retrieve all the data points inside the circle, i.e., points with index values 11, 32, and 51. Finally, the point 32 is identified as the nearest neighbor. The search algorithm for CNN adopts a similar approach. It approximates a search range which is guaranteed to bound all the answer objects, issues a window query to retrieve all the objects inside the search range, and finally filters out those unqualified.

All the indexes mentioned above are based on a central tree-based structure, like R-tree and B-tree. However, employing a tree-based index on a linear broadcast channel to support spatial queries results in several deficiencies. First, clients can only start the search when they re- trieve the root node in the channel. Replicating the index tree in multiple places in the broadcast channel provides multiple search starting points, shortening the initial root-probing time. However, a prolonged broadcast cycle leads to a long ac- cess latency experienced by the clients. Second, wireless broadcast media is not error-free. In case of losing intermediate nodes during the search process, the clients are forced to either restart the search upon an upcoming root node or scan the subsequential broadcast for other possible nodes in order to resume the search, thus ex- tending the tuning time. *Distributed spatial index* (*DSI*), a fully distributed spatial index structure, is motivated by these observations [5]. A similar distributed structure was proposed in [3] as well to support access to spatial data on air.

DSI is very different from tree-based indexes, and is not a hierarchical structure. Index infor- mation of spatial objects is fully distributed in DSI, instead of simply replicated in the broad- cast. With DSI, the clients do not need to wait for a root node to start the search. The search process launches immediately after a client tunes into the broadcast channel and hence the initial probe time for index information is minimized. Furthermore, in the event of data loss, clients resume the search quickly.

Like HCI, DSI also adopts Hilbert curve to determine broadcast order of data objects. Data objects, mapped to point locations in a 2-D space, are broadcast in the ascending order of their HC index values. Suppose there are N objects in total, DSI chunks them into n_F *frames*, with each hav- ing n_o objects ($n_F = \lceil N/n_o \rceil$). The space covered by Hilbert Curve shown in Fig. 5 is used as a running example, with solid dots representing the locations of data objects (i.e., $N = 8$), Figure 6 demonstrates a DSI structure with n_o set to 1, i.e., each frame contains only one object.

In addition to objects, each frame also has an index table as its header, which maintains infor- mation regarding to the HC values of data objects to be broadcast with specific waiting interval from the current frame. This waiting interval can be denoted by delivery time difference or number of data frames apart, with respect to the current frame. Every index table keeps n_i entries, each of which, τ_j, is expressed in the form of $\langle HC'_j, P_j \rangle$, $j \in [0, n_i)$. P_j is a pointer to the r^j -th frame after the current frame, where r (>1) is an exponential base (i.e., a system-wide parameter), and HC'_j is the HC value of the first object inside the frame pointed by P_j. In addition to τ_j, an index table also keeps the HC values HC_k ($k \in [1, n_o]$) of all the objects obj_k that are contained in the current frame. This extra information, although occupying litter extra bandwidth, can provide a more precise image of all the objects inside current frame. During the retrieval, a client can compare HC_k s of the objects against the one she has interest in, so the retrieval of unnecessary object whose size is much larger than an HC value can be avoided.

Refer to the example shown in Fig. 5, with corresponding DSI depicted in Fig. 6. Suppose $r = 2, n_o = 1, n_F = 8$, and $n_i = 3$. The index tables corresponding to frames of data objects O_6 and O_{32} are shown in the figure. Take the index table for frame O_6 as an example: τ_0 contains a pointer to the next upcoming (2^0-th) frame whose first object's HC value is 11, τ_1 contains a pointer to the second (2^1-th) frame with HC value for the first object (the only object) 17, and the last entry τ_2 points to the fourth (2^2-th) frame. It also keeps the HC value 6 of the object O_6 in the current frame. Search algorithm for window queries and kNN searches are proposed.

Key Applications

Location-Based Service

Wireless broadcast systems, because of the scalability, provide an alternative to disseminate location-based information to a large number of users. Efficient air indexes enable clients to selectively tune into the channel and hence the power consumption is reduced.

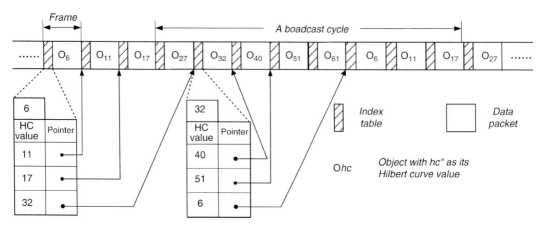

Air Indexes for Spatial Databases, Fig. 6 Distributed spatial index

Moving Objects Monitoring

Many moving objects monitoring applications are interested in finding out all the objects that currently satisfy certain conditions specified by the users. In many cases, the number of moving objects is much larger than the number of submitted queries. As a result, wireless broadcast provides an ideal way to deliver subscribed queries to the objects, and those objects that might affect the queries can then report their current locations.

Cross-References

▶ Nearest Neighbor Query
▶ Space-Filling Curves for Query Processing
▶ Spatial Indexing Techniques
▶ Voronoi Diagrams

Recommended Reading

1. Acharya D, Kumar V. Location based indexing scheme for days. In: Proceedings of the 4th ACM International Workshop on Data Engineering for Wireless and Mobile Access; 2005. p. 17–24.
2. Gedik B, Singh A, Liu L. Energy efficient exact knn search in wireless broadcast environments. In: Proceedings of the 12th ACM International Symposium on Geographic Information Systems; 2004. p. 137–46.
3. Im S, Song M, Hwang C. An error-resilient cell-based distributed index for location-based wireless broadcast services. In: Proceedings of the 5th ACM International Workshop on Data Engineering for Wireless and Mobile Access; 2006. p. 59–66.
4. Imielinski T, Viswanathan S, Badrinath BR. Data on air – organization and access. IEEE Trans Knowl Data Eng. 1997;9(3):353.
5. Lee WC, Zheng B. Dsi: a fully distributed spatial index for wireless data broadcast. In: Proceedings of the 23rd International Conference on Distributed Computing Systems; 2005. p. 349–58.
6. Xu J, Zheng B, Lee W-C, Lee DL. The d-tree: an index structure for location-dependent data in wireless services. IEEE Trans Knowl Data Eng. 2002;16(12):1526–42.
7. Zheng B, Lee W-C, Lee DL. Spatial queries in wireless broadcast systems. ACM/Kluwer J Wirel Netw. 2004;10(6):723–36.
8. Zheng B, Xu J, Lee W-C, Lee L. Grid-partition index: a hybrid method for nearest-neighbor queries in wireless location-based services. VLDB J. 2006;15(1):21–39.
9. Zheng B, Lee W-C, Lee DL. On searching continuous k nearest neighbors in wireless data broadcast systems. IEEE Trans Mobile Comput. 2007;6(7):748–61.

AJAX

Alex Wun
University of Toronto, Toronto, ON, Canada

Definition

AJAX is an acronym for "Asynchronous JavaScript and XML" and refers to a collection of web development technologies used together to create highly dynamic web applications.

Key Points

AJAX does not refer to a specific technology, but instead refers to a collection of technologies used in conjunction to develop dynamic and interactive web applications. The two main technologies comprising AJAX are the JavaScript scripting language and the W3C open standard XMLHttpRequest object API. While the use of XML and DOM are important for standardized data representation, using neither XML nor DOM is required for an application to be considered AJAX-enabled since the XMLHttpRequest API actually supports any text format.

Using the XMLHttpRequest API, web applications can fetch data asynchronously while registering a callback function to be invoked once the fetched data is available. More concretely, the XMLHttpRequest object issues a standard HTTP POST or GET request to a web server but returns control to the calling application immediately after issuing the request. The calling application is then free to continue execution while the HTTP request is being handled on the server. When the HTTP response is received, the XMLHttpRequest object calls back into the function that was supplied by the calling application so that the response can be processed. The asynchronous callback model used in AJAX applications is analogous to the Operating System technique of using interrupt handlers to avoid blocking on I/O. As such, development using AJAX necessarily requires an understanding of multi-threaded programming.

There are three main benefits to using AJAX in web applications:

1. *Performance*: Since XMLHttpRequest calls are asynchronous, client-side scripts can continue execution after issuing a request without being blocked by potentially lengthy data transfers. Consequently, web pages can be easily populated with data fetched in small increments in the background.
2. *Interactivity*: By maintaining long-lived data transfer requests, an application can closely approximate real-time event-driven behavior without resorting to periodic polling, which can only be as responsive as the polling frequency.
3. *Data Composition*: Web applications can easily pull data from multiple sources for aggregation and processing on the client-side without any dependence on HTML form elements. Data composition is also facilitated by having data adhere to standard XML and DOM formats.

The functionality provided by AJAX allows web applications to appear and behave much more like traditional desktop applications. The main difference is that data consumed by the application resides primarily out on the Internet – one of the concepts behind applications that are labeled as being representative "Web 2.0" applications.

Cross-References

▶ MashUp
▶ Web 2.0/3.0
▶ XML

Recommended Reading

1. The document object model: W3C working draft. Available at: http://www.w3.org/DOM/
2. The XMLHttpRequest object: W3C working draft. Available at: http://www.w3.org/TR/XMLHttpRequest/

Allen's Relations

Peter Revesz[1] and Paolo Terenziani[2]
[1]University of Nebraska-Lincoln, Lincoln, NE, USA
[2]Dipartimento di Scienze e Innovazione Tecnologica (DiSIT), Università del Piemonte Orientale, "Amedeo Avogadro", Alessandria, Italy

Synonyms

Qualitative relations between time intervals; Qualitative temporal constraints between time intervals

Definition

A (convex) *time interval* I is the set of all time points between a starting point (usually denoted by I⁻) and an ending point (I⁺). Allen's relations

Allen's Relations, Table 1 Translation of Allen's interval relations between two intervals I and J into conjunctions of point relations between I−, I+, J−, and J+

	I⁻ J⁻	I⁻ J⁺	I⁺ J⁻	I⁺ J⁺
After		>		
Before			<	
Meets			=	
Met_by		=		
During	>			<
Contains	<			>
Equal	=			=
Finishes	>			=
Finished_by	<			=
Starts	=			<
Started_by	=			>
Overlaps	<		>	<
Overlapped_by	>	<	>	

model all possible relative positions between two time intervals [1]. There are 13 different possibilities, depending on the relative positions of the endpoints of the intervals (Table 1).

For example, "there will be a guest speaker during the database system class" can be represented by Allen's relation I_{Guest} *During* $I_{Database}$ (or by $I^-_{Guest} > I^-_{Database} \land I^+_{Guest} < I^+_{Database}$ considering the relative position of the endpoints). Moreover, any subset of the 13 relations, excluding the empty subset, is a relation in Allen's interval algebra (therefore, there are $2^{13}-1$ relations in Allen's algebra). Such subsets are used in order to denote ambiguous cases, in which the relative position of two intervals is only partially known. For instance, I_1 (before, meets, overlaps) I_2 represents the fact that I_1 is before *or* meets *or* overlaps I_2.

Key Points

In many cases, the exact time interval when facts occur is not known, but (possibly imprecise) information on the relative temporal location of

Allen's Relations, Fig. 1
Visualization of Allen's interval relations

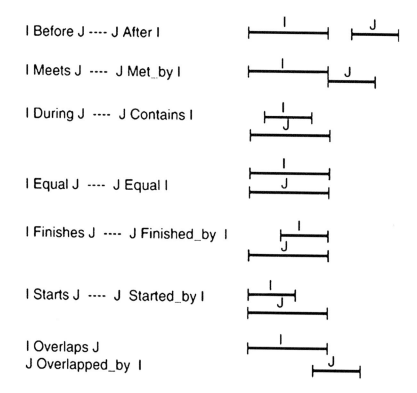

I Before J ---- J After I

I Meets J ---- J Met_by I

I During J ---- J Contains I

I Equal J ---- J Equal I

I Finishes J ---- J Finished_by I

I Starts J ---- J Started_by I

I Overlaps J
J Overlapped_by I

facts is available. Allen's relations allow one to represent such cases of *temporal indeterminacy*. For instance, planning in artificial intelligence is the first application of Allen's relations. A graphical representation of the basic 13 Allen's relations is shown in Fig. 1.

Allen's relations are specific cases of *temporal constraints*. Namely, they are qualitative temporal constraints between time intervals. Given a set of such constraints, *qualitative temporal reasoning* can be used in order to make inferences (e.g., to check whether the set of constraints is consistent).

Finally, notice that, in many entries of this encyclopedia, the term *(time) period* has been used with the same meaning of *(time) interval* in this entry.

Cross-References

▶ Qualitative Temporal Reasoning
▶ Temporal Constraints
▶ Temporal Indeterminacy

Recommended Reading

1. Allen JF. Maintaining knowledge about temporal intervals. Commun ACM. 1983;26(11):832–43.

measure for diversified search, where it is assumed that a topic q has multiple possible intents $\{i\}$, and that *intent probabilities $Pr(i|q)$* and *per-intent relevance assessments* are available. For a given diversified ranked list, let $I_i(r) = 1$ if the document at rank r is relevant to intent i and let $I_i(r) = 0$ otherwise; let $C_i(r) = \sum_{k=1}^{r} I_i(k)$. For a given parameter α (typically set to 0.5), let the *novelty-biased gain* at r be $ng(r) = \sum_i I_i(r)(1 - \alpha)^{C_i(r-1)}$. This reflects the view that (a) the value of a document is basically given by the number of intents it covers; but that (b) for each intent, whenever a relevant document is found, the value of the next relevant document should be discounted (the *diminishing return* property). α-nDCG at *measurement depth* (or *document cutoff*) d is defined as:

$$\alpha\text{-}nDCG = \frac{\sum_{r=1}^{d} ng(r)/\log(r+1)}{\sum_{r=1}^{d} ng^*(r)/\log(r+1)}$$

where $ng^*(r)$ is the novelty-biased gain at rank r of an *ideal ranked list*. Unfortunately, finding the ideal list is an *NP*-complete problem, so a greedy approximation is used in practice.

α-nDCG

Tetsuya Sakai
Waseda University, Tokyo, Japan

Synonyms

Alpha-nDCG

Definition

α-nDCG is a variant of Normalised Discounted Cumulative Gain (nDCG). It is an evaluation

Main Text

α-nDCG penalises retrieval of many relevant documents for the same intent, and thereby encourages retrieval of a few relevant documents for many intents. However, note that, unlike some other evaluation measures for diversified search (e.g., D-measure and ERR-IA), α-nDCG cannot utilise per-intent *graded* relevance assessments: the novelty-biased gain only considers whether each intent is covered by the document or not. Also, the above formulation of α-nDCG does not consider the intent probabilities $Pr(i|q)$, although there is a version of α-nDCG that does incorporate them just like D-measure and ERR-IA.

Cross-References

▶ Advanced Information Retrieval Measures
▶ Discounted Cumulated Gain
▶ D-Measure
▶ ERR-IA

AMOSQL

Peter M. D. Gray
University of Aberdeen, Aberdeen, UK

Definition

AMOSQL [1] is a functional language having its roots in the functional query languages OSQL [2] and DAPLEX [3] with extensions of mediation primitives, multi-directional foreign functions, late binding, *active rules*, etc. Queries are specified using the select-from-where construct as in SQL. Furthermore, AMOSQL has aggregation operators, nested subqueries, disjunctive queries, quantifiers, and is relationally complete.

AMOSQL is a functional query language operating within the environment of Amos II, which is an open, light-weight, and extensible database management system (DBMS) with a functional data model. Each Amos II server contains all the traditional database facilities, such as a storage manager, a recovery manager, a transaction manager, and a query language. The system can be used as a single-user database or as a multi-user server to applications and to other Amos II peers. It has mainly been used for experiments with Mediators [4].

Key Points

AMOSQL is often used within a distributed mediator system where several mediator peers communicate over the Internet. Functional views provide transparent access to data sources from clients and other mediator peers. Amos II mediators are composable since a mediator peer can regard other mediator peers as data sources. In AMOSQL [1] the top-level language uses function application within an SQL-like syntax, for example to list children of a particular parent who like sailing:

```
create function sailch(person p) ->
string as
select name(c) from person c
where parent(c) = p and hobby(c) =
'sailing';
```

This is turned internally into a typed object comprehension.

$sailch(p) == [name(c) \mid c <- person; parent(c)= p; hobby(c) = 'sailing']$

Some functions may be defined as views on other databases accessed through mediators. In the comprehension form, such functions may easily be substituted because of referential transparency, leading to a longer conjunction with extra clauses [5].

Suppose information about *hobby(c)* was held in connection with *sports- person*, a subclass of person with a *surname* attribute:

$hobby(c) == [recreation(x) \mid x <- sportsperson; surname(x) = name(c)]$

These comprehensions merge into the following which can be further simplified:

$sailch(p) == [name(c) \mid c <- person; parent(c)=p; x <- sportsperson;$
$surname(x) = name(c); recreation(x) = 'sailing']$

Note that the internal form of comprehension does not explicitly distinguish the use of generators but simply use equality as in a filter. The conceptual advantage of this is that generators are not always a fixed role and some optimizations may reverse the role of filter and generator (systems with explicit generators use rewrite rules to do this).

Cross-References

▶ Comprehensions
▶ Functional Query Language

Recommended Reading

1. Fahl G, Risch T, Sköld M. 1AMOS – an architecture for active mediators. In: Proceedings of the Workshop on Next Generation Information Technologies and Systems; 1993.
2. Beech D. A foundation of evolution from relational to object databases. In: Advances in Database Technology, Proceedings of the 1st International Conference on Extending Database Technology; 1988. p. 251–70.
3. Shipman DW. The functional data model and the data language DAPLEX. ACM Trans Database Syst. 1981;6(1):140–73.
4. Risch T, Josifovski V, Katchaounov T. Chapter 9, Functional data integration in a distributed mediator system. In: Gray PMD, Kerschberg L, King PJH, Poulovassilis A, editors. The functional approach to data management. Berlin/Heidelberg/New York: Springer; 2004.
5. Josifovski V, Risch T. Functional query optimization over object-oriented views for data integration. J Intell Inf Syst. 1999;12(2–3):165–90.

AMS Sketch

Alin Dobra
University of Florida, Gainesville, FL, USA

Synonyms

AGMS sketch; Sketch; Tug-of-war sketch

Definition

AMS sketches are randomized summaries of the data that can be used to compute aggregates such as the second frequency moment (the self-join size) and sizes of joins. AMS sketches can be viewed as random projections of the data in the frequency domain on ± 1 pseudo-random vectors. The key property of AMS sketches is that the product of projections on the same random vector of frequencies of the join attribute of two relations is an unbiased estimate of the size of join of the relations. While a single AMS sketch is inaccurate, multiple such sketches can be computed and combined using averages and medians to obtain an estimate of any desired precision.

Historical Background

The AMS sketches were introduced in 1996 by Noga Alon, Yossi Matias, and Mario Szegedy as part of a suit of randomized algorithms for approximate computation of frequency moments. The same authors, together with Phillip Gibbons, extended the second frequency moment application of AMS sketches to the computation of the size of join of two relations, a more relevant database application. The initial work on AMS sketches fostered a large amount of subsequent work on data streaming algorithms including generalizations and extensions of AMS sketches. Alon, Matias, and Szegedy received the Gödel Prize in 2005 for their work on AMS sketches.

Foundations

While the AMS sketches were initially introduced to compute the second frequency moment, since the reader might be more familiar with database terminology, the problem of estimating the size of join of two relations will be considered here instead. Notice that the size of the self join size of a relation coincides with the second frequency moment of the relation thus the treatment here is slightly more general but not more complicated.

Problem Setup

To set up the problem, assume access is provided to two relations F and G each with a single attribute a. Since it is convenient, denote with f_i and g_i the frequency of value i of attribute a in relation F and G, respectively. Assume that elements of F and G are streamed the result of the query: $COUNT(F \bowtie_a G)$ needs to be computed or estimated. Consider the following example:

| Stream F: | a | 1 | 1 | 2 | 3 | 1 | 3 |, |

frequency vector **f**:

i	1	2	3
f_i	3	1	2

Stream G: $\boxed{a \mid 3 \quad 1 \quad 3 \quad 1 \quad 1}$,

frequency vector $\mathbf{g}$:

i	1	2	3
g_i	3	0	2

Elements of F and G are assumed to arrive one by one (i.e., are streamed). If the frequency vectors $\mathbf{f}$ and $\mathbf{g}$ can be maintained, than the result of the query $COUNT(F \bowtie_a G)$ can be computed clearly in the following example:

$$\text{COUNT}(F \bowtie_\alpha G) = \mathbf{fg}^T$$
$$= [3 \ 1 \ 2] \begin{bmatrix} 3 \ 0 \ 2 \end{bmatrix}^T$$
$$= 3 \cdot 3 + 1 \cdot 0 + 2 \cdot 2$$
$$= 13$$

Observe that the size of join can be written as the dot product of the frequency vectors of the two relations. Expressing the size of the join in terms of frequencies of the join attribute is key for AMS sketch based approximation.

Main Idea

Assume now that the estimate $COUNT(F \bowtie_a G)$ needs to be computed but only less than linear space, in terms of the size of the frequency vectors, is available. As it turns out, exact computation is not possible with less space (in an asymptotic sense), but approximate computation is possible. The AMS sketches prove that they allow the approximation of the size of join using sub-linear space.

The main idea behind AMS sketches is to summarize the entire frequency table by projecting it on a random vector. The value thus obtained will be referred to as an elementary sketch. Then, use the two elementary sketches, one for each relation, to recover approximately the result of the query. Interestingly, a random vector $\xi = [\xi_1 \ldots \xi_n]$ of ± 1 values suffices to obtain projections with the desired properties. For simplicity, random vectors for which $\forall i, E[\xi_i] = 0$ are preferred. With this:

- Sketch of F, $X_F = \mathbf{f}\xi^T$
- Sketch of G, $X_G = \mathbf{g}\xi^T$
- $X = X_F X_G$ estimates $COUNT(F \bowtie_a G)$ since

$$E[X] = E[\mathbf{f}\xi^T \xi \mathbf{g}^T] = \mathbf{f}E[\xi^T \xi]\mathbf{g}^T = \mathbf{f}I\mathbf{g}^T = \mathbf{fg}^T$$

- if $E[\xi^T \xi] = I$. To ensure this, property distinct elements of ξ must be pair-wise independent, i.e., $\forall i \neq i'$, $\xi_i^2 = 1$, $E[\xi_i \xi_{i'}] = 0$

For the particular random vector $\xi = [\xi_1 \ \xi_2 \ \xi_3] = [-1 \ +1 \ -1]$, the value of the elementary sketches and the overall estimate will be:

$$X_F = \mathbf{f}\xi^T = -4$$
$$X_G = \mathbf{g}\xi^T = -5$$
$$X = X_F X_G = (-4)(-5) = 20 \approx 13$$

The error of the estimate X is due to its variance that can be shown to have the property:

$$\text{Var}(X) \leq 2\mathbf{ff}^T \mathbf{gg}^T = 2 \ \text{SJ}(F) \ \text{SJ}(G)$$

as long as the random vector ξ is 4-wise independent, i.e., $\forall i_1 \neq i_2 \neq i_3 \neq i_4$, $E[\xi_{i_1} \xi_{i_2}] = 0$, $E[\xi_{i_1} \xi_{i_2} \xi_{i_3} \xi_{i_4}] = 0$.

Since a higher degree of independence would not make the sketch more precise, 4-wise independence suffices. This is important since 4-wise independent ± 1 random vectors can be generated *on the fly* by combining a small seed s and the index of the entry using $\xi_i(s) = h(s, i)$ with h a special hash function that guarantees the 4-wise independence of the components of ξ. The fact that elements of ξ can be generated on the fly is important since space can be saved and, more importantly, because sketches X_F and X_G can be computed using constant storage. This is how this can be accomplished using the previous example:

$$X_F = \mathbf{f}\xi^T = \sum_i f_i \xi_i$$

$$= \sum_{t \in F} \xi_{t.a} = \xi_1 + \xi_1 + \xi_2 + \xi_3 + \xi_1 + \xi_3$$

$$= h(s,1) + h(s,1) + h(s,2)$$
$$+ h(s,3) + h(s,1) + h(s,3)$$

$$X_G = \mathbf{g}\xi^T = \sum_i g_i \xi_i$$

$$= \sum_{t \in G} \xi_{t.a} = \xi_3 + \xi_1 + \xi_3 + \xi_1 + \xi_1$$

$$= h(s,3) + h(s,1) + h(s,3)$$
$$+ h(s,1) + h(s,1)$$

From this example, it can be observed that, to maintain the elementary sketches over the streams F and G, the only operation needed is to increment X_F and X_G by the value of $\xi_{t.a}$ using the function $h(\cdot)$ and the seed s where $t.a$ is the value of attribute a of the current tuple t arriving on the data stream. The fact that the elementary sketches can be computed so easily by considering one element at the time in an arbitrary order is what makes the AMS sketches appealing as an approximation technique.

Improving the Basic Schema

Since the streams F and G are summarized by a single number X_F and X_G, respectively, it is not expected that the estimate will be very precise (this is suggested as well by the above example). In order to improve the accuracy of X, a standard technique in randomized algorithms can be used (i.e., generating multiple independent copies of random variable X). Copies of X are averaged in order to decrease the variance (thus the error). The median of such averaged values of X is used to estimate $COUNT(F \bowtie_a G)$ since medians improve confidence. Multiple copies of X can

be obtained using multiple seeds as depicted in Fig. 1.

It can be shown that:

- Average $\frac{8\operatorname{Var}(X)}{\epsilon^2 E^2[X]}$ independent copies of X to reduce error to ϵ
- Median of $2 \log 1/\delta$ such averages increases the confidence to $1 - \delta$

Key Applications

AMS sketches are particularly well suited for computing aggregates when data is either streamed (or a single pass over the data is allowed/desirable) or distributed at multiple sites. Thus, AMS sketches are relevant for processing large amount of data, as is the case in data warehousing, or processing distributed/streaming data, as is the case for computing networking statistics.

Experimental Results

To get an understanding of how the AMS sketches perform in the problem of estimating the self join size of a relation, consider the following setup. The domain of the attribute on which the self join size is computed is set to 16,384. The seize of the relation is fixed at 100,000 tuples. The distribution of the frequencies of join attribute values are generated according to a Zipf distribution with a varying Zipf coefficient. The number of medians is set to 1 (no median computation) and the number of elementary sketches averaged is set to 1,024.

The relative error, both theoretical and empirical, of AMS sketches is depicted in

AMS Sketch, Fig. 1 Combining elementary sketches to estimate $COUNT(F \bowtie_a G)$ with relative error at most ϵ with probability at least $1 - \delta$

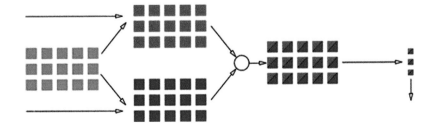

AMS Sketch, Fig. 2
Relative error of AMS
sketches as a function of
Zipf coefficient

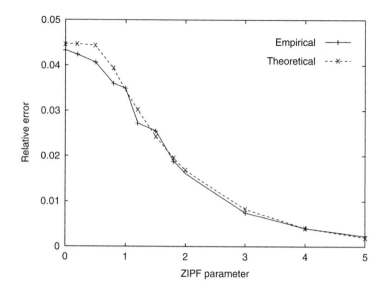

Fig. 2. The following observations confirm the intuition based on theory for the behavior of AMS sketches: (i) on the self join size problem the error is acceptable for sketches of size in the order of 2,000 words, (ii) the error decreases somewhat as the skew increases, and (iii) the theoretical prediction follows entirely the empirical behavior.

URL to Code

http://www.cs.rutgers.edu/~muthu/massdal-code-index.html
http://www.cise.ufl.edu/~adobra/AQP/code.html

Cross-References

▶ Approximate Query Processing
▶ Data Stream

Recommended Reading

1. Alon N, Gibbons PB, Matias Y, Szegedy M. Tracking join and self-join sizes in limited storage. J Comput Syst Sci. 2002;64(3):719–47.
2. Alon N., Matias Y., and Szegedy M. The space complexity of approximating the frequency moments. In: Proceedings of the 28th Annual ACM Symposium on Theory of Computing; 1996. p. 20–9.
3. Charikar M., Chen K., and Farach-Colton M. Finding frequent items in data streams. In: Proceedings of the 29th International Colloquium on Automata, Languages and Programming; 2002. p. 693–703.
4. Cormode G. and Garofalakis M. Sketching streams through the net: distributed approximate query tracking. In: Proceedings of the 31st International Conference on Very Large Data Bases; 2005. p. 13–24.
5. Das A., Gehrke J., and Riedewald M. Approximation techniques for spatial data. In: Proceedings of the ACM SIGMOD International Conference on Management of Data; 2004. p. 695–706.
6. Dobra A., Garofalakis M., Gehrke J., and Rastogi R. Processing complex aggregate queries over data streams. In: Proceedings of the ACM SIGMOD International Conference on Management of Data; 2002. p. 61–72.
7. Rusu F, Dobra A. Pseudo-random number generation for sketch-based estimations. ACM Trans Database Syst. 2007;32(2):11.
8. Rusu F. and Dobra A. Statistical analysis of sketch estimators. In: Proceedings of the ACM SIGMOD International Conference on Management of Data; 2007. p. 187–98.

Anchor Text

Vassilis Plachouras
Yahoo! Research, Barcelona, Spain

Synonyms

Anchor; Anchor text surrogate

Definition

Anchor text is the text associated with a hyperlink pointing from one Web document to another. Anchor text provides a concise description of a document, not necessarily written by the author of the document. It is very effective in Web search retrieval tasks, and in particular, in tasks where the aim is to find the home page of a given website.

Key Points

Hyperlink analysis algorithms explicitly employ the hyperlinks between Web documents to find high quality or authoritative Web documents. A form of implicit use of the hyperlinks in combination with content analysis is the use anchor text associated with the incoming hyperlinks of documents. Web documents can be represented by an anchor text surrogate, which is formed by collecting the anchor text associated with the hyperlinks pointing to the document.

The anchor text of the incoming hyperlinks provides a concise description for a Web document. The used terms in the anchor text may be different from the ones which occur in the document itself, because the author of the anchor text is not necessarily the author of the document. Eiron and McCurley [2] found similarities in the distribution of terms between the anchor text of Web documents and the queries submitted to an intranet search engine by users. Similarities were also found in the use of abbreviations and technical terms.

Craswell et al. [1] show that anchor text is effective for navigational search tasks and more specifically for finding home pages of Web sites. They report that searching for home pages of websites using an index of anchor text performs better than using an index of the document contents. Upstill et al. [3] also suggest that the anchor text of the incoming hyperlinks from documents outside a corpus of documents enhances the retrieval effectiveness for homepage finding.

Cross-References

▶ Document Links and Hyperlinks
▶ Field-Based Information Retrieval Models

Recommended Reading

1. Craswell N, Hawking D, Robertson S. Effective site finding using link anchor information. In: Proceedings of the 24th Annual International ACM SIGIR Conference on Research and Development in Information Retrieval; 2001. p. 250–7.
2. Eiron N, McCurley KS. Analysis of anchor text for web search. In: Proceedings of the 26th Annual International ACM SIGIR Conference on Research and Development in Informaion Retrieval; 2003. p. 459–60.
3. Upstill T, Craswell N, Hawking D. Query-independent evidence in home page finding. ACM Trans Inform Syst. 2003;21(3):286–313.

Annotation

Amarnath Gupta
San Diego Supercomputer Center, University of California San Diego, La Jolla, CA, USA

Definition

An annotation is any form of additional information "superposed" on any existing data or document.

Example: If a scientist records her experimental data in a relational database and then marks some "cells" of a table with the comment "consistent with previous findings," this additionally "marked" information is an annotation.

Key Points

Often annotations are not originally intended to be part of the collected data, and hence no data or schema structure was designed to hold it. Annotating data is a very common practice in science, where scientists would literally "mark" experimental observation with comments and often use annotations to share their opinions in a collaborative study. One can annotate data at the level of whole data sets, groups of data elements (like columns), or values. As larger-scale experiments are conducted and larger collaborations are formed, management of the annotated data becomes a serious challenge. In recent times, the emerging importance of annotation in scientific data management has been recognized by the Information Management Community, leading to a variety of research in annotation management.

Cross-References

▶ Annotation-Based Image Retrieval
▶ Biomedical Scientific Textual Data Types and Processing

Recommended Reading

1. Bhagwat D, Chiticariu L, Tan WC, Vijayvargiya G. An annotation management system for relational databases. In: Proceedings of the 30th International Conference on Very Large Data Bases; 2004. p. 900–11.
2. Buneman P, Khanna S, Tan W-C. On propagation of deletions and annotations through views. In: Proceedings of the 21st ACM SIGACT-SIGMOD-SIGART Symposium on Principles of Database Systems; 2002. p. 150–8.
3. Geerts F, Kementsiesidis A, Milano D. MONDRIAN: annotating and querying databases through colors and blocks. In: Proceedings of the 22nd International Conference on Data Engineering; 2006. p. 82.
4. Gertz M, Sattler K-U. Integrating scientific data through external, concept-based annotations. In: Proceedings of the Workshop on Efficiency and Effectiveness of XML Tools and Techniques and Data Integration Over the Web; 2002. p. 220–40.
5. Murthy S, Maier D, Delcambre LML. Querying bi-level information. In: Proceedings of the 7th International Workshop on the World Wide Web and Databases; 2004. p. 7–12.
6. Srivastava D, Velegrakis Y. Intensional associations between data and metadata. In: Proceedings of the ACM SIGMOD International Conference on Management of Data; 2007. p. 401–12.

Annotation-Based Image Retrieval

Xin-Jing Wang[1,2] and Lei Zhang[1,3]
[1]Microsoft Research Asia, Beijing, China
[2]Micros Facebook, CA, USA
[3]Microsoft Research, Redmond, WA, USA

Synonyms

Semantic image retrieval; Tag-based image retrieval; Tag-based image search; Text-based image retrieval

Definition

Given (i) a textual query and (ii) a set of images and their annotations (phrases or keywords), annotation-based image retrieval systems retrieve images according to the matching score of the query and the corresponding annotations. There are three levels of queries according to Eakins [1]:

- Level 1: Retrieval by primitive features such as color, texture, shape, or the spatial location of image elements, typically querying by an example, i.e., "find pictures like this."
- Level 2: Retrieval by derived features, with some degree of logical inference. For example, "find a picture of a flower."
- Level 3: Retrieval by abstract attributes, involving a significant amount of high-level reasoning about the purpose of the objects or scenes depicted. This includes retrieval of named events, of pictures with emotional or religious significance, etc., e.g., "find pictures of a joyful crowd."

Together, levels 2 and 3 are referred to as semantic image retrieval, which can also be regarded as annotation-based image retrieval. A comprehensive review and analysis on image search in the past 20 years was written by Zhang and Rui [2], which details the system framework, feature extraction and image representation, indexing, and big data's potential.

Historical Background

There are two frameworks of image retrieval [3]: annotation based (or more popularly, text based) and content based. The annotation-based approach can be tracked back to the 1970s. In such systems, the images are manually annotated by text descriptors, which are used by a database management system (DBMS) to perform image retrieval. There are two disadvantages with this approach. The first is that a considerable level of human labor is required for manual annotation. The second is that because of the subjectivity of human perception, the manually labeled annotations may not converge. To overcome the above disadvantages, content-based image retrieval (CBIR) was introduced in the early 1980s. In CBIR, images are indexed by their visual content, such as color, texture, and shapes. In the past decade, several commercial products and experimental prototype systems were developed, such as QBIC, Photobook, Virage, VisualSEEK, Netra, and SIMPLIcity. Comprehensive surveys in CBIR can be found in [1, 4].

However, the discrepancy between the limited descriptive power of low-level image features and the richness of user semantics, which is referred to as the "semantic gap," bounds the performance of CBIR. On the other hand, due to the explosive growth of visual data (both online and offline) and the phenomenal success in Web search, there has been increasing expectation for image search technologies. Because of these reasons, the main challenge of image retrieval is understanding media by bridging the semantic gap between the bit stream and the visual content interpretation by humans [5]. Hence, the focus is on automatic image annotation techniques.

Foundations

The state-of-the-art image auto-annotation techniques include four main categories [3, 5]:

(i) using machine learning tools to map low-level features to concepts, (ii) exploring the relations among image content and the textual terms in the associated metadata, (iii) generating semantic template (ST) to support high-level image retrieval, and (iv) making use of both the visual content of images and the textual information obtained from the Web to learn the annotations.

Machine Learning Approaches

A typical approach is using support vector machine (SVM) as a discriminative classifier over image low-level features. Though straightforward, it has been shown effective in detecting a number of visual concepts.

Recently there is a surge of interest in leveraging and handling relational data, e.g., images and their surrounding texts. Blei et al. [6] extend the latent Dirichlet allocation (LDA) model to the mix of words and images and proposed a correlation LDA model. This model assumes that there is a hidden layer of topics, which are a set of latent factors and obey the Dirichlet distribution, and words and regions are conditionally independent on the topics, i.e., generated by the topics. This work used 7,000 Corel photos and a vocabulary of 168 words for annotation.

Relation Exploring Approaches

Another notable direction for annotating image visual content is exploring the relations among image content and the textual terms in the associated metadata. Such metadata are abundant but are often incomplete and noisy. By exploring the co-occurrence relations among the images and the words, the initial labels may be filtered and propagated from initial labeled images to additional relevant ones in the same collection [5].

Jeon et al. [7] proposed a cross-media relevance model to learn the joint probabilistic distributions of the words and the visual tokens in each image, which are then used to estimate the likelihood of detecting a specific semantic concept in a new image.

Semantic Template Approaches

Though it is not yet widely used in the above-mentioned techniques, semantic template (ST) is a promising approach in annotation-based image retrieval (a map between high-level concept and low-level visual features).

Chang and Chen [8] show a typical example of ST, in which a visual template is a set of icons or example scenes/objects denoting a personalized view of concepts such as meetings and sunset. The generation of a ST is based on user definition. For a concept, the objects, their spatial and temporal constraints, and the weights of each feature of each object are specified. This initial query scenario is provided to the system, and then through the interaction with users, the system finally converges to a small set of exemplar queries that "best" match (maximize the recall) the concept in the user's mind.

In contrast, Zhuang et al. [9] generate ST automatically in the process of relevance feedback, whose basic idea is to refine retrieval outputs based on interactions with the user. A semantic lexicon called WordNet is used in this system to construct a network of ST. During the retrieval process, once the user submits a query concept (keyword), the system can find a corresponding ST and thus target similar images.

Large-Scale Web Data-Supported Approaches

Good scalability to a large set of concepts is required in ensuring the practicability of image annotation. On the other hand, images from the Web repositories, e.g., Web search engines or photo sharing sites, come with free but less reliable labels. In [10], a novel search-based annotation framework was proposed to explore such Web-based resources. Fundamentally, it is to automatically expand the text labels of an image of interest, using its initial keyword and image content.

The process of [10] is shown in Fig. 1. It contains three stages, the text-based search stage, the content-based search stage, and the annotation learning stage, which are differentiated using different colors (black, brown, blue) and labels (A, B, C). When a user submits a query

image as well as a query keyword, the system first uses the keyword to search a large-scale Web image database (2.4 million images crawled from several Web photo forums), in which images are associated with meaningful but noisy descriptions, as tagged by "A." in Fig. 1. The intention of this step is to select a semantically relevant image subset from the original pool. Visual feature-based search is then applied to further filter the subset and save only those visually similar images (the path labeled by "B" in Fig. 1). By these means, a group of image search results which are both semantically and visually similar to the query image are obtained. To speed up the visual feature-based search procedure, a hash encoding algorithm is adopted to map the visual features into hash codes, by which inverted indexing technique in text retrieval area can be applied for fast retrieval. At last, based on the search results, the system collects their associated textual descriptions and applies the search result clustering (SRC) algorithm to group the images into clusters. The reason of using SRC algorithm is that (i) it is proved to be significantly effective in grouping documents semantically, and (ii) more attractively, it is capable of learning a name for each cluster that best represents the common topics of a cluster's member documents. By ranking these clusters according to a ranking function and setting a certain threshold, the system selects a group of clusters and merges their names as the final learnt annotations for the query image, which ends the entire process (C in Fig. 1).

The technique has been advanced in the following aspects. Firstly, [11] discussed the special case that the query is only an image, instead of both an image and a text. In this case, no text-based search stage is available, so that it is more challenging for image representation, image indexing (to support real-time search), image retrieval metric, and annotation detection (i.e., mining salient words or phrases from surrounding texts of retrieved images) techniques. Dai et al. [12] proposed a Bayesian model as a better annotation detection model than in [10], and [13] presented novel image representation and

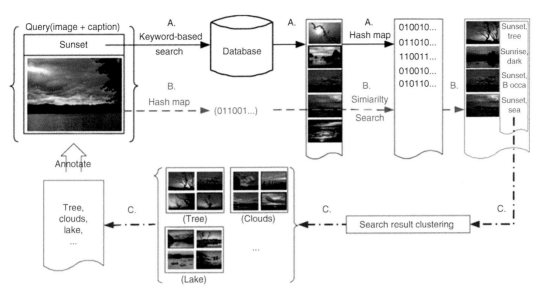

Annotation-Based Image Retrieval, Fig. 1 Framework of the search-based annotation system

image indexing techniques. Secondly, to annotate images that have duplicated versions on the Web is shown to have great research and commercial potentials [14]. Such images are called duplicated images. The reasons are (1) duplicate image detection is a well-defined research problem, compared to general image similarity retrieval, since there is no clear semantic definition of image similarity; (2) how to effectively and efficiently discover duplicate images (i.e., require both retrieval precision and recall) from a Web-scale dataset is a challenging research problem [15]; and (3) duplicate images tend to belong to Web user-interested categories such as celebrities, famous locations, movie posters, funny pictures, etc., and to annotate such images has great commercial values for image search, e.g., to improve search relevance and user experience and to reduce image index [13].

Key Applications

Due to the explosive growth of visual data (both online and offline), effective annotation-based image search becomes a highly supportive technique for many multimedia applications.

Long et al. [1] applied the annotation-based image search technique in [10] to understand user images for monetization. The image annotations help ad providers to select related advertisements for targeted advertising.

El-Saban et al. [16] adopted annotation-based image search to automatically caption mobile captured videos. Such a technique helps save human labor in annotating their videos, and the auto-captions can be directly used to facilitate video search, which has great commercial value.

Zhang et al. [17] utilized annotation-based image search to collect celebrity face images, which shows an example of how to automatically collect a large-scale dataset for computer vision research.

Cross-References

▶ Cross-Modal Multimedia Information Retrieval
▶ Hash-Based Indexing
▶ Image Querying
▶ Indexing and Similarity Search
▶ Information Retrieval
▶ Object Recognition
▶ Multimedia Information Retrieval Model

Recommended Readings

1. Long F, Zhang HJ, Feng DD. Fundamentals of content-based image retrieval. In: Feng D, editor. Multimedia information retrieval and management: technological fundamentals and applications. Berlin/Heidelberg: Springer; 2013; Part I. p. 1–26. ISSN: 1860-4862.
2. Zhang L, Rui Y. Image search from thousands to billions in 20 years. ACM TOMCCAP 2013;9(1s): Article No. 36. Special Issue on the 20th Anniversary of the ACM MM Conference; New York.
3. Liu Y, Zhang D, Lu G, Ma W-Y. A survey of content-based image retrieval with high-level semantics. Pattern Recogn. 2007;40(1):262–82.
4. Rui Y, Huang TS, Chang S-F. Image retrieval: current techniques, promising directions, and open issues. J Vis Commun Image Represent. 1999;10(4):39–62.
5. Chang S-F, Ma W-Y, Smeulders A. Recent advances and challenges of semantic image/video search. In: Proceedings of the IEEE International Conference on Acoustics, Speech, and Signal Processing; 2007. p. 1205–8.
6. Blei D, Jordan MI. Modeling annotated data. In: Proceedings of the 26th Annual International ACM SIGIR Conference on Research and Development in Information Retrieval; 2003. p. 127–34.
7. Jeon J, Lavrenko V, Manmatha R. Automatic image annotation and retrieval using cross-media relevance models. In: Proceedings of the 26th Annual International ACM SIGIR Conference on Research and Development in Information Retrieval; 2003. p. 119–26.
8. Chang S-F, Chen W, Sundaram H. Semantic visual templates: linking visual features to semantics. In: Proceedings of the International Conference on Image Processing; 1998. p. 531–4.
9. Zhuang Y, Liu X, Pan Y. Apply semantic template to support content-based image retrieval. In: Proceedings of the SPIE, Storage and Retrieval for Media Databases; 1999. p. 442–9.
10. Wang X-J, Zhang L, Jing F, Ma W-Y. AnnoSearch: image auto-annotation by search. In: Proceedings of the IEEE International Conference on Computer Vision and Pattern Recognition; 2006. p. 1483–90.
11. Wang X-J, Zhang L, Li X, Ma W-Y. Annotating images by mining image search results. IEEE Trans Pattern Anal Mach Intell. 2008;30(11):1919–32.
12. Dai LC, Wang X-J, Zhang L, Yu NH. Efficient tag mining via mixture modeling for real-time search-based image annotation. In: Proceedings of the IEEE International Conference on Multimedia and Expo; 2012.
13. Wang X-J, Zhang L, Ma W-Y. Duplicate-search-based image annotation using web-scale data. Proc IEEE. 2012;100(9):2705–21.
14. Wang X-J, Zhang L, Liu M, Li Y, Ma W-Y. ARISTA – image search to annotation on billions of web photos. In: Proceedings of the IEEE International Conference on Computer Vision and Pattern Recognition; 2010.
15. Wang X-J, Zhang L, Liu C. Duplicate discovery on 2 billion internet images. IEEE Conference on Computer Vision and Pattern Recognition; 2013.
16. El-Saban M, Wang X-J, Hasan N, Bassiouny M, Refaat M. Seamless annotation and enrichment of mobile captured video streams in real-time. In: Proceedings of the IEEE International Conference on Multimedia and Expo; 2011.
17. Zhang X, Zhang L, Wang X-J, Shum H-Y. Finding celebrities in billions of webpages. IEEE Transactions on Multimedia. 2012;14(4):995–1007.
18. Eakins J, Graham M. Content-based image retrieval. Technical report. Tyne: University of Northumbria at Newcastle; 1999.
19. Wang X-J, Yu M, Zhang L, Ma W-Y. Advertising based on users' photos. In: Proceedings of the IEEE International Conference on Multimedia and Expo; 2009.

Anomaly Detection on Streams

Spiros Papadimitriou
IBM T.J. Watson Research Center, Hawthorne, NY, USA

Definition

Anomaly detection generally refers to the process of automatically detecting events or behaviors which deviate from those considered normal. It is an unsupervised process, and can thus detect anomalies which have not been previously encountered. It is based on estimating a model of typical behavior from past observations and consequently comparing current observations against this model. It can be performed either on a single stream or among multiple streams. Anomaly detection encompasses outlier detection as well as change detection and therefore is closely related to forecasting and clustering methods.

Historical Background

Anomaly detection in streams has close connections to traditional outlier detection, as well as to change detection. The former is a common and widely studied topic in statistics [11]. The latter emerged in the context of statistical monitoring and control for continuous processes and the widely used CUSUM algorithm was proposed as early as 1954 [9]. With the emergence of data stream management systems, anomaly detection in this setting has received significant attention, with applications in network management and intrusion detection, environmental monitoring, and surveillance, to mention a few.

Foundations

Anomaly detection is closely related to outlier detection and change detection. After a review of the main ideas, the streaming case is presented.

Outlier Detection
The existing approaches to outlier detection can be broadly classified into the following categories. Typically, outlier detection relies on a model for the data. Model parameters are estimated based on appropriately chosen historical data. As new observations arrive, they are either compared directly against the model and are declared outliers if the fit is poor. Alternatively, a second set of model parameters may be estimated from recent observations. If there is a statistically significant difference among the two sets of parameters, the new observations are declared as outliers.

Clustering-Based and Forecasting-Based Approaches
Many clustering and forecasting algorithms detect outliers as by-products. However, not all clustering or forecasting procedures can be easily turned into outlier detection procedures.

Distribution-Based Approaches
Methods in this category are typically found in statistics textbooks. They deploy some standard distribution model (e.g., Gaussian) and flag as outliers those objects which deviate from the model. These work well in many occasions, but may be unsuitable for high-dimensional data sets,

or when reasonable assumptions about the distribution of data points cannot be made.

Distance-Based and Density-Based Approaches
A point in a data set is a distance-based outlier if at least a fraction β of all other points are further than r from it. This outlier definition is based on a single, global criterion determined by the parameters r and β. This can lead to problems when the data set has both dense and sparse regions. Density-based approaches aim to remedy this problem, by relying on the local density of each point's neighborhood.

Change Detection
Sequential hypothesis testing and sequential change detection arose out of problems in statistical process control. Assume a collected sequence of observations, modeled as random variables $X_1, X_2, \ldots, X_t, \ldots$. Additionally, assume that X_t are drawn from a distribution with parameter θ and that a test of whether the true parameter is θ_0 or θ_1 is desired.

The Sequential Likelihood Ratio Test (SLRT) relies on the logarithm of likelihood ratios $z_t := \log(p(x_t\,;\,\theta_0)/p(x_t\,;\,\theta_1))$ and tests the cumulative sum $z_1 + \ldots + z_t$ to decide upon the true parameter.

This can be extended to other settings, such as detecting changes in other distribution parameters. For example, in its simplest form, CUSUM tests for a shift in the mean by essentially applying SLRT, assuming points independently drawn from a Gaussian distribution with known variance. Many other versions have appeared since the CUSUM test was first proposed [9], relaxing or modifying some of these assumptions.

In general, change detection is closely related to outlier detection; in fact, change detection may also be viewed as outlier detection along the time axis.

Streaming Algorithms
In a streaming setting, there are two key challenges that need to be addressed:

1. *Limited resources.* In a streaming setting, a large number of observations arrives over time and the total volume of data grows indefinitely. However processing and storage capacity are limited, in comparison to the amount of data. Therefore, data summarization or *sketching* techniques need to be applied, in order to extract a few, relevant features from the raw data.

2. *Concept drift.* In an indefinitely growing collection of observations, changes in the underlying features (e.g., distribution parameters) may not necessarily correspond to anomalies, but rather be part of normal changes in the behavior of the system. Thus, mechanisms to handle such non-stationarity or *concept drift* and adapt to changing behavior are necessary [12].

Next, several of the approaches that have been studied in the literature are reviewed.

Sketching Techniques

In the past several year, a number of techniques for *sketch* or *synopsis* construction have appeared, with applications to many stream processing problems. Some examples include CM sketches, AMS sketches, FM sketches, and Bloom filters [2]. Other summarization techniques specifically for data clustering on streams have appeared, such as those in [1, 4], which can be easily extended for outlier detection on streams.

Burst Detection

In many applications, the appearance of sudden bursts in the data often signifies an anomaly. For example, in a network monitoring application, a burst in the traffic volume to a particular destination may signify a denial of service (DoS) attack. Thus, *burst detection* on streams has received significant attention. Examples of such work include [7] and [14].

Correlation Analysis

Often a collection of multiple streams is available and measurements from different streams may be highly correlated with each other. If the strength of correlations changes over time [13] or the number of correlated components varies [10], this often signifies changes in the underlying data-generating process that may be due to anomalies.

Change Analysis

More generally, detecting significant changes has been studied in the context of stream processing [3].

Key Applications

Intrusion Detection

With the widespread adoption of the internet, various forms of malware (e.g., viruses, worms, trojans, and botnets) have become a serious and costly issue. Most intrusion detection systems (IDS) rely on known signatures to identify malicious payloads or behaviors. However, there are several efforts underway for automatic detection of suspicious activity on the fly, as well as for automating the signature extraction process.

System Monitoring

Maintenance costs for large computer clusters or networks is traditionally labor-intensive and contributes a large fraction of total cost of ownership. Hence, autonomic computing initiatives aim at automating this process. An important first step is the automatic, unsupervised detection of abnormal events (e.g., node or link failures) based on continuously collected system metrics. Streaming anomaly detection methods are used to address this problem.

Process Control

Applications in quality control and industrial process control have traditionally provided much of the impetus for the development of change detection methods. Machinery used in a production chain (e.g., food preparation or chip fabrication) typically monitor a large number of process parameters at each step. Early detection of sudden changes in those parameters is important to identify potential flaws in the process which can severely affect end product quality.

Pervasive Healthcare

Small and cheap sensors which can continuously monitor patient physiological data (e.g., temperature, blood pressure, heart rate, ECG measurements, glucose levels, etc.) are becoming widely available. Anomaly detection methods can prove essential in enabling early diagnosis of potential life-threatening conditions, as well as preventive healthcare.

Civil Infrastructure

Early detection of faults by continuously monitoring civil infrastructure components (e.g., bridges, buildings, and roadways) can reduce maintenance costs and increase safety. Similarly, surveillance systems on urban environments rely on anomaly detection methods to spot suspicious activities and increase security.

Future Directions

Certain anomalies can be detected only by taking into account information collected from a large number of different sources. Even if data ownership issues are resolved, collecting all this information at a central site is often infeasible due to its large volume. A number of efforts have tackled this problem in the past few years, but much remains to be done, especially as the scale of information collected increases. Also related to this trend is anomaly detection on more complex data, such as time-evolving graphs.

Cross-References

▶ Change Detection on Streams
▶ Outlier Detection

Recommended Reading

1. Aggarwal CC, Han J, Wang J, and Yu PS. A Framework for clustering evolving data streams. In: Proceedings of the 29th International Conference on Very Large Data Bases; 2003. p. 81–92.
2. Aggarwal CC, Yu PS. A survey of synopsis construction in data streams. In: Data streams: models and algorithms. New York: Springer; 2007.
3. Cormode G, Muthukrishnan S. What's new: finding significant differences in network data streams. IEEE/ACM Trans Netw. 2005;13(6):1219–32.
4. Guha S, Meyerson A, Mishra N, Motwani R, O'Callaghan L. Clustering data streams: theory and practice. IEEE Trans Knowl Data Eng. 2003;15(3):515–28.
5. Hulten G, Spencer L, and Domingos P. Mining time-changing data streams. In: Proceedings of the 7th ACM SIGKDD International Conference on Knowledge Discovery and Data Mining; 2001. p. 97–106.
6. Jain AK, Murty MN, Flynn PJ. Data clustering: a review. ACM Comput Surv. 1999;31(3): 264–323.
7. Kleinberg J. Bursty and hierarchical structure in streams. In: Proceedings of the 8th ACM SIGKDD International Conference on Knowledge Discovery and Data Mining; 2002. p. 91–101.
8. Lee W, Stolfo SJ, Mok KW. Adaptive intrusion detection: a data mining approach. Artif Intell Rev. 2000;14(6):533–67.
9. Page ES. Continuous inspection schemes. Biometrika. 1954;41(1):100–15.
10. Papadimitriou S, Sun J, and Faloutsos C. Streaming pattern discovery in multiple time-series. In: Proceedings of the 31st International Conference on Very Large Data Bases; 2005. p. 697–708.
11. Peter JR, Annick ML. Robust regression and outlier detection. New York: Wiley; 1987.
12. Wang H, Fan W, Yu PS, and Han J. Mining concept-drifting data streams using ensemble classifiers. In: Proceedings of the 9th ACM SIGKDD International Conference on Knowledge Discovery and Data Mining; 2003. p. 226–35.
13. Zhu Y and Shasha D. StatStream: statistical monitoring of thousands of data streams in real time. In: Proceedings of the 28th International Conference on Very Large Data Bases; 2002. p. 358–69.
14. Zhu Y and Shasha D. Efficient elastic burst detection in data streams. In: Proceedings of the 9th ACM SIGKDD International Conference on Knowledge Discovery and Data Mining; 2003. p. 336–45.

Anonymity

Simone Fischer-Hübner
Karlstad University, Karlstad, Sweden

Synonyms

Namelessness; Nonidentifiability

Definition

The term anonymity originates from the Greek word "anonymia," which means "without a name."

In the context of computing, anonymity has been defined in [2] as follows: "Anonymity of a subject means that the subject is not identifiable within a set of subjects, the anonymity set." The anonymity set is the set of all possible subjects, e.g., the set of all possible senders of a message or the set of all possible recipients of a message (dependent on the knowledge of an attacker). *Sender anonymity* means that a message cannot be linked to the sender, while *receiver anonymity* implies that a certain message cannot be linked to the receiver of that message. Relationship anonymity of a sender and recipient means that even though a sender and a recipient can be identified as participating in some communication, they cannot be identified as communicating with each other, i.e., sender and recipient are unlinkable. The definition above corresponds to the definition of anonymity in [1]: "Anonymity of a user means that the user may use a resource or service without disclosing the user's identity."

To reflect the possibility to quantify anonymity, a slightly modified definition has also provided by [2]: "Anonymity of a subject from the attacker's perspective means that the attacker cannot sufficiently identify the subject within a set of subjects, the anonymity set."

Data protection legislation usually defines data as anonymous if information concerning personal or material circumstances can no longer (or with disappropriate amount of time, expense, and labor) be attributed to an individual.

Key Points

Providing anonymity is the best strategy for achieving privacy. If individuals can act anonymously and if their data are kept in an anonymous form, their privacy is not affected (and consequently, data protection legislation is not applicable).

For providing anonymity, however, it has to be guaranteed that potential attackers cannot or not sufficiently identify the individuals.

A simple anonymity metric is the anonymity set size, which relates to k-anonymity of databases requiring that a search for identifying information results in a group of at least k candidate records [3]. Further proposals for measuring the degree of anonymity are based on Shannon's entropy for taking the distribution of probabilities that link the subjects of an anonymity set to the item of interest into account.

Cross-References

▶ Privacy
▶ Privacy-Enhancing Technologies
▶ Privacy Metrics

Recommended Reading

1. Common Criteria Project, common criteria for information technology security evaluation, part 2: security functional requirements, Jan 2010, ISO/IEC 15408-1:2009. www.commoncriteriaportal.org.
2. Pfitzmann A. Hansen M. A terminology for talking about privacy by data minimization: anonymity, unlinkability, unobservability, pseudonymity, and identity management version 0.34. http://dud.inf.tu-dresden.de/Anon_Terminology.shtml. 10 Aug 2010.
3. Sweeney L. k-anonymity: a model for protecting privacy. Int J Uncertain Fuzziness Knowl-Based Syst. 2002;10(5):557–70.

ANSI/INCITS RBAC Standard

Yue Zhang and James B. D. Joshi
University of Pittsburgh, Pittsburgh, PA, USA

Synonyms

RBAC standard

Definition

The ANSI/INCITS RBAC standard includes the definition of the *RBAC Reference Model* and the *RBAC System and Administrative Functional Specification*.

The *RBAC Reference Model* constitutes four model components: *core RBAC model, hierarchical RBAC model, RBAC with Static Separation of Duty* and *RBAC with Dynamic Separation of Duty* [1]. The RBAC models with Separation of Duty are also called *Constrained RBAC*. The *core RBAC* includes the following entities:

- *USERS, ROLES, OPS,* and *OBS* are the sets of users, roles, operations and objects, respectively.
- *UA* ⊆ *USERS* × *ROLES* is a many-to-many mapping from *USERS* to *ROLES*.
- *Assigned_users*: (*r:ROLES*) → 2^{USERS}, is mapping of role *r* onto a set of users. Formally: *assigned_users*(*r*) = {*u*ε*USERS* | (*u, r*) ε *UA*}
- *PRMS* = $2^{(OBS × OPS)}$ is the set of permissions.
- *PA* ⊆ *PERMS* × *ROLES* is a many-to-many mapping from *PERMISSIONS* to *ROLES*.
- *Assigned_permissions*(*r:ROLES*) → 2^{PRMS}, is mapping of role *r* onto a set of permissions. Formally: *assigned_permissions*(*r*) = {*p*ε*PRMS* | (*p, r*) ε *PA*}
- *Op*(*p: PRMS*) → {*op*⊆*OPS*}, is the permission to operation mapping.
- *Ob*(*p: PRMS*) → {*ob*⊆*OBS*}, is the permission to object mapping.
- *SESSIONS* = the set of sessions
- *Session_users* (*s:SESSIONS*) → *USERS* is mapping of session *s* onto the corresponding user.
- *Session_roles* (*s:SESSIONS*) → 2^{ROLES} is mapping of session *s* onto a set of roles.
- *Avail_session_perms*(*s:SESSIONS*) → 2^{PRMS} gives the permissions available to a user in a session =

$$\left[\bigcup_{r \in session\_roles} assigned\_permissions(r) \right]$$

The *hierarchical RBAC* model extends the *core* RBAC model with the *general* and *limited* role hierarchies. The *general role hierarchy* is defined below:

- *RH* ⊆ *ROLES* × *ROLES* is a partial order on *ROLES*, written as ≥, where $r_1 ≥ r_2$ only if all permissions of r_2 are also permissions of r_1, and all users who can assume r_1 can also assume r_2.
- *Authorized_users*(*r: ROLES*) → 2^{USERS}, is mapping of role *r* onto a set of users in the presence of a role hierarchy. Formally: *authorized_users*(*r*) = {*u*∈*USERS* | *r′*; ≥ *r*, (*u, r′*) ∈ *UA*}.
- *Authorized_permissions*(*r: ROLES*) → 2^{PRMS}, is mapping of role *r* onto a set of permissions in the presence of a role hierarchy. Formally: *authorized_permissions*(*r*) = {*p*∈*PRMS* | *r′* ≥ *r*, (*p, r′*) ∈ *PA*}

The *limited hierarchy* is similar to the *general role hierarchies* with the following limitation:

$$\forall r, r_1, r_2 \in ROLES, \quad r \geq r_1 \wedge r \geq r_2 \Rightarrow r_1 = r_2$$

The *constrained RBAC* model extends the hierarchical RBAC with *Static Separation of Duty* (SSoD) and *Dynamic SoD* (DSoD) constraints.

SSoD Constraint: SSD ⊆ (2^{ROLES} × N) is a collection of pairs (*rs, n*) in *Static Separation of Duty*, where each *rs* is a role set, and *n* is a natural number ≥2, with the property that no user is assigned (or authorized if hierarchy is present) to *n* or more roles from the set *rs* in each (*rs, n*) ∈ *SSD*.

DSoD Constraint: DSD ⊆ (2^{ROLES} × N) is collection of pairs (*rs, n*) in *Dynamic Separation of Duty*, where each *rs* is a role set and *n* is a natural number ≥2, with the property that no user may activate *n* or more roles from the set *rs* in each *dsd* ∈ *DSD*.

The RBAC System and Administrative Function Specification details various features required in a RBAC system. These features are divided into three categories: *administrative operations, administrative reviews,* and *system level functionality* [1]. The administrative operations include functions that are important for administrative interface that enable administrative activities of creating, deleting and maintaining RBAC model elements and relations. The administrative reviews feature include interfaces for performing queries on RBAC elements and relations. The system level functionality provides support creating user session creation, activating/deactivating roles, making access decision and enforcement of constraints.

ANSI RBAC also suggests a methodology of creating functional packages, which allows systems to select one of the four RBAC reference models depending upon the access control needs and add optional features [1].

Key Points

The ANSI/INCTIS RBAC standard essentially evolved from the original RBAC96 model [2], and its predecessor NIST RBAC standard [3]. Several advanced features of RBAC such as cardinality constraints, hybrid hierarchy, much fine-grained SoD constraints, features related to the administration of RBAC, etc., are not covered in the standard. The link (http://csrc.nist.gov/groups/SNS/rbac/standards.html) includes the most updated information on ANSI RBAC.

Cross-References

▶ Role-Based Access Control

Recommended Reading

1. ANSI. American National Standard for Information Technology – role based access control. ANSI INCITS, 359-2004, February 2004.
2. Sandhu RS, Coyne EJ, Feinstein HL, Youman CE. Role-based access control models. IEEE Comput. 1996;29(2):38–47.
3. Ferraiolo DF, Sandhu R, Gavrila S, Kuhn DR, Chandramouli R. Proposed NIST standard for role-based access control. ACM Trans Inf Syst Secur. 2001;4(3):224–74. https://doi.org/10.1145/501978.501980. http://doi.acm.org/10.1145/501978.501980.

Answering Queries Using Views

Vasilis Vassalos
Athens University of Economics and Business, Athens, Greece

Synonyms

Rewriting Queries Using Views

Definition

Answering queries using views refers to a data management problem and the set of related techniques, algorithms and other results to address it. The basic formulation of the problem is the following: Given a query Q over a database schema Σ, expressed in a query language L_Q and a set of views $V_1, V_2, \ldots, V_n$ over the same schema, expressed in a query language L_V, is it possible to answer the query Q using (only) the views $V_1, V_2, \ldots, V_n$?

The problem has a number of related formulations: What is the *maximal set* of tuples in the answer of Q that can be obtained from the views? If it is possible to access both the views and the database relations, what is the cheapest query execution plan for answering Q?

From the above, it is clear that this is more generally a family of problems. Problems of different complexity, often admitting different (or no) solutions and solution techniques, result from making choices about various "parameters" of the original problem. For example, one can choose the languages L_Q and L_V, the language L_R used to answer Q from the views, the ability to access the relations of the schema Σ, the possible existence of constraints, e.g., equality-generating or tuple-generating dependencies, as well as the semantics of the views (sound, complete, or both).

The problem is also referred to as query rewriting using views.

Historical Background

The problem of putting together information from multiple sources has a relatively long history within the database community. In the 1970s, distributed databases offered a controlled solution to the problem: the data are structured with a single schema and are put under the control of a single, albeit distributed, database system. In the early 1980s, the challenges of integrating full-fledged relational databases was studied in the context of multidatabases. An important direction of multidatabase research was static schema integration, i.e., creating in advance a single new schema with as much of the information of the original schemas as possible. Query processing could then be performed on the integrated schema. "Reusing" the original database tables also received some attention in the multidatabase context. In the early 1990s, techniques were developed in order to make use of existing materialized views to speed up the processing of queries that did not mention them explicitly. These techniques were precursors to the more general techniques developed a few years later for the problem of answering queries using views. The problem was cast in its current form in a seminal paper by Levy, Mendelzon, Sagiv, and Srivastava in 1995.

Foundations

Preliminaries

To fully specify the problem of answering queries using views one needs to decide on the languages used for the queries, the views and, in some cases, the rewritings. This entry presents the basic case of answering *conjunctive queries* using *conjunctive views*, possibly in the presence of constraints. The relevant definitions are presented next. Techniques developed for different languages or data models, such as bag queries (and views), XML queries, queries with aggregates, etc., can be found in the Recommended Reading list. Moreover, detailed presentations of the techniques discussed in the article, including exceptions, optimizations, and other technical is-

sues, can be found in the research articles on the Recommended Reading list.

Conjunctive Queries

A *conjunctive* query can be represented as:

$$q\left(\overline{X}\right) \leftarrow s_1\left(\overline{X}_1\right) \wedge \cdots \wedge s_n\left(\overline{X}_n\right)$$

where q and $s_1, \ldots, s_n$ are predicate names. In general, $s_1, \ldots, s_n$ refer to database relations. The atoms $s_1(\overline{X}_1), \ldots, s_n\ (\overline{X}_n)$ that appear in the body of the query are called *subgoals* of the query, The atom $q(\overline{X})$ is called the *head* of the query and defines the answer relation. The tuples $\overline{X}, \overline{X}_1, \ldots, \overline{X}_n$ contain either variables or constants. The variables in $\overline{X}$ are the *distinguished* variables of the query, the rest of the variables are the *existential* variables. An important condition for a conjunctive query is *safety*: A query is safe if $\overline{X} \subseteq \overline{X}_1 \cup \ldots \cup \overline{X}_n$, i.e., every distinguished variable must also appear in a query subgoal.

Use *Vars(Q)* and *Subg(Q)* to refer to the set of variables (and constants) in Q and subgoals of Q respectively. $Q(D)$ refers to the result of evaluating the query Q over the database D.

Views

A *view* is a query whose head defines a new database relation. If this relation is not stored, the view is called a *virtual* view. If the results of executing the view are stored, it is called a *materialized* view and the relation is the *extension* of the view. Denote by D_V the database D extended with the extensions of the views belonging to a view set V.

Containment & Equivalence

The concepts of query containment and equivalence are central to query rewriting theory (see entry "▶ Query Rewriting") as they are used to test the correctness of a rewriting of a query using a set of views.

Definition 1

A query Q_1 is contained in a query Q_2, denoted $Q_1 \sqsubseteq Q_2$, if for all database instances D, the set of tuples computed for Q_1 is a subset of those computed for Q_2, i.e., $Q_1(D) \subseteq Q_2(D)$. The two queries are equivalent if $Q_1 \sqsubseteq Q_2$ and $Q_2 \sqsubseteq Q_1$.

Containment mappings provide a necessary and sufficient syntactic condition for testing query containment of conjunctive queries.

Definition 2

A mapping τ from Vars(Q_2) to Vars(Q_1) is a containment mapping if

- *It is the identity on constants*
- *It maps every subgoal inSubg(Q_2) to a subgoal inSubg(Q_1), and*
- *It maps the head of Q_2 to the head of Q_1.*

A seminal result in semantic query optimization is that a query Q_1 is contained in Q_2 if and only if there is a containment mapping from Q_2 to Q_1.

Rewritings

Given a query Q and a set of views $V_1, V_2, \ldots, V_n$ over the same database schema, the goal is to find an *equivalent rewriting Q'* of the query, i.e., a query Q' that uses one or more of the views in the body.

Definition 3

A query Q' is an equivalent rewriting of query Q that uses the set of views $V = \{V_1, V_2, \ldots, V_n\}$ if

- *Q and Q' are equivalent, and*
- *Q' refers to the views in V.*

For every database D, Q' is evaluated over D_V. A rewriting Q' is *locally minimal* if no literals from Q' can be removed while retaining equivalence to Q. A rewriting is *globally minimal* if there is no other rewriting with fewer literals. If the rewriting refers *only* to the views in V and to no other relations then the rewriting is called *complete*. When looking for rewritings, it is preferable to find those that are cheaper to evaluate than the original query. Below is an example of a query and a rewriting using views.

Example 1

Consider the following query Q and view V

$$Q : q(X, U) \leftarrow p(X, Y) \wedge p_0(Y, Z) \wedge p_1$$
$$(X, W) \wedge p_2(W, U).$$

$$V : v(A, B) \leftarrow p(A, C) \wedge p_0$$
$$(C, B) \wedge p_1(A, D).$$

Q can be rewritten using V as follows:

$$Q' : q(X, U) \leftarrow v(X, Z) \wedge p_1$$
$$(X, W) \wedge p_2(W, U).$$

By substituting the view, the first two literals of the query can be removed. However, although the third literal in Q is guaranteed to be satisfied by V, it cannot be removed from the query, as the variable D is projected out in the head of V. Hence, if p_1 were removed from the query, the join condition between p_1 and p_2 could not be enforced.

In several settings, *maximally contained* rewritings need to be considered. Unlike equivalent rewritings, maximally contained rewritings may differ depending on the language used to express the rewritings. Therefore, the following definition depends on a particular query language:

Definition 3

Let Q be a query and $V = \{V_1, \ldots, V_m\}$ be a set of views over the same database schema, and L. be a query language. The query Q' is a maximally contained rewriting of Q using V with respect to L. if:

- *Q' is a query in L. that refers only to the views in V,*
- *$Q' \sqsubseteq Q$, and*
- *for every rewriting $Q_1 \in L.$, such that $Q_1 \sqsubseteq Q$, $Q_1 \sqsubseteq Q'$.*

Characterizing Rewritings

Answering queries using views is closely related to the problem of query containment. The following proposition provides a necessary and sufficient condition for the existence of a rewriting of Q that includes a view V.

Proposition 1

Let Q and V be conjunctive queries with built-in predicates. There is a rewriting of Q using V if and only if $\pi_\varnothing (Q) \sqsubseteq \pi_\varnothing (V)$, i.e., the projection of Q onto the empty set of columns is contained in the projection of V onto the empty set of columns.

This proposition provides a complete characterization of the problem of using views for query answering. Two other important characteristics of the problem are that, for conjunctive queries and views, a rewriting that does not introduce new variables and does not include database relations that do not appear in the original query, can always be found. These characteristics allow significant pruning of the search space for a minimal rewriting of Q, but do not always hold for more expressive settings of the problem. Finally, a minimal rewriting of a query Q using a set of views V without built-in predicates does not need more than the number of literals in the query. In particular, *if the body of Q has p literals and Q' is a locally minimal and complete rewriting of Q using V, then Q' has at most p literals.*

The bound provided above does not hold when the database relations have functional dependencies. In such a case the size of a minimal rewriting is at most $p + d$ literals, where d is the sum of the arities of the literals in Q. In the presence of built-in predicates, the size of the rewritten query is at most exponential in the size of Q.

The above properties determine the complexity of the problem. In particular, if Q is a conjunctive query with built-in predicates and V is a set of conjunctive views *without* built-in predicates, then the problem of determining whether there exists a rewriting of Q that uses V is NP-complete. If the views in V have built-in predicates, the problem is Π^P_2-complete. If neither the query nor the views have built-in predicates, then finding a rewriting with at most k literals, where k is the number of literals in the body of Q, is NP-complete. Finally, if the query and the views have built-in predicates, then finding a rewriting with at most k literals is in Σ^P_3.

Techniques for Answering Queries Using Views

The above characterization suggests a two-step algorithm for finding rewritings of a query Q using a set of views V. At first, find some containment mapping from V to Q and add to Q the appropriate atoms of V, resulting in a new query Q'. Then, minimize Q' by removing literals from Q that are redundant. An algorithm needs to consider every possible conjunction of k or fewer view atoms, where k is the number of subgoals in the query. Algorithms that attempt to explore more effectively the search space to produce maximally contained rewritings include the Bucket Algorithm, the Inverse-Rules Algorithm and the MiniCon Algorithm.

The Bucket Algorithm

The main idea underlying the algorithm is that the query rewritings that need to be considered can be significantly reduced if for each subgoal in the query the relevant views are identified.

To demonstrate the algorithm, the following query and views are used (The example used is from [10]):

$$Q_1(x) \leftarrow cites\ (x,y) \wedge cites\ (y,x)$$
$$\wedge sameTopic\ (x,y)$$
$$V_1(a) \leftarrow cites\ (a,b) \wedge cites\ (b,a)$$
$$V_2(c,d) \leftarrow sameTopic\ (c,d)$$
$$V_3(f,h) \leftarrow cites\ (f,g) \wedge cites\ (g,h)$$
$$\wedge sameTopic\ (f,g)$$

Given a query Q, the Bucket Algorithm proceeds in two steps. In the first step, the algorithm creates a bucket for each subgoal in Q that is not in $C(Q)$, where by $C(Q)$ refers to the subgoals of comparison predicates of Q. Each bucket contains the views that are relevant to answering the particular subgoal. For the example, the following buckets are created:

cites(x, y)	cites(y, x)	sameTopic(x, y)
$V_1(x)$	$V_1(x)$	$V_2(x, y)$
$V_3(x, y)$	$V_3(x, y)$	$V_3(x, y)$

The algorithm also requires that every distinguished variable in the query should be mapped to a distinguished variable in the view. Hence, the algorithm does not include the entry $V_1(y)$ even though it is possible to map the subgoal $cites(x, y)$ in the query to the subgoal $cites(b, a)$ in V_1.

In the second step, for each element of the Cartesian product of the buckets, the algorithm constructs a conjunctive rewriting and checks whether it is contained (or can be made to be contained) in the query. If so, the rewriting is added to the answer. Therefore, the result of the Bucket Algorithm is a union of conjunctive rewritings.

In the example, the algorithm will try to combine V_1 with the other views and fail. Then it will consider the rewritings involving V_3 and V_2, and discover a contained rewriting. Finally, it will consider the rewritings involving only V_3, and find a contained rewriting by equating the variables in the head of V_3. It is possible to add an additional check that will determine whether a resulting rewriting will be redundant, as is the case here with the rewriting combining V_3 and V_2. Therefore, the only minimal rewriting in this example is: $Q_1'(x) \leftarrow V_3(x, x)$. This rewriting is in fact an equivalent one.

The Bucket Algorithm misses important interactions between view subgoals by considering each subgoal separately. Consequently, the buckets can contain unusable views and, as a result, the second step of the algorithm can become very expensive.

The Inverse Rules Algorithm

The main idea of the Inverse-Rules Algorithm is to construct a set of rules that *invert* the view definitions, i.e., rules that compute tuples for the database relations from tuples of the views. The inverse rules together with a conjunctive query

Q constitute a maximally contained rewriting for Q that is represented as a union of conjunctive queries. Considering the views of the previous example, the algorithm would construct the following inverse rules:

$$R_1 : cites\ \big(a, f_{V_1}(a)\big) \leftarrow V_1(a)$$
$$R_2 : cites\ \big(f_{V_1}(a), a\big) \leftarrow V_1(a)$$
$$R_3 : sameTopic\ (c, d) \leftarrow V_2(c, d)$$
$$R_4 : cites\ \big(f, f_{V_3}(f, h)\big) \leftarrow V_3(f, h)$$
$$R_5 : cites\ \big(f_{V_3}(f, h), h\big) \leftarrow V_3(f, h)$$
$$R_6 : sameTopic\ \big(f, f_{V_3}(f, h)\big) \leftarrow V_3(f, h)$$

For every view relation V with an existential variable Z_i, a function symbol $f_{V,i}$ is introduced. The above rules provide all the information about the database relations that can be extracted from the view extension. Intuitively, a tuple of the form (A) in the extension of the view V_1 is a *witness* of two tuples in the relation $cites$: a tuple of the form (A, Z), for some value of Z, and a tuple of the form (Z, A), for the same value of Z. The rules will generate from a tuple $V_1(A)$ two such tuples that will contain the unique functional term $f_{V_1}(A)$: it is a syntactic stand-in for the unknown (but known to exist) common value in the two tuples.

The rewriting of a query Q using the set of views V is simply the query consisting of Q and the inverse rules for V (There is a systematic way to eliminate the function symbols). Two important advantages of the algorithm are that the inverse rules can be constructed ahead of time, independent of a particular query, in polynomial time, and that the technique is also applicable virtually without change to recursive queries.

The rewritings produced by the Inverse-Rules Algorithm have some shortcomings when used for query evaluation. First, applying the inverse rules to the extension of the views may invert some of the computation done to produce the extent of the view. Second, they may access views that have no relevance to the query.

The MiniCon Algorithm

The MiniCon Algorithm starts by considering, for each subgoal in the query, which view specializations contain subgoals that can "cover" it.

A view specialization is created from a view by possibly equating some head variables. Once the algorithm finds a partial mapping ϕ from a subgoal g in the query to a subgoal g_1 in the view specialization V, it extends it to a minimal additional set of subgoals G of the query that must also be mapped to V. In particular, G includes all subgoals of the query that contain some variable of g mapped by φ to an existential variable of V. This set of subgoals and mapping information is call a *MiniCon Description* (MCD), and can be viewed as a generalized bucket. Having considered in advance how each of the variables in the query can interact with the available views, the second phase of the MiniCon Algorithm needs to consider significantly fewer combinations of these generalized buckets.

Using the query and views of the previous example, the MCDs formed during the first phase of the MiniCon Algorithm are:

$V(Y)$	h	ϕ	G
$V_2(c, d)$	$c \to c, d \to d$	$x \to c, y \to d$	3
$V_3(f, f)$	$f \to f, h \to f$	$x \to f, y \to f$	1, 2, 3

where h is a head homomorphism on V_i, V ($\overline{Y}$) is the result of applying h to V_i to create a view specialization, ϕ is a partial mapping from $Vars(Q_1)$ to $h(Vars(V_i))$ and G is a subset of the subgoals in Q_1 that are covered by some subgoal in $h(V_i)$ under the mapping ϕ.

In the second phase, the MCDs are combined to produce the query rewritings. In this phase the algorithm considers combinations of MCDs, and for each valid combination it creates a conjunctive rewriting of the query. There is no need for containment testing, as opposed to the Bucket algorithm. The combinations of MCDs considered by the MiniCon Algorithm are those that cover all the subgoals of the query with pairwise disjoint subsets of subgoals. The final maximally contained rewriting is a union of conjunctive queries.

As of 2008, MiniCon has been shown to be the most efficient current algorithm for answering conjunctive queries using conjunctive views.

Chase and Backchase

In the presence of a set C of constraints on the relations and the views, Chase and Backchase is a powerful technique for finding *equivalent* rewritings using the views.

The chase (see entry "▶ Chase") is a well-known technique that can be used to decide containment of queries under constraints. If the chase of Q_1 with C terminates producing a query Q_c then $Q_1 \sqsubseteq_C Q_2$ if and only if $Q_C \sqsubseteq Q_2$. $Q_1 \sqsubseteq_C Q_2$ means that $Q_1 \sqsubseteq Q_2$ on every database instance that satisfies the constraint set C.

For the equivalent rewriting problem under constraints C, Q_1 is given and must effectively find Q_2 such that $Q_1 \equiv_C Q_2$. The algorithm proceeds in two steps. In the *chase* step, it chases Q_1 with C until no more chase steps are possible. This results in a query U called the *universal plan*. The universal plan is a query over the database relations and the views that conceptually incorporate all possible alternative ways to answer Q_1 in the presence of the constraints C. In particular, any minimal conjunctive query equivalent to Q_1 under C is isomorphic to a subquery of U. Hence, to find all minimal rewritings of Q_1, the backchase step of the algorithm searches the finite space of subqueries of U. Specifically, for each subquery Q_s it checks for equivalence with Q_1 by chasing Q_s with C. Q_s is equivalent to Q_1 if and only if there is a containment mapping from Q_1 into an intermediate chase result of Q_s.

Chase and Backchase applies if the constraint set C includes tuple-generating and equality-generating dependencies. It is sound and complete if C is *weakly acyclic*.

Key Applications

The conceptual framework and techniques developed for query rewriting using views have had significant impact in a number of areas of information systems, especially (but not exclusively) in the areas of information integration and data integration (see entry "▶ Information Integration" and "Data Integration"). In particular, as of 2007, data integration theory and systems follow two main architectural approaches for defining and processing queries in a virtual (In a virtual integration system, as opposed to a data warehouse (see entry "▸ Data Warehouse"), data resides only in their original locations/sources. The integration system accesses them every time it needs to answer a query) integration system. In *global-as-view* the integrated *global view* (or views) of the disparate data sources is defined (using a standard query language or an ad hoc specification system) in terms of the *local* data sources, and queries are directly expressed in terms only of the global view(s). In this case, processing a query over this global view involves *view expansion*. In *local-as-view* data integration on the other hand, local data sources are expressed in terms of an a priori given *global schema*, i.e., they are expressed as views over the relations of the global schema. Queries are also expressed in terms of the global schema. Since in an integration system the only available data reside in the local sources, answering a query requires rewriting it in terms of the available sources, i.e., rewriting it using the views.

In the above context, sources describe their contents in terms of the local schema. In addition, for sources with limited processing power, their capabilities need to also be taken into account. Since integrated queries are answered by retrieving data from local sources, the data retrieval requests (i.e., the queries) submitted to the sources need to conform to their capabilities. Query rewriting using views has provided the conceptual framework and tools to address the *capability-based rewriting* problem.

Another area of application of the ideas of answering queries using views is query optimization. When the set of relations mentioned in queries overlaps with those mentioned in one or more views, and the views select one or more attributes selected by the query, the view may be *usable* by the query. Moreover, certain indices can also be modeled as materialized views. Using the materialized views may lower the cost of query processing, depending on the available access methods and the selectivity of the involved predicates. Answering queries using views is used to decide view usability and generate the appropriate query plans.

Finally, answering queries using views, especially in the presence of constraints, has influenced the theory of *data exchange*. Data ex-

change, or data translation, involves taking data described in one, *source*, schema and translating, without loss of information, into data structured under a different, *target*, schema.

Future Directions

Even though significant progress has been made in this area, the applicability and/or efficiency of the developed techniques is often limited. Future work is needed on rewriting techniques that can perform correctly and efficiently for "real" SQL views on real SQL queries, including queries with aggregates, nesting, bag semantics, and user-defined functions. Moreover, the conceptual framework of rewriting using views applies to problems in graph querying, Web service composition, and rewriting of XQuery.

Cross-References

▸ Chase
▸ Data Warehouse
▸ Information Integration
▸ Query Containment
▸ Query Rewriting
▸ View Definition
▸ Views

Recommended Reading

1. Abiteboul S, Duschka O. Complexity of answering queries using materialized views. In: Proceedings of the 17th ACM SIGACT-SIGMOD-SIGART Symposium on Principles of Database Systems; 1998. p. 254–63.
2. Cohen S, Nutt W, Sagiv Y. Containment of aggregate queries. In: Proceedings of the 9th International Conference on Database Theory; 2003. p. 111–25.
3. Deutsch A, Popa L, Tannen V. Query reformulation with constraints. ACM SIGMOD Rec. 2006;35(4)
4. Duschka O, Geneserth M. Answering recursive queries using views. In: Proceedings of the 16th ACM SIGACT-SIGMOD-SIGART Symposium on Principles of Database Systems; 1997. p. 109–16.
5. Halevy A. Answering queries using views: a survey. VLDB J. 2001;10(4):270–94.
6. Lenzerini M. Data integration: a theoretical perspective. In: Proceedings of the 21st ACM SIGACT-SIGMOD-SIGART Symposium on Principles of Database Systems; 2002. p. 233–46.
7. Levy AY, Mendelzon AO, Sagiv Y, Srivastava D. Answering queries using views. In: Proceedings of the

14th ACM SIGACT-SIGMOD-SIGART Symposium on Principles of Database Systems; 1995. p. 95–104.

8. Lin V, Vassalos V, Malakasiotis P. MiniCount: efficient rewriting of COUNT-Queries using views. In: Proceedings of the 22nd International Conference on Data Engineering; 2006. p. 1.

9. Papakonstantinou Y, Vassalos V. Query Rewriting using semistructured views. In: Proceedings of the ACM SIGMOD International Conference on Management of Data; 1999. p. 455–66.

10. Pottinger R, Halevy A. Minicon: A scalable algorithm for answering queries using views. VLDB J. 2001;10(2–3):182–98.

11. Xu A, Meral OZ. Rewriting XPath queries using materialized views. In: Proceedings of the 31st International Conference on Very Large Data Bases; 2005. p. 121–32.

Anti-monotone Constraints

Carson Kai-Sang Leung
Department of Computer Science, University of Manitoba, Winnipeg, MB, Canada

Synonyms

Anti-monotonic constraints

Definition

A constraint C is *anti-monotone* if and only if for all itemsets S and S':

if $S \supseteq S'$ and S satisfies C, then S' satisfies C.

Key Points

Anti-monotone constraints [1, 2] possess the following nice property. If an itemset S satisfies an anti-monotone constraint C, then all of its subsets also satisfy C (i.e., C is downward closed). Equivalently, any superset of an itemset violating an anti-monotone constraint C also violates C. By exploiting this property, anti-monotone constraints can be used for pruning in frequent itemset mining with constraints. As frequent itemset mining with constraints aims to find itemsets that are frequent and satisfy the constraints, if an itemset violates an anti-monotone constraint C, all its supersets (which

would also violate C) can be pruned away and their frequencies do not need to be counted. Examples of anti-monotone constraints include $min(S.Price) \geq \$20$ (which expresses that the minimum price of all items in an itemset S is at least \$20) and the usual frequency constraint $support(S) \geq minsup$ (i.e., $frequency(S) \geq minsup$). For the former, if the minimum price of all items in S is less than \$20, adding more items to S would not increase its minimum price (i.e., supersets of S would not satisfy such an anti-monotone constraint). For the latter, it is widely used in frequent itemset mining, with or without constraints. It states that (i) all subsets of a frequent itemset are frequent and (ii) any superset of an infrequent itemset is also infrequent. This is also known as the *Apriori property*.

Cross-References

▶ Frequent Itemset Mining with Constraints

Recommended Reading

1. Lakshmanan LVS, Leung CK-S, Ng RT. Efficient dynamic mining of constrained frequent sets. ACM Trans Database Syst. 2003;28(4):337–89. https://doi.org/10.1145/958942.958944.

2. Ng RT, Lakshmanan LVS, Han J, Pang A. Exploratory mining and pruning optimizations of constrained associations rules. In: Proceedings of the ACM SIGMOD International Conference on Management of Data; 1998. p. 13–24.

Applicability Period

Christian S. Jensen[1] and Richard T. Snodgrass[2,3]
[1]Department of Computer Science, Aalborg University, Aalborg, Denmark
[2]Department of Computer Science, University of Arizona, Tucson, AZ, USA
[3]Dataware Ventures, Tucson, AZ, USA

Definition

The *applicability period* (or *period of applicability*) for a modification (generally an insertion, deletion, or update) is the time period that modification is to be applied. Generally the modification is a sequenced modification and the period applies to valid time. This period should be distinguished from *lifespan*.

Key Points

The applicability period is specified within a modification statement. In constrast, the lifespan is an aspect of a stored fact.

This illustration uses the TSQL2 language, which has an explicit *VALID* clause to specify the applicability period within an *INSERT*, *DELETE*, or *UPDATE* statement.

For insertions, the applicability period is the valid time of the fact being inserted. The following states that Ben is in the book department for 1 month in 2007.

* INSERT INTO EMPLOYEE
* VALUES ('Ben', 'Book')
* VALID PERIOD '[15 Feb 2007, 15 Mar 2007]'

For a deletion, the applicability period states for what period of time the deletion is to be applied. The following modification states that Ben in fact was not in the book department during March.

* DELETE FROM EMPLOYEE
* WHERE Name = 'Ben'
* VALID PERIOD '[1 Mar 2007, 31 Mar 2007]'

After this modification, the lifespan would be February 15 through February 28.

Similarly, the applicability period for an *UPDATE* statement would affect the stored state just for the applicability period.

A *current modification* has a default applicability period that either extends from the time the statement is executed to forever, or when now-relative time is supported from the time of execution to the ever-increasing current time.

Cross-References

▶ Current Semantics
▶ Lifespan
▶ Now in Temporal Databases
▶ Sequenced Semantics
▶ Temporal Database
▶ Time Period
▶ TSQL2
▶ Valid Time

Recommended Reading

1. Snodgrass RT, editor. The TSQL2 temporal query language. Kluwer; 1995.
2. Snodgrass RT. Developing time-oriented database applications in SQL. San Francisco: Morgan Kaufmann; 1999.

Application Benchmark

Denilson Barbosa[1], Ioana Manolescu[2], and Jeffrey Xu Yu[3]
[1]University of Alberta, Edmonton, AL, Canada
[2]INRIA Saclay–Île de France, Orsay, France
[3]Department of Systems Engineering and Engineering Management, The Chinese University of Hong Kong, Hong Kong, China

Synonyms

Benchmark; Performance benchmark

Definition

An application benchmark is a suite of tasks that are representative of typical workloads in an application domain.

Key Points

Unlike a MICROBENCHMARK, an application benchmark specifies broader tasks that are aimed at exercising most components of a system or tool. Each individual task in the benchmark is assigned a relative *weight*, usually reflecting its frequency or importance in the application being modeled. A meaningful interpretation of the benchmark results has to take these weights into account.

The Transaction Processing Performance Council (TPC) is a body with a long history of defining and published benchmarks for database systems. For instance, it has defined benchmarks for Online Transaction Processing applications (TPC-C and TPC-E), Decision Support applications (TPC-H), and for an Application Server setting (TPC-App).

Other examples of application benchmarks are: the OO1 and OO7 benchmarks, developed for object-oriented databases, and XMark, XBench and TPoX, developed for XML applications.

Cross-References

▶ Microbenchmark
▶ XML Benchmarks

Recommended Reading

1. Carey MJ, DeWitt DJ, Naughton JF. The OO7 benchmark. In: Proceedings of the ACM SIGMOD International Conference on Management of Data; 1993. p. 12–21.
2. Gray J, editor. The benchmark handbook for database and transaction systems. 2nd ed. San Mateo: Morgan Kaufmann; 1993.
3. Nicola M, Kogan I, Schiefer B. An XML transaction processing benchmark. In: Proceedings of the ACM SIGMOD International Conference on Management of Data; 2007. p. 937–48.
4. Schmidt A, Waas F, Kersten ML, Carey MJ, Manolescu I, Busse R. XMark: a benchmark for XML data management. In: Proceedings of the 28th International Conference on Very Large Data Bases; 2002. p. 974–85.
5. Transaction Processing Performance Council. Available at: http://www.tpc.org/default.asp
6. Yao BB, Özsu MT, Khandelwal N. XBench benchmark and performance testing of XML DBMSs. In: Proceedings of the 20th International Conference on Data Engineering; 2004. p. 621–33.

Application Recovery

David Lomet
Microsoft Research, Redmond, WA, USA

Synonyms

Exactly-once execution; Fault-tolerant applications; Persistent applications; Recovery guarantees; Transaction processing

Definition

Systems implement application recovery to enable applications to survive system crashes and provide "exactly-once execution" in which the result of executing the application is equivalent to a single execution where no system crashes or failures occur.

Historical Background

Application recovery was first commercially provided by IBM's CICS (Customer Information Control System). Generically, these kinds of systems became known as transaction processing monitors (TP monitors) [5, 9]. With a TP monitor, applications are decomposed into a series of steps. Each step is executed within a transaction. A step typically consists of reading input state from a database or transactional queue, executing some business logic, perhaps processing user input or reading and writing to a database, and, finally, writing state for the next step into database or queue [4]. If a step failure occurs, its transaction is aborted. Since the prior step results are stably stored, the step can be re-executed after system recovery.

Application recovery does not always involve transactions, however, as the early work by Borg et al. demonstrates [6]. More recently, in the web context, TP monitors have been renamed as application servers (app servers). App servers are similar to TP monitors, where state is explicitly managed frequently by using transactions.

Ongoing interest in application recovery is illustrated by the "recovery-oriented computing" (ROC) project [3] and the Phoenix project [1].

Foundations

Introduction

Exactly-Once Applications: While application programmers are usually familiar with the problem area for which an application is written, they are frequently also faced with having to deal with "system problems" of reliability and availability. The system goal is to permit applications to achieve "exactly-once execution" [2]. For example, an airline reservation system wants to issue exactly the number of tickets a customer requests, instead of no tickets or twice the number.

Types of Failures: Applications cannot survive all forms of failures. If an exactly-once execution produces a "deterministic" failure, then every execution of the application will lead to that same failure. Such deterministic failures are called "hard failures." The failures that software can deal with are called "soft failures." A subsequent execution of the system will usually avoid the state leading to the system failure. Soft failures arise in a number of ways.

1. *Software non-determinism*: A software system is non-deterministic if, when re-executed, it results in a different execution path than a prior execution. Non-determinism can arise when, for example, paths are determined by relative processor speed or the sequence of external events. Such software bugs have been called "Heisenbugs" (hard failures being "Bohrbugs").
2. *Soft hardware failures*: Hardware can also suffer from "Heisenbugs." For example, a transient hardware failure may be triggered by an environmental cause, such as a cosmic ray changing a memory bit, etc.
3. *Operator failures*: Systems occasionally require operator intervention. Operators, being human, make mistakes. An operator is un-

likely to make the same mistake at the same point in a subsequent execution.

Recovery is effective because failures are usually "soft" [3, 9], which is why database recovery is so successful.

Support for Application Persistence: Applications which guarantee "exactly-once" execution are called "persistent applications" because application state persists across or despite system failures. The traditional method for providing application persistence has been to use a TP monitor. However, new approaches permit implicit state management, in which the application programmer delegates the problem of managing application state to the system infrastructure.

Implicit state management does not require application steps to execute within transactions. It permits applications to be coded more "naturally" (though restrictions exist). Application state is made stable "under the covers." Phoenix [1] does this via logging application non-determinism and replaying the captured non-determinism after a failure to recover application state. CORBA's approach [11], having shorter downtime but more costly normal execution, replicates application state so that if part of the system fails, a copy of the state is available elsewhere for continued execution. Implicit state management is discussed here.

Application persistence (recovery) requires different techniques than database recovery: (i) Applications are usually single threaded and "piecewise deterministic," while databases execute highly concurrent code. (ii) Application state is frequently distributed, so dealing with distribution is essential. (iii) Databases change state entirely within a transaction, while application state may frequently change outside of a transaction.

Persisting Application State

Implicit state management requires the infrastructure supporting an application to capture application state and make it stable in some way, relieving the programmer of this task. Two approaches have been developed. The approaches are not as different as they appear. Both require that applications be piecewise deterministic, with clearly identified non-deterministic events that can be "applied" at an application instance to be used in deterministic re-execution that can generate the same state as the original execution.

Recovery: Recovery technology usually assumes that a failing application will be reactivated after a system crash and its activities recovered on the original system. Thus recovery-oriented approaches usually capture application state stably (e.g., "on disk") so that it survives a system failure. There are two parts to this:

1. Writing a "snapshot" of the state to stable storage from time to time. It matters when this snapshot, also called a checkpoint, occurs.
2. Stably logging non-deterministic events encountered by the application in arrival order.

Recovery, in which the state is re-created following a system failure, consists of re-installing the latest captured state (checkpoint) followed by re-executing the application by "feeding" events from the log to re-create application state as of the last logged event.

Logging can be either optimistic or pessimistic [7]. Pessimistic logging eagerly makes log records stable, usually as events occur in order to ensure that execution is "exactly once." Optimistic logging defers for a time making events stable, frequently sacrificing exactly-once execution. Pessimistic logging can be greatly optimized. Optimistic logging can be constrained so that exactly-once execution is assured. This brings these techniques toward some middle ground in performance, though important differences remain.

Replication: Replication technology usually assumes that an application executing on a failing system A will continue execution on a separate system B where a replica of its state has been maintained. The problem for replica-oriented approaches is to capture and keep in sync an application's state on these separate systems. There are two generic approaches to this:

1. Designate one system as the primary system and the others as secondaries. Non-deterministic events are sent to the primary system, which then relays them to the secondary in the order that it received them. Replicas thus execute in response to the same sequence of events. A secondary becomes the primary on failure of the primary. If a secondary fails, the primary needs to know this and re-create another secondary.
2. Use atomic broadcast to send non-deterministic events to all systems maintaining replicas "simultaneously" [1]. Atomic broadcast guarantees that replicas receive all messages and in the same order. If there is a failure, then the remaining replicas participating in the atomic broadcast need to know the members of the new group so that the atomic broadcast protocol continues to work correctly, possibly including the creation of a new replica.

Intermediate Approaches: The line between replication and recovery approaches is not always crisp: (i) A recovery system can maintain its log by sending the log records to another system, hence using a second system as stable storage instead of a disk. This second site might then become the site where the application is subsequently executed should the original site fail. (ii) A second system for a replica might not execute the events forwarded to it immediately, but rather simply "log" them. Then, only if a primary replica fails would it perhaps then "catch up" by executing the stored events.

Thus a second system might serve as a cold standby system (retaining a log of events) or as a hot standby, immediately executing the application in response to the events. Warm standbys are also possible in which an application is executed at a secondary site in a lazy fashion, lagging the state of a primary site, but not by too much. Whether this should be viewed as recovery or replication is not really important.

Distributed Applications

Many applications are distributed; e.g., a web application might consist of a client component providing the user interface, one or more middle-tier components executing business logic, and back-end components that typically provide transactions and database functionality. The new problem is to coordinate the states of multiple software components executing different parts of the application [6, 7].

1. The state of the set of application components needs to be "causal," i.e., every message receive reflected in the state of some component must always be accompanied by a sender in a state where the message has been sent.
2. Messages (non-deterministic events in a distributed system) must result in an "exactly-once" execution by the receiving component.
3. The application must be able to interact with users or other elements outside of the persistence infrastructure, which may not obey the required protocols.
4. The application must be able to interact with transactional elements like databases.
5. Different strategies have different costs and impact the balance between normal runtime costs and recovery costs and time to recovery.

Contracts for Persistent Components: Providing persistence for components of a distributed application requires an agreed-upon set of protocols or "contracts" [1, 2] involving component state and message stability, repeated sending of messages, eliminating duplicate messages, etc. The basic contract between persistent components, called the "committed interaction contract" or CIC, places burdens on both sender and receiver of a message at the time it is sent.

The *sender of a message* ensures causality. The sender ensures that its state as of the time of message send and the message will be persistent earlier than the receiver's state is persisted. Further, the sender must continue to send the message until the receiver acknowledges the message, in order to deal with unreliable networks and crashed receivers, etc. The CIC does not specify how to do this, but the recovery approach usually writes non-deterministic events to a log and flushes the log when a message is sent should the state include previously received messages since the last log flush.

The *receiver of the message* ensures exactly-once execution. The receiver eliminates duplicate messages the sender may send in its effort to provide reliable delivery. The receiver executes in response to a message only if receiver state does not already reflect having received the message and bypasses execution otherwise, returning the same result as produced by the original execution. Finally, the receiver ensures that the state resulting from the message receipt is stable at an appropriate moment. The CIC contract does not specify how to do this, but the receiver might use a table of messages received, or some high water mark for messages when the sender is known, to which it compares each incoming message.

External and Transactional Components: Many web applications involve users entering information at a keyboard and a database that stores the results of business dealings, e.g., the purchase of a plane ticket. Hence, persistent components must interact with elements outside of the boundaries of the application and its supporting system. This requires new forms of "contracts" in which the main burden is placed on the persistent components to enable an ensemble of elements to achieve exactly-once execution, e.g., one purchase of the plane ticket.

External components, including users, may not obey CIC requirements. Hence, external interactions must be limited to ensure exactly-once execution. For users, both reads and writes might be exposed as having multiple occurrences. A failure in the middle of a user interaction may require entering data more than once or repeating an output if the system fails between the event and its stable logging. Systems typically minimize the problem window by immediate logging, with a log flush, in response to external events.

A transactional component only occurs at the edge of an application system, responding to requests for and updates to its data. Transactions can abort, with the transactional component's state reset to remove all effects of the aborted transaction. The persistent component must be prepared to handle transaction aborts at any time. Transactions actually reduce the burden for a persistent component as it need not ensure that its state is stable at each interaction within a transaction. A system crash will eventually lead to the abort of any interrupted transaction. However, at the point where the persistent component requests a transaction commit, it must obey the usual requirements for a sender in a CIC.

Optimizations: Optimizations can reduce the normal runtime cost of providing persistent applications. These exploit additional information known to the application programmer when the application is either written or being deployed.

1. Some components may be read-only, producing no external side effects. So a read-only component need do no logging itself, while a calling persistent component can log lazily because the read can be repeated if needed. A functional call component in which the result depends only on the arguments of the call requires no logging as the call can be replayed idempotently.
2. If called components extend the time they stably remember prior calls and results for idempotence, then the calling component need not log to make its state stable prior to each call. It can depend upon the called components in a sequence of calls to capture the call results to enable its deterministic replay.
3. When a component is a "server" for exactly one client, the client can capture what would be non-deterministic calls and their order for the server component, relieving the server from needing to log calls messages. Combining this with item 2 enables persistent components without logging [8, 10]. A logless component's state is regenerated by its client replaying the sequence of calls and it re-executing its own series of calls to other components that have captured the call results. Logless components are easily deployed anywhere, can be freely replicated, and hence make persistence simple, flexible, and low cost.

Checkpoints: Recovery time is shortened via checkpoints. Component state consists of its variables and its execution time stack. By checkpointing component state when there is no execution within the component, the checkpoint can be accomplished solely by capturing its variable values; e.g., checkpoint might be taken transparently during the midst of a return from a call to a component. The cost of the checkpoint and the need for fast recovery time dictate checkpoint frequency.

Discussion

Making declarative the requirements for application persistence, as represented by interaction contracts, makes it easier to provide and optimize exactly-once execution by expanding the range of implementation options, including interesting optimizations.

The level of the contract is also important. A CIC could have been derived from reliable messaging instead of more explicitly as repeated sending of messages with duplicate elimination. However, reliable messaging does not describe what happens to messages once they are delivered. By describing the requirements for state stability, etc., it becomes clear that delivery is not, by itself, sufficient. Further, where "persistent reliable messaging" is used, this would require extra log forces that can frequently be avoided with the "end-to-end" application persistence protocol.

System-provided transparent application persistence is an example of delegating a serious problem to the "system," similar to how transactions delegate concurrency control and recovery to database systems. There is no need for special application programmer consideration, simplifying the application logic and improving programmer productivity.

Key Applications

Application recovery (persistence) is used wherever exactly-once semantics is required. Traditionally, this has been within transaction processing systems, but now more commonly, this involves web applications for e-business, whether for end-user customers or business-to-business dealings. If there are financial or legal requirements for applications, they will be built using some form of application recovery to ensure exactly-once execution.

Cross-References

▶ Transaction

Recommended Reading

1. Barga R, Chen S, Lomet D. Improving logging and recovery performance in phoenix/App. In: Proceedings of the 20th International Conference on Data Engineering; 2004.
2. Barga R, Lomet D, Shegalov G, Weikum G. Recovery guarantees for internet applications. ACM Trans Internet Technol. 2004;4(3): 289–328.
3. Berkeley/Stanford Recovery-Oriented Computing (ROC) Project. http://roc.cs.berkeley.edu. 10 Oct 2008.
4. Bernstein P, Hsu M, Mann B. Implementing recoverable requests using queues. In: Proceedings of the ACM SIGMOD International Conference on Management of Data; 1990. p. 112–22.
5. Bernstein P, Newcomer E. Principles of transaction processing. Morgan Kaufmann; 1997.
6. Borg A, Baumbach J, Glazer S. A message system supporting fault tolerance. In: Proceedings of the 9th ACM Symposium on Operating System Principles; 1983. p. 90–9.
7. Elnozahy EN, Alvisi L, Wang Y, Johnson DB. A survey of rollback-recovery protocols in message-passing systems. ACM Comput Surv. 2002;34(3):375–408.
8. Frølund S, Guerraoui R. 2000. A pragmatic implementation of e-Transactions. In: Proceedings of the 19th Symposium on Reliable Distributed Systems; 2000. p. 186–95.
9. Gray J, Reuter A. Transaction processing: concepts and techniques. San Mateo: Morgan Kaufmann; 1993.
10. Lomet D. Persistent middle tier components without logging. In: Proceedings of the International Conference on Database Engineering and Applications; 2005. p. 37–46.
11. Narasimhan P, Moser L, Melliar-Smith PM. Lessons learned in building a fault-tolerant CORBA System. In: Proceedings of the International Conference on Dependable Systems and Networks; 2002. p. 39–44.

Application Server

Heiko Schuldt
Department of Mathematics and Computer Science, Databases and Information Systems Research Group, University of Basel, Basel, Switzerland

Synonyms

Java application server; Web application server

Definition

An *Application Server* is a dedicated software component in a three-tier or multi-tier architecture which provides application logic (business logic) and which allows for the separation of application logic from user interface functionality (client layer), delivery of data (Web server), and data management (database server).

Key Points

Modern information systems, especially information systems on the Web, follow an architectural paradigm that is based on a separation of concerns. In contrast to monolithic (single-tier) architectures or two-tier client/server architectures where business logic is bundled with other functionality, three-tier or multi-tier architectures consider dedicated application servers which exclusively focus on providing business logic.

In three-tier or multi-tier architectures, application servers typically make use of several middleware services which enable the communication within and between layers and which coordinate the access to shared data or services (transactional middleware, database middleware). Application servers usually provide the basis for the execution of distributed applications with transactional guarantees on top of persistent data. In large-scale deployments, systems might encompass several instances of application servers (application server clusters). This allows for the distribution of client requests across application server instances for the purpose of load balancing (replication in multi-tier architectures).

Early application servers evolved from distributed TP Monitors. Over time, a large variety of application servers has emerged. The most prominent class consists of Java application servers (Java EE) that follow a service-oriented architecture where functionality is provided by means of services with well-defined interfaces (e.g., Web services).

Cross-References

- ▶ Database Middleware
- ▶ Enterprise Application Integration
- ▶ Java Enterprise Edition
- ▶ Message Queuing Systems
- ▶ Middleware Support for Database Replication and Caching
- ▶ Multitier Architecture
- ▶ Replication in Multitier Architectures
- ▶ Transactional Middleware
- ▶ Web Services

Recommended Reading

1. Jacobs D. Data management in application servers. Datenbank-Spektrum. 2004;8:5–11.
2. Gupta A. Java EE 7 Essentials. Sebastopol, CA, USA: O'Reilly; 2013.
3. Raghavachari M, Reimer D, Johnson RD. The Deployer's problem: configuring application servers for performance and reliability. In: Proceedings of the 25th International Conference on Software Engineering; 2003. p. 484–9.

Application-Level Tuning

Philippe Bonnet[1] and Dennis Shasha[2]
[1]Department of Computer Science, IT University of Copenhagen, Copenhagen, Denmark
[2]Department of Computer Science, New York University, New York, NY, USA

Synonyms

Query tuning; Tuning the application interface

Definition

An under-appreciated tuning principle asserts *start-up costs are high; running costs are low*. When applied to the application level, this principle suggests performance of a few bulk operations that manipulate and transport a lot of data rather than many small operations that act on small amounts of data. To make this concrete, this entry discusses several examples and draws lessons from each.

Historical Background

Application-level tuning is about changing the way a task is performed. This entails finding a better algorithm or finding a better way to handle the database. The first is difficult to automate, but the latter goes back to the very first use of the relational databases. Whether on disk or in main memory, databases have generally always performed best when a single statement accesses all and exactly the data needed for a task.

Foundations

Application-level tuning has the nice property that it often is a pure win. Whereas adding or removing indexes often entails a trade-off between insert/delete/update performance and query performance, application-level tuning for the most part improves the performance of certain queries without hurting the performance of others. Rewriting queries to use resources more efficiently also often gives a greater benefit than physical changes.

Assemble Object Collections in Bulk

Object-oriented encapsulation is the principle of shielding the user of an object from the object's implementation. Encapsulation sometimes is interpreted as *the specification is all that counts*. That interpretation can, unfortunately, lead to horrible performance.

The reason is simple. The first design that seems to occur to object-oriented implementers is to make relational records (or sometimes fields) into objects. This has the virtue of generality. Fetching one of these objects then translates to a fetch of a record or a field. So far, so good. But then the temptation is to build bulk fetches from fetches on little objects (the so-called "encapsulation imperative"). The net result is a proliferation of small queries instead of one large query.

Consider for example a system that delivers and stores trade information. Each document type (e.g., a report on a customer account) is produced according to a certain schedule that may differ from one trade type to another. "Focus" information relates trade types to risk analysts. That is, risk analysts may focus on one trade type or another. This gives a pair of tables of the form:

- Focus(analyst, tradetype)
- Tradeinstance(id, tradetype, tradedetail)

When an analyst logs in, the system gives information about the trade instances in which he or she would be interested. This can easily be done with the join:

- select tradeinstance.id, tradeinstance. tradedetail
- from tradeinstance, focus
- where tradeinstance.tradetype = focus. tradetype
- and focus.analyst = {input analyst name}

But if each trade type is an object and each trade instance is another object, then one may be tempted to write the following code:

- Focus focustypes = new Focus();
- Focus.init({*input analyst name*});
- for (Enumeration e = focustypes.elements(); e.hasMoreElements();)
- {
- TradeInstance tradeinst = new TradeInstance();
- tradeinst.init(e.nextElement());
- tradeinst.print();
- }

This application program will first issue one query to find all the trade types for the analyst (within the init method of Focus class):

- select tradeinstance.tradetype
- from focus
- where focus.analyst = {input analyst name}

and then for each such type *t* to issue the query (within the init method of TradeInstance class):

- select tradeinstance.id, tradeinstance.tradedetail
- from tradeinstance
- where tradeinstance.tradetype = t

This is much slower than the previous SQL formulation. The join is performed in the application and not in the database server.

The point is not that object-orientation is bad-encapsulation contributes to maintainability. The point is that programmers should keep their minds open to the possibility that accessing a bulk object (e.g., a collection of documents) should be done directly rather than by forming the member objects individually (incurring a start-up cost each time) and then grouping them into a bulk object on the application side.

The Art of Insertion
This entry has discussed retrieving data so far. Inserting data rapidly requires understanding the sources of overhead of putting a record into the database:

- As in the retrieval case, the first source of overhead is an excessive number of round trips across the database interface. This occurs if the batch size of inserts is too small. A more radical approach is to assemble all the data to be inserted into a file, load that file into a temporary table, and then insert from that temporary table into the target table. This can improve performance by a factor of 100 or more when tables are large.
- The second issue has to do with the ancillary overhead that an insert causes: updating all the indexes on the table. Even a single index can hurt performance. For this reason, it is often a good idea to add indexes after loading the data.

Application-Level Tuning, Fig. 1

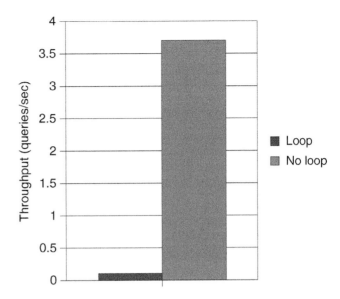

Key Applications

Application-level tuning has proved useful in every setting where performance is an issue. Whereas index and physical layer tuning can be more easily automated, the benefits of application-level tuning are often greater.

Experimental Results

Looping Hurts

This experiment illustrates the overhead of crossing the application interface. Two hundred tuples are fetched (these are small tuples composed of two integers) using a range query (no loop), or looping on the client-side to fetch one tuple at a time.

Figure 1 traces the throughput observed for this experiment on MySQL 6.0 with a warm buffer. There is an almost two orders of magnitude penalty when looping through the tuples. This is due to the overhead of crossing the application interface 200 times.

Url to Code and Data Sets

Loop experiment: http://www.databasetuning.org/?sec=loop

Cross-References

▶ Application server
▶ DBMS Interface
▶ Java Database Connectivity
▶ Open Database Connectivity
▶ SQL

Recommended Reading

1. Celko J. Joe celko's SQL for smarties: advanced SQL programming. 3rd ed. San Francisco: Morgan Kaufmann; 2005.
2. Shasha D, Bonnet P. Database tuning: principles, experiments and troubleshooting techniques. San Francisco: Morgan Kaufmann; 2002.
3. Tow D. SQL tuning. Beijing: O'Reilly; 2003.

Applications of Emerging Patterns for Microarray Gene Expression Data Analysis

Guozhu Dong[1] and Jinyan Li[2]
[1]Wright State University, Dayton, OH, USA
[2]Nanyang Technological University, Singapore, Singapore

Definition

This topic is related to applications of emerging patterns in the bioinformatics field, in particular for in-silico cancer diagnosis by mining emerging patterns from large scale microarray gene expression data.

Key Points

The contemporary gene expression profiling technologies such as cDNA microarray chips and Affymetrics DNA microarry chips can measure the expression levels of thousands even tens of thousands of genes simultaneously. This provides a great opportunity to identify specific genes or gene groups that are responsible for a particular disease, for example, the subtypes of childhood leukemia disease. Reference [3] proposed to use emerging patterns to capture the signature patterns between the gene expression profiles of colon tumor cells and normal cells. This was the first bioinformatics work studying how gene groups and their expression intervals signify the difference between diseased and normal cells. That paper also proposed to design treatment plans to cure the diseased cells by adjusting certain genes' expression level based on the discovered emerging patterns. Reference [2] reported simple rules underlying gene expression profiles of more than six subtypes of acute lymphoblastic leukemia (ALL) patients. The rules are converted from the mined emerging patterns. The rules are also used to construct a classifier (PCL) that reached a benchmark diagnosis accuracy of 96% on an independent test data set. As gene expression data sets contain many attributes, the discovery of emerging patterns by border-differential based algorithms is sometimes slow. To tackle this problem, reference [1] proposed a CART-based approach to discover a proportion of emerging patterns from the high-dimensional gene expression profiling data with a high speed. The ZBDD based approach discussed in the emerging patterns entry of this volume is also fast and can handle large number of genes.

Cross-References

▶ Emerging Patterns
▶ Emerging Pattern Based Classification

Recommended Reading

1. Boulesteix A-L, Tutz G, Strimmer K. A CART-based approach to discover emerging patterns in microarray data. Bioinformatics. 2003;19(18):2465–72.
2. Li J, Liu H, Downing JR, Eng-Juh YA, Wong L. Simple rules underlying gene expression profiles of more than six subtypes of acute lymphoblastic leukemia (ALL) patients. Bioinformatics. 2003;19(1):71–8.
3. Li J, Wong L. Identifying good diagnostic genes or genes groups from gene expression data by using the concept of emerging patterns. Bioinformatics. 2002;18(5):725–34.

Applications of Sensor Network Data Management

Farnoush Banaei-Kashani[1] and Cyrus Shahabi[2]
[1]Computer Science and Engineering, University of Colorado Denver, Denver, CO, USA
[2]University of Southern California, Los Angeles, CA, USA

Synonyms

Applications of sensor databases; Applications of sensor network databases; Applications of sensor networks

Definition

Sensor networks allow for micro-monitoring of different phenomena of interest in arbitrary physical environments. With this unique capability, sensor networks can capture the events in the real world as they happen in the form of high-resolution spatiotemporal data of various modalities and provide the opportunity for real-time querying and analysis of the data for immediate response and control. Such functionality is desirable in many classic applications while enabling numerous other novel applications. From the data management perspective, there is a consensus among database researchers that management and analysis of the massive, dynamic, distributed, and uncertain data in sensor network applications are going to be one of the new grand challenges for the database community:

> Sensor information processing will raise many of the most interesting database issues in a new environment, with a new set of constraints and opportunities.

– Excerpt from the Lowell Database Research Self-Assessment (By a group of 30 senior database researchers, "*The Lowell database research self-assessment*," Communications of the ACM, Volume 48, Issue 5, May 2005.)

Historical Background

Shortly after the introduction of the sensor networks and their potential applications about a decade ago [7], management of the sensor data was recognized as one of the main challenges in realizing the sensor network applications [2].

Foundations

Sensor Databases

One can think of a sensor network as a distributed database that collects, stores, and indexes the sensor data to answer the queries received from external users/applications as well as internal system entities. By considering a sensor network as a database, one envisions some of the benefits of the traditional databases potentially for sensor databases, e.g., reduced application development time, convenient multiuser data access and querying with a well-defined generic interface, efficient data reuse, and most importantly data independence. Physical data independence is a particularly beneficial advantage of the sensor database approach, because as compared to the physical layer of the traditional databases, the physical infrastructure of the sensor networks is much more sophisticated. With physical data independence in sensor databases, the logical schema of the data exposed to the users is separated from the physical schema that defines the complex and probably changing implementation of the data structures and operations on the physical network. By separating the logical and physical schemas, users/applications are isolated from the typical complications of the distributed data processing in the volatile sensor networks and can focus on designing the logical structure of their queries. Hence, the use of the sensor data is significantly facilitated.

Sensor Database Distinctions

Sensor databases are different from traditional distributed databases in both physical specifications and data characteristics. At the physical level, nodes of the database (i.e., sensor nodes) are severely constrained in resources, such as memory space, storage space, CPU power, and, most importantly, energy. Moreover, in sensor databases nodes and links of the network are both highly volatile. On the other hand, with sensor devices continuously collecting measurements from the environment, sensor data is naturally very dynamic. Besides, due to inaccuracy of the sensor devices, signal interference, noise, etc., uncertainty is also an inherent characteristic of the sensor data. With such physical and data characteristics, maintaining the illusion of a database is arguably a more difficult objective with sensor databases as compared to that of the traditional distributed databases and, accordingly, requires new data management solutions:

- Database operators should be delay tolerant and tolerant to frequent updates of the data.

- Query execution should be performed *in network* for energy efficiency; similarly, data storage and access should be designed for energy efficiency.
- Data acquisition plan is required to determine what data to collect [15].
- Participatory and opportunistic data transfer in sensor networks [16].
- Sensor query language should be augmented with new operators to specify duration and sampling rate of the data acquisition.
- Query execution plan should be dynamically optimized to account for variable access delay and uncertain data availability.
- Data uncertainty should be accounted for.
- Volatility of the sensor network should be hidden to provide the illusion of a stable database.
- Continuous queries should be supported, as sensor networks are primarily used for long-term monitoring.
- Meaningful data digests should be maintained to allow for answering historical queries, since data is continuously collected despite the limited space for storage.
- Aggregate spatiotemporal queries and range queries should be supported, for energy efficiency [18, 19].
- Approximate queries should be supported, as they are more meaningful with sensor data.
- Triggers should be supported for the event-driven monitoring applications.

One approach to implement sensor databases is to transfer all data to one or a small number of external base stations, where a traditional database system can be exploited. Alternatively, the data can be stored within the network itself with a balanced and optimal data storage plan. Although with the first approach one can more conveniently employ and extend the data management solutions applicable with the traditional databases, the second approach, termed *in-network storage*, allows for tighter coupling between query processing, on the one hand, and networking and application semantics, on the other hand. Tight coupling can potentially enable more energy-efficient query processing in sensor databases. To evaluate the query processing

performance with a particular sensor database implemented with either of these approaches, one can use the standard distributed database performance metrics such as incurred communication cost, query time, indexing time, throughput, load balance among nodes, data update overhead, and storage requirements.

Key Applications

As compared to the traditional wireline sensor networks that have been in use for decades, the more recent wireless sensor networks enable low cost and rapid deployment of the sensing network while supporting mobility. With these desirable characteristics due to the wireless technology, recently the standard applications of the sensing networks are revived, and new applications that were otherwise unthinkable are identified. The key classes of applications for sensor databases/networks are discussed below.

Environmental Monitoring

Environmental monitoring applications, specifically habitat monitoring [4], are among the earliest applications of the sensor networks. With the habitat monitoring applications, sensors are deployed to monitor animals or plants in their original habitats with most convenience for the scientists and least disturbance for the wildlife. With other environmental applications, sensors can be used to collect earth science and atmospheric data for environmental explorations, such as the study of the air pollution, global warming, etc., and also early detection and prediction of the natural and man-made disasters, such as hurricanes, wildfires, earthquakes, and biological hazards.

Military Intelligence

With rapid deployment, and inexpensive and untethered sensors, wireless sensor networks are well positioned as the tool to collect battlefield data for real-time battlefield intelligence [11]. For instance, wireless sensors can be utilized for geofencing (i.e., deploying a sensor network as a transparent fence to protect an area against unauthorized trespassing), enemy tracking, and battlefield exploration and condition assessment particularly in hazardous environments. In military intelligence applications, the small form factor, reliability, interoperability, and durability of the sensor nodes under severe environmental conditions are particularly critical requirements.

Asset Management

Businesses with large and high-turnover inventories of assets (such as construction companies, utility companies, and trucking companies) can benefit from automated asset management systems in improving the utilization of their resources [12]. With automated asset management, sensor networks are deployed to collect real-time data about exact location and condition of an inventory of assets automatically. The collected data provides the opportunity for real-time analysis of the resource usage, which in turn enables timely and optimal decision-making on handling, supplying, delivering, storing, and other asset management tasks. Various types of sensing devices, such as GPS devices and passive radio-frequency identification (RFID) tags, are applicable with the sensor networks used for asset monitoring.

Building Monitoring

The recent attempt aiming at optimizing the energy performance of the buildings by deep sensing of the building conditions, dubbed Building Information Modeling (BIM) (Federal BIM Program. URL: http://www.gsa.gov/bim/), heavily relies on the sensor network technology. With BIM energy tools enabled by sensing networks, one can monitor, e.g., the temperature and lighting conditions in the building, and accordingly regulate the heating and cooling systems, ventilators, and lights dynamically for best energy performance [9, 13]. Also as a safety tool, building sensor networks can detect and report threats, such as existence of the biological agents in the environment as well as physical intrusions.

Automotive

With the new standards such as dedicated short-range communication (DSRC) designated for ve-

hicle communications, in the near future, cars will be able to communicate information to each other and to the roadside infrastructures. With this capability, while in traffic cars can form a so-called vehicular sensor network, where each car equipped with sensing devices (e.g., camera, thermometer, etc.) acts as a mobile sensor node. In a vehicular sensor network, cars can share information and analyze the aggregate information about the road conditions, congestions, nearby emergencies, etc., for applications such as collision prevention, congestion avoidance, and flow optimization [10].

Healthcare

Sensor networks can effectively improve the accuracy of the patient care and, consequently, the safety of the patients when they become physically incapacitated and require immediate medical attention [1, 8]. Sensor networks allow this by enabling close and automated monitoring of the patient's vital signs. When monitoring is coupled with real-time analysis of the signs, the sensor-enabled healthcare system can alert the right person at the right time to attend to the patient. Such healthcare systems are applicable both at homes of the elderly and at the hospitals.

Industrial Monitoring

Sensors can be used to monitor industrial processes for safety as well as manufacturing optimization [5, 17]. One can also deploy sensors to monitor the condition of the industrial equipments for preventative maintenance and also safety of the operators. Wireline sensors have been in use for a long time in various industry sectors such as oil companies (both upstream and downstream) and chemical plants. Wireless technology and inexpensive sensors have greatly facilitated and extended the use of the sensing networks for process and equipment monitoring, encouraging oil companies, e.g., to develop smart oil fields by equipping the oil wells and other assets with wireless sensors (see, e.g., [3]).

Future Directions

With the current trend, sensor networks are being applied with increasingly more complex, large-scale, and distributed systems (e.g., the federal intelligent transportation system (Federal ITS Program. URL: http://www.its.dot.gov/)). Such applications demand deployment of large-scale sensor networks and, accordingly, require fully decentralized solutions for sensor data management to achieve scalability.

Data Sets

- Intel lab data set [6]. URL: http://db.csail.mit.edu/labdata/labdata.html
- Precipitation data set [19]. URL: http://www.jisao.washington.edu/data_sets/widmann/
- IHOP data set. URL: http://www.eol.ucar.edu/rtf/projects/ihop_2002/spol/

Cross-References

▶ Continuous Queries in Sensor Networks
▶ Data Acquisition and Dissemination in Sensor Networks
▶ Data Aggregation in Sensor Networks
▶ Data Storage and Indexing in Sensor Networks
▶ Data Uncertainty Management in Sensor Networks
▶ Database Languages for Sensor Networks
▶ In-Network Query Processing
▶ Sensor Networks

Recommended Reading

1. Banaei-Kashani F, et al. Monitoring mobility disorders at home using 3D visual sensors and mobile sensors (demo paper). Wireless Health. 2013.
2. Bonnet P, Gehrke J, Seshadri P. Towards sensor database systems. In: Proceedings of the 2nd International Conference on Mobile Data Management; 2001. p. 3–14.
3. The Center for Interactive Smart Oilfield Technologies. http://cisoft.usc.edu/
4. Cerpa A, Elson J, Estrin D, Girod L, Hamilton M, Zhao J. Habitat monitoring: application driver for wireless communications technology. In: Proceed-

ings of the SIGCOMM Workshop on Data Communications in Latin America and the Caribbean; 2001.
5. Chong C, Kumar SP. Sensor networks: evolution, opportunities, and challenges. Proc IEEE. 2003; 91(8):1247–1256.
6. Culler D, Estrin D, Srivastava M. Sensor network applications (cover feature). IEEE Comput. 2004;37(8):41–49.
7. Estrin D, Govindan R, Heidemann J, Kumar S. Next century challenges: scalable coordination in sensor networks. In: Proceedings of the 5th Annual International Conference on Mobile Computing and Networking; 1999. p. 263–70.
8. Ho L, Moh M, Walker Z, Hamada T, Su C. A prototype on RFID and sensor networks for elder healthcare. In: Proceedings of the 2005 ACM SIGCOMM Workshop on Experimental Approaches to Wireless Network Design and Analysis; 2005. p. 70–5.
9. Hu X, Wang B, Ji H. A wireless sensor network-based structural health monitoring system for highway bridges. Comput Aided Civ Infrastruct Eng Spec Issue: Health Monit Struct. 2013;28(3):193–209.
10. Lee U, Magistretti E, Zhou B, Gerla M, Bellavista P, Corradi A. MobEyes: smart mobs for urban monitoring with a vehicular sensor network. IEEE Wirel Commun. 2006;13(5):52–57.
11. Nemeroff J, Garcia L, Hampel D, Di Pierro S. Application of sensor network communications. In: Proceedings of the Military Communications Conference; 2001. p. 336–41.
12. RFID Journal. URL: http://www.rfidjournal.com/.
13. Schmid T, Dubois-Ferrière H, Vetterli M. SensorScope: experiences with a wireless building monitoring sensor network. In: Proceedings of the Workshop on Real-World Wireless Sensor Networks; 2005.
14. Sharifzadeh M, Shahabi C. Utilizing voronoi cells of location data streams for accurate computation of aggregate functions in sensor networks. GeoInformatica. 2006;10(1):1–26.
15. Shirani-Mehr H, Banaei-Kashani F, Shahabi C. Efficient viewpoint selection for urban texture documentation. GSN 2009.
16. Shirani-Mehr H, Banaei-Kashani F, Shahabi C. Using location-based social networks for quality-aware participatory data transfer. LBSN 2010 in conjunction with ACMGIS 2010.
17. Yamaji M, Ishii Y, Shimamura T, Yamamoto S. Wireless sensor network for industrial automation. In: Proceedings of the 5th International Conference on Networked Sensing Systems; 2008.
18. Yoon S, Shahabi C. The Clustered AGgregation (CAG) technique leveraging spatial and temporal correlations in wireless sensor networks. ACM Trans Sensor Netw. 2007;3(1):1–39.
19. Zhao F, Guibas L. Wireless sensor networks: an information processing approach. 1st ed. San Francisco: Morgan Kaufmann; 2004.
20. Center for Embedded Networked Sensing. http://www.cens.ucla.edu/

Approximate Queries in Peer-to-Peer Systems

Wolf Siberski and Wolfgang Nejdl
L3S Research Center, University of Hannover,
Hannover, Germany

Synonyms

Aggregate queries in P2P systems; Top-k queries in P2P systems.

Definition

Peer-to-peer (P2P) networks enable the interconnection of a huge amount of information sources without imposing costs for a central coordination infrastructure. Due to the dynamic and self-organizing nature of such networks, it is not feasible to guarantee completeness and correctness as in traditional distributed databases. Therefore, P2P systems are usually applied in areas where approximate query evaluation, i.e., the computation of a nearly complete and correct answer set, is sufficient. As the most frequent application of querying in P2P is search, many of these algorithms fall into the class of top-k query algorithms. Another important case is the approximation of aggregate query results.

Historical Background

P2P networks use approximate querying from the outset. In Gnutella, an unstructured network, the query is distributed in a limited neighborhood only, thus the result is usually not complete. The early top-k query algorithms for P2P are based on such unstructured networks. PlanetP [6], a P2P network for information retrieval, employs gossiping to replicate index information among all peers and sends queries to the best peers according to this replicated index. This approach has been extended to a super-peer network in [11]. While the first Distributed Hash Table (DHT) systems such as CHORD and Pastry aim at complete and correct answers, later

structured networks approximate the result set, especially P2P networks for information retrieval [2, 15]. Frequently, approximate P2P networks build upon algorithms for distributed databases or distributed IR. For example, for source selection PlanetP relies on an extended version of GlOSS [7], Minerva [15] on CORI [4]. Odissea [13] uses either Fagins Algorithm (FA) or a Threshold Algorithm (TA) as top-k algorithm, KLEE [9] extends TA, etc. (▶ Top-k Selection Queries on Multimedia Datasets).

Foundations

The main challenge in P2P approximate query processing is to select the optimal subset of peers to which the query is forwarded. However, the selection criteria for this subset are completely different for top-k and aggregate queries. While in top-k the goal is to identify the peers holding top objects, approximate aggregate querying algorithms need to find a representative sample of peers. In both cases, each selected peer evaluates the query locally; the respective responses are collected and merged to compute the final result set.

Approximation for Top-k Queries

Regardless of the chosen network topology, all distributed top-k algorithms consist of the following elements [14]:

Indexing. Determines what is indexed and how index information is collected within the network.

Source selection. Determines the peers a query is sent to.

Result merging. Determines how local result lists are merged to form the final result set.

Unstructured Networks

In an unstructured P2P network peers form a random graph. In these networks, the only available routing strategy is *filtered flooding*, i.e., forwarding the query to selected neighbors. For effective filtering, the peers maintain content indexes. For each neighbor peer, the index allows to look up which kind of content is reachable via this peer. In the case of information retrieval, the index holds term frequencies of these subnets. This index information is built by gossiping: each peer periodically sends the content summaries it holds to its neighbors, where they are merged with the index. Thus, over time each peer gathers more and more accurate information about the whole network. Frequently, bloom filters or hash sketches are used to represent the content summaries. The typical representative for this approach is PlanetP [6]. Evaluations have shown that this algorithm type only scales to several thousands of peers, due to the limitations of filtered flooding. Even with complete index information, the query still usually has to be sent to a high fraction of all peers to reach the peers holding the top-k objects. Also, gossiping induces rather high index maintenance costs.

Hierarchical Networks

Hierarchical topologies can overcome some of these limitations. In these topologies, particularly powerful peers form a super-peer backbone. Information sources are not connected with each other, but always assigned to one of the super-peers. This topology is especially suited for adaptation of traditional distributed top-k algorithms: each super-peer acts as coordinating node for its peers. In some systems the peers form a tree-shaped network; this has the advantage that the same aggregation algorithm can be used up to the root peer, but at the price of extremely uneven load distribution. Therefore, the usual approach is to restrict the hierarchy to two levels. In this case, filtered flooding is used to distribute queries within the super-peer backbone. Maintaining the index independently from the actual queries can impose a high overhead; this can be avoided by building a *query-driven index* [2].

Structured Networks

Arguably the most efficient peer-to-peer networks are DHTs, where peers form highly structured network topologies. However, they only provide the usual hash table feature, storage and retrieval by key. The DHT maintains lists of top peers for each feature [3, 13]. As in the case of hierarchical peers, the index can be improved by considering query statistics [12]. For a query, first these lists are retrieved, and then an established algorithm such as CORI or GlOSS is used for source selection. To retrieve the top-k objects, the TA family of algorithms are the state of the art. TPUT is especially suitable because it limits the retrieval process to three phases: first, the query initiator determines a lower bound for the object score by requesting scores of the top-k objects at each peer. Second, the initiator requests all objects having at least the threshold score. It is guaranteed that all top-k objects are in the returned sets. Finally, the initiator determines these objects and requests the actual content. TPUT has been evolved to an approximate algorithm with probabilistic guarantees in [9].

Approximation for Aggregate Queries

Efficient approximation of aggregate queries in unstructured P2P networks can be done by sampling. Starting at the peer issuing the query, the query travels along a random path to gather the sample. The challenge is to choose this path such that the query is indeed received by a uniform sample of the network. Note that the standard approach (*Markov-Chain random walk*) of selecting each outgoing edge with equal probability does not yield a uniform sample, but favors nodes with high degrees. This can be approximately compensated by scaling down the local peer value with the probability of this peer being selected. To reduce errors due to clustering within the network, the random walk can be modified such that only each ith peer on the path is considered for the sample. [1] shows how usual aggregate queries (COUNT, SUM, AVG) can be computed in this way with low error rates. While gossiping also has be proposed to compute aggregates in unstructured networks, this method does not scale to large networks [10].

An efficient method to compute COUNT queries in DHT networks has been proposed in [10]. This approach relies on locally computed hash sketches which are inserted into the DHT. For hash sketches of length k the actual counting requires $O(k)$ DHT lookups, resulting in $O(k \cdot log n)$ messages.

Key Applications

Top-k queries are used in distributed information retrieval scenarios, such as digital library networks.

An important application for aggregate queries in massively distributed networks is the gathering of network statistics, e.g., to identify security risks or to monitor performance [8]. Approximate aggregate queries are also gaining importance in the area of sensor network [5], where limitations of the sensor hardware (processor, memory, power supply) are key factors for query algorithm design.

Cross-References

▶ Peer Data Management System
▶ Top-K Selection Queries on Multimedia Datasets

Recommended Reading

1. Arai B, Das G, Gunopulos D, Kalogeraki V. Efficient approximate query processing in peer-to-peer networks. IEEE Trans Knowl Data Eng. 2007;19(7):919–33.
2. Balke WT, Nejdl W, Siberski W, Thaden U. Progressive distributed top-k retrieval in peer-to-peer networks. In: Proceedings of the 21st International Conference on Data Engineering; 2005. p. 174–85.
3. Bender M, Michel S, Triantafillou P, Weikum G, Zimmer C. MINERVA: collaborative P2P search. In: Proceedings of the 31st International Conference on Very Large Data Bases; 2005. p. 1263–6.
4. Callan JP, Lu Z, Croft WB. Searching distributed collections with inference networks. In: Proceedings of the 18th Annual International ACM SIGIR Conference on Research and Development in Information Retrieval; 1995. p. 21–8.
5. Chu D, Deshpande A, Hellerstein JM, Hong W. Approximate data collection in sensor networks using probabilistic models. In: Proceedings of the 22nd International Conference on Data Engineering; 2006. p. 48.
6. Cuenca-Acuna FM, Peery C, Martin RP, Nguyen RD. Planet P: using gossiping to build content addressable peer-to-peer information sharing communities. In: Proceedings of the 12th IEEE International Symposium on High Performance Distributed Computing; 2003. p. 236–46.
7. Gravano L, Garcia-Molina H, Tomasic A. GlOSS: Text-source discovery over the internet. ACM Trans Database Syst. 1999;24(2):229–64.
8. Hellerstein JM, Condie T, Garofalakis MN, Loo BT, Maniatis P, Roscoe T, Taft N. Public health for the internet (PHI). In: Proceedings of the 3rd Biennial Conference on Innovative Data Systems Research; 2007. p. 332–40.
9. Michel S, Triantafillou P, Weikum G. Klee: a framework for distributed top-k query algorithms. In: Proceedings of the 31st International Conference on Very Large Data Bases; 2005. p. 637–48.
10. Ntarmos N, Triantafillou P, Weikum G. Counting at large: efficient cardinality estimation in internet-scale data networks. In: Proceedings of the 22nd International Conference on Data Engineering; 2006. p. 40.
11. Seshadri S, Cooper BF. Routing queries through a peer-to-peer infobeacons network using information retrieval techniques. IEEE Trans Parallel Distrib Syst. 2007;18(12):1754–65.
12. Skobeltsyn G, Luu T, Podnar ZI, Rajman M, Aberer K. Web text retrieval with a P2P query-driven index. In: Proceedings of the 33rd Annual International ACM SIGIR Conference on Research and Development in Information Retrieval; 2007. p. 679–86.
13. Suel T, Mathur C, Wu J, Zhang J, Delis A, Kharrazi M, Long X, Shanmugasundaram K. ODISSEA: a peer-to-peer architecture for scalable web search and information retrieval. In: Proceedings of the 6th International Workshop on the World Wide Web and Databases; 2003. p. 67–72.
14. Yu C, Philip G, Meng W. Distributed top-n query processing with possibly uncooperative local systems. In: Proceedings of the 29th International Conference on Very Large Data Bases; 2003. p. 117–28.
15. Zimmer C, Tryfonopoulos C, Weikum G. MinervaDL: an architecture for information retrieval and filtering in distributed digital libraries. In: Proceedings of the 11th European Conference on Research and Advanced Technology for Digital Libraries; 2007. p. 148–60.

Approximate Query Processing

Qing Liu
CSIRO, Hobart, TAS, Australia

Synonyms

Approximate query answering

Definition

Query processing in a database context is the process that deduces information that is available in the database. Due to the huge amount of data available, one of the main issues of query processing is how to process queries efficiently. In many cases, it is impossible or too expensive for users to get exact answers in the short query response time. Approximate query processing (AQP) is an alternative way that returns approximate answer using information which is similar to the one from which the query would be answered. It is designed primarily for aggregate queries such as count, sum and avg, etc. Given a SQL aggregate query Q, the accurate answer is y while the approximate answer is y'. The relative error of query Q can be quantified as:

$$Error(Q) = \mid \frac{y - y'}{y} \mid .\qquad(1)$$

The goal of approximate query processing is to provide approximate answers with acceptable accuracy in orders of magnitude less query response time than that for the exact query processing.

Historical Background

The earliest work on approximate answers to decision support queries appears in Morgenstein's dissertation from Berkeley [1]. And the approximate query processing problem has been studied extensively in the last 10 years. The main motivations [2] which drive the techniques being developed are summarized as follows.

First, with the advanced data collection and management technologies, nowadays there are a large number of applications with data sets about gigabytes, terabytes or even petabytes. Such massive data sets necessarily reside on disks or tapes, making even a few accesses of the base data sets comparably slow. In many cases, precision to "last decimal" is not required for a query answer. Quick approximation with some error guarantee (e.g., the resident population in Australia $21,126,700 +/- 200$) is adequate to provide insights about the data.

Second, decision support system (DSS) and data mining are popular approaches to analyzing large databases for decision making. The main characteristic of the DSS is that aggregation queries (e.g., count, sum, avg, etc.) are executed on large portion of the databases, which can be very expensive and resource intensive even for a single analysis query. Due to the exploratory nature of decision making, iterative process involves multiple query attempts. Approximate answers with fast response time gives users the ability to focus on their explorations and quickly identify truly interesting data. It provides a great scalability of the decision support applications.

Third, approximate query processing is also used to provide query preview. In most cases, users are only interested in a subset of the entire database. Given a trial query, query preview provides an overview about the data distribution. The users can preview the number of hits and refine the queries accordingly. This prevents users from fruitless queries such as zero-hits or mega-hits. Figure 1 shows an example of query preview interface of NASA EOSDIS (Earth Observing System Data and Information System) project, which is attempting to provide online access to a rapidly growing archive of scientific earth data about the earth's land, water, and air. In the query preview, users select rough ranges for three attributes: area, topic (a menu list of parameters such as atmosphere, land surface, or oceans) and temporal coverage. The number of data sets for each topic, year, and area is shown on preview bars. The result preview bar, at the bottom of the interface, displays the total approximate number of data sets which satisfy the query.

Finally, sometimes network limitation or disk storage failure would cause the exact answers unaffordable or unavailable. An alternative solution is to provide an approximate answer based on the local cached data synopsis.

Foundations

Due to the acceptability of approximate answers coupled with the necessity for quick query response time, approximate query processing has

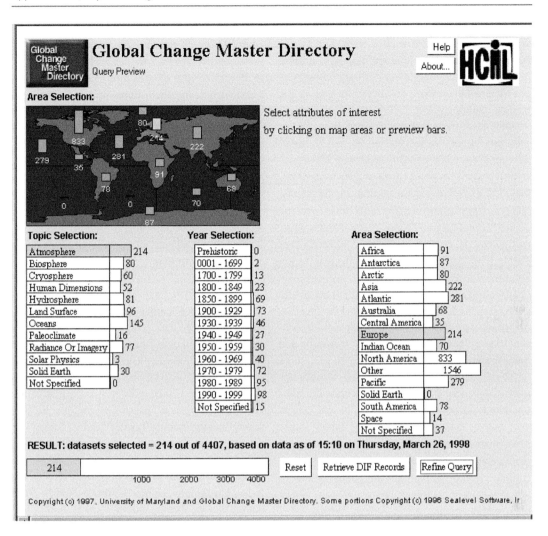

Approximate Query Processing, Fig. 1 NASA EOSDIS interface of query preview

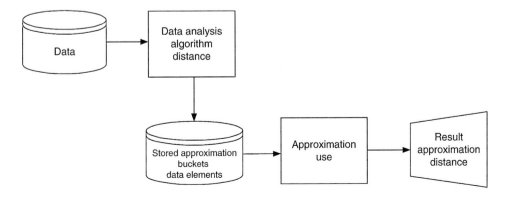

Approximate Query Processing, Fig. 2 Generic flow of approximation process

emerged as a cost effective approach for dealing with the large amount of data. This speed-up is achieved by answering queries based on samples or other synopses (summary) of data whose size is orders of magnitude smaller than that of the original data sets.

Ioannidis presented the generic flow of approximate query process (Fig. 2) in [3]. Here, data analysis is the overall approach which derives the synopsis from original data to approximate the underlying data distribution. Typically, the *algorithm* partitions the data based on the *distance* function into groups of similar elements, called *buckets*, clusters, patterns, or several other names. The *data elements* that fall in each bucket are then represented by the synopses for *approximation use* which corresponds to the actual purpose of the whole data approximation process. The quality of approximation result can be measured by the *distance* between the synopses and the original data.

There are two basic approaches to achieve approximate query processing: pre-computed synopsis and online query processing.

Pre-computed Synopsis

Approximate query processing using pre-computed synopsis includes two steps: construct synopsis prior to query time, answer query approximately using synopsis at query time.

To provide high accurate query answer, the key issue to construct synopsis is how to represent the underlying data distribution precisely and compactly. Generally, the data distribution can be classified into two groups: uniform distribution and non-uniform distribution. The synopsis for uniform data distribution assumes the objects are distributed uniformly in the data space. For point objects locating in two-dimensional space $[U^x_{min}, U^y_{min}] \times [U^x_{max}, U^y_{max}]$, the query result size is estimated as $N \times area(Q)/((U^x_{max} - U^x_{min}) \times (U^y_{max} - U^y_{min}))$, where N is the data set size and $area(Q)$ is the area of window query Q.

There are various techniques developed for non-uniform data distribution. They can also be divided into two groups: parametric and non-parametric. The parametric techniques try to use

parameters to catch the original data distributions. Although the models can summarize data distributions with a few descriptive parameters, if the underlying data do not follow any known distributions, or their linear combinations, the model fitting techniques produce inferior results.

The non-parametric techniques use different approaches to summarize the data distributions. Generally, it is possible to classify these techniques into three categories according to the strategies adopted:

1. Sampling techniques
2. Histogram techniques
3. Wavelet techniques

Sampling The basic idea of sampling is that a small random sample of the data often well-represent all the data. Therefore, query would be answered based on the pre-sampled small amount of data and then scaled up based on the sample rate. Figure 3 shows an example where 50% of data are sampled during the pre-computed stage. Given a query "how many Sony laptops are sold in R", the approximate result is "select 2 * sum(*) from S where $S.product =$ '*SonyLaptop*'", which is 12. In R, the exact answer is 11.

The main issue of sampling method is to decide what sample criteria should be used to select data. The sampling techniques are classified into the following groups [4]:

1. *Uniform sampling*. Data is sampled uniformly
2. *Biased sampling*. A non-uniform random sample is pre-computed such that parts of the database deemed "more important" than the rest
3. *Icicles*. A biased sampling technique that is based on known workload information
4. *Outlier indexing*. Indexing outliers and biased sampling the remaining data
5. *Congressional sampling*. Targeting group by queries with aggregation and trying to maximize the accuracy for all groups (large or small) in each group-by query
6. *Stratified sampling*. Generalization of outlier indexing, Icicles and congressional sampling.

Approximate Query Processing, Fig. 3
Example of sampling

R: Sales

Product	Amount
Sony laptop	1
Sony laptop	1
Sony laptop	2
Sony laptop	3
Sony laptop	4
Dell printer	1
Dell printer	2
LCD monitor	1

S: Sampled sales

Product	Amount
Sony laptop	1
Sony laptop	2
Sony laptop	3
Dell printer	2

It targets minimizing error in estimation of aggregates for the given workload

Sample-based procedures are robust in the presence of correlated and nonuniform data. Most importantly, sampling-based procedures permit both assessment and control of estimation errors. The main disadvantage of this approach is the overhead it adds to query optimization. Furthermore, join operation could lead to significant quality degradations because join operator applied on two uniform random sample can result in a non-uniform sample of the join result which contains very few tuples.

Histograms Histogram techniques are the most commonly used form of statistics in practice (e.g., they are used in DB2, Oracle and Microsoft SQL Server). This is because they incur almost no run-time overhead and produce low-error estimates while occupying reasonably small space.

The basic idea is to partition attribute value domain into a set of buckets and query is answered based on the buckets. The main issues of histogram construction and query are as follows:

1. How to partition data into bucket
2. How to represent data in each bucket
3. How to estimate answer using the histogram

For one-dimensional space, a histogram on an attribute X is constructed by partitioning the data distribution of X into B ($B \geq 1$) buckets and approximating the frequencies and values in each bucket. Figure 4a is an example of original data set and Fig. 4b shows its data distribution. Figure 4c is an example of histogram constructed accordingly, where $B = 3$.

If there are several attributes involved in a query, a multi-dimensional histogram is needed to approximate the data distribution and answer such a query. A multi-dimensional histogram on a set of attributes is constructed by partitioning the joint data distribution of the attributes. They have the exact same characteristics as one-dimensional histograms, except that the partition rule needs to be more intricate and cannot always be clearly analyzed because there cannot be ordering in multiple dimensions [5].

To represent data in each bucket, it includes value approximation and frequency approximation. Value approximation captures how attribute values are approximated within a bucket. And frequency approximation captures how frequencies are approximated within a bucket.

The two main approaches for value approximation are *continuous value assumption* and *uniform spread assumption* [6]. Continuous value assumption only maintains min and max value without indication of how many values there are or where they might be. Under the uniform

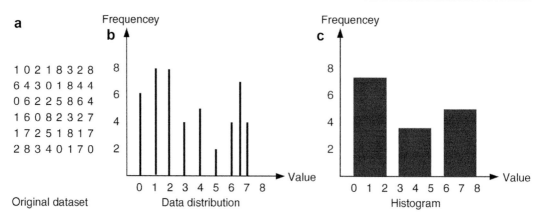

Approximate Query Processing, Fig. 4 Example of histogram

spread assumption, one also maintain the number of values within each bucket and approximates the actual value set by the set that is formed by (virtually) placing the same number of values at equal distances between the min and max value in multi-dimensional space [7].

With respect to frequency approximation, almost all work deal with *uniform distribution assumption*.

The benefit of a histogram synopsis is that it can be easily used to answer many query types, including the aggregate and non-aggregate queries. However, one of the issues of histogram approach is it is hard to calculate a theoretical error bound. Thus the evaluations on the histogram synopsis usually rely heavily on the experiment results. Further more, histogram-based approaches become problematic when dealing with the high-dimensional data sets that are typical for modern decision support applications. This is because as the dimensionality of the data increases, both the storage overhead (i.e., number of buckets) and the construction cost of histograms that can achieve reasonable error rates increase in an explosive manner.

Wavelet Wavelet is a mathematical tool for hierarchical decomposition of functions using recursive pairwise averaging and differencing at different resolutions. It represents a function in terms of a coarse overall shape, plus details that range from broad to narrow. It is widely used in the signal and image processing.

Matias et al. [8] first proposed the use of Haar-wavelet coefficients as synopsis for estimating the selectivity of window queries. The basic idea is to apply wavelet decomposition to the input data collection to obtain a compact data synopsis that comprises a select small collection of wavelet coefficients. Figure 5 shows an example of hierarchical decomposition tree of Haar-wavelet. The leaf nodes are original data and non-leaf nodes are wavelet coefficients generated by averaging and differencing from their two children.

Later, the wavelet concept was extended to answer more general approximate queries. The results of recent studies have clearly shown that wavelets can be very effective in handling aggregates over high-dimensional online analytical processing (OLAP) cubes, while avoiding the high construction costs and storage overheads of histogram techniques.

Another important part of the above three technologies for approximate query processing is synopsis maintenance. If the data distribution is not changed significantly, the data synopsis would be updated accordingly to reflect such change. Otherwise, a new data synopsis will be constructed and discard the old one. Refer to the specific techniques for more details.

There are a few other work that do not belong to the above three categories to approximate the underlying data distribution. For example, recently Das et al. [9] present a framework that is based on randomized projections. This is the first

**Approximate Query
Processing, Fig. 5**
Example of Haar-wavelet
hierarchical decomposition

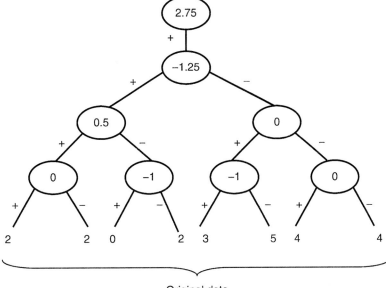

Original data

work in the context of spatial database which provides probability quality guarantees with respect to query result size approximation.

Online Query Processing

The motivation of online query processing is that the data analysis is to extract unknown information from data. It is an iterative process starting by asking broad questions and continually refining them based on approximate or partial results. Therefore, instead of optimizing for query response time, it needs to balance two conflicting performance goals: minimizing uneventful "dead time" between updates for the users, while simultaneously maximizing the rate at which partial or approximate answers approach a correct answer. Refer to [10] for details.

Compared with pre-computed synopses approach, the advantages of online query processing approach is it does not require pre-processing and progressive refinement of approximate results at runtime can quickly lead to a satisfied results. However, the main obstacles for this technique to be practical is it significantly changes the query processor of current commercial query processing system which is not desirable.

Key Applications

AQP in Relational Data Management
The AQUA system proposed by the Bell lab can support various kinds of approximate queries over the relational database.

AQP in Spatial Data Management
Alexandria Digital Library allows user to define spatial queries and returns the approximate number of hits quickly as the initial result and then user can refine the query accordingly.

AQP in Stream Data Management
MIT proposed Aurora as a data stream management system, which virtually supports answering approximate queries over the data stream.

AQP in Sensor Network
The techniques of TinyDB system, proposed by Massachusetts Institute of Technology and UC Berkeley, can lead to orders of magnitude improvements in power consumption and increased accuracy of query results over non-acquisitional systems that do not actively control when and where data is collected.

AQP in Semantic Web Search

For semantic web search, the viewpoints of users performing a web search, ontology designers and annotation designers do not match. This leads to missed answers. Research studies have shown AQP is of prime importance for efficiently searching the Semantic Web.

Cross-References

▶ Approximate Queries in Peer-to-Peer Systems
▶ Clinical Decision Support
▶ Data Mining
▶ Histogram
▶ Online Analytical Processing
▶ Query Optimization
▶ Wavelets on Streams

Recommended Reading

1. Morgenstein J. Computer based management information systems embodying answer accuracy as a user parameter. PhD thesis, University of California, Berkeley; 1980.
2. Garofalakis M, Gibbons P. Approximate query processing: taming the teraBytes: a tutorial. In: Proceedings of the 27th Interantional Conference on Very Large Data Bases; 2001.
3. Ioannidis Y. Approximation in database systems. In: Proceedings of the 9th International Conference on Database Theory; 2003. p. 16–30.
4. Das G. Sampling methods in approximate query answering systems. In: Wang J, editors. Invited Book Chapter, Encyclopedia of data warehousing and mining. Information Science Publishing; 2005.
5. Poosala V, Ioannidis Y. Selectivity estimation without the attribute value independence assumption. In: Proceedings of the 23th International Conference on Very Large Data Bases; 1997. p. 466–75.
6. Poosala V, Ioannidis Y, Haas P, Shekita E. Improved histograms for selectivity estimation of range predicates. In: Proceedings of the ACM SIGMOD International Conference on Management of Data; 1996. p. 294–305.
7. Ioannidis Y. The history of histograms (abridged). In: Proceedings of the 29th International Conference on Very Large Data Bases; 2003. p. 19–30.
8. Matias Y, Vitter J, Wang M. Wavelet based histograms for selectivity estimation. In: Proceedings of the ACM SIGMOD International Conference on Management of Data; 1998. p. 448–59.
9. Das A, Gehrke J, Riedewald M. Approximation techniques for spatial data. In: Proceedings of the ACM SIGMOD International Conference on Management of Data; 2004. p. 695–700.
10. Hellerstein J, Haas P, Wang H. Online aggregation. In: Proceedings of the ACM SIGMOD International Conference on Management of Data; 1997. p. 171–82.

Approximate Reasoning

Vilém Novák
Institute for Research and Applications of Fuzzy Modeling, University of Ostrava, Ostrava, Czech Republic

Definition

Approximate reasoning is a deduction method which makes it possible to derive a conclusion on the basis of imprecisely characterized situation (quite often using linguistically specified fuzzy IF-THEN rules) and a new information that can also be imprecise. The basic scheme of approximate reasoning is the following:

$$
\begin{aligned}
\text{Condition} : &\ \text{IF } X \text{ is } A_1 \text{ THEN } Y \text{ is } B_1 \\
&\ \cdots\cdots\cdots\cdots\cdots\cdots\cdots \\
&\ \text{IF } X \text{ is } A_m \text{ THEN } Y \text{ is } B_m \\
\text{Premise} : &\ \quad X \text{ is } A' \\
&\ \cdots\cdots\cdots\cdots\cdots\cdots\cdots \\
\text{Conclusion} : &\ \quad Y \text{ is } B'
\end{aligned}
\tag{1}
$$

where "Condition" is a linguistic description consisting of a set of fuzzy/linguistic IF-THEN rules and A' is a possible modification of antecedent of some of the former rules. For example, "X is small" can be replaced by "X is very small."

Key Points

The mathematical model of approximate reasoning depends on the way how the linguistic description forming the condition is interpreted (see Fuzzy/Linguistic If-Then Rules and Linguistic Descriptions).

Let X is A' be interpreted by a fuzzy set $A' \subseteq U$. If "Condition" is interpreted by a fuzzy relation $R \subseteq U \times V$ then the result of approximate reasoning is a fuzzy set $B' \subseteq$ which interprets "Y is B'" and which is obtained using the formula

$$B'(y) = \bigvee_{x \in U} (A'(x) \otimes R(x, y)) \qquad (2)$$

where $\otimes$ is a t-norm (product in residuated lattice).

Alternative approximate reasoning method is *perception-based logical deduction*. Its idea consists of finding *perception* of the given measured value of the input $X = x_0$. The perception is an evaluative expression occurring among $A_1, \ldots, A_m$ that fits x_0 in the best way. Then the corresponding fuzzy IF-THEN rule is fired and the proper output is derived. More details can be found in [2, 3].

Cross-References

▶ Fuzzy Relation
▶ Fuzzy Set
▶ Fuzzy/Linguistic IF-THEN Rules and Linguistic Descriptions
▶ Triangular Norms

Recommended Reading

1. Klir GJ, Yuan B. Fuzzy sets and fuzzy logic: theory and applications. New York: Prentice-Hall; 1995.
2. Novák V, Lehmke S. Logical structure of fuzzy IF-THEN rules. Fuzzy Sets Syst. 2006;157(15):2003–29.
3. Novák V, Perfilieva I. On the semantics of perception-based fuzzy logic deduction. Int J Intell Syst. 2004;19:1007–31.
4. Novák V., Perfilieva I., and Močkoř J . Mathematical principles of fuzzy logic. Kluwer, Boston/Dordrecht, 1999.

Approximation of Frequent Itemsets

Jinze Liu
University of Kentucky, Lexington, KY, USA

Synonyms

AFI

Definition

Consider an $n \times m$ binary matrix D. Each row of D corresponds to a transaction t and each column of D corresponds to an item i. The (t, i)-element of D, denoted $D(t, i)$, is 1 if transaction t contains item i, and 0 otherwise. Let $T_0 = \{t_1, t_2, \ldots, t_n\}$ and $I_0 = \{i_1, i_2, \ldots, i_m\}$ be the set of transactions and items associated with D, respectively.

Let D be as above, and let $\varepsilon_r, \varepsilon_c \in [0, 1]$. An itemset $I \subseteq I_0$ is an approximate frequent itemset AFI($\varepsilon_r, \varepsilon_c$), if there exists a set of transactions $T \subseteq T_0$ with $|T| \geq minsup\,|T_0|$ such that the following two conditions hold:

1. $\forall i \in T, \frac{1}{|I|} \sum_{j \in I} D(i, j) \geq (1 - \epsilon_r)$;
2. $\forall j \in I, \frac{1}{|T|} \sum_{i \in T} D(i, j) \geq (1 - \epsilon_c)$;

Historical Background

Relational databases are ubiquitous, cataloging everything from market-basket data [1] to genomic data collected in biological experiments [2]. A binary matrix is one common representation of relational databases. Rows in the matrix often correspond to the objects, while columns represent attributes of the objects. The binary value of each matrix entry indicates the presence (1) or absence (0) of an attribute in the object. For example, in a market-basket database, rows represent transactions, columns represent product items, and a binary entry indicates whether an item is contained in the transaction [1]. Frequent itemset mining [1] is a key technique in the analysis of such data. In the binary representation, a *frequent itemset* corresponds to a sub-matrix of 1s, where the itemset (the set of columns) are supported by a sufficiently large number of transactions (set of rows).

While frequent itemsets and the algorithms to generate them have been well studied, the problem is that the data in real application is often imperfect. In a transaction database, frequent itemsets might be obscured by the vagaries of the market and human behaviors. Items expected to be purchased together by a customer might not appear together in some transactions when one of them is out of stock in the market or overstocked by the customer. In addition, empirical data is subject to measurement noise. For example, Microarray data is often error-prone due to variations in the experimental technology and the stochastic nature of biological processes.

The noise recorded in real applications undermines the ultimate goal of the classical frequent itemset algorithms, i.e., revealing the itemset that is present in a sufficient fraction of transactions (Fig. 1). In fact, when noise is present, the classical frequent itemset algorithms may discover multiple small fragments of the true itemset while missing the true itemset itself. The problem worsens for the most interesting large itemsets since they are more vulnerable to noise.

The approximate frequent itemset AFI(ε_r, ε_c) denotes a collection of submatrices of D where the row-wise and column-wise noise levels within each submatrix are below ε_r and ε_c respectively. The classical exact frequent itemset (EFI) is a member of AFI(ε_r, ε_c) where ε_r and ε_c are set to be 0 (Fig. 2). The noise thresholds ε_r and ε_c are usually below 30%. In cases when the noise in either row or column is not restricted, AFI(ε_r, $*$) or AFI($*$, ε_c) will be used to denote the corresponding families. AFI(ε_r, $*$) corresponds

to the same family of itemsets, namely *Error Tolerant Itemset (ETI)*, defined by Yang et al. [3].

Foundations

The AFI-mining algorithm [4, 5] generalizes the framework of level-wise itemset enumeration. First, the Apriori property of exact frequent itemset mining doesn't hold for AFI when noise is allowed. Instead, conditions under which candidate itemsets can be pruned are established and employed in the AFI algorithm. Secondly, methods that systematically enumerate candidate itemsets without multiple scans of the database are also developed.

Noise-Tolerant Support Pruning

The anti-monotone property of exact frequent itemsets is the key to eliminating the exponential search space in frequent itemset mining [1]. In particular, the anti-monotone property ensures that a $(k + 1)$ exact itemset can be pruned if anyone of its k sub-itemsets is not sufficiently supported. However, the allowance of noise may lower the support necessary for the sub-itemsets of a noise-tolerant itemset. The following theorem suggests a lower bound of support for pruning the candidate itemsets in generating AFIs.

Theorem 1

Given a minsup threshold, If a length $(k + 1)$-itemset I' is an AFI(ε_r, ε_c), for any of its k-sub-itemset $I \subseteq I'$, the number of transactions containing no more than ε_r fraction of noise must be at least

$$n \cdot minsup \cdot \left(1 - \frac{k\epsilon_c}{\lfloor k\epsilon_r \rfloor + 1}\right) \quad (1)$$

The noise-tolerant pruning support is defined as the following:

Definition Given ε_c, ε_r and *minsup*, the **noise-tolerant support** for a length-k itemset.

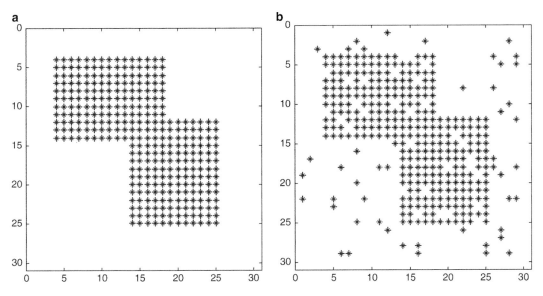

Approximation of Frequent Itemsets, Fig. 1 Patterns with and without noise

$$minsup_I^k = minsup \cdot \left(1 - \frac{k\epsilon_c}{\lfloor k\epsilon_r \rfloor + 1}\right) + \tag{2}$$

Here $(a)_+ = \max\{a, 0\}$.

The noise-tolerant support threshold is used as the basis of a pruning strategy for AFI mining. The strategy removes supersets of a given I from further consideration when I has support less than $minsup_I{}^k$. In the special case that $\varepsilon_r = \varepsilon_c = 0$, $minsup_I{}^k = minsup$, which is consistent with the anti-monotone property of exact frequent itemsets [1].

0/1 Extensions A transaction t supports a k-itemset I if t contains at least a fraction $1 - \varepsilon_r$ of the items in I. The transaction set of I consists of all the transactions supporting I. Starting with singleton itemsets, the AFI algorithm generates $(k + 1)$-itemsets from k-itemsets in a breadth-first fashion. For each candidate itemset I, transactions t supporting I are generated. The transaction set of a $(k + 1)$ itemset is constructed from the transaction sets of its k-item subsets in one of two different ways, depending on the value of k and ε_r.

0-extension and 1-extension are the two basic steps to be taken for the efficient collection of the supporting transactions. They obtain the supports

based on the support of its sub-itemset while avoiding the repeated database scans plaguing the algorithms proposed by [3, 6].

Lemma 1

(1-Extension) If $\lfloor k \cdot \varepsilon_r \rfloor = \lfloor (k + 1) \cdot \varepsilon_r \rfloor$ *then any transaction that does not support a k-itemset will not support its $(k + 1)$ superset.*

The Lemma is based on the fact that if no additional noise is allowed when generating $(k + 1)$ itemset, a transaction does not support a (k-itemset won't support its $(k + 1)$ superset since the number of 1s it contains is always smaller than or equal or $\frac{\lfloor k_* \epsilon \rfloor - 1 + 1}{k+1} < \epsilon$. Thus if $\lfloor k \cdot \varepsilon_r \rfloor = \lfloor (k + 1) \cdot \varepsilon_r \rfloor$ then the transaction set of a ($k + 1$) itemset I is the intersection of transaction sets of its length k subsets. This is called a *1-extension*.

Lemma 2

(0-Extension) If $\lfloor k \cdot \varepsilon_r \rfloor + 1 = \lfloor (k + 1) \cdot \varepsilon_r \rfloor$ *then any transaction supporting a (k-itemset also supports its $(k + 1)$ supersets.*

The procedure of 0-extension embodies how noise can be encompassed into a frequent itemset. If additional noise is allowed in a $(k + 1)$-itemset, it is intuitive that a transaction that supports a k-itemset will also support its $k + 1$-item superset,

Approximation of Frequent Itemsets, Fig. 2 Relationships of various AFI criteria

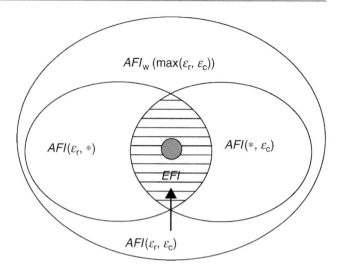

no matter whether the $k + 1$ entry is 1 or 0. To reflect this property, if $\lfloor k \cdot \varepsilon_r \rfloor + 1 = \lfloor (k + 1) \cdot \varepsilon_r \rfloor$, the transaction set of a $(k + 1)$ itemset I is the union of the transaction sets of its length k subsets. This is called a *0-extension*.

Experimental Results

In order to test the quality of the algorithm, data with an embedded pattern and overlaid random errors are generated. The discovered patterns are evaluated against the true patterns which are known in apriori. The methods, exact frequent itemset (EFI), ETI and AFI, are compared in terms of their capabilities in discovering the true patterns in the presence of noise.

Two measures jointly describing the quality are employed. They are "recoverability" and "spuriousness." Recoverability is the fraction of the embedded patterns recovered by an algorithm, while spuriousness is the fraction of the mined results that fail to correspond to any planted cluster. A truly useful data mining algorithm should achieve high recoverability with little spuriousness to dilute the results. A detailed description of the two measures is given in [4].

Multiple data sets were created and analyzed to explore the relationship between increasing noise levels and the quality of the result. Noise was introduced by bit-flipping each entry of the full matrix with a probability equal to p. The probability p was varied over different runs from 0.01 to 0.2. The number of pattern blocks embedded also varied, but the results were consistent across this parameter. The results when one or three blocks were embedded in the data matrix are presented in Fig. 3a, b, respectively.

In both cases, the exact method performed poorly as noise increased. Beyond $p = 0.05$ the original pattern could not be recovered, and all of the discovered patterns were spurious. In contrast, the error-tolerant algorithms, ETI and AFI, were much better at recovering the embedded matrices at the higher error rates. However, the ETI algorithm reported many more spurious results than AFI. Although it may discover the embedded patterns, ETI generates many more patterns that are not of interest, which may overshadow the real patterns of interest. The AFI algorithm consistently demonstrated higher recoverability of the embedded pattern while maintaining a lower level of spuriousness.

Application

AFI mining algorithm can be generally used to find dense 1s blocks in a large binary matrix. The following are example applications.

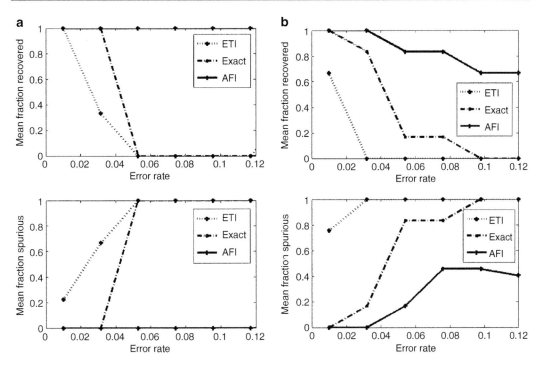

Approximation of Frequent Itemsets, Fig. 3   Algorithm quality versus noise level

- E-commerce application: discover approximate frequent itemset in large transactional databases [4, 5].
- Biogeographic application: discover regions associated with migration in biogeographic research [4].
- Microarray analysis: find noise tolerant co-expressed patterns.

Cross-References

▶ Association Rule Mining on Streams
▶ Data Mining

Recommended Reading

1. Agrawal R, Imielinski T, Swami A. Mining association rules between sets of items in large databases. In: Proceedings of the ACM SIGMOD International Conference on Management of Data; 1993. p. 207–16.

2. Creighton C, Hanash S. Mining gene expression databases for association rules. Bioinformatics. 2003;19(1):79–86.
3. Yang C, Fayyad U, Bradley PS. Efficient discovery of error-tolerant frequent itemsets in high dimensions. In: Proceedings of the 7th ACM SIGKDD International Conference on Knowledge Discovery and Data Mining; 2001. p. 194–203.
4. Liu J, Paulsen S, Wang W, Nobel A, Prins J. Mining approximate frequent itemset from noisy data. In: Proceedings of the 5th IEEE International Conference on Data Mining; 2005. p. 721–4.
5. Seppanen JK, Mannila H. Dense itemsets. In: Proceedings of the 10th ACM SIGKDD International Conference on Knowledge Discovery, and Data Mining; 2004. p. 683–8.
6. Liu J, Paulsen S, Sun X, Wang W, Nobel A, Prins J. Mining Approximate frequent itemset in the presence of noise: algorithm and analysis. In: Proceedings of the SIAM International Conference on Data Mining; 2006. p. 405–11.
7. Pei J, Tung AK, Han J. Fault-tolerant frequent pattern mining: problems and challenges. In: Proceedings of the Workshop on Research Issues in Data Mining and Knowledge Discovery; 2001.
8. UCI machine learning repository. http://www.ics.uci.edu/ mlearn/MLSummary.html

Apriori Property and Breadth-First Search Algorithms

Bart Goethals
University of Antwerp, Antwerp, Belgium

Synonyms

Downward closure property; Levelwise search; Monotonicity property

Definition

Given a large database of sets of items, called transactions, the goal of frequent itemset mining is to find all subsets of items, called itemsets, occurring frequently in the database, i.e., occurring in a given minimum number of transactions.

The search space of all itemsets is exponential in the number of different items occurring in the database. Hence, the naive approach to generate and count the frequency of all itemsets over the database can not be achieved within reasonable time. Also, the given databases could be massive, containing millions of transactions, making frequency counting a tough problem in itself.

Therefore, numerous solutions have been proposed to perform a more directed search through the search space, almost all relying on the well known Apriori-property. These solutions can be divided into breadth-first search and depth-first search, of which the first is discussed here.

Historical Background

The original motivation for searching frequent itemsets came from the need to analyze so called supermarket transaction data, that is, to examine customer behavior in terms of the purchased products. Frequent sets of products describe how often items are purchased together. In 1993, Agrawal, Imilienski, and Swami introduced this problem, and proposed a first algorithm to solve it. Shortly after that, in 1994, the algorithm

was improved and named *Apriori*. The main improvement was to exploit the monotonicity property of the frequency of itemsets, later referred to as the *Apriori property*. The same technique was independently proposed by Mannila, Toivonen, and Verkamo, after which both works were combined in one book chapter [1]. Since then, hundreds of improvements and new algorithms have been developed, many of them relying on the breadth-first search strategy as proposed in the Apriori algorithm.

Foundations

A set of items I and a database D of subsets of I, called *transactions*, is given. An *itemset* $I \subseteq I$ is some set of items; its *support* in D is defined as the number of transactions in D that contain all items in I. An itemset is called *frequent* in D if its support in D is greater than or equal to a given minimum support threshold σ. The goal is now, given a minimal support threshold σ and a database D, to find all frequent itemsets in D.

Instead of naively generating and counting all possible itemsets, several collections of *candidate itemsets* are generated iteratively, and their supports computed until all frequent itemsets have been generated. Obviously, the size of a collection of candidate itemsets must not exceed the size of available main memory. Moreover, it is important to generate as few candidate itemsets as possible, since computing the supports of a collection of itemsets is a time consuming procedure. In the best case, only the frequent itemsets are generated and counted. Unfortunately, this ideal is impossible generally, which will be shown later in this section.

The main underlying property exploited by most algorithms is that support is monotone decreasing with respect to extension of an itemset.

Property 1 (Apriori Property) *Given a transaction database D over I, let $X, Y \subseteq I$ be two itemsets. Then,*

$$X \subseteq Y \Rightarrow support(Y) \leq support(X).$$

Input: $\mathcal{D}, \sigma$
Output: $\mathcal{F}(\mathcal{D}, \sigma)$
```
 1:  C_1 := {{i} | i ∈ I}
 2:  k := 1
 3:  while C_k ≠ {} do
 4:      for all transactions (tid,I) ∈ D ← do
 5:          for all candidate sets X ∈ C_k do
 6:              if X ⊆ I then
 7:                  Increment X. support by 1
 8:              end if
 9:          end for
10:      end for
11:      F_k := {X ∈ C_k | X.support ≥ σ}
12:      C_{k+1} := {}
13:      for all X,Y ∈ F_k, such that X[i] = Y[i]
14:          for 1 ≤ i ≤ k − 1, and X[k] < Y[k] do
15:              I := X ∪ {Y[k]}
16:              if ∀ J ⊂ I, |J| = k: J ∈ F_k then
17:                  Add I to C_{k+1}
18:              end if
19:          end for
20:      Increment k by 1
21:  end while
```

Apriori Property and Breadth-First Search Algorithms, Fig. 1 Apriori

Hence, if an itemset is infrequent, all of its supersets must be infrequent. In the literature, this property is also called the monotonicity property, or also the downward closure property, since the set of frequent itemsets is closed with respect to set inclusion. This property is of crucial importance for all frequent itemset mining algorithms as it allows for pruning large parts of the search space. As soon as an itemset is known to be infrequent, none of its supersets has to be considered anymore.

The Apriori Algorithm

For simplicity, assume that items in transactions and itemsets are kept sorted in their lexicographic order unless stated otherwise.

The itemset mining phase of the Apriori algorithm is given in Fig. 1. The notation $X[i]$ is used to represent the ith item in X; the k-*prefix* of a set X is the k-set $\{X[1], \ldots, X[k]\}$, and $\mathcal{F}_k$ denotes frequent k-sets.

The algorithm performs a breadth-first (level-wise) search through the search space of all sets by iteratively generating and counting a collection of candidate sets. More specifically, a set is candidate if all of its subsets are counted and frequent. In each iteration, the collection C_{k+1} of candidate sets of size $k + 1$ is generated, starting with $k = 0$. Obviously, the initial set C_1 consists of all items in $\mathcal{I}$ (line 1). At a certain level k, all candidate sets of size $k + 1$ are generated. This is done in two steps. First, in the *join* step, the union $X \cup Y$ of sets $X, Y \subset \mathcal{F}_k$ is generated if they have the same $k - 1$-prefix (lines 13–15). In the *prune* step, $X \cup Y$ is inserted into C_{k+1} only if all of its k-subsets are frequent and thus, occur in $\mathcal{F}_k$ (lines 16–17).

To count the supports of all candidate k-sets, the database is scanned one transaction at a time, and the supports of all candidate sets that are included in that transaction are incremented (lines 4–7). All sets that turn out to be frequent are inserted into $\mathcal{F}_k$ (line 11).

If the number of candidate sets is too large to remain in main memory, the algorithm can be easily modified as follows. The candidate generation procedure stops and the supports of all generated candidates are counted. In the next iteration, instead of generating candidate sets of size $k + 2$, the remaining candidate $k + 1$-sets are generated and counted repeatedly until all frequent sets of size $k + 1$ are generated and counted.

Although this is a very efficient and robust algorithm, its main drawback lies in its inefficient support counting mechanism. Fortunately, a lot of counting optimizations have been proposed for many different situations.

Optimizations

A lot of other algorithms proposed after the introduction of Apriori retain the same general structure, adding several techniques to optimize certain steps within the algorithm. Since the performance of the Apriori algorithm is almost completely dictated by its support counting procedure, most research has focused on that aspect of the Apriori algorithm. Here, only four out of more than hundreds of improvement proposals are outlined, but at least, these four represent the most influential and largest jumps forward.

Item Reordering

One of the most important optimizations which can be effectively exploited by almost any frequent set mining algorithm, is the reordering of items.

The underlying intuition is to assume statistical independence of all items. Then, items with high frequency tend to occur in more frequent sets, while low frequent items are more likely to occur in only very few sets.

For example, in the case of Apriori, sorting the items in support ascending order improves the distribution of the candidate sets within the used data structure [4]. Also, the number of candidate sets generated during the join step can be reduced in this way. Unfortunately, until now, no results have been presented on an optimal ordering of all items for any given algorithm and only vague intuitions and heuristics are given, supported by practical experiments.

Partition

As the main drawback of Apriori is its slow and iterative support counting mechanism, Savasere et al. [12] proposed the Partition algorithm.

The main novelty in the Partition algorithm, compared to Apriori, is that the database is partitioned into several disjoint parts and the algorithm generates for every part all sets that are relatively frequent within that part. The parts of the database are chosen in such a way that each part fits into main memory, allowing for much more efficient counting mechanisms. Then, the algorithm merges all relatively frequent sets of every part together. This results in a superset of all frequent sets over the complete database, since a set that is frequent in the complete database must be relatively frequent in one of the parts. Finally, the actual supports of all sets are computed during a second scan through the complete database.

Sampling

Another technique to solve Apriori's slow counting is to use sampling as proposed by Toivonen [13].

The presented Sampling algorithm picks a random sample from the database that fits in main memory, then finds all relatively frequent patterns in that sample, and finally verifies the results with the rest of the database. In the cases where the sampling method does not produce all frequent sets, the missing sets can be found by generating all remaining potentially frequent sets and verifying their supports during a second pass through the database. The probability of such a failure can be kept small by decreasing the minimal support threshold. However, for a reasonably small probability of failure, the threshold must be drastically decreased, which can cause a combinatorial explosion of the number of candidate patterns. Nevertheless, in practice, finding all frequent patterns within a small sample of the database can be done very fast using fast in-memory support counting techniques. In the next step, all true supports of these patterns must be counted after which the standard levelwise algorithm could finish finding all other frequent patterns by generating and counting all candidate patterns iteratively. It has been shown that this technique usually needs only one more scan resulting in a significant performance improvement [13].

Concise Representations

If the number of frequent sets for a given database is large, it could become infeasible to generate them all. Moreover, if the database is dense, or the minimal support threshold is set too low, then there could exist a lot of very large frequent sets, which would make sending them all to the output infeasible to begin with. Indeed, a frequent set of size k includes the existence of at least $2^k - 1$ frequent sets, i.e., all of its subsets. To overcome this problem, several proposals have been made to generate only a concise representation of all frequent sets for a given database such that, if necessary, the frequency of a set, or the support of a set not in that representation can be efficiently determined or estimated [2, 5–11].

Key Applications

The Apriori property and the breadth-first search algorithms have broad applications in mining frequent itemsets and association rules. Please refer to the entries of frequent itemset mining and association rules. The Apriori property and the breadth-first search algorithms can be extended to mine sequential patterns. Refer to the entry of sequential patterns. Moreover, they can also be used to tackle other data mining problems such as density-based subspace clustering.

Experimental Results

So far, several hundreds of scientific papers present different techniques and optimizations for frequent set mining and it seems that this trend is keeping its pace. For a fair comparison of some of these algorithms, a contest was organized to find the best implementations in order to understand precisely why and under what conditions one algorithm would outperform another [3]. Although there were no clear winners, one of the rather surprising results of that contest was that the original Apriori algorithm often still performs among the best.

Cross-References

▶ Association Rule Mining on Streams
▶ Closed Itemset Mining and Nonredundant Association Rule Mining
▶ Frequent Itemset Mining with Constraints
▶ Frequent Itemsets and Association Rules

Recommended Reading

1. Agrawal R, Mannila H, Srikant R, Toivonen H, Verkamo A. Fast discovery of association rules. In: Fayyad U, Piatetsky-Shapiro G, Smyth P, Uthurusamy R, editors. Advances in knowledge discovery and data mining. Cambridge, MA: MIT; 1996. p. 307–28.
2. Bayardo J, Roberto J. Efficiently mining long patterns from databases. In: Proceedings of the ACM SIGMOD International Conference on Management of Data; 1998. p. 85–93.
3. Bodon F. A fast apriori implementation. In: Proceedings of the ICDM Workshop on Frequent Itemset Mining Implementations; 2003.
4. Borgelt C, Kruse R. Induction of association rules: apriori implementation. In: Proceedings of the 15th Conference on Computational Statistics; 2002. p. 395–400.
5. Boulicaut JF, Bykowski A, Rigotti C. Free-sets: a condensed representation of boolean data for the approximation of frequency queries. Data Min Knowl Discov. 2003;7(1):5–22.
6. Bykowski A, Rigitti C. A condensed representation to find frequent patterns. In: Proceedings of the 20th ACM SIGACT-SIGMOD-SIGART Symposium on Principles of Database Systems; 2001. p. 267–73.
7. Calders T, Goethals B. Mining all non-derivable frequent itemsets. In: Principles of Data Mining and Knowledge Discovery, 6th European Conference; 2002. p. 74–85.
8. Calders T, Goethals B. Minimal k-free representations of frequent sets. In: Principles of Data Mining and Knowledge Discovery, 7th European Conference; 2003. p. 71–82.
9. Gunopulos D, Khardon R, Mannila H, Saluja S, Toivonen H, Sharma R. Discovering all most specific sentences. ACM Trans Database Syst. 2003;28(2):140–74.
10. Mannila H. Inductive databases and condensed representations for data mining. In: Proceedings of the 14th International Conference on Logic Programming; 1997. p. 21–30.
11. Pasquier N, Bastide Y, Taouil R, Lakhal L. Discovering frequent closed itemsets for association rules. In: Proceedings of the 7th International Conference on Database Theory; 1999. p. 398–416.
12. Savasere A, Omiecinski E, Navathe S. An efficient algorithm for mining association rules in large databases. In: Proceedings of the 21th International Conference on Very Large Data Bases; 1995. p. 432–44.
13. Toivonen H. Sampling large databases for association rules. In: Proceedings of the 22th International Conference on Very Large Data Bases; 1996. p. 134–45.

Architecture-Conscious Database System

John Cieslewicz[1] and Kenneth A. Ross[2]
[1]Google Inc., Mountain View, CA, USA
[2]Columbia University, New York, NY, USA

Synonyms

Architecture-aware database system; Architecture-sensitive database system; Hardware-conscious database system

Definition

Database systems designed with awareness of and a sensitivity to the underlying computer hardware are "architecture-conscious." In an architecture-conscious database system implementation, the performance characteristics of computer hardware guide algorithm and system design.

Historical Background

Database system implementation has been, in varying ways, architecture conscious from the advent of the relational database. For instance, System R [2], an early relational database system prototype included the number of I/Os as a cost metric in its optimizer. At a very high level, the implementers of System R included the characteristics of the underlying hardware in their analysis. This trend has continued with growing attention paid by the database research community to the effects of hardware technology on database performance. Architecture-conscious design took on greater importance as processor speeds improved by four orders of magnitude between 1980 and 2005, while memory latency improved by less than a single order of magnitude. Because of this performance gap, memory accesses, which are central to any database workload, became relatively expensive, requiring database researchers to design database algorithms with this hardware limitation in mind. New computational features have been added to microprocessors, including SIMD instructions, branch prediction, and memory prefetching. These techniques help to improve single threaded performance and instruction level parallelism (ILP), which is the simultaneous processing of instructions from the same thread. In order to take maximal advantage of many of these features, databases must be designed with them in mind.

The introduction of chip multiprocessors (CMP), in particular, has created new challenges and opportunities for improving database performance. Due to problems with power usage,

heat dissipation, and diminishing single threaded performance returns from increasingly complex logic to exploit additional ILP, beginning early in the 2000s chip architects switched their design emphasis from faster clock rates to increased on chip parallelism. It is expected that the degree of parallelism supported by CMPs will continue to increase, making it imperative to design database operators for effective on-chip parallelism. In addition to single threaded performance, achieving good performance on chip multiprocessors requires awareness of thread level parallelism (TLP), the simultaneous processing of instructions from multiple instruction streams. All of these features provide additional opportunities for architecture-conscious database system design.

Foundations

The underlying hardware architecture of which an architecture-conscious database system must be aware includes all computer components. For database systems, the most critical components are persistent storage, the memory hierarchy, and the microprocessor. Research in this field is conducted using architecture simulators and with real hardware. Both are valuable tools, as simulators can test new techniques on hypothetical future hardware designs and using real hardware yields results that are applicable to systems in production now. This article focuses on identifying and explaining the architectural features that have performance implications for databases, and providing high-level design guidelines for database systems implementers using these architectures. A more detailed description of specific database implementation techniques is given in [6].

Persistent Storage

As mentioned above, the impact of storage subsystems has been part of database implementation from the beginning, and many database textbooks describe the costs of various relational operators in terms of the number of I/Os required. Accessing secondary storage incurs significant latency compared with accessing data in memory.

For magnetic disk based storage systems, sequential access is favored because each read and write requires a large, fixed cost (rotational and seek latency) before the relatively fast reading or writing. Because of this property, databases are often optimized for sequential reads and writes to storage, e.g., reading entire buffer pages rather than individual records. Magnetic disks, however, are not the only type of persistent storage. Flash memory storage devices also provide high density persistent storage, but have different properties than magnetic disks. As of 2007, flash memory does not have density comparable to magnetic disks, but the technology is improving. Flash memory, unlike a magnetic disk, has no mechanical parts, uses less power, and supports a higher number of random I/Os per second. These different characteristics require new thinking about the way a database system uses persistent storage. For instance, the read, write, and erase properties of flash memory make is less suitable for update-in-place database operations than magnetic disks. Changes to page layout and logging have been introduced to overcome this difference [11]. Flash memory's support for more random I/Os per second, also leads to changes in cost-performance metrics such as the five-min rule [7].

Main Memory and Cache Optimizations

Memory density has increased while prices have fallen, resulting in affordable systems in which a database's entire working set can be held in memory. For such systems, I/O latency no longer dominates. At the same time, processor speeds have increased at a much faster rate than memory latency, making a data load from main memory relatively more expensive. Computer architects have combated this problem by adding caches to processors. A cache is a small, but very fast memory on the same chip as the processor. The cache provides a processor with fast access to frequently used data (temporal locality) or data that resides near recently used data (spatial locality). When data is found in the cache, it is called a *cache hit*, but when data is not found in the cache this is a *cache miss* and main memory must be accessed. Accessing main memory is a longer latency operation that can stall the processor's pipeline. On database workloads, cache misses for data and instructions have been shown to account for a significant portion of execution time [1]. Therefore, data structures and query execution techniques that make more efficient use of limited cache resources are complementary improvements that reduce cache misses and improve performance.

In addition to improving cache use, the challenge posed by the memory bottleneck can be overcome by overlapping memory latency with computation. Prefetching facilitates that overlapping by attempting to load data into the cache before it is needed. Prefetching can be controlled both in hardware and software. In hardware prefetching, the processor attempts to identify access patters, such as a sequential scan, and load data ahead of when it is needed. In software prefetching, the programmer inserts prefetch instructions into a program to provide the hardware with hints about what data should be loaded into the cache. Software prefetching must be used when there is no pattern to the memory accesses. Chen et al. have improved the performance of tree-based index searches and hash joins by using prefetch instructions [3, 4].

Microarchitecture Optimizations

Conditional branches are a performance hazard on many microarchitectures. A conditional branch represents a control dependency, after which the processor does not immediately know which instruction to execute. On architectures with long pipelines, the outcome of a conditional test is not known for many cycles. In the interim, the processor must either stop issuing instructions or guess the outcome of the condition. This control dependency can limit instruction level parallelism (ILP) because the compiler is limited in its ability to reorder instructions around a branch and the processor is unsure which instructions to issue after a branch instruction. Many modern processors attempt to guess the outcome of the branch using hardware branch prediction, so that instructions continue to be

Architecture-Conscious Database System, Fig. 1 Number of branch mispredictions and execution time for a one-sided range selection with varying selectivity. The *red*, "Conditional Branch," line is an implementation using a conditional branch, while the *green*, "No-Branch," line is an implementation where the conditional branch's control dependency has been replaced with a data dependency [8]

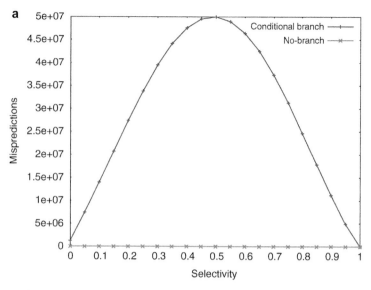

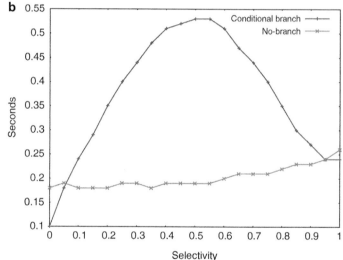

issued in spite of the branch. Branch prediction attempts to identify patterns in the recent history of branch outcomes in order to guess the present branch's outcome correctly. If the branch is predicted correctly, the branch causes no delay in the pipeline. When the prediction is incorrect, the pipeline must be flushed and restarted because invalid instructions have been issued. This cost is the "branch misprediction penalty" and is related to the depth of the pipeline. When there is no pattern to the branching, branch prediction fails up to half the time, and the misprediction penalty degrades performance.

In database workloads, where branches are often data dependent, branch misprediction has a noticeable impact on performance [1]. Figure 1a shows the number of mispredictions of performing a simple range selection on an array of 100 million uniformly distributed four byte integers on a 2.4 GHz Intel Core 2 Duo processor. The graphs in Fig. 1 show data for a selection implementation using a conditional branch and a "no-branch" implementation in which the conditional branch's control dependency has been changed to a data dependency [12]. The maximum number of mispredictions occur in the conditional branch implementation when the selectivity is 0.5 because the branch prediction is wrong half of the time. This high rate of branch misprediction impacts performance, as shown in Fig. 1b. In contrast, the implementation with no conditional branch is less sensitive to the selectivity, with its modest rise in execution time (Fig. 1b) attributed to the cost of writing more output as the selectivity increases. Ross extends the query optimizer's cost model to take the cost of branch misprediction into account and demonstrates techniques for evaluating selection conditions using varying numbers of conditional branches [12].

Modern microarchitectures feature special vector operations, also known as *single instruction multiple data* (SIMD) instructions. These instructions operate on wide registers containing multiple variables, enabling data parallelism. For example, a 128 bit SIMD register can contain 16 *char* data types or four 32-bit integers. A wide range of mathematical, logical, and data movement operations can be applied in parallel to those variables using a single instruction. SIMD instructions can be used to improve the data parallelism present in database execution, resulting in higher performance [14]. For some architectures, such as the Cell Broadband Engine, which are highly optimized for SIMD processing, SIMD-izing database operations is crucial to achieving good performance [9]. SIMD instructions can be used to transform control dependencies into data dependencies. For instance, when using scalar instructions, selection is commonly performed with a comparison operation followed by a conditional branch. In contrast, with SIMD instructions, a selection is performed by applying a SIMD comparison operation to multiple elements in parallel. The result of this operation is a bit-mask that can be used to select the elements that pass the condition using other instructions. Transforming a control dependency into a data dependency can also be accomplished with predicated scalar instructions such as a *conditional move* instruction. Regardless of how the control dependency is eliminated, doing so will improve ILP, particularly if the branch is difficult to predict, as is the case in the example shown in Fig. 1.

On-Chip Parallelism

A chip multiprocessor (CMP) provides multiple on chip hardware thread contexts, often in the form of multiple processor cores on a single die. A processor core's pipeline may be shared by different instruction streams, a design known as simultaneous multithreading (SMT). Some CMPs have both multiple cores and multiple threads per core. As of 2007, the most cores available on a commodity processor was eight, with each core supporting four threads for a total of 32 threads per CMP.

Hardware support for thread level parallelism (TLP) in a CMP differs from a symmetric multiprocessor (SMP) in that a CMP is one processor die with multiple processor cores and an SMP is a system composed of multiple processors sharing one main memory. This difference has two important implications. First, when data is shared among two or more processors in an SMP system, modifications to that data by any processor is coordinated by a cache coherency protocol, which uses the system bus to communicate information about changes to shared data. The system bus is significantly slower than the processors and if a large amount of data is shared among the processors or it is frequently updated, cache coherency overhead can significantly impact performance. In contrast, because all of the cores of a CMP are on the same die, communication between the cores can occur at chip speeds, greatly reducing the coherency overhead of shared data. Second, in a SMP system, although the main memory is shared, each processors' cache hierarchy is separate. In contrast, many CMPs share some cache levels among the processor cores or threads. Because some cache resources may be shared, designing threads to share data in the cache can be beneficial provided that coordinating the sharing does no introduce large amounts of overhead. These differences have implications for architecture-conscious database systems such as the challenge of designing core database operations that take advantage of the on-chip parallelism afforded by chip multiprocessors in a manner that will scale as CMPs provide more hardware thread contexts.

Chip multiprocessors present opportunities and challenges for improving database performance [8]. Finding parallelism in database systems is not difficult. Most database systems have long supported concurrent queries from different users. Other forms of parallelism can be found within a single query or within a single operator. Parallelism alone does not guarantee optimal performance on a CMP. An architecture-conscious database system on a CMP must exploit the on-chip parallelism in a manner that uses the parallel resources to maximum effect. For instance, Cieslewicz and Ross find that balancing parallelism, cache sharing, and thread communication is key to achieving good performance when computing aggregates using a CMP [5]. Chip multiprocessors can also help alleviate some of the problems associated with the memory bottleneck described previously. By having multiple concurrently executing thread contexts on one chip, a stall due to a cache miss in one thread does not stall the entire processor, leading to higher overall processor utilization. Similar to techniques using prefetch instructions described earlier, an on-chip thread can also be used to explicitly load data into the cache for other another thread [13].

Techniques that perform well on a uniprocessor may not be appropriate for a CMP. For instance, Johnson et al. found that work-sharing schemes for in-memory workloads on CMPs actually reduced performance [10]. This was because work-sharing created a bottleneck at the shared work, limiting the total parallelism. An important lesson is that in order to achieve optimal database performance on a CMP, database implementers must carefully evaluate design decisions that may limit parallelism.

Key Applications

Databases are crucial to a wide range of data storage and analysis activities. Optimizing core database operations to take advantage of all of the features and resources available on modern hardware will result in better performance, which in turn has the ability to directly improve the experience of all users of databases.

Future Directions

Adapting database operators to increased on-chip parallelism afforded by CMPs is one open problem. As the amount of on-chip parallelism increases, new bottlenecks and scaling challenges will emerge. Another area for future research includes using non-traditional architectures, such as the Cell Broadband Engine [9] and other future heterogeneous processors, for database operations. As long as computer architecture continues to evolve, architecture-conscious databases will need to adapt to new technologies.

Cross-References

► Cache-Conscious Query Processing

Recommended Reading

1. Ailamaki A, DeWitt DJ, Hill MD, Wood DA. DBMSs on a modern processor: where does time go? In: Proceedings of the 25th International Conference on Very Large Data Bases; 1999. p. 266–77.
2. Chamberlin DD, Astrahan MM, Blasgen MW, Gray JN, King WF, Lindsay BG, Lorie R, Mehl JW, Price TG, Putzolu F, Selinger PG, Schkolnick M, Slutz DR, Traiger IL, Wade BW, Yost RA. A history and evaluation of system R. Commun ACM. 1981;24(10):632–46.
3. Chen S, Ailamaki A, Gibbons PB, Mowry TC. Improving hash join performance through prefetching. ACM Trans Database Syst. 2007;32(3):17.
4. Chen S, Gibbons PB, Mowry TC, Valentin G. Fractal prefetching B+ trees: optimizing both cache and disk performance. In: Proceedings of the ACM SIGMOD International Conference on Management of Data; 2002. p. 157–68.
5. Cieslewicz J, Ross KA. Adaptive aggregation on chip multiprocessors. In: Proceedings of the 33rd International Conference on Very Large Data Bases; 2007. p. 339–50.

6. Cieslewicz J, Ross KA. Database optimizations for modern hardware. Proc IEEE. 2008;96(5):863–78.
7. Graefe G. The five-minute rule twenty years later, and how flash memory changes the rules. In: Proceedings of the Workshop on Data Management on New Hardware; 2007.
8. Hardavellas N, Pandis I, Johnson R, Mancheril N, Ailamaki A, Falsafi B. Database servers on chip multiprocessors: limitations and opportunities. In: Proceedings of the 3rd Biennial Conference on Innovative Data Systems Research; 2007. p. 79–87.
9. Héman S, Nes N, Zukowski M, Boncz P. Vectorized data processing on the cell broadband engine. In: Proceedings of the 3rd Workshop on Data Management on New Hardware; 2007.
10. Johnson R, Hardavellas N, Pandis I, Mancheril N, Harizopoulos S, Sabirli K, Ailamaki A, Falsafi B. To share or not to share? In: Proceedings of the 33rd International Conference on Very Large Data Bases; 2007. p. 351–62.
11. Lee SW, Moon B. Design of flash-based DBMS: an in-page logging approach. In: Proceedings of the ACM SIGMOD International Conference on Management of Data; 2007. p. 55–66.
12. Ross KA. Selection conditions in main memory. ACM Trans Database Syst. 2004(1);29:132–61.
13. Zhou J, Cieslewicz J, Ross KA, Shah M. Improving database performance on simultaneous multithreading processors. In: Proceedings of the 31st International Conference on Very Large Data Bases; 2005. p. 49–60.
14. Zhou J, Ross KA. Implementing database operations using SIMD instructions. In: Proceedings of the ACM SIGMOD International Conference on Management of Data; 2002. p. 145–56.

Archiving Experimental Data

Reagan W. Moore
School of Information and Library Science,
University of North Carolina at Chapel Hill,
Chapel Hill, NC, USA

Synonyms

Preservation; Reference collections

Definition

The archiving of data is the process by which a project preserves both data and representation information that is needed to interpret the data.

Historical Background

An increasing fraction of science research is based on the analysis of large data collections. Experimental data are collected from measurements by sensors from repeatable experiments conducted in laboratories. Observational data are collected from sensors that measure properties of the natural environment. Simulation output files are generated by applications that run on supercomputers. The size of each of these types of research data can be massive, measured in petabytes (thousands of terabytes), such that there is physically not enough room to keep the data on high-speed disk file systems.

The archiving of experimental data traditionally focused on the migration of data from high-performance (more costly) storage systems to the lowest cost media that were available (originally tape-based systems). Files were manually copied from disk onto the tape archive. The IEEE Mass Storage System Reference Model version 5 defines the properties that a tape archive should provide [4]. The properties include the ability to automate retrieval of the file, manage the name space used to identify the file, and manage migration of files between tapes. While these capabilities enable the long-term management of the files, each project was required to manage independently any representation information needed to interpret the file and the meaning of the data. The OAIS model defines representation information that should be provided for preservation of data [6]. This includes information about the source of the data, descriptions of the structures present within the data file, identification of the application that can manipulate the data structure, descriptions of the meaning of the data, and identification of a knowledge community that can interpret the meaning. Archiving of experimental data is successful when the data can be retrieved, interpreted, and manipulated by the project at an unspecified future time.

Foundations

The archiving of experimental data is facilitated through the development of standards for the descriptive terms used to describe meaning and provenance, standards for describing the structures present in the file, and standards for the services that manipulate the data structures:

- Standard semantics. The terms used to describe physical phenomena are created by each discipline. The meaning of these terms evolves over time as a better understanding is developed of the physical phenomena. A preservation environment will need to support evolution of the semantic meaning, through the mapping of prior vocabulary sets to the new vocabulary. This is typically done through the use of ontologies that define logical mappings between terms, spatial correlation of terms to maps, assignment of temporal coordinates to processes, and specification of functional relationships between physical quantities.
- Standard formats. Each file is a linear string of bits, on which structure is imposed. A file format describes how to parse the bit string into structures that can be named. Since the types of observational or experimental or simulation data vary extensively, each community chooses a standard format for specific physical phenomena. As new phenomena are measured, new data formats are created. Thus, a preservation environment needs to manage either characterization of the file formats or migration of the file formats onto new standards. A promising approach to handle data format migration is based on Data Format Description Languages that enable the description of the structure of the file using an XML syntax.
- Standard access methods. Each scientific community defines standard processing steps for manipulating the structures present within the standard data formats, using the standard semantic terms. These standard processing mechanisms can be ported on

top of data grid technology to enable their use within the preservation environment [5]. An alternate approach is to port the required display application to each new operating system technology. This assumes the data can be retrieved from the preservation environment, and a copy is placed on the computer where the emulated application is executed.

The size of science data collections poses a major scalability challenge and strongly impacts the implementation of the data archiving process. The large number of files and the large size of an individual file require support mechanisms that may not be provided by a specific storage system. The science research projects listed below use data grid technology to overcome limitations in current storage technology and to simplify incorporation of new technology within the preservation environment. Data grids are software middleware that insulate the scientific collections from dependencies on current storage technology through the implementation of infrastructure independence [2]. The names used to describe the science data are managed by the data grid independently of the choice of storage system. This makes it possible to ensure consistent naming even when data are distributed across multiple storage systems or migrated from old technology to new technology.

Data grids provide standard operations for manipulating science data. The operations are executed at the remote storage system through interoperability mechanisms that can be applied on any choice of storage. Data grids also provide standard interfaces that enable the use of a preferred access mechanism. The result is the ability to use a particular research group's access interface to manipulate data stored on multiple types of remote storage systems.

The mechanisms provided by data grids to manage science data explicitly handle issues of scale:

- Large file size. An individual file can be terabytes in size. The time required to move such a file over a network to a storage system can be excessive. By using parallel I/O streams, the time can be decreased by multiple factors of two. Effectively, the data file is separated into multiple segments, each of which is transmitted over two to four parallel I/O streams.

- Large file size. The retrieval of a large file, even with parallel I/O streams, may take multiple minutes. For cases where a small subset of the large file is needed, it is more efficient to filter the file at the remote storage system and send only the desired subset back over the network. Data grids support remote manipulation of files through the execution of remote procedures at the location where the file resides.

- Large number of files. Each storage system has a maximum number of files that it is designed to handle. Large collections can exceed the number of files that can be written to a single storage system. To overcome this limitation, data grids aggregate small files into a single larger file called a container. The container is written to the storage system. The data grid maintains information that allows it to track the location in the container of each file. For example, the 2-Micron All Sky Survey [8] astronomy image collection containing five million images was archived on tape by aggregating the images into 147,000 containers. If the images in each container are from the same area on the sky, then retrieval of the container can result in multiple related images becoming accessible, improving bulk access to the collection.

- Large number of files. For the archived files to be useful, descriptive information is needed about each file to support discovery. The descriptive information may be extractable from the file. However, if a discovery request is issued that requires the parsing of every file in the collection, the time to satisfy the request may be exceptionally long. Data grids manage metadata for each file in a metadata catalog. The metadata catalog is stored as tables in a relational database, enabling efficient searches. Data grids can use remote procedures to parse the descriptive metadata from the file and bulk load the metadata into the metadata catalog.

- Large size of collections. The management of integrity across large collections may be viewed as an intractable problem. File integrity can be verified by reading each file, calculating a checksum, and then comparing the checksum with a previous value that was stored in the metadata catalog. If the file has become corrupted, the two checksums will not be equal. If a single tape drive is used to read a petabyte collection, a sustained data transfer rate of 33 MB/s is needed to read the entire collection in a year. Thus, integrity checking of large collections can require the dedication of significant hardware resources. If a problem is detected, a second copy is required to be able to repair the corruption. Data grids support the replication of data onto multiple storage systems that may be located at geographically remote locations. The geographic separation is needed to ensure recovery from natural disasters. Synchronization of replicas is done to verify integrity of the files.

Key Applications

Science disciplines are generating massive collections of experimental, observational, and simulation data. Single projects such as the Southern California Earthquake Center [7] generate and store from a single simulation more than 0.5 petabytes of data describing seismic wave propagation from earthquakes on the San Andreas Fault. The Large Synoptic Survey Telescope (LSST) plans to capture more than 130 petabytes of observational data [3]. The LSST project takes photographs of the sky to track near-earth objects, supernovae, and microlensing events that can provide information on the structure of the Universe. The BaBar high-energy physics experiment moved more than 2 petabytes of experimental data from the Stanford Linear Accelerator in Palo Alto, California, to Lyon, France, for analysis by collaborating physicists [1]. In each case, the data are archived for comparison with future research results.

Future Directions

The ability to validate the trustworthiness of a digital repository is becoming an essential requirement for the archiving of experimental data. When scientific collections are archived, each community defines assessment criteria that they expect the preservation environment to maintain. The assessment criteria may be related to retention and disposition policies, or to time-dependent access controls, or to specifications of required descriptive metadata, or to required access mechanisms for data display and manipulation. An emerging requirement for scientific collections is the characterization of the management policies under which the desired collection properties are enforced.

Rule-based data grids provide the mechanisms needed not only to enforce the application of the collection management policies but also to automate the execution of data management policies. As scientific collections grow to the petabyte size, the labor required to administer the collections can become onerous. This is driven by the use of distributed storage systems to manage the collections. Once the collection resides on multiple types of storage systems, located on multiple administrative domains at geographically remote sites, it becomes very hard to control what is happening. A network router or storage system may be taken down for maintenance, or an operational procedure may change, causing an unexpected result. Data grids provide mechanisms to recover from such problems through the use of replicas, checksums, and synchronization. If problems are detected and repaired through administrator-initiated actions, when the number of detected problems becomes too large, the administrator is no longer able to keep up with the workload.

Rule-based data grids minimize the labor required by the administrator by automating execution of management policies. Management policies are defined that control the preserva-

tion processes that are applied to the collection (e.g., validate checksum, verify presence of required metadata, implement the retention policy). Assessment criteria are specified that evaluate whether the management policies have been correctly applied. In a rule-based data grid, the preservation processes are expressed as sets of micro-services that are executed at the remote storage system. Each micro-service in turn is composed from standard operations that the data grid implements for each type of storage system. Management policies are expressed as sets of rules that control the execution of the micro-services. A rule engine is installed at each remote storage system to ensure that the policies are enforced, no matter which access mechanism is used to interact with the data grid. The assessment criteria are mapped to queries on persistent state information that is generated after the execution of each micro-service. Such a rule-based data grid is capable of monitoring its own operations and verifying the trustworthiness of the digital repository that holds the archived scientific collection [2].

Cross-References

► Data Warehouse
► Disaster Recovery
► Information Lifecycle Management
► Meta data Repository
► Replication

Recommended Reading

1. BaBar B meson high energy physics project. Available at: http://www.slac.stanford.edu/BFROOT/
2. Integrated Rule-Oriented Data System (iRODS). Available at: http://irods.org/
3. Large Synoptic Survey Telescope (LSST). Available at: http://www.lsst.org/lsst//
4. Miller SW. Mass storage reference model special topics. In: Proceedings of the 9th IEEE Symposium on Mass Storage Systems; 1988. p. 3–7.
5. Moore R. Building preservation environments with data grid technology. Am Arch. 2006;69(1):139–58.
6. OAIS, reference model for an open archival information system, ISO standard ISO 14721:2003. Available at: http://nost.gsfc.nasa.gov/isoas/ref_model.html

7. Southern California Earthquake Center (SCEC). Available at: http://www.scec.org/
8. 8-micron all sky survey. Available at: http://www.ipac.caltech.edu/2mass/

Armstrong Axioms

Solmaz Kolahi
University of British Columbia, Vancouver, BC, Canada

Definition

The term *Armstrong axioms* refers to the sound and complete set of inference rules or axioms, introduced by William W. Armstrong [2], that is used to test logical implication of functional dependencies.

Given a relation schema $R[U]$ and a set of functional dependencies Σ over attributes in U, a functional dependency f is logically implied by Σ, denoted by $\Sigma \models f$, if for every instance I of R satisfying all functional dependencies in Σ, I satisfies f. The set of all functional dependencies implied by Σ is called the *closure* of Σ, denoted by Σ^+.

Key Points

Armstrong axioms consist of the following three rules:

- *Reflexivity*: If $Y \subseteq X$, then $X \to Y$.
- *Augmentation*: If $X \to Y$, then $XZ \to YZ$.
- *Transitivity*: If $X \to Y$ and $Y \to Z$, then $X \to Z$.

Note that in the above rules XZ refers to the union of two attribute sets X and Z. Armstrong axioms are *sound* and *complete*: a functional dependency f is derivable from a set of functional dependencies Σ by applying the axioms if and only if $\Sigma \models f$ (refer to [1] for more information).

Cross-References

▶ Functional Dependency
▶ Implication of Constraints

Recommended Reading

1. Abiteboul S, Hull R, Vianu V. Foundations of databases. Reading: Addison-Wesley; 1995.
2. Armstrong W. Dependency structures of data base relationships. In: Proceedings of the IFIP Congress; 1974.

Array Databases

Peter Baumann
Jacobs University, Bremen, Germany

Synonyms

Raster databases

Definition

Array (also called raster or grid): a collection of data items sharing the same data type where each item has a coordinate associated which sits at grid points in a rectangular, axis-parallel subset of the Euclidean space $\mathbf{Z}^d$ for some d > 0 (same as arrays in programming languages).

Array database system: a database system with modeling and query support for multidimensional arrays.

Array query language: a query language allowing declarative retrieval on multidimensional arrays.

Historical Background

Traditionally, all data not tractable with relational tables have been considered "unstructured"; this has long included multidimensional ("n-D") ar-

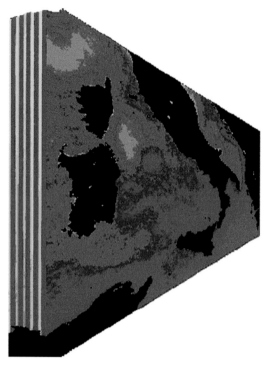

Array Databases, Fig. 1 Retrieval result from a 3-D x/y/t satellite image time series data cube on sea surface temperature; data cube queried contains about 10,000 satellite images (image: rasdaman screenshot, data: NASA, service: DLR)

(integer) pixel coordinates and r, g, and b represent the color values of a pixel. However, SQL does not allow to conveniently express imaging operations, and the linearization of arrays on disk does not preserve spatial proximity of neighboring pixels (not to speak about the tuple maintenance overhead of traditional databases). Effectively, this leads to a performance that is worse by orders of magnitude as the SS-DB [9, 34] benchmark shows (Fig. 2).

The only exception is when the data space (often referred to as "data cube," irrespective of the number of dimensions) is populated very sparsely, i.e., most of the positions contain null values; in this case, enumeration of the few existing data points together with their coordinates comprises an effective compression. While this works well for statistics data (and is being utilized by ROLAP systems), scientific and engineering data often are dense so that such relational storage is not an option.

Storage as BLOBs is not a viable solution either. Again, n-D neighborhood is not preserved, and as the DBMS knows no semantics of the linearized byte string, no operational support is available. Even worse, typically one would want to compress the string thereby effectively disabling any computation of array element positions – the complete array has to be decompressed in main memory prior to accessing its elements.

A first early step beyond BLOBs was accomplished with PICDMS [10] where 2-D retrieval, still procedural and without suitable storage support, has been introduced. A declarative n-D array query language with an algebraic foundation and a suitable architecture, which is implemented and in operational use, has been introduced in [1, 2]. Marathe and Salem present a 2-D database array language [18]. In [3] nested relational calculus is extended with multidimensional arrays, obtaining a model called NCRA and a query language derived from it, AQL. Both seem to be more theoretically motivated and aim in particular at complexity studies for array addressing. Mennis et al. [20] introduce 3-D map algebra, a framework suitable for handling 2-D and 3-D geo raster data.

rays although these have a very regular structure. Arrays form an important, widespread information structure appearing in virtually all domains and effectively make up for a large part of today's "Big Data" as spatiotemporal sensor, image, simulation, and statistics data in science, engineering, business, and beyond.

Due to the inability of traditional database systems to manage such data appropriately database systems have not been used in these domains for handling the mass data, but only the small, metadata which can be mapped conveniently to tables.

Array Databases aim at changing this by offering the same versatile query support on n-D arrays as is available on sets (Fig. 1).

While arrays are not directly supported by SQL, arrays can be mapped to tables on principle. For example, an RGB image can be represented as a table R(x,y,r,g,b) where x and y store the

Array Databases, Fig. 2
Relational representation
(*left and center*) of a 2-D
image matrix (*right*):
reading tuples along *x*
accesses contiguous tuple
lists, whereas cutouts along
y lead to fragmented, slow
tuple access

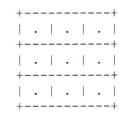

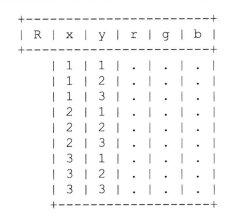

```
+------------------------+
| R  |  x |  y |  r |  g |  b |
+------------------------+
|  1 |  1 |  . |  . |  . |
|  1 |  2 |  . |  . |  . |
|  1 |  3 |  . |  . |  . |
|  2 |  1 |  . |  . |  . |
|  2 |  2 |  . |  . |  . |
|  2 |  3 |  . |  . |  . |
|  3 |  1 |  . |  . |  . |
|  3 |  2 |  . |  . |  . |
|  3 |  3 |  . |  . |  . |
+------------------------+
```

```
+----------+
|  . |  . |  . |
+----------+
|  . |  . |  . |
+----------+
|  . |  . |  . |
+----------+
```

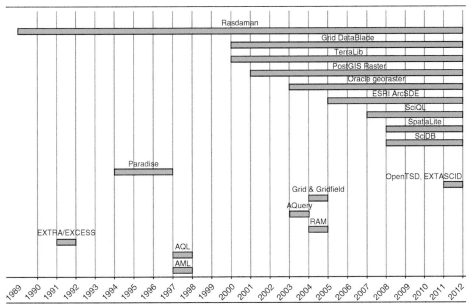

Array Databases, Fig. 3 Early history of Array DBMSs

Only recently Array Databases have found acceptance as a research field within the database community, and support for multidimensional ("n-D") arrays has been recognized as a research challenge. Array DBMSs offer data definition of arrays (usually embedded in a relational context) and operations embedded in a declarative query language (usually SQL). Important implementations, at different stages of evolution, include (in chronological order) rasdaman, PostGIS Raster, SciQL, and SciDB. Relevant non-database approaches include SciHadoop [7], an array front-end to Hadoop, as well as SciSpark, analogously building on

Spark and specialized towards climate analysis. Today, we see first Array Databases in practical use, gradually permeating into science, agencies, and industry; one example is the EarthServer initiative with several services based on data assets exceeding 500 TB at the time of this writing [13].

Figure 3 gives a brief timeline of early Array DBMS research. The rasdaman approach effectively has pioneered the field. Several approaches have emerged subsequently, but have seized to develop further, possibly due to researchers leaving academic world. Later on, it can be observed that an increasing number of array engines have started to emerge – a trend which is continuing today.

Scientific Fundamentals

The collection paradigm in databases encompasses sets, bags, lists, and arrays. Each of these concepts forms a separate, distinct information category – for example, arrays might be emulated, e.g., by nested lists, but this will not lead to usable query concepts (e.g., for slicing an array along an arbitrary axis), nor can this be a basis for efficient, scalable implementations.

Consequently, all database aspects need to be reconsidered for array support, including conceptual modeling (what are appropriate operations?), storage management (how to manage objects spanning many disk blocks, if not several media?), and query evaluation (what are efficient and optimizable processing strategies?).

Algebraic Modeling

Formally, an array a is a function $a: X \rightarrow F$ where X is a finite axis-parallel hypercube in Euclidean space $\mathbf{Z}^d$ for some $d > 0$ and F is some algebra. Let X be the array's *domain*, F its *cell type*, and the individual locations $x \in X$ carrying some value $f \in F$ the array's *cells*.

Following good practice in databases, an array query language should be declarative and safe in evaluation. Declarative in this context means that there is no explicit iteration sequence over an array (or part of it) during evaluation – conceptually, all cells should be inspected simultaneously. This opens up avenues for efficient storage and evaluation patterns, and query optimization in general (see below). Avoiding explicit iterations also keeps queries *safe in evaluation*, i.e., every query is evaluated in a finite number of finite-time steps.

Several algebraic models exist, such as [2, 16, 18]; an attempt to integrate three important models (rasdaman, SciQL, and SciDB) has been started, but discontinued. Array Algebra has been influential as basis of the rasdaman system [2], the OGC WCPS standard [3], and the forthcoming ISO SQL/MDA standard [19, 22]; we

adopt the latter for our discussion. The algebra requires two primitives to express a wide range of practically relevant operations; we inspect them in turn.

The *array* operator creates an array over some given domain, specified as the cross product of axis extents, and defines a position variable takes on each cell location in turn, without any specified iteration order. In the *values* clause, cell values are provided as an expression evaluated for each value of the iteration variable.

Example: "A cutout of array a specified by x interval (100,200) and y interval (100,300)."

```
mdarray   [ x(100:200), y(100:300) ]
elements  a[x,y]
```

In most array query languages this can be abbreviated as

```
a[ x(100:200), y(100:300) ]
```

Figure 4 illustrates graphically subsetting operations from some sample 3-D array.

Aside from copying cells as above, the resulting array can also have new values assigned which may or may not depend on other arrays.

Example: "Array a (of size 1024×768), with intensity reduced by a factor of 2":

```
mdarray   [ x(0:1023), y(0:767) ]
elements  a[x,y] / 2
```

Such operations are frequently abbreviated by applying the operation on hand directly to the array, in the example obtaining:

```
a/2
```

All unary and binary operations on cells can thus be lifted to become array operations. The following example performs a traffic light classification of array values, based on thresholds $t1$ and $t2$:

```
case
   when a > t2 then {255,0,0}
   when a > t1 then {0,255,255}
   else        {0,255,0}
end
```

The aggregate operator combines cell values into one scalar result, similar to set aggregates. Like the array constructor, an iteration domain is

Array Databases, Fig. 4
Various types of subsetting
from an array: trimming
(*left*, which keeps the
original dimension) and
slicing (which reduces the
number of dimensions,
right)

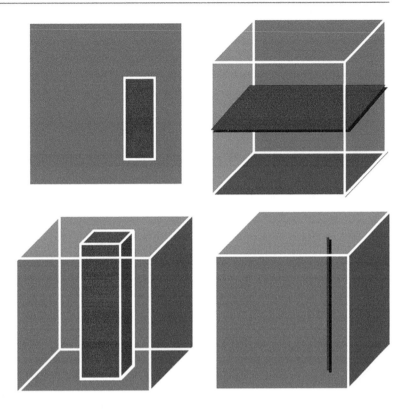

specified as the cross product of axis intervals,
and cell values are inspected in nondeterministic
sequence by evaluating a given expression which
may contain position iterator occurrences. Additionally, the aggregating operation is indicated
which is required to be commutative and associative.

Example: "The sum of all values in *a* (with
same extent, 1024,768, as before)"

```
aggregate +
over  [ x(0:1023), y(0:767) ]
using a[x,y]
```

Common shorthands include *count_cells()*,
add _cells(*a*), minimum, maximum, and
quantifiers.

Often, both primitives appear in combination.
Let us formulate a histogram of 256 buckets over
some 8-bit grayscale array *a*:

```
mdarray  [ b(0:255) ]
elements add_cells( a = b )
```

The inner expression, *a = n*, is an application
of the formerly introduced "lifted" operation: For
every bucket value *b*, array *a* is compared against

it cell by cell, yielding a Boolean array. The result
value is obtained by summing up all equality
occurrences flagged in the Boolean matrix and assigning it to the histogram array cell at position *b*.

Such languages allow formulating statistical
and imaging operations which can be written
analytically without using loops. In [16] it has
been proven that the expressive power of such
array languages is equivalent to relational query
languages with ranking.

Array Query Languages

As arrays in applications typically do not occur
freestanding, but in conjunction with "metadata"
describing geo location, simulation runs which
generated the data set, etc., it makes sense to
combine array management with some "metadata
management" based on conventional technology
like relations with some data management approach, such as relational technology; the only
known exception is SciDB [32] which is a freestanding Array DBMS.

Orchestration of arrays into a relational
database schema can be done in two ways:

adding arrays as a new column type ("*array-as-attribute*") or as table-like constructs ("*array-as-table*"). Most systems – such as rasdaman, Teradata, Oracle, PostGIS Raster – follow an *array-as-attribute* approach, and likewise does the existing array support in ISO SQL [24]. Two models, SciQL [35] and SciDB [32], deviate and follow an *array-as-table* approach (in analogy to ROLAP modeling); for example, in SciQL an RGB image is defined as follows:

```
create array R(
x int dimension [0:1023],
y int dimension[0:767],
r int, g int, b int )
```

Retrieving a 100 × 100 subset is done via a standard set syntax where brackets indicate that a particular column is to be interpreted as array index:

```
select [x], [y], r, g, b
from R
where x between 0 and 99
and y between 0 and 99
```

For comparison, the equivalent rasdaman query is as follows, assuming array attribute *a* in R:

```
select a[ x(0:99), y(0:767) ] from R
```

Disadvantages of the *array-as-table* approach come with scale. It remains unclear how array-as-table scales to millions of arrays, such as with satellite image archives, given that SQL does not foresee iteration over table sets. Also, given that arrays and tables are twins, such huge numbers of arrays will lead to a significant burden for the data dictionary. This conflicts with the generally accepted best practice that the schema should remain stable in face of update traffic; otherwise, clients doing updates would need schema modification rights in the database, which clearly is not desirable.

Finally, an *array-as-attribute* example shows how array joins can be expressed: "the difference between all images in table A and B":

```
select encode( a.img - b.img,
      "image/jpeg" )
from a, b
```

On the fly we have introduced the *encode()* function which delivers the result array in some

(suitable) data format. Dual function *decode()* reads in data coming in some data format. In rasdaman, Media Type (formerly known as MIME Type) identifiers [25] are used to identify the format.

To achieve a tight coupling between relational and array world, a pair of *nest* and *unnest* operations is provided. An array is converted ("unnested") into a table by defining a table where each dimension maps to a table attribute, together with the array attributes. Conversely, a table can be converted into an array through a nest expression where the array columns that will become array dimensions are indicated.

Finally, a language implementation aspect needs to be mentioned. Typically, object-relational capabilities are used to implement arrays. Only two systems (rasdaman and SciQL) undertake the significant effort of a deep integration which requires extending the QL parser and implementing a dedicated query processor. However, such effort pays off in more convenient syntax, more optimization potential due to the better query understanding of the evaluation engine, and more power. The reason is that arrays are not data types (as supported by object-relational extensions), but data type constructors (also called templates) which need to be instantiated with a concrete dimension, domain, and cell type. For example, Oracle GeoRaster and PostGIS Raster are limited to 2-D arrays and a given set of cell types. SciDB, on the other hand, extensively relies on User-Defined Data Types (UDFs) for array operators.

Physical Modeling

In almost all practically relevant scenarios array objects are by orders of magnitude larger than disk blocks. Access patterns are strongly linked to the Euclidean neighborhood of array cells (), therefore it is a main goal to preserve proximity on persistent storage through some suitable spatial clustering. It is common, therefore, to partition n-D arrays into n-D sub-arrays called tiles [1] or chunks [30] which then form the unit of access to persistent storage (Fig. 5).

Obviously, the concrete partitioning chosen greatly affects disk traffic and, hence, overall

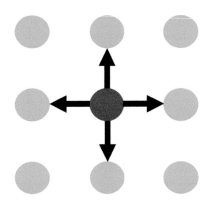

Array Databases, Fig. 5   n-D Euclidean neighborhood of array cells

query performance. By adjusting the partitioning – statically in advance or dynamically at query time – to the workloads, the number of partitions fetched from persistent storage can be minimized (ideally: 1). Challenge is to find a partitioning which supports a given workload. For example, when building x/y/t remote sensing data cubes imagery comes in x/y slices with a thickness of 1 along t. Time series analysis, on the contrary requires cutouts with long t intervals and (possibly) limited spatial x/y extent.

While this principle is generally accepted partitioning techniques vary to some extent. PostGIS Raster [27] allows only 2D x/y tiles and suggests small sizes, like 100×100 pixels. Teradata arrays are limited to less than 64 kB [33]. SciDB offers a two-level partitioning where smaller partitions can be gathered in container partitions. Further, SciDB allows overlapping partitions so that queries requiring adjacent pixels (link in convolution operations) do not require reading the neighboring partitions [31]. In rasdaman, a storage layout sublanguage allows to define partitioning along several strategies [5]. For example, in "directional tiling" ratios of partition edge extents are indicated, rather than absolute sizes; this allows to balance mixed workloads containing, e.g., spatial timeslice extraction and temporal timeseries analysis. In the "area of interest tiling" strategy, hot spots are indicated and the system automatically determines an optimal partitioning (Fig. 6).

To quickly determine the partitions required – a typical range query – some spatial index, such as the R-Tree, proves advantageous. As opposed to spatial (i.e., vector) databases the situation with arrays is relatively simple: the target objects, which have a box structure (as opposed to general polygons), partition a space of known extent. Hence, most spatial indexes can be expected to perform decently.

Often, compression of tiles is advantageous [11]. Still, in face of very large array databases tertiary storage may be required, such as tape robots [29, 30].

Query Evaluation

We briefly sketch the rasdaman query processor as one example of an array engine (Fig. 7). The (relational) set tree is unchanged over standard SQL; the only extension is that the *select* and *where* clause can contain array subexpressions; these are evaluated in separate array subtrees which, from the relational set tree view, act as element processing subtrees (similar to other column data types). Hence, following the *array-as-attribute* paradigm, array processing does not interfere with the overall set tree. The array subtree is evaluated through tile streaming. Leaf nodes read the tiles required for result processing and pass them upwards.

Query Optimization

The above presented separation of concerns leaves much freedom for optimization, parallelization, and, generally, distributed query processing [12].

Algebraic query rewriting gives good results. In rasdaman, 150 rules are applied to every incoming query to see whether a rewrite can gain efficiency. For example, for the query

```
select avg_elements( a.array + b.array )
from a, b
```

an equivalence can be found as shown in Fig. 8. There, the red double arrows indicate expensive tile streams whereas the single-line blue arrows represent negligibly inexpensive scalar transport. The left-hand side requires three expensive operations whereas the right-hand side needs only

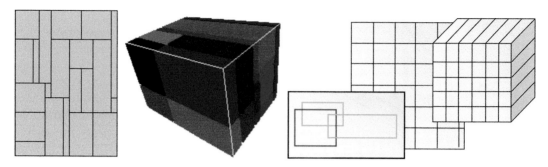

Array Databases, Fig. 6 Sample tiling of 2-D and 3-D arrays (*left* and *center*) and rasdaman tiling strategies *area-of-interest*, *regular*, and *directional* (*right*)

Array Databases, Fig. 7
Sketch of query processing in rasdaman: array subtree embedded in set tree processing

```
select a.img < avg_cells( b.img + c.img )
from    a, b, c
```

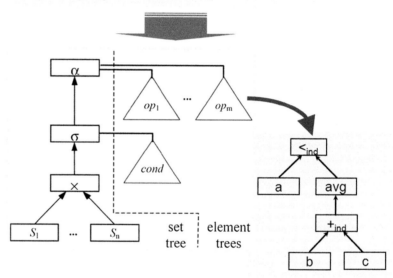

two, so substitution increases performance by 33%. Translated back, the rewritten query reads as follows, obviously more efficient as only scalar numbers are added instead of images:

```
select avg_elements( a.array )
     + avg_elements( b.array )
from    a, b
```

Many array operations are nonblocking, hence streaming can go far up the tree, allowing optimizations like tile-parallel evaluation or sending subtrees to different processing hardware [15]. Blocking operators form natural boundaries, e.g., for subtree shipping to different cores or nodes. Code shipping is important as array queries typ-

ically are CPU-bound, in contrast to the mostly IO-bound SQL table queries.

Architecture

Handling of arrays has been implemented by crafting dedicated architectures from scratch [1, 14], by emulating arrays through relations [35], or through User-Defined Functions (UDFs – i.e., external code linked into a standard relational kernel [32]). Parallel query evaluation is a highly desirable feature due to the massive sizes of arrays, and is supported through various techniques, including a centralized master with parallelized UDFs [32] as in SciDB, dedicated parallelizing algorithms as in EXTASCID [8],

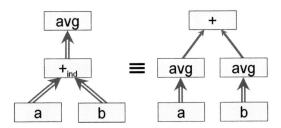

Array Databases, Fig. 8 Algebraic equivalence rule "*avg_elements*(a + b) ≡ *avg_elements*(a) + *avg_elements*(b)"

or peer-to-peer federations of independent nodes without a single point of failure as is the case with rasdaman.

Some systems need to import data in a fixed ASCII format, such as CSV [32], others allow scientific data import directly [1, 14]. Further, it is possible to directly work on scientific archives without prior import of data, which saves data centers from duplicating massive array data sets [13].

Key Applications

As sensor, image, simulation output, and statistics data, arrays are at the heart of science and engineering. Coming both as sampled data (such as satellite, telescope, and microscopy images) and computer generated data sets (such as weather and cosmological simulations) they regularly represent the Big Data challenging current technology. We briefly inspect Earth and Life Sciences to show examples.

Earth Sciences

By far the most important application domain for Array Databases in terms of both citizen impact and commercial relevance is geospatial raster data. These come, among others, as 1-D sensor timeseries, 2-D x/y satellite imagery, 3-D x/y/image timeseries and x/y/z geophysical voxel models, and 4-D x/y/z/t weather data.

Due to the importance of this application domain and due to the fact that it is understood well enough that even domain-specific query languages exist (cf. the OGC Web Coverage Processing Service (WCPS) [4]), a separate entry has

been devoted in this Encyclopedia – see "Geo Raster Databases."

Life Science

In human brain imaging, the research goal is to understand the relations between brain structure and its function. In experiments, human subjects have to perform some mental task while activity parameters such as brain temperature, electrical activity, and oxygen consumption are measured by PET or fMRI CAT scans. The resulting 3-D x/y/z data cubes have, e.g., a resolution of 1 mm and an overall volume in the Megabyte size; these sizing parameters are continuously growing as instruments improve. In a computationally expensive warping operation the brain images get normalized against some chosen standard brain so that organs always sit at known voxel coordinates (Fig. 9 left). In the end a brain image set as large as possible is desired to achieve high statistical significance.

Traditionally, feature search is the only property through standard SQL on structured and semi-structured data; specialized tools can perform search on single images or small sets thereof. With array databases, thousands of experiments each with large image sets can be searched by brain feature. The brain organs can be registered as voxel masks (Fig. 9 right). The query types arising mainly perform standard statistics per brain.

Example: "*A parasagittal view of all scans containing critical Hippocampus activations, TIFF-coded*," whereby positional parameter \$1 denotes the slicing position, \$2 the intensity threshold value, \$3 the confidence, as chosen by a user through some interactive interface:

```
select    encode(
    ht.img[ x($1) ],
    "image/tiff" )
from   HeadTomograms as ht,
    Hippocampus as mask
where   count_true_elements(
    ht.img > $2 and mask.img )
    / count_elements ( mask.img )
> $3
```

Another Life Science example is the analysis of gene expression, i.e., the activity of reading out genes for reproduction in a living body. The

Array Databases, Fig. 9
Human brain activation
map (*left*) and brain organ
masks (*right*)

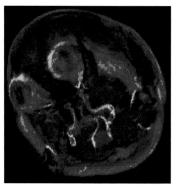
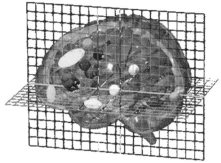

research goal in gene expression analysis is to understand how and when genes express to become manifest in the phenotype (i.e., the organism).

The staining process delivers activity patterns (Fig. 10 left). It is common to combine three gene images into one by randomly assigning them to the red, green, and blue channel, respectively (Fig. 10 top right). Traditionally, diagrams like Fig. 10 bottom right have been constructed by researchers.

Array queries allow searching the resulting 4-D x/y/z/t activity cube and generate, among others, the views researchers are used to. A gene expression database might contain 4-D objects where the first dimension lists the fruit fly genes in some chosen order and the other three spatial dimensions allow addressing into the fruit fly body. Then, a query can slice these 4-D cubes and recombine these slices into the aforementioned RGB overlay.

Example: "*Genes $1, $2, and $3 at age $4, as JPEG images*" where the $ parameters get substituted by actual user input (e.g., through interactive selection menus):

```
select    encode(
  {1c,0c,0c} * e.img[$1,*:*,*:*,$4]
+ {0c,1c,0c} * e.img[$2,*:*,*:*,$4]
+ {0c,0c,1c} * e.img[$3,*:*,*:*,$4],
  "image/jpeg" )
from    EmbryoImages as e
where   e.id=...
```

Future Directions

On conceptual level, a convergence can be seen on array languages towards the array-as-attribute approach. Based on this, standardization of array query languages is seen worthwhile: at the time of this writing, ISO is finalizing SQL/MDA which extends the array stub in the current SQL standard with n-D arrays and declarative operations. The examples given in this entry follow the current SQL/MDA proposal [19]; an overview is available in [22].

In the geo service field, the Open Geospatial Consortium (OGC) Web Coverage Processing Service (WCPS) standard, adopted in 2010, establishes a spatiotemporal query language for regular and irregular grids [3, 4]. Its syntax is aligned with XQuery to allow a later integration, as in Earth Observation metadata typically are provided as XML documents [23].

On implementation side, distributed query processing has highest priority (given that arrays represent "Big Data"). Likewise, optimization and new hardware support is being investigated actively. Physical tuning (such as adaptive array partitioning and distribution) based on approaches like [14] have a high potential for performance improvement, too. Finally, canonical benchmarks are desirable to allow a neutral comparison of systems.

Beyond performance and scalability enhancements of Array DBMSs, two main future research thrusts are currently visible: Tensor Algebra and safe array iterations. Support for general Linear Algebra/Tensor Algebra operations on massive arrays is a requirement coming from statistics, Machine Learning, and other highly relevant Big Data Analytics techniques. As many of these algorithms are inherently iterative, query languages should support safe iterations. In contrast

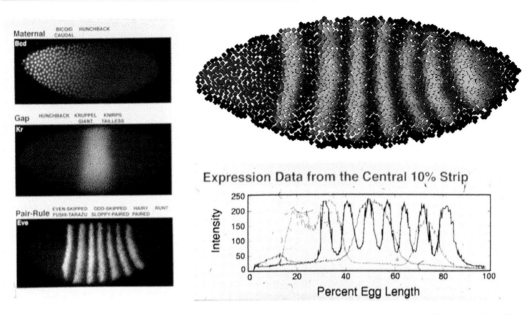

Array Databases, Fig. 10 Gene activity maps for a fruit fly embryo (*left*), overlay into RGB (*top right*), slice aggregating activity over the 10% central strip of a snapshot (*bottom right*) [26]

to known recursion schemes like Magic Sets, these iterations are directed by – often computationally very expensive – numerical criteria. While such capabilities are highly desirable for manifold applications it is a wide open research field as of today how to provide iterative queries – in particular: how to express termination criteria meaningfully – on conceptual level and how to evaluate such queries efficiently.

Experimental Results

As a young discipline, array databases currently lack comprehensive, generally accepted benchmarks; establishing these between Array DBMSs and also comparing with related technology is a field of active research. However, it is safe to say that Array DBMSs can offer enhanced functionality and performance over classic technology – for example, in a direct comparison the SS-DB [9, 34] benchmark finds array databases up to 200x faster than relational technology. While this benchmark is relying on a few handpicked examples, work on a systematic benchmark is ongoing [6, 21]. A comprehensive overview on Array Databases and related technology is being undertaken by the Research Data Alliance [28].

In terms of installations, rasdaman databases today exceed 500 TB and are growing towards crossing the Petabyte frontier [13].

URL to Code

Several projects make their source code freely available; the following is a nonexhaustive list:

- rasdaman: http://www.rasdaman.org (free download, no registration);
- SciDB: http://www.scidb.org (free download, registered users only);
- PostGIS Raster: http://trac.osgeo.org/postgis/wiki/WKTRaster (free download, no registration)

Cross-References

▶ Biomedical Image Data Types and Processing
▶ Digital Elevation Models
▶ Geographic Information System
▶ Geo Raster Data Management

► Image Management for Biological Data
► Online Analytical Processing
► Query Optimization
► Query Rewriting
► Range Query
► Spatial and Spatiotemporal Data Models and Languages
► Spatial Data Analysis
► Spatial Data Mining
► Spatial Data Types
► Spatial Indexing Techniques

Recommended Reading

1. Baumann P. On the management of multidimensional discrete data. VLDB J. 1994;4(3):401–44. Special Issue on Spatial Database Systems.
2. Baumann P. A database array algebra for spatio-temporal data and beyond. In: Proceedings of the 5th Workshop on Next Generation Information Technologies and Systems; 1999. p. 76–93.
3. Baumann P. The OGC web coverage processing service (WCPS) standard. GeoInformatica. 2010;14(4):447–79.
4. Baumann P. OGC web coverage processing service (WCPS) language interface standard. OGC document 08-068r2; 2010a.
5. Baumann P, Feyzabadi S, Jucovschi C. Putting pixels in place: a storage layout language for scientific data. In: Proceedings of the IEEE ICDM Workshop on Spatial and Spatiotemporal Data Mining; 2010b. p. 194–201.
6. Baumann P, Stamerjohanns H. Benchmarking large arrays in databases. In: Proceedings of the Workshop on Big Data Benchmarking; 2012. p. 94–102.
7. Buck J, Watkins N, LeFevre J, Ioannidou K, Maltzahn C, Polyzotis N, Brandt SA. SciHadoop: array-based query processing in Hadoop. In: Proceedings of the High Performance Computing, Networking, Storage and Analysis Super Computing; 2011. p. 66:1–66:11.
8. Cheng Y, Rusu F. Astronomical data processing in EXTASCID. In: Szalay A, Budavari T, Balazinska M, Meliou A, Sacan A editors. Proceedings of the 25th International Conference on Scientific and Statistical Database Management; 2013. Article 47. https://doi.org/10.1145/2484838.2484875.
9. Cheng Y, Rusu F. Formal representation of the SS-DB benchmark and experimental evaluation in EX-TASCID. Distrib Parallel Databases. 2013;33(3):277. https://doi.org/10.1007/s10619-014-7149-7.
10. Chock M, Cardenas A, Klinger A. Database structure and manipulation capabilities of a picture database management system (PICDMS). IEEE ToPAMI. 1984;6(4):484–92.
11. Dehmel A. A compression engine for multidimensional array database systems. PhD thesis, TU München; 2002.
12. Dumitru A, Merticariu V, Baumann P. Exploring cloud opportunities from an array database perspective. In: Proceedings of the ACM SIGMOD Workshop on Data Analytics in the Cloud; 2014.
13. EarthServer: The EarthServer Initiative. www.earthserver.eu. Seen 12 Apr 2017.
14. Furtado P, Baumann P. Storage of multidimensional arrays based on arbitrary tiling. In: Proceedings of the International Conference on Data Engineering; 1999. p. 328–36.
15. Hahn K, Reiner B. Parallel query support for multidimensional data: inter-object parallelism. In: Proceedings of the 13th International Conference on Database and Expert Systems Applications; 2002.
16. Libkin L, Machlin R, Wong L. A query language for multidimensional arrays: design, implementation and optimization techniques. In: Proceedings of the ACM SIGMOD International Conference on Management of Data; 1996. p. 228–39.
17. Machlin R. Index-based multidimensional array queries: safety and equivalence. In: Proceedings of the 26th ACM SIGACT-SIGMOD-SIGART Symposium on Principles of Database Systems; 2007.
18. Marathe A, Salem K. A language for manipulating arrays. In: Proceedings of the 23th International Conference on Very Large Data Bases; 1997. p. 46–55.
19. Melton J, Baumann P, Misev D. ISO/IEC 9075–15 SQL MDA (multi-dimensional arrays).
20. Mennis J, Viger R, Tomlin CD. Cubic map algebra functions for spatio-temporal analysis. Cartogr Geogr Inf Sci. 2005;32(1):17–32.
21. Merticariu G, Misev D, Baumann P. Measuring storage access performance in array databases. In: Proceedings of the 7th Workshop on Big Data Benchmarking; 2016.
22. Misev D, Baumann P. Extending the SQL array concept to support scientific analytics. In: Proceedings of the Scientific and Statistical Database Management; 2014. p. 10:1–10:11.
23. N.n.: ISO/IEC 19139 XML schema, http://www.isotc211.org/2005/gmd/. Seen 29 July 2014.
24. N.n.: ISO/IEC 9075–1 SQL Foundation.
25. N.n.: Multipurpose internet mail extensions (MIME) part one: format of internet message bodies, https://tools.ietf.org/html/rfc2045. Seen 12 Apr 2017.
26. Pisarev A, Poustelnikova E, Samsonova M, Baumann P. Mooshka: a system for the management of multidimensional gene expression data in situ. Inf Syst. 2003;28(4):269–85.
27. PostGIS: PostGIS Raster manual. Seen 29 July 2014.
28. RDA: Array Database Assessment Working Group. https://www.rd-alliance.org/groups/array-database-working-group.html. Seen 12 Apr 2017.
29. Reiner B, Hahn K. Hierarchical storage support and management for large-scale multidimensional array database management systems. In: Proceedings of

the 13th International Conference on Database and Expert Systems Applications; 2002.

30. Sarawagi S, Stonebraker M. Efficient organization of large multidimensional arrays. In: Proceedings of the International Conference on Data Engineering; 1994. p. 328–36.

31. Soroush E, Balazinska M, Wang D. ArrayStore: a storage manager for complex parallel array processing. In: Proceedings of the ACM SIGMOD International Conference on Management of Data; 2011. p. 253–64.

32. Stonebraker M, Brown P, Poliakov A, Raman S. The architecture of SciDB. In: Proceedings of the 23rd International Conference on Scientific and Statistical Database Management; 2011. p. 1–16.

33. Teradata: User-Defined Data Type, ARRAY Data Type, and VARRAY Data Type Limits. Seen 29 July 2014.

34. XLDB: Science Benchmark. http://www.xldb.org/science-benchmark/. Seen 12 Apr 2017.

35. Zhang Y, Kersten M L, Ivanova M, Nes, N. SciQL, bridging the gap between science and relational DBMS. In: Desai BC, Cruz IF, Bernardino J, editors. Proceedings of the 15th Symposium on International Database Engineering and Applications; 2011. p. 124–33.

Association Rule Mining on Streams

Philip S. Yu[1] and Yun Chi[2]
[1]Computer Science Department, University of Illinois at Chicago, Chicago, IL, USA
[2]NEC Laboratories America, Cupertino, CA, USA

Definition

Let $I = \{i_1, \ldots, i_m\}$ be a set of items. Let S be a stream of transactions in a sequential order where each transaction is a subset of I. For an *itemset* X, which is a subset of I, a transaction T in S is said to *contain* the itemset X if $X \subseteq T$. The *support* of X is defined as the fraction of transactions in S that contain X. For a given support threshold $s\%$, X is *frequent* if the support of X is greater than or equal to $s\%$, i.e., if at least $s\%$ transactions in S contain X. For a given *confidence* threshold $c\%$, an *association rule* $X \Rightarrow Y$ holds if $X \cup Y$ is frequent and at least $c\%$ of transactions in S that contain X also contain Y. The problem of association rule mining on streams is to discover all association rules that hold in a stream of transactions.

Historical Background

In 1993, Rakesh Agrawal et al. [1] proposed the framework for association rule mining. Since this seminal work, a lot of research work has been done to improve the efficiency of association rule mining algorithms, to extend the definition of associations rule, and to apply association rule mining to other types of data such as time sequence data and structured data such as trees and graphs. On the other hand, research on data streams started around 2000 when several data stream management systems were originated (e.g., the Brandeis AURORA project, the Cornell COUGAR project, and the Stanford STREAM project) to solve new challenges in applications such as network traffic monitoring, transaction data management, Web click streams monitoring, sensor networks, etc. Because association rule mining plays an important role in these data stream applications, along with the development of data stream management systems, developing association rule mining algorithms for data streams has become an important research topic.

Foundations

Two Subproblems

Algorithms for association rule mining usually consist of two steps. The first step is to discover frequent itemsets where all frequent itemsets that meet the support threshold are discovered. The second step is to derive association rules where based on the frequent itemsets discovered in the first step, the association rules that meet the confidence criterion are derived. Because the second step, deriving association rules, can be solved efficiently in a straightforward manner, most of researches focus mainly on the first step, i.e., how to efficiently discover all frequent itemsets in data streams. Therefore, in the rest of this entry, the focus will be on frequent itemset mining in data streams.

Key Challenges

Frequent itemset mining in a non-streaming setting is already a challenging problem. For example, due to combinatorial explosion, there may be a huge number of frequent itemsets, and the main challenge is how to efficiently enumerate, discover, and store frequent itemsets. Data streams, because of their unique features, pose many further challenges to frequent itemset mining. Some of these new challenges are described as the following:

- *Single access of data* In data streams, data are arriving continuously with high speed and in large volume. As a consequence, in many cases it is impractical to store all data in persistent media; in other cases, it is too expensive to (randomly) access data multiple times. The challenge is to discover frequent itemsets while the data can only be assessed once.
- *Unbounded data* Another feature of data streams is that data are unbounded. In comparison, storage that can be used to discover or maintain the frequent itemsets is limited. A consequence of unbounded data is that whether an itemset is frequent depends on time. The challenge is to use limited storage to discover dynamic frequent itemsets from unbounded data.
- *Real-time response* Because data stream applications are usually time critical, there are requirements on response time. For some restricted scenarios, algorithms that are slower than the data arriving rate are useless. The challenge is therefore to efficiently mine frequent itemsets in real time.

Data Models

Compared with finite data in traditional databases, data streams are unbounded and the number of transactions increases with time. Due to these characteristics, different data models have been proposed.

The first data model is an accumulative model. In this model, all data in the range from the beginning to the current time are considered in an equal fashion. Therefore, frequent itemsets are defined on the accumulated data where new data are appended continuously as time grows.

The second data model is based on a sliding window. That is, although the whole data stream is unbounded, the frequent itemsets are defined based on the most recent data that fall within a temporal sliding window whose ending time is the current time. One justification for such a sliding-window model is that due to concept drifts, data distribution in streams may change with time, and very often people are interested in the most recent patterns.

The third data model falls in between the first two models – while all data are considered in frequent itemset mining, they are weighted differently according to a predefined weighting function. A very commonly used weighting function is the exponentially decaying function. That is, for a transaction at time τ, its weight is $\alpha(\tau) = \exp.(\tau - t)$, where t is the current time.

Algorithm Types

Based on the mining results, existing frequent itemsets mining algorithms on data streams can be roughly divided into two categories: *exact* mining algorithms and *approximate* mining algorithms.

Exact mining algorithms provide as results all the frequent itemsets in the data streams together with their accurate supports. Usually the focus of exact algorithms is to efficiently update frequent itemsets when new transactions arrive and (in the sliding-window data model) when old transactions expire.

Approximate mining algorithms, on the other hand, focus more on finite memory usage and single access of data, at the cost of the accuracy of the mining results. Approximate algorithms target low false-positive rate, zero, or low false-negative rate, and tight error bounds on the estimation for the supports of the frequent itemsets.

Representative Algorithms

Cheung et al. [6, 7] proposed algorithms *FUP* and FUP_2 for incrementally updating frequent itemsets. Thomas et al. [13] presented a similar algorithm. Both Cheung's and Thomas's algorithms assume batch updates and take advantage of the relationship between the original database (*DB*) and the incrementally changed transactions (*db*). *FUP* is similar to the well-known Apriori algorithm [1], which is a multiple-step algorithm. The key observation of *FUP* is that by adding *db* to *DB*, some previously frequent itemsets will remain frequent and some previously infrequent itemsets will become frequent (these itemsets are called *winners*); at the same time, some previously frequent itemsets will become infrequent (these itemsets are called *losers*). The key technique of *FUP* is to use information in *db* to filter out some winners and losers and therefore reduce the size of candidate set in the Apriori algorithm. Because the performance of the Apriori algorithm relies heavily on the size of candidate set, *FUP* improves the performance of Apriori greatly. FUP_2 extended *FUP* by allowing deleting old transactions from a database as well. Therefore, *FUP* is restricted to the accumulative data model while FUP_2 can be used on the sliding-window data model as well. The algorithm proposed by Thomas et al. is similar to FUP_2 except that in addition to frequent itemsets, a negative border is maintained. In the algorithm, the frequent itemsets in *db* are mined first. At the same time, the counts of frequent itemsets (and itemsets on the negative border) in *DB* are updated. Then based on the change of the frequent itemsets in *DB*, the negative border in *DB*, and the frequent itemsets in *db*, the frequent itemsets in the updated database are computed with a possible scan of the updated database. Because the updated database is scanned at most once, Thomas's algorithm has very good performance. Thomas's algorithm can be used for both the accumulative and the sliding-window data models. In addition, *FUP*, FUP_2, and Thomas's algorithm all fall into the category of exact mining algorithms.

Veloso et al. [14] proposed an algorithm *ZIGZAG* for mining frequent itemsets in evolving databases. Later, Otey et al. [11] extended *ZIGZAG* into parallel and distributed algorithms. *ZIGZAG* is similar to Cheung's and Thomas's algorithms in that it achieves its speedup by using the relationship between *DB* and *db*. However, *ZIGZAG* has many distinct features. First, *ZIGZAG* mainly used *db* to speed up the support counting of frequent itemsets in the updated database, and it does not discover the frequent itemsets in *db* itself. As a result, for a given minimum support, *ZIGZAG* can handle batch update with arbitrary block size. Second, *ZIGZAG* adapts the techniques proposed in the *GENMAX* algorithm [9] and in each update only maintains *maximal* frequent itemsets. Because the information on maximal frequent itemsets and their supports is not enough to generate association rules (because the support information of some non-maximal frequent itemsets may be missing), a second step is used in *ZIGZAG* in which the updated database is scanned to discover all frequent itemsets and their supports.

Chi et al. [5] developed an algorithm, *Moment*, to mine *closed* frequent itemsets over data stream sliding windows. In this work the authors introduced a compact data structure, called *closed enumeration tree* (CET), to maintain a dynamically selected set of itemsets over a sliding window. The selected itemsets contain a boundary between closed frequent itemsets and the rest of the itemsets. Concept drifts in a data stream are reflected by boundary movements in the CET and can be efficiently captured. Both *ZIGZAG* and *Moment* are exact mining algorithms that use the sliding-window data model.

Charikar et al. [3] presented a one-pass algorithm, *Count Sketch*, that returns most frequent *items* whose frequencies satisfy a threshold with high probabilities. Manku et al. [10] developed a randomized algorithm, the *Sticky Sampling* algorithm, and a deterministic algorithm, the *Lossy Counting* algorithm, for maintaining frequent *items* over a data stream where for a given time *t*, the frequent items are defined over the *entire* data stream up to *t*. The algorithms guarantee no false negative and a bound on the error of estimated frequency (the guarantees are in a probabilistic sense for the randomized algo-

Association Rule Mining on Streams, Table 1 The categorization of the representative algorithms according to their data models and algorithm types

	Data model		
	Accumulative	Sliding window	Weighted
Exact mining algorithm	FUP [6]	FUP_2 [7] Thomas's [13] $ZIGZAG$ [11, 14] $Moment$ [4]	
Algorithm mining algorithm	$Count\ Sketch$ [3] $Sticky$ $Sampline$ [10] $Lossy$ $Counting$ [10]	$FTP\text{-}DS$ [12]	$estDec$ [2] Giannella's [8]

rithm). The *Lossy Counting* algorithm is extended to handle frequent *itemsets*, where a trie is used to maintain all frequent itemsets and the trie is updated by batches of transactions in the data stream. The algorithms of Manku et al. offer a tunable compromise between memory usage and error bounds. *Count Sketch*, *Sticky Sampling*, and *Lossy Counting* are all approximate mining algorithms that use the accumulative data model.

Teng et al. [12] presented an algorithm, *FTP-DS*, that mines frequent temporal patterns from data streams of itemsets. *FTP-DS* is an approximate mining algorithm that uses the sliding-window data model. Chang et al. [2] presented an algorithm, *estDec*, that mines recent frequent itemsets where the frequency is defined by an aging function. Giannella et al. [8] proposed an approximate algorithm for mining frequent itemsets in data streams during arbitrary time intervals. An in-memory data structure, *FP-stream*, is used to store and update historic information about frequent itemsets and their frequency over time, and an aging function is used to update the entries so that more recent entries are weighted more. Both *estDec* and Giannella's algorithm are approximate mining algorithms on weighted transactions. However, Giannella's algorithm further provides different error levels for data at multiple time granularities.

The above representative algorithms are summarized in Table 1. For a more detailed survey on algorithms for frequent itemset mining over data streams, refer to a comprehensive survey by Cheng et al. [4].

Key Applications

For most data stream applications, there are needs for mining frequent patterns and association rules from data streams. Some key applications in various areas are listed in the following:

- *Performance monitoring* Monitors network traffic and performance and detects abnormality and intrusion
- *Transaction monitoring* Monitors transactions in retail stores, ATM machines, and financial markets
- *Log record mining* Mines patterns from telecommunication calling records, Web server log, etc.
- *Sensor network mining* Mines patterns in streams coming from sensor networks or surveillance cameras

Experimental Results

For all the algorithms presented in this entry, there are supporting experimental studies in the corresponding references. Commonly compared performance metrics are memory usage, speed of the mining process, and (for the approximate algorithms) errors such as false-positive rate, false-negative rate, and support errors for the frequent itemsets.

Data Sets

A Linux version of the synthetic data generator originally developed by Agrawal et al. [1] is available at http://miles.cnuce.cnr.it/~palmeri/datam/DCI/datasets.php.

Some other synthetic and real-life data sets are available from the Frequent Itemset Mining Dataset Repository at http://fimi.cs.helsinki.fi/data/.

Furthermore, pointers to some related data sets are available at the KDnuggets Web site http://www.kdnuggets.com/datasets/.

Cross-References

▶ Approximation of Frequent Itemsets
▶ Association Rule Mining on Streams
▶ Change Detection on Streams
▶ Closed Itemset Mining and Nonredundant Association Rule Mining
▶ Continuous Queries in Sensor Networks
▶ Data Aggregation in Sensor Networks
▶ Data Estimation in Sensor Networks
▶ Data Mining
▶ Data Sketch/Synopsis
▶ Data Stream
▶ Frequent Items on Streams
▶ Frequent Itemsets and Association Rules
▶ Incremental Computation of Queries
▶ One-Pass Algorithm
▶ Pattern-Growth Methods
▶ Randomization Methods to Ensure Data Privacy
▶ Real-Time Transaction Processing
▶ Sensor Networks
▶ Stream Mining
▶ Stream Models
▶ Streaming Applications
▶ Trie
▶ Web Services
▶ Web Transactions

Recommended Reading

 1. Agrawal R, Imielinski T, Swami A. Mining association rules between sets of items in large databases. In: Proceedings of the ACM SIGMOD International Conference on Management of Data; 1993. p. 207–16.
 2. Chang JH, Lee WS. Finding recent frequent itemsets adaptively over online data streams. In: Proceedings of the 9th ACM SIGKDD International Conference on Knowledge Discovery and Data Mining; 2003. p. 487–92.
 3. Charikar M, Chen K, Farach-Colton M. Finding frequent items in data streams. In: Proceedings of the 29th International Colloquium on Automata, Languages and Programming; 2002. p. 693–703.
 4. Cheng J, Ke Y, Ng W. A survey on algorithms for mining frequent itemsets over data streams. Knowl Int Syst. 2008;16(1):1–27.
 5. Chi Y, Wang H, Yu PS, Muntz RR. Catch the moment: maintaining closed frequent itemsets in a data stream sliding window. Knowl Inf Syst. 2006;10(3): 265–94.
 6. Cheung DW, Han J, Ng V, Wong CY. Maintenance of discovered association rules in large databases: an incremental updating technique. In: Proceedings of the 12th International Conference on Data Engineering; 1996. p. 106–14.
 7. Cheung DW, Lee SD, Kao B. A general incremental technique for maintaining discovered association rules. In: Proceedings of the 5th Intereational Conference on Database Systems for Advanced Applications; 1997. p. 185–94.
 8. Giannella C, Han J, Pei J, Yan X, Yu PS. Mining frequent patterns in data streams at multiple time granularities. In: Kargupta H, Joshi A, Sivakumar K, Yesha Y, editors. Data mining: next generation challenges and future directions. AAAI; 2004.
 9. Gouda K, Zaki MJ. Efficiently mining maximal frequent itemsets. In: Proceedings of the 1st IEEE Intereational Conference on Data Mining; 2001. p. 163–70.
10. Manku G, Motwani R. Approximate frequency counts over data streams. In: Proceedings of the 28th International Conference on Very Large Data Bases; 2002. p. 346–57.
11. Otey ME, Parthasarathy S, Wang C, Veloso A, Meira W Jr. Parallel and distributed methods for incremental frequent itemset mining. IEEE Trans Syst Man Cybern B. 2004;34(6):2439–50.
12. Teng W-G, Chen M-S, Yu PS. A regression-based temporal pattern mining scheme for data streams. In: Proceedings of the 29th International Conference on Very Large Data Bases; 2003. p. 98–104.
13. Thomas S, Bodagala S, Alsabti K, Ranka S. An efficient algorithm for the incremental updation of association rules in large databases. In: Proceedings of the 3rd International Conference on Knowledge Discovery and Data Mining; 1997. p. 263–6.
14. Veloso A, Meira Jr W, de Carvalho M, Pôssas B, Parthasarathy S, Zaki MJ. Mining frequent itemsets in evolving databases. In: Proceedings of the SIAM International Conference on Data Mining; 2002.

Association Rules

Jian Pei
School of Computing Science, Simon Fraser
University, Burnaby, BC, Canada

Definition

Let I be a set of *items*, where each item is a literal.
A *transaction* $T \subseteq I$ is a subset of I. Association
rules are defined on a set of transactions T.

An association rule R is in the form of $X \to Y$,
where X and Y are two sets of items, that is,
$X, Y \subseteq I$. R is associated with two measures, the
support sup(R) and the confidence *conf (R)*. The
support *sup(R)* is the probability that X appears in
a transaction in T. The confidence *conf (R)* is the
conditional probability that when X appears in a
transaction, Y also appears.

Historical Background

The concept of association rules was firstly proposed by Agrawal et al. [1] for market basket
analysis. A well-known illustrative example of
association rules is "Diaper $\to$ Beer" which can
be explained by the fact that when dads buy
diapers for their babies, they also buy beer at the
same time for their weekends game watching.

Apriori, an efficient algorithm for mining association rules, was developed by Agrawal and
Srikant [2], while the similar idea was explored
by Mannila et al. [3]. There have been many studies trying to improve the efficiency of Apriori (refer to entry "▶ Apriori Property and Breadth-First
Search Algorithms"). A pattern-growth approach
for mining frequent itemsets without candidate
generation was developed by Han et al. [4] and
has been further improved by many studies since
2001 (see entry "▶ Pattern-Growth Methods").

To remove redundancy in association rules,
Pasquier et al. [5] proposed the notion of frequent
closed itemsets using formal concept analysis.
Several efficient algorithms have been developed

(refer to entry "▶ Closed Itemset Mining and
Nonredundant Association Rule Mining").

Association rules have been extended in
several ways, such as sequential patterns and
sequential association rules (refer to entry
"▶ Sequential Patterns"), spatial association
rules [6], cyclic association rules [7], negative
association rules [8], intertransaction association
rules [9], multilevel generalized association rules
[10, 11], and quantitative association rules (refer
to entry "▶ Quantitative Association Rules").

In addition to support and confidence, some
other interestingness measures for association
rules were explored, such as [12, 13].

Foundations

Let I be a set of *items*, where each item is a literal.
An *itemset X* is a subset of items, that is, $X \subseteq I$. A
transaction $T \subseteq I$ consists of a *transaction id* and
an itemset. A *transaction database* T is a multiset
of transactions.

An association rule R is in the form of $X \to Y$,
where X and Y are two itemsets, that is, $X, Y \subseteq I$.
R is associated with two measures: the *support*
sup(R) and the confidence *conf (R)*. The support
sup(R), given by $sup(R) = Pr(X)$, is the probability that X appears in a transaction in T. The confidence *conf(R)*, given by $conf(R) = \frac{\sup(X \cup Y)}{\sup(X)} =$
$Pr(Y|X)$, is the conditional probability that when
X appears in a transaction, Y also appears.

Given a transaction database T, a minimum
support threshold *minsup*, and a minimum confidence threshold *minconf*, the problem of *association rule mining* is to find the complete
set of association rules whose supports are at
least *minsup* and whose confidences are at least
minconf.

Association rules can be mined in two steps. In
the first step, the complete set of frequent itemsets
are identified. An itemset is called *frequent* if
$Pr(X) \geq minsup$. In the second step, frequent
itemsets are used to generate association rules.

More often than not, to make association rule
mining interesting, a user may specify a *minimum support threshold min_sup* and a *minimum
confidence threshold min_conf*. Then, only the

association rules whose supports and confidence pass those thresholds, respectively, should be returned. Alternatively, in some situations, a user may want to find the top-k association rules with the largest support and/or confidence. Some other kinds of constraints can also be specified. Such thresholds and constraints may be used by some association rule mining methods to speed up the mining procedure.

Key Applications

Association rules have been extensively mined and used in many applications. For example, mining association rules about customers' market baskets helps to identify the products that customers like to purchase together, or some products that may trigger the purchases of some other products. Such information can help business in one way or another. For example, knowing that the purchase of diapers may potentially lead to the purchase of beer, a store can put beer beside diapers so that the sales of beer can be boosted.

Association rules are also mined on biological data and clinic data. For example, mining the association rules among symptoms and disease can help to diagnose diseases.

An important application of association rules is to construct classifiers using association rules. Technically, association rules among features and the target class labels can be mined and the rules can be used to make prediction on cases with unknown class labels. Research has found that associative classifiers, classifiers using association rules, are accurate and highly understandable in a few applications such as those with many features [14–16].

Cross-References

▶ Approximation of Frequent Itemsets
▶ Apriori Property and Breadth-First Search Algorithms
▶ Closed Itemset Mining and Nonredundant Association Rule Mining
▶ Data Mining

▶ Emerging Patterns
▶ Frequent Itemset Mining with Constraints
▶ Frequent Itemsets and Association Rules
▶ Pattern-Growth Methods
▶ Quantitative Association Rules
▶ Sequential Patterns

Recommended Reading

1. Agrawal R, Imielinski T, Swami A. Mining association rules between sets of items in large databases. In: Proceedings of the ACM SIGMOD Interntional Conference on Management of Data; 1993. p. 207–16.
2. Agrawal R, Srikant R. Fast algorithms for mining association rules. In: Proceedings of the 20th International Conference on Very Large Data Bases; 1994. p. 487–99.
3. Mannila H, Toivonen H, Verkamo AI. Efficient algorithms for discovering association rules. In: Proceedings of the AAAI 1994 Workshop Knowledge Discovery in Databases; 1994. p. 181–92.
4. Han J, Pei J, Yin Y. Mining frequent patterns without candidate generation. In: Proceedings of the ACM SIGMOD International Conference on Management of Data; 2000. p. 1–12.
5. Pasquier N, Bastide Y, Taouil R, Lakhal L. Discovering frequent closed itemsets for association rules. In: Proceedings of the 7th International Conference on Database Theory; 1999. p. 398–416.
6. Koperski K, Han J. Discovery of spatial association rules in geographic information databases. In: Proceedings of the ACM SIGMOD International Conference on Management of Data; 1995. p. 47–66.
7. Özden B, Ramaswamy S, Silberschatz A. Cyclic association rules. In: Proceedings of the 14th International Conference on Data Engineering; 1998. p. 412–21.
8. Savasere A, Omiecinski E, Navathe S. Mining for strong negative associations in a large database of customer transactions. In: Proceedings of the 14th International Conference on Data Engineering; 1998. p. 494–502.
9. Lu H, Han J, Feng L. Stock movement and n-dimensional inter-transaction association rules. In: Proceedings of the ACM SIGMOD Workshop on Research Issues in Data Mining and Knowledge Discovery; 1998. p. 1201–17.
10. Han J, Fu Y. Discovery of multiple-level association rules from large databases. In: Proceedings of the 21th International Conference on Very Large Data Bases; 1995. p. 420–31.
11. Srikant R, Agrawal R. Mining generalized association rules. In: Proceedings of the 21th International Conference on Very Large Data Bases; 1995. p. 407–19.

12. Brin S, Motwani R, Silverstein C. Beyond market basket: generalizing association rules to correlations. In: Proceedings of the ACM SIGMOD International Conference on Management of Data; 1997. p. 265–76.
13. Piatetsky-Shapiro G. Discovery, analysis, and presentation of strong rules. In: Piatetsky-Shapiro G, Frawley W, editors. Knowledge discovery in databases. Menlo Park/Cambridge, MA: AAAI/MIT; 1991. p. 229–38.
14. Dong G, Li J. Efficient mining of emerging patterns: discovering trends and differences. In: Proceedings of the 5th ACM SIGKDD International Conference on Knowledge Discovery and Data Mining; 1999. p. 43–52.
15. Li W, Han J, Pei J. CMAR: accurate and efficient classification based on multiple class-association rules. In: Proceedings of the 1st IEEE International Conference on Data Mining; 2001. p. 369–76.
16. Liu B, Hsu W, Ma Y. Discovering the set of fundamental rule changes. In: Proceedings of the 7th ACM SIGKDD International Conference on Knowledge Discovery and Data Mining; 2001. p. 335–40.
17. Agrawal R, Mannila H, Srikant R, Toivonen H, Verkamo AI. Fast discovery of association rules. In: Fayyad U, Piatetsky-Shapiro G, Smyth P, Uthurusamy R, editors. Advances in knowledge discovery and data mining. Menlo Park/Cambridge: AAAI/MIT; 1996. p. 307–28.
18. Agrawal R, Srikant R. Mining sequential patterns. In: Proceedings of the 11th International Conference on Data Engineering; 1995. p. 3–14.
19. Pei J, Han J, Lu H, Nishio S, Tang S, Yang D. H-Mine: hyper-structure mining of frequent patterns in large databases. In: Proceedings of the 1st IEEE International Conference on Data Mining; 2001. p. 441–8.
20. Pei J, Han J, Mao R. CLOSET: an efficient algorithm for mining frequent closed itemsets. In: Proceedings of the ACM SIGMOD Workshop on Research Issues in Data Mining and Knowledge Discovery; 2000. p. 11–20.

Asymmetric Encryption

Ninghui Li
Purdue University, West Lafayette, IN, USA

Synonyms

Public-key encryption

Definition

Asymmetric encryption, also known as public-key encryption, is a form of data encryption where the encryption key (also called the public key) and the corresponding decryption key (also called the private key) are different. A message encrypted with the public key can be decrypted only with the corresponding private key. The public key and the private key are related mathematically, but it is computationally infeasible to derive the private key from the public key. Therefore, a recipient could distribute the public key widely. Anyone can use the public key to encrypt messages for the recipient and only the recipient can decrypt them.

Key Points

A public-key encryption algorithm requires a trapdoor one-way function, i.e., a function that is easy to compute but hard to invert unless one knows some secret trapdoor (i.e., the private key). Existing public-key encryption algorithms are based on computational problems in number theory.

The most well-known public-key encryption algorithms include RSA and El Gamal. RSA uses exponentiation modulo a product of two large primes to encrypt and decrypt, and its security is connected to the presumed difficulty of factoring large integers. The El Gamal cryptosystem relies on the difficulty of the discrete logarithm problem. The introduction of elliptic curve cryptography in the mid 1980s has yielded a new family of analogous public-key algorithms. Elliptic curves appear to provide a more efficient way to leverage the discrete logarithm problem, particularly with respect to key size.

Cross-References

▶ Data Encryption
▶ Symmetric Encryption

Recommended Reading

1. Diffie W, Hellman ME. New directions in cryptography. IEEE Trans Inf Theory. 1976;22(6):644–54.
2. El Gamal T. A public key cryptosystem and a signature scheme based on discrete logarithms. In: Advances in cryptology: proceedings of CRYPTO '84. LNCS; 1985. p. 10–8.
3. Rivest RL, Shamir A, Adleman LM. A method for obtaining digital signatures and public-key cryptosystems. Commun ACM. 1978;21(2):120–6.

Atelic Data

Vijay Khatri[1], Richard T. Snodgrass[2,3], and Paolo Terenziani[4]
[1]Operations and Decision Technologies Department, Kelley School of Business, Indiana University, Bloomington, IN, USA
[2]Department of Computer Science, University of Arizona, Tucson, AZ, USA
[3]Dataware Ventures, Tucson, AZ, USA
[4]Dipartimento di Scienze e Innovazione Tecnologica (DiSIT), Università del Piemonte Orientale "Amedeo Avogadro", Alessandria, Italy

Synonyms

Point-based temporal data; Snapshot data

Definition

Atelic data is temporal data describing facts that do not involve a goal or culmination. In ancient Greek, *telos* means "goal," and α is used as prefix to denote negation. In the context of temporal databases, atelic data is that data for which both *upward* and *downward* (temporal) *inheritance* hold. Specifically,

- *Downward inheritance*. The *downward inheritance* property implies that one can infer from temporal data d that holds at valid time t (where t is a time period) and that d holds in any subperiod (and sub-point) of t.
- *Upward inheritance*. The *upward inheritance* property implies that one can infer from temporal data d that holds at two consecutive or overlapping time periods t_1 and t_2 and that d holds in the union time period $t_1 \cup t_2$.

Atelic data is differentiated from telic data, in which neither upward nor downward inheritance holds.

Key Points

Starting from Aristotle [1], researchers in areas such as philosophy, linguistics, cognitive science, and computer science have noticed that different types of facts can be distinguished according to their temporal behavior. Specifically, since atelic facts do not have any specific goal or culmination, they can be seen as "temporally homogeneous" facts, so that both upward and downward inheritance hold for them. For example, the fact that an employee (say, John) works for a company (say, ACME) would be considered atelic because of lack of goal or accomplishment. As a consequence, if John has worked for ACME from January 20, 2007 to September 23, 2007, it can be correctly inferred that John was working for ACME in May 2007 (or at any specific time point in May; therefore, downward inheritance does hold); furthermore, from the additional fact that John has also worked for ACME from September 23, 2007 to February 2, 2008, it can be correctly inferred that John worked for ACME from January 20, 2007 to February 2, 2008 (therefore, upward inheritance does hold).

In Aristotle's categorization, all possible facts are divided into two categories, telic and atelic; a telic fact, e.g., "John built a house," has goal or culmination [1].

Since both upward and downward inheritance properties hold for atelic data, such data supports the conventional "snapshot-by-snapshot" (i.e., point-based) viewpoint: the intended semantics of a temporal database is the set of conventional (atemporal) databases holding at each time point. As a consequence, most approaches to temporal databases support (only) atelic facts even if, in several cases, the representation allows, as a syntactic sugar, the use of periods to denote convex sets of time points. An integrated temporal database model that supports both telic and atelic data semantics has been developed [2].

Cross-References

► Period-Stamped Temporal Models
► Point-Stamped Temporal Models
► Telic Distinction in Temporal Databases

Recommended Reading

1. Aristotle. The categories on interpretation prior analytics. Cambridge, MA: Harvard University Press; 2002.
2. Terenziani P, Snodgrass RT. Reconciling point-based and interval-based semantics in temporal relational databases: a proper treatment of the telic/atelic distinction. IEEE Trans Knowl Data Eng. 2004;16(4): 540–51.

Atomic Event

Jonas Mellin and Mikael Berndtsson
University of Skövde, The Informatics Research Centre, Skövde, Sweden
University of Skövde, School of Informatics, Skövde, Sweden

Synonyms

Primitive event

Definition

An atomic event is considered to be indivisible and instantaneous.

Key Points

If an event is non-instantaneous, then it is possible to divide into a beginning of this event (an initiator) and an ending of this event (a terminator). Therefore, atomic events must be instantaneous. In active database literature, the concept primitive event is typically used instead of atomic event. The major reason is probably that system primitives and application primitives are not distinguished.

Cross-References

► Composite Event
► Event Detection
► Event Specification

Atomicity

Gerhard Weikum
Department 5: Databases and Information Systems, Max-Planck-Institut für Informatik, Saarbrücken, Germany

Definition

The *atomicity of actions* on a database is a fundamental guarantee that database systems provide to application programs. Whatever state modifications an atomic action may perform are guaranteed to be executed in an *all-or-nothing* manner: either all state changes caused by the action will be installed in the database or none. This property is important in the potential presence of failures that could interrupt the atomic action. The database system prepares itself for this case by *logging* state modifications and providing automated *recovery* as part of the failure handling or system restart. These implementation aspects are transparent to the application program and are thus a major relief for the programs' failure handling and boost the application development productivity.

Historical Background

Since the early 1970s (or even earlier), transaction processing systems for airline reservations and debit/credit banking had means for recovery and concurrency control that were similar to atomic actions. However, these implementation techniques were hardly documented in publicly available literature and still far from a principled, universal solution. The major credit for the modern concept of atomicity (and transactions) belongs to the 1998 Turing Award winner Jim Gray [6–8]. Closely related notions of atomic actions have been proposed by other authors at around the same time, including [11–13], which in turn were inspired by the informal work on "spheres of control" by Bjork and Davies [3, 4].

Foundations

As an example for the importance of atomic actions, consider a sequence of steps that read and write two bank-account records, x and y, in order to transfer some amount of money from account x to account y:

$$T1 : R (x) W (x) \qquad R (y) W (y)$$
$$T2 : \qquad R (x) R (y)$$

If there is a system failure after writing x but before writing y, the underlying database becomes inconsistent, with the transferred money seemingly lost in "mid-flight." Even worse, the application program may not even know if this problem actually occurred or not (the system may have succeeded in writing y just before it crashed but could not send a return code anymore). If, on the other hand, the database system executes the entire step sequence as an atomic action, the failure handling for the application program becomes much simpler as it can always restart on a clean, consistent database.

In database systems the atomic actions themselves can be flexibly defined by the application programs, by demarcating the begin and end of a *transaction* with explicit interface calls. For example, an entire sequence of SQL command invocations can be made atomic. By default, usually every individual SQL operation is guaranteed to be atomic. These guarantees are part of the *transactional ACID properites*: atomicity, consistency-preservation, isolation, and durability [8]. In some scientific communities outside of database systems research, atomicity is meant to include the *isolation* guarantee. Atomic actions are then (alternatively) defined to be state-modification sequences whose effects are ensured (by the underlying run-time environment) to be equivalent to *indivisible actions* with all effect appearing to be instantaneous upon the completion of the entire sequence. In the presence of concurrent accesses to the same shared data, this combined atomicity/isolation property provides the illusion, despite the fact that in reality accesses by different programs are interleaved. Defining this principle in formal terms leads to the concept of *serializability* and methods for *concurrency control* as part of the database (or other run-time) system [2, 6, 14].

As an example for the importance of isolation (as an additional property of atomic actions), consider an extended variant of the earlier example. Assume to the money-transfer process – now viewed as a transaction T1 – a second transaction T2 reads both bank-account records x and y in order to analyze financial portfolios and perform some kind of risk assessment. The following concurrent execution, with time proceeding from left to right, could be possible:

$$T1 : R (x) W (x) \qquad R (y) W (y)$$

$$T2 : \qquad R (x) R (y)$$

With this step interleaving, transaction T2 would see an inconsistent database, namely, the state of x after money is withdrawn from x and the state of y before money is deposited there. This may lead to a distorted analysis and false conclusions in the decision-making of a financial broker. Running both T1 and T2 as atomic and isolated transactions would prevent this particular interleaving and guarantees that only such executions are allowed that are provably equivalent to a sequential execution where such an anomaly is impossible.

The atomicity guarantee includes all "side effects" of an action as far as the database state is concerned. For example, the effects of a database trigger are covered by the guarantee, but effects outside of the database such as sending a message are outside the scope of the database system guarantees. Modern application servers and message brokers, on the other hand, may provide such guarantees about messages (e.g., atomic multicasts) and application state (e.g., specific program variables) beyond the database. A traditional implementation technique to this end is to support failure-resilient queues. Modern database systems have integrated such message queues and application state management and extend their atomic actions to them, providing more comprehensive *application recovery*. A new research trend in programming languages is to provide atomicity guarantees to arbitrary programs, not just database applications, in order to simplify exception handling and generally ease programmers' work. For example, method invocations in an object-oriented language could be made atomic by means of an underlying *transactional memory* as part of the language's run-time system (and possibly even hardware architecture).

The atomicity concept simplifies failure handling at the application program level, but it does not mask failures. Rather a typical approach is that the program notices the failing of an atomic action by a corresponding system return code (for the atomic action invocation itself, for the end-of-action demarcation call, or upon the next interaction with the database system if the previous call simply timed out without any response) and then has to retry the action. This paradigm still requires explicit coding for the retrying, and this may require special care about nonidempotent effects or additional system guarantees and state testing for ensuring *idempotence*. Some advanced methods for application recovery can automate these retrials and testing for nonidempotence, thus strengthening the all-or-nothing guarantee for atomicity into an *exactly once execution* guarantee with complete failure masking [1].

On the other hand, for some data-intensive applications outside of database systems, atomicity may be an overly strong property if applied to entire processes; this holds particularly for long-lived workflows and cooperative work. Although these applications still benefit from atomic actions for smaller-grained operations, additional forms of *relaxed atomicity* or extended atomicity would be desirable. The database research community has developed a variety of such models, most notably the model of open nested transactions and the ACTA framework [5, 9].

Future Directions

Atomicity is a ground-breaking, fundamental contribution that first emerged in database systems but is increasingly pursued also by other research communities like programming languages, operating systems, dependable system design, and also formal reasoning and program verification [10]. There are several strategic reasons for this growing interest and extended application of atomic actions:

- Web services, long-running workflows across organizations, large scale peer-to-peer platforms, and ambient-intelligence environments with huge numbers of mobile and embedded sensor/actor devices critically need support for handling or even masking concurrency and component failures and may mandate rethinking the traditional atomicity concept.
- There is a proliferation of open systems where applications are constructed from preexisting components. The components and their configurations are not known in advance, and they can change on the fly. Thus, it is crucial that atomicity properties of components are composable and that one can predict and reason about the behavior of the composite system.

- Modern applications and languages like Java lead millions of developers into concurrent programming. This is a drastic change from the classical situation where only a few hundred "five-star wizard" system programmers and a few thousand programmers working in scientific computing on parallel supercomputers would have to cope with the inherently complex issues of concurrency and advanced failure handling.

- On an even broader scale, the drastically increasing complexity of the new and anticipated applications will require enormous efforts and care towards dependable systems or it may lead into a major "dependability crisis." Atomicity is an elegant basic asset to build on in the design, implementation, and composition of complex systems and the reasoning about system behavior and guaranteed properties.

Cross-References

► ACID Properties
► Open Nested Transaction Models
► Software Transactional Memory
► Transaction

Recommended Reading

1. Barga RS, Lomet DB, Shegalov G, Weikum G. Recovery guarantees for internet applications. ACM Trans Internet Technol. 2004;4(3):289–328.
2. Bernstein PA, Hadzilacos V, Goodman N. Concurrency control and recovery in database systems. Reading: Addison-Wesley, MS; 1987.
3. Bjork LA. Recovery scenario for a DB/DC system. In: Proceedings of the ACM Annual Conference; 1973. p. 142–6.
4. Davies CT. Recovery semantics for a DB/DC system. In: Proceedings of 1st ACM Annual Conference; 1973. p. 136–41.
5. Elmagarmid AK, editor. Database transaction models for advanced applications. San Fransisco: Morgan Kaufmann; 1992.
6. Eswaran KP, Gray J, Lorie RA, Traiger IL. The notions of consistency and predicate locks in a database system. Commun ACM. 1976;19(11):624–33.
7. Gray J. Notes on database operating systems. In: Operating systems – an advanced course. London: Springer; 1978.
8. Gray J, Reuter A. Transaction processing: concepts and techniques. San Fransisco: Morgan Kaufmann; 1993.
9. Jajodia S, Kerschberg L, editors. Advanced transaction models and architectures. Noewell: Kluwer; 1997.
10. Jones CB, Lomet DB, Romanovsky AB, Weikum G, Fekete A, Gaudel M-C, Korth HF, Rogério de Lemos J, Moss EB, Rajwar R, Ramamritham K, Randell B, Rodrigues L. The atomic manifesto: a story in four quarks. ACM SIGMOD Rec. 2005;34(1):63–9.
11. Lampson B. Atomic transactions. In: Distributed systems – \architecture and implementation. New York: Springer; 1981.
12. Lomet DB. Process structuring, synchronization, and recovery using atomic actions. In: Proceedings of the ACM Conference on Language Design for Reliable Software; 1977. p. 128–37.
13. Randell B. System structure for software fault-tolerance. IEEE Trans Softw Eng. 1975;1(2):221–32.
14. Weikum G, Vossen G. Transactional information systems: theory, algorithms, and the practice of concurrency control and recovery. San Fransisco: Morgan Kaufmann; 2002.

Audio

Lie Lu[1] and Alan Hanjalic[2]
[1]Microsoft Research Asia, Beijing, China
[2]Delft University of Technology, Delft, The Netherlands

Synonyms

Audible sound

Definition

Audio refers to audible sound – the sound perceivable by the human hearing system, or the sound of a frequency belonging to the *audible frequency range* (20–20,000 Hz). Audio can be generated from various sources and perceived as speech, music, voices, noise, or any combinations of these. The perception of an audible sound starts by the sound pressure waves hitting the eardrum of the outer ear. The generated vibrations are transmitted to the cochlea of the inner ear to produce mechanical displacements along the basilar membrane. These displacements are further transduced into electrical activity along the auditory nerve fibers, and finally "analyzed" and "understood" in the central auditory system [4, 7].

Historical Background

The step from the fundamental definition of audio towards the concept of *audio signal* can be seen as a step towards the birth of the modern consumer electronics. An audio signal is a signal that contains audio information in the audible frequency range. The technology for generating, processing, recording, broadcasting and retrieving audio signals, first *analog* and later on *digital* ones, has rapidly grown for over a century, from the pioneering radio broadcasting and telephony systems to advanced mobile communication infrastructures, music players, speech recognition and synthesis tools, and audio content analysis, indexing and retrieval solutions. This growth may have been initiated by the research in the field of signal processing, but it has been maintained and has continuously gained in strength through an extensive interdisciplinary effort involving signal processing, information theory, human-computer interaction, psychoacoustics, psychology, natural language processing, network and wireless technology, and information retrieval.

Foundations

Digital Audio
An audio signal is an analog signal, which can be represented as a one-dimensional function $x(t)$, where t is a continuous variable representing time. To facilitate storage and processing of such signals in computers, they can be transformed into *digital signals* by *sampling* and *quantization*.

Sampling is the process in which one audio signal value (*sample*) is taken for each time interval (*sampling period*) T. This results in a *discrete audio* signal $x(n) = x(nT)$, where n is a numeric sequence. The sampling period T determines the

sampling frequency that can be defined as $f = 1/T$. Typical sampling frequencies of digital audio are 8, 16, 32, 48, 11.025, 22.05, and 44.1 kHz (Hz represents the number of samples per second). Based on the Nyquist-Shannon sampling theorem, the sampling frequency must be at least two times larger than the band limit of the audio signal in order to be able to reconstruct the original analog signal back from its discrete representation. In the next step, each sample in the discrete audio signal is *quantized* with a bit resolution, which makes each sample be represented by a fixed limited number of bits. Common bit resolution is 8-bit or 16-bit per sample. The overall result is a digital representation of the original audio signal, that is referred to as *digital audio signal* or, if it is just considered as a set of bits, for instance for the purpose of storage and compression, as *digital audio data*.

Audio Coding and Compression
The digitization process described above leads to the basic standard of digital audio representation or *coding* named *Pulse Code Modulation* (PCM), which was developed in 1930–1940s. PCM is also the standard digital audio format in computers and Compact Disc (CD). PCM can be integrated into a widely used WAV format, which consists of the digital audio data and a *header* specifying the sampling frequency, bits per sample, and the number of audio channels.

As a basic audio coding format, PCM keeps all samples obtained from the original audio signal and all bits representing the samples. This format is therefore also referred to as *raw* or *uncompressed*. While it preserves all the information contained in the original analog signal, it is also rather expensive to store. For example, a one-hour *stereo* (A Cambridge Dictionary definition of stereo: a way of recording or playing sound so that it is separated into two signals and produces more natural sound) audio signal with 44.1 kHz sampling rate and 16 bits per sample requires 635MB of digital storage space. To save storage in computers and improve the efficiency of audio transmission, processing and management, *compression* theory and algorithms can be applied to decrease the size of a digital audio signal while still keeping the quality of the signal and communicated information at the acceptable level.

Starting with the variants of PCM, such as *Differential Pulse Code Modulation* (DPCM) and *Adaptive Differential Pulse Code Modulation* (ADPCM), a large number of audio compression approaches have been developed [5]. Some most commonly used approaches include MP3/ACC defined in the MPEG-1/2 standard [2, 3], Windows Media Audio (WMA) developed by Microsoft, and RealAudio (RA) developed by RealNetworks. These approaches typically lead to a compressed audio signal being about 1/5 to 1/10 of the size of the PCM format.

Audio Content Analysis

Audio content analysis aims at extracting descriptors or *metadata* related to audio content and allowing content-based search, retrieval, management and other user actions performed on audio data. The research in the field of audio content analysis has built on the synergy of many scientific disciplines, such as signal processing, pattern recognition, machine learning, information retrieval, and information theory, and has been conducted in three main directions, namely *audio representation*, *audio segmentation*, and *audio classification*.

Audio representation refers to the extraction of audio signal properties, or *features*, that are representative of the audio signal composition (both in temporal and spectral domain) and audio signal behavior over time. The extracted features then serve as input into audio segmentation and audio classification. Audio segmentation aims at automatically revealing semantically meaningful temporal segments in an audio signal, which can then be grouped together (using e.g., a *clustering* algorithm) to facilitate search and browsing. Finally, an audio classification algorithm classifies a piece of audio signal into a pre-defined semantic class, and assigns the corresponding label (e.g., "applause," "action," "highlight," "music") to it for the purpose of text-based search and retrieval.

Audio Retrieval

Audio retrieval aims at retrieving sound samples from a large corpus based on their relation to an input query. Here, the query can be of different types and the expected results may vary depending on the application context. For example, in the *content-based retrieval* scenario, a user may use the text term "applause" to search for the audio clips containing the audio effect "applause." Clearly, the results obtained from audio classification can help annotate the corresponding audio samples, audio segments or audio tracks, and thus facilitate this search and retrieval strategy. However, audio retrieval can also be done by using an audio data stream as a query, i.e., by performing *query-by-example* [6]. For instance, one could aim at retrieving a song and all its variants by simply singing or humming its melody line.

In another retrieval scenario, the user may want to retrieve the exact match to the query or some information related to it. This typically falls into the application domain of *audio fingerprinting* [1]. An audio fingerprint is a highly compact feature-based representation of an audio signal enabling extremely fast search for a match between the signal and a large-scale audio database for the purpose of audio signal identification.

Key Applications

Audio technology is widely used in diverse applications areas, such as broadcasting, telephony, mobile communications, entertainment, gaming, hearing aids, and the management of large-scale audio/music collections.

Cross-References

► Audio Classification
► Audio Content Analysis
► Audio Representation
► Audio Segmentation
► Multimedia Data

Recommended Reading

1. Haitsma J, Kalker T. A highly robust audio fingerprinting system with an efficient search strategy. J New Music Res. 2003;32(2):211–21.

2. ISO/IEC 11172-3:1993. Information technology – coding of moving pictures and associated audio for digital storage media at up to about 1,5 Mbit/s – part 3: Audio; 1993.
3. ISO/IEC 13818-3:1998. Information technology – generic coding of moving pictures and associated audio information – part 3: Audio; 1998.
4. Pickles JO. An introduction to the physiology of hearing. London: Academic; 1988.
5. Spanias A, Painter T, Atti V. Audio signal processing and coding. Hoboken: Wiley; 2007.
6. Wold E, Blum T, Wheaton J. Content-based classification, search and retrieval of audio. IEEE Multimedia. 1996;3(3):27–36.
7. Yang X, Wang K, Shamma SA. Auditory representations of acoustic signals. IEEE Trans Inform Theory. 1992;38(2):824–39.

Audio Classification

Lie Lu[1] and Alan Hanjalic[2]
[1]Microsoft Research Asia, Beijing, China
[2]Delft University of Technology, Delft, The Netherlands

Synonyms

Audio categorization; Audio indexing; Audio recognition

Definition

Audio classification aims at classifying a piece of audio signal into one of the pre-defined *semantic classes*. It is typically realized as a combination of a *learning* step to learn a statistical model of each semantic class, and an *inference* step to estimate which semantic class is closest to the given piece of audio signal.

Historical Background

Audio classification associates *semantic labels* with audio signals, and can also be referred to as *audio indexing, audio categorization* or *audio recognition*. As such, audio classification plays an important role in facilitating search and retrieval in large-scale audio collections (databases). Semantic labels are used to represent semantic classes or *semantic concepts*, which can be defined at different abstraction and complexity levels. Typical examples of basic semantic audio classes are *speech, music, environmental sounds*, and *silence*, which can be detected rather effectively using the methods like [8, 9, 10, 17]. Examples of mid-level semantic concepts are *key audio effects*, like *applause, cheer, ball-hit, whistling, car-racing, siren, gun-shot*, and *explosion*. Finally, the detection of higher-level semantic concepts, such as sport highlights [1, 14, 15, 16] and action scenes in movies [3, 11], is usually performed by analyzing the sequence of detected key audio effects.

Foundations

Figure 1 shows a general classification scheme, which is typically composed of two main steps: *learning* and *inference*. In the learning step, a model of each semantic class is built based on a set of training data, and with a specific learning scheme. Then, in the inference step, a new, unseen collection of data is associated with a semantic label, the model of which best resembles the properties of the data. Various schemes have been employed so far for realizing both the learning and inference steps. These schemes include sets of heuristic rules, vector Quantization (VQ), k-nearest neighbor (kNN), decision tree, Gaussian mixture model (GMM), support vector machine (SVM), boosting, Bayesian decision, hidden Markov model (HMM), and neural network. More information about these schemes can be found in [4, 5].

While directly applying the previously mentioned learning and inference algorithms works well for straightforward classification tasks, there are a number of issues which should be taken into account when applying this scheme for audio classification. In the following sections, some critical issues are addressed that need to be considered when designing audio classification algorithms, and it is shown on the example of an existing approach how these issues can effectively be resolved. The classification cases discussed in the sections below concern the mid-level semantic concepts (key audio effects) and hierarchies of higher-level semantic concepts.

Key Audio Effect Detection

Several issues play a role in key audio effect detection in a continuous audio signal, and need to be resolved in order to secure reliable classification. The most important issues can be described as follows:

1. Key audio effect detection in a long, continuous audio signal, is typically approached by applying a sliding window of a given length (e.g., 0.5 s) to the signal. The audio segment captured by the window at a given time stamp is then used as the basic unit to associate with a key audio effect. An important implicit assumption here is that each segment corresponds to one and only one semantic class. However, a sliding window is often too short to capture one complete effect, which leads to over-segmentation. The sliding window could also be too long and capture several effects within one segment.
2. The targeted audio effects are usually sparsely distributed over the signal, and there are plenty of non-target sounds that are to be rejected. Many existing approaches assume having a complete set of semantic classes available, and classify any audio segment into one of these semantic classes. Other methods use thresholds to discard the sounds with low classification confidence [3]. However, the threshold

setting becomes troublesome for a large number of key effects.
3. Audio effects are usually related to each other. For example, some audio effects such as *applause* and *laughter* are likely to happen together, while others are not. Taking into account the transition relationships between audio effects is therefore likely to improve the detection of each individual effect.

To investigate the possibilities for effectively resolving the abovementioned issues when designing algorithms for key audio effect detection, the approach proposed in [2] will be discussed as an example. In this hierarchical, probabilistic framework, as illustrated in Fig. 2, an HMM model is first built for each key audio effect based on the complete set of audio samples, and the defined models are then used to compose the *Key Audio Effect Pool*. Then, comprehensive *background* models are also established to cover all non-target sounds that complement the targeted key effects. Thus, the non-target sounds would be detected as background sounds and excluded from the target audio effect sequence. Moreover, a higher-level probabilistic model is used to connect these individual models with a *Grammar Network*, in which the transition probabilities among various audio effects and background sounds are taken into account for finding the optimal audio effect sequence. Then, for a given input audio stream, the optimal audio effect sequence is found among the candidate paths using the Viterbi algorithm, and the location and duration of each key audio effect in the stream are determined simultaneously, without the need to pre-segment the audio stream into audio segments. In the following both the learning and

Audio Classification,
Fig. 1 An illustration of a general classification scheme

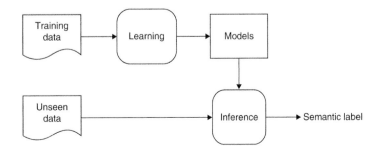

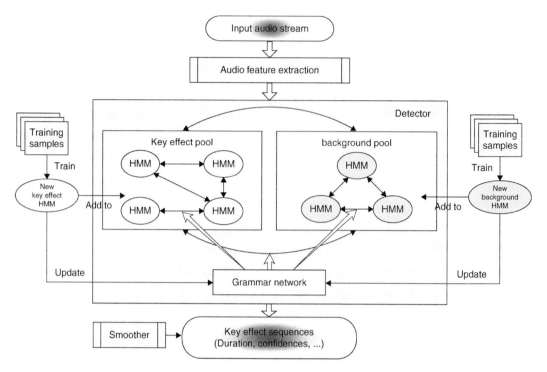

Audio Classification, Fig. 2 The framework for audio effect detection [13], consisting of three main parts: key audio effect pool, background sound pool, and grammar network

inference step from [2] are discussed in more detail.

Classifier Learning

In the approach from [2], each key audio effect and background sound are modeled using HMMs, since HMM provides a natural and flexible way for modeling time-varying process [12]. The main issue that needs to be resolved for an HMM is the parameter selection, which includes (i) the optimal model size (the number of states); (ii) the number of Gaussian mixtures for each state, and (iii) the topology of the model.

To select the model size for a key audio effect category, one needs to balance the number of hidden states in the HMM and the computational complexity in the learning and inference processes. In general, a sufficient number of states are required to describe all the significant behavioral characteristics of a signal over time. However, when the number of states increases, the computational complexity grows dramatically and more training samples are required. Unlike

speech modeling, in which the basic units such as tri-phones could be adopted to specify the number of states, general key audio effects lack such basic units, and thus make the choice of the state numbers difficult. As an example of an approach in this direction, a clustering-based method was proposed in [2, 13, 17] to estimate a reasonable number of states (model size) per audio effect. The clustering step was realized through an improved, unsupervised k-means algorithm, and the resulting number of clusters is taken as the model size.

The number of Gaussian mixtures per state is usually determined experimentally. For instance, the method from [2] adopts 32 Gaussian mixtures for each state in the HMM. This number is larger than those used in other related methods in order to secure a sufficient discriminative ability of the models to identify a large diversity of audio effects in general audio streams.

The most popular HMM topology is the left-to-right or the fully connected one. The left-to-right structure only permits transitions between adjacent states; while the fully connected structure allows transitions between any two states in the model. Different topologies can be used to model audio effects with different properties. For instance, for key audio effects with obvious time-progressive signal behavior, such as *car-crash* and *explosion*, the left-to-right structure should be adopted, while for audio effects without distinct evolution phases, such as *applause* and *cheer*, the fully connected structure is more suitable.

Regarding the background sound modeling, a straightforward approach is to build a large HMM, and train it with as many samples as possible. However, background sounds are very complex and diverse, and their feature vectors are typically widely scattered in the feature space, so that both the number of states and the Gaussian mixtures per state of such a HMM must be particularly large to secure a representation of all possible background sounds. As an alternative, the method from [2] modeled the background sounds as a set of subsets of basic audio classes. It is namely so that in most practical applications the background sounds can be further classified into a few basic categories, such as *speech*, *music*, and other *noise*. Thus, if background models could be trained from all these respective subsets, the training data would be relatively concentrated, and the training time would be reduced. Another advantage of building these subset models is that they could provide additional useful information in high-level semantic inference. For example, *music* is usually used in the background of movies, and *speech* is the most dominant component in talk shows. Following the discussion from above, three background models are built in [2] using the fully connected HMMs for *speech*, *music*, and *noise*. Here, *noise* is referred to as all background sounds except *speech* and *music*. To provide comprehensive descriptions of the background, 10 states and 128 Gaussian mixtures in each state are used in modeling.

The Grammar Network in Fig. 2 is an analogy to a language model in speech processing. It organizes all the HMM models for continuous recognition. Two models are connected in the Grammar Network if the corresponding sounds

are likely to occur after each other, both within and between the key audio effect pool and the background sound pool. For each connection, the corresponding transition probability is set and taken into account when finding the optimal effect sequence from the input stream.

The transition probabilities between two models can be statistically learned from a set of training data. If no sufficient training data are available, a heuristic approach can be deployed. For instance, the approach presented in [2] is based on the concept of *Audio Effect Groups*, and assumes that (i) only audio effects in the same group can happen subsequently, (ii) there should be background sounds between any two key audio effects belonging to different groups, and (iii) the transition probability is uniformly distributed per group. An audio effect group can be seen as a set of audio effects that usually occur together. An example Grammar Network with audio effect groups indicated as G_1-G_k is illustrated in Fig. 3.

Probabilistic Inference

Using the learned classification framework setup described above, the *Viterbi* algorithm can be used to choose the optimal state sequence from the continuous audio stream, as:

$$S_{\text{optimal}} = \arg\max_{s} \Pr(s|M, O). \qquad (1)$$

Here, s is the candidate state sequence, M represents the hierarchical structure, and O is the observation vector sequence of the input audio stream. In terms of practical realization of this classification scheme, the corresponding state and log-probability are obtained first for each audio frame. Then, a complete audio effect or background sound can be detected by merging adjacent frames belonging to the same sound model. Before this merging step, a smoothing filter can be applied to remove the classification outliers in the sequences of consecutive frames. The final classification confidence can be measured by averaging the log-probabilities of the classified audio frames. In addition, the

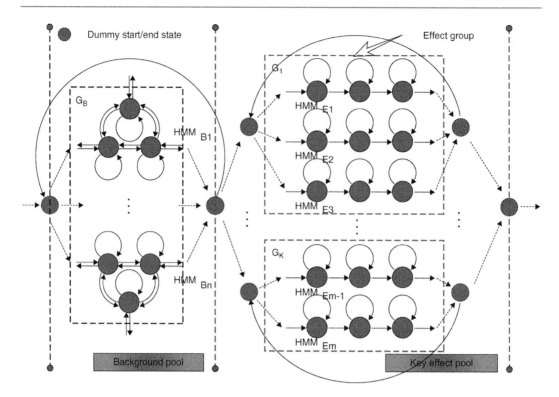

Audio Classification, Fig. 3 An illustration of the *Grammar Network* with *Audio Effect Groups*, where G_k is the *k*th *Effect Group* and G_B is the *Background Sound Pool*. For convenience, all the key audio effect models are presented as 3-state left-to-right HMMs, and all the background models are denoted as 3-state fully connected HMMs. The dummy start and end states are used to link models, and make the structure more clear [2]

starting time stamp and duration of each sound occurrence can be obtained simultaneously.

From Key Audio Effects to a Hierarchy of Semantic Concepts

Based on the obtained key audio effect sequence, methods can be developed to perform audio classification at a higher level, such as the level of audio events and scenes. While high-level semantic concepts can generally also be detected using the general scheme from Fig. 1, using key audio effects as an intermediate classification level has proved to be a more effective way to perform indexing at this level.

To infer high-level semantics from audio effects, most existing methods are rule-based [1, 16], or employ a statistical classification [3, 11]. Heuristic inference is straightforward and can be easily applied in practice. However, it is usually laborious to find a proper rule set if the situation is complex. For example, the rules usually involve many thresholds which are difficult to set; some rules may be in conflict with others, and some cases may not be well-covered. People are used to designing rules from a positive view but ignoring those negative instances, thus many false alarms are introduced although high recall can be achieved. Classification-based methods provide solutions from the view of statistical learning. However, the inference performance relies highly on the completeness and the size of the training samples. Without sufficient data, a positive instance not included in the training set will usually be misclassified. Thus these approaches are usually prone to high precision but low recall. Furthermore, it is inconvenient to combine prior knowledge into the classification process in these algorithms.

To integrate the advantages of heuristic and statistical learning methods, a Bayesian network-based approach is proposed in [2]. A Bayesian network [6] is a directed acyclic graphical model

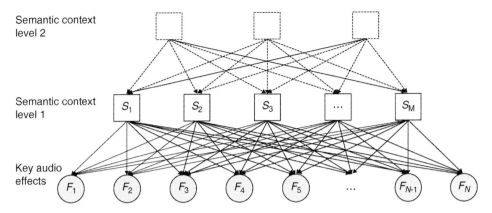

Audio Classification, Fig. 4 An example of a Bayesian network for inference of a hierarchy of semantic concepts. Arcs are drawn from cause to effect. Following the convention, discrete variables are represented as squares while continuous variables are indicated as circles. Furthermore, observed variables are shaded, while hidden variables are not

that encodes probabilistic relationships among nodes which denote random variables related to semantic concepts. A Bayesian network can handle situations where some data entries are missing, as well as avoid the overfitting of training data [6]. Thus, it weakens the influence from unbalanced training samples. Furthermore, a Bayesian network can also integrate prior knowledge by specifying its graphic structure.

Figure 4 illustrates the topology of an example Bayesian network with three layers. Nodes in the bottom layer are the observed audio effects. Nodes in the second layer denote high-level semantic categories such as scenes, while those in the top layer denote much higher semantic concepts. In Fig. 4, the nodes in adjacent layers can be fully connected, or partially connected based on the prior knowledge of the application domain. For instance, if it is known *a priori* that some key effects have no relationships with a semantic category, the arcs from that category node to those key effect nodes could be removed. A Bayesian network with a manually specified topology utilizes human knowledge in representing the conditional dependencies among nodes, thus it can describe some cases which are not covered in the training samples.

The nodes in the upper layers are usually assumed to be discrete binaries which represent the presence or absence of a corresponding semantic category, while the nodes in the bottom

layer produce continuous values of a Gaussian distribution

$$P\left(F_i | \mathbf{pa}_i\right) \sim N\left(\boldsymbol{\mu}_i, \boldsymbol{\Sigma}_i\right)(1 \leq i \leq N) \quad (2)$$

where F_i is a 2-dimensional observation vector of the ith audio effect and is composed of the normalized duration of an effect and its detection confidence in a given semantic segment. The conditional argument $\mathbf{pa}_i$ denotes a possible assignment of values to the parent nodes of F_i, while $\boldsymbol{\mu}_i$ and $\boldsymbol{\Sigma}_i$ are the mean and covariance of the corresponding Gaussian distribution respectively. In the training phase, all these conditional probability distributions are uniformly initialized and then updated by maximum likelihood estimation using the EM algorithm. In the inference process, the junction tree algorithm [7] can be used to calculate the occurrence probability of each semantic category. Thus, given the information on the audio effects, the semantics in each layer can be inferred based on the posterior probabilities. A semantic segment can be classified into the cth semantic category with the maximum marginal posterior probability:

$$c = \arg\max_j \Pr\left(S_j | F\right) \quad 1 \leq j \leq M$$
$$\text{where } \mathbf{F} = \{F_1, F_2, \cdots, F_N\} \quad (3)$$

With this scheme, human knowledge and machine learning are effectively combined to perform high-level semantic inference. In other words, the topology of the network can be designed according to the prior knowledge of an application domain, and the optimized model parameters can then be estimated by statistical learning.

Key Applications

Audio classification is typically applied for content-based search and retrieval in and management of large-scale audio collections (databases).

Cross-References

▶ Audio Content Analysis
▶ Audio Representation
▶ Audio Segmentation

Recommended Reading

1. Baillie M, Jose JM. Audio-based event detection for sports video. In: Proceedings of the 2nd International Conference on Image and Video Retrieval; 2003. p. 300–9.
2. Cai R, Lu L, Hanjalic A, Zhang HJ, Cai LH. A flexible framework for key audio effects detection and Auditory context inference. IEEE Trans Audio Speech Lang Process. 2006;14(3):1026–39.
3. Cheng WH, Chu WT, Wu, JL. Semantic context detection based on hierarchical audio models. In: Proceedings of the 5th ACM SIGMM International Workshop on Multimedia Information Retrieval; 2003. p. 109–15.
4. Duda RO, Hart PE, Stork DG. Pattern classification. 2nd ed. New York: Wiley; 2000.
5. Hastie T, Tibshirani R, Friedman J. The elements of statistical learning: data mining, inference, and prediction. New York: Springer; 2001.
6. Heckerman D. A tutorial on learning with Bayesian networks. Microsoft Research, Redmond, Washington, Tech. Rep. MSR-TR-95-06; 1995.
7. Huang C, Darwiche A. Inference in belief networks: a procedural guide. Int J Approx Reason. 1996;15(3):225–63.
8. Liu Z, Wang Y, Chen T. Audio feature extraction and analysis for scene segmentation and classification. J VLSI Signal Process Syst Signal Image Video Technol. 1998;20(1–2):61–79.
9. Lu L, Zhang HJ, Jiang H. Content analysis for audio classification and segmentation. IEEE Trans Speech Audio Process. 2002;10(7):504–16.
10. Lu L, Zhang HJ, Li S. Content-based audio classification and segmentation by using support vector machines. ACM Multimed Syst J. 2003;8(6): 482–92.
11. Moncrieff S, Dorai C, Venkatesh S. Detecting indexical signs in film audio for scene interpretation. In: Proceedings of the IEEE International Conference on Multimedia and Expo; 2001. p. 1192–5.
12. Rabiner LR. A tutorial on hidden Markov models and selected applications in speech recognition. Proc IEEE. 1989;77(2):257–86.
13. Reyes-Gomez MJ, Ellis DPW. Selection, parameter estimation, and discriminative training of hidden Markov models for general audio modeling. In: Proceedings of the IEEE International Conference on Multimedia and Expo; 2003. p. 73–6.
14. Rui Y, Gupta A, Acero A. Automatically extracting highlights for TV baseball programs. In: Proceedings of the 8th ACM International Conference on Multimedia; 2000. p. 105–15.
15. Xiong Z, Radhakrishnan R, Divakaran A, Huang TS. Audio events detection based highlights extraction from baseball, golf and soccer games in a unified framework. In: Proceedings of the IEEE International Conference on Multimedia and Expo; 2003. p. 401–4.
16. Xu M, Maddage N, Xu CS, Kankanhalli M, Tian Q. Creating audio keywords for event detection in soccer video. In: Proceedings of the IEEE International Conference on Multimedia and Expo; 2003. p. 281–4.
17. Zhang T, Jay Kuo CC. Hierarchical system for content-based audio classification and retrieval. In: Proceedings of the SPIE: Multimedia Storage and Archiving Systems III; 1998. p. 398–409.

Audio Content Analysis

Lie Lu[1] and Alan Hanjalic[2]
[1]Microsoft Research Asia, Beijing, China
[2]Delft University of Technology, Delft, The Netherlands

Synonyms

Audio information retrieval; Semantic inference in audio

Definition

An *audio signal* is a signal that contains information in the audible frequency range. Audio content analysis refers to a set of theories, algorithms and systems that aim at extracting descriptors or metadata related to audio content and allowing search, retrieval and other user actions performed on audio signals.

Historical Background

Multimedia content analysis has been one of the most booming research directions in the past years. With the objective of providing fast, natural, intuitive and personalized content-based access to vast multimedia data collections, and building on the synergy of many scientific disciplines, such as signal processing, pattern recognition, machine learning, information retrieval, information theory, natural language processing and psychology, the research initiative born around the end of the 1980s has succeeded in inspiring and mobilizing enormous number of researchers worldwide. This initiative turned into a broad research effort that has continuously gained in strength ever since. As an integrated part of multimedia (multimodal) documents, audio plays an important role in multimedia content analysis. In particular, audio content analysis can be combined with content analysis of visual information to jointly address semantic inference from *multimedia data* streams [5]. However, also taken separately, audio content analysis has been recognized as the key technology for providing easy access to rapidly growing audio archives, and in particular to music collections [3].

Foundations

The underlying problem of audio content analysis is to infer the information about the semantics of the content carried by an audio signal. Here, the audio signal can be a rather simple one consisting of one audio type only (e.g., pure speech, mu-

sic), or a more complex audio signal (*composite audio*) resulting from a superposition of several audio types, like music, speech, audio effects, and noise. The audio content semantics includes the information on audio types (e.g., speech, music, noise, or any combination of these), on audio semantic classes (e.g., applause, solo instrument, guitar, speaker X or Y), and on audio content structure (e.g., the breaks between semantically coherent audio segments, clusters of *auditory scenes*).

The structure of a general audio content analysis system is illustrated in Fig. 1. Depending on the levels at which prior knowledge is specified and the system is trained, this inference system can be realized by employing various approaches ranging from purely supervised to fully unsupervised ones. For example, if the scheme in Fig. 1 is seen as a speech recognition system, the prior knowledge, such as the labeled audio data, dictionary and grammar, need to be pre-collected to train both an acoustic model and a language model in a supervised fashion [6]. Similarly, trained models of semantic classes such as car-racing, siren, gun-shot, and explosion can be used to detect the occurrences of these sounds in movie soundtracks [2, 4]. Compared to these supervised realizations, an unsupervised approach can be employed to find clusters in audio data corresponding to auditory scenes [1, 7], or to find "unusual" events in the sound track of a surveillance signal [9].

While the realizations of the Scheme in Fig. 1 can be very different with respect to the types of input audio signals they handle, the prior knowledge and trained models they use, the types of inference techniques they employ (e.g., supervised versus unsupervised), and the inference results they are expected to provide (e.g., semantic categories versus clusters of data), the research targeting these realizations can be said to follow three main directions that also roughly define the scope of the audio content analysis research field. These directions are *audio representation*, *audio segmentation*, and *audio classification*.

- Audio representation refers to the extraction of audio signal properties, or *features*, that

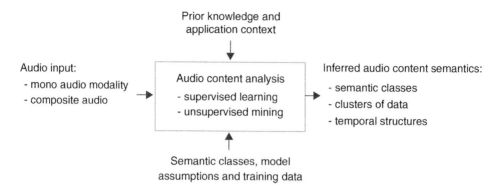

Audio Content Analysis, Fig. 1 A general audio content analysis scheme, specifying various possible types of input audio signals, prior knowledge and application context, the types of inference techniques and inference results

are representative of the audio signal, such as short time energy, zero-crossing rate, pitch, or spectrum. Obtaining a compact feature-based representation of an audio signal can improve the efficiency of audio processing and benefit many applications based on such processing (e.g., audio retrieval).

• Audio segmentation aims to automatically reveal coherent and semantically meaningful temporal segments in an audio signal. Examples of such segments are those containing pure speech or music, or those corresponding to *auditory scenes* [7]. The segments discovered through segmentation can also be clustered together into semantically coherent groups [1] to provide the possibility for an easy content access (e.g., through browsing).

• While the above mentioned segmentation and clustering processes are typically *unsupervised*, audio classification classifies a piece of audio signal into one of the predefined semantic classes using *supervised* methods of machine learning and pattern classification. The semantic classes can be defined at the level of basic audio types (e.g., speech, music) [8], audio effects (e.g., applause, gun-shot, car chasing) [2, 4], or auditory scenes (e.g., action, highlights, romance).

The term "audio" in the context of audio content analysis typically stands for general or *composite audio* signals [1]. When dealing with

specific audio types, such as speech and music, and related applications, dedicated research has been deployed and led to a number of specific research directions like music information retrieval [3] and speech recognition [6].

Key Applications

Audio content analysis is typically applied for content-based search and retrieval in the management of large-scale audio (or multimedia) collections (databases).

Cross-References

▸ Audio Classification
▸ Audio Representation
▸ Audio Segmentation
▸ Multimedia Data
▸ Video Content Analysis
▸ Video Scene and Event Detection

Recommended Reading

1. Cai R, Lu L, Hanjalic A. Unsupervised content discovery in composite audio. In: Proceedings of the IEEE International Conference on Multimedia and Expo; 2005. p. 628–37.
2. Cai R, Lu L, Hanjalic A, Zhang H-J, Cai L-H. A flexible framework for key audio effects detection and auditory context inference. IEEE Trans Audio Speech Lang Process. 2006;14(3):1026–39.

3. Casey M, et al. Content-based music information retrieval: current directions and future challenges. In: Proceedings of the IEEE, Special Issue on Advances in Multimedia Information Retrieval. 2008;96(4): 668–96.
4. Cheng W-H, Chu W-T, Wu J-L. Semantic context detection based on hierarchical audio models. In: Proceedings of the 5th ACM SIGMM International Workshop on Multimedia Information Retrieval; 2003. p. 109–15.
5. Hanjalic A. Content-based analysis of digital video. Norwell: Kluwer; 2004.
6. Huang X, Acero A, Hon HW. Spoken language processing: a guide to theory, algorithm, and system development. Upper Saddle River: Prentice; 2001.
7. Lu L, Cai R, Hanjalic A. Audio elements based auditory scene segmentation. In: Proceedings of the IEEE International Conference on Acoustics, Speech and Signal Processing; 2006. p. 17–20.
8. Lu L, Zhang H-J, Jiang H. Content analysis for audio classification and segmentation. IEEE Trans Speech Audio Process. 2002;10(7):504–16.
9. Radhakrishnan R, Divakaran A, Xiong Z. A time series clustering based framework for multimedia mining and summarization using audio features. In: Proceedings of the 6th ACM SIGMM International Workshop on Multimedia Information Retrieval; 2004. p. 157–64.

Audio Metadata

Werner Kriechbaum
IBM Development Lab, Böblingen, Germany

Synonyms

Music metadata

Definition

Audio, first used in 1934 refers to "Sound, esp. recorded or transmitted sound ... and signals representing this" [Oxford English Dictionary, Oxford 2005, Vol. 1, p. 780].

Metadata is data about data of any sort in any media, describing an individual datum, content item, or a collection of data including multiple content items. In that way metadata facilitates the understanding, characterization, use and management of data.

Audio metadata is structured, encoded data that describes content and representation characteristics of audio entities to facilitate the automatic or semiautomatic identification, discovery, assessment, interpretation, and management of the described entities, as well as their generation, manipulation, and distribution.

Historical Background

Audio metadata predate audio data by centuries. Since antiquity artists and theoreticians alike were interested to classify the effects that could be produced by the combination of different tones and to devise rules for the proper composition of music [12]. The baroque *Affektenlehre* (doctrine of the affections) provided the composers of music with rules and schemes to express affects like love, fear, or hate, and on the other hand gave the listeners a reference system to decode the emotional content of a piece of music. Falling into oblivion with the end of the baroque area similar classification schemes resurfaced with the advent of silent films. Score snippets classified this way enabled the pianist accompanying the film to select music appropriate to the mood of the film scene on the fly.

Like many other technological innovations the modern history of audio metadata started in Bell Labs with the vocoder, developed by Homer Dudley during the late 20s and early 30s of the last century [2]. As part of a speech analysis/synthesis system the vocoder was used to reduce the amount of storage needed to store human speech. The analysis part of the system computed a spectrogram (a variant of a short-term power spectrum) and extracted the fundamental frequency of the speech signal – the first low-level audio metadata. This paved the way for the development of an ever increasing stream of audio analysis techniques that can be traced for example in the "Transactions of the IRE Professional Group on Audio" and its IEEE follow-on and spin-off journals (Available at: http://ieeexplore.ieee.org/xpl/RecentIssue.jsp?punumber = 8340).

MPEG-7 Audio standard [7], finalized in 2002, selected and standardized a set of low- and high-level descriptors from the plethora of available signal processing techniques and tools. MPEG-7 low-level audio descriptors are derived from the time-frequency analysis of the audio signal and include:

- Basic descriptors: e.g., the audio waveform envelope
- Basic spectral descriptors: e.g., the centroid of the audio spectrum
- Signal parameters: e.g., the fundamental frequency
- Timbral temporal descriptors: e.g., the logarithmic attack time
- Timbral spectral descriptors: e.g., the harmonic spectral centroid
- Spectral basis descriptors: which are low-dimensional projections of the spectrum
- Silence

The MPEG-7 high-level descriptors, which are usually collections of further low-level descriptors with additional context, comprise:

- Audio signature
- Musical instrument timbre
- Monophonic melody
- HMM sound model
- Spoken content model

Markup languages and among those most notably HyTime [6], provided the second major force that influenced the development of audio metadata. The linking concepts introduced by HyTime provided mechanisms to connect (textual) symbolic representations and (non-textual) realizations of a media entity without embedding the link in either. The Standard Music Description Language (SMDL) [5] – developed as a HyTime application and up to now not promoted from a draft standard to an international standard – applied these techniques to standardize architecture for the representation of music. From an SMDL point of view a single musical work is comprised of four domains:

- Logical domain: "The logical domain is the basic musical content – the essence from which all performances and editions of the work are derived, including virtual time values, nominal pitches, etc. The logical domain is describable as 'the composer's intentions with respect to pitches, rhythms, harmonies, dynamics, tempi, articulations, accents, etc.,'" [5, p. 5]
- Gestural domain: Any number of performances of the logical domain, e.g., a digital audio recording capturing a concert
- Visual domain: Any visual rendering of the logical domain, e.g., a score
- Analytical domain: Any number of music-theoretical analyses like e.g., Schenkerian analysis [A. Cadwallader/D. Gagné, Analysis of Tonal Music: A Schenkerian Approach, Oxford 1998]

There is only a faint echo of this concept in the MPEG-7 Audio standard but the ideas are taken up again in the IEEE P1599 Recommended Practice for Definition of a Commonly Acceptable Musical Application using the XML Language (MX) which is as the time of this writing still a draft but should undergo ballot soon. In the terminology of MX, the SMDL domains are called layers and the aspects of music addressed are refined to six:

- General: Information that applies to the piece of music as a whole, e.g., the opus number
- Logic: The symbolic description of the music (equivalent to the SMDL logical domain)
- Structural: Description of music objects and their causal relationship
- Notational: The score (similar to the SMDL visual domain)
- Performance: An audible rendering of the piece of music like e.g., MIDI, CSound, etc.
- Audio: Links to and description of digital audio

The great success of the compact disc and the almost ubiquitous availability of tools to create own copies or compilations from digital audio in MP3 format spawned a further community-based audio metadata "standardization" initiative. Since according to the original Red Book specification [IEC 60908 Ed. 2.0 b:1999 Audio recording – Compact disc digital audio system] the compact discs did not include metadata like disc name, track names or author information the need arose to add this information from a supplemental database and to embed at least part of this information in an MP3 file generated from the audio disc. The basic idea to bind the audio disc to its database entry used by all such systems is to compute a hash based on the track information available in the CD's table of contents. This hash is used as key for the metadata stored in the database. An entry in the open-source GPL-licensed audio metadata database freedb [http://www.freedb.org/] contains the following information:

- DISCID: the hash key
- DTITLE: Artist and disc title separated by "/"
- DYEAR: Year the compact disc was released
- DGENRE: The genre in textual form. In addition to this genre entry the database is split into eleven different categories (blues, classical, country, data, folk, jazz, newage, reggae, rock, soundtrack, misc) to minimize hash collisions. The information in the genre field can conflict with the category and will do so when two discs from the same category produce the same hash key. The recommended resolution for the key collision is to put the second disc in another category.
- TTITLEN: The title of track N
- EXTD: Extended data (i.e., any interesting information) for the audio disc
- EXTTN: Extended data for track N

Like the original audio disc the MP3 standard [ISO/IEC 11172-3:1993 Information technology – Coding of moving pictures and associated audio for digital storage media at up to about 1,5 Mbit/s – Part 3: Audio] did not provide for the inclusion of textual material as part of the encoded audio. This did not deter the user community from enhancing MP3 files with textual information. One result of these efforts is the (informal) ID3 standard [http://www.id3.org/id3v2.4.0-structure and http://www.id3.org/id3v2.4.0-frames] which specifies how to prepend metadata to an MP3 file. There is a rich set of predefined ID3 frames for example:

- Text information frames: Information like album, author, artist(s), etc.
- URL link frames: Links to e.g., the web page of a performer.
- Event timing codes: Time stamps for events in the audio like e.g., verse start, refrain start, theme start, key change, theme end, profanity, profanity end.
- Unsynchronized lyrics or text transcription
- Synchronized lyrics or text.
- Equalization: Equalizer settings for the encoded audio.
- Attached picture

The tag set can be extended since ID3 parsers – like HTML parsers – have to ignore unknown tags.

Foundations

From a metadata point of view, audio is rather ill-defined. All but the basic technical metadata describing the recording setup and the physical metadata characterizing the recorded signal, are specific for the recorded material and require content specific expertise. What makes sense for the description of the recording of a starting airplane is quite different from the data needed to describe birdsong or an opera recording. And even the description of the audio signal on the physical level is not without problems when the metadata are collected for human usage. On their way from the eardrum to the brain, auditory signals undergo a non-linear transformation caused by one's sound perception system [3, 9] and all the concepts our cognition forms about audio events are based on this transformed signal. The perceived intensity (loudness) of a signal, for example, is quite different from the physical intensity of the signal. Low-level descriptors of audio signals have to take these differences into account, especially when they are used to derive more complex cognitive descriptors.

Since the domain of the recorded material has a marked influence on the metadata useful and necessary to describe the audio recording, the following discussion is restricted to the recording of music. Almost all music, except monophonic works, is realized by the co-operation of musicians that perform in parallel different subsets of the music. But in many cases, grouping mechanisms in one's auditory perception transform even monophonic music in two or more perceived separate streams [1]. Therefore even pure music not accompanying a stage play can be understood and described as multimedia data and the scientific fundamentals discussed in the entry on multimedia metadata apply to music as well. Furthermore all music can exist in two different forms: a symbolic representation (the score), and a realization (the audio recording). Information like metrum or key that is hard to derive from the analysis of the recorded audio is readily available from the score. In addition a score is accessible for music theoretical analysis that leads to further metadata [8]. Whenever one of the two, score or realization, is missing, it can, at least in principle, be derived from the other. The realizations generated by performing a score vary to a considerable degree. First of all most scores do not give an absolute reference point for the frequency. To map the note A4 (the A above the middle C) to a frequency of 440 Hertz, now an ISO standard [ISO 16:1975 Acoustics – Standard tuning frequency (Standard musical pitch)], is a rather new convention established by an international conference in 1936. Throughout history a variety of reference frequencies for A4 have been used, varying from as low as 392 Hertz up to 466 Hertz. Prior to the acceptance of equal temperament as tuning standard in the second half of the eighteenth century, a variety of tuning standards were in use and rendering baroque music on historic instruments with a historic temperament leads to a quite different performance than the one produced by a modern orchestra with

well-tempered tuning. Similar variability exists for the global tempo (at least prior to the use of the metronome ticks to specify absolute time values) and the dynamic. And of course each individual performer or orchestra has its personal style of phrasings and embellishments. Therefore one score gives rise to a variety of realizations and at least when one is interested in identifying music all metadata derived from the realization of the score should in the end lead to the same piece of music. But scores themselves are by no means static; throughout history they have been transformed to adapt them to different needs. Examples of such transformations are transpositions (shifts in pitch) to adapt the score to the ambitus of an instrument, or piano reductions where an orchestral piece is simplified in such a way that its "essence" can be rendered on a piano. Like with the variation in realizations metadata for the description of music should be able to cope with this variability. Both types of variation exist not only in classical western music but in popular music or non-western music like Indonesian gamelan or Indian ragas as well.

Music has a complex temporal organisation and a rich semantic structure that defines a natural segmentation for the audio stream. Like in images, in music many features are characteristic for segments, vary from segment to segment, and become meaningless when averaged over all segments. Musical structure is not arbitrary but conforms to a set of possible patterns: The sonata form [10] for example is one of the most influential structural patterns during the classical era of western music. The structure of the sonata form is built from three to five pieces: an optional introduction, an exposition, a middle part (*Durchführung*), a repeat, and an optional coda. Usually these five elements are not atomic but further structured and some segments of this substructure are derived from each other by transformations like e.g., transposition, inversion, reflection, or tempo changes. Besides being metadata in its own right, structural information linked with the audio material allows a natural navigation of the recorded performance. As outlined above, the approach to document structure is prescriptive. As in nineteenth century music theory it is assumed that there is an ideal architecture for a sonata form, and that a piece of music not conforming to these rules is in error. In the twentieth century this concept has come under considerable criticism [e.g., 11] and a descriptive approach has been advocated. As a consequence, each musical work is likely to have more than one semantic segmentation, depending on the analysis approaches chosen. To be of any use, both a controlled vocabulary describing the structures and an ontology describing the relationships among them are needed. But this is by no means specific to the structure of music. For many other audio metadata standardized controlled vocabularies and ontologies are still lacking. For example without amendment the Dublin Core [http://www.dublincore.org/] term creator, "An entity primarily responsible for making the resource" [http://purl.org/dc/terms/creator], attached to a piece of music makes it hard if not impossible to recognize the specific role of the creator. Felix Weingartner as creator could refer to his role as conductor (he conducted what is believed to be the first complete recording of Beethoven's symphonies), or his role as composer (symphonies, string quartets, operas) or his role as editor (he edited the complete works of Berlioz).

Key Applications

Audio metadata are essential for any search for audio data. In addition music metadata help to reveal similarities in style or structure between different compositions or different realisations of a piece of music. There are many frameworks for the analysis and manipulation of music, two GPLed packages that allow easy experimentation with different metadata concepts for music are CLAM [http://www.clam.iua.upf.edu/] and RU-BATO® [http://www.rubato.org/].

Cross-References

▶ Image Metadata
▶ Multimedia Metadata
▶ Video Metadata

Recommended Reading

1. Deutsch D, editor. The psychology of music. 2nd ed. San Diego: Academic; 1999.
2. Dudley HW. The vocoder. Bell Labs Rec. 1939;17(2):122–6.
3. Handel S. Listening. Cambridge, MA: MIT; 1989.
4. IEEE P1599/D5.0. Draft recommended practice for definition of a commonly acceptable musical application using the XML Language. New York; 2008.
5. ISO/IEC DIS 10743. Standard music description language (SMDL). 1995 July.
6. ISO/IEC 10744:1997. Information technology – Hypermedia/Time-based structuring language (HyTime). Geneva; 1997.
7. ISO/IEC 15938-4:2002. Information technology – Multimedia content description interface – Part 4: Audio. Geneva; 2002.
8. Mazzola G. The topos of music. Basel: Birkhäuser; 2002.
9. Moore B, editor. Hearing. San Diego: Academic; 1995.
10. Mauser S, editor. Handbuch der musikalischen Gattungen. Laaber: Laaber Verlag; 1993.
11. Rosen C. Sonata forms. 2nd ed. New York: Norton; 1980.
12. Zaminer F. Geschichte der Musiktheorie. Darmstadt; 1984.

Audio Representation

Lie Lu[1] and Alan Hanjalic[2]
[1] Microsoft Research Asia, Beijing, China
[2] Delft University of Technology, Delft, The Netherlands

Synonyms

Audio characterization; Audio feature extraction

Definition

An *audio signal* is a signal that contains information in the audible frequency range. Audio representation refers to the extraction of audio signal properties, or *features*, that are representative of the audio signal composition (both in temporal and spectral domain) and audio signal behavior over time. *Feature extraction* is typically combined with *feature selection*, through which the best set of features for the intended operation on the audio signal is defined.

Historical Background

Audio feature extraction typically leads to a strongly reduced audio signal representation. Obtaining such representation can improve the efficiency of audio processing and benefit many applications based on such processing. For example, a compact representation of an audio signal in the form of a *fingerprint* can enable extremely fast search for a match between this signal and a large-scale audio database for the purpose of audio signal identification. Further, if audio features are carefully chosen, they can capture the information from the original data that is relevant to subsequent audio signal analysis and processing steps, while leaving out the redundant and irrelevant (noisy) information parts. This possibility to simultaneously improve the efficiency and robustness of audio signal analysis and processing indicates the importance of the *feature selection* step as the basis step in *audio content analysis*, and in particular in *audio segmentation* and *audio classification*.

Foundations

Audio features can be divided into *temporal* and *spectral* features that capture the temporal and spectral characteristics of an audio signal, respectively. In terms of the way the features are extracted, a division into *frame-level* and *window-level* features can be made. An audio *frame* is the elementary temporal segment of the signal, from which features are extracted. The length of an audio frame typically varies between 10 and 50 ms. Due to its short duration, a frame can be said to contain (close-to) stationary signal behavior. The window-level features are extracted from a longer audio segment, comprising a number of consecutive frames, and are typically marked by applying a sliding window to the signal. While most audio features are extracted at the frame level, window-level features are mainly derived from the frame-level features by investigating their variation along the frames within the window, e.g., the mean and standard deviation of frame-level features. This expansion of the frame-level feature consideration from an individual frame to a series of consecutive frames proved to be useful in many applications, which indicates the importance of window-level features.

Table 1 gives an overview of the typical features proposed in literature to perform various operations on audio, and in particular, the audio segmentation and classification. The features in the table are ranked according to the frequency of their usage in literature. Multiple names indicated per row of the table stand for one and the same feature and/or its variants. Also, the notation *t*, *s*, *fl* and *wl* is used to indicate whether a feature is temporal, spectral, frame-level or a window-level feature. The table is followed by detailed descriptions of a subset of the most prominent frame-level and window-level features. Finally, information is provided about the processes of feature normalization and selection to generate optimal feature sets (vectors) for audio representation.

Zero-Crossing Rate

Zero-crossing rate (*ZCR*) is defined as the relative number of times the audio signal crosses the zero-line within a frame. It can be computed using the following expression:

$$ZCR = \frac{1}{2(N-1)} \sum_{m=1}^{N-1} \left| \text{sgn}\left[x(m+1)\right] \right. \tag{1}$$
$$\left. - \text{sgn}\left[x(m)\right] \right|$$

Here, sgn[] is a sign function, *x(m)* is the discrete audio signal, *m = 1...N*, and *N* is the frame length.

The *ZCR* is a computationally simple measure of the general frequency content of a signal, and as such it is particularly useful in characterizing audio signals in terms of the *voiced* and *unvoiced* sound categories. As speech signals are generally

Audio Representation, Table 1 An overview of audio features most frequently used in literature for the purpose of audio segmentation and classification

Features	Level	Temporal/spectral
Short time energy, root mean square (RMS), spectrum power, volume, loudness	*fl*	*t, s*
Zero crossing rate (ZCR)	*fl*	*t*
Mel-frequency cepstral coefficient (MFCC)	*fl*	*s*
Spectral centroid, brightness, frequency centroid	*fl*	*s*
Bandwidth	*fl*	*s*
Sub-band energy (distribution), sub-band power, band-energy ratio	*fl*	*s*
Short time fundamental frequency, pitch, harmonic frequency	*fl*	*s*
LPC-derived cepstral coefficients (LPCC)	*fl*	*s*
Linear predictive coding (LPC)	*fl*	*s*
Spectral rolloff	*fl*	*s*
Spectral peak	*fl*	*s*
Spectral moments	*fl*	*s*
Spectral flatness	*fl*	*s*
Harmonicity	*fl*	*s*
Harmonicity prominence	*fl*	*s*
Sub-band partial prominence	*fl*	*s*
Wavelet decomposition	*fl*	*s*
MPEG-7 audio features	*fl*	*s*
Spectrum flux	*wl*	*s*
Percentage of low-energy frames, low short-time energy ratio (LSTER), non-silence ratio	*wl*	*t, s*
High ZCR ratio (HZCRR)	*wl*	*t*
Noise frame ratio, noise or non-voice ratio	*wl*	*t*
4 Hz modulation energy	*wl*	*t, s*
Pulse metric	*wl*	*t*

composed of alternating voiced and unvoiced sounds, which is not the case in music signals, the variation in the *ZCR* values is expected to be larger for speech signals than for music signals. Due to its discriminative power in separating speech, music and various audio effects, *ZCR* is often employed in audio content analysis algorithms. An illustration of its practical usage can be found in [7, 8, 10–12, 15].

Short Time Energy

Short Time Energy (*STE*) is the total spectral power of a frame. It can be computed from the audio signal directly, as

$$STE = \frac{1}{N} \sum_{m=1}^{N} |x(m)|^2, \qquad (2)$$

or from its *Discrete Fourier Transform* (*DFT*) coefficients, as

$$STE = \sum_{k=0}^{K/2} |F(k)|^2 \qquad (3)$$

Here, $F(k)$ denotes the *DFT* coefficients, $|F(k)|^2$ is the signal power at the discrete frequency k, and K is the order of *DFT*. In [7, 14], this energy is computed using the logarithmic expression, to get a measure in (or similar to) decibels.

Similar to *ZCR, STE* is also an effective feature for discriminating between speech and music signals. For example, there are more silence (or unvoiced) frames in speech than in music. As a result, the variation of *STE* in speech is in general much higher than in music. An illustration of the practical usage of this feature can be found in [7, 10, 12, 14, 15].

Sub-band Energy Distribution

To further exploit the energy information based on the *STE* feature defined above, the *sub-band energy distribution* (*SBED*) can be computed. This distribution can be obtained by dividing the frequency spectrum into sub-bands, and by computing for each sub-band j the ratio D_j between the energy contained in that sub-band and the total spectral power of the frame:

$$D_j = \frac{1}{STE} \sum_{L_j}^{H_j} |F(k)|^2. \tag{4}$$

Here, L_j and H_j are the lower and upper bound of sub-band j respectively. The sub-band division can be done in various ways, such as in octave-scale [7] or in mel-scale [1].

Since the spectrum characteristics are rather different for sounds produced by different sources (e.g., human voice, music, environmental noise) the *SBED* feature has often been used for general audio classification [5, 10], and, in particular, for discriminating between different sound effects [1, 14].

Brightness and Bandwidth

Brightness and *bandwidth* are related to the first- and second-order statistics of the spectrum, respectively. The brightness is the centroid of the spectrum of a frame, and can be defined as:

$$w_c = \frac{\sum_{k=0}^{K/2} k \, |F(k)|^2}{\sum_{k=0}^{K/2} |F(k)|^2} \tag{5}$$

Bandwidth is the square root of the power-weighted average of the squared difference between the spectral components and the centroid:

$$B = \sqrt{\frac{\sum_{k=0}^{K/2} (k - w_c)^2 |F(k)|^2}{\sum_{k=0}^{K/2} |F(k)|^2}} \tag{6}$$

Brightness and Bandwidth characterize the shape of the spectrum, and roughly indicate the timbre quality of a sound. From this perspective, brightness and bandwidth can provide useful information for audio classification processes [11, 14].

Mel-Frequency Cepstral Coefficient (MFCC)

The set of *Mel-Frequency Cepstral Coefficients* [10] is a cepstral representation of the audio signal obtained based on the mel-scaled spectrum. The log spectral amplitudes are first mapped onto the perceptual, logarithmic *mel-scale*, using a triangular band-pass filter bank. Then, the mel-scaled spectrum is transformed into MFCC using the Discrete Cosine Transform (*DCT*).

$$c_x = \sqrt{\frac{2}{K}} \sum_{k=1}^{K} (\log S_k) \cos \left[n \, (k - 0.5) \, \pi / K \right]$$

$$n = 1, 2, ..., L \tag{7}$$

Here, c_n is the n-th MFCC, K is the number of band-pass filters, S_k is the mel-scaled spectrum after passing the k-th triangular band-pass filter, and L is the order of the cepstrum.

MFCC is commonly used in speech recognition and speaker recognition systems. However, *MFCC* also proved to be useful in discriminating between speech and other sound classes, such as music, which explains its wide usage in the audio analysis and processing literature [3, 6, 12].

Sub-band Partial Prominence and Harmonicity Prominence

The *Sub-Band Partial Prominence* (*SBPP*) is used to measure whether there are salient frequency components (i.e., *partials*) in a sub-band. In other words, the *SBPP* estimates the existence of prominent partials in sub-bands [1].

It is computed by accumulating the variation between adjacent frequency bins in each sub-band, that is

$$S_p(i) = \frac{1}{H_i - L_i} \sum_{j=L_i}^{H_i-1} \left| \hat{F}(k+1) - \hat{F}(k) \right| \tag{8}$$

Here, L_i and H_i are the lower and upper boundaries of the ith sub-band respectively, and the value of $S_p(i)$ indicates the corresponding prominence of salient partial components. The *SBPP* value $S_p(i)$ for sub-bands containing salient components is expected to be large. In order to reduce the impact induced by the energy variation over time, the original *DFT* spectral coefficient vector F is first converted to the decibel scale and constrained to the unit L_2-norm [2] to yield the new spectral coefficient vector used on (8):

$$\hat{F} = \frac{10\log_{10}(F)}{\|10\log_{10}(F)\|} \tag{9}$$

If now the property of an ideally harmonic sound (with one dominant fundamental frequency f_0) is considered, its spectral energy is highly concentrated and precisely located at those predicted harmonic positions which are the multiples of the fundamental frequency f_0. To detect this situation, the following three factors could be measured: (i) the energy ratio between the detected harmonics and the whole spectrum; (ii) the deviation between the detected harmonics and predicted positions; and (iii) the concentration degree of the harmonic energy. Based on the above, the *Harmonicity Prominence* (*HP*) was defined in [14] to estimate the harmonic degree of a sound. The *HP* measure takes into account the above three factors and can be defined as

$$H_p$$
$$= \frac{\sum_{n=1}^{N} E^{(n)} \left(1 - |B_r^{(n)} - f_n|/0.5f_0\right)\left(1 - B_w^{(n)}/B\right)}{E} \tag{10}$$

Here, $E^{(n)}$ is the energy of the detected nth harmonic contour in the range of $[f_n - f_0/2, f_n +$

$f_0/2]$ and the denominator E is the total spectral energy. The ratio between $E^{(n)}$ and E stands for the first of the three factors identified above. Further, f_n is the nth predicted harmonic position and is defined as

$$f_n = nf_0 \sqrt{1 + \beta(n^2 - 1)} \tag{11}$$

where β is the *inharmonicity modification factor*, and $B_y^{(n)}$ and $B_w^{(n)}$ are the brightness and bandwidth of the nth harmonic contour, respectively. The brightness $B_y^{(n)}$ is used instead of the detected harmonic peak in order to estimate a more accurate frequency center. The bandwidth $B_y^{(n)}$ describes the concentration degree of the nth harmonic. It is normalized by a constant B, which is defined as the bandwidth of an instance where the energy is uniformly distributed in the search range. Thus, the components $\left(1 - |B_y^{(n)} - f_n|/0.5f_0\right)$ and $\left(1 - B_w^{(n)}/B\right)$ in the numerator of (10) represent the second and the third factor defined above.

An illustration of the definition of *harmonicity prominence* is shown in Fig. 1. Detailed explanation of the computation and usage of this feature can be found in [1]. The harmonic audio analysis using the *SBPP* and *HP* features enables sophisticated audio classification, like for instance, the discrimination between *cheer* and *laughter*. This is possible because *laughter*, as opposed to *cheer*, usually contains prominent harmonic partials.

High ZCR Ratio

High ZCR Ratio (*HZCRR*) [6] is defined as the fraction of frames in the analysis window, whose *ZCR* values are 50% higher than the average *ZCR* computed in the window, that is

$$HZCRR = \frac{1}{2N} \sum_{n=0}^{N-1}$$
$$[\text{sgn}(ZCR(n) - 1.5avZCR) + 1] \tag{12}$$

Here, n is the frame index, $ZCR(n)$ is the zero-crossing rate at the n-th frame, N is the total number of frames, $avZCR$ is the average ZCR in the analysis window, and sgn[] is a sign function.

Audio Representation, Fig. 1 Definition of *harmonicity prominence.* The *horizontal* axis represents the frequency, and the *vertical* axis denotes the energy. The harmonic contour is the segment between the adjacent valleys separating the harmonic peaks. Based on the harmonic contour, three factors, that is, the peak energy, energy centroid (*brightness*) and degree of concentration (*bandwidth*), are computed to estimate the *harmonicity prominence*, as illustrated at the second harmonic in this example.

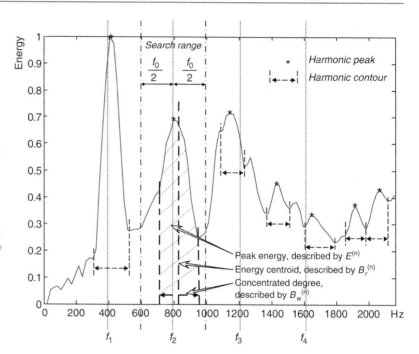

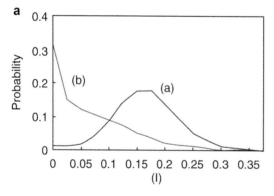

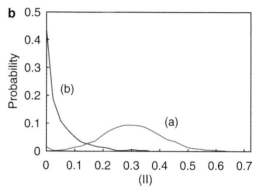

Audio Representation, Fig. 2 An illustration of probability distribution curves of **a**$HZCRR$ (a) speech and (b) music, and **b**$LSTER$ (a) speech and (b) music

The variation of $HZCRR$ is expected to be higher in speech signals than in music. Figure 2a shows the probability distribution curves of $HZCRR$ computed for a large number of speech and music signals. Using the cross-point of two displayed $HZCRR$ curves as the threshold to discriminate speech and music leads to the classification error of 19.36%, as shown in [6].

Low Short-Time Energy Ratio

As an analogy for selecting $HZCRR$ to model the variations of the ZCR within the analysis window,

the *low short-time energy ratio* ($LSTER$) [6, 11] can be defined to model the variation of the *short-time energy* (STE) in this window. $LSTER$ is defined as the fraction of the frames within the analysis window, whose STE values are less than a half of the average STE measured in the window, that is,

$$LSTER = \frac{1}{2N} \sum_{n=0}^{N-1}$$

$$[\text{sgn}\,(0.5 av STE - STE(n)) + 1] \tag{13}$$

Here, N is the total number of frames in the analysis window, $STE(n)$ is the short time energy at the n-th frame, and $avSTE$ is the average STE in the window. The $LSTER$ measure of speech is expected to be much higher than that of music. This can be seen clearly from the probability distribution curves of $LSTER$ obtained for a large number of speech and music signals, as illustrated in the Fig. 2b. Using the cross-point of two displayed $LSTER$ curves as the threshold to discriminate speech and music leads to the classification error of 8.27%, as presented in [6].

Spectrum Flux

Spectrum Flux (SF) is defined as the average variation of the spectrum between adjacent two frames in the analysis window, that is

$$SF = \frac{1}{(N-1)(K-1)} \sum_{n=1}^{N-1} \sum_{k=1}^{K-1}$$

$$[\log(A(n,k) + \delta) - \log(A(n-1,k) + \delta]^2 \tag{14}$$

Here, $A(n, k)$ is the absolute value of the k-th DFT coefficient of the n-th frame, K is the order of DFT, δ is a very small value used to avoid computation overflow, and N is the number of frames in the analysis window. The SF of speech is expected to be larger than that of music. It was also found that the spectrum flux of environmental sounds is generally very high, and that it changes more dynamically than for speech and music [6]. To illustrate this, Fig. 3a shows the SF computed for an audio segment consisting of speech (0–200 s), music (201–350 s) and environmental sounds (351–450 s). Usage examples of this feature are provided in [8, 6, 11].

Noise Frame Ratio

Noise frame ratio (NFR) is defined as the ratio of noise frames in a given audio clip. A frame is considered as a noise frame if the maximum local peak of its normalized correlation function is lower than a pre-set threshold. The NFR is usually used to discriminate environmental sound from music and speech, and to detect noisy sounds.

For example, the NFR value of a noise-like environmental sound is higher than that of music, because it contains many more noise frames. As can be observed in Fig. 3b, considering higher NFR values can prove quite discriminative in separating these two types of audio. An illustration of the usage of this feature can be found in [5, 6].

Feature Vector Generation

After they are extracted, features need to be combined together to form a complete audio representation and to provide input into audio analysis and processing steps. Since the values and dynamics of these features may vary considerably over the feature set, simply concatenating them all into a long feature vector is not likely to lead to good results. Therefore, a *normalization* needs to be performed on the features first to equalize their scales. The normalization is usually performed using the mean and standard deviation per feature, namely as $x_i' = (x_i - \mu_i)/\sigma_i$, where x_i is the i-th feature, and where the corresponding mean μ_i and standard deviation σ_i can be obtained from the analyzed data set. Next to the normalization, *feature selection* is usually performed to improve the effectiveness of the feature vector while minimizing its dimension. While feature selection can be realized in many ways [4], a typical approach involves the *principle component analysis* (PCA) [13]. Technically, PCA is an orthogonal linear transformation that transforms the data to a new coordinate system to reveal the main characteristics (*principal components*) of the data that contribute most to the variance in data, and therefore best explain the data. The principal components can be obtained by performing a covariance analysis or *singular value decomposition* (SVD) [13]. If X' is a set of N-dimensional normalized feature vectors from M segments (usually $M \gg N$), then X' can be written as an $M \times N$ matrix, where each row corresponds to a feature vector of one audio segment. By applying the SVD, the matrix X' can be written as

$$X' = USV^T \tag{15}$$

In terms of SVD, V and U are, respectively, an $N \times N$ and $M \times N$ matrix containing the right

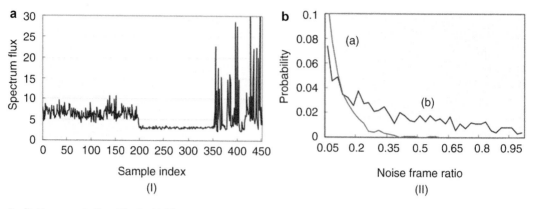

Audio Representation, Fig. 3 (**a**) The spectrum flux curve of speech (0–200 s), music (201–350 s) and environmental sounds (351–450 s). (**b**) The probability distribution curves of *NFR*: (*a*) music and (*b*) environmental sound

and left singular vectors, while the diagonal $N{\times}N$ matrix $S = \text{diag}\{\lambda_1 , \ldots, \lambda_N ,\}$ contains singular values, with $\lambda_1 \geq \lambda_2 \geq \ldots \geq \lambda_N$. In terms of *PCA*, singular vectors (columns) of the matrix V can be seen as principal components of X', each of which has its corresponding singular value. The larger this singular value is, the more principal (or more important) the component is. Assuming that V_m is a matrix keeping the first m principal components (by keeping the first m columns from V), the original feature set X' can be replaced by a reduced, PCA-transformed feature set

$$X'' = X' V_m \qquad (16)$$

which only preserves those features that are relevant to subsequent audio signal analysis and processing steps, while leaving out the redundant and irrelevant (noisy) features.

Key Applications

Audio representation provides fundamentals for audio classification and audio segmentation.

Cross-References

▶ Audio Classification
▶ Audio Content Analysis
▶ Audio Segmentation

Recommended Reading

1. Cai R, Lu L, Hanjalic A, Zhang H-J, Cai L-H. A flexible framework for key audio effects detection and auditory context inference. IEEE Trans Audio Speech Lang Process. 2006;14(3):1026–39.
2. Casey MA. MPEG-7 sound-recognition tools. IEEE Trans Circuits Syst Video Technol. 1997;11(6):737–47.
3. Foote J. Content-based retrieval of music and audio. In: Proceedings of the SPIE Multimedia Storage and Archiving Systems II; 1997. p. 138–47.
4. Guyon I, Elisseeff A. An introduction to variable and feature selection. J Mach Learn Res. 2003;3(7/8):1157–82.
5. Liu Z, Wang Y, Chen T. Audio feature extraction and analysis for scene segmentation and classification. J VLSI Signal Process Syst. 1998;20(1–2):61–79.
6. Lu L, Zhang H-J, Jiang H. Content analysis for audio classification and segmentation. IEEE Trans Speech Audio Process. 2002;10(7):504–16.
7. Lu L, Zhang H-J, Li S. Content-based audio classification and segmentation by using support vector machines. ACM Multimed Syst J. 2003;8(6):482–92.
8. Peltonen V, Tuomi J, Klapuri AP, Huopaniemi J, Sorsa T. Computational auditory scene recognition. In: Proceedings of the IEEE International Conference on Acoustics, Speech and Signal Processing; 2002. p. 1941–4.
9. Rabiner L, Juang BH. Fundamentals of speech recognition. Englewood Cliffs: Prentice-Hall; 1993.
10. Saunders J. Real-time discrimination of broadcast speech/music. In: Proceedings of the IEEE International Conference on Acoustics, Speech and Signal Processing; 1996. p. 993–6.
11. Scheirer E, Slaney M. Construction and evaluation of a robust multifeature music/speech discriminator.

In: Proceedings of the IEEE International Conference on Acoustics, Speech and Signal Processing; 1997. p. 1331–4.

12. Tzanetakis G, Cook P. Marsyas: a framework for audio analysis. Organ Sound. 2000;4(3).

13. Wall ME, Rechtsteiner A, Rocha LM. Singular value decomposition and principal component analysis. In: Berrar DP, Dubitzky W, Granzow M, editors. A practical approach to microarray data analysis. Norwell: Kluwer; 2003. p. 91–109. LANL LA-UR-02-4001.

14. Wold E, Blum T, Wheaton J. Content-based classification, search and retrieval of audio. IEEE Multimedia. 1996;3(3):27–36.

15. Zhang T, Kuo C-CJ. Video content parsing based on combined audio and visual information. In: Proceedings of the SPIE: Multimedia Storage and Archiving Systems IV; 1999. p. 78–89.

Audio Segmentation

Lie Lu[1] and Alan Hanjalic[2]
[1]Microsoft Research Asia, Beijing, China
[2]Delft University of Technology, Delft, The Netherlands

Synonyms

Audio parsing; Auditory scene detection

Definition

Audio segmentation refers to the class of theories and algorithms designed to automatically reveal semantically meaningful temporal segments in an audio signal, also referred to as *auditory scenes* [7]. These scenes can be seen as equivalents of paragraphs in text, and can serve as input into audio categorization processes, either supervised (audio classification) or unsupervised (audio clustering). Through these processes, semantically similar auditory scenes can be grouped together and/or labeled using semantic indexes to provide multi-level, non-linear content-based access to large audio documents and collections.

Historical Background

Automatic detection of *auditory scenes* is an important step in enabling high-level semantic inference from general audio signals, and can benefit various content-based applications involving both audio and multimodal (multimedia) data sets. Traditional approaches to audio segmentation usually rely on a direct analysis of low-level audio features, that is, the targeted audio segments were often defined to coincide with a consistent low-level feature behavior [2, 10, 11]. This idea served as a basis for numerous approaches for audio segmentation. For example, in [10], a method for scene segmentation was presented that uses low-level features, such as cepstral and cochlear decomposition, combined with the listener model and various time scales. Motivated by the known limitations of traditional low-level feature based approaches, an approach was proposed in [7] to discover auditory scenes based on an analysis of *audio elements*, which can be seen as equivalents to the words in a text document. In this approach that draws an analogy to text document analysis, an audio track is described as a sequence of audio elements, and auditory scenes are segmented based on the semantic affinity among these audio elements and their co-occurrence.

Foundations

Traditional approaches to audio parsing relying directly on audio features have proved effective for many applications, and in particular for those where knowledge on the basic audio modalities (speech, music, and noise) is critical. However, for other applications, like those where higher-level content categories, e.g., semantic concepts, become interesting, the low-level feature based approaches have shown deficiencies due to their incapability of capturing the entire content diversity of a typical semantic concept. The audio segments obtained by typical feature-based approaches are short and of no higher semantic meaning, if compared to the true semantic segments, like, for instance, *logical*

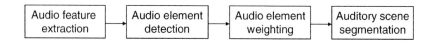

Audio Segmentation, Fig. 1 The framework for audio segmentation based on audio elements [6]

story units targeted by the algorithms of high-level video parsing [3], or the paragraphs in a text document.

To come closer to the level of auditory scenes, a promising alternative is to design and employ suitable mid-level audio content representations. Figure 1 shows the framework for audio segmentation [6] where the input audio is first decomposed into various *audio elements* such as speech, music, various audio effects and any combination of these. Then, *audio element weighting* is performed to reveal the importance of an audio element to represent an audio document or any of its parts. The audio elements with highest weights can be adopted as the *key audio elements*, being the most characteristic for the semantics of the analyzed audio data [1, 4, 5, 8, 9, 12–14]. Finally, auditory scenes can be characterized and detected based on the audio elements they contain, just as the paragraphs of a text document can be characterized and detected using a vector of words and their weights. As shown in [7], introducing the mid-level audio content representation in the form of audio elements enables splitting the semantics inference process into two steps, which leads to more robustness compared to inferring the semantics from low-level features directly.

The usefulness of audio elements for audio content analysis was already recognized, e.g., in [13], where the audio elements such as *applause*, *ball-hit*, and *whistling*, are extracted and used to detect the highlights in sports videos. However, this and similar methods usually adopted supervised data analysis and classification methods. There, the scene categories and the corresponding audio elements need to be predefined, which is usually difficult to do for general audio documents. Further, the effectiveness of supervised approaches relies heavily on the quality and quantity of the training data. This makes such approaches difficult to generalize. In view of this, a number of unsupervised approaches were proposed, including the approach to audio element discovery [1], and an approach to auditory scene segmentation [7]. The latter exploits the co-occurrence phenomena among audio elements to realize the segmentation scheme from Fig. 1. This is based on the rationale that, in general, some audio elements will rarely occur together in the same semantic context. On the other hand, the auditory scenes with similar semantics usually contain similar sets of typical audio elements. For example, many action scenes may contain *gunshots* and *explosions*, while a typical scene in a situation comedy may be characterized by a combination of *applause, laughter, speech,* and *light music.*

Audio Elements Detection and Weighting

As proposed in[1], an iterative *spectral clustering* method can be used to decompose an audio document into audio elements. Spectral clustering can be seen as an optimization problem of grouping similar data based on eigenvectors of a (possibly normalized) affinity matrix. Ng et al. [9] proposed a method to use k eigenvectors simultaneously to partition the data into k clusters, and successfully applied this to a number of complicated clustering problems. To further improve the robustness of the clustering process, the self-tuning strategy [14] can be adopted to set the context-based scaling factors for different data densities. This removes the need for the assumption that each cluster in the input data has a similar distribution density in the feature space, which is inherent in the standard spectral clustering algorithm, but usually not satisfied in complex audio data. Using this clustering method, short audio segments (e.g., one second

Audio Segmentation, Fig. 2 An illustration of previous approaches to auditory scene segmentation, where a vertical line indicates a detected scene boundary: (a) using time interval between key audio elements; (b) using semantic affinity between neighboring key audio elements; and (c) investigating the relationship of (key) audio elements on a large temporal scale

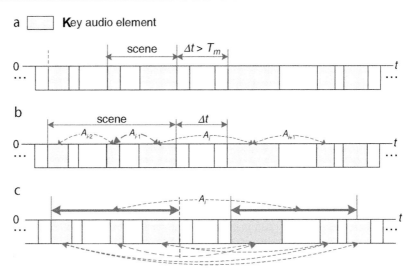

in length [1]) can be grouped into natural semantic clusters that can then be adopted as audio elements.

In the next step, the obtained audio elements are assigned the importance weights to indicate their prominence in characterizing the content of audio data. Here, two cases can be considered. The first case assumes that only one audio document is available for weight computation. Then, a number of heuristic importance indicators, including *Element Frequency, Element Duration,* and *Average Element Length,* are proposed in [1] to compute the weight. The second case assumes that multiple audio documents are available to learn the weights. In this case, inspired by the effectiveness of *term frequency (TF)* and *inverse document frequency (IDF)* used for word weighting in text document analysis, the equivalents of these measures can be defined and employed for the case of audio segmentation. As described in [1, 4, 5, 8, 9, 12–14], four factors, including *expected term frequency (ETF),* e*xpected inverse document frequency (EIDF), expected term duration (ETD),* and *expected inverse document duration (EIDD),* can be defined and combined together to give the importance weight of each audio element. These factors take into account the discriminative power of the occurrence frequency and the duration of a particular audio element to characterize the semantics of an audio document.

Auditory Scene Segmentation

In view of the way it is defined, an auditory scene may consist of multiple, concatenated and semantically related audio elements. An example of such an auditory scene is a *humor* scene consisting of several interleaved segments of *speech, laughter, cheer,* and possibly also some *light music.* In [6], a simple segmentation scheme was presented that employs crisply defined key audio elements. As shown in Fig. 2a, two adjacent key audio elements are assumed to be in the same auditory scene if the time interval between them is sufficiently short. Then the scene boundaries are aligned to the key audio elements, while the background audio elements between two scenes are discarded. Clearly, the algorithm is quite naive and does not fully exploit the relationship between audio elements and auditory scenes. To improve the detection performance, the notion of *semantic affinity* between two contiguous key audio elements was introduced in [1]. This affinity takes into account both the co-occurrence statistics of these key audio elements and the time interval between them, and was employed in [1] to locate the auditory scene boundaries. As shown in Fig. 2b, auditory scene boundaries are found between two key audio elements if their semantic affinity is low.

The performance of the segmentation methods discussed above strongly depends on the definition of a key audio element and the reliability of

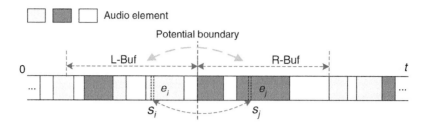

Audio Segmentation, Fig. 3 An illustration of an approach to audio segmentation [7], where s_i and s_j are two audio segments, and e_i and e_j are their corresponding audio element identities

its detection. Crisply defining key audio elements and detecting them in composite audio documents may be rather difficult due to multiple superimposed audio modalities. Therefore, a more reliable solution would be to work with all audio elements instead, and rely on their importance weights. This idea also follows the analogy to the classical text [4] and video scene segmentation approaches [3, 5]. As illustrated in Fig. 2c, an approach in this direction would decide about the presence of a scene boundary at the observed time stamp based on an investigation of the semantic affinity between audio elements taken from a broader range and surrounding this time stamp.

The possibilities for realizing the audio segmentation idea from Fig. 2c are now illustrated on the example of the method proposed in [7]. Here, just like in text document analysis, the measure for semantic affinity is not based on the feature-based similarity between two audio segments, but on their joint ability to represent a semantically coherent piece of audio. With this in mind, the definition of semantic affinity in [7] is based on the following intuitive assumptions:

- Affinity between two audio segments is high if the corresponding audio elements usually occur together.
- The larger the time interval between two audio segments, the lower their affinity.
- The higher the importance weights of the corresponding audio elements, the more important is the role these elements will play in the auditory scene segmentation process, and therefore the more significant the computed semantic affinity value will be.

Figure 3 shows an example audio element sequence, where each temporal block belongs to an audio element and where different classes of audio elements are represented by different colors/grayscales. The semantic affinity between the segments s_i and s_j can now be computed as a function consisting of three components, each of which reflects one of the assumptions stated above. The following measure is proposed in [1]:

$$A\left(S_i, S_j\right) = Co\left(e_i, e_j\right) e^{-T\left(S_i, S_j\right)/T_m} P_{e_i} \, P_{e_j} \tag{1}$$

Here, the notation e_i and e_j is used to indicate the audio element identities of the segments s_i and s_j, that is, to describe their content (e.g., speech, music, noise, or any combination of these). P_{ei} and P_{ej} are the importance weights of audio elements e_i and e_j, while $T(s_i, s_j)$ is the time interval between the audio segments s_i and s_j. Further, T_m is a scaling factor which can be set to 16 s, following the discussions on human memory limit [3]. The exponential expression in Eq. 1 is inspired by the content coherence computation formula introduced in [12]. Further, $Co(e_i, e_j)$ stands for the co-occurrence between two audio elements, e_i and e_j, in the entire observed audio document, and measures their joint ability to characterize a semantically coherent auditory scene.

To estimate the co-occurrence between two audio elements, one can rely on the average time interval between two audio elements. The shorter the time interval, the higher the co-occurrence probability is. The procedure for estimating the

Audio Segmentation, Fig. 4 An example of the smoothed confidence curve and the auditory scene segmentation scheme, where $S_1^* \sim S_5^*$ are five obtained auditory scenes and Th and Th_2 are two thresholds

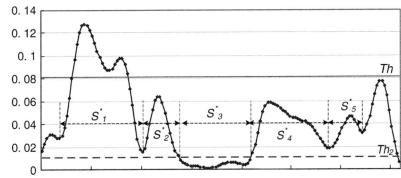

value $Co(e_i, e_j)$ can then be summarized in the following three steps [1]:

1. First, compute D_{ij}, the average time interval between audio elements e_i and e_j, which is obtained by investigating the co-occurrences of the observed audio elements in the audio signal. For each segment belonging to audio element e_i, the nearest segment corresponding to e_j is located, and then D_{ij} is obtained as the average temporal distance between e_i and e_j.
2. D_{ji} is computed as an analogy to D_{ij}. One should note that D_{ij} is not always equal to D_{ji}.
3. The co-occurrence value can now be found as

$$Co\left(e_i, e_j\right) = e^{-\frac{D_{ij} + D_{ji}}{2\mu_D}} \qquad (2)$$

where μ_D is the average of all D_{ij} and D_{ji} values. The choice for an exponential formula in Eq. 2 is made to keep the influence of audio element co-occurrence on the overall semantic affinity comparable with the influence of the time interval between the audio segments Eq. 1.

Based on the semantic affinity Eq. 1, the confidence of being within an auditory scene at the time stamp t can now be computed by averaging the affinity values obtained for all pairs of segments s_i and s_j surrounding the t, that is,

$$C(t) = \frac{1}{N_l N_r} \sum_{i=1}^{N_l} \sum_{j=1}^{N_r} A\left(S_i, S_j\right)$$

$$= \frac{1}{N_l N_r} \sum_{i=1}^{N_l} \sum_{j=1}^{N_r} Co\left(e_1, e_j\right) e^{-T\left(S_i, S_j\right)/T_m} P_{e_i} P_{e_j}$$

$$(3)$$

where N_l and N_r are the numbers of audio segments considered left and right from the potential boundary (as captured by the intervals L-Buf and R-Buf in Fig. 3). Using this expression, a confidence curve can be obtained over the timeslots of potential boundaries, as illustrated in Fig. 4. The boundaries of auditory scenes can now be detected simply by searching for local minima of the curve. In the approach from [1], the curve is first smoothed by using a median filter and then the auditory scene boundaries are found at places at which the following criteria are fulfilled:

$$C(t) < C(t + 1);$$

$$C(t) < C(t - 1); \quad C(t) < Th \qquad (4)$$

Here, the first two conditions secure a local valley, while the last condition prevents high valleys from being detected. The threshold Th is set experimentally as $\mu_a + \sigma_a$, where μ_a and σ_a are the mean and standard deviation of the curve, respectively.

The obtained confidence curve is likely to contain long sequences of low confidence values, as shown by the segment S_3^* in Fig. 4. These sequences typically consist of the background audio elements which are weakly related to each other and also have low importance weights. Since it is not reasonable to divide such a sequence into smaller segments, or to merge them into neighboring auditory scenes, one could choose to isolate these sequences by including all consecutive audio segments with low affinity values into a separate auditory scene. Detecting such scenes is an analogy to detecting pauses in speech. Inspired by this, the corresponding threshold (Th_2 in Fig. 4) can be set by using an approach similar to background noise level detection in speech analysis [12].

Key Applications

Audio segmentation is typically applied for content-based search and retrieval in and management of large-scale audio collections (databases).

Cross-References

▶ Audio Classification
▶ Audio Representation
▶ Video Content Structure

Recommended Reading

1. Cai R, Lu L, Hanjalic A Unsupervised content discovery in composite audio. In: Proceedings of the 13th ACM International Conference on Multimedia; 2005. p. 628–37.
2. Foote J. Automatic audio segmentation using a measure of audio novelty. In: Proceedings of the IEEE International Conference on Multimedia and Expo; 2000. p. 452–5.
3. Hanjalic A, Lagendijk RL, Biemond J. Automated high-level movie segmentation for advanced video-retrieval systems. IEEE Trans Circuits Syst Video Technol. 1999;9(4):580–8.
4. Kozima H. Text segmentation based on similarity between words. In: Proceedings of the 31st Annual Meeting on Association for Computational Linguistics; 1993. p. 286–8.
5. Kender JR, Yeo B-L. Video scene segmentation via continuous video coherence. In: Proceedings of the IEEE International Conference on Computer Vision and Pattern Recognition; 1998. p. 367–73.
6. Lu L, Cai R, Hanjalic A. Towards a unified framework for content-based audio analysis. In: Proceedings of the IEEE International Conference on Acoustics, Speech and Signal Processing; 2005. p. 1069–72.
7. Lu L, Cai R, Hanjalic A. Audio elements based auditory scene segmentation. In: Proceedings of the IEEE International Conference on Acoustics, Speech and Signal Processing; 2006. p. 17–20.
8. Lu L., Hanjalic A. Towards optimal audio keywords detection for audio content analysis and discovery. In: Proceedings of the 14th ACM International Conference on Multimedia; 2006. p. 825–34.
9. Ng AY, Jordan MI, Weis Y. On spectral clustering: analysis and an algorithm. In: Proceedings of the Advances in Neural Information Processing Systems; 2001. p. 849–56.
10. Sundaram H, Chang S-F. Audio scene segmentation using multiple features, models and timescales. In: Proceedings of the IEEE International Conference on Acoustics, Speech and Signal Processing; 2000. p. 2441–4.
11. Tzanetakis G, Cook P. Multifeature audio segmentation for browsing and annotation. In: Proceedings of the IEEE Workshop on Applications of Signal Processing to Audio and Acoustics; 1999. p. 103–6.
12. Wang D, Lu L, Zhang H-J. Speech segmentation without speech recognition. In: Proceedings of the IEEE International Conference on Acoustics, Speech and Signal Processing; 2003. p. 468–71.
13. Xu M, Maddage N, Xu C-S, Kankanhalli M, Tian Q. Creating audio keywords for event detection in soccer video. In: Proceedings of the IEEE International Conference on Multimedia and Expo; 2003. p. 281–4.
14. Zelnik-Manor L, Perona P. Self-tuning spectral clustering. In: Proceedings of the Advances in Neural Information Processing Systems; 2004. p. 1601–8.

Auditing and Forensic Analysis

Brian Levine and Gerome Miklau
University of Massachusetts, Amherst, MA, USA

Synonyms

Accountability; Monitoring

Definition

The goal of *database auditing* is to retain a secure record of database operations that can be used to verify compliance with desired security policies, to trace policy violations, or to detect anomalous patterns of access. An audit log can contain the authorization ID and time stamp of read and write operations in the database, as well as a record of server connections, login attempts and authorization changes. Government and institutional regulations for the management of sensitive information often require auditing of data disclosure and data modification.

Database forensics is the analysis of the state of a database system to validate hypotheses about past events that are relevant to an alleged crime or violation of policy. Evidence supporting a forensic analysis may be found in an audit log (if available) but may also be recovered from any other component of a database system including table storage, the transaction log, temporary caches, or backup media. A challenge of working with forensically recovered evidence is that it is typically provides only a partial record of an event, possibly based on remnants of deleted information or on inference from incomplete information. On the other hand, it can be difficult to completely eradicate digital evidence from databases, which may be critical for preventing disclosure and complying with policy mandating limited data retention.

While auditing is focused on the preservation and analysis of specific data as required by law or internal corporate policy, digital forensics is broader in scope, potentially relating to any criminal or civil proceeding. Auditors and forensic analysts share some common goals, however the auditor usually relies on information retained intentionally by the system. The forensic analyst is more likely to rely on unintended remnants and inference about past events.

Historical Background

Auditing has been a common practice in settings where sensitive data is managed by computer systems, such as financial and military applications. Auditing has grown in importance as institutions are increasingly required to prove compliance with privacy regulations, or are mandated to discover and publicly respond to exploited vulnerabilities in their systems. Database forensics is an emerging subfield of digital forensics. Computer crime and investigations began receiving attention the late 1970s, but the realization that digital evidence is relevant to a spectrum of crimes has occurred within the last decade. Databases are common components of operating systems, web browers, and email programs and are an important focus of digital investigations.

Foundations

Database Auditing

The auditing component of a database system must support the collection, storage, and protection of sufficient historical data to enable desired auditing inquiries. Typical auditing queries might include the following:

- Display the query expression for all operations which modified more than ten rows.
- List the authorization IDs of users or client programs who performed SELECT queries on the *Patients* table between 10 p.m. and 5 a.m. last week.
- List any records in the *Employee* table whose *salary* field has been modified more than twice in the past 12 months.

In misuse detection or intrusion detection, the audit log may be used to assess more complex behavior such as: *Is today's workload of update operations similar to "normal" patterns of database usage?*

To support such inquiries, an audit log records the client programs and users who are executing operations, the data objects modified or disclosed, and the context of those operations. For each operation performed, an audit log could contain the SQL query string along with contextual information such as time of day, authorization ID, and network connection information. For update and deletion operations, the audit log may contain the removed values, and the previous values of modified data.

Analyzing database disclosure resulting from a sequence of SELECT queries can be more complex than analyzing the history of database modifications (inserts, updates, and deletes). Auditing disclosure has been the subject of intense research [1, 2, 4], but faces subtle challenges because a user may be able to infer information not directly released through an executed query. For example, a user's access to an individual database record can be hidden in a sequence of aggregate queries.

In establishing auditing policies, the credibility and completeness of the audit log must be carefully considered to ensure that all relevant events in the system are preserved. For example, the effects of an aborted transaction will be removed from the system for database consistency and atomicity. But a record of aborted transactions, and the reason for abort, could be important to an audit analysis [4].

Naturally, data that will never be relevant to the audit queries under consideration need not be recorded. Further, the retention of recorded audit data must be carefully determined and enforced. The log should be available for appropriate audit inquiries when they arise, but it also contains highly sensitive information and should be destroyed once the period of legitimate inquiry has passed.

Systems Issues and Performance Considerations To support auditing, database systems must efficiently collect required data and permit analysis. Modern commercial databases contain a range of native auditing features which usually include more than one type of log. Common features include a system log to record all connections to the database, and an query log to record each query expression submitted to the database. An auditing policy can be chosen to specify the level of detail that should be logged (e.g., all accesses to relations, or individual tuples; logging of first access in a session or all accesses, etc.).

Logged data may be written to files outside database storage, or to system tables within the database. In the former case, protection depends on operating system access control, while in the latter case protection of audit data depends on the access controls of the database. In particular, the DBA often has privileges to read or alter system tables. When audit data is stored in tables it can have a substantial performance impact on normal database operations.

Database users can implement their own auditing through user-level triggers. A trigger is a user-defined procedure that executes before or after designated events in the database. The triggering event can be an insert, delete, or update command, and most systems allow for tuple-level execution (in which the rule executes once for each tuple affected) or statement level (in which rule executes once for each statement). User-level triggers can be inefficient (especially tuple-level triggers) and the scope of events that can act as trigger events may be limited.

The database transaction log is designed to support critical ACID properties for the concurrent execution of transactions. The transaction log typically includes the before and after images of all database modifications to allow for rollback of aborted transactions and the redo of changes lost due to system failure. While the transaction log contains a wealth of information relevant to auditing, it has a number of limitations when used as an auditing mechanism. First, key data is missing from the transaction log: namely a record of read accesses to the database, as well as some operational context information. In addition, performing an audit analysis using the transaction log could require recovering the state of the database as of a past moment in time. Although a number of current systems provide such point-in-time recovery by reinstating backups and rolling transactions forward, using this mechanism for audit analysis is very inefficient. Finally, the retention period of transaction log data and audit log data may be substantially different. Transaction logs are often implemented as circular files in which old log records no longer needed for recovery are overwritten. Retention periods for auditing data may be much longer.

In a persistent database (also known as an archiving, or transaction-time database) the historical state of the database is purposely retained as modifications are applied. In such systems a deletion never destroys data and an update merely creates a new version of a tuple. It is possible to pose queries "as-of" any past point in time. Persistent databases have received considerable attention from the research community [5], motivated by both auditing and other applications.

Combined with query logs and system logs, a persistent database can offer the most complete audit collection along with efficient audit analysis since the historical state of the database can be queried.

Support for persistence has not been widely implemented, and is not usually used to support auditing in commercial systems. A number of research projects have built persistent, temporal or transaction-time databases, many as extensions to existing systems like MySQL [8], BerkeleyDB [7], and SQL Server.

Protecting the Audit Log It is essential that the audit log be protected from unauthorized modification so that it accurately reflects history. The database administrator, and other privileged parties, should not be capable of tampering with the audit log. Typically there is no legitimate reason for records in an audit log to be modified. The log should be append-only and may be implemented using write-once media. Deletion of the audit log should occur only when it is clear that the log is no longer needed for audit inquiries. Cryptographic techniques have been proposed for detecting tampering of database audit logs and for efficiently tracing the location of illegally modified records once tampering has been discovered.

It is equally critical that the confidentiality of the audit log be protected. The audit log poses multiple privacy threats: the audit log contains records from database that may be sensitive, as well as a history of how the database was used (the users of the database, the queries that were executed, and times of day can all violate the privacy of individuals). Viewing audit logs is a highly privileged operation in most systems. Recent research has investigated the use of cryptography to permit searching over encrypted audit logs to minimize the disclosed data during an investigation [10].

Database Forensics

The goal of database forensics is the analysis of a database system's contents to validate hypotheses about past events that are relevant to an alleged crime or violation of policy. This is a challenge since recovered evidence is typically only a par-

tial record of past events. The goal of the analysis is a formal presentation of recovered data in a court of law, and therefore it is critical for the investigator to also understand legal concepts, including evidence handling. Unlike auditing, there is no limitation on the type of data that can be of interest. Broadly, there are two types of evidence.

- Database evidence can be the direct subject of a crime. For example, records can store contraband, such as images of child pornography, copyrighted media, and stolen intellectual property.
- Evidence can be indirectly related to a crime, for example data from a log that verifies that a relationship exists between two users or computers. Or a log of database query terms can corroborate the notion that a user had knowledge of and intent to possess particular contraband content found in their file system.

Typically investigations cover not only databases but other computer systems (e.g., file systems, email stores, web history) as well as aspects of a crime scene beyond the computer. Evidence from the entire scope of a crime scene must be synthesized and reported as testimony that is persuasive to an adjudicator.

Harvesting Database Evidence Databases are among the most complicated systems found in modern computers. Investigators can harvest evidence from table storage, indexes, transaction and audit logs, temporary caches, database catalogs, or archived copies of records on backed up media.

Data values have a complex lifetime with a database system. When input to a database, records begin in an active state, which means a database and its services need the record in order to function properly. Database operations can change or create active records or remove the purpose of active records. In the latter case, records become expired. For example, a record becomes expired when it is deleted and when there is no combination of operations left that would ever use the record again. Forensic investigators are interested in both active and recoverable expired records.

The lifetime of data is illustrated in Fig. 1 for the simple case of records in table storage. An insertion creates an active record, a deletion changes a record from being active to being expired and eventually the data may be overwritten by another database operation. After expiration, a tuple can be either removed, or it can continue to exist as recoverable slack data. Methods of removal are discussed below.

In the case of table storage, each paged file of storage is shared by many records. The database API enables users or investigators to retrieve all active values. When data is deleted by users, typically a single bit is flipped in the page file to indicate removal of the data. However, the record can be retrieved outside the mechanisms of the API.

Other database mechanisms can also leave recoverable records. Updates to records with variable lengths can replace one or more attribute values with smaller attribute values; the tail-end of the old record will remain partially recoverable until it is overwritten. Or an administrator may initiate a *vacuum* command to improve storage performance. When vacuum executes, on many systems, the reorganization is not performed completely in place. In addition to reorganizing records within and across pages, the size of the file used for table storage may be reduced, returning space to the file system, creating the possibility that the database records can be recovered through file system forensics.

Expired data can also be found in stored indexes if, for example, entries in B+tree nodes are deleted but not overwritten immediately. Temporary relations, materialized for improved query processing or used for external sorting, also contain data recoverable as filesystem slack. Data can be recovered from transaction logs, which are typically implemented sequentially written circular files where the newest data overwrites the oldest portion of the log. The amount of recoverable data is dependent on the file system space allocated to the log and characteristics of the database, including the rate and size of updates and checkpoints. Similarly, the amount of data stored in backups varies with policy, storage capacity, and use.

System Transparency and Privacy Forensic analysis is not restricted to active tuples because database designs do not strive to eliminate unintended retention of data accessible through interfaces that are not controlled by the database. This incongruence between what the database presents to users and what is actually stored represents a threat to privacy and confidentiality.

For example, as stated above, businesses can unintentionally violate privacy regulations when deleted data is left in table or file storage. Adversaries that investigate databases recovered from lost or stolen computers can reveal sensitive information that was thought to be deleted.

From this point of view, it is desirable for a database to operate in a forensically *transparent* way [9]. Stahlberg et al. have proposed a set of desiderata to determine the extent to which a database system is forensically transparent. A database system is forensically transparent if it satisfies all three desiderata.

- *Clarity*: The impact of each operation on the state of records, whether active or expired, is clear to the user.

Auditing and Forensic Analysis, Fig. 1 The flow of data during its lifetime. It begins in the active state. Before it is deleted and becomes expired it will often be retained as database slack or filesystem slack [9]

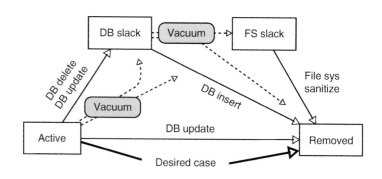

- *Purposeful retention*: Only *active* records should be retained by the database.
- *Complete removal*: Expired records must be *removed* by the system within a short, fixed time from when they become expired. In other words, there must be a small upper bound on the time that slack data exists in the database.

Databases that satisfy these desiderata provide a reliable interface to the user in terms of what data is actually stored in the system.

Removal of Data Ensuring that removed data is unrecoverable from a database system is difficult. Records can be removed by overwriting storage with a standard pattern (e.g., all zeros) or through the use of encryption.

Deletion through overwriting can be costly in terms of performance and therefore should be limited to small records or files. Asynchronous overwriting during idle periods can ameliorate performance degradation while opening a window of opportunity for recovering data. Secure removal of data can also be accomplished by storing data in encrypted form and using overwriting to remove the decryption key. This technique was first proposed for efficient simultaneous removal of data from files and backup logs. Stahlberg et al. [9] present an extended discussion of the use of encryption keys and overwriting in the different internal database mechanisms.

Key Applications

Auditing and forensic analysis are critical operations in any setting where sensitive or high-value data is exposed to untrusted users. For example, database applications that manage financial, national intelligence, or medical information must maintain audit records of past data and operations, and may be the subject of forensic analysis if policy violations occur.

Future Directions

A frequent outcome of auditing and forensic analysis is the detection of a malicious or mistaken operation in the past. Correcting the corrupted database, especially after many valid transactions have been applied, is a significant challenge receiving recent attention by the research community [3, 6].

Also note that the above discussion has focused on a single database server. In modern applications, a comprehensive audit or forensic analysis is likely to involve many integrated systems where data objects are derived from diverse sources and follow a complex workflow. In such settings auditing inquiries share some similarities with data provenance (or lineage), which is a record of how a data item has come to exist in a database or view.

Cross-References

▸ Database Security
▸ Inference Control in Statistical Databases
▸ Intrusion Detection Technology

Recommended Reading

1. Adam NR, Wortmann JC. Security-control methods for statistical databases: a comparative study. ACM Comput Surv. 1989;21(4):515–56.
2. Agrawal R, Bayardo RJ, Faloutsos C, Kiernan J, Rantzau R, Srikant R. Auditing compliance with a hippocratic database. In: Proceedings of the 30th International Conference on Very Large Data Bases; 2004. p. 516–27.
3. Ammann P, Jajodia S, Liu P. Recovery from malicious transactions. IEEE Trans Knowl Data Eng. 2002;14(5):1167–85.
4. Castano S, Fugini MG, Martella G, Samarati P. Database security. New York: ACM/Addison-Wesley; 1994.
5. Jensen CS, Mark L, Roussopoulos N. Incremental implementation model for relational databases with transaction time. IEEE Trans Knowl Data Eng. 1991;3(4):461–73.
6. Lomet D, Vagena Z, Barga R. Recovery from "bad" user transactions. In: Proceedings of the ACM SIGMOD International Conference on Management of Data; 2006. p. 337–46.
7. Snodgrass RT, Collberg CS. The τ-BerkeleyDB temporal subsystem. Available at www.cs.arizona.edu/tau/tbdb/
8. Snodgrass RT, Collberg CS. The τ-MySQL transaction time support. Available at www.cs.arizona.edu/tau/tmysql

9. Stahlberg P, Miklau G, Levine B. Threats to privacy in the forensic analysis of database systems. In: Proceedings of the ACM SIGMOD International Conference on Management of Data; 2007. p. 91–102.
10. Waters B, Balfanz D, Durfee G, Smetters D. Building an encrypted and searchable audit log. In: Proceedings of the Network and Distributed System Security Symposium; 2004. p. 91–102.

Authentication

Marina Blanton
University of Notre Dame, Notre Dame, IN, USA

Definition

Authentication is a broad term, which is normally referred to mechanisms of ensuring that entities are who they claim to be or that data has not been manipulated by unauthorized parties. Thus, *entity authentication* or *identification* refers to the means of verifying user identity, after which the user will be granted appropriate privileges. *Data origin authentication* refers to the means of ensuring that the data comes from an authentic source and has not been tampered with during the transmission.

Historical Background

The need for user authentication in early computer systems arose once it became possible to support multi-user environments. Similarly, data base systems that can be accessed by multiple users with different privileges have to rely on user authentication to enforce proper access control. There is a variety of mechanisms that allow users to authenticate themselves, but password-based authentication is currently the most widely used form of identification.

Data origin authentication (or *data authentication* for short) is also an old concept which gained importance with the adoption of inter-computer communications. With respect to data base systems, data authentication is crucial in distributed environments and when data bases are disseminated to other entities.

Foundations

Identification

The purpose of *entity authentication* or *identification* is to allow one party (the *verifier*) to gather evidence that the identity of another party (the *claimant*) is as claimed. Thus, authentication protocols should permit honest parties to successfully finish the protocol with the claimant's identity being accepted as authentic and make it difficult for dishonest parties to impersonate an identity of another user. Impersonation must remain difficult even for an adversary who can observe a large number of successful executions of the authentication protocol by another entity.

Identification mechanisms can normally be divided into the following types depending on how the identity evidence is gathered:

1. *The user knows a secret.* Such types of identification include passwords, personal identification numbers (PINs), or secret keys.
2. *The user possesses a token.* This is normally based on a hardware token such as magnetic-striped cards (or smartcards) or other custom-designed devices that generate time-variant passwords.
3. *The user has a physical characteristic.* Identification can be based on characteristics inherent to the user being authenticated such as biometrics, handwritten signatures, keystroke dynamics, facial and hand geometries, voice, and others.

Note that for added security often different mechanisms can be combined together. For example, PIN-based authentication is almost always used in conjunction with a physical device that stores information about its owner (e.g., user ID, credit card number, etc.). Likewise, biometric-based recognition can be used in combination with a password or a physical token.

Before a user will be able to engage in an authentication protocol, she needs to *register* with the system and store the data (e.g., a password, keys, biometric data) that will thereafter aid the system in the authentication process. In what follows, we first treat password-based authentication, followed by stronger forms of entity authentication.

Password-Based Authentication

Identification based on conventional time-invariant passwords is the most widely used form of identification even though such approaches do not provide strong authentication. A password is a string of (normally eight or more) characters associated with a certain user, which serves the purpose of a shared secret between the user and the system. When a user initiates the identification process, she supplies the system with the pair (*userid*, *password*), where *userid* identifies the user and *password* provides the necessary evidence that the user possesses the secret.

The most straightforward approach for the system to store passwords is in the clear text. This, however, allows the system administrator, or an adversary in case of system compromise, to recover passwords of individual users leading to security concerns. Thus, most systems first apply a one-way *hash function* to each password and store the output in the system. Then when a user supplies the password during the identification process, the password is first hashed and then compared to the string stored in the system. This security measure no longer allows cleartext passwords to be recovered, but does not mitigate many other attacks on passwords such as:

- *Replay of passwords.* Because passwords are reusable, an adversary who obtains password information (by either seeing the user type the password, using a keylogger program, or capturing the password in transit from the user to the system) will be able to reuse it and successfully impersonate the user.
- *Exhaustive search.* An adversary might attempt to guess user passwords by trying all possible strings as potential passwords (on the

verifier itself or by obtaining a copy of the password file and performing this attack off-line). Generally it is infeasible for an adversary to try all passwords if they are chosen from a sufficiently large space, but it is possible to exhaust short passwords (e.g., of length six characters or less). Specific rules on what consititutes a valid password (e.g., a password must start with a letter) narrow the search space.

- *Dictionary attack.* It is well known that users tend to choose passwords that they can easily remember but which are considered weak from the security point of view. Thus, an adversary might try to guess a user password using words from a dictionary and variations thereof. If a user makes a poor password choice, such an attack can have a high probability of success. Dictionary attacks become increasingly complicated, testing for combinations of words, common substitutions and misspellings, insertion of additional symbols, words from foreign languages, etc.

To decrease the vulnerability of the system to these attacks, additional measures are normally employed, some of which are:

- *Salting passwords.* In order to make guessing attacks less effective, many systems use an additional random string, called *salt*, with each password. Before a password is stored, it is augmented with a random salt, hashed, and then stored in the system along with the salt. This prevents an attacker who is in possession of a file with many user passwords to launch a dictionary attack against all of them at the same time, and requires each user's password to be tested individually.
- *Slowing down password verification.* To defeat against attacks that perform trials of a large number of passwords, the function that computes the hash of the password can be made more computationally extensive. This, for example, can be done by iterating the hash computation n times. When increasing the computation for password verification, a

care, however, must be taken not to impose a burden on legitimate users.

- *Limiting the number of unsuccessful password guesses.* It is common for a user account to be locked after the number of unsuccessful authentication attempts exceeds a certain threshold. The owner of a locked account must then contact an administrator and have the account re-activated. Limiting the number of unsuccessful password guesses is an effective mechanism for mitigating exhaustive or dictionary search attacks performed on the verifier.
- *Password rules.* To protect against guessing attacks, often certain rules are imposed on user choice of passwords such as the minimal password length and/or usage of capital letters, numbers, and special symbols. Such rules normally strengthen the password choices but they also limit the password search space. Also, a technique called *password aging* is often employed to force users to choose a new password after a certain period of time. If such a period is rather short, however, users will be unable to choose and remember a new strong password at each time period, thus weakening the security of the system with bad password choices or choosing obvious modifications to the previously used passwords.

It is always a challenge to find a balance between memorability of passwords (passwords that are hard to remember tend to be written down) and their resistance to dictionary attacks (i.e., passwords with enough randomness in them). Thus to aid users in choosing less predictable passwords which they can remember, techniques to create computer-generated *pronounceable passwords* exist, which are based on mnemonics. Also, recently various solutions have been developed to use images for authentication (a user is given a number of images and is asked to identify the set of pre-selected images), graphical interfaces (a user draws a pattern on a grid that has to match a previously chosen pattern), etc. Such systems, however, have not undergone an extensive amount of evaluation and it is difficult to determine the level of security they provide.

Since a major security concern with fixed passwords is the possibility of replaying them, a natural way to improve their security is to consider *one-time passwords*. As the name suggests, in such systems each password is used only once and there are different ways to realize them, which are briefly outlined next.

- The user and the system initially agree on a sequence of secret passwords. Each time the user authenticates to the system, a new password is used. This solution is simple but requires maintenance of the shared list and synchronization of which password will be used next (due to unreliability of communication channels).
- The user updates her password with each instance of the authentication protocol. For instance, the user might send the new password encrypted under a key derived from her current password. This method crucially relies on the correct communication of the new password to the system.
- The new password is derived with each instance of the authentication protocol using a one-way *hash function*. As an example, consider the scheme called S/Key due to Lamport [2]. The user begins with a secret k and applies a one-way *hash function* h to produce a sequence of values $k, h(k), h(h(k)), \ldots, h^t(k)$. The password for ith identification session is $k_i = h^{t-i}(k)$. When the user authenticates $(i + 1)$st time with k_{i+1}, the server (which has k_i for that user stored) checks whether $h(k_{i+1}) = k_i$ and, if so, accepts the authentication and replaces k_i with k_{i+1}. This check convinces the server because the function h is considered to be infeasible to invert and only the legitimate user will be able to construct k_{i+1} that passes the check. Synchronization in presence of errors may also be required.

Challenge-Response Identification

Challenge-response techniques provide a strong form of entity authentication as they are not vulnerable to replay attacks. The main idea behind such protocols is that the claimant possesses a secret. During an identification session, the server sends to the claimant a *challenge* randomly chosen from a large space. The claimant computes a *response*, which is a function of her secret and the server's challenge, and sends it to the server. It is important to note that the response does not provide information about the user secret, and cannot be used to successfully compute responses to server's challenges in the future by someone who monitors the message exchange.

A variety of cryptographic challenge-response techniques exist. They can be based on (i) symmetric encryption, (ii) one-way *hash functions*, or (iii) public-key encryption. A more detailed explanation of such techniques is beyond the scope of this article and can be found in standard textbooks on cryptography such as [3, 7].

Data Origin Authentication

The purpose of *data origin authentication* is to ensure that the data comes from a trusted source and has not been tampered with during the transmission. Techniques that permit verifying data authenticity can be divided in two categories: (i) the communicating parties *share a common secret* and (ii) the communicating parties *do not share a secret*.

Consider that in a distributed database environment two systems communicate often and there is a need to ensure data integrity. Then such systems can share a secret S that permits them to use *message authentication codes* (MAC) for data authentication. That is, prior to transmitting the data, the sender constructs a MAC using S and sends it along with the data. After obtaining the data, the receiver uses the shared secret to compute a MAC of the received message and compare it with the MAC received. If the check succeeds, the data is accepted as authentic, and it is discarded otherwise.

In cases when the sender and the receiver do not already have a secret known to both of them, *digital signatures* can be used to verify the authenticity of the sender and integrity of the data. This mechanism assumes that the sender has a public-private key pair, which is used to sign the data. Then the sender uses her private key to sign the message and the receiver uses the corresponding public key to check whether the message arrived intact. The standard way of producing a digital signature on data is to first apply a one-way *hash function* on the data to compute its digest and then sign the digest. The purpose of computing the digest is to compress the data to a short string, which then can be efficiently signed to produce a fixed-size signature.

In cases when integrity of a database needs to be verified with some regularity while only parts of it change, more advanced techniques can be used. In particular, the technique called a Merkle hash tree is commonly used to produce a digital signature on a hierarchically structured set of documents (e.g., an XML tree of documents). In such a tree, digests of individual nodes are computed and then combined in a bottom-up fashion to result in a single short digest of the tree. The owner of the data produces a signature on the root node only. Verification of data integrity in such trees can be done faster than recomputing digests of the entire tree if the user would like to verify the integrity of only a part of the tree. In particular, verification of integrity of a number of adjacent documents is proportional to the number of the documents being verified plus the height of the tree.

Key Applications

The main application of authentication is access control. Namely, in multi-user systems a user authenticates to the system and is granted access to specific resources determined by her access privileges. The mechanisms used to determine user access rights vary drastically from one system to another and are based on the type of access control (e.g., role-based, discretionary, etc.,) and access control policies.

Also, authentication applications and services such as Kerberos, X.509 Authentication Service, or Public-Key Infrastructure (PKI) can be used to aid in the authentication process.

In case of data authentication, verification of data integrity and authenticity is essential in determining trustworthiness of the data. For example, updated database records that are coming from a trusted source can be safely used to modify the current contents of the database. If, on the other hand, data integrity and authenticity of the modifications cannot be verified, in many cases such data will not be trusted.

Cross-References

► Access Control
► Digital Signatures
► Hash Functions
► Merkle Trees
► Message Authentication Codes
► Security Services
► Storage Security

Recommended Reading

1. Bishop M. Computer security: art and science. Boston: Addison Wesley Professional; 2002.
2. Haller N. The S/Key one-time password system. In: Proceedings of the Network and Distributed Systems Security Symposium; 1994. p. 151–7.
3. Menezes A, van Oorschot P, Vanstone S. Handbook of applied cryptography. Boca Raton: CRC; 1996.
4. Pfleeger C, Pfleeger S. Security in computing. 3rd ed. Englewood Cliffs: Prentice-Hall; 2003.
5. Schneier B. Applied cryptography: protocols, algorithms, and source code in C. 2nd ed. New York: Wiley; 1996.
6. Stallings W. Cryptography and network security: principles and practices. 4th ed. New Delhi: Pearson Prentice Hall; 2006.
7. Stinson D. Cryptography: theory and practice. 3rd ed. Boca Raton: Chapman & Hall/CRC; 2006.

Automatic Image Annotation

Nicolas Hervé and Nozha Boujemaa
INRIA Paris-Rocquencourt, Le Chesnay, France

Synonyms

Auto-annotation; Image classification; Multimedia content enrichment; Object detection and recognition.

Definition

The widespread search engines, in the professional as well as the personal context, used to work on the basis of textual information associated or extracted from indexed documents. Nowadays, most of the exchanged or stored documents have multimedia content. To reduce the technological gap so that these engines still can work on multimedia content, it is very convenient developing methods capable to generate automatically textual annotations and metadata. These methods will then allow to enrich the upcoming new content or to post-annotate the existing content with additional information extracted automatically if ever this existing content is partly or not annotated.

A broad diversity in the typology of manual annotation is usually found in image databases. Part of them is representing contextual information. The author, date, place or technical shooting conditions are quite frequent. Some semantic or subjective annotations, like emotions that flow out from images, can be found. Some other annotations could be related to the visual content of images. They provide information on a given image such as indicating whether it is a drawing, a map or a photograph ... For photographs, the global aspect is often specified (vertical/horizontal, color/black and white, indoor/outdoor, landscape, portrait ...), as well as the presence of remarkable objects or persons.

The aim of automatic image annotation approaches is to provide efficient methods that extract automatically the visual content of pictures allowing semantic labeling of images. This is generally achieved by learning algorithms that, once being trained on annotated subcorpora, are able to suggest keywords to the archivist through object detection/recognition and image classification methods.

Historical Background

The exploration of visual content databases and their querying to retrieve some specific content usually rely on textual annotations that have been previously provided manually by human operators. The outcome of the tremendous improvements in digitization and acquisition devices is the availability of exponentially growing content. Usual annotation techniques then became more and more difficult to apply because they are time and cost consuming. Moreover, manual annotations are far from being perfect. They are often focused on the context, subjective, partial and driven by the needs of the end-users at the time they are produced. As these needs are evolving, part of the existing annotations becomes irrelevant and others are missing. This is especially true with the arising of Internet and the availability of all kind of databases online. An other issue lies in the lack of controlled vocabularies for most of the databases making difficult for the end-user to guess what query words he has to use in order to retrieve the content he has in mind.

Visual content indexing and retrieval community have achieved significant progress in the recent years [1] toward efficient approaches for visual features extraction and visual appearance modeling together with developing advanced mechanism for interactive visual information retrieval. One of the major issues was and remains the semantic gap [2, 3].

Two main types of images databases could be distinguished. Specific databases are focused on a given restricted field. In the scientific domain, one can cite satellite images for weather forecast or cultivation study, medical images or botanical databases for species recognition. They are also found in the cultural heritage domain (e.g., paintings databases) or the military and security domain (e.g., fingerprints and faces databases). On the other side, generic databases contain very different images, without any *a priori* on their content. This is usually the case for professional news agencies, illustration photo stock collections and personal family and holiday photo albums. Only methods for generic content databases labeling will be addressed.

By analyzing automatically images and characterizing them with low-level features (mainly colors, textures and shapes) CBIR systems [4] provided new query paradigms that enable users to express their needs. The main one is "query by example" where the system retrieves images of the database that are the most similar to a given example. The scientific community has been facing the well known semantic gap problem for a while which remain the major concern of the research community. Since the late 90s, relevance feedback mechanism is one of possible solutions to this difficult problem.

The early papers on automatic annotation that have been published tackled image orientation detection or the classical indoor versus outdoor and city versus landscape classifications of photographs [5, 6]. Recently, relevance feedback allows moreover helping for interactive mass-annotation of image collections. This approach is often referred to as semi-automatic image annotation.

Foundations

Despite some of its drawbacks, the query by keyword is still very useful and quite natural for the end-user [7]. Automatic annotation generates such keywords to enrich the images semantic descriptions and ease further querying. Because of the computational costs of all current approaches, the existing systems are always composed of two parts. An offline part is in charge of indexing the visual content and generating the annotations. Eventually, a human operator can help the system during the process or after it to validate/invalidate the produced annotations. In such cases, one talks of semi-automatic annotation systems. The second part, online and real-time, is a query by keywords module.

As the main purpose is to describe the visual content, the term "visual concept" is preferred over keyword to describe the labels a system has to discover in images. As previously mentioned, these visual concepts could be related to either global appearance of the image or presence of some objects. Objects detection can also be refined in generic object class detection or specific object instance detection. For example, one can ask a system to label only the "vehicle" concept, or more precisely to distinguish cars, motorbikes, boats and airplanes, and, at a very specific level, being able to recognize different makes of cars. This is the same problem with annotating persons. Being able to detect the presence of a person in an image is a different procedure and result than recognizing him. As face recognition is a well studied problem which is tackled by a specific research community. The ability to generalize from a few examples and to reach higher abstraction levels is natural for humans but it is very challenging task to achieve with current state-of-the-art's annotation systems [8–11].

One of the fundamental hypotheses of automatic annotation is that what looks similar is probably semantically similar. Most of the approaches rely on this assumption. The main generic steps of automatic annotation are described below. First, visual features are extracted automatically from images in order to obtain representations in a visual space. The second step is to build models that will link the visual concepts to the relevant information in the visual space. When new content is proposed, models are then able to predict the corresponding visual concepts.

The performances evaluation of such methods may rely on the usage of the annotations by the final users. As in most of information retrieval systems, precision and recall measures are used. Precision emphasizes the retrieval of relevant documents earlier and recall focuses on the retrieval of the full set of relevant documents. Precision and recall are complementary to judge the quality of a system. But for some applications, precision is the only important measure. This is especially the case when a huge image database is available (like Internet). When doing a query, a user is more interested in the first satisfying results than in the complete relevant result set.

Images Description with Low-Level Features

The visual description of images is of great importance as it is the raw material on which further models are built. There is not a universally good low-level features extractor. In specific databases, *a priori* on the content of images

can be used to extract specialized features that will better describe their special nature. For example, numerous features can be found in the literature for faces or fingerprints description. In generic databases, compromises have to be made between exhaustiveness, fidelity to the content, ability to generalize and different invariance degrees (illumination changes, rotations, scales, occlusions . . .). The use of inappropriate features leading to poor performances of a system has often been described as semantic gap. In this case, one rather faces the numerical gap, meaning that the visual information is present in images but it has not been extracted correctly.

Due to their ability to generalize to content in different conditions, statistical features are often used. They gather color, shape and texture information in histograms, separately or jointly. Color histograms are among the first features used to describe images. They vary depending on the underlying color space that is used, the quantization parameters, different weighting schemes or the use of co-occurrences of colors. Shapes can be described by properties of edges found in images like their types, orientations or lengths. Textures are focusing on the analysis of frequencies in images. They often rely on Fourier transform, Gabor filter banks or wavelets. Some features also combine different types of information, mixing for example color and texture in a single representation. Typically, these visual features are represented by vectors in high-dimensional spaces (generally between a few tens and a few hundreds dimensions) [6, 12].

Initially, the features were extracted over the full image. This approach is well suited to describe the global aspect of the content but is too coarse to represent small details and objects. Features need to be extracted locally. First, a support region has to be determined. Once its location, shape and size are known, features are computed on this small portion of the image. These features can be of the same types as those extracted at a global level or they can be specialized according to the nature of the support regions. Several strategies are used to select the support regions. Segmentation algorithms try to find the boundaries between homogeneous regions in images [13]. Segmentation is a difficult problem in itself that is not well defined. Unfortunately, the general trend has always been to focus on segmentation that detects objects, which is already a highly semantic task and, thus, not really achievable through automatic processes. Alternative approaches consist on sliding windows and fixed grid, with varying sizes and spacing, are common ways of obtaining dense sampling of the visual content [5, 12]. Another popular region selection approach is based on local features detectors via point-of-interest. They were originally designed for image registration. These detectors are generally attracted to specific areas of images that have high variation in the visual signal, such as the vicinity of edges and corners of regions. They allow the selection of a very small proportion of image locations having the highest visual variance [14, 15]. Typically, when using dense sampling, point-of-interest or when mixing them, between a few hundreds and a few thousands features are extracted per image. The computational cost is then much higher than with global features. Some representations also try to carry other information, like geometrical relations between features locations or contextual information [16].

With global features, the image representation is straightforward. However, even when local features are to be used, learning algorithms may sometimes require a global image representation that encompasses all the local visual information. The bag-of-visual-words representation, very much inspired by the classical bag-of-words representation for text, is one of the most popular for images. A visual vocabulary composed of visual words (some representative features) is generated. An image is then represented by a coordinate vector, each value of which expresses the degree of importance of a feature with respect to the image and/or the database as a whole. The creation of a visual vocabulary is an important step in the full process. The selected visual words (a few hundreds to a few thousands) have to be representative of the database content as they will serve as a basis for further representation. Creating a good vocabulary will avoid the loss of too much local information. Generally, clustering

algorithms either supervised or not, are used with a sample of the database. This step can be seen as a quantization of the local features.

Whatever the selected representation, the similarity between images is measured by a distance functional in the visual space. A broad variety has been developed: classical Euclidean distance (L2), L1, earth mover distance (EMD), chi-squared (χ^2), vector angle, histogram intersection, ...

Learning and Models

Although several formulations have been proposed, the main purpose of building models for visual concepts is to associate them with the visual space regions that best represent them. This problem is at the cross-roads of computer vision, data mining and machine learning. Usually, the models are built through a supervised learning process. For given visual concepts, an algorithm is fed with a training dataset containing both positive and negative images regarding the concepts to learn. This algorithm has to find the discriminant information from the visual space that best models the concepts. Generally, the available annotations for the training set are not localized. The presence of a visual concept for an image is known, but its exact location is not provided. This is the case for almost all professional and personal databases. In the same way, annotating new images does not require to locate exactly the visual concept, but only to predict its presence. This is the main distinction that can be made with object detection tasks.

Two main learning algorithm families are used:

- *Generative*: the system tries to estimate density distribution of concepts in the visual space or other hidden variables [13]. Popular examples include Gaussian Mixture Models, Hidden Markov Models, Bayesian networks, Latent Semantic Analysis and translation models from the text processing community. The Expectation-Maximization algorithm is often used to train these models.

Automatic Image Annotation, Fig. 1 Pictures annotated with the "tennis" keyword

• *Discriminative*: instead of trying to model the distributions, discriminative approaches are focusing on detecting the boundaries between classes. For each visual concept, the annotation process is then often formalized as a two-class classification problem (present/not present). Although appearing to be a little bit more effective, discriminative approaches do not have the elegance of generative ones. They act more as black boxes and relationships between the different variables are not explicit, and thus difficult to analyze. The most famous algorithms are Support Vector Machine (SVM) [12, 15], boosting (e.g., adaboost) [14] and all flavors of discriminant analysis (linear – LDA, biased – BDA, multiple – MDA, Fisher – FDA).

Both approaches may use global or local representations. Methods using bag-of-words

Automatic Image Annotation, Fig. 2 Pictures displayed after two iterations

representations are also called "multiple instance learning." Sometimes, pre-processes may also be used to prepare the data in order to enhance the performances or to reduce the computational costs: feature selection, dimensionality reduction, scaling or normalization. Current systems are often composed of several components, using different low-level features, combining them according to different schemes and training models with multiple learning strategies.

Once the models have been learned, they can be used to predict the visual concepts. Two types of predictions are possible. Hard decision simply indicates the presence or absence of the concept. Soft decision also provides a degree of confidence in the prediction, allowing ranking more easily the results when answering an end-user query and

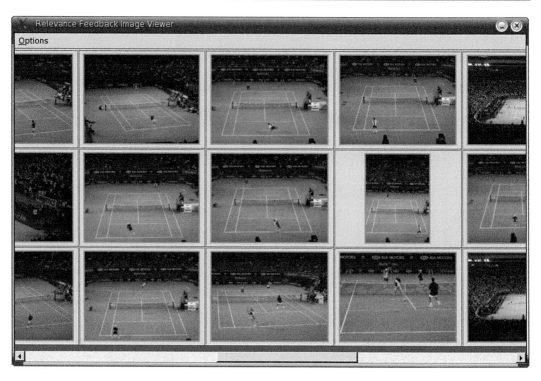

Automatic Image Annotation, Fig. 3   All positive pictures basket allowing mass-annotation

thus improving the retrieval of pertinent images earlier.

Current Results

The different methods are actually mature enough to predict global visual concepts like image types and scene categories. Regarding local concepts, huge improvements still need to be made in order to provide useful applications to real users. Both dense sampling and point-of-interest have shown to perform quite well on research databases, but results on real databases are quite poor [11, 12]. A good indication of state-of-the-art performances can be obtained in the results of official benchmark campaigns like ImagEval, Pascal VOC, Imageclef or Trecvid. In all cases, the contextual information (visual or from the existing metadata) has shown to be of great importance in the results.

When the availability of correctly annotated images for a given visual concept is not guaranteed, the offline learning approach is not possible.

One of the solutions is then to interact with a user through relevance feedback, also called interactive learning. In a few iterations, the user will provide the system positive and negative examples and guide it to recognize the visual concept. Part of the mechanism involved are the same as offline learning, but the training labels are provided online by a user. As an example, the following screen captures from Ikona [4] are showing the IAPR-TC12 database, used for the Imageclef benchmark. The first screen displays all the images annotated with the "tennis" keyword. The user is only interested in pictures where a tennis court is visible. He indicates positive (green border) and negative (red border) examples. After two iterations, one can see on the second screen that a lot of tennis court pictures have been retrieved. None of them was annotated with the "tennis" keyword. After a few more iterations, when no more correct pictures are retrieved from the database, the user is able to annotate massively all the pictures gathered through the iterations that were kept in a specific basket (third screen) (Figs. 1, 2, and 3).

Key Issues and Future Research

The lack of generalization ability for both visual features and learning algorithms has to be compensated by a huge number of training examples. Depending on the complexity of the visual concept, a good estimation is around a hundred positive examples and ten times more negatives examples for the training set. Paradoxically, despite the tremendous amount of images available nowadays, finding content that has been reliably annotated for training dataset is hard.

The computational complexity is also still too high for real-time annotation when dealing with several thousands of visual concepts. Research is made on scalability issues in machine learning and is linked to existing high-dimensional data indexing structures.

Progresses for better description and integration of all types of available information need to be achieved.

There are also some questions arising: will the problem be solved with more computational power when one is able to process images at every scale and location in real-time? Are massive collaborative annotation websites, like Flickr, going to change the annotation paradigm by transforming Internet in a giant common repository? What is the impact of GPS metadata, and more generally all the new information captured directly when a photograph is taken?

Key Applications

- Professional content owners: post-editing
- Personal family and holiday photo albums
- Web image search
- Searching into poorly human-made annotated corpora enhancing the quality of search results

Cross-References

▶ Annotation-Based Image Retrieval
▶ Boosting
▶ Image Database
▶ Image Retrieval and Relevance Feedback
▶ Object Recognition
▶ Support Vector Machine
▶ Visual Content Analysis

Recommended Reading

1. Smeulders AWM, Worring M, Santini S, Gupta A, Jain R. Content-based image retrieval at the end of the early years. IEEE Trans Pattern Analy Machine Intell. 2000;22(12):1349–80.
2. Boujemaa N., Fauqueur J., and Gouet V. What's beyond query by example? Technical report, INRIA; 2003.
3. Hare JS, Lewis PH, Enser PGB, Sandom CJ. Mind the gap: another look at the problem of the semantic gap in image retrieval. In: Proceedings of the SPIE: Multimedia Content Analysis, Management, and Retrieval; 2006.
4. Boujemaa N, Fauqueur J, Ferecatu M, Fleuret F, Gouet V, Le Saux B, Sahbi H. Ikona: interactive specific and generic image retrieval. In: Proceedings of the International Workshop on Multimedia Content-Based Indexing and Retrieval; 2001. Available at: http://www-rocg.inria.fr/imedia/mmcbirzod.html.
5. Szummer M, Picard RW. Indoor-outdoor image classification. In: Proceedings of the Workshop on Content-Based Access to Image and Video Databases; 1998.
6. Vailaya A, Jain A, Zhang H-J. On image classification: city images vs. landscapes. Pattern Recogn J. 1998;31(12):1921–35.
7. Enser PGB, Sandom CJ, Lewis PH. Automatic annotation of images from the practitioner perspective. In: Proceedings of the 4th International Conference. Image and Video Retrieval; 2005. p. 497–506.
8. Datta R, Li J, Wang JZ. Content-based image retrieval – approaches and trends of the new age. In: Proceedings of the 7th ACM SIGMM International Workshop on Multimedia Information Retrieval; 2005. p. 253–62.
9. Hanjalic A, Sebe N, Chang E. Multimedia content analysis, management and retrieval: trends and challenges. In: Proceedings of the SPIE: Multimedia Content Analysis, Management, and Retrieval; 2006.
10. Lew MS, Sebe N, Djeraba C, Jain R. Content-based multimedia information retrieval: state of the art and challenges. ACM Trans Multimedia Comp Comm Appl. 2006;2(1):1–19.
11. Ponce J, Hebert M, Schmid C, Zisserman A, editors. Toward category-level object recognition, Lecture notes in computer science. New York: Springer; 2006.
12. Hervé N, Boujemaa N. Image annotation: which approach for realistic databases? In: Proceedings of the 6th ACM International Conference Image and Video Retrieval; 2007. p. 170–7.
13. Barnard K, Duygulu P, Forsyth D, de Freitas N, Blei DM, Jordan MI. Matching words and pictures. J Mach Learn Res. 2003;3(6):1107–35.

14. Opelt A, Pinz A, Fussenegger M, Auer P. Generic object recognition with boosting. Pattern Anal Mach Intell. 2006;28(3):416–31.
15. Zhang J, Marszalek M, Lazebnik S, Schmid C. Local features and kernels for classification of texture and object categories: a comprehensive study. Int J Comput Vis. 2007;73(2):213–38.
16. Amores J, Sebe N, Radeva P. Context-based object-class recognition and retrieval by generalized correlograms. IEEE Trans Pattern Anal Mach Intell. 2007;29(10):1818–33.

Autonomous Replication

Cristiana Amza[1] and Jin Chen[2]
[1]Department of Electrical and Computer Engineering, University of Toronto, Toronto, ON, Canada
[2]Computer Engineering Research Group, University of Toronto, Toronto, ON, Canada

Synonyms

Adaptive database replication; Autonomic database replica allocation; Database provisioning

Definition

Autonomic database replication refers to dynamic allocation of servers to applications in shared server clusters, in such a way to meet per-application performance requirements. Autonomic database replication enables the service provider to efficiently multiplex data center resources across applications in order to save per-server costs related to human management, power, and cooling.

Historical Background

The concept of autonomic computing and the associated research area of automated, adaptive self-management in data centers were introduced by IBM as a grand challenge project in the early 2000s. Other companies, which have responded or have had similar proposals of their own, include Microsoft, Intel, Sun, and HP. Related industry efforts in this area have been on developing open standards for resource monitoring tools, e.g., as available on IBM's Alphaworks (http://www.alphaworks.ibm.com), and academic or collaborative industry-academia efforts related to applying machine learning techniques for automated adaptation in cluster systems [6, 11, 18].

Scientific Fundamentals

Dynamic content servers, such as Amazon.com and eBay.com, commonly use a three-tier architecture (see Fig. 1) that consists of a front-end web server tier, an application server tier that implements the business logic, and a back-end database tier that stores the dynamic content of the site.

Large data centers may host multiple applications concurrently, such as e-commerce, auctions, news, and games. The cooling and power costs of gross hardware over-provisioning for each application's estimated peak load are making efficient resource usage crucial. Furthermore, the excessive personnel costs involved in server management motivate an automated approach to resource allocation for applications in large sites.

Dynamic resource allocation techniques, i.e., dynamic provisioning of servers to multiple applications in each tier of the dynamic content site, have been recently introduced to address the increasing costs of ownership for large dynamic content server clusters. These automatic solutions add servers to an application's allocation based on perceived or predicted performance bottlenecks caused by either load spikes or component failures; they remove resources from a application's allocation when in underload.

If services experience daily patterns with peak loads for each service type at a different time (e.g., daytime for e-commerce, evening for auctions, morning for news sites, night for gaming), there are opportunities for reassigning hardware resources from one service to another. Thus, instead of gross hardware over-provisioning for each application's estimated peak load, dynamic

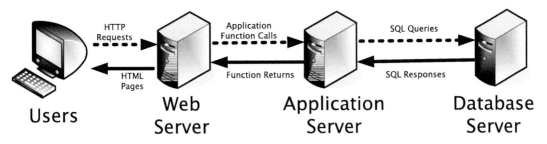

Autonomous Replication, Fig. 1 Three-tier architecture

resource provisioning techniques enable the service provider to efficiently multiplex data center resources across applications.

Common Architecture for Database Replica Provisioning

Figure 2 shows the common architecture of sites with dynamic provisioning in the database server tier. A resource manager makes the replica allocation decisions for each application hosted on the site based on the application requirements and the current system state. The requirements are expressed in terms of a service level agreement (SLA) that consists of a latency requirement on the application's queries. The current system state includes the current performance of this application and the system capacity. The resource manager operates in two modes, underload and overload. During underload, the number of replicas exceeds overall demand, and allocation decisions per application are made independently. During overload, the manager uses either a utility-based scheme, e.g., profit based, or a fairness scheme, e.g., equal share, to allocate replicas to applications.

The allocation decisions are communicated to a set of schedulers, one per application, interposed between the application and the database tiers. Each scheduler fulfills the following functions: (i) It keeps track of the current *database set* allocated to its application and allocates or removes replicas from its managed set according to the resource manager's decisions, (ii) It distributes the corresponding incoming requests onto the respective database replicas, (iii) It samples various system and application metrics, e.g., the average application latency, periodically, from its *database set* in order to perceive or predict resource bottlenecks, (iv) It provides consistent replication, i.e., one-copy serializability [5], at all the replicas allocated to the application it manages.

Strong consistency is desirable for dynamic provisioning in the database tier of a multi-tier data center, for transparency reasons; the replicated nature of the back-end database and its configuration adaptations are thereby hidden from the application server, which interacts with the replicated back end as with a single database. Any eager replication scheme with one-copy serializability [5] can be used within each application's replica set. However, existing dynamic provisioning schemes typically leverage the presence of the scheduler and the synchronous, one query at a time, nature of the communication between the application and database tiers in a data center to implement a middleware replication solution in the scheduler itself. Upon receiving a query from the application server, the scheduler sends the query using a read-one, write-all replication scheme to the replica set allocated to the application. The scheduler assigns a global serialization order to all transactions pertaining to its application based on conservatively perceived table-level conflicts between transactions and ensures that transactions execute in this order at all the corresponding database replicas. Table-level concurrency control affords optimizations based on conflict awareness in the scheduler [2, 3], which offset any penalties due to the coarse-grain control for read-intensive e-commerce applications [15].

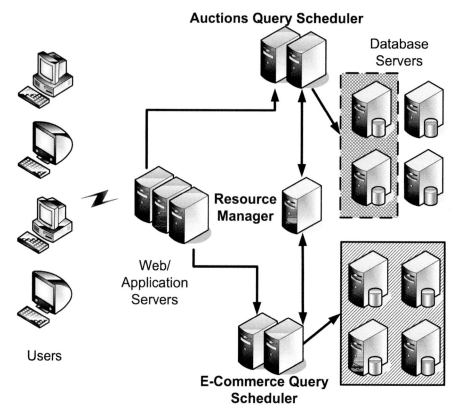

Autonomous Replication, Fig. 2 Cluster architecture

The scheduler is also in charge of bringing a new replica up to date by a process called *data migration*, during which all missing updates are applied on that replica. Integrating a stale replica into an application's allocation occurs through an online reconfiguration technique [15,21] without stalling ongoing transactions on already active replicas for that application. Finally, each scheduler may itself be replicated for availability [2,3].

Overview of Dynamic Provisioning Solutions

Several fully transparent provisioning solutions [4, 13, 17] have been recently introduced to address the increasing cost of management problem. Many of these approaches [4,13,17] investigate dynamic provisioning of resources within the (mostly) stateless web server and application server tiers. In this article, we focus on the dynamic resource allocation solutions within the stateful database tier, which commonly becomes the bottleneck [1].

Regardless of the tier it applies to, a dynamic provisioning solution can be either *proactive* or *reactive* depending on its capabilities for predicting performance bottlenecks in the dynamic content server. *Proactive* solutions use sophisticated system models for prediction, such as queuing models [4], utility models [17], machine learning models [6,11,18], or marketplace approaches [8]. *Proactive* provisioning techniques use the system model predictions for triggering allocations in advance of expected need. This is especially important for tiers with a higher adaptation delay, such as the database tier. In contrast, *reactive* approaches do not use prediction but rather detect and react to existing resource bottlenecks. They rely on predefined thresholds for application-level or system metrics, such as latency or CPU usage for triggering changes in application allocation.

Proactive approaches typically do not model multiple database replicas as part of their solution search space [4, 14].

Challenges for Database Replica Provisioning

Adapting to a bottleneck caused by either a workload spike, or a failure, by allocating additional replicas to the application poses a set of challenges, which we detail next.

Adaptation Delay

Adding a database replica to an application's allocation is not an immediate process. The database state of the new replica(s) for that application will be stale and must be brought up to date, or a new instance of that application may need to be installed before it can be used. Furthermore, load balancing for the old and new replicas needs to occur, and the buffer pool at the new replica(s) needs to be warm before the new replica(s) can be used effectively.

Oscillations in Allocation

Oscillations in database allocations to applications may occur during system instability induced by adaptations. As discussed earlier, the replica addition process can be long. *During* the adaptation phases, i.e., data migration, buffer pool warm-up, and load stabilization, the latency will remain high or may even temporarily continue to increase as shown in Fig. 3. Latency sampling during this potentially long time is thus not necessarily reflective of a continued increase in load but of system instability after an adaptation is triggered. If the system takes further decisions based on sampling latency during the stabilization time, it may continue to add further replicas which are unnecessary and hence will need to be removed later. This is an oscillation in allocation which carries performance penalties for other applications running on the system due to potential interference.

Either a reactive or proactive policy that measures application-level metrics, periodically, including during the replica addition process, can suffer from allocation instability. Allocation oscillations, in their turn, cause cross application

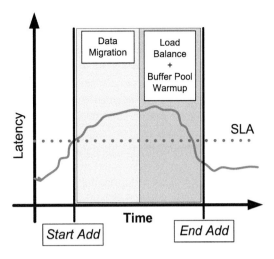

Autonomous Replication, Fig. 3 Latency instability during replica addition

interference due to the price paid for warming up the buffer pool as part of the "context switch" between applications on the machines involved.

While rapid load fluctuations may induce similar behavior, simple smoothing or filtering techniques can offer some protection to very brief load spikes. All existing dynamic provisioning schemes use some form of smoothing or filtering, to dampen brief load fluctuations.

Design Choices and Trade-Offs for Database Provisioning

In designing a dynamic provisioning solution for the database back-end, we need to consider the design trade-off between allocating replicas to applications in a *disjoint* versus *overlapping* manner. Consider the case where we allocate a *disjoint* set of machines to host the replicas of each application and we dynamically adjust each application's cluster partition. When an application requires an additional replica, it must use an unallocated machine or a machine allocated to another application. In either case, we incur the full *adaptation delay* for replica addition to the application.

Replica addition delay can be avoided altogether with *fully overlapped* replicas, where all the database applications are replicated across all the available cluster machines. In this case, there is no replica addition delay because replicas do not have to be added or removed. However, this approach causes interference due to resource sharing. For example, when multiple database applications run on the same machine or inside the same, their performance can degrade due to buffer pool interference. This discussion shows that there is a trade-off between using disjoint and fully overlapped replica allocation strategies. Disjoint allocation reduces interference and thus improves steady-state performance. Fully overlapped allocation avoids replica addition delay and thus can speed up the system's response to load spikes and failures.

A practical solution is to use a *partial overlap* strategy [15], where the application allocations are disjoint, but, for each application, batched updates are periodically executed on a set of database machines outside of that application's allocation, thus keeping them partially up to date.

Performance Modeling for Resource Allocation

Many performance modeling-related techniques have been proposed to support proactive resource allocation (e.g., resource provisioning for database replicas) in data centers. A *performance model* is a mathematical function that calculates an estimate of the application performance for a range of resource configurations.

Automatic machine learning-based performance model [9, 10, 20] building iterates through two steps: (1) gathering experimental samples and (2) modeling computation. Gathering experimental samples means actuating the experimental system into a given resource configuration, running a specific application workload on the live system (or equivalent), and measuring the application latency. Modeling computation involves mathematical interpolation for building the model on existing sampling data. While modeling computation is typically on the order of fractions of seconds, experimental sampling may take months for mapping out the entire resource configuration space of an application with sufficient statistical accuracy. This is due to dynamic effects, for instance, cache warm-up time, which make reliable

actuation and sampling expensive even for a single configuration point.

At the other end of the spectrum is using analytical models that rely on sysadmin or analyst's semantic knowledge of the system and application [12, 16, 19]. However, these analytical models are precise only for restricted parts of the system, specific application workload mix, or resource configurations. They are brittle to dynamic changes and require too much domain expertise.

A compromise ensemble approach [7] leverages automated black-box long-term learning of the system itself, coupled with administrator semantic awareness and expertise, wherever available to build an ensemble/mixture of models. It validates model templates using monitoring data for the resources the model is sensitive to. In this way, it incrementally constructs an ensemble of models for the purposes of capacity planning, performance inquiry, or resource allocation.

Key Applications

Autonomic database replication is used for dynamic resource allocation in the back-end database of dynamic content web sites hosting e-commerce applications. Write queries in dynamic content applications are typically more lightweight and have a much lower memory footprint compared to read queries [2]. For instance, in e-commerce applications, an update query typically updates only the record pertaining to a particular customer or product, while read queries caused by browsing involve expensive database joins as a result of complex search criteria. Moreover, read queries are much more frequent than write queries. Hence, a *partial overlap* replication solution, which causes minimal resource interference, is typically used in order to reduce the adaptation delay [15].

Recommended Reading

1. Amza C, Cecchet E, Chanda A, Cox A, Elnikety S, Gil R, et al. Specification and implementation of dynamic web site benchmarks. In: Proceedings of the

5th IEEE Workshop on Workload Characterization; 2002.

2. Amza C, Cox AL, Zwaenepoel W. Conflict-aware scheduling for dynamic content applications. In: Proceedings of the 4th USENIX Symposium on Internet Technologies and Systems; 2003. p. 6–6.

3. Amza C, Cox AL, Zwaenepoel W. Distributed versioning: consistent replication for scaling back-end databases of dynamic content web sites. In: Proceedings of the ACM/IFIP/USENIX International Middleware Conference; 2003. p. 282–304.

4. Bennani MN, Menasce DA. Resource allocation for autonomic data centers using analytic performance models. In: Proceedings of the 2nd International Conference on Autonomic Computing; 2005. p. 229–40.

5. Bernstein PA, Hadzilacos V, Goodman N. Concurrency control and recovery in database systems. Reading: Addison-Wesley; 1987.

6. Chen J, Soundararajan G, Amza C. Autonomic provisioning of backend databases in dynamic content web servers. In: Proceedings of the 3rd International Conference on Autonomic Computing; 2006. p. 123–33.

7. Chen J, Soundararajan G, Ghanbari S, Amza C. Model ensemble tools for self-management in data centers. In: Proceedings of the 8th International Workshop on Self Managing Database Systems, ICDE Workshop; 2013. p. 36–43.

8. Coleman K, Norris J, Candea G, Fox A. Oncall: defeating spikes with a free-market server cluster. In: Proceedings of the 1st International Conference on Autonomic Computing; 2004.

9. Duan S, Thummala V, Babu S. Tuning database configuration parameters with iTuned. Proc VLDB Endowment. 2009;2(1):1246–57.

10. Ganapathi A, Kuno HA, Dayal U, Wiener JL, Fox A, Jordan MI, et al. Predicting multiple metrics for queries: better decisions enabled by machine learning. In: Proceedings of the 25th International Conference on Data Engineering; 2009. p. 592–603.

11. Ghanbari S, Soundararajan G, Chen J, Amza C. Adaptive learning of metric correlations for temperature-aware database provisioning. In: Proceedings of the 4th International Conference on Autonomic Computing; 2007. p. 26.

12. Gulati A, Kumar C, Ahmad I, Kumar K. BASIL: automated IO load balancing across storage devices. In: Proceedings of the 8th USENIX Conference on File and Storage Technologies; 2010. p. 169–82.

13. IBM Corporation: Automated provisioning of resources for data center environments. http://www-306.ibm.com/software/tivoli/solutions/provisioning/ (2003).

14. Karve A, Kimbrel T, Pacifici G, Spreitzer M, Steinder M, Sviridenko M, et al. Dynamic placement for clustered web applications. In: Proceedings of the 15th International World Wide Web Conference; 2006. p. 595–604.

15. Soundararajan G, Amza C. Reactive provisioning of backend databases in shared dynamic content server clusters. ACM Trans Auton Adapt Syst. 2006;1(2):151–88.

16. Soundararajan G, Lupei D, Ghanbari S, Popescu AD, Chen J, Amza C. Dynamic resource allocation for database servers running on virtual storage. In: Proceedings of the 7th USENIX Conference on File and Storage Technologies; 2009. p. 71–84.

17. Tesauro G, Das R, Walsh WE, Kephart JO. Utility-function-driven resource allocation in autonomic systems. In: Proceedings of the 2nd International Conference on Autonomic Computing; 2005. p. 70–7.

18. Tesauro G, Jong NK, Das R, Bennani MN. On the use of hybrid reinforcement learning for autonomic resource allocation. Clust Comput. 2007;10(3):287–99.

19. Urgaonkar B, Pacifici G, Shenoy PJ, Spreitzer M, Tantawi AN. An analytical model for multi-tier internet services and its applications. In: Proceedings of the 2005 ACM SIGMETRICS International Conference on Measurement and Modeling of Computer Systems; 2005. p. 291–302.

20. Wang M, Au K, Ailamaki A, Brockwell A, Faloutsos C, Ganger GR. Storage device performance prediction with CART models. In: Proceedings of the 2004 ACM SIGMETRICS International Conference on Measurement and Modeling of Computer Systems; 2004. p. 412–13.

21. Wu S, Kemme B. Postgres-R(SI): combining replica control with concurrency control based on snapshot isolation. In: Proceedings of the 21st International Conference on Data Engineering; 2005. p. 422–33.

Average Precision

Ethan Zhang[1] and Yi Zhang[2]
[1]University of California, Santa Cruz, CA, USA
[2]Yahoo! Inc., Santa Clara, CA, USA

Definition

Average precision is a measure that combines recall and precision for ranked retrieval results. For one information need, the average precision is the mean of the precision scores after each relevant document is retrieved.

$$\text{Average Precision} = \frac{\sum_r P@r}{R}$$

where r is the rank of each relevant document, R is the total number of relevant documents, and $P@r$ is the precision of the top-r retrieved documents.

Key Points

The average precision is very sensitive to the ranking of retrieval results. The relevant documents that are ranked higher contribute more to the average than the relevant documents that are ranked lower. Changes to the ranking of relevant documents have a significant impact on the average precision score. Average precision is considered a reasonable evaluation measure for emphasizing returning more relevant documents earlier.

Cross-References

▶ MAP
▶ Precision
▶ Precision at n
▶ Recall
▶ Standard Effectiveness Measures

Average Precision at n

Nick Craswell and Stephen Robertson
Microsoft Research Cambridge, Cambridge, UK

Synonyms

AP@n

Definition

Average Precision at n is a variant of Average Precision (AP) where only the top n ranked doc-

uments are considered (please see the entry on ▶ Average Precision for its definition). AP is already a top-heavy measure, but has a recall component because it is normalized according to R, the number of relevant documents for a query. In AP@n there are a number of options for normalization, for example, normalize by n or normalize by min(n,R).

Key Points

The well-known measure Average Precision has a number of lesser-known variants, used in TREC [3] and elsewhere. Before and during TREC-1, it was usual to calculate an 11-point interpolated Precision-Recall curve, and take the average of these 11 precision values, giving an "interpolated AP." In TREC-2 and beyond, the modern non-interpolated AP was introduced. It calculates precision at each relevant document.

A number of other AP variants arise in a precision-oriented setting, where it is possible to calculate Average Precision at n. AP@n takes into account both the number of relevant documents in the top n and the positions of those documents. This is in contrast to Precision at n (P@n), which ignores position. It is defined as:

$$AP@n = \sum_{i=1}^{n} \frac{rel(i) \times P@i}{NF}$$

where $rel(i) = 1$ if the ith retrieved document is relevant and $rel(i) = 0$ otherwise, and NF is the normalization factor. In Average Precision it is usual to normalize by the number of relevant documents, i.e., NF = R. In a precision-oriented setting, this presents two problems. First, in precision-oriented evaluation one may not have judged enough documents to know R, or estimate it accurately. Second, if R is greater than n, a ceiling of n/R is imposed on the measure. For example, if one knows R = 100 relevant documents, then the best possible top-20, containing 20 relevant documents, will score an AP@20 of 0.2.

Three alternate normalization factors (NF) have been considered, only one of which has been used in TREC. Here, r is the number of relevant documents retrieved and n is the number of documents retrieved.

- *Normalize by r*: Baeza-Yates and Ribeiro-Neto [1] includes this variant, calling it "Average Precision at Seen Relevant Documents." This measure has the property that it may decrease when a relevant document is promoted into the top-n, because this increases the normalization factor. This seems counter-intuitive.
- *Normalize by n*: This variant was introduced for use in Web search evaluation [2]. In cases where R is less than n, this variant has a ceiling of less than 1. The scale of the measure is $0 \leq AP@n \leq min(R/n,1)$. It may be considered undesirable that the scale of the measure varies from query to query, for example when measuring the mean.
- *Normalize by min(n,R)*: This is the "modified average precision" from the TREC-7 Very Large Collection track [3]. The ceiling is always 1 for this variant.

The normalization NF = min(n,R) allows AP@n scores to have the full range of 0,1, while retaining the property that promoting a relevant document always increases the score.

The arithmetic mean of AP@n over a set of queries can be called Mean Average Precision at n (MAP@n), in the same way that Average Precision relates to Mean Average Precision.

Cross-References

▶ Average Precision
▶ Precision at n
▶ Precision-Oriented Effectiveness Measures

Recommended Reading

1. Baeza-Yates RA, Ribeiro-Neto B. Modern information retrieval. Reading: Addison-Wesley; 1999.
2. Hawking D, Craswell N, Bailey P, Griffiths K. Measuring search engine quality. Inf Retr. 2001;4(1): 33–59.
3. Voorhees EM, Harman DK. TREC: experiment and evaluation in information retrieval. Cambridge, MA: MIT Press; 2005.

Average Precision Histogram

Steven M. Beitzel[1], Eric C. Jensen[2], and Ophir Friede[3]
[1]Telcordia Technologies, Piscataway, NJ, USA
[2]Twitter, Inc., San Francisco, CA, USA
[3]Georgetown University, Washington, DC, USA

Definition

The average precision histogram plots the performance of a single run produced by an information retrieval system on a per-query basis. Each data point on the abscissa represents a query used in the evaluation process. The corresponding points on the ordinate measures the difference in this run's performance on the given topic relative to the median average precision of other runs on that topic.

Key Points

Average precision histograms are often used to illustrate the performance differences that an information retrieval system may exhibit across different queries in an evaluation set in relation to other systems or runs that are evaluated on the same set of queries. These histograms can be used to identify the queries in an evaluation set which pose the most difficulty for information retrieval systems. Various tracks in the NIST Text Retrieval Conference (TREC) have used average precision histograms in the post-competition analysis of submitted runs [1]. An example average precision histogram for 25 queries is shown in Fig. 1.

Average Precision Histogram, Fig. 1
Example average precision histogram

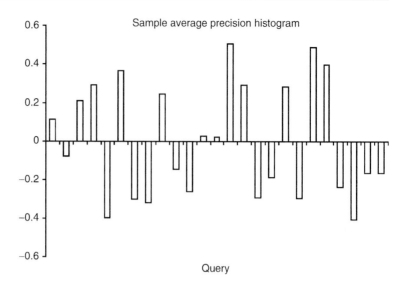

Sample average precision histogram

Query

Cross-References

▶ Average Precision
▶ Effectiveness Involving Multiple Queries

Recommended Reading

1. National Institute of Standards and Technology. TREC-2003 common evaluation metrics. Available online at: http://trec.nist.gov/pubs/trec12/appendices/measures.ps. 2003. Retrieved on 27 Aug 2007.

$$\text{ARP} = \frac{1}{n} \sum_n \text{RP}_n$$

where RP represents the R-Precision value for a given topic from the evaluation set of n topics. R-Precision is defined as the precision after R documents have been retrieved by the system, where R is also the total number of judged relevant documents for the given topic. Precision is defined as the portion of retrieved documents that are truly relevant to the given query topic.

Average R-Precision

Steven M. Beitzel[1], Eric C. Jensen[2], and Ophir Frieder[3]
[1]Telcordia Technologies, Piscataway, NJ, USA
[2]Twitter, Inc., San Francisco, CA, USA
[3]Georgetown University, Washington, DC, USA

Definition

The Average R-precision is the arithmetic mean of the R-precision values for an information retrieval system over a set of n query topics. It can be expressed as follows:

Key Points

R-precision places lower emphasis on the exact ranking of the relevant documents returned by an information retrieval system. This can be useful when a topic has a large number of judged relevant documents, or when an evaluator is more interested in measuring aggregate performance as opposed to the fine-grained quality of the ranking provided by the system.

As an example, consider two query topics: topic A has ten relevant documents, and topic B has six relevant documents. Suppose further that an information retrieval system returns five relevant documents in the top ten retrieved for topic A, two relevant documents in the top six retrieved for topic B. For this case, Average R-precision for this run would be:

$$\mathrm{ARP} = \frac{\frac{5}{10} + \frac{2}{6}}{2} \approx 0.4167$$

Cross-References

▶ Average Precision
▶ Effectiveness Involving Multiple Queries
▶ R-Precision

Recommended Reading

1. National Institute of Standards and Technology. TREC-2004 common evaluation measures. Available online at: http://trec.nist.gov/pubs/trec14/appendices/CE.MEASURES05.pdf. 2005. Retrieved on 27 Aug 2007.

A

B

B+-Tree

Donghui Zhang[1], Kenneth Paul Baclawski[2], and Vassilis J. Tsotras[3]
[1]Paradigm4, Inc., Waltham, MA, USA
[2]Northeastern University, Boston, MA, USA
[3]University of California-Riverside, Riverside, CA, USA

Synonyms

B-tree

Definition

The B+-tree is a disk-based, paginated, dynamically updateable, balanced, and treelike index structure. It supports the exact match query as well as insertion/deletion operations in $O(\log_p n)$ I/Os, where n is the number of records in the tree and p is the page capacity in number of records. It also supports the range searches in $O(\log_p n + t/p)$ I/Os, where t is the number of records in the query result.

Historical Background

The binary search tree is a well-known data structure. When the data volume is so large that the tree does not fit in main memory, a disk-based search tree is necessary. The most commonly used disk-based search trees are the B-tree and its variations. Originally invented by Bayer and McCreight [2], the B-tree may be regarded as an extension of the balanced binary tree, since a B-tree is always balanced (i.e., all leaf nodes are on the same level). Since each disk access retrieves or updates an entire block of information between memory and disk rather than a few bytes, a node of the B-tree is expanded to hold more than two child pointers, up to the block capacity. To guarantee worst-case performance, the B-tree requires that every node (except the root) has to be at least half full. Because of this requirement, an exact match query, insertion or deletion operation must access at most $O(\log_p n)$ nodes, where p is the page capacity in number of child pointers, and n is the number of objects. The most popular variation of the B-tree is the B+-tree [3, 4]. In a B+-tree, objects are stored only at the leaf level, and the leaf nodes are organized into a double linked list. As such, the B+-tree can be seen as an extension of the Indexed Sequential Access Method (ISAM), a static (and thus possibly unbalanced if updates take place) disk-based search tree proposed by IBM in the mid 1960s.

Foundations

Structure
The B+-tree is a tree structure where every node corresponds to a disk block and which satisfies the following properties:

- The tree is balanced, i.e., every leaf node has the same depth.
- An internal node stores a list of keys and a list of pointers. The number of pointers is one more than the number of keys. Every node corresponds to a key range. The key range of an internal node with k keys is partitioned into $k + 1$ subranges, one for each child node. For instance, suppose that the root node has exactly two keys, 100 and 200. The key range of the root node is divided into three subranges $(-\infty, 100)$, $(100, 200)$, and $(200, +\infty)$. Note that a key in an internal node does not need to occur as the key of any leaf record. Such a key serves only as a means of defining a subrange.
- A leaf node stores a list of records, each having a key and some value.
- Every node except the root node is at least half full. For example suppose that an internal node can hold up to p child pointers (and $p-1$ keys, of course) and a leaf node can hold up to r records. The half full requirement says any internal node (except the root) must contain at least $\lceil p/2 \rceil$ child pointers and any leaf node (except the root) must contain at least $\lceil r/2 \rceil$ records.
- If the root node is an internal node, it must have at least two child pointers.
- All the leaf nodes are organized, in increasing key order, into a double linked list.

An example B+-tree is given in Fig. 1. It is assumed that every node has between two and four entries. In a leaf node, an entry is simply a record. In an internal node, an entry is a pair of (key, child pointer), where the key for the first entry is NULL. To differentiate a leaf entry

(which corresponds to an actual record) from a key in an index entry, each leaf entry is followed by a "*".

Query Processing

The B+-tree efficiently supports not only *exact-match queries*, which find the record with a given key, but also *range queries*, which find the records whose keys are in a given range. To perform an exact-match query, the B+-tree follows a single path from the root to a leaf. In the root node, there is a single child pointer whose key range contains the specified key. If one follows the child pointer to the corresponding child node, inside the child node there is also a single child pointer whose key range contains the desired key. Eventually, one reaches a leaf node. The desired record, if it exists, must be located in this node. As an example, Fig. 1 shows the search path if one searches for the record with key = 41. Besides exact-match queries, the B+-tree also supports range queries. That is, one can efficiently find all records whose keys belong to a range R. In order to do so, all the leaf nodes of a B+-tree are linked together. To search for all records whose keys are in the range $R = [low, high]$, one performs an exact match query for key = *low*. This leads to a leaf node. One examines all records in this leaf node, and then follows the sibling link to the next leaf node, and so on. The algorithm stops when a record with key > *high* is encountered. An example is shown in Fig. 2.

Insertion

To insert a new record, the B+-tree first performs an exact-match query to locate the leaf node

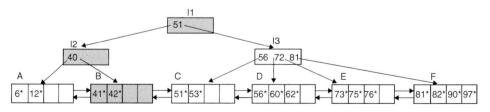

B+-Tree, Fig. 1 Illustration of the B+-tree and exact-match query processing. To search for a record with key = 41, nodes I_1, I_2 and B are examined

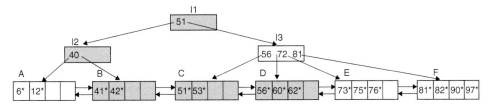

B+-Tree, Fig. 2 Illustration of the range query algorithm in the B+-tree. To search for all records with keys in the range [41, 60], the first step is to find the leaf node containing 41* (I_1, I_2 and B are examined). The second step is to follow the right-sibling pointers between leaf nodes and examine nodes C and D. The algorithm stops at D because a record with key > 60 is found

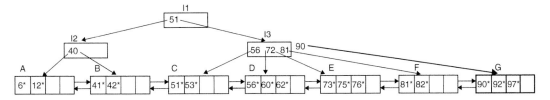

B+-Tree, Fig. 3 Intermediate result of inserting record 92* into Fig. 1. The leaf node F is split. The smallest key in the new node G, which is 90, is *copied up* to the parent node

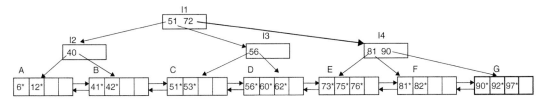

B+-Tree, Fig. 4 Continued from Fig. 3. Final result of inserting record 92*. The overflowing internal node I_3 is split. The middle key, 72, is *pushed up* to the parent node

where the record should be stored, then the record is stored in the leaf node if there is enough space available. If there is not enough space, the leaf node is split. A new node is allocated, and half of the records, the ones with the larger keys in the overflowing node, are moved to the new node. A new index entry (the smallest key in the new node and a pointer to the new node) is inserted into the parent node. This may, in turn, cause the parent node to overflow, and so on. In the worst case, all nodes along the insertion path are split. If the root node is split into two, a new root node is allocated and therefore the height of the tree increases by one.

As an example, Fig. 3 shows an intermediate result of inserting record 92* into Fig. 1. In particular, the example illustrates that splitting a leaf node results in a "*copy up*" operation. The result is intermediate because the parent node I_3 will also be split.

The complete result after inserting 92* is shown in Fig. 4. Here the overflowing internal node I_3 is split. In particular, the example illustrates that splitting an internal node can result in a "*push up*" operation.

Deletion

To delete a record from the B+-tree, one first uses the exact-match query algorithm to locate the leaf node that contains the record, and then the record is removed from the leaf node. If the node is at least half full, the algorithm finishes. Otherwise, the algorithm tries to redistribute records between an immediate sibling node and the underflowing node. If redistribution is not possible, the underflowing node is merged with an immediate sibling node. Note that this merge is always possible.

As an example, Fig. 5 shows the intermediate result of deleting record 41* from the B+-tree shown in Fig. 4. Note that when merging two

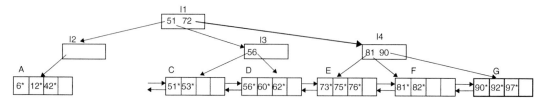

B+-Tree, Fig. 5 Intermediate result of deleting 41* from Fig. 4. Node *B* is merged with node *A*. Key 40 (as well as the pointer to node *B*) is *discarded* from the parent node

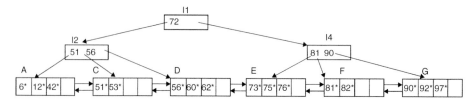

B+-Tree, Fig. 6 Continued from Fig. 5. Final result of deleting 41*. Node I_2 is merged with node I_3. Key 51 from the parent node is *dragged down*

leaf nodes, a key in the parent node is *discarded*, which is the reverse operation of *copy up*. The result is intermediate because node I_2 is also underflowing.

Figure 6 illustrates the final result of deleting 41*. The underflow of node I_2 is handled by merging I_2 with I_3. This merge causes key 51 from the parent node to be *dragged down* to I_2. It is the reverse operation of *push up*.

Comparison with Some Other Index Structures

Compared with the B-tree, the B+-tree stores all records at the leaf level and organizes the leaf nodes into a double linked list. This enables efficient range queries. Compared with ISAM, the B+-tree is a fully dynamic structure that balances itself nicely as records are inserted or deleted, without the need for overflow pages. Compared with external hashing schemes such as Linear Hashing and Extendible Hashing, the B+-tree can guarantee logarithmic query cost in the worst case (while hashing schemes have linear worst-case cost, although this is very unlikely), and can efficiently support range queries (whereas hashing schemes do not support range queries).

Key Applications

The B+-tree index has been implemented in most, if not all, relational database management systems such as Oracle, Microsoft SQL Server, IBM DB2, Informix, Sybase, and MySQL. Further, it is implemented in many filesystems including the NTFS filesystem (for Microsoft Windows), ReiserFS filesystem (for Unix and Linux), XFS filesystem (for IRIX and Linux), and the JFS2 filesystem (for AIX, OS/2, and Linux).

Future Directions

The impact of the B+-tree index is very significant. Many disk-based index structures, such as the R-tree [4] and k-d-B-tree [5] or their variants, are extensions of the B+-tree. Concurrency in B+-trees was studied in [6]. The Universal B-tree, which extends the B+-tree to index multidimensional objects, was studied in [1].

Cross-References

► Extendible Hashing
► Indexed Sequential Access Method

▶ Linear Hashing
▶ R-Tree (and Family)
▶ Tree-Based Indexing

Recommended Reading

1. Bayer R. The universal B-tree for multidimensional indexing: general concepts. In: Proceedings of the International Conference on Worldwide Computing and Its Applications; 1997. p. 198–209.
2. Bayer R, McCreight EM. Organization and maintenance of large ordered indices. Acta Inf. 1972;1(3): 173–89.
3. Comer D. The ubiquitous B-tree. ACM Comput Surv. 1979;11(2):121–37.
4. Knuth D. The art of computer programming. Sorting and searching, vol. 3. Reading: Addison Wesley; 1973.
5. Robinson JT. K-D-B tree: a search structure for large multidimensional dynamic indexes. In: Proceedings of the ACM SIGMOD International Conference on Management of Data; 1981. p. 10–8.
6. Srinivasan V, Carey MJ. Performance of B+ tree concurrency algorithms. VLDB J. 1993;2(4):361–406.
7. Theodoridis Y. The R-tree-portal. 2003. http://www. rtreeportal.org

Backup and Restore

Kazuhisa Fujimoto
Hitachi Ltd., Tokyo, Japan

Synonyms

Backup copy

Definition

Backup is the action of collecting data stored on non-volatile storage media to aid recovery in case the original data are lost or becomes inaccessible due to corruption or damage caused by, mainly, a failure in storage component hardware (such as a hard disk drive or controller), a disastrous event, an operator's mistake, or intentional alteration or erasure of the data. In addition, the data collected by this action are called a backup. Restore is the action of copying a backup to on-line storage for use by applications in a recovery. The term "back up and restore" indicates both actions.

Key Points

The purpose of backup includes recovering data from storage media failure. For this reason, it is common to make backup to different storage media. Removable storage media is used because it is relatively less expensive than online storage media like hard disk drive, and it is easy to bring it offsite in case of site failure. Hard disk drives are becoming popular as backup storage media, because of its decreasing bit cost and quick restore capability.

There are many ways to take backup. A full backup is the way to make a full copy of data. It requires equal capacity of original data, unless data compression is not used. If full backup is taken daily from Monday to Friday and they are kept for the following week, five times capacity is required for backup.

The following two kinds of backup are used to reduce the size of backup data:

Incremental backup A backup that copies all data modified since the last full backup. Modified data can be copied many times until next full backup is taken. To restore data when incremental backups are in use, the latest full backup and the latest incremental backup are required.

Differential backup A backup that copies all data modified since the last backup. Modified data are not copied more than once. To restore data when differential backups are used, the latest full backup and all differential backups or all backups are required.

Cross-References

▶ Disaster Recovery
▶ Replication

Bag Semantics

Todd J. Green
University of Pennsylvania, Philadelphia, PA,
USA

Synonyms

Duplicate Semantics; Multiset Semantics

Definition

In the ordinary relational model, relations are sets of tuples, which by definition do not contain "duplicate" entries. However, RDBMSs typically implement a variation of this model where relations are *bags* (or *multisets*) of tuples, with duplicates allowed. Formally, a bag is a mapping of tuples to natural number *multiplicities*; a set can be viewed as a special case of a bag where all tuple multiplicities are 0 or 1. The operations of the relational algebra are extended to operate on bags by defining their action on tuple multiplicities. RDBMSs based on bags rather than sets are said to implement *bag semantics* (rather than *set semantics*). Duplicates may occur at multiple levels: in source relations, in materialized views, or in query answers. A variation of bag semantics called *bag-set semantics* is obtained by requiring source relations to be sets, while allowing views and query answers to contain duplicates. Bag-set semantics represents a practical compromise between bag semantics and set semantics.

Historical Background

In the ordinary relational model, relations are sets of tuples, which by definition do not contain "duplicate" entries. However, RDBMS implementations, beginning with System R and INGRES, deviated from the "pure" relational model by allowing duplicate tuples in query answers, thus making bags (multisets), rather than sets, the primary collection type in query processing. The primary motivation for allowing duplicate tuples was the need to avoid expensive duplicate elimination in query processing. Instead, duplicate elimination would only be performed if explicitly requested by the user, e.g., via the SQL *DISTINCT* keyword. In their use of bag relations, RDBMS implementations were ahead of theory, and the precise semantics of relational algebra on bags was not studied until the 1980s [6, 11, 12]. Klug [12] pointed out the failure of certain algebraic identities under bag semantics, and also gave a precise semantics of aggregate functions that did not depend on bag semantics. Dayal et al. initiated the study of query optimization under bag semantics in [6]. Bag containment of unions of conjunctive queries (UCQs) was shown to be undecidable by Ioannidis and Ramakrishan [9], while bag equivalence of conjunctive queries (CQs) was shown to be decidable, in fact the same as isomorphism, by Chaudhuri and Vardi [3]. The Π_2^p -hardness of checking bag containment of CQs was also established in [3], but the decidability of the problem remains open (see Future Directions). The same paper introduced the terminology "bag-set semantics" to describe the semantics obtained by requiring source relations to be sets while allowing views and query answers to contain duplicates, and established the decidability of bag-set equivalence of CQs. For conjunctive queries with inequalities (CQ$^{\neq}$), bag containment and bag-set containment were both shown to be undecidable by Jayram et al. [10]. Bag equivalence of UCQs was shown to be decidable (and like CQs, the same as isomorphism) by Cohen et al. [5]. Cohen [4] studied query equivalence for a generalization of bag-semantics called *combined semantics*, which captures user-specified elimination of duplicates at intermediate stages of query processing. A bag semantics for Datalog was proposed by Mumick et al. [14], and Mumick and Shmueli [15] studied computational problems related to infinite multiplicities (which may occur in query answers because of recursion). Going beyond the relational model, several papers have studied bag semantics in the context of query languages for nested relations; see [2, 8] and references therein.

Foundations

Bag relational algebra. For simplicity, the definitions here assume the unnamed perspective [1] of the relational model. Fix a countable domain $\mathbb{D}$ of database values. A *bag relation* of arity k is a mapping $R : \mathbb{D}^k \to$ associating *tuples* with their *multiplicities.* Conceptually, a tuple not present in the bag relation is mapped to zero. It is typically required that bag relations have *finite support*, i.e., R is zero on all but finitely many tuples. A *duplicate tuple* is a tuple whose multiplicity is greater than 1. From the perspective of bag semantics, an ordinary set relation of arity k is a mapping $R : \mathbb{D}^k \to \{0, 1\}$ (again of finite support), in other words, a bag relation with no duplicate tuples. The relational algebra can be extended to bag relations by defining the action of the relational operators on tuple multiplicities:

- *Selection.* If R is a bag relation of arity k and the selection predicate P maps each k-tuple to either 0 or 1 then $\sigma_P R$ is the bag relation of arity k defined by

$$(\sigma_P R)(t) \stackrel{\text{def}}{=} R(t) \cdot P(t).$$

- *Projection.* If R is a bag relation of arity k and $V = (v_1, \ldots, v_n)$ is a list of indices, $1 \le v_i \le k$, then $\pi_V R$ is a bag relation of arity n, defined by

$$(\pi_V R)(t) \stackrel{\text{def}}{=} \sum_{t' \text{ s.t.} t' = \pi_V(t)} R(t').$$

- *Cross product.* If R_1 is a bag relation of arity k_1 and R_2 is a bag relation of arity k_2, then $R_1 \times R_2$ is a bag relation of arity $k_1 + k_2$, defined by

$$(R_1 \times R_2)(t) \stackrel{\text{def}}{=} R_1(t_1) \cdot R_2(t_2),$$

- where t is a $(k_1 + k_2)$-tuple obtained by concatenating t_1 and t_2.
- *Union.* If R_1 and R_2 are bag relations of arity k, then $R_1 \cup R_2$ is a bag relation of arity k, defined by

$$(R_1 \cup R_2)(t) \stackrel{\text{def}}{=} R_1(t) + R_2(t).$$

- *Intersection.* If R_1 and R_2 are bag relations of arity k, then $R_1 \cap R_2$ is a bag relation of arity k, defined by

$$(R_1 \cap R_2)(t) \stackrel{\text{def}}{=} \min(R_1(t), R_2(t)).$$

- *Difference.* If R_1 and R_2 are bag relations of arity k, then $R_1 - R_2$ is a bag relation of arity k, defined by

$$(R_1 - R_2)(t) \stackrel{\text{def}}{=} \max(R_1(t) - R_2(t), 0),$$

- in other words, the calculation uses *proper subtraction* of natural numbers.

Example 1

$$R = \begin{array}{|cc|c|} \hline a & b & 2 \\ c & b & 3 \\ c & d & 1 \\ \hline \end{array}, \quad S = \begin{array}{|cc|c|} \hline b & c & 5 \\ b & d & 1 \\ d & d & 2 \\ \hline \end{array}, \quad R \circ S = \begin{array}{|cc|c|} \hline a & c & 2 \cdot 5 = 10 \\ a & d & 2 \cdot 1 = 2 \\ c & c & 1 \cdot 5 = 5 \\ c & d & 3 \cdot 1 + 1 \cdot 2 = 5 \\ \hline \end{array}$$

where $R \circ S$ denotes relational composition, i.e., $R \circ S \stackrel{\text{def}}{=} \pi_{1,4} \, \sigma_{2=3}(R \times S)$.

Example 2

$$R = \begin{array}{|cc|c|} a & b & 1 \\ c & b & 1 \\ c & d & 1 \end{array}, \quad S = \begin{array}{|cc|c|} b & c & 1 \\ b & d & 1 \\ d & d & 1 \end{array},$$

$$\pi_{1,3}(R \times S) = \begin{array}{|cc|l|} a & b & 1 \cdot 1 + 1 \cdot 1 = 2 \\ a & d & 1 \cdot 1 = 1 \\ c & b & 1 \cdot 1 + 1 \cdot 1 + 1 \cdot 1 + 1 \cdot 1 = 4 \\ c & d & 1 \cdot 1 + 1 \cdot 1 = 2 \end{array}$$

Note that the use of projection results in duplicate tuples in the query output, even though the source tables R and S are duplicate-free. Duplicates may also be introduced by the union operator.

Bags and SQL. Practical query languages such as SQL are based on bag semantics rather than set semantics, since eliminating duplicates typically requires an expensive sort. However, SQL provides a way to emulate set semantics by the use of an explicit *duplicate elimination* operator, which is specified using the *DISTINCT* keyword. This operator converts a bag relation into the corresponding set relation. For example, to execute the SQL query

```
SELECT DISTINCT R.A, S.C
FROM R, S
WHERE R.B = S.B
```

the RDBMS will compute the bag join of R and S, then eliminate duplicates to produce a set relation. The result is the same as the join of R and S under set semantics.

SQL also provides alternate versions of the union, intersection, and difference operators which differ in their handling of duplicates. In particular, *UNION ALL*, *INTERSECT ALL*, and *EXCEPT ALL* correspond to the bag operations as defined above, i.e., they retain duplicates and produce bag relations as output. Meanwhile, *UNION*, *INTERSECT*, and *EXCEPT* perform duplicate elimination and produce set relations as output. For *UNION*, duplicate elimination is performed after the bag union, while for *INTERSECT* and *EXCEPT*, the duplicate elimination is performed beforehand on the operands.

Query reformulation with bag semantics. The context of bag semantics poses unique challenges in query reformulation and optimization, as classical optimization techniques such as query minimization do not necessarily transfer to bag semantics. For example, under set semantics, the query $R \bowtie R$ can be minimized by removing the redundant self-join to produce the equivalent query R. However, under bag semantics, the resulting query is not equivalent to the original, as the "redundant" self-join may increase the multiplicity of some output tuples. The rest of this section summarizes some of the theoretical results which are known regarding containment and equivalence of queries under bag semantics. In studying query reformulation, it is convenient to assume queries are expressed in a Datalog-style syntax. Thus, the algebraic query $R \bowtie R$ corresponds to the CQ

$$Q(x, y) : -R(x, y), R(x, y),$$

while the query R corresponds to the CQ

$$Q'(x, y) : -R(x, y).$$

As the example illustrates, repetitions of an atom in the body of a CQ are significant. As with set semantics, the bag semantics of CQs is defined in terms of *valuations* which assign the variables in the CQ values from the domain $\mathbb{D}$. The multiplicity of an output tuple is computed by summing over all valuations which produce that output tuple, and for each valuation, taking the product of the source multiplicities of the tuples corresponding to the body. Thus, for a CQ

$$Q(\overline{x}) : -A_1(\overline{z}_1), \ldots, A_n(\overline{z}_n),$$

the multiplicity of an output tuple t is given by

$$Q(t) \overset{\text{def}}{=} \sum_{v} \prod_{i=1}^{n} A_i(v(\overline{z}_i)), \qquad (1)$$

where the sum is over valuations $v : vars(Q) \to \mathbb{D}$ such that $v(\overline{x}) = t$. Note that this definition agrees with the bag relational algebra definition for SPJ queries (using selection, projection, and cross product).

Example 3

The relational algebra query from Example 1 corresponds to the CQ

$$Q(x, z) : -R(x, y), S(y, z).$$

Applying Q to the bag relations R and S from Example 1 yields $Q(c, d) = 3 \cdot 1 + 1 \cdot 2 = 5$, where the $3 \cdot 1$ is produced by the valuation v which sends $x \mapsto c, y \mapsto b,$ and $z \mapsto d$, and the $1 \cdot 2$ is produced by the valuation v' which sends $x \mapsto c, y \mapsto d,$ and $z \mapsto d$.

For a UCQ $\overline{Q} = (Q_1, \ldots, Q_n)$ the multiplicity of an output tuple is computed by summing over all the CQs in $\overline{Q}$:

$$\overline{Q}(t) \overset{\text{def}}{=} \sum_{i}^{n} Q_i(t). \qquad (2)$$

This definition agrees with the bag relational algebra definition for SPJU queries (using selection, projection, cross product, and union).

Example 4

Applying the UCQ Q' defined by the CQs

$$Q'(x, z) : -R(x, y), S(y, z)$$

$$Q'(z, y) : -R(x, y), S(x, z)$$

to the set relations R and S from Example 2, the multiplicity of (c, d) in the output is $1 \cdot 1 +$

$1 \cdot 1 = 2$, *with each of the CQs contributing a term to the sum.*

A query P is *bag-contained* in a query Q, denoted $P \sqsubseteq_b Q$, if for all bag instances I, for every output tuple t, $P(I)(t) \le Q(I)(t)$. When $P \sqsubseteq_b Q$ and $Q \sqsubseteq_b P$, P and Q are said to be *bag-equivalent*, denoted $P \equiv_b Q$. The analogous notions of set-containment and set-equivalence are denoted $P \sqsubseteq_s Q$ and $P \equiv_s Q$, respectively. With set semantics, it is well-known that a CQ Q can be optimized by deleting atoms in the body of Q, producing a minimal *core* of Q which is set-equivalent to Q. However, under bag semantics, this optimization no longer works, as deleting an atom in the body of a CQ always produces an inequivalent query, in fact:

Theorem 1 ([3])

For CQs P, Q, $P \equiv_b Q$ iff $P \cong Q$.

Here, $P \cong Q$ denotes that P and Q are *isomorphic*. Two CQs P and Q are isomorphic if there is a (bijective) renaming of variables $\theta : vars(P) \to vars(Q)$ which transforms P into Q. Checking isomorphism of directed graphs (i.e., Boolean CQs over a single binary predicate), is known to be in *NP*, but is not known or believed to be either in *P* or *NP*-complete. This holds also for general relational structures (and arbitrary CQs). Thus, although there are fewer opportunities for optimization of CQs under bag semantics, checking bag-equivalence is presumably computationally easier than checking set-equivalence (which is known to be *NP*-complete). It turns out that Theorem 1 can be generalized to UCQs:

Theorem 2 ([5])

For UCQs $\overline{P}, \overline{Q}$, $\overline{P} \equiv_b \overline{Q}$ iff $\overline{P} \cong \overline{Q}$.

Bag-equivalence of unions of conjunctive queries with inequalities (UCQ$^{\ne}$s) is also known to be decidable:

Theorem 3 ([5])

For UCQ$^{\ne}$ s P, Q, checking $P \equiv_b Q$ is in PSPACE.

The same result was obtained earlier for CQ$^{\ne}$s by Nutt et al. [16]. The exact complexity in each case (CQ$^{\ne}$s or UCQ$^{\ne}$s) seems to be open.

(Theorem 1 implies a graph isomorphism-hard lower bound).

In contrast to equivalence, which seems to only become easier to check for bag semantics than for set semantics, containment can become much harder. For CQs, the following lower bound on the complexity of bag containment is known:

Theorem 4 ([3])

For CQs P, Q, checking $P \sqsubseteq_b Q$ is Π_2^p -hard.

However, this is only a lower bound, and as of 2008, it is not known whether the problem is even decidable. For UCQs, the question has been resolved negatively:

Theorem 5 ([9])

For UCQs $\overline{P}, \overline{Q}$, checking $\overline{P} \sqsubseteq_b \overline{Q}$ is undecidable.

The reduction used to prove this result highlights a close connection between checking bag containment of UCQs and Hilbert's Tenth Problem, which concerns checking for the existence of solutions to Diophantine equations. Another (non-trivial) reduction from the same problem was used to establish a similar result for CQ$^{\neq}$s:

Theorem 6 ([10])

For CQ$^{\neq}$ s P, Q, checking $P \sqsubseteq_b Q$ is undecidable.

Query reformulation with bag-set semantics. Recall that with bag-set semantics, all source tuples are assumed to have cardinality 1 (or 0, if not present). For query optimization, the main ramification compared to bag semantics is that redundant (repeated) atoms in the body of a CQ are immaterial under bag-set semantics and can be simply deleted from the body, producing an *irredundant* CQ. For example, the CQs

$$Q_1(x,z) : -R(x,y), S(y,z)$$

$$Q_2(x,z) : -R(x,y), R(x,y), S(y,z)$$

are bag-set equivalent, denoted $Q_1 \equiv_{bs} Q_2$, and Q_1 is irredundant. The following result states that eliminating repeated atoms is essentially the

only optimization possible for CQs under bag-set semantics:

Theorem 7 ([3])

For irredundant CQs P, Q, $P \equiv_{bs} Q$ iff $P \equiv_b Q$ (hence $P \equiv_{bs} Q$ iff $P \cong Q$).

This result was essentially a rediscovery of a well-known result in graph theory due to Lovász [13], who showed that for finite relational structures F, G, if $| \text{Hom}(F, H)| = | \text{Hom}(G, H)|$ for all finite relational structures H, where $\text{Hom}(A, B)$ is the set of homomorphisms $h : A \rightarrow B$, then $F \cong G$. In database terminology, this says that bag-set equivalence of irredundant Boolean CQs is the same as isomorphism.

Theorem 7 was extended to UCQs in [5], and the PSPACE upper bound of Theorem 3 was also shown to hold for bag-set equivalence of UCQ$^{\neq}$s in [5].

In general, results on bag-set semantics correspond to results on bag semantics via the following *transfer lemma*:

Lemma 1 ([7])

There exists a mapping $\phi : CQ \rightarrow CQ$ (which extends to UCQs by applying it componentwise on CQs), a mapping f from bag instances to set instances, and a mapping g from set instances to bag instances, such that for any UCQ $\overline{Q}$, bag instance I, and set instance J:

1. $\overline{Q}(I) = \phi(\overline{Q})(f(I))$
2. $\phi(\overline{Q})(J) = \overline{P}(g(J))$

In particular this lemma shows that bag-containment of CQs (UCQs) is polynomial time reducible to bag-set containment, as shown for CQs in [3]. Thus Theorem 5 also implies that bag-set containment of UCQs is undecidable. Also, Lemma 1 generalizes to queries with inequality predicates, hence Theorem [10] also implies also the undecidability of bag-set containment of CQ$^{\neq}$s, as shown in [10]. Finally, note that for UCQs $\overline{P}, \overline{Q}$ the transformation φ can be defined such that $\overline{P} \cong \overline{Q}$ iff $\phi(\overline{P}) \cong \phi(\overline{Q})$. Theorem 2 thus follows from a similar result for bag-set equivalence, also shown in [5]:

Theorem 8 ([5])

For unions of irredundant CQs $\overline{P}, \overline{Q}$, we have $\overline{P} \equiv_{bs} \overline{Q}$ iff $\overline{P} \cong \overline{Q}$.

Bag semantics of datalog queries. As with the relational algebra, the semantics of Datalog queries can be extended operate on bag instances. This is done by defining how *derivation trees* for an output tuple contribute to the multiplicity of the tuple. The *fringe* of a derivation tree τ of an output tuple t, denoted *fringe*(τ), is the bag of leaves (source tuples) in the derivation tree. The count of an output tuple is computed by summing over all derivation trees, and taking the product of the multiplicities of the source tuples in the fringe. If a source tuple appears several times in the fringe, it is counted that many times in the product. Formally, for aDatalog program Q and

source bag instance I, the multiplicity of a tuple t in the output is defined by

$$Q(t) \stackrel{\text{def}}{=} \sum_{\tau} \prod_{A(t') \in \text{fringe}(\tau)} A(t') \qquad (3)$$

where the sum is over all derivation trees for t. Note that when Q is a CQ or UCQ, the above definition agrees with definitions (1.1) and (1.2). However, a basic difference is that for recursive queries, there may be *infinitely many* derivation trees for an output tuple. In this case, its count is defined to be ∞.

Example 5

For the transitive closure query TC and bag relation R defined by

$$
\begin{array}{ll}
TC(x, y) & :- \quad R(x, y) \\
TC(x, z) & :- \quad TC(x, y), R(y, z)
\end{array}
\qquad
R =
\begin{array}{|ccc|}
\hline
a & b & 1 \\
a & c & 2 \\
c & b & 1 \\
c & d & 1 \\
d & d & 3 \\
\hline
\end{array}
$$

evaluating the query yields TC(a, b) = 1 + 2 · 1 = 3 but TC(c, d) = ∞.

Computational problems related to infinite counts are therefore of central interest for Datalog programs with bag semantics. The most basic problem is to check whether or not a count for a given output tuple is ∞. Clearly, computing the set of all derivation trees, and then checking whether or not the set is finite, is not feasible. However, it turns out that the problem is decidable in polynomial time (data complexity), even for Datalog extended with safe stratified negation [15]. The related problem of checking statically whether answer counts are finite for all possible source bag instances is also decidable, even for Datalog extended with negation on unary edb's [15]. However, checking this property is undecidable for Datalog extended with safe stratified negation [15].

Key Applications

The use of bag semantics in practical RDBMS implementations has led to a reexamination of fundamental issues in query optimization, namely, query containment and query equivalence. The switch from classical set semantics to bag semantics turns out to have a radical impact on these issues. Bag semantics also poses challenges for processing of recursive Datalog programs, where infinite multiplicities may arise in the output.

Future Directions

The most salient open problem involving bag semantics concerns containment of conjunctive queries. This was shown to be Π_2^p -hard in [3], but the decidability of the problem remains open. In contrast, for set semantics, the problem is known to be *NP*-complete.

Recommended Reading

1. Abiteboul S, Hull R, Vianu V. Foundations of databases. Reading: Addison-Wesley; 1995.
2. Buneman P, Naqvi S, Tannen V, Wong L. Principles of programming with complex objects and collection types. Theor Comput Sci. 1995;149(1):3–48.
3. Chaudhuri S, Vardi MY. Optimization of *real* conjunctive queries. In: Proceedings of the 12th ACM SIGACT-SIGMOD-SIGART Symposium on Principles of Database Systems; 1993.
4. Cohen S. Equivalence of queries combining set and bag-set semantics. In: Proceedings of the 25th ACM SIGACT-SIGMOD-SIGART Symposium on Principles of Database Systems; 2006, p. 70–9.
5. Cohen S, Nutt W, Serebrenik A. Rewriting aggregate queries using views. In: Proceedings of the 18th ACM SIGACT-SIGMOD-SIGART Symposium on Principles of Database Systems; 1999.
6. Dayal U, Goodman N, Katz, RH. An extended relational algebra with control over duplicate elimination. In: Proceedings of the 1st ACM SIGACT-SIGMOD Symposium on Principles of Database Systems; 1982, p. 117–23.
7. Green TJ. Containment of conjunctive queries on annotated relations. In: Proceedings of the 12th International Conference on Database Theory; 2009.
8. Grumbach S, Libkin L, Milo T, Wong L. Query languages for bags: expressive power and complexity. SIGACT News; 1996, p. 27.
9. Ioannidis YE, Ramakrishnan R. Containment of conjunctive queries: beyond relations as sets. ACM Trans Database Syst. 1995;20(3):288–324.
10. Jayram TS, Kolaitis PG, Vee E. The containment problem for *real* conjunctive queries with inequalities. In: Proceedings of the 25th ACM SIGACT-SIGMOD-SIGART Symposium on Principles of Database Systems; 2006.
11. Klausner A, Goodman N. Multirelations – semantics and languages. In Proceedings of the 11th International Conference on Very Large Data Bases; 1985.
12. Klug AC. Equivalence of relational algebra and relational calculus query languages having aggregate functions. J ACM. 1982;29(3):699–717.
13. Lovász L. Operations with structures. Acta Math Hungarica. 1967;18(3–4):321–8.
14. Mumick IS, Pirahesh H, Ramakrishnan R. The magic of duplicates and aggregates. In: Proceedings of the 16th International Conference on Very Large Data Bases; 1990, p. 264–77.
15. Mumick IS, Shmueli O. Finiteness properties of database queries. In: Proceedings of the 4th Australian Database Conference 1993.
16. Nutt W, Sagiv Y, Shurin S. Deciding equivalences among aggregate queries. In: Proceedings of the 17th ACM SIGACT-SIGMOD-SIGART Symposium on Principles of Database Systems; 1998, p. 214–23.

Bagging

Wei Fan[1] and Kun Zhang[2]
[1]IBM T.J. Watson Research, Hawthorne, NY, USA
[2]Xavier University of Louisiana, New Orleans, LA, USA

Synonyms

Bootstrap aggregating

Definition

Bagging (**B**ootstrap **Agg**regating) uses "majority voting" to combine the output of different inductive models, constructed from bootstrap samples of the same training set. A bootstrap has the same size as the training data, and is uniformly sampled from the original training set with replacement. That is, after an example is selected from the training set, it is still kept in the training set for subsequent sampling and the same example could be selected multiple times into the same bootstrap sample. When the training set is sufficiently large, on average, a bootstrap sample has 63.2 % unique examples from the original training set, and the rest are duplicates. In order to make full use of bagging, typically, one need to generate at least 50 bootstrap samples and construct 50 classifiers using these samples. During prediction, the class label receiving the most votes or most predictions from the base level 50 classifiers will be the final prediction. Normally, Bagging is more accurate than a single classifier trained from the original training set. Statistically, this is due to multiple classifiers' power to reduce the effect of overfitting on the training data and statistical variance. Bagging works the best for non-stable inductive models, such as decision trees, and its advantage is limited for stable methods such as logistic regression, SVM, naive Bayes, etc.

Historical Background

Bagging was originally proposed by Leo Breiman in 1996 [3]. The motivation comes from the fact that unstable inductive learners such as decision trees tend to generate very different models with just a slight change in the training data, for example, the inclusion or exclusion of one example. This is due to the "greedy-based" search heuristics of these methods to find the best hypothesis to fit the labeled training data. Statistically, the effect is that the variance in bias and variance decomposition of error is large. Bias is the systematic error of the chosen learning technique, and variance is due to variations in the training data, i.e., different training data produce different models. Bagging was proposed to remedy these problems. In order to overcome variations of a single training set, multiple bootstrap samples are used to replace the single original training set. Since each bootstrap sample is randomly selected from the original training data, it still "preserves" the main concept to be modeled. However, each bootstrap sample is different, thus offsetting the "individuality" of any single training set that contributes to variance. In the same time, any possible over-sample or "non-sample" of particular labeled examples from the training data (recall that 63.2 % are unique examples and the rest are duplicates) that could reduce learning accuracy is resolved by majority voting or choosing the predicted labels with the most number of votes from base line models. The simple observation is that the probability for most models to make the same mistakes is much less than any single model itself.

Foundations

The key idea of Bagging can be illustrated by a simple synthetic example. In Fig. 1, the true decision boundary that separates two classes of examples is a simple linear function, $y = x$, where examples above the line belongs to one class and examples below the line belongs to another class. The labeled training examples are randomly sampled from the universe of examples and is obviously not exhaustive. When a decision tree

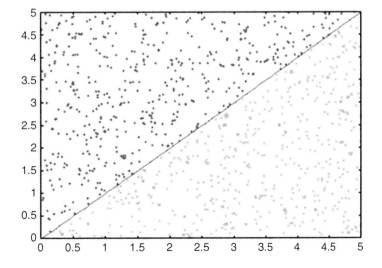

Bagging, Fig. 1 Training data for a simple linear function

Bagging, Fig. 2 Decision
boundary constructed by a
single tree

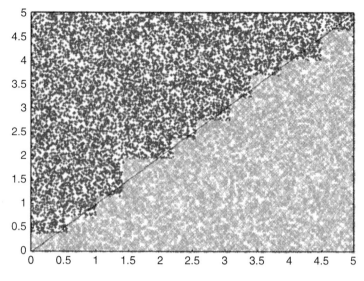

Bagging, Fig. 3 Decision
boundary constructed by
bagging

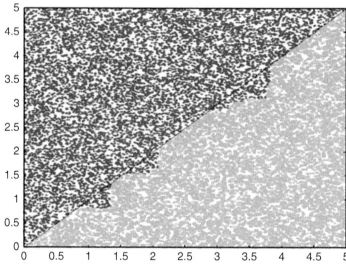

algorithm is applied to the sampled training data,
it will construct a tree model particular to that
sample set. Figure 2 shows the decision boundary
of a single tree constructed by C4.5 algorithm
[12]. As observed, the "stair-step" shape of the
decision boundary is due to "bias" or systematic
error of decision tree algorithm, where the splits
are always perpendicular to the axes. Once an
algorithm is chosen, its bias is hard to avoid. At
the same time, the exact position and size of each
step is due to variations in individually sampled
training data, and it contributes to "variance"
term in the prediction error. Next, this entry
studies how "variance" can be effectively reduced

by applying Bagging. Figure 3 demonstrated the
decision boundary of Bagging generated from 50
bagged C4.5 trees. It is clear that the decision
boundary is much closer to the perfect "$y = x$"
line, and only three steps are left.

Formally, the reduction in variance can be
approximated by the following equation:

$$\epsilon = \beta + \frac{\sigma}{\sqrt{n}}, \qquad (1)$$

where the error ε is decomposed into bias β and
variance σ. The reduction in variance is scaled
by $\sqrt{n}$ where n is the number of bagging models.

Additionally, the idea of Bagging can be applied to both classification problem or discrete variable prediction (such as the synthetic example given above) and regression problems or continuous variable prediction. In regression problems, the estimated value of different models are averaged as the final prediction.

For further reading and understanding of Bagging and its various applications, refer to the 15 papers in the "Recommended Reading" section. For theories to explain how and why Bagging works, the best source is the original paper by Leo Breiman [3], that explains the statistical fundamentals that Bagging is built upon. Additionally [5], explains the appropriate understanding of overfitting and how overfitting plays a role in classifers' generalization power. In essence, Bagging helps to correct the overfitting problem of unstable learners, such as decision trees. For various information on how to build an accurate decision tree, the base model where Bagging normally combines, the following list of papers contain solid information [2, 4, 9, 11–13]. For state-of-the-art and future works of Bagging, one of the most novel ideas is "randomization" and different treatment on this subject can be found in various papers [1, 6, 7, 15]. During the time when Bagging was proposed, an alternative method called "Boosting" was proposed to resolve the instability of inductive learners and some papers that theoretically describe its motivation and practice can be found in [8, 10]. For a detailed and systematic comparison of Bagging and many other state-of-the art algorithms on a difficult application problem "skewed and stochastic ozone day forecasting", the works in [14] contain useful information.

Key Applications

Bagging is most suitable for applications where both the training set is not too big, i.e., the number of training examples can fit into main memory of the machine, and the chosen inductive learner is unstable, such as decision trees. This is mainly because generating multiple bootstrap sample involves multiple scans of the training data and can

be a bottleneck if most operations rely on disk. At the same time, if the number of examples does not fit into main memory of the machine, learning can incur swapping cost and take a long time to complete. As discussed previously, bagging's success is mainly limited to unstable learners that normally have large variance.

As a good example, "ozone day prediction" [14] illustrates how well Bagging can perform in practice. What makes this problem interesting is that there are 72 continuous features in the collected dataset, only about 3 % of 2,500+ collected days are positive. It is important to understand that for a problem with 72 dimensions, 2,500 training examples are trivial in size. Even if these 72 features are binary in values, the total problem size is an astronomical number 2^{72}, and obviously, 2,500 samples from this space is "really nothing".

Applying non-stable inductive learning method, such as decision tree, on this problem is unlikely to obtain satisfactory result. The simple reason is that these methods tend to "overfit" or build unnecessarily detailed models to fit well on the given 2,500 examples, but do not generalize well on unseen testing data in the problem space of 2^{72}. As illustrated in a precision-recall plot in Fig. 4, Bagging consistently obtains higher "precision" than its comparable single tree methods. Recall is the percentage of "true ozone days" that are correctly predicted as ozone days, and precision is the percentage of "predicted ozone days" that are actually ozone days. By reading the consistently higher precision numbers by Bagging on the same recall number, it clearly demonstrates that Bagging is more accurate than single decision trees.

Future Directions

Most recently, a completely counter-intuitive method called "Random Decision Tree" [6, 7, 15] has been proposed to reduce both bias and variance in inductive learning. Each tree in a random tree ensemble is used as a structure to summarize the training data, and importantly, the structure of the tree is semi-randomly

Bagging, Fig. 4 Bagging versus single decision tree

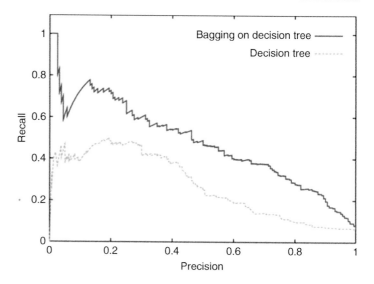

constructed, and data are "filled-in" the tree structure to summarize statistics of the training data. During prediction, the estimated probability from each tree is averaged as the final conditional probability $P(\mathbf{y}|\mathbf{x})$ and the class label with the highest probability is chosen as the predicted label. Since it does not incur any overhead of producing bootstraps or using information gain to perform feature selection as Bagging does, random decision tree can be highly efficient. One application on difficult skewed and high dimensional ozone day forecast can be found in [14]. Similar to Bagging, Random Decision Tree is applicable to both classification and regression problems.

Cross-References

▶ Boosting
▶ Decision Trees

Recommended Reading

1. Amit Y, Geman D. Shape quantization and recognition with randomized trees. Neural Comput. 1997;9(7):1545–88.
2. Bradford JP, Kunz C, Kohavi R, Brunk C, Brodley CE. Pruning decision trees with misclassification costs. In: Proceedings of the European Conference on Machine Learning; 1998. p. 131–6.
3. Breiman L. Bagging predictors. Mach Learn. 1996;24(2):123–40.
4. Buntine W. Learning classification trees. In: Hand DJ, editor. Artificial intelligence frontiers in statistics. London: Chapman & Hall; 1993. p. 182–201.
5. Domingos P. Occam's two razors: the sharp and the blunt. In: Proceedings of the 4th International Conference on Knowledge Discovery and Data Mining; 1998.
6. Fan W, Wang H, Yu PS, Ma S. Is random model better? On its accuracy and efficiency. In: Proceedings of the 19th International Conference on Data Engineering; 2003.
7. Fan W, Greengrass E, McCloskey J, Yu PS, Drummey K. Effective estimation of posterior probabilities: explaining the accuracy of randomized decision tree approaches. In: Proceedings of the IEEE International Conference on Data Mining; 2005. p. 154–61.
8. Freund Y, Schapire R. A decision-theoretic generalization of on-line learning and an application to boosting. Comput Syst Sci. 1997;55(1):119–39.
9. Gehrke J, Ganti V, Ramakrishnan R, Loh W-Y. BOAT-optimistic decision tree construction. In: Proceedings of the ACM SIGMOD International Conference on Management of Data; 1999.
10. Kearns M, Mansour Y. On the boosting ability of top-down decision tree learning algorithms. In: Proceedings of the Annual ACM Symposium on the Theory of Computing; 1996. p. 459–68.
11. Mehta M, Rissanen J, Agrawal R. MDL-based decision tree pruning. In: Proceedings of the 1st International Conference on Knowledge Discovery and Data Mining; 1995. p. 216–21.
12. Quinlan R. C4.5: programs for machine learning. Los Altos: Morgan Kaufmann; 1993.
13. Shawe-Taylor J, Cristianini N. Data-dependent structural risk minimisation for perceptron decision trees. In: Jordan M, Kearns M, Solla S, editors. Ad-

vances in neural information processing systems 10. Cambridge, MA: MIT Press; 1998. p. 336–42.

14. Zhang K, Fan W. Forecasting skewed biased stochastic ozone days: analyses, solutions and beyond. Knowl Inf Syst. 2008;14(3):299–326.

15. Zhang K, Xu Z, Peng J, Buckles BP. Learning through changes: an empirical study of dynamic behaviors of probability estimation trees. In: Proceedings of the IEEE International Conference on Data Mining; 2005. p. 817–20.

Bayesian Classification

Wynne Hsu
National University of Singapore, Singapore, Singapore

Synonyms

Bayes classifier

Definition

In classification, the objective is to build a classifier that takes an unlabeled example and assigns it to a class. Bayesian classification does this by modeling the probabilistic relationships between the attribute set and the class variable. Based on the modeled relationships, it estimates the class membership probability of the unseen example.

Historical Background

The foundation of Bayesian classification goes back to Reverend Bayes himself [2]. The origin of Bayesian belief nets can be traced back to [15]. In 1965, Good [4] combined the independence assumption with the Bayes formula to define the Naïve Bayes Classifier. Duda and Hart [14] introduced the basic notion of Bayesian classification and the naïve Bayes representation of joint distribution. The modern treatment and development of Bayesian belief networks is attributed to Pearl [8]. Heckerman [13] later reformulated the

Bayes results and defined the probabilistic similarity networks that demonstrated the practicality of Bayesian classification in complex diagnostic problems.

Foundations

Bayesian classification is based on Bayes Theorem. It provides the basis for probabilistic learning that accommodates prior knowledge and takes into account the observed data.

Let X be a data sample whose class label is unknown. Suppose H is a hypothesis that X belongs to class Y. The goal is to estimate the probability that hypothesis H is true given the observed data sample X, that is, $P(Y|X)$.

Consider the example of a dataset with the following attributes: Home Owner, Marital Status, and Annual Income as shown in Fig. 1. Credit Risks are Low for those who have never defaulted on their payments and credit risks are High for those who have previously defaulted on their payments.

Assume that a new data arrives with the following attribute set: X = (Home Owner = Yes, Marital Status = Married, Annual Income = High). To determine the credit risk of this record, it is noted that the Bayes classifier combines the predictions of all alternative hypotheses to determine the most probable classification of a new instance. In the example, this involves computing $P(High|X)$ and $P(Low|X)$ and to determine whether $P(High|X) > P(Low|X)$?

However, estimating these probabilities is difficult, since it requires a very large training set that covers every possible combination of the class label and attribute values. Instead, Bayes theorem is applied and it resulted in the following equations:

$$P(High|X) = P(X|High) P(High) /P(X) \text{ and}$$

$$P(Low|X) = P(X|Low) P(Low) /P(X)$$

$P(High)$, $P(Low)$, and $P(X)$ can be estimated from the given dataset and prior knowledge. To estimate the class-conditional probabilities

Customer ID	Home owner	Marital status	Annual income	Credit risks
1	Yes	Married	High	Low
2	No	Single	Medium	High
3	No	Married	Medium	High
4	Yes	Divorced	Low	High
5	No	Married	High	Low
6	No	Divorced	High	Low
7	Yes	Single	Low	Low
8	No	Single	High	Low
9	No	Married	Medium	High
10	Yes	Single	Medium	Low

Bayesian Classification, Fig. 1 Dataset example

P(X|High), P(X|Low), there are two implementations: the Naïve Bayesian classifier and the Bayesian Belief Networks.

In the Naïve Bayesian classifier [13], the attributes are assumed to be conditionally independent given the class label y. In other words, for an n-attribute set X = $(X_1, X_2,...,X_n)$, the class-conditional probability can be estimated as follows:

$$P(Y|X) = \alpha P(Y) \prod_i P(X_i|Y)$$

In the example,

$(X|\text{Low}) = P\,(\text{Home Owner} = \text{Yes}|\text{Credit Risk}$

$= \text{Low}) \times P\,(\text{Marital Status}$

$= \text{Married}|\text{Credit Risk} = \text{Low})$

$\times P\,(\text{Annual Income}$

$= \text{High}|\,\text{Credit Risk} = \text{Low}$

$= 3/4 \times 2/4 \times 4/4 = 3/8\,P\,(X|\text{High})$

$= P\,(\text{Home Owner} = \text{Yes}|\text{Credit Risk}$

$= \text{High}) \times P\,(\text{Marital Status} = \text{High})$

$\times P\,(\text{Annual Income} = \text{High}|\text{Credit}$

$\text{Risk} = \text{High}) = 0$

Putting them together, P(High|X) = P(X|High) P(High)/P(X) = 0

P(*Low*|X) = P(X|*Low*) P(*Low*)/P(X) > 0

Since P(Low|X) > P(High|X), X is classified as having Credit Risk = Low.

In other words,

$$Classifiy(X) = \arg\max_y P\,(Y = y)$$

$$\prod_i P\,(X_i|Y = y)$$

In general, Naïve Bayes classifiers are robust to isolated noise points and irrelevant attributes. However, the presence of correlated attributes can degrade the performance of naïve Bayes classifiers as they violate the conditional independence assumption. Fortunately, Domingos and Pazzani [3] showed that even when the independence assumption is violated in some situations, the naïve Bayesian classifier can still be optimal. This has led to a wide spread use of naïve Bayesian classifiers in many applications. Jaeger [9] also further clarifies and distinguishes the concepts that can be recognized by naïve Bayes classifiers and the theoretical limits on learning the concepts from data.

There are many extensions to the naïve Bayes classifier that impose limited dependencies among the feature/attribute nodes, such as tree-augmented naïve Bayes [6], and forest-augmented naïve Bayes [11].

Bayesian belief network [10] overcomes the rigidity imposed by this assumption by allowing the dependence relationships among a set of attributes to be modeled as a directed acyclic graph. Associated with each node in the directed acyclic graph is a probability table. Note that a node in the Bayesian network is conditionally independent of its non-descendants, if its parents are known.

Refer to the running example. Suppose the probabilistic relationships among Home Owner, Marital Status, Annual Income, and Credit Risks are shown in Fig. 2. Associated with each node is the corresponding conditional probability table relating the node to its parent node(s).

In the Bayesian belief network, the probabilities are estimated as follows:

$$P(Risk|X) = \alpha \sum_I P(Risk|Income)$$

$$\sum_O P(Owner) \sum_S P(Status)$$

$$P(Income|Owner, Status)$$

In the above example,

$$P(Low|X) = P(Low|Income = High)$$
$$* P(Income = High|Owner$$
$$= Yes, Status = Married)$$
$$= 1 * 1 = 1$$

Alternatively, by recognizing that given Annual Income, Credit Risks is conditionally independent of Home Owner and Marital status, then

$$P(Low|X) = P(Low|Income = High) = 1$$

This example illustrates that the classification problem in Bayesian networks is a special case of belief updating for any node (target class) Y in the network, given evidence X.

While Bayesian belief network provides an approach to capture dependencies among variables using a graphical model, constructing the network is time consuming and costly. Substantial research has been and is still continuing to address the inference as well as the automated construction of Bayesian networks by learning from data. Much progress has been made recently. So applying Bayesian networks for classification is no longer as time consuming and costly as before,

and the approach is gaining headway into the mainstream applications.

Key Applications

Bayesian classification techniques have been applied in many applications. Here, a few more common applications of Bayesian classifiers are mentioned.

Text Document Classification

Text classification refers to the grouping of texts into several clusters so as to improve the efficiency and effectiveness of text retrieval. Typically, the text documents are pre-processed and the key words chosen. Based on the selected keywords of the documents, probabilistic classifiers are built. Dumais et al. [5] show naïve Bayes classifier yields surprisingly good classifications for text documents.

Image Pattern Recognition

In image pattern recognition, a set of elementary or low level image features are selected which describe some characteristics of the objects. Data extracted based on this feature set are used to train Bayesian classifiers for subsequent object recognition. Aggarwal et al. [1] did a comparative study of three paradigms for object recognition - Bayesian Statistics, Neural Networks and Expert Systems.

Medical Diagnostic and Decision Support Systems

Large amounts of medical data are available for analysis. Knowledge derived from analyzing these data can be used to assist the physician in subsequent diagnosis. In this area, naïve Bayesian classifiers performed exceptionally well. Kononenko et al. [12] showed that the naïve Bayesian classifier outperformed other classification algorithms on five out of the eight medical diagnostic problems.

Email Spam Filtering

With the growing problem of junk email, it is desirable to have an automatic email spam filters

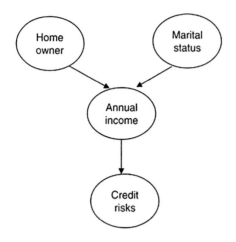

Owner = Yes
0.4

(Table associated with node owner)

Status	
Married	0.4
Single	0.4
Divorced	0.2

(Table associated with node status)

	Income = High	Income = Medium	Income = Low
HO = Yes, MS = Married	1	0	0
HO = Yes, MS = Divorced	0	0	1
HO = Yes, MS = Single	0	0.5	0.5
HO = No, MS = Married	0.33	0.67	0
HO = No, MS = Divorced	1	0	0
HO = No, MS = Single	0.5	0.5	0

(Table associated with node income)

	Risk = High	Risk = Low
Income = High	0	1
Income = Medium	0.75	0.25
Income = Low	0.5	0.5

(Table associated with node credit risks)

Bayesian Classification, Fig. 2 Bayesian belief network of the credit risks dataset

to eliminate unwanted messages from a user's mail stream. Bayesian classifiers that take into consideration domain-specific features for classifying emails are now accurate enough for real world usage.

Data Sets

http://archive.ics.uci.edu/beta/datasets.html
http://spamassassin.apache.org/publiccorpus/

URL to Code

More recent lists of Bayesian Networks software can be found at:
  Kevin Murphy's website:
  http://www.cs.ubc.ca/~murphyk/Bayes/bnsoft.html
  Google directory:
  http://directory.google.com/Top/Computers/Artificial_Intelligence/Belief_Networks/Software/

Specialized naïve Bayes classification software:

jBNC – a java toolkit for variants of naïve Bayesian classifiers, with WEKA interface

Cross-References

▶ Probabilistic Retrieval Models and Binary Independence Retrieval (BIR) Model

References

1. Aggarwal JK, Ghosh J, Nair D, Taha I. A comparative study of three paradigms for object recognition – Bayesian statistics, neural networks, and expert systems. In: Advances in image understanding: a festschrift for Azriel Rosenfeld. Washington, DC: IEEE Computer Society Press; 1996. p. 241–62.
2. Bayes T. An essay towards solving a problem in the doctrine of chances. Philos Trans R Soc. 1763;53:370–418.
3. Domingos P, Pazzani M. Beyond independence: conditions for the optimality of the simple Bayesian classifier. In: Proceedings of the 13th International Conference on Machine Learning; 1996. p. 105–12.
4. Duda RO, Hart PE. Pattern classification and scene analysis. New York: Wiley; 1973.
5. Dumais S, Platt J, Heckerman D, Sahami M. Inductive learning algorithms and representations for text categorization. In: Proceedings of the International Conference on Information and Knowledge Management; 1998.
6. Friedman N, Geiger D, Goldszmidt M. Bayesian network classifiers. Mach Learn. 1997;29:131–63.
7. Good IJ. The estimation of probabilities: an essay on modern Bayesian methods. Cambridge, MA: MIT Press; 1965.
8. Heckerman D. Probabilistic similarity networks. ACM doctoral dissertation award series. Cambridge, MA: MIT Press; 1991.
9. Jaeger M. Probabilistic classifiers and the concepts they recognize. In: Proceedings of the 20th International Conference on Machine Learning; 2003. p. 266–73.
10. Jensen FV. An introduction to Bayesian networks. New York: Springer; 1996.
11. Keogh E, Pazzani M. Learning augmented Bayesian classifiers: a comparison of distribution-based and classification-based approaches. In: Proceedings of the 7th International Workshop on Artificial Intelligence and Statistics; 1999.
12. Kononenko I, Bratko I, Kukar M. Application of machine learning to medical diagnosis. In: Machine learning, data mining and knowledge discovery: methods and applications. New York: Wiley; 1998.
13. Langley P, Iba W, Thompson K. An analysis of Bayesian classifiers. In: Proceedings of the 10th National Conference on Artificial Intelligence. 1992. p. 3–8.
14. Pearl J. Probabilistic reasoning in intelligenet systems: networks of plausible inference. San Mateo: Morgan Kaufmann; 1988.
15. Wright S. Correlation and causation. J Agric Res. 1921;20(7):557–85.

Benchmark Frameworks

Philippe Bonnet[1] and Dennis Shasha[2]
[1]Department of Computer Science, IT University of Copenhagen, Copenhagen, Denmark
[2]Department of Computer Science, New York University, New York, NY, USA

Definition

A benchmark framework is a collection of tools for generating and executing benchmarks. Examples include HammerDB and YCSB;

HammerDB is a tool for executing benchmarks on relational database systems. It supports a range of existing relational database systems. It is configured to run TPC-C and TPC-H benchmarks (including schema creation, data generation, measurements, and workload execution). It can also capture and replay traces on Oracle databases. HammerDB provides a visual interface and a command line interface for additional scripting and automation.

YCSB is a tool for characterizing the performance of cloud-based data serving systems. It supports a range of existing data platforms including document databases, key-value stores, full-text search engines, BigTable databaes, block stores, and relational databases. It is configured to run latency benchmarks (with variable load on the data serving system) as well as throughput benchmarks (with a variable number of servers attached to the data serving system). Workloads represent a mix of read/write operations, data sizes, and request distributions. A core package is provided and users can develop their own (by changing existing configurations or writing new code).

URL to Code and Data Sets

HammerDB: http://www.hammerdb.com/
YCSB: https://github.com/brianfrankcooper/YCSB

Recommended Reading

1. Cooper BF, Silberstein A, Tam E, Ramakrishnan R, Sears R. Benchmarking cloud serving systems with YCSB. In: Proceedings of the 1st ACM Symposium on Cloud Computing; 2010. p. 143–154.

Big Data Platforms for Data Analytics

Volker Markl[1], Vineyak Borkar[2],
Matei Zaharia[3], Till Westmann[4], and
Alexander Alexandrov[5]
[1]IBM Almaden Research Center, San Jose, CA, USA
[2]CTO and VP of Engineering, X15 Software, San Francisco, CA, USA
[3]Douglas T. Ross Career Development Professor of Software Technology, MIT CSAIL, Cambridge, MA, USA
[4]Oracle Labs, Redwood City, CA, USA
[5]Database and Information Management (DIMA), Institute of Software Engineering and Theoretical Computer Science, Berlin, Germany

Synonyms

Big data management systems; Data intensive computing software; Predictive analytics platforms

Definition

Due to the volume, velocity, and variety of data now coming from the Web, social media, and personal devices, the analysis of "Big Data" has become a priority. A number of software platforms have been developed to support the analysis of massive data sets using clusters of computers working in parallel. These platforms fall into two categories: those based on the relational data model and its SQL query language, and those with more flexible data models and query languages tailored to less rigidly structured data. The latter category is referred to here as Big Data Platforms. (SQL analytics on Big Data are covered separately.)

Historical Background

Today's platforms for Big Data Analytics are the result of technical work carried out in two computer systems software fields: database systems and distributed systems.

Parallel Databases

In the field of database systems, the need to manage, query, and analyze enterprise data volumes that exceed the capabilities of a single server led to the development of parallel database systems starting in the mid-1980s [6]. Most are architecturally "shared nothing," being based on the use of a networked cluster of individual machines, each with their own private processors, main memories, and disks; inter-machine coordination and data communication are accomplished via message passing. Notable first-generation parallel database systems were developed at the University of Wisconsin (Gamma) and the University of Tokyo (GRACE), and the first major commercial system was developed by Teradata (which remains a market leader today). Parallel database systems spread relational data over such a cluster based on a partitioning strategy, such as hash or range partitioning, and queries are processed by employing partitioned-parallel divide-and-conquer techniques. Parallel database systems are fronted by a high-level, declarative language, SQL, so users are shielded from the complexities of parallel programming. Parallel database systems have been an extremely successful application of parallel computing, and a number of commercial products exist today.

Scalable Data Processing

In the distributed systems world, the rise of the Web brought a need to index and query its huge content. SQL and parallel relational databases were not suited for this purpose, although shared-nothing clusters again emerged as the hardware platform of choice. In the early 2000s, Google developed the Google File System (GFS) [7] and MapReduce programming model [5] to allow programmers to store and process large volumes of data by writing a few user-defined functions that operate on one data item or one group of items. The MapReduce framework applies these functions in parallel to data instances in distributed files (map) and to sorted groups of instances sharing a common key (reduce) – not unlike the partitioned parallelism in parallel database systems. Apache's Hadoop MapReduce platform is the most prominent implementation of this paradigm for the rest of the Big Data community. Shortly thereafter came relationally inspired declarative languages (like Pig and

Hive, covered later) that are layered on top of MapReduce and HDFS and compile down to Hadoop MapReduce jobs. These languages have greatly boosted productivity for developers for many Big Data use cases.

Convergence

Today there is a convergence of these two fields. New platforms for Big Data Analytics, as well as new Big Data Management platforms (which store and manage data as well as supporting its analysis), are being developed based on combinations of ideas drawn from both worlds. There is also a movement toward platform support for more complex forms of data analysis (e.g., iterative algorithms in addition to dataflows).

Scientific Fundamentals

System Layers

Big Data platforms for data analytics can be organized in a layered architecture; Fig. 1 characterizes a number of today's platforms in this manner. The heart of the layered architecture consists of a **low-level API** for dataflow assembly that targets a (usually unique) parallel execution engine. In parallel database systems, this corresponds to the APIs used for physical plan description and

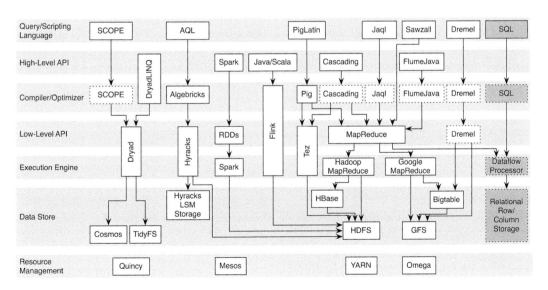

Big Data Platforms for Data Analytics, Fig. 1 Platforms and layers

the runtime systems that evaluate them. Low-level APIs vary with respect to the family of supported operators but usually include physical operators for collection processing (e.g., `sort`, `hashjoin`), as well as first-class support for UDFs through second-order functions (`map` f, `reduce` g, `filter` p). The low-level API creates a dataflow execution graph which describes a set of bulk data processing operators to be executed (vertices) and the data exchange links between them (edges).

The **Execution Engine** evaluates the dataflow graph in a data-parallel manner. This is achieved by spawning a set of parallel instances for each operator and wiring the channels between adjacent operators based on the configured link type (e.g., all-to-all, point-to-point). Data exchange between worker processes can be initiated by the client (push) or by the receiver (pull) [9].

Internally, the engine assumes a model for the element type of the processed collections whose transparency varies between the different systems. Element members used for collection processing in particular can be represented either explicitly as data fields (in a record/tuple model) or implicitly as functions (in a black-box object model with key selectors).

For pure dataflows, scheduling can be realized via lazy deployment of the spawned task instances in topological order. Native support for recursive dataflows at the engine level, however, requires a model that incorporates some notion of state (for the iteratively refined results) and repeated execution (for iterative processing) at the task level.

Failure tolerance (for continued forward progress) relies on maintaining lineage information about the computed collections. It can be realized pessimistically, via intermediate result materialization or, optimistically, via partial reevaluation or constraint-based random re-initialization of the lost partitions.

High-level APIs are embedded as libraries in a programming language that is typically identical to or compatible with the low-level API (e.g., Java, Scala). Their main goal is to abstract away certain physical execution aspects, most notably algorithm and degree of parallelism se-

lection. The APIs advocate a functional style of dataflow assembly characterized by fluent syntax via chaining of functions (usually called *transformations*) over a core data type representing a parallel collection (RDD, DataSet). UDFs can be integrated seamlessly into this style of assembly through in-place parameter definition, either as lambda expressions (in Scala, Java 8) or anonymous inner classes (in Java).

Various standalone **Query and Scripting Languages** exist on top of the APIs discussed above. Characteristic for these languages is a more concise, declarative dataflow syntax, often akin to SQL, as well as integrated support for named expressions that facilitate a scripting approach to program assembly.

Translating the scripting language or high-level API expressions to the low-level APIs expected by the execution engines involves a language compiler/optimizer and/or a dataflow execution planner. The range of optimizations supported by the language compiler/optimizer is typically constrained by the degree of semantic information available at the source layer.

Influential Systems

Hadoop Universe Hadoop MapReduce is an open-source implementation of the MapReduce programming model and execution engine architecture proposed by Google [5]. It offers a Java API for writing scalable data-processing jobs. A MapReduce job consists of a *map* phase (formally a *flatMap*):

$$\texttt{map}(f : I \Rightarrow (K, V)^*) : I^* \Rightarrow (K, V)^*$$

in which a UDF f is applied per element on an input collection I^* and the resulting collections are flattened as an intermediate collection of key-value pairs $(K, V)^*$. The intermediate result is then grouped by key and passed to a *reduce* phase

$$\texttt{reduce}(h : (K, V^*) \Rightarrow O^*) : (K, V^*)^* \Rightarrow O^*$$

in which a UDF h is applied per group. The grouping phase (also called *shuffle*) is realized as a highly scalable parallel sort where the data is partitioned on the group key and shuffled through the network using a lazy "pull" model. Fault tolerance utilizes the fact that the intermediate result is materialized in order to restart only the current phase (*map* or *reduce*) in case of failure, while load balancing can be achieved through work stealing in the *map* phase. A fundamental limitation of MapReduce is the hard-coded form of the assembled execution plans and the lack of explicit support for multiple inputs. Due to this limitation, various systems optimize and translate a more generic dataflow API with higher-order functions (Cascading, Scalding) or a declarative language (Hive, Pig) into a series of MapReduce jobs.

Dryad/SCOPE Microsoft has developed its own suite of Big Data processing technologies. Dryad [10] is a parallel runtime platform that enables a programmer (or program) to build a dataflow execution graph out of sequential programs and then connect them using one-way channels. Programs are graph vertices, while the channels are the edges. In Dryad, a job is actually a graph generator that can synthesize any directed acyclic graph; a Dryad graph can even change during execution in response to situations encountered as the computation progresses. Several higher-level Microsoft programming models use Dryad as their execution engine. DryadLINQ [12] generates Dryad computations from LINQ (Language-Integrated Query) query expressions in C#. SCOPE [15] is an SQL-inspired scripting language targeted for large-scale data analysis. SCOPE data is modeled as sets of rows composed of typed columns, and each rowset has a well-defined schema. The SCOPE runtime implements many standard physical operators, thereby saving Dryad users from implementing similar functionality repetitively.

AsterixDB/Hyracks Hyracks [4] is a push-based data-parallel runtime at the same level as Hadoop and Dryad. Hyracks Jobs are DAGs consisting of operators and connectors. Operators consume partitions of input data

and produce output partitions, while connectors perform various redistributions of data between operators. The Hyracks scheduler analyzes each job DAG to identify groups of operators (stages) that can be co-scheduled while adhering to the operators' blocking requirements. Job stages are parallelized and executed in data dependency order. Hyracks is extensible, so users can define new operators and connectors if desired. AsterixDB [3] is a Big Data Management System for storing, indexing, and querying large collections of semistructured data. AsterixDB's data model (ADM) is JSON-like, with extensions in terms of extra data types, and it has a query language (AQL) that provides full query capabilities over one or multiple ADM datasets.

Spark extended the DAG execution model to allow storing intermediate results, called *resilient distributed datasets* (RDDs), in a fault-tolerant manner [13]. This enables it to support iteration (e.g., repeated jobs that access the same dataset) and incremental stream processing via mini-batches [14]. RDDs provide fault tolerance at low cost by using a lineage graph to reconstruct missing partitions instead of data replication; thus they can efficiently utilize memory and storage media on a large cluster. Spark includes high-level APIs in Java, Scala, and Python, as well as higher-level libraries for relational queries (Spark SQL), machine learning (MLlib), graph processing (GraphX [8]), and stream processing [14]. The relational layer performs traditional relational query optimization for the portions of the program expressed in it.

Flink Akin to Spark, Flink (initially developed as Stratosphere [2]) exposes a Java or Scala dataflow API based around higher-order collection transformations. In contrast to Spark or Hadoop MapReduce, and similar to Hyracks, Flink supports key-based operations (e.g., joins, grouping, aggregations) on generic objects as it models keys implicitly through key selector functions rather than explicitly through (key, value) tuples. Flink performs cost-based algorithm and data shipment strategy selection in order to translate the dataflow expressions into an optimal execution graph. The runtime is Java based, but it manages memory off-heap by keeping a serialized version of the processed objects. Operator implementations try to mitigate the overhead of deserialization by working on the serialized data whenever possible. In addition to that, the runtime natively supports bulk synchronous iterative dataflows in two flavors (batch and delta-centric).

Dremel is a low-latency query system at Google that runs interactive queries in an SQL-like language [11]. Unlike prior SQL-based systems, it is optimized for nested data, offering a special columnar storage format that can represent such data and operators that run on it. Dremel runs queries as a tree of operators with a "push" model and can return partial results early to mitigate stragglers.

Key Applications

Key applications for Big Data analysis include marketing, finance, public safety, public health, national security, law enforcement, medicine, political science, and governmental policy-making.

Cross-References

▶ Distributed Machine Learning
▶ Graph Mining
▶ Parallel Database Management
▶ SQL Analytics on Big Data

Recommended Reading

1. Data, data everywhere. The Economist; 25 Feb 2010.
2. Alexandrov A, Bergmann R, Ewen S, Freytag J-C, Hueske F, Heise A, Kao O, Leich M, Leser U, Markl V, Naumann F, Peters M, Rheinländer A, Sax M, Schelter S, Höger M, Tzoumas K, Warneke D. The stratosphere platform for big data analytics. VLDB J. 2014;(6):1–26.
3. Alsubaiee S, Altowim Y, Altwaijry H, Behm A, Borkar VR, Bu Y, Carey MJ, Cetindil I, Cheelangi M, Faraaz K, Gabrielova E, Grover R, Heilbron Z, Kim Y, Li C, Li G, Ok JM, Onose N, Pirzadeh P, Tsotras VJ, Vernica R, Wen J, Westmann T. Asterixdb: a scalable, open source BDMS. Proc VLDB Endow. 2014;7(14):1905–16.
4. Borkar VR, Carey MJ, Grover R, Onose N, Vernica R. Hyracks: a flexible and extensible foundation for data-intensive computing. In: Abiteboul S, Böhm K, Koch C, Tan K-L, editors. Proceedings of the 27th International Conference on Data Engineering; 2011. p. 1151–62.
5. Dean J, Ghemawat S. Mapreduce: a flexible data processing tool. Commun ACM. 2010;53(1):72–77.
6. DeWitt DJ, Gray J. Parallel database systems: the future of high performance database systems. Commun ACM. 1992;35(6):85–98.
7. Ghemawat S, Gobioff H, Leung S. The Google File System. In: Scott ML, Peterson LL, editors. Proceedings of the 19th ACM Symposium on Operating Systems Principles; 2003. p. 29–43.
8. Gonzalez JE, Xin RS, Dave A, Crankshaw D, Franklin MJ, Stoica I. Graphx: graph processing in a distributed dataflow framework. In: Proceedings of the 11th USENIX Symposium on Operating System Design and Implementation; 2014.
9. Graefe G. Query evaluation techniques for large databases. ACM Comput Surv. 1993;25(2):73–169.
10. Isard M, Budiu M, Yu Y, Birrell A, Fetterly D. Dryad: distributed data-parallel programs from sequential building blocks. In: Ferreira P, Gross TR, Veiga L, editors. Proceedings of the 2007 EuroSys Conference; 2007. p. 59–72.
11. Melnik S, Gubarev A, Long JJ, Romer G, Shivakumar S, Tolton M, Vassilakis T. Dremel: interactive analysis of web-scale datasets. Proc VLDB Endow. 2010;3(1):330–39.
12. Yu Y, Isard M, Fetterly D, Budiu M, Erlingsson Ú, Gunda PK, Currey J. Dryadlinq: a system for general-purpose distributed data-parallel computing using a high-level language. In: Draves R, van Renesse R, editors. Proceedings of the 8th USENIX Symposium on Operating Systems Design and Implementation; 2008. p. 1–14.
13. Zaharia M, Chowdhury M, Das T, Dave A, Ma J, McCauly M, Franklin MJ, Shenker S, Stoica I. Resilient distributed datasets: a fault-tolerant abstraction for in-memory cluster computing. In: Gribble SD, Katabi D, editors. Proceedings of the 9th USENIX Symposium on Networked Systems Design and Implementation; 2012. p. 15–28.
14. Zaharia M, Das T, Li H, Hunter T, Shenker S, Stoica I. Discretized streams: fault-tolerant streaming computation at scale. In: Proceedings of the 24th ACM Symposium on Operating System Principles; 2013.
15. Zhou J, Bruno N, Wu M, Larson P, Chaiken R, Shakib D. SCOPE: parallel databases meet MapReduce. VLDB J. 2012;21(5):611–36.

Big Stream Systems

Nathan Backman
Computer Science, Buena Vista University,
Storm Lake, IA, USA

Synonyms

Continuous workflow execution frameworks;
Distributed stream processing

Definition

Big stream systems aim to bring the scalability of batch processing frameworks to stream applications. Stream processing systems have different constraints than batch processing systems as well as a different set of challenges. The unbounded and potentially high-volume nature of streams require stream applications to execute continuously and to limit the role of disk-based storage. The throughput of high-volume streams can exceed the throughput of disks, and the stream data may not have any lasting value beyond the meaning that can be extracted from them. Big stream systems address the challenge of achieving high scalability in stream processing by (1) keeping data moving and off of disks, (2) implementing fault-tolerant strategies to allow stream data to persist in the event of faults, and (3) spreading computational workloads across many nodes while preserving the integrity and order of the stream amidst its distributed execution.

Historical Background

Database management systems have provided the traditional and lasting framework upon which structured data analysis has been performed. In the early 2000s, traditional database techniques were applied to streams to provide temporal analyses of the unbounded data streams [1]. Streams, by their nature, are temporal entities such that the order and temporal relations between data

within the stream carry meaning. Therefore it is not just the data within the stream that are processed but it is the stream itself hence the name "stream" processing. This area opened up new topics for research, applied to streams, in the context of distributed processing, fault tolerant designs to preserve stream and result integrity in the midst of failures or corrupt data, and new declarative language interfaces more suitable for stream analysis.

In the mid-2000s, the MapReduce [5] parallel programming model and associated data processing framework provided a means to process extremely large amounts of unstructured data on clusters of incredible size and with relative ease in a fault-tolerant manner. This model expressed parallel computations as a series of *map* and *reduce* operations. The map operation performs data transformation and filtering and was often followed by a reduce operation that performed like-key aggregation. An open source implementation of MapReduce, Hadoop [2], was quickly developed and has since been the predominant implementation of MapReduce.

Big stream systems sit at the intersection of traditional stream processing frameworks and large-scale, batch-processing frameworks. Their goal is to address the challenge of processing streams that require more computational power and throughput than can be provided by a small set of computing nodes. They are expected to scale to dozens and even hundreds of computing nodes.

One of the earliest of these systems is the Hadoop Online Prototype [4]. It augmented Hadoop to process unbounded streams by pipelining data, in real time, between the maps and reduce processes running on the participating computing nodes. However, HOP was a simple prototype and did not support the full range of requirements of stream processing applications.

In the early 2010s, more sophisticated big stream systems were unveiled which included S4, Storm, and Spark Streaming. These frameworks were designed to achieve scalable and continuous processing while utilizing large clusters of computing nodes. These frameworks will be discussed in more detail in the following section.

Scientific Fundamentals

Stream applications can be represented as a directed acyclic graph of operators. Such workflows receive their input from unbounded data sources that feed data to downstream workflow operators. Operators are capable of consuming one or many streams and will produce intermediate streams to be consumed by additional downstream operators. Data produced from the workflow outputs can be utilized by the application programmer.

Stream processing occurs by transforming, filtering, or aggregating subsets (*windows*) of a stream. A window is denoted by a *size* and a *slide* to identify the relative amount of the stream to evaluate and to identify how frequently it should be evaluated. The size of a window is commonly bounded by time or by record count. The slide denotes how frequently the window should be evaluated. For example, a time-based window may have a size of 1 min and a slide of 10 s and a count-based window may have a size of 100 records and a slide of 20 records.

Hadoop Online Prototype

The *Hadoop Online Prototype* (HOP) [4] achieved rudimentary continuous query execution by augmenting Hadoop. Data are pipelined from continuously executing map processes to periodically executed reduce processes. The prototype supports only window evaluation based on elapsed wall-clock time such that the reduce operation is triggered when a timer surpasses a threshold. This strategy enforces simultaneous aggregation of all concurrent windows (one for each key). This precludes windowing strategies which evaluate windows only once they are fully materialized. For example, it may be desirable to evaluate the window for a key once it has accrued *size* records or one may wish to utilize embedded application data to determine if a temporal boundary has passed.

HOP is also unable to ensure that streams maintain their order. If we suppose that each map task consumes and produces an ordered stream, then we should expect that any merged stream at a downstream reduce task input will be out of order. Regardless of the strategy used to form windows, it would be possible for records to end up in the incorrect window when being processed if the order of the merged stream is not reconstructed. For those continuous applications that do not rely on stream order, this is not a significant issue.

The programming interface for HOP is identical to that of Hadoop regarding the application logic. A programmer can enable pipelining by setting a flag and can then specifying the temporal window size for periodic evaluation of reduce tasks. If a stream application workflow requires more than a map operation followed by a reduce operation, then it is the application programmer's burden to communicate data between adjacent Hadoop jobs. They must themselves pipeline data from the reduce task of the upstream Hadoop job to the continuously running map tasks of a "downstream" Hadoop job. It is then also the programmer's burden to specify the number of map and reduce tasks for each running Hadoop job in the programmer's workflow and to migrate data between them accordingly.

Apache S4

Apache S4 [6] debuted in 2010 as a stream processing framework, built from the ground up to be distributed, scalable, and fault tolerant. The MapReduce programming model inspired the authors to also utilize keyed data. As with in HOP, the benefit to this is twofold; it provides a simple mechanism to facilitate widely distributed processing via simple key-based partitioning, and it enabled a slim Java programming interface for application programmers who can focus on application logic as simple key-value processing. The authors did relax standard stream processing constraints by permitting some amount of data loss (in the event of failures and excessive stream volumes) and also do not allow the dynamic addition or removal of computing nodes. While preventing data loss is a nonnegotiable requirement for many stream applications, there is still a large class of applications that *can* tolerate some degree of loss.

The fundamental unit of computation in S4 is represented by a *processing element*. A processing element can be considered an operator in a stream application's workflow, and its task is to process keyed data. S4 achieves parallelism with scalability by allocating new sub-instances of a logical processing element for each distinct key observed in the stream and assigns the resulting processing element to a *processing node*. The processing nodes are the logical hosts to processing elements and facilitate communication between processing elements.

Directing keyed data to the appropriate processing element is accomplished via consistent hashing (as in MapReduce). The key is hashed, and the resulting value is mapped onto the processing node that hosts the relevant processing element. The authors of S4 have stated that continuous applications generating a large number of distinct keys may be required to implement their own form of resource management if the number of processing elements approaches the resource limitations of the nodes. The programmer may therefore be required to generate a policy to release infrequently utilized processing elements so that resources may be reclaimed.

The state of a processing node (which is actually the collection of internal state of the processing elements it hosts) is periodically stored as a checkpoint on remote storage for recovery purposes. In the event of a failure, the most recent checkpoint can be loaded onto a spare node waiting on standby to be inserted back into the application workflow. Any state produced between the time of a checkpoint and a failure is lost. If there are no nodes available on standby, then the checkpoint data will be unable to be restored onto a node. This means that the workload and state also cannot be absorbed by any of the active processing nodes. An additional consequence of the inability to shuffle processing elements between active nodes is that there is no current means to perform load balancing across processing nodes. Thus, processing elements are statically placed on these nodes. It is unclear how this assignment is performed, but its static nature results in an inability to make adjustments in the event that there is skew in the stream that affects the workloads of individual processing elements and nodes.

Stream applications built atop S4 are programmed in Java. The data objects passed within the framework and through the intermediate streams may be any Java Object that implements the java.io.Serializable interface. Constructing a stream application workflow consists of defining and instantiating *ProcessingElement* subclasses as workflow operators and connecting them together via named streams. Also, there is an interface for defining timer-based windows (similar to HOP). When the timer elapses for a processing element object, all of its processing element sub-instances (one for each key) will be processed sequentially. Similar to HOP, S4 also makes no mention of preserving stream order after parallel computation to ensure that windows contain temporally adjacent data.

As an application programmer launches their application, they will have an opportunity to specify the cluster and number of nodes to utilize. This decision is made statically at launch time and cannot be dynamically modified at a later time.

Apache Storm

Apache Storm [3] is another stream processing framework that was developed with scalability and fault tolerance in mind. It originated at Back-Type, was further developed at Twitter (after Twitter acquired BackType), and now lives within the Apache Incubator project.

Stream application workflows are defined in Storm as *stream topologies* and are represented by a set of *spouts* and *bolts*. A spout defines a stream input, and a bolt defines a stream operator that processes data within a stream. Bolts therefore consume input or intermediate streams (one or many) and produce streams as their output. Spouts and bolts are executed as many tasks throughout a cluster. A node in a cluster facilitates a worker process for each stream topology that it manages. The worker process may then host a number of single-threaded *executors* – one or more for each spout and bolt allocated to the worker process. Each executor can periodically and sequentially execute one or more tasks for the spout or bolt it supports as data arrive. This model allows for the parallelization of stream workflows onto clusters of nodes.

Data communicated within a Storm topology are guaranteed delivery. They will either complete their journey through the pipeline within a timeout threshold or they will be replayed from the relevant input spout. This requires spouts to be responsible for caching input data (locally or externally) to be made available for replay.

Data may be communicated to downstream bolts in predefined ways. The *shuffle* method allows data to be forwarded to any task associated with the downstream bolt. The *grouping* method utilizes consistent hashing (similar to MapReduce and S4) to send like-keyed data to a common task associated with the downstream bolt.

The programming interface for specifying the actions a bolt should take is slightly more complicated than doing the same in MapReduce or in S4. In Storm, programmers must maintain their own state for any form of aggregation or analysis that may take place across multiple keys. Therefore the classic word count example would require the programmer to maintain their own dictionary or map of words with their respective counts. Allowing the programming to manage state in this way also allows the programmer to restore state in the event of faults.

Perhaps one of the biggest drawbacks of Storm is that it does not support temporal window-based stream analysis nor does their documentation provide any mention of maintaining stream order. The absence of such features makes it impossible to implement stream applications that analyze the temporal relationships of data within the stream. At this point in time, Storm remains more suitable for continuous applications in which temporal data relationships have little or no value.

Spark Streaming

Spark Streaming [7] takes a different approach to processing streams than is seen in more traditional systems. Instead of representing a continuous workflow of operators, streams are discretized into small temporally bound batches that can be processed using the methods of a batch-processing framework. The focus of this platform is to provide scalable and fault-tolerant

stream processing without replication or disk storage by using in-memory Resilient Distributed Datasets (RDDs) to recompute recently lost data. This strategy does therefore require a certain amount of caching, but data can be quickly recomputed without requiring reinsertion into an upstream point in a stream workflow. RDDs contribute to scalability as they represent contiguous batches of data in the stream. Manipulating data in batches requires much less overhead than to manipulate the individual records within them. Fine-tuning an appropriate batch size, however, requires an attentive application programmer to do exploratory testing on a per-application basis. Batches are specified by temporal bounds and appear to be fixed in their temporal size throughout the stream application's lifetime. Performing stream processing in this batch-oriented manner certainly improves scalability but does come at the cost of increasing latency. Therefore Spark Streaming achieves high scalability by sacrificing sub-second latencies.

In addition to providing a fast and in-memory approach to achieving fault tolerance, Spark Streaming preserves stream order and supports true windowed stream computations. Windows in Spark Streaming can be defined temporally (by a window length and a slide interval) or by record count. Spark Streaming has an API that facilitates such windows for a number of operation types. It is unclear if Spark Streaming supports windows based on timestamps embedded within application data or whether application timestamps can be used to form and identify batches as data are inserted into the system.

In implementing windowing, Spark Streaming supports two sliding window evaluation models. The first, naïve evaluation, computes the result of a window by processing all of the batches that contribute to the window. If a batch participates in multiple windows, then it would be reprocessed for each of those windows. The second method, incremental evaluation, allows for the preservation of window state that can be reused for adjacent and overlapping windows. The newest batch to arrive will *increment* the window state by adding its contributions while the oldest (and no longer relevant) batch will *decrement* the window state by removing its contributions. Instead of a batch being processed for each window it participates in, batches will only be invoked on window state twice in order to increment and decrement that state.

Key Applications

High frequency trading
Online log processing
Serving targeted web-based advertisements

Cross-References

▶ Fault Tolerance and High Availability in Data Stream Management Systems
▶ MapReduce
▶ Stream Processing
▶ Window-Based Query Processing

Recommended Reading

1. Abadi D, Ahmad Y, Balazinska M, Çetintemel U, Cherniack M, Hwang J, Lindner W, Maskey A, Rasin A, Ryvkina E, Tatbul N, Xing Y, Zdonik S. The design of the borealis stream processing engine. In: Proceedings of the 2nd Biennial Conference on Innovative Data Systems Research; 2005. p. 277–89.
2. Apache Hadoop. The Apache Software Foundation. 2014. http://hadoop.apache.org. Accessed 1 June 2014.
3. Apache Storm. The Apache Software Foundation. 2014. http://storm.incubator.apache.org. Accessed 1 June 2014.
4. Condie T, Conway N, Alvaro P, Hellerstein J, Elmeleegy K, Sears R. MapReduce Online. In: Proceedings of the 7th USENIX Symposium on Networked Systems Design & Implementation; 2010.
5. Dean J, Ghemawat S. MapReduce: simplified data processing on large cluster. In: Proceedings of the 6th USENIX Symp. on Operating System Design and Implementation; 2004.
6. Neumeyer L, Robbins B, Nair A, Kesari A. S4: distributed stream computing platform. In: Proceedings of the 10th IEEE International Conference on Data Mining Workshops; 2010.
7. Zaharia M, Das T, Li H, Hunter T, Shenker S, Stoica, I. Discretized streams: a fault-tolerant model for scalable stream processing. In: Proceedings of the 24th ACM Symposium on Operating System Principles; 2013.

Biological Metadata Management

Zoé Lacroix[1], Cartik R. Kothari[2], Peter Mork[3], Mark D. Wilkinson[4], and Sarah Cohen-Boulakia[5]
[1]Arizona State University, Tempe, AZ, USA
[2]Biomedical Informatics, Ohio State University, College of Medicine, Columbus, OH, USA
[3]The MITRE Corporation, McLean, VA, USA
[4]University of British Columbia, Vancouver, BC, Canada
[5]University Paris-Sud, Orsay Cedex, France

Definition

Metadata characterize biological resources by core information including a name, a description of its input and its output (parameters or format), its address, and various additional properties. Resources are organized with respect to metadata that characterize their content (for data sources), their semantics (in terms of ontological classes and relationships), their characteristics (syntactical properties), their performance (with metrics and benchmarks), their quality (curation, reliability, trust), etc.

Historical Background

Digital resources for the Life Sciences include a variety of data sources and applications whose number increases dramatically every year [4]. Although this rich and valuable offering provides scientists with multiple options to implement and execute their scientific protocols (i.e., pipelines, dataflows, workflows), selecting the resources suitable for implementing each scientific step remains a difficult task. Scientific protocols are typically implemented using the resources a scientist is most familiar with, instead of the resources that may best meet the protocol's needs. The number of resources a scientist uses regularly, knowing their structure, the quality of data and annotations they offer, the capabilities made available by the provider to access, analyze, and display the data are ridiculously small compared to the thousands of resources made available on the Web [2]. Metadata are not only critical in selecting suitable resources for implement scientific protocols, but they are essential to the proper composition and integration of resources in a platform such as a workflow system or wrapped into a database mediation system. They also play a decisive role in the analysis of data, in particular to track data provenance. Finally, they contribute to data curation and the management of Life Sciences resources in a global and linked digital biological maze.

Foundations

Metadata management relies on the description of resources including the resource name, identification, and all additional information that may be relevant to locating, evaluating, and using the resource. A *resource identifier* is a sequence of characters that uniquely identifies a resource and is globally shared and understood over a network. A resource is analogous to a node on the Web. The ubiquitous Uniform Resource Locator (URL) is an example of a resource identifier, which uses the location, the local directory path, and the local file name of the resource to locate it on the Web. Unique Resource Identifiers (URIs) include URLs that not only identify the resource but describe its primary access mechanism or network location, and Uniform Resource Names (URN) that identify a resource by name in a particular namespace. The unique identification of resources is an unresolved problem in the life sciences community. Different protein, gene, and molecular interaction databases often assign separate identifiers to the same resource, a phenomenon known as *coreference*. Leveraging the information from all these databases becomes problematic, leading to duplicate records and inconsistency. To alleviate this problem, many Life Sciences databases cross reference their identifiers.

Metadata are data that describe a resource. Metadata include a wide range of information from attribution metadata, such as those attributes

defined in the Dublin Core, to detailed policy metadata indicating who can access the resource under what conditions. *Semantic metadata* include the description of a resource with respect to the domain knowledge (e.g., a data source provides information about proteins, a tool computes the translation of a RNA sequence into an AA sequence). *Syntactic metadata* provide the description of the resource interface. *Summary metadata* describe the actual contents of the resource. These metadata include free text summaries and statistical summaries of the instances (values) contained in the database. Summary metadata can be classified along several axes: (i) textual versus quantitative, (ii) structured versus unstructured, and (iii) manually generated versus automatically generated. By far the most common type of summary metadata are textual. For example, NAR [4] maintains a listing of hundreds of biomedical resources. For each resource, they provide a brief description of the contents of that resource. Textual metadata allow an application developer or end-user to search for resources using keywords or phrases. The success of existing approaches seems to show that it is a familiar and intuitive operation, which works well when searching for reasonably well-defined concepts. Textual metadata are unstructured (i.e., free text) and manually curated. Alternatively, summary metadata can take the form of keywords drawn from a *controlled vocabulary*, such as Medical Subject Headings (MeSH) terms. A controlled vocabulary makes it easier to search for resources, assuming the vocabulary is sufficiently expressive and used consistently to annotate the resources. In most cases, textual metadata are generated manually, although there is some research in automatically extracting keywords from a resource for its annotation.

Quantitative metadata describe resources in terms of numeric datatypes. In the simplest case, these metadata specify the range of values that can be found in the resource. For example, all of the subjects in a pediatric database would be younger than 18. More detailed summaries are also possible. In the case of quantitative metadata, unstructured metadata make little sense. The end user needs to know what a given number represents, including relevant units. Quantitative metadata can still be generated manually or automatically. The former is required if the resource does not contain the necessary raw data. For example, if a pediatric database does not contain the ages of its subjects, the relevant age range must be specified manually. However, when the resource does contain the necessary raw data, quantitative metadata can be generated automatically. Moreover, the amount of detail in the metadata can vary depending on the needs of the resource owners and community members searching for resources. A current research challenge is determining the appropriate granularity for quantitative metadata and using these metadata to estimate the extent to which a given resource matches the end user's search criteria. *Statistical metadata* and benchmarks provide an additional layer exploited when the scientist wishes to predict the outcome of an execution. These metadata are particularly useful when several resources are combined to evaluate alternative evaluation strategies and select the most efficient one with respect to the protocols' aim [5]. The domain and range of resources, as well as resource overlaps contribute to the statistical description of resources. Similarly, information related to the quality of the resource (e.g., curation) may be exploited to optimize the quality of the execution.

Structural metadata describe the resource interface and the intention of the resource provider. These metadata can take on many forms including database schemata, Unified Modeling Language (UML) diagrams or Web service descriptions. What structural metadata provide are a description of how the resource provider intends to organize and deliver data. Controlled vocabularies capture domain knowledge and clarify resource descriptions (e.g., identical concepts). Controlled vocabularies are naturally extended by logical or conceptual representations such as expressed in a domain ontology. More generally, a metadata registry containing structural metadata allows an application developer to search for resources that are intended to contain particular types of data. Moreover, once a developer discovers a useful resource, he has a good idea of how to interact with that resource, both in terms

of formulating queries and processing results. Structural metadata only indicate what sorts of queries can be posed, not whether those queries will return meaningful results. For example, the structural metadata for a card catalog might indicate that it includes, for each entry, a list of authors. Thus, one could reasonably query the card catalog for all books authored by John Grisham. However, if the card catalog supports a medical (non-fiction) library, it is unlikely that this query will return any record.

Bioinformatics resources may be represented with formats and standards developed by various communities driven by disparate motivations including business, library, Web, etc. The Resource Description Framework (RDF) and the RDF Schema (RDFS) were the earliest adopted standards for representing metadata about Web resources. The Dublin Core Metadata Elements Set (DCMES) is a standard set of metadata elements that can be used to describe a generic resource to facilitate its discovery and use. RDF specifically provides a very simple "triples" syntax or Subject-Predicate-Object syntax to capture resource metadata. Universal Description, Discovery and Integration (UDDI) is the XML-based format to register businesses on the Web proposed by OASIS. Dublin Core is a standard (NISO Standard Z39.85-2007) for cross-domain information resource description. The Web Ontology Language OWL, based on earlier languages OIL and DAML+OIL, is a W3C recommendation that extends XML, RDF, and RDF Schema (RDF-S) by providing additional vocabulary along with a formal semantics with descriptions of classes, along with their related properties and instances. The Web Service Description Language (WSDL) and its extension WSDL-S are used respectively as resource description and semantic annotation. Resource description and registration developed for the life science include BioMoby, PISE/Mobyle, caBIG, and SOAPlab.

The Life Sciences community has a number of metadata annotation standards. The Darwin Core (DwC) is a metadata standard from the National Biological Information Infrastructure (NBII) for annotating the objects contained within natural history specimen collections and species observation databases. These annotations are used to retrieve records of natural history specimens and observation records from local libraries, integrating them with other collections across the United States and making them available on the Web. The Access to Biological Collections Data Schema (ABCD Schema) is a complementary, hierarchical metadata standard for the annotation of biological specimens. Mappings exist from the terms of Darwin Core to the ABCD Schema that illustrate the overlap between the two standards and the few differences. Organizations such as NBII, the Integrated Taxonomic Information System and the Global Biodiversity Information Facility leverage metadata standards such as ABCD Schema and DwC to discover and utilize information pertaining to species and natural history specimens that is distributed around the world.

The Minimum Information About a Microarray Experiment (MIAME) [1] from the Microarray Gene Expression Data (MGED) Society, is used to describe sufficient information about a microarray experiment to reproduce it unambiguously. An increasing number of data providers are embracing the MIAME standard and several journals including NAR, Cell, and Nature require MIAME compliant data as a condition for publishing microarray based papers. The Minimum Information About a Proteomics Experiment (MIAPE) [7] is a minimum information reporting requirement for proteomics experiments. It is analogous to the MIAME standard for transcriptomics data. MIAPE is distributed across several modules, each of which is useful to describe a different proteomics experiment. Example modules are the MIAPE Gel Electrophoresis module, the MIAPE Mass Spectrometry module and the MIAPE Column Chromatography module. The Minimum Information required for reporting a Molecular Interaction Experiment (MIMIx) [6] has been developed as a framework to capture metadata about molecular interaction experiments. MIAME, MIAPE and MIMIx are being developed under the auspices of the Proteomics Standards Initiative at the Human Proteomics Organization.

Metadata standards vary in complexity from the simple format of the Dublin Core to the complex requirements of MIAME. A number of organizations are currently involved in the development of metadata standards in various fields of study. The Federal Geographic Data Committee (FGDC) is involved in the creation of several metadata standards such as the Spatial Data Transfer Standard for the exchange of spatial data, the National Vegetation Classification Standard to support a national vegetation classification system, the Biological Data Profile for the documentation of biological data, and the Utilities Data Content Standard to standardize geospatial information for utility systems. In addition the FGDC publishes the Content Standard for Digital Geospatial Metadata (CSDGM) for the annotation of geospatial data in the form of maps, atlases and satellite images. The Ocean Biogeographic Information System as an initiative to make marine biogeographic data available to a worldwide audience. OBIS has developed a taxonomic hierarchy of metadata elements, called the OBIS taxonomy for annotation of marine species observations. GeoConnections is a Canadian initiative to make geospatial data in the form of maps and satellite images easily available to a worldwide audience. GeoConnections uses the CSDGM published by the FGDC, to annotate its geospatial data.

Key Applications

Resource discovery systems exploit metadata in order to map the user requirements to the resource characteristics. For example, caCORE is an n-tier data management and integration infrastructure that combines several interconnected software and services. The Cancer Bioinformatics Infrastructure Objects (caBIO) model is at the heart of caCORE. The caBIO model contains definitions of concepts and inter-concept relationships that are common to biomedical research. These concept definitions in caBIO are the basis upon which data from distributed repositories are integrated. These repositories include gene and homolog databases (e.g., UniGene and Homologene), pathway databases (e.g., BioCarta), vocabulary and terminology repositories (e.g., the

National Cancer Institute (NCI) Thesaurus, NCI Metathesaurus, and the Gene Ontology). The PathPort framework developed at The Virginia Bioinformatics Institute, presents a Web based interface that makes it possible for end users to invoke local and distributed biological Web services that are described in WSDL, in a location and platform independent manner.

Metadata offer metrics that may be used to predict the outcome of the execution of a scientific protocol on selected resources [5]. Metadata provide resource characterization that allows their comparison with similar resources, thus addressing the need of combining multiple complementary resources to implement completely a single task (e.g., data coverage). Path-based systems such as BioNavigation exploit statistical metadata to predict the resources that are likely to return the most entries, the best characterized (most attributes) entries, etc. [2].

Although some degree of transparency is often needed in queries, scientists also expect to be aware of the provenance of the answers. In order to analyze the results obtained from the execution of their scientific protocols, they often need to understand the process that produced the dataset. In particular, they need to know which resources have been used, and how entries have been linked to generate the answer to their question. Data traceability is related to the degree of information pertaining to the resources used to implement the process as well as the integration used to combine them. Because of this, reasoning on data provenance may exploit scientific resource metadata repositories. For example, BioGuideSRS allows the user to visualize the correspondence between the graph of entities and the graph of sources-entities [3]. By selecting an entity, the user visualizes the sources which provide information about this entity; similarly, by selecting a relationship, the user visualizes the links between sources which achieve this relationship. Second, the data obtained as a result yielded by BioGuideSRS to the user is systematically associated with the path which has been used to obtain it. In this way, the user knows the exact sequence of sources and links used. This approach was demonstrated with the ZOOM*UserViews system.

Future Directions

Metadata management remains a critical issue for the Life Sciences. First, the community has not agreed on common metadata to publish together with a resource so that it is properly identified, located, and used by scientists. Multiple discussions related to the representation of scientific objects generate the design of a large number of ontologies as published by the Open Biomedical Ontology (OBO) group. Although this effort contributes significantly to the better understanding of scientific information, it produces ontologies that may overlap and that are difficult to integrate. This semantic gap is aggravated by the diversity of models and formats used by biological data providers. Moreover, the community shows reluctance to adopting recommendations from the W3C Semantic Web for the specification of resources. The lack of a common publishing process for resources affects their impact significantly. In particular, it challenges the development of resource repositories to support resource discovery. Consequently, it affects the ability for scientists to select resources suitable to implement the scientific tasks involved in their protocols. The development of adequate technology, still in its infancy, is rather limited by the lack of a *franca lingua* for metadata. Future developments include the identification of metrics that adequately capture the characteristics of resources, the design of benchmarks to evaluate and compare similar resources, automated data curation approaches that exploit and update resource metadata, automated classification of resources, data provenance analysis, etc.

Data Sets

NAR http://nar.oxfordjournals.org/

BMC Source Code for Biology and Medicine http://www.scfbm.org/home

Bioinformatics Links Directory http://bioinformatics.ca/links_directory/

Open Biomedical Ontologies (OBO) http://obofoundry.org/

Next Generation Biology Workbench (Swami) http://www.ngbw.org

URL to Code

Medical Subjects Headings (MeSH) http://www.nlm. nih.gov/mesh

UDDI http://www.uddi.org/

OWL Web Ontology Language http://www.w3.org/TR/owl-features/

DAML+OIL http://www.w3.org/TR/daml+oil-reference

BioMOBY http://biomoby.org/

PISE http://www.pasteur.fr/recherche/unites/sis/Pise/

Mobyle http://www.pasteur.fr/recherche/unites/sis/Pise/mobyle.html

caBIG http://cabig.nci.nih.gov/

SOAPlab http://www.ebi.ac.uk/Tools/webservices/soaplab/overview

National Biological Information Infrastructure (NBII) http://www.nbii.gov/

Microarray Gene Expression Data (MGED) Society http://www.mged.org

Federal Geographic Data Committee (FGDC) http://www.fgdc.gov

Ocean Biogeographic Information System (OBIS) http://www.iobis.org

GeoConnections http://www.geoconnections.org

caCORE http://ncicb.nci.nih.gov/infrastructure/cacoresdk

ZOOM*UserViews system http://db.cis.upenn.edu/research/provwf.html

BioNavigation http://bioinformatics.eas.asu.edu/

BioGuide http://bioguide-project.net/

Cross-References

▶ Biological Resource Discovery
▶ Dublin Core
▶ Graph Management in the Life Sciences
▶ Metadata

▶ Ontology
▶ Resource Description Framework
▶ Unified Modeling Language
▶ Web Services
▶ Web Services and the Semantic Web for Life Science Data
▶ XML

Recommended Reading

1. Brazama A, Hingamp P, Quackenbush J, Sherlock G, Spellman P, Stoeckert C, Aach J, Ansorge W, Ball CA, Causton HC, Gaasterland T, Holstege FCP, Kim IF, Markowitz V, Matese JC, Parkinson H, Robinson A, Sarkans U, Schulze-Kremer S, Stewart J, Taylor R, Vilo J, Vingron M. Minimum information about a microarray experiment (MIAME) – toward standards for microarray data. Nat Genet. 2001;29(4): 365–71.
2. Cohen-Boulakia S, Davidson S, Froidevaux C, Lacroix Z, Vidal ME. Path-based systems to guide scientists in the maze of biological resources. J Bioinform Comput Biol. 2006;4(5):1069–95.
3. Cohen-Boulakia S, Biton O, Davidson S, Froidevaux C. BioGuideSRS: querying multiple sources with a user-centric perspective. Bioinformatics. 2007;23(10):1301–3.
4. Galperin MY. The molecular biology database collection: 2007 update. Nucleic Acids Res. 2007;35(Database issue): D3–4.
5. Lacroix Z, Raschid L, Eckman B. Techniques for optimization of queries on integrated biological resources. J Bioinform Comput Biol. 2004;2(2): 375–411.
6. Orchard S, Salwinski L, Kerrien S, Montecchi-Palazzi L, Oesterheld M, Stmpflen V, Ceol A, Chatraryamontri A, Armstrong J, Woollard P, Salama JJ, Moore S, Wojcik J, Bader GD, Vidal M, Cusick ME, Gerstein M, Gavin AC, Superti-Furga G, Greenblatt J, Bader J, Uetz P, Tyers M, Legrain P, Fields S, Mulder N, Gilson M, Niepmann M, Burgoon L, De Las Rivas J, Prieto C, Perreau VM, Hogue C, Mewes HW, Apweiler R, Xenarios I, Eisenberg D, Cesareni G, Hermjakob H. The minimum information required for reporting a molecular interaction experiment (MIMIx). Nat Biotechnol. 2007;25(8):894–8.
7. Taylor CF, Paton NW, Lilley KS, Binz PA, Julian RK, Jones AR, Zhu W, Apweiler R, Aebersold R, Deutsch EW, Dunn MJ, Heck AJR, Leitner A, Macht M, Mann M, Martens L, Neubert TA, Patterson SD, Ping P, Seymour SL, Souda P, Tsugita A, Vandekerckhove J, Vondriska TM, Whitelegge JP, Wilkins MR, Xenarios I, Yates JR, Hermjakob H. The minimum information about a proteomics experiment (MIAPE). Nat Biotechnol. 2007;25(8):887–93.

Biological Networks

Amarnath Gupta
San Diego Supercomputer Center, University of
California San Diego, La Jolla, CA, USA

Synonyms

Biological pathways; Molecular interaction
graphs; Protein-protein interaction networks;
Signal transduction networks; Transcriptional
networks

Definition

A biological network is a graph-structured representation of binarized interactions among biological objects. Typically, the nodes in such
a graph represent biological molecules, and the
edges are labeled to represent different forms of
interactions between molecules.

Example: A transcriptional network is a directed graph where a node represents either a
protein (a transcription factor) or a region of the
chromosome such that the edges can be constructed from the protein node to the chromoso-
mal region. The edge in the graph represents that
the protein can initiate the transcription (production of messenger RNA) process.

Key Points

A biological network is typically a node and
edge attributed graph, where the edges can have
different semantics depending on the kind of
network. In some networks, the edges may be
weighted, denoting, for instance, the probability
of the interaction taking place. In some networks,
like the protein-protein interaction graph, the
edges are undirected. In some cases, like signal
transduction networks, the edges represent
the flow of time. Querying, integrating, and
simulating are typical operations performed on
biological networks.

Cross-References

► Graph Data Management in Scientific Applications

Recommended Reading

1. Baitaluk M, Qian X, Godbole S, Raval A, Ray
   A, Gupta A. PathSys: integrating molecular interaction graphs for systems biology. BMC Bioinf.
   2006;7(1):55.
2. Eckman BA, Brown PG. Graph data management
   for molecular and cell biology. IBM J Res Dev.
   2006;50(6):545–60.
3. Leser U. A query language for biological networks.
   Bioinformatics. 2005;21(Suppl 2):ii33–9.

Biological Resource Discovery

Zoé Lacroix[1], Cartik R. Kothari[2], Peter Mork[3],
Rami Rifaieh[4], Mark D. Wilkinson[5], Juliana
Freire[6,7,8], and Sarah Cohen-Boulakia[9]
[1]Arizona State University, Tempe, AZ, USA
[2]Biomedical Informatics, Ohio State University,
College of Medicine, Columbus, OH, USA
[3]The MITRE Corporation, McLean, VA, USA
[4]University of California-San Diego, San Diego,
CA, USA
[5]University of British Columbia, Vancouver,
BC, Canada
[6]NYU Tandon School of Engineering, Brooklyn,
NY, USA
[7]NYU Center for Data Science, New York, NY,
USA
[8]New York University, New York, NY, USA
[9]University Paris-Sud, Orsay Cedex, France

Definition

Resources for the Life Sciences include various
expedients including (access to) data stored in flat
files or databases (e.g., a query form or a textual
search engine), links between resources (index
or hyperlink), or services such as applications

or tools. *Resource discovery* is the process of identifying and locating existing resources that have a particular property. Machine-based resource discovery relies on crawling, clustering, and classifying resources discovered on the Web automatically. Resource discovery systems allow the expression of queries to identify and locate resources that implement scientific tasks and have properties of interest.

Historical Background

Resource selection relies on the identification of the resources suitable to achieve each task and the ability to compose the selected resources into a meaningful and efficient executable protocol. Metadata constitute the core information requisite to evaluate the suitability of Life Sciences resources to achieve a scientific task. Metadata critical to resource discovery include (i) resource publication, identification, and location and (ii) semantic and (iii) syntactic descriptions. First scientists need to be aware of existing resources. If academic publications such as Nucleic Acids Research (NAR) [7] or BMC Source Code for Biology and Medicine have provided valuable media where bioinformaticians may publish their resources, they require significant manpower to identify and evaluate the potential of each resource and compile and record their location and description for future use. Core resource description in a unified format accessible to scientists and machines alike and resource repositories contribute greatly to ease the problem of resource identification and location.

Foundations

Resource discovery relates to the activity of identifying a resource suitable to implement a particular task. Resource discovery relies on various metadata that specify the characteristics of resources thus allowing the mapping of the requirements to the resource specifications. The type of metadata chosen to represent resources will constrain resource discovery. For example, textual metadata drawn from a controlled, hierarchical vocabulary support resource discovery by automatically expanding search terms to include more specific terms. However, textual metadata are not sufficient when searching on specific criteria; for example, consider a researcher interested in finding datasets of "MRI images for subjects between the ages of 18 and 24." Syntactic (formats) and semantic (concepts) metadata describe how a resource is organized and provide some insight into what type of information might be found in the resource while summary metadata specify the content of the resource. Although structural metadata are normally generated to help application developers understand how to interact with the resource, they can also be collected to support resource discovery. For example, in the caBIG framework, structural metadata are represented as common data elements. Each common data element references a common terminology (the NCI thesaurus in this case) and may also contain free-text documentation describing that data element both providing a semantic representation of the resource. The metadata registry also maps common data elements to resources that provide instances of that element. An application developer searches the metadata registry by providing a collection of keywords; the registry returns a list of data elements that contain those keywords.

Resource discovery is the interface between resource metadata on one side and resource integration to implement complex scientific protocols (or workflows, queries, pipelines) on the other. Indeed the motivation for discovering a resource is drawn from the need to implement a scientific task. Most approaches to support resource discovery only locate one resource at a time, regardless of their future composition to implement complex workflows. In contrast, path-based guiding systems such as BioNavigation and BioGuide provide the ability to express resource discovery queries to identify resources that can be composed to express scientific protocols expressed as connected scientific tasks [4].

Key Applications

BioMoby is an open source, extensible framework that enables the representation, discovery, retrieval, and integration of biological data from distributed data repositories and analysis services. By registering their analysis and data access services with BioMoby, service providers agree to use and provide service specifications in a shared semantic space. The BioMoby Central registry now hosts more than a thousand services in the United States, Canada, and several other countries across the world. BioMoby uses a datatype hierarchy to facilitate the automated discovery of Web services capable of handling specific input datatypes. As a minimal Web-based interface, the Gbrowse Moby service browser can be used by biologists to discover and invoke biological Web services from the Moby registry and seamlessly chain these services to compose multi-step analytical workflows. The process is data centric, relying on input and output datatype specifications of the services. The Seahawk client interface can infer the datatype of the input data files directly and immediately presents the biologist with a list of Web services that can process the input file. Users of the Seahawk interface are relieved of the necessity to familiarize themselves with datatype hierarchies and instead are free to concentrate on the analytical aspects of their work. The BioMoby service encyclopedia provides a query interface to the repository of services. MOBY-S Web Service Browser retrieves bioinformatics resources with respect to a data type. Additional interfaces to BioMoby services include registry browsers that provide access to the complete list of registered BioMoby services organized in a HTML page.

These interfaces are convenient when searching for services with respect to a specific data format (input), but they are not suitable when searching for services with respect to their scientific meaning rather than their format. Another critical limitation of the approaches occurs when no single service achieves the task. In order to allow the discovery of the services that can be used to express scientific protocols, combinations of services must be retrieved. Scientific data integration systems such as workflow systems [6] enable the composition and execution of bioinformatics services. Combining a workflow approach with a service representation that guarantees compatibility of data formats offers a great value to the scientist who has selected the services to use and wishes to combine them in an executable workflow. A resource is selected because it uses or produces the expected format (e.g., *FASTA*) rather than because it implements the expected scientific aim or because it is efficient. This is illustrated by the BioMoby plug-in in Taverna. When a BioMoby service, e.g., *DragonDB_TBlastN*, is included in a workflow, its output format, e.g., *NCBI_BLAST_TEXT*, can be searched (brief search) against available formats to determine if it is an input to any other service registered in BioMoby [9]. The characterization of a resource provided by existing formats such as Web services does not include the level of metadata necessary to evaluate the suitability of resources beyond the description of its input and output. More advanced resource formats such as OWL-S, WSDL-S, SAWSDL, and BioMoby aim at providing a semantic layer to capture better what the resource does in addition to its input and output data formats. The semantic part of the resource registration allows the classification of resources into a hierarchy of classes thus enhancing resource discovery. For example, the semantic search method of the BioMoby plug-in in Taverna traverses the object ontology and recursively extracts the parent nodes of that particular output object [9]. However, existing approaches do not offer an interface that allows the discovery of services with respect to their scientific meaning expressed in an ontology. To overcome this difficulty, path-based guiding systems such as SemanticMap and BioGuide can be combined with integration platforms to allow the discovery of resources suitable to implement scientific workflows. For example, BioGuide extends SRS [5] to allow a unique interface to discover resources and express queries over integrated data sources [3].

Resource discovery can exploit further resource metadata to predict the outcome of a workflow execution and select the resource more likely to produce the expected output. For example, BioNavigation [4] exploits various statistical metadata combined with semantic data to rank resources with respect to the users' criteria. Resources are ranked with respect to their cardinality, the characterization of their entries (number of attributes), etc. A user interested in *retrieving as many genes involved in a particular disease* selects a path in a domain ontology together with the corresponding ranking criteria. BioNavigation returns a ranking of all implementations of the conceptual path [10].

BioSpider is a system that integrates biological and chemical online databases. Given a biological or chemical identifier, BioSpider produces a report containing physicochemical, biochemical, and genetic information about the identifier. Ngu et al. [11] proposed an approach to classify search interfaces by probing these interfaces and trying to match the control flow of the interface against a standard control flow. InfoSpiders is a multi-agent focused crawler specialized for biomedical information whose goal is to fetch information about diseases when given information about genes. The Adaptive Crawler for Hidden-Web Entry (ACHE) points is a focused crawler specialized for locating searchable Web forms that serve as entry points to online databases and Web services. Context-aware form clustering (CAFC) is a clustering approach that models Web forms as a set of hyperlinked objects and considers visible information in the form context – both within and in the neighborhood of forms – as the basis for similarity comparison. A repository of scientific resources was automatically compiled using the approach [1].

Future Directions

Biological resource discovery remains a critical issue for the Life Sciences. The development of a system to support resource discovery is directly constrained by the information pertaining to scientific resources made available to the users as well as the formats designed to represent these rich metadata. For these reasons research on resource discovery for Life Sciences still is in its infancy. Scientists dramatically need assistance at each level of the process from the identification of the resources that best would meet the experimental requirements to the actual composition of the resources in an executable workflow. Future developments include the design of systems that combine various orthogonal aims for resource selection such as semantics (what the resource does), statistics (prediction of the result), syntax (schema mapping for resource composition), performance (efficiency), quality, etc.

Data Sets

NAR http://nar.oxfordjournals.org/
BMC Source Code for Biology and Medicine http://www.scfbm.org/home
Bioinformatics Links Directory http://bioinformatics.ca/links_directory/
Semantic Map for Structural Bioinformatics http://bioserv.rpbs.jussieu.fr/SBMap/
Automatically compiled list of biological resources http://formsearch.cs.utah.edu
Open Biomedical Ontologies (OBO) http://obofoundry.org/
Next Generation Biology Workbench (Swami) http://www.ngbw.org

URL to Code

caBIG http://cabig.nci.nih.gov/
BioMOBY http://biomoby.org/
SOAPlab http://www.ebi.ac.uk/Tools/webservices/soaplab/overview
SemanticMap http://bioinformatics.eas.asu.edu/
myGRID http://www.mygrid.org.uk/
Taverna http://taverna.sourceforge.net/
MOBY-S http://mobycentral.icapture.ubc.ca/
BioNavigation http://bioinformatics.eas.asu.edu/
BioGuide http://bioguide-project.net/
Seahawk http://biomoby.open-bio.org/
Remora http://lipm-bioinfo.toulouse.inra.fr/remora/cgi/remora.cgi
Kepler http://www.kepler-project.org/
BioSpider http://biospider.ca
InfoSpiders http://www.informatics.indiana.edu/fil/IS/

Cross-References

► Biological Metadata Management
► Dublin Core
► Graph Management in the Life Sciences
► Metadata
► Ontology
► Resource Description Framework
► Unified Modeling Language
► Web Services
► Web Services and the Semantic Web for Life Science Data
► XML

Recommended Reading

1. Barbosa L, Tandon S, Freire J. Automatically constructing a directory of molecular biology databases. In: Data integration in the life sciences. DILS 2007. LNCS, vol. 4544. Berlin: Springer. p. 6–16.
2. Clark T, Martin S, Liefeld T. Graphically distributed object identification for biological knowledge bases. Brief Bioinform. 2004;5(1):59–70.
3. Cohen-Boulakia S, Biton O, Davidson S, Froidevaux C. BioGuideSRS: querying multiple sources with a user-centric perspective. Bioinformatics. 2006;23(10):1301–3.
4. Cohen-Boulakia S, Davidson S, Froidevaux C, Lacroix Z, Vidal ME. Path-based systems to guide scientists in the maze of biological resources. J Bioinforma Comput Biol. 2006;4(5):1069–95.
5. Etsold T, Harris H, Beaulah S. Chapter 5 – SRS: an integration platform for databanks and analysis tools in bioinformatics. In: Lacroix Z, Critchlow T, editors. Bioinformatics: managing scientific data. Los Altos: Morgan Kaufmann; 2003. p. 109–46.
6. Fox GC, Gannon D, editors. Concurrency and Computation: Practice and Experience, Special Issue: Workflow in Grid Systems, vol. 18(10). Chichester: Wiley; 2006.
7. Galperin MY. The molecular biology database collection: 2007 update. Nucleic Acids Res. 2007;35(Database issue):D3–4.
8. Good BM, Wilkinson MD. The life sciences semantic web is full of creeps! Brief Bioinform. 2006;7(3):275–86.
9. Kawa EA, Senger M, Wilkinson MD. BioMoby extensions to the Taverna workflow management and enactment software. BMC Bioinf. 2006;7(1):523.
10. Lacroix Z, Raschid L, Eckman B. Techniques for optimization of queries on integrated biological resources. J Bioinforma Comput Biol. 2004;2(2):375–411.
11. Ngu AHH, Rocco D, Critchlow T, Buttler D. Automatic discovery and inferencing of complex bioinformatics web interfaces. World Wide Web. 2005;8(4):463–93.
12. Wolstencroft K, Alper P, Hull D, Wroe C, Lord P, Stevens R, Goble C. The myGrid ontology: bioinformatics service discovery. Int J Bioinforma Res Appl. 2007;3(3):303–25.

Biological Sequences

Amarnath Gupta
San Diego Supercomputer Center, University of California San Diego, La Jolla, CA, USA

Synonyms

DNA sequences; Protein sequence

Definition

A biological sequence is a sequence with a small fixed alphabet, and represents a naturally occurring or experimental generated fragment of genetic or protein material or any intermediate product (like the messenger RNA).

Example: A DNA fragment has the 4 character alphabet 'A', 'C', 'T', 'G'. Chromosomes are long strings over this alphabet.

Key Points

Biological sequences can be long. A full chromosome may have millions of characters. Therefore development of proper storing and indexing strategies is very important for fast retrieval. Suffix tree based indexes have been used successfully for long biological sequences. Further, approximate string matching techniques with potential deletions and insertions are important for biological sequences. BLAST is a well known algorithm used for approximate matching and ranking of biological sequences.

Cross-References

▶ Index Structures for Biological Sequences
▶ Query Languages and Evaluation Techniques for Biological Sequence Data
▶ Query Languages for the Life Sciences

Recommended Reading

1. Brown AL. Constructing genome scale suffix trees. In: Proceedings of the 2nd Asia-Pacific Bioinformatics Conference; 2004.
2. Hunt E, Atkinson MP, Irving RW. A database index to large biological sequences. In: Proceedings of the 27th International Conference on Very Large Data Bases; 2001. p. 139–48.
3. Phoophakdee B, Zaki MJ. TRELLIS +: an effective approach for indexing genome-scale sequences using suffix trees. In: Proceedings of the Pacific Symposium on Biocomputing (online proceedings); 2008. p. 90–101.
4. Tian Y, Tata S, Hankins RA, Patel JM. Practical methods for constructing suffix trees. VLDB J. 2005;14(3):281–99.

Biomedical Data/Content Acquisition, Curation

Nigam Shah
Stanford University, Stanford, CA, USA

Synonyms

Biomedical data annotation

Definition

The largest source of biomedical knowledge is the published literature, where results of experimental studies are reported in natural language. Published literature is hard to query, integrate computationally or to reason over. The task of reading published papers (or other forms of experimental results such as pharmacogenomics datasets) and distilling them down into structured knowledge that can be stored in databases as well as knowledgebases is called curation. The statements comprising the structured knowledge are called annotations. The level of structure in annotation statements can vary from loose declarations of "associations" between concepts (such as associating a paper with the concept "colon cancer") to statements that declare a precisely defined relationship between concepts with explicit semantics. There is an inherent tradeoff between the level of detail of the structured annotations and the time and effort required to create them. Curation to create highly structured and computable annotations requires PhD level individuals to curate the literature. In the molecular biology research community, this task is performed primarily by curators employed by genome databases such as the saccharomyces genome database [7]. In the biomedical research community this task is performed by curators employed by community portals such as AlzForum for Alzheimer's research [9] and PharmGKB for pharmacogenomics [31]. In the medical community such curation is still an ignored task, with some groups, such as RCTBank [26], pioneering the effort to curate clinical trial reports.

Historical Background

In the biomedical domain, curation began with the formation of cDNA, EST and gene sequence databases such as GenBank. Initially, curation was restricted to the task of assigning a functional annotation (usually in free text) to a sequence being submitted to GenBank. Scientists performing the experiments and submitting the data performed the task on their own. With the rise in the amount of sequence data and subsequently data on the function, structure and cellular locations of gene products along with the formation of communities of researchers around specific model organisms, the task of curation gradually became centralized in the role of a curator at model organism databases. Interaction amongst the curators and leading scientists led to the creation of projects such as the gene ontology project [1] in 1998, which led to a systematic basis for creating annotations about the molecular function, biological process and cellular locations of gene products. In subsequent years, user groups formed around other kinds of data, such as microarray gene expression data, resulting in the creation of information models for structuring the metadata pertaining to high throughput experiments. Individual research groups, such as Ecocyc, have already maintained a high level of curation effort, particularly for Information about biological pathways; although it was the success of the gene ontology project that resulted in the widespread appreciation for the need of curated content. With the continued rise in the amount and diversity of biomedical data, the need for curation continues to increase; both in terms of the number of man-hours required and in the level detail desired in the resulting annotations.

Foundations

In the course of their work, biomedical investigators must integrate a growing amount of diverse information. It is not possible for scientists to bring together this large amount of information without the aid of computers. Researchers have turned to ontologies – which allow representation of experimental results in a structured form – to facilitate interoperability among databases by indexing them with standard terms as well as to create knowledge bases that store large amounts of knowledge in a structured manner. Ontologies provide researchers with both the structure into which experimental results, facts and findings have to be put into as well as the words (or terms) to be used in populating the structure with instances [8, 14]. If the ontologies are well-designed, then the resulting knowledge bases can be used to retrieve relevant facts, to organize and interpret disparate knowledge, to infer non-

obvious relationships, and to evaluate hypotheses posited by scientists.

Assertion annotations – assertions or statements about the relationships among biological entities and the processes in which they participate – are a crucial link between abstractions of experimental results and the theory (or theories) that explain the underlying results. The national center for biomedical ontology develops methods and tools that enable the easy creation of such assertion annotations [11]. However, even with tool support, the creation of assertion annotations is manual, hard and expensive. In addition to this annotation as assertion viewpoint, another predominant use of annotations is to provide metadata for datasets stored in databases. In this case, these metadata-annotations are not assertions about any biological entity but instead provide additional information about the experiment or dataset examining the biological entity. Such metadata-annotations provide information about experimental conditions, the disease that the dataset pertains to, the perturbation applied during the experiment, and so on. Metadata-annotations do not state a biological fact like an assertion based on interpretation of experimental results. These two kinds of annotations: (i) assertion annotations and (ii) metadata annotations, are highlighted in Fig. 1.

Curation is the process by which annotations (either assertions or metadata) are created. (The word "annotation" is also used by some as a verb to describe this process of curation to create annotations; this has lead to wide-spread confusion in the community about the meaning of annotation.) Until now curation has been largely a manual process requiring highly qualified individuals to read and interpret the text in published papers to create the annotations. It is important to note that even though curation is carried out by skilled personnel, different curators have different opinions on what "knowledge" is being reported in the paper. Increasingly, automated methods are being employed to assist in the curation task because of the fact that manual curation is unlikely to scale and keep pace with the growth of biomedical data and literature [3]. Curation is typically carried out using a tool that allows the curator to select

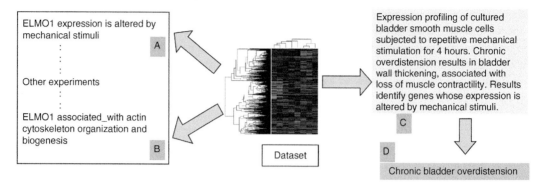

ELMO1 expression is altered by
mechanical stimuli

A

⋮

Other experiments

⋮

ELMO1 associated_with actin
cytoskeleton organization and
biogenesis

B

Dataset

Expression profiling of cultured
bladder smooth muscle cells
subjected to repetitive mechanical
stimulation for 4 hours. Chronic
overdistension results in bladder
wall thickening, associated with
loss of muscle contractility. Results
identify genes whose expression is
altered by mechanical stimuli.

C

D

Chronic bladder overdistension

Biomedical Data/Content Acquisition, Curation, Fig. 1 Shows the relationship of the assertion-annotations and metadata-annotations with datasets. The dataset is analyzed by a researcher to make a fine grained statement (**a**), which states the particular observation made in the dataset and reported in a publication. Several such statements get published in scientific papers. A curator, after reading a multitude of these papers creates an *assertion-annotation* (such as a GO annotation) as a summary statement (**b**) based on the fine grained statements. The dataset is also described by the researcher in terms of the disease studied, the cell lines used, the experimental conditions that existed etc. This description (**c**), comprises the *metadata-annotation* of the dataset, and is usually in natural language; although at times it is done using a CV. (**d**) shows the "tag" from a controlled vocabulary or ontology that can be assigned to this dataset upon processing the text description computationally

relevant ontology terms and associate them with the entity being annotated. The same tool writes out the resulting annotations in a custom format.

Technical Issues

The Different Types of Expressivity of Ontologies/Vocabularies Used to Create the Annotations

As discussed, annotations can range from simple terms that are "associated" with a particular resource to structured assertions that use explicit logical relationships. Depending on the required use – that of creating assertion-annotations or creating metadata-annotations – the ontologies used in the annotation process need to have adequate expressivity in terms of the different relationships the user can use during the curation process. A detailed discussion on the kinds of relationships that are available in biomedical ontologies is can be found in [27]. The most widely used artifacts for annotation are controlled vocabularies (CVs). A CV provides a list of terms whose meanings are specifically defined. Terms from a CV are usually used for indexing records in a database. The Gene Ontology (GO) is the most widely used CV in databases serving biomedical researchers

[1]. The GO provides terms that are "associated" with particular gene products for describing their molecular function (MF), biological process (BP) and cellular component (CC). Arguably, CVs provide the most return-on-effort in terms of facilitating database search and interoperability.

Storage Schemes and Data Models to Store These Annotations in Underlying Databases

Most annotations when created initially are stored as flat text files. However in order for the annotations to be useful to researchers, they need to be stored in database systems that support efficient storage and querying. Naturally the database schema and the data model to which the annotations conform to becomes an important issue. Until recently the trend was to create a relational schema corresponding to the annotation model used for a particular curation workflow and each group created its own annotation model as well as schema. This lead to various "silo" databases that need to be mapped to one another. The need for such mapping lead to the creation of groups, such as BioPAX, which proposed "exchange formats" to map silo databases to one another [4]. Recently, semantic web technology is receiving a lot of attention in the biomedical community because of the promise of "automatic" interoperability if different groups use consistent identifier (URIs) as well as the Resource Description Framework (RDF) format to describe entities and resources in their annotations [22].

Techniques for Indexing the Curated Annotation for Retrieval

Assuming the issue of creating annotations (manually or computationally) is adequately addressed, special attention needs to be paid to the appropriate indexing of annotations. For example, once a publication is annotated by associating it with the term *melanoma*, in order to ensure appropriate retrieval when someone searches for *skin neoplasms* it is essential to index the same paper with terms such as *skin neoplasm* (because *melanoma* is a kind of *skin neoplasm*). This can be accomplished by pre-computing all such inferred annotations or by real time query expansion using the hierarchy among the terms *melanoma* and *skin neoplasm*.

Workflow Aspects of the Curation Process

As noted before, different curators can have different opinions on what "knowledge" is being reported in the paper. The level of this agreement is quantified by calculating inter-curator agreement using a variety of methods [6]. Using detailed curation guidelines, many projects achieve inter-curator agreement in the range of 85–90 % and some as high as 94 % [6]. Currently, curation is typically carried out using a tool, such as Phenote, (www.phenote.org.) that allows the curator to select relevant ontology terms and associate them with the entity being annotated. The same tool writes out the resulting annotations in a custom format. Usually the workflow for curation differs by organization and the kinds of source (such as published papers or medical records or clinical trials) being curated. Currently, there are no off-shelf workflow systems that provide a generic curation workflow. Increasingly curation is becoming web based and the tools used for curation are tied to a database which stores the annotations (See the Alzforum and SWAN projects for an example). There are also

efforts to make curation collaborative, and several wiki-based projects such as wikipathways (www.wikipathways.org) are underway in the field.

Key Applications

The discovery process in biomedical research is cyclical; Scientists examine existing data to formulate models that explain the data, design experiments to test the hypotheses and develop new hypotheses that incorporate the data generated during experimentation. Currently, in order to advance this cycle, the experimentalist must perform several tasks: (i) gather information of many different types about the biological entities that participate in a BP, (ii) formulate hypotheses (or models) about the relationships among these entities, (iii) examine the different data to evaluate the degree to which his/her hypothesis is supported, and (iv) refine the hypotheses to achieve the best possible match with the data. In today's data-rich environment, this is a very difficult, time-consuming, and tedious task.

If existing data, information and knowledge are curated to create knowledge bases that store large amounts of knowledge in a structured manner [14, 15] the resulting knowledge bases can be used to retrieve relevant facts, to organize and interpret disparate knowledge and to computationally evaluate hypotheses and model posited by scientists [2, 18, 19]. For example, EcoCyc is a comprehensive source of structured knowledge on metabolic pathways in E. Coli and can be used to reason about E. Coli metabolism. Reactome is a source of structured knowledge on BPs related to signal transduction, gene regulation and metabolism in eukaryotic organisms [13]. The creation of such knowledgebases requires that the task of curation be carried out with great detail and that the tradeoffs between the complexity of the annotation structure required and the curation overhead entailed by that be balanced. Understanding the curation cost is a significant factor in determining the feasibility of proposed knowledge-driven applications [12].

There are several public as well as private groups that curate biomedical literature and other data to create highly structured knowledge bases. A majority of these knowledge bases are centered on biological pathways and Ecocyc and Reactome are the leading examples. In Pharmacogenomics, PharmGKB is a resource that provides curated knowledge on the interactions between genotype and pharmacological effects of drugs [31]. The Semantic Web Applications in Neuroscience (SWAN) project is a resource providing curated knowledge on Alzheimer's research with a focus on capturing the evolving scientific discourse as the research progresses [16]. In the commercial sector, companies such as Ingenuity offer subscription access to curated literature content as well as curation-for-fee services.

Such curated and structured content is primarily used to interpret the results of high-throughput datasets in the light of prior knowledge. At the simplest level, coloring nodes of a pathway according the increase or decrease in their expression level in a particular assay is a widely used approach. Another widely used approach is that of counting the annotations, such as the association with a particular BP, assigned to a set of biological entities, such as genes deemed significant for a particular cancer, and analyzing for a statistically significant difference in the distribution of the annotation counts as compared to a reference such as the set of all the genes assayed.

The other key use of curated and structured content is to support computer aided reasoning; with the goal of inferring possible explanations for biological phenomena [25], for evaluating alternative explanations for biological phenomena [19], for automated question answering [29] and automatically constructing as well as extending existing structured descriptions of BPs such as pathways [23].

Future Directions

As the amount and diversity of data, information and knowledge rise in the biomedical domain, there is a recognized need to be able to compute with the existing knowledge [5]. As the use of ontology rises in the biomedical domain [30], the appreciation for the need of curated content is also rapidly increasing; along with the realization that manual curation is unlikely to keep pace with the needs of the community [3].

These trends have led several groups, such as the BioAI group at Arizona State University and the SWAN group, to propose the use of distributed and collaborative curation in an attempt to leverage the "wisdom of the masses" [10, 17]. Collaborative curation holds tremendous promise for the field if the community can arrive at an agreed upon platform and formalism using which researchers can contribute structured content. The other clear future direction is the use of text-mining in the curation pipeline as a "force multiplier" to increase the productivity of existing curation efforts. The computational pharmacology group at University of Colorado is conducting exciting research in this direction. Both community-based collaborative curation as well as the use of text-mining to increase the efficiency of curation tools will be activities to follow closely for those interested in biomedical data acquisition and curation.

Cross-References

▶ Annotation
▶ Biological Metadata Management

Recommended Reading

1. Ashburner M et al. Geneontology: tool for the unification of biology. Nat Genet. 2000;25(1):25–9.
2. Baral C et al. A knowledge based approach for representing and reasoning about signaling networks. Bioinformatics. 2004;20(1):15–22.
3. Baumgartner Jr WA et al. Manual curation is not sufficient for annotation of genomic databases. Bioinformatics. 2007;23(13):41–8.
4. BioPax-Consortium. BioPAX: biological pathways exchange, 2006. Available from: http://www.biopax.org/2006
5. Bodenreider O, Stevens R. Bio-ontologies: current trends and future directions. Brief. Bioinform. 2006;7(3):256–74.
6. Camon EB et al. An evaluation of GO annotation retrieval for BioCreAtIvE and GOA. BMC Bioinf. 2005;6(Suppl 1):S17.

7. Cherry JM et al. SGD: saccharomyceas genome database. Nucl Acids Res. 1998;26(1):73–9.
8. Ciccaresse P, Wu E, Clark T. An overview of the SWAN 1.0 ontology of scientific discourse. In: Proceeding of 16th International World Wide Web Conference; 2007.
9. Clark T, Kinoshita J. Alzforum and SWAN: the present and future of scientific web communities. Brief Bioinform. 2007;8(3):163–71.
10. Gao Y et al. SWAN: a distributed knowledge infrastructure for Alzheimer disease research. J Web Semant. 2006;4(3):222–8.
11. Gibson M. Phenote. Berkeley Bioinformatics and Ontology Project (BBOP), National Center for Biomedical Ontology, Lawrence Berkeley National Laboratory; 2007.
12. Hunter L, Cohen KB. Biomedical language processing: what's beyond PubMed? Mol Cell. 2006;21(5):589–94.
13. Joshi-Tope G et al. Reactome: a knowledge base of biological pathways. Nucl Acids Res. 2005;33(Database Issue):D428–32.
14. Karp PD. An ontology for biological function based on molecular interactions. Bioinformatics. 2000;16(3):269–85.
15. Karp PD. Pathway databases: a case study in computational symbolic theories. Science. 2001;293(5537):2040–4.
16. Katz AE et al. Molecular staging of genitourinary malignancies. Urology. 1996;47(6):948–58.
17. Leslie M. Netwatch. Science. 2006;312(5781):1721.
18. Massar JP et al. BioLingua: a programmable knowledge environment for biologists. Bioinformatics. 2004;21(2):199–207.
19. Racunas SA et al. HyBrow: a prototype system for computer-aided hypothesis evaluation. Bioinformatics. 2004;20(Suppl 1):257–64.
20. Reactome Curator Guide. http://wiki.reactome.org/index.php/Reactome_Curator_Guide
21. Rise of the Bio-Librarian - the field of biocuration expands as the data grow. http://www.the-scientist.com/article/display/23316/
22. Ruttenberg A et al. Advancing translational research with the Semantic Web. BMC Bioinf. 2007;8(Suppl 3):S2.
23. Rzhetsky A et al. GeneWays: a system for extracting, analyzing, visualizing, and integrating molecular pathway data. J Biomed Inform. 2004;37(1):43–53.
24. Second International Biocuration Meeting, San Jose, 25–28 Oct, 2007. http://biocurator.org/Mtg2007/index.html
25. Shrager J et al. Deductive biocomputing. PLoS One. 2007;2(4):e339.
26. Sim I, Olasov B, Carini S. The Trial Bank system: capturing randomized trials for evidence-based medicine. In: American Medical Informatics Association Annual Symposium Proceedings; 2003. p. 1076
27. Smith B et al. Relations in biomedical ontologies. Genome Biol. 2005;6(5):R46.
28. Spasic I, Ananiadou S, McNaught J, Kumar A. Text mining and ontologies in biomedicine: making sense of raw text. Brief Bioinform. 2005;6(3):239–51.
29. Tari L., et al. BioQA. 2007. http://cbioc.eas.asu.edu/bioQA/v2/index.html
30. The National Center for Biomedical Ontology. 2006. Available at: www.biontology.org
31. Thorn CF, Klein TE, Altman RB. PharmGKB: the pharmacogenetics and pharmacogenomics knowledge base. In: Methods in molecular biology, vol. 311. Springer. p. 179–91.

Biomedical Image Data Types and Processing

Sameer Antani
National Institutes of Health, Bethesda, MD, USA

Synonyms

Data types: Image, Video, Pixel, Voxel, Frame; Conceptual data types: Pixel, Point, Edge, Volume, Region of interest, Shape, Color, Texture, Feature; Format: Joint photographic experts group (JPEG), Digital imaging and communications in medicine (DICOM), JPEG2000; Imaging Technique: X-Ray, Magnetic resonance imaging (MRI), Computerized tomography (CT), Ultrasound, Positron emission tomography (PET), Nuclear magnetic resonance (NMR), Microscopy, Single photon emission computerized tomography (SPECT), Fluoroscopy; Image processing: Compression, Wavelet compression, Functional mapping, Image reconstruction, 2D image processing, Texture analysis, Edge detection, 3D image processing, Surface detection, Image content analysis; Storage and retrieval: Image databases, Content-based image retrieval (CBIR), Visual similarity, Feature indexing, Multimedia information retrieval

Definition

The entry term describes biomedical image types (X-Ray, CT, MR, PET) stored in a particular format (DICOM, JPEG) that can be processed for visual enhancement (windowing, leveling) or extraction of features for further processing as needed in specific applications (generate 3D volumes from 2D slices, Content-Based Image Retrieval (CBIR)).

Historical Background

Both image processing and databases have been studied for over four decades. Biomedical image and data processing and storage systems have received significant attention within the last two decades. Image acquisition, image processing, and image analysis has gained significant importance in clinical medicine, biomedical research and education. Correspondingly, biomedical image databases have also found increasing use in recent years. Images stored as flat files, or in compressed formats on file servers and made accessible via links stored in relevant database records. Significant progress has been made in image types, formats, and content being computed and stored in these databases. Radiological images are now available through Web-based access to Picture Archiving and Communications Systems (PACS). This information can help in processing and further use of these data. Image and image feature indexing is a topic of significant research interest for visual retrieval. Visual image retrieval is rapidly becoming a key component of imaging informatics applications. Some specialized medical systems and some with some frequently used Web search engines are already providing visual search capability.

Foundations

Imaging has taken on a very important role in clinical medicine, biomedical research, and education. Biomedical visual data are acquired using a variety of techniques: single frame images; 3D volumes composed of single frame images; and made and as time-synchronized multiple frames as video data. In addition, these data are acquired at varying scales ranging from gross anatomy to the cellular level. Each image data type has specific acquisition methods, set of image processing methods for feature extraction that aid in analysis for targeted purposes, compression and storage methods, and particular data handling methods [1]. The image database primarily serves as a file storage mechanism with various processes for analysis and retrieval traditionally included in utility applications. Image databases are typically found in practical use as "multimedia databases" or "multimedia information systems" in the form of Radiological Information Systems (RIS), Hospital Information Systems (HIS), and Picture Archiving and Communication Systems (PACS). Such systems link textual data to image data through file links stored in database records. Image databases imply use of image feature indexing strategies such as metric index trees, multidimensional data trees, spatial databases, and R-trees, for specialized use such as Content-Based Image Retrieval (CBIR) [2, 3].

Image data types are challenging to define in standard terms such as integers, characters, strings, etc. An image is composed of pixels or in case of 3D images may be considered to composed of a set of elements of conceptual data type called voxels. Wikipedia (http://www.wikipedia.org) defines voxel as a portmanteau of words volumetric and pixel representing a unit element on a 3D image. Each such element (pixel or voxel) can be considered a complex data type as its content may be expressed using n-bits where n may be 8, 12, 16, 24, or 32. Typically 8-, 12-, and 16-bit images are gray scale images. A color pixel is typically 24-bits in depth comprising of three 8-bit channels for the additive color primaries (RED, GREEN, BLUE), though it is possible to have color images with other bit depths. This information is not natively stored in the image but needs to be exposed to the application through image metadata that may be stored in particular formats, such as a DICOM (http://dicom.nema.org/) on JPEG image header [4].

The images can be generated using a variety of techniques. Radiographic or X-Ray images, Computerized Tomography (CT), Magnetic Resonance images (MRI), Positron Emission Tomography (PET) images are examples of various imaging techniques. Techniques such as CT and MRI image the desired anatomical region in closely spaced sections. These sectional images can then be processed to create views along desired axes (axial, coronal, sagittal, oblique) as well as generate 3D volumetric data rendering.

Image processing is a term that includes functions and methods that focus on enhancement of images for improved human visualization or computer analysis, such as windowing, leveling, object edge detection, among others [5, 6]. It also is synonymous with application of methods whereby features such as edges, textures, and surfaces, among others, can be computed for making measurements, computer-aided diagnosis, visual enhancement, identifying anatomical structures, determining unique image content signature, etc. For instance, using the above example of generating 3D volumetric data from 2D image slices, for a data set of MRI slices of the brain, it would be necessary to segment the edges from each 2D MRI image slice and register them with corresponding edges from the same anatomy in other slices. The next step would be to convert these edges into surfaces formed across these slices in order to generate 3D volumetric data.

With the increasing use of images in medical care and research, it becomes necessary make this data connect with other image and non-image data. The resulting database systems have evolved as PACS, Electronic Health Record (EHR) and Electronic Medical Record (EMR) systems. Such systems are increasingly common npw in hospitals and medical centers. These database systems are capable of storing and retrieving text data, for example, a patient record containing test results and other medical history, along with image data. The systems may exist on a single computer, a local network of computers, or distributed over a wide area network. Variants of these systems developed for medical research studies can also correlate between different study participant information and keep track of longitudinal information.

In database processing it is often necessary to define the image to be of a particular type. This can assist in data and type verification as well as communication of semantics to other applications that may be using the data. Some database systems require storage of images in their native form as undefined BLOB data types while others, including most PACS, prefer to maintain references to image data files that are stored in traditional directory (folder) file structures. The choice between these approaches is largely determined by storage and computational efficiency and dependent on particular applications and solutions. In either scenario it is efficient to store the image metadata as database records. While this information may be available in image header files, it requires the additional step of accessing and opening each image file for any database operation involving images.

Image processing steps often result in features extracted from images. These features could be pixels, or image regions, measurements of color, texture, edges forming a shape, surfaces, etc. Each of these features could be standardized to be a data type or could use standard data types, e.g., a 3-channel color histogram could be represented using 3D arrays that could represent 3D histograms. Often there are mentions of 4D images. These include the element of time, as observation of the same image region during image acquisition. Each such conceptual type may be defined as a predefined data type for purposes of image analysis or visualization. Other operations could include transforming the image or extracted characteristics from the spatial domain several into other domains through frequency analysis, such as Fourier analysis or other transforms. Selecting these or other image processing methods is heavily dependent on the nature of the images and several methods are covered in [1, 5, 6]. Further, it may be necessary to compress the images in order to minimize data storage requirements or improve transmission efficiency over networks. These decisions must be made carefully in light of possible data loss found in typical implementations of common image compression methods such as JPEG or JPEG2000 [4].

In summary, biomedical image processing is critical to analysis and use of biomedical images for clinical medicine, research and education. These images may also have other associated images as well as text data. All this information is stored in biomedical databases that use a combination of image types, header information, image units, and content through the extracted features as data types. Images may be indexed through multidimensional indexing trees or be linked to flat files stored in a folder or accessed via a file server.

Key Applications

Medical Imaging and Informatics Systems: Medical Information Systems, like the PACS, hold medical data about patients, medical research study participants, etc. This medical data is typically heterogeneous comprising of linkages to other databases, such as electronic medical records containing the medical history, clinical notes, lab reports, and any acquired images. The text data in a DICOM image that is typically stored in a PACS is fielded. The radiology readings can also be fielded, exist as a block of free text, or in some older systems are converted into an bitmap image. The image data can be from various sources and in a variety of formats. The radiological image repositories are increasingly considered as rich resources for data mining. This has created the field of study called medical imaging informatics. Image and text information are used to inform a learning engine that can then be trained to assist in clinical or research applications. Such engines are in their infancy and their description and definition is an active subject of research [7]. As reported in the article, mining of such image and text data could serve as key component of evidence-based medicine and significantly improve clinical care.

Content-Based Image Retrieval (CBIR): While the PACS systems allow querying images by searching fielded text using text keywords and structured SQL-like queries, an alternative complementary form of image-based querying has gained significant research interest in recent years [2, 3]. This approach, called Content-Based Image Retrieval (CBIR), uses distinguishing features extracted from images to serve as indices. These features are then used to find images similar to an image query. For example, in the OpenI system [3] developed by the U.S. National Library of Medicine (NLM), part of the National Institutes of Health, the user is able to search a collection of Open Access biomedical research articles from PubMed Central repository using text keywords and images. The image features are matched with the collection of over 2.3 million images derived from approximately 750,000 articles. The system also contains a collection of 8,000 X-rays and a collection of teaching images from the the University of Southern California, Norris Medical Library [8]. The image matching is done using a combination of color, texture, and meta-features that are converted into visual words for rapid retrieval using NLM's Essie Search Engine. This is an example of a Multimodal Information Retrieval System. Technologies such as these could be used for medical case-based retrieval that are being studied for improving clinical care and research.

Finding similar images tends to be a very subjective matter and is heavily dependent on the extracted features from the images. Additionally, it is also dependent on the level of detail extracted. For example, one measure of image similarity is overall appearance of the image, which in case of X-Ray images, can be characterized by a histogram of pixel intensity levels in the image. While this *global* measure may be sufficient for overall similarity, it is insufficient in expressing *local* pathology that can only be captured by feature extraction within the region of interest. An intelligently implemented hierarchical strategy works better in a heterogeneous collection of images.

Given the subjective nature of image content and human perception, it is challenging to evaluate systems through system characteristics or reported performance measures alone. These results are sensitive to the kinds of image data that the system is operating on, extracted features, query capability, and several other gaps that need to be overcome for developing an "ideal" system. A framework for these gaps is discussed in [9]. Other medical systems are reviewed in [10]. The Cross Language Evaluation Forum (CLEF) benchmarking competition has evaluates image classification and image retrieval in a biomedical setting on an annual basis (CLEF-Campaign, http://www.clef-campaign.org) and permits use of text data commonly found with medical images to improve usability. Due to the transient nature of academic systems, and lack of detail available in commercial systems, such a venue provides valuable metrics for comparing various systems and assessing the state-of-the-art.

Cross-References

- ▶ Annotation-Based Image Retrieval
- ▶ Feature-Based 3D Object Retrieval
- ▶ Feature Extraction for Content-Based Image Retrieval
- ▶ Image
- ▶ Image Content Modeling
- ▶ Image Database
- ▶ Image Management for Biological Data
- ▶ Image Metadata
- ▶ Image Representation
- ▶ Image Retrieval and Relevance Feedback
- ▶ Image Segmentation
- ▶ Indexing and Similarity Search
- ▶ Indexing Metric Spaces
- ▶ Multimedia Data
- ▶ Multimedia Databases
- ▶ Multimedia Data Indexing
- ▶ Multimedia Data Storage
- ▶ Relevance Feedback for Content-Based Information Retrieval
- ▶ Spatial Data Types
- ▶ Two-Dimensional Shape Retrieval
- ▶ Visual Content Analysis

Recommended Reading

1. Beutel J, Kundel HL, Van Metter RL, editors. Handbook of medical imaging, vols. 1, 2, and 3. Bellingham: SPIE Press.
2. Samet H. Foundations of multidimensional and metric data structures. San Francisco: Morgan Kaufman; 2006.
3. Demner-Fushman D, Antani SK, Simpson M, Thoma GR. Design and development of a multimodal biomedical information retrieval system. J Comput Sci Eng. 2012;6(2):168–77.
4. Joint Photographic Experts Group (JPEG). http://www.jpeg.org/. American Medical Information Association; 2007. p. 826–30.
5. Gonzales RC, Woods RE, editors. Digital image processing. 2nd ed. Upper Saddle River: Prentice Hall.
6. Sonka M, Hlavac V, Boyle R, editors. Image processing, analysis, and machine vision. 2nd ed. Washington, DC: PWS Publishing.
7. Reiner BI, Siegel EL. The clinical imperative of medical imaging informatics. J Digit Imaging. 2009;22(4):345–7.
8. Rehman I, Smith CF. Orthopaedic surgical anatomy teaching collection. 2002. http://cdm15799.contentdm.oclc.org/cdm/landingpage/collection/p15799coll50.
9. Deserno TM, Antani S, Long R. Ontology of gaps in content-based image retrieval. J Digit Imaging. 2008;22(2):202–15.
10. Müller H, Michoux N, Bandon D, Geissbuhler A. A review of content-based image retrieval systems in medical applications – clinical benefits and future directions. Int J Med Inform. 2004;73(1):1–23.

Biomedical Scientific Textual Data Types and Processing

Li Zhou[1] and Hua Xu[2]
[1]Partners HealthCare System Inc., Boston, MA, USA
[2]Columbia University, New York, NY, USA

Synonyms

Annotation; Biomedical literature; Curation; Indexing; Information retrieval; Information retrieval models/metrics/operations; MEDLINE/PubMed; Scientific knowledge bases; Semi-structured text retrieval; Text extraction; Text mining; Web search and crawling

Definition

Vast amounts of biomedical scientific information and knowledge are recorded in text [1, 7]. Various scientific textual data in the biomedical domain may generally be disseminated through the following resources [7, 11]: biomedical literature (e.g., original reports and summaries of research in journals, books, reports, and guidelines), biological databases (e.g., annotations in gene/protein databases), patient records (e.g., clinical narrative reports), and web content.

A variety of techniques have been applied to identify, extract, manage, integrate and exploit knowledge from biomedical text. Some researchers [11] divide biomedical scientific textual data processing into three major activities as shown in Fig. 1: information retrieval (IR), information extraction (IE), and text mining (TM).

Information retrieval [2, 11] is the science of indexing and searching for information particularly in text or other unstructured forms. The aim of IR is to identify relevant documents in response to a particular query, which forms the basis of any knowledge discovery process.

Information extraction [11] aims to identify and extract categorized or semantically well-defined data (entities, relations or events) from text documents in a certain domain, as well as to create structured knowledge bases that can be accessed by other informatics applications. Typical subtasks of IE are named entity, relation and event recognition (e.g., recognition of protein names and interactions between proteins), coreference (e.g., identifying whether a chain of noun phrases refer to the same object), and terminology extraction (e.g., finding the relevant terms for a given corpus).

Text mining (TM) [2, 6] is the process of discovering and extracting interesting and non-trivial patterns and knowledge from unstructured text data. The primary goal of TM is to retrieve knowledge that is hidden in text, and to present the distilled knowledge to users in a concise form. Typical subtasks of TM may include pattern discovery, hypothesis generation, correlation discovery, etc. However, some researchers give a broader definition of TM which overlaps with IR and IE on certain tasks such as text classification, text clustering and named entity recognition.

Historical Background

Before the invention of computers, results of biomedical research have been published as journal or conference prints for a long time. Bibliographic databases that typically contained references to literature on library shelves was the first application of using computers to improve library service. MEDLINE (Medical Literature Analysis and Retrieval System Online) is the U.S. National Library of Medicine's (NLM) premier bibliographic database that contains over 16 million references to journal articles in life sciences with a concentration on biomedicine. As an online interactive searchable bibliographic database, MEDLINE was introduced in 1971 by NLM, to replace its previous version called MEDLARS (Medical Literature Analysis and Retrieval System). In 1997, PubMed was developed by the National Center for Biotechnology Information (NCBI) at the NLM, to provide free and efficient access to MEDLINE through the World Wide Web. Currently, MEDLINE/PubMed is probably the best-known biomedical literature reference database. The use of high-throughput experimental technologies has dramatically increased the pace of biomedical knowledge discovery. In 2006, over 623,000 references to published articles were added to the MEDLINE database. A large amount of effort has been spent on improving the performance of IR on the MEDLINE database. A distinctive feature of MEDLINE is that the records are indexed with NLM's Medical Subject Headings (MeSH).

During the past decade, more and more full-text biomedical publications have become available on the Internet, though most of them have restricted access. In 1999, a bold new initiative called PubMed Central (PMC) was designed at the U.S. National Institutes of Health (NIH) to provide a central repository for literature in the life sciences with open access. To date, there are more than 300 journals that have joined PMC and they provide free access to their publications. BioMed Central, a commercial publisher, also provides free access to papers published in their journals.

Various text processing methods, such as natural language processing (NLP) and machine learning (ML) technologies, have been extensively studied in the domain of computer and information science. However, they have not been widely applied to biomedical text before the 1990s, largely due to the lack of available biomedical text. Starting at the mid 1990s, text processing technologies have been gradually applied to biomedical text on different tasks, such as information retrieval, biomedical entity recognition, text clustering and classification, and knowledge discovery.

Foundations

As mentioned above, biomedical scientific textual data processing applies methods and technologies from multiple disciplines, including linguistics, computer science, statistics, and so on. In general, the major stages of processing scientific textual data to exploit rich knowledge include retrieval of relevant documents, extraction of named entities and relations, and discovery of new knowledge. However, some processes may not follow the exact steps. This entry adopts a classification by Natarajan et al. [11] on major constituent technologies for knowledge discovery in text (see Fig. 1). Scientific fundamentals for each stage will be discussed in the following sections.

Information Retrieval
Conventional IR methods are often based on keyword queries, using Boolean logic, vector space models or probabilistic models [1, 7]. One of the simplest forms of IR is to search keywords in documents that are indexed by a set of keywords. Search algorithms are used to identify the relevant documents based on the number of index keywords that match query keywords. One disadvantage of this approach is that the documents are determined either relevant or irrelevant. There is no further ranking. Vector space model is an algebraic model for representing text documents. When applying the model to IR, both the query and documents are represented as vectors, whose dimensions correspond to different terms in the query or documents. There are different methods to compute the weight of terms in the vectors and the tf-idf weighting is one of the best known schemes. Relevancy rankings of documents to a query can be determined by calculating the document similarities between the query and documents, via measurements such as cosine similarity of two vectors. Probabilistic models treat the process of document retrieval as a probabilistic inference and similarities are computed as probabilities that a document is relevant for a given query. An advantage of probabilistic model is that documents are ranked in decreasing order of their probability of being relevant to the query.

Information Extraction
Approaches to named entity recognition generally fall into three categories: lexicon based, rule based and statistically-based [1]. For example, part-of-speech tagging, inductive rule learning, decision trees, Bayesian model, support vector machines, as well as combined methods have

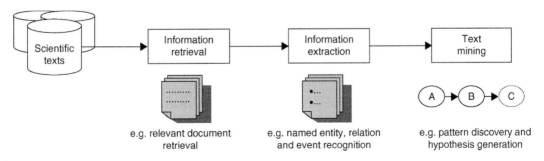

Biomedical Scientific Textual Data Types and Processing, Fig. 1 Major stages of processing biomedical scientific textual data and relevant subtasks [11]

been applied to this problem. For discovering relationships among entities, variant techniques have been used. Shallow parsing is often used to focus on specific parts of the text to analyze predefined words such as verbs and nouns. Some systems combine natural language processing and co-occurrence techniques, while others apply machine learning techniques.

Text Mining

Text classification and clustering are the most widely used techniques in biomedical text mining. While text classification is a form of learning from pre-classified examples, text clustering is referred to as unsupervised learning. Bayesian models were widely used in the early days. In recent years, more advanced machine learning methods, such as k-nearest neighbors, artificial neural networks, support vector machines, expectation maximization, and fuzzy clustering have been used. Logical inference models [13] have been applied to hypothesis generation which attempts to uncover relationships that are not present in the text but instead are inferred by the other existing relationships. There are a variety of techniques for knowledge discovery from biomedical text using graphs and knowledge models.

Key Applications

Information retrieval technologies have been used extensively to help users to find relevant articles that they are interested in. MEDLINE/PubMed has used various methods to improve the performance of searches. It provides keyword-based Boolean search to allow users to search by keywords, as well as document-based search, which implements the vector space model and could find documents for similar topics. With the availability of full-text articles online, more IR applications have tried to search full-text articles for detailed information. For example, the focus of the 2006 genomics track of the Text Retrieval Conference (TREC) [8] was to retrieve answers for biological questions from full text articles. The European Bioinformatics Institute (EBI)

at the European Molecular Biology Laboratory (EMBL) has developed a biomedical information retrieval system called "CiteXplore," which combines literature search with text mining tools for biology. It also links biomedical literature sources to existing bioinformatics databases, such as SwissProt.

Although most of biomedical text processing tools are still in the research stage, some of them have shown potential uses. Many information extraction systems have been used to build knowledge bases from biomedical literature. Different approaches have been reported to extract relations among biomedical entities of interest (e.g., gene/protein). GENIES (GENomic Information Extraction System) [4] is an NLP-based system that extracts molecular pathways from literature. It semantically parses sentences into a structured form for relation extraction. PASTA (Protein Active Site Template Acquisition) [5] is a system that uses manually created templates to extract relationships between amino acid residues and their functions within a protein. The PreBIND [3] system uses Support Vector Machine (SVM) technology to locate protein-protein interaction data in the literature, thus to facilitate the curation process for protein databases. IR and IE systems are also combined to build more sophisticated systems to help specific tasks in biology, such as biological database curation tools that can help curators find related articles and identify critical findings from biological articles [14]. The iHOP [9] system extracts protein-relationship from the literature. It also includes advanced search modes for discovery and visualization of protein-protein-interaction network [9].

Another potential application of text mining tools is to discover new knowledge from literature, for example, helping biomedical researchers to generate new research hypotheses. ARROW-SMITH and BITOLA [9, 14] are two online tools that provide the function of literature-based knowledge discovery. ARROWSMITH detects indirect associations between concepts that are not directly linked in the literature. BITOLA is designed for disease candidate gene discovery by mining the bibliographic database MEDLINE.

Cross-References

► Data Mining
► Data, Text, and Web Mining in Healthcare
► Information Retrieval
► Text Mining
► Text Mining of Biological Resources

Recommended Reading

1. Chen H, Friedman W, Hersh SS, editors. Fuller medical informatics: knowledge management and data mining in biomedicine. Secaucus: Springer; 2005.
2. Cohen AM, Hersh WR. A survey of current work in biomedical text mining. Brief Bioinform. 2005;6(1):57–71.
3. Donaldson I, Martin J, deBruijn B, Wolting C, Lay V, Tuekam B, Zhang S, Baskin B, Bader G, Michalickova K, et al. PreBIND and textomy – mining the biomedical literature for protein-protein interactions using a support vector machine. BMC Bioinf. 2003;4(1):11.
4. Friedman C, Kra P, Yu H, Krauthammer M, Rzhetsky A. GENIES: a natural-language processing system for the extraction of molecular pathways from journal articles. Bioinformatics. 2001;17(Suppl 1):S74–82.
5. Gaizauskas R, Demetriou G, Artymiuk PJ, Willett P. Protein structures and information extraction from biological texts: the PASTA system. Bioinformatics. 2003;19(1):135–43.
6. Hearst M. Untangling text data mining. In: Proceedings of the 27th Annual Meeting of the Association for Computational Linguistics; 1999.
7. Hersh W. Information retrieval: a health and biomedical perspective. New York: Springer; 2003.
8. Hersh W, Cohen A, Roberts P, Rekapalli HK. TREC 2006 genomics track overview. In: Proceedings of the Text Retrieval Conference; 2006. Available at: http://trec.nist.gov/pubs/trec15/papers/GEO06.OVERVIEW.pdf
9. Hoffmann R, Valencia A. A gene network for navigating the literature. Nat Genet. 2004;36(7):664.
10. Hristovski D, Peterlin B. Literature-based disease candidate gene discovery. In: Proceedings of the Medinfo; 2004. p. 1649.
11. Natarajan J, Berrar D, Hack CJ, Dubizky W. Knowledge discovery in biology and biotechnology texts: a review of techniques, evaluation strategies, and applications. Crit Rev Biotechnol. 2005;25(1–2):31–52.
12. Smalheiser N, Swanson D. Using ARROWSMITH: a computer-assisted approach to formulating and assessing scientific hypotheses. Comput Methods Prog Biomed. 1998;57(3):149–53.
13. Swanson DR Complementary structure in disjoint science literatures. In: Proceedings of the 23rd Annual International ACM SIGIR Conference on Research and Development in Information Retrieval; 1990. p. 280–9.
14. Yeh AS, Hirschman L, Morgan AA. Evaluation of text data mining for database curation: lessons learned from the KDD Challenge Cup. Bioinformatics. 2003;19(Suppl 1):i331–9.

Biostatistics and Data Analysis

Mehmet M. Dalkiliç
Indiana University, Bloomington, IN, USA

Definition

Biostatistics is the application of probability and statistical techniques to the biological sciences. Probability has played a significant role in areas like genetics where combinatorics validate conjectures about the relationships of genes and the environment. Recently, combinatorics has become one of the main approaches to solving problems, e.g., motif discovery in bioinformatics. In the nineteenth century, well-known biologists like Herman von Helmholtz advocated that biological phenomenon could be understood using techniques in the physical sciences (remnants of this view still are present today). That approach, together with the "vitalism" movement [3], impeded the use of statistics. By the beginning of this century, however, statistics has become *de rigueur* in virtually all biological publications.

Historical Background

Although statistics as a mathematical area can be traced further, biostatistics is often associated with the work of Francis Galton (1822–1911). His major contribution was demonstrating that statistical methods could be beneficial in biology. Carrying on this tradition were Karl Pearson (1857–1936) and R.A. Fisher (1890–1962). While a number of standard applications and techniques have not changed for more than half-a-century, the availability of computing has made some (heretofore infeasible) techniques available. The classic text in this area is by Sokal and Rohlf, "Biometry" [5] now in its third edition. It should be pointed out that the demarcation often cited between "frequentists" [6] and "Bayesians" [1] is fairly well evident in biostatistics. In the former, physical, repeatable, identical, and random experiments can be associated to a mathematical limit, e.g., P (heads) $= \lim_{flips \to \infty} \frac{\#heads}{flips} = 1/2$ is the probability of getting heads when flipping a coin. Bayesians, on the other hand, presume probability to be a degree of belief (or subjective probability). This can be written as *posterior* $\propto$ *prior x likelihood*, in symbols $P(H|E) \propto P(H)P(E|H)$. The distinctive feature is that Bayesian statisticians will associate values with the *posterior*, *hypothesis*, and *prior* (likelihood), whereas non-Bayesian statisticians will only consider the hypotheses in constrained settings. Classical statistics is often the primary choice. However, Bayesian techniques are becoming increasingly popular in systems biology-biology that integrates and evaluates disparate data usually in the form of very large graphs. In these graphs, nodes are typically genes and edges evidence of relationship.

Foundations

As explained eloquently in [5], there is a deeper philosophical debate centered on whether biological phenomena can ultimately be modeled using deterministic approaches. That debate aside, the use of statistics is continuing to grow as the amounts of biological data grow. Indeed, recent technologies (high-throughput) are producing several orders of magnitude more data in comparison to traditional approaches. Statistical tools, from this perspective, become a necessity. Furthermore, a growing number of biologists now believe that there is benefit in examining data normally not studied within one's own specialized domain-this is called a systems biology. While the foundations for systems biology are still under development, it is clear that it will rely heavily on Bayesian reasoning. Typically

the disparate experimental, textual, etc., data are structured as a directed, acyclic graph where nodes are random variables and edges are effects. The basic attempt is to discover sets of independent variables and form a better understanding of the joint probability. For example, one might take the some 14,000 *Drosophila m.* genes and presume a joint distribution $P(X_1, X_2, ..., X_{14000})$ where X_i represents the probability that gene i has some expression level. Bayesians build large graphs relying on conditional probabilities and so-called "separations" that expose which random variables are independent of one another. They then examine the behavior of the graph under particular conditions. The interested reader is guided to [4]. Topics studied in biostatistics are too numerous to list (e.g., multivariate regression, analysis of covariance, linear discriminant analysis, principal component analysis, and so forth); therefore, a sample that reflects the kind of tools that are used and most prevalent techniques will be given. Most of the biostatistics used is parametric; the models result from human expertise. This is opposed to nonparametric models (or data-driven) that rely on the data itself. With the availability of cheap, fast computation, however, the use of nonparametric models has exploded. Given a set of data $\mathbf{x} = x_1, x_2, ..., xn$, the probability $P(x|\theta_1, \theta_2, ..., \theta_m)$ is parametric if m is not dependent on n; otherwise it is nonparametric. Linear regression is a parametric model that presumes a linear relationship between two random variables (rv) both having Gaussian distributed noise. It is presumed the variables are real valued. If rv Y is a function of rv X, then the phrase, "A regression of Y on X," is used. One is determining the optimal coefficients β_0, β_1 given data $D = \{\langle x_1, y_1 \rangle, ..., \langle x_n, y_n \rangle\}$ on function $Y = \beta_1 X + \beta_0$. Regression, though simple, provides a first step in understanding the relationship between to rvs. Regression can be used to predict, adjust, explain, etc., and is common in biostatistics. Regression produces an optimal linear relationship between rvs Y and X. Typically, one minimizes the squares of the residuals – the difference between the hypothetical point and observed point. The use is so ubiquitous that virtually every mathematical or statistical package has this available. As pointed out many times in the area, because two rvs are *not* linearly related, does not mean they are *not* functionally related.

Correlation, related to Linear Regression, is as often used. It is so similar in so many ways to linear regression that the literature will often have the use of one, when in fact, it is the other that is warranted or even makes sense. Succinctly, regression examines how one rv depends on another; Correlation examines how two rvs behave together-in concert or more formally *covary*. Correlation is related to moments and is often called product-moments. One of the most popular is the Pearson product-moment $p = \frac{\sigma_{XY}}{\sigma_X \sigma_Y}$ where σ_i is the variance of the joint (numerator) and individual rvs (denominator). The value $\rho \in [-1, 1]$ where as $|\rho| \approx 1$, the variables appear to have an association. As ρ approaches 0, this indicates they do not. The positive and negative score reflects the direction of the association. Another popular correlation examines how pairs of pairs of values behave called ranks. Kendall's τ examines two pairs of pairs $\langle x_i, y_y \rangle, \langle x_j, y_j \rangle$ observing if *both* values of one pair are greater or smaller than the other pair; if so, then the pair of pairs is called concordant. If this is not satisfied, then the pair of pairs is discordant. The formula for Kendall's $\tau = \frac{N}{n(n-1)}$ where N is the count of ranks and n the sample size. It should be pointed out this is a nonparametric statistic and value $\tau \in (-1, 1)$.

A good deal of analysis is done with simple hypothesis testing of sample statistics. Because an entire population can be seldom checked, to establish a degree of certainty about a property we use sampling techniques (statistics). Sampling is done randomly to hopefully reflect the underlying distribution, especially if it is not known. If the sample is consistent with the conjecture about the existence of a property (called a hypothesis), then the hypothesis is said to be *accepted*; if not, it is *rejected*. There exist two disparate hypotheses that are simultaneously proposed when doing this analysis. The Null hypothesis, typically denoted H_0, is that the property was observed likely through chance. The Alternative hypothesis, typically denoted by $H a$, is that the property is *not* due to chance. While recapitulating this process in its entirety is not appropriate here, we summarize the following sequence of standard steps that are typically followed for such analysis: (**0**) State the two hypotheses H_0 and *Ha*. It must be the case that only one can be true; (**1**) Establish the protocol for acquiring and using sample data. The decision usually depends on a chosen test statistic (a real-valued function of the sample); (**3**) Calculate the test statistic's value; (**4**) Check whether the statistic's value is likely to have occurred by chance or not typically using tables [2].

There are two types of errors that can result in this procedure. A Type I error will reject a null hypothesis when it is actually true. The significance level is the probability of committing this error and is often denoted by α. A Type II error is accepting a null hypothesis when it is actually false. The power of the test is the probability of not committing a Type II error. A *P*-value is, in a sense, an extreme case in examining H_0. It will reflect the strength of the evidence. *P*-values are very useful, since one can apply any number of significance levels reflecting confidence in the data.

ANOVA (analysis of variance), developed by Fisher almost a century ago remains a popular tool that examines differences in populations means $\mu = \sum_{i=1}^{N} \frac{X_i}{N}$, for values $X_1, X_2, \ldots, X_N$ one of the measures of central tendency of a population (e.g., median, mode). Population variance, $\sigma^2 = \sum_{i=1}^{N} \frac{(X_i - \mu)^2}{N}$ measures the spread of values. There are two models of ANOVA, Model I and Model II, which form H_0 and *Ha* based on mean and variance, respectively. The interested reader is guided to [5] for an in-depth presentation.

Key Applications

The initial application of biostatistics was to study evolution and natural selection. It now plays an essential role in sequence and genome analysis, protein structure prediction, proteomics, phylogenetics, etc. A current challenge for data analysis is to consider extensions to relational data, to include probabilistic data and uncertainty, as they their roles in biological inquiry. For database researchers, one of the challenges is to move in thinking from a Boolean model (relational) to one that involves probability and uncertainty, conflicting data, nonreplicable data, and data that have orders of magnitude more attributes than tuples. Data in biology are often unnormalized and may lack primary key identifiers normalization.

Future Directions

Biologists and computer scientists use different paradigms when considering data and analysis. Biologists are reductionists and focus on tightly constrained problems, whereas computer scientists pursue generic technological solutions that can be applied to multiple problems. Researchers in data management and biostatistics must keep these differences in mind as they move towards developing useful yet generic tools to serve biologists into the future. The future of biostatistics will depend directly on the computational resources available. Typically, two different approaches to problems are taken: combinatorial (or enumerative) or statistical. The former is usually very fast, but lacks the ability to discover nuanced relationships. Statistical approaches are computationally expensive, e.g., Expectation-Maximization, but with increasingly powerful computers and clusters, this computational impediment is slowly eroding.

Cross-References

▶ Annotation
▶ Biomedical Data/Content Acquisition, Curation
▶ Data Quality Assessment
▶ F-Measure
▶ Principal Component Analysis
▶ Probabilistic Databases
▶ Spectral Clustering
▶ Taxonomy: Biomedical Health Informatics
▶ Term Weighting
▶ Text Analytics

▶ Text Compression
▶ Two-Poisson Model
▶ Uncertainty in Events

Recommended Reading

1. Lee PM. Bayesian statistics. 2nd ed. Arnold; 2003.
2. Lindley DV, Scott WF. New Cambridge statistics tables. 2nd ed. Cambridge University Press; 1995.
3. Myers CS. Vitalism: a brief historical and critical review. Mind. 1900;9(35):319–31.
4. Neapolitan RE. Learning Bayesian networks. Prentice Hall; 2003.
5. Sokal R, Rohlf FJ. Biometry. 3rd ed. New York: W.H. Freeman and Company; 1995.
6. von Mises R. Probability, statistics, and truth (trans: Neyman J, Scholl D, Rabinowitsch E.). 1939.

Bitemporal Indexing

Mirella M. Moro[1] and Vassilis J. Tsotras[2]
[1]Departamento de Ciencia da Computaçao, Universidade Federal de Minas Gerais – UFMG, Belo Horizonte, MG, Brazil
[2]University of California-Riverside, Riverside, CA, USA

Synonyms

Bi-temporal access methods

Definition

A bi-temporal index is a data structure that supports both temporal time dimensions, namely, transaction time (the time when a fact is stored in the database) and valid time (the time when a fact becomes valid in reality). The characteristics of the time dimensions supported imply various properties that the bi-temporal index should have to be efficient. As traditional indices, the performance of a temporal index is described by three costs: (i) storage cost (i.e., the number of pages the index occupies on the disk), (ii) update cost (the number of pages accessed to perform an update on the index, e.g., when adding, deleting, or updating a record), and (iii) query cost (the number of pages accessed for the index to answer a query).

Historical Background

Most of the early work on temporal indexing has concentrated on providing solutions for transaction-time databases. A basic property of transaction time is that it always *increases*. Each newly recorded piece of data is time-stamped with a new, larger transaction time. The immediate implication of this property is that previous transaction times *cannot* be changed. Hence, a transaction-time database can "roll back" to, or answer queries for, any of its previous states.

On the other hand, a valid-time database maintains the entire temporal behavior of an enterprise as best known now. It stores the current knowledge about the enterprise's past, current, or even future behavior. If errors are discovered about this temporal behavior, they are corrected by modifying the database. In general, if the knowledge about the enterprise is updated, the new knowledge modifies the existing one. When a correction or an update is applied, previous values are not retained. It is thus not possible to view the database as it was before the correction/update.

By supporting both valid and transaction time, a bi-temporal database combines the features of the other temporal database types. While it keeps its past states, it also supports changes anywhere in the valid-time domain. Hence, the overlapping and persistent methodologies proposed for transaction-time indexing can be applied [5, 6, 9]. The difference with transaction-time indexing is that the underlying access method should be able to dynamically manage intervals (like an R-tree, a quad-tree, etc.). For a worst-case comparison of temporal access methods, the reader is referred to Ref. [7].

Foundations

When considering temporal indexing, it is important to realize that the valid and transaction time dimensions are *orthogonal* [3]. While in various scenarios it may be assumed that data about a fact is entered in the database at the same time as when it happens in the real world (i.e., valid and transaction time coincide), in practice, there are many applications where this assumption does not hold. For example, data records about the sales that occurred during a given day are recorded in the database at the end of the day (when batch processing of all data collected during the day is performed). Moreover, a recorded valid time may represent a later time instant than the transaction time when it was recorded. For example, a contract may be valid for an interval that is later than the (transaction) time when this information was entered in the database. Such properties are critical in the design of a bi-temporal access method since the support of both valid and transaction time affects directly the way records are created or updated. Note that the term "interval" is used here to mean a "convex subset of the time domain" (and not a "directed duration"). This concept has also been named a "period"; in this discussion, however, only the term "interval" is used.

The reader is referred to the entry on ▶ Transaction-Time Indexing, in which a transaction-time database was abstracted as an evolving collection of objects; updates arrive in increasing transaction-time order and are always applied on the latest state of this set. In other words, previous states cannot be changed. Thus, a transaction-time database represents and stores the database activity; objects are associated with intervals based on this database activity. In contrast, in the entry on ▶ Valid-Time Indexing, a valid-time database was abstracted as an evolving collection of interval objects, where each interval represents the validity interval of an object. The allowable changes in this environment are the addition/deletion/modification of an interval object. A difference with the transaction-time abstraction is that the collection's evolution (past states) is *not* kept. Note that when considering the valid time dimension, changes do not necessarily come in increasing time order; rather, they can affect *any* interval in the collection. This implies that a valid-time database can correct errors in previously recorded data. However, only a single data state is kept, the one resulting after the correction is applied.

A bi-temporal database has the characteristics of both approaches. Its abstraction maintains the evolution (through the support of transaction time) of a dynamic collection of (valid-time) interval objects. Figure 1 offers a conceptual view of a bi-temporal database. Instead of maintaining a single collection of interval objects (as a valid-time database does), a bi-temporal database maintains a sequence of such collections $C(t_i)$ indexed by transaction time. Assume that each interval I represents the validity interval of a contract in a company. In this environment, the user can represent how the knowledge about company contracts evolved. In Fig. 1, the t-axis (v-axis) corresponds to transaction (valid) times. At transaction time t_1, the database starts with interval objects I_x and I_y. At t_2, a new interval object I_z is recorded, etc. At t_5 the valid-time interval of object I_x is modified to a new length.

When an interval object I_j is inserted in the database at transaction time t, a record is created with the object's surrogate (contract_no I_j), a valid-time interval (contract duration), and an initial transaction-time interval $[t, UC]$. When an object is inserted, it is not yet known if it (ever) will be updated. Therefore, the right endpoint of the transaction-time interval is filled

with the variable until changed (UC), which will be changed to another transaction time if this object is later updated. For example, the record for interval object I_z has transaction-time interval $[t_2, t_4]$, because it was inserted in the database at transaction time t_2 and was "deleted" at t_4. Note that the collections $C(t_3)$ and $C(t_4)$ correspond to the collections C_a and C_b in Fig. 1 in the ▶ Valid-Time Indexing entry, assuming that at transaction time t_4, the erroneous contract I_z was deleted from the database.

Based on the previous discussion, an index for a bi-temporal database should (i) store past states, (ii) support addition/deletion/modification changes on the interval objects of its current logical state, and (iii) efficiently access and query the interval objects on any state.

Figure 1 summarizes the differences among the various database types. Each collection $C(t_i)$ can be thought of on its own, as a separate valid-time database. A valid-time database differs from a bi-temporal database since it keeps *only one* collection of interval objects (the latest). A transaction-time database differs from a bi-temporal database in that it maintains the history of an evolving set of *plain* objects instead of *interval* objects. A transaction-time database differs from a conventional (nontemporal) database in that it also keeps its *past* states instead of only the latest state. Finally, the difference between a valid time and a conventional database is that the former keeps *interval* objects (and these intervals can be queried).

There are three approaches that can be used for indexing bi-temporal databases.

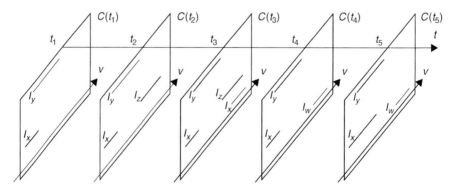

Bitemporal Indexing, Fig. 1 A bi-temporal database

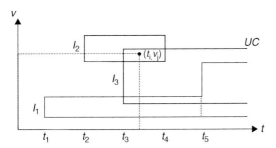

Bitemporal Indexing, Fig. 2 The bounding-rectangle approach for bi-temporal objects

Approach 1: The first one is to have each bi-temporal object represented by a "bounding rectangle" created by the object's valid- and transaction-time intervals and to store it in a conventional multidimensional structure like the R-tree. This approach has the advantage of using a single index to support both time dimensions, but the characteristics of transaction time create a serious overlapping problem [5]. A bi-temporal object with valid-time interval I that is inserted in the database at transaction time t is represented by a rectangle with a transaction-time interval of the form $[t, UC]$. All bi-temporal objects that have not been deleted (in the transaction sense) will share the common transaction-time endpoint UC (which, in a typical implementation, could be represented by the largest possible transaction time). Furthermore, intervals that remain unchanged will create long (in the transaction-time axis) rectangles, a reason for further overlapping. A simple bi-temporal query that asks for all valid-time intervals that at transaction time t_i contained valid time v_j corresponds to finding all rectangles that contain point (t_i, v_j). Figure 2 illustrates the bounding-rectangle approach; only the valid and transaction axis are shown. At t_5, the valid-time interval I_1 is modified (enlarged). As a result, the initial rectangle for I_1 ends at t_5, and a new enlarged rectangle is inserted ranging from t_5 to UC.

Approach 2: To avoid overlapping, the use of two R-trees has also been proposed [5]. When a bi-temporal object with valid-time interval I is added in the database at transaction time t, it is inserted at the *front* R-tree. This tree keeps bi-temporal objects whose right transaction endpoint is unknown. If a bi-temporal object is later deleted at some time $t' > t$, it is physically deleted from the front R-tree and inserted as a rectangle of height I and width from t to t' in the *back* R-tree. The back R-tree keeps bi-temporal objects with known transaction-time interval. At any given time, all bi-temporal objects stored in the front R-tree share the property that they are alive in the transaction-time sense. The temporal information of every such object is thus represented simply by a vertical (valid-time) interval that "cuts" the transaction axis at the transaction time when this object was inserted in the database. Insertions in the front R-tree objects are in increasing transaction time, while physical deletions can happen anywhere on the transaction axis.

In Fig. 3, the two R-tree methodologies for bi-temporal data are divided according to whether their right transaction endpoint is known. The scenario of Fig. 2 is presented here (i.e., after time t_5 has elapsed). The query is then translated into an interval intersection and a point enclosure problem. A simple bi-temporal query that asks for all valid-time intervals which contained valid time v_j at transaction time t_i is answered with two searches. The back R-tree is searched for all rectangles that contain point (t_i, v_j). The front R-tree is searched for all vertical intervals that intersect a horizontal interval H that starts from the beginning of transaction time and extends until point t_i at height v_j.

When an R-tree is used to index bi-temporal data, overlapping may also incur if the valid-time intervals extend to the ever-increasing *now*. One approach could be to use the largest possible valid-time time stamp to represent the variable *now*. In Ref. [2] the problem of addressing both the *now* and UC variables is addressed by using bounding rectangles/regions that increase as the time proceeds. A variation of the R-tree, the GR-tree, is presented. The index leaf nodes capture the exact geometry of the bi-temporal regions of data. Bi-temporal regions can be static or growing

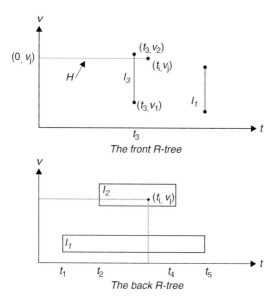

Bitemporal Indexing, Fig. 3 The two R-tree methodologies for bi-temporal data

and rectangles or stair shapes. Two versions of the GR-tree are explored, one using minimum bounding rectangles in non-leaf nodes and one using minimum bounding regions in non-leaf nodes. Details appear in Ref. [2].

Approach 3: Another approach to address bi-temporal problems is to use the notion of partial persistence [1, 4]. This solution emanates from the abstraction of a bi-temporal database as a sequence of collections $C(t)$ (in Fig. 1) and has two steps. First, a good index is chosen to represent each $C(t)$. This index must support dynamic addition/deletion of (valid-time) interval objects. Second, this index is made partially persistent. The collection of queries supported by the interval index structure implies which queries are answered by the bi-temporal structure. Using this approach, the bi-temporal R-tree that takes an R-tree and makes it partially persistent was introduced in Ref. [5].

Similar to the transaction-time databases, one can use the "overlapping" approach [3] to create an index for bi-temporal databases. It is necessary to use an index that can handle the valid-time intervals and an overlapping approach to provide

the transaction-time support. Multidimensional indexes can be used for supporting intervals, for example, an R-tree or a quad-tree. The overlapping R-tree was proposed in Ref. [6], where an R-tree maintains the valid-time intervals at each transaction time instant. As intervals are added/deleted or updated, overlapping is used to share common paths in the relevant R-trees. Likewise, [9] proposes the use of quad-trees (which can also be used for spatiotemporal queries).

There are two advantages in "viewing" a bi-temporal query as a "partial persistence" or "overlapping" problem. First, the valid-time requirements are disassociated from the transaction-time ones. More specifically, the valid time support is provided from the properties of the R-tree, while the transaction-time support is achieved by making this structure "partially persistent" or "overlapping." Conceptually, this methodology provides fast access to the $C(t)$ of interest on which the valid-time query is then performed. Second, changes are always applied to the most current state of the structure and last until updated (if ever) at a later transaction time, thus avoiding the explicit representation of variable UC. Considering the two approaches, overlapping has the advantage of simpler implementation, while the partial persistence approach avoids the possible logarithmic space overhead.

Key Applications

The importance of temporal indexing emanates from the many applications that maintain temporal data. The ever-increasing nature of time imposes the need for many applications to store large amounts of temporal data. Accessing such data specialized indexing techniques is necessary. Temporal indexing has offered many such solutions that enable fast access.

Cross-References

▸ B+-Tree
▸ R-Tree (and Family)

► Temporal Database
► Transaction-Time Indexing
► Valid-Time Indexing

Recommended Reading

1. Becker B, Gschwind S, Ohler T, Seeger B, Widmayer P. An asymptotically optimal multiversion B-tree. VLDB J. 1996;5(4):264–75.
2. Bliujute R, Jensen CS, Saltenis S, Slivinskas G. R-tree based indexing of now-relative bitemporal data. In: Proceedings of the 24th International Conference on Very Large Data Bases; 1998. p. 345–56.
3. Burton FW, Huntbach MM, Kollias JG. Multiple generation text files using overlapping tree structures. Comput J. 1985;28(4):414–6.
4. Driscoll JR, Sarnak N, Sleator DD, Tarjan RE. Making data structures persistent. J Comput Syst Sci. 1989;38(1):86–124.
5. Kumar A, Tsotras VJ, Faloutsos C. Designing access methods for bitemporal databases. IEEE Trans Knowl Data Eng. 1998;10(1):1–20.
6. Nascimento MA, Silva JRO. Towards historical R-trees. In: Proceedings of the 1998 ACM Symposium on Applied Computing; 1998. p. 235–40.
7. Salzberg B, Tsotras VJ. A comparison of access methods for time-evolving data. ACM Comput Surv. 1999;31(2):158–221.
8. Snodgrass RT, Ahn I. Temporal databases. IEEE Comput. 1986;19(9):35–42.
9. Tzouramanis T, Vassilakopoulos M, Manolopoulos Y. Overlapping linear quadtrees and spatio-temporal query processing. Comput J. 2000;43(4):325–43.

Bitemporal Interval

Christian S. Jensen[1] and Richard T. Snodgrass[2,3]
[1]Department of Computer Science, Aalborg University, Aalborg, Denmark
[2]Department of Computer Science, University of Arizona, Tucson, AZ, USA
[3]Dataware Ventures, Tucson, AZ, USA

Definition

Facts in a bitemporal database may be timestamped by time values that are products of time intervals drawn from two orthogonal time domains that model valid time and transaction time, respectively. A *bitemporal interval* then is given by an interval from the valid-time domain and an interval from the transaction-time domain, and denotes a rectangle in the two-dimensional space spanned by valid and transaction time.

When associated with a fact, a bitemporal interval then identifies an interval (valid time) during which that fact held (or holds or will hold) true in reality, as well as identifies an interval (transaction time) when that belief (that the fact was true during the specified valid-time interval) was held, i.e., was part of the current database state.

Key Points

In this definition, a time interval denotes a convex subset of the time domain. Assuming a discrete time domain, a bitemporal interval can be represented with a non-empty set of bitemporal chronons or granules.

Cross-References

► Bitemporal Relation
► Chronon
► Temporal Database
► Temporal Granularity
► Time Domain
► Time Interval
► Transaction Time

Recommended Reading

1. Bettini C, Dyreson CE, Evans WS, Snodgrass RT, Wang XS. A glossary of time granularity concepts. In: Etzion O, Jajodia S, Sripada S, editors. Temporal databases: research and practice. LNCS, vol. 1399. Berlin: Springer; 1998. p. 406–13.
2. Jensen CS, Dyreson CE, editors. A consensus glossary of temporal database concepts-February 1998 version. In: Etzion O, Jajodia S, Sripada S, editors. Temporal databases: research and practice. LNCS, vol. 1399. Berlin: Springer; 1998. p. 367–405.

Bitemporal Relation

Christian S. Jensen[1] and Richard T. Snodgrass[2,3]
[1]Department of Computer Science, Aalborg
University, Aalborg, Denmark
[2]Department of Computer Science, University of
Arizona, Tucson, AZ, USA
[3]Dataware Ventures, Tucson, AZ, USA

Synonyms

Fully temporal relation; Temporal relation; Valid-time and transaction-time relation

Definition

A *bitemporal relation* captures exactly one valid time aspect and one transaction time aspect of the data it contains. This relation inherits its properties from valid-time relations and transaction-time relations. There are no restrictions as to how either of these temporal aspects may be incorporated into the tuples.

Key Points

In this definition, "bi" refers to the capture of exactly two temporal aspects. An alternative definition states that a bitemporal relation captures one or more valid times and one or more transaction times. In this definition, "bi" refers to the existence of exactly two types of times.

One may adopt the view that the data in a relation represents a collection of logical statements, i.e., statements that can be assigned a truth values. The valid times of these so-called facts are the times when these are true in the reality modeled by the relation. In cases where multiple realities are perceived, a single fact may have multiple, different valid times. This might occur in a relation capturing archaeological facts for which there no agreements among the archaeologists. In effect, different archaeologists perceive different realities.

Transaction times capture when database objects are current in a database. In case an object migrates from one database to another, the object may carry along its transaction times from the predecessor databases, termed *temporal generalization*. This then calls for relations that capture multiple transaction times.

The definition of bitemporal is used as the basis for applying bitemporal as a modifier to other concepts such as "query language." A query language is bitemporal if and only if it supports any bitemporal relation. Hence, most query languages involving both valid and transaction time may be characterized as bitemporal.

Relations are named as opposed to databases because a database may contain several types of relations. Most relations involving both valid and transaction time are bitemporal according to both definitions.

Concerning synonyms, the term "temporal relation" is commonly used. However, it is also used in a generic and less strict sense, simply meaning any relation with time-referenced data.

Next, the term "fully temporal relation" was originally proposed because a bitemporal relation is capable of modeling both the intrinsic and the extrinsic time aspects of facts, thus providing the "full story." However, this term is no longer used.

The term "valid-time and transaction-time relation" is precise and consistent with the other terms, but is also lengthy.

Cross-References

▶ Bitemporal Interval
▶ Temporal Database
▶ Temporal Generalization
▶ Transaction Time
▶ Valid Time

Recommended Reading

1. Jensen CS, Dyreson CE. A consensus glossary of temporal database concepts-February 1998 version. In: Etzion O, Jajodia S, Sripada S, editors. Temporal databases: research and practice. LNCS, vol. 1399. Berlin/New York: Springer; 1998. p. 367–405.

Bitmap Index

Chee-Yong Chan
National University of Singapore, Singapore,
Singapore

Definition

An index on an attribute provides an efficient way to access data records associated with a given range of values for the indexed attribute. Typically, an index stores a list of RIDs (called a RID-list) of all the records associated with each distinct value v of the indexed attribute. In a *bitmap index*, each RID-list is represented in the form of a bit vector (i.e., bitmap) where the size of each bitmap is equal to the cardinality of the indexed relation, and the i-th bit in each bitmap corresponds to the i-th record in the indexed relation. The simplest bitmap index design is the *Value-List index*, which is illustrated in Fig. 1b for an attribute A of a 12-record relation R in Fig. 1a. In this bitmap index, there is one bitmap E^v associated with each attribute value $v \in [0,9]$ such that the i-th bit of E^v is set to 1 if and only if the i-th record has a value v for the indexed attribute.

Historical Background

The idea of using bitmap indexes to speed up selection predicate evaluation has been recognized since the early 1970s [4]. Some early implementations of bitmap processing techniques include PC DBMSs (e.g., FoxPro, Interbase), a scientific/statistical database application developed at Lawrence Berkeley Laboratory [11], and Model 204, which is a commercial DBMS for the IBM mainframe [7].

The main advantage of using a bitmap index is the CPU efficiency of bitmap operations (AND, OR, XOR, NOT). Furthermore, compared to RID-based indexes, bitmap indexes are more space-efficient for attributes with low cardinality and more I/O-efficient for evaluating selection predicates with low selectivities. For example, assuming each RID requires four bytes of storage and ignoring any compression, bitmap indexes are more space-efficient if the attribute cardinality is less than 32, and reading a bitmap is more I/O-efficient than reading a RID-list if the selectivity factor of the selection predicate is more than $\frac{1}{32}$ ($\approx 3.2\%$). Another advantage of bitmap indexes is that they are very amenable to parallelization due to the equal-sized bitmaps and the nature of the bitwise operations.

A variety of bitmap index designs have been proposed since the early days. Besides the simple Value-List index illustrated in Fig. 1b, another early bitmap index design is the *Bit-Sliced index (BSI)* which is implemented in Model 204 and Sybase IQ [7]. A BSI for an attribute with a cardinality of C consists of $k = \lceil \log_2(C) \rceil$ bitmaps, with one bitmap associated with each bit in the binary representation of C. Compared to the Value-List index, the BSI is more space-efficient with an attribute value v being encoded by a string of k bits corresponding to its binary representation. The BSI design can be generalized to use a nonbinary base b such that it consists of $k(b - 1)$ bitmaps $B_i^j : 1 \leq k, 0 \leq jb$, where $k = \lceil \log_b(C) \rceil$. Each attribute value v is expressed in base b as a sequence of k base-b digits $v_k v_{k-1}...v_2 v_1$, and each bitmap B_i^j represents the set of records with $v_i \leq j$. Using a larger base number improves the index's performance for evaluating range predicates at the cost of an increased space cost. An example of a base-10 BSI is shown in Fig. 1c. Both Model 204 and Sybase IQ implemented base-10 Bit-Sliced indexes [7].

Several bitmap index designs have also been implemented in a scientific/statistical database application at Lawrence Berkeley Laboratory [11]. These bitmap indexes include binary encoded indexes (equivalent to binary BSI), unary encoded indexes (equivalent to non-binary BSI), K-of-N encoded indexes (generalizations of Value-List indexes where each attribute value is encoded by a N-bit string with exactly K bits set to 1), and superimposed encoded indexes based on superimposed encoding which is useful for indexing set-valued attributes.

Bitmap Index, Fig. 1 Examples of bitmap indexes. (**a**) indexed attribute A, (**b**) equality-encoded index (or value-list index), (**c**) range-encoded index (or base-10 bit-sliced index), (**d**) interval-encoded index

a / **b** — indexed attribute and equality-encoded index

	A	E^9	E^8	E^7	E^6	E^5	E^4	E^3	E^2	E^1	E^0
1	3	0	0	0	0	0	0	1	0	0	0
2	2	0	0	0	0	0	0	0	1	0	0
3	1	0	0	0	0	0	0	0	0	1	0
4	2	0	0	0	0	0	0	0	1	0	0
5	8	0	1	0	0	0	0	0	0	0	0
6	2	0	0	0	0	0	0	0	1	0	0
7	9	1	0	0	0	0	0	0	0	0	0
8	0	0	0	0	0	0	0	0	0	0	1
9	7	0	0	1	0	0	0	0	0	0	0
10	5	0	0	0	0	1	0	0	0	0	0
11	6	0	0	0	1	0	0	0	0	0	0
12	4	0	0	0	0	0	1	0	0	0	0

c — range-encoded index

R^8	R^7	R^6	R^5	R^4	R^3	R^2	R^1	R^0
1	1	1	1	1	1	0	0	0
1	1	1	1	1	1	1	0	0
1	1	1	1	1	1	1	1	0
1	1	1	1	1	1	1	0	0
1	0	0	0	0	0	0	0	0
1	1	1	1	1	1	1	0	0
0	0	0	0	0	0	0	0	0
1	1	1	1	1	1	1	1	1
1	1	0	0	0	0	0	0	0
1	1	1	1	0	0	0	0	0
1	1	1	0	0	0	0	0	0
1	1	1	1	1	0	0	0	0

d — interval-encoded index

I^4_1	I^3_1	I^2_1	I^1_1	I^0_1
0	1	1	1	1
0	0	1	1	1
0	0	0	1	1
0	0	1	1	1
1	0	0	0	0
0	0	1	1	1
0	0	0	0	0
0	0	0	0	1
1	1	0	0	0
1	1	1	1	0
1	1	1	0	0
1	1	1	1	1

Bitmap Index, Fig. 2 example of base-$<3,4>$ indexes. (**a**) Indexed attribute A (**b**) equality-encoded index (**c**) range-encoded index (**d**) interval-encoded index

a — Indexed attribute A

	A	decomposition
1	3	$0 \times 4 + 3$
2	2	$0 \times 4 + 2$
3	1	$0 \times 4 + 1$
4	2	$0 \times 4 + 2$
5	8	$2 \times 4 + 0$
6	2	$0 \times 4 + 2$
7	9	$2 \times 4 + 1$
8	0	$0 \times 4 + 0$
9	7	$1 \times 4 + 3$
10	5	$1 \times 4 + 1$
11	6	$1 \times 4 + 2$
12	4	$1 \times 4 + 0$

b — equality-encoded index

E^2_2	E^1_2	E^0_2		E^3_1	E^2_1	E^1_1	E^0_1
0	0	1		1	0	0	0
0	0	1		0	1	0	0
0	0	1		0	0	1	0
0	0	1		0	1	0	0
1	0	0		0	0	0	1
0	0	1		0	1	0	0
1	0	0		0	0	1	0
0	0	1		0	0	0	1
0	1	0		1	0	0	0
0	1	0		0	0	1	0
0	1	0		0	1	0	0
0	1	0		0	0	0	1

c — range-encoded index

R^1_2	R^0_2		R^2_1	R^1_1	R^0_1
1	1		0	0	0
1	1		1	0	0
1	1		1	1	0
1	1		1	0	0
0	0		1	1	1
1	1		1	0	0
0	0		1	1	0
1	1		1	1	1
1	0		0	0	0
1	0		1	1	0
1	0		1	0	0
1	0		1	1	1

d — interval-encoded index

I^1_2	I^0_2		I^1_1	I^0_1
0	1		0	0
0	1		1	0
0	1		1	1
0	1		1	0
0	0		0	1
0	1		1	0
0	0		1	1
0	1		0	1
1	0		0	0
1	0		1	1
1	0		1	0
1	0		0	1

Interest in bitmap indexes was revived in the mid-1990s due to the emergence of data warehousing applications which are characterized by read-mostly query workloads dominated by large, complex ad hoc queries [7]. All the major DBMS vendors (IBM, Microsoft, Oracle, and Sybase) also started to support bitmap indexes in their products around this time.

Foundations

- Chan and Ioannidis proposed a two-dimensional framework to characterize the design space of bitmap indexes [2]. The two orthogonal parameters identified for bitmap indexes (with an attribute cardinality of C) are (i) the arithmetic used to represent attribute values; i.e., how an attribute value is decomposed into digits according to some base (e.g., base-C arithmetic is used in a Value-List index); and (ii) the encoding scheme of each decomposed digit in bits (e.g., each attribute value in a Value-List index is encoded by turning on exactly one out of C bits). Consider an attribute value $v \in [0, C)$ and a sequence of n base numbers $B = b_n, b_{n-1}, \ldots, b_1>$, where

$$b_n = \left\lceil \frac{C}{\prod_{i-1}^{n-1} b_i} \right\rceil \text{ and } b_i \geq 2, \ i \in [1, n].$$

Using B, v can be decomposed into a unique sequence of n digits $<v_n, v_{n-1}, \ldots, v_1>$ as follows: $v_i = V_i \bmod b_i$, where $V_1 = v$ and $V_i = \left\lfloor \frac{v_{i-1}}{b_{i-1}} \right\rfloor$, for $1 < i \leq n$. Thus, $v = v_n \left(\prod_{j=1}^{n-1} b_j \right) + \cdots + v_i \left(\prod_{j=1}^{i-1} b_j \right) + \cdots + v_2 b_1 + v_1$. Note that each $v i$ is a base-$b i$ digit (i.e., $0 \leq v i < b i$). Each choice of n and base-sequence B gives a different representation of attribute values and therefore a different index (known as a Base-B). The index consists of n *components* (i.e., one component per digit) where each component individually is now a collection of bitmaps. Figure 2b shows a base-$<3,4>$ Value-List index that consists of two components: the first component has four bitmaps $\{E_1^3, E_1^2, E_1^1, E_1^0\}$ and the second component has three bitmaps $\{E_2^2, E_2^1, E_2^0\}$. Note that the k-th bit in each bitmap E_i^j is set to 1 if and only if $v_j = j$, where $<v_2, v_1>$ is the $<3,4>$-decomposition of the k-th record's indexed attribute value. For the encoding scheme dimension, there are two basic encoding schemes: equality encoding and range encoding. Consider the i-th component of an index with base b_i, and a value $v_i \in [0, b_i - 1]$. In the *equality encoding scheme*, v_i is encoded by b_i bits, where all the bits are set to 0 except for the bit corresponding to v_i, which is set to 1. Thus, an equality-encoded component (with base b_i) consists of b_i bitmaps $\{E_i^{b_i-1}, \ldots, E_i^0\}$

such that the k-th bit in each bitmap E_i^j is set to 1 if and only if $v_i = j$, where v_i is the i-th digit of the decomposition of the k-th record's indexed attribute value. In the *range encoding scheme*, v_i is encoded again by b_i bits, with the v_i rightmost bits set to 0 and the remaining bits (starting from the one corresponding to v_i and to the left) set to 1. The k-th bit in each bitmap $R_i^{b_i-1}$ is set to 1 if and only if $v_i \leq j$, where v_i is the i-th digit of the decomposition of the k-th record's indexed attribute value. Since the bitmap $R_i^{b_i-1}$ has all bits set to 1, it does not need to be stored, so a range-encoded component consists of $(b_i - 1)$ bitmaps $\{R_i^{b_i-2}, \ldots, R_i^0\}$. Value-List and Bit-Sliced indexes therefore correspond to equality-encoded and range-encoded indexes, respectively. Figures 1c and 2c shows the range-encoded indexes corresponding to the equality-encoded indexes in Figs. 1b and 2b. Details of query processing algorithms and space-time tradeoffs of equality/range-encoded, multicomponent bitmap indexes are given in [4].

A new encoding scheme, called *interval encoding*, was proposed in [5]. For an attribute with a cardinality of C, a value $v \in [0, C)$ is encoded using $\lceil \frac{C}{2} \rceil$ bits such that if $v \lceil \frac{C}{2} \rceil$, then v is encoded by setting the $(v + 1)$ rightmost bits to 1 and the remaining bits to 0; otherwise, v is encoded by setting the $(C - 1 - v)$ leftmost bits to 1 and remaining bits to 0. Thus, the interval encoding scheme consists of $\lceil \frac{C}{2} \rceil$ bitmaps $\{I^{\lceil \frac{C}{2} \rceil - 1}, \ldots, I^0\}$, where each bitmap I^i is associated with a range of $(m + 1)$ values $[j, j + m]$, $m = \lfloor \frac{C}{2} \rfloor - 1$, such that the k-th bit in a bitmap $I j$ is set to 1 if and only if the k-th record's indexed attribute value is in $[j, j + m]$. Figures 1d and 2d shows the interval-encoded indexes corresponding to the equality-encoded indexes in Figs. 1b and 2b. Note that interval encoding has better space-time tradeoff than range encoding: it has the same worst-case evaluation cost of two bitmap scans as range encoding, but its space requirement is about half that of range encoding.

Wu and Bachmann proposed a variant of binary BSI called *encoded bitmap index (EBI)* [12, 13]. Instead of encoding each attribute value simply in terms of its binary representation, an EBI uses a lookup table to map each attribute value to a distinct bit string; this flexibility enables optimization of the value-to-bit-string mapping, by exploiting knowledge of the query workload, to reduce the number of bitmap scans for query evaluation. Thus, binary BSI is a special case of EBI. Another similar index design called *Encoded-Vector index (EVI)* is used in IBM DB2. Instead of storing the index as a collection of $\lceil log_b(C) \rceil$ bitmaps as in EBI, EVI is organized as a single vector of $\lceil log_b(C) \rceil$-bit strings, and the purpose of the lookup table optimization is to reduce the CPU cost of bit string comparisons when evaluating selection queries of the form "A $\in \{v_1, v_2, \ldots, v_n\}$." Another related index is the *Projection index* [7], which is implemented in Sybase IQ.

Complex, multitable join queries (such as star-join queries) can also be evaluated very efficiently using *bitmapped join indexes* [6], which are indexes that combine the advantages of *join indexes* and bitmap representation. A *join index* for the join between two relations R and S is a precomputation of their join result defined by $\Pi_{R.rid, S.rid}(R \bowtie_p S)$ where p is the join predicate between R and S. Thus, a join index on $R \bowtie S$ can be thought of as a conventional index on the table R, where the attribute being indexed is the "virtual" attribute S.rid; i.e., each distinct S.rid value v is associated with a list of all R.rid values that are related to v via the join. A *bitmapped join index* [6] is simply a join index with the RID-lists represented using bitmaps. Bitmapped join indexes are implemented in Informix Red Brick Warehouse and Oracle. Bitmap indexes can also be applied to evaluate queries that involve aggregate functions (e.g., SUM, MIN/MAX, MEDIAN); evaluation algorithms for Value-List and Bit-Sliced indexes are discussed in a paper by O'Neil and Quass [7]. Efficient algorithms for performing arithmetic operations (addition and subtraction) on binary BSIs are proposed in [8].

As bitmap indexes become less efficient for larger attribute cardinality, a number of approaches have been developed to reduce their space requirement. Besides using multicomponent bitmap indexes [2, 11], another common space-reduction technique is to apply compression. In Model 204 [7], the bitmaps are compressed by using a hybrid representation; specifically, each individual bitmap is partitioned into a number of fixed-size segments, and segments that are dense are stored as verbatim bitmaps, while sparse segments are converted into RID-lists. While generic compression techniques (e.g., LZ77) are effective in reducing both the disk storage and retrieval cost of bitmap indexes, the savings in I/O cost can be offset by the high CPU incurred for decompressing the compressed bitmaps before they can be operated with other bitmaps. A number of specialized compression techniques that enable bitmaps to be operated on without a complete decompression have been proposed: Byte-aligned Bitmap Code (BBC) (which is used by Oracle) and Word-Aligned Hybrid code (WAH) [15]. Some performance studies of compressed bitmap indexes are reported in [1–3, 15].

A different approach proposed to reduce the size of bitmaps is to use *range-based bitmaps (RBB)* [14], which have been applied to index large data sets in tertiary storage systems as well as large, multidimensional data sets in scientific applications [15]. Unlike Value-List indexes where there is one bitmap for each distinct attribute value, the RBB approach partitions the attribute domain into a number of disjoint ranges and constructs one bitmap for each range of values (this is also known as *binning*). Thus, RBB provides a form of lossy compression which requires additional postprocessing to filter out false positives. Koudas [5] has examined space-optimal RBBs for equality queries when both the attribute and query distributions are known. More recently, Sinha and Winslett have proposed *multiresolution bitmap indexes* to avoid the cost of filtering out false positives for RBB [10]. For example, in a two-resolution bitmap index, it has a lower resolution index consisting of RBB (i.e., with each bitmap representing a range of attribute values) and a higher resolution index consisting of one bitmap for each distinct attribute value. By combining the efficiency of lower resolution indexes and the precision of higher resolution indexes, queries can be evaluated efficiently without false positive filtering.

Key Applications

Today, bitmap indexes are supported by all major database systems, and they are particularly suitable for data warehousing applications [7]. Bitmap indexes have also been used in scientific databases (e.g., [10, 11, 15]), indexing data on tertiary storage systems (e.g., [1]), and data mining applications.

Future Directions

The design space for bitmap indexes is characterized by four key parameters: levels of resolution (which affects the number of levels of bitmap indexes and the index granularity at each level), attribute value representation (which affects the number and size of the index components at each level) encoding scheme (which affects how the bitmaps in each component are encoded), and storage format (i.e., uncompressed, compressed, or a combination of compressed and uncompressed). While there are several performance studies that have examined various combinations of the above parameter space (e.g., [16]), a comprehensive investigation into the space-time tradeoffs of the entire design space is, however, still lacking and deserves to be further explored. The result of such a study can be applied to further enhance automated physical database tuning tools.

Cross-References

▶ Access Path
▶ Bitmap-Based Index Structures
▶ Data Warehouse

Recommended Reading

1. Amer-Yahia S, Johnson T. Optimizing queries on compressed bitmaps. In: Proceedings of the 26th International Conference on Very Large Data Bases; 2000. p. 329–38.
2. Chan CY, Ioannidis YE. Bitmap index design and evaluation. In: Proceedings of the ACM SIGMOD International Conference on Management of Data; 1998. p. 355–66.
3. Chan CY, Ioannidis YE. An efficient bitmap encoding scheme for selection queries. In: Proceedings of the ACM SIGMOD International Conference on Management of Data; 1999. p. 215–26.
4. Knuth DE. Retrieval on secondary keys. Chap. 6. In: The art of computer programming: sorting and searching, vol. 3. Reading: Addison-Wesley; 1973. p. 550–67.
5. Koudas N. Space efficient bitmap indexing. In: Proceedings of the International Conference on Information and Knowledge Management; 2000. p. 194–201.
6. O'Neil P, Graefe G. Multi-table joins through bitmapped join indices. ACM SIGMOD Record. 1995;24(3)8–11.
7. O'Neil P, Quass D. Improved query performance with variant indexes. In: Proceedings of the ACM SIGMOD International Conference on Management of Data; 1997. p. 38–49.
8. Reinfret D, O'Neil P, O'Neil E. Bit-sliced index arithmetic. In: Proceedings of the ACM SIGMOD International Conference on Management of Data; 2001. p. 47–57.
9. Sinha R, Winslett M. Multi-resolution bitmap indexes for scientific data. ACM Trans Database Syst. 2007;32(3):1–39.
10. Wong HKT, Liu H-F, Olken F, Rotem D, Wong L. Bit transposed files. In: Proceedings of the 11th International Conference on Very Large Data Bases; 1985. p. 448–57.
11. Wu MC. Query optimization for selections using nitmaps. In: Proceedings of the ACM SIGMOD International Conference on Management of Data; 1999. p. 227–38.
12. Wu MC, Buchmann AP. Encoded bitmap indexing for data warehouses. In: Proceedings of the 14th International Conference on Data Engineering; 1998. p. 220–30.
13. Wu KL, Yu PS. Range-based bitmap indexing for high cardinality attributes with skew. Technical report, IBM Watson Research Center. 1996.
14. Wu K, Otoo EJ, Shoshani A. Optimizing bitmap indices with efficient compression. ACM Trans Database Syst. 2006;31(1):1–38.
15. Wu K, Shoshani A, Stockinger K. Analyses of multi-level and multi-component compressed bitmap indexes. ACM Trans Database Syst. 2010;35(1): 1–52.

B

Bitmap-Based Index Structures

Guadalupe Canahuate and
Hakan Ferhatosmanoglu
The Ohio State University, Columbus, OH, USA

Synonyms

Bitmap Index; Projection Index

Definition

A bitmap-based index is a binary vector that represents an interesting property and indicates which objects in the dataset satisfy the given property. The vector has a 1 in position i if the i-th data object satisfies the property, and 0 otherwise. Queries are executed using fast bitwise logical operations supported by hardware over the binary vectors.

Historical Background

Bitmap-based indexing was first implemented in Computer Corporation of America's Model 204 in the mid-1980s by Dr. Patrick O'Neil. The bitmap index from Model 204 was a hybrid between verbatim (uncompressed) bitmaps and RID lists. Originally, a bitmap was created for each value in the attribute domain. The entire bitmap index is smaller than the original data as long as the number of distinct values is less than the number of bits used to represent the attribute in the original data. For example, if an integer attribute has cardinality 10 and integers are stored using 32 bits, then the bitmap index for such an attribute, which only has 10 bit vectors, is 3.2 times smaller than the original data. For floating point attributes, the bitsliced index (BSI) [8] stores each bit of the binary representation of the attribute independently. Bitsliced indexes are never more than the size of the original data. However, in general, all bitslices need to be accessed to answer a query.

In order to reduce the number of bit vectors that needed to be read to answer a query, more complex encodings for bitmap-based indexes have been proposed, such as range encoded bitmaps [15, 8], encoded bitmaps [16], and interval encoded bitmaps [4, 5]. For attributes with high cardinality including floating point attributes, binning [7, 13] was proposed to reduce the number of values in the domain, and therefore the number of bit vectors in the index. In addition, special compression techniques were developed to improve the performance of the bitmap indexes. The two most popular techniques are Byte-aligned Bitmap Code (BBC) [1], and Word Aligned Hybrid (WAH) [18] compression method. These compression techniques allow query execution over the compressed bitmaps. With compression, bitmap indexes have been proven to perform well with high cardinality attributes [17].

Foundations

Bitmap tables are a special type of binary matrices. Each binary row in the bitmap table represents one tuple in the database. The bitmap columns are produced by quantizing the attributes in the database into categories or bins. Each tuple in the database is then encoded based on which bin each attribute value falls into. Bitmap index encoding is based on the properties of physical row identifiers and there is a one-to-one correspondence between the data objects and the bits in the bitmap vector. Therefore, given a bit position, the location of its corresponding table row can be computed by simple arithmetic operations. For tables with a clustered-index or index-ordered tables, a mapping table is used to map bit positions to row locations.

Bitmap Encoding

For equality encoded bitmaps (also called simple encoding or projection index) [8], if a value falls into a bin, this bin is marked "1", otherwise "0". Since a value can only fall into a single bin, only a *single* "1" can exist for each row of each attribute. After binning, the whole database is converted into a large 0–1 bitmap, where rows correspond to tuples and columns correspond to bins. Figure 1 shows an example using a table with one attribute with cardinality 4. Columns 3–6 of Fig. 1 show the equality encoded bitmap for this table. The first tuple t_1 has value 1, therefore only the corresponding bit in the bitmap $=1$ is set. Columns 7 and 8, show the bitmaps for equality encoding with binning. In this example, the attribute was quantized into two bins.

For range encoded bitmaps [4], a bin is marked "1" if the value falls into it or a smaller bin, and "0" otherwise. Using this encoding, the last bin for each attribute is all 1s. Thus, this column is not explicitly stored. Columns 9–11 in Fig. 1 show the range encoding for the attribute. The first tuple t_1 has the smallest value 1, therefore all the bitmaps have the first bit set. For interval encoded bitmaps, every bitmap represents a range of $\lceil \frac{C}{2} \rceil$. values, where C is the cardinality of the attribute. This encoding allows to answer any range or point query on one attribute by reading at most two bit vectors [4, 5]. Columns 12–14 in Fig. 1 show the interval encoded bitmap. The fifth tuple t_5 has value 2, therefore the fifth bit is set for the [1, 2] and [2, 3] interval bitmaps. Bit-sliced indexes (BSI) [8] can be considered as a special case of the encoded bitmaps [16]. With the bit-sliced index the bitmaps encode the binary representation of the attribute value. Columns 15–17 of Fig. 1 show the bit-sliced index of the table. The third tuple t_3 has value 3 which is represented by the binary number 011, therefore the corresponding bit in slices 2^1 and 2^0 are set. Encoded bitmaps encode the binary representation of the attribute bins. Therefore, only $\lceil \log_2 bins \rceil$ bitmaps are needed to represent all values.

With binning, several values are encoded in the same bitmap. However, the results from the queries are supersets of the actual results, therefore additional disk access may be needed to evaluate the candidates and retrieve the exact results. Several binning strategies, based on data distribution and query workloads, and several strategies for candidate evaluation have been proposed [10, 11]. A binned bitmap index augmented with an auxiliary order-preserving bin-based Clustering (OrBiC) [19] can significantly reduce the I/O cost of candidate evaluation.

Query Execution

Bitmap indexes can provide efficient performance for point and range queries thanks to fast bitwise operations (AND, OR, NOT) which are supported by hardware. With equality encoded bitmaps a point query is executed by ANDing together the bit vectors corresponding to the values specified in the query conditions. For example, finding the data points that correspond to a query where Attribute 1 is equal to 3 and Attribute 2 is equal to 5 is only a matter of ANDing the two bitmaps together. Equality Encoded Bitmaps are optimal for point queries. Range queries are executed by first ORing together all the bit vectors specified by each range in the query conditions and then ANDing the answers together. If the query range for an attribute queried includes more than half of the cardinality then one executes the query by taking the complement of the ORed bitmaps that are not included in the query condition. The performance of queries with large-range query conditions can be improved by using multi-resolution bitmap indexes [12], i.e., bitmaps with different interval size: at the highest level one

Tuple	Value	Equality				Eq w/binning		Range			Interval			Bitsliced		
		= 1	= 2	= 3	= 4	{1,2}	{3,4}	≤3	≤2	≤1	[1,2]	[2,3]	[3,4]	2^2	2^1	2^0
t_1	1	1	0	0	0	1	0	1	1	1	1	0	0	0	0	1
t_2	2	0	1	0	0	1	0	1	1	0	1	1	0	0	1	0
t_3	3	0	0	1	0	0	1	1	0	0	0	1	1	0	1	1
t_4	1	1	0	0	0	1	0	1	1	1	1	0	0	0	0	1
t_5	2	0	1	0	0	1	0	0	1	1	1	1	0	0	1	0
t_6	4	0	0	0	1	0	1	0	0	0	0	0	1	1	0	0
t_7	1	1	0	0	0	1	0	1	1	1	1	0	0	0	0	1
t_8	3	0	0	1	0	0	1	1	0	0	0	1	1	0	1	1

Bitmap-Based Index Structures, Fig. 1 Bitmap index examples for a table with one attribute with cardinality 4

bit vector corresponds to an individual value and at lower levels each bit vector corresponds to a bin of values. Queries are executed using combined resolution bitmaps that minimize both the number of bitwise operations and the number of candidate points. With more complex encodings, the query evaluation strategy depends on the encoding and the range being queried.

Bitmap Compression

The goal of the bitmap index compression is twofold: to reduce the space requirement of the index and to maintain query execution performance by executing the query over the compressed index. The two most popular run-length compression techniques are the Byte-aligned Bitmap Code (BBC) [1] and the Word-Aligned Hybrid (WAH) compression method [18]. BBC stores the compressed data in bytes while WAH stores it in words. WAH is simpler because it only has two types of words, while BBC has four. Both techniques are based on the idea of run length encoding that represents consecutive bits of the same symbol (also called a fill or a gap) by their bit value and their length. The bit value of a fill is called the fill bit. BBC first divides the bit sequence into bytes and then group bytes into runs. A run consists of a fill word followed by a tail of literal bytes. A run always contains a number of whole bytes as it represents the fill length as number of bytes rather than number of bits. The byte alignment property limits a fill length to be an integer multiple of bytes. This ensures that during any bitwise logical operation a tail byte is never broken into individual bits. Similar to BBC, WAH is a hybrid between the run length encoding and the literal scheme. WAH stores compressed data in words rather than in bytes. The most significant bit of a word distinguishes between a literal word(0) and a fill word(1). Lower bits of a literal word contain the bit values from the bit sequence. The second most significant bit of a fill word is the fill bit and the lower (w-2) bits store the fill length. Imposing word alignment ensures that the logical operation functions only need to access words not bytes or bits. In general, bit operations over

the compressed WAH bitmap file are faster than BBC while BBC gives better compression ratio.

Typically, complex encoded bitmaps do not compress well as the bit density of such bitmaps is relatively high. Recently, reordering has been proposed as a preprocessing step for improving the compression of bitmaps. The objective with reordering is to increase the performance of run length encoding. By reordering the data, the compression ratio of large Boolean matrices can be improved [6]. However, optimal matrix reordering is NP-hard and the authors use traveling salesman heuristics to compute the new order. The idea of reordering is also applied to compression of bitmap indexes [9]. The authors show that the tuple reordering problem is NP-complete and propose a Gray code ordering heuristic.

Bitmap update is an expensive operation. The common practice is to drop the index, do a batch update, and recreate the index. For improved update performance, it is possible to add to each compressed bitmap a pad-word that encodes all possible rows using non-set bits [3]. For insertions, only the updated column needs to be accessed as opposed to all the bitmap columns. This technique is suitable for maintaining an online index when the update rate is not very high. An alternative to run-length compression proposed in the literature is the Approximate Bitmap Encoding [2], where hashing is used to encode the data objects in a bloom filter. This technique can offer improved performance for selection queries over small number of rows.

Key Applications

Bitmap indexing has been traditionally used in data warehouses to index large amount of data that are infrequently updated. More recently, bitmaps have found many other applications such as visualization and indexing of scientific data. Other works have applied bitmaps for term matching and similarity searches. In general, one could claim that bitmaps would outperform most approaches in the domains where full scan of the data are unavoidable and computations can be done using parallel bitwise operations.

Bitmap-based indexes are successfully implemented in commercial Database Management Systems. Oracle implements a version of the simple bitmap encoding compressed using BBC. Sybase IQ has two different types of bitwise indexes. IBM DB2 dynamically builds bitmaps from a single-column B-tree to join tables. Informix uses bitmapped indexes since version 8.21 released in 1998. PostgreSQL supports bitmap index scans since version 8.1 where dynamic bitmaps are created to combine the results of multiple index conditions using bitwise operations.

Cross-References

▶ Query Processing and Optimization in Object Relational Databases

Recommended Reading

1. Antoshenkov G. Byte-aligned bitmap compression. In: Data Compression Conference, Oracle Corp; 1995.
2. Apaydin T, Canahuate G, Ferhatosmanoglu H, Tosun A. Approximate encoding for direct access and query processing over compressed bitmaps. In: Proceedings of the 32nd International Conference on Very Large Data Bases; 2006.
3. Canahuate G, Gibas M, Ferhatosmanoglu H. Update conscious bitmap indices. In: Proceedings of the 19th International Conference on Scientific and Statistical Database Management; 2007.
4. Chan CY, Ioannidis YE. Bitmap index design and evaluation. In: Proceedings of the ACM SIGMOD International Conference on Management of Data; 1998.
5. Chan CY, Ioannidis YE. An efficient bitmap encoding scheme for selection queries. ACM SIGMOD Rec. 1999.
6. Johnson D, Krishnan S, Chhugani J, Kumar S, Venkatasubramanian S. Compressing large boolean matrices using reordering techniques. In: Proceedings of the 30th International Conference on Very Large Data Bases; 2004.
7. Koudas N. Space efficient bitmap indexing. In: Proceedings of the International Conference on Information and Knowledge Management; 2000.
8. O'Neil P, Quass D. Improved query performance with variant indexes. In: Proceedings of the ACM SIGMOD International Conference on Management of Data; 1997.
9. Pinar A, Tao T, Ferhatosmanoglu H. Compressing bitmap indices by data reorganization. In: Proceedings of the 21st International Conference on Data Engineering; 2005.
10. Rotem D, Stockinger K, Wu K. Optimizing candidate check costs for bitmap indices. In: Proceedings of the International on Information and Knowledge Management; 2005.
11. Rotem D, Stockinger K, Wu K. Minimizing I/O costs of multi-dimensional queries with bitmap indices. In: Proceedings of the 18th International Conference on Scientific and Statistical Database Management; 2006.
12. Sinha R, Winslett M. Multi-resolution bitmap indexes for scientific data. ACM Trans Database Syst. 2007;32(3):16.
13. Stockinger K. Design and implementation of bitmap indices for scientific data. In: Proceedings of the International Conference on Database Engineering and Applications; 2001.
14. Stockinger K, Wu K, Shoshani A. Evaluation strategies for bitmap indices with binning. In: Proceedings of the 15th International Conference on Database and Expert Systems Applications; 2004.
15. Wong HK, Liu H, Olken F, Rotem D, Wong L. Bit transposed files. In: Proceedings of the 11th International Conference on Very Large Data Bases; 1985.
16. Wu M-C, Buchmann A. Encoded bitmap indexing for data warehouses. In: Proceedings of the 14th International Conference on Data Engineering; 1998.
17. Wu K, Otoo EJ, Shoshani A. On the performance of bitmap indices for high cardinality attributes. In: Proceedings of the 30th International Conference on Very Large Data Bases; 2004.
18. Wu K, Otoo EJ, Shoshani A. Optimizing bitmap indexes with efficient compression. ACM Trans Database Syst. 2006;31:1–38.
19. Wu K, Stockinger K, Shoshani A. Breaking the curse of cardinality on bitmap indexes. In: Proceedings of the 20th International Conference on Scientific and Statistical Database Management; 2008.

Blind Signatures

Barbara Carminati
Department of Theoretical and Applied Science,
University of Insubria, Varese, Italy

Definition

The key feature of blind signature schemes is that the sender and the signer of the message are two distinguished entities. In particular, given a message m sent by a sender, blind signature has the nice property that the signer digitally signs a blinded version of m, i.e., m', without the disclosure of any information about the original message. The obtained blind signature can be then verified by using the public key of the signer and the original message m, instead of m'.

Key Points

In 1982, David Chaum introduced the concept of blind signatures for protecting user privacy during electronic payment transactions [1]. This scheme has been devised for scenarios where the sender and the signer of the message are two distinguished entities, with the aim of preventing the signer from observing the message he or she signs. More precisely, given a message m generated by A, a signer B is able to digitally sign a blinded version of m', i.e., $DS_B(m')$, without the disclosure of any information about the original message. The key property of blind signatures is that the obtained signature $DS_B(m')$ can be verified by using the public key of B and the original message m, instead of m'. This property makes blind signatures particularly suitable for applications where sender privacy is the main concern, like for instance, in digital cash protocols and electronic voting systems.

Several digital signature schemes can be used to obtain a blind signature, like for instance RSA and DSA schemes. To explain the generation and verification of a blind signature, in the following the RSA algorithm is considered. Let $PK_B = (n, e)$ and $SK_B = (n, d)$ be the public and private keys of signer B.

Blinding. Let k be a random integer chosen by the sender A, such that $0 \leq k \leq n - 1$ and $gcd(n, k) = 1$. Thus, given a message m, the sender generates the blinded version of it, i.e., m', by combing m with k. According to RSA, the blinded version of m is computed as $m' = (mk^e) \bmod n$.

Blind signature generation. The signer digitally signs the blinded version m' using the RSA digital signature algorithm; that is, $DS_B(m') = (m')^d \bmod n$.

Signature unblinding. The intended verifier can obtain the signature of m, i.e., $DS_B(m)$, as follows $DS_B(m) = k^{-1} DS_B(m')$. This signature can be validated by the RSA verification algorithm, according to the standard process.

Cross-References

▶ Digital Signatures
▶ Privacy-Enhancing Technologies

Recommended Reading

1. Chaum D. Blind signatures for untraceable payments. In: Advances in Cryptology Proceedings of Crypto 82; 1983. p. 199–203.

Bloom Filters

Michael Mitzenmacher
Harvard University, Boston, MA, USA

Synonyms

Hash filter

Definition

A Bloom filter is a simple, space-efficient randomized data structure based on hashing that represents a set in a way that allows membership queries to determine whether an element is a member of the set. False positives are possible, but not false negatives. In many applications, the space savings afforded by Bloom filters outweigh the drawbacks of a small probability for a false positive. Various extensions of Bloom filters can be used to handle alternative settings, such as when elements can be inserted and deleted from the set, and more complex queries, such as when each element has an associated function value that should be returned.

Historical Background

Burton Bloom introduced what is now called a Bloom filter in his 1970 paper [1], where he described the technique as an extension of hash-coding methods for applications where error-free methods require too much space and were not strictly necessary. The specific application he considered involved hyphenation: a subset of words from a standard dictionary require specialized hyphenation, while the rest could be handled by a few simple and standard rules. Keeping a Bloom filter of the words requiring specialized hyphenation dramatically cut down disk accesses. Here, false positives caused words that could be handled by the simple rules to be treated as special cases. Bloom filters were also used in early UNIX spell-checkers [2, 3], where a false positive would allow a misspelled word to be ignored, and were suggested as a way of succinctly storing a dictionary of insecure passwords, where a false positive would disallow a potentially secure password [4].

Bloom filters were applied in databases to reduce the amount of communication and computation for join operations, especially distributed join operations [5, 6, 7]. The term Bloomjoin is sometimes used to describe a semijoin operation that utilizes Bloom filters. The Bloom filter is used to represent join column values, so that matching values can be found by querying the Bloom filter. False positives cause false matches of tuples that must later be removed. However, the communication and computation gains from using the filter can yield significant advantages even with these false positives.

Foundations

A Bloom filter for representing a set $S = \{x_1, x_2, \ldots, x_n\}$ of n elements from a universe U consists of an array of m bits, initially all set to 0. The filter uses k independent hash functions $h_1, \ldots, h_k$ with range $\{1, \ldots, m\}$. For each element $x \in S$, the bits $h_i(x)$ are set to 1 for $1 \le i \le k$. (A location can be set to 1 multiple times.) To check if an item y is in S, one checks whether all $h_i(y)$ are set to 1. If not, then clearly y is not a member of S. If all $h_i(y)$ are set to 1, the data structure returns that y is in S. There is no possibility of a false negative, but it is possible that for $y \notin S$, all $h_i(y)$ are set to 1, in which case the data structure gives a false positive.

The fundamental issue in the analysis of the Bloom filter is the *false positive probability* for an element not in the set. In the analysis, generally it is assumed that the hash functions map each element in the universe to a random number independently and uniformly over the range. While this is clearly an optimistic assumption, it appears to be suitable for practical implementations. With this assumption, after all the elements of S are hashed into the Bloom filter, the probability that a specific bit is still 0 is

$$p' = (1 - 1/m)^{kn} \approx e^{-kn/m}.$$

It is convenient to use the approximation $p = e^{-kn/m}$ in place of p'. If ρ is the proportion of 0 bits after all the n elements are inserted in the table, then conditioned on ρ the probability f of a false positive is

$$f = (1 - \rho)^k \approx (1 - p')^k \approx (1 - p)^k$$
$$= \left(1 - e^{-kn/m}\right)^k.$$

These approximations follow since $E[\rho] = p'$, and ρ can be shown to be highly concentrated around p' using standard techniques [8].

The optimal number of hash function can be found by finding where $(1 - e^{-kn/m})^k$ is minimized as a function of k. Simple calculus reveals the optimum occurs when $k = \ln2 \cdot (m/n)$, giving a false positive probability f of

$$f = \left(1 - e^{-kn/m}\right)^k = (1/2)^k \approx (0.6185)^{m/n}.$$

In practice, k must be an integer, and the choice of k might depend on application-specific questions.

Key Applications

Essentially any application that requires membership checks against a list or set of objects, and for which space is at a premium, is a candidate for a Bloom filter. In some applications, Bloom filters may also save computational resources. The consequences of false positives, however, need to be carefully considered in all circumstances.

Bloom filters enjoyed a recent resurgence in the networking community after their use in a paper by Fan et al. on distributed Web caches [9]. Instead of having caches distribute lists of URLs (or lists of their 16-byte MD5 hashes) corresponding to the cache contents, the authors demonstrate a system that saves in communication costs by sharing Bloom filters of URLs. False positives may cause a server to request a page from another nearby server, even when that server does not hold the page. In this case, the page must subsequently be retrieved from the Web. If the false positive probability is sufficiently low, the penalty from these occasional failures is dominated by the improvement in overall network traffic. Countless further applications in databases, peer-to-peer networks, overlay networks, and router architectures have arisen, and the use of Bloom filters and their many variants and extensions has expanded dramatically [10].

In databases, applications include the previously mentioned Bloomjoin variation of the semijoin. In a similar manner, Bloom filters can be effectively used to estimate the size of semijoin operations, using the fact that Bloom filters can be used to estimate the size of a set (and the size of set unions and intersections) [11]. Another early database application utilizes differential files [12]. In this setting, all changes to a database that occur during a given time period are processed as a batch job, and a differential file tracks changes that occur until the batched update occurs. To read a record then requires determining if the record has been changed by some transaction in the differential file. If not, the record can be read directly from the database, which is generally much quicker and more efficient than performing the necessary processing on the differential file. Instead of keeping a list of all records that have changed since the last update, a Bloom filter of the records that have been changed can be kept. Here, a false positive would force the database to check both the differential file and the database on a read even when a record has not been changed. If false positives are sufficiently rare, this cost is minimal.

Future Directions

Recent research related to Bloom filters has followed three major directions: alternative constructions, improved analysis, and extensions to more challenging questions. All three of these directions are likely to continue to grow.

Research in alternative constructions has focused on building data structures with the same capabilities as a Bloom filter or related data structures with improved efficiency, sometimes particularly for a specialized domain, such as router hardware. Improved analysis has considered reducing for example the amount of randomness required for a Bloom filter, or its performance under specific classes of hash functions.

The largest area of research has come from extending the basic idea of the Bloom filter to more general and more difficult problems. Arguably, the success of Bloom filters relies on their flexibility, and numerous variations on the theme have arisen. For example, counting Bloom filters extend Bloom filters by allowing sets to change dynamically via insertions and deletions of elements [9, 13]. Spectral Bloom filters extend Bloom filters to handle multi-sets [14]. Count-min sketches extend Bloom filters to approximately track counts associated to items in a data stream, such as byte-counts for network flows [8]. Bloomier filters extend Bloom filters to the situation where each element of S is associated with a function value from a discrete, finite set of values, and the data structure should return that value [15]. Approximate concurrent state machines further extend Bloom filter to the setting where both the set can change due to insertions and deletions and the function values associated with set elements can change dynamically [16]. The list of variations continues to grow steadily as more applications are found.

URL to Code

- en.wikipedia.org/wiki/Bloom_filter
- www.patrow.net/programming/hashfunctions/index.html
- search.cpan.org/~mceglows/Bloom-Filter-1.0/Filter.pm

Cross-References

▶ Join
▶ Semijoin

Recommended Reading

1. Bloom B. Space/time tradeoffs in hash coding with allowable errors. Commun ACM. 1970;13(7):422–6.
2. McIlroy MD. Development of a spelling list. IEEE Trans Commun. 1982;30(1):91–9.
3. Mullin JK, Margoliash DJ. A tale of three spelling checkers. Software Pract Exp. 1990;20(6):625–30.
4. Spafford EH. Opus: preventing weak password choices. Comp Sec. 1992;11(3):273–8.
5. Babb E. Implementing a relational database by means of specialized hardware. ACM Trans Database Syst. 1979;4(1):1–29.
6. Bratbergsengen K. Hashing methods and relational algebra operations. In: Proceedings of the 10th International Conference on Very Large Data Bases; 1984. p. 323–33.
7. Mackett LF, Lohman GM. R* optimizer validation and performance evaluation for distributed queries. In: Proceedings 27th International Conference on Very Large Data Bases; 1986. p. 149–59.
8. Cormode G, Muthukrishnan S. An improved data stream summary: the count-min sketch and its applications. J Algorithms. 2003;55(1):58–75.
9. Fan L, Cao P, Almeida J, Broder AZ. Summary cache: a scalable wide-area Web cache sharing protocol. IEEE/ACM Trans Network. 2000;8(3):281–93.
10. Broder A, Mitzenmacher M. Network applications of Bloom filters: a survey. Internet Math. 2005;1(4):485–509.
11. Mullin JK. Estimating the size of a relational join. Inf Syst. 1993;18(3):189–96.
12. Gremilion LL. Designing a Bloom filter for differential file access. Commun ACM. 1982;25(9):600–4.
13. Mitzenmacher M. Compressed Bloom filters. IEEE/ACM Trans Network. 2002;10(5):604–12.
14. Cohen S, Matias Y. Spectral Bloom filters. In: Proceedings of the ACM SIGMOD International Conference on Management of Data; 2003. p. 241–52.
15. Chazelle B, Kilian J, Rubinfeld R, Tal A. The Bloomier filter: an efficient data structure for static support lookup tables. In: Proceedings of the 15th Annual ACM-SIAM Symposium on Discrete Algorithms; 2004. p. 30–9.
16. Bonomi F, Mitzenmacher M, Panigrahy R, Singh S, Varghese G. Beyond Bloom filters: from approximate membership checks to approximate state machines. Comput Commun Rev. 2006;36(4):315–26.

BM25

Giambattista Amati
Fondazione Ugo Bordoni, Rome, Italy

Synonyms

OKAPI retrieval function; Probabilistic model

Definition

BM25 is a ranking function that ranks a set of documents based on the query terms appearing in each document, regardless of the interrelationship between the query terms within a document (e.g., their relative proximity). It is not a single function, but actually a whole family of scoring functions, with slightly different components and parameters. It is used by search engines to rank matching documents according to their relevance to a given search query and is often referred to as "Okapi *BM25*," since the Okapi information retrieval system was the first system implementing this function. The BM25 retrieval formula belongs to the *BM* family of retrieval models (*BM* stands for Best Match) that is the weight of a term *t* in a document *d* is

$$\frac{tf}{k+tf} \ln \frac{(r_t + 0.5) \cdot (N - R - n_t + r_t + 0.5)}{(n_t - r_t + 0.5) \cdot (R - r_t + 0.5)}$$

$$\left[BM's \ \text{family} \right]$$

where

- R is the number of documents known to be relevant to a specific topic
- r_t is the number of relevant documents containing the term
- N is the number of documents of the collection
- n_t is the document frequency of the term
- tf is the frequency of the term in the document
- k is a parameter

The BM25 document-query matching function is

$$\sum_{t \in q} w_t \cdot \ln \frac{(r_t + 0.5) \cdot (N - R - n_t + r_t + 0.5)}{(n_t - r_t + 0.5) \cdot (R - r_t + 0.5)}$$

$$[BM25] \qquad (1)$$

where

- q is the query,
- $w_t = (k_1 + 1) \frac{tf}{k+tf} \cdot (k_3 + 1) \frac{tf_q}{(k_3 + tf_q)}$
- tf_q is the frequency of the term within the topic from which q was derived
- l and avg$_l$ are, respectively, the document length and average document length.
- k is $k_1 \left((1-b) + b \left(\frac{l}{\text{avg}_l} \right) \right)$
- k_1, b, and k_3 are parameters which depend on the nature of the queries and possibly on the database
- k_1 and b are set by default to 1.2 and 0.75, respectively, k_3 to 1000

By using these default parameters, the unexpanded BM25 ranking function, that is the BM25 applied in the absence of information about relevance ($R = r = 0$), is

$$\sum_{t \in q} \frac{2.2 \cdot tf}{0.3 + 0.9 \frac{l}{\text{avg}_l} + tf} \cdot \frac{1001 \cdot tf_q}{1000 + tf_q} \ln$$

$$\frac{N - n_t + 0.5}{n_t + 0.5} \qquad (2)$$

Historical Background

The successful formula of the BM family, the BM25, was introduced by Robertson and Walker in 1994 [1, 3] and it has its root in Harter's 2-Poisson model of eliteness for indexing. Before BM25, a direct exploitation of eliteness for document retrieval was explored by Robertson, Van Rijsbergen, Porter, Williams, and Walker [2, 3] who plugged the Harter 2-Poisson model into the standard probabilistic model of relevance of Robertson and Spark Jones [4]. The evolution of the 2-Poisson model as designed by Robertson, Van Rijsbergen, and Porter has thus motivated the birth of the BM family of term-weighting forms. The BM25 formula is the matching function of the Okapi information retrieval system of City University in London.

Foundations

BM25 derives from both the 2-Poisson model and the probabilistic binary independence model of relevance that is as a combination of the probabilistic model

$$\text{Prob}\,(rel, d|q) \propto \prod_{t \in q \cap d} \frac{\text{Prob}\,(t|rel) \cdot \text{Prob}\,(\bar{t}|\overline{rel})}{\text{Prob}\,(t|\overline{rel}) \cdot \text{Prob}\,(\bar{t}|\overline{rel})}$$

$$(3)$$

and the 2-Poisson model

$$\text{Prob}\,(X = tf) = p \cdot \frac{e^{-\lambda_{E_t}} \lambda_{E_t}{}^{tf}}{tf!} + (1-p) \cdot \frac{e^{-\lambda_{\overline{E_t}}} \lambda_{\overline{E_t}}{}^{tf}}{tf!}$$

$$(4)$$

where p is the prior probability of a document to belong to the elite set of the term.

However, the 2-Poisson has three parameters: the mean term frequency of the term in the elite set $\lambda_{\overline{E_t}}$ (a set of documents with a large number of occurrences of the term), the mean term frequency of the term in the rest of the collection, $\lambda_{\overline{E_t}}$, and the mixing parameter p. Therefore, the 2-Poisson model needs reasonable approximations in order to make the probabilistic mixture a workable retrieval model.

The combination of the notion of eliteness with that of relevance generates the Robertson, Van Rijsbergen, and Porter's query term-document matching function:

$$w = \ln \frac{\left(p_1 \lambda_{E_t}{}^{tf} e^{-\lambda_{E_t}} + (1 - p_1) \lambda_{\overline{E_t}}{}^{tf} e^{-\lambda_{\overline{E_t}}} \right)}{\left(p_2 \lambda_{E_t}{}^{tf} e^{-\lambda_{E_t}} + (1 - p_2) \lambda_{\overline{E_t}}{}^{tf} e^{-\lambda_{\overline{E_t}}} \right)}$$

$$\frac{\left(p_2 e^{-\lambda_{E_t}} + (1 - p_2) e^{-\lambda_{\overline{E_t}}} \right)}{\left(p_1 e^{-\lambda_{E_t}} + (1 - p_1) e^{-\lambda_{\overline{E_t}}} \right)} \tag{5}$$

where p_1 and p_2 are the conditional probabilities of a document to be or not in the elite set, respectively, given the set of relevant documents. Since elite set (documents with high query term frequency) and relevance (documents relevant to the query) are highly correlated, then it can be assumed that $p_1 > p_2$.

With little algebra, Eq. 5 is equivalent to

$$w = \ln \frac{\left(p_1 + (1 - p_1) \left(\frac{\lambda_{\overline{E_t}}}{\lambda_{E_t}} \right)^{tf} e^{\lambda_{E_t} - \lambda_{\overline{E_t}}} \right)}{\left(p_2 + (1 - p_2) \left(\frac{\lambda_{\overline{E_t}}}{\lambda_{E_t}} \right)^{tf} e^{\lambda_{E_t} - \lambda_{\overline{E_t}}} \right)}$$

$$\frac{\left(p_2 e^{-\lambda_{E_t} + \lambda_{\overline{E_t}}} + (1 - p_2) \right)}{\left(p_1 e^{-\lambda_{E_t} + \lambda_{\overline{E_t}}} + (1 - p_1) \right)} \tag{6}$$

Equation 6 can be rewritten as

$$w\,(\mathrm{tf}) = \ln C\,(\mathrm{tf}) + \ln C_0$$

where C_0 is the ratio of the two components of the cross-product ratio not containing the variable tf. The first derivative with respect to the variable tf is

$$\frac{dw}{d\,\mathrm{tf}} = \frac{(p_1 - p_2) \cdot e^{\lambda_{E_t} - \lambda_{\overline{E_t}}} \cdot \left(\frac{\lambda_{\overline{E_t}}}{\lambda_{E_t}} \right)^{tf} \cdot \ln \left(\frac{\lambda_{E_t}}{\lambda_{\overline{E_t}}} \right)}{C\,(\mathrm{tf})}$$

The derivative is always positive because $p_1 > p_2$, $\lambda_{\overline{E_t}} \ll \lambda_{E_t}$ and thus $\ln \left(\frac{\lambda_{E_t}}{\lambda_{\overline{E_t}}} \right) > 0$ in the 2-Poisson model. Therefore $w(tf)$ is a monotonically increasing function. The limiting form of Eq. 6 for $tf \to \infty$ is

$$w = \ln \frac{p_1 \left(p_2 e^{-\lambda_{E_t} + \lambda_{\overline{E_t}}} + (1 - p_2) \right)}{p_2 \left(p_1 e^{-\lambda_{E_t} + \lambda_{\overline{E_t}}} + (1 - p_1) \right)}$$

Since $e^{-\lambda_{E_t} + \lambda_{\overline{E_t}}} \sim 0$, this limit is very close to

$$w \sim \ln \frac{p_1 (1 - p_2)}{p_2 (1 - p_1)} \tag{7}$$

Because Eq. 6 is monotonic with respect to the within-document term frequency tf, document ranking is obtained by decreasing order of the tf value. Hence because for the highest values of the term frequency tf (i.e., for $tf \to \infty$), the limiting form of Eq. 7 can be taken as the actual score of the topmost documents.

In 1994, Robertson and Walker defined as an approximation of w of Eq. 7 the product

$$w = \frac{tf}{tf + K} \cdot \ln \frac{\mathrm{Prob}\,(t|rel)\,\mathrm{Prob}\,\left(\bar{t}|\overline{rel} \right)}{\mathrm{Prob}\,(t|\overline{rec})\,\mathrm{Prob}\,\left(\bar{t}|rel \right)} \tag{8}$$

Indeed, both Eqs. 6 and 8 have Formula 7 as their limit for large tf. Varying the parameter K and using equation

$$\frac{\mathrm{Prob}\,(t|rel)\,\mathrm{Prob}\,\left(\bar{t}|\overline{rel} \right)}{\mathrm{Prob}\,\left(t|\overline{rel} \right)\,\mathrm{Prob}\,\left(\bar{t}|rel \right)} =$$

$$\frac{(r_t + 0.5) \cdot (N - R - n_t + r_t + 0.5)}{(n_t - r_t + 0.5) \cdot (R - r_t + 0.5)} \tag{9}$$

the *BM* weighting formulas are derived.

Cross-References

▶ Divergence-from-Randomness Models
▶ Information Retrieval

▶ Probabilistic Retrieval Models and Binary Independence Retrieval (BIR) Model
▶ Relevance
▶ Term Weighting
▶ TF*IDF
▶ Two-Poisson Model

Recommended Reading

1. Robertson SE, Walker S, Beaulieu MM, Gatford M, Payne A. Okapi at trec-4. In: Harman DK, editor. NIST special publication 500-236. Proceedings of the 4th Text Retrieval Conference; 1996.
2. Robertson SE, Walker S. Some simple approximations to the 2-poisson model for probabilistic weighted retrieval. In: Proceedings of the 17th Annual International ACM SIGIR Conference on Research and Development in Information Retrieval; 1994. p. 232–41.
3. Robertson SE, Van Rijsbergen CJ, Porter M. Probabilistic models of indexing and searching, Chapter 4. In: Robertson SE, Van Rijsbergen CJ, Williams PW, editors. Information retrieval research. London: Butterworths; 1981. p. 35–56.
4. Robertson SE, Sparck-Jones K. Relevance weighting of search terms. J Am Soc Inf Sci. 1976;27(3):129–46.

Boolean Model

Massimo Melucci
University of Padua, Padua, Italy

Definition

In the Boolean model for Information Retrieval (**IR**), a document collection is a set of documents, and an index term is the subset of documents indexed by the term itself. An index term can also be seen as a proposition which asserts whether the term is a property of a document, that is, if the term occurs in the document or, in other words, if the document is about the concept represented by the term.

The interpretation of a query is set-theoretical. In practice, a query is a Boolean expression where the set operators are the usual intersec-

tion, union, and complement, and the operands are index terms. The document subsets which correspond to the index terms of the query are combined through the set operators. The system returns the documents which belong to the subset expressed by the query and make the query true.

Historical Background

The Boolean model for **IR** was proposed as a paradigm for accessing large-scale systems since the 1950s. The idea of composing queries as Boolean propositions was at that time considered advantageous for the end user and simple to implement, so that its use rapidly spread in the 1960s. The main reason for the rapid advent of the Boolean model was the presence of experienced end users who also were expert of the domain of the searched documents. These users were expected to be able to quite effectively express their own information needs by composing index terms and operators as propositions. More recently, some user studies suggested that even experienced users prefer to employ the less Boolean operators as possible and to submit quite simple queries. It is at later step that a user may refine the query by adding operators to retrieve more relevant documents or to exclude nonrelevant documents.

A possible reason for the success of the Boolean model was the controlled, coordinated, and very often manual indexing of the collections. Term that are manually assigned to documents are often quite homogeneous terms, namely, referred to a restricted domain and without any ambiguity (e.g., synonymy or polysemy). In this way, manual indexing carefully avoids ambiguous terms and correctly associated related terms. Since then, the Boolean model was adopted by various other information management systems, e.g., office information systems, library catalogue systems, and database systems.

Currently, the Boolean model is quite ubiquitous since the advanced search functions constitute a constant of almost every information management systems, World Wide Web (WWW) search engines included. However, it is a matter of fact that the success of this model decreased over time because of the decreasing proportion of experienced end users, the increasing heterogeneity of the document collections, and the lack of a support for ranking documents. The most glaring example of type of information retrieval system which is less based on the Boolean model than in the past is the search engine for the World Wide Web (**WWW**). The indexes of a search engine for the **WWW** are fueled with more and more heterogenous **WWW** pages and is increasingly and mostly accessed by unexperienced end users. Various research works showed that the percentage of users who use advanced search functions is very limited, even when they are quite expert of the domain or in using the search engine graphical interface.

Foundations

In the Boolean model for information retrieval (IR), a document collection is a set of documents

$$D = \{d_1, \ldots, d_N\}$$

where each document is considered as an element without specific properties other than the membership to D.

An index term t is the subset of documents

$$T \subseteq D$$

such that each $d \in T$ is a document indexed by T. An index term can also be seen as a proposition which asserts whether the term is a property of a document, that is, if the term occurs in the document or, in other words, if the document is about the concept represented by the term. It follows that a proposition

$$d \in T$$

can be either true or false depending on whether d is indexed by T or not, respectively.

The interpretation of a query is set-theoretical. A query is a Boolean expression in conjunctive normal form (CNF)

$$Q = (T_{1,1} \wedge \cdots \wedge T_{1,n_1}) \vee \cdots \vee (T_{k,1} \wedge \cdots \wedge T_{k,n_k})$$

A query in Conjunctive Normal Form (**CNF**) is a conjunction of k disjunctions where the j-th disjunction is among n_j terms (i.e., document subsets). Each query can be converted into an equivalent **CNF** using the Boolean logic laws such as the distributive law.

The artificial query languages that implement the Boolean logic operators $\wedge, \vee, \neg$ utilize the reserved keywords AND, OR, NOT, respectively.

The **IR** system that is based on the Boolean model returns the documents which belong to the subset expressed by the query, that is, it returns the documents which make the query Q true.

Suppose, for example, some authors are writing their own documents, say, d_1, d_2, d_3 using, say, three terms, namely, A, B, C. Suppose, for example, that the collection stores the documents d_1, d_2, d_3, which, respectively, contains the following index terms: $\{A\}, \{A, B\}, \{B, C\}$. According to the Boolean model, three subsets $A = \{d_1, d_2\}$, $B = \{d_2, d_3\}$, $C = \{d_3\}$ are defined after indexing the collection. The query

$$Q = A \wedge B$$

is true when document d_2 is returned; the documents d_1, d_3 are not returned by an **IR** system that implements the Boolean model because they do not make Q true.

The terms *unambiguously* describe one concept each – the absence of ambiguity of the natural language is an important assumption so as to make this model effective. As a document may be about one or two concepts, a document may belong to more subsets.

The example shows that a concept is addressed by one or more documents and a term is associated to a single document subset. These document subsets are the extensional expressions of the concept described by an index term. In other words, the enumeration of the document subset is the description of the concept labeled by the index term. Indeed, the Boolean model was thought as a means for describing concepts without recurring to ontological description, but to a simple, yet powerful logical language.

The index terms and the associated document subsets are decided at indexing time by a human indexer or an automated process driven by an indexing algorithm. If the set operators were not available, the end user could use one index term at a time for formulating his own query thus having a very few degrees of freedom for formulating his own information need. Thanks to the set-based view imposed by the Boolean model, the end user can utilize the Boolean algebra for constructing new subsets. These new subsets are what results from Boolean expressions based on the classical operators, namely, intersection, union, and complement. Suppose, for example, A, B, C are the document subsets associated to three index terms; let $A = \{d_1, d_2\}$, $B = \{d_2, d_3\}$, $C = \{d_3\}$ and $A \wedge \neg C$ be a new subset. This subset only contains d_2 thus expressing the fact that d_2 is about $A \wedge \neg C$.

What is here important is that the construction of document subsets through the Boolean algebra allows the end user for expressing new concepts which were not explicitly thought by the authors – indeed, the author of d_2 did not think that d_2 was about $A \wedge \neg C$. With this respect, the Boolean model is a powerful language because it would permit to represent the knowledge stored in a document collection by means of an algebra. A system based on the Boolean model can efficiently answer these queries by performing some simple algorithms which implement set operations and process sets.

Despite its strengths and simplicity, the Boolean model has some weaknesses. Understanding the main weaknesses of the Boolean model is crucial for making the application of this model to realistic contexts effective. The weaknesses of the Boolean model can be summarized as follows:

- Set operator confusion
- Expressiveness gap
- Index term ambiguity
- Null output
- Output overload

Null output and output overload are due to the lack of support for ranking documents.

Set operator confusion occurs whenever intersection and union, namely, AND and OR, are exchanged when composing a query. Because of the imprecision of the natural language, humans often use "or" for saying "and" and vice versa. This confusion causes a similar exchange when using the artificial Boolean language. Suppose, for example, the collection is $\{d_4, d_5\}$, d_4 is about A and d_5 is about B. Using the natural language, end user's information need might be "the documents about A *and* B" to mean the whole document collection. To obtain the whole collection as a result, the user should use $A \vee B$. However, if the user translated the "and" of his own natural language request into an AND operator, the translation of the request into an artificial language would be $A \wedge B$, which returns the empty set. Things are as more difficult as the query is more complex.

Expressiveness gap is the loss of expressiveness encountered whenever an information need has to be translated to an artificial language like the Boolean language from a more expressive natural language. Suppose, for example, the end user wants to retrieve documents about the Boolean language in **IR** or database systems. The use of the Boolean query language requires that each word or group of words is corresponded to an index term, namely, to a document subsets, and that some set operators are applied to these sets; for example, a query might be (boolean AND information retrieval) OR database systems. The problem of the expressiveness gap is amplified by the multiplicity of queries which can be formulated as a representation of the same information need. As a consequence, a variety of document subsets not all being the same can be retrieved. Suppose, for example, the end user wants to retrieve documents about the Boolean language in **IR** or database systems. Potential Boolean queries are, for example:

- (boolean AND information retrieval) OR database systems
- boolean AND (information retrieval OR database systems)

The expressiveness gap occurs because the request expressed in natural language does not indicate if and where the parentheses are located, whereas the location of the parentheses is crucial when using the Boolean algebra. Although both are valid expressions, they provide different results.

Index term ambiguity occurs whenever a term has two distinct meanings (polysemy) or two terms have the same meaning (synonymy). When a term is polysemous, documents about two distinct concepts co-occur in the same subset. As a consequence, irrelevant documents may be retrieved when the end user employs the term in his own query. Suppose, for example, bank $= \{d_6, d_7\}$, but d_6 is about bank as river bank and d_7 is about bank as bank branch. When the user expresses his own Boolean query using bank as term, both documents are retrieved; if, however, the user needs information about river banks, one irrelevant document, i.e., d_7, is retrieved. In this event, precision is lower than the precision measured if ambiguity does not occur.

When two terms are synonyms, two distinct document subsets are defined, yet the documents are about the same concept. As a consequence, relevant documents may be missed when the end user employs only one of the two terms in his own query. Suppose, for example, personal computer $= \{d_8, d_9\}$ and PC $= \{d_{10}\}$, but PC is the acronym of personal computer and therefore d_{10} should be retrieved, but it is not. When the user expresses his own Boolean query using PC as term, only d_{10} is retrieved; if, however, the use needs information about personal computers, two relevant documents, i.e., d_8, d_9, are missed. In this event, recall decreases.

Null output occurs whenever the end user submits a very restrictive query to the system to an extent to make the returned document subset very small, if not even empty. Limit examples are terms being absent from the index, which are associated to empty document subsets or two index terms whose intersection is empty because there are no common documents. The event of intersecting disjoint document subsets is very probable since the document subsets are usually small if compared with the document collection. Moreover, Boolean systems do not usually keep track of the semantic relationships between terms, and therefore connecting by AND two semantically related terms whose document subsets are disjoint would give the empty set as result. To reduce null output, some query terms can be related by OR or some AND should be removed.

Output overload occurs whenever the result document subset is too large for being effectively browsed by the end user. This drawback often happens when too many ORs or too few ANDs are utilized for formulating the query. Even though a document subset is small if compared with the document collection, a subset of, say, 1,000 documents is very large for the end user who is required to inspect all of them. A strategy for avoiding output overload is reducing the number of OR or adding some AND.

Overall, the side effects of null output and output overload are hardly controlled by the end user who is requested to add and delete terms or Boolean operators so as to make the query an effective description of his own information need. This task, which is called query expansion, would overload the cognitive effort of the end user, if performed using a Boolean systems.

To overcome the problems of null output and output overload, the notion of level of coordination was introduced. The level of coordination is a measure of the degree to which each returned document matches the query. In this way, the level of coordination provides a score for ranking the documents. This ranking allows the user for deciding how many documents to inspect and the system for cutting the bottom-ranked documents off the list.

The level of coordination is calculated as follows. A Boolean query is transcribed in conjunctive normal form, namely, as a list of propositions related by AND where each proposition is a disjunction of elementary propositions. The level of coordination of a document is the number of conjoined propositions satisfied by the document. Suppose, for example, A, B, C are the document subsets associated to three index terms; let $A = \{d_1, d_2\}$, $B = \{d_2, d_3\}$, $C = \{d_3\}$

and A AND $(C$ OR $B)$ be the query; the level of coordination of d_1 is 1 because only the first proposition, i.e., A is satisfied by d_1, the level of d_2 is 2 because d_2 belongs to both A and C OR B, and the level of d_3 is $1 - d_1, d_3$ would have not been retrieved if the level of coordination were not been calculated because they do not satisfy the query.

A variation of the level of coordination was introduced for taking the variable size of document subsets into account – indeed, the document subsets are of arbitrary size and therefore a small subset may be treated as a large subset. This variation has been called weighted level of coordination: instead of assigning a constant weight to each proposition made true by a document, a different weighted is assigned depending on the proposition; for example, an Inverse Document Frequency (**IDF**) may be used. The weighed level of coordination is then the sum of the weights assigned to every proposition. In practice, each conjoined proposition is assigned a weight function which increases when the concept's importance increases; for example, an **IDF** may be used since rare concepts are likely to be important. Each document is assigned the weight function values of the concepts occurring in the document. The weight function depends on the disjoined concepts that are conjoined together. The weight function value of a disjunction may be, for example, the average, the maximum, or other functions of the weight function values of the disjoined concepts.

The level of coordination was a simple yet effective way to dealing with the situations in which a document does not make true a **CNF** query; it assigns a weight to each disjunction and sums the weights according the truth value of a disjunction with respect a document. Other approaches to extending the Boolean model to deal with uncertainty of inference and with lack of implication were implemented by a probabilistic datalog system connecting logic and probability (see [7]) and by a common theoretical framework encompassing logic, probability, and vector spaces (see [19] and [16]).

At query processing, an **IR** system leverages the **CNF** of a query. The disjunctions are pro-

cessed to build k posting lists. For each disjunction $T_{j,1} \vee \cdots \vee T_{j,n_j}$, the n_j posting lists that correspond to $T_{j,1}, \ldots, T_{j,n_j}$ are retrieved and merged. The k posting lists that result from merging are then processed and are conjoined together by a series of k conjunction operators (i.e., $\wedge$ or AND). **CNF** query processing is efficient because (i) a document must occur in every posting list built after merging the posting lists of the disjunctions and (ii) the **IR** system processes the k posting lists in ascending order of size, that is, from the least frequent term to the most frequent term. In this way, the system can skip over the posting list of frequent terms to find the documents that also contain the infrequent terms.

Key Applications

The research papers and articles on the Boolean model flourished until when other models, e.g., the probabilistic models and the vector space model, appeared on the scene of the **IR** theatre, and the **WWW** search engines drastically changed the audience and the document collections. The knowledge of this model is however a basic element of an **IR** system because the Boolean query language is provided by many if not all information management systems. As a matter of fact, **WWW** search engines still provide Boolean operators together with other operators such as those filtering results by file type or Internet domain. All the operators are crucial to improve retrieval effectiveness, especially recall.

Cross-References

▶ Indexing Units of Structured Text Retrieval
▶ Logical Models of Information Retrieval
▶ Precision
▶ Probabilistic Retrieval Models and Binary Independence Retrieval (BIR) Model
▶ Query Expansion for Information Retrieval
▶ Query Expansion Models
▶ Recall
▶ Vector-Space Model

Recommended Reading

1. Bar-Hillel Y. Language and information. Reading: Addison-Wesley; 1964.
2. Belkin NJ, Cool C, Croft WB, Callan JP. The effect of multiple query representations on information retrieval system performance. In: Proceedings of the 16th Annual International ACM SIGIR Conference on Research and Development in Information Retrieval; 1993. p. 339–46.
3. Blair D. Language and representation in information retrieval. Amsterdam: Elsevier; 1990.
4. Cooper W. Getting beyond Boole. Inform Process Manage. 1988;24:243–48.
5. Croft W, Turtle H, Lewis D. The use of phrases and structured queries in information retrieval. In: Proceedings of the 14th Annual International ACM SIGIR Conference on Research and Development in Information Retrieval; 1991. p. 32–45.
6. Croft W, Metzler D, Strohman T. Search engines: information retrieval in practice. Boston: Addison Wesley; 2009.
7. Fuhr N. Probabilistic datalog – a logic for powerful retrieval methods. In: Proceedings of the ACM International Conference on Research and Development in Information Retrieval; 1995.
8. Grefenstette G, editor. Cross-language information retrieval. International series on information retrieval. Dordecht: Kluwer Academic; 1998.
9. Hersh W, Hickam D. An evaluation of interactive boolean and natural language searching with an online medical textbook. J Am Soc Inform Sci. 1995;46(7):478–89.
10. Hull D. A weighted boolean model for cross language text retrieval. In: Grefenstette G, editor. Cross-language information retrieval. Boston: Kluwer. p. 119–36.
11. Korfhage R. Information storage and retrieval. New York: Wiley; 1997.
12. Kowalski G, Maybury M. Information retrieval systems: theory and implementation. Dordecht: Kluwer; 2000.
13. Lancaster F, Warner A. Information retrieval today. Arlington: Information Resources; 1993.
14. Lee J. Properties of extended boolean models in information retrieval. In: Proceedings of the 17th Annual International ACM SIGIR Conference on Research and Development in Information Retrieval; 1994. p. 182–90.
15. Lee J, Kim W, Kim M, Lee Y. On the evaluation of boolean operators in the extended boolean retrieval framework. In: Proceedings of the 16th Annual International ACM SIGIR Conference on Research and Development in Information Retrieval; 1993. p. 291–97.
16. Melucci M. Introduction to information retrieval and quantum mechanics. Berlin/Heidelberg: Springer; 2015.
17. Nie JY, Lepage F. Toward a broader logical model for information retrieval. In: Crestani F, Lalmas M,
18. van Rijsbergen CJ, editors. Uncertainty and logics: advanced models for the representation and retrieval of information. Boston/Dordrecht: Kluwer Academic Press; 1998. p. 17–38.
19. Radecki T. Generalized boolean methods of information retrieval. Int J Man Mach Stud. 1983;18(5):407–39.
20. van Rijsbergen C. The geometry of information retrieval. Cambridge: Cambridge University Press; 2004.
21. Wong S, Ziarko W, Raghavan V, Wong P. Extended boolean query processing in the generalized vector space model. Inform Syst. 1989;14(1):47–63.

Boosting

Zhi-Hua Zhou
National Key Lab for Novel Software
Technology, Nanjing University, Nanjing, China

Definition

Boosting is a kind of ensemble methods [13] which produces a strong learner that is capable of making very accurate predictions by combining rough and moderately inaccurate learners (which are called as *base learners* or *weak learners*). In particular, boosting sequentially trains a series of base learners by using a *base learning algorithm*, where the training examples wrongly predicted by a base learner will receive more attention from the successive base learner. After that, it generates a final strong learner through a weighted combination of these base learners.

Historical Background

In 1989, Kearns and Valiant posed an interesting theoretical question, i.e., whether two complexity classes, *weakly learnable* and *strongly learnable* problems, are equal. In other words, whether a *weak* learning algorithm that performs just slightly better than random guess can be boosted into an arbitrarily accurate *strong* learning algorithm. In 1990, Schapire [9] proved that the answer to the question is "yes," and the proof is a construction, which is the first boosting algorithm. One year later, Freund developed a more efficient algorithm. However, both algorithms suffered from some practical deficiencies. Later, in 1995, Freund and Schapire [3] developed the AdaBoost algorithm which raised the hot wave of boosting research.

Foundations

AdaBoost is the most influential boosting algorithm. Let X and Y denote the instance space and the set of class labels, respectively, and assume $Y = \{-1, +1\}$. Given a training set $D = \{(x_1, y_1), (x_2, y_2), \ldots, (x_m, y_m)\}$ where $x_i \in X$ and $y_i \in Y$ $(i = 1, \ldots, m)$ and a base learning algorithm which can be decision tree, neural networks, or any other learning algorithms, the AdaBoost algorithm works as follows:

First, it assigns equal weights to all the training examples (x_i, y_i) $(i \in \{1, \ldots, m\})$. Denote the distribution of the weights at the tth learning round as D_t. From the training set and D_t, the algorithm generates a base learner $h_t: X \rightarrow Y$ by calling the base learning algorithm. Then, it uses the training examples to test h_t, and the weights of the incorrectly classified examples will be increased. Thus, an updated weight distribution D_{t+1} is obtained. From the training set and D_{t+1}, AdaBoost generates another base learner by calling the base learning algorithm again. Such a process is repeated for T times, each of which is called a *round*, and the final learner is derived by weighted majority voting of the T base learners, where the weights of the learners are determined during the training process. In practice, the base learning algorithm may be a learning algorithm which can use weighted training examples directly; otherwise the weights can be exploited by sampling the training examples according to the weight distribution D_t. The pseudo-code of AdaBoost is shown in Fig. 1.

Freund and Schapire [3] proved that the training error of the final learner H is upper bounded by

$$\epsilon_D = \Pr_{i \sim D}[H(x_i) \neq y_i] \leq 2^T \prod_{t=1}^{T} \sqrt{\epsilon_t (1 - \epsilon_t)},$$

which can be written as

$$\epsilon_D \leq \prod_{t=1}^{T} \sqrt{1 - 4\gamma_t^2} \leq \exp\left(-2\sum_{t=1}^{T} \gamma_t^2\right),$$

where $\gamma_t = 1/2 - \epsilon_t$. Thus, if each base learner is slightly better than random so that $\gamma_t \geq \gamma$ for some $\gamma > 0$, the training error will drop exponentially fast in T since the upper bound is at most $e^{-2T\gamma^2}$.

Freund and Schapire [3] also gave a generalization error bound of H in terms of its training error ϵ_D, the size m of the training set, the VC-dimension d of the base learner space, and the number of rounds T, by

$$\epsilon \leq \epsilon_D + \tilde{O}\left(\sqrt{\frac{Td}{m}}\right)$$

with high probability, where $\tilde{O}(\cdot)$ is used to hide all logarithmic and constant factors instead of using $O(\cdot)$ which hides only constant factors.

The above generalization error bound suggests that AdaBoost will overfit if it runs for many rounds since T is in the numerator. However, empirical observations show that AdaBoost often does *not* overfit even after a large number of rounds, and sometimes it is even able to reduce the generalization error after the training error has already reached zero. Thus, later, Schapire et al. [10] presented another generalization error bound:

$$\epsilon \leq \Pr_{i \sim D}\left[\text{margin}_f (x_i, y_i) \leq \theta\right] + \tilde{O}\left(\sqrt{\frac{d}{m\theta^2}}\right)$$

for any $\theta > 0$ with high probability, where the *margin* of f on (x_i, y_i) was defined as

---

Input: Data set $\mathcal{D} = \{(x_1, y_1), (x_2, y_2), \cdots, (x_m, y_m)\}$;
     Base learning algorithm $\mathcal{L}$;
     Number of learning rounds T.
Process:
   $D_1(i) = 1/m.$ % Initialize the weight distribution
   for $t = 1, \cdots, T$:
     $h_t = \mathcal{L}(\mathcal{D}, D_t)$;   % Train a base learner h_t from $\mathcal{D}$ using distribution D_t
     $\epsilon_t = Pr_{i \sim D_i}[h_t(x_i \neq y_i)]$;   % Measure the error of h_t

   $\alpha_t = \frac{1}{2} \ln \left(\frac{1-\epsilon_t}{\epsilon_t} \right)$; % Determine the weight of h_t

   $D_{t+1}(i) = \dfrac{D_t(i)}{Z_t} \times \begin{cases} \exp(-\alpha_t) & \text{if } h_t(x_i) = y_i \\ \exp(\alpha_t) & \text{if } h_t(x_i) \neq y_i \end{cases}$

   $\qquad = \dfrac{D_t(i)\exp(-\alpha_t y_i h_t(x_i))}{Z_t}$ % Update the distribution, where Z_t is

   $\qquad\qquad\qquad$ % a normalization factor which enables D_{t+1} be a distribution
   end.
Output: $H(x) = \text{sign}(f(x)) = \text{sign}\left(\Sigma_{t=1}^T \alpha_t h_t(x) \right)$

Boosting, Fig. 1 The AdaBoost algorithm

$$\text{margin}_f(x_i, y_i) = \frac{y_i f(x_i)}{\sum_{t=1}^T |\alpha_t|}$$

$$= \frac{y_i \sum_{i=1}^T |\alpha_t h_t(x)|}{\sum_{t=1}^T |\alpha_t|},$$

whose value is in $[-1, +1]$ and is positive only if H classifies (x_i, y_i) correctly. In fact, the magnitude of margin can be explained as a measure of confidence in prediction. The larger the magnitude of margin, the higher confidence of prediction. Note that when $H(x_i) = y_i$, $\text{margin}_f(x_i, y_i)$ can still be increased as t increases. Thus, the above margin-based generalization bound gives an answer to the question of why AdaBoost is able to reduce the generalization error even after the training error reaches zero, that is, the confidence in prediction can be increased further.

Breiman [2] indicated that the minimum margin is crucial in the margin theory. He, however, theoretically and empirically showed that an increased minimum margin does not necessarily lead to an increased generalization performance;

this almost sentenced the margin theory to death. Seven years later, Reyzin and Schapire [8] found that the complexity of base learners in Breiman's experiments was not controlled well, and they argued that rather than minimum margin, the average margin or median margin is crucial. Later on, great efforts have been devoted to this line of research [5, 12]; finally the importance of margin distribution was theoretically proved, and it was disclosed that both the margin mean and variance are crucial [5].

In addition to the margin theory, there are some other theoretical efforts for understanding AdaBoost, such as the statistical view [4], etc. Many variants or extensions of AdaBoost have been developed [6, 13], which make boosting become a big family of ensemble methods.

In contrast to another famous ensemble method, bagging (which reduces *variance* significantly but has little effect on *bias*), boosting can significantly reduce bias in addition to reducing variance [13]. So, on weak learners such as decision stumps, which are one-level decision trees, Boosting is usually more effective.

Key Applications

The first application of boosting was on optical character recognition by Drucker et al. Later, boosting was applied to diverse tasks such as text categorization, speech recognition, image retrieval, medical diagnosis, etc. [6, 13]. It is worth mentioning that AdaBoost has been combined with a cascade process for face detection [11], and the resulting face detector was 15 times faster than state-of-the-art face detectors at that time but with comparable accuracy, which was recognized as one of the major breakthroughs in computer vision (in particular, face detection) during the past two decades.

Future Directions

The defense of margin theory of boosting [5] has much implications, among which is the establishment of a unified theoretical framework for the two powerful learning paradigms, i.e., boosting and support vector machines. The novel insight that, rather than the long believed minimum margin, the margin mean and variance are crucial to the generalization performance leads to a new and promising direction, i.e., large margin distribution learning [14].

It has been observed that boosting performs poorly when abundant noise exists. Making boosting more robust to noise is an important task. Moreover, boosting suffers from some general deficiencies of ensemble methods [13], such as the lack of comprehensibility, i.e., the knowledge learned by boosting is not understandable to the user. Trying to overcome those deficiencies is an important future direction.

Experimental Results

Empirical studies on boosting have been reported in many papers, such as [1, 7].

Cross-References

► Bagging
► Decision Trees
► Ensemble
► Neural Networks
► Support Vector Machine

Recommended Reading

1. Bauer E, Kohavi R. An empirical comparison of voting classification algorithms: bagging, boosting, and variants. Mach Learn. 1999;36(1–2):105–39.
2. Breiman L. Prediction games and arcing classifiers. Neural Comput. 1999;11(7):1493–517.
3. Freund Y, Schapire RE. A decision-theoretic generalization of on-line learning and an application to boosting. J Comput Syst Sci. 1997;55(1):119–39 (A short version appeared in the Proceedings of Euro-COLT'95).
4. Friedman J, Hastie T, Tibshirani R. Additive logistic regression: a statistical view of boosting with discussions. Ann Stat. 2000;28(2):337–407.
5. Gao W, Zhou Z-H. On the doubt about margin explanation of boosting. Artif Intell. 2013;203:1–18.
6. Meir R, Rätsch G. An introduction to boosting and leveraging. In: Mendelson S, Smola AJ, editors. Advanced lectures in machine learning. LNCS vol. 2600. Berlin: Springer; 2003. p. 118–83.
7. Opitz D, Maclin R. Popular ensemble methods: an empirical study. J Artif Intell Res. 1999;11(1): 169–98.
8. Reyzin L, Schapire RE. How boosting the margin can also boost classifier complexity. In: Proceedings of the 23rd International Conference on Machine Learning, Pittsburgh; 2006. p. 753–60.
9. Schapire RE. The strength of weak learn ability. Mach Learn. 1990;5(2):197–227.
10. Schapire RE, Freund Y, Bartlett P, Lee WS. Boosting the margin: a new explanation for the effectiveness of voting methods. Ann Stat. 1998;26(5):1651–86.
11. Viola P, Jones M. Rapid object detection using a boosted cascade of simple features. In: Proceedings of the IEEE Computer Society Conference on Computer Vision and Pattern Recognition; 2001. p. 511–8.
12. Wang L, Sugiyama M, Yang C, Zhou Z.H, Feng J. On the margin explanation of boosting algorithm. In: Proceedings of the 21st Annual Conference on Learning Theory; 2008. p. 479–90.
13. Zhou Z-H. Ensemble methods: foundations and algorithms. Boca Raton: CRC Press; 2012.
14. Zhou Z-H. Large margin distribution learning. In: Proceedings of Artificial Neural Networks in Pattern Recognition; 2014.

Bootstrap

Hwanjo Yu
University of Iowa, Iowa City, IA, USA

Synonyms

Bootstrap estimation; Bootstrap sampling

Definition

The bootstrap is a statistical method for estimating the performance (e.g., accuracy) of classification or regression methods. The bootstrap is based on the statistical procedure of sampling *with replacement*. Unlike other estimation methods such as cross-validation, the same object or tuple can be selected for the training set more than once in the boostrap. That is, each time a tuple is selected, it is equally likely to be selected again and re-added to the training set.

Historical Background

The bootstrap sampling was developed by Bradley Efron in 1979, and mainly used for estimating the statistical parameters such as mean, standard errors, etc. [2]. A meta-classification method using the bootstrap called *bootstrap aggregating* (or *bagging*) was proposed by Leo Breiman in 1994 to improve the classification by combining classifications of randomly generated training sets [1].

Foundations

This section discusses a commonly used bootstrap method, *0.632 bootstrap*. Given a dataset of N tuples, the dataset is sampled N times, with replacement, resulting in a *bootstrap sample* or training set of N tuples. It is very likely that some of the original data tuples will occur more than once in the training set. The data tuples that were not sampled into the training set end up forming the test set. If this process is repeated multiple times, on average 63.2 % of the original data tuples will end up in the training set and the remaining 36.8 % will form the test set (hence, the name, 0.632 bootstrap).

The figure, 63.2 %, comes from the fact that a tuple will not be chosen with probability of 36.8 %. Each tuple has a probability of $1/N$ of being selected, so the probability of not being chosen is $(1 - 1/N)$. The selection is done N times, so the probability that a tuple will not be chosen during the whole time is $(1 - 1/N)^N$. If N is large, the probability approaches $e^{-1} = 0.368$. Thus, 36.8 % of tuples will not be selected for training and thereby end up in the test set, and the remaining 63.2 % will form the training set.

The above procedure can be repeated k times, where in each iteration, the current test set is used to obtain an accuracy estimate of the model obtained from the current bootstrap sample. The overall accuracy of the model is then estimated as

$$Acc(M) = \sum_{i=1}^{k} \left(0.632 \times cc(M_i)_{test\_set} \right. \\ \left. +0.368 \times Acc(M_i)_{train\_set} \right), \quad (1)$$

where $Acc(M_i)_{train\_set}$ and $Acc(M_i)_{test\_set}$ are the accuracy of the model obtained with bootstrap sample i when it is applied to training set and test set respectively in sample i.

Key Applications

The bootstrap method is preferably used for estimating the performance when the size of dataset is relatively small.

Cross-References

▶ Cross-Validation

Recommended Reading

1. Breiman L. Bagging predictors. Machine Learning; 1996.
2. Efron B, Tibshirani RJ. An introduction to the bootstrap. Boca Raton: CRC Press; 1994.

Boyce-Codd Normal Form

Marcelo Arenas
Pontifical Catholic University of Chile, Santiago, Chile

Synonyms

BCNF

Definition

Let $R(A_1,...,A_n)$ be a relation schema and Σ a set of functional dependencies over $R(A_1, \ldots, A_n)$. Then (R, Σ) is said to be in Boyce-Codd normal form (BCNF) if for every nontrivial functional dependency $X \rightarrow A$ implied by Σ, it holds that X is a superkey for R.

Key Points

In order to avoid update anomalies in database schemas containing functional dependencies, BCNF was introduced by Codd in [1] (Codd pointed out in [1] that this normal form was developed by Raymond F. Boyce and himself.). This normal form is defined in terms of the notion of superkey as shown above. For example, given a relation schema $R(A, B, C)$ and a set of functional dependencies $\Sigma = \{AB \rightarrow C, C \rightarrow B\}$, it does not hold that $(R(A, B, C), \Sigma)$ is in BCNF since C is not a superkey for R. On the other hand, $(S(A,$ $B, C), \Gamma)$ is in BCNF if $\Gamma = \{A \rightarrow BC\}$, since A is a superkey for S in this case.

It should be noticed that relation schema $R(A, B, C)$ above is in 3NF if $\Sigma = \{AB \rightarrow C, C \rightarrow B\}$, although this schema is not in BCNF. In fact, BCNF is strictly stronger than 3NF; every schema in BCNF is in 3NF, but there exist schemas (as the one shown above) that are in 3NF but not in BCNF.

For every normal form, two problems have to be addressed: how to decide whether a schema is in that normal form and how to transform a schema into an equivalent one in that normal form. As opposed to the case of 3NF, it can be tested efficiently whether a relation schema is in BCNF. A relation schema (R, Σ) is in BCNF if and only if for every nontrivial functional dependency $X \rightarrow Y \in \Sigma$, it holds that X is a superkey. Thus, it is possible to check efficiently whether (R, Σ) is in BCNF by using the linear-time algorithm for functional dependency implication developed by Beeri and Bernstein [2]. On the negative side, given a relation schema S, it is not always possible to find a database schema S' such that S' is in BCNF and S' is a lossless and dependency preserving decomposition of S. In fact, relation schema $R(A, B, C)$ and set of functional dependencies $\{AB \rightarrow C, C \rightarrow B\}$ do not admit a dependency-preserving decomposition in BCNF.

Cross-References

▶ Fourth Normal Form
▶ Normal Forms and Normalization
▶ Second Normal Form (2NF)
▶ Third Normal Form

Recommended Reading

1. Codd EF. Recent investigations in relational data base systems. In: Proceedings of the IFIP Congress, Information Processing 74; 1974, p. 1017–21.
2. Beeri C, Bernstein P. Computational problems related to the design of normal form relational schemas. ACM Trans Database Sys. 1979;4(1):30–59.

BP-Completeness

Dirk Van Gucht
Indiana University, Bloomington, IN, USA

Synonyms

Instance-completeness; Relation-completeness

Definition

A relational query language Q is BP-complete if for each relational database D, the set of all relations defined by the queries of Q on D is equal to the set of all first-order definable relations over D. More formally, fix some infinite universe $\mathbf{U}$ of atomic data elements. A *relational database schema* S is a finite set of relation names, each with an associated arity. A *relational database* D with schema S assigns to each relation name of S a *finite* relation over $\mathbf{U}$ of its arity. The domain of D, $dom(D)$, is the set of all atomic data elements occurring in the tuples of its relations. Let FO^S be the set of first-order formulas over signature S and the equality predicate, and let $FO^S(D) = \{\phi(D)|\phi \in FO^S\}$. (For a formula $\phi \in FO^S$ with free variables $(x_1, \ldots, x_m)$, $\phi(D)$ denotes the m-ary relation over $dom(D)$ defined by ϕ, where the variables in ϕ are assumed to range over $dom(D)$.) Let Q^S denote those queries of Q defined over schema S, and let $Q^S(D) = \{q(D)|q \in Q^S\}$, i.e., the set of relations defined by queries of Q applied to D. Then, Q is *BP- complete* if for each relational database D over schema S,

$$Q^S(D) = FO^S(D).$$

In the words of Chandra and Harel, "BP-completeness can be seen to be a measure of the power of a language to express relations and *not* of its power to express functions having relations as outputs, i.e., queries." In fact, there exist BP-complete languages that do not express the same queries.

Key Points

Chandra and Harel introduced the concept of BP-completeness and attributed it to Bancilhon and Paredaens who were the first to study it. Bancilhon and Paredaens considered the following decision problem: given a relational database D and a relation R defined over $dom(D)$, does there exists a first-order formula ϕ such that $\phi(D) = R$? (Paredaens considered this problem for the relational algebra, but by Codd's theorem on the equivalence of first-order logic and the relational algebra, these decision problems are the same.) They gave an algebraic, language-independent characterization of this problem by showing that such a first-order formula exists if and only if for each bijection h: $dom(D) \rightarrow dom(D)$, if $h(D) = D$ then $h(R) = R$. Equivalently, if h is an automorphism of D then it is also an automorphism of R. (Here, $h(D)$ and $h(R)$ are the natural extensions of h to D and R, respectively.) For a relational database D over schema S and a relation R, denote by $Aut(D)$ and $Aut(R)$ the sets of automorphisms of D and R, respectively, and let $R^S(D) = \{R \mid R$ is a relation over $dom(D)$ such that $Aut(D) \subseteq Aut(R)\}$. Then, an alternative characterization for the BP-completeness of Q is to require that for each relational database D over schema S,

$$Q^S(D) = R^S(D).$$

Van den Bussche showed that this characterization follows from Beth's Theorem about the explicit and implicit definability of first-order logic. The concept of BP-completeness has been generalized as well as specialized to query languages over other database models.

Cross-References

▶ Computationally Complete Relational Query Languages
▶ Query Language
▶ Relational Calculus
▶ Relational Algebra

Recommended Reading

1. Bancilhon F. On the completeness of query languages for relational databases. In: Proceedings of the Mathematical Foundations of Computer Science; 1978.
2. Chandra A, Harel D. Computable queries for relational databases. J Comput Syst Sci. 1980;21(2):156–78.
3. Paredaens J. On the expressive power of the relational algebra. Inform Process Lett. 1978;7(2):107–11.
4. Van den Bussche. J. Applications of Alfred Tarski's ideas in database theory. In: Proceedings of the 15th International Workshop on Computer Science Logic; 2001. p. 20–37.

Bpref

Nick Craswell
Microsoft Research Cambridge, Cambridge, UK

Definition

Bpref is a preference-based information retrieval measure that considers whether relevant documents are ranked above irrelevant ones. It is designed to be robust to missing relevance judgments, such that it gives the same experimental outcome with incomplete judgments that Mean Average Precision would with complete judgments.

Key Points

In a test collection where all relevant documents have been identified, experiments using bpref and MAP should give the same outcome, for example both systems should agree that system A is better than system B. However, if the relevance judgments are incomplete, for example where only half the pool has been judged, MAP becomes unstable and may incorrectly show that system B is better than system A. The bpref measure was developed to maintain the correct ordering of systems (A better than B) even with incomplete judgments.

Given a ranked list of search results and a set of R known relevant documents and N known irrelevant documents, bpref first identifies the top-R irrelevant documents in the list. For these irrelevant documents n and relevant documents r the measure is:

$$\text{bpref} = \frac{1}{R}\sum_{r}\left(1 - \frac{|n \text{ ranked higher than } r|}{\min(R, N)}\right)$$

The bpref paper [1] found good agreement between full-judgment MAP and reduced-judgment bpref even if the judged set was reduced by 80 %. Later work has introduced other measures with similar or better properties, notably Inferred Average Precision [2].

Cross-References

▶ Average Precision
▶ Precision-Oriented Effectiveness Measures

Recommended Reading

1. Buckley C, Voorhees EM. Retrieval evaluation with incomplete information. In: Proceedings of 30th Annual International ACM SIGIR Conference on Research and Development in Information Retrieval; 2004. p. 25–32.
2. Yilmaz E, Aslam JA. Estimating average precision with incomplete and imperfect judgments. In: Proceedings of International Conference on Information and Knowledge Management; 2006. p. 102–11.

Browsing

Kent Wittenburg
Mitsubishi Electric Research Laboratories, Inc., Cambridge, MA, USA

Synonyms

Perusal; Scanning

Definition

"Browsing" has two definitions in the context of visual interfaces for database systems:

1. The human activity of visual perception and interpretation of electronic content when there is no specific target object being sought
2. The human activity of clicking or tapping on a sequence of elements in an information display that results in a sequence of screens of information state

Definition (1) implies something about the mental intent of the user performing the act of information processing. The intent can be to learn the gist of "what is there." It may be to take in information in order to be entertained or informed. Or it may simply be a part of involuntary visual scanning within an environment. In the physical world, examples include flipping through the pages of a book, scanning quickly through a menu, or glancing down a store aisle.

Definition (2) is used in the context of contrasting the human-computer interaction paradigm of link-following versus searching in World Wide Web or digital library applications. "Browsing" in this context refers to a sequence of clicking or tapping behaviors on highlighted elements (hyperlinks) in order to go directly to a subsequent state of the information display. In contrast, "searching" entails specifying and then executing a query. The interaction paradigms of browsing and searching are independent of the intentional state of the user – the individual may or may not be looking for something specific in either case.

Key Points

A branch of research in visual interfaces for browsing in the first sense is rapid serial visual presentation (RSVP) [2]. The human psychology of visual perception, specifically models of short-term visual memory, can inform the design of visual interfaces for browsing. Visual information can be presented in space or in time or in some combination thereof. For image presentation specifically, is it better to utilize screen real estate to present multiple images concurrently or to present them over time? If over time, should images move in some specified path or remain motionless? What are the best controls to give to the user in order to support the tasks of rapid perusal in order to get a gist of the content within that information space or to help in making a selection for further detailed investigation? These are some of the questions being asked by researchers and designers in the context of RSVP interfaces.

An important aspect of browsing as link-following has to do with information "scent." Just as animals' scent can be used as an indicator of their territory or former presence, the presentation of information links provides hints of what a user would find if he or she were to follow the link. The theory of information foraging [1] provides insight into human behavior in information seeking that can help inform the design for interfaces and electronic documents. According to Pirolli, humans tend to make decisions about what links to follow and how long to continue along certain paths based on resource-limited assessments of the costs of such activities relative to a judgment of payoff.

Cross-References

▶ Browsing in Digital Libraries
▶ Discovery
▶ Human-Computer Interaction
▶ Information Foraging
▶ Information Navigation
▶ Information Retrieval
▶ Searching Digital Libraries
▶ Visual Interaction
▶ Visual Interfaces
▶ Web Information Retrieval Models

Recommended Reading

1. Pirolli P. Information foraging theory: adaptive inter-action with information. New York: Oxford University Press; 2007.
2. Spence R, Witkowski M. Rapid serial visual presentation: design for cognition. London: Springer; 2013.

Browsing in Digital Libraries

Rao Shen[1] and Edward A. Fox[2]
[1] Yahoo!, Sunnyvale, CA, USA
[2] Virginia Tech, Blacksburg, VA, USA

Synonyms

Exploring; Looking over/through; Surfing

Definition

Informally, browsing is a process that involves looking through a collection of information. Thus, in a traditional library, one may wander about the stacks, glancing at titles of works in regions where one expects to find interesting material. In the broadest sense, browsing is considered as one type of exploration, typically less directed or purposeful than searching, in that the goal or result is not always precisely known in advance. In the World Wide Web, this is very common, making use of "browsers" like Firefox or Internet Explorer. Sometimes the colorful term "surfing" is used when it seems appropriate to emphasize the excitement that some feel when exploring new Web content or the thrill of getting closer to some desirable result. Yet, the WWW is but one example of a hypertext (or, if multiple media types are considered, hypermedia) environment. Accordingly, in a digital library, which many believe must include a hypertext [4], browsing involves moving from one information object to another according to some connections or links or structures.

Historical Background

A hypertext is a high-level interactive navigational structure which allows nonsequential exploration of and access to information [3]. In the most general case, it consists of nodes connected by directed links in a graph structure. A directed link in the hypertext graph is called a hyperlink. Today's most famous hypertext system is the World Wide Web. The Web's structure contains hyperlinks connecting nodes that each can be associated with either a complete document (e.g., HTML page) or with a location in a document (e.g., a start tag in an HTML page).

Three models for Web browsing defined by Baeza-Yates and Ribeiro-Neto [1] are flat browsing, structure-guided browsing, and utilizing the hypertext model. If a document collection is organized as a one-dimensional list, exploring the document space is an example of flat browsing; if the organization is hierarchical instead of flat, then browsing is structure guided. The first two models have a hypertext graph that is made up of a disconnected bipartite graph and a tree, respectively, so they are special cases of the third model.

Browsing and searching are two paradigms for finding things on the Web. Some Web search engines provide Web directories used for browsing (and also for searching). Web directories are hierarchical taxonomies that classify human knowledge. Although a taxonomy can be considered as a tree, there can be cross-references. One of the advantages of browsing a Web directory, in most cases, is that if a user finds what she is looking for in the taxonomy, then the answer that is found classified under that category almost certainly will be useful. In Web directories, a search can be reduced to a sub-tree of the taxonomy. However, searching may miss related pages that are not in that part of the taxonomy, if a user cannot formulate her information need sufficiently broadly.

Foundations

Browsing and searching are two methods of pursuing information on the Web; they also are two

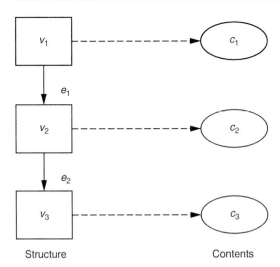

Structure   Contents

Browsing in Digital Libraries, Fig. 1 A simple hypertext

understood as a traversal of a directed linked graph.

Returning to the previous example, one can observe that the left part of Fig. 1 is a directed linked graph (V, E), where each right part is a set of contents $C = \{c_1, c_2, c_3\}$ and where c_i ($1 <= i <= 3$) is the contents of a Web page associated with node v_i. An event of traversing along edge $e_1 = (v_1, v_2)$ is associated with a function:

$E \rightarrow C$, which retrieves the contents of node v_2, i.e., c_2.

Note that (sorted) lists and hierarchical trees can be represented as hypertexts, so browsing can be over common organizations, such as alphabetical lists of authors or indented lists of categories and subcategories; these are familiar from books with author indexes or tables of contents.

major paradigms for exploring digital libraries. The Web has no maintenance organization. Individuals add and delete pages at will. On the other hand, digital libraries require proper collection maintenance, and they will succeed only if their content is well organized [2].

The 5S (streams, structures, scenarios, spaces, and societies) framework [4], which is used as a formal base upon which to describe digital libraries, defines structures to specify the way in which parts of a whole are managed or organized. In such digital libraries, structures can represent hypertexts.

For example, Fig. 1 illustrates a simple hypertext designed to provide access to objects in a digital library, in chronological order. It is made up of structural hyperlinks that follow chronological order (see solid arrows in Fig. 1) and external reference links (see dashed arrows in Fig. 1).

Thus, a formal and precise definition of browsing is illustrated in Fig. 1: Given a hypertext of a digital library, with vertex set V and edge set E, browsing is a set of sequences of traverse link events over the hypertext, such that event e of traversing from node v_k to v_t is associated with a function which retrieves the contents of node v_t. Hence, browsing a digital library is the procedure of navigating the hypertext, and it also can be

Key Applications

Modeling and analyzing browsing in digital libraries helps ease development and maintenance in those digital libraries. Some domain-specific digital libraries have heterogeneous data that should be organized using several schemes. Thus, digital objects in an archaeological digital library may fit into various categories of archaeological data such as figurine images, bone records, locus sheets, and site plans. They can be organized according to different hierarchical structures (e.g., animal bone records are organized based on sites where they are excavated, temporal sequence, and animal names). These hierarchical structures contain one or more hierarchically arranged categories. In addition, they can be refined based on taxonomies existing in botany and zoology or through classification and description of artifacts by archaeologists.

Thus, an archaeological digital library may provide multidimensional browsing (see Fig. 2) to allow users to move along any of the navigational dimensions or a combination thereof. Navigational dimensions correspond to hierarchical structures used to browse digital objects, as mentioned above.

For example, an archaeologist might browse through three dimensions: space, object, and time. She can start from any of these dimensions and move along by clicking. The scenario shown in Fig. 2 tells that she is interested in the artifact records from the tomb numbered 056 in area A of the Bab edh-Dhra site. The clickstream representing her navigation path is denoted "Site=Bab edh-Dhra >>PARTITION=A>>SUBPARTITION = 056." While this navigation path is within the first dimension, it also is associated with the other dimensions. The second dimension shows there is only one type of object, i.e., pottery, from that particular location. The third dimension presents the two time periods associated with those pottery records. Hence, the dynamic coverage and hierarchical structure of those dimensions yields a tool supporting learning and exploring. The user can navigate across dimensions. By clicking "EARLY BRONZE II" in the third dimension, she can view all of the interesting artifact records from the EARLY BRONZE II period.

Browsing may present a useful starting point for active exploration of an answer space. Subsequent browsing and searching, in any combination, also can be employed to refine or enhance users' initial, possibly under-specified, information needs.

Browsing context is associated with a user's navigation path. Browsing results within a certain browsing context typically are a set of records (web pages), e.g., there are 35 pottery records within the browsing context represented by the

navigation path denoted "Site=Bab edh-Dhra >>PARTITION=A>>SUBPARTITION = 056." For example, assume a user wants to find saucer records in the set of thirty five pottery records. She types "saucer" in the search box as shown in Fig. 3. She switches from browsing to searching, so searching then is a natural extension of browsing.

Browsing may be provided as a post-retrieval service to organize searching results hierarchically in digital libraries. For example, in Fig. 4 one can see that 88 equus records are retrieved through the basic searching service, in response to a query "equus." They are organized into three dimensions after the user clicks the button "View search results hierarchically" (see Fig. 5).

Browsing may be supported by visualization to provide a starting point for users. Graphic overviews of a digital library collection can display category labels hierarchically based on the facets. Categories can be visualized as a hyperbolic tree [5] as well as through a traditional node-link representation of a tree.

A hyperbolic tree in Fig. 6 shows hierarchical relationships among excavation data in an archaeological digital library based on spatial, temporal, and artifact-related taxonomies. A node name represents a category, and a bubble attached to a node represents a set of archaeological records. The size of a bubble attached to a node reflects the number of records belonging to that category. The hyperbolic tree supports "focus + context" navigation; it also provides an overview of records organized in the archaeological digital library. It shows that the records are from seven

---

You are in: Main >> SITE=Bab edh-Dhra >> PARTITION=A >> SUBPARTITION=056   Save this Navigation Path

Search within this context for [   ] [Go]

View Records for the Context Below

Browse by space:: SITE=Bab edh-Dhra::PARTITION=A::SUBPARTITION=056:: LOCUS
    Unclassified

Browse by object:: :: OBJECTTYPE
    Pottery

Browse by time:: :: Period
    EARLY BRONZE II   EARLY BRONZE III

Browsing in Digital Libraries, Fig. 2 Multidimensional browsing

Browsing in Digital Libraries, Fig. 3 Search saucer records

archaeological sites (the Megiddo site has the most) and are of twelve different types.

Future Directions

Browsing and searching are often provided by digital libraries as separate services. Developers commonly see these functions as having different underlying mechanisms, and they follow a functional, rather than a task-oriented, approach to interaction design. While exhibiting complementary advantages, neither paradigm alone is adequate for complex information needs. Searching is popular because of its ability to identify information quickly. On the other hand, browsing is useful when appropriate search keywords are unknown or unavailable to users. Browsing also is appropriate when a great deal of contextual information is obtained along the navigation path. Therefore, a synergy between searching and browsing is required to support users' information seeking goals. Browsing and searching can be converted and switched to each other under certain conditions [5]. This suggests some new possibilities for blurring the dividing line between browsing and searching. If these two services are not considered to have different underlying mechanisms, they will not be provided as separated functions in digital libraries and may be better integrated.

Text mining and visualization techniques provide digital libraries additional powerful exploring services, with possible beneficial effects on browsing and searching. Digital library exploring services such as browsing, searching, clustering, and visualization can be generalized in the context of a formal digital library framework [5]. The theoretical approach may provide a systematic and functional method to design and implement exploring services for domain focused digital libraries.

Experimental Results

See Figs. 2, 3, 4, 5 and 6 above and the corresponding explanation.

Data Sets

See Figs. 2, 3, 4, 5 and 6 above and the ETANA Digital Library [5].

Cross-References

▶ Digital Libraries

Recommended Reading

1. Baeza-Yates R, Ribeiro-Neto B. Modern information retrieval. Reading: Addison-Wesley; 1999.
2. Fox EA, Urs SR. Digital libraries, chap. 12. In: Cronin B, editor. Annual review of information science and technology, vol. 36. Medford: Information Today, Inc.; 2002. p. 503–89.

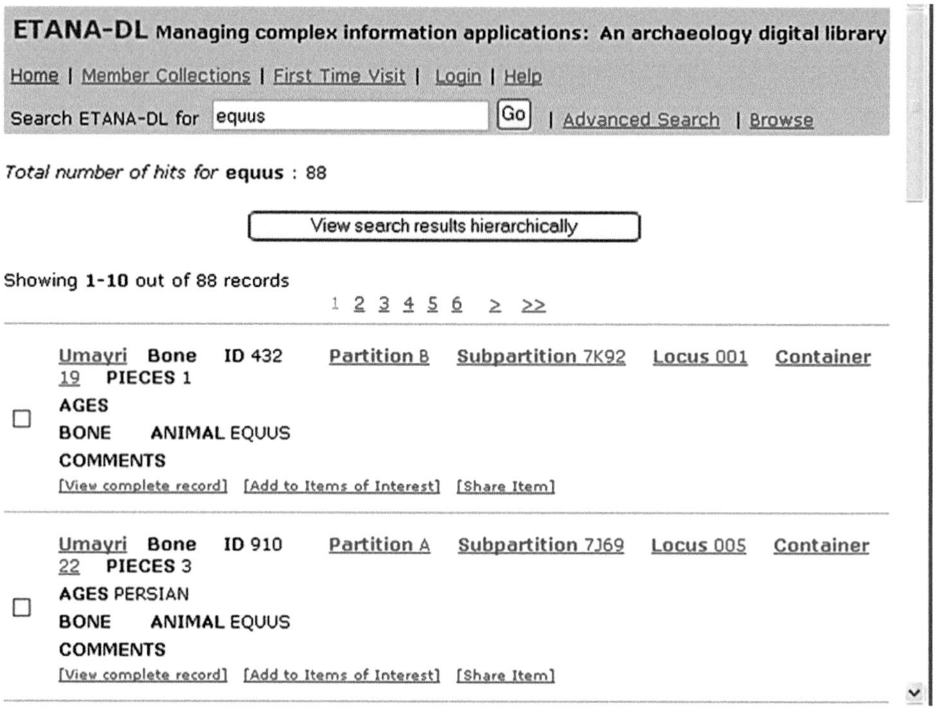

Browsing in Digital Libraries, Fig. 4 Equus records are retrieved through basic searching

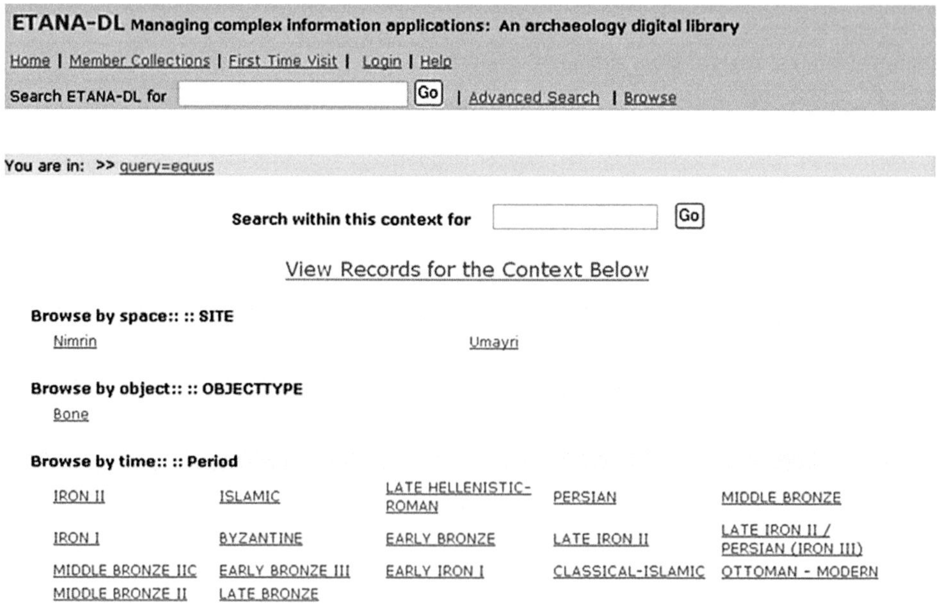

Browsing in Digital Libraries, Fig. 5 Retrieved equus records are organized into three dimensions

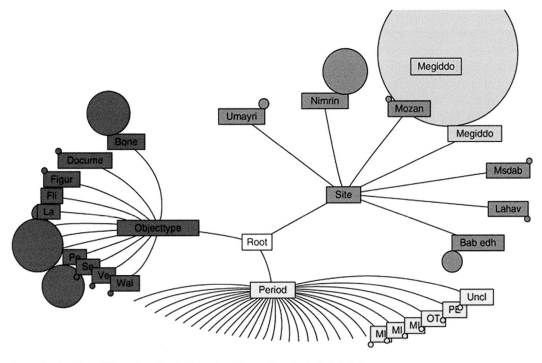

Browsing in Digital Libraries, Fig. 6 Overview of an archaeological digital library

3. Fox EA, Rous B, Marchionini G. ACM's hypertext and hypermedia publishing projects. In: Berk E, Devlin J, editors. Hypertext/hypermedia handbook. New York: McGraw-Hill; 1991. p. 465–7.
4. Goncalves M, Fox EA, Watson L, Kipp N. Streams, structures, spaces, scenarios, societies (5S): a formal model for digital libraries. ACM Trans Inf Syst. 2004;22(2):270–312.
5. Shen R., Vemuri N., Fan W., Torres R., and Fox E.A. Exploring digital libraries: integrating browsing, searching, and visualization. In: Proceedings 6th ACM/IEEE-CS joint Conference on Digital Libraries. 2006, p. 1–10.

B-Tree Locking

Goetz Graefe
Google, Inc., Mountain View, CA, USA

Synonyms

B-tree concurrency control; Crabbing; Key range locking; Key value locking; Latch coupling; Latching; Lock coupling; Row-level locking

Definition

B-tree locking controls concurrent searches and updates in B-trees. It separates transactions in order to protect the B-tree contents and it separates threads in order to protect the B-tree data structure. Nowadays, the latter is usually called latching rather than locking.

Historical Background

Bayer and Schkolnick [1] presented multiple locking (latching) protocols for B*-trees (all data records in the leaves, merely separator keys or "reference keys" in upper nodes) that combined high concurrency with deadlock avoidance. Their approach for insertion and deletion is based on deciding during a root-to-leaf traversal whether a node is "safe" from splitting (during an insertion) or merging (during a deletion), and on reserving appropriate locks (latches) for ancestors of unsafe nodes.

Lehman and Yao defined B^{link}-trees by relaxing the B-tree structure in favor of higher concurrency [8]. Srinivasan and Carey demonstrated their high performance using detailed simulations [13]. Jaluta et al. recently presented a detailed design for latching in B^{link}-trees, including a technique to avoid excessive link chains and thus poor search performance [7].

IBM's System R project explored multiple transaction management techniques, including transaction isolation levels and lock duration, predicate locking and key value locking, multi-granularity and hierarchical locking, etc. These techniques have been adapted and refined in many research and product efforts since then. Research into multi-level transactions [14] and into open nested transactions [3, 12] enables crisp separation of locks and latches – the former protecting database contents against conflicts among transactions and the latter protecting in-memory data structures against conflicts among concurrent threads.

Mohan's ARIES/KVL design [10, 11] explicitly separates locks and latches, i.e., logical database contents versus "structure maintenance" in a B-tree. A key value lock covers both a gap between two B-tree keys and the upper boundary key. In non-unique indexes, an intention lock on a key value permits operations on separate rows with the same value in the indexed column. In contrast, other designs include the row identifier in the unique lock identifier and thus do not need to distinguish between unique and non-unique indexes.

Lomet's design for key range locking [4] attempts to adapt hierarchical and multi-granularity locking to keys and half-open intervals but requires additional lock modes, e.g., a "range insert" mode, to achieve the desired concurrency. Graefe's design [9] applies traditional hierarchical locking to keys and gaps (open intervals) between keys, employs ghost (pseudo-deleted) records during insertion as well as during deletion, and permits more concurrency with fewer special cases. The same paper also outlines hierarchical locking exploiting B-trees' hierarchical structure or multi-field B-tree keys.

Foundations

The foundations of B-tree locking are the well-known transaction concepts, including multi-level transactions and open nested transactions, and pessimistic concurrency control, i.e., locking. Multiple locking concepts and techniques are employed, including two-phase locking, phantom protection, predicate locks, precision locks, key value locking, key range locking, multi-granularity locking, hierarchical locking, and intention locks.

Preliminaries

Most work on concurrency control and recovery in B-trees assumes what Bayer and Schkolnick call B^*-trees [1] and what Comer calls B^+-trees [2], i.e., all data records are in leaf nodes and keys in non-leaf or "interior" nodes act merely as separators enabling search and other operations but not carrying logical database contents. Following this tradition, this entry ignores the original design of B-trees with data records in interior nodes.

Also ignored are many other variations of B-trees here. This includes what Comer, following Knuth, calls B^*-trees, i.e., attempting to merge an overflowing node with a sibling rather than splitting it immediately. Among the ignored techniques are whether or not underflow is recognized and acted upon by load balancing and merging nodes, whether or not empty nodes are removed immediately or ever, whether or not leaf nodes form a singly or doubly linked list using physical pointers (page identifiers) or logical boundaries (fence keys equal to separators posted in the parent node during a split), whether suffix truncation is employed when posting a separator key, whether prefix truncation or any other compression is employed on each page, and the type of information associated with B-tree keys. Most of these issues have little or no bearing on locking in B-trees, with the exception of sibling pointers, as indicated below where appropriate.

Two Forms of B-Tree Locking

B-tree locking, or locking in B-tree indexes, means two things. First, it means concurrency control among concurrent database transactions querying or modifying database contents and its representation in B-tree indexes. Second, it means concurrency control among concurrent threads modifying the B-tree data structure in memory, including in particular images of disk-based B-tree nodes in the buffer pool.

These two aspects have not always been separated cleanly. Their difference becomes very apparent when a single database request is processed by multiple parallel threads. Specifically, two threads within the same transaction must "see" the same database contents, the same count of rows in a table, etc. This includes one thread "seeing" updates applied by the other thread. While one thread splits a B-tree node, however, the other thread should not observe intermediate and incomplete data structures. The difference also becomes apparent in the opposite case when a single operating system thread is multiplexed to serve all user transactions.

These two purposes are usually accomplished by two different mechanisms, locks and latches. Unfortunately, the literature on operating systems and programming environments usually uses the term locks for the mechanisms that in database systems are called latches, which can be confusing.

Locks separate transactions using read and write locks on pages, on B-tree keys, or even gaps (open intervals) between keys. The latter two methods are called key value locking and key range locking. Key range locking is a form of predicate locking that uses actual key values in the B-tree and the B-tree's sort order to define predicates. By default, locks participate in deadlock detection and are held until end-of-transaction. Locks also support sophisticated scheduling, e.g., using queues for pending lock requests and delaying new lock acquisitions for lock conversions, e.g., an existing shared lock to an exclusive lock. This level of sophistication makes lock acquisition and release fairly expensive, often thousands of CPU cycles, some of those due to cache faults in the lock manager's hash table.

Latches separate threads accessing B-tree pages, the buffer pool's management tables, and all other in-memory data structures shared among multiple threads. Since the lock manager's hash table is one of the data structures shared by many threads, latches are required while inspecting or modifying a database system's lock information. With respect to shared data structures, even threads of the same user transaction conflict if one thread requires a write latch. Latches are held only during a critical section, i.e., while a data structure is read or updated. Deadlocks are avoided by appropriate coding disciplines, e.g., requesting multiple latches in carefully designed sequences. Deadlock resolution requires a facility to roll back prior actions, whereas deadlock avoidance does not. Thus, deadlock avoidance is more appropriate for latches, which are designed for minimal overhead and maximal performance and scalability. Latch acquisition and release may require tens of instructions only, usually with no additional cache faults since a latch can be embedded in the data structure it protects.

Latch Coupling and B^{link}-Trees

Latches coordinate multiple concurrent threads accessing shared in-memory data structures, including images of on-disk storage structures while in the buffer pool. In the context of B-trees, latches solve several problems that are similar to each other but nonetheless lend themselves to different solutions.

First, a page image in the buffer pool must not be modified (written) by one thread while it is interpreted (read) by another thread. For this issue, database systems employ latches that differ from the simplest implementations of critical sections and mutual exclusion only by the distinction between read-only latches and read-write latches, i.e., shared or exclusive access. Latches are useful not only for pages in the buffer pool but also for the buffer pool's table of contents or the lock manager's hash table.

Second, while following a pointer (page identifier) from one page to another, e.g., from a parent node to a child node in a B-tree index, the pointer must not be invalidated by another thread, e.g., by deallocating a child page or balancing the load among neighboring pages. This issue requires retaining the latch on the parent node until the child node is latched. This technique is traditionally called "lock coupling" or better "latch coupling."

Third, "pointer chasing" applies not only to parent-child pointers but also to neighbor pointers, e.g., in a chain of leaf pages during a scan. This issue is similar to the previous, with two differences. On the positive side, asynchronous read-ahead may alleviate the frequency of buffer faults. On the negative side, deadlock avoidance among scans in opposite directions requires that latch acquisition code provides an immediate failure mode.

Fourth, during a B-tree insertion, a child node may overflow and require an insertion into its parent node, which may thereupon also overflow and require an insertion into the child's grandparent node. In the most extreme case, the B-tree's root node splits and a new root node is added. Going back from the leaf towards the B-tree root works well in single-threaded B-tree implementations, but in multi-threaded code it introduces the danger of dead-locks. This issue affects all updates, including insertion, deletion, and even record updates, the latter if length changes in variable-length records can lead to nodes splitting or merging. The most naïve approach, latching an entire B-tree with a single exclusive latch, is obviously not practical in multi-threaded servers.

One approach latches all nodes in exclusive mode during the root-to-leaf traversal. The obvious problem in this approach is the potential concurrency bottleneck, particularly at a B-tree's root. Another approach performs the root-to-leaf search using shared latches and attempts an upgrade to an exclusive latch when necessary. A third approach reserves nodes using "update" or "upgrade" latches. A refinement of these three approaches retains latches on nodes along its root-to-leaf search only until a lower, less-than-full node guarantees that split operations will not propagate up the tree beyond the lower node.

Since most nodes are less than full, most insertion operations will latch no nodes in addition to the current one.

A fourth approach splits nodes proactively during a root-to-leaf traversal for an insertion. This method avoids both the bottleneck of the first approach and the failure point (upgrading a latch) of the second approach. Its disadvantage is that it wastes some space by splitting earlier than truly required. A fifth approach protects its initial root-to-leaf search with shared latches, aborts this search when a node requires splitting, restarts a new one, and upon reaching the node requiring a split, acquires an exclusive latch and performs the split.

An entirely different approach relaxes the data structure constraints of B-tress and divides a node split into two independent steps. Each node has a high fence key and a pointer to its right neighbor, thus the name B^{link}-trees. The right neighbor might not yet be referenced in the node's parent and a root-to-leaf search might need to proceed to the node's right neighbor. The first step of splitting a node creates the high fence key and a new right neighbor. The second, independent step posts the high fence key in the parent. The second step should happen as soon as possible yet it may be delayed beyond a system reboot or even a crash. The advantage of B^{link}-trees is that allocation of a new node and its initial intro-duction into the B-tree is a local step, affecting only one preexisting node. The disadvantages are that search may be a bit less efficient, a solution is needed to prevent long linked lists among neighbor nodes during periods of high insertion rates, and verification of a B-tree's structural consistency is more complex and perhaps less efficient.

Key Range Locking

Locks separate transactions reading and modi-fying database contents. For serializability, read locks are retained until end-of-transaction. Write locks are always retained until end-of-transaction in order to ensure the ability to roll back all changes if the transaction aborts. High concur-rency requires a fine granularity of locking, e.g., locking individual keys in B-tree indexes. The terms key value locking and key range locking are often used interchangeably.

Key range locking is a special form of predicate locking. The predicates are defined by intervals in the sort order of the B-tree. Interval boundaries are the key values currently existing in the B-tree, which form half-open intervals including the gap between two neighboring keys and one of the end points.

In the simplest form of key range locking, a key and the gap to the neighbor are locked as a unit. An exclusive lock is required for any form of update of this unit, including modifying non-key fields of the record, deletion of the key, insertion of a new key into the gap, etc. Deletion of a key requires a lock on both the old key and its neighbor; the latter is required to ensure the ability to re-insert the key in case of transaction rollback.

High rates of insertion can create a hotspot at the "right edge" of a B-tree index on an attribute correlated with time. With next-key locking, one solution verifies the ability to acquire a lock on $+\infty$ but does not actually retain it. Such "instant locks" violate two-phase locking but work correctly if a single acquisition of the page latch protects both verification of the lock and creation of the new key on the page.

In those B-tree implementations in which a deletion does not actually erase the record and instead merely marks the record as invalid, "pseudo-deleted," or a "ghost" record, each ghost record's key participates in key range locking just like a valid record's key. Another technique to increase concurrency models a key, the appropriate neighboring open interval, and the combination of key and open interval as three separate items [9]. These items form a hierarchy amenable to multi-granularity locking. Moreover, since key, open interval, and their combination are all identified by the key value, additional lock modes can replace multiple invocations of the lock manager by a single one, thus eliminating the execution costs of this hierarchy.

Multi-granularity locking also applies keys and individual rows in a non-unique index, whether such rows are represented using multiple copies of the key, a list of row identifiers

associated with a single copy of the key, or even a bitmap. Multi-granularity locking techniques exploiting a B-tree's tree structure or a B-tree's compound (multi-column) key have also been proposed. Finally, "increment" locks may be very beneficial for B-tree indexes on materialized summary views [5].

Both proposals need many details worked out, e.g., appropriate organization of the lock manager's hash table to ensure efficient search for conflicting locks and adaptation during structure changes in the B-tree (node splits, load balancing among neighboring nodes, etc.).

Key Applications

B-tree indexes have been called ubiquitous more than a quarter of a century ago [2], and they have become ever more ubiquitous since. Even for single-threaded applications, concurrent threads for maintenance and tuning require concurrency control in B-tree indexes, not to mention online utilities such as online backup. The applications of B-trees and B-tree locking are simply too numerous to enumerate them.

Future Directions

Perhaps the most urgently needed future direction is simplification - concurrency control and recovery functionality and code are too complex to design, implement, test, tune, explain, and maintain. Elimination of any special cases without a severe drop in performance or scalability would be welcome to all database development and test teams.

At the same time, B-trees are employed in new areas, e.g., Z-order UB-trees for spatial and temporal information, various indexes for unstructured data and XML documents, in-memory and on-disk indexes for data streams and as caches of reusable intermediate query results. It is unclear whether these application areas require new concepts or techniques in B-tree concurrency control.

Online operations – load and query, incremental online index creation, reorganization & optimization, consistency check, trickle load and zero latency in data warehousing including specialized B-tree structures.

Scalability – granularities of locking between page and index based on compound keys or on B-tree structure; shared scans and sort-based operations including "group by," merge join, and poor man's merge join (index nested loops join); delegate locking (e.g., locks on orders cover order details) including hierarchical delegate locking.

B-tree underpinnings for non-traditional database indexes, e.g., blobs, column stores, bitmap indexes, and master-detail clustering.

Confusion about transaction isolation levels in plans with multiple tables, indexes, materialized and indexed views, replicas, etc.

URL to Code

Gray and Reuter's book [6] shows various examples of sample code. In addition, the source of various open-source database systems is readily available.

Cross-References

▶ Database Benchmarks
▶ Locking Granularity and Lock Types
▶ Two-Phase Commit
▶ Two-Phase Locking

Recommended Reading

1. Bayer R, Schkolnick M. Concurrency of operations on B-trees. Acta Inf. 1977;9(1):1–21.
2. Comer D. The ubiquitous B-tree. ACM Comput Surv. 1979;11(2):121–37.
3. Eliot J, Moss B. Open nested transactions: semantics and support. In: Proceedings of Workshop on Memory Performance Issues; 2006.
4. Graefe G. Hierarchical locking in B-tree indexes. In: The Conference for Database Systems for Business, Technology, and Web; 2007. p. 18–42.
5. Graefe G, Zwilling MJ. Transaction support for indexed views. In: Proceedings of ACM SIGMOD International Conference on Management of Data; 2004.
6. Gray J, Reuter A. Transaction processing: concepts and techniques. San Francisco: Morgan Kaufmann; 1993.
7. Jaluta I, Sippu S, Soisalon-Soininen E. Concurrency control and recovery for balanced B-link trees. VLDB J. 2005;14(2):257–77.
8. Lehman PL, Yao SB. Efficient locking for concurrent operations on B-trees. ACM Trans Database Syst. 1981;6(4):650–70.
9. Lomet DB. Key range locking strategies for improved concurrency. In: Proceedings of 19th International Conference on Very Large Data Bases; 1993. p. 655–64.
10. Mohan C. ARIES/KVL: a key-value locking method for concurrency control of multiaction transactions operating on B-tree indexes. In: Proceedings of 16th International Conference on Very Large Data Bases; 1990. p. 392–405.
11. Mohan C, Haderle DJ, Lindsay BG, Pirahesh H, Schwarz PM. ARIES: a transaction recovery method supporting fine-granularity locking and partial rollbacks using write-ahead logging. ACM Trans Database Syst. 1992;17(1):94–162.
12. Ni Y, Menon V, Adl-Tabatabai A-R, Hosking AL, Hudson RL, Moss JEB, Saha B, Shpeisman T. Open nesting in software transactional memory. In: Proceedings of 12th ACM SIGPLAN Symposium on Principles and Practice of Parallel Programming; 2007. p. 68–78.
13. Srinivasan V, Carey M.J. Performance of B-tree concurrency algorithms. In: Proceedings of ACM SIGMOD International Conference on Management of Data; 1991. pp. 416–25.
14. Weikum G. Principles and realization strategies of multilevel transaction management. ACM Trans Database Syst. 1991;16(1):132–80.

Buffer Management

Giovanni Maria Sacco
Dipartimento di Informatica, Università di Torino, Torino, Italy

Definition

The database buffer is a main-memory area used to cache database pages. Database processes request pages from the buffer manager, whose responsibility is to minimize the number of secondary memory accesses by keeping needed pages in the buffer. Because typical database workloads are I/O-bound, the effectiveness of buffer management is critical for system performance.

Historical Background

Buffer management was initially introduced in the 1970s, following the results in virtual memory systems. One of the first systems to implement it was IBM System-R. The high cost of main-memory in the early days forced the use of very small buffers and consequently moderate performance improvements.

Scientific Fundamentals

The buffer is a main-memory area subdivided into frames, and each frame can contain a page from a secondary storage database file. Database pages are requested from the buffer manager. If the requested page is in the buffer, it is immediately returned to the requesting process with no secondary memory access. Otherwise, a fault occurs and the page is read into a free frame. If no free frames are available, a "victim" page is selected and its frame is freed by clearing its content, after writing it to secondary storage if the page was modified. Usually any page can be selected as a victim, but some systems allow processes actively using a page to fix or pin it, in order to prevent the buffer manager from discarding it [1]. Asynchronous buffered write operations have an impact on the recovery subsystem and require specific protocols not discussed here.

There are obvious similarities between buffer management and virtual memory (VM) systems [2]. In both cases, the caching system tries to keep needed pages in main-memory in order to minimize secondary memory accesses and hence speed up execution. As in VM systems, buffer management is characterized by two policies: the **admission policy**, which determines when pages are loaded into the buffer, and the **replacement policy**, which selects the page to be replaced

when no empty frames are available. The admission policy normally used is demand paging (i.e., pages are read into the buffer when requested by an application), although prefetching (pages are read before applications request them) was studied (e.g., [3]). Since the interaction with the caching system is orders of magnitude less frequent in database systems than in VM systems, "intelligent" replacement policies such as LRU [2] (the Least Recently Used page is selected for replacement) can be implemented in software, with no performance degradation. Inverted page tables are used because their space requirement is proportional to the buffer size rather than to the entire database space as in normal page tables. Finally, database pages in the buffer can be shared among different processes, whereas the amount of sharing in VM systems is usually negligible.

The VM Approach

Besides minor architectural differences, the buffer manager can be used exactly as a virtual memory system for database pages, so that buffer management is transparent for the database system. In this VM approach, research is focused on effective replacement policies, which include GCLOCK and LRD (Least Reference Density) [1], LRU, and, more recently, efficient policies that account for page popularity, such as LRU-K [4].

LRU does not discriminate between frequently and infrequently referenced pages, and once a page is admitted in a buffer of B frames, it will stay there for at least B-1 references, even if not referenced again. Therefore, a potentially large portion of the buffer can be wasted by caching useless (i.e., infrequent) pages. **LRU-K** dynamically estimates the interreference distance for each page in the buffer by keeping the history of the last K references for each page in the buffer. A shorter interreference distance means a more frequently accessed page. Consequently, the page selected for replacement is the one with the largest interreference distance, i.e., the one with the maximum backward K-distance. The backward K-distance $b_t(p, K)$

at time t is defined as the distance backward to the Kth most recent reference the page p. When K = 1, only the last reference is considered, and LRU-1 is equivalent to LRU. As K grows, so do space and time overheads, because longer histories must be stored and kept ordered. At the same time, larger values of K improve the estimate of the interreference distance, but make the algorithm less responsive to dynamic changes in page popularity, so that, in practice, LRU-2 is normally used. The **2Q** (two queues) [5] algorithm provides an efficient, constant-time algorithm equivalent to LRU-2 replacement.

Although LRU-K improves LRU replacement through additional information about page access frequency, it is fundamentally different from the Least Frequently Used (LFU) replacement, because LFU does not adapt its estimate to evolving access patterns. The idea of exploiting both recency and frequency of access (also present in LRD replacement [1]) is extended by **LRFU** (least recently/frequently used) replacement [6], to model a parametric continuum of replacement policies ranging from LRU to LFU. The Combined Recency and Frequency (CRF) information is associated to each page in the buffer. CRF weights page references giving higher weights to more recent references and combines frequency and recency information through a parameter λ ($0 \leq \lambda \leq 1$). The page to be replaced is the one with minimum CRF. When $\lambda = 0$, LRFU becomes LFU; when $\lambda = 1$, it becomes LRU. The optimal λ depends on the actual workload, but a self-tuning strategy can be used. An efficient implementation of LRFU exists, and experiments show this replacement strategy to outperform LRU-K with small buffer sizes.

A known problem in the VM approach is that applications with fast sequential scans over large relations tend to fill the buffer with useless pages accessed in the scan and flush the active pages of other applications out of memory, thereby significantly increasing the overall fault rate [7]. In a sequential scan, the current page is (a) rapidly accessed several times in order to read all the tuples on it and (b) once the last tuple in the page is read, the page will never be reaccessed

again. Reaccess in (a) makes VM strategies to incorrectly estimate that the current page is likely to be reaccessed in the future and therefore to keep it in the buffer at the expense of other pages. In order to avoid this problem, some strategies use parametric correctives. LRU-K, for instance, uses the Correlated Reference Period. During this period, a page freshly admitted to the buffer cannot be replaced because additional references are expected. At the same time, subsequent references during this period are not tracked because they are expected to be correlated and do not give a reliable indication of future behavior. The length of such period is at the same time critical and difficult to estimate.

The Predictive Approach

A completely different, predictive approach was proposed with the **hot set model** [7, 8]. In database systems with nonprocedural data manipulation languages (such as relational or object-relational database systems), it is not the programmer but the system query optimizer that determines the access plan for a query. Such access plan is based on a small number of primitives, such as sequential scans, nested loop joins, among others, whose reference string is known or can be estimated before actual execution. As an example, consider the execution of a tuplewise nested loop join between relation R (3 tuples/page, 40 pages) and relation S (2 tuples/page, 30 pages). Assuming S inner and indicating by Xi the i-th page of relation X, and by $\{a\}^N$ N repetitions of string a, the reference string is exactly $\{R1, \{S1\}^2, \ldots, \{S30\}^2\}^3, \{R2, \{S1\}^2, \ldots, \{S30\}^2\}^3, \ldots, \{R40, \{S1\}^2, \ldots, \{S30\}^2\}^3$. Figure 1 plots the number of faults as a function of the available buffer space, under LRU replacement.

Figure 1 shows that differently from VM systems where fault curves tend to be smooth, the fault curve here is discontinuous. If the buffer is not large enough to contain all the pages in the inner relation plus one frame for the current page in the outer one, no reusal occurs and the fault rate is the same as for 1 frame. If sufficient space exists, all the needed pages are kept in the buffer, and the cost becomes linear in the number of pages of the two relations. In this case, the entire behavior of the plan in which S is inner is completely characterized by two *hot points* and their corresponding fault rate:

- $hp1 = 1$, $faults(hp1) = |R|(1 + pages(S))$
- $hp2 = 1 + pages(S)$, $faults(hp2) = pages(R) + pages(S)$

The number of frames needed by a plan is called its hot set size and usually is the largest hot point no larger than the total buffer size. Note that the length of the loop on the inner relation is exactly known at the database level, but it is very hard to discover in the VM approach.

The basic idea of the hot set model is that given an access plan and a replacement policy, the number of faults as a function of the available buffer space can be predicted. Three main types of reusal, modeled upon primitive strategies, are considered:

1. Simple reusal. It models the sequential scan of a relation: the current page is accessed several times in order to read all the tuples on it, but once the last tuple in the page is read, the page will never be reaccessed again. There is only one hot point at 1 frame;

2. Loop reusal. Used in the example in Fig. 1. There are as many hot points as there are loops;
3. Index reusal. It includes both clustered and unclustered index access. Reusal can occur when indices are used in a join, or when an index is accessed by several processes. In unclustered index access, data pages are accessed randomly, and the hot set is estimated through Yao's function.

It must be stressed that LRU replacement is not required and that other replacement strategies can be adopted [8, 9]. As a matter of fact, both simple and loop reusal do not benefit from LRU replacement at all. The **query locality set model** (QLSM) [9] extends the hot set model by determining the hot set size on a file-instance basis rather than on a primitive operation basis and by adopting replacement policies appropriate for each type of reusal. As an example, a nested loop join between R and S (S inner) is characterized by two different access patterns: a simple reusal (sequential scan) on R, requiring 1 frame, and a loop reusal on S, requiring 1 to pages(S) frames. MRU (most recently used) replacement can be used for loop reusal, since it is more efficient than LRU for this type of reference string.

In fact, since the reference string of the application is known or can be estimated, the investigation of a plurality of replacement policies is not

Buffer Management, Fig. 1 Number of faults as a function of the available buffer space, under LRU replacement

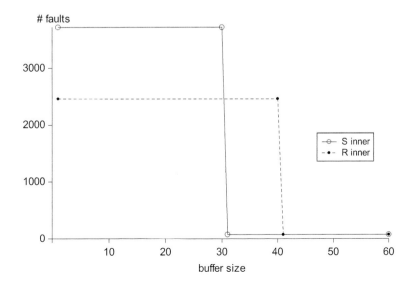

needed and optimal replacement policies (such as Belady's OPT [10]) can be used. This opportunity was first exploited in **OLRU** [11], which derives an optimal replacement policy for clustered index reusal, e.g., repeated access to clustered B+ trees. In this case, the Independent Reference Model (IRM) [2] is used. IRM was originally proposed as a theoretical evaluation model for VM systems and assumes, in extreme synthesis, that the probability of reaccess for each page is known and stationary. Under this assumption, it can be proved that the optimal buffering strategy for a buffer of B frames consists of ordering the pages by decreasing reaccess probabilities, fixing the first B-1 high-probability pages in the buffer and using the remaining frame to access all the other low-probability pages. Since the optimal strategy only requires a ranking among pages, it can be straightforwardly applied to a B+ tree of order m and height h by assuming a uniform distribution of access to the leaf level. In this case, the probability of reaccessing a page at level j ($0 \leq j < h$) is $1/m^j$. The optimal policy is then an allocation by levels, i.e., fixing the first levels in the tree in the buffer. LRU replacement allocates the buffer by traversal stacks and is therefore suboptimal, unless severe deviations from uniformity in data page access occur, in which case OLRU and LRU are similar. LRU-K, which assumes IRM as well, can be seen as a runtime approximation of OLRU.

IRM can also be used for loop reusal. In this case, the reaccess probability is the same for any page in the loop, and optimal replacement consists of fixing B-1 pages in the loop in the buffer. Any subset of cardinality B-1 can be fixed in the buffer. If the subset consisting of the first B-1 pages in the loop is selected, the resulting replacement strategy is equivalent to MRU and OPT.

The use of optimal replacement strategies is no substitute for research on efficient implementation of database operations. Optimal replacement (e. g., OPT) does not guarantee per se the optimality of an evaluation method. Consider nested scans [12] vs. nested loops: for insufficient buffer sizes, nested scans are more efficient than nested loops with OPT replacement. In fact, OPT guarantees the minimum number of faults for a given reference string only (e.g., the one produced by nested loops). A different reference string (e.g., the one produced by nested scans) can in fact produce a lower number of faults.

The difference in performance among different replacement strategies (including optimal replacement) becomes negligible or null when the buffer allocated to a query is equal to its largest hot point. When the largest hot point exceeds the available buffer, its size can be reduced by increasing data locality in the buffer through preprocessing. Known methods for join operations by sequential scans include sorting, fragmentation [13] (*aka* recursive hash partitioning), and semijoins. As an example, the basic version of fragmentation recursively partitions both relations on the value of the join attribute until the smallest fits in B-1 frames: the largest hot point is now B, all required data fit in the buffer, and the cost is minimum. Data locality is increased by partitioning: in fact, if K total values of the join attributes are uniformly partitioned among N partitions, each partition P only contains a subset S_P of join values whose cardinality is K/N and each relation fragment in P only the tuples whose join value is in S_P.

A complete characterization of query buffer requirements and corresponding access costs is required for two major reasons. First, query optimizers need to have a precise **cost estimate** as a function of the available buffer size, in order to discriminate execution plans. This is especially important because (a) cost curves can intersect as shown in Fig. 1: R inner is cheaper in the interval [1, 30], whereas S inner is better or no worse for buffer sizes of at least 31 frames; (b) many query evaluation strategies (e.g., nested loops) do not explicitly account for available memory, and they must be compared with other strategies (such as fragmentation/recursive hash partitioning), in which available buffer space is fully accounted for and directly managed.

Second, **thrashing** phenomena [2] can occur in database buffers as in virtual memory systems and become potentially more frequent as the number of concurrent active users increases. Thrashing occurs when the available memory is insufficient to keep all the pages each active process needs. Consequently, processes steal pages from each other and, if the available memory is severely overcommitted, the system collapses because all activity is devoted to swapping pages to and from main-memory. In VM systems, thrashing can be avoided by (a) monitoring the pagination device for excessive utilization or (b) using the working set model [2] in order to estimate the memory requirements of active processes. In buffer management, there is no pagination device and accesses may be directed to a high number of secondary devices. In addition, the working set model, which is an expensive run-time estimator, cannot be efficiently used in database systems [8].

Thrashing avoidance, and the more general problem of **scheduling** queries for execution in order to optimally use buffer resources, is considerably simplified in the predictive approach because buffer requirements for each query are known before execution. The simplest policy [8] schedules queries for execution in such a way that the sum of their hot set sizes does not exceed the total buffer space. Each query can then be run in isolation in its own buffer partition, and this, by definition, avoids thrashing. However, page sharing, which is relatively frequent in database systems, is not accounted for: if two different active queries request the same page, two faults occur. If sharing is considered these additional faults can be avoided, and the actual required buffer size can decrease because a page shared by several processes requires only one frame. For these reasons, hot set scheduling [7, 8] maintains an additional global LRU chain to manage free pages and, on a local page fault, scans the entire buffer for potential shared pages. In addition, a measure of buffer consumption is used in order to avoid unnecessarily inflating hot set sizes, which would result in serializing small queries behind queries with high buffer requirements. Variations of this scheduling algorithm include DBMIN [9] (where different local replacement policies can be used), scheduling with marginal gains [14], and scheduling with prefetching [3].

The VM Versus the Predictive Approach

When compared to the predictive approach, the VM approach has the advantage of placing all concerns on buffering into a single system component that can be seen as a black box from the rest of the system. In addition, the VM approach is more generally applicable since it does not require nonprocedural interactions and inherently implements page sharing.

However, for nonprocedural systems, the predictive approach solves a number of important problems:

- Uniformity of estimated costs for query optimization. Query plans can be compared, regardless of whether methods directly manage memory or not.
- Thrashing avoidance, and efficient, low cost query scheduling. The predictive approach characterizes requirements before execution and does not require expensive run-time estimators.
- Resource planning for self-tuning databases. Predictive characterization of buffer requirements can be used to determine the optimal buffer size for actual workloads.
- Better performance. Since reference strings are known, optimal replacement policies can be used and buffer requirements carefully tuned.

There is a duality between detecting sequential scans and detecting page sharing. The VM approach has no problems in implementing page sharing. VM strategies work reasonably well for index access, but despite correctives, they tend to break on simple and loop reusal. Conversely, sequential scan detection and inner loop size detection is trivial in the predictive approach, but since prediction is based on a query run in isolation, run-time corrections are required for page sharing. This duality suggests that a combination of the two approaches might be beneficial.

Key Applications

The trend towards cheaper and larger main-memories does not make buffer management less important. In fact, the increase in available main-memory has been so far matched by a larger increase in secondary storage capacity and in the amount of data to be managed, in the complexity of queries, and in the number of users. Consequently, buffer management continues to be a fundamental topic for database systems, and indeed its results carry over to different areas, such as web caching.

Current research areas in database systems include automatic buffer sizing in self-tuning databases [15], extendibility of buffer management strategies [16], specific buffering strategies for object databases [17], XML databases and P2P data architectures, multidimensional databases, real-time databases where process priorities must be considered in scheduling, and, of course, new index structures or evaluation strategies for which buffer analysis is required [18]. In addition, challenging applications managing very large amounts of data such as sensor data, search engines, large digital libraries, among others require high-performance buffer management.

Interestingly, some of the results of the predictive approach carry back to VM systems. A new research direction in VM management is to exploit access hints from processes [19]. Thus, the knowledge of the reference string which was shown to be central to efficient buffer management in database systems is now being considered in VM systems as well.

Cross-References

▶ Evaluation of Relational Operators

Recommended Reading

1. Effelsberg W, Haerder T. Principles of database buffer management. ACM Trans Database Syst. 1984;9(4):560–95.

2. Coffman EG Jr, Denning PJ. Operating systems theory. Englewood Cliffs: Prentice-Hall; 1973.

3. Cai FF, Hull MEC, Bell DA. Buffer management for high performance database systems. In: Proceedings of the High-Performance Computing on the Information Superhighway; 1997. p. 633–8.

4. O'Neil EJ, O'Neil PE, Weikum G. The LRU-K page replacement algorithm for database disk buffering. In: Proceedings of ACM SIGMOD Conference; 1993. p. 297–306.

5. Johnson T, Shasha D. 2Q: a low overhead high performance buffer management replacement algorithm. In: Proceedings of the 20th International Conference on Very Large Data Bases; 1994. p. 439–50.

6. Lee D, Choi J, Kim J-H, Noh SH, Min SL, Cho Y, Kim CS. On the existence of a spectrum of policies that subsumes the least recently used (LRU) and least frequently used (LFU) policies. In: Proceedings of ACM SIGMETRICS Conference; 1999. p. 134–43.

7. Sacco GM, Schkolnick M. A mechanism for managing the buffer pool in a relational database system using the hot set model. In: Proceedings of the 8th International Conference on Very Data Bases; 1982. p. 257–62.

8. Sacco GM, Schkolnick M. Buffer management in relational database systems. ACM Trans Database Syst. 1986;11(4):473–98.

9. Chou H-T, DeWitt DJ. An evaluation of buffer management strategies for relational database systems. In: Proceedings of the 11th International Conference on Very Large Data Bases; 1985. p. 174–88.

10. Belady LA. A study of replacement algorithms for a virtual-storage computer. IBM Syst J. 1966;5(2):78–101.

11. Sacco GM. Index access with a finite buffer. In: Proceedings of the 13th International Conference on Very Large Data Bases; 1987. p. 301–9.

12. Kim W. A new way to compute the product and join of relations. In: Proceedings of 1980 ACM SIGMOD Conference; 1980. p. 179–87.

13. Sacco GM. Fragmentation: a technique for efficient query processing. ACM Trans Database Syst. 1986;11(2):113–33.

14. Ng R, Faloutsos C, Sellis T. Flexible buffer allocation based on marginal gains. In: Proceedings of ACM SIGMOD Conference; 1991. p. 387–96.

15. Storm AJ, Garcia-Arellano C, Lightstone SS, Diao Y, Surendra M. Adaptive self-tuning memory in DB2. In: Proceedings of the 32nd International Conference on Very Large Data Bases; 2006. p. 1081–92.

16. Goh L, Shu Y, Huang Z, Ooi C. Dynamic buffer management with extensible replacement policies. VLDB J. 2006;15(2):99–120.

17. Kemper A, Kossmann D. Dual-buffering strategies in object bases. In: Proceedings of the 20th International Conference on Very Large Data Bases; 1994. p. 427–38.

18. Corral A, Vassilakopoulos M, Manolopoulos Y. The impact of buffering on closest pairs queries using R-trees. In: Proceedings of the 5th East European

Conference on Advances in Databases and Information Systems; 2001. p. 41–54.

19. Gu X, Ding C. A generalized theory of collaborative caching. SIGPLAN Not. 2012;47(11):109–20.

Buffer Manager

Goetz Graefe
Google, Inc., Mountain View, CA, USA

Synonyms

Cache manager

Definition

If a buffer pool is employed in a database management system, the associated software must provide appropriate services for efficient query processing, correct transaction execution, and effective sharing and reuse of database pages. It must provide interfaces for page access including pinning and latching pages, and it must invoke primitives for disk I/O and synchronization.

Historical Background

Database buffer pool management was studied heavily in the 1970s and 1980s as the new relational database management system posed new challenges, in particular non-procedural queries with range scans. Reliance of virtual memory and file system buffer pool was investigated but rejected due to performance issues (read-ahead, prefetch) and correctness issues (transaction management, write-behind, write-through).

Buffer pool management is currently not a very active research area. It may be revived in order to serve deep storage hierarchies, e.g., slow disk, fast disk, flash memory, main memory, and CPU cache. It may also be revived in the context of very large memories, e.g., query optimization that considers residence in the buffer pool or physical database design that includes temporary or partial indexes that exist only in the buffer pool.

Foundations

This section describes a database system's buffer pool and its management by focusing on management of individual pages in the buffer pool, replacement policies, asynchronous I/O, and the requirements imposed by concurrency control and recovery.

Buffer Pool Interfaces

The principal methods provided by the buffer pool manager request and release a page. A page request fixes or pins a page, i.e., it protects it from replacement as well as movement within the buffer pool, thus permitting access to the page by in-memory pointers. Variants of requesting and releasing a page apply to disk pages immediately after allocation (no need to read the page contents from disk) and immediately after deallocation (no need to save the page contents).

In addition, a buffer pool may provide methods for concurrency control among threads in order to protect page contents, also known as latching. Pragmatically, the methods to pin and to release a page include parameters that control latching.

Concurrency control among transactions is usually not provided by the buffer pool. In aid of logging and recovery, a buffer pool must support a method to force a page to disk and may support a method to retain one page until another page has been written.

For performance, a buffer pool may support methods to hint asynchronous prefetch, read-ahead, and write-behind. If the buffer pool serves a memory pool for other software layers, it must support methods for memory allocation and deallocation.

The principal methods upon which a buffer pool manager relies are disk reading and writing, including asynchronous operations, and scheduling primitives to implement latching.

If the buffer pool size is dynamic, memory allocation and deallocation are also required. If the buffer pool supports multiple page sizes, it requires fast methods for moving (copying) page contents from one memory address to another.

Replacement Policies

The goal of the replacement policy (or retention policy) is to speed up future page accesses. Prediction of future accesses can be based on past accesses (how recent, how frequent, whether read or write) or on hints from higher software layers within the database management system. Standard policies include LRU (least recently used, implemented using a doubly-linked list), LRU-K (least recent K uses), LFU (least frequently used), second chance (usually implemented following a clock metaphor), generalized clock (using counters instead of a single bit per page frame), and combinations of those. Many combinations have been proposed, including the hot set model, the query locality set model, adaptive replacement cache, etc.

They differ in their heuristics to separate pages used only once, e.g., in a large sequential scan, from pages likely to be reused, e.g., pages containing the database catalog or root pages of B-tree indexes. Alternative designs let higher software layers hint the likelihood of reuse, e.g., love/hate hints or keep/toss hints.

Dirty pages (containing recent updates) may be retained longer than clean ones because their replacement cost and delay are twice as high (write plus read instead of merely a read operation) and because correct preparation for recovery may impose restrictions on the order in which pages are written. For example, write-ahead logging requires writing the relevant log page to stable storage before overwriting old database contents. On the other hand, non-logged operations (e.g., index creation) require flushing dirty pages as part of transaction commit.

Asynchronous I/O

Rather than merely responding to requests from higher software layers, a buffer pool may employ or enable asynchronous I/O, in three forms:

Write-behind cleans the buffer pool of dirty pages in order to complete update transactions as fast as possible yet enable quick page replacement without needing a write operation prior to a read operation. However, most write-behind is driven by checkpoints rather than page faults, based on typical checkpoint intervals (very few minutes) and retention intervals as calculated or optimized using the five-minute rule (many minutes or even hours). A write-behind operation may leverage a disk seek forced by another read or write operation. Alternatively, a write-behind operation may move the data to achieve this effect, e.g., in log-structured file systems and write-optimized database indexes.

Read-ahead speeds up large scans, e.g., a range scan in a B-tree or a complete scan of a heap structure. The appropriate amount of read-ahead is the product of bandwidth and latency, i.e., the smallest of I/O bandwidth and processing bandwidth multiplied with the delay from initiation to completion of a read operation. Read-ahead may be triggered by observation of the access pattern or by a hint from a higher software layer, e.g., a table scan in a query execution plan.

Prefetch accelerates fetch operations by loading into the buffer pool precisely those pages that contain needed data records. Prefetch in heap structures is quite straightforward. In B-tree indexes, prefetch may apply to all tree levels or merely to the leaf level, combined with synchronous read operations or large read-ahead for interior B-tree nodes. Prefetch speeds up not only to ordinary forward processing but also to transaction rollback as well as system recovery.

Concurrency Control and Recovery

The buffer pool may participate in concurrency control, e.g., if multi-version concurrency control requires multiple versions of individual pages. The mechanisms for pinning and latching are essential for coordination of multiple threads accessing the same in-memory data structures, including in-memory images of on-disk pages.

The buffer pool always participates in the preparation for recovery including transaction rollback, media recovery, and system recovery. For example, by retaining all modified pages in the buffer pool until transaction commit, one can avoid logging *undo* information in the persistent log – this is called a "no steal" policy. By forcing merely log pages to stable storage, one can avoid writing all modified database pages back to disk – this is called a "no force" policy. Most database management systems use "steal – no force" by default. The log volume of index creation and similar operations is often minimized using a "force" policy.

Logging volume can be reduced by ensuring specific write sequences. For example, when a B-tree node is split and some records are moved to a new node, one can avoid logging the moved records by writing the new page before writing the modified old page.

The buffer pool also participates in checkpoint processing. In addition to recording active transactions, a checkpoint must write all dirty database pages to the log or to the database. Proactive asynchronous write-behind may lessen the number of write operations during the checkpoint.

Cooperative Buffer Pool Management

Multiple buffer pools may cooperate. Prototypical examples include client-server operation (in which the buffer pools form a hierarchy) and shared-disk database systems (in which the buffer pools are peers). Those environments require optimizations for both data traffic among buffer pools and control messages, in particular for concurrency control and lock management.

Key Applications

A buffer pool and its management software are required in any database management system that employ multiple levels in a memory hierarchy, process and store data at different levels, and do not rely other means for moving data between those levels. The main example is main memory and disks – the buffer pool manager manages which data pages are immediately available for access, e.g., from the query execution engine. A CPU cache is a level in the memory hierarchy above the main memory, but data movement between main memory and CPU cache are automatic. A database management system could rely on a file system and its buffer pool manager but usually does not due to performance issues (e.g., read-ahead, prefetch) and due to correctness issues (write-behind, write-through). A database management system could rely on virtual memory provided by the operating system but typically does not for the same reasons.

Future Directions

While basic buffer pool management in traditional database management systems is well understood, there are many developments that build upon it. For example, integration of database cache, mid-tier cache, and web cache may become imperative in order to maximize efficiency and thus minimize costs for hardware, management, power, and cooling.

Buffer pools and buffer management will become more pervasive with the increased virtualization of storage and processing. At the same time, it will become more complex due to missing information on the true cost and location of data. It will also become more pervasive and complex due to increasing use of peer-to-peer storage, communication, and processing.

Any buffer pool becomes more effective with data compression and co-location. Compression reduces the space required locally (in the buffer pool) and remotely. Co-location techniques such as master-detail clustering enable access to multiple related records or pieces of information within a single frame in the buffer pool and with a single access to the remote location.

In a very large buffer pool, one might create temporary on-disk structures that never even exist on disk. For example, a temporary index on a permanent table may be created yet retained in the buffer pool. During contention in the buffer pool, the index is dropped. Ideally, such an index is partitioned, created and dropped incrementally, and left behind as a free side effect of query execution.

The data structures that manage a buffer pool, both its contents descriptors and its replacement policy, can be complex yet require very high concurrency, in particular in forthcoming many-core processors. Hardware-assisted transactional memory may simplify the software implementation effort as well as increase concurrency and performance.

In deep memory hierarchies, e.g., a three-level hierarchy of traditional memory, flash memory, and disk, contents descriptors and data structures in aid of the replacement policy may be separate. For example, the contents descriptors may need to be persistent if the flash memory is part of the persistent database, but all data structures for the replacement policy (between flash memory and disk) might be in the traditional memory.

Cross-References

- ▶ B-tree locking
- ▶ Buffer pool
- ▶ Hierarchy
- ▶ Storage manager

Recommended Reading

1. Bansal S, Modha DS. CAR: clock with adaptive replacement. In: Proceedings of the 3rd USENIX Conference on File and Storage Technologies; 2004. p. 187–200.
2. Chou H-T, DeWitt DJ. An evaluation of buffer management strategies for relational database systems. Algorithmica. 1986;1(3):311–36.
3. Effelsberg W, Härder T. Principles of database buffer management. ACM Trans Database Syst. 1984;9(4):560–95.
4. Gray J, Putzolu GR. The 5 minute rule for trading memory for disk accesses and the 10 byte rule for trading memory for CPU Time. In: Proceedings of the ACM SIGMOD International Conference on Management of Data; 1987. p. 395–8.
5. Ramamurthy R, DeWitt DJ. Buffer-pool Aware Query Optimization. In: Proceedings of the 2nd Biennial Conference on Innovative Data Systems Research; 2005. p. 250–61.
6. Stonebraker M. Operating system support for database management. Commun ACM. 1981;24(7):412–8.

Buffer Pool

Goetz Graefe
Google, Inc., Mountain View, CA, USA

Synonyms

I/O cache; Page cache

Definition

Cost constraints (dollars per gigabyte) prohibit in-memory databases in most cases, but processors can access and manipulate data only while it is in memory. The in-memory buffer pool holds database pages currently in use and retains those deemed likely to be used again soon.

The buffer pool and the buffer management component within the storage layer of a database management system provide fast access and fast recall of on-disk pages using in-memory images of those pages. In addition to the size of individual pages and of the entire buffer pool, key issues are (i) page replacement in response to buffer faults, and (ii) page retention and update in aid of database recovery.

A database buffer pool differs from virtual memory as it contributes to correctness and efficiency of query and update processing, e.g., by pinning pages while in use and by ensuring a write sequence that guarantees the ability to recover. In some systems, the buffer pool permits "stealing" memory for query processing operations such as sorting and hash join, for utilities such as reorganization and consistency checks, etc.

Historical Background

Gray and Putzolo's paper introducing the five-minute rule makes a strong case for using multiple levels of the memory hierarchy for data collections that include both hot and cold data, i.e., data that are accessed with different frequencies. Hot data are kept at a high level of the memory hierarchy, whereas cold data are fetched as needed and buffered temporarily.

A buffer pool is also needed if the provided granularity of access is too coarse. Disks permit random page accesses, whereas query execution, predicate evaluation, etc. require accesses to individual records, fields, and bytes.

Early research demonstrated that virtual memory is not appropriate for use in database management systems, as observed by Härder, Stonebraker, Traiger, and their research teams for early relational database management systems. Reasons include both correctness, specifically with respect to recovery, and performance, specifically asynchronous read-ahead and write-behind.

In order to avoid double page faults, a buffer pool should not be subject to page replacement by virtual memory provided by the operating system. One means to achieve this is to vary the size of the buffer pool in response to paging rates of the virtual memory. The five-minute rule gives guidance for the appropriate sizing of memory and buffer pool.

Techniques to map on-disk database contents into virtual memory advanced in the context of object-oriented databases and persistent programming languages but did not result in wide adoption.

Flash memory still requires an in-memory buffer pool for access to bytes within pages, yet it can also serve as a buffer pool for pages that require faster access than rotating disks can provide. In fact, buffer pool management techniques such as replacement policies apply to all levels in a multi-level memory hierarchy, e.g., CPU caches, RAM, flash memory, disk caches, rotating disk, tape media, etc. In an extreme case, a fast disk may serve a buffer pool for a slower disk; alternatively, both disks may serve as permanent storage and buffer replacement policies may be adapted for page placement on those disks, possibly including frequent page migration.

Foundations

This section describes a database system's buffer pool, replacement policies, and the requirements imposed by concurrency control and recovery.

Buffer Frames

A buffer pool contains many frames, each capable of holding the image of an on-disk page. Pages can be fixed-length or variable-length, i.e., multiple base pages, typically a power of 2. Space management is complex for variable-length pages; one technique employed commercially relies on multiple buffer pools with a single page size in each, i.e., a fixed-length page frame in each buffer pool.

In addition to being idle or unused, buffer frames can be pinned to protect the page from replacement, latched (locked) to protect against concurrent readers or writers, or in transit from or to disk. Pinning and latching are important for performance; they enable the database's query execution software to inspect or update records directly in the buffer pool.

A buffer pool may hold multiple versions of the same on-disk page in aid of transaction isolation, compression, asynchronous write-behind, or protection from partial writes. Write-behind with a single copy inhibits further updates until the write operation completes. The copy step may compact free space between variable-length records.

A small descriptor data structure is used for each frame in the buffer pool. It identifies the on-disk page and its status with respect to pinning, latching, versioning, etc. The descriptor also participates in a look-up scheme, typically a hash table, and in data structures used for page replacement.

Buffer Pool Data Structures

In addition to page frames and their descriptors, a buffer pool needs data structures to locate buffered pages and to manage page replacement including page frames currently unused. Locating a buffered page, i.e., mapping from a disk address (page number) to an in-memory address, usually is implemented with a hash table.

B

The data structures needed for page replacement depend on the page replacement policy – one method that is simple but not ideal is to employ a doubly linked list of page descriptors with pages ordered the time since they were last used. When a page is pinned, it is removed from the list. When it is unpinned, it is inserted at the head of the list. When a page is needed for replacement, the page at the tail of the list is chosen.

Replacement Policies

The goal of the replacement policy (or retention policy) is to speed up future page accesses. Prediction of future accesses can be based on past accesses (how recent, how frequent, whether read or write) or on hints from higher software layers within the database management system. Standard policies include LRU (least recently used, implemented using a doubly-linked list), LRU-K (least recent K uses), LFU (least frequently used), second chance (usually implemented following a clock metaphor), generalized clock (using counters instead of a single bit per page frame), and combinations of those. Many combinations have been proposed, including the hot set model, the query locality set model, adaptive replacement cache, etc.

They differ in their heuristics to separate pages used only once, e.g., in a large sequential scan, from pages likely to be reused, e.g., pages containing the database catalog or root pages of B-tree indexes. Alternative designs let higher software layers hint the likelihood of reuse, e.g., love/hate hints or keep/toss hints.

Dirty pages (containing recent updates) may be retained longer than clean ones because their replacement cost and delay are twice as high (write plus read instead of merely a read operation) and because correct preparation for recovery may impose restrictions on the order in which pages are written. For example, write-ahead logging requires writing the relevant log page to stable storage before overwriting old database contents. On the other hand, non-logged operations (e.g., index creation) require flushing dirty pages as part of transaction commit.

Key Applications

A buffer pool and its management software are required in any database management system that employ multiple levels in a memory hierarchy, process and store data at different levels, and do not rely other means for moving data between those levels. The main example is main memory and disks – the buffer pool manager manages which data pages are immediately available for access, e.g., from the query execution engine. A CPU cache is a level in the memory hierarchy above the main memory, but data movement between main memory and CPU cache are automatic. A database management system could rely on a file system and its buffer pool manager but usually does not due to performance issues (e.g., read-ahead, prefetch) and due to correctness issues (write-behind, write-through). A database management system could rely on virtual memory provided by the operating system but typically does not for the same reasons.

Future Directions

In deep memory hierarchies, e.g., a three-level hierarchy of traditional memory, flash memory, and disk, contents descriptors and data structures in aid of the replacement policy may be separate. For example, the contents descriptors may need to be persistent if the flash memory is part of the persistent database, but all data structures for the replacement policy (between flash memory and disk) might be in the traditional memory.

Cross-References

▶ B-tree Locking
▶ Storage Manager

Recommended Reading

1. Bansal S, Modha DS. CAR: clock with adaptive replacement. In: Proceedings of the 3rd USENIX Conference on File and Storage Technologies; 2004. p. 187–200.

2. Chou H-T, DeWitt DJ. An evaluation of buffer management strategies for relational database systems. Algorithmica. 1986;1(3):311–36.
3. Effelsberg W, Härder T. Principles of database buffer management. ACM Trans Database Syst. 1984;9(4):560–95.
4. Gray J, Putzolu GR. The 5 minute rule for trading memory for disk accesses and the 10 byte rule for trading memory for CPU time. In: Proceedings of the ACM SIGMOD International Conference on Management of Data; 1987. p. 395–8.
5. Ramamurthy R, DeWitt DJ. Buffer-pool aware query optimization. In: Proceedings of the 2nd Biennial Conference on Innovative Data Systems Research; 2005. p. 250–61.
6. Stonebraker M. Operating system support for database management. Commun ACM. 1981;24(7):412–8.

Business Intelligence

Stefano Rizzi
DISI, University of Bologna, Bologna, Italy

Definition

Business intelligence (often referred to as *BI*) is a business management term that indicates the capability of adding more intelligence to the way business is done by companies. More precisely, it refers to a set of tools and techniques that enable a company to transform its business data into timely and accurate information for the decisional process, to be made available to the right persons in the most suitable form. Business intelligence systems are used by decision makers to get a comprehensive knowledge of the business and of the factors that affect it, as well as to define and support their business strategies. The goal is to enable data-based decisions aimed at gaining competitive advantage, improving operative performance, responding more quickly to changes, increasing profitability and, in general, creating added value for the company.

Historical Background

Business intelligence was coined as a term in 1958 by Hans P. Luhn, to indicate a system capable of processing documents aimed at extracting information to be disseminated to the various sections of an enterprise. Within the industrial world the term was then revived in the early 1990s, to denote a set of technologies (mainly rooted in reporting systems and in Decision Support Systems) aimed at satisfying the managers' request for efficiently and effectively analyzing the enterprise data in order to better understand the situation of their business and improving the decision process. In the mid-1990s business intelligence became an object of interest for the academic world, and 10 years of research managed to transform a bundle of naive techniques into a well-founded approach to information extraction and processing that led to defining the modern architectures of data warehousing systems. Currently, business intelligence includes not only the tools to gather, provide access to, and analyze data and information about company operations, but also a wide array of technologies used to support a closed decisional loop (known as *Business Performance Management*) where the company performance is measured by a set of indicators (commonly called *Key Performance Indicators, KPIs*) whose target values are determined by the company strategy, and where the actions taken are aimed at matching current and target values for these indicators [1].

Scientific Fundamentals

The decisional process includes four macrophases: *analyzing* the problem and obtaining useful information from data, *understanding* the problem and transforming information into knowledge, *deciding* and acting accordingly, and *measuring* the performances following the action taken [2]. In general, we can say that the goal of business intelligence is to support decision makers in closing this loop.

Broadly speaking, business intelligence applications require an infrastructure that includes dedicated hardware, network facilities, DBMSs, back-end software, and front-end software. More specifically, the enabling technologies must provide capability of storing huge volumes of data, computational power to ensure satisfactory

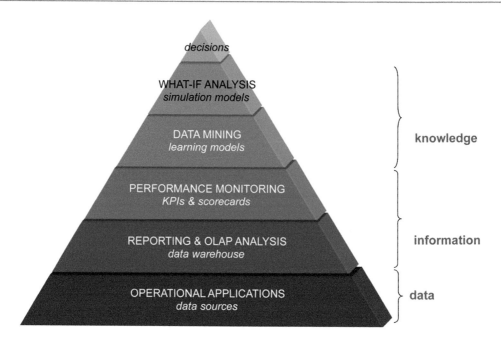

Business Intelligence, Fig. 1 The BI pyramid

querying performance, sophisticated techniques for visualizing complex information, and effective software interoperability. On the human side, mental agility, inclination to accept changes, and creativity are also needed.

The different levels of business intelligence can be intuitively summarized by the so-called *BI pyramid*, depicted in Fig. 1, that also illustrates the data-information-knowledge transformation.

Architectures

From an architectural point of view, the core of a business intelligence system is usually a data warehouse, i.e., a repository that stores the corporate historical data in a consistent and integrated form. The data warehouse is periodically refreshed by an ETL process that extracts data from operational data sources, transforms them to put them in multidimensional form, and finally loads them into the data warehouse. A number of applications may be built around the data warehouse, for instance aimed at supporting OLAP (On-Line Analytical Processing) analysis, data mining, what-if analysis, forecasting, balanced scorecards preparation, geospatial analysis, and click-stream analysis.

Different architectures have been proposed in the literature for specific types of applications. For instance, the architecture in [1] is completed by a reactive data flow, more suited for monitoring the time-critical operational processes by supporting real-time applications (see Fig. 2). The architecture in [3] emphasizes real-time data integration, distribution of the data warehouse in the cloud, and information interchange between decision makers. Information interchange is also the goal of the architecture outlined in [4], that supports the run-time rewriting of analysis queries across a network of business intelligence systems. The architecture proposed in [5], sketched in Fig. 3, integrates enterprise data with social data extracted by crawling processes and semantically enriched aimed at giving decision makers a timely perception of the market mood and help them explain the phenomena of business and society. Finally, the architecture in [6] builds on a spatial data warehouse to closely integrate a geographical dimension with enterprise data, aimed at enabling a more effective monitoring and interpretation of business events with specific reference to their territorial distribution.

Business Intelligence,
Fig. 2 An architecture to
support business activity
monitoring

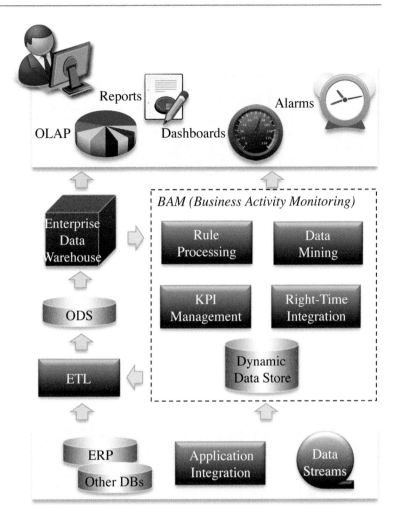

Functions

From a functional point of view, the advent of so-called *BI 2.0* [3] has significantly enriched the scope of business intelligence in different directions:

- The basic idea of **pervasive BI** is that information should be easily and timely accessed through devices with different computation and visualization capabilities, and with sophisticated and customizable presentations, by everyone in the organization. The research works in this direction mainly focus either on innovative techniques for compact yet expressive visualization of data or on approaches for personalizing the user experience based, for instance, on query preferences and recommendations [7].

- **Real-time BI** is about delivering information about business operations as they occur, with near-0 latency. This typically requires moving from traditional data warehouse architectures, based on a periodic refresh of data via ETL processes, to more sophisticated architectures based for instance on *trickle & flip* or *direct trickle feed* approaches, or even to *data stream warehouses*.

- In **service-oriented BI**, business intelligence is hosted as a service provided to users across the Internet. As discussed in [8], services can be provided mainly at four levels: *infrastructure-as-a-service*, where only the hardware is virtualized; *platform-*

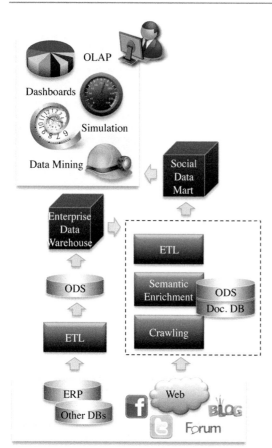

Business Intelligence, Fig. 3 An architecture for social business intelligence

and the capability of dynamically integrating new, useful data in the decision making process (which is made faster by the adoption of agile design methodologies).

Methodologies

On the methodological side, we remark that the different architectures and specific functions discussed above often require ad-hoc approaches to be designed and maintained. For instance, creating a social BI application is not so much about designing data and reports, while it mostly has to do with providing effective and efficient support to quick maintenance iterations – this is because of the huge dynamism of user-generated contents and of the pressing need of immediately perceiving and timely reacting to changes in the environment [5]. Similarly, in the case of collaborative BI a large methodological effort is normally required to design the (virtual or physical) integration between the different data warehouses by setting mappings between schemata and instances.

Tools

A wide array of tools are available, with either commercial and open-source license, to cover one or more phases of the data-intensive processes involved in business intelligence or to deliver different applications. The main categories of tools are listed below:

- *Data storage*: tools used to efficiently store data for the decision support based either on relational technologies (e.g., Oracle DBMS) or on NoSQL technologies (e.g., MongoDB).
- *ETL*: tools supporting the extraction, transformation, and loading of data from various sources into a data warehouse (e.g., Talend).
- *Reporting and dashboards*: tools that generate static (i.e., mainly not interactive) reports (e.g., Crystal Reports).
- *OLAP*: tools that enable users to analyze multidimensional data interactively from multiple perspectives using intuitive operations like roll-up, drill-down, and slice-and-dice (e.g., MicroStrategy).
- *Data warehouse appliances*: integrated high-performance architectures for data warehous-

as-a-service, where also basic software (like DBMSs) is provided; *software-as-a-service*, where the whole business intelligence application is installed and maintained by the provider; and *business process-as-a-service*, where even part of the company business is externalized, thus introducing a complete dependency on the provider.
- **Self-service BI** is an approach where business users can proactively create custom analysis reports and queries without involving the IT department. While OLAP can be seen as a basic form of self-service BI because it enables users to explore multidimensional cubes, more sophisticated implementations require the sharing of business glossaries and metadata dictionaries (which is made easier by relying on conceptual documentation for cubes)

ing – including servers, storage, operating systems, and DBMS – specifically targeted for big data analytics and requiring little configuration effort (e.g., Netezza).

- *Data mining*: tools that support automatic or semi-automatic analysis of large quantities of data to extract previously unknown interesting patterns (e.g., R).
- *Simulation and what-if analysis*: tools that enable users to create simulation models and run them to build predictions of future business scenarios (e.g., PowerSim).
- *Text analysis and semantic enrichment*: tools that analyze and enrich textual content by identifying its structured parts and/or assigning each sentence a polarity using sentiment analysis techniques (e.g., Synthema).

Key Applications

Business intelligence techniques are fruitfully applied in all areas of business, in both private and public companies and organizations, whenever managers need support to analyze performances, understand the market trends and the underlying reasons, evaluate the impact of the business strategies they apply [9]. More specifically, common applications of business intelligence tools are reporting, OLAP, data mining and knowledge discovery, business performance management, benchmarking, predictive and prescriptive analytics, and alerting.

Among the emerging applications of business intelligence techniques we mention **social BI**, that aims at effectively and efficiently combining corporate data with user-generated content to let decision-makers analyze and improve their business based on the trends and moods perceived from the environment. In this context, the most widely used category of user-generated content is the one coming in the form of textual clips; among the research challenges we mention delivering effective techniques for *sentiment analysis*, aimed at interpreting each sentence and if possible assign a *polarity* (i.e., positive, negative, or neutral) to it, and modeling hierarchies of topics (i.e., specific concepts of interest within

the subject area), whose complexity overcomes the possibilities encompassed by traditional hierarchy modeling techniques.

Future Directions

As mentioned in the previous sections, the "new era" of business intelligence points well beyond mere data and reports. Among the challenges and goals not attained yet, we mention two relevant areas that have been investigated by the research community only to some extent so far:

- In **collaborative BI**, the information assets of a company are empowered thanks to cooperation and data sharing with other companies and organizations, so that the decision-making process is extended beyond the company boundaries [10]. Data warehouse integration is an enabling technique for collaborative BI. Two categories of approaches were mainly devised: *warehousing approaches*, where the integrated data are physically materialized, and *federated approaches*, where integration is virtual. In both cases, it is assumed that all components to be integrated share the same schema, or at least that a global schema is given. In contexts where the different parties have a common interest in collaborating but each of them wants to preserve its autonomy and view of business, defining a global schema is often unfeasible; to cope with this, *peer-to-peer approaches* have been emerging to enable each peer to transparently formulate queries also involving the other peers, typically based on a set of mappings that establish semantic relationships between the peers' schemata [4].
- Dynamically integrating data into the decisional process is the goal of **on-demand BI**, where information in an enterprise data warehouse is completed by correlating it on-the-fly with external information that may come from the corporate intranet, be acquired from some external vendor, or be derived from the internet. On-demand BI is also called **situational BI** to emphasize that the decision process is

enriched with situational data, i.e., data that have a narrow focus on a specific business problem and, typically, a short lifespan for a small group of users [11]. Often, these data are not owned and controlled by the decision maker; their search, extraction, integration, and storage for reuse or sharing should be accomplished by decision makers without any intervention by designers or programmers.

We close this section by observing that, especially with applications that involve huge volumes of unstructured data like on-demand BI and social BI, the separation between the two neighboring areas of business intelligence and big data analytics seems to blur. See [12] for a discussion of the basic differences between the two areas.

Cross-References

▶ Active, Real-Time, and Intellective Data Warehousing
▶ Data Mining
▶ Data Warehousing in Cloud Environments
▶ Data Warehousing Systems: Foundations and Architectures
▶ OLAM
▶ OLAP Personalization and Recommendation
▶ Online Analytical Processing
▶ Predictive Analytics
▶ Prescriptive Analytics
▶ Visual Online Analytical Processing (OLAP)
▶ What-If Analysis

Recommended Reading

1. Golfarelli M, Rizzi S, Cella I. Beyond data warehousing: what's next in business intelligence? In: Proceedings ACM 7th International Workshop on Data Warehousing and OLAP; 2004. p. 1–6.
2. Vitt E, Luckevich M, Misner S. Business intelligence: making better decisions faster. Microsoft Press; 2002.
3. Trujillo J, Maté A. Business intelligence 2.0: a general overview. In: Aufaure M-A, Zimanyi E, editors. eBISS 2011. LNBIP 96. Paris: Springer; 2012, p. 98–116.
4. Golfarelli M, Mandreoli F, Penzo W, Rizzi S, Turricchia E. OLAP query reformulation in peer-to-peer data warehousing. Inf Syst. 2012;37(5):393–411.
5. Gallinucci E, Golfarelli M, Rizzi S. Meta-stars: multidimensional modeling for social business intelligence. In: Proceedings 16th International Workshop on Data Warehousing and OLAP; 2013. p. 11–8.
6. Golfarelli M, Mantovani M, Ravaldi F, Rizzi S. Lily: a geo-enhanced library for location intelligence. In: Proceedings 15th International Conference on Data Warehousing and Knowledge Discovery; 2013. p. 72–83.
7. Kozmina N, Niedrite L. Research directions of OLAP personalization. In: Proceedings 19th International Conference on Information Systems Development; 2010. p. 345–56.
8. Abelló A, Romero O. Service-oriented business intelligence. In: Aufaure M-A, Zimanyi E, editors. eBISS 2011. LNBIP 96; 2012. p. 156–85.
9. Chen H, Chiang R, Storey V. Business intelligence and analytics: from big data to big impact. MIS Q. 2012;36(4):1165–88.
10. Rizzi S. Collaborative business intelligence. In: Aufaure M-A, Zimanyi E, editors. eBISS 2011. LNBIP 96. Paris: Springer; 2012. p. 186–205.
11. Abelló A, et al. Fusion cubes: towards self-service business intelligence. Int J Data Warehouse Min. 2013;9(2):66–88.
12. Pedersen, TB. Managing Big Multidimensional Data. In: Proceedings 15ème Conférence Internationale sur l'Extraction et la Gestion des Connaissances. 2015. p. 3–6.

Business Process Execution Language

W. M. P. van der Aalst
Eindhoven University of Technology,
Eindhoven, The Netherlands

Synonyms

BPEL; BPEL4WS

Definition

The *Business Process Execution Language for Web Services* (BPEL) has emerged as a standard for specifying and executing processes. It is supported by many vendors and positioned as the "process language of the Internet." BPEL is XML based and aims to enable "programming in the large," i.e., using BPEL new services can be composed from other services.

Key Points

BPEL [2, 3] supports the modeling of two types of processes: executable and abstract processes. An *abstract*, (not executable) *process* is a business protocol, specifying the message exchange behavior between different parties without revealing the internal behavior for any one of them. This abstract process views the outside world from the perspective of a single organization or (composite) service. An *executable process* views the world in a similar manner. However, things are specified in more detail such that the process becomes executable, i.e., an executable BPEL process specifies the execution order of a number of *activities* constituting the process, the *partners* involved in the process, the *messages* exchanged between these partners, and the *fault* and *exception handling* required in cases of errors and exceptions.

A BPEL process itself is a kind of flow-chart, where each element in the process is called an *activity*. An activity is either a primitive or a structured activity. The set of *primitive activities* contains: invoke, invoking an operation on a web service; receive, waiting for a message from an external source; reply, replying to an external source; wait, pausing for a specified time; assign, copying data from one place to another; throw, indicating errors in the execution; terminate, terminating the entire service instance; and empty, doing nothing.

To enable the presentation of complex structures the following *structured activities* are defined: sequence, for defining an execution order; switch, for conditional routing; while, for looping; pick, for race conditions based on timing or external triggers; flow, for parallel routing; and scope, for grouping activities to be treated by the same fault-handler. Structured activities can be

nested and combined in arbitrary ways. Within activities executed in parallel the execution order can further be controlled by the usage of links (sometimes also called control links, or guarded links), which allows the definition of directed graphs. The graphs too can be nested but must be acyclic.

The terminology above is based on BPEL 1.1 which was introduced in 2003 [3]. A new version of the standard [2] was published in 2007. This version has been approved as an OASIS Standard. This new version resolves many semantical issues [1, 5]. Moreover, new activity types such as repeatUntil, validate, forEach (parallel and sequential), rethrow, extensionActivity, and compensateScope, have been added and some of the existing activities have been renamed (switch/-case renamed to if/else and terminate renamed to exit). Many extensions have been proposed, including BPEL4People which enables BPEL activities to be executed by human resources [4]. However, more graphical oriented languages like BPMN are taking over the role of BPEL.

Cross-References

▶ Business Process Modeling Notation
▶ Business Process Management
▶ Choreography
▶ Composition
▶ Orchestration
▶ Web Services
▶ Workflow Management
▶ Workflow Patterns

Recommended Reading

1. van der Aalst WMP, Dumas M, ter Hofstede AHM, Russell N, Verbeek HMW, Wohed P. Life after BPEL? In: Web Services and Formal Methods; 2005. p. 35–50.
2. Alves A, Arkin A, Askary S, Barreto C, Bloch B, Curbera F, Ford M, Goland Y, Guzar A, Kartha N, Liu CK, Khalaf R, Koenig D, Marin M, Mehta V, Thatte S, Rijn D, Yendluri P, Yiu A. Web services business process execution language, version 2.0 (OASIS Standard). WS-BPEL TC OASIS. 2007. http://docs.oasis-open.org/wsbpel/2.0/wsbpel-v2.0.html.

B

3. Andrews T, Curbera F, Dholakia H, Goland Y, Klein J, Leymann F, Liu K, Roller D, Smith D, Thatte S, Trickovic I, Weerawarana S. Business process execution language for web services, version 1.1. Standards Proposal by BEA Systems, International Business Machines Corporation, and Microsoft Corporation; 2003.
4. Kloppmann M, Koenig D, Leymann F, Pfau G, Rickayzen A, von Riegen C, Schmidt P, Trickovic I. WS-BPEL extension for people BPEL4People. In: Proceedings of 22nd International Conference on Conceptual Modeling; 2005.
5. Wohed P, van der Aalst WMP, Dumas M, ter Hofstede AHM. Analysis of web services composition languages: the case of BPEL4WS. In: Proceedings of 22nd International Conference on Conceptual Modeling; 2003. p. 200–15.

Business Process Management

W. M. P. van der Aalst
Eindhoven University of Technology,
Eindhoven, The Netherlands

Synonyms

Case handling; Process management; Workflow management

Definition

Information technology has changed business processes within and between enterprises. Traditionally, information technology was mainly used to support individual tasks ("type a letter") and to store information. However, today business processes and their information systems are intertwined. Processes heavily depend on information systems and information systems are driven by the processes they support [1]. *Business Process Management* (BPM) is concerned with the interactions between processes and information systems. An important element of BPM is the modeling and analysis of processes. Processes can be designed using a wide variety of languages ranging from BPMN to Petri nets. Some of these languages allow for analysis techniques (e.g., model checking and simulation) to answer questions related to correctness and performance. Models can be used to configure generic software tools, e.g., middleware, workflow management systems, ERP systems, etc. These systems, also referred to as *Business Process Management Systems* (BPMS), are used to enact relevant business processes. A business process management system can be defined as: *a generic software system that is driven by explicit process designs to enact and manage operational business processes.* The system should be "process-aware" and "generic" in the sense that it is possible to modify the processes it supports. The process designs are often graphical and the focus is on structured processes that need to handle many cases. Workflow management systems are typical examples of such "process-aware" systems. An important technological enabler for business process management systems is the Service Oriented Architecture (SOA). The partitioning of processes into services makes it easier to isolate the process-logic.

Historical Background

Traditionally, information systems are viewed from either a *process-centric* or an *information-centric* perspective. The information-centric view focuses on the information managed by the system. Database management systems provide the functionality required to store and retrieve data. Since the 1970s, there have been consensus on the modeling of data. Although there are different languages and different types of database management systems, the fundamental concepts are quite stable for the information-centric view of information systems. The process-centric view on information systems on the other hand can be characterized by the term "divergence." There is little consensus on the fundamental concepts. Despite the availability of established formal languages (e.g., Petri nets and process calculi) industry has been pushing ad-hoc/domain-specific languages. As a result there is a plethora of systems and languages available today.

An good starting point from a scientific perspective is the early work on office information systems. In the 1970s, people like Skip Ellis, Anatol Holt, and Michael Zisman already worked on so-called office information systems, which were driven by explicit process models [2]. It is interesting to see that the three pioneers in this area independently used Petri-net variants to model office procedures. During the 1970s and 1980s, there was great optimism about the applicability of office information systems. Unfortunately, few applications succeeded. As a result of these experiences, both the application of this technology and research almost stopped for a decade. Consequently, hardly any advances were made in the 1980s. In the 1990s, there again was a huge interest in these systems [3]. The number of workflow management systems developed in the period 1995–2005 and the many papers on workflow technology illustrate the revival of office information systems. Today workflow management systems are readily available. However, their application is still limited to specific industries such as banking and insurance. In fact, workflow technology is often hidden inside other systems. For example, ERP systems like SAP and Oracle provide workflow engines. Many other platforms include workflow-like functionality. For example, integration and application infrastructure software such as IBMs Websphere provides extensive process support.

When comparing today's business process management systems to the workflow management systems of the nineties two things can be noted. First of all, the focus is no longer exclusively on automation and enactment, e.g., process analysis (simulation, process mining, verification, etc.) is increasingly important. Second, the use of web technology makes it easier to realize such systems even if processes are scattered over multiple organizations.

Foundations

Business process management looks at the relationships between business processes and information systems. Using information systems in an innovative way enables new types of business processes. For example, making paper documents electronic may enable the concurrent execution of tasks thus shortening flow times. Moreover, characteristics of business processes lead to requirements for business process management systems (cf. ▸ Workflow Patterns). Business process management is *not* limited to the automation of business processes. For example, it is vital to *analyze* processes before and after they are enacted. During the design phase it is vital to use verification techniques to assess the correctness of the process design. Moreover, simulation techniques can be used to estimate the performance of the process once it is realized. While the processes is running, the information system needs to record information about actual events and realized performance. Using process mining techniques and other types of business intelligence, the event logs of systems can be analyzed. Based on such a diagnosis, the process can be improved.

The modeling and analysis of processes plays a central role in business process management. Therefore, the choice of language to represent an organization's processes is essential. Three types of languages can be identified:

1. *Formal languages*: Processes have been studied using theoretical models. Mathematicians have been using Markov chains, queueing networks, etc. to model processes. Computer scientists have been using Turing machines, transition systems, Petri nets, and process algebras to model processes. All of these languages have in common that they have *unambiguous semantics* and allow for *analysis*.
2. *Conceptual languages*: Users in practice have problems using formal languages. They prefer to use higher-level languages. Examples are BPMN (Business Process Modeling Notation), EPCs (Event-Driven Process Chains), UML activity diagrams, etc. These language are typically *informal*, i.e., they do not have a well-defined semantics and do not allow for analysis. Moreover, the lack of semantics makes it impossible to directly execute them.
3. *Execution languages*: Formal language typically abstract from "implementation

details" (e.g., data structures) and conceptual languages only provide an approximate description of the desired behavior. Therefore, more technical languages are needed for enactment. An example is the BPEL (Business Process Execution Language) language. Most vendors provide a proprietary execution language.

The existence and parallel use of these three types of languages causes many problems. The lack of consensus makes it difficult to exchange models. The gap between conceptual languages and execution languages leads to re-work and a disconnect between users and implementers. Moreover, both types of languages are not supported by advanced analysis tools.

Figure 1 shows the typical architecture of a business process management system. The figure also shows three roles of people involved: management, designer, and worker. The "heart" of the business process management system is the enactment service also known as "workflow engine" [3, 4]. This engine is offering the right pieces of work (work-items) to workers at the right point in time. In order to do this, it needs to have detailed descriptions of the processes and

organizations involved. Using design tools one can model processes and organizations. Note that Fig. 1 presents an idealized view. As indicated before there may be different languages (formal, conceptual, and execution languages) involved. The designer may first model the process in an informal manner. This model is then converted into a model that can be enacted or analyzed. The enactment service is driven by models in order to offer the right piece of work to the right persons at the right time. Moreover, the enactment service is starting applications and provides access to case data. During execution all kinds of information are recorded and at any time there is a (partial) history (i.e., audit trails, event logs, etc.) and a current state (run-time data). This information can be used for all kinds of analysis. Some types of analysis focus on the process design (e.g., verification and simulation). Other types of analysis focus on the actual behavior of the process (e.g., process mining).

As indicated before, different types of people are involved (management, designers, and workers). Moreover, business process management systems have a characteristic *life-cycle* Figure 2 shows the four phases of such a life-cycle [3]. In the *design phase*, the processes

Business Process Management, Fig. 1
Architecture of a business process management system

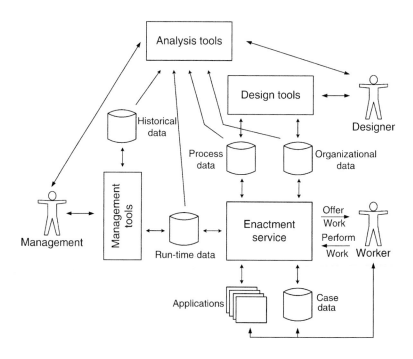

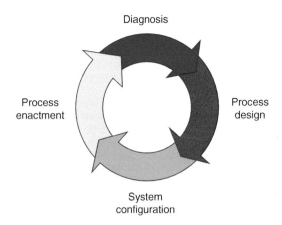

Business Process Management, Fig. 2 BPM life-cycle

are (re)designed. In the *configuration phase*, designs are implemented by configuring a process aware information system (e.g., a BPMS). After configuration, the *enactment phase* starts where the operational business processes are executed using the system configured. In the *diagnosis phase*, the operational processes are analyzed to identify problems and to find things that can be improved. The focus of traditional workflow management (systems) is on the lower half of the BPM life-cycle. As a result there is little support for the diagnosis phase. Moreover, support in the design phase is limited to providing an editor while analysis and real design support are missing. It is remarkable that few systems provide good support simulation, verification, and validation of process designs. Another problem of conventional workflow systems is the lack of flexibility. Fortunately, the emphasis is shifting from automation of highly structured processes to issues such as flexibility and analysis. Case handling systems such as FLOWer and academic prototypes such as DECLARE, YAWL/worklets, and ADEPT offer innovative ways of supporting flexible processes. Process mining tools such as ProM, ARIS PPM, etc. allow for the analysis of actual behavior. This supports the diagnosis phase and triggers process improvement.

The architecture demonstrated in Fig. 1 does not show any organizational boundaries. In the traditional setting it was very difficult to support inter organizational processes. Web services and the Service Oriented Architecture (SOA) simplify the distribution of processes over different organizations [5]. Moreover, the paradigm shift towards services has also changed the architecture within a single organization. When focusing on processes, two terms are important: (i) choreography and (ii) orchestration.

Choreography is concerned with the exchange of messages between those services. *Orchestration* is concerned with the interactions of a single service with its environment. While choreography can be characterized by reaching an agreement and the monitoring of the overall progress, the focus of orchestration is more on the implementation of a particular service by describing the process logic and linking this to neighboring services. Orchestration languages are close to traditional workflow languages (BPMN, BPEL, Petri nets, etc.). An important characteristic of such languages is the ability to compose a service by using other services. The role of choreography languages (e.g., WS-CDL) is less clear.

Business process management is clearly related to management science. For example, topics such as operations research, operations management, business process re-engineering are highly relevant [6]. There are also clear links with coordination languages and theory. Coordination can be defined as "managing dependencies between activities." Process modeling languages and concepts such as choreography and orchestration are obviously related to coordination.

Key Applications

Banking

The financial industry has changed dramatically using both business process re-engineering and workflow-like technologies. Many processes have been rationalized using business process management techniques. The rise of e-banking led to a dramatic reduction of people and offices. Banking processes are supported and monitored by business process management systems.

Government

Government organizations need to react quickly to new legislation, i.e., the corresponding processes need to be modified based on changes in tax laws, customs procedures, immigration laws, corporate governance, safety regulations, etc. Business process management assists in dealing with these changes and further improving the processes.

Business-to-Business

Mergers and virtual enterprises trigger the need for cross-organizational workflows. Web services and business process management techniques can assist in connecting process fragments from different organizations. The SOA combined with languages like BPEL provides a good basis for cross-organizational workflows.

Health-Care

Business process management techniques have mainly been applied to structured processes. Given the nature of care processes, it is not easy to streamline these processes and to support them with workflow-like systems. However, they only way to reduce costs and improve effectively is to provide better support for such processes. Hence, the health-care domain poses an interesting and relevant challenge for business process management.

Cross-References

▶ Business Process Execution Language
▶ Business Process Modeling Notation
▶ Composition
▶ Choreography
▶ Orchestration
▶ Process Mining
▶ Web Services
▶ Workflow Management
▶ Workflow Management and Workflow Management System
▶ Workflow Model Analysis
▶ Workflow Patterns

Recommended Reading

1. Dumas M, van der Aalst WMP, ter Hofstede AHM. Process-aware information systems: bridging people and software through process technology. New York: Wiley; 2005.
2. van der Aalst WMP. Business process management demystified: a tutorial on models, systems and standards for workflow management. In: Desel J, Reisig W, Rozenberg G, editors. Lectures on concurrency and petri nets. LNCS, Vol. 3098. Berlin/Heidelberg/New York: Springer; 2004. p. 1–65.
3. van der Aalst WMP, van Hee KM. Workflow management: models, methods, and systems. Cambridge, MA: MIT; 2004.
4. Leymann F, Roller D. Production workflow: concepts and techniques. Upper Saddle River: Prentice Hall PTR; 1999.
5. Weske M. Business process management: concepts, languages, architectures. Berlin/Heidelberg/New York: Springer; 2007.
6. van der Aalst WMP. Business process management: a comprehensive survey. ISRN Softw Eng. 2013:1–37.
7. Georgakopoulos D, Hornick M, Sheth A. An overview of workflow management: from process modeling to workflow automation infrastructure. Distrib Parallel Databases. 1995;3:119–53.
8. Jablonski S, Bussler C. Workflow management: modeling concepts, architecture, and implementation. London: International Thomson Computer; 1996.
9. Reijers H. Design and control of workflow processes: business process management for the service industry. LNCS, Vol. 2617. Berlin/Heidelberg/New York: Springer; 2003.
10. Dumas M, La Rosa M, Mendling J, Reijers H. Fundamentals of business process management. Berlin/New York: Springer; 2013.

Business Process Modeling

Marlon Dumas
University of Tartu, Tartu, Estonia

Synonyms

Workflow modeling

Definition

A business process model is a representation of the way an organization operates to achieve a goal, such as delivering a product or a service. Business process models may be given as input to a workflow management system to automatically coordinate the tasks composing the business process model. However, business process modeling may be conducted purely for documentation purposes or to analyze and improve the operations of an organization, without this improvement effort implying automation by means of a workflow system.

A typical business process model is a graph consisting of at least two types of nodes: task nodes and control nodes. Task nodes describe units of work that may be performed by humans or software applications or a combination thereof. Control nodes capture the flow of execution between tasks, therefore establishing which tasks should be enabled or performed after completion of a given task. Business process models, especially when they are intended for automation, may also include object nodes denoting inputs and outputs of tasks. Object nodes may correspond to data required or produced by a task. But in some notations, they may correspond to physical documents or other artifacts that need to be made available or that are produced or modified by a task. Additionally, business process models may include elements for capturing resources that are involved in the performance of tasks. For example, in two notations for business process modeling, the Unified Modeling Language (UML) Activity Diagrams and the Business Process Model and Notation (BPMN), resource types are captured as *swimlanes*, with each task belonging to one swimlane (or multiple in the case of UML). In other notations, such as event-driven process chains (EPC), this is represented by attaching resource types to each task.

Historical Background

The idea of documenting business processes in order to improve customer satisfaction and internal efficiency dates back to at least the 1960s when it was raised, among others, by Levitt [9]. Levitt contended that organizations should avoid focusing exclusively on goods manufacturing and should view their "entire business process as consisting of a tightly integrated effort to discover, create, arouse, and satisfy customer needs." However, it is not until the 1970s that the idea of using computer systems to coordinate business operations based on process models emerged. Early scientific work on business process modeling was undertaken in the context of *office information systems* by Ellis [3] among others. During the 1970s and 1980s, there was optimism in the IT community regarding the applicability of process-oriented office information systems. Unfortunately, few deployments of this concept succeeded, partly due to the lack of maturity of the technology but also due to the fact that the structure and operations of organizations were centered around the fulfillment of individual functions rather than end-to-end processes. Following these early negative experiences, the idea of process modeling lost ground during the next decade.

Toward the early 1990s, however, there was a renewed interest in business process modeling. Instrumental in this revival was the popularity gained in the management community by the concept of *Business Process Reengineering* (BPR) advocated by Hammer [4] and Davenport [1] among others. BPR contended that overspecialized tasks carried across multiple organizational units should be unified into coherent and globally visible processes. This management trend fed another wave of business process technology known as workflow management. Workflow management, in its original form, focused on capturing business processes composed of tasks performed by human actors and requiring data, documents, and forms to be transferred between them. Workflow management was successfully applied to the automation of routine and high-volume administrative processes such as order-to-cash, procure-to-pay, and insurance claim handling. However it was less successful in application scenarios requiring the integration of heterogeneous information systems and those involving high levels of evolution and change. Also, standardization efforts in the field of business process modeling and management, led by the Workflow Management Coalition (WfMC), failed to gain wide adoption.

The BPR trend also led to the emergence of business process modeling tools that support the analysis and simulation of business processes, without targeting their automation. The ARIS platform, based on the EPC notation, is an example of this family of tools. Other competing tools supported alternative notations such as IDEF3 and several variants of flowcharts.

By the late 1990s, the management community had moved from the concept of BPR to that of *Business Process Improvement* (BPI), which advocates an incremental and continuous approach to adopting process orientation. This change of trend, combined with the limited success of workflow management and its failure to reach standardization, led to the term *workflow* acquiring a negative connotation. Nonetheless, the key concepts underpinning workflow management reemerged at the wake of the twenty-first century under the umbrella of *Business Process Management* (BPM). BPM adopts a more holistic view than workflow management, covering not only the automated coordination of processes but also their monitoring and continuous analysis and improvement. Also, BPM does not only deal with the coordination of human tasks but also the integration of heterogeneous information and software systems.

With BPM came the realization that business processes should be analyzed and designed both from a business viewpoint and from a technical viewpoint and that methods are required to bridge these viewpoints. This realization led to a convergence of modeling languages. Until the mid-2000s, two rather distinct types of modeling languages coexisted: (i) languages for modeling processes for the purpose of documentation and analysis, such as UML Activity Diagrams and EPCs, and (ii) languages for specifying business processes for the purpose of automated coordination, such as WS-BPEL and YAWL (Yet Another Workflow Language) [18]. The late 2000s witnessed a wide convergence toward a single process modeling languages, namely, BPMN, which is designed to cover the needs of

the entire lifecycle of a business process, from discovery and analysis to implementation and execution.

Foundations

Business process modeling can be approached from a number of perspectives [5]. The *control-flow perspective* describes the ordering and causality relationships between tasks. The *data perspective* (or information perspective) describes the data that are taken as input and produced by the activities in the process. The *resource perspective* describes the structure of the organization and identifies resources, roles, and groups. The *task perspective* describes individual steps in the processes and thus connects the other three perspectives.

Many business process modeling notations emphasize the control-flow perspective. This is the case, for example, of process modeling notation based on *flowcharts*. Flowcharts consist of actions (also called activities) and decision nodes connected by arcs that denote sequential execution. In their basic form, flowcharts do not make a separation between actions performed by different actors or resources. However, an extension of flowcharts, known as cross-functional flowcharts, allows one to perform this partitioning. Flowcharts inspired many other contemporary process modeling notations, including UML Activity Diagrams and later BPMN.

A simplified business process model for credit approval in BPMN is given in Fig. 1. An execution of this process model is triggered by the receipt of a credit application, denoted by the leftmost element in the model. Following this, the application is checked for completeness. Two outcomes are possible: either the application is marked as complete or as incomplete. This choice is represented by the *decision gateway* depicted by the diamond labeled by an "X" sign. If the application is incomplete, additional information is sought from the applicant. Otherwise, two checks are performed in parallel: a credit history check and an income check. This is denoted by a

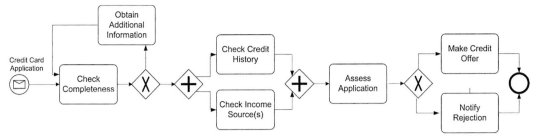

Business Process Modeling, Fig. 1 Example of a business process model in BPMN

parallel split gateway (the first diamond labeled by a "+" sign). The two parallel checks then converge into a *synchronization gateway* (the second diamond labeled by a "+"). The credit history check may be performed automatically, while the income source check may require an officer to contact the applicant's employer by phone. After these checks, the application is assessed, and it is either accepted or rejected.

A second family of notations for business process modeling are those based on state machines. State machines provide a simple approach to modeling behavior. In their basic form, they consists of states connected by transitions. When used for business process modeling, it is common for the transitions in a state machine to be labeled by event-condition-action (ECA) rules. The semantics of a transition labeled by an action is the following: if the execution of the state machine is in the source state, an occurrence of the event (type) has been observed, and the condition holds and then the transition may be taken. If the transition is taken, the action associated with the transition is performed, and the execution moves to the target state. A transition may contain any combination of these three elements: e.g., only an event, only a condition, only an action, an event and a condition but no action, etc. In some variants of state machines, it is possible to attach activities to any given state. The semantics is that when the state is entered, the activity in question may be started, and the state can only be exited once this activity has completed. Several commercial business process management systems support the execution of process models defined as state machines. Examples include the "Business State Machine" models supported by

IBM Websphere and the "Workflow State Machine" models supported by Microsoft Windows Workflow Foundation.

A disadvantage of using basic state machines for process modeling is their tendency to lead to state explosion. Basic state machines represent sequential behavior: the state machine can only be in one state at a time. This, when representing a business process model with parallel execution threads, one essentially needs to enumerate all possible permutations of the activities that may be executed concurrently. Another source of state explosion is exception handling. Statecharts have been proposed as a way to address this state explosion problem. Statecharts include a notion of concurrent components as well as transitions that may interrupt the execution of an entire component of the model. The potential use of statecharts for business process modeling has been studied in the research literature, for example, in the context of the Mentor research prototype [12], although it has not been adopted in commercial products.

Figure 2 shows a statechart process model intended to be equivalent to the BPMN "Credit Approval" process model of Fig. 1. It can be noted that decision points are represented by multiple transitions sharing the same source state and labeled with different conditions (conditions are written between square brackets). Parallel execution is represented by means of a composite state divided into two concurrent components. Each of these components contains one single state. In the general case, each concurrent component may contain an arbitrarily complex state machine.

Finally, a third family of notations for control-flow modeling of business processes are those based on *Petri nets*. In the research literature, business process modeling notations based on Petri nets have been used mainly for studying expressiveness and verification problems (e.g., detecting deadlocks in business process models). Also, certain Petri net-based notations are supported by business process modeling tools and workflow management systems. A notable example is YAWL. YAWL is an extension of Petri nets designed on the basis of the *workflow patterns*. A key design goal of YAWL is to support as many workflow patterns as possible in a direct manner while retaining the theoretical foundation of Petri nets. Like Petri nets, YAWL is based on the concepts of place and transition (Transitions are called *tasks* in YAWL, while places are called *conditions*.). YAWL also incorporates a concept of *decorator*, which is akin to the concept of gateway in BPMN. In addition, it supports the concepts of subprocess (called *composite task*) and cancelation region. The latter is used to capture the fact that the execution of a set of tasks may be interrupted by the execution of another task and can be used, for example, for fault handling. Figure 3 shows a YAWL process model intended to be equivalent to the BPMN "Credit Approval"

process model of Fig. 1 and the statechart process model of Fig. 2. The symbols attached to the left and the right of each task are the decorators. The decorator on the left of task "Check Completeness" is an XOR-merge decorator, meaning that the task can be reached from either of its incoming flows. The decorator of the right of this same task is an XOR-split, which corresponds to the decision gateway in BPMN. The decorator just before "Check Credit History" is an AND-split (akin to a parallel split gateway) and the one attached to the left of "Assess Application" is an AND-join and it is used to denote a full synchronization.

The data perspective of process modeling can be captured using notations derived from dataflow diagrams. Dataflow diagrams are directed graphs composed of three types of nodes: *processes*, *entities*, and *data stores*. Entities denote actors, whether internal or external to an organization. Processes denote units of work that transform data. Data stores denote units of data storage. A data store is typically used to store data objects of a designated type. The arcs in a dataflow diagram are called

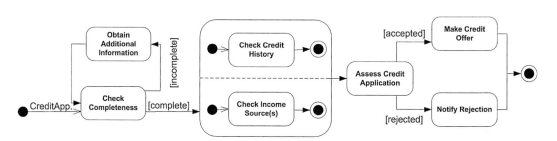

Business Process Modeling, Fig. 2 Example of a business process model captured as a statechart

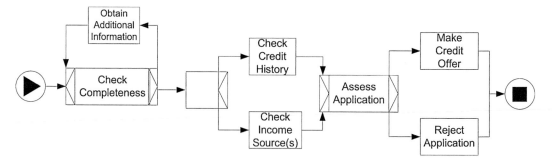

Business Process Modeling, Fig. 3 Example of a business process model captured in YAWL

dataflows. Dataflows denote the transfer of data between entities, processes, and data stores. They may be labeled with the type of data being transferred.

Dataflow diagrams are often used very liberally, with almost no universally accepted syntactic constraints. However, best practice dictates that each process should have at least one incoming dataflow and one outgoing dataflow. Dataflows may connect entities to processes and vice versa but cannot be used to connect entities to data stores or to connect one entity to another directly. It is possible to connect two processes directly by means of a dataflow, in which case it means that the output of the first process is the input of the second. It is also possible to assign ordinal numbers to processes in order to capture their execution order. Dataflow diagrams also support decomposition: Any process can be decomposed into an entire dataflow diagram consisting of several subprocesses.

Dataflow diagrams are quite restricted insofar as they do not capture decision points and they are not suitable for capturing parallel execution or synchronization. Because of this, dataflow diagrams are not used to capture processes in details; instead, they are used as a high-level notation.

On the other hand, some elements from dataflow diagrams have found their way into other process modeling notations such as EPCs, which incorporate constructs for capturing input and output data of functions. Similarly, Activity Diagrams have a notion of object nodes, used to represent inputs (outputs) consumed (produced) by activities. Also, more sophisticated variants of dataflow languages have been adopted in the field of *scientific workflow* [10], which is concerned by the automation of processes involving large datasets and complex computational steps in fields such as biology, astrophysics, and environmental sciences.

When modeling business processes for execution (e.g., by a workflow system), it is important to capture not only the flow of data into and out of tasks but also data transformation, extraction, and aggregation steps. At this level of detail, it may become cumbersome to use simple dataflows. Instead, executable process modeling languages

tend to rely on *scoped variables.* Variables may be defined, for example, at the level of the process model. Tasks may take as input several input and output parameters, and data mappings are used to connect the data available in the process model variables and the input and output parameters. Data mappings may be inbound (i.e., expressions that extract data from variables in the task's encompassing scope and assign values to the task's input parameter) or outbound (from the output parameters to the variables in the encompassing scope). This is the approach used, for example, in YAWL, Mentor, and AdeptFlex [14]. It is also the approach adopted by many commercial workflow products and by WS-BPEL.

Scientific Fundamentals

One area of research in the area of business process modeling is the analysis of their expressiveness [6]. Several classes of workflow notations have been identified from this perspective: structured business process models, synchronizing business process models, and standard business process models. Structured business process models are those that can be composed out of four basic control-flow constructs, one for choosing one among multiple blocks of activities, another for executing multiple blocks of activities in parallel, a third one for executing multiple blocks of activities in sequence, and finally a fourth one for repeating a block of activities multiple times. An activity by itself can be a block, and more complex blocks can be formed by composing smaller blocks using one of the four constructs. It has been shown that structured business process models are well behaved, meaning in particular that their execution cannot result in deadlocks [13]. On the other hand, this class of models has relatively low expressive power. There exist unstructured business process models that cannot be translated to equivalent structured ones under reasonable notions of equivalence [13]. Synchronizing business process models are acyclic business process models composed of activities that are connected by transitions. These transitions correspond to control links in the WS-BPEL terminology. They impose precedence relations and they can be labeled with conditions. An activity can be executed if all its predecessors have been either completed or skipped. Like structured business process models, this class of models are well behaved. However, they are not very expressive.

Finally, standard business process models correspond to the subset of BPMN composed of tasks, decision gateways, merge gateways (where multiple alternative branches converge), parallel split gateways, and synchronization gateways. This class of models is more expressive than the former two, but on the other hand, some of these models may contain deadlocks. Other classes of business process models are those corresponding to Workflow Nets, which are more expressive than standard business process models as they can capture different types of choices. YAWL models are even more expressive due to the presence of the cancelation region construct and another construct (called the "OR-join") for synchronization of parallel branches, where some of these branches may never complete. A significant amount of literature in the area of workflow verification has dealt with studying properties of modeling languages incorporating a notion of OR-join [7].

Another area of research in business process modeling relates to quality of process models and factors that determine the degree of understandability of process models. This research has led to a number of complexity metrics for process models that have been empirically shown to be correlated with understandability [15]. Also, a number of guidelines have been proposed to help modelers ensure that their process models are understandable by a broad range of users [11]. A more comprehensive set of criteria for assessing the quality of process models is provided by SEQUAL (Semiotic Quality Framework) [8]. This framework goes beyond internal (physical) quality properties of process models by taking into account the goal of process modeling and the organizational and social context in which a process model is intended to be used.

Key Applications

Business process modeling can be applied across the entire lifecycle of business processes [2], all the way from the so-called discovery (documentation) phase to the analysis, redesign, automation, and monitoring phases. In some cases, organizations undertake business process modeling projects for the sake of documenting key activities and using this documentation for employee induction and knowledge transfer. Business process models are sometimes required for compliance reasons, for example, as a result of legislation in the style of the Sarbanes-Oxley Act in the United States which requires public companies to have internal control processes in place to detect and prevent frauds. A process model that captures how an organization currently works is called an "as is" process model. Such "as is" models are useful in understanding bottlenecks and generating ideas for improvement using process analysis techniques, including simulation and activity-based costing. The integration of improvement ideas in a business process leads to a redesigned "to be" process model. Some of these improvement ideas may include automating (parts of) the business process. At this stage, business process models need to be made more detailed. Once they have been made executable, business process models may be deployed in a business process execution engine (also called a workflow engine). Process models also play a central role during the monitoring phase of the business process, for example, in the context of performance monitoring and in the context of *process mining*, where enables users to analyze the actual execution of a business process as recorded in event logs.

Future Directions

Standardization efforts in the field of business process modeling have led to wide consensus around BPMN. It is recognized however that while BPMN is suitable for modeling standardized business processes, it is less suitable when it comes to capturing business processes with high levels of variability, such as healthcare processes, product development processes, or other knowledge-intensive and creative processes. For such processes, a number of paradigms have been proposed. One such paradigm explored in the research literature is that of *declarative process modeling*, which relies on the principle that "everything is allowed except what is forbidden." In other words, rather than specifying the allowed flows of a process, declarative process modeling focuses on specifying constraints that must be fulfilled during the execution of a process. A prototypical example of a declarative process modeling language is Declare [17].

Another complementary paradigm for modeling processes with high variability is *artifact-centric process modeling*, which relies on two tenets. The first tenet is that processes should be decomposed not in terms of subprocesses as in BPMN but rather in terms of interacting entity types (also called artifact types), each with its own lifecycle. The second tenet is that artifact lifecycles should be modeled in a data-driven approach, which is based on rules that capture what is possible in a given state of a business entity. An example of an artifact-centric process modeling language is GSM (Guard-Stage-Milestone) [16], where processes are decomposed into artifact types whose lifecycle is captured in terms of stages that are opened when certain guards (data conditions) become true and closed when certain milestones (also captured as data conditions) are fulfilled. Concepts behind the GSM language have made their way into another modeling standard, namely, CMMN (Case Management Model and Notation). CMMN is positioned as an alternative to BPMN. Whereas BPMN is intended to capture normative processes, CMMN is targeted at processes with high levels of run-time variability. At present however, adoption of CMMN remains limited, and the question of how to capture processes with high variability remains largely open.

Cross-References

- ▶ Activity Diagrams
- ▶ Business Process Execution Language
- ▶ Business Process Management
- ▶ Business Process Modeling Notation
- ▶ Business Process Reengineering
- ▶ Petri Nets
- ▶ Workflow Model Analysis
- ▶ Workflow Patterns

Recommended Reading

1. Davenport TH. Process innovation: reengineering work through information technology. Boston: Harvard Business School; 1992.
2. Dumas M, La Rosa M, Reijers HA, Mendling J. Fundamentals of business process management. Berlin/New York: Springer; 2013.
3. Ellis CA. Information control nets: a mathematical model of office information flow. In: Proceedings Conference on Simulation, Measurement and Modeling of Computer Systems; 1979. p. 225–40.
4. Hammer M. Reengineering work: don't automate, obliterate. Harvard Bus Rev. 1990;68(4):104–12.
5. Jablonski S, Bussler C. Workflow management: modeling concepts, architecture, and implementation. London: International Thomson Computer; 1996.
6. Kiepuszewski B, Ter Hofstede AHM, van der Aalst WMP. Fundamentals of control flow in workflows. Acta Inform. 2003;39(3):143–209.
7. Kindler E. On the semantics of EPCs: resolving the vicious circle. Data Knowl Eng. 2006;56(1):23–40.
8. Krogstie J. Quality in business process modeling. Cham: Springer; 2016.
9. Levitt T. Marketing myopia. Boston: Harvard Business Press; 1960. p. 45–56.
10. Ludäscher B, Altintas I, Berkley C, Higgins D, Jaeger E, Jones M, Lee EA, Tao J, Zhao Y. Scientific workflow management and the Kepler system. Concurr Comput–Pract Exp. 2006;18(10):1039–65.
11. Mendling J, Reijers HA, van der Aalst WMP. Seven process modeling guidelines (7PMG). Inf Softw Technol. 2010;52(2):127–36.
12. Muth P, Wodtke D, Weissenfels J, Dittrich A, Weikum G. From centralized workflow specification to distributed workflow execution. J Intell Inform Syst. 1998;10(2):159–84.
13. Polyvyanyy A, García-Bañuelos L, Dumas M. Structuring acyclic process models. Inf Syst. 2012;37(6):518–38.
14. Reichert M, Dadam P. ADEPTflex: supporting dynamic changes of workflow without loosing control. J Intell Inform Syst. 1998;10(2):93–129.
15. Reijers HA, Mendling J. A study into the factors that influence the understandability of business process models. IEEE Trans Syst Man Cybern Part A. 2011;41(3):449–62.

16. Vaculín R, Hull R, Heath T, Cochran C, Nigam A, Sukaviriya P. Declarative business artifact centric modeling of decision and knowledge intensive business processes. In: Proceedings of the 15th IEEE International Enterprise Distributed Object Computing Conference; 2011. p. 151–60.
17. van der Aalst WMP, Pesic M, Schonenberg H. Declarative workflows: balancing between flexibility and support. Comput Sci R&D 2009;23(2):99–113.
18. van der Aalst WMP, ter Hofstede AHM. YAWL: yet another workflow language. Inform Syst. 2004;30(4):245–75.

Business Process Modeling Notation

W. M. P. van der Aalst
Eindhoven University of Technology, Eindhoven, The Netherlands

Synonyms

BPMN

Definition

The Business Process Modeling Notation (BPMN), also known as the Business Process Model and Notation, is a graphical notation for drawing business processes. It is proposed as a standard notation for drawing models understandable by different business users. BPMN aims to bridge the communication gap that frequently occurs between business process design and implementation. The language is similar to other informal notations such as UML activity diagrams and extended event-driven process chains.

Key Points

BPMN was initially developed by Business Process Management Initiative (BPMI). It is now being maintained by the Object Management Group (OMG) who released Version 2.0 in 2011 [1]. The intent of BPMN is to standardize a business process modeling notation in the face of many different modeling notations and viewpoints.

A model expressed in terms of BPMN is called a Business Process Diagram (BPD). A BPD is essentially a flowchart composed of different elements. There are four basic categories of elements: (i) *Flow Objects*, (ii) *Connecting Objects*, (iii) *Swimlanes*, and (iv) *Artifacts* [1]. Flow objects are the main graphical elements to define the behavior of a process. There are three types of flow objects: *Events*, *Activities*, and *Gateways*. Events are comparable to places in a Petri net, i.e., they are used to trigger and/or connect activities. There are different types of activities. Atomic activities are referred to as tasks. Gateways are used to model splits and joins. The flow objects can be connected to establish a control-flow. Swimlanes are just a means to structure processes. Artifacts are used to add data or to further annotate process models.

Figure 1 shows the basic set of symbols used by BPMN. These symbols can be combined to construct a BPD. On the left hand side, four types of events are shown. Three type of splits are shown. The data-based XOR gateway split passes control to exactly one of its output arcs. The parallel fork gateway passes control to all output arcs. The event-based XOR gateway selects one output arc based on the occurrence of the corresponding event. Note that Fig. 1 shows only a subset of all possible notations. The complete language is rather complex. The specification itself [1] is 308 pages without providing any formal semantics.

BPMN is an informal language aiming at the communication and not directly at execution. Therefore, the language is positioned as the design language for BPEL, i.e., BPMN diagrams are gradually refined into BPEL specifications. Given the lack of formal semantics this is not a trivial task [2, 3]. Therefore, several attempts have been made to provide semantics for a subset of BPMN [4].

Cross-References

▶ Business Process Execution Language
▶ Business Process Management
▶ Composition

**Business Process
Modeling Notation,
Fig. 1** BPMN notation

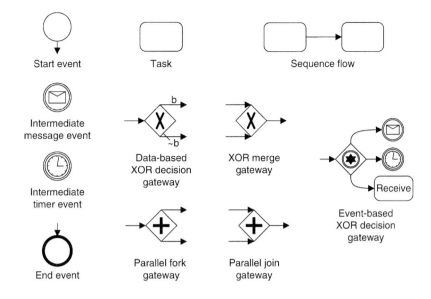

▶ Orchestration
▶ Petri Nets
▶ Web Services
▶ Workflow Management
▶ Workflow Patterns

Recommended Reading

1. OMG Business Process Model and Notation (Version 2.0). 2011. http://www.omg.org/spec/BPMN/2.0.
2. White S. Using BPMN to model a BPEL process. BPTrends. 2005;3(3):1–18.
3. Wohed P, van der Aalst WWP, Dumas M, ter Hofstede AHM, Russell N. On the suitability of BPMN for business process modelling. In: Proceedings of International Conference on Business Process Management; 2006. p. 161–76.
4. Weske M. Business process management: concepts, languages, architectures. Berlin/Heidelberg/New York: Springer; 2007.

Business Process Reengineering

Chiara Francalanci
Politecnico di Milano University, Milan, Italy

Synonyms

Business process redesign

Definition

Business process reengineering refers to a substantial change of a company's organizational processes that (i) is enabled by the implementation of new information technologies that were not previously used by the company, (ii) takes an interfunctional (or interorganizational) perspective, i.e., involves multiple organizational functions (or organizations) that cooperate along processes (iii) considers end-to-end processes, i.e., processes that deliver a service to a company's customers, (iv) emphasizes the integration of information and related information technologies to obtain seamless technological support along processes.

Historical Background

In 1993, Hammer and Champy [3] introduced the concept of business process reengineering as a radical and fast change of organizational processes that leverages information technology. Their work was the first of a wave of research contributions that analyzed the role and impact of information technologies in business process reengineering. This research was accompanied by a widespread use of the term business process reengineering in the industry to indicate large projects that involved the implementation of client-server architectures and the concurrent redesign of core business processes. Client-server enabled a more extensive sharing of organizational data within organizations by making server data accessible to personal computers through more user friendly client applications. In turn, this enabled a redesign of end-to-end business processes towards greater inter-functional cooperation. These changes primarily involved service companies, such as insurance companies, banks, research institutions, and so on, since client server had much broader application in service industries due to a more widespread use of personal computers by all employees and in all activities, both clerical and operating. More recently, the term has been extended to indicate any substantial redesign of business processes that is enabled by information technologies.

Foundations

The interfunctional integration of business processes has been recognized as a fundamental lever to improve organizational performance ever since the early studies within the information perspective of organizational theory. Going back to Galbraith's analysis [5], two complementary methods for increasing the information processing capacity of an organization are proposed. The first improves the communication of information within organizations along hierarchies of authority through *vertical* information systems. Information is gathered from lower hierarchical levels, consolidated, and conveyed to higher decision-making centers. While very effective for routine decision making, vertical information systems can quickly become overloaded in conditions of increasing uncertainty. In this case, firms can resort to direct *lateral* communication, Galbraith's second method. Liaison roles, task forces, project teams, and matrix structures are increasingly powerful mechanisms to facilitate this lateral interchange of information. While the first method reinforces the communication of information across levels of authority, the second consists of a set of organizational solutions al-

lowing information exchanges orthogonal to the hierarchy.

Solutions for direct lateral communication permit higher organizational efficiency when hierarchies become inadequate information processors because of increasing coordination needs. This happens when dependencies between agents are difficult to foresee, as they continually change over time. In principle, within a hierarchy, lateral communication can take place only through levels of authority. Any two agents communicating through hierarchical levels experience an efficiency that is inversely proportional to their distance in the hierarchy. If dependencies between agents are stable and cause recurring paths of lateral communication, hierarchies of authority can be built by grouping agents who require most frequent interaction. By minimizing the distance between interacting agents, hierarchical coordination can be efficient. On the contrary, when organizations operate in conditions of high uncertainty, lateral information exchange becomes essential. Since demands for responsiveness and flexibility in today's business environment are raised as a result of increased uncertainty, the lateral exchange of information becomes critical to support overall information processing needs. Business process reengineering involves the redesign of organizational processes towards a higher degree of lateral communication.

Traditional Intra-organizational Reengineering

Historically, mainframe-based architectures could support the implementation of vertical information systems, but were inherently inadequate for lateral communication. In centralized architectures, no distinction was made between data and applications. Data belonged to the application creating them and could be accessed only through that same application. Consistent with Galbraith's recommendations for the use of IT to support vertical communication, design methodologies witnessed a focus on the hierarchical conception and implementation of information systems. For example, the Normative

Application Portfolio [9] adopted a layered view of organizations, following Anthony's hierarchical framework of planning and control systems. Three main classes of applications were distinguished, supporting operations, management, and strategy, respectively. The application portfolio was created by building vertically from one level to the next, in order to guarantee that activities on lower levels supported higher level activities.

In the mid-1970s, databases and database management systems (DBMS) introduced significant changes in the design of IT architectures. Based on database technology, a new type of IT architecture was implemented, which can be broadly categorized as *centralized*. Centralized architectures allow the *logical separation* between data and applications. The layer of software services constituting the DBMS logically separated data and applications and permitted their independent design. Data common to different applications could be designed and managed as a unified resource. Database management systems exported data manipulation services that could be accessed by any application.

The management of data as a unified resource favored information sharing by integrating data common to different applications. Users running different applications could exchange information by storing and retrieving data in the central database. Conversely, designers could conceive new applications taking advantage of previously gathered and integrated information. A classical example is the use of accounting data by financial applications, allowing more precise financial analyses through detailed information on a firm's cash flows. Likewise, replenishment could be optimized through integration with order fulfillment data.

In the 1990s, client-server architectures allowed the *physical separation* between data and applications. Unlike the logical separation provided by databases in centralized architectures, the physical separation allowed the storage of data and the execution of applications on any computer. Within a client-server environment, shared data were typically stored on the server, but applications could be stored and run on local servers or personal computers.

The DBMS layer of centralized architectures was complemented by an additional layer of network services achieving the physical separation between data and applications. This made the number and the location of resources transparent to individual nodes in the architecture. By relying on peripheral processing and storage capacities of individual nodes, distributed architectures could grow incrementally. This provided the flexibility to continuously adjust to requirements and to implement a variety of applications according to individual needs. This greater flexibility enabled lateral organizational solutions that allowed different functions to take full advantage of organizational information with a variety of functionalities that could be more easily adjusted to changing requirements. These functionalities are now incorporated inside ERP (Enterprise Resource Planning) systems representing fully integrated software solutions that embed laterally integrated organizational processes.

Supply Chain Management Process Reengineering

More recently, the Web service paradigm is shifting reengineering activities towards interorganizational and interpersonal processes. Web service platforms are designed to wrap a company's information system and make selected functionalities available as web services to both internal and external users. This opens up a number of new opportunities. First, the Web service paradigm is causing a radical redesign of supply chain management processes (see Fig. 1). Supply chain management applications (SCM) support the integration of suppliers into a company's information system. This integration allows concurrent and efficient planning of production activities along the value chain. SCM involve a learning process, as shown in Fig. 1. This learning process starts from the monitoring of suppliers to measure their performance and, hence, optimize procurement activities. Then, a subset of efficient and reliable suppliers is selected for tighter integration, ranging from electronic orders to re-

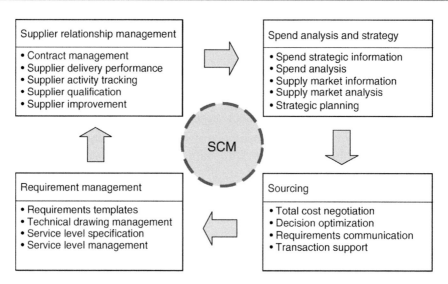

Business Process Reengineering, Fig. 1 The supply chain management (SCM) learning cycle

quirement management and electronic requirement management and codesign. If the supplier management process is effective, a company can build and evaluation and qualification system that can lead to official certifications that suppliers themselves can leverage as part of their brand equity. Overall, SCM involves a deep reengineering of supply management processes that represents the objective of a number of current projects.

Knowledge Management Process Reengineering

Knowledge management processes constitute a second important objective of current business process reengineering activities enabled by the Web and by the more recent Web service paradigm. Knowledge management systems (KMS) are "IT-based systems developed to support and enhance the organizational processes of knowledge creation, storage/retrieval, transfer, and application" [1, p. 114]. KMS span a large and complex spectrum from help desk and customer care applications to those designed to develop employee skills. Virtual communities and collaborative environments are forms of KMS and KMS can also serve as corporate knowledge repositories and maps of expertise.

The literature on Knowledge Management Systems (KMS) largely assumes that an individual's knowledge can be captured and converted into group or organization-available knowledge. When individuals do not contribute to such systems, the knowledge creation capability of the firm is adversely affected. However, there is little clarity in the information systems literature on which mechanisms are necessary for conversion to take place. Many in the information systems literature (e.g., [6]) have argued for social influences (such as culture), hierarchical authority, and/or economic incentives, all of which are external influences that rely on the broader social context outside the KMS system. An alternative view to these external influences is internal or intrinsic motivation. This view assumes that there is little that a broader context outside of the person and the person's interactions with KMS can do to enhance contributions. "Creating and sharing knowledge are intangible activities that can neither be supervised nor forced out of people. They only happen when people cooperate voluntarily" [6]. Intrinsic motivation implies that the activity is performed because of the immediate satisfaction it provides in terms of flow, self-defined goal, or obligations of personal and social identity rather than some external factor or goal. The external factors can even undermine knowledge contributions if they interfere with intrinsic motivation. Although both views are likely to play a role in knowledge sharing, the internal view is important in organizations that take on characteristics of knowledge era (or postmodern) organizations. Such organizations require a more participative and self-managing knowledge worker compared to industrial era organizations. Knowledge-era organizations are associated with increased egalitarianism among positions, increased availability of information and knowledge resources, and increased self-management of knowledge workers. Wikipedia.org represents a typical example of postmodern self-organizing KMS, in which everyone in the world can contribute spontaneously with personal expertise and where control over contributions appropriateness is not based on any hierarchical structure, but is completely peer-based.

In a postmodern organization, an employee's commitment to his or her organization results from autonomous forms of organizing and the resulting self-expression and feelings of responsibility and control of the outputs of work. With knowledge workers, where work is associated with flows of knowledge and information, rather than flows of materials, self-expression, responsibility, and control are often targeted to knowledge outputs. Knowledge workers want to share their knowledge across the organization while preserving their association with these contributions. This type of psychological attachment, or emotional connection between the person and knowledge is well documented in open source initiatives. When given a choice to require or not require attribution to one's creative works (permit others to copy, distribute, display, modify the work but only if given credit without any economic implications), individuals invariably choose the requirement of future users of their knowledge to attribute knowledge to them (http://Creativecommons.org/). The importance of psychological attachment to knowledge is no less important in commercial contexts.

KMS do not necessarily harness psychological attachment between knowledge embedded in the system and the individual who is the source of that knowledge. On one side, it is risky for the organization to let knowledge reside within the minds of individuals, in a form that can easily leak across the firm's boundaries and lead to a loss of competitive advantage. On the other hand, individual employees may perceive their personal goals to be poorly served by sharing knowledge unless the system helps to manage the knowledge worker's personal attachment to knowledge and make this attachment known in relevant organizational communities. In knowledge intensive organizations, people's distinctiveness depends upon their possessed knowledge. KMS that do not help construct, communicate, and defend the psychological attachment between the knowledge and the knowledge worker, and the rest of the organization can reduce the motivation to contribute knowledge to KMS. Fostering psychological attachment is likely to increase the knowledge workers quality of contributions, not only the quantity. In fact, the more complex and tacit the knowledge contributed to KMS, the higher the effort that knowledge workers are likely to put in their contributing activities. When the KMS fosters psychological attachment to contributions, knowledge workers increase their likelihood to engage also those knowledge sharing activities that require greater effort in order to be carried out.

Process Modeling Languages and Techniques

Reengineering initiatives typically involve a process modeling phase that supports the analysis of existing processes and the specification of new processes. UML (www.uml.org) represents a quasi-standard for process modeling and is widely used in business process reengineering.

Over the years, the scope of business processes and BPM has broadened. Initially, BPM, or workflow, was a technique that helped design largely human-based, paper-driven processes within a corporate department. For example, to handle a claim, an insurance claims process, taking as input a scanned image of a paper claims form, would pass the form electronically from the mailbox (or worklist) of one claims specialist to that of another, mimicking the traditional movement of interoffice mail from desk to desk. The contemporary process orchestrates complex system interactions, and is itself a service capable of communicating and conversing with the processes of other companies according to well-defined technical contracts. A retailer's process to handle a purchase order, for example, is a service that uses XML messages to converse with the service-based processes of consumers and warehouses.

A number of new modeling languages have recently been proposed to accommodate this complexity. For example, the i* model and its subsequent developments within the TROPOS project allow the representation of strategic relationships between organizations and their relationships with goals, resources and system components (cf. [10]). The impact of the structure of cooperation forms on the interactions among actors are modeled in the field of Multi-Agent Systems to determine an architectural solution based on typical software non-functional requirements (i.e., security, integrity, modularity, etc.). Moreover, the i* model supports the analysis of high-level goals together with non-functional requirements. The ability to move from high-level goals to sub-goals is also provided by KAOS (cf. [4]). KAOS is a formal approach for analyzing goals and for transforming goals into requirements for the software system. Finally, GBRAM (Goal Based Requirements Analysis Method, cf. [2]) proposes a method supporting the initial identification of high-level goals. Different from i* and KAOS, GBRAM does not assume that high-level goals are previously identified and provides a set of strategies to elicit goals from all available sources of information.

Key Applications

Enterprise Resource Planning (ERP), Web Services (WS), Knowledge Management Systems (KMS).

Cross-References

▶ Database Management System
▶ Unified Modeling Language
▶ Web Services

Recommended Reading

1. Alavi M, Leidner DE. Knowledge management and knowledge management systems: conceptual foundations and research issues. MIS Q. 2001;25(1):107–36.
2. Anton A. Goal identification and refinement in the specification of software-based information systems, Ph.D. Dissertation. Georgia Institute of Technology, Atlanta, 1997.
3. Champy J, Hammer M. Reengineering the corporation. New York: Harper Collins; 1993.
4. Dardenne A, Lamsweerde A, van Fickas S. Goal-directed requirements acquisition. Sci Comput Program. 1993;20(1–2):3–50.
5. Galbraith JR. Organization design. Reading: Addison-Wesley; 1977.
6. Jarvenpaa SL, Staples DS. The use of collaborative electronic media for information sharing: an exploratory study of determinants. J Strateg Inf Syst. 2000;9(2–3):129–54.
7. Kim WC, Mauborgne R. Procedural justice, strategic decision making, and the knowledge economy. Strateg Manag J. 1988;19(4):323–38.
8. Malone TW, Crowston K. The interdisciplinary study of coordination. ACM Comput Surv. 1994;26(1):87–119.
9. Nolan RL. Managing the data resource function. St. Paul: West Publishing Company; 1982.
10. Yu E, Mylopoulos J. Using goal, rules and methods to support reasoning in business process reengineering. Int J Intell Syst Account Financ Manag. 1996;5(1):1–13.

C

Cache-Conscious Query Processing

Kenneth A. Ross
Columbia University, New York, NY, USA

Synonyms

Cache-aware query processing; Cache-sensitive query processing

Definition

Query processing algorithms that are designed to efficiently exploit the available cache units in the memory hierarchy. Cache-conscious algorithms typically employ knowledge of architectural parameters such as cache size and latency. This knowledge can be used to ensure that the algorithms have suitable temporal and/or spatial locality on the target platform.

Historical Background

Between 1980 and 2005, processing speeds improved by roughly four orders of magnitude, while memory speeds improved by less than a single order of magnitude. As of 2017, it is common for data accesses to RAM to require several hundred CPU cycles to resolve. Many database workloads have shifted from being I/O bound to being memory/CPU-bound as the amount of memory per machine has been increasing. For such workloads, improving the locality of data-intensive operations can have a direct impact on the system's overall performance.

Scientific Fundamentals

A cache is a hardware unit that speeds up access to data. Several cache units may be present at various levels of the memory hierarchy, depending on the processor architecture. For example, a processor may have a small but fast Level-1 (L1) cache for data and another L1 cache for instructions. The same processor may have a larger but slower L2 cache storing both data and instructions. Many processors also have an L3 cache. On multicore processors, the lower level caches are typically shared among groups of cores. A special kind of cache for mapping virtual memory to physical memory is known as the translation lookaside buffer (TLB).

On a system with multiple CPUs, the caches of the different CPUs interact to ensure coherent access to data. Accessing data from a remote CPU cache is slower than accessing the corresponding local cache. Similarly, accessing data resident in remote CPU RAM is slower than accessing data from local RAM, a phenomenon known as nonuniform memory access (NUMA).

Some initial analysis would typically be performed to determine the performance characteristics of a workload. For example, Ailamaki et al. [2] used hardware performance counters to demonstrate that several commercial systems

© Springer Science+Business Media, LLC, part of Springer Nature 2018
L. Liu, M. T. Özsu (eds.), *Encyclopedia of Database Systems*,
https://doi.org/10.1007/978-1-4614-8265-9

were, at that time, suffering many L2 data cache misses and L1 instruction cache misses. Based on such observations, one can determine that the L2 data cache and L1 instruction cache are targets for performance tuning.

If the operating system does not provide direct access to system parameters such as the cache size, a database system can run a calibration test to estimate the relevant parameters [14].

To get good cache performance, algorithm designers typically utilize one or more of the following general approaches:

- Improve spatial locality, so that data items that are often accessed together are in the same cache lines.
- Improve temporal locality for data, so that after an initial cache miss, subsequent data item accesses occur while the item is still cache-resident.
- Improve temporal locality for instructions, so that code that needs to be applied to many data items is executed many times while the instructions reside in the cache.
- Hide the memory latency. Latency can be hidden in several ways. If the data access pattern is predictable, prefetching data elements into the cache can overlap the memory latency with other work. On architectures that support multiple simultaneous outstanding memory requests, cache miss latencies can be overlapped with one another.
- Avoid cache pollution. Use nontemporal loads and stores to prevent caching data items that are not going to be reused in the near future.
- Avoid thrashing the TLB. Do not perform random accesses to a large number of active memory regions.
- Sample the data to predict the cache behavior, and choose an algorithm accordingly.

Examples of each of these approaches are given below.

Spatial Locality

In many query-processing contexts, only a few columns for a table are needed. In such cases, it pays to organize the table column-wise, so that column values from consecutive records are contiguous. Cache lines then contain many useful data elements. A row-wise organization would require more cache-line accesses, since each cache line would contain some data from unneeded columns. Examples of systems with column-wise storage are Sybase IQ [13], MonetDB [4], and C-Store [21]. The PAX storage model [1] allows for column-wise storage within each disk page. The main advantage of such an approach is that existing page-oriented database systems can improve cache behavior with limited changes to the whole system.

Chilimbi et al. [8] advocate placing multiple levels of a binary tree within a cache line, to reduce the number of cache misses per traversal. Chilimbi et al. also use cache coloring to place frequently accessed items in certain ranges of physical memory. The idea is to reduce the number of conflict misses by making sure that the low-order bits of the addresses of certain items cannot be the same.

To reduce the number of cache lines needed to search for an item in an index, Rao and Ross proposed CSS-trees [18] and CSB+-trees [19]. CSS-trees eliminate pointers; a node contains only keys. Nodes are of fixed size, typically one cache line, aligned to the cache-line boundaries. Nodes are stored in an array in such a way that the children of a node can be determined using simple arithmetic operations, making pointers unnecessary. CSB+-trees extend this idea, allowing just one pointer per node and requiring that all sibling nodes be contiguous. CSB+-trees have better update performance than CSS-trees while retaining almost all of the cache efficiency. The diagram below shows a CSB+-Tree of Order 1. Note that each node has only one child pointer and that each node's children are allocated contiguously.

Several other ways to compress B+-tree nodes for cache performance, such as key compression and key-prefix truncation, are discussed in [12].

Temporal Locality

Blocking is a general technique for ensuring temporal locality. Data is processed in cache-sized units, so that all data within the block stays

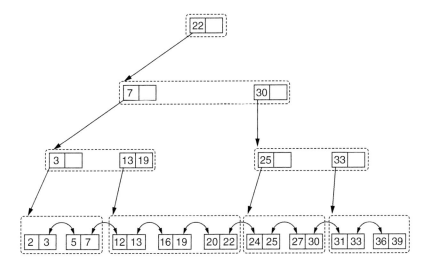

cache-resident. The blocks are then combined in a later phase. Alpha-Sort [15] is an example of such a method: cache-sized units of input data are read and quick-sorted into runs. These runs are merged in a later phase. Padmanabhan et al. [16] modified a commercial database system to pass data between certain operators in blocks and demonstrated improved cache behavior.

Buffering is a related strategy to improve temporal locality. Zhou and Ross [24] propose buffering to speed up bulk B-tree index lookups. By sending probes only one level at a time through the tree, in batches, one can amortize the cost of reading an index node over many probes. The savings in data cache misses usually outweigh the extra cost of reading and writing to intermediate buffers. Zhou and Ross also examine the code size of database operators and propose to buffer data to ensure that the footprint of the active code is smaller than the size of the L1 instruction cache [25]. Again, the savings in instruction cache misses usually outweigh the cost of buffering.

Partitioning the data into cache-sized units for later processing is the dual of blocking. Examples include the partitioned hash join [20] and radix-join [4].

When multiple processors or multiple threads within a single processor access a shared cache, cache interference can result. Even if each individual thread is cache-conscious, the total cache resources may be insufficient for all threads, and cache thrashing may result. To counter this interference, one could design cooperative threads that work together on a common task using common cache-conscious data structures. Multithreaded join operators [11, 23] and aggregation operators [9] have been proposed.

Many divide-and-conquer style algorithms generate temporal locality at recursively smaller granularities. Such algorithms have been termed *cache oblivious* because they can achieve locality at multiple levels of the memory hierarchy without explicit knowledge of the cache parameters [10]. A distinct but related concept is a *cache adaptive* algorithm, which is an algorithm that exhibits good cache performance even if the available cache memory changes during the course of the computation [3].

Prefetching

Prefetching involves reading data into the cache ahead of when it is to be used. When access patterns can be predicted in advance and when memory bandwidth is not saturated, prefetching can effectively hide the memory latency. Some hardware platforms automatically recognize certain access patterns, such as regular fixed-stride access to memory. The hardware then automatically prefetches ahead in the access sequence. For access patterns that are not so easily recognized or for hardware platforms that do not support hardware prefetching, one can explicitly prefetch memory locations using software.

Chen et al. [7] show how to prefetch parts of a B+-tree node or CSB+-tree node to get a bigger effective node size. For example, if the memory system can support n outstanding memory requests, then a node consisting of n cache lines could be retrieved in only slightly more time than a single cache line. Since wider nodes result in shallower trees, the optimal node size might be several cache lines wide.

Chen et al. [5] use prefetching to speed up hash joins. The internal steps for processing records are divided into stages. A memory access is typically required between stages. Stages for multiple records are scheduled, so that while data for a forthcoming operation is being prefetched, useful work is being performed on other records.

Zhou et al. [23] define the notion of a work-ahead set, a data structure that describes a memory location and a computation stage for some data-intensive operation. One thread of a two-threaded system is devoted purely to prefetching the data into the cache, while the other thread does the algorithmic work.

Avoiding Cache Pollution and TLB Thrashing

As previously mentioned, partitioning can improve cache behavior by generating cache-resident subproblems. Nevertheless, the partitioning process itself is vulnerable to cache performance pitfalls, both cache misses and TLB thrashing [14, 22]. To overcome these difficulties, a combination of performance optimizations has been employed. Partitioning first into a cache-resident buffer and then flushing the buffer one cache line at a time have several benefits [22]. TLB misses can be amortized over a cache line's worth of data items. Nontemporal store instructions can prevent the caching of the written data in some of the CPU cache levels. Write-combining registers available on modern CPUs allow writing of a whole cache line without reading its previous contents. Even in-place partitioning can be cache-efficient [17].

Sampling

Inspector joins sample the data during an initial partitioning phase [6]. This information is used to accelerate a cache-optimized join algorithm for processing the partitions. Cieslewicz et al. [9] sample a stream of tuples for aggregation to estimate (among other things) the locality of reference of group-by values. Based on that information, an appropriate aggregation algorithm is chosen for the remainder of the stream.

Key Applications

Data-intensive operators such as sorts, joins, aggregates, and index lookups, form the "assembly language" into which complex queries are compiled. By making these operators as efficient as possible on modern hardware, all database system users can effectively exploit the available resources.

Future Directions

Future processors are likely to scale by placing many cores on a chip, with only a modest increase in clock frequency. As a result, the amount of cache memory per processor may actually decrease over time, making cache optimization even more critical. For chips with shared caches, interference between cores will be a significant performance hazard. While locality is good for cache behavior, it can be bad for concurrency due to hot-spots of contention [9]. Cache performance will need to be considered together with parallelism to find appropriate performance trade-offs.

Cross-References

▶ Architecture-Conscious Database System

Recommended Reading

1. Ailamaki A, DeWitt DJ, Hill MD, Skounakis M. Weaving relations for cache performance. In: Proceedings of the 27th International Conference on Very Large Data Bases; 2001.
2. Ailamaki A, et al. DBMSs on a modern processor: where does time go? In: Proceedings of the 25th

International Conference on Very Large Data Bases; 1999.

3. Bender MA, Ebrahimi R, Fineman JT, Ghasemiesfeh G, Johnson R, McCauley S. Cache-adaptive algorithms. In: Proceedings of the 25th Annual ACM-SIAM Symposium on Discrete Algorithms; 2014. p. 958–71.

4. Boncz PA, Manegold S, Kersten ML. Database architecture optimized for the new bottleneck: memory access. In: Proceedings of the 25th International Conference on Very Large Data Bases; 1999.

5. Chen S, Ailamaki A, Gibbons PB, Mowry TC. Improving hash join performance through prefetching. In: Proceedings of the 20th International Conference on Data Engineering; 2004.

6. Chen S, et al. Inspector joins. In: Proceedings of the 31st International Conference on Very Large Data Bases; 2005. p. 817–28.

7. Chen S, Gibbons PB, Mowry TC. Improving index performance through prefetching. In: Proceedings of the ACM SIGMOD International Conference on Management of Data; 2001.

8. Chilimbi TM, Hill MD, Larus JR. Cache-conscious structure layout. In: Proceedings of the ACM SIGPLAN 1999 Conference on Programming Language Design and Implementation; 1999.

9. Cieslewicz J, Ross KA. Adaptive aggregation on chip multiprocessors. In: Proceedings of the 33rd International Conference on Very Large Data Bases; 2007. p. 339–50.

10. Frigo M, Leiserson CE, Prokop H, Ramachandran S. Cache-oblivious algorithms. In: Proceedings of the 40th Annual Symposium on Foundations of Computer Science; 1999. p. 285–98.

11. Garcia P, Korth HF. Database hash-join algorithms on multithreaded computer architectures. In: Proceedings of the 3rd Conference on Computing Frontiers; 2006. p. 241–51.

12. Graefe G, Larson P. B-tree indexes and CPU caches. In: Proceedings of the 17th International Conference on Data Engineering; 2001.

13. MacNicol R, French B. Sybase IQ multiplex – designed for analytics. In: Proceedings of the 30th International Conference on Very Large Data Bases; 2004. p. 1227–30.

14. Manegold S, et al. What happens during a join? Dissecting CPU and memory optimization effects. In: Proceedings of the 26th International Conference on Very Large Data Bases; 2000.

15. Nyberg C, Barclay T, Cvetanovic Z, Gray J, Lomet DB. Alphasort: a cache-sensitive parallel external sort. VLDB J. 1995;4(4):603–27.

16. Padmanabhan S, Malkemus T, Agarwal R, Jhingran A. Block oriented processing of relational database operations in modern computer architectures. In: Proceedings of the 17th International Conference on Data Engineering; 2001.

17. Polychroniou O, Ross KA. A comprehensive study of main-memory partitioning and its application to large-scale comparison- and radix-sort. In: Proceed-ings of the ACM SIGMOD International Conference on Management of Data; 2014.

18. Rao J, Ross KA. Cache conscious indexing for decision-support in main memory. In: Proceedings of the 25th International Conference on Very Large Data Bases; 1999.

19. Rao J, Ross KA. Making B+ trees cache conscious in main memory. In: Proceedings of the ACM SIGMOD International Conference on Management of Data; 2000.

20. Shatdal A, Kant C, Naughton JF. Cache conscious algorithms for relational query processing. In: Proceedings of the 20th International Conference on Very Large Data Bases; 1994. p. 510–21.

21. Stonebraker M, Abadi DJ, Batkin A, Chen X, Cherniack M, Ferreira M, Lau E, Lin A, Madden S, O'Neil EJ, O'Neil PE, Rasin A, Tran N, Zdonik SB. C-store: a column-oriented DBMS. In: Proceedings of the 31st International Conference on Very Large Data Bases; 2005.

22. Wassenberg J, Sanders P. Engineering a multi-core radix sort. In: Proceedings of the 17th International Euro-Par Conference; 2011. p. 160–9.

23. Zhou J, Cieslewicz J, Ross KA, Shah M. Improving database performance on simultaneous multithreading processors. In: Proceedings of the 31st International Conference on Very Large Data Bases; 2005. p. 49–60.

24. Zhou J, Ross KA. Buffering accesses to memory-resident index structures. In: Proceedings of the 29th International Conference on Very Large Data Bases; 2003.

25. Zhou J, Ross KA. Buffering database operations for enhanced instruction cache performance. In: Proceedings of the ACM SIGMOD International Conference on Management of Data; 2004.

Calendar

Christian S. Jensen[1] and Richard T. Snodgrass[2,3]
[1] Department of Computer Science, Aalborg University, Aalborg, Denmark
[2] Department of Computer Science, University of Arizona, Tucson, AZ, USA
[3] Dataware Ventures, Tucson, AZ, USA

Definition

A *calendar* provides a human interpretation of time. As such, calendars ascribe meaning to temporal values such that the particular meaning or interpretation provided is relevant to its users.

In particular, calendars determine the mapping between human-meaningful time values and an underlying time line.

Key Points

Calendars are most often cyclic, allowing human-meaningful time values to be expressed succinctly. For example, dates in the common Gregorian calendar may be expressed in the form <*month, day, year*> where the month and day fields cycle as time passes.

The concept of calendar defined here subsumes commonly used calendars such as the Gregorian calendar, the Hebrew calendar, and the Lunar calendar, though the given definition is much more general. This usage is consistent with the conventional English meaning of the word.

Dershowitz and Reingold's book presents complete algorithms for fourteen prominent calendars: the present civil calendar (Gregorian), the recent ISO commercial calendar, the old civil calendar (Julian), the Coptic an Ethiopic calendars, the Islamic (Muslim) calendar, the modern Persian (solar) calendar, the Bahá'í calendar, the Hebrew (Jewish) calendar, the Mayan calendars, the French Revolutionary calendar, the Chinese calendar, and both the old (mean) and new (true) Hindu (Indian) calendars. One could also envision more specific calendars, such as an academic calendar particular to a school, or a fiscal calendar particular to a company.

Cross-References

▶ Calendric System
▶ SQL
▶ Temporal Database

Recommended Reading

1. Bettini C, Dyreson CE, Evans WS, Snodgrass RT, Wang XS. A glossary of time granularity concepts. In: Etzion O, Jajodia S, Sripada S, editors. Temporal databases: research and practice, LNCS, vol. 1399. Berlin: Springer; 1998. p. 406–13.
2. Dershowitz N, Reingold EM. Calendrical calculations. Cambridge: Cambridge University Press; 1977.
3. Jensen CS, Dyreson CE. editors. Böhlen M, Clifford J, Elmasri R, Gadia SK, Grandi F, Hayes P, Jajodia S, Käfer W, Kline N, Lorentzos N, Mitsopoulos Y, Montanari A, Nonen D, Peressi E, Pernici B, Roddick JF, Sarda NL, Scalas MR, Segev A, Snodgrass RT, Soo MD, Tansel A, Tiberio R, Wiederhold G. A consensus glossary of temporal database concepts – February 1998 Version. In: Etzion O, Jajodia S, Sripada S. editors. Temporal databases: research and practice. LNCS, vol. 1399. Berlin: Springer; 1998. p. 367–405.
4. Urgun B, Dyreson CE, Snodgrass RT, Miller JK, Kline N, Soo MD, Jensen CS. Integrating multiple calendars using τZaman. Softw Pract Exper. 2007;37(3): 267–308.

Calendric System

Curtis E. Dyreson[1], Christian S. Jensen[2], and Richard T. Snodgrass[3,4]
[1]Utah State University, Logan, UT, USA
[2]Department of Computer Science, Aalborg University, Aalborg, Denmark
[3]Department of Computer Science, University of Arizona, Tucson, AZ, USA
[4]Dataware Ventures, Tucson, AZ, USA

Definition

A calendric system is a collection of calendars. The calendars in a calendric system are defined over contiguous and non-overlapping intervals of an underlying time-line. Calendric systems define the human interpretation of time for a particular locale as different calendars may be employed during different intervals.

Key Points

A calendric system is the abstraction of time available at the conceptual and logical (query language) levels. As an example, a Russian calendric system could be constructed by considering the sequence of six different calendars used in that region of the world. In prehistoric epochs, the Geologic calendar and Carbon-14 dating (another form of calendar) are used to measure time. Later, during the Roman empire, the lunar calendar developed by the Roman republic was used. Pope Julius, in the first century b.c., introduced a solar calendar, the Julian calendar. This calendar was in use until the 1917 Bolshevik revolution when the Gregorian calendar, first introduced by Pope Gregory XIII in 1572, was adopted. In 1929, the Soviets introduced a continuous schedule work week based on 4 days of work followed by 1 day of rest, in an attempt to break tradition with the 7-day week. This new calendar, the Communist calendar, had the failing that only 80% of the work force was active on any day, and it was abandoned after only 2 years in favor of the Gregorian calendar, which is still in use today.

The term "calendric system" has been used to describe the calculation of events within a single calendar. However, the given definition generalizes that usage to multiple calendars in a very natural way.

Cross-References

► Calendar
► Temporal Database
► Time Interval

Recommended Reading

1. Jensen CS, Dyreson CE, editors. Böhlen M, Clifford J, Elmasri R, Gadia SK, Grandi F, Hayes P, Jajodia S, Käfer W, Kline N, Lorentzos N, Mitsopoulos Y, Montanari A, Nonen D, Peressi E, Pernici B, Roddick JF, Sarda NL, Scalas MR, Segev A, Snodgrass RT, Soo MD, Tansel A, Tiberio R, Wiederhold G. A consensus glossary of temporal database concepts – February 1998 version. In: Etzion O, Jajodia S, Sripada S. editors. Temporal databases: research and practice. LNCS, vol. 1399. Berlin: Springer; 1998, p. 367–405.

CAP Theorem

Alan Fekete
University of Sydney, Sydney, NSW, Australia

Definition

The CAP theorem states that it is impossible to design a distributed data management platform that provides always consistent (C) data accessed through always available (A) operations if there is the possibility that the set of nodes may partition (P), that is, where there may be nodes which are unable to communicate with each other.

Key Points

The CAP theorem was stated in 1999 by Fox and Brewer [2], and it was proved mathematically by Gilbert and Lynch in 2002 [3]. The CAP theorem is often used to justify NoSQL data platforms that do not keep data consistent, but rather provide eventual consistency (q.v.) or other weak consistency models for replicated data (q.v.).

A common description of the CAP theorem is that a system can have at most two out of the three properties: consistency (C), availability (A), and partition tolerance (P). However, consistency and availability are properties that the system might choose to offer to users, or not; but P is a property of the communication environment, and so it is not up to the system designer to decide about this.

The main idea of the theorem is to suppose (for contradiction) that a system has C, A, and P. Now consider a write (changing x from 0 to 1, say) that is done on one side of a partition; because the system is always available, this write will succeed. Now consider a read of x which is requested on the other side of the partition; this must also succeed (by A), but there is no way to the information from the write to get to the response; so the read will not be able to return 1 but can only report on the previous value 0. Thus the execution doesn't reflect a view of a single consistent data item, contradicting the assumption of C.

Even when communication between nodes is possible, it can take a relatively long time compared to local computation. Abadi [1] extended the CAP idea to PACELC: "if there is a partition (P), how does the system trade off availability and consistency (A and C); else (E), when the system is running normally in the absence of partitions, how does the system trade off latency (L) and consistency (C)?"

Recommended Reading

1. Abadi D. Consistency tradeoffs in modern distributed database system design: CAP is only part of the story. IEEE Comput. 2012;45(2):37–42.
2. Fox A, Brewer E. Harvest, yield and scalable tolerant systems. In: Proceedings of Workshop on Hot Topics in Operating Systems; 1999. p. 174–8.
3. Gilbert S, Lynch N. Brewer's conjecture and the feasibility of consistent, available, partition-tolerant web services. SIGACT News. 2002;33(2):51–9.

Cardinal Direction Relationships

Spiros Skiadopoulos
University of Peloponnese, Tripoli, Greece

Synonyms

Directional relationships; Orientation relationships

Definition

Cardinal direction relationships are qualitative spatial relations that describe how an object is placed relative to other objects utilizing a coordinate system. This knowledge is expressed using symbolic (qualitative) and not numerical (quantitative) methods. For instance, *north* and *southeast* are cardinal direction relationships. Such relationships are used to describe and constrain the relative positions of objects and can be used to pose queries such as "Find all objects a, b, and c such that a is *north of* b and b is *southeast of* c."

Historical Background

Qualitative spatial relationships (QSR) approach common sense knowledge and reasoning about space using symbolic rather than numerical methods [5]. QSR has found applications in many diverse scientific areas such as geographic information systems, artificial intelligence, databases, and multimedia. Most researchers in QSR have concentrated on the three main aspects of space, namely, topology, distance, and direction. The uttermost aim in these lines of research is to define new and more expressive categories of spatial operators, as well as to build efficient algorithms for the automatic processing of queries involving these operators.

Foundations

Several models capturing cardinal direction relationships have been proposed in the literature. Typically, a cardinal direction relationship is a binary relation that describes how a *primary object a* is placed relative to a *reference object b* utilizing a coordinate system (e.g., object a is *north of* object b). Early models for cardinal direction relationships approximate an extended spatial object by a representative point [3,6]. For instance, objects in Fig. 1 are approximated by their centroid. Typically, such models partition the space around the reference object b into a number of mutually exclusive areas. For instance, the *projection* model partitions the space using lines parallel to the axes (Fig. 1a), while the *cone* model partitions the space using lines with an origin angle φ (Fig. 1b). Depending on the adopted model, the relation between two objects may change. For instance, consider Fig. 1c. According to the projection model, a is northeast of b, while according to the cone model, a is north of b. Point-based approximations may be crude [4]; thus, later models more finely approximate an object using a representative area (instead of a point) and express directions on these approximations [9, 10]. Most commonly, these methods use the minimum bounding box as a representative area (the minimum bounding box of an object a is the smallest rectangle, aligned with the axis, that encloses a). Unfortunately, even with such finer approximations, models that approximate both the primary and the reference object may give misleading directional relationships when objects are overlapping, intertwined, or horseshoe-shaped [4].

Cardinal Direction Relationships, Fig. 1
Projection and cone models for point approximations

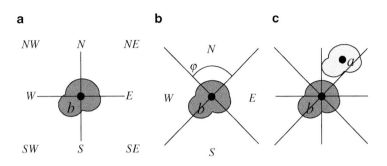

Cardinal Direction Relationships, Fig. 2
Extending the projection model

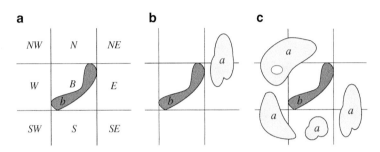

Recently, more precise models for cardinal direction relationships have been proposed. Such models define directions on the exact shape of the primary object and only approximate the reference object (using its minimum bounding box). The *projection-based directional relationships* (*PDR*) model is the first model of this category [4, 12, 13]. The *PDR* model partitions the plane around the reference object into nine areas similarly to the projection model (Fig. 2a). These areas correspond to the minimum bounding box (*B*) and the eight cardinal directions (*N*, *NE*, *E*, etc.). Intuitively, the cardinal direction relationship is characterized by the names of the reference areas occupied by the primary object. For instance, in Fig. 2b, object *a* is partly *NE* and partly *E* of object *b*. This is denoted by *a NE:E b*. Similarly in Fig. 2c, *a B:S:SW:W:NW:N:E:SE b* holds. In total, the *PDR* model identifies 511 ($= 2^9 - 1$) relationships.

Clearly, the *PDR* model offers a more precise and expressive model than previous approaches that approximate objects using points or rectangles [4]. The *PDR* model adopts a projection-based partition using lines parallel to the axes (Fig. 2). Typically, most people find it more natural to organize the surrounding space using lines with an origin angle similarly to the cone model

(Fig. 3). This partition of space is adopted by the *cone-based directional relationships* (CDR) model. Similarly to the *PDR* model, the CDR model uses the exact shape of the primary object and only approximates the reference object using its minimum bounding box [15]. Interestingly, for the CDR model, the space around the reference object is partitioned into five areas using a cone-like partition (Fig. 3a). The cardinal direction relationship is formed by the areas that the primary object falls in. For instance, in Fig. 3b, *a* is south of *b*. This is denoted by *a S b*. Similarly in Fig. 3c, *a B:W:N b* holds. In total, the CDR model identifies 31 ($= 2^5 - 1$) relationships.

In another line of research, cardinal direction relationships are modeled as ternary relationships [2]. Given three objects *a*, *b*, and *c*, the ternary model expresses the direction relation of the primary object *a* with respect to a reference frame constructed by objects *b* and *c*. Specifically, the convex hull, the internal and the external tangents of objects *b* and *c* divide the space into five areas as in Fig. 4a. These areas correspond to the following directions: right side (*RS*), before (*BF*), left side (*LS*), after (*AF*), and between (*BT*). Similarly to *PDR* and CDR, the name of the areas that *a* falls into determines the relation. For instance, in Fig. 4b, *a* is before and to the left side of

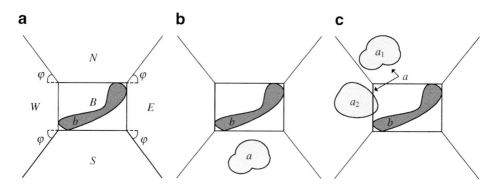

Cardinal Direction Relationships, Fig. 3 Extending the cone model

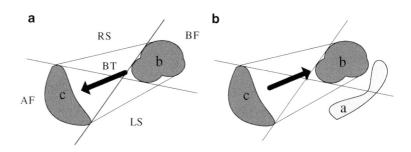

Cardinal Direction Relationships, Fig. 4 Ternary cardinal direction relationships

b and c. This is denoted by $LS{:}BF(a,b,c)$. Notice that, if the order of the reference objects changes, the relationship also changes. For instance, in Fig. 4b, $RS{:}AF(a,c,b)$ also holds.

For all the above models of cardinal direction relationships, research has focused on four interesting operators: (i) efficiently determining the relationships that hold between a set of objects, (ii) calculating the inverse of a relationship, (iii) computing the composition of two relationships, and (iv) checking the consistency of a set of relationships. These operators are used as mechanisms that compute and infer cardinal direction relationships. Such mechanisms are important as they are in the heart of any system that retrieves collections of objects similarly related to each other using spatial relations. Table 1 summarizes current research on the aforementioned problems.

Key Applications

Cardinal direction relationships intuitively describe the relative position of objects and can be used to constrain and query spatial configurations. This information is very useful in several applications like geographic information systems, spatial databases, spatial arrangement and planning, etc.

Future Directions

There are several open and important problems concerning cardinal direction relationships. For the models discussed in the previous section, as presented in Table 1, there are four operators that have not been studied (two for the CDR and two for the ternary model). Another open issue is the integration of cardinal direction relationships with existing spatial query answering algorithms and data indexing structures (like the R-tree). Finally, with respect to the modeling aspect, even the most expressive cardinal direction relationship models define directions by approximating the reference objects. Currently, there is not an intuitive, simple, and easy to use model that defines cardinal direction relationships on the exact shape of the involved objects.

Cardinal Direction Relationships, Table 1 Operations for cardinal direction relationships

Model	Computation	Inverse	Composition	Consistency
Point approximations	[11]	[6]	[6]	[6]
Rectangle approximations	[10]	[10]	[10]	[10]
$\mathcal{P}$DR	[14]	[1]	[12]	[7, 8, 13]
$\mathcal{C}$DR	Open problem	[15]	[15]	Open problem
Ternary	[2]	[2]	Open problem	Open problem

Cross-References

▶ Geographic Information System
▶ Spatial Operations and Map Operations
▶ Topological Relationships

Recommended Reading

1. Cicerone S, Di Felice P. Cardinal directions between spatial objects: the pairwise-consistency problem. Inf Sci. 2004;164(1–4):165–88.
2. Clementini E, Billen R. Modeling and computing ternary projective relations between regions. IEEE Trans Knowl Data Eng. 2006;18(6):799–814.
3. Freksa C. Using orientation information for qualitative spatial reasoning. In: Proceedings of the International Conference on Spatial Information Theory; 1992. p. 162–78.
4. Goyal R. Similarity assessment for cardinal directions between extended spatial objects. PhD Thesis, Department of Spatial Information Science and Engineering, University of Maine; 2000.
5. Hernández D. Qualitative representation of spatial knowledge. LNCS, vol. 804. Berlin: Springer; 1994.
6. Ligozat G. Reasoning about cardinal directions. J Visual Lang Comput. 1998;9(1):23–44.
7. Liu W, Li S. Reasoning about cardinal directions between extended objects: the NP-hardness result. Artif Intell. 2011;175(18): 2155–69.
8. Liu W, Zhang X, Li S, Ying M. Reasoning about cardinal directions between extended objects. Artif Intell. 2010;174(12–13):951–83
9. Mukerjee A, Joe G. A qualitative model for space. In: Proceedings of 7th National Conference on AI; 1990. p. 721–7.
10. Papadias D. Relation-based representation of spatial knowledge. PhD Thesis, Department of Electrical and Computer Engineering, National Technical University of Athens; 1994.
11. Peuquet DJ, Ci-Xiang Z. An algorithm to determine the directional relationship between arbitrarily-shaped polygons in the plane. Pattern Recognit. 1987;20(1):65–74.
12. Skiadopoulos S, Koubarakis M. Composing cardinal direction relations. Artif Intell. 2004;152(2):143–71
13. Skiadopoulos S, Koubarakis M. On the consistency of cardinal directions constraints. Artif Intell. 2005;163(1):91–135.
14. Skiadopoulos S, Giannoukos C, Sarkas N, Vassiliadis P, Sellis T, Koubarakis M. Computing and managing cardinal direction relations. IEEE Trans Knowl Data Eng. 2005;17(12):1610–23.
15. Skiadopoulos S, Sarkas N, Sellis T, Koubarakis M. A family of directional relation models for extended objects. IEEE Trans Knowl Data Eng. 2007;19(8):1116–30

Cartesian Product

Cristina Sirangelo
IRIF, Paris Diderot University, Paris, France

Synonyms

Cross product

Definition

Given two relation instances R_1, over set of attributes U_1, and R_2, over set of attributes U_2 – with U_1 and U_2 disjoint – the Cartesian product $R_1 \times R_2$ returns a new relation, over set of attributes $U_1 \cup U_2$, consisting of tuples $\{t | t(U_1) \in R_1 \text{ and } t(U_2) \in R_2\}$. Here t($U$) denotes the restriction of the tuple t to attributes in the set U.

Key Points

The Cartesian product is an operator of the relational algebra which extends to relations the usual notion of Cartesian product of sets.

Since the sets of attributes of the input relations are disjoint, in $R_1 \times R_2$, each tuple of R_1 is combined with each tuple of R_2; moreover the arity of the output relation is the sum of the arities of R_1 and R_2.

As an example, consider a relation *Students* over attributes (*student-number, student-name*), containing tuples {(1001, *Black*), (1002, *White*)}, and a relation *Courses* over attributes (*course-number, course-name*), containing tuples {(*EH*1, *Databases*), (*GH*5, *Logic*)}. Then *Students* × *Courses* is a relation over attributes (*student-number, student-name, course-number, course-name*) containing tuples (1001, *Black*, *EH*1, *Databases*), (1001, *Black*, *GH*5, *Logic*), (1002, *White*, *EH*1, *Databases*), and (1002, *White*, *GH*5, *Logic*).

The Cartesian product can also be viewed as a special case of natural join, arising when the set of attributes of the operands are disjoint. However, relations over non-disjoint sets of attributes can also be combined by the Cartesian product, provided that the renaming operator is used to rename common attributes in one of the two relations.

In the presence of attribute names, the Cartesian product is commutative. In the case that relation schemas do not come with attribute names, but are specified by a relation name and arity, the Cartesian product $R_1 \times R_2$ returns the concatenation $t_1 \ t_2$ of all pairs of tuples such that $t_1 \in R_1$ and $t_2 \in R_2$. Moreover the output schema is specified by the sum of the arities of the input schemas. In this case the Cartesian product is a noncommutative operator.

Cross-References

▶ Join
▶ Relational Algebra

Cataloging in Digital Libraries

Mary Lynette Larsgaard
University of California-Santa Barbara, Santa Barbara, CA, USA

Synonyms

Cataloging; Classification

Definition

Cataloging is using standard rules to create a mainly text surrogate that describes an object sufficiently in detail so that the object is uniquely differentiated from all other objects. Without looking at the object, a user can quickly and efficiently learn enough about the object to know if it suits the user's needs. It is generally considered to include bibliographic description and the application of subjects, both as words and as classification. Metadata creation using standard rules may be considered to be a form of cataloging; both cataloging and metadata creation require software with the ability for the user to do searching, browsing, navigation, and display and the ability for the agency doing the work to manage digital objects; to provide preservation and security of metadata, data, and user privacy; and to provide at some level interoperability with other software and other agencies doing the same work.

Historical Background

Devising and using methods of arranging and describing information – termed, within the standard library world, classification and cataloging, respectively – have been primary concerns of libraries ever since libraries began, in the ancient world of the Greeks and the Romans. A collection of information without classification and cataloging is not a library. The point of classification and cataloging is to make access quick and easy for users; it was discovered very early that putting like objects together (classification) and creating text or relatively speaking much smaller surrogates to describe an information object (cataloging) made finding information much quicker for the user.

Experiments in using digital records in libraries started in approximately the late 1960s. But it was only in the mid-1970s, with the success of what is called "shared cataloging" – many libraries using a catalog record contributed as "original cataloging" by the first library to catalog the item – that using digital systems for cataloging came into its own. This sharing of bibliographic records in online form began with OCLC, initially as a consortium of college libraries in Ohio (starting in 1967), but growing rapidly to become the most successful such library utility, currently with about 17,000 participating libraries in 170 countries and territories (http://www.oclc.org). The development of integrated library systems (ILS) – software, or a combination of software and hardware, that permits a library to perform multiple functions, such as acquisitions/ordering, cataloging and classification, circulation, preservation, and reference, using digital files in large databases with many tables – has continued to the present. The use of authorized records – that is, always using the same form of an author, subject, title, or name/title – with cross-references from terms not used is crucial to enabling users to find all resources on a given topic, author, or author title (see "Library of Congress Authorities" online authority system, http://authorities.loc.gov).

The inception and speedy growth of the World Wide Web (Web) since the mid-1990s have spread this interest in arrangement and description of, and access to, information objects to nonlibrary communities and within the library world to how specifically arrange and describe digital objects made available over the Web in digital libraries. There have been many standards for the description of information objects. They are most often intended either to apply at least in theory to all materials (e.g., Resource Description and Access, hereafter referred to as RDA; Anglo-American Cataloging Rules, hereafter referred to as AACR; Dublin Core, which began in 1995, http://www.dublincore.org/documents/dces/) or to apply to the description of a specific body of information (e.g., for digital geospatial data, "Content Standard for Digital Geospatial Metadata"; second edition, 1998, http://www.fgdc.gov/standards/projects/FGDC-standards-projects/metadata/base-metadata/v2_698.pdf and ISO Standard 19,115, "Geographic Information, metadata; Information géographique, métadonnées," 2014).

Foundations

Classification

Classification and cataloging complement each other. Classification is a form of subject cataloging, which is where the major overlap between the two occurs. Classification tends to be hierarchical, breaking a given world of information or knowledge into broad divisions (e.g., Law) and then breaking that into smaller divisions (e.g., education for law, law of various countries, etc.). Classification is most often placing like items about like subjects (e.g., works by Shakespeare) and like genres or formats (e.g., maps) together. It also provides a physical location within a library for each object, be it digital or hard copy. The most prominent systems used are the Dewey Decimal Classification (often used by public libraries and smaller libraries generally), and the Library of Congress Classification (most often used by university and other research libraries in the United States). There are many more systems, such as the Bliss Classification and the Colon Classification, and special and research libraries (such as the New York Public Library) devise and maintain their own systems. Devising a classification system is easy; maintaining it is difficult, expensive, and time-consuming. A library maintaining a classification system unique to itself will find that it is far less work to convert to a rigorously maintained system – maintained by some central agency – that is used by many libraries than it is to maintain a unique system not used by any other collections.

Classification of digital objects at first glance seems unnecessary. Why not just assign an arbitrary number (e.g., a unique identifier that database software assigns to each separate catalog record) and be done with it? Libraries of digital objects have found that, for several reasons, it is very practical to assign classification to digital objects just as one would to hard-copy objects. The main one is that very often one needs to move around or to perform the same operation (e.g., create distribution forms) on large numbers of digital objects in groups of items that are, e.g., the same file type. For example, the Alexandria Digital Library (ADL), a collection of digital geospatial data that began in the mid-1990s (http://webclient.alexandria. ucsb.edu), gives a digital geospatial data object (e.g., a scan of a paper map; a born-digital object) the same classification number as the hard-copy geospatial data equivalent (e.g., a paper map). Classification of digital objects allows one file to be accessible from multiple classification numbers, an action that few libraries of hard-copy items have ever been able to afford. For example, a digital map of California and Nevada may have two classification numbers (either pointing toward one file or storing the same file two separate places), one for California and one for Nevada. Few libraries of hard-copy items have financial resources available to buy multiple copies of an item and store a copy at each applicable classification number.

Cataloging

A catalog record provides information on the thematic and physical nature of a cataloged resource (whether hard copy or digital). Libraries first used hard-copy catalogs, generally book-format catalogs, then cards, and then beginning in the late 1960s the use of databases as catalogs. This started with in 1969 the Library of Congress' MAchine-Readable Catalog (MARC) format for the transmission of catalog-card information in digital form (http://www.loc.gov/marc/). While MARC may be used as the machine format for bibliographic records formulated using any set of cataloging rules, it is most often used for records based on the cataloging rules of the Anglo-American library community, AACR (Anglo-American Cataloging Rules) in its various editions, first issued in 1967 and used through early 2013, followed by RDA.

The International Federation of Library Associations (IFLA) has been a prime agent in the move toward an international cataloging standard. In the early 1990s, IFLA's study group on the Functional Requirements for Bibliographic Records (FRBR) began work on its report to recommend basic functionalities of a catalog and bibliographic-record data requirements. The group's final report was issued in 1998 (Functional Requirements for Bibliographic Records (FRBR) (http://www.ifla. org/files/assets/cataloguing/frbr/frbr_2008.pdf).

While FRBR put forward several ideas, the one that most engaged the interest and discussion of the library community, which is a major part of the structure of RDA, was the importance of incorporating into the bibliographic description the concept of the relationship between the work, the expression, the manifestation, and the item (called group 1 entities). A work is a distinct intellectual or artistic creation but a concept rather than an actual physical object. An expression is the intellectual or artistic realization of a work but still not generally an actual physical object. The manifestation is all copies of a published object (e.g., all copies of a printed map; all copies of a DVD of a specific piece of music), and the item is one copy of a manifestation. For example, all versions (i.e., all editions) of a given map are a work, while the 1973 edition of all physical forms of the map is an expression, with each hard-copy or digital issuance of that map being a manifestation, and each copy of any one physical form being an item.

Key Applications

Key applications of classification and cataloging:

- Classification: Arrangement of digital objects in digital libraries. See previous section.
- Cataloging: Creation of metadata for digital objects in digital libraries.

While full-text searching is extremely useful and a major step forward in the history of information retrieval, it does have the following main problems: it may give the user very large numbers of hits; it is not as efficient as searching well-constructed text surrogates, and it does not work for what are primarily non-text materials (e.g., music, maps, etc.). The reason for creating metadata records for digital objects is the same as that for performing standard cataloging – constructing a surrogate for the item so that users may quickly and efficiently find resources that suit the users' needs. Metadata is constructed by both nonlibrary entities (e.g., federal government agencies) and by libraries. Metadata records include but are not limited to the information contained in what the standard library cataloging world terms "bibliographic records."

Metadata records may be longer, shorter, or the same length as catalog records; they may have relatively few fields (e.g., Dublin Core has 15 fields) or much longer because they contain far more and generally more detailed technical information than a catalog record would. The latter is much more likely to include a URL that points to an online version of that technical information, quite possibly to a metadata record. Metadata records may be constructed for general audiences (which may mean shorter records) or, very focused, often technically skilled audiences; for example, a geographic digital dataset (such as a geographic information system, more commonly known as a GIS) is often quite large and complicated, with many layers of information, and therefore is by no means as easily browsed – in order to determine its suitability for use – as is a hard-copy map. For example, the catalog record for digital orthophoto quarter quadrangles (DOQQs; mosaics of rectified aerial photographs) is relatively brief (see Fig. 1) – about one standard printed page – when compared with the metadata record for DOQQs, which is seven pages. For example, see http://gisdata.usgs.gov/metadata/doqq.htm. What makes a metadata record longer is the many different kinds of metadata that may be included in one record – technical, administrative, descriptive, preservation, transport, and usage. Even for metadata records, controlled vocabulary – what the library world calls authority control – is essential for subjects and authors. Agencies that are under pressure to get as many digital resources available online as quickly as possible may be tempted to create brief metadata records, formulated without a schema or rules. This is a considerable mistake and leads to far more work later on, when the metadata rules need to be set up and the records have to be recreated. Especially is this the case when a project expands far beyond its original size (e.g., from 100 items to several thousand), or when there are several different sets of metadata that would be far more useful if the records could all be searched simultaneously, rather than the user having to search each metadata set separately [13].

A library generating metadata records has two major options: load the records into the library's ILS (integrated library system) online catalog or create what is in effect another ILS, or at the very least an online catalog for the metadata records, with links to the digital resource being described. The first technique generally requires that the records be in MARC format, since the alternative is that the online catalog software must be capable of searching over multiple catalog-record databases in different formats, with each set of records possibly formulated according to different rules; this latter may be termed "federated searching."

For the second technique, the following is required: software (UNIX; a database manager; a user interface; a digitization operation if other than born-digital resources are involved; middleware to connect inquiries on the user interface with the data and return results; what is sometimes called software for digital asset management); hardware, and computer technical staff/programmers (for a digital library of any size, a minimum of three computer programmers to deal with adding new data and metadata and maintaining and improving the system, including the interface, plus one more programmer to deal with the operating system, disk storage, and manipulation and maintenance of the directories of digital material). There are many softwares for digital access management. One that has been used in libraries is CONTENTdm, a product of OCLC (http://www.oclc.org/contentdm.en/html), used at the time of writing of this entry by about 2000 organizations worldwide. It is attractive to libraries for several reasons; one main one is that one may use standard cataloging records, created by a given library on OCLC's WorldCat (an international online catalog, with hundreds of millions of catalog records), for these digital resources, with a link in the resource to the given library's URL for the scan. This means the library would not need to set up a metadata schema and

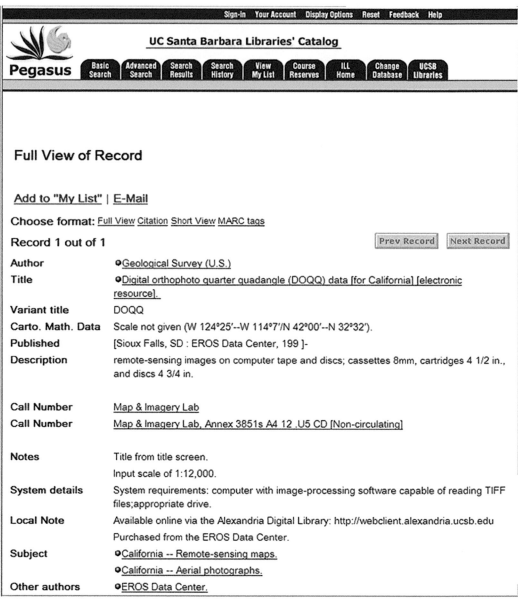

Cataloging in Digital Libraries, Fig. 1 Record from a library online catalog

software for the creation and searching of the metadata records, but rather could use RDA and the usual method of entering or loading bibliographic records on WorldCat (http://www. oclc.org/worldcat.en.html). Setting up and using CONTENTdm will require considerable work by computer programmers, to do all the software configuration and back-end work.

An example of this is the ADL (Alexandria Digital Library) catalog, http://webclient. alexandria.ucsb.edu, initially constructed between the mid-1990s and the early 2000s. ADL (http://www.alexandria.ucsb.edu) has changed as of April 2014 to ADRL, the Alexandria Digital Research Library; the first collection is theses and dissertations, in electronic form, of the University of California, Santa Barbara. Because of the popularity over the last 10 years of libraries digitizing their rare or unique resources, there are many more, for example, the University of Nevada, Las Vegas's many digital collections (http://digital.library.unlv.edu/).

As previously indicated, there are numerous metadata standards. The following are major standards that digital libraries creating metadata records will probably need to deal with, at least in the United States: Dublin Core, XML, METS, and MODS.

Dublin Core

Dublin Core (DC) (http://dublincore.org/) is used by both library and nonlibrary agencies for cataloging digital content. Its adaptability to any form or type of digital data and its brevity (15 fields, what libraries term minimal-level cataloging) with no fields required and all fields repeatable make it very flexible. While DC may be used either as "qualified" (each of the 15 elements may be qualified in some way to make the information clear, e.g., for coverage, one might state that the geographic area is given in decimal degrees), the experience in libraries over the nearly 25 years since DC was made available for use is that it is strongly advised only qualified DC be used. This is because unqualified DC results in metadata records that are so unstructured as to be nearly useless [12]. For digital libraries needing to keep at least one foot solidly in the traditional library world, there is a DC-to-MARC2 crosswalk at http://www.loc.gov/marc/dccross.html and also one the other direction.

XML

XML has achieved primacy as the format of choice for metadata for digital libraries and is of considerable importance to the standard library world, as evidenced by the Library of Congress having MARC21 in XML available over the Web at http://loc.gov/marc/marcxml.html. ISO 2709 (Information and Documentation – Format for Information Exchange; http://www.iso.org/iso/iso_catalogue/catalogue_tc/catalogue_detail.htm?csnumber=41319, 2008) has been used for many years and has worked well, but the library community needed a standard exclusively for MARC records (of which there are over 100 million worldwide) in XML. The standard's most current edition was published in 2013; it is ISO 25577, with the short name of MarcXchange (http://www.iso.org/iso/home/store/catalogue_tc/catalogue_detail.htm?csnumber=62878).

METS and MODS

METS (Metadata Encoding and Transmission Standard) is a standard for encoding descriptive, administrative, and structural metadata of objects in a digital library (http://www.loc.gov/standards/mets/).

The "Metadata Object Description Schema" (MODS) is intended both to carry metadata from existing MARC21 records and to be used to create new catalog records. It has a subset of MARC fields – unlike MARC21. It uses language-based tags rather than numeric tags. It occasionally regroups elements from the MARC21 bibliographic format (http://www.loc.gov/standards/mods/). METS and MODS are both expressed using XML and are maintained by the Library of Congress's Network Development and MARC Standards Office.

Conclusion

Organizations creating metadata records and arranging digital files are best advised to follow well-maintained national or international standards. In no case should organizations just starting out on this work create their own standards. Instead, the use of sturdy standards – some of which have an extensions feature, to enable customization of the records to the library users' needs – is recommended.

Creating metadata records is relatively easy compared with the difficult and expensive work of setting up what is in effect an ILS online catalog. Agencies need to consider this very carefully. If the agency cannot sustain the programming effort required to develop and then to maintain and add to the catalog, then the agency should not begin the project. Digital libraries are at least as expensive and time-consuming to develop and maintain as are hard-copy libraries.

Cross-References

► Annotation
► Audio Metadata
► Biomedical Data/Content Acquisition, Curation
► Browsing in Digital Libraries
► Classification by Association Rule Analysis
► Classification in Streams
► Clinical Data Acquisition, Storage, and Management
► Cross-Modal Multimedia Information Retrieval
► Data Warehouse Metadata
► Digital Libraries
► Discovery
► Dublin Core
► Field-Based Information Retrieval Models
► Geographical Information Retrieval
► Image Metadata
► Indexing Historical Spatiotemporal Data
► Information Retrieval
► Information Retrieval Models
► Metadata Registry, ISO/IEC 11179
► Metadata
► Metadata Interchange Specification
► Metadata Registry, ISO/IEC 11179
► Meta Data Repository
► Metasearch Engines
► LOC METS
► Multimedia Metadata
► Ontology
► Ontology
► Schema Mapping
► Searching Digital Libraries
► Text Indexing and Retrieval
► XML
► XML Metadata Interchange

Recommended Reading

1. American Library Association. Anglo-American cataloging rules. Chicago: American Library Association; 1967. In various editions.
2. Borgman CL. From Gutenberg to the global information infrastructure: access to information in the networked world. Cambridge: MIT Press; 2000.
3. Chan LM. Cataloging and classification: an introduction. Blue Ridge Summit: Scarecrow Press; 2007.
4. Foulonneau M, Riley J. Metadata for digital resources: implementation, systems design and interoperability. Oxford: Chandos; 2008.
5. IFLA Study Group. Functional requirements for bibliographic records (FRBR). UBCIM publications, new series, vol. 19. Munchen: K.G. Saur; 1998. Available online at: http://www.ifla.org/VII/s13/frbr/frbr.htm
6. Libraries. Encyclopedia Britannica. Micropedia. 2002;7:333–4; Macropedia 22:947–63. Chicago: Encyclopedia Britannica. Available online at: http://search.eb.com/
7. Library of Congress. 1969? MARC21 Concise Bibliographic. Washington, DC: Library of Congress; 1969. Available online at: http://www.loc.gov/marc/
8. Miller SJ. Metadata for digital collections: a how-to-do-it manual. London: Facet; 2011.
9. Reitz JM. Dictionary for library and information science. Westport: Libraries Unlimited; 2004.
10. American Library Association. Resource description and access. Chicago: American Library Association; 2010.
11. Svenonius E. The intellectual Foundation of Information Organization. Cambridge: MIT Press; 2000.
12. Tennant R. Bitter harvest: problems and suggested solutions for OAI-PMH data and service providers. Oakland: California Digital Library; 2004. Available online at: http://www.cdlib.org/inside/projects/harvesting/bitter_harvest.html
13. Woodward E. Metadata for image collections. Am Libr. 2014;45(6):42–4.

Causal Consistency

Alejandro Z. Tomsic[1,2] and Marc Shapiro[2,3]
[1]Sorbonne-Universités-UPMC-LIP6, Paris, France
[2]Inria Paris, Paris, France
[3]Sorbonne-Universités-UPMC-LIP6, Paris, France

Synonyms

Causal consistency (CC); Causal memory; Causal-plus consistency (Causal+ consistency, C+C)

Definition

Causal consistency is a "Data Consistency Model" (*q.v.*) initially introduced for message passing (distributed) systems and later for shared memory systems. It ensures that, writes are observed by every party in the system in *potential-causality order*.

Potential-causality order is a global partial order of operations that tracks whether some operation may be influenced (caused) by some other operation. Formally, operation u *potentially causally precedes* operation v or, equivalently, operation v *potentially causally depends* on operation u if any of the following conditions hold (For simplicity, we drop "potentially" and shorten to "causally precedes" and to "causally depends" respectively. Similarly, "potential-causality order" will be shortened to "causality order."):

1. Thread of Execution: u and v are two operations in a single thread of execution, and operation u executed before operation v.
2. Reads-From: u is a write operation, and v is a read operation that returns the value written by u.
3. Transitivity: There exists an operation w, such that u causally precedes w, and w causally precedes v.

If neither u causally precedes v nor v causally precedes u, we say that u and v are concurrent.

Under *causal consistency*, if two writes are concurrent, then parties in the system may observe them in either relative order. If write u causally precedes write v, then, if any party p observes (reads) v, i.e., v is "visible" to p, then u must also be visible to p.

Key Points

Causal consistency is an important correctness criterion for replicated systems that require low latency and/or that need to remain available under network partitions between replicas (see "▶ CAP Theorem"). Indeed, it is the strongest consistency model that does not require cross-replica synchronization in the critical path of client operations [2,6].

In a causally-consistent replicated database, an update u is first applied to its origin replica R, then propagated in the background to other replicas. An implementation must ensure that a receiving replica makes this update visible respecting causality order. FIFO propagation is not sufficient to ensure this, since u may depend on updates originating at replicas other than R.

Therefore, an implementation generally piggy-backs, upon update u, metadata that identifies the updates that causally precede u. A replica R' receiving u checks the metadata. Different systems use different metadata representations and mechanisms for making updates visible respecting causality order.

There is a fundamental trade-off in the size of metadata. Tracking u's predecessors precisely requires the metadata to either contain the entire set of updates that precede u, or to consist of a vector clock with one entry per master replica. Although it is often possible to optimise it, in the worst case the metadata can inflate bandwidth and/or processing costs tremendously.

A compact representation [3,7] avoids these costs, but introduces *false dependencies*, which increase the delay between the moment the update is sent from the origin replica, and the time it is made visible to read operations at other replicas.

Causal consistency does not impose an order between independent writes; in particular, two replicas may update the same data item concurrently. If these updates are made visible to different replicas in differing orders and do not commute, the replicas may diverge. We refer the reader to the entries on "Conflict resolution and reconciliation" and on "Conflict-Free Replicated Data Types". These techniques can be used to ensure that all eventually converge to the same value. Causal Consistency with automatic conflict resolution is known as *Causal+ Consistency*.

Historical Notes

Lamport [4] originally defined the Happened-Before relation, referring to sending and receiving messages in point 2; causality order is based on this definition. Ahamad et al. [1] extended the definition to reads and writes in shared memory; our definition of Causal Consistency is closely patterned after their Causal Memory. Lloyd et al. [5] define Causal+ Consistency. Mahajan et al. [6] claim that a specific variant of Causal Consistency is the strongest consistency model that is compatible with availability in failure-prone networks; this result is formalised and refined by [2].

Cross-References

- ► CAP Theorem
- ► Data Replication
- ► Eventual Consistency
- ► Weak Consistency Models for Replicated Data

Recommended Reading

1. Ahamad M, Neiger G, Burns JE, et al. Causal memory: definitions, implementation, and programming. Distrib Comput. 1995;9(1):37–49.
2. Attiya H, Ellen F, Morrison A. Limitations of highly-available eventually-consistent data stores. In: Proceedings of the ACM SIGACT-SIGOPS 34th Symposium on the Principles of Distributed Computing; 2015. p. 385–94.
3. Du J, Iorgulescu C, Roy A, et al. GentleRain: cheap and scalable causal consistency with physical clocks. In: Proceedings of the 5th ACM Symposium on Cloud Computing; 2014. p. 4:1–4:13.
4. Lamport L. Time, clocks, and the ordering of events in a distributed system. Commun ACM. 1978;21(7):558–65.
5. Lloyd W, Freedman MJ, Kaminsky M, et al. Don't settle for eventual: scalable causal consistency for wide-area storage with COPS. In: Proceedings of the 23rd ACM Symposium on Operating System Principles; 2011. p. 401–16.
6. Mahajan P, Alvisi L, Dahlin M. Consistency, availability, and convergence. Technical Report UTCS TR-11-22, Department of Computer Science, The University of Texas at Austin. Austin; 2011.
7. Zawirski M, Preguiça N, Duarte S, et al. Write fast, read in the past: causal consistency for client-side applications. In: Proceedings of the ACM/IFIP/USENIX 14th International Middleware Conference; 2015. p. 75–87.

Certain (and Possible) Answers

Gösta Grahne
Concordia University, Montreal, QC, Canada

Synonyms

True answer (Maybe answer); Validity (Satisfiability)

Definition

Let T be a finite theory expressed in a language L, and φ an L-sentence. Then T finitely entails φ, in notation $T \models \varphi$, if all finite models of T also are models of φ. A theory T is said to be *complete in the finite* if for each L-sentence φ either $T \models \varphi$ or $T \models \neg\varphi$. In particular, if T is incomplete (not complete in the finite), then there is an L-sentence φ, such that $T \nvDash \varphi$ and $T \nvDash \neg\varphi$. It follows from classical logic that a first order theory is complete in the finite if and only if all its finite models are isomorphic. Consider now a theory

$$T_1 = \begin{cases} R(a,b) \wedge R(a,c), \\ \forall x, y\colon R(x,y) \rightarrow (x,y) = (a,b) \vee (a,c), \\ a \neq b, a \neq c, b \neq c. \end{cases}$$

where a, b, and c are constants. This theory is complete, and clearly for instance $T \models R(a, b)$, $T \models R(a, c)$, and $T \nvDash R(d, c)$, for all constants d different from a and b. Consider then the theory

$$T_2 = \begin{cases} R(a,b) \vee R(a,c), \\ \forall x, y\colon R(x,y) \rightarrow (a,c) = (a,c) \vee (a,c), \\ a \neq b, a \neq c, b \neq c. \end{cases}$$

This theory is incomplete, since for instance $T_2 \nvDash R(a, b)$, and $T_2 \nvDash \neg R(a, b)$. If "finitely entails" is equated with "certainly holds," it is possible to say that $R(a, b)$ and $R(a, c)$ *certainly* hold in T_1. Dually, it is possible to say that $R(a, b)$ *possibly* holds in T_2, since $T_2 \nvDash \neg R(a, b)$, and similarly that $R(a, c)$ possibly holds in T_2.

Key Points

An *incomplete database* is similar to a logical theory: it is defined using a finite specification, usually a *table T* (relation with nulls and conditions) of some sort, and a function *Rep* that associates a set of complete (ordinary, finite) databases *Rep(T)* with *T*. Then each instance $I \in Rep(T)$ represents one isomorphism class (isomorphism up to renaming of the constants) of the finite models of the table *T* regarded as a logical theory. Depending on the interpretation of facts missing from *T*, either the *closed world assumption* is made [9], which postulates or axiomatizes (as in the middle "row" in T_1 and T_2) that any facts not deducible from *T* are *false*, or the *open world assumption* (omit the middle rows), in which there are certain and possible facts, but no false ones. There is actually a spectrum of closed world assumptions, ranging up to semantics best axiomatized in third order logic [4].

Having settled on a representation *T*, and an interpretation *Rep*, the *certain answer* to a query *Q* on an incomplete database *T*, is now defined as $\bigcap_{I \in Rep(T)} Q(I)$, sometimes also denoted $\bigcap Q(Rep(T))$. In database parlance the certain answer consists of those facts that are true in *every* possible database instance *I* that *T* represents. Likewise, the *possible answer* to a query *Q* on an incomplete database *T*, consists of those facts which are true in *some* possible database, i.e., $\bigcup_{I \in Rep(T)} Q(I)$. Needless to say, the possible answer $\bigcup Q(Rep(T))$ is interesting only under a closed world assumption, since otherwise every fact is possible.

These definitions are clear and crisp, but unfortunately it doesn't mean that they always have tractable computational properties. Consider the membership problem for the set

$$CERT(Q) = \{(t, T) : t \in \cap Q (Rep(T))\}.$$

If *T* actually is a complete instance *I*, it is well known that CERT(*Q*) has polynomial time complexity, for any first order (relational algebra) or datalog query *Q*. Likewise, the set

$$POSS(Q) = \{(t, T) : t \in \cup Q(Rep(T)).$$

has PTIME complexity for first order and datalog queries *Q*, and tables *T* that actually are complete databases.

A table *T* with *unmarked nulls* is a classical instance containing existentially quantified variables (nulls), such that each existential quantifier has only one variable in its scope. This means that each occurrence of a null can be independently substituted by a constant for obtaining one possible database in *Rep(T)*.

If only simple incomplete databases with unmarked nulls are allowed, only existential first order queries *Q* need to be admitted, or alternatively algebraic expressions with operators from $\{\pi, \sigma, \cup, \bowtie\}$, in order for CERT(*Q*) to become coNP-complete, and POSS(*Q*) to become NP-complete [2]. The use of inequalities $\neq$ or disjunctions $\vee$ in *Q* is essential. If the use of inequalities and disjunctions is denied, CERT(*Q*) and POSS(*Q*) remain in PTIME. If one admits arbitrary first order or full relational queries *Q*, along with an open world assumption, CERT(*Q*) and POSS(*Q*) become undecidable. This follows from validity and satisfiability of a variant of first order logic known to be undecidable [3]. (Note that under the open world assumption POSS(*Q*) equals all possible databases, assuming POSS(*Q*) $\neq \varnothing$. The problem then becomes to decide whether POSS(*Q*) is non-empty or not.)

On the other hand, if the representation mechanism allowed for *T* is more powerful that the simple incomplete databases above, CERT(*Q*) and POSS(*Q*) again become coNP and NP complete, respectively, already with *Q* being the identity query. For this, the *conditional tables* of [6] are needed. As observed in [5], conditional tables can be obtained as a closure of simple incomplete databases by requiring that the exact result $\{q(I) : I \in Rep(T)\}$ of any relational algebra query on a any table *T* is representable by conditional table.

In other words, for each T conditional table and Q relational algebra query, there exists a conditional table U, such that $Rep(U) = Q(Rep(T))$.

Another way of representing incomplete databases, is to consider an *information integration* scenario, where the basic facts are stored in *views* of a virtual *global schema*. For instance, in the integration scenario

$$T_3 = \begin{cases} V(a), \\ \forall x, y : R(a, c) \to V(x), \\ \forall x : V(x) \to x = a, \\ \forall x, y : R(a, c) \to x = a \end{cases}$$

gives (closed world) $Rep(T_3) = \{I : V^I = \pi_1(R^I) = \{(a)\}\}$. ($R^I$ means the value (interpretation) of predicate symbol R in instance/model I. The meaning of V^I is similar.) Open world (omit in T_3 the third and fourth rows) $Rep(T_3)$ would be defined as $\{I : V^I \supseteq \{(a)\}, \pi_1(R^I) \supseteq \{(a)\}$. If T is allowed to use conjunctive queries (such as the second row of T_3) to express the views V in terms of the global relations R, then $\text{CERT}(Q)$ is in PTIME for existential first order and datalog queries under the open world assumption, and coNP complete under the closed world assumption [1]. The latter is due to the negation implicit in the closed world assumption. If one allows inequalities $\neq$ in the query, $\text{CERT}(Q)$ is coNP complete also under the open world assumption. Undecidability of $\text{CERT}(Q)$ is achieved by allowing negation in the query or the view definitions, or, under the open world assumption by allowing view definitions in (recursive) datalog.

Finally, one can see the *data exchange* problem [7, 8] as a variation of the integration problem. The data exchange problem consists of importing the data from a source database R_s to a target database R_t, using data dependencies (implicational sentences) to express the translation. For example, a (closed world) exchange scenario could be

$$T_4 = \{R_s(a), \forall x : [R_s(x) \leftrightarrow \exists y : R_t(x, y)].$$

The base facts are in a source database R_s, and the user query is expressed against the target database R_t. As T_4 obviously is incomplete, the certain answer of Q on T_4 is defined as $\cap Q(Rep(T_4))$, and the possible answer as $\cup Q(Rep(T_4))$. It is perhaps no big surprise that essentially the same complexity landscape for $\text{CERT}(Q)$ and $\text{POSS}(Q)$ as in the previous table- and integration-scenarios emerges: the boundaries between undecidability, intractability, and polynomial time depend on similar restrictions on the use of negation, of inequalities or unions in the exchange mappings, and on the open or closed world assumptions.

Cross-References

▸ Conditional Tables
▸ Incomplete Information
▸ Naive Tables
▸ Null Values

Recommended Reading

1. Abiteboul S, Duschka OM. Complexity of answering queries using materialized views. In: Proceedings of the 17th ACM SIGACT-SIGMOD-SIGART Symposium on Principles of Database Systems; 1998. p. 254–63.
2. Abiteboul S, Kanellakis PC, Grahne G. On the representation and querying of sets of possible worlds. Theor Comput Sci. 1991;78(1):158–87.
3. Di Paola RA. The recursive unsolvability of the decision problem for the class of definite formulas. J ACM. 1969;16(2):324–7.
4. Eiter T, Gottlob G, Gurevich Y. Curb your theory! a circumspective approach for inclusive interpretation of disjunctive information. In: Proceedings of the 13th International Joint Conference on AI; 1993. pp. 634–39.
5. Green TJ, Tannen V. Models for incomplete and probabilistic information. In: Advances in Database Technology, Proceedings of the 10th International Conference on Extending Database Technology; 2006.
6. Imielinski T, Lipski W. Incomplete information in relational databases. J ACM. 1984;31(4):761–91.
7. Kolaitis PG. Schema mappings, data exchange, and metadata management. In: Proceedings of the 24th ACM SIGACT-SIGMOD-SIGART Symposium on Principles of Database Systems; 2005. p. 61–75.
8. Libkin L. Data exchange and incomplete information. In: Proceedings of the 25th ACM SIGACT-SIGMOD-SIGART Symposium on Principles of Database Systems; 2006. p. 60–9.
9. Reiter R. On closed world data bases. In Logic and Data Bases. 1977. p. 55–76.

Change Detection on Streams

Daniel Kifer
Yahoo! Research, Santa Clara, CA, USA

Synonyms

Change detection and explanation on streams

Definition

A *data stream* is a (potentially infinite) sequence of data items $x_1, x_2, \ldots$. As opposed to traditional data analysis, it is not assumed that the data items are generated independently from the same probability distribution. Thus *change detection* is an important part of data stream mining. It consists of two tasks: determining when there is a change in the characteristics of the data stream (preferably as quickly as possible) and explaining what is the nature of the change.

The nature of the data stream model means that it may be infeasible to store all of the data or to make several passes over it. For this reason, change detection algorithms should satisfy the following desiderata: the memory requirements should be constant or increase logarithmically, and the algorithm should require only one pass over the data.

Historical Background

There has been a lot of work on detecting change in time series data *after* all of the data has been collected. Change point analysis [5] is a statistical field devoted to detecting the point in time where the distribution of the data has changed. The description of the change is often concerned with how the parameters of the distributions (such as the mean) have changed. Scan statistics [11] can be used to detect and describe changes (not just along the time dimension) by identifying regions where the probability mass has changed the most. For examples of offline analysis of change in terms of data mining models see [6] for itemset mining, [10] for itemset mining and decision trees, and [14] for burst detection. Offline algorithms for describing change include [1] for hierarchical numerical data and [7, 8] for semi-structured data.

The offline methods are useful for data analysis, but as data acquisition becomes easier and easier, the data stream model becomes more and more relevant. The assumptions behind this model are that there is so much data that it cannot all be stored on disk (let alone kept in memory) and that the data arrives at such a rate that expensive online computations are impractical. Often results are expected in real-time – for example, notification of change should occur as soon as possible. The data stream model is discussed in detail in [2].

In the data stream model, one can predefine a certain set of stochastic processes. One of these processes is initially active and the goal is to determine when a different process in that set becomes active [3]. The description of the change is then the identity of the new process. Alternative approaches [9, 13] avoid making distributional assumptions by using ideas from nonparametric statistics. In these approaches the main idea is to divide the domain of the data into (possibly overlapping) regions, estimate the probability of the regions, and then determine whether the changes in probability in any of the regions are statistically significant. Alternate approaches for handling change include testing if the data are exchangeable (i.e., any permutation of the data is equally likely) [15] and developing data mining algorithms (such as decision tree construction) that adapt to change [12].

Foundations

Let $x_1, x_2, \ldots$, be a potentially infinite sequence of data points. To detect changes in the data stream, one first has to determine a plausible framework that describes how the data can be generated. In the simplest case, data are generated from one of two probability distributions S_1 and S_2 (for example a Gaussian with mean 0 and variance 1, and a Gaussian with mean 10 and variance 1). Initially, the data points are generated independently from S_1 and after some time the data are generated independently from S_2. The celebrated CUSUM algorithm by Page [4] can be used to detect that a change from S_1 to S_2 occurred by comparing the likelihoods that parts of the data were generated by S_1 or S_2. Suppose S_1 has density f_1 and S_2 has density f_2, and let $\delta > 0$ be a threshold.

A user of the change-detection system is interested in the first time k where. $\sum_{i=k}^{now} \log(f_2(x_i)/f_1(x_i)) > \delta$. When this happens, the system signals that a change has occurred and can return k as a plausible estimate of the change point. This test can be done in an online fashion by defining $T_0 = 0$ and $T_k = \max(T_{k-1} + \log(f_2(x_k)/f_1(x_k)),0)$ and signaling a change if $T_{\text{now}} > \delta$. Typically S_1 is chosen based on an initial sample of the data stream and S_2 is then chosen to represent the smallest change whose detection is desired. For example, suppose that S_1 is chosen to be a Gaussian with mean 0 and variance 1, and suppose that for the current application it is desirable to detect any change in mean greater than 10. Then a natural choice for S_2 is a Gaussian with mean 10 and variance 1.

This framework has been generalized by Bansal and Papantoni-Kazakos [3] to the case where S_1 and S_2 are stochastic processes that need not generate each point independently. Additional generalizations of the CUSUM algorithm, including the case of multiple data generating distributions, are discussed in [4].

The framework of the CUSUM algorithm is an example of a *parametric* framework: the set of possible data-generating distributions has been prespecified and elements in that set can be identified using a small number of parameters. Parametric approaches are powerful in cases where they can accurately model the data. In cases where the data is not well modeled by a parametric framework, performance may deteriorate in terms of more false change reports and/or fewer detection of changes.

Kifer et al. [13] showed how to avoid problems with parametric approaches by using a nonpara-

metric framework. In this framework the data points $x_1, x_2, \ldots,$ are k-dimensional vectors of real numbers. Point x_1 is generated by some (arbitrary) probability distribution F_1, x_2 is generated by F_2 (independently of x_1), x_3 is generated by probability distribution F_3 (independently of x_1 and x_2), etc. A change is defined as a change in the data-generating distribution; if the first change occurs at time n_1 then $F_1 = F_2 = \ldots = F_{n_1-1} \neq F_{n_1}$; if the second change occurs at time n_2 then $F_{n_1} = F_{n_1+1} = \ldots = F_{n_2-1} \neq F_{n_2}$, etc. This framework for detecting change consists of three parts: a collection of regions of interests, a method for estimating probabilities, and a statistical test.

Regions of Interest

A collection of regions of interest serves two purposes: to restrict attention to changes that are considered meaningful, and to provide a means for describing the change.

Ideally a change is said to have occurred whenever the data-generating distribution changes. However, for many practical applications not all changes are meaningful. For example, consider the case where F_1 is a probability distribution that assigns probability 1 to the set of real numbers between 0 and 1 whose seventeenth significant digit is odd and furthermore suppose that F_1 is uniform over this set. From time 1 up to $n - 1$ the data are generated independently from the distribution F_1. At time n a change occurs and from that point the data are generated independently from the distribution F_n defined as follows: F_n assigns probability 1 to the set of real numbers between 0 and 1 whose seventeenth significant digit is even and is uniform over this set. Letting f_1 be the probability density function for F_1 and f_n be the probability density function for F_n it can be seen that f_1 and f_n are very different according to some common similarity measures. Indeed, the L^1 distance between f_1 and f_n (i.e., $\int_0^1 |f_1(x) - f_n(x)| \, dx$), is as large as possible. However, in many applications it is of no practical consequence whether the true distribution is F_1 or F_n. This is because one may be interested only in questions such as "what is the probability that the next point is larger than 0.75" or "what is the probability that the next point is within the safety range of 0.14–0.95." For these types of questions one would only be interested in the probabilities of various intervals, so instead of receiving notification of arbitrary types of change, one would be happy to know only when some interval has become more or less probable. In this case, the intervals are said to be the *regions of interest*. In general, if the domain of each data element is D then the set of regions of interest is a collection of subsets of D. Note that regions of interest can be overlapping, as in the case of intervals, or they can form a partition of the domain. Dasu et al. [9] also proposed to partition the domain based on an initial sample of the data.

Once the regions of interest have been specified, the goal is to report a change whenever the system detects that the probability of a region has changed. The region (or regions) with the largest change in probability are then given as the description of the change.

Estimating Probabilities

Since the true distributions F_i are unknown, it is necessary to estimate them. A *window of size m* is a set of m consecutive data points. The initial distribution is estimated using a window that contains the first m data points and the most recent distribution is estimated using a window containing the most recent m data points (in practice, several change detection algorithms can be run in parallel, each using a different value of m). In each window W_i, the probability of a particular region of interest R can be estimated by $\frac{|W_i \cap R|}{|W_i|}$, the fraction of points in the window that occur in the region. Alternatively, if the regions of interest form a partition of the domain into k regions, then a Bayesian-style correction $\frac{|W_i \cap R| + \alpha}{|W_i| + \alpha k}$ can also be used [9].

Statistical Testing

In this setting, a *statistic f* is a function that assigns a number to a pair of windows (W_1, W_2). The larger this number is, the more likely it is that the points in one window were generated from one distribution and the points in the other window were generated from a different distribution.

A *statistical test* is a statistic f and a real number τ that serves as a threshold; when $f(W_1, W_2) \geq \tau$ then one can conclude that the points in W_1 were generated from a different distribution than the points in W_2.

For each $i \geq 1$ let W_i be the window that contains the points $x_i, \ldots, x_{i+m-1}$, so that W_1 is the set of the first m data points. To use a statistical test f with threshold τ, one computes the values $f(W_1, W_m), f(W_1, W_{m+2}), f(W_1, W_{m+3}), \ldots$, and signals a change the first time i such that $f(W_1, W_i) \geq \tau$. At this time, the current window W_i is considered to be a set of m points generated from the new distribution. The distribution may change again in the future, so one proceeds by computing the values of $f(W_i, W_{i+m}), f(W_i, W_{i+m+1}), f(W_i, W_{i+m+2})$, etc., until another change is detected.

Note that in order for it to be useful, a statistic should be easy to compute since a new value must be computed every time a new data point arrives. The value of the threshold τ should also be carefully chosen to avoid incorrect detections of change. A *false positive* is said to have occurred if the algorithm reports a change when the underlying distribution has not changed. Since a stream is a potentially unbounded source of data, false positives will occur and so instead of bounding the probability of a false positive, the goal is to choose a value of τ that bounds the *expected rate* of false positives. Several statistics for detecting change and a method for choosing τ are presented next.

Let A be the collection of regions of interest. Let W and W' be two windows and let P and P' be the corresponding probability estimates: for any $A \in A$, $P(A)$ is the estimated probability of region A based on window W and $P'(A)$ is the estimated probability of A based on window W'. The following statistics can be used in the change detection framework [13]:

$$d_A(W, W') = \sup_{A \in A} |P(A) - P'(A)|$$

$$\phi_A(W, W') = \sup_{A \in A} \frac{|P(A) - P'(A)|}{\sqrt{\min\left\{\frac{P(A) + P'(A)}{2}, 1 - \frac{P(A) + P'(A)}{2}\right\}}}$$

$$\Xi_A(W, W') = \sup_{A \in A} \frac{|P(A) - P'(A)|}{\sqrt{\frac{P(A) + P'(A)}{2}\left(1 - \frac{P(A) + P'(A)}{2}\right)}}$$

Note that when A is the set of intervals of the form $(-\infty, b)$ then d_A is also known as the Kolmogorov-Smirnov statistic. For any one of these statistics, the region $A \in$ A which maximizes the statistic is the region where the change in observed probability is the most statistically significant; this region (or the ℓ most significant regions, depending on user preferences) and its change in probability is therefore the description of the change.

To use these statistics, one must determine the value of the threshold τ and the corresponding expected rate of false positives. To do this one can take advantage of the fact that for one-dimensional data, the worst-case behavior of the d_A, ϕ_A, and Ξ_A statistics occur when the data are generated by continuous distributions and that the statistics behave in the same way for all continuous distributions [13]. This means that one can perform an offline computationally-intensive simulation to determine τ and then use this value for *any* one-dimensional stream afterwards.

To perform the simulation, a user must specify a test statistic f and four parameters: a window size m, a real number p between 0 and $\frac{1}{2}$, a large integer $q >= 2m$ (e.g., 1 million), and the number of repetitions B. For each repetition i, generate q points independently from any continuous distribution (e.g., from a Gaussian distribution with mean 0 and variance 1). Compute the value $\tau_i \equiv \max_{m \le j \le q-m+1} f(W_1, W_j)$ (this represents the largest value of the statistic f that would have encountered if this were the real data). After B repetitions, choose a value for the threshold τ such that τ is greater than $(1 - p)B$ of the values $\tau_1, \ldots, \tau_B$. This value of τ guarantees that the probability of a false positive in the first q points is approximately p.

To compute the expected rate of false positives corresponding to τ, one first notes that false reports of change should occur in pairs for the following reason. Once a false positive has occurred, one has a window with points that are considered anomalous (since they caused a change to be reported); as new data points arrive, one compares the m most recent points (which are still generated from the original distribution) with this anomalous window using the chosen test statistic and therefore a second report of change should soon occur. Thus one can upper bound the expected number H of false positives in the first q points using the following probability distribution: $P(H = 2) = p$, $P(H = 4) = p^2$, etc., and $P(H = 0) = \frac{1-2p}{1-p}$. The expected value is $\frac{2p}{(1-p)^2}$ and one can use this as an upper bound on the number of false positives in the first q points. One can approximate the expected number of errors in the next q points also by $\frac{2p}{(1-p)^2}$ so that the expected rate of false positives is approximated by $\frac{2p}{q(1-p)^2}$.

When the regions of interest form a partition of the domain into k regions, other statistics, such as the KL-distance can be used [9]:

$$KL_A\left(W, W'\right) = \sum_{A \in A} P(A) \log \frac{P(A)}{P'(A)}$$

where the probabilities $P(A)$ and $P'(A)$ are estimated using the Bayesian correction (i.e., $P(A) = \frac{|W \cap A| + \alpha}{|W| + \alpha k}$). Dasu et al. propose using the KL-distance with the following scheme (which uses a user-defined parameter γ): initially collect m data points for the window W_1 and use these points to create the regions of interest which partition the (possibly high-dimensional) domain; then compute $KL_A(W_1, W_m), KL_A(W_1, W_{m+1}), KL_A(W_1, W_{m+2})$, etc., and report a change whenever $n\gamma$ consecutive values of the statistic exceed a threshold τ. The value of τ depends on the points in W_1 and must be recomputed every time a change is detected. As before, τ is estimated via simulation.

To determine the value of τ, choose a parameter p $(0 < p < 1/2)$ and number of repetitions B. For each repetition i, use the probability distribution P estimated from W_1 to generate two windows V_1 and V_2 of m points each. Define τ_i to be $KL_A(V_1, V_2)$. After B repetitions, choose τ so that it is greater than $(1 - p)B$ of the $\tau_1, \ldots, \tau_B$.

Key Applications

Data mining, network monitoring.

Future Directions

Key open problems include efficiently detecting change in high-dimensional spaces (see also [9]) and detecting change in streams where data points are not generated independently.

Cross-References

▶ Stream Mining

Recommended Reading

1. Agarwal D, Barman D, Gunopulos D, Korn F, Srivastava D, Young N. Efficient and effective explanation of change in hierarchical summaries. In: Proceedings of the 13th ACM SIGKDD International Conference on Knowledge Discovery and Data Mining; 2007. p. 6–15.
2. Babcock B, Babu S, Datar M, Motwani R, Wisdom J. Models and issues in data stream systems. In: Proceedings of the 21st ACM SIGACT-SIGMOD-SIGART Symposium on Principles of Database Systems; 2002. p. 1–16.
3. Bansal RK, Papantoni-Kazakos P. An algorithm for detecting a change in a stochastic process. IEEE Trans Inf Theor. 1986;32(2):227–35.
4. Basseville M, Nikiforov IV. Detection of abrupt changes: theory and application. Englewood Cliffs, NJ: Prentice-Hall; 1993.
5. Carlstein E, H-G M, Siegmund D, editors. Change-point problems. Hayward: Institute of Mathematical Statistics; 1994.
6. Chahrabarti S, Sarawagi S, Dom B. Mining surprising patterns using temporal description length. In: Proceedings of the 24th International Conference on Very Large Data Bases; 1998. p. 606–17.
7. Chawathe SS, Abiteboul S, Widom J. Representing and querying changes in semi-structured data. In: Proceedings of the 14th International Conference on Data Engineering; 1998. p. 4–13.
8. Chawathe SS, Garcia-Molina H. Meaningful change detection in structured data. In: Proceedings of the ACM SIGMOD International Conference on Management of Data; 1997. p. 26–37.
9. Dasu T, Krishnan S, Venkatasubramanian S, Yi K. An information-theoretic approach to detecting changes in multi-dimensional data streams. In: Proceedings of the 38th Symposium on the Interface of Statistics, Computing Science, and Applications; 2006.
10. Ganti V, Gehrke J, Ramakrishnan R. Mining data streams under block evolution. SIGKDD Explorations. 2002;3(2):1–10.
11. Glaz J, Balakrishnan N. Scan statistics and applications. Boston: Birkhäuser; 1999.
12. Hulten G, Spencer L, Domingos P. Mining time-changing data streams. In: Proceedings of the 7th ACM SIGKDD International Conference on Knowledge Discovery and Data Mining; 2001. p. 97–106.
13. Kifer D, Ben-David S, Gehrke J. Detecting change in data streams. In: Proceedings of the 30th International Conference on Very Large Data Bases; 2004. p. 180–91.
14. Kleinberg JM Bursty and hierarchical structure in streams. In: Proceedings of the 8th ACM SIGKDD International Conference on Knowledge Discovery and Data Mining; 2002. p. 91–101.
15. Vovk V, Nouretdinov I, Gammerman A. Testing exchangeability on-line. In: Proceedings of the 20th International Conference on Machine Learning; 2003. p. 768–75.

Channel-Based Publish/Subscribe

Hans-Arno Jacobsen
Department of Electrical and Computer Engineering, University of Toronto, Toronto, ON, Canada

Synonyms

Event channel; Event service

Definition

Channel-based publish/subscribe is a communication abstraction that supports data dissemination among many sources and many sinks. It is an instance of the more general publish/subscribe concept. The communication channel mediates between publishing data sources and subscribing data sinks and decouples their interaction.

Key Points

Publishing data sources submit messages to the channel and subscribing data sinks listen to the channel. All messages published to the channel are received by all subscribers listening on the channel. The channel broadcasts a publication message to all listening subscribers.

The channel decouples the interaction among publishing data sources and subscribing data sinks. The same decoupling characteristics as discussed under the general publish/subscribe concept apply here as well. Realizations of this model found in practice vary in the exact decoupling offered. To properly qualify as publish/subscribe, at least the anonymous communication style must exist. That is publishing clients must not be aware of who the subscribing clients are and how many subscribing clients exist, and vice versa. Thus, channel-based publish/subscribe enables the decoupled interaction of n sources with m sinks for $n, m \geq 1$.

Channel-based publish/subscribe systems often allow the application developer to create multiple logical channels, where each channel can be configured to offer different qualities-of-service to an application. Furthermore, a channel can be dedicated to the dissemination of messages pertaining to a specific subject or type. The channel-based publish/subscribe model does not support message filtering, except through the use of various channels to partition the publication space. It is the clients' responsibility to select the right channel for the dissemination of messages, which are sent to all listeners on the channel. This enables a limited form of filtering by constraining messages to be disseminated on one channel to a given message type. Finally, channel-based publish/subscribe is often coupled with client-side filtering, where messages are still broadcast throughout the channel, but filtered upon arrival at the data sink before passing to the application. More fine-grained filtering functionalities are provided by the other publish/subscribe models, such as the topic-based model and the content-based model.

In channel-based publish/subscribe, the publication data model is defined by the type of message the channel-based communication abstraction supports. This is often closely tied to the programming language or the library that implements the model.

Similarly, the subscription language model is defined by the programming language or library that allows the application developer to select channels to listen to, unless special provisions for subscriber-side filtering are offered. If supported, subscriber-side filtering can be arbitrarily complex, even selecting messages based on their content, as offered by the content-based publish/-subscribe model.

Matching in the sense of evaluating a publication message against a set of subscriptions, as is common in the other publish/subscribe instantiations, does not occur in channel-based publish/-subscribe.

Channel-based publish/subscribe systems are often coupled with different client interaction styles. These are the *push-style* and the *pull-style*. In the push-style, data sources initiate the transfer of messages to the channel, which delivers the messages to all listening data sinks. In the pull-style, data sinks initiate the message transfer by requesting messages from the channel, which requests any available messages from all connected data sources. Both interaction styles can also be combined. That is on one channel some clients can connect to the channel through the push-style, while others connect via the pull-style.

Channel-based publish/subscribe systems are distinguished by the qualities-of-service the channel offers to its clients, such as various degrees of reliability, persistence, real-time constraints, and message delivery guarantees. Channel-based publish/subscribe relates to topic-based publish/subscribe in that publishing a message to a channel is similar to associating a message with a topic, which could be the name or identity of the channel. However, in topic-based publish/subscribe this association is reflected in the message itself, while in channel-based publish/subscribe the association is indirect, reflected by selecting a channel, not part of the message. Also, topics can go far beyond channel identities, as discussed under the topic-based publish/subscribe concept. Examples that follow the channel-based publish/subscribe model are the CORBA Event Service [2], IP multicast [?], Usenet newsgroups [?], mailing lists, and

Chart   417

group communication [?]. Elements of channel-based publish/subscribe can also be found in the Java Messaging Service [1], the OMG Data Dissemination Service [3], and other messaging middleware. However, these approaches are not directly following the channel-based model as described above; rather these approaches are enriched with elements of message queuing, topic-based publish/subscribe, and content-based publish/subscribe.

There are many applications of channel-based publish/subscribe. Examples include change notification, update propagation, information dissemination, newsgroups, email lists, and system management. Channel-based publish/-subscribe serves well, if one or more entities have to communicate date to an anonymous group of receivers that may change over time, without the need of filtering messages within the channel.

In the literature the term channel-based publish/subscribe is not used uniformly. Abstractions that exhibit the above described functionality are also often referred to as event services, event channels, and simply channels. Messages disseminated to listeners are also often referred to as events. Publishing data sources are often referred to as publishers, producers or suppliers, and subscribing data sinks are often referred to as subscribers, consumers or listeners.

Cross-References

▶ Content-Based Publish/Subscribe
▶ Publish/Subscribe
▶ Topic-Based Publish/Subscribe

Recommended Reading

1. Hapner M, Burridge R, Sharma R. Java message service. Sun Microsystems, version 1.0.2 edition. 9 November 1999.
2. OMG. Event service specification, version 1.2, formal/04-10-02 edition. October 2004.
3. OMG. Data distribution service for real-time systems, version 1.2, formal/07-01-01 edition. January 2007.

Chart

Hans Hinterberger
Department of Computer Science, ETH Zurich, Zurich, Switzerland

Synonyms

Diagram; Graph; Information graphic; Map

Definition

A chart is an instrument to consolidate and display information.

The term is applied to virtually any graphic that displays information, be it a map used for navigation, a plan for military operations, a musical arrangement, barometric pressure, genealogical data, and even lists of tunes that are most popular at a given time.

Definitions of specialized charts typically include the graphical method on which the chart is based (e.g., bar chart) and or its application area (e.g., CPM chart), but it does not specify design principles. Tufte [2] introduces the notion of "chartjunk" and defines it to be that part of the chart which functions only as decoration, all of which is considered to redundant data ink.

Sometimes charts are reduced to refer to maps and diagrams, excluding graphs and tables.

Key Points

Because charts are used for different purposes, there exist almost as many different types of charts as there are applications. Some maps, however, are more general in their use and therefore assigned to one of the following four categories: graphs, maps, diagrams, and tables. Each of these categories is broken down further into subcategories. In the category diagrams, one therefore finds pie charts as well as flow charts or organization charts. A detailed categorization can be found in [1].

Charts are not only used to visualize data. They serve as a useful tool for many tasks where information is the main ingredient such as planning, presentation, analysis, and monitoring.

Cross-References

▸ Data Visualization
▸ Diagram
▸ Graph
▸ Map
▸ Table
▸ Thematic Map

Recommended Reading

1. Harris RL. Information graphics: a comprehensive illustrated reference. New York/Oxford: Oxford University Press; 1999.
2. Tufte ER. The visual display of quantitative information. Cheshire: Graphics Press; 1983.

Chase

Alin Deutsch[1] and Alan Nash[2]
[1]University of California-San Diego, La Jolla, CA, USA
[2]Aleph One LLC, La Jolla, CA, USA

Definition

The chase is a procedure that takes as input a set Σ of constraints and an instance I. The chase does not always terminate, but if it does it produces as output an instance U with the following properties:

1. $U \models \Sigma$; that is, U satisfies Σ.
2. $I \to U$; that is, there is a homomorphism from I to U.
3. For every instance J (finite or infinite), if $J \models \Sigma$ and $I \to J$, then $U \to J$.

In [7], an instance that satisfies (1) and (2) above is called a *model of Σ and I* and an instance that satisfies (3) above is called *strongly universal*.

In summary, the chase is a procedure which – whenever it terminates – yields a strongly-universal model.

Comments

1. The set Σ of constraints is usually a set of tuple-generating dependencies (tgds) and equality-generating dependencies (egds) [5], or, equivalently, embedded dependencies [5, 10]. However, the chase has been extended to wider classes of constraints and to universality under functions other than homomorphisms [6, 7, 9]. In this case, the chase often produces a strongly-universal model *set* (see below), instead of a single model.
2. It was noted in [7] that in database applications, *weak universality* (condition 3 above restricted to finite instances) would suffice. Nevertheless, the chase gives strong universality.

Historical Background

The term "chase" was coined in [14], where it was used to test the logical implication of dependencies (i.e., whether all databases satisfying a set Σ of dependencies must also satisfy a given dependency σ). The implication problem was one of the key concerns of dependency theory, with applications to automatic schema design. The chase was defined in [14] for the classes of functional, join and multivalued dependencies. Related chase formulations for various kinds of dependencies were introduced in [15, 17]. The work [5] unified the treatment of the implication problem for various dependency classes by introducing and defining the chase for tuple-generating and equality-generating dependencies (sufficiently expressive to capture all prior dependencies).

Ancestors of the chase (introduced as unnamed algorithms) appear in [2–4]. Aho et al.

[4] introduces tableaux, a pattern-based representation for relational queries, and shows how to check the equivalence of tableau queries in the presence of functional dependencies, with applications to query optimization. To this end, the tableaux are modified using an algorithm that coincides with the chase with functional dependencies. The same algorithm is used in [3] for minimization of tableaux under functional dependencies. This algorithm is extended in [2] to include also multivalued dependencies, for the purpose of checking whether the join of several relations is lossless (i.e., the original relations can be retrieved as projections of the join result).

The chase was extended to include disjunction and inequality in [9], and to arbitrary $\forall\exists$-sentences in [6]. Independently, [13] extended the chase to a particular case of disjunctive dependencies incorporating disjunctions of equalities between variables and constants (see also [12]). There are also extensions of the chase to deal with more complex data models beyond relational. The chase (and the language of embedded dependencies) is extended in [16] to work over complex values and dictionaries. For an excellent survey of the history of the chase prior to 1995, consult [1].

Foundations

A *tuple-generating dependency (tgd)* is a constraint σ of the form

$$\forall \overline{x}, \overline{y} \ (\alpha \, (\overline{x}, \overline{y}) \rightarrow \exists \overline{z} \beta \, (\overline{x}, \overline{z}))$$

where α and β are conjunctions of relational atoms. Furthermore, every variable in $\overline{x}$ appears in both α and β. The $\forall \overline{x}, \overline{y}$ prefix of universal quantifiers is usually omitted. If $\overline{z}$ is empty, then σ is *full*.

An *equality-generating dependency (egd)* is a constraint φ of the form

$$\forall x_1, x_2, \overline{y} \ (\alpha \, (x_1, x_2, \overline{y}) \rightarrow x_1 = x_2)$$

where α is a conjunction of relational atoms.

The chase is used on instances whose active domain consists of constants and *labeled nulls*. A *homomorphism* from A to B is denoted $A \rightarrow B$. It is a mapping h on the constants and nulls in A that (i) preserves constants (i.e., $h(c) = c$ for every constant c) and preserves relationships (i.e., for every tuple $R(x_1, \ldots, x_n) \in A$, that is $R(h(x_1), \ldots, h(x_n)) \in B$). Two instances A and B are *homomorphically equivalent* if $A \rightarrow B$ and $B \rightarrow A$.

The chase is a natural procedure for building strong universal models. Indeed, it turns out that checking for strong universality is undecidable as shown in [7]). In contrast, checking whether an instance is a model can be done efficiently. Therefore, it is natural to define any procedure for constructing strong universal models by steps which always preserve strong universality while attempting to obtain a model and then to check whether a model was indeed obtained. This is precisely what the chase does.

A tgd $\sigma \in \Sigma$ *fails* (or *applies*) on instance A and tuple $\overline{a}$ if there is tuple $\overline{b}$ in A such that the premise α of σ satisfies $A \vDash \alpha(\overline{a}, \overline{b})$, yet there is no tuple $\overline{c}$ in A such that the conclusion β of σ satisfies $A \vDash \beta(\overline{a}, \overline{c})$. Assume that the instance A' is obtained by adding to A the tuples in $\beta(\overline{a}, \overline{n})$ where $\overline{n}$ is a tuple of new nulls. Then A' is the result of *firing* σ on A, $\overline{a}$. Notice that $A \subseteq A'$ and that σ does not fail on A', $\overline{a}$. It is easy to verify that if A is strongly universal for Σ and I, then so is A' (towards this, it is essential that all the nulls in $\overline{n}$ be new and distinct).

An egd $\sigma \in \Sigma$ *fails* (or *applies*) on instance A and values a_1, a_2 if there is tuple $\overline{b}$ in A such that the premise α of σ satisfies $A \vDash \alpha(a_1, a_2, \overline{b})$, yet $a_1 \neq a_2$. If a_2 is a null a_2 is replaced everywhere in A with a_1 to obtain A', then say that A' is the result of *firing* σ on A, a_1, a_2. Notice that $A \rightarrow A'$ and that σ does not fail on A', a_1, a_1. It is easy to verify that if A is strongly universal for Σ and I, then so is A'. If a_2 is a constant, but a_1 is null, then it is possible to replace a_1 everywhere in A with a_2 instead. However, if both a_1 and a_2 are constants, then it is not possible to satisfy σ and preserve strong universality and the chase *fails*.

The standard chase procedure proceeds as follows.

1. Set $A_0 = I$.
2. Repeat the following:
   1. If A_n is a model of Σ and I, stop and return A_n.
   2. Otherwise, there must be either
      1. a tgd σ and $\overline{a}$ such that σ fails on A, $\overline{a}$, or
      2. an egd σ' and a_1, a_2 such that σ' fails on A, a_1, a_2.

      Obtain A_{n+1} by picking one such σ and $\overline{a}$ and firing σ on A_n, $\overline{a}$, or by picking one such σ' and a_1, a_2 and firing σ' on A, a_1, a_2. (This is one *chase step* of the standard chase.)

Notice that, at every chase step, there may be a choice of σ and $\overline{a}$, respectively σ' and a_1, a_2. How these choices are picked is often left unspecified and in that case the standard chase is non-deterministic. The chase *terminates* if A_n is a model of Σ and I for some n.

The chase of instance I with tgds Σ produces a sequence of instances $I = A_0 \subseteq A_1 \subseteq A_2 \subseteq \ldots$ such that every A_i is strongly universal for Σ and I. The chase with tgds and egds produces a sequence $I = A_0 \rightarrow A_1 \rightarrow A_2 \rightarrow \ldots$ such that every A_i is strongly universal for Σ and I. In the presence of egds, it is no longer the case that $A_i \subseteq A_j$ for $i \leq j$ and there is the additional complication that a chase step may fail. The chase for tgds and egds is described in more detail in [1].

Example 1 Consider the schema consisting of two relations:

1. employee Emp(ss#, name, dept#), with social security, name, and dept. number, and
2. department Dept(dept#, name, location, mgr#), with dept. number, name, location, and its manager's social security number.

Assume that Σ consists of the constraints

- σ_1: dept# is a foreign key in Emp,
- σ_2: mgr# is a foreign key in Dept, and
- σ_3: every manager manages his own department.

(This omits the constraints that say that ss# is a key for Emp and that dept# is a key for Dept to keep the example simple.) These constraints can be written as follows (where σ_1 and σ_2 are tgds and σ_3 is an egd):

- σ_1: Dept$(d, e, \ell, m) \rightarrow \exists n, d'Emp(m, n, d')$,
- σ_2: Emp$(s, n, d) \rightarrow \exists e, \ell, m$Dept$(d, e, \ell, m)$, and
- σ_3: Dept(d, e, ℓ, m), Emp$(m, n, d') \rightarrow d = d'$.

Consider the initial instance

$$I_0 = \text{Dept}(1, \text{"HR"}, \text{"somewhere"}, 333 - 33 - 3333)$$

containing a single tuple. Then in the first step of the chase, σ_1 fires, giving

$$I_1 = \{\text{Dept}(1, \}\text{HR}\}, \}\text{somewhere}\}, 33),$$
$$\text{Emp}(33, \alpha, \beta)\}$$

where α and β are labeled nulls. In the second step, both σ_2 and σ_3 apply. If σ_3 fires, then β is set to 1 and yields

$$I_2 = \{\text{Dept}(1, \text{"HR"}, \text{"somewhere"}, 33),$$
$$\text{Emp}(33, \alpha, 1)\}.$$

Since I_2 satisfies Σ, the chase terminates. However, if instead at the second step σ_2 fires, then it gives

$$I_2' = \{\text{Dept}(1, \text{"HR"}, \text{"somewhere"}, 33),$$
$$\text{Emp}(33, \alpha, \beta), \text{Dept}(\beta, \gamma, \delta, \epsilon)\}$$

where γ, δ, and ε are new nulls. In this case, it is possible to continue firing σ_1, σ_2, and σ_3 in such a way as to obtain a chase that does not terminate, perpetually introducing new nulls.

If the standard chase (or any other chase listed below) terminates, it yields a strongly-universal model of Σ and I and it is straightforward to verify that all such models are homomorphically equivalent. Therefore the result of the standard chase is unique up to homomorphic equivalence. However, the choice of what constraint to fire and on what tuple may affect whether the chase terminates or not.

There are several variations of the chase, which shall be called here the *standard* chase, the *parallel* chase, and the *core* chase. The standard chase was described above. In the *parallel chase*, at every chase step σ is fired on $A_n, \overline{a}$ for all pairs $(\sigma, \overline{a})$ such that σ fails on $A, \overline{a}$.

One writes I^{Σ} for the result of the chase on Σ and I, if the chase terminates. In that case, one says that I^{Σ} is defined. In general, it holds that if $A \rightarrow B$, then $A^{\Sigma} \rightarrow B^{\Sigma}$, whenever the latter are defined.

It was shown in [7] that the standard chase is incomplete, in the following sense: it may be that Σ and I have a strongly-universal model, yet the standard chase does not terminate. The parallel chase is also incomplete in this sense. In contrast, the *core chase* introduced in [7] is complete: if a strongly universal-model exists, the core chase terminates and yields such a model. A chase step of the core chase consists of one chase step of the parallel chase, followed by computing the core of the resulting instance.

Any of the above mentioned variations of the chase can be applied to sets of constraints which consist of

1. tgds only
2. tgds and egds
3. tgds and egds with disjunctions
4. tgds and egds with disjunctions and negation which are equivalent to general $\forall\exists$ sentences

The chase with tgds and egds has been described above. The chase has been extended to handle disjunction and negation. In this case, it gives not a single model, but a set S of models which is strongly universal, in the sense that for any model J (finite or infinite) of Σ and I, there is a model $A \in S$ such that $A \rightarrow J$. Such a set arises from a single initial model by branching due to disjunction. For example, consider the set Σ with the single disjunctive tgd

$$\sigma : R(x) \rightarrow S(x) \vee T(x)$$

and the instance I containing the single fact $R(1)$. Clearly every model of Σ and I, must contain either $S(1)$ or $T(1)$. It is easy to verify that the set $S = \{I_1, I_2\}$ where $I_1 = \{R(1), S(1)\}$ and $I_2 = \{R(1), T(1)\}$ is strongly universal for I and Σ, but no proper subset of S is. The disjunctive chase with Σ on I consists of a single step, which produces not a single model, but the set S of models. Intuitively, whenever a disjunctive tgd fires on a set W of models, it produces, for every instance $A \in W$, one instance for every disjunct in its conclusion. For details, to see how negation is handled, and to see how universality for functions other than homomorphisms is achieved, see [6, 7].

It was shown in [7] that it is undecidable whether the standard, parallel, or core chase with a set of tgds terminates. A widely-applicable, efficiently-checkable condition on a set Σ of tgds, which is sufficient to guarantee that the chase with Σ on any instance I terminates, was introduced in [9, 11]. A set of tgds satisfying this condition is called *weakly acyclic* in [11] and is said to have *stratified witnesses* in [9]. A more widely-applicable condition, also sufficient for chase termination, was introduced in [7], where a set of tgds satisfying this condition is called *stratified*.

Key Applications

The chase has been used in many applications, including

- Checking containment of queries under constraints (which in turn is used in such query rewriting tasks as minimization, rewriting using views, and semantic optimization)
- Rewriting queries using views
- Checking implication of constraints
- Computing solutions to data exchange problems
- Computing certain answers in data integration settings

To check whether a query P is contained in a query Q under constraints Σ, written $P \sqsubseteq_{\Sigma} Q$, it is sufficient to (1) treat P as if it was an instance in which the free variables are constants and the bound variables are nulls (this is known as the "frozen instance" or "canonical database" [1] corresponding to P) (2) chase it with Σ, and if this chase terminates to yield P^Σ (3) check whether the result of this chase is contained in Q, written $P^\Sigma \sqsubseteq Q$. In symbols, if the chase described above terminates, then

$$P \sqsubseteq_\Sigma Q \text{ iff } P^\Sigma \sqsubseteq Q.$$

That is, the chase reduces the problem of query containment under constraints to one of query containment without constraints.

To check whether a set Σ of tgds implies a tgd σ of the form

$$\forall \overline{x}, \overline{y} \, (\alpha \, (\overline{x}, \overline{y}) \rightarrow \exists \overline{z} \beta \, (\overline{x}, \overline{z}))$$

which is logically equivalent to

$$\forall \overline{x} (\underbrace{\exists \overline{y} \alpha \, (\overline{x}, \overline{y})}_{Q_\alpha(\overline{x})} \rightarrow \underbrace{\exists \overline{z} \beta \, (\overline{x}, \overline{z})}_{Q_\beta(\overline{x})}),$$

it suffices to check whether the query Q_α in the premise of σ is contained under the constraints Σ in the query Q_β in the conclusion of σ. That is, if $Q_\alpha{}^\Sigma$ is defined, then

$$\Sigma \models \sigma \text{ iff } Q_\alpha \sqsubseteq_\Sigma Q_\beta \text{ iff } Q_\alpha^\Sigma \sqsubseteq Q_\beta.$$

The chase was also employed to find equivalent rewritings of conjunctive queries using conjunctive query views, in the presence of constraints. Given a set v of conjunctive query views and a conjunctive query Q, one can construct, using the chase, a query R expressed in terms of v, such that Q has some equivalent rewriting using v if and only if R is itself such a rewriting. Moreover, every minimal rewriting of Q is guaranteed to be a sub-query of R. The algorithm for constructing R and exploring all its sub-queries is called the *Chase&Backchase (CB)* [8], and it is sound and complete for finding all minimal rewritings under a set Σ of embedded dependencies, provided the chase with Σ terminates

[9]. The CB algorithm constructs R by simply (i) constructing a set Σ_v of tgds extracted from the view definitions, and (ii) chasing Q with $\Sigma \cup \Sigma_v$ and restricting the resulting query to only the atoms using views in v.

In [11] is was shown that the certain answers to a union Q of conjunctive queries on a ground instance I under a set Σ of source-to-target tgds and target tgds and egds can be obtained by computing $Q(U)$ – where U is a *universal solution* for I under Σ – then discarding any tuples with nulls. Universal solutions, which are the preferred solutions to materialize in data exchange, are closely related to strongly-universal models [7] and it was shown in [11] that they can be obtained using the chase.

Cross-References

- ▶ Data Exchange
- ▶ Database Dependencies
- ▶ Equality-Generating Dependencies
- ▶ Query Containment
- ▶ Query Optimization
- ▶ Query Rewriting
- ▶ Rewriting Queries Using Views
- ▶ Tuple-Generating Dependencies

Recommended Reading

1. Abiteboul S, Hull R, Vianu V. Foundations of databases. Reading: Addison Wesley; 1995.
2. Aho AV, Beeri C, Ullman JD. The theory of joins in relational databases. ACM Trans Database Syst. 1979a;4(3):297–314.
3. Aho AV, Sagiv Y, Ullman JD. Efficient optimization of a class of relational expressions. ACM Trans Database Syst. 1979b;4(4):435–54.
4. Aho AV, Sagiv Y, Ullman JD. Equivalence of relational expressions. SIAM J Comput. 1979c;8(2):218–46.
5. Beeri C, Vardi MY. A proof procedure for data dependencies. J ACM. 1984;31(4):718–41.
6. Deutsch A, Ludaescher B, Nash A. Rewriting queries using views with access patterns under integrity constraints. In: Proceedings of the 10th International Conference on Database Theory; 2005. p. 352–67.

7. Deutsch A, Nash A, Remmel J. The chase revisited. In: Proceedings of the 27th ACM SIGACT-SIGMOD-SIGART Symposium on Principles of Database Systems; 2008. p. 149–58.
8. Deutsch A, Popa L, Tannen V. Physical data independence, constraints, and optimization with universal plans. In: Proceedings of the 25th International Conference on Very Large Data Bases; 1999. p. 459–70.
9. Deutsch A, Tannen V. XML queries and constraints, containment and reformulation. Theor Comput Sci. 2005;336(1):57–87, preliminary version in ICDT 2003.
10. Fagin R. Horn clauses and database dependencies. J ACM. 1982;29(4):952–85.
11. Fagin R, Kolaitis PG, Miller RJ, Popa L. Data exchange: semantics and query answering. Theor Comput Sci. 2005;336(1):89–124, preliminary version in PODS 2005.
12. Fuxman A, Kolaitis PG, Miller RJ, Tan WC. Peer data exchange. ACM Trans Database Syst. 2006;31(4):1454–98, preliminary version in PODS 2005.
13. Grahne G, Mendelzon AO. Tableau techniques for querying information sources through global schemas. In: Proceedings of the 7th International Conference on Database Theory; 1999. p. 332–47.
14. Maier D, Mendelzon AO, Sagiv Y. Testing implications of data dependencies. ACM Trans Database Syst. 1979;4(4):455–69.
15. Maier D, Sagiv Y, Yannakakis M. On the complexity of testing implication of functional and join dependencies. J ACM. 1981;28(4):680–95.
16. Popa L, Tannen V. An equational chase for path-conjunctive queries, constraints, and views. In: Proceedings of the 7th International Conference on Database Theory; 1999. p. 39–57.
17. Vardi M. Inferring multivalued dependencies from functional and join dependencies. Acta Informatica. 1983;19:305–24.

Checksum and Cyclic Redundancy Check Mechanism

Kazuhisa Fujimoto
Hitachi Ltd., Tokyo, Japan

Synonyms

Cyclic redundancy check (CRC)

Definition

Checksum and CRC are schemes for detecting the errors of data which occur during transmission or storage. The data computed and appended to original data in order to detect errors are also referred as checksum and CRC.

A checksum consists of a fixed number of bits computed as a function of the data to be protected, and is appended to the data. To detect errors, the function is recomputed, and the result is compared to that appended to the data. Simple implementation of checksum is to divide the data into same length bits chunk and to make exclusive-or of all chunks. Cyclic redundancy check mechanism exploits mathematical properties of cyclic codes. Specifically, CRC uses polynomial devisor circuits with a given generator polynomial so as to obtain the remainder polynomial. The remainder is similarly appended to the original data for transmission and storage, and then utilized for error detection. CRC can be used as a kind of checksum.

Key Points

CRC is usually expressed by the use of binary polynomials due to mathematical convenience. When original data $M(x)$ is given, basic CRC mechanism calculates redundancy data $R(x)$ by using a pre-defined generator polynomial $G(x)$. That is, supposing the degree of $G(x)$ is m, a polynomial $M(x) * x^m$ is divided by $G(x)$ and the remainder is used for R(x) such that a concatenated polynomial $T(x) = M(x) * x^m + R(x)$ is divisible by $G(x)$. The obtained $T(x)$ is used for transmission or storage. For error detection, CRC mechanism similarly checks the divisibility of $T(x)$ by $G(x)$. These encoding and detection processes can be implemented by using multi-level shift register circuits.

Given below is an example of CRC calculation. Assume that a generator polynomial and original data are given as follows.

$$G(x) = x^3 + x + 1$$

$$(binary\ expression : 1011)$$

$$M(x) = x^4 + 1 \quad (10001)$$

In this case, a remainder polynomial $R(x)$ can be obtained by dividing $M(x) * x^3$ by $G(x)$.

$$R(x) = x \quad (010)$$

Therefore, the resulting data $T(x)$ can be obtained as follows.

$$T(x) = x^7 + x^3 + x \quad (10001010)$$

Theoretically, CRC is capable of detecting m-bit long or shorter bust errors. This property is suitable for communication infrastructure and storage infrastructure, which often introduce burst errors rather than random errors.

Cross-References

▶ Disk

Recommended Reading

1. Houghton A. Error coding for engineers. Dordrecht: Kluwer; 2001.
2. Sweeney P. Error control coding from theory to practice. New York: Wiley; 2002.

Choreography

W. M. P. van der Aalst
Eindhoven University of Technology,
Eindhoven, The Netherlands

Definition

In a service oriented architecture (SOA) services are interacting by exchanging messages, i.e., by combining services more complex services are created. Choreography is concerned with the composition of such services seen from a global viewpoint focusing on the common and complementary observable behavior. Choreography is particularly relevant in a setting where there is not a single coordinator.

Key Points

The terms orchestration and choreography describe two aspects of integrating services to create business processes [1, 3]. The two terms overlap somewhat and the distinction is subject to discussion. Orchestration and choreography can be seen as different "perspectives." Choreography is concerned with the exchange of messages between those services. Orchestration is concerned with the interactions of a single service with its environment.

Figure 1 illustrates the notion of choreography. The dashed area shows the focal point of choreography, i.e., the aim is to establish a "contract" containing a "global" definition of the constraints under which messages are exchanged. Unlike orchestration, the viewpoint is not limited to a single service. The Web Services Choreography Description Language (WS-CDL, cf. [2]) and the Web Service Choreography Interface (WSCI) are two languages aiming at choreography. Since the focus is on agreement rather than enactment, choreography is quite different from traditional workflow languages. The goal is not to control and enact but to coordinate autonomous parties. Some characterize choreography as "Dancers dance following a global

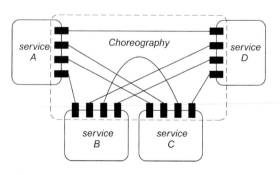

Choreography, Fig. 1 Choreography

scenario without a single point of control" to emphasize this distinction.

Cross-References

▶ Business Process Execution Language
▶ Business Process Management
▶ Orchestration
▶ Web Services
▶ Workflow Management

Recommended Reading

1. Dumas M, van der Aalst WMP, ter Hofstede AHM. Process-aware information systems: bridging people and software through process technology. New York: Wiley; 2005.
2. Kavantzas N, Burdett D, Ritzinger G, Fletcher T, Lafon Y. Web Services Choreography Description Language Version 1.0 (W3C Candidate Recommendation). 2005. http://www.w3.org/TR/2005/CR-ws-cdl-10-20051109/.
3. Weske M. Business process management: concepts, languages, architectures. Berlin: Springer; 2007.

Chronon

Curtis E. Dyreson
Utah State University, Logan, UT, USA

Synonyms

Instant; Moment; Time quantum; Time unit

Definition

A chronon is the smallest, discrete, non-decomposable unit of time in a temporal data model. In a one-dimensional model, a chronon is a *time interval* or *period*, while in an *n*-dimensional model it is a non-decomposable region in *n*-dimensional time. Important special types of chronons include valid-time, transaction-time, and bitemporal chronons.

Key Points

Data models often represent a time line by a sequence of non-decomposable, consecutive time periods of identical duration. These periods are termed chronons. A data model will typically leave the size of each particular chronon unspecified. The size (e.g., 1 μs) will be fixed later by an individual application or by a database management system, within the restrictions posed by the implementation of the data model. The number of chronons is finite in a bounded model (i.e., a model with a minimum and maximum chronon), or countably infinite otherwise. Consecutive chronons may be grouped into larger intervals or segments, termed *granules*; a chronon is a granule at the lowest possible granularity.

Cross-References

▶ Temporal Granularity
▶ Time Domain
▶ Time Instant
▶ Time Interval

Recommended Reading

1. Dyreson CE, Snodgrass RT. The base-line clock. In: The TSQLZ temporal query language. Kluwer; 1987. p. 73–92.
2. Dyreson CE, Snodgrass RT. Timestamp semantics and representation. Inf Syst. 1993;18(3):143–66.

Citation

Prasenjit Mitra
The Pennsylvania State University, University Park, PA, USA

Synonyms

Bibliography; Reference

Definition

A citation is a reference from one article to another article. A citation is a record that consists of the names of the authors, the title of the referred article, the time and place of publication, as well as various other fields. The fields in the citation should collectively specify unambiguously where the full text of the referred article could be obtained. Typically, all citations are presented at the end of the referring article. However, articles in certain domains list the citations as footnotes in the pages where the reference occurs. Citations can range from references to be to single articles or to entire books.

Key Points

Often, authors have to refer to knowledge that is derived from another work. For example, when quoting text from another article or book, the author must specify from which article or book the quotation is obtained. Authors need to refer to other works in order to point out preliminary information on the shoulders of which the current treatise stands, to refer to related work and contrast the current work with previous works, etc. A citation is used for primarily two purposes: (i) to provide a reference to an article or book such that the reader can retrieve the article or book easily and read the article to gain additional knowledge, and (ii) to provide credit (or discredit for "negative" citations) to the authors of the works that are being cited.

There are various widely used formats for citations. Citation formats vary by discipline; typically a discipline adheres to one (or a few) "style-guide" that indicates what fields should be mentioned in a citation and how the fields should be formatted and presented. Recently, with the proliferation of electronic documents published over the World-Wide-Web, citations to Uniform Resource Locators (URLs) of websites are increasingly common. Unlike printed articles and books, websites are dynamic and can change frequently. Therefore, in order to specify precisely which version of the webpage was being referred, apart from the publication date, authors usually provide the date on which the website was accessed.

Citations analysis has been performed to identify the impact of published articles. Because different authors use different formats, automatic analysis of citations requires *citation matching*. Citation matching helps identify which different citations formatted differently refer to the same article or book. The term *bibliometrics* is used to refer to metrics designed based on citation analysis. Citation indexing for academic journals was popularized by Eugene Garfield [1, 2]. A citation index contains the information about which document cites which. The term *co-citation* refers to the frequency with which two documents are cited together [3]. Today, Google Scholar (http://scholar.google.com) provides a readily-available collection of indexed citations on the web.

Cross-References

▶ Digital Libraries

Recommended Reading

1. Garfield E. Citation indexing: its theory and application in science, technology, and humanities. New York: Wiley; 1979.
2. Garfield E. Citation analysis as a tool in journal evaluation: journals can be ranked by frequency and impact of citations for science policy studies. Science. 1972;178(4060):471–9.
3. Small H. Co-citation in the scientific literature: a new measure of the relationship between two documents. J Am Soc Inf Sci. 1973;24(4):265–9. Wiley Periodicals.

Classification

Ian H. Witten
University of Waikato, Hamilton, New Zealand

Synonyms

Classification learning; Concept learning; Learning with a teacher; Statistical decision techniques; Supervised learning

Definition

In *Classification learning*, an algorithm is presented with a set of classified examples or "instances" from which it is expected to infer a way of classifying unseen instances into one of several "classes". Instances have a set of features or "attributes" whose values define that particular instance. Numeric prediction, or "regression," is a variant of classification learning in which the class attribute is numeric rather than categorical. Classification learning is sometimes called *supervised* because the method operates under supervision by being provided with the actual outcome for each of the training instances. This contrasts with clustering where the classes are not given, and with association learning which seeks any association – not just one that predicts the class.

Historical Background

Classification learning grew out of two strands of work that began in the 1950s and were actively pursued throughout the 1960s: statistical decision techniques and the Perceptron model of neural networks. In 1955, statisticians Bush and Mosteller published a seminal book *Stochastic Models for Learning* which modeled in mathematical terms the psychologist B. F. Skinner's experimental analyses of animal behavior using reinforcement learning [2]. The "perceptron" was a one-level linear classification scheme developed by Rosenblatt around 1957 and published in his book *Principles of Neurodynamics: Perceptrons and the Theory of Brain Mechanisms* [10]. In a response published in 1969, Minsky and Papert argued that perceptrons were simplistic in terms of their representational capability and had been greatly over-hyped as potentially universal learning machines [6]. This scathing response by widely-respected artificial intelligence pioneers dampened research in neural nets and machine learning in general. Meanwhile, in 1957 others were investigating the application of Bayesian decision schemes to pattern recognition; the general

conclusion was that full Bayesian models were prohibitively expensive. In 1960 Maron investigated in the context of information retrieval what has since become known as the "naïve Bayes" approach, which assumes independence between attributes notwithstanding overwhelming evidence to the contrary [5]. Other early machine learning work was buried in cybernetics, the study of feedback and derived concepts such as communication and control in living and artificial organisms. Throughout the 1960s classification learning applied to pattern recognition was the central thread of the embryo field of machine learning, as underlined by the subtitle of Nilsson's 1965 landmark book *Learning Machines – Foundations of Trainable Pattern-Classifying Systems* [7].

Symbolic learning techniques began to recover from the doldrums in the late 1970s, with influential and almost simultaneous publications by Breiman et al. on *classification and regression trees* (the CART system) [1] and Quinlan on *decision tree induction* (the ID3 and later C4.5 systems) [8, 9]. Whereas Breiman was a statistician, Quinlan was an experimental computer scientist who first used decision trees not to generalize but to condense large collections of chess endgames. Their work proceeded independently, and the similarities remained unnoticed until years later. CART (by default) produces multivariate trees whose tests can involve more than one attribute: these are more accurate and smaller than the univariate trees produced by Quinlan's systems, but take longer to generate.

The first workshop devoted to machine learning was held in 1980 at Carnegie-Mellon University. Further workshops followed in 1983 and 1985. These invitation-only events became an open conference in 1988. Meanwhile the journal *Machine Learning* was established in 1986. By the 1990s the subject had become the poster child of artificial intelligence – a successful, burgeoning, practical technology that eschewed the classical topics of general knowledge representation, logical deduction, theorem proving, search techniques, computational linguistics, expert systems and philosophical foundations that still characterize the field today. Classification learning, which forms the core of machine learning, outgrew its behaviorist and neurological roots and moved into the practical realm of database systems.

Early work focused on the *process* of learning – learning curves, the possibility of sustained learning, and the like – rather than the *results* of learning. However, with the new emphasis on applications, objective techniques of empirical testing began to supplant the scenario-based style of evaluation that characterized the early days. A major breakthrough came during the 1980s, when researchers finally realized that evaluating a learning system on its training data gave misleading results, and instead put the subject on a secure statistical footing.

Foundations

One of the most instructive lessons learned since the renaissance of classification in the 1980s is that simple schemes often work very well. Today, practitioners strongly recommend the adoption of a "simplicity-first" methodology when analyzing practical datasets. There are many different kinds of simple structure that datasets can exhibit. One dataset might have a single attribute that does all the work, the others being irrelevant or redundant. Alternatively, the attributes might contribute independently and equally to the final outcome. Underlying a third dataset might be a simple contingent structure involving just a few attributes. In a fourth, a few independent rules may govern the assignment of instances to classes. In a fifth, classifications appropriate to particular regions of instance space might depend on the distance between the instances themselves. A sixth might exhibit dependence among numeric attributes, determined by a sum of attribute values with appropriately chosen weights. This sum might represent the final output for numeric prediction, or be compared to a fixed threshold in a binary decision setting. Each of these examples leads to a different style of method suited to discovering that kind of structure.

Rules Based on a Single Attribute
Even when instances have several attributes, the classification decision may rest on the value of just one of them. Such a structure constitutes a set of rules that all test the same attribute (or, equivalently, a one-level decision tree). It can be found by evaluating the success, in terms of the total number of errors on the training data, of testing each attribute in turn, predicting the most prevalent class for each value of that attribute. If an attribute has many possible values – and particularly if it has numeric values – this may "overfit" the training data by generating a rule that has almost as many branches as there are instances. Minor modifications to the scheme overcome this problem.

A startling discovery published in 1993 was that "very simple classification rules perform well on most commonly used datasets" [3]. In an empirical investigation of the accuracy of rules that classify instances on the basis of a single attribute, on most standard datasets the resulting rule was found to be as accurate as the structures induced by the majority of machine learning systems – which are far more complicated. The moral? – always compare new methods with simple baseline schemes.

Statistical Modeling (See Entry ▸ Bayesian Classification)
Another simple technique is to use all attributes and allow them to make contributions to the decision that are *equally important* and *independent* of one another, given the class. Although grossly unrealistic – what makes real-life datasets interesting is that the attributes are certainly not equally important or independent – it leads to a statistically-based scheme that works surprisingly well in practice. Employed in information retrieval as early as 1960 [5], the idea was rediscovered, dubbed "naïve Bayes," and introduced into machine learning 30 years later [4]. Despite the disparaging moniker it works well on many actual datasets. Over-reliance on the independence of attributes can be countered by applying attribute selection techniques.

Divide and Conquer Technique (See Entry ▸ Decision Tree Classification)
The process of constructing a decision tree can be expressed recursively. First, select an attribute

to use at the root, and make a branch for each possible value. This splits the instance set into subsets, one for every value of the attribute. Now repeat the process recursively for each branch, using only those instances that actually reach the branch. If all instances at a node have the same classification, stop developing that part of the tree. This method of "top-down induction of decision trees" was explored and popularized by Quinlan [8, 9]. The nub of the problem is to select an appropriate attribute at each stage. Of many heuristics that have been investigated, the dominant one is to measure the expected amount of information gained by knowing that attribute's actual value. Having generated the tree, it is selectively pruned back from the leaves to avoid over-fitting. A series of improvements include ways of dealing with numeric attributes, missing values, and noisy data; and generating rules from trees.

Covering Algorithms (See Entry ▶ Rule-Based Classification)

Classification rules can be produced by taking each class in turn and seeking a rule that covers all its instances, at the same time excluding instances not in the class. This bottom-up approach is called *covering* because at each stage a rule is identified that "covers" some of the instances. Although trees can always be converted into an equivalent rule set, and vice versa, the perspicuity of the representation often differs. Rules can be symmetric whereas trees must select one attribute to split on first, which can produce trees that are much larger than an equivalent set of rules. In the multiclass case a decision tree split takes account of all classes and maximizes the information gained, whereas many rule generation methods concentrate on one class at a time, disregarding what happens to the others.

Instance-Based Learning (See Entry ▶ Nearest Neighbor Classification)

Another approach is to store training instances verbatim and, given an unknown test instance, use a distance function to determine the closest training instance and predict its class for the test instance. Suitable distance functions are the Euclidean or Manhattan (city-block) metric; attributes should be normalized to lie between 0 and 1 to compensate for scaling effects. For nominal attributes that assume symbolic rather than numeric values, the distance between two values is 1 if they are not the same and 0 otherwise. In the k-nearest neighbor strategy, some fixed number of nearest neighbors – say five – are located and used together to determine the class of the test instance by majority vote. Another way of proofing the database against noise is to selectively and judiciously choose the exemplars that are added. Nearest-neighbor classification was notoriously slow until advanced data structures like kD-trees were applied in the early 1990s.

Linear Models (See Entry ▶ Linear Regression)

When the outcome and all attributes are numeric, linear regression can be used. This expresses the class as a linear combination of the attributes, with weights that are calculated from the training data. Linear regression has been popular in statistical applications for decades. If the data exhibits a nonlinear dependency, the best-fitting straight line will be found, where "best" is interpreted in the least-mean-squared-difference sense. Although this line may fit poorly, linear models can serve as building blocks for more complex learning schemes.

Linear Classification (See Entry ▶ Neural Networks)

The idea of linear classification is to find a hyperplane in instance space that separates two classes. (In the multi-class case, a binary decision can be learned for each pair of classes). If the linear sum exceeds zero the first class is predicted; otherwise the second is predicted. If the data is linearly separable – that is, it can be separated perfectly using a hyperplane – the perceptron learning rule espoused by Rosenblatt is guaranteed to find a separating hyperplane [10]. This rule adjusts the weight vector whenever the prediction for a particular instance is erroneous: if the first class is predicted the instance (expressed as a vector) is added to the weight vector (making it more likely that the result will be positive next time around); otherwise the instance is subtracted.

There have been many powerful extensions of this basic idea. Support vector machines use linear decisions to implement nonlinear class boundaries by transforming the input using a nonlinear mapping. Multilayer perceptrons connect many linear models in a hierarchical arrangement that can represent nonlinear decision boundaries, and use a technique called "back-propagation" to distribute the effect of errors through this hierarchy during training.

Missing Values

Most datasets encountered in practice contain missing values. Sometimes different kinds are distinguished (e.g., unknown vs. unrecorded vs. irrelevant values). They may occur for a variety of reasons. There may be some significance in the fact that a certain instance has an attribute value missing – perhaps a decision was taken not to perform some test – and that might convey information about the instance other than the mere absence of the value. If this is the case, *not tested* should be recorded as another possible value for this attribute. Only someone familiar with the data can make an informed judgment as to whether a particular value being missing has some significance or should simply be coded as an ordinary missing value. For example, researchers analyzing medical databases have noticed that cases may, in some circumstances, be diagnosable strictly from the tests that a doctor decides to make, regardless of the outcome of the tests. Then a record of which values are "missing" is all that is needed for a complete diagnosis – the actual measurements can be ignored entirely!

Meta-learning

Decisions can often be improved by combining the output of several different models. Over the past decade or so the techniques of *bagging*, *boosting*, and *stacking* have been developed that learn an ensemble of models and deploy them together. Their performance is often astonishingly good. Researchers have struggled to understand why, and during that struggle new methods have

emerged that are sometimes even better. For example, whereas human committees rarely benefit from noisy distractions, shaking up bagging by adding random variants of classifiers can improve performance. Boosting – perhaps the most powerful of the three methods – is related to the established statistical technique of additive models, and this realization has led to improved procedures.

Combined models share the disadvantage of being rather hard to analyze: they can comprise dozens or even hundreds of individual learners and it is not easy to understand in intuitive terms what factors are contributing to the improved decisions. In the last few years methods have been developed that combine the performance benefits of committees with comprehensible models. Some produce standard decision tree models; others introduce new variants of trees that provide optional paths.

Evaluation

For classification problems, performance is naturally measured in terms of the *error rate*. The classifier predicts the class of each test instance: if it is correct, that is counted as a success; if not, it is an error. The error rate is the proportion of errors made over a whole set of instances, and reflects the overall performance of the classifier. Performance on the training set is definitely *not* a good indicator of expected performance on an independent test set. A classifier is *overfitted* to a dataset if its structure reflects that particular set to an excessive degree. For example, the classifier might be generated by rote learning without any generalization whatsoever. An overfitted classifier usually exhibits performance on the training set which is excellent but far from representative of performance on other datasets from the same source.

In practice, one must predict performance bounds based on experiments with whatever data is available. Labeled data is required for both training and testing, and is often hard to obtain. A single data set can be partitioned for training and testing in various different ways. In a popular statistical technique called *cross-validation* the experimenter first decides on a fixed number of "folds," or partitions of the data – say three. The data is split into three approximately equal portions, and each in turn is used for testing while the remainder serves for training. The procedure is repeated three times so that in the end every instance has been used exactly once for testing. This is called *threefold cross-validation*. "Stratification" is the idea of ensuring that all classes are represented in all folds in approximately the right proportions. *Stratified tenfold cross-validation* has become a common standard for estimating the error rate of a classification learning scheme. Alternatives include *leave-one-out* cross-validation, which is effectively n-fold cross-validation where n is the size of the data set; and the *bootstrap*, which takes a carefully-judged number of random samples from the data with replacement and uses these for training, combining the error rate on the training data (an optimistic estimate) with that on the test data (a pessimistic estimate, since the classifier has only been trained on a subset of the full data) to get an overall estimate.

Key Applications

Classification learning is one of the flagship triumphs of research in artificial intelligence. It has been used for problems that range from selecting promising embryos to implant in a human womb during in vitro fertilization to the selection of which cows in a herd to sell off to an abattoir. Fielded applications are legion. They include decisions involving judgment, such as whether a credit company should make a loan to a particular person; screening images, such as the detection of oil slicks from satellite images; load forecasting, such as combining historical load information with current weather conditions and other events to predict hourly demand for electricity; diagnosis, such as fault finding and preventative maintenance of electromechanical devices; marketing and sales, such as detecting customers who are likely to switch to a competitor.

URL to Code

The Weka machine learning workbench is a popular tool for experimental investigation and comparison of classification learning techniques, as well as other machine learning methods. It is described in [11] and available for download from http://www.cs.waikato.ac.nz/ml/weka.

Cross-References

▶ Abstraction
▶ Association Rules
▶ Bagging
▶ Bayesian Classification
▶ Boosting
▶ Bootstrap
▶ Cataloging in Digital Libraries
▶ Classification by Association Rule Analysis
▶ Clustering Overview and Applications
▶ Cross-Validation
▶ Data Mining
▶ Decision Rule Mining in Rough Set Theory
▶ Decision Tree Classification
▶ Fuzzy Set Approach
▶ Genetic Algorithms
▶ Linear Regression
▶ Log-Linear Regression
▶ Nearest Neighbor Classification
▶ Neural Networks
▶ Receiver Operating Characteristic
▶ Rule-Based classification
▶ Support Vector Machine

Recommended Reading

1. Breiman L, Friedman JH, Olshen RA, Stone CJ. Classification and regression trees. Pacific Grove: Wadsworth; 1984.
2. Bush RR, Mosteller F. Stochastic models for learning. New York: Wiley; 1955.
3. Holte RC. Very simple classification rules perform well on most commonly used datasets. Mach Learn. 1993;11:63–91.
4. Kononebko I. ID3, sequential Bayes, naïve Bayes and Bayesian neural networks. In: Proceedings of the 4th European Working Session on Learning; 1989. p. 91–8.

5. Maron ME, Kuhns JL. On relevance, probabilistic indexing and information retrieval. J ACM. 1960;7(3):216–44.
6. Minsky ML, Papert S. Perceptrons. Cambridge: MIT Press; 1969.
7. Nilsson NJ. Learning machines. New York: McGraw-Hill; 1965.
8. Quinlan JR. Induction of decision trees. Mach Learn. 1986;1(1):81–106.
9. Quinlan JR. C4.5: programs for machine learning. San Francisco: Morgan Kaufmann; 1993.
10. Rosenblatt F. Principles of neurodynamics. Washington, DC: Spartan; 1961.
11. Witten IH, Frank E. Data mining: practical machine learning tools and techniques. 2nd ed. San Francisco: Morgan Kaufmann; 2003.

Classification by Association Rule Analysis

Bing Liu
University of Illinois at Chicago, Chicago, IL, USA

Synonyms

Associative classification

Definition

Given a training dataset D, build a classifier (or a classification model) from D using an association rule mining algorithm. The model can be used to classify future or test cases.

Historical Background

In the previous section, it was shown that a list of rules can be induced or mined from the data for classification. A decision tree may also be converted to a set of rules. It is thus only natural to expect that association rules [1] be used for classification as well. Yes, indeed! Since the first classification system (called CBA) that used association rules was reported in [10], many techniques and systems have been proposed by researchers [2–4, 6–8, 13, 15, 16]. CBA is based on class association rules (CAR), which are a special type of association rules with only a class label on the right-hand-side of each rule. Thus, syntactically or semantically there is no difference between a rule generated by a class association rule miner and a rule generated by a rule induction system (or a decision tree system for that matter). However, class association rule mining inherits the completeness property of association rule mining [1]. That is, all rules that satisfy the user-specified minimum support and minimum conference are generated. Other classification algorithms only generate a small subset of rules existing in data for classification [9, 10].

Most existing classification systems based on association rules (also called *associative classifiers*) employ CARs directly for classification, although their ways of using CARs can be quite different [3, 7, 8, 10, 15, 16]. To deal with unbalanced class distributions, the multiple minimum class supports approach is proposed in [9, 11], which gives each class a different minimum support based on its relative frequency in the data. In [2, 4, 6, 13], the authors also proposed to use rules as features or attributes to augment the original data or even to replace the original data. That is, in these techniques, CARs are not directly used for classification, but are used only to expand or to replace the original data. Any classification technique can be used subsequently to build the final classifier based on the expanded data, e.g., naïve Bayesian and SVM. Since the number of class association rules can be huge, closed rule sets have been proposed for classification in [3]. This approach helps solve the problem that in many data sets the complete sets of CARs cannot be generated due to combinatorial explosion. The closed rule set is a smaller, lossless and concise representation of all rules. Thus, long rules (rules with many conditions) may be used in classification, which otherwise may not be generated but can be crucial for accurate classification. Finally, normal association rules may be used for prediction or classification as well.

This section thus introduces the following three approaches to using association rules for classification:

1. Using class association rules for classification
2. Using class association rules as features or attributes
3. Using normal association rules for classification

The first two approaches can be applied to tabular data or transactional data. The last approach is usually employed for transactional data only. Transactional data sets are difficult to handle by traditional classification techniques, but are very natural for association rules. Below, the three approaches are described in turn. Note that various sequential rules can be used for classification in similar ways as well if sequential data sets are involved [6].

Foundations

Classification Using Class Association Rules

As mentioned above, a class association rule (CAR) is an association rule with only a class label on the right-hand side of the rule. Any association rule mining algorithm can be adapted for mining CARs. For example, the Apriori algorithm [1] for association rule mining was adapted to mine CARs in [10].

There is basically no difference between rules generated from a decision tree (or a rule induction system) and CARs if only categorical (or discrete) attributes (more on this later) are considered. The differences are in the mining processes and the final rule sets. CAR mining finds all rules in data that satisfy the user-specified minimum support (minsup) and minimum confidence (minconf) constraints. A decision tree or a rule induction system finds only a subset of the rules (expressed as a tree or a list of rules) for classification. In many cases, rules that are not in the decision tree (or the rule list) may be able to perform the classification more accurately. Empirical comparisons reported by several

researchers have shown that classification using CARs can perform more accurately on many data sets than decision trees and rule induction systems [7, 8, 10, 15, 16].

The complete set of rules from CAR mining is also beneficial from the rule usage point of view. In many applications, the user wants to act on some interesting rules. For example, in an application for finding causes of product problems in a manufacturing company, more rules are preferred to fewer rules because with more rules, the user is more likely to find rules that indicate causes of problems. Such rules may not be found by a decision tree or a rule induction system. A deployed data mining system based on CARs is reported in [12] for finding actionable knowledge from manufacturing and engineering data sets.

One should, however, also bear in mind of the following differences between CAR mining and decision tree construction (or rule induction):

1. Decision tree learning and rule induction do not use the minsup or minconf constraint. Thus, some rules that they find can have very low supports, which, of course, are likely to be pruned because the chance that they overfit the training data is high. Although a low minsup for CAR mining can be used, it may cause combinatorial explosion. In practice, in addition to minsup and minconf, a limit on the total number of rules to be generated may be used to further control the CAR generation process. When the number of generated rules reaches the limit, the algorithm stops. However, with this limit, long rules (with many conditions) may not be generated. Recall that the Apriori algorithm works in a level-wise fashion, i.e., short rules are generated before long rules. In some applications, this may not be an issue as short rules are often preferred and are sufficient for classification or for action. Long rules normally have very low supports and tend to overfit the data. However, in some other applications, long rules can be useful.
2. CAR mining does not use continuous (numeric) attributes, while decision trees deal with continuous attributes naturally. Rule induction can use continuous attributes as well.

There is still no satisfactory method to deal with such attributes directly in association rule mining. Fortunately, many attribute discretization algorithms exist that can automatically discretize the value range of a continuous attribute into suitable intervals (e.g., [5]), which are then considered as discrete values.

Mining Class Association Rules for Classification

There are many techniques that use CARs to build classifiers. Before describing them, it is useful to first discuss some issues related to CAR mining for classification.

Rule pruning: CAR rules are highly redundant, and many of them are not statistically significant (which can cause overfitting). Rule pruning is thus needed. The idea of pruning CARs is basically the same as tree pruning in decision tree building or rule pruning in rule induction. Thus, it will not be discussed further (see [8, 10] for some of the pruning methods).

Multiple minimum class supports: A single minsup may be inadequate for mining CARs because many practical classification data sets have uneven class distributions, i.e., some classes cover a large proportion of the data, while others cover only a very small proportion (which are called *rare* or *infrequent classes*).

For example, there is a data set with two classes, Y and N. 99% of the data belong to the Y class, and only 1% of the data belong to the N class. If the minsup is set to 1.5%, no rule for class N will be found. To solve the problem, the minsup needs to be lowered. Suppose the minsup is set to 0.2%. Then, a huge number of overfitting rules for class Y may be found because minsup = 0.2% is too low for class Y.

Multiple minimum class supports can be applied to deal with the problem. A different *minimum class support* $minsup_i$ for each class c_i can be assigned, i.e., all the rules of class c_i must satisfy $minsup_i$. Alternatively, one single total minsup can be provided, denoted by t_minsup, which is then distributed to each class according to the class distribution:

$$minsup_i = t\_minsup \times sup(c_i)$$

where $sup(c_i)$ is the support of class c_i in the training data. The formula gives frequent classes higher minsups and infrequent classes lower minsups. There is also a general algorithm for mining normal association rules using multiple minimum supports in [9, 11].

Parameter selection: The parameters used in CAR mining are the minimum supports and the minimum confidences. Note that a different minimum confidence may also be used for each class. However, minimum confidences do not affect the classification much because classifiers tend to use high confidence rules. One minimum confidence is sufficient as long as it is not set too high. To determine the best $minsup_i$ for each class c_i, a range of values can be tried to build classifiers and then use a validation set to select the final value. Cross-validation may be used as well.

Classifier Building

After all CAR rules are found, a classifier is built using the rules. There are many existing approaches, which can be grouped into three categories.

Use the strongest rule: This is perhaps the simplest strategy. It simply uses CARs directly for classification. For each test instance, it finds the strongest rule that covers the instance. A rule *covers* an instance if the instance satisfies the conditions of the rule. The class of the strongest rule is then assigned as the class of the test instance. The strength of a rule can be measured in various ways, e.g., based on confidence, χ^2 test, or a combination of both support and confidence values.

Select a subset of the rules to build a classifier: The representative method of this category is the one used in the CBA system [10]. The method is similar to the sequential covering method, but applied to class association rules with additional enhancements as discussed above.

Let the set of all discovered CARs be S. Let the training data set be D. The basic idea is to select a subset L ($\subseteq S$) of high confidence rules to cover D. The set of selected rules, including a default class, is then used as the classifier. The selection of rules is based on a total order defined on the rules in S.

C

Classification by Association Rule Analysis, Fig. 1 A simple classifier building algorithm

Algorithm CBA(S, D)
1   S = sort(S);   // sorting is done according to the precedence>
2   $RuleList = \varnothing$;   // the rule list classifier
3   **for** each rule $r \in S$ in sequence **do**
4   **if** $D \neq \varnothing$ AND r classifies at least one example in D correctly **then**
5   delete from D all training examples covered by r;
6   add r at the end of $RuleList$
7   **endif**
8   **endfor**
9   add the majority class as the default class at the end of $RuleList$

Definition: *Given two rules, r_i and r_j, $r_i \succ r_j$ (called r_i precedes r_j, or r_i has a higher precedence than r_j) if*

1. *The confidence of r_i is greater than that of r_j, or*
2. *Their confidences are the same, but the support of r_i is greater than that of r_j, or*
3. *Both the confidences and supports of r_i and r_j are the same, but r_i is generated earlier than r_j.*

A CBA classifier L is of the form:

$$L = <r_1, r_2, \ldots, r_k, \text{default} - \text{class}>$$

where $r_i \in S$, $r_a \succ r_b$ if $b > a$. In classifying a test case, the first rule that satisfies the case classifies it. If no rule applies to the case, it takes the default class (*default-class*). A simplified version of the algorithm for building such a classifier is given in Fig. 1. The classifier is the *RuleList*.

This algorithm can be easily implemented by making one pass through the training data for every rule. However, this is extremely inefficient for large data sets. An efficient algorithm that makes at most two passes over the data is given in [10].

Combine multiple rules: Like the first approach, this approach does not have an additional step to build a classifier. At the classification time, for each test instance, the system first finds the subset of rules that covers the instance. If all the rules in the subset have the same class, the class is assigned to the test instance. If the rules have different classes, the system divides the rules into groups according to their classes, i.e., all rules of the same class are in the same group. The system

then compares the aggregated effects of the rule groups and finds the strongest group. The class label of the strongest group is assigned to the test instance. To measure the strength of a rule group, there again can be many possible techniques. For example, the CMAR system uses a weighted χ^2 measure [8].

Class Association Rules as Features

In the above two approaches, rules are directly used for classification. In this approach, rules are used as features to augment the original data or simply form a new data set, which is then fed to a traditional classification algorithm, e.g., decision trees or the naïve Bayesian algorithm.

To use CARs as features, only the conditional part of each rule is needed, and it is often treated as a Boolean feature/attribute. If a data instance in the original data contains the conditional part, the value of the feature/attribute is set to 1, and 0 otherwise. Several applications of this method have been reported [2, 4, 6, 13]. The reason that this approach is helpful is that CARs capture multi-attribute or multi-item correlations with class labels. Many classification algorithms do not find such correlations (e.g., the naïve Bayesian method), but they can be quite useful.

Classification Using Normal Association Rules

Not only can class association rules be used for classification, but also normal association rules. For example, association rules are commonly used in e-commerce Web sites for product recommendations, which work as follows: When a customer purchases some products, the system recommends him/her some other related products based on what he/she has already purchased.

Recommendation is essentially a prediction problem. It predicts what a customer is likely to buy. Association rules are naturally applicable to such applications. The classification process is as follows:

1. The system first uses previous purchase transactions (the same as market basket transactions) to mine association rules. In this case, there are no fixed classes. Any item can appear on the left-hand side or the right-hand side of a rule. For recommendation purposes, usually only one item appears on the right-hand side of a rule.
2. At the prediction (e.g., recommendation) time, given a transaction (e.g., a set of items already purchased by a customer), all the rules that cover the transaction are selected. The strongest rule is chosen and the item on the right-hand side of the rule (i.e., the consequent) is the predicted item and is recommended to the user. If multiple rules are very strong, multiple items can be recommended.

This method is basically the same as the "use the strongest rule" method described earlier. Again, the rule strength can be measured in various ways, e.g., confidence, χ^2 test, or a combination of both support and confidence. Clearly, the other two classification methods discussed earlier can be applied here as well.

The key advantage of using association rules for recommendation is that they can predict any item since any item can be the class item on the right-hand side. Traditional classification algorithms only work with a single fixed class attribute, and are not easily applicable to recommendations.

Finally, it should be noted that multiple minimum supports in rule mining [11] can be of significant help. Otherwise, *rare items* will never be recommended, which is called the *coverage* problem [14]. It is shown in [14] that using multiple minimum supports can dramatically increase the coverage.

Key Applications

The applications of associative classifiers are very wide. Three main scenarios are briefly described below.

1. Since classification using class association rules is a supervised learning technique, it can be (and has been) used as a classification algorithm just like any other classification algorithm from machine learning, e.g., decision trees, naïve Bayesian classifiers, SVM, and rule induction. In many cases, an associative classifier performs better than these classic machine learning techniques.
2. Apart from classification, individual class association rules themselves are very useful in practice due to the completeness property. In many practical applications (especially diagnostic data mining applications), the user wants to find interesting rules that are actionable. As discussed earlier, traditional classification algorithms (e.g., rule induction or any other technique) are not suitable for such applications because they only find a small subset of rules that exist in data. Many interesting or actionable rules are not discovered. A deployed data mining system, called Opportunity Map, for Motorola Corporation was based on class association rules [12]. When this entry was written, the system had been in use in Motorola for more than 2 years and further improvements were still being made. Although the system was originally designed for finding rules that indicate causes of phone call failures, it had been used in a variety of other applications in Motorola.
3. Using normal association rules for classification or prediction is also very common, especially for the transaction type of data. For such kind of data, as described above, traditional classification techniques are not easily applicable because they can only predict some fixed class items (or labels).

Cross-References

▶ Association Rule Mining on Streams
▶ Decision Trees

Recommended Reading

1. Agrawal R, Srikant R. Fast algorithms for mining association rules. In: Proceedings of the 20th International Conference on Very Large Data Bases; 1994. p. 487–99.
2. Antonie ML, Zaiane O. Text document categorization by term association. In: Proceedings of the 2nd IEEE International Conference on Data Mining; 2002. p. 19–26.
3. Baralis E, Chiusano S. Essential classification rule sets. ACM Trans Database Syst. 2004;29(4): 635–74.
4. Cheng H, Yan X, Han J, Hsu C-W. Discriminative frequent pattern analysis for effective classification. In: Proceedings of the 23rd International Conference on Data Engineering; 2007. p. 706–15.
5. Dougherty J, Kohavi R, Sahami M. Supervised and unsupervised discretization of continuous features. In: Proceedings of the 12th International Conference on Machine Learning; 1995. p. 194–202.
6. Jindal N, Liu B. Identifying comparative sentences in text documents. In: Proceedings of the 32nd Annual International ACM SIGIR Conference on Research and Development in Information Retrieval; 2006. p. 244–51.
7. Li J, Dong G, Ramamohanarao K. Making use of the most expressive jumping emerging patterns for classification. In: Advances in Knowledge Discovery and Data Mining, 4th Pacific-Asia Conference; 2000. p. 220–32.
8. Li W, Han J, Pei J. CMAR: accurate and efficient classification based on multiple class-association rules. In: Proceedings of the 2001 IEEE International Conference on Data Mining; 2001. p. 369–76.
9. Liu B. Web data mining: exploring hyperlinks, contents and usage data. Berlin: Springer; 2007.
10. Liu B, Hsu W, Ma Y. Integrating classification and association rule mining. In: Proceedings of the 4th International Conference on Knowledge Discovery and Data Mining; 1998. p. 80–6.
11. Liu B, Hsu W, Ma Y. Mining association rules with multiple minimum supports. In: Proceedings of the 5th ACM SIGKDD International Conference on Knowledge Discovery and Data Mining; 1999. p. 337–41.
12. Liu B, Zhao K, Benkler J, Xiao W. Rule interestingness analysis using OLAP operations. In: Proceedings of the 12th ACM SIGKDD International Conference on Knowledge Discovery and Data Mining; 2006. p. 297–306.
13. Meretakis D, Wüthrich B. Extending naïve Bayes classifiers using long itemsets. In: Proceedings of the 5th ACM SIGKDD International Conference on Knowledge Discovery and Data Mining; 1999. p. 165–74.
14. Mobasher B, Dai H, Luo T, Nakagawa N. Effective personalization based on association rule discovery from web usage data. In: Proceedings of the 3rd ACM Workshop on Web Information and Data Management; 2001. p. 9–15.
15. Wang K, Zhou S, He Y. Growing decision trees on support-less association rules. In: Proceedings of the 6th ACM SIGKDD International Conference on Knowledge Discovery and Data Mining; 2000. p. 265–9.
16. Yin X, Han J. CPAR: classification based on predictive association rules. In: Proceedings of the 2003 SIAM International Conference on Data Mining; 2003.

Classification in Streams

Charu C. Aggarwal
IBM T. J. Watson Research Center, Yorktown Heights, NY, USA

Synonyms

Knowledge discovery in streams; Learning in streams

Definition

The classification problem is a well defined problem in the data mining domain, in which a training data set is supplied, which contains several feature attributes, and a special attribute known as the class attribute. The class attribute is specified in the training data, which is used to model the relationship between the feature attributes and the class attribute. This model is used in order to predict the unknown class label value for the test instance.

A data stream is defined as a large volume of continuously incoming data. The classification problem has traditionally been defined on a static training or test data set, but in the stream scenario, either the training or test data may be in the form of a stream.

Historical Background

The problem of classification has been studied so widely in the classification literature, that a single source for the problem cannot be identified. Most likely, the problem was frequently encountered in practical commercial scenarios as a statistical problem, long before the field of machine learning was defined. With advances in hardware technology, data streams became more common, and most data mining problems such as clustering and association rule mining were applied to the data stream domain. Domingos and Hulten [2] were the first to model the problem in the context of data streams.

Foundations

There are numerous techniques available for classification in the classical literature [3]. However, most of these techniques cannot be used directly for the stream scenario. This is because the stream scenario creates a number of special constraints which are as follows:

- The data stream typically contains a large volume of continuously incoming data. Therefore the techniques for training or testing need to be very efficient. Furthermore, a data point may be examined only once over the course of the entire computation. This imposes hard constraints on the nature of the algorithms which may be used for stream classification. This constraint is generally true of almost all data mining algorithms.
- Often the patterns in the underlying data may evolve continuously over time. As a result, the model may soon become stale for data mining purposes. It is therefore important to keep the models current even when the patterns in the underlying data may change. This issue is known as concept drift.

- Many stream classification methods have considerable memory requirements in order to improve computational efficiency. The stream case is particularly resource constrained, since the memory may sometimes be limited, while the computational efficiency requirements continue to be very high.
- In many cases, the rate of incoming data cannot be controlled easily. Therefore, the classification process needs to be nimble enough in order to provide effective tradeoffs between accuracy and efficiency.

Most of the known classification methods can be made to work in the data stream scenario with a few modifications. These modifications are generally designed to deal with either the one-pass constraint, or the stream evolution scenario. The different types of classifiers which can be modified for the data stream scenario are as follows:

- Nearest Neighbor Classifiers: In these techniques, the class label of the nearest neighbor to the target record is used in order to perform the classification. Since the nearest neighbor cannot be defined easily over the entire stream, a stream sample is used in order to perform the classification. This stream sample can be dynamically maintained with the one-pass constraint with the use of a technique called *reservoir sampling*. In order to deal with issues of stream evolution, one can used *biased* reservoir sampling. In biased sampling, a time decay function is used in order to maintain a sample which is biased towards more recent data points.
- Decision Tree Classifiers: In this techniques, decision trees need to be built in one pass of the stream. A method known as Very Fast Decision Trees (VFDT) was proposed in [2] which uses probabilistic split methods in order to create decision trees with predictable accuracy. In particular, the Hoeffding inequality is used in order to ensure that the generated tree produces the same tree as a conventional learner. Several other techniques were proposed by the same authors subsequently, which deal with the time-changing aspect of the data streams.

- Cluster-based Classifiers: An on-demand stream classification model was proposed which uses clustering techniques in order to build the optimal model for a classifier on demand. In this technique, a micro-clustering technique is used in order to compress the underlying data into clusters. The data belonging to different classes are compressed into different clusters. For a given test example, the class of the closest cluster is used in order to predict the class label. One key aspect of this classifier is that it assumes that both the training and the test data are in the form of a stream. The technique calculates the optimal horizon for using the cluster statistics.
- Ensemble Classifiers: In this case, a combination of different models is used in order to deal with the issue of concept drift. This is because different kinds of models work better with different kinds of data patterns. Therefore, an optimal model is picked depending upon the current data pattern. The idea is that different classifiers are more effective for different kinds of data patterns. Therefore, by making an optimal choice of the classifier from the ensemble, it is possible to improve the classification accuracy significantly.
- Bayes Classifiers: The naive Bayes classifier computes the Bayes a-posteriori probabilities of a test instance belonging to a particular class using the inter-attribute independence assumption. The key in adapting such classifiers is to be able to effectively maintain the statistics used to compute conditional probabilities in one pass. In the case of an evolving data stream, the statistics need to be maintained over particular user-specific horizons.

A number of other methods for stream classification also exist which cannot be discussed within the scope of this entry. A detailed survey on classification methods may be found in [3].

Key Applications

Stream classification finds application to numerous data domains such as network intrusion detection, target marketing and credit card fraud detection. In many of these cases, the incoming data clearly has very large volume. For example, typical intrusion scenarios have a large volume of incoming data. Similarly, in the case of target-marketing, super-store transactions may have very large volumes of incoming data.

Many of the traditional classification applications are still used in the batch mode, since the stream technology is still in its infancy, and it is sometimes simpler to collect a sample of the data set and run a batch process on it. Most of the traditional problems for the classification domain will eventually be transformed to the data stream scenario. This is because more and more data domains are being converted to the stream scenario with advances in hardware technology.

Cross-References

▶ Association Rule Mining on Streams
▶ Clustering on Streams
▶ Data Stream

Recommended Reading

1. Aggarwal CC, editor. Data streams: models and algorithms. Berlin/Heidelberg/New York: Springer; 2007.
2. Domingos P, Hulten G. Mining high speed data streams. In: Proceedings of the 6th ACM SIGKDD International Conference on Knowledge Discovery and Data Mining; 2000. p. 71–80.
3. James M. Classification algorithms. New York: Wiley; 1985.

Client-Server Architecture

M. Tamer Özsu
Cheriton School of Computer Science,
University of Waterloo, Waterloo, ON, Canada

Definition

Client-server DBMS (database management system) refers to an architectural paradigm that separates database functionality between client machines and servers.

Historical Background

The original idea, which is to offload the database management functions to a special server, dates back to the early 1970s [1]. At the time, the computer on which the database system was run was called the *database machine*, *database computer*, or *backend computer*, while the computer that ran the applications was called the *host computer*. More recent terms for these are the *database server* and *application server*, respectively.

The client-server architecture, as it appears today, has become a popular architecture around the beginning of 1990s [2]. Prior to that, the distribution of database functionality assumed that there was no functional difference between the client machines and servers (i.e., an earlier form of today's peer-to-peer architecture). Client-server architectures are believed to be easier to manage than peer-to-peer systems, which has increased their popularity.

Foundations

Client-server DBMS architecture involves a number of database client machines accessing one or more database server machines. The general idea is very simple and elegant: distinguish the functionality that needs to be provided and divide these functions into two classes: server functions and client functions. This provides a *two-level architecture* that makes it easier to manage the complexity of modern DBMSs and the complexity of distribution.

In client-server DBMSs, the database management functionality is shared between the clients and the server(s) (Fig. 1). The server is responsible for the bulk of the data management tasks as it handles the storage, query optimization, and transaction management (locking and recovery). The client, in addition to the application and the user interface, has a *DBMS client* module that is responsible for managing the data that are

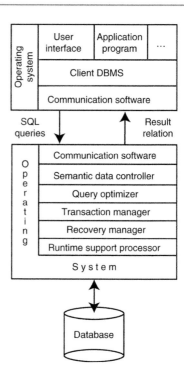

Client-Server Architecture, Fig. 1 Client-server reference architecture

cached to the client, and (sometimes) managing the transaction locks that may have been cached as well. It is also possible to place consistency checking of user queries at the client side, but this is not common since it requires the replication of the system catalog at the client machines. The communication between the clients and the server(s) is at the level of SQL statements: the clients pass SQL queries to the server without trying to understand or optimize them; the server executes these queries and returns the result relation to the client. The communication between clients and servers are typically over a computer network.

In the model discussed above, there is only one server which is accessed by multiple clients. This is referred to as *multiple client-single server* architecture [3]. There are a number of advantages of this model. As indicated above, they are simple; the simplicity is primarily due to the fact that data management responsibility is delegated to one server. Therefore, from a data management perspective, this architecture is similar to centralized databases although there are some (important) differences from centralized systems in the way transactions are executed and caches are managed. A second advantage is that they provide predictable performance. This is due to the movement of non-database functions to the clients, allowing the server to focus entirely on data management. This, however, is also the cause of the major disadvantage of client-server systems. Since the data management functionality is centralized at one server, the server becomes a bottleneck and these systems cannot scale very well.

The disadvantage of the simple client-server systems are partially alleviated by a more sophisticated architecture where there are multiple servers in the system (the so-called *multiple client-multiple server* approach). In this case, two alternative management strategies are possible: either each client manages its own connection to the appropriate server or each client knows of only its "home server", which then communicates with other servers as required. The former approach simplifies server code, but loads the client machines with additional responsibilities, leading to what has been called "heavy client" systems. The latter approach, on the other hand, concentrates the data management functionality at the servers. Thus, the transparency of data access is provided at the server interface, leading to "light clients."

The integration of workstations in a distributed environment enables an extension of the client-server architecture and provides for a more efficient function distribution. Application programs run on workstations, called *application servers*, while database functions are handled by dedicated computers, called *database servers*. The *clients* run the user interface. This leads to the present trend in three-tier distributed system architecture, where sites are organized as specialized servers rather than as general-purpose computers (Fig. 2).

The application server approach (indeed, a n-tier distributed approach) can be extended by the introduction of multiple database servers and multiple application servers, as can be done in client-server architectures. In this case, it is common for each application server to be dedicated to one or a few applications, while database servers

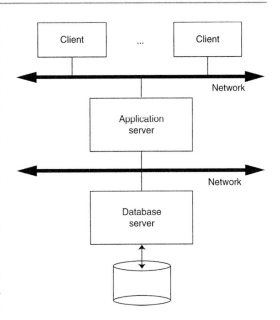

Client-Server Architecture, Fig. 2 Database server approach

operate in the multiple server fashion discussed above.

Key Applications

Many of the current database applications employ either a two-layer client-server architecture of the three-layer application-server approach.

Cross-References

▶ Data Stream Management Architectures and Prototypes

Recommended Reading

1. Canaday RH, Harrisson RD, Ivie EL, Rydery JL, Wehr LA. A back-end computer for data base management. Commun ACM. 1974;17(10):575–82.
2. Orfali R, Harkey D, Edwards J. Essential client/server survival guide. New York: Wiley; 1994.
3. Özsu MT, Valduriez P. Principles of distributed database systems. 2nd ed. Englewood Cliffs: Prentice-Hall; 1999.

Clinical Data Acquisition, Storage, and Management

Chimezie Ogbuji
Cleveland Clinic Foundation, Cleveland, OH, USA

Synonyms

Case report forms; Clinical data management systems; Electronic data capture

Definition

The management of clinical data for supporting patient care and for supporting retrospective clinical research requires a means to acquire the clinical data and a repository that stores the data and provides the functions necessary for managing them over their lifetime. Typically, patient data are collected "at the point of care" (i.e., onsite where health care is being provided) and entered into a patient record system [2]. Data entry is typically the first line of precaution for maintaining a certain amount of quality on the data collected. Subsequently, a representative from an externally sponsoring organization or authorized personnel from within the health care institution then extracts a select set of medical record data into a Clinical Data Management System (*CDMS*). The entries in such systems are often referred to as secondary patient records since they are derived from a primary patient record and are used by personnel who are not involved in direct patient care [2]. These systems are also often referred to as *patient registries*.

The study committee of the Institute of Medicine (IOM) defined [2] a computer-based patient record (*CPR*) as an electronic patient record that resides in a system specifically designed to support users by providing accessibility to complete and accurate data, alerts, reminders, clinical decision support systems, links to medical knowledge, and other aids.

Such systems are also often referred to as Electronic Health Records (*EHR*s). The committee also defined a *primary* patient record as one used by health care professionals while providing patient care services to review patient data or document their own observations, actions, or instructions [2]. Finally, the committee emphasized the distinction between clinical data and the systems that capture and process this data by defining a patient record *system* as the set of components that form the mechanism by which patient records are created, used, stored, and retrieved.

A CDMS is the repository for the management of the data used for clinical studies or trials. One of the core services provided by a CDMS is to facilitate the identification (and correction) of errors due to human entry as well as errors that existed in the source from which the data was gathered. Data completeness implies that CDMS will accommodate an expected range and complexity for the data in the system [2]. In addition, the CDMS can employ the use and enforcement of one or more vocabulary standards.

Finally, a CDMS will also provide services for querying the data as well as generate reports from the data. The generated reports and results of such queries are typically transmitted to a centralized authority or normalized for use by statistical analysis tools.

Historical Background

Virtually every person in the United States who has received health care in the United States since 1918 has a patient record [2]. Most of these records consist of structured paper forms with sections that consist solely of narrative text. However, conventional patient records can also appear in other forms such as scanned media, microfilm, optical disk, etc.

They are created and used most frequently in health care provider settings. However, their use also extends to other facilities such as correctional institutions, the armed forces, occupational health programs, and universities [2].

The process of recording patient care information has primarily consisted of entry into a paper patient record. For the purpose of a clinical trial or study, data are manually transcribed into a paper Case Report Form (*CRF*). CRFs are typically in the form of a questionnaire formulated to collect information specific to a particular clinical trial. The ICH Guidelines for Good Clinical Practice define [4] the CRF as:

A printed, optical, or electronic document designed to record all of the protocol required information to be reported to the sponsor on each trial subject.

CRFs are then collected by a representative from the sponsoring organization and manually entered into a CDMS. This secondary transcription is often called *double data entry*. In some cases, Optical Character Recognition (*OCR*) is used to semi-automate the transcription from a CRF into a CDMS [1].

Traditionally, clinical data management systems consist of infrastructure built on top of relational database management systems. Depending on the nature of the requirements for the creation of analysis data sets for biostatisticians, accuracy of the data, and speed of data entry, a wide spectrum of database management or spreadsheet systems are used as the underlying medium of storage for the CDMS.

Good database design and proper application of relational model theory for normalizing the data is typically used to ensure data accuracy. Traditional relational query languages such as SQL are used to identify and extract relevant variables for subsequent analysis or reporting purposes.

Foundations

Electronic Data Capture

There is a slow, but steady move by pharmaceutical companies towards the adoption of an electronic means of capturing patient record information directly from the source and at the point of care into an electronic system that submits the data relevant to the trial to the sponsor or to other consumers of electronic patient record data. This new shift of emphasis from paper to a direct electronic system is often referred to (in the health care industry) as Electronic Data Capture (*EDC*) [3].

Infrastructure and Standards for Data Exchange

Once patient record data are collected and stored in an electronic information system, the increasing need to transfer the machine-readable data to external systems emphasizes the importance of standardized formats for communication between these disparate systems [2]. Efforts to standardize a common format for communication between health care systems and other external consumers of health care data have settled on the adoption of Extensible Markup Language (*XML*) as the primary data format for the transmission of Health Level 7 (*HL7*) messages.

HL7 is an organization with a mission to develop standards that improve the delivery of care, optimize the management of workflow, reduce ambiguity in clinical terminology and facilitate efficient transfer of knowledge between the major stakeholders.

Document Models and Management Systems

As its name implies, XML is a markup language for describing structured data (or *documents*) in a manner that is both human- and machine-readable. It can be extended to support users who wish to define their own vocabularies. It is meant to be highly reusable across different information systems for a variety of purposes. It is recommended by the World Wide Web Consortium (*W3C*) and is a free and open standard.

XML is at the core of an entire suite of technologies produced by the W3C that includes languages for querying XML documents as well as describing their structure for the purpose of validating their content. This suite of technologies is meant to serve as infrastructure for a contemporary set of information systems each known more broadly as a Document Management System (*DMS*).

Common Components of Information Systems

Like most information systems, document management systems are comprised of a particular data model (XML in this case), one or more query languages, and a formal processing model for systems that wish to support queries written in language against the underlying data. Document management systems (and information systems in general) typically also offer security services that ensure limited access to the data. This is particularly important for clinical data management systems.

With respect to the kind of services they provide for the systems that are built on top of them (such as clinical data management systems), document management systems are very much like relational database management systems. However, whereas relational database management systems have an underlying relational model that is tabular and rigid, document management systems have a data model that is hierarchical with data elements that can be extended to support new terminology over the life of the data.

Text-Oriented Information Systems

Most modern computer-based patient record systems mainly adopt information systems with hierarchical, relational, or text-oriented data models. Text-based information systems typically store their content primarily as narrative text and often employ natural language processing for extracting structured data for transcription into a clinical data management system. Querying such systems usually involves keyword-based searches that use text indexes that are used to associate words with the sections of narrative in which they can be found.

Key Applications

Electronic data capture and clinical data management systems constitute the majority of the infrastructure necessary in the overall process of clinical research from the point of interaction with primary patient records all the way to the analysis of the curated clinical research data. The sections below describe the major areas where their application makes a significant difference.

Electronic Data Collection Options

There are a variety of ways in which data can be acquired electronically for transcription into a clinical data management system. The most desired means is one where the data are directly retrieved electronically from an existing source such as the primary patient record. This method is often referred to as *single entry* [1]. It requires that the primary patient record adopt or align with a set of consistent format standards such that they can facilitate the support of primary care as well as reuse for the purpose of (unanticipated) clinical research. Unfortunately the lack of adoption of computer-based patient records remains a primary impediment to this more direct means of acquiring clinical data [2, 3].

Alternatively, clinical data can be transcribed from a primary patient record into a secondary patient record using some form of an electronic user interface on a particular device. Typically, such user interfaces are either web browser-based (i.e., they are built on top of an existing web browser such as Internet Explorer or Firefox) or they are written as independent applications that are launched separately. The latter approach is often referred to as a *thick-client system* [6].

Patient Registries

The set of functions associated with a secondary computer-based patient record system is often adopted from the underlying information system. Modern document and relational database management systems are equipped with a wide spectrum of capabilities each of which is directly relevant to the needs of users of these systems. This includes: content organization, archival, creation of documents, security, query services, disaster recovery, and support for web-based user interfaces.

Clinical Workflow Management

Equally important to the clinical data is the management of the pattern of activity, responsibilities, and resources associated with a particular clinical study or trial. These patterns are often referred to as *workflow*. Orchestrating the overall process can also have a significant impact on the success of a clinical study. Clinical data management systems sometimes have off-the-shelf capabilities for managing workflow. These usually support some amount of automation of the workflow process. Document management systems that include capabilities for building customized application are well-suited for supporting workflows that are either specific to a particular study protocol or capable of supporting multiple (or arbitrary) protocols.

Quality Management, Report Generation, and Analysis

Finally, document management and relational database management systems include capabilities for monitoring error in the data collected. This is often supported through the application of a set of common constraints that are relevant to the research protocol. Typically, these systems have an automated mechanism for indicating when the underlying data does not adhere to the constraints specified.

In addition, document management and relational database management systems include services for generating reports and extracting variables for statistical analysis.

Future Directions

Modern information management systems are adopting standards for representation formats that push the envelope of machine-readability. In particular, the W3C has recently been developing a suite of technologies that build on the standards associated with the World Wide Web and introduce a formal model for capturing knowledge in a manner that emphasizes the meaning of terms rather than their structure. Such approaches to modeling information are often referred to as knowledge representation or conceptual models. This particular collection of standards is commonly referred to as *semantic web technologies* [5].

Semantic web technologies are built on a graph-based data model known as the Resource Description Framework (*RDF*) as well as a language for describing conceptual models for RDF data known as Ontology Web Language (*OWL*). RDF leverages a highly-distributable addressing and naming mechanism known as Uniform Resource Identifiers (*URI*s) that is the foundation of the current web infrastructure.

Semantic web technologies also include a formal mechanism for rendering or transforming XML document dialects into RDF known as Gleaning Resources Descriptions from Dialects of Languages (*GRDDL*). Finally a common query language has been defined for accessing data expressed in RDF known as *SPARQL*.

The Institute of Medicine has indicated [2] that the flexibility of computer-based patient records is primarily due to their adoption of a data dictionary that can be expanded to accommodate new elements. In addition, the IOM has identified [2] the following as crucial to the evolution of content and standard formats in computer-based patient record systems:

- The content of CPRs must be defined and contain a uniform core set of data elements.
- Data elements must be named consistently via the enforcement of some form of vocabulary control.
- Format standards for data exchange must be developed and used.

In addition, the IOM's study committee identified the ability for CPRs to be linked with other clinical records as a critical attribute of a comprehensive computer-based patient record. The combination of these observations is a strong indication that in the future, clinical data management systems will be built on information management systems that adopt semantic web technologies in order to better meet the growing needs of the management of clinical research data.

Finally, a new generation of technologies for building declarative web applications will lower the technological barrier associated with the kind of user interfaces necessary for the adoption of electronic data capture methods at health care institutions. In particular, an XML-based technology known as *XForms* is well positioned to have a significant impact on the front end of the clinical data pipeline (data collection).

XForm applications are web form-based, independent of the device on which they are deployed and use XML as the data model of the underlying content. This approach has strong correspondence with the current direction of clinical data exchange standards with the adoption of XML as the format for communication between health care systems.

In the near future, lightweight devices (such as Tablet PCs) will connect to remote, distributed computer-based patient record systems over a secure web-based network protocol. Electronic data capture will be implemented by XForm applications that run in a browser and compose XML documents that represent sections of a computer-based patient record. These documents will adhere to a standard document format for the exchange of medical records such as the HL7 Clinical Document Architecture (*CDA*). The HL7 CDA is an XML-based document markup standard that specifies the structure and semantics of clinical documents for the purpose of exchange.

These documents will be securely transmitted directly into a primary computer-based patient record which employs XML as its core data model and uses GRDDL to also store an RDF representation of the document that conforms to a formal, standard ontology (expressed in OWL) that describes the meaning of the terms. This ontology provides a certain degree of logical consistency that facilitates ad hoc analysis through the use of logical inference.

Patients that meet the criteria for a particular research protocol will be identified by a SPARQL query that is dispatched against the patient record system, which uses terminology easily understood by the investigators themselves (rather than an intermediary database administrator). These patient records will then be transmitted directly into a clinical data management system (or patient registry) that will include the facilities for managing the workflow associated with the relevant research protocol. These facilities will be implemented as web applications built on the same underlying information management systems as those used by the primary computer-based patient records.

Cross-References

▶ Clinical Data Quality and Validation
▶ Electronic Health Record

Recommended Reading

1. Anisfeld MH, Prokscha S. Practical guide to clinical data management. Boca Raton: CRC Press; 1999.
2. Committee on Improving the Patient Record, Institute of Medicine. The computer-based patient record: an essential technology for health care. Revised edition. Washington, DC: National Academies Press; 1997.
3. Lori A, Nesbitt. Clinical research: What it is and how it works. Sudbury: Jones and Bartlett Publishers; 2003.
4. Rondel RK, Varley SA, Webb CF. Clinical data management. Chichester: Wiley; 2000.
5. Ruttenberg A, Clark T, Bug W, Samwald M, Bodenreider O, Chen H, Doherty D, Forsberg K, Gao Y, Kashyap V, Kinoshita J, Luciano J, Marshall MS, Ogbuji C, Rees J, Stephens S, Wong GT, Elizabeth W, Zaccagnini D, Hongsermeier T, Neumann E, Herman I, Cheung KH. Advancing translational research with the Semantic Web. BMC Bioinf. 2007;8(Suppl 3):S2.
6. Wilson D, Pace MD, Elizabeth W, Staton MSTC. Electronic data collection options for practice-based research networks. Ann Fam Med. 2005;3(Suppl 1):S2–4.

Clinical Data and Information Models

Chintan Patel and Chunhua Weng
Columbia University, New York, NY, USA

Definition

A formal representation of the clinical data using entities, types, relationships and attributes. The abstraction of clinical data into an information model enables reusability and extensibility of the database to satisfy different application needs and accommodate changes in the underlying data.

Key Points

The clinical domain is a data rich environment with multitude of different data entities ranging from several thousands of laboratory tests, procedures or medications that change often with new ones getting added almost every day. Furthermore these data are generated from different information systems or devices (often from different vendors) in the hospital. Integrating such wide variety of data streams into a common information model is a challenging task.

Most healthcare databases use generic information models [3, 4] such as event-component models with an Entity-Attribute-Value [5] (EAV) schema to represent the data (see Fig. 1). The advantage of using a generic information model is to accommodate the data heterogeneity and extensibility. Generally, an external terminology or vocabulary is used in conjunction with a generic information model to represent the clinical domain (laboratory tests, medications and so on) and the healthcare activities, for example, LOINC is a standard vocabulary for representing laboratory data or SNOMED CT for healthcare activities.

Various information models have been proposed towards standardizing the representation of clinical data. The goal of standardizing the information model is to facilitate exchange, sharing and reuse of clinical data by different systems locally as well as nationally. Following are some current standardized models:

HL7 Reference Information Model: The HL7 standards organization [2] has developed a Reference Information Model (RIM) to share consistent meaning of healthcare data beyond local context. The RIM specifies a set of abstract bases classes Entity, Role, Participation and Act, which contain specific classes/attributes such as Person, Organization, Patient, Provider, Intent, Observation and so on. This model is used to create concrete concepts by combining the RIM types, for example, *elevated blood pressure* would be represented in RIM as class = Observation with code = Finding of increased blood pressure (SNOMED#241842005), mood = Event, interpretation code = abnormal (HL7#A), target site = heart (LOINC#LP7289). Note that standardized terminology codes (SNOMED CT and LOINC) are used to represent specific findings and modifiers. An implementation of HL7 RIM based model over a relational database schema is described here [1].

openEHR Reference Model: The openEHR specification [6] (developed largely by the institutions in EU and Australia) provides information models for the electronic health record (EHR), demographics, data structures, integration and so on. The openEHR EHR model represents various facets of EHR such as clinician/patient interaction, audit-trailing, technology/data format independence and supporting secondary uses. The openEHR project uses the notion of archetypes that enable domain experts to formally model a domain concept (or an aggregation of concepts), corresponding constraints and other compositions, for example, an archetype on *blood pressure measurement* consists of systolic, diastolic measurements and units with other clinically relevant information such as history.

Recommended Reading

1. Eggebraaten TJ, Tenner JW, Dubbels JC. A health-care data model based on the HL7 reference information model. IBM Syst J. 2007;46(1):5–18.
2. HL7 Reference Information Model. 2008.http://www.hl7.org/. Accessed 18 Apr 2008.
3. Huff S, Rocha R, Bray B, Warner H, Haug P. An event model of medical information representation. J Am Med Inform Assoc. 1995;2(2):116–34.
4. Johnson S. Generic data modeling for clinical repositories. J Am Med Inform Assoc. 1996;3(5):328–67.
5. Nadkarni P, Marenco L, Chen R, Skoufos E, Shepherd G, Miller P. Organization of heterogeneous scientific data using the EAV/CR representation. J Am Med Inform Assoc. 1999;6(6):478–571.
6. OpenEHR. Reference information model. 2008.http://www.openehr.org/. Accessed 18 Apr 2008.

Clinical Data Quality and Validation

Chintan Patel and Chunhua Weng
Columbia University, New York, NY, USA

Definition

Clinical data quality is defined as the accuracy and completeness of the clinical data for the purposes of clinical care, health services and

**Clinical Data
and Information Models,
Fig. 1** The event
component information
model

Event (Chest X-Ray*) -> **Attribute1** = **Value1** (on-date* = mm/dd/yyyy)
-> **Attribute2** = **Value2** (finding-site* = breast*)
-> **Attribute3** = **Value3** (has-morphology* = neoplasm*)
...
*coded using an external terminology

other secondary uses such as decision support and clinical research. The quality of clinical data can be achieved by the standardization, inspection and evaluation of the data generating processes and tools [2].

Key Points

The term data quality can potentially have different meanings or interpretation based on the domain or the application using the data [1]. Even within the context of clinical databases, there exists a multitude of different data types (administrative data, procedure data, laboratory data and so on) that may be used for several different applications such as clinical report generation, billing or research. The major components of clinical data quality can be broadly characterized as follows:

Accuracy

Clinical data are often generated by automated systems (such as lab equipment) or manually entered by clinicians (notes). These data generating processes are prone to errors that result in incorrect data being stored in the database. The severity of errors can vary significantly, for example, a minor misspelling in patient history note versus a prescription error in drug dosage order can lead to drastically different outcomes in terms of patient care. The accuracy of clinical data is defined as the proportion of correct data (truly representing the actual patient condition or measurement) in the clinical database. The accuracy of clinical data depends on the enforcement of well-defined data entry standards and protocols.

Completeness

It is defined as the availability of data elements in a clinical database that are necessary to accomplish a given task, for example, a clinical trial recruitment application with detailed eligibility criteria would require information from the clinical notes in addition to coded problem list data. The completeness of a patient record is critical for a clinician to choose a most appropriate treatment plan for the patient. The availability of complete patient information is critical during an emergency condition. In the case of unavailability of data elements, some applications tend to substitute data sources, which can lead to suboptimal results. Consider for example, a clinical decision support application reusing coarse ICD (International Classification of Disease) classification to generate decisions.

Reliability

The notion of "repeatability" – to determine whether the clinical data generation processes produce consistent data at different times or settings. Hospitals are a chaotic environment with multiple care providers taking care of a single patient. It becomes critical to develop data entry protocols to ensure consistent representation of patient information in the clinical database. Often to eliminate the variations across different users the data entry software systems such as the EMR (electronic medical record) contain various checks to ensure the correctness and completeness of the data elements [3]. The coding of clinical data using terminologies such as ICD has to be done in a consistent fashion to facilitate applications that require data integration or comparative analysis.

Maintaining quality in clinical databases is a continuous process requiring strong commitment from different stakeholders involved. The amount of electronic biomedical data generated is growing at an exponential rate. Developing high quality clinical databases can have significant implications for the applications reusing the data.

Cross-References

▶ Quality and Trust of Information Content and Credentialing

Recommended Reading

1. Arts D, De Keizer N, Scheffer G. Defining and improving data quality in medical registries: a literature review, case study, and generic framework. J Am Med Inform Assoc. 2002;9(6):600–11.
2. Black N. High-quality clinical databases: breaking down barriers. Lancet. 2006;353(9160):1205–11.
3. Hogan W, Wagner M. Accuracy of data in computer-based patient records. J Am Med Inform Assoc. 1997;4(5):342–97.

Clinical Decision Support

Adam Wright
Partners HealthCare, Boston, MA, USA

Synonyms

CDS; Decision support

Definition

Clinical Decision Support systems are computer systems which assist humans in making optimal clinical decisions. While clinical decision support systems are most often designed for clinicians, they can also be developed to assist patients or caregivers. Common examples of clinical decision support systems include drug-drug interaction checks, dose range checking for medication and preventive care reminders.

Historical Background

The first clinical decision support system was described in 1959 by Robert Ledley and Lee Lusted [6] in their paper "Reasoning foundations of medical diagnosis; symbolic logic, probability, and value theory aid our understanding of how physicians reason." Ledley and Lusted described an analog computer used to sort cards containing a diagnosis and a series of punches which represented symptoms. By selecting the cards which matched the symptoms present in a given case a clinician could develop a possible set of diagnosis.

In 1961, Homer Warner [15] described a clinical decision support system for diagnosing congenital heart defects. The system was developed around a contingency table that mapped clinical symptoms to forms of congenital heart disease. A physician would input the patient's symptoms and findings from the clinical exam and other studies into the system, which would then proceed to suggest the most probable diagnoses based on the contingency table.

In the 1970s, Edward Shortliffe developed the well-known MYCIN system for antibiotic therapy. MYCIN was an expert system with a large knowledge base of clinical rules [12]. Users of MYCIN would input known facts about their patient, and MYCIN would apply them to the rule base using backward chaining to yield a probable causative agent for infections as well as suggestions for antibiotic therapy.

While the systems described so far all focused on a specific area of medicine, the INTERNIST-I system, developed by Randy Miller, Harry Pople and Jack Myers [8] targeted the broad domain of diagnosis in internal medicine. The INTERNIST-I knowledge base consisted of a large set of mappings between symptoms and diagnoses. These links were scored along three axes: evoking strength (the likelihood that a patient has a diagnosis given a particular symptom), frequency (how often a symptom is present given a particular diagnosis) and import (how critical it is that a particular diagnosis be considered given that it is possible or probable based on a set of symptoms). Octo Barnett's DXplain system for diagnostic decision support was developed around the same time as INTERNIST-I.

The earliest decision support systems were standalone, but the second wave in clinical decision support, beginning in the 1970s, was the integration of decision support systems into broader clinical information systems. The first two examples of this integration were the Health Evaluation through Logical Processing (HELP) system at the University of Utah and LDS Hospital, and the Regenstrief Medical Records System (RMRS) developed at the Regenstrief Institute in Indianapolis. The HELP system, which was used for many facets of patient care, had support for the development of a variety of kinds of decision support, and was especially well known for its Bayesian reasoning modules. The RMRS was developed, from the ground up, with a large knowledge base of clinical care rules. Both HELP and RMRS are in active use today.

Most current commercially available clinical information systems have some support for clinical decision support, and efforts to standardize representation and enable the sharing of decision support content are ongoing.

Foundations

Development of clinical decision support systems entails a variety of issues. The first step in developing any clinical information system is to identify an important clinical target, and then consider interventions. The most critical database systems related issues are knowledge representation, storage and standards.

Issues of Knowledge Representation

Once a desired clinical decision support target has been identified and relevant medical knowledge has been collected, the knowledge must somehow be represented. Knowledge in clinical decision support systems has been represented in a variety of ways, the most common being if-then rules, expert systems, probabilistic and Bayesian systems and reference content.

Perhaps the simplest form of knowledge is if-then rules. Much of clinical decision support con-

tent can be represented this way (for example "if the acetaminophen dose is 10 g per day, alert the user that this is too high" or "if the patient is over 50 years of age and has not had a sigmoidoscopy, recommend one"). These rules are frequently designed to be chained together, although generally in a fixed and predetermined pattern.

More complex than simple if-then rules are expert systems. These systems are composed of large knowledge bases which contain many intermediate states and assertions. Like if-then rules, these rules are composed of an antecedent, a consequent and an implication. However, expert systems are generally designed to elicit emergent behavior from extensive chaining including, in many cases, goal-directed backward chaining.

Probabilistic and Bayesian systems share much in common with if-then rules. However, instead of modeling knowledge and clinical states as deterministic values, they use probabilities. By combining these probabilities with knowledge provided by the user, these systems can estimate the likelihood of various diagnostic possibilities, or the relative utility of different therapeutic modalities. It is important to note that many expert systems employ probabilistic or Bayesian reasoning.

A simpler form of knowledge representation is reference knowledge designed to be read by a human. This form of decision support provides information to the user but expects him or her to formulate a plan of action on his or her own. In many cases knowledge, such as clinical guidelines, can be equivalently modeled as rules or as reference content. Reference content is simpler to construct, but it sometimes can not be as proactive as rule-based content.

Storage of Clinical Knowledge in Database Systems

A key challenge for developers of database systems for clinical decision support is selecting the optimal strategy for storing clinical knowledge in a database. This selection has many tradeoffs among performance, space, maintainability and human readability.

Rule based decision support content is often stored as compiled or interpreted code and, when properly integrated into clinical information support systems, can be very efficient. However, many systems instead choose to store rules in some intermediate form, often indexed according to their trigger (a clinical event, such as a new prescription, which causes decision support rules to fire). A chained hash table with these triggers as its keys and decision support rules to invoke as values can be a particularly efficient representation.

In cases of particularly high transaction volume, where performance is important and the number of rules to evaluate is large, more sophisticated storage and processing mechanisms can be used. One of the most effective in terms of performance (although not necessarily in terms of space) is Charles Fogarty's Rete algorithm. The Rete algorithm is an efficient network-based method for pattern matching in rule-based systems.

Because it is not rule based, reference knowledge requires a different set of storage and retrieval strategies, based largely on the principles of information retrieval. In general, these strategies employ one or some combination of full-text search and metadata queries.

Standards for Sharing Clinical Decision Support Content Between Database Systems

In addition to the aforementioned issues of internal representation of clinical knowledge, there are also issues relating to the sharing of clinical decision support content between systems. Several standards for sharing such content have been proposed, beginning with Arden Syntax, a standard for event-driven rule-based decision support content. Other standards, such as Guideline Interchange Format (GLIF) and the related expression language GELLO exist to represent more complex forms of clinical knowledge. While construction of standards for representing clinical knowledge may seem straightforward, issues relating to terminology and a reference model for patient information have proven formidable.

An alternate approach to strict structured knowledge representation formalisms for sharing clinical decision support content is the use

of services. Several recent efforts, including SEBASTIAN and SANDS have defined a set of interfaces and, in the case of SANDS, patient data models to help overcome prior difficulties in sharing decision support content.

Key Applications

Applications of clinical decision support can be categorized along a variety of axes, including intervention type (alert, reminder, reference information, etc.), clinical purpose (diagnosis, therapy, prevention), disease target (diabetes, hypertension, cancer, etc.) and user (physician, nurse, patient, etc.).

Several clinical decision support systems have been described in the historical background section. Additional significant systems include:

- Morris Collen's system for "Automated Multiphasic Screening And Diagnosis."
- Howard Bleich's system for diagnosis and treatment of acid-base disorders.
- A system for the diagnosis and management of abdominal complaints developed by F.T. de Dombal.
- The ATTENDING system developed by Perry Miller and designed to critique and suggest improvements to anesthesia plans.
- A system for ventilator management by Dean Sittig.
- A blood product ordering critiquing system by Reed Gardner.
- An antibiotic advising system by Scott Evans.

Experimental Results

There is a long experimental tradition in the field of clinical decision support, and many systems have shown strong results, even for the earliest systems. Warner's system for congenital heart defects was compared favorably to experienced cardiologists, MYCIN proposed clinically appropriate antibiotic therapy 75% of the time (and got better as more rules were added) and

INTERNIST performed about as well as average doctors at diagnosis.

Just as significant is the effect that such systems have on physician practice. In a landmark paper, Clem McDonald described the results of an experimental trial performed within the RMRS system. In the trial, half of the physician users of RMRS received patient care suggestions based on the knowledge base of rules, while half did not. Physicians who received the suggestions carried them out 51% of the time, while physicians who did not receive suggestions performed the actions that would have been suggested only 22% of the time. When the reminder system was turned off, physician performance returned almost immediately to baseline.

There have been several significant systematic reviews of clinical decision support systems. A 2005 review by Amit Garg [2] found that decision support systems were associated with improved provider performance in 64% of the controlled trials reviewed. Another systematic review by Ken Kawamoto found that decision support systems improved performance in 68% of trials, and that systems designed to the highest criteria improved performance in 94% of trials.

Cross-References

► Clinical Data and Information Models

Recommended Reading

1. Bates DW, Kuperman GJ, Wang S, et al. Ten commandments for effective clinical decision support: making the practice of evidence-based medicine a reality. J Am Med Inform Assoc. 2003;10(6):523–30.
2. Garg AX, Adhikari NK, McDonald H, et al. Effects of computerized clinical decision support systems on practitioner performance and patient outcomes: a systematic review. JAMA. 2005;293(10):1223–38.
3. Kawamoto K, Houlihan CA, Balas EA, Lobach DF. Improving clinical practice using clinical decision support systems: a systematic review of trials to identify features critical to success. BMJ. 2005;330(7494):765.
4. Kawamoto K , Lobach DF. Design, implementation, use, and preliminary evaluation of SEBASTIAN, a standards-based web service for clinical decision support. In: Proceedings of the AMIA Symposium; 2005. p. 380–4.
5. Kuperman GJ, Gardner RM, Pryor TA. HELP: a dynamic hospital information system. New York: Springer; 1991.
6. Ledley RS, Lusted LB. Reasoning foundations of medical diagnosis; symbolic logic, probability, and value theory aid our understanding of how physicians reason. Science. 1959;130(3366):9–21.
7. McDonald CJ. Protocol-based computer reminders, the quality of care and the non-perfectability of man. N Engl J Med. 1976;295(24):1351–5.
8. Miller RA, Pople HE, Myers JD. Internist-1, an experimental computer-based diagnostic consultant for general internal medicine. N Engl J Med. 1982;307(8):468–76.
9. Osheroff JA, Pifer EA, Sittig DF, Jenders RA, Teich JM. Improving outcomes with clinical decision support: an implementers' guide. Chicago: HIMSS; 2005.
10. Osheroff JA, Teich JM, Middleton B, Steen EB, Wright A, Detmer DE. A roadmap for national action on clinical decision support. J Am Med Inform Assoc. 2007;14(2):141–5.
11. Sittig DF, Wright A, Osheroff JA, et al. Grand challenges in clinical decision support. J Biomed Inform. 2007;41(2):387–92.
12. Shortliffe EH, Davis R, Axline SG, Buchanan BG, Green CC, Cohen SN. Computer-based consultations in clinical therapeutics: explanation and rule acquisition capabilities of the MYCIN system. Comput Biomed Res. 1975;8(4):303–20.
13. Wright A, Goldberg H, Hongsermeier T, Middleton B. A description and functional taxonomy of rule-based decision support content at a large integrated delivery network. J Am Med Inform Assoc. 2007;14(4):489–96.
14. Wright A, Sittig DF. SANDS: a service-oriented architecture for clinical decision support in a National Health Information Network. J Biomed Inform. 2008;41(6). https://doi.org/10.1016/j.jbi.2008.03.001.
15. Warner HR, Toronto AF, Veasey LG, Stephenson R. A mathematical approach to medical diagnosis. Application to congenital heart disease. JAMA. 1961;177(3):177–83.

Clinical Document Architecture

Amnon Shabo (Shvo)
University of Haifa, Haifa, Israel

Synonyms

CDA; CDA R1; CDA R2

Definition

The Clinical Document Architecture (CDA) is a document markup standard that specifies the structure and semantics of clinical documents for the purpose of exchange and share of patient data. The standard is developed by Health Level Seven (HL7) – a Standards Development Organization [2] focused on the area of healthcare. At the time of writing this entry, two releases of CDA were approved: CDA R1 was approved in 2000 and CDA R2 in 2005. Both releases are part of the HL7 new generation of standards (V3), all derived from a core reference information model (RIM) that assures semantic consistency across the various standards such as laboratory, medications, care provision and so forth. The RIM is based on common data types and vocabularies, and together these components constitute the HL7 V3 Foundation that is an inherent part of the CDA standard specification.

Key Points

Clinical documents such as discharge summaries, operative notes and referral letters are ubiquitous in healthcare and currently exist mostly in paper. The computerized clinical document is similar in purpose to its paper counterpart and the clinician's narratives are a key component of both versions. Narratives are compositions based on the natural language of the writer, while computerized structuring of a document is limited to some computer language. The design of the CDA standard strives to bridge the gap between these "languages" especially when it comes to the mixture of structured and unstructured data intertwined to describe the same phenomena, while addressing two important goals: human readability and machine-processability. The drive to structure medical narratives is also challenging the thin line between art and craftsmanship in the medical practice [3].

The basic structure of a CDA document consists of a header and a body. The header represents an extensive set of metadata about the document such as time stamps, the type of docu-

ment, encounter details and of course the identification of the patient and those who participated in the documented encounter or service. While the header is a structured part of the document and is similar in the two releases of CDA, the body consists of clinical data organized in sections and only in CDA R2 it enables the formal representation of structured data along with narratives [1]. Data is structured in clinical statements based on entries such as observations, medication administrations, or adverse events where several entries are associated into a compound clinical statement. Nevertheless, only the narrative parts of the CDA body are mandatory, which makes CDA easy to adopt if structured data is not yet available. It is even possible to simply wrap a non-XML document with the CDA header or create a document with a structured header and sections containing only narrative content. The purpose of this design is to encourage widespread adoption, while providing an information infrastructure to incrementally move toward structured documents, serving the goal of semantic interoperability between disparate health information systems.

Beside text, CDA can also accommodate images, sounds, and other multimedia content. It can be transferred within a message and can be understood independently, outside the relaying message and its sending and receiving systems. CDA documents are encoded in Extensible Markup Language (XML), and they derive their machine processable meaning from the RIM, coupled with specific vocabularies.

A CDA document is a collection of information that is intended to be legally authenticated and has to be maintained by an organization entrusted with its care (stewardship). Inherent in the HL7 CDA standard are mechanisms for dealing with the authentication and versioning of documents so that it can be used in medical records enterprise repositories as well as in cross-institutional sharing of personal health information to facilitate continuity of care.

Cross-References

▶ Electronic Health Record

Recommended Reading

1. Dolin RH, Alschuler L, Boyer S, Beebe C, Behlen FM, Biron PV, Shabo A. HL7 Clinical Document Architecture, Release 2. J Am Med Inform Assoc. 2006;13(1):30–9.
2. Health Level Seven (HL7). http://www.hl7.org
3. Shabo A. Synopsis of the patient records section: structuring the medical narrative in patient records – a further step towards a multi-accessible EHR. The IMIA 2004 yearbook of medical informatics: towards clinical bioinformatics. 2004.

Clinical Event

Dan Russler
Oracle Health Sciences, Redwood Shores, CA, USA
Georgia Tech Research Institute, Atlanta, Georgia, USA

Definition

Vernacular Definition

1. In event planning circles, a "clinical event" is an event, e.g., meeting or party, attended by clinicians as opposed to administrative or financial personnel.

Technical Definitions

1. A state transition, normally a "create" or "update" state transition, targeting a record in an electronic medical record system or one of the systems associated with an electronic medical record system.
2. A report generated within a clinical trial that is subsequently evaluated for the presence of an adverse event by a clinical trial Clinical Event Committee.

Words often confused by use of the term "Clinical Event" include: Clinical Event (multiple definitions); Adverse Event; Clinical Act; Patient Event, Information Event.

The primary technical definition of "clinical event" includes the kind of "events" that are monitored by a "clinical event monitor" [1, 2, 3] used in synchronous or asynchronous decision support functions. Examples of these events include clinical orders, electronic medical record entries, admission, transfer and discharge notifications, lab results, and patient safety reports. These events trigger state transitions in an electronic medical record system or related system.

Typically, once the clinical event monitoring system, such as an HL7 Arden Syntax-based system, discovers a state transition, in the electronic medical record system, a decision support rule is applied to the clinical event and related data in order to determine whether a notification of a person or another system is required.

Key Points

"Event" or "Action" analysis traces its roots to the work of Aristotle on propositions. Propositions usually follow the form of Subject-Predicate and, upon analysis, may be found to be "true" or "false." The classic example of a proposition is "Socrates is a man." "Socrates" is the "Subject" and "is a man" is the "Predicate." An analogous proposition in healthcare is "Peter has a potassium level of 5.5 mg/dl." Clinical Events are propositions in healthcare that may be evaluated themselves by clinicians as "true" or "false" or may be applied in rules that evaluate to true or false.

For example, the creation of a record asserting that "Peter has a potassium level of 5.5 mg/dl" might trigger a clinical event monitoring system to implement the rule: "If potassium level record created, then evaluate if ("record value" >5.0); if "true," then notify Dr. X."

"Event-driven programming" as opposed to "procedural programming" utilizes the same kinds of predicate logic in evaluating state transitions or triggers to state transitions in a modern computer-programming environment. Consequently, Clinical Events drive programming logic in many modern systems.

The HL7 Reference Information Model (RIM) describes clinical events; the term "Act" in the RIM identifies objects that are instantiated in XML communications between systems or in records within the electronic healthcare systems themselves. These "Acts" correspond to "clinical events" used for monitoring systems in healthcare. However, in the RIM, "Event" is defined narrowly as an instance of an Act that has been completed or is in the process of being completed. Clinical event monitoring systems may also evaluate HL7 "Orders or Requests" or other kinds of "Act" instances as events of interest (www.hl7.org).

Cross-References

► Clinical Observation
► Clinical Order
► Event Driven Architecture
► Interface Engines in Healthcare

Recommended Reading

1. Glaser J, et al. Impact of information events on medical care. In: Proceedings of the 1996 HIMSS Annual Conference; 1996.
2. Hripisak G et al. Design of a clinical event monitor. Comput Biomed Res. 1996;29(3):194–221.
3. McDonald C. Action-oriented decisions in ambulatory medicine. Chicago: Yearbook Medical Publishers; 1981.

Clinical Knowledge Repository

Roberto A. Rocha
Partners eCare, Partners HealthCare System,
Wellesley, MA, USA

Synonyms

Clinical content database; Clinical content registry; Clinical content repository; Clinical knowledge base; Clinical knowledge directory; Clinical knowledge management repository

Definition

A clinical knowledge repository (CKR) is a multipurpose storehouse for clinical knowledge assets. "Clinical knowledge asset" is a generic term that describes any type of human or machine-readable electronic content used for computerized clinical decision support. A CKR is normally implemented as an enterprise resource that centralizes a large quantity and wide variety of clinical knowledge assets. A CKR provides integrated support to all asset lifecycle phases such as authoring, review, activation, revision, and eventual inactivation. A CKR routinely provides services to search, retrieve, transform, merge, upload, and download clinical knowledge assets. From a content curation perspective, a CKR has to ensure proper asset provenance, integrity, and versioning, along with effective access and utilization constraints compatible with collaborative development and deployment activities. A CKR can be considered a specialized content management system, designed specifically to support clinical information systems. Within the context of clinical decision support systems, a CKR can be considered a special kind of knowledge base designed to manage multiple types of human and machine-readable clinical knowledge assets.

Key Points

In recent years, multiple initiatives have attempted to better organize, filter, and apply the ever-growing biomedical knowledge. Among these initiatives, one of the most promising is the utilization of computerized clinical decision support systems. Computerized clinical decision support can be defined as computer systems that provide the correct amount of relevant knowledge at the appropriate time and context, contributing to improved clinical care and outcomes. A wide variety of knowledge-driven tools and methods have resulted in multiple modalities of clinical decision support, including information selection and retrieval, information aggregation and presentation, data entry assistance, event monitors, care workflow assistance, and descriptive or predictive modeling. A CKR provides an integrated storage platform that enables the creation and maintenance of multiple types of knowledge assets. A CKR ensures that different modalities of decision support can be combined to properly support the activities of clinical workers. Core requirements guiding the implementation of a CKR include clinical knowledge asset provenance (metadata), versioning, and integrity. Other essential requirements include the proper representation of access and utilization constraints, taking into account the collaborative nature of asset development processes and deployment environments. Another fundamental requirement is to aptly represent multiple types of knowledge assets, where each type might require specialized storage and handling. The CKR core requirements are similar to those specified for other types of repositories used for storage and management of machine-readable assets.

Historical Background

Biomedical knowledge has always been in constant expansion, but unprecedented growth is being observed during the last decade. Over 32% of the 22.3 million citations accumulated by MEDLINE until January of 2014 were created in the last 10 years, with an average of over 708,400 new citations per year [1]. The number of articles published each year is commonly used as an indicator of how much new knowledge the scientific community is creating. However, from a clinical perspective, particularly for those involved with direct patient care, the vast amount of new knowledge represents an ever-growing gap between what is known and what is routinely practiced. Multiple initiatives in recent years have attempted to better organize, filter, and apply the knowledge being generated. Among these various initiatives, one of the most promising is the utilization of computerized clinical decision support systems [2]. In fact, some authors avow that clinical care currently mandates a degree

of individualization that is inconceivable without computerized decision support [3].

Computerized clinical decision support can be defined as computer systems that provide the correct amount of relevant knowledge at the appropriate time and context, ultimately contributing to improved clinical care and outcomes [4]. Computerized clinical decision support has been an active area of informatics research and development for the last three decades [5]. A wide variety of knowledge-driven tools and methods have resulted in multiple modalities of clinical decision support, including information selection and retrieval (e.g., infobuttons, crawlers), information aggregation and presentation (e.g., summaries, reports, dashboards), data entry assistance (e.g., forcing functions, calculations, evidence-based templates for ordering and documentation), event monitors (e.g., alerts, reminders, alarms), care workflow assistance (e.g., protocols, care pathways, practice guidelines), and descriptive or predictive modeling (e.g., diagnosis, prognosis, treatment planning, treatment outcomes). Each modality requires specific types of knowledge assets, ranging from production rules to mathematical formulas, and from automated workflows to machine learning models. A CKR provides an integrated storage platform that enables the creation and maintenance of multiple types of assets using knowledge management best practices [6].

The systematic application of knowledge management processes and best practices to the biomedical domain is a relatively recent endeavor [5]. Consequently, a CKR should be seen as an evolving concept that is progressively being recognized as a fundamental component for the acquisition, storage, and maintenance of clinical knowledge assets. Most clinical decision support systems currently in use still rely on traditional knowledge bases that handle a single type of knowledge asset and do not provide direct support for a complete asset lifecycle. Another important principle is the recognition that different modalities of decision support have to be combined and subsequently integrated with information systems to properly support the activities of clinical workers. The premise of integrating multiple modalities of clinical decision support reinforces the need for knowledge management processes supported by a CKR.

Foundations

Core requirements guiding the implementation of a CKR include clinical knowledge asset provenance (metadata), versioning, and integrity. Requirements associated with proper access and utilization constraints are also essential, particularly considering the collaborative nature of most asset development processes and deployment environments. Another fundamental requirement is to aptly represent multiple types of knowledge assets, where each type might require specialized storage and handling. The CKR core requirements are generally similar to those specified for other types of repositories used for storage and management of diverse machine-readable assets.

Requirements associated with asset provenance can be implemented using a rich set of metadata properties that describe the origin, purpose, evolution, and status of each clinical knowledge asset. The metadata properties should reflect the information that needs to be captured during each phase of the knowledge asset lifecycle process, taking into account multiple iterative authoring and review cycles, followed by a possibly long period of clinical use that might require multiple periodic revisions (updates). Despite the diversity of asset types, each with a potentially distinct lifecycle process, a portion of the metadata properties should be consistently implemented, enabling basic searching and retrieval services across asset types. Ideally, the shared metadata should be based on metadata standards (e.g., "Dublin Core Metadata Element Set" (http://dublincore.org/documents/dces/)). The adoption of standard metadata properties also simplifies the integration of external collections of clinical knowledge assets in a CKR. In addition to a shared set of properties, a CKR should also accommodate extended sets of properties specific for each clinical knowledge asset type and its respective lifecycle process. Discrete namespaces are commonly used to represent type-specific extended metadata properties.

Asset version and status, along with detailed change tracking, are vital requirements for a CKR. Different versioning strategies can be used, but as a general rule there should be only one clinically active version of any given knowledge asset. This general rule is easily observed if the type and purpose of the clinical knowledge asset remains the same throughout its lifecycle. However, a competing goal is created with the very desirable evolution of human-readable assets to become machine-readable. Such evolution invariably requires the creation of new knowledge assets of different types and potentially narrower purposes. In order to support this "natural" evolution, a CKR should implement the concept of asset generations, while preserving the change history that links one generation to the next. Also within a clinical setting, it is not uncommon to have to ensure that knowledge assets comply with, or directly implement, different norms and regulations. As a result, the change history of a clinical knowledge asset should identify the standardization and compliance aspects considered, enabling subsequent auditing and/or eventual certification.

Ensuring the integrity of clinical knowledge assets is yet another vital requirement for a CKR. Proper integrity guarantees that each asset is unique within a specific type and purpose, and that all its required properties are accurately defined. Integrity requirements also take into account the definition and preservation of dependencies between clinical knowledge assets. These dependencies can be manifested as simple hyperlinks, or as integral content defined as another independent asset. Creating clinical knowledge assets from separate components or modules (i.e., modularity) is a very desirable feature in a CKR - one that ultimately contributes to the overall maintainability of the various asset collections. However, modularity introduces important integrity challenges, particularly when a new knowledge asset is being activated for clinical use. Activation for clinical use requires a close examination of all separate components, sometimes triggering unplanned revisions of components already in routine use. Another important integrity requirement is the ability to validate the structure and the content of a clinical knowledge asset against predefined templates (schemas) and dictionaries (ontologies). Asset content validation is essential for optimal integration with clinical information systems. Ideally, within a given healthcare organization all clinical information systems and the CKR should utilize the same reference ontologies.

Contextual characteristics of the care delivery process establish the requirements associated with proper access, utilization, and presentation of the clinical knowledge assets. The care delivery context is a multidimensional constraint that includes characteristics of the patient (e.g., gender, age group, language, clinical condition), the clinical worker (e.g., discipline, specialty, role), the clinical setting (e.g., inpatient, outpatient, ICU, Emergency Department), and the information system being used (e.g., order entry, documentation, monitoring), among others. The care delivery context normally applies to the entire clinical knowledge asset, directly influencing search, retrieval, and presentation services. The care delivery context can also be used to constrain specific portions of a knowledge asset, including links to other embedded assets, making them accessible only if the constraints are satisfied. An important integrity challenge created by the systematic use of the care delivery context is the need for reconciling conflicts caused by incompatible asset constraints, particularly when different teams maintain the assets being combined. In this scenario, competing requirements are frequently present, namely the intention to maximize modularity and reusability versus the need to maximize clinical specificity and ease or use.

The accurate selection, retrieval, and presentation of unstructured assets is generally perceived as a simple but very useful modality of clinical decision support, particularly if the information presented to the clinical worker is concise and appropriate to the care being delivered. However, the appropriateness of the information is largely defined by the constraints imposed by the aforementioned care delivery context. Moreover, the

extent of indexing ("retrievability") of most collections of unstructured clinical knowledge assets is not sufficient to fully recognize detailed care delivery context expressions. Ultimately, the care delivery context provides an extensible mechanism for defining the appropriateness of a given clinical knowledge asset in response to a wide variety of CKR service requests.

The requirements just described are totally or partially implemented as part of general-purpose (enterprise) content management systems. However, content management systems have been traditionally constructed for managing primarily human-readable electronic content. Human-readable content, more properly characterized as unstructured knowledge assets, include narrative text, diagrams, and multimedia objects. When combined, these unstructured assets likely represent the largest portion of the inventory of clinical knowledge assets of any healthcare institution. As a result, in recent years different healthcare organizations have deployed CKRs using enterprise content management systems, despite their inability to manage machine-readable content.

Key Applications

Computerized Clinical Decision Support, Clinical Knowledge Engineering, Clinical Information Systems.

Cross-References

▶ Biomedical Data/Content Acquisition, Curation
▶ Clinical Data Acquisition, Storage, and Management
▶ Clinical Decision Support
▶ Dublin Core
▶ Evidence-Based Medicine
▶ Executable Knowledge
▶ Metadata
▶ Reference Knowledge

Recommended Reading

1. Statistical Reports on MEDLINE®/PubMed® Baseline Data, National Library of Medicine, Department of Health and Human Services [Online]. Available at: http://www.nlm.nih.gov/bsd/licensee/baselinestats.html. Accessed 29 Jun 2014.
2. Wyatt JC. Decision support systems. J R Soc Med. 2000;93(12):629–33.
3. Bates DW, Gawande AA. Improving safety with information technology. N Engl J Med. 2003;348(25):2526–34.
4. Berner E. Clinical decision support systems: state of the art. Rockville: Agency for Healthcare Research and Quality; 2009. http://healthit.ahrq.gov/sites/default/files/docs/page/09-0069-EF_1.pdf. Accessed 29 Jun 2014.
5. Greenes RA, editor. clinical decision support: the road to broad adoption. 2nd ed. Burlington: Academic; 2014.
6. Rocha RA, Maviglia SM, Sordo M, Rocha BH. Clinical knowledge management program. In: Greenes RA, editor. Clinical decision support – the road to broad adoption. 2nd ed. Burlington: Academic; 2014. p. 773–817.

Clinical Observation

Dan Russler
Oracle Health Sciences, Redwood Shores, CA, USA
Georgia Tech Research Institute, Atlanta, Georgia, USA

Synonyms

Clinical judgment; Clinical result; Clinical test; Finding of observation

Definition

1. The act of measuring, questioning, evaluating, or otherwise observing a patient or a specimen from a patient in healthcare; the act of making a clinical judgment.
2. The result, answer, judgment, or knowledge gained from the act of observing a patient or a specimen from a patient in healthcare.

These two definitions of "observation" have caused confusion in clinical communications, especially when applying the term to the rigor of standardized terminologies. When developing a list of observations, the terminologists have differed on whether the list of terms should refer to the "act of observing" or the "result of the observation."

Logical Observation Identifiers Names and Codes (LOINC) (www.loinc.org) focus on observation as the "act of observing." Systematized Nomenclature of Medicine (SNOMED) (www.ihtsdo.org) asserts that "General finding of observation of patient" is a synonym for "General observation of patient." Of note is the analysis in HL7 that identifies many shared attributes between descriptions of the act of observing and the result obtained. As a consequence, in HL7 Reference Information Model (RIM), both the act of observing and the result of the observation are contained in the same Observation Class (www.hl7.org).

Key Points

The topic of clinical observation has been central to the study of medicine since medicine began. Early physicians focused on the use of all five senses in order to make judgments about the current condition of the patient, i.e., diagnosis, or to make judgments about the future of patients, i.e., prognosis. Physical exam included sight, touch, listening, and smell. Physicians diagnosed diabetes by tasting the urine for sweetness.

As more tests on bodily fluids and tissues were discovered and used, the opportunity for better diagnosis and prognosis increased. Philosophy of science through the centuries often included the study of clinical observation in addition to the study of other observations in nature.

During the last century, the study of rigorous testing techniques that improve the reproducibility and interpretation of results has included the development of extensive nomenclatures for naming the acts of observation and observation results, e.g., LOINC and SNOMED. These terminologies were developed in part to support the safe application of expert system rules to information recorded in the electronic health care record.

The development of the HL7 Reference Information Model (RIM) was based on analysis of the "act of observing" and the "result of the act of observing" [1]. Today, new Entity attributes proposed for the HL7 RIM are evaluated for inclusion based partly on whether the information is best communicated in a new attribute for an HL7 Entity or best communicated in an HL7 Observation Act.

Improved standardization of clinical observation techniques, both in the practice of bedside care and the recording of clinical observations in electronic healthcare systems is thought to be essential to the continuing improvement of healthcare and patient safety.

Cross-References

▶ Clinical Event
▶ Clinical Order
▶ Interface Engines in Healthcare

Recommended Reading

1. Russler D, et al. Influences of the unified service action model on the HL7 reference information model. In: JAMIA Symposium Supplement, Proceedings SCAMC; 1999. p. 930–4.

Clinical Ontologies

Yves A. Lussier and James L. Chen
University of Chicago, Chicago, IL, USA

Synonyms

Clinical classifications; Clinical nomenclatures; Clinical terminologies

Definition

An ontology is a formal representation of a set of heterogeneous concepts. However, in the life sciences, the term clinical ontology has also been more broadly defined as also comprising all forms of classified terminologies, including classifications and nomenclatures. Clinical ontologies provide not only a controlled vocabulary but also relationships among concepts allowing computer reasoning such that different parties, like physicians and insurers, can efficiently answer complex queries.

Historical Background

As the life sciences integrates increasingly sophisticated systems of patient management, different means of data representation have had to keep pace to support user systems. Simultaneously, the explosion of genetic information from breakthroughs from the Human Genome Project and gene chip technology have further expedited the need for robust, scalable platforms for handling heterogeneous data. Multiple solutions have been developed by the scientific community to answer these challenges at all different levels of biology.

This growing field of "systems medicine" starts humbly at the question – how can one best capture and represent complex data in a means that can be understood globally without ambiguity? In other words, does the data captured have the same semantic validity after retrieval as it did prior? These knowledgebases are in of themselves organic. They need to be able to expand, shrink, and rearrange themselves based on user or system needs. This entry will touch upon existing clinical ontologies used in a variety of applications.

Foundations

The complexity of biological data cannot be understated. Issues generally fall into challenges with (i) definition, (ii) context, (iii) composition, and (iv) scale. One cannot even take for granted that the term "genome" is well-understood. Mahner found five different characterizations for the term "genome" [8]. Ontologies then provide a means of providing representational consistency through their structure and equally important provide the ability to connect these terms together in a semantically informative and computationally elegant manner [9]. This has led to their ubiquity in the life sciences. Formal ontologies are designated using frames or description logics [5]. However, few life science knowledgebases are represented completely in this manner due to difficulties with achieving consensus on definitions regarding the terms and the effort required to give context to the terms. Thus, this article defines well-organized nomenclatures and terminologies as clinical ontologies – regardless if their terms adhere to strict formalism.

Looking at elevations in gene expression, it matters what organism and under what experimental conditions the experiment was conducted. Clinical context changes the meaning of terms. The term "cortex" can either indicate a part of the kidney or that of the brain. Generalized or "essential hypertension" can be what is known colloquially as "high blood pressure" or localized to the lungs as "pulmonary hypertension." One can have pulmonary hypertension but not essential hypertension. This leads to the next representational challenge – that of composition. Should hypertension be represented implicitly as "essential hypertension" and as "pulmonary hypertension"? Or should it be stored explicitly as "hypertension" with a location attribute?

These representational decisions are driven by the queries that may be asked. The difficulty arises in anticipating the queries and in post-processing of the query to split the terminological components of the overall concept. Finally, the knowledge model needs to be able to scale upward. The same decision logic that was relevant when the knowledgebase contained 100 concepts needs to still be relevant at 1,000,000 concepts.

Properties of Clinical Ontologies

Ontologies vary widely in their degree of formalism and design. With this comes differing computability. In 1998, Cimino proposed desirable properties for purposes of clinical computation [3, 4]. Table 1 summarizes the overall properties of the commonly used clinical ontologies.

1. Concept-oriented: a single concept is the preferred unit
2. Formal semantic definition: well-defined terms
3. Nonredundancy: each concept needs to be unique
4. Nonambiguity: different concepts should not overlap or be conflated
5. Relationships: the structure of connections between concepts differentiate ontologies:

- Monohierarchy (tree): each concept only has one parent concept
- Polyhierarchy: each concept may multiply inherit from multiple parents
- Directed Acycle Graph (DAG): there are no cycles in the graph – in other words, children concepts may not point to parent terms

Clinical Ontologies, Table 1 Properties of clinical ontologies

| Ontology | Architecture | | | | | |
	Concept oriented	Formal semantic definition	Concept permanence	Nonredundancy	Uniqueness	Relationship
ICD-9	+		±	+	+	M
LOINC	±		+	+		P
CPT	±		±			M
SNOMED	+	+	+	+	+	DAG
UMLS	+		+	+	+	CG

M monohierarchy/tree, *P* polyhierarchy, *DAG* directed acyclic graph, *CG* cyclic graph

Clinical Ontologies, Table 2 Coverage of classification, nomenclatures and ontologies

| Ontology | Content | | | | | | Number of concepts (order of magnitude) |
	Diseases	Anatomy	Morphology	Labs	Procedures	Drugs	
ICD-9	X						10^4
LOINC				X			10^5
CPT					X		10^4
SNOMED	X	X	X	X	X	X	10^5
UMLS	X	X	X	X	X	X	10^6

Key Applications

This section reviews different, well-used life science ontologies used to annotate datasets. First, this discussion summarizes a select number of archetypal clinical ontologies that comprise one or several types of clinical entities such as diseases, clinical findings, procedures, laboratory measurements, and medications. Table 2 below summarizes the content coverage of each of archetypal health ontologies.

Prototypical Clinical Ontologies

(a) **The Systematized Nomenclature of Medicine (SNOMED CT)**

SNOMED CT is the most extensive set of publically available collection of clinical concepts. It is organized as a directed acyclic graph (DAG) and contains class/subclass relationships and partonomy relationships. It is maintained by the College of American Pathologists and is available in the United States through a license from the National Library of Medicine in perpetuity. SNOMED CT is one of the designated data standards for use in U.S. Federal Government systems for the electronic exchange of clinical health information. SNOMED CT is now owned by the International Healthcare Terminology Standards Development Organization [6].

(b) **International Statistical Classification of Diseases (ICD-9, ICD-10, ICD-CM)**

ICD-9 and ICD 10 are detailed ontologies of disease and symptomatology used ubiquitously for reimbursement systems (i.e.,

Medicare/Medicaid) and automated decision support in medicine. ICD-10 is used worldwide for morbidity and mortality statistics. Owned by the World Health Organization (WHO), licenses are available generally free for research. ICD-9 CM is a subtype of ICD-9 with clinical modifiers for billing purposes [11].

(c) **Medical Subject Headings (MeSH)**

MeSH grew out of an effort by the NLM for indexing life science journal articles and books. {Nelson S.J., 2001 #6}. The extensive controlled vocabulary MeSH serves as the backbone of the MEDLINE/PubMed article database. MeSH can be browsed and downloaded free of charge on the Internet [10].

(d) **International Classification of Primary Care (ICPC-2, ICPC-2-E)**

ICPC is a primary care encounter classification system [12]. It has a biaxial structure of 17 clinical systems and 7 types of data. It allows for the classification of the patient's reason for encounter (RFE), the problems/diagnosis managed, primary care interventions, and the ordering of the data. of the primary care session in an episode of care structure. ICPC-2-E refers to a revised electronic version.

(e) **Diagnostic and Statistical Manual of Mental Disorders (DSM-IV, DSM-V)**

The DSM is edited and published by the American Psychiatric Association provides categories of and diagnosis criteria for mental disorders [2]. It is used extensively by clinicians, policy makers and insurers. The original version of the DSM was published in 1962. DSM-V is due for publication in May 2012. The diagnosis codes are developed to be compatible with ICD-9.

(f) Logical Observation Identifiers Names and Codes (LOINC)

LOINC is a database protocol aimed at standardizing laboratory and clinical codes. The Regenstrief Institute, Inc, maintains the LOINC database and supporting documentation. LOINC is endorsed by the American Clinical Laboratory Association and College of American Pathologist and is one of the accepted standards by the US Federal Government for information exchange [7].

(g) Current Procedural Terminology (CPT)

The CPT code set is owned and maintained by the American Medical Association through the CPT Editorial Panel [1]. The CPT code set is used extensively to communicate medical and diagnostic services that were rendered among physicians and payers. The current version is the CPT 2008.

Cross-References

► Anchor text
► Annotation
► Archiving Experimental Data
► Biomedical Data/Content Acquisition, Curation
► Classification
► Clinical Data Acquisition, Storage, and Management
► Clinical Data and Information Models
► Clinical Decision Support
► Data Integration Architectures and Methodology for the Life Sciences
► Data Types in Scientific Data Management
► Data Warehousing for Clinical Research
► Digital Curation
► Electronic Health Record
► Fully Automatic Web Data Extraction

► Information Integration Techniques for Scientific Data
► Integration of Rules and Ontologies
► Logical Models of Information Retrieval
► Ontology
► Ontologies and Life Science Data Management
► Ontology
► Ontology Elicitation
► Ontology Engineering
► Ontology Visual Querying
► OWL: Web Ontology Language
► Semantic Data Integration for Life Science Entities
► Semantic Web
► Storage Management
► Taxonomy: Biomedical Health Informatics
► Web Information Extraction

Recommended Reading

1. American Medical Association. Cited; Available at: http://www.cptnetwork.com
2. American Psychiatric Association. Cited; Available at: http://www.psych.org/MainMenu/Research/DSMIV.aspx
3. Cimino JJ. Desiderata for controlled medical vocabularies in the twenty-first century. Methods Inf Med. 1998;37(4–5):394–403.
4. Cimino JJ. In defense of the Desiderata. [comment]. J Biomed Inform. 2006;39(3):299–306.
5. Gruber TR. Toward principles for the design of ontologies used for knowledge sharing. Int J Hum Comput Stud. 1995;43(4–5):907–28.
6. I.H.T.S.D. Cited; Available from: http://www.ihtsdo.org/our-standards/snomed-ct
7. Khan AN et al. Standardizing laboratory data by mapping to LOINC. J Am Med Inform Assoc. 2006;13(3):353–5.
8. Mahner M, Kary M. What exactly are genomes, genotypes and phenotypes? and what about phenomes? J Theor Biol. 1997;186(1):55–63.
9. Musen MA et al. PROTEGE-II: computer support for development of intelligent systems from libraries of components. Medinfo. 1995;8(Pt 1):766–70.
10. Nelson SJ, Johnston D, Humphreys BL. Relationships in medicical subject headings. In: Carol AB, Rebecca G, editors. Relationships in the organization of knowledge. Dordecht: Kluwer; 2001. p. 171–84.
11. World Health Organization. Cited; Available at: http://www.who.int/classifications/icd/en/
12. World Organization of National Colleges, Academies, and Academic Associations of General Practitioners/Family Physicians, ICPC. International classification of primary care. Oxford: Oxford University Press; 1987.

Clinical Order

Dan Russler
Oracle Health Sciences, Redwood Shores, CA,
USA
Georgia Tech Research Institute, Atlanta,
Georgia, USA

Synonyms

Order item; Procedure order; Procedure request;
Service item; Service order; Service request

Definition

The act of requesting that a service be performed
for a patient.

Clinical orders in healthcare share many characteristics with purchase orders in other industries. Both clinical orders and purchase orders establish a customer-provider relationship between the person placing the request for a service to be provided and the person or organization filling the request. In both cases, the clinical order and purchase order are followed by either a promise or intent to fill the request, a decline to fill the request, or a counter-proposal to provide an alternate service. In both scenarios, an authorization step such as an insurance company authorization or a credit company authorization may be required. Therefore, the dynamic flow of communications between a placer and filler in a clinical order management system and a purchase order management system are very similar.

Both clinical order and purchase order management systems maintain a catalog of items that may be requested. These items in both kinds of systems may represent physical items from supply or services from a service provider. Each of these items in both kinds of systems is associated with an internally unique identifier, a text description, and often a code. Dates, status codes, delivery locations, and other attributes of a clinical order and purchase order are also similar. Therefore, in addition to similarities in the dynamic flow of order communications, the structure of the content in clinical orders and purchase orders is similar.

Logical Observation Identifiers Names and Codes (LOINC) (www.loinc.org) describe many of the requested services in healthcare, especially in laboratory systems. Other procedural terminologies exist for healthcare, either independently in terminologies like LOINC or included in more comprehensive terminologies such as Systematized Nomenclature of Medicine (SNOMED) (www.ihtsdo.org).

Key Points

Clinical orders exist in the context of a larger clinical management, process. The order management business process of an organization, that includes defining a catalog of services to be provided and then allowing people to select from the catalog of services, is common in many industries. However, the decision support opportunities for helping providers select the optimum set of services for a patient are often more complex in healthcare than occurs in other industries. The outcomes of this selection process are studied in clinical research, clinical trials on medications and devices, and in organizational quality improvement initiatives. Finally, the outcomes of the service selection process are used to improve the clinical decision support processes utilized by providers selecting services for patients. This business process in healthcare as well as in many other industries describes a circular feedback loop defined by the offering of services, the selection of services, the delivery of services, the outcome of services, and finally, the modification of service selection opportunities and decision support.

In the HL7 Reference Information Model (RIM), "ACT" classes sub-typed with the moodCode attribute support the healthcare improvement process (www.hl7.org). These objects with process "moods" support the sequence of objects created during the execution of a process defined in Business Process Execution Language (BPEL) in a service oriented architecture that begins with an "order", evolves into an "appointment", which then is completed as an "event". The reason the term "mood" is used is that the values of the moodcode attribute are analogous to the models of verbs in many languages, e.g., the "Definition mood" used to define service catalogs corresponds to the "infinitive" verbal mood, i.e., a possible action; the "Request or Order mood" corresponds to the "imperative" verbal mood; the "Event mood" corresponds to the "indicative" verbal mood; and the "Goal mood," which describes the desired outcome of the selected service, corresponds to the "subjunctive" verbal mood.

Cross-References

▶ Clinical Event
▶ Clinical Observation
▶ Interface Engines in Healthcare

Closed Itemset Mining and Nonredundant Association Rule Mining

Mohammed J. Zaki
Rensselaer Polytechnic Institute, Troy, NY, USA

Synonyms

Frequent concepts; Rule bases

Definition

Let I be a set of binary-valued attributes, called *items*. A set $X \subseteq I$ is called an *itemset*. A transaction database D is a multiset of itemsets, where each itemset, called a transaction, has a unique identifier, called a tid. The *support* of an itemset X in a dataset D, denoted $sup(X)$, is the fraction of transactions in D where X appears as a subset.

X is said to be a *frequent* itemset in D if $sup(X) \geq minsup$, where $minsup$ is a user defined minimum support threshold. An (frequent) itemset is called *closed* if it has no (frequent) superset having the same support.

An *association rule* is an expression $A \Rightarrow B$, where A and B are itemsets, and $A \cap B = \emptyset$. The *support* of the rule is the joint probability of a transaction containing both A and B, given as $sup(A \Rightarrow B) = P(A \wedge B) = sup(A \cup B)$. The *confidence* of a rule is the conditional probability that a transaction contains B, given that it contains A, given as: $conf(A \Rightarrow B) = P(B|A) = \frac{P(A \wedge B)}{P(A)} = \frac{sup(A \cup B)}{sup(A)}$. A rule is frequent if the itemset $A \cup B$ is frequent. A rule is confident if $conf \geq minconf$, where $minconf$ is a user-specified minimum threshold. The aim of non-redundant association rule mining is to generate a *rule basis*, a small, non-redundant set of rules, from which all other association rules can be derived.

Historical Background

The notion of closed itemsets has its origins in the elegant mathematical framework of Formal Concept Analysis (FCA) [3], where they are called *concepts*. The task of mining frequent closed itemsets was independently proposed in [7, 11]. Approaches for non-redundant association rule mining were also independently proposed in [1, 9]. These approaches rely heavily on the seminal work on rule bases in [5, 6]. Efficient algorithms for mining frequent closed itemsets include CHARM [10], CLOSET [8] and several new approaches described in the Frequent Itemset Mining Implementations workshops [4].

Foundations

Let $I = \{i_1, i_2, \ldots, i_m\}$ be the set of items, and let $T = \{t_1, t_2, \ldots, t_n\}$ be the set of tids, the transaction identifiers. Just as a subset of items is called an itemset, a subset of tids is called a tidset. Let $\mathbf{t} : 2^I \rightarrow 2^T$ be a function, defined as follows:

$$\mathbf{t}(X) = \{t \in T | X \subseteq \mathbf{i}(t)\}$$

That is, $\mathbf{t}(X)$ is the set of transactions that contain *all* the items in the itemset X. Let $\mathbf{i} : 2^T \rightarrow 2^I$ be a function, defined as follows:

$$\mathbf{t}(Y) = \{i \in I \,|\, \forall t \in Y, t \text{ contains } x\}$$

That is, $\mathbf{i}(T)$ is the set of items that are contained in *all* the tids in the tidset Y. Formally, an itemset X is closed if $\mathbf{i} \circ \mathbf{t}(X) = X$, i.e., if X is a fixed-point of the closure operator $\mathbf{c} = \mathbf{i} \circ \mathbf{t}$. From the properties of the closure operator, one can derive that X is the maximal itemset that is contained in all the transactions $\mathbf{t}(X)$, which gives the simple definition of a closed itemset, namely, a closed itemset is one that has no superset that has the same support.

Based on the discussion above, three main families of itemsets can be distinguished. Let $\mathcal{F}$ denote the set of all frequent itemsets, given as

$$\mathcal{F} = \{X \,|\, X \subseteq I \text{ and } sup(X) \geq minsup\}$$

Let $\mathcal{C}$ denote the set of all closed frequent itemsets, given as

$$\mathcal{C} = \{X \,|\, X \in \mathcal{F} \text{ and } \nexists Y \supset X \text{ with } sup(X) = sup(Y)\}$$

Finally, let $\mathcal{M}$ denote the set of all *maximal* frequent itemsets, given as

$$\mathcal{M} = \{X \,|\, X \in \mathcal{F} \text{ and } \nexists Y \supset X \text{ such that } Y \in \mathcal{F}\}$$

The following relationship holds between these sets: $\mathcal{M} \subseteq \mathcal{C} \subseteq \mathcal{F}$, which is illustrated in Fig. 1, based on the example dataset shown in Table 1 and using minimum support $minsup = 3$. The *equivalence classes* of itemsets that have the same tidsets have been shown clearly; the largest itemset in each equivalence class is a closed itemset. The figure also shows that the maximal itemsets are a subset of the closed itemsets.

Mining Closed Frequent Itemsets

CHARM[8] is an efficient algorithm for mining closed itemsets. Define two itemsets X,Y of length k as belonging to the same *prefix equivalence class*, $[P]$, if they share the $k-1$ length prefix P, i.e., $X = Px$ and $Y = Py$, where $x,y \in I$. More formally, $[P] = \{Px_i \mid x_i \in I\}$, is the class of all itemsets sharing P as a common prefix. In CHARM there is no distinct candidate generation and support counting phase. Rather, counting is simultaneous with candidate generation. For a given prefix class, one performs intersections of the tidsets of all pairs of itemsets in the class, and checks if the resulting tidsets have cardinality at least *minsup*. Each resulting frequent itemset generates a new class which will be expanded in the next step. That is, for a given class of itemsets with prefix P, $[P] = \{Px_1, Px_2, \ldots, Px_n\}$, one performs the intersection of Px_i with all Px_j with $j > i$ to obtain a new class $[Px_i] = [P']$ with elements $P'x_j$ provided the itemset $Px_i x_j$ is frequent. The computation progresses recursively

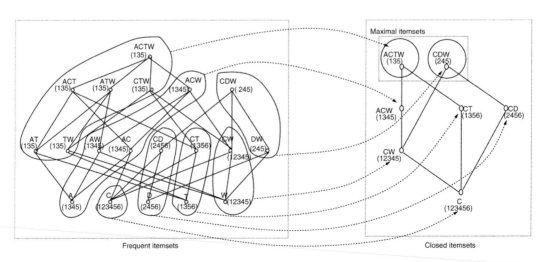

Closed Itemset Mining and Nonredundant Association Rule Mining, Fig. 1 Frequent, closed frequent and maximal frequent itemsets

until no more frequent itemsets are produced. The initial invocation is with the class of frequent single items (the class $[\varnothing]$). All tidset intersections for pairs of class elements are computed. However in addition to checking for frequency, CHARM eliminates branches that cannot lead to closed sets, and grows closed itemsets using subset relationships among tidsets. There are four cases: if $\mathbf{t}(X_i) \subset \mathbf{t}(X_j)$ or if $\mathbf{t}(X_i) = \mathbf{t}(X_j)$, then replace every occurrence of X_i with $X_i \cup X_j$, since whenever X_i occurs X_j also occurs, which implies that $\mathbf{c}(X_i) \subseteq \mathbf{c}(X_i \cup X_j)$. If $\mathbf{t}(X_i) \supset \mathbf{t}(X_j)$ then replace X_j for the same reason. Finally, further recursion is required if $\mathbf{t}(X_i) \neq \mathbf{t}(X_j)$. These four properties allow CHARM to efficiently prune the search tree (for additional details see [10]).

Figure 2 shows how CHARM works on the example database shown in Table 1. First, CHARM sorts the items in increasing order of support, and initializes the root class as $[\varnothing] = \{D \times 2456, T \times 1356, A \times 1345, W \times 12345,$

$C \times 123456\}$. The notation $D \times 2456$ stands for the itemset D and its tidset $\mathbf{t}(D) = \{2,4,5,6\}$. CHARM first processes the node $D \times 2456$; it will be combined with the sibling elements. DT and DA are not frequent and are thus pruned. Looking at W, since $\mathbf{t}(D) \neq \mathbf{t}(W)$, W is inserted in the new equivalence class $[D]$. For C, since $\mathbf{t}(D) \subset \mathbf{t}(C)$, all occurrences of D are replaced with DC, which means that $[D]$ is also changed to $[DC]$, and the element DW to DWC. A recursive call with class $[DC]$ is then made and since there is only a single itemset DWC, it is added to the set of closed itemsets C. When the call returns to D (i.e., DC) all elements in the class have been processed, so DC itself is added to C.

When processing T, $\mathbf{t}(T) \neq \mathbf{t}(A)$, and thus CHARM inserts A in the new class $[T]$. Next it finds that $\mathbf{t}(T) \neq \mathbf{t}(W)$ and updates $[T] = \{A,W\}$. When it finds $\mathbf{t}(T) \subset \mathbf{t}(C)$ it updates all occurrences of T with TC. The class $[T]$ becomes $[TC] = \{A,W\}$. CHARM then makes a recursive call to process $[TC]$. When combining TAC with TWC it finds $\mathbf{t}(TAC) = \mathbf{t}(TWC)$, and thus replaces TAC with $TACW$, deleting TWC at the same time. Since $TACW$ cannot be extended further, it is inserted in C. Finally, when it is done processing the branch TC, it too is added to C. Since $\mathbf{t}(A) \subset \mathbf{t}(W) \subset \mathbf{t}(C)$ no new recursion is made; the final set of closed itemsets C consists of the uncrossed itemsets shown in Fig. 2.

Closed Itemset Mining and Nonredundant Association Rule Mining, Table 1 Example transaction dataset

	$i(t)$
1	ACTW
2	CDW
3	ACTW
4	ACDW
5	ACDTW
6	CDT

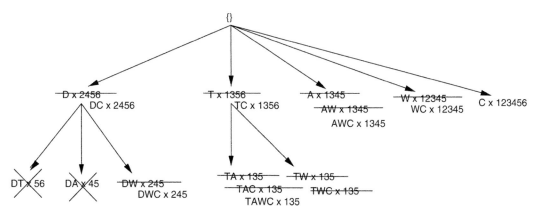

Closed Itemset Mining and Nonredundant Association Rule Mining, Fig. 2 CHARM: mining closed frequent itemsets

Non-redundant Association Rules

Given the set of closed frequent itemsets $\mathcal{C}$, one can generate all non-redundant association rules. There are two main classes of rules: (i) those that have 100% confidence, and (ii) those that have less than 100% confidence [9]. Let X_1 and X_2 be closed frequent itemsets. The 100% confidence rules are equivalent to those directed from X_1 to X_2, where $X_2 \subseteq X_1$, i.e., from a superset to a subset (not necessarily proper subset). For example, the rule $C \Rightarrow W$ is equivalent to the rule between the closed itemsets $\mathbf{c}(W) \Rightarrow \mathbf{c}(C) \equiv CW \Rightarrow C$. Its support is $sup(CW) = 5/6$, and its confidence is $\frac{sup(CW)}{sup(W)} = 5/5 = 1$, i.e., 100%. The less than 100% confidence rules are equivalent to those from X_1 to X_2 where $X_1 \subset X_2$, i.e., from a subset to a proper superset. For example, the rule $W \Rightarrow T$ is equivalent to the rule $\mathbf{c}(W) \Rightarrow \mathbf{c}(W \cup T) \equiv CW \Rightarrow ACTW$. Its support is $sup(TW) = 3/6 = 0.5$, and its confidence is $\frac{sup(TW)}{sup(W)} = 3/5 = 0.6$ or 60%. More details on how to generate these non-redundant rules appears in [9].

Key Applications

Closed itemsets provide a loss-less representation of the set of all frequent itemsets; they allow one to determine not only the frequent sets but also their exact support. At the same time they can be orders of magnitude fewer. Likewise, the non-redundant rules provide a much smaller, and manageable, set of rules, from which all other rules can be derived. There are numerous applications of these methods, such as market basket analysis, web usage mining, gene expression pattern mining, and so on.

Future Directions

Closed itemset mining has inspired a lot of subsequent research in mining compressed representations or summaries of the set of frequent patterns; see [2] for a survey of these approaches. Mining compressed pattern bases remains an active area of study.

Experimental Results

A number of algorithms have been proposed to mine frequent closed itemsets, and to extract nonredundant rule bases. The Frequent Itemset Mining Implementations (FIMI) Repository contains links to many of the latest implementations for mining closed itemsets. A report on the comparison of these methods also appears in [4]. Other implementations can be obtained from individual author's websites.

Data Sets

The FIMI repository has a number of real and synthetic datasets used in various studies on closed itemset mining.

Url to Code

The main FIMI website is at http://fimi.cs. helsinki.fi/, which is also mirrored at: http:// www.cs.rpi.edu/~zaki/FIMI/

Cross-References

▶ Association Rule Mining on Streams
▶ Data Mining

Recommended Reading

1. Bastide Y, Pasquier N, Taouil R, Stumme G, Lakhal L. Mining minimal non-redundant association rules using frequent closed itemsets. In: Proceedings of the 1st International Conference on Computational Logic; 2000. p. 972–86.
2. Calders T, Rigotti C, Boulicaut J-F. A survey on condensed representation for frequent sets. In: Boulicaut J-F, De Raedt L, Mannila H, editors. Constraint-based mining and inductive databases, LNCS, vol. 3848. Berlin: Springer; 2005. p. 64–80.
3. Ganter B, Wille R. Formal concept analysis: mathematical foundations. Berlin/Heidelberg/New York: Springer; 1999.
4. Goethals B, Zaki MJ. Advances in frequent itemset mining implementations: report on FIMI'03. SIGKDD Explor. 2003;6(1):109–17.

5. Guigues JL, Duquenne V. Familles minimales d'implications informatives resultant d'un tableau de donnees binaires. Math Sci Hum. 1986;24(95):5–18.
6. Luxenburger M. Implications partielles dans un contexte. Math Inf Sci Hum. 1991;29(113):35–55.
7. Pasquier N, Bastide Y, Taouil R, Lakhal L. Discovering frequent closed itemsets for association rules. In: Proceedings of the 7th International Conference on Database Theory; 1999. p. 398–416.
8. Pei J, Han J, Mao R. Closet: an efficient algorithm for mining frequent closed itemsets. In: Proceedings of the ACM SIGMOD Workshop on Research Issues in Data Mining and Knowledge Discovery; 2000. p. 21–30.
9. Zaki MJ. Generating non-redundant association rules. In: Proceedings of the 6th ACM SIGKDD International Conference on Knowledge Discovery and Data Mining; 2000. p. 34–43.
10. Zaki MJ, Hsiao CJ. CHARM: an efficient algorithm for closed itemset mining. In: Proceedings of the SIAM International Conference on Data Mining; 2002. p. 457–73.
11. Zaki MJ, Ogihara M. Theoretical foundations of association rules. In: Proceedings of the ACM SIGMOD Workshop on Research Issues in Data Mining and Knowledge Discovery; 1998.

Closest-Pair Query

Antonio Corral[1] and Michael Vassilakopoulos[2]
[1]University of Almeria, Almeria, Spain
[2]University of Thessaly, Volos, Greece

Synonyms

Closest pairs; Incremental k-distance join; k-Closest pair join; k-Closest pair query; k-Distance join

Definition

Given two sets P and Q of objects, a closest pair (CP) query discovers the pair of objects (p, q) with a distance that is the smallest among all object pairs in the Cartesian product P × Q. Similarly, a k closest pair query (k-CPQ) retrieves k pairs of objects from P and Q with the minimum distances among all the object pairs. In spatial databases, the distance is usually defined according to the Euclidean metric, and the set of objects P and Q are disk-resident. Query algorithms aim at minimizing the processing cost and the number of I/O operations, by using several optimization techniques for pruning the search space.

Historical Background

The closest pair query, has been widely studied in computational geometry. More recently, this problem has been approached in the context of spatial databases [4, 8, 12, 14]. In spatial databases, existing algorithms assume that P and Q are indexed by a spatial access method (usually an R-tree [1]) and utilize some pruning bounds and heuristics to restrict the search space.

Reference [8] was the first to address this issue, and proposed the following distance-based algorithms: incremental distance join, k distance join and k distance semijoin between two R-tree indices. The incremental processing reports one-by-one the desired elements of the result in ascending order of distance (k is unknown in advance and the user can stop when he/she is satisfied by the result). The algorithms follow the Best-First (BF) traversal policy, which keeps a heap with the entries of the nodes visited so far (it maintains a priority queue which contains pairs of index entries and objects, and pop out the closest pair and process it). BF is near-optimal for CP queries; i.e., it only visits the pairs of nodes necessary for obtaining the result with a high probability. In [12] several modifications to the algorithms of [8] had been proposed in order to improve performance. Mainly, a method was proposed for selecting the sweep axis and direction for the plane sweep technique in bidirectional node expansion which minimizes the computational overhead of [8].

Later, an improved version of BF and several algorithms that follow Depth-First (DF) traversal ordering from the non-incremental point of view (which assumes that k is known in advance and reports the k elements of the result all together at the end of the algorithm) was proposed in [4]. In general, a DF algorithm visit the roots of the two R-trees and recursively follows the pair of entries $< E_P, E_Q >$, $E_P \in R_P$ and $E_Q \in R_Q$, whose MINMINDIST is the minimum distance among all pairs. At the opposite of BF, DF is sub-optimal, i.e., it accesses more nodes than necessary. The main disadvantage of BF with respect to DF is that it may suffer from buffer thrashing if the available memory is not enough for the heap (it is space-consuming), when a great quantity of elements of the result is required. In this case, part of the heap must be migrated to disk, incurring frequent I/O accesses. The implementation of DF is by recursion, which is available in most of the programming languages, and linear-space consuming with respect to the height of the R-trees. Moreover, BF is not favored by page replacement policies (e.g., LRU), as it does not exhibit locality between I/O accesses.

Another interesting contribution to the CP query was proposed by [14], in which a new structure called the b-Rdnn tree was presented, along with a better solution to the k-CP query when there is high overlap between the two datasets. The main idea is to find k objects from each dataset which are the closest to the other dataset.

There are a lot of papers related to k-CP query, like buffer query [3], iceberg distance join query [13], multi-way distance join query [6], k-nearest neighbor join [2], closest pair query with spatial constraints [11], etc. For example, a buffer query [3] involves two spatial datasets and a distance threshold ρ; the answer to this query is a set of pairs of spatial objects, one from each input dataset, that are within distance ρ of each other.

Foundations

In spatial databases, existing algorithms assume that sets of spatial objects are indexed by a spatial access method (usually an R-tree [1]) and utilize some pruning bounds to restrict the search space. An R-tree is a hierarchical, height balanced multi-dimensional data structure, designed to be used in secondary storage based on B-trees. The R-trees are considered as excellent choices for indexing various kinds of spatial data (points, rectangles, line-segments, polygons, etc.). They are used for the dynamic organization of a set of spatial objects approximated by their Minimum Bounding Rectangles (MBRs). These MBRs are characterized by *min* and *max* points of rectangles with faces parallel to the coordinate axis. Using the MBR instead of the exact geometrical representation of the object, its representational complexity is reduced to two points where the most important features of the spatial object (position and extension) are maintained. The R-trees belong to the category of data-driven access methods, since their structure adapts itself to the MBRs distribution in the space (i.e., the partitioning adapts to the object distribution in the embedding space). Figure 1a shows two points sets P and Q (and the node extents), where the closest pair

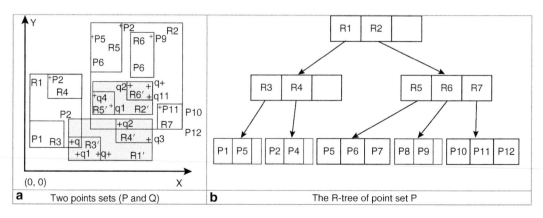

Closest-Pair Query, Fig. 1 Example of an R-tree and a point CP query

is (p_8, q_8), and Fig. 1b is the R-tree for the point set $P = \{p_1, p_2, \ldots, p_{12}\}$ with a capacity of three entries per node (branching factor or fan-out).

Assuming that the spatial datasets are indexed on any spatial tree-like structure belonging to the R-tree family, then the main objective while answering these types of spatial queries is to reduce the search space. In [5], three MBR-based distance functions to be used in algorithms for CP queries were formally defined, as an extension of the work presented in [4]. These metrics are MINMINDIST, MINMAXDIST and MAXMAXDIST. MINMINDIST (M_1, M_2) between two MBRs is the minimum possible distance between any point in the first MBR and any point in the second MBR. Maxmaxdist between two MBRs (M_1, M_2) is the maximum possible distance between any point in the first MBR and any point in the second MBR. Finally, MINMAXDIST between two MBRs (M_1, M_2) is the minimum of the maximum distance values of all the pairs of orthogonal faces to each dimension. Formally, they are defined as follows:

- Given two MBRs $M_1 = (a, b)$ and $M_2 = (c, d)$, in the d-dimensional Euclidean space,
- $M_1 = (a, b)$, where $a = (a_1, a_2, \ldots, a_d)$ and $b = (b_1, b_2, \ldots, b_d)$ such that $a_i \leq b_i\ 1 \leq i \leq d$
- $M_2 = (c, d)$, where $c = (c_1, c_2, \ldots, c_d)$ and $d = (d_1, d_2, \ldots, d_d)$ such that $a_i \leq b_i\ 1 \leq i \leq d$

the MBR-based distance functions are defined as follows:

$$MINMINDIST\ (M_1, M_2)$$

$$= \sqrt{\sum_{i=1}^{d} \begin{cases} (c_i - b_i)^2, & c_i > b_i \\ (a_i - d_i)^2, & a_i > d_i \\ 0, & \text{otherwise} \end{cases}}$$

$$MAXMAXDIST\ (M_1, M_2)$$

$$= \sqrt{\sum_{i=1}^{d} \begin{cases} (d_i - a_i)^2, & c_i > b_i \\ (b_i - c_i)^2, & a_i > d_i \\ \max\left\{(d_i - a_i)^2, (b_i - c_i)^2\right\}, & \text{otherwise} \end{cases}}$$

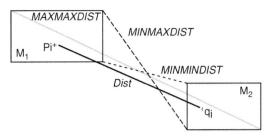

Closest-Pair Query, Fig. 2 MBR-based distance functions in 2-dimensional Euclidean space

$$MINMAXDIST\ (M_1, M_2)$$

$$= \sqrt{\min_{1 \leq j \leq d} \left\{ x_j^2 + \sum_{i=1, i \neq j}^{d} y_i^2 \right\}}$$

where

$$x_j = \min\left\{ |a_j - c_j|, |a_j - d_j|, |b_j - c_j|, |b_j - d_j| \right\} \text{ and}$$
$$y_i = \max\left\{ |a_i - d_i|, |b_i - c_i| \right\}$$

To illustrate the distance functions MIN-MINDIST, MINMAXDIST and MAXMAXDIST which are the basis of query algorithms for CPQ, in Fig. 2, two MBRs and their MBR-based distance functions and their relation with the distance (dist) between two points (p_i, q_j) are depicted in 2-dimensional Euclidean space.

According to [5], MINMINDIST(M_1, M_2) is monotonically non-decreasing with the R-tree heights. MINMINDIST(M_1, M_2) and MAXMAXDIST(M_1, M_2) serve respectively as lower and upper bounding functions of the Euclidean distance from the k closest pairs of spatial objects within the MBRs M_1, and M_2. In the same sense, MINMAXDIST(M_1, M_2) serves as an upper bounding function of the Euclidean distance from the closest pair of spatial objects enclosed by the MBRs M_1 and M_2. As long as the distance functions are consistent, the branch-bound algorithms based on them will work correctly [5].

Moreover, the general pruning mechanism for k-CP queries over R-tree nodes using branch-and-bound algorithms is the following: *if MIN-MINDIST$(M_1, M_2) > z$, then the pair of MBRs*

(M_1, M_2) *will be discarded*, where z is the distance value of the k-th closest pair that has been found so far (during the processing of the algorithm), or the distance value of the k-th largest MAXMAXDIST found so far (z is also called as the pruning distance).

Branch-and-bound algorithms can be designed following DF or BF traversal ordering (Breadth-First traversal order (level-by-level) can also be implemented, but the processing of each level must follow a BF order) to report k closest pairs in non-incremental way (for incremental processing the ordering of traversal must be BF [8]).

As an example, Fig. 3 shows the BF k-CPQ algorithm for two R-trees, for the non-incremental processing version. This algorithm needs to keep a minimum binary heap (H) with the references to pairs of internal nodes (characterized by their MBRs) accessed so far from the two different R-trees and their minimum distance ($<$MINMINDIST, $Addr_{MPi}$, $Addr_{MQj}>$). It visits the pair of MBRs (nodes) with the minimum MINMINDIST in H, until it becomes empty or the MINMINDIST value of the pair of MBRs located in the root of H is larger than the distance value of the k-th closest pair that has been found so far (z). To keep track of z, an additional data structure that stores the k closest pairs discovered during the processing of the algorithm is needed. This data structure is organized as a maximum binary heap (k-heap) and holds pairs of objects according to their minimum distance (the pair with the largest distance resides in the root). In the implementation of k-CPQ algorithm, the following cases must be considered: (i) initially the k-heap is empty (z is initialized to ∞), (ii) the pairs of objects reached at the leaf level are inserted in the k-heap until it gets full (z keeps the value of ∞), (iii) if the distance of a new pair of

Closest-Pair Query, Fig. 3
Best-First k-CPQ
Algorithm using R-trees

```
Create and initialize H, k-heap, z = ∞;
for each pair of MBRs <M_Pi, M_Qj> in <root_RP, root_RQ> do
        H.insert(MINMINDIST(M_Pi, M_Qj), Addr_MPi, Addr_MQj);
enddo

While (not H.isEmpty()) and (H.MinimumDistance() ≤ z) do
        minimum = H.deleteMinimum();
        node_RP = RP.readNode(minimum.Addr_MP);
        node_RQ = RQ.readNode(minimum.Addr_MQ);
        if (node_RP and node_RQ are internal nodes) then
                for each pair of MBRs <M_Pi, M_Qj> in < node_RP, node_RQ> do
                        if (MINMINDIST(M_Pi, M_Qj) ≤ z) then
                                H.insert(MINMINDIST(M_Pi, M_Qj), Addr_MPi, Addr_MQj);
                        endif
                enddo
        else
                for each pair of points <p_i, q_j> in <node_RP, node_RQ> do
                        distance = dist(p_i, q_j);
                        if (k-heap.isFull()) then
                                if (distance ≤ z) then
                                        k-heap.delete Maximum();
                                        k-heap.insert(distance, p_i, q_j);
                                        z = k-heap.get Maximum();
                                endif
                        else
                                k-heap.insert(distance, p_i, q_j);
                        endif
                enddo
        endif
endod
Destroy H;
```

Closest-Pair Query, Fig. 4
Using plane-sweep
technique over MBRs from
two R-tree nodes

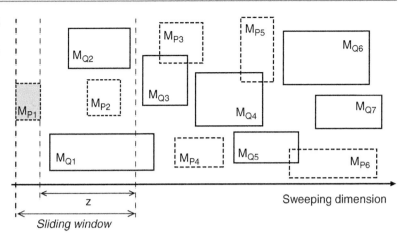

objects discovered at the leaf level is smaller than the distance of the pair residing in the k-heap root, then the root is extracted and the new pair is inserted in the k-heap, updating this data structure and z (distance of the pair of objects residing in the k-heap root).

Several optimizations had been proposed in order to improve performance, mainly with respect to the CPU cost. For instance, a method for selecting the sweep axis and direction for the plane sweep technique has been proposed [12]. But the most important optimization is the use of the plane-sweep technique for k-CPQ [5, 12], which is a common technique for computing intersections. The basic idea is to move a sweep-line perpendicular to one of the dimensions, so-called the sweeping dimension, from left to right. This technique is applied for restricting all possible combinations of pairs of MBRs from two R-tree nodes from R_P and R_Q. If this technique is not used, then a set with all possible combinations of pairs of MBRs from two R-tree nodes must be created. In general, the technique consists in sorting the MBRs of the two current R-tree nodes, based on the coordinates of one of the lower left corners of the MBRs in increasing order. Each MBR encountered during a plane sweep is selected as a pivot, and it is paired up with the non-processed MBRs in the other R-tree node from left to right. The pairs of MBRs with MINMINDIST on the sweeping dimension that are less than or equal to z (pruning distance) are selected for processing. After all possible pairs

of MBRs that contain the pivot have been found, the pivot is updated with the MBR of the next smallest value of a lower left corner of MBRs on the sweeping dimension, and the process is repeated. In summary, the application of this technique can be viewed as a *sliding window* on the sweeping dimension with a width of z starting in the lower end of the pivot MBR, where all possible pairs of MBRs that can be formed using the MBR of the pivot and the other MBRs from the remainder entries of the other R-tree node that fall into the current sliding window are chosen. For example, in Fig. 4, a set of MBRs from two R-tree nodes ($\{M_{P1}, M_{P2}, M_{P3}, M_{P4}, M_{P5}, M_{P6}\}$ and $\{M_{Q1}, M_{Q2}, M_{Q3}, M_{Q4}, M_{Q5}, M_{Q6}, M_{Q7}\}$) is shown. Without plane-sweep, $6*7 = 42$ pairs of MBRs must be generated. If the plane-sweep technique is applied over the X axis (sweeping dimension) and taking into account the distance value of z (pruning distance), this number of possible pairs will reduced considerably (the number of selected pairs of MBRs using the plane sweep technique is only 29).

Key Applications

Geographical Information Systems
Closest pair is a common distance-based query in the spatial database context, and it has only recently received special attention. Efficient algorithms are important for dealing with the large amount of spatial data in several GIS applica-

tions. For example, k-CPQ can discover the K closest pairs of cities and cultural landmarks providing an increase order based on its distances.

Data Analysis

Closest pair queries have been considered as a core module of clustering. For example, a proposed clustering algorithm [10] owes its efficiency to the use of closest pair query, as opposed to previous quadratic-cost approaches.

Decision Making

A number of decision support tasks can be modeled as closest pairs query. For instance, find the top k factory-house pairs ordered by the closeness to one another. This gives us a measure of the effect of individual factory on individual household, and can give workers a priority to which factory to address first.

Future Directions

k-closest pair query is a useful type of query in many practical applications involving spatial data, and the traditional technique to handle this spatial query generally assumes that the objects are static. Objects represented as a function of time have been studied in other domains, as in spatial semijoin [9]. For this reason, closest pair query in spatio-temporal databases could be an interesting line of research.

Another interesting problem to study is the monitoring of k-closest pairs over moving objects. It aims at maintaining closest pairs results while the underlying objects change the positions [15]. For example, return k pairs of taxi stands and taxies that have the smallest distances.

Other interesting topics to consider (from the static point of view) are to study k-CPQ between different spatial data structures (Linear Region Quadtrees for raster and R-trees for vector data), and to investigate k-CPQ in non-Euclidean spaces (e.g., road networks).

Experimental Results

In general, for every presented method, there is an accompanying experimental evaluation in the corresponding reference. References [4, 5, 8] compare BF and DF traversal order for conventional k-CPQ (from the incremental and non-incremental point of view). In [7], a cost model for k-CPQ using R-trees was proposed, evaluating their accuracy. Moreover, experimental results on k-closest pair queries to support the fact that b-Rdnn tree is a better alternative with respect to the R*-trees, when there is high overlap between the two datasets, were presented in [14].

Data Sets

A large collection of real datasets, commonly used for experiments, can be found at: http://www.rtreeportal.org/

URL to Code

R-tree portal (see above) contains the code for most common spatial access methods (mainly R-tree and variations), as well as data generators and several useful links for researchers and practitioners in spatial databases.

The sources in C + + of k-CPQ are in: http://www.ual.es/~acorral/DescripcionTesis.htm

Cross-References

▶ Multi step Query Processing
▶ Nearest Neighbor Query
▶ R-Tree (and Family)
▶ Spatial Indexing Techniques
▶ Spatial Join

Recommended Reading

1. Beckmann N, Kriegel HP, Schneider R, Seeger B. The R*-tree: an efficient and robust

access method for points and rectangles. In: Proceedings of the ACM SIGMOD International Conference on Management of Data; 1990. p. 322–31.

2. Böhm C, Krebs F. The k-nearest neighbour join: turbo charging the KDD process. Knowl Inform Syst. 2004;6(6):728–49.

3. Chan EPF. Buffer queries. IEEE Trans Knowl Data Eng. 2003;15(4):895–910.

4. Corral A, Manolopoulos Y, Theodoridis Y, Vassilakopoulos M. Closest pair queries in spatial databases. In: Proceedings of the ACM SIGMOD International Conference on Management of Data; 2000. p. 189–200.

5. Corral A, Manolopoulos Y, Theodoridis Y, Vassilakopoulos M. Algorithms for processing K-closest-pair queries in spatial databases. Data Knowl Eng. 2004;49(1):67–104.

6. Corral A, Manolopoulos Y, Theodoridis Y, Vassilakopoulos M. Multi-way distance join queries in spatial databases. GeoInformatica. 2004;8(4):373–402.

7. Corral A, Manolopoulos Y, Theodoridis Y, Vassilakopoulos M. Cost models for distance joins queries using R-trees. Data Knowl Eng. 2006;57(1):1–36.

8. Hjaltason GR, Samet H. Incremental distance join algorithms for spatial databases. In: Proceedings of the ACM SIGMOD International Conference on Management of Data; 1998. p. 237–48.

9. Iwerks GS, Samet H, Smith K. Maintenance of spatial semijoin queries on moving points. In: Proceedings of the 30th International Conference on Very Large Data Bases; 2004. p. 828–39.

10. Nanopoulos A, Theodoridis Y, Manolopoulos Y. C^2P: clustering based on closest pairs. In: Proceedings of the 27th International Conference on Very Large Data Bases; 2001. p. 331–40.

11. Papadopoulos AN, Nanopoulos A, Manolopoulos Y. Processing distance join queries with constraints. Comput J. 2006;49(3):281–96.

12. Shin H, Moon B, Lee S. Adaptive multi-stage distance join processing. In: Proceedings of the ACM SIGMOD International Conference on Management of Data; 2000. p. 343–54.

13. Shou Y, Mamoulis N, Cao H, Papadias D, Cheung D.W. Evaluation of iceberg distance joins. In: Proceedings of the 8th International Symposium Advances in Spatial and Temporal Databases; 2003. p. 270–88.

14. Yang C, Lin K. An index structure for improving closest pairs and related join queries in spatial databases. In: Proceedings of the International Conference on Database Engineering and Applications; 2002. p. 140–49.

15. Zhu M, Lee DL, Zhang J. k-closest pair query monitoring over moving objects. In: Proceedings of the 3rd International Conference on Mobile Data Management; 2002. p. 14–14.

Cloud Computing

Chandra Krintz
Department of Computer Science, University of California, Santa Barbara, CA, USA

Definition

Cloud computing refers to a service-oriented computing methodology in which computational, networking, and storage capabilities, and software services are virtualized and contracted for via Service-Level Agreements (SLAs), through well-defined interfaces that are exercised via a network connection.

Historical Background

Cloud computing has its origins in two separate but related technological developments that emerged in the commercial sector during the 1990s. The first is very-large-scale e-commerce as evinced by "Internet-scale" companies such as Google, eBay, and Yahoo! These companies fielded proprietary web-facing infrastructure designed to support millions of simultaneous, anonymized users via the common carrier Internet.

The second development, pioneered by the company Salesforce.com, was "Software-as-a-Service" (SaaS) – the credentialed, and for-fee access to proprietary software via programmatic web-facing Application Programming Interfaces (APIs). Until that time, software was typically licensed and shipped (via persistent media) to consumers for installation on locally maintained hardware. Salesforce.com offered Customer Relationship Management (CRM) software hosted on infrastructure maintained by the company and accessed via network-accessible interfaces only. Thus, customers would purchase access to the software over the Internet. This model of network-facing services accessed via APIs relieves users of the installation and maintenance burden of downloading or purchasing "shrink wrapped" software and installing or updating it directly on their computers. Perhaps more importantly, this model also facilitates collection and analysis of user activity, purchasing behavior, and personal information with which vendors can tailor application features, user experiences, and marketing and advertisement activities.

The scale at which these network-facing software venues operate (often termed web scale or Internet scale) mandates the heavy use of automation to manage the infrastructure that is used to host them. Large-scale e-commerce and SaaS vendors have proprietary internal automation systems that monitor and maintain the infrastructure that hosts their respective web-facing services. In late 2003 at Amazon.com, Chris Pinkham and Benjamin Black proposed to standardize the Amazon-internal infrastructure automation around curated web services. As a result, Amazon Web Services (AWS) would serve Amazon-internal needs, but would also be capable of providing rental access to virtualized infrastructure to Amazon customers. Launched in 2006, AWS offered web-service access to data center infrastructure components (virtual servers, network connectivity, disk volumes) and infrastructure services (storage, message queuing, domain name service, etc.), thereby creating the first publicly accessible, general purpose cloud.

The term "cloud," coined in August of 2007 by Eric Schmidt at Google and borrowed from networking nomenclature, refers to the location-obscured service delivery mechanism inherent in the model. In a networking context, the networking infrastructure provides connectivity as a service. That is, users of the network cannot "see" or manipulate the array of infrastructure components that must be employed to implement l network connectivity. Instead, the connectivity is provided as a user-opaque service that is often depicted as a cloud in many high-level schematic representations.

"Cloud computing" refers to the implementation of this abstraction for generalized computing, storage, and higher-level software services. Cloud users gain access to these capabilities as services without the ability to "see" or know how they are implemented. To allow users to reason about the quality of the cloud resources they employ as services, each service is typically made available with a Service Level Agreement (SLA) defining minimum performance and reliability levels that the cloud provider attempts to maintain for the service. Cloud computing is also typically implemented using vast collections of commodity hardware components located in multiple data centers (often termed warehouse-scale computers). The organization and automation used to manage this hardware is usually proprietary, but the components themselves are commodities both to ensure scalable management and also to minimize capital investment.

Scientific Fundamentals

In general, cloud computing offers users access to computational capabilities via a set of network-facing web services. Web services and the separation between usage and implementation that they enforce make it possible for the consumers of cloud services to buy or rent them from a provider in a customer-vendor relationship. Cloud automation is also useful as a technology for supporting user communities within an organization that share a common set of computing resources that the organization has purchased and manages locally.

Cloud Abstractions

Cloud services are often decomposed into three categories.

- **Infrastructure as a Service (IaaS)** provides the ability to allocate and release virtualized data center infrastructure components via a set of web service APIs. These components include virtual machines with a variety of capabilities and operating systems, ephemeral and persistent storage volumes, network topologies and firewall rules, load balancers, etc.
- **Platform as a Service (PaaS)** exports via APIs shared, scalable, preinstalled, and automatically managed software resources for use

by applications that have been uploaded to the platform. PaaS resources typically include application servers, databases, security and authentication support, task queue services, messaging systems, and analytics technologies, among others.

- **Software as a Service (SaaS)** exports a web-facing API for a specific application. Consumers of a SaaS service access a customizable instance of the application that shares backend services and infrastructure resources with other instances and users.

Both IaaS and PaaS provide services that are designed to support applications written and installed by their users. SaaS refers to the notion that the application itself is accessed as a web-facing service.

Architecturally, it is possible to layer the different cloud abstractions. A SaaS application can use PaaS services that are implemented using resources that are culled from an IaaS cloud. This layering is increasingly common; however, currently public cloud vendors typically specialize either in IaaS or PaaS and their customers use these services to implement their own SaaS. The predominant public IaaS vendor is Amazon.com with Rackspace, IBM, Hewlett-Packard, Google, and Terremark offering competitive alternatives. The leading public PaaS vendors are Google and Microsoft, but there are emerging competitive offerings available from EngineYard, Heroku, AppFog, and Red Hat, among others.

Both IaaS and PaaS vendors are increasingly offering services across abstraction layers. Amazon's AWS, originally strictly an IaaS cloud, now offers a variety of database and content management services. Similarly, Google's original PaaS offering, Google App Engine, is now coupled with an IaaS offering called Compute Engine (also with other storage and analytics services) as part of Google Cloud Platform. Microsoft Azure was originally a PaaS, but now supports many IaaS features. Thus, future cloud products will likely include both IaaS and PaaS services so that users can choose the right combination of abstractions for a particular application.

Public and Private Clouds

IT organizations often wish to offer their respective user communities the self-service and automated access to computational capabilities that public clouds provide, using their own internal hardware resources. These clouds that are operated by organizations for their constituencies are often termed private clouds to differentiate them from retail clouds that are operated by vendors for public consumption.

Architecturally, private clouds typically differ from public clouds in two key ways. First, they are quota controlled so that users can acquire resource quantities that do not exceed the quotas set for them by the cloud administrators. The reason for this difference stems from the way in which the expense associated with the infrastructure is amortized.

In a public cloud setting, the retail relationship directly monetizes the investment in the infrastructure and its management costs. Because access is fully automated and self-service, the marginal costs associated with additional workload diminish, but charging is typically linear. Thus, profits increase with usage scale. To ensure that SLAs can be met, users are often restricted in the number of resources that can acquire at any one time, but not in the duration over which they may retain an allocation. That is, a public cloud will usually rent a resource to a user for a logically indefinite period of time (until the lease is terminated or payment for the lease is no longer received). To ensure that users do not need to queue waiting for resources (i.e., the cloud appears to be always responsive and available), the total number of resources a single user can acquire is limited to an insignificant fraction of the resources available in the cloud. Because no user can substantially drain the cloud of resources, the cloud administrator need only predict growth or shrinkage of the user population in order to understand how to provision the cloud itself.

In a private cloud setting, the resources are often budgeted with an explicit quota in mind, either per user or per project. That is, the organization purchases and sites resources that are sufficient to accomplish a specific purpose to support a specific user community. The organization wishes to allow the resources to be used in an automated, self-service manner (to streamline the IT support burden, typically), but users are given fixed quotas on the number and duration of usage they are entitled to consume from the cloud. In this way, organizations implement the speed and customizability that clouds offer while aligning allocation proportions with budget planning and conformance. As a result, users of private clouds are provided with a slightly different abstraction in terms of scale and usage governance than users of public clouds even if the APIs are similar or the same.

A second way in which private and public clouds differ is with respect to the degree to which the infrastructure and the cloud abstractions are managed separately. The software abstractions and hardware configurations that implement public clouds are typically co-designed. That is, the cloud abstraction designers "know" what hardware capabilities and configurations will be available to implement their abstractions. Similarly, the hardware and configuration are chosen specifically to support a set of cloud abstractions.

In a private cloud setting, the cloud software is usually deployed as a control and management facility for an existing, predefined hardware deployment. That is, the hardware and its configuration are typically first chosen to meet organizational goals with respect to budgetary allocation, product delivery timing, security policies, etc. Once the infrastructure and its configuration are determined, the private cloud software then functions as an overlaying operating system that exports the cloud APIs, thereby hiding the underlying infrastructure details.

For this reason, private cloud software must be designed in a way that makes it portable between a large variety of hardware components and deployment configurations. While the public cloud software designer can build into the code dependencies on specific hardware components or hardware configurations (e.g., network connectivity), the designer of private cloud software must attempt to support the largest possible range of "unseen" hardware component and deployment scenarios.

Also for this reason, most (but not all) of the predominant private cloud software systems are available as open source including Open-Stack, Eucalyptus, CloudStack, Open Nebula, OpenShift, AppScale, and Cloud Foundry. The technical transparency, configurability, and ability to modify and customize the software to suit site-specific needs have made open source platforms dominant in private cloud settings. Notable nonopen source alternatives include vCloud Server (a VMWare product) and Azure Enterprise (a version of Microsoft Azure that can be deployed as a private cloud).

Cloud Architectures

The precise implementations of public clouds are closely held trade secrets. Each public cloud vendor's profit margin depends on proprietary administration and management techniques that can provide a competitive advantage. Alternatively, because private clouds are often built from open source components using existing data center configurations, more is known about the state of the art in private cloud implementation.

IaaS

IaaS clouds must allow users to provision four principle data center functionalities:

- Virtual Machines – The computing elements of IaaS clouds are typically implemented by virtualized machine instances that are hosted by physical hardware using hypervisor technologies. Each virtual machine is characterized by a number of computing elements (cores), a memory size, and an initial storage capacity (which may or may not be deallocated when the virtual machine is decommissioned).
- Persistent Storage – Users must be able to allocate storage elements that persist beyond the lifetime of their virtual machines. While the precise semantics associated with this storage vary from cloud to cloud, most IaaS clouds include facilities for allocating persistent disk volumes (that can be attached and detached from different virtual machines) and eventu-

ally consistent storage objects (that are accessed via a web service interface).

- Network Capabilities – IaaS clouds allow users to control network connectivity (in a limited way) between virtual machines and between virtual machines and the network connected to the cloud itself (i.e., the "outside" world). They also often provide dynamically allocated and programmatically controlled load balancers and firewalls. Finally, most IaaS clouds provide support for configuring Domain Name Service (DNS) entries for the resources they host.
- Identity – IaaS clouds typically implement role-based access control necessitating the ability to provision user identities. This functionality is either strictly under the control of the cloud administrator or includes a delegation capacity in which users provisioned by the administrator can be given the rights necessary to provision other users.

The technologies that are available to implement these four functionalities vary widely. For example, there are several production-quality hypervisors available that can implement virtual machines in a cloud setting including Xen, KVM, VMWare ESX, and Oracle VirtualBox. Persistent storage volumes can be allocated and attached via a network connection using most Storage Area Networking (SAN) products as well as a number of storage overlay systems. Network provisioning can make use of dynamically assigned VLANs, software defined networking (e.g., OpenFlow enabled switches), and hardware and/or software load balancers. Finally, identity provisioning may be implemented using LDAP, Active Directory, and/or certificate-based web service policies.

While the specific technologies used to implement any given cloud may not be known, all clouds must allow their users to manipulate these technologies via web-service interfaces that are both scalable and self-service. Thus, at some level, all cloud implementations must include technologies that can field and respond to web service requests (using either the SOAP or REST messaging formats). Web service technologies also vary widely, but most include facilities for user authentication, HTTP connection management, argument marshaling and unmarshaling, and scalable database access.

Thus, an IaaS cloud communicates with its users via authenticated web-service requests and responses that are implemented via a web service "front end." User requests are translated into one or more internal commands to manipulate the various component technologies internal to the cloud. These commands may take the form of subsequent internal web service requests (if the cloud implementation uses web services as its internal messaging system) or some other messaging paradigm that is hidden from the external cloud users.

For scalability reasons, IaaS clouds typically report request status via an asynchronous polling interface. That is, when a user makes a request, the cloud acknowledges that it has received the request in a reply, but the status of the request (including the time at which it will be fulfilled) is not immediately returned. Instead, the user must poll the cloud (using a request ID) to determine if and when the request has been fulfilled or has been terminated due to an error (e.g., an authentication failure).

In addition, the cloud must track the status of all internal commands and resources in a way that is scalable and highly available. To implement the necessary data integrity associated with cloud state, most IaaS clouds use an internal, scalable database deployment with some form of high-availability capability. Without a scalable database, user requests (each of which may trigger multiple commands to manipulate infrastructure components that must also be tracked) may be delayed by database performance. Moreover, the integrity of the system depends on the ability to determine, unambiguously, the status of each request, and resource at all times. If the system does not implement high availability, then an internal failure or partial failure (e.g., network partition) must cause the cloud to stop. As the cloud scales, the probability of a (possibly transient) internal failure escalates. Thus, clouds implement high availability in order to achieve functionality at scale.

PaaS

PaaS clouds must allow users to provision the software infrastructure for an application. This includes:

- Application Execution Engines – The computing elements of PaaS clouds are programmatic responses to web service requests and background tasks. Configuration, deployment, management, and scaling of application server instances can be user self-service or automated by the platform. The latter are also referred to as application PaaS systems (aPaaS).
- Application Services – PaaS systems also export services via application programming interfaces (APIs) from which scalable, web-based applications can be constructed and managed. The platform typically deploys, manages, and scales these services automatically.

PaaS systems make these components available to users via APIs, which enable and hide their shared use and implementation. Users upload their applications, typically written in high level languages, e.g., Java, Python, PHP, and Ruby, to the platform for execution. The platform forwards HTTP requests to instances of the application and executes programmer-defined background tasks. The platform can provision, withdraw, and load-balance data center resources for such execution according to load or as directed by users. PaaS systems isolate and sandbox concurrently executing applications via system-level virtualization or other operating system and programmatic mechanisms (e.g., managed programming languages, containers, and namespaces).

PaaS systems support multiple programming languages and models through the use of different application execution engines. PaaS systems also make implementations of common application components available as scalable, shared services via network-facing APIs. Thus, the user of the cloud is relieved of the administrative burden associated with installing and configuring these frequently used services. Typically, the PaaS service implementations and configurations

are highly scalable, reliable, and fault tolerant so that the cloud can support large numbers of applications, users, requests, and tasks concurrently. Moreover, the separation between the applications and the services via APIs allows the underlying platform and service implementation to change and evolve without significantly impacting usage. Thus, users get access to a high-quality and cloud-administered set of services for their applications automatically. Common PaaS application services include structured and blob data storage and caching, queuing systems, and user authentication, among others.

Data Center Maintenance and Power Optimization

Because the cloud obscures from the user (via different abstractions) the specific way in which each request is fulfilled, it offers the opportunity for automated power optimization and infrastructure maintenance in the data center. That is, the user cannot "see" which specific machine, storage device, or network ensemble is being used to implement her or his request and, instead, is given an SLA that defines the minimum performance that will be delivered. Thus, the cloud automation system is free to optimize power consumption subject to the constraints of the SLAs it guarantees.

There are two primary methodologies for optimizing power usage. The first is to implement multitenancy for the purposes of consolidation so that idle components may be powered down. Because cloud allocations are isolated on a per-user basis, it is possible to map the requests from unrelated users to the same resources as long as the SLA guarantees can still be met. Thus, each resource can host multiple "tenants" who (because of the cloud abstractions) do not know their usage is co-located. By implementing multitenancy, the cloud can ensure that resource utilization is maximized on the resources that are in use, thereby allowing the others to be powered down. When request load exceeds the capacity (including the effects of multi-tenancy on SLA guarantees) of the in-use resources, the cloud must be able to power up additional resources automatically.

The second related technique is to migrate workload so that resources can be idled and powered down. Virtual machines, for example, can be migrated "live" between machines with minimal interruption of availability and/or network connectivity. Similarly, many SANs provide the ability to move volumes between storage pools. Thus, again because the cloud obscures the actual physical hardware and its location within the cloud, it is possible to move workload within the cloud so that resources can be idled and powered down until needed.

Key Applications

Cloud computing systems have been designed and specialized for Internet-accessible applications and services. As such, there are a large number of toolkits and support technologies to facilitate the development, configuration, deployment, and maintenance of web-based applications and services. Moreover, cloud systems automate the dynamic addition and removal (i.e., elastic use) of resources in response to load and use. In addition, public cloud systems give users access to vast resources at very low cost. Examples of application domains that require and benefit significantly from such capabilities include e-commerce, social, business internal, collaboration, data storage and sharing, data analytics, and gaming applications. Because these capabilities are exported via APIs (web and programmatic), cloud systems are similarly ideal for deploying and supporting the "backend" software that mobile and resource-constrained devices can use to extend their capabilities and coordinate.

Cross-References

▶ Data Migration Management
▶ Infrastructure-as-a-Service (IaaS)
▶ Multitenancy
▶ Platform-as-a-Service (PaaS)
▶ Software-as-a-Service (SaaS)

Recommended Reading

1. http://aws.amazon.com/whitepapers/
2. http://dl.acm.org/citation.cfm?id=1294281
3. http://research.google.com/archive/bigtable.html
4. http://research.google.com/archive/mapreduce.html
5. http://research.google.com/pubs/DistributedSystemsandParallelComputing.html
6. http://research.google.com/pubs/pub36971.html
7. http://research.microsoft.com/apps/pubs/default.aspx?id=189249

Cloud Intelligence

Torben Bach Pedersen
Department of Computer Science, Aalborg University, Aalborg, Denmark

Definition

Cloud Intelligence (CI) is a collection of technologies emerging from the migration of business intelligence (BI) and analytics technologies to a cloud computing environment combined with exploiting the massive range of new intelligence opportunities opened up by cloud computing and Big Data.

Key Points

Cloud Intelligence can be characterized as BI and analytics *in*, *for*, and *with* the cloud.

In the cloud refers to the fact that cloud intelligence solutions will be offered "as-a-service", running in the cloud rather than at user sites. The cloud intelligence services should be dynamically scalable to a global level. Thus, massively parallel computing techniques such as MapReduce, and beyond are the standard underlying computing platform. Another aspect of running on a cloud platform is the fundamentally new economic model needed for cloud intelligence. In traditional BI, the (large) cost of building a BI system is initially covered by an enterprise investment which must later be paid back through savings or new earnings in the enterprise. In cloud intelligence, there will typically no longer be a central entity paying the bill. Instead, pay-as-you-go models that allow users to pay (a small amount) per use, e.g., of a data set, in return for a one-time advantage, will become the norm in combination with open source inspired models, supporting that cloud intelligence systems will be *grown* over time (rather than built) through a collaborative community effort.

For the cloud means that cloud intelligence solutions should be able to handle the full multitude of complex "Big Data" available in the cloud *and* the complex analyses that should be performed on that data. The types of data available in the cloud include structured data, text data, e.g., web pages, documents, or tweets, semi-structured data, e.g., XML or Semantic Web, graph data, e.g., from social networks or scientific applications, social network status updates, sensor data, e.g., RFID, and geo data, e.g., positions from GPS or indoor positioning technologies. The analyses to perform include both type-specific analyses, e.g., sentiment analysis for tweets, as well as more advanced analyses combining several data types. Two particular types of emerging analytics applications try to predict the future (*Predictive Analytics*), and further use the predictions to suggest (prescribe) optimal decisions/actions in certain scenarios (*Prescriptive Analytics*). With all this, potentially sensitive, data available in the cloud, built-in protection of *data privacy* is needed.

With the cloud refers to fully utilizing the new *use* opportunities offered by the cloud. Cloud intelligence, as opposed to traditional BI, will be much more bottom-up and user-driven as opposed to the top-down enterprise-driven approach of traditional BI. The users will increasingly be private citizens or other types of more independent actors collaborating in ever-changing constellations to achieve a temporary common goal. Cloud Intelligence is collaborative by nature. Many people, knowing each other or not, share data and results, and thus contribute, consciously or not, to achieve a goal. Additionally, the data "belonging" to a user with a certain (open-ended) question is typically not enough to answer the question. Thus, new data, typically available on the Internet, must be discovered, selected, acquired, integrated, and presented. However, current BI/analytics tools do not support this process, as they are designed for a closed-world setting, where data and questions are more or less known in advance. The recent proposal of Fusion Cubes [1] aims to support exactly these scenarios.

A true cloud intelligence solution should support *all* three aspects (in, for, with the cloud). Here, the concept of a *cloud warehouse* [4] has been proposed as a unified platform for cloud intelligence.

In addition to Cloud BI service offerings from the big vendors, several companies now offer dedicated cloud intelligence products and services [5, 6]. New research events have also occurred, e.g., the VLDB workshop series on Cloud Intelligence (2014).

Cross-References

▶ Big Data Platforms for Data Analytics
▶ Business Intelligence
▶ Cloud Computing
▶ Data Warehousing in Cloud Environments
▶ Data Warehousing on Nonconventional Data
▶ Predictive Analytics
▶ Prescriptive Analytics
▶ Privacy

Recommended Reading

1. Abelló A, Darmont J, Etcheverry L, Golfarelli M, Mazón J-N, Naumann F, Pedersen TB, Rizzi S, Trujillo J, Vassiliadis P, Vossen G. Fusion cubes: towards self-service business intelligence. Int J Data Warehous Min. 2013;9(2):66–88.c.
2. Cloud Intelligence Workshop series. 2013. http://eric.univ-lyon2.fr/cloud-i/. Accessed 26 Aug 2014.
3. Pedersen TB. Research challenges for cloud intelligence: invited talk. In: Proceedings of the BEWEB 2010, Part of EDBT/ICDT Workshops Proceedings; 2010.
4. Pedersen TB, Pedersen D, Riis K. On-demand multidimensional data integration: toward a semantic foundation for cloud intelligence. J Supercomput. 2013;65(1):217–57.

5. VLDB Solutions. Cloud Intelligence data warehouse platform. 2014. http://www.cloudintelligence.co.uk. Accessed 26 Aug 2014.
6. Sonian. Sonian Cloud Intelligence. 2014. http://sonian.com/cloud-intelligence/. Accessed 26 Aug 2014.
7. Darmont J, Pedersen TB, Middelfart M. Cloud intelligence: what is REALLY new? In: Darmont J, Pedersen TB, editors. Proceedings of the 1st International Workshop on Cloud Intelligence; 2012.
8. Darmont J, Pedersen TB. Cloud intelligence: challenges for research and industry. In: Darmont J, Pedersen TB, editors. Proceedings of the 2nd International Workshop on Cloud Intelligence; 2013.

Cluster and Distance Measure

Dimitrios Gunopulos
Department of Computer Science and Engineering, The University of California at Riverside, Bourns College of Engineering, Riverside, CA, USA

Synonyms

Segmentation; Unsupervised learning

Definition

Clustering

Clustering is the assignment of objects to groups of similar objects (clusters). The objects are typically described as vectors of features (also called attributes). So if one has n attributes, object x is described as a vector $(x_1,..,x_n)$. Attributes can be numerical (scalar) or categorical. The assignment can be hard, where each object belongs to one cluster, or fuzzy, where an object can belong to several clusters with a probability. The clusters can be overlapping, though typically they are disjoint. Fundamental in the clustering process is the use of a distance measure.

Distance Measure

In the clustering setting, a distance (or equivalently a similarity) measure is a function that quantifies the similarity between two objects.

Key Points

The choice of a distance measure depends on the nature of the data, and the expected outcome of the clustering process. The most important consideration is the type of the features of the objects. One first focuses on distance measures when the features are all numerical. This includes features with continuous values (real numbers) or discrete values (integers). In this case, typical choices include:

1. The L_p norm. It is defined as $D(x, y) = \left(\sum_{1 \le i \le n} (X_1 - Y_1)^p \right)^{1/p}$. Typically p is 2 (the intuitive and therefore widely used Euclidean distance), or 1 (the Manhattan or city block distance), or infinity (the Maximum distance).
2. The Mahalanobis distance. It is defined as $D(x, y) = (x - y) \sum^{-1} (x - y)^T$ which generalizes the Euclidean and allows the assignment of different weights to different features.
3. The angle between two vectors, computed using the inner product of two vectors $x \cdot y$.
4. The Hamming distance, which measures the number of disagreements between two binary vectors.

In different settings different distance measures can be used. The edit, or Levenshtein, distance, is an extension of the Hamming distance, and is typically used for measuring the distance between two strings of characters. The edit distance is defined as the minimum number of insertions, deletions or substitutions that it takes to transform one sting to another.

When two time series are compared, the Dynamic Time Warping distance measure is often used to quantify their distance. The length of the Longest Common Subsequence (LCSS) of two time series is also frequently used to provide a similarity measure between the time series. LCSS is a similarity measure because the longest common subsequence becomes longer when two time series are more similar. To create a distance measure, LCSS is typically normalized by dividing by the length of the longest of the two sequences, and then subtracting the ratio from one.

Finally, when sets of objects are compared, the Jaccard coefficient is typically used to compute their distance. The Jaccard coefficient of sets A and B is defined as $J(A, B) = |A \cap B| / |A \cup B|$, that is, the fraction of the common elements over the union of the two sets.

The majority of the distance measures used in practice, and indeed most of the ones described above are metrics. Formally, a distance measure D is a metric if it obeys the following properties:

For objects A, B, (i) D(A,B) $\geq$ 0, (ii) D(A,B) = 0 if and only if A = B, and (iii) D(A,B) = D(B,A), and (iv) for any objects A,B,C, D(A,B) + D(B,C) $\geq$ D(A,C) (triangle inequality).

Most distance measures can be trivially shown to observe the first three properties, but do not necessarily observe the triangle inequality. For example, the constrained Dynamic Time Warping distance, a typically used measure to compute the similarity between time series which does not allow arbitrary stretching of a time series, is not a metric because it does not satisfy the triangle inequality. Experimental results have shown that the constrained Dynamic Time Warping distance performs at least as good as the unconstrained one and it is also faster to compute, thus justifying its use although it is not a metric. Note however that, if it is so required, any distance measure can be converted into a metric by taking the shortest path between objects A and B in the complete graph where each object is a node and each edge is weighted by the distance between the two nodes.

Cross-References

▶ Clustering Overview and Applications
▶ Data Mining

Recommended Reading

1. Everitt BS, Landau S, Leese M. Cluster analysis. Chichester: Wiley; 2001.
2. Jain AK, Murty MN, Flyn PJ. Data clustering: a review. ACM Comput Surv. 1999;31(3):264.
3. Theodoridis S, Koutroubas K. Pattern recognition. Academic; 1999.

Clustering for Post Hoc Information Retrieval

Dietmar Wolfram
University of Wisconsin-Milwaukee,
Milwaukee, WI, USA

Synonyms

Document clustering

Definition

Clustering is a technique that allows similar objects to be grouped together based on common attributes. It has been used in information retrieval for different retrieval process tasks and objects of interest (e.g., documents, authors, index terms). Attributes used for clustering may include assigned terms within documents and their co-occurrences, the documents themselves if the focus is on index terms, or linkages (e.g., hypertext links of Web documents, citations or co-citations within documents, documents accessed). Clustering in IR facilitates browsing and assessment of retrieved documents for relevance and may reveal unexpected relationships among the clustered objects.

Historical Background

A fundamental challenge of information retrieval (IR) that continues today is how to best match user queries with documents in a queried collection. Many mathematical models have been developed over the years to facilitate the matching process. The details of the matching process are usually hidden from the user, who is only presented with an outcome. Once candidate documents have been identified, they are presented to the user for perusal. Traditional approaches have relied on ordered linear lists of documents based on calculated relevance or another sequencing criterion (e.g., date, alphabetical by title or author). The resulting linear list addresses the assessed relationship of documents to queries, but not the relationships between the documents. Clustering techniques can reduce this limitation by creating groups of documents (or other objects of interest) to facilitate more efficient retrieval or perusal and evaluation of retrieved sets.

The application of clustering techniques to IR extends back to some of the earliest experimental IR systems including Gerard Salton's SMART system, which relied on document cluster identification within a vector space as a means of quickly identifying sets of relevant documents. The rationale for applying clustering was formalized as the "cluster hypothesis," proposed by Jardine and van Rijsbergen [8]. This hypothesis proposes that documents that are relevant to a query are more similar to each other than to documents that are not relevant to the query. The manifestation of this relationship can be represented in different ways by grouping like documents or, more recently, visualizing the relationships and resulting proximities in a multi-dimensional space.

Early applications of clustering emphasized its use to more efficiently identify groups of related, relevant documents and to improve search techniques. The computational burden associated with real-time cluster identification during searches on increasingly larger data corpora and the resulting lackluster performance improvements have caused clustering to lose favor as a primary mechanism for retrieval. However, clustering methods continue to be studied and used (see, for example, [10]). Much of the recent research into clustering for information retrieval has focused on other areas that support the retrieval process. For instance, clustering has been used to assist in query expansion, where additional terms for retrieval

may be identified. Clustering of similar terms also can be used to construct thesauri, which can be used to index documents [4].

Recent research on clustering has highlighted its benefits for post hoc retrieval tasks, in particular for the presentation of search results to better model user and usage behavior. The focus of applications presented here is on these post hoc IR tasks, dealing with effective representation of groups of objects once identified to support exploratory browsing and to provide a greater understanding of users and system usage for future IR system development.

Scientific Fundamentals

Methods used to identify clusters are based on cluster analysis, a multivariate exploratory statistical technique. Cluster analysis relies on similarities or differences in object attributes and their values. The granularity of the analysis and the validity of the resulting groups are dependent on the range of attributes and values associated with objects of interest. For IR applications, clusters are based on common occurrences and weights of assigned terms for documents, the use of query terms, or linkages between objects of interest represented as hypertext linkages or citations/co-citations.

Clustering techniques can be divided into hierarchical and non-hierarchical approaches. Non-hierarchical clustering methods require that a priori assumptions be made about the nature and number of clusters, but can be useful if specific cluster parameters are sought. The k-means clustering family of algorithms, which partition a set of observations into k clusters, has been widely used. Recent studies have proposed more effective methods that extend the k-means approach [9], integrate k-means with other methods such as harmony clustering [6], or use other approaches such as a latent semantic indexing subspace signature model [14]. Hierarchical clustering, which is more commonly used, begins with many small groups of objects that serve as initial clusters. Existing groups are clustered into larger groups until only one cluster remains. Visually, the structure and relationship of clusters may be represented as a dendrogram, with different cluster agglomerations at different levels on the dendrogram representing the strength of relationship between clusters. Other visualization techniques may be applied and are covered elsewhere. In hierarchical methods, the shorter the agglomerative distance, the closer the relationship and the more similar the clusters are. As an exploratory technique, there is no universally accepted algorithm to conduct the analysis, but the general steps for conducting the analysis are similar. First, a similarity measure is applied to the object attributes, which serves as the basis for pairwise comparisons. Standard similarity or distance measures applied in IR research such as the Euclidean distance, cosine measure, Jaccard coefficient, and Dice coefficient can be used. Next, a method for cluster determination is selected. Common methods include: single complete linkage, average linkage, nearest neighbor, furthest neighbor, centroid clustering (representing the average characteristics of objects within a cluster), and Ward's method. Each method uses a different algorithm to assess cluster membership and may be found to be more appropriate in given circumstances. Outcomes can vary significantly depending on the method used. This flexibility underscores one of the challenges for effectively implementing cluster analysis. With no one correct or accepted way to conduct the analysis, outcomes are open to interpretation, but may be viewed as equally valid. For example, single linkage clustering, which links pairs of objects that most closely resemble one another, is comparatively simple to implement and has been widely used, but can result in lengthy linear chains of clusters. Parameters may need to be specified that dictate the minimum size of clusters to avoid situations where there are large orders of difference in cluster membership. Another challenge inherent in clustering is that different clustering algorithms can produce similar numbers of clusters, but if some clusters contain few members, this does little to disambiguate the members within large clusters. The number of clusters that partition the object set can be variable in hierarchical clustering. More clusters result in

fewer objects per cluster with greater inter-object similarity, but with potentially more groups to assess. It is possible to test for an optimal number of clusters using various measures that calculate how differing numbers of clusters affect cluster cohesiveness.

Clustering may be implemented in dynamic environments by referencing routines based on specific clustering algorithms developed by researchers or through specialty clustering packages. Details on clustering algorithms for information retrieval can be found in Rasmussen [11]. Standard statistical and mathematical software packages such as SAS and SPSS also support a range of clustering algorithms. Special algorithms may need to be applied to very large datasets to reduce computational overhead, which can be substantial for some algorithms.

Key Applications

In addition to early applications of clustering for improving retrieval efficiency, clustering techniques in IR have included retrieval results presentation, and modeling of IR user and usage characteristics based on transactions logs. Although largely a topic of research interest, some applications have found their way into commercial systems. Clustering of search results has been applied by several Web-based search services since the late 1990s, some of which are no longer available. Most notable of the current generation is Yippy (yippy.com), formerly known as Clusty, which organizes retrieval results from several search services around topical themes.

The application of clustering to support interactive browsing has been an active area of investigation in recent years. Among the earliest demonstrations for this purpose was the Scatter/Gather method outlined by Cutting et al. [5], in which the authors demonstrated how clustering of retrieved items can facilitate browsing for vaguely defined information needs. This approach was developed to serve as a complement to more focused techniques for retrieval assessment. In application, the method presents users with a set of clusters that serves as the starting point for browsing. The user selects the clusters of greatest interest. The contents of those clusters are then gathered into a single cluster, which now serves as the corpus for a new round of clustering, into which the new smaller corpus of items is scattered. The process continues until the user's information need is met or the user abandons the search. To support real time clustering of datasets, the authors developed an efficient clustering algorithm, called buckshot, plus a more accurate algorithm, named fractionation, to permit more detailed clustering in offline environments where a timely response is less critical. Another algorithm, called cluster digest, was used to encapsulate the topicality of a given cluster based on the highest weighted terms within the cluster. Hearst and Pedersen [7] evaluated the efficacy of Scatter/Gather on the top-ranked retrieval outcomes of a large dataset, and tested the validity of the cluster hypothesis. The authors compared the number of known relevant items to those appearing in the generated clusters. A user study was also conducted, which demonstrated that participants were able to effectively navigate and interact with the system incorporating Scatter/Gather.

Mobile computing devices have made mobile IR possible, but the small display on many mobile devices for presenting a linear list does not scale well. Carpineto, Mizzaro, Romano and Snidero [1] tested the use of clustering of search results on mobile devices by using mobile clustering engines in comparison to plain search linear lists. The authors suggested the use of search results clustering may make mobile, goal-directed searching feasible and faster.

Increasingly, IR systems provide access to heterogeneous collections of documents. The question arises whether the cluster hypothesis and the benefits of capitalizing on its attributes extend to the distributed IR environment, where additional challenges include the merger of different representations of documents and identification of multiple occurrences of documents across the federated datasets. Crestani and Wu [3] conducted an experimental study to determine whether the cluster hypothesis holds in a distributed environment. They simulated a

distributed environment by using different combinations of retrieval environments and document representation heterogeneity, with the most sophisticated implementation representing three different IR environments with three different collections. Results of the different collections and systems were clustered and compared. The authors concluded that the cluster hypothesis largely holds true in distributed environments, but fails when brief surrogates of full text documents are used.

With the growing availability of large IR system transaction logs, clustering methods have been used to identify user and usage patterns. By better understanding patterns in usage behavior, IR systems may be able to identify types of behaviors and accommodate those behaviors through context-sensitive assistance or through integration of system features that accommodate identified behaviors. Chen and Cooper [2] relied on a rich dataset of user sessions collected from the University of California MELVYL online public access catalog system. Based on 47 variables associated with each user session (e.g., session length in seconds, average number of items retrieved, average number of search modifications), their analysis identified six clusters representing different types of user behaviors during search sessions. These included help-intensive searching, knowledgeable usage, and known-item searching. Similarly, Wen et al. [12] focused on clustering of user queries in an online encyclopedia environment to determine whether queries could be effectively clustered to direct users to appropriate frequently asked questions topics. IR environments that cater to a broad range of users are well-known for short query submissions by users, which make clustering applications based solely on query term co-occurrence unreliable. In addition to the query content, the authors based their analysis on common retrieved documents viewed by users. By combining query content with common document selections, a link was established between queries that might not share search terms. The authors demonstrated how the application of their clustering method, which was reportedly adopted by the encyclopedia studied, could effectively guide users to appropriate frequently asked questions.

The previous examples represent only a sample of clustering applications in an IR context. Additional recent research developments and applications using clustering may be found in Wu et al. [13].

Cross-References

▶ Data Mining
▶ Text Mining
▶ Visualization for Information Retrieval

Recommended Reading

1. Carpineto C, Mizzaro S, Romano G, Snidero M. Mobile information retrieval with search results clustering: prototypes and evaluations. J Am Soc Inf Sci Technol. 2009;60(5):877–95.
2. Chen HM, Cooper MD. Using clustering techniques to detect usage patterns in a web-based information system. J Am Soc Inf Sci Technol. 2001;52(11): 888–904.
3. Crestani F, Wu S. Testing the cluster hypothesis in distributed information retrieval. Inf Process Manage. 2006;42(5):1137–50.
4. Crouch CJ. A cluster-based approach to thesaurus construction. In: Proceedings of the 11th Annual International ACM SIGIR Conference on Research and Development in Information Retrieval; 1988. p. 309–20.
5. Cutting DR, Karger DR, Pedersen JO, Tukey JW. Scatter/Gather: a cluster-based approach to browsing large document collections. In: Proceedings of the 15th Annual International ACM SIGIR Conference on Research and Development in Information Retrieval; 1992. p. 318–29.
6. Forsati R, Mahdavi M, Shamsfard M, Meybodi MR. Efficient stochastic algorithms for document clustering. Inform Sci. 2013(Jan);220:269–91.
7. Hearst MA, Pedersen JO. Reexamining the cluster hypothesis: scatter/gather on retrieval results. In: Proceedings of the 19th Annual International ACM SIGIR Conference on Research and Development in Information Retrieval; 1996. p. 76–84.
8. Jardine N, van Rijsbergen C. The use of hierarchic clustering in information retrieval. Inf Storage Retr. 1971;7(5):217–40.
9. Kalogeratos A, Likas A. Document clustering using synthetic cluster prototypes. Data Knowl Eng. 2011;70(3):284–306.
10. Liu X, Croft WB. Cluster-based retrieval using language models. In: Proceedings of the 30th Annual International ACM SIGIR Conference on Research and Development in Information Retrieval; 2004. p. 186–93.
11. Rasmussen E. Clustering algorithms. In: Frakes WB, Baeza-Yates R, editors. Information retrieval data structures & algorithms. Englewood Cliffs: Prentice Hall; 1992. p. 419–42.
12. Wen JR, Nie JY, Zhang HJ. Query clustering using user logs. ACM Trans Inf Syst. 2002;20(1):59–81.
13. Wu W, Xiong H, Shekhar S, editors. Clustering and information retrieval. Norwell: Kluwer; 2004.
14. Zhu WZ, Allen RB. Document clustering using the LSI subspace signature model. J Am Soc Inf Sci Technol. 2013;64(4):844–60.

Clustering on Streams

Suresh Venkatasubramanian
University of Utah, Salt Lake City, UT, USA

Definition

An instance of a clustering problem (see clustering) consists of a collection of points in a distance space, a measure of the *cost* of a clustering, and a measure of the *size* of a clustering. The goal is to compute a partitioning of the points into clusters such that the cost of this clustering is minimized, while the size is kept under some predefined threshold. Less commonly, a threshold for the cost is specified, while the goal is to minimize the size of the clustering.

A data stream (see data streams) is a sequence of data presented to an algorithm one item at a time. A stream algorithm, upon reading an item, must perform some action based on this item and the contents of its working space, which is sublinear in the size of the data sequence. After this action is performed (which might include copying the item to its working space), the item is discarded.

Clustering on streams refers to the problem of clustering a data set presented as a data stream.

Historical Background

Clustering (see clustering) algorithms typically require access to the entire data to produce an effective clustering. This is a problem for large data sets, where random access to the data, or repeated access to the entire data set, is a costly operation. For example, the well-known k-means heuristic is an iterative procedure that in each iteration must read the entire data set twice. One set of approaches to performing clustering on large data involves sampling: a small subset of data is extracted from the input and clustered, and then this clustering is extrapolated to the entire data set.

The *data stream* paradigm [1] came about in two ways: first, as a way to model access to large streaming sources (network traffic, satellite imagery) that by virtue of their sheer volume, cannot be archived for offline processing and need to be aggregated, summarized and then discarded in real time. Second, the streaming paradigm has shown itself to be the most effective way of accessing large databases: Google's Map Reduce [2] computational framework is one example of the efficacy of stream processing.

Designing clustering algorithms for stream data requires different algorithmic ideas than those useful for traditional clustering algorithms. The online computational paradigm [3] is a potential solution: in this paradigm, an algorithm is presented with items one by one, and using only information learned up to the current time, must make a prediction or estimate on the new item being presented. Although the online computing paradigm captures the sequential aspect of stream processing, it does not capture the additional constraint that only a small portion of the history may be stored. In fact, an online algorithm is permitted to use the entirety of the history of the stream, and is usually not limited computationally in any way. Thus, new ideas are needed to perform clustering in a stream setting.

Foundations

Preliminaries

Let X be a domain and d be a distance function defined between pairs of elements in X. Typically,

it is assumed that d is a metric (i.e., it satisfies the triangle inequality $d(x, y) + d(y, z) \geq d(x, z) \forall x, y, z \in X$). One of the more common measures of the cost of a cluster is the so-called *median cost*: the cost of a cluster $C \subseteq X$ is the function

$$\text{cost}(C) = \sum_{x \in C} d\left(x, c^*\right)$$

where $c^* \in X$, the *cluster center*, is the point that minimizes $\text{cost}(C)$. The *k-median problem* is to find a collection of k disjoint clusters, the sum of whose costs is minimized.

An equally important cost function is the *mean cost*: the cost of a cluster $C \subseteq X$ is the function

$$\text{cost}(C) = \sum_{x \in C} d^2\left(x, c^*\right)$$

where c^* is defined as before. The *k-means problem* is to find a collection of clusters whose total *mean cost* is minimized. It is useful to note that the median cost is more robust to outliers in the data; however, the mean cost function, especially for points in Euclidean spaces, yields a very simple definition for c^*: it is merely the centroid of the set of points in the cluster. Other measures that are often considered are the *k-center cost*, where the goal is to minimize the maximum radius of a cluster, and the diameter cost, where the goal is to minimize the maximum diameter of a cluster (note that the diameter measure does not require one to define a cluster center).

A data stream problem consists of a sequence of items $x_1, x_2, \ldots, x_n$, and a function $f(x_1, \ldots, x_n)$ that one wishes to compute. The limitation here is that the algorithm is only permitted to store a *sublinear* number of items in memory, because n is typically too large for all the items to fit in memory. Further, even random access to the data is prohibitive, and so the algorithm is limited to accessing the data in sequential order.

Since most standard clustering problems (including the ones described above) are NP-hard in general, one cannot expect solutions that minimize the cost of a clustering. However, one can often show that an algorithm comes close to being optimal: formally, one can show that the cost achieved by an algorithm is within some multiplicative factor c of the optimal solution. Such an algorithm is said to be a c-approximation algorithm. Many of the methods presented here will provide such guarantees on the quality of their output. As usual, one should keep in mind that these guarantees are *worst-case*, and thus apply to any possible input the algorithm may encounter. In practice, these algorithms will often perform far better than promised.

General Principles

Stream clustering is a relatively new topic within the larger area of stream algorithms and data analysis. However, there are some general techniques that have proven their usefulness both theoretically as well as practically, and are good starting points for the design and analysis of stream clustering methods. This section reviews these ideas, as well as pointing to examples of how they have been used in various settings.

Incremental Clustering

The simplest way to think about a clustering algorithm on stream data is to imagine the stream data arriving in chunks of elements. Prior to the arrival of the current chunk, the clustering algorithm has computed a set of clusters for all the data seen so far. Upon encountering the new chunk, the algorithm must update the clusters, possibly expanding some and contracting others, merging some clusters and splitting others. It then requests the next chunk, discarding the current one. Thus, a core component of any stream clustering algorithm is a routine to incrementally update a clustering when new data arrives. Such an approach was developed by Charikar et al. [4] for maintaining clusterings of data in a metric space using a diameter cost function. Although their scheme was phrased in terms of incremental clusterings, rather than stream clusterings, their approach generalizes well to streams. They show that their scheme yields a provable approximation to the optimal diameter of a k-clustering.

Representations

One of the problems with clustering data streams is choosing a representation for a cluster. At the very least, any stream clustering algorithm stores the location of a cluster center, and possibly the number of items currently associated with this cluster. This representation can be viewed as a *weighted point*, and can be treated as a single point in further iterations of the clustering process. However, this representation loses information about the geometric size and distribution of a cluster. Thus, another standard representation of a cluster consists of the center and the number of points augmented with the sum of squared distances from the points in the cluster to the center. This last term informally measures the variation of points within a cluster, and when viewed in the context of density estimation via Gaussians, is in fact the sample variance of the cluster.

Clusters reduced in this way can be treated as weighted points (or weighted balls), and clustering algorithms should be able to handle such generalized points. One notable example of the use of such a representation is the one-pass clustering algorithm of Bradley et al. [5], which was simplified and improved by Farnstrom et al. [6]. Built around the well known k-means algorithm (that iteratively seeks to minimize the k-means measure described above), this technique proceeds as follows.

Algorithm 1 Clustering with representations

Initialize cluster centers randomly
While chunk of data remains to be read **do**
Read a chunk of data (as much as will fit in memory), and cluster it using the k-means algorithm.
For each cluster, divide the points contained within it into the core (points that are very close to the center under various measures), and the periphery.
Replace the set of points in the core by a summary as described above. Discard all remaining points.
Use the current cluster list as the set of centers for the next chunk.

It is important that representations be *linear*. Specifically, given two chunks of data c, c', and their representations r, r', it should be the case that the representation of $c \cup c'$ be formed from a linear combination of r and r'. This relates to the idea of *sketching* in stream algorithms, and is important because it allows the clustering algorithm to work in the (reduced) space of representations, rather than in the original space of data. Representations like the one described above are linear, and this is a crucial factor in the effectiveness of these algorithms.

Hierarchical Clustering

Viewing a cluster as a weighted point in a new clustering problem quickly leads to the idea of *hierarchical clustering*: by thinking of a point as a single-element cluster, and connecting a cluster and its elements in a parent-child relationship, a clustering algorithm can represent multiple levels of merges as a tree of clusters, with the root node being a single cluster containing all the data, and each leaf being a single item. Such a tree is called a Hierarchical Agglomerative Clustering (HAC), since it can be viewed bottom-up as a series of agglomerations. Building such a hierarchy yields more general information about the relationship between clusters, and the ability to make better judgments about how to merge clusters.

The well-known clustering algorithm BIRCH [7] makes use of a hierarchy of cluster representations to cluster a large database in a few passes. In a first pass, a tree called the CF-tree is constructed, where each internal node represents a cluster of clusters, and each leaf represents a cluster of items. This tree is controlled by two parameters: B, the branching factor, and T, a diameter threshold that limits the size of leaf clusters. In further passes, more analysis is performed on the CF-tree to compress clusters further. The tree is built much in the way a B+-tree is built: new items are inserted in the deepest cluster possible, and if the threshold constraint is violated, the cluster is split, and updates are propagated up the tree.

BIRCH is one of the best-known large-data clustering algorithms, and is generally viewed as a benchmark to compare other clustering algorithms against. However, BIRCH does not provide formal guarantees on the quality of the

clusterings thus produced. The first algorithm that computes a hierarchical clustering on a stream while providing formal performance guarantees is a method for solving the k-median problem developed by Guha et al. [8, 9]. This algorithm is best described by first presenting it in a non-streaming context:

Algorithm 2 Small space

Divide the input into l disjoint parts.

Cluster each part into $O(k)$ clusters. Assign each point to its nearest cluster center.

cluster the $O(lk)$ cluster centers, where each center is weighted by the number of points assigned to it.

Note that the total space required by this algorithm is $O(\ell k + n/\ell)$. The value of this algorithm is that it propagates good clusterings: specifically, if the intermediate clusterings are computed by algorithms that yield constant-factor approximations to the best clustering (under the k-median cost measure), then the final output will also be a (larger) constant factor approximation to the best clustering. Also note that the final clustering step may itself be replaced by a recursive call to the algorithm, yielding a hierarchical scheme.

Converting this to a stream algorithm is not too difficult. Consider each chunk of data as one of the disjoint parts the input is broken into. Suppose each part is of size m, and there exists a clustering procedure that can cluster these points into $2k$ centers with reasonable accuracy. The algorithm reads enough data to obtain m centers ($m^2/2k$ points). These m "points" can be viewed as the input to a second level streaming process, which performs the same operations. In general, the ith-level stream process takes $m^2/2k$ points from the $(i-1)$th-level stream process and clusters them into m points, which are appended to the stream for the next level.

The guarantees provided by the method rely on having accurate clustering algorithms for the intermediate steps. However, the general paradigm itself is useful as a heuristic: the authors show that using the k-means algorithm as the intermediate clustering step yields reasonable clustering results in practice, even though the method comes with no formal guarantees.

On Relaxing the Number of Clusters

If one wishes to obtain guarantees on the quality of a clustering, using at least k clusters is critical; it is easy to design examples where the cost of a $(k - 1)$-clustering is much larger than the cost of a k-clustering. One interesting aspect of the above scheme is how it uses weaker clustering algorithms (that output $O(k)$ rather than k clusters) as intermediate steps on the way to computing a k-clustering. In fact, this idea has been shown to be useful in a formal sense: subsequent work by Charikar et al. [10] showed that if one were to use an extremely weak clustering algorithm (in fact, one that produces $O(k \log n)$ clusters), then this output can be fed into a clustering algorithm that produces k clusters, while maintaining overall quality bounds that are better than those described above. This idea is useful especially if one has a fast algorithm that produces a larger number of clusters, and a more expensive algorithm that produces k clusters: the expensive algorithm can be run on the (small) output of the fast algorithm to produce the desired answer.

Clustering Evolving Data

Stream data is often temporal. Typical data analysis questions are therefore often limited to ranges of time ("in the last three days," "over the past week," "for the period between Jan 1 and Feb 1," and so on). All of the above methods for clustering streams assume that the goal is to cluster the entire data stream, and the only constraint is the space needed to store the data. Although they are almost always incremental, in that the stream can be stopped at any time and the resulting clustering will be accurate *for all data seen upto that point*, they cannot correctly output clusterings on *windows* of data, or allow the influence of past data to gradually wane over time. Even with non-temporal data, it may be important to allow the data analysis to operate on a subset of the data to capture the notion of *concept drift* [11], a term that is used to describe a scenario when natural data characteristics change as the stream evolves.

Sliding Windows

A popular model of stream analysis is the *sliding window* model, which introduces a new parameter W. The goal of the stream analysis is to produce summary statistics (a clustering, variance estimates or other statistics), on the *most recent W items only*, while using space that is sublinear in W. This model can be thought of as represented by a *sliding window* of length W with one end (the sliding end) anchored to the current element being read. The challenge of dealing with sliding windows is the problem of deletion. Although not as general as a fully dynamic data model where arbitrary elements can be inserted and deleted, the sliding window model introduces with the problem of updating a cluster representation under deletions, and requires new ideas.

One such idea is the *exponential histogram*, first introduced by Datar et al. [12] to estimate certain statistical properties of sliding windows on streams, and used by Babcock et al. [13] to compute an approximate k-median clustering in the sliding window model. The idea here is to maintain a set of buckets that together partition all data in the current window. For each bucket, relevant summary statistics are maintained. Intuitively, the smaller the number of items assigned to a bucket, the more accurate the summary statistics (in the limit, the trivial histogram has one bucket for each of the W items in the window). The larger this number, the fewer the number of buckets needed. Balancing these two conflicting requirements yields a scheme where each bucket stores the items between two timestamps, and the bucket sizes increase exponentially as they store items further in the past. It requires more detailed analysis to demonstrate that such a scheme will provide accurate answers to queries over windows, but the use of such exponentially increasing bucket sizes allows the algorithm to use a few buckets, while still maintaining a reasonable approximation to the desired estimate.

Hierarchies of Windows

The sliding window model introduces an extra parameter W whose value must be justified by external considerations. One way of getting around this problem is to maintain statistics for multiple values of W (typically an exponentially increasing family). Another approach, used by Aggarwal et al. [14] is to maintain *snapshots* (summary representations of the clusterings) at time steps at different levels of resolution. For example, a simple two level snapshot scheme might store the cluster representations computed after times t, $t + 1, \ldots t + W$, as well as $t, t + 2, t + 4, \ldots t + 2W$ (eliminating duplicate summaries as necessary). Using the linear structure of representations will allow the algorithm to extract summaries for time intervals: they show that such a scheme uses space efficiently while still being able to detect evolution in data streams at different scales.

Decaying Data

For scenarios where such a justification might be elusive, another model of evolving data is the *decay model*, in which one can think of a data item's influence waning (typically exponentially) with time. In other words, the value of the ith item, instead of being fixed at x_i, is a function of time $x_i(t) = x_i(0)\exp(-c(t - i))$. This reduces the problem to the standard setting of computing statistics over the entire stream, while using the decay function to decide which items to remove from the limited local storage when computing statistics. The use of exponentially decaying data is quite common in temporal data analysis: one specific example of its application in the clustering of data streams is the work on HPStream by Aggarwal et al. [15].

Key Applications

Systems that manage large data sets and perform data analysis will require stream clustering methods. Many modern *data cleaning systems* require such tools, as well as large scientific databases. Another application of stream clustering is for network traffic analysis: such algorithms might be situated at routers, operating on packet streams.

Experimental Results

Most of the papers cited above are accompanied by experimental evaluations and comparisons to prior work. BIRCH, as mentioned before, is a common benchmarking tool.

Cross-References

▶ Information Retrieval

Recommended Reading

1. Muthukrishnan S. Data streams: algorithms and applications. Found Trend Theor Comput Sci. 2005;1(2):117–236.
2. Dean J, Ghemaway S. MapReduce: simplified data processing on large clusters. In: Proceedings of the 6th USENIX Symposium on Operating System Design and Implementation; 2004. p. 137–50.
3. Borodin A, El-Yaniv R. Online computation and competitive analysis. New York: Cambridge University Press; 1998.
4. Charikar M, Chekuri C, Feder T, Motwani R. Incremental clustering and dynamic information retrieval. SIAM J Comput. 2004;33(6):1417–40.
5. Bradley PS, Fayyad UM, Reina C. Scaling clustering algorithms to large databases. In: Proceedings of the 4th International Conference on Knowledge Discovery and Data Mining; 1998. p. 9–15.
6. Farnstrom F, Lewis J, Elkan C. Scalability for clustering algorithms revisited. SIGKDD Explor. 2000;2(1):51–7.
7. Zhang T, Ramakrishnan R, Livny M. BIRCH: A new data clustering algorithm and its applications. Data Min Knowl Discov. 1997;1(2):141–82.
8. Guha S, Meyerson A, Mishra N, Motwani R, O'Callaghan L. Clustering data streams: theory and practice. IEEE Trans Knowl Data Eng. 2003;15(3):515–28.
9. Guha S, Mishra N, Motwani R, O'Callaghan L. Clustering data streams. In: Proceedings of the 41st Annual Symposium on Foundations of Computer Science; 2000. p. 359.
10. Charikar M, O'Callaghan L, Panigrahy R. Better streaming algorithms for clustering problems. In: Proceedings of the 35th Annual ACM Symposium on Theory of Computing; 2003. p. 30–9.
11. Domingos P, Hulten G. Mining high-speed data streams. In: Proceedings of the 6th ACM SIGKDD International Conference on Knowledge Discovery and Data Mining; 2000. p. 71–80.
12. Datar M, Gionis A, Indyk P, Motwani R. Maintaining stream statistics over sliding windows: (extended

abstract). In: Proceedings of the 13th Annual ACM
- SIAM Symposium on Discrete Algorithms; 2002.
p. 635–44.

13. Babcock B, Datar M, Motwani R, O'Callaghan L.
    Maintaining variance and k-medians over data
    stream windows. In: Proceedings of the 22nd ACM
    SIGACT-SIGMOD-SIGART Symposium on Princi-
    ples of Database Systems; 2003. p. 234–43.

14. Aggarwal CC, Han J, Wang J, Yu PS. A framework
    for clustering evolving data streams. In: Proceedings
    of the 29th International Conference on Very Large
    Data Bases; 2003. p. 81–92.

15. Aggarwal CC, Han J, Wang J, Yu PS. A frame-
    work for projected clustering of high dimensional
    data streams. In: Proceedings of the 30th Interna-
    tional Conference on Very Large Data Bases; 2004.
    p. 852–63.

Clustering Overview and Applications

Dimitrios Gunopulos
Department of Computer Science and
Engineering, The University of California at
Riverside, Bourns College of Engineering,
Riverside, CA, USA

Synonyms

Unsupervised learning

Definition

Clustering is the assignment of objects to groups
of similar objects (clusters). The objects are typically described as vectors of features (also called
attributes). Attributes can be numerical (scalar) or
categorical. The assignment can be hard, where
each object belongs to one cluster, or fuzzy,
where an object can belong to several clusters
with a probability. The clusters can be overlapping, though typically they are disjoint. A distance measure is a function that quantifies the
similarity of two objects.

Historical Background

Clustering is one of the most useful tasks in data
analysis. The goal of clustering is to discover
groups of similar objects and to identify interesting patterns in the data. Typically, the clustering
problem is about partitioning a given data set into
groups (clusters) such that the data points in a
cluster are more similar to each other than points
in different clusters [4, 8]. For example, consider
a retail database where each record contains items
purchased at the same time by a customer. A
clustering procedure could group the customers
in such a way that customers with similar buying
patterns are in the same cluster. Thus, the main
concern in the clustering process is to reveal the
organization of patterns into "sensible" groups,
which allow one to discover similarities and differences, as well as to derive useful conclusions
about them. This idea is applicable to many fields,
such as life sciences, medical sciences and engineering. Clustering may be found under different
names in different contexts, such as unsupervised
learning (in pattern recognition), numerical taxonomy (in biology, ecology), typology (in social
sciences) and partition (in graph theory) [13].

The clustering problem comes up in so many
domains due to the prevalence of large datasets
for which labels are not available. In one or two
dimensions, humans can perform clustering very
effectively visually, however in higher dimensions automated procedures are necessary. The
lack of training examples makes it very difficult
to evaluate the results of the clustering process. In
fact, the clustering process may result in different
partitioning of a data set, depending on the specific algorithm, criterion, or choice of parameters
used for clustering.

Foundations

The Clustering Process
In the clustering process, there are no predefined
classes and no examples that would show what
kind of desirable relations should be valid among
the data. That is the main difference from the task
of classification: Classification is the procedure

of assigning an object to a predefined set of categories [FSSU96]. Clustering produces initial categories in which values of a data set are classified during the classification process. For this reason, clustering is described as "unsupervised learning"; in contrast to classification, which is considered as "supervised learning." Typically, the clustering process will include at least the following steps:

1. Feature selection: Typically, the objects or observations to be clustered are described using a set of features. The goal is to appropriately select the features on which clustering is to be performed so as to encode as much information as possible concerning the task of interest. Thus, a pre-processing step may be necessary before using the data.
2. Choice of the clustering algorithm. In this step the user chooses the algorithm that is more appropriate for the data at hand, and therefore is more likely to result to a good clustering scheme. In addition, a similarity (or distance) measure and a clustering criterion are selected in tandem
    - The distance measure is a function that quantifies how "similar" two objects are. In most of the cases, one has to ensure that all selected features contribute equally to the computation of the proximity measure and there are no features that dominate others.
    - The clustering criterion is typically a cost function that the clustering algorithm has to optimize. The choice of clustering criterion has to take into account the type of clusters that are expected to occur.
3. Validation and interpretation of the results. The correctness of the results of the clustering algorithm is verified using appropriate criteria and techniques. Since clustering algorithms define clusters that are not known a priori, irrespective of the clustering methods, the final partition of the data typically requires some kind of evaluation. In many cases, the experts in the application area have to integrate the clustering results with other experimental evidence and analysis in order to draw the right conclusion.

After the third phase the user may elect to use the clustering results obtained, or may start the process from the beginning, perhaps using different clustering algorithms or parameters.

Clustering Algorithms Taxonomy

With clustering being a useful tool in diverse research communities, a multitude of clustering methods has been proposed in the literature. Occasionally similar techniques have been proposed and used in different communities. Clustering algorithms can be classified according to:

1. The type of data input to the algorithm (for example, objects described with numerical features or categorical features) and the choice of similarity function between two objects.
2. The clustering criterion optimized by the algorithm.
3. The theory and fundamental concepts on which clustering analysis techniques are based (e.g., fuzzy theory, statistics).

A broad classification of clustering algorithms is the following [8, 14]:

1. *Partitional clustering algorithms*: here the algorithm attempts to directly decompose the data set into a set of (typically) disjoint clusters. More specifically, the algorithm attempts to determine an integer number of partitions that optimize a certain criterion function.
2. *Hierarchical clustering algorithms*: here the algorithm proceeds successively by either merging smaller clusters into larger ones, or by splitting larger clusters. The result of the algorithm is a tree of clusters, called dendrogram, which shows how the clusters are related. By cutting the dendrogram at a desired level, a clustering of the data items into disjoint groups is obtained.
3. *Density-based clustering*: The key idea of this type of clustering is to group neighbouring objects of a data set into clusters based on density conditions. This includes grid-based algorithms that quantise the space into a finite number of cells and then do operations in the quantised space.

For each of above categories there is a wealth of subtypes and different algorithms for finding the clusters. Thus, according to the type of variables allowed in the data set additional categorizations include [14]: (i) Statistical algorithms, which are based on statistical analysis concepts and use similarity measures to partition objects and they are limited to numeric data. (ii) Conceptual algorithms that are used to cluster categorical data. (iii) Fuzzy clustering algorithms, which use fuzzy techniques to cluster data and allow objects to be classified into more than one clusters. Such algorithms lead to clustering schemes that are compatible with everyday life experience as they handle the uncertainty of real data. (iv) Crisp clustering techniques, that consider non-overlapping partitions so that a data point either belongs to a class or not. Most of the clustering algorithms result in crisp clusters, and thus can be categorized in crisp clustering. (v) Kohonen net clustering, which is based on the concepts of neural networks.

In the remaining discussion, partitional clustering algorithms will be described in more detail; other techniques will be dealt with separately.

Partitional Algorithms

In general terms, the clustering algorithms are based on a criterion for assessing the quality of a given partitioning. More specifically, they take as input some parameters (e.g., number of clusters, density of clusters) and attempt to define the best partitioning of a data set for the given parameters. Thus, they define a partitioning of a data set based on certain assumptions and not necessarily the "best" one that fits the data set.

In this category, K-Means is a commonly used algorithm [10]. The aim of K-Means clustering is the optimisation of an objective function that is described by the equation:

$$E = \sum_{i=1}^{c} \sum_{x \in C_i} d(x, m_i)$$

In the above equation, m_i is the center of cluster C_i, while $d(x, m_i)$ is the Euclidean distance between a point x and m_i. Thus, the criterion function E attempts to minimize the distance of every point from the center of the cluster to which the point belongs.

It should be noted that optimizing E is a combinatorial problem that is NP-Complete and thus any practical algorithm to optimize it cannot guarantee optimality. The K-means algorithm is the first practical and effective heuristic that was suggested to optimize this criterion, and owes its popularity to its good performance in practice. The K-means algorithm begins by initialising a set of c cluster centers. Then, it assigns each object of the dataset to the cluster whose center is the nearest, and re-computes the centers. The process continues until the centers of the clusters stop changing.

Another algorithm of this category is PAM (Partitioning Around Medoids). The objective of PAM is to determine a representative object (medoid) for each cluster, that is, to find the most centrally located objects within the clusters. The algorithm begins by selecting an object as medoid for each of c clusters. Then, each of the non-selected objects is grouped with the medoid to which it is the most similar. PAM swaps medoids with other non-selected objects until all objects qualify as medoid. It is clear that PAM is an expensive algorithm with respect to finding the medoids, as it compares an object with the entire dataset [12].

CLARA (Clustering Large Applications), is an implementation of PAM in a subset of the dataset. It draws multiple samples of the dataset, applies PAM on samples, and then outputs the best clustering out of these samples [12]. CLARANS (Clustering Large Applications based on Randomized Search), combines the sampling techniques with PAM. The clustering process can be presented as searching a graph where every node is a potential solution, that is, a set of k medoids. The clustering obtained after replacing a medoid is called the neighbour of the current clustering. CLARANS selects a node and compares it to a user-defined number of their neighbours searching for a local minimum. If a better neighbor is found (i.e., having lower-square error), CLARANS moves to the neighbour's node and the process starts again; otherwise the current clustering is a local optimum. If the local optimum is found, CLARANS starts with a new randomly selected node in search for a new local optimum.

The algorithms described above result in crisp clusters, meaning that a data point either belongs to a cluster or not. The clusters are non-overlapping and this kind of partitioning is further called crisp clustering. The issue of uncertainty support in the clustering task leads to the introduction of algorithms that use fuzzy logic concepts in their procedure. A common fuzzy clustering algorithm is the Fuzzy C-Means (FCM), an extension of classical C-Means algorithm for fuzzy applications [2]. FCM attempts to find the most characteristic point in each cluster, which can be considered as the "center" of the cluster and, then, the grade of membership for each object in the clusters.

Another approach proposed in the literature to solve the problems of crisp clustering is based on probabilistic models. The basis of this type of clustering algorithms is the EM algorithm, which provides a quite general approach to learning in presence of unobservable variables [11]. A common algorithm is the probabilistic variant of K-Means, which is based on the mixture of Gaussian distributions. This approach of K-Means uses probability density rather than distance to associate records with clusters. More specifically, it regards the centers of clusters as means of Gaussian distributions. Then, it estimates the probability that a data point is generated by the jth Gaussian (i.e., belongs to jth cluster). This approach is based on Gaussian model to extract clusters and assigns the data points to clusters assuming that they are generated by normal distribution. Also, this approach is implemented only in the case of algorithms based on the EM (Expectation Maximization) algorithm.

Another type of clustering algorithms combine graph partitioning and hierarchical clustering algorithms characteristics. Such algorithms include CHAMELEON [9], which measures the similarity among clusters based on a dynamic model contrary to the clustering algorithms discussed above. Moreover in the clustering process both the inter-connectivity and closeness between two clusters are taken into account to decide how to merge the clusters. The merge process based on the dynamic model facilitates the discovery of natural and homogeneous clusters. Also it is applicable to all types of data as long as a similarity function is specified. Finally, BIRCH [ZRL99] uses a data structure called CF-Tree for partitioning the incoming data points in an incremental and dynamic way, thus providing an effective way to cluster very large datasets.

Partitional algorithms are applicable mainly to numerical data sets. However, there are some variants of K-Means such as K-prototypes, and K-mode [7] that are based on the K-Means algorithm, but they aim at clustering categorical data. K-mode discovers clusters while it adopts new concepts in order to handle categorical data. Thus, the cluster centers are replaced with "modes," a new dissimilarity measure used to deal with categorical objects.

The K-means algorithm and related techniques tend to produce spherical clusters due to the use of a symmetric objective function. They require the user to set only one parameter, the desirable number of clusters K. However, since the objective function gets smaller monotonically as K increases, it is not clear how to define what is the best number of clusters for a given dataset. Although several approaches have been proposed to address this shortcoming [14], this is one of the main disadvantages of partitional algorithms. Another characteristic of the partitional algorithms is that they are unable to handle noise and outliers and they are not suitable to discover clusters with non-convex shapes. Another characteristic of K-means is that the algorithm does not display a monotone behavior with respect to K. For example, if a dataset is clustered into M and 2 M clusters, it is intuitive to expect that the smaller clusters in the second clustering will be subsets of the larger clusters in the first; however this is typically not the case.

Key Applications

Cluster analysis is very useful task in exploratory data analysis and a major tool in a very wide spectrum of applications in many fields of business and science. Clustering applications include:

1. *Data reduction.* Cluster analysis can contribute to the compression of the information included in the data. In several cases, the amount of the available data is very large and its processing becomes very demanding. Clustering can be used to partition the data set into a number of "interesting" clusters. Then, instead of processing the data set as an entity, the representatives of the defined clusters are adopted in the process. Thus, data compression is achieved.

2. *Hypothesis generation.* Cluster analysis is used here in order to infer some hypotheses concerning the data. For instance, one may find in a retail database that there are two significant groups of customers based on their age and the time of purchases. Then, one may infer some hypotheses for the data, that it, "young people go shopping in the evening," "old people go shopping in the morning."

3. *Hypothesis testing.* In this case, the cluster analysis is used for the verification of the validity of a specific hypothesis. For example, consider the following hypothesis: "Young people go shopping in the evening." One way to verify whether this is true is to apply cluster analysis to a representative set of stores. Suppose that each store is represented by its customer's details (age, job, etc.) and the time of transactions. If, after applying cluster analysis, a cluster that corresponds to "young people buy in the evening" is formed, then the hypothesis is supported by cluster analysis.

4. *Prediction based on groups.* Cluster analysis is applied to the data set and the resulting clusters are characterized by the features of the patterns that belong to these clusters. Then, unknown patterns can be classified into specified clusters based on their similarity to the clusters' features. In such cases, useful knowledge related to this data can be extracted. Assume, for example, that the cluster analysis is applied to a data set concerning patients infected by the same disease. The result is a number of clusters of patients, according to their reaction to specific drugs. Then, for a new patient, one identifies the cluster in which he/she can be classified and based on this decision his/her medication can be made.

5. *Business Applications and Market Research.* In business, clustering may help marketers discover significant groups in their customers' database and characterize them based on purchasing patterns.

6. *Biology and Bioinformatics.* In biology, it can be used to define taxonomies, categorize genes with similar functionality and gain insights into structures inherent in populations.

7. *Spatial data analysis.* Due to the huge amounts of spatial data that may be obtained from satellite images, medical equipment, Geographical Information Systems (GIS), image database exploration etc., it is expensive and difficult for the users to examine spatial data in detail. Clustering may help to automate the process of analysing and understanding spatial data. It is used to identify and extract interesting characteristics and patterns that may exist in large spatial databases.

8. *Web mining.* Clustering is used to discover significant groups of documents on the Web huge collection of semi-structured documents. This classification of Web documents assists in information discovery. Another application of clustering is discovering groups in social networks.

In addition, clustering can be used as a preprocessing step for other algorithms, such as classification, which would then operate on the detected clusters.

Cross-References

▶ Dimension Reduction Techniques for Clustering
▶ Document Clustering
▶ Feature Selection for Clustering
▶ Hierarchical Clustering
▶ Semi-supervised Learning
▶ Spectral Clustering
▶ Subspace Clustering Techniques
▶ Text Clustering
▶ Visual Clustering
▶ Visualizing Clustering Results

Recommended Reading

1. Agrawal R, Gehrke J, Gunopulos D, Raghavan P. Automatic subspace clustering of high dimensional data for data mining applications. In: Proceedings of the ACM SIGMOD International Conference on Management of Data; 1998. p. 94–105.
2. Bezdeck JC, Ehrlich R, Full W. FCM: Fuzzy C-Means algorithm. Comput Geosci. 1984;10(2–3): 191–203.
3. Ester M, Kriegel H.-Peter, Sander J, Xu X. A density-based algorithm for discovering clusters in large spatial databases with noise. In: Proceedings of the 2nd International Conference on Knowledge Discovery and Data Mining; 1996. p. 226–31.
4. Everitt BS, Landau S, Leese M. Cluster analysis. London: Hodder Arnold; 2001.
5. Fayyad UM, Piatesky-Shapiro G, Smuth P, Uthurusamy R. Advances in knowledge discovery and data mining. Menlo Park: AAAI Press; 1996.
6. Han J, Kamber M. Data mining: concepts and techniques. San Fransisco: Morgan Kaufmann Publishers; 2001.
7. Huang Z. A fast clustering algorithm to cluster very large categorical data sets in data mining. In: Proceedings of the ACM SIGMOD Workshop on Research Issues in Data Mining and Knowledge Discovery; 1997.
8. Jain AK, Murty MN, Flyn PJ. Data clustering: a review. ACM Comput Surv. 1999;31(3):264–323.
9. Karypis G, Han E-H, Kumar V. CHAMELEON: a hierarchical clustering algorithm using dynamic modeling. IEEE Comput. 1999;32(8):68–75.
10. MacQueen JB Some methods for classification and analysis of multivariate observations. In: Proceedings of the 5th Berkeley Symposium on Mathematical Statistics and Probability; 1967. p. 281–97.
11. Mitchell T. Machine learning. New York: McGraw-Hill; 1997.
12. Ng R, Han J. Efficient and effective clustering methods for spatial data mining. In: Proceedings of the 20th International Conference on Very Large Data Bases; 1994. p. 144–55.
13. Theodoridis S, Koutroubas K. Pattern recognition. New York: Academic; 1999.
14. Vazirgiannis M, Halkidi M, Gunopulos D. Uncertainty handling and quality assessment in data mining. New York: Springer; 2003.
15. Wang W, Yang J, Muntz R. STING: a statistical information grid approach to spatial data mining. In: Proceedings of the 23th Internationa Conference on Very Large Data Bases; 1997. p. 186–95.
16. Zhang T, Ramakrishnman R, Linvy M. BIRCH: an efficient method for very large databases. In: Proceedings of the ACM SIGMOD International Conference on Management of Data; 1996. p. 103–14.

Clustering Validity

Michalis Vazirgiannis
Athens University of Economics and Business, Athens, Greece

Synonyms

Cluster stability; Cluster validation; Quality assessment; Stability-based validation of clustering

Definition

A problem one faces in clustering is to decide the optimal partitioning of the data into clusters. In this context visualization of the data set is a crucial verification of the clustering results. In the case of large multidimensional data sets (e.g., more than three dimensions) effective visualization of the data set is cumbersome. Moreover the perception of clusters using available visualization tools is a difficult task for humans that are not accustomed to higher dimensional spaces. The procedure of evaluating the results of a clustering algorithm is known under the term *cluster validity*. Cluster validity consists of a set of techniques for finding a set of clusters that best fits natural partitions (of given datasets) without any a priori class information. The outcome of the clustering process is validated by a cluster validity index.

Historical Background

Clustering is a major task in the data mining process for discovering groups and identifying interesting distributions and patterns in the underlying data. In the literature a wide variety of algorithms for different applications and sizes of data sets. The application of an algorithm to a data set, assuming that the data set offers a clustering tendency, aims at discovering its inherent partitions. However, the clustering process is an unsupervised process, since there are no predefined classes or examples. Then, the various clustering algorithms are based on some assumptions in order to define a partitioning of a data set. As a consequence, they may behave in a different way depending on: (i) the features of the data set (geometry and density distribution of clusters) and (ii) the input parameter values.

One of the most important issues in cluster analysis is the evaluation of clustering results to find the partitioning that best fits the underlying data. This is the main subject of cluster validity. If clustering algorithm parameters are assigned an improper value, the clustering method results in a partitioning scheme that is not optimal for the specific data set leading to wrong decisions. The problems of deciding the number of clusters better fitting a data set as well as the evaluation of the clustering results has been subject of several research efforts. The procedure of evaluating the results of a clustering algorithm is known under the term cluster validity. In general terms, there are three approaches to investigate cluster validity. The first is based on *external criteria*. This implies that the results of a clustering algorithm are evaluated based on a pre-specified structure, which is imposed on a data set and reflects one's intuition about the clustering structure of the data set. The second approach is based on *internal criteria*. The results of a clustering algorithm may be evaluated in terms of quantities that involve the vectors of the data set themselves (e.g., proximity matrix). The third approach of clustering validity is based on *relative criteria*. Here the basic idea is the evaluation of a clustering structure by comparing it to other clustering schemes, resulting by the same algorithm but with different parameter values. There are two criteria proposed for clustering evaluation and selection of an optimal clustering scheme: (i) Compactness, the members of each cluster should be as close to each other as possible. A common measure of compactness is the variance, which should be minimized. (ii) Separation, the clusters themselves should be widely spaced.

Foundations

This section discusses methods suitable for the quantitative evaluation of the clustering results, known as cluster validity methods. However, these methods give an indication of the quality of the resulting partitioning and thus they can only be considered as a tool at the disposal of the experts in order to evaluate the clustering results. The cluster validity approaches based on external and internal criteria rely on statistical hypothesis testing. In the following section, an introduction to the fundamental concepts of hypothesis testing in cluster validity is presented.

In cluster validity the basic idea is to test whether the points of a data set are randomly structured or not. This analysis is based on the *Null Hypothesis*, denoted as *Ho*, expressed as a statement of random structure of a data set X. To test this hypothesis, statistical tests are used, which lead to a computationally complex procedure. Monte Carlo techniques are used as a solution to this problem.

External Criteria
Based on external criteria, one can work in two different ways. First, one can evaluate the resulting clustering structure **C**, by comparing it to an independent partition of the data **P** built according to one's intuition about the clustering structure of the data set. Second, one can compare the proximity matrix P to the partition **P**.

Comparison of C with Partition P (Non-hierarchical Clustering)
Let $C = \{C_1 \ldots C_m\}$ be a clustering structure of a data set X and $P = \{P_1 \ldots P_s\}$ be a defined

partition of the data. Refer to a pair of points (x_v, x_u) from the data set using the following terms:

- **SS**: if both points belong to the same cluster of the clustering structure **C** and to the same group of partition **P**.
- **SD**: if points belong to the same cluster of **C** and to different groups of **P**.
- **DS**: if points belong to different clusters of **C** and to the same group of **P**.
- **DD**: if both points belong to different clusters of **C** and to different groups of **P**.

Assuming now that **a**, **b**, **c** and **d** are the number of SS, SD, DS and DD pairs respectively, then a + b + c + d = **M** which is the maximum number of all pairs in the data set (meaning, M = N(N − 1)/2 where N is the total number of points in the data set).

Now define the following indices to measure the degree of similarity between **C** and **P**:

1. *Rand Statistic*: R = (a + d)/M
2. *Jaccard Coefficient*: J = a/(a + b + c)

     The above two indices range between 0 and 1, and are maximized when m=s. Another known index is the:
3. *Folkes and Mallows index*:

$$FM = a/\sqrt{m_1 m_2} = \sqrt{\frac{a}{a+b} \cdot \frac{a}{a+c}} \quad (1)$$

where $m_1 = (a + b)$, $m_2 = (a + c)$.

For the previous three indices it has been proven that the higher the values of these indices are the more similar **C** and **P** are. Other indices are:

4. *Huberts Γ statistic*:

$$\Gamma = (1/M) \sum_{i=1}^{N-1} \sum_{j=i+1}^{N} X(i,j) Y(i,j) \quad (2)$$

High values of this index indicate a strong similarity between the matrices X and Y.

5. *Normalized Γ statistic*:

     where $X(i,j)$ and $Y(i,j)$ are the (i,j) element of the matrices X, Y respectively that one

wants to compare. Also μ_x, μ_y, σ_x, σ_y are the respective means and variances of X, Y matrices. This index takes values between −1 and 1.

$$\hat{\Gamma} = \frac{\left[(1/M) \sum_{i=1}^{N-1} \sum_{j=i+1}^{N} (X(i,j) - \mu_X)(Y(i,j) - \mu_Y)\right]}{\sigma_X \sigma_Y}$$

All these statistics have right-tailed probability density functions, under the random hypothesis. In order to use these indices in statistical tests, one must know their respective probability density function under the Null Hypothesis, H_o, which is the hypothesis of random structure of the data set. Thus, if one accepts the Null Hypothesis, the data are randomly distributed. However, the computation of the probability density function of these indices is computationally expensive. A solution to this problem is to use Monte Carlo techniques.

After having plotted the approximation of the probability density function of the defined statistic index, its value, denoted by q, is compared to the $q(C_i)$ values, further referred to as q_i. The indices R, J, FM, Γ defined previously are used as the q index mentioned in the above procedure.

Internal Criteria

Using this approach of cluster validity the goal is to evaluate the clustering result of an algorithm using only quantities and features inherited from the data set. There are two cases in which one applies internal criteria of cluster validity depending on the clustering structure: (i) hierarchy of clustering schemes, and (ii) single clustering scheme.

Validating Hierarchy of Clustering Schemes

A matrix called cophenetic matrix, P_c, can represent the hierarchy diagram that is produced by a hierarchical algorithm. The element $P_c(i, j)$ of cophenetic matrix represents the proximity level at which the two vectors x_i and x_j are found in the same cluster for the first time. A statistical index can be defined to measure the degree of similarity between P_c and P (proximity matrix) matrices. This index is called *Cophenetic Correlation Coefficient* and defined as:

$$CPCC =$$

$$\frac{(1/M) \sum_{i=1}^{N-1} \sum_{j=i+1}^{N} d_{ij}c_{ij} - \mu_P\mu_C}{\sqrt{\left[(1/M) \sum_{i=1}^{N-1} \sum_{j=i+1}^{N} d_{ij}^2 - \mu_P^2\right]\left[(1/M) \sum_{i=1}^{N-1} \sum_{j=i+1}^{N} c_{ij}^2 - \mu_C^2\right]}},$$

(4)

where $M = N \times (N-1)/2$ and N is the number of points in a data set. Also, μ_p and μ_c are the means of matrices P and P_c respectively, and are defined in the (Eq. 5):

$$\mu_P = (1/M) \sum_{i=1}^{N-1} \sum_{j=i+1}^{N} P(i,j),$$

$$\mu_C = (1/M) \sum_{i=1}^{N-1} \sum_{j=i+1}^{N} P_c(i,j)$$

(5)

Moreover, d_{ij}, c_{ij} are the (i,j) elements of P and P_c matrices respectively. The CPCC values range in $[-1, 1]$. A value of the index close to 1 is an indication of a significant similarity between the two matrices.

Validating a Single Clustering Scheme

The goal here is to find the degree of match between a given clustering scheme C, consisting of n_c clusters, and the proximity matrix P. The defined index for this approach is Hubert's Γ statistic (or normalized Γ statistic). An additional matrix for the computation of the index is used, that is

$$Y(i,j)$$

$$= \begin{cases} 1, & \text{if } x_i \text{ and } x_j \text{ belong to different clusters} \\ 0, & \text{otherwise array.where } i,j = 1, \ 1/4, \ N. \end{cases}$$

The application of Monte Carlo techniques is also a means to test the random hypothesis in a given data set.

Relative Criteria

The major drawback of techniques based on internal or external criteria is their high computational complexity. A different validation approach is discussed in this section. The fundamental idea of the relative criteria is to choose the best clustering scheme of a set of defined schemes according to a pre-specified criterion. More specifically, the problem can be stated as follows:

Let P_{alg} be the set of parameters associated with a specific clustering algorithm (e.g., the number of clusters n_c). Among the clustering schemes C_i, $i = 1, \ldots, n_c$, is defined by a specific algorithm. For different values of the parameters in P_{alg}, choose the one that best fits the data set.

Then, consider the following cases of the problem:

1. *P_{alg} does not contain the number of clusters, n_c, as a parameter.* In this case, the choice of the optimal parameter values are described as follows: The algorithm runs for a wide range of its parameters' values and the largest range for which n_c remains constant is selected (usually $n_c \ll N$ (number of tuples)). Then the values that correspond to the middle of this range are chosen as appropriate values of the P_{alg} parameters. Also, this procedure identifies the number of clusters that underlie the data set.

2. *P_{alg} contains n_c as a parameter.* The procedure of identifying the best clustering scheme is based on a validity index. Selecting a suitable performance index, q, one proceeds with the following steps:

   • Clustering runs for all values of n_c between n_{cmin} and n_{cmax} defined a priori by the user.

   • For each of n_c values, the algorithm runs r times, using different sets of values for the other parameters of the algorithm (e.g., different initial conditions).

   • The best values of the index q obtained by each n_c are plotted as the function of n_c.

Based on this plot, the best clustering schemes are identified. There are two approaches for defining the best clustering depending on the behavior of q with respect to n_c. Thus, if the validity index does not exhibit an increasing or decreasing trend as n_c increases, one seeks the max (min) of the plot. On the other hand, for indices that increase (decrease) as the number of clusters increase, one searches for the values of n_c at which a significant local change in value of the index occurs. This change appears as a "knee" in the plot and it is an indication of the number of clusters underlying the data set. The absence of a knee is an indication that the data set possesses no clustering structure. Below, some representative relative validity indices are presented.

The Modified Hubert Γ Statistic

The definition of the modified Hubert Γ statistic is given by the equation

$$\Gamma = (1/M) \sum_{i=1}^{N-1} \sum_{j=i+1}^{N} P(i,j) \cdot Q(i,j) \quad (6)$$

where N is the number of objects in a data set, $M = N(N-1)/2$, P is the proximity matrix of the data set and Q is an $N \times N$ matrix whose (i, j) element is equal to the distance between the representative points (v_{ci}, v_{cj}) of the clusters where the objects x_i and x_j belong.

Similarly, one can define the normalized Hubert Γ statistic, given by equation

$$\widehat{\Gamma} = \frac{\left[(1/M) \sum_{i=1}^{N-1} \sum_{j=i+1}^{N} (P(i,j) - \mu_P)(Q(i,j) - \mu_Q) \right]}{\sigma_P \sigma_Q}. \quad (7)$$

where μ_P, μ_Q, σ_P, σ_Q are the respective means and variances of P, Q matrices.

If the $d(v_{ci}, v_{cj})$ is close to $d(x_i, x_j)$ for $i, j = 1, 2,\ldots,N$, P and Q will be in close agreement and the values of Γ and $\widehat{\Gamma}$ (normalized Γ) will be high. Conversely, a high value of Γ ($\widehat{\Gamma}$) indicates the existence of compact clusters. Thus, in the plot of normalized Γ versus n_c, one seeks a significant knee that corresponds to a significant increase of normalized Γ. The number of clusters at which the knee occurs is an indication of the number

of clusters that occurs in the data. Note that for $n_c = 1$ and $n_c = N$, the index is not defined.

Dunn Family of Indices

A cluster validity index for crisp clustering proposed by Dunn (1974), aims at the identification of "compact and well separated clusters". The index is defined in the following equation for a specific number of clusters

$$D_{n_c} = \min_{i=1,\ldots,n_c} \left\{ \min_{j=i+1,\ldots,n_c} \left(\frac{d(c_i, c_j)}{\max_{k=1,\ldots,n_c} (\operatorname{diam}(c_k))} \right) \right\} \quad (8)$$

where $d(c_i, c_j)$ is the dissimilarity function between two clusters c_i and c_j defined as $d(c_i, c_j) = \min_{x \in C_i, y \in C_j} d(x, y)$, and diam(c) is the diameter of a cluster, which may be considered as a measure of clusters' dispersion. The diameter of a cluster C can be defined as follows:

$$\operatorname{diam}(C) = \max_{x,y \in C} \{d(x, y)\} \quad (9)$$

If the data set contains compact and well-separated clusters, the distance between the clusters is expected large and the diameter of the clusters is expected small. Based on the Dunn's index definition, one concludes that large values of the index indicate the presence of compact and well-separated clusters.

The problems of the Dunn index are: (i) its considerable time complexity, and (ii) its sensitivity to the presence of noise in data sets, since these are likely to increase the values of the diameter.

RMSSDT, SPR, RS, CD

This family of validity indices is applicable in the cases that hierarchical algorithms are used to cluster the data sets. Hereafter the discussion refers to the definitions of four validity indices, which have to be used simultaneously to determine the number of clusters existing in the data set. These four indices are applied to each step of a *hierarchical* clustering algorithm and they are known as:

- *Root-mean-square standard deviation (RMS STD) of the new cluster,*
- *Semi-partial R-squared (SPR),*
- *R-squared (RS),*
- *Distance between two clusters (CD).*

Getting into a more detailed description of them, one can say that:

RMSSTD of a new clustering scheme defined at a level of a clustering hierarchy is the square root of the variance of all the variables (attributes used in the clustering process). This index measures the homogeneity of the formed clusters at each step of the hierarchical algorithm. Since the objective of cluster analysis is to form homogeneous groups the RMSSTD of a cluster should be as small as possible. Where the values of RMSSTD are higher than the ones of the previous step, one has an indication that the new clustering scheme is worse.

In the following definitions, the term SS is used, which means *Sum of Squares* and refers to the equation:

$$SS = \sum_{i=1}^{n} \left(X_i - \overline{X} \right)^2 \qquad (10)$$

Along with this, additional terms will be used, such as:

1. SS_w referring to the sum of squares within group,
2. SS_b referring to the sum of squares between groups,
3. SS_t referring to the total sum of squares, of the whole data set.

In the case cluster join to form a new one, *SPR*- for the new cluster – is defined as the difference between SS_w of the new cluster and the sum of the SS_w's values of clusters joined to obtain the new cluster (*loss of homogeneity*), divided by the SS_t for the whole data set. This index measures the loss of homogeneity after merging the two

clusters of a single algorithm step. If the index value is zero then the new cluster is obtained by merging two perfectly homogeneous clusters. If its value is high then the new cluster is obtained by merging two heterogeneous clusters.

RS of the new cluster is the ratio of SS_b over SS_t. SS_b is a measure of difference between groups. Since $SS_t = SS_b + SS_w$, the greater the SS_b the smaller the SS_w and vice versa. As a result, the greater the differences between groups, the more homogenous each group is and vice versa. Thus, RS may be considered as a measure of dissimilarity between clusters. Furthermore, it measures the degree of homogeneity between groups. The values of RS range between 0 and 1. Where the value of RS is zero, there is an indication that no difference exists among groups. On the other hand, when RS equals 1 there is an indication of significant difference among groups.

Key Applications

There is a certain cross disciplinary interest for clustering validity method and indices. A prominent area where cluster validity measures apply is the area of biological data [2, 6]. Patterns hidden in gene expression data offer a tremendous opportunity for an enhanced understanding of functional genomics. However, the large number of genes and the complexity of biological networks greatly increase the challenges of comprehending and interpreting the resulting mass of data, which often consists of millions of measurements. The data mining process aims to reveal natural structures and identify interesting patterns in the underlying data. Clustering techniques constitute a first essential step toward addressing this challenge. Moreover recent research effort papers in the area of image segmentation [3, 13].

The area is fertile as the clustering issue is a fundamental problem and the application domains are still widening. Challenging relevant research directions [9] follow:

- Is there a principled way to measure the quality of a clustering on particular data set?

- Can every clustering task be expressed as an optimization of some explicit, readily computable, objective cost function?
- Can stability be considered a first principle for meaningful clustering?
- How should the similarity between different clusterings be measured?
- Can one distinguish clusterable data from structureless data?
- What are the tools that should be imported from other relevant areas of research?

Cross-References

Recommended Reading

1. Bezdek JC, Pal NR. Some new indexes of cluster validity. IEEE Trans Syst Man Cybern Part B. 1998;28(3):301–15.
2. Datta S, Datta S. Comparisons and validation of statistical clustering techniques for microarray gene expression data. Bioinformatics. 2003;19(4):459–66.
3. El-Melegy MT, Zanaty EA, Abd-Elhafiez WM, Farag AA. On cluster validity indexes in fuzzy and hard clustering algorithms for image segmentation. In: Proceedings of the International Conference on Image Processing; 2007. p. 5–8.
4. Halkidi M, Batistakis Y, Vazirgiannis M. On clustering validation techniques. J Intell Inf Syst. 2001;17(2–3):107–45.
5. Halkidi M, Gunopulos D, Vazirgiannis M, Kumar N, Domeniconi C. A clustering framework based on subjective and objective validity criteria. ACM Trans Knowl Discov Data. 2008;1(4):1 25
6. Jiang D, Tang C, Zhang A. Cluster analysis for gene expression data: a survey. IEEE Trans Knowl Data Eng. 2004;16(11):1370–86.
7. Kim M, Ramakrishna RS. New indices for cluster validity assessment. Pattern Recogn Lett. 2005;26(15):2353–63.
8. Maulik U, Bandyopadhyay S. Performance evaluation of some clustering algorithms and validity indices. IEEE Trans Pattern Anal Mach Intell. 2002;24(12):1650–4.
9. NIPS 2005 workshop on theoretical foundations of clustering, Saturday, December 10th, 2005. Available at: http://www.kyb.tuebingen.mpg.de/bs/people/ule/clustering_workshop_nips05/clustering_workshop_nips05.htm_
10. Pal NR, Bezdek JC. On cluster validity for the fuzzy c-means model. IEEE Trans Fuzzy Syst. 1995;3(3):370–9.
11. Rand WM. Objective criteria for the evaluation of clustering methods. J Am Stat Assoc. 1971;66(336):846–50.
12. Wang J-S, Chiang J-C. A cluster validity measure with a hybrid parameter search method for the support vector clustering algorithm. Pattern Recog. 2008;41(2):506–20.
13. Zhang J, Modestino JW. A model-fitting approach to cluster validation with application to stochastic model-based image segmentation. IEEE Trans Pattern Anal Mach Intell. 1990;12(10):1009–17.

Clustering with Constraints

Ian Davidson
University of California-Davis, Davis, CA, USA

Synonyms

Semi-supervised clustering

Definition

The area of clustering with constraints makes use of hints or advice in the form of constraints to aid or bias the clustering process. The most prevalent form of advice are conjunctions of pair-wise instance level constraints of the form must-link (ML) and cannot-link (CL) which state that pairs of instances should be in the same or different clusters respectively. Given a set of points P to cluster and a set of constraints C, the aim of clustering with constraints is to use the constraints to improve the clustering results. Constraints have so far being used in two main ways: (i) Writing algorithms that use a standard distance metric but attempt to satisfy all or as many constraints as possible and (ii) Using the constraints to learn a distance function that is then used in the clustering algorithm.

Historical Background

The idea of using constraints to guide clustering was first introduced by Wagstaff and Cardie in their seminal paper ICML 2000 [13] with a modified COBWEB-style algorithm that attempts to satisfy all constraints. Later [14] they introduced

constraints to the k-means algorithms. Their algorithms (as most algorithms now do) look at satisfying a conjunction of must-link and cannot-link constraints. Independently, Cohn et al. [3, 4] introduced constraints as a user feedback mechanism to guide the clustering algorithm to a more useful result.

In 2002 Xing and collaborators [15] (NIPS 2002) and Klein and collaborators (ICML 2002) [12] explored making use of constraints by learning a distance function for non-hierarchical clustering and a distance matrix for hierarchical clustering respectively.

Basu and collaborators more recently have looked at key issues such as which are the most informative sets of constraints [2] and seeding algorithms using constraints [1]. Gondek has explored using constraints to find orthogonal/alternative clusterings of data [3, 11]. Davidson

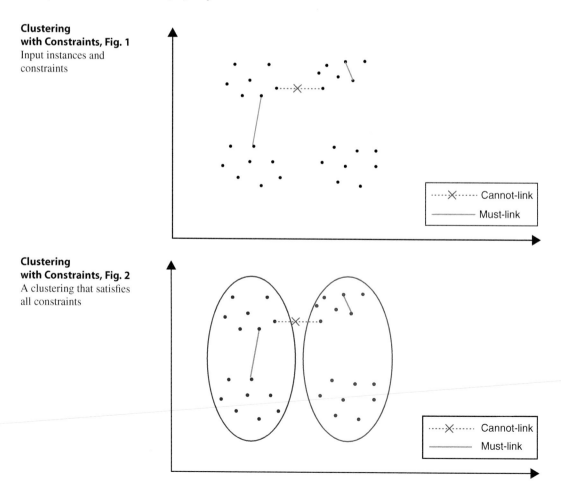

Clustering with Constraints, Fig. 1 Input instances and constraints

Clustering with Constraints, Fig. 2 A clustering that satisfies all constraints

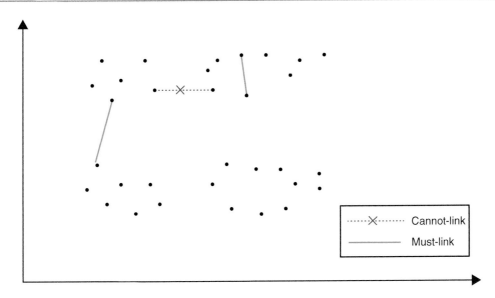

Clustering with Constraints, Fig. 3 Input instances and constraints

and Ravi explored the intractability issues of clustering under constraints for non-hierarchical clustering [6], hierarchical clustering [5] and non-hierarchical clustering with feedback [9].

Foundations

Clustering has many successful applications in a variety of domains where the objective function of the clustering algorithm finds a novel and useful clustering. However, in some application domains the typical objective functions may lead to well-known or non-actionable clusterings of the data. This could be overcome by an ad hoc approach such as manipulating the data. The introduction of constraints into clustering allows a principled approach to incorporate user preferences or domain expertise into the clustering process so as to guide the algorithm to a desirable solution or away from an undesirable solution. The typical semi-supervised learning situations involves having a label associated with a subset of the available instances. However in many domains, knowledge of the relevant categories is incomplete and it is easier to obtain pairwise constraints either automatically or from domain experts.

Types of Constraints. Must-link and cannot-link constraints are typically used since they can be easily generated from small amounts of labeled data (generate a must-link between two instances if the labels agree, cannot-link if they disagree) or from domain experts. They can be used to represent geometric properties [6, 14] by noting that for instance, making the maximum cluster diameter be α is equivalent to enforcing a conjunction of cannot-link constraints between all points whose distance is greater than α. Similarly, clusters can be separated by distance at least δ by enforcing a conjunction of must-link constraints between all points whose distance is less than δ. Both types of instance-level constraints have interesting properties that can be used to effectively generate many additional constraints. Must-link constraints are transitive: $ML(x,y)$, $ML(y,z) \rightarrow ML(x,z)$ and cannot link constraints have an entailment property: $ML(a,b), ML(x,y)$, $CL(a,x) \rightarrow CL(a,y)$, $CL(b,x)$, $CL(b,y)$.

How Constraints Are Used. Constraints have typically been used in clustering algorithms in two ways. Constraints can be used to modify the cluster assignment stage of the cluster algorithm [4, 14], to enforce satisfaction of the constraints or as many as possible [2, 6]. These approaches typically use a standard distance or likelihood

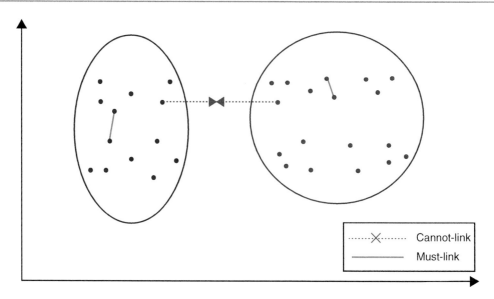

Clustering with Constraints, Fig. 4 A learnt distance space respective of the constraints

function. Alternatively, the distance function of the clustering algorithm can also be trained either before or after the clustering actually occurs using the constraints [12, 15]. The former are called constraint-based approaches and the later distance based approaches.

Constraint-Based Methods. In constraint-based approaches, the clustering algorithm itself (typically the assignment step) is modified so that the available constraints are used to bias the search for an appropriate clustering of the data. Figure 2 shows how though two clusterings exist (a horizontal and vertical clustering) just three constraints can rule out the former.

Constraint-based clustering is typically achieved using one of the following approaches:

1. Enforcing constraints to be satisfied during the cluster assignment in the clustering algorithm [5, 13].
2. Modifying the clustering objective function so that it includes a term for satisfying specified constraints. Penalties for violating constraints have been explored in the maximum likelihood framework [2] and distance framework [6].
3. Initializing clusters and inferring clustering constraints based on neighborhoods derived from labeled examples [1].

Each of the above approaches provides a simple method of modifying existing partitional and agglomerative style hierarchical algorithms to incorporate constraints. For more recent advances in algorithm design such as the use of variational techniques for constrained clustering see [3].

Distance-Based Methods. In distance-based approaches, an existing clustering algorithm that uses a distance measure is employed. However, rather than use the Euclidean distance metric, the distance measure is first trained to "satisfy" the given constraints. The approach of Xing and collaborators [15] casts the problem of learning a distance *metric* from the constraints so that the points (and surrounding points) that are part of the must-link (cannot-link) constraints are close together (far apart). They consider two formulations: firstly learning a generalized Mahanabolis distance metric which essentially stretches or compresses each axis as appropriate. Figure 4 gives an example where the constraints can be satisfied by stretching the x-axis and compressing the y-axis and then applying a clustering algorithm to the new data space. The second formulation allows a more complex transformation on the space of points.

Klein and collaborators [12] explore learning a distance *matrix* from constraints for agglomerative clustering. Only points that are directly involved in the constraints are brought closer together or far apart using a multi-step approach of making must-linked points have a distance of 0 and cannot-linked points having the greatest distance.

There have been some algorithms that try to both enforce constraints and learn distance functions from constraints [2].

Key Applications

Key application areas include images, video, biology, text, web pages, audio (speaker identification) [3] and GPS trace information [14].

URL to Code

http://www.constrained-clustering.org

Cross-References

▶ Semi-supervised Learning

Recommended Reading

1. Basu S, Banerjee A, Mooney R. Semi-supervised clustering by seeding. In: Proceedings of the 19th International Conference on Machine Learning; 2002. p. 27–34.
2. Basu S, Banerjee A, Mooney RJ. Active semi-supervision for pairwise constrained clustering. In: Proceedings of the SIAM International Conference on Data Mining; 2004.
3. Basu S, Davidson I, Wagstaff K, editors. Constrained clustering: advances in algorithms, theory and applications. New York: Chapman & Hall/CRC Press; 2008.
4. Cohn D, Caruana R, McCallum A. Semi-supervised clustering with user feedback. Technical Report 2003–1892. Cornell University; 2003.
5. Davidson I, Ravi SS. Agglomerative hierarchical clustering with constraints: theoretical and empirical results. In: Principles of Data Mining and Knowledge Discovery, 9th European Conference; 2005. p. 59–70.
6. Davidson I, Ravi SS. Clustering with constraints: feasibility issues and the k-means algorithm. In: Proceedings of the SIAM International Conference on Data Mining; 2005.
7. Davidson I, Ravi SS. Identifying and generating easy sets of constraints for clustering. In: Proceedings of the 15th National Conference on AI; 2006.
8. Davidson I, Ester M, Ravi SS. Efficient incremental clustering with constraints. In: Proceedings of the 13th ACM SIGKDD International Conference on Knowledge Discovery and Data Mining; 2007. p. 204–49.
9. Davidson I, Ravi SS. Intractability and clustering with constraints. In: Proceedings of the 24th International Conference on Machine Learning; 2007. p. 201–8.
10. Davidson I, Ravi SS. The complexity of non-hierarchical clustering with instance and cluster level constraints. Data Mining Knowl Discov. 2007;14(1):25–61.
11. Gondek D, Hofmann T. Non-redundant data clustering. In: Proceedings of the 2004 IEEE International Conference on Data Mining; 2004. p. 75–82.
12. Klein D, Kamvar SD, Manning CD. From instance-level constraints to space-level constraints: making the most of prior knowledge in data clustering. In: Proceedings of the 19th International Conference on Machine Learning; 2002. p. 307–14.
13. Wagstaff K, Cardie C. Clustering with instance-level constraints. In: Proceedings of the 17th International Conference on Machine Learning; 2000. p. 1103–10.
14. Wagstaff K, Cardie C, Rogers S, Schroedl S. Constrained K-means clustering with background knowledge. In: Proceedings of the 18th International Conference on Machine Learning; 2001. p. 577–84.
15. Xing E, Ng A, Jordan M, Russell S. Distance metric learning, with application to clustering with side-information. Adv Neural Inf Process Syst. 2002;15:505.

Collaborative Filtering

Mohamed Sarwat[1] and Mohamed F. Mokbel[2]
[1]School of Computing, Informatics, and Decision Systems Engineering, Arizona State University, Tempe, AZ, USA
[2]Department of Computer Science and Engineering, University of Minnesota-Twin Cities, Minneapolis, MN, USA

Synonyms

Social filtering

Definition

Collaborative filtering assumes a set of n users $\mathcal{U} = \{u_1, \ldots, u_n\}$ and a set of m items $\mathcal{I} = \{i_1, \ldots, i_m\}$. Each user u_j expresses opinions about a set of items $\mathcal{I}_{u_j} \subseteq \mathcal{I}$. Many applications assume opinions are expressed through an explicit numeric rating (e.g., one through five stars), but other methods are possible (e.g., hyperlink clicks, Facebook "likes"). For an active user u_a, collaborative filtering predicts the rating $\mathcal{F}(u_a, i_r)$ that u_a would give to item i_r such that $i_r \in \mathcal{I}_r$ and $\mathcal{I}_{u_a} \cap \mathcal{I}_r = \emptyset$, i.e., the user has not rated the suggested items.

Historical Background

In the early 1990s, collaborative filtering has emerged as one of the many ways to recommend useful information that are relevant to users. Collaborative filters predict how much a user would like a specific item based upon other users who showed similar (or dissimilar) tastes in the past. In the early 2000s, collaborative filtering (as a recommendation technique) became widely adopted by internet companies and online retail stores like Amazon [9]. Such companies employed collaborative filtering to recommend relevant products (e.g., books) to their users based on their historical preferences. The popularity of collaborative filtering mainly stems from the fact that it is relatively simple and yet exhibits very high accuracy in predicting users' preferences.

A collaborative filtering algorithm takes as input a set of users $\mathcal{U}$, items $\mathcal{I}$, and ratings (history of users' opinions over items) $\mathcal{R}$. It then analyzes the historical preferences (tastes) of many users to predict how much a specific user would like a certain item. Computer scientists have come up with several collaborative filtering algorithms to generate relevant recommendations to users [1, 5]. From one perspective, collaborative filtering recommenders fall into two main categories: (a) neighborhood based [10, 11] that calculates the similarity between system users or items and leverages that to estimate how much a user like an item and (b) matrix factorization [2, 6] that uses linear algebra techniques to analyze the user/item ratings data and hence predict how much a user would like an unseen item.

Scientific Fundamentals

The collaborative filtering process is usually broken into two phases: (1) a *model generation* phase that creates a model storing correlations between items and users and (2) a *rating prediction* phase that uses the model to predict items' ratings. There are several methods to perform collaborative filtering including item based [11], user based [10], regression based [11], or approaches that use more sophisticated probabilistic models (e.g., Bayesian Networks [2]). Below we describe three popular collaborative filtering methods: item-based [11] and user-based [10] collaborative filtering and singular value decomposition (a matrix factorization method [6]) in use today (e.g., Amazon [9]).

Item-Based Collaborative Filtering

The item-based collaborative filtering technique builds, for each of the m items $\mathcal{I}$ in the database, a list $\mathcal{L}$ of *similar* items. Given two items i_p and i_q, we can derive their similarity score $sim(i_p, i_q)$ by representing each as a vector in the user-rating space and then use a similarity function over the two vectors to compute a numeric value representing the strength of their relationship. Figure 1 depicts this item-item model-building process. Conceptually, we can represent the ratings data as a matrix, with users and items each representing a dimension, as depicted on the left side of Fig. 1. The similarity function, $sim(i_p, i_q)$, computes the similarity of vectors i_p and i_q using *only* their co-rated dimensions. In our example u_j and u_k represent the co-rated dimensions. Finally, we store i_p, i_q, and $sim(i_p, i_q)$ in our model, as depicted on the right side of Fig. 1. The similarity measure need not be symmetric, i.e., it is possible that $sim(i_p, i_q) \neq sim(i_q, i_p)$. Many similarity measures have been proposed in the literature [11]. One of the most popular measures used is the cosine distance, calculated as in Eq. 1.

$$sim(i_p, i_q) = k \frac{\vec{i_p} \cdot \vec{i_q}}{\|\vec{i_p}\| \|\vec{i_q}\|} \qquad (1)$$

Here, items i_p and i_q are represented as vectors in the user-rating space, and k represents a dampening factor that discounts the influence of item pairs having high scores but only a *few* common ratings; given the co-rating count between two items as $corate(i_p, i_q)$, k is defined as

$$k = \begin{cases} 1 & corate(i_p, i_q) \geq 50 \\ corate(i_q, i_q)/50 & \text{otherwise} \end{cases} \qquad (2)$$

Model Truncation. It is common practice in collaborative filtering to reduce the model size by truncating the similarity list $\mathcal{L}$ for each object [11] (e.g., item or user). For the item-item model, truncation means storing in $\mathcal{L}$ only a small fraction of similar items for each of the m items in the database. Such a practice has positive performance implications, as a smaller $\mathcal{L}$ implies a more efficient rating prediction process (per Eq. 3). Also, for the item-item method, it has been observed that truncating $\mathcal{L}$ has minimal impact on the *quality* of recommendations [11]. Truncation is also beneficial to the user-user method from both an efficiency and quality standpoints. In general, the criteria used for truncating $\mathcal{L}$ is unique to each collaborative filtering technique. However, two common approaches are: (1) store the k most similar items to an item i, where ($k \ll m$), and (2) store only items l that have a similarity score (i.e., $sim(i, l)$) greater than a threshold $\mathcal{T}$.

Rating prediction in the item-based cosine method produces the top-n items based on predicted score using two steps. (1) *Reduction*: cut down the model such that each item i left in the model is an item *not* rated by user u_a, while i's similarity list $\mathcal{L}$ contains only items l already rated by u_a. (2) *Compute*: the predicted rating $F_{(u_a, i)}$ for an item i and user u_a is calculated as a weighted sum [11]:

$$F_{(u_a, i)} = \frac{\sum_{l \in \mathcal{L}} sim(i, l) * r_{u_a, l}}{\sum_{l \in \mathcal{L}} sim(i, l)} \qquad (3)$$

The prediction is the sum of the user's rating for a related item l, $r_{u_a, l}$, weighted by the similarity to the candidate item i. The prediction is normalized by the sum of scores between i and l.

User-Based Collaborative Filtering

The user-user model is similar in nature to the item-item paradigm, except that the model calculates similarity between users (instead of items). This calculation is performed by comparing user vectors in the item-rating space. For example, in Fig. 1, focusing on the user/item matrix, users u_j and u_k can be represented as vectors in item space and compared based on the items they have co-rated (i.e., i_p and i_q). The user-user model primarily uses cosine distance and Pearson correlation as similarity measures [2], much like that of the item-item paradigm with the exception that similarity is measured in item space rather than user space.

Rating prediction in the user-based collaborative filtering paradigm is similar in spirit to the item-based method. Recall that the similarity list

Collaborative Filtering,
Fig. 1 Item-based model generation

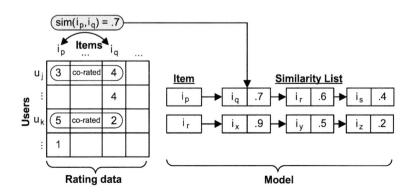

$\mathcal{L}$ in the user-user paradigm is a list of similar users to a particular user u. A prediction $F_{(u_a,i)}$ for an item i given user u_a is calculated as in Eq. 4. This value is the weighted average of deviations from a related user u_l's mean. In this equation, $r_{u_l,i}$ represents a user u_l's (non-zero) rating for item i, while $\bar{r}_{u_a}$ and $\bar{r}_{u_l}$ represent the average rating values for users u_a and u_l, respectively.

$$F_{(u_a,i)} = \bar{r}_{u_a} + \frac{\sum_{l \in \mathcal{L}}(r_{u_l,i} - \bar{r}_{u_l}) * sim(u_a, u_l)}{\sum_{l \in \mathcal{L}} |sim(u_a, u_l)|}$$
$$(4)$$

Matrix Factorization

Matrix factorization reduces the user/item rating space into two latent factor space matrices: (1) user factor matrix (p), contains a set of user vectors such that each user vector $p_u \in p$ denotes the weights that each user would assign to a set of item features (latent factors), and (2) item factor matrix (q), consists of a set of item vectors such that each item vector $q_i \in q$ denotes the weights that qualify how much each item belongs to a set of features (latent factors).

$$\min_{q*,p*} \sum_{(u,i) \in k} (r_{ui} - q_i^T.p_u)^2 + \lambda(||q_i||^2 + ||p_u||^2)$$
$$(5)$$
$$F(u,i) = q_i^T.p_u \qquad (6)$$

To learn the matrix factorization model, the system uses techniques like singular value decomposition (SVD), stochastic gradient descent, and alternating least square to minimize the regularized squared error (see Eq. 5). For matrix factorization, the predicted rating valued $F(u,i)$ for each item i not rated by u is calculated as the dot product of both the user feature vector p_u and the item feature vector transpose (q_i^T) (see Eq. 6).

Key Applications

Recommendation

Personalized recommendation is by far the main application that uses collaborative filtering. For instance, collaborative filtering has been leveraged to recommend movies (e.g., Netflix), books (e.g., Amazon), and goods (e.g., eBay).

Personalized News Feed

A variation of collaborative filtering has been also used to filter news feed [3] (and social media) to generate relevant news feed to users of social media websites (e.g., Facebook, Twitter).

Experimental Results

The authors in [4] present a comprehensive accuracy evaluation of several collaborative filtering algorithms that include user-based, item-based, and singular value decomposition. The authors in [7] present experimental evaluation that measures the query processing efficiency and scalability of item-based collaborative filtering techniques.

Data Sets

MovieLens

Three data sets (of different sizes) are extracted from the MovieLens system (a real movie recommendation website). The largest MovieLens data set consists of 10 million ratings (scale from 1 to 5) assigned by 72,000 users to 10,000 movies. The MovieLens data sets also consist of users' demographic information like age and gender as well as movies' attributes like genre and release year.

Netflix Prize

A data set is provided by Netflix to participants in the Netflix prize. The data set consists of 100 million ratings (scale from 1 to 5) assigned by 480,000 anonymous users to over 17,000 movies.

Yahoo! Music

This data set is collected by Yahoo! Music services between 2002 and 2006. It contains over 717 million ratings of 136,000 songs given by 1.8 million users of Yahoo! Music services. The data set also provides attributes for each song like the artist, album, and genre.

URL to Code

LensKit

LensKit (http://lenskit.org) is a Java-based open-source software that provides a generic (out-of-the-box) implementation of various collaborative filtering algorithms.

Apache Mahout

Mahout (https://mahout.apache.org) is an open-source software that implements machine learning and data mining algorithms on top of scalable cluster computing platforms (currently implemented on-top-of Hadoop). Mahout provides support for major collaborative filtering algorithms that include user-based collaborative filtering, item-based collaborative filtering, and matrix factorization.

RecDB

RecDB (http://www-users.cs.umn.edu/~sarwat/RecDB/) [8,12] is a recommendation engine built entirely inside PostgreSQL (relational DBMS) to answer recommendation-based queries. RecDB provides support for major collaborative filtering algorithms that includes user-/item-based and singular value decomposition.

Cross-References

▶ Recommender Systems

Recommended Reading

1. Adomavicius G, Tuzhilin A. Toward the next generation of recommender systems: a survey of the state-of-the-art and possible extensions. IEEE Trans Knowl Data Eng TKDE. 2005;17(6):734–49.
2. Breese JS, Heckerman D, Kadie C. Empirical analysis of predictive algorithms for collaborative filtering. In: Proceedings of the 14th Conference on Uncertainty in Artificial Intelligence; 1998.
3. Das A, Datar M, Garg A, Rajaram S. Google news personalization: scalable online collaborative filtering. In: Proceedings of the 16th International World Wide Web Conference; 2007.
4. Ekstrand MD, Ludwig M, Konstan JA, Riedl J. Rethinking the recommender research ecosystem: reproducibility, openness, and lenskit. In: Proceedings of the 5th ACM Conference on Recommender Systems; 2011.
5. Koren Y, Bell RM. Advances in collaborative filtering. In: Recommender systems handbook. Springer; 2011. p. 145–86. https://link.springer.com/book/10.1007/978-0-387-85820-3
6. Koren Y, Bell RM, Volinsky C. Matrix factorization techniques for recommender systems. IEEE Comput. 2009;42(8):30–7.
7. Levandoski JJ, Ekstrand MD, Ludwig M, Eldawy A, Mokbel MF, Riedl J. Recbench: benchmarks for evaluating performance of recommender system architectures. Proc VLDB Endowment. 2011;4(11):911–20.
8. Levandoski JJ, Sarwat M, Mokbel MF, Ekstrand MD. RecStore: an extensible and adaptive framework for online recommender queries inside the database engine. In: Proceedings of the 15th International Conference on Extending Database Technology; 2012.
9. Linden G, et al. Amazon.com recommendations: item-to-item collaborative filtering. IEEE Internet Comput. 2003;7(1):76–80.
10. Resnick P, Iacovou N, Suchak M, Bergstrom P, Riedl J. GroupLens: an open architecture for collaborative filtering of netnews. In: Proceedings of the 1994 Conference on Computer Supported Cooperative Work; 1994.
11. Sarwar B, Karypis G, Konstan J, Riedl J. Item-based collaborative filtering recommendation algorithms. In: Proceedings of the 10th International World Wide Web Conference; 2001.
12. Sarwat M, Avery J, Mokbel MF. RecDB in action: recommendation made easy in relational databases. Proc VLDB Endowment. 2013;6(12):1242–5.

Column Segmentation

Sunita Sarawagi
IIT Bombay, Mumbai, India

Synonyms

Information extraction; Record extraction; Text segmentation.

Definition

The term column segmentation refers to the segmentation of an unstructured text string into segments such that each segment is a column of a structured record.

As an example, consider a text string S = *"18100 New Hampshire Ave. Silver Spring,*

MD 20861" representing an unstructured form of an Address record. Let the columns of this record be House number, Street name, City name, State, Zip and Country. In column segmentation, the goal is to segment S and assign a column label to each segment so as to get an output of the form:

```
House Number    :   18100
Street Name     :   New Hampshire Ave.
City            :   Silver Spring
State           :   MD
Zip             :   20861
Country         :   --
```

Historical Background

The column segmentation problem is a special case of a more general problem of Information Extraction (IE) that refers to the extraction of structure from unstructured text. Column segmentation is typically performed on short text strings where most of the tokens belong to one of a fixed set of columns. In the more general IE problem, the unstructured text could be an arbitrary paragraph or an HTML document where the structured entities of interest form a small part of the entire string.

There is a long history of work on information extraction [5]. Most of the early work in the area was in the context of natural language processing, for example for extracting named entities like people names, organization names, and location names from news articles. The early systems were based on hand-coded set of rules and relied heavily on dictionaries of known records. Later systems were based on statistical methods like maximum entropy taggers [9], Hidden Markov Models [11] and Conditional Random Fields (CRFs) [7].

In the database research community, interest in column segmentation arose in the late 1990s as a step in the process of cleaning text data for data warehousing. Many commercial tools were developed purely for the purposes of cleaning names and addresses. These were based on hand-coded, rule-based, database driven methods that work only for the region that they are developed for and do not extend to other domains. Much manual work has to be done to rewrite these rules when shifting the domain from one locality to another. This led to the adoption of statistical techniques [1, 3] which proved to be more robust to noisy inputs.

Foundations

A formal definition of column segmentation follows. Let $Y = \{y_1, \ldots, y_m\}$ denote the set of column types of the structured record. Given any unstructured text string $\mathbf{x}$, column segmentation finds segments of $\mathbf{x}$ and labels each with one of the columns in Y. The input $\mathbf{x}$ is typically treated as a sequence of tokens obtained by splitting $\mathbf{x}$ along a set of delimiters. Let $x_1, \ldots, x_n$ denote such a sequence of tokens. A segmentation of $\mathbf{x}$ is a sequence of segments $s_1 \ldots s_p$. Each segment s_j consists of a *start position* t_j, an *end position* u_j, and a *label* $y_j \in Y \cup \{\text{"Other"}\}$. The special label "Other" is used to label tokens not belonging to any of the columns. The segments are assumed to be contiguous, that is, segment s_{j+1} begins right after segment s_j ends. Also, the last segment ends at n and the first segment starts at 1.

As a second example consider a citation String $T = P.P.Wangikar, T.P. Graycar, D.A. Estell, D.S. Clark, J.S. Dordick (1993) Protein and Solvent Engineering of Subtilising BPN'$ in

```
Title     :   Protein and Solvent Engineering
              of Subtilising BPN
Authors   :   P.P.Wangikar, T.P.Graycar,
              D.A. Estell, D.S. Clark, J.S. Dordick
Year      :   1993
Venue     :   Nearly Anhydrous Organic Media
              J.Amer. Chem. Soc.
Volume    :   115
Number    :   --
Pages     :   12231-12237
```

Nearly Anhydrous Organic Media J.Amer. Chem. Soc. 115, 12231–12237. and a set of columns: Author names, title, year, publication venue, volume, number. A segmentation of this string is:

In this example, the tokens "in", "("and")" of the input have been assigned label "Other".

Challenges

The problem of column segmentation is challenging because of the presence of various kinds of noise in the unstructured string.

- The same column might be represented in many different forms, for example "Street" might be abbreviated as "St." or "st".
- The order in which columns appear might be different in different strings: for example, in some citations authors could be before title, and after title in others.
- Columns might be missing: some addresses might contain a country name, others may not.
- Strings from different sources might be formatted differently: for example, some citations might use a comma to separate fields whereas others might have no regular delimiter between fields.

Main Techniques

A column segmentation technique needs to combine information from multiple different sources of evidence to be able to correctly recognize segmentations in noisy strings. One source is the characteristic words in each elements, for example the word "street" appears in road-names. A second source is the limited partial ordering between its element. Often the first element is a house number, then a possible building name and so on and the last few elements are zipcode and state-name. A third source is the typical number of words in each element. For example, state names usually have one or two words whereas road names are longer. Even within a field, some words are more likely to appear in the beginning of the field rather than towards its end. Often, there is a pre-existing database of known values of columns. Match of a substring of the text to an existing database column, can be a valuable clue for segmentation. The format of

the entry, presence of certain regular expression, capitalization, and punctuation patterns can be useful when word-level matches are absent. A good column segmentation technique would combine evidence from all of these clues in performing the final segmentation.

The three main types of column segmentation techniques are:

Rule-Based Systems

A rule-based technique, as the name suggests, encodes one or more of the above clues as rules. These are applied in a specified order and when more than two rules conflict, another set of rule resolution mechanisms are used to decide which one wins.

Here are some examples of rules that can be used to extract columns from citation records:

- Punctuation CapsWord {2–10} Dot → Title
- CapsWord Comma Initial Dot Initial Dot → Author name
- Initial Dot CapsWord Comma → Author name
- AllCaps Words {1–2} Journal → Journal

For example, the first rule marks as title any substring of two to ten capitalized words appearing between a punctuation and a full-stop. The second rule marks as an author name any substring consisting of a capitalized word followed by comma and two initials. This would identify strings of the form "Gandhi, M. K." as author names. Whereas the third rule would mark strings like "V. Ganti," as author names. The fourth rule would mark string like "ACM computing Journal" as journal names.

Such rules could be either hand-coded or learnt from example datasets [2, 6]. Existing rule-based techniques are able to concentrate only on a subset of the above mentioned clues to limit the complexity of the learnt rules. They provide high precision segmentation in uniform settings where the amount of noise is limited. When the input becomes noisy, rule-based systems tend to lose on recall.

Hidden Markov Models

Hidden Markov Models (HMMs) provide an intuitive statistical method for combining many of the above clues in a unified model. A HMM is a probabilistic finite state automata where the states represent the fields to be extracted, directed edges between edges are attached with probability values indicating probability of transitioning from one state to another, and states are attached with a distribution over the words that can be generated from the state. A segmentation of a string S is achieved by finding the sequence of states for which the product of the probability of generating the words in S and following the transitions in state sequences is maximized. Such a path can be found efficiently using a dynamic programming algorithm. The parameters controlling the transition and word distributions of states are learnt using examples of correctly segmented strings.

An example, of a Hidden Markov Model trained to recognize Indian addresses appears in Fig. 1. The number of states is 10 and the edge labels depict the state transition probabilities. For example, the probability of an address beginning with House Number is 0.92 and that of seeing a City after Road is 0.22. The dictionary and the emission probabilities are not shown for compactness.

For more details on the use of HMMs in column segmentation see [1, 3, 11].

Conditional Models

A limitation of HMMs is that the distribution that controls the generation of words within a state is generative, and can therefore capture only a limited set of properties of the words it can generate. For example, it is complicated to account for various orthographic properties of words, like its capitalization pattern, or the delimiter following that word. These limitations are removed by recently proposed formalisms like Conditional Random Fields (CRFs) that capture the conditional distribution of column sequence *given* the sequence of words in a string S. This enables the incorporation of any arbitrary set of clues derived from a word and the words in its neighborhood. Also it becomes easy to incorporate clues derived from the degree of match of a proposed column with pre-existing values in the database.

A CRF models the conditional probability distribution over segmentations $\mathbf{s}$ for a given input sequence $\mathbf{x}$ as follows:

$$\Pr\left(\mathbf{s}\,\middle|\,\mathbf{x},\,\mathbf{W}\right) = \frac{1}{Z\,(x)}\exp\left(\mathbf{W}\cdot\sum_{j}\mathbf{f}\,(j,\,\mathbf{x},\,\mathbf{s})\right)$$

(1)

Where $\mathbf{f}\,(j,\mathbf{x},\mathbf{s})$ is a vector of local feature functions $f_1 \ldots f_N$ of $\mathbf{s}$ at the jth segment and $\mathbf{W} = (W_1, W_2, \ldots, W_N)$ is a weight vector that en-

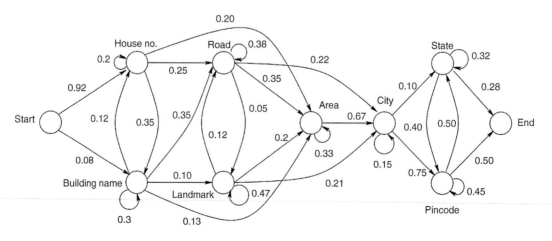

Column Segmentation, Fig. 1 An example of trained HMM for segmenting addresses

codes the importance of each feature function in $\mathbf{f}$. $Z\left(\mathbf{x}\right) = \sum_{s'} \exp\left(\mathbf{W} \cdot \sum_j \mathbf{f}\left(j, \mathbf{x}, \mathbf{s}'\right)\right)$ is a normalization factor. The label of a segment depends on the label of the previous segment and the properties of the tokens comprising this segment and the neighboring tokens. Thus a feature for segment $s_j = \left(t_j, u_j, y_j\right)$ is a function of the form $f\left(y_j, y_{j-1}, \mathbf{x}, t_j, u_j\right)$ that returns a numeric value. Example of such features are:

$f_8\ \left(y_i,\ y_{i-1},\ \mathbf{x},\ 3,\ 5\right) = [x_3\ x_4\ x_5$ appears in a journal list] . $[y_i = $ journal]

$f_{12}\ \left(y_i,\ y_{i-1},\ \mathbf{x},\ 19,\ 19\right) = [x_{19}$ is an integer] . $[y_i = $ year] . $[y_{i-1} = $ month]

The weight vector $\mathbf{W}$ is learnt during training via a variety of methods, such as likelihood maximization [7]. During *segmentation*, thegoal is to find a $\mathbf{s} = s_1 \ldots s_p$ for the input sequence $\mathbf{x} = x_1 \ldots x_n$ such that $\Pr\left(\mathbf{s} \middle| \mathbf{x}, \mathbf{W}\right)$ (as defined by (1) is maximized.

$$\arg\max_s \Pr\left(\mathbf{s} \middle| \mathbf{x},\ \mathbf{W}\right) = \arg\max_s \mathbf{W} \cdot \sum_j \mathbf{f}\left(y_i,\ y_{i-1},\ \mathbf{x},\ t_j,\ u_j\right)$$

The right hand side can be efficiently computed using dynamic programming. Let L be an upper bound on segment length. Let $\mathbf{s}_{i:y}$ denote set of all partial segmentation starting from 1 (the first index of the sequence) to i, such that the last segment has the label y and ending position i. Let $V\ (i,\ y)$ denote the largest value of $\mathbf{W} \cdot \sum_j \mathbf{f}\left(j, \mathbf{x}, \mathbf{s}'\right)$ for any $\mathbf{s}' \in \mathbf{s}_{i:y}$. The following recursive calculation finds the best segmentation:

$$V\ (i,\ y) = \begin{cases} \max_{y',\ i'=i-L\ldots i-1}\ V\ (i',\ y') + \mathbf{W} \cdot \mathbf{f}(y,\ y',\ \mathbf{x},\ i'+1,\ i) & \text{if } i > 0 \\ 0 & \text{if } i = 0 \\ -\infty & \text{if } i < 0 \end{cases}$$

The best segmentation then corresponds to the path traced by $\max_y V(|\mathbf{x}|, y)$.

More details on CRFs can be found in [7] and the extension of CRFs for segmentation can be found in [8, 10]. Reports an empirical evaluation of CRFs with HMMs for segmenting paper citations [4]. Shows how to perform efficient segmentation using CRFs in the presence of a large pre-existing database of known values.

Key Applications

Column segmentation has many applications, including,

Cleaning of text fields during warehouse construction: In operational datasets, text fields like addresses are often recorded as single strings. When warehousing such datasets for decision support, it is often useful to identify structured elements of the address. This not only allows for richer structured queries, it also serves as a useful pre-processing step for duplicate elimination.

Creation of citation databases: A key step in the creation of citation databases like Citeseer and Google Scholar, is to resolve for each citation, which paper it refers to in the database. Citations as extracted from papers are unstructured text strings. These have to be segmented into component author names, titles, years, and publication venue before they can be correctly resolved to a paper entry in the database.

Extraction of product information from product descriptions: Comparison shopping websites often need to parse structured fields representing various attributes of product from unstructured HTML sources.

URL to Code

Java packages for column segmentation using conditional random fields are available via Source Forge at http://crf.sf.net and as part of the Mallet package at http://mallet.cs.umass.edu

Cross-References

► Data Cleaning

Recommended Reading

1. Agichtein E, Ganti V. Mining reference tables for automatic text segmentation. In: Proceedings of the 10th ACM SIGKDD International Conference on Knowledge Discovery and Data Mining; 2004. p. 20–9.
2. Aldelberg B. Nodose: a tool for semi-automatically extracting structured and semi-structured data from text documents. In: Proceedings of the ACM SIGMOD International Conference on Management of Data; 1998. p. 283–94.
3. Borkar VR, Deshmukh K, Sarawagi S. Automatic text segmentation for extracting structured records. In: Proceedings of the ACM SIGMOD International Conference on Management of Data; 2001. p. 175–86.
4. Chandel A, Nagesh PC, Sarawagi S. Efficient batch top-k search for dictionary-based entity recognition. In: Proceedings of the 22nd International Conference on Data Engineering; 2006.
5. Cunningham H. Information extraction, automatic. In: Encyclopedia of Language and Linguistics. 2nd ed. 2005.
6. Kushmerick N, Weld DS, Doorenbos R. Wrapper induction for information extraction. In: Proceedings of the 15th International Joint Conference on AI; 1997. p. 729–37.
7. Lafferty J, McCallum A, Pereira F. Conditional random fields: Probabilistic models for segmenting and labeling sequence data. In: Proceedings of the 18th International Conference on Machine Learning; 2001. p. 282–9.
8. Peng F, McCallum A. Accurate information extraction from research papers using conditional random fields. In: Proceedings of the Human Language Technology Conference and North American Chapter of the Association for Computational Linguistics; 2004. p. 329–36.
9. Ratnaparkhi A. Learning to parse natural language with maximum entropy models. Mach Learn. 1999;34:151.
10. Sarawagi S, Cohen WW. Semi-markov conditional random fields for information extraction. In: Advances in Neural Information Processing Systems. 17, 2004.
11. Seymore K, McCallum A, Rosenfeld R. Learning Hidden Markov Model structure for information extraction. In: Papers from the AAAI-99 Workshop on Machine Learning for Information Extraction; 1999. p. 37–42.

Column Stores

Pingpeng Yuan and Hai Jin
Service Computing Technology and System Lab, Cluster and Grid Computing Lab, School of Computer Science and Technology, Huazhong University of Science and Technology, Wuhan, China

Synonyms

Column-oriented DBMS; Columnar DBMS

Definition

A column store is a database powered with column-oriented storage and access mechanisms. At the conceptual schema level, a column store consists of multiple columns. Some related columns can be grouped to a column family. Furthermore, several column families can form a super column, which can be seen as a "view" on a number of tables [8]. Super column can also be viewed as a map of tables.

At physical storage level, a column store places all values of a column in a sequential order on a storage media, then the values of the next column, and so on. Storing data column by column makes it possible to retrieve data in a column without fetching other columns. The column-oriented approach is in contrast to row-oriented databases or row stores and can significantly speed up column-based access.

The column store is made popular by Google's Bigtable [7], which is also considered as a column family store.

Historical Background

Column store DBMSs can be traced back to the 1970s, when transposed files have been implemented from the early days of DBMS development [2]. Copeland GP et al first showed the advantages of a fully decomposed storage model

(DSM) over n-ary storage model (NSM) [9]. Subsequent research on join and projection indices further strengthened the advantages of DSM over NSM. KDB [19] was the first commercially available column-oriented database developed in 1993. KDB has proven itself fast and capable of holding massive amounts of data, widely used in the financial industry. In 1995, Sybase IQ [19] was matured as a column-oriented database with many deployments and good tooling support.

Although the idea of the DSM layout to improve performance has been around for a long time, it was not a major alternative to row-oriented stores until the 2000s when many column-oriented open-source and commercial implementations were made available. MonetDB [15], MonetDB/X100 [6], and C-Store [18] are some of the pioneers in the design of modern column-oriented database systems and vectorized query execution [2]. The technology developed in MonetDB has been emulated by others and directly leads to the vectorwise commercial product. C-Store is produced as a joint research project optimized for reads [18]. Vertica [23] was a commercial spin-off of C-Store development effort, while MonetDB spawned Vectorwise [15]. All these column-oriented systems apply similar techniques such as column-oriented storage, multicore execution, and automatic storage pruning for analytical workloads. Another interesting column store-related effort is the SAP HANA system [22], which takes a different approach to analytic workloads and focuses on columnar in-memory storage and tight integration with other business applications.

The success of column stores has led legacy RDBMS vendors to add columnar storage options to their existing engines [13]. For example, InfiniDB [21] was MySQL compatible warehouse columnar engine that is multi-terabyte capable. Greenplum [20] was a hybrid column-/row-oriented database based on PostgreSQL with several enhancements to allow efficient parallel execution over multiple machines.

The increasing amount of semi-structured and unstructured data in the Web poses huge challenges for relational database systems (RDBMS). Many companies and organizations developed their own non-relational storage systems, which are coined today as NoSQL (Not only SQL) databases. From the data model perspective, NoSQL includes key-value stores, document-oriented stores, column family stores, and graph stores [16]. Many key-value stores in the NoSQL category are actually hybrid row/column store unlike pure relational column databases. Examples of such column family NoSQL stores include HBase [4], Hypertable [11], and Cassandra [3].

Scientific Fundamentals

In the standard relational data model, a database consists of named tables, each with M columns ($M > 1$). Each row in a table describes a record, and each column corresponds to a field in the record. A record is a logical structure assembled from the M number of fields. For example, let us assume we have six students and thus there are six records, each of which contains five fields, (Table 1). The first column serves as the surrogate attribute for each tuple of the relation. Its value of a tuple is unique.

There are several ways to implement the tables. Most stores implement the tables by using a row-oriented approach based on an n-ary storage model (NSM) [9]. This approach stores data as seen in the conceptual schema. An alternative to implementing physical tables is projection. That is, a new relation is formed by projection on one or more attributes and its surrogate. When the columns in a projection are related, the projection is similar to column family or super column. Considering the example (Table 1), we project

Column Stores, Table 1 Academic record

No.	Name	Subject number	Credit points	Mark
1	Bob	COM1010	6	65
2	Alice	HSY1010	6	78
3	John	JRN2905	6	79
4	Jane	JRN1903	6	70
5	Joe	JRN2906	6	79
6	Mike	JRN3903	6	72

Column Stores, Table 2
Decomposing academic record

(a) Students

No.	Name
1	Bob
2	Alice
3	John
4	Jane
5	Joe
6	Mike

(b) Subjects

No.	Subject number	Credit points
1	COM1010	6
2	HSY1010	6
3	JRN2905	6
4	JRN1903	6
5	JRN2906	6
6	JRN3903	6

(c) Mark

No.	Mark
1	65
2	78
3	79
4	70
5	79
6	72

Column Stores, Table 3
Further decomposing academic record

(a) Subject name

No.	Subject number
1	COM1010
2	HSY1010
3	JRN2905
4	JRN1903
3	JRN2906
6	JRN3903

(b) Credit point

No.	Credit points
1	6
2	6
3	6
4	6
5	6
6	6

Column Stores, Fig. 1
Row store vs. column store

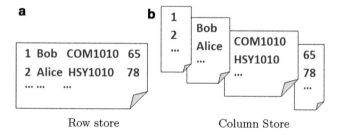

Row store   Column Store

the table into three sub-tables (Table 2). We can proceed toward a decomposition storage model (DSM) so that each decomposed relation is binary, containing one attribute value and one surrogate [9]. Considering the example again, since Tables 2a, c are two-dimensional tables, they will not be decomposed. Table 2b will be decomposed into two binary relations (Table 3a, b). DSM pairs each attribute value with the surrogate of its conceptual schema record in a binary relation [9]. Thus, any number of conceptual schema attributes can be supported without additional complexity. Changing an attribute in the conceptual schema causes no change in the storage structure because the storage structure is still the simple binary relation [9].

The above model provides a separation between the conceptual and storage schema. However, data is not physically stored in storage media using this logical data model. When storing the data into storage media, most RDBMSs store data in rows. In this layout, a DBMS stores every attribute of a given row in sequence, with the last entry of one row followed by the first entry of the next row. For example, the above relation would be stored using row-based format as shown in Fig. 1a. In contrast to row-oriented stores, an alternative approach is to store raw tables in column-based format. In a column store architecture, the values for each single column (or attribute) are stored contiguously [18]. Each attribute of a relational table is physically stored as a separate column. Figure 1b is the column-oriented storage of our running example in Table 1. This column-oriented approach allows queries to load only the required attributes.

There is much debate on performance of row store and column store. Whether or not a column-oriented system will be more efficient than row-based stores depends heavily on the workload.

The reason is that different workloads will execute operations over different pieces of data [10]. Comparisons between row-oriented and column-oriented data layouts are typically concerned with the efficiency of secondary storage I/O for a given workload. For example, if we want to access data row by row, then it will cost less storage I/O for a row store than a column store as shown in Fig. 2. However, reading a column will cost less storage I/O over a column store than a row store.

In practice, column-oriented storage layouts are well-suited for OLAP-like workloads which typically involve aggregates over a small subset of all columns of data. For example, when an aggregate is computed over a small subset of all columns of data, column stores are more efficient. OLAP-like workloads are generally Read-mostly applications which include customer relationship management (CRM) systems, electronic library, to name a few. Row-oriented storage layouts are more efficient when many columns of a single row are required for computation at the same time and the entire row can be retrieved with sequential storage I/O. The typical workloads are OLTP-like workloads.

To truly take advantage of a column store architecture, the query optimizer must be deeply aware of columnar storage and optimization techniques. Maintaining indexes for block address, record keys, and column value would enable faster access to multiple blocks of column store. Generally, the adjacent data of a column is similar, and same value may have multiple appearances. Thus, compression schemes, such as run-length encoding, can be used to optimize the storage size of column store [14]. Since on-the-fly tuple reconstruction for multi-attribute queries is necessary in column store, partial sideways cracking can minimize the tuple reconstruction cost in a self-organizing way [12]. Because data is stored in columns, the operations on data can be executed in a highly parallelized manner. Late materialization and invisible join are examples of optimization technologies that can significantly speed up joins in a column store [2]. The column-based organization also makes it easy to access memory sequentially. This presents huge optimizations opportunity for in-memory database systems.

Key Applications

A number of applications can directly benefit from column store systems, such as data warehouse, RDFs, graphs, and search engines.

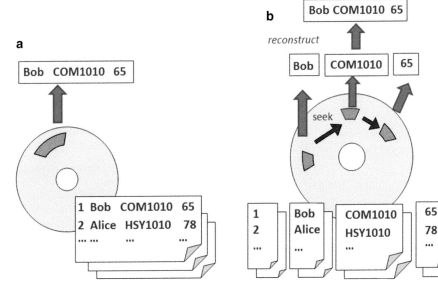

Column Stores, Fig. 2 Row store vs. column store. (**a**) Read a record on row store. (**b**) Read a record on column store

Data warehouse: A data warehouse is a database that is designed for facilitating querying and analysis. In data warehouses, typical tasks are OLAP queries which involve aggregates performed over only a few columns of large numbers of rows. Hence, Column stores are widely regarded as the preferred architectures for data warehousing because they can only load target columns instead of all columns of large number of rows. In such environments, conventional RDBMs vendors also incorporated columnar storage and columnar compression into their row-oriented DBMS.

RDF and Graph: Column store has been used to store RDF data [1], where a fat table is vertically partitioned into multiple two-column tables, one for each unique property (predicate). The first column is for subject, whereas the other column is for object. Although those tables can be stored using either row-oriented or column-oriented DBMS, the column store is a more popular storage solution [1]. This approach is easy to implement and can provide superior performance for queries with value-based restrictions on properties [24]. However, this approach may suffer from scalability problems when the size of tables varied significantly [17].

Search engine: Search engines and decision support systems serve similar purpose in general. In addition, there are trends toward a confluence of column stores and search engines. For example, MonetDB/X100 is considered as a backend for a search engine [5].

Cross-References

▶ Database Management System
▶ Graph Database
▶ NoSQL Stores
▶ Resource Description Framework

Recommended Reading

1. Abadi DJ, Marcus A, Madden SR, Hollenbach K. Scalable semantic web data management using vertical partitioning. In: Proceedings of the 33rd International Conference on Very Large Data Bases; 2007. p. 411–22.

2. Abadi DJ, Madden SR, Hachem N. Column-stores vs. row-stores: how different are they really? In: Proceedings of the ACM SIGMOD International Conference on Management of Data; 2008. p. 967–80.

3. Apache Cassandra. Cassandra. http://cassandra.apache.org/ (2015).

4. Apache HBase. Hbase. http://hbase.apache.org/ (2015).

5. Bjørklund TA, Gehrke J, Torbjørnsen Ø. A confluence of column stores and search engines: opportunities and challenges. In: Proceedings of the 35th International Conference on Very Large Data Bases; 2009. p. 1–12

6. Boncz P, Zukowski M, Nes N. Monetdb/x100: hyper-pipelining query execution. In: Proceedings of the 2nd Biennial Conference on Innovative Data Systems Research; 2005, p. 225–37.

7. Chang F, Dean J, Ghemawat S, Hsieh WC, Wallach DA, Burrows M, Chandra T, Fikes A, Gruber RE. Bigtable: a distributed storage system for structured data. In: Proceedings of the 7th USENIX Symposium on Operating System Design and Implementation; 2006. p. 205–18.

8. Column family. https://en.wikipedia.org/wiki/Column_family.

9. Copeland GP, Khoshafian SN. A decomposition storage model. In: Proceedings of the ACM SIGMOD International Conference on Management of Data; 1985. p. 268–79.

10. Halverson A, Beckmann JL, Naughton JF, DeWitt DJ. A comparison of C-store and row-store in a common framework. In: Proceedings of the 32nd International Conference on Very Large Data Bases; 2006. p. 553–64.

11. Hypertable. Hypertable. http://hypertable.org/ (2015).

12. Idreos S, Kersten ML, Manegold S. Self-organizing tuple reconstruction in column-stores. In: Proceedings of the ACM SIGMOD International Conference on Management of Data; 2009. p. 297–308.

13. Lamb A, Fuller M, Varadarajan R, Tran N, Vandiver B, Doshi L, Bear C. The vertica analytic database: C-store 7 years later. Proc VLDB Endowment. 2012;5(12):1790–801.

14. Mohapatra A, Genesereth M. Incrementally maintaining run-length encoded attributes in column stores. In: Proceedings of the International Conference on Database Engineering and Applications; 2012. p. 146–54.

15. MonetDB. MonetDB, 2010. https://www.monetdb.org/Home.

16. Nosql. https://en.wikipedia.org/wiki/NoSQL.

17. Sidirourgos L, Goncalves R, Kersten M, Nes N, Manegold S. Column-store support for RDF data management: not all swans are white. Proc VLDB Endowment. 2008;1(2):1553–63.

18. Stonebraker M, Abadi DJ, Batkin A, Chen X, Cherniack M, Ferreira M, Lau E, Lin A, Madden S, ÓNeil E, ÓNeil P, Rasin A, Tran N, Zdonik S. C-store: a column-oriented DBMS. In: Proceedings

of the 31st International Conference on Very Large Data Bases; 2005. p. 553–64.

19. Wikipedia. Column-oriented DBMS. https://en.wikipedia.org/wiki/Column-oriented_DBMS (2015).
20. Wikipedia. Greenplum. https://en.wikipedia.org/wiki/Greenplum (2015).
21. Wikipedia. Infinidb. https://en.wikipedia.org/wiki/InfiniDB (2015).
22. Wikipedia. Sap hana. https://en.wikipedia.org/wiki/SAP_HANA (2015).
23. Wikipedia. Vertica. https://en.wikipedia.org/wiki/Vertica (2015).
24. Yuan PP, Liu P, Wu BW, Liu L, Jin H, Zhang WY. TripleBit: a fast and compact system for large scale RDF data. Proc VLDB Endowment. 2013;6(7):517–28.

Common Warehouse Metamodel

Liam Peyton
University of Ottawa, Ottawa, ON, Canada

Synonyms

Common Warehouse Metadata Interchange (CWMI); CWM

Definition

The Common Warehouse Metamodel (CWM™) is an adopted specification from the OMG (Object Management Group) standards body. It defines standard interfaces that can be used to enable easy interchange of data warehouse and business intelligence metadata between data warehouse tools, data warehouse platforms and data warehouse metadata repositories in distributed heterogeneous environments. It supports relational, non-relational, multi-dimensional, and most other objects found in a data warehousing environment. It leverages three other standards from OMG:

- UML – Unified Modeling Language
- MOF – Meta Object Facility
- XMI – XML Metadata Interchange

The Object Management Group has been an international, open membership, not-for-profit computer industry consortium since 1989 with over 700 member organizations.

Historical Background

An initial Request For Proposal (RFP) for a common warehouse metadata interchange (CWMI) was issued by the OMG (Object Management Group) in 1998. A joint submission was received by the OMG in 1999 from Dimension EDI, Genesis Development Corporation, Hyperion Solutions, International Business Machines, NCR, Oracle, UBS AG, and Unisys.

At the time, there was a competing initiative from the Meta Data Coalition (MDC) which was supported by Microsoft and others. In 2000, however the two initiatives merged when the MDC joined OMG [5]. In 2001, version 1.0 of the specification was adopted with the name: Common Warehouse Metamodel (CWM™). The currently adopted version is 1.1 [1] (March, 2003).

The purpose of the Common Warehouse Metamodel specification was to make it possible for large organizations to have a metadata repository with a single metamodel. In practice this was not possible to achieve, since every data management and analysis tool requires different metadata and different metadata models [3]. Instead, the CWM specification defines interfaces that facilitate the interchange of data warehouse metadata between tools. In particular, the OMG Meta-Object Facility (MOF™) bridges the gap between dissimilar meta-models by providing a common basis for meta-models. If two different meta-models are both MOF-conformant, then models based on them can reside in the same repository.

However, compliance with the CWM specification does not guarantee tools from different vendors will integrate well, even when they are "CWM-compliant." The OMG addressed some of these issues by releasing patterns and best practices to correct these problems in a supplementary specification, the Common Warehouse Metamodel (CWM™)) Metadata

Interchange Patterns (MIP) Specification. Version 1.0 [2] was released in March 2004.

Foundations

The Common Warehouse Metamodel enables organizations and tool vendors to define and represent their metadata, metadata models and the processes which manipulate them in a common format so that the information can be streamed between tools and accessed programmatically [4].

The basic architecture and key technologies supporting the Common Warehouse Metamodel are shown in Fig. 1, on the next page. Metadata in a variety of formats, and from a variety of sources (Tools, Repositories, Databases, Files, etc.) is defined and represented in UML notation, based on the objects and classes that are defined in the Common Warehouse Metamodel. That representation is persisted in an XML notation that can be streamed to other tools, repositories, databases or files based on the XMI protocol. Finally, MOF is used to provide a broker facility that supports the ability to define and manipulate metamodels programmatically using fine grained CORBA interfaces. Using this architecture, organizations can create a single common repository which stores all the CWM-modeled descriptions of metadata and metamodels.

An example of a CWM description of a table from a relational database is shown below, along with the metadata description of the type of one of its columns (**type="22"**);
<CWMRDB:Table xmi.id="_15" name="MyTableName">
<CWM:Classifier.feature>
<CWMRDB:Column xmi.id="_16" name="my PrimaryKeyID" precision="4" type="_17"/>
<CWMRDB:Column xmi.id="_18" name="my ForeignKey1ID" precision="4" type="_17"/>
<CWMRDB:Column xmi.id="_19" name="my ForeignKey2ID" precision="4" type="_17"/>
<CWMRDB:Column xmi.id="_20" name="my ForeignKey3ID" precision="4" type="_17"/>
<CWMRDB:Column xmi.id="_21" name= "description" length="200" type="_22"/>
</CWM:Classifier.feature>

<CWM:Namespace.ownedElement>
<CWMRDB:ForeignKey xmi.id="_23" name= "unnamed_23" namespace="_15" feature="_19" uniqueKey=
"_24"/>
</CWM:Namespace.ownedElement>
</CWMRDB:Table>
<CWMRDB:SQLSimpleType
xmi.id="_22"
name="VARCHAR2"
visibility="public"characterMaximumLength= "200"
characterOctetLength="1" type Number="12"/>

The CWM specifications consists of a collection of metamodels (defined in UML) that capture all the elements of metadata, metamodels, and their processing that can be expressed when exchanging information between tools. These are what is identified in Fig. 1 as the Common Warehouse Metamodel. It is organized into five layers of abstraction.

Object Model

The Object Model layer is the base layer of the Common Warehouse Metamodel. The metamodels in the Object Model layer define the subset of UML that is used for creating and describing the CWM. They are the building blocks used by all the metamodels in the upper layers. This enables CWM to leverage UML's concepts without requiring implementations to support all full of UML's capabilities.

- Core
  The Core metamodel contains basic classes and associations used by all other CWM metamodels like Namespace, Constraint, Attribute, ModeledElement etc.
- Behavioral
  The Behavioral metamodel describe the behavior of CWM types and how that behavior is invoked with classes like Event, Parameter, CallAction etc.
- Relationships
  The Relationships metamodel describes two types of relationships between object

Common Warehouse Metamodel, Fig. 1
Common warehouse metamodel architecture

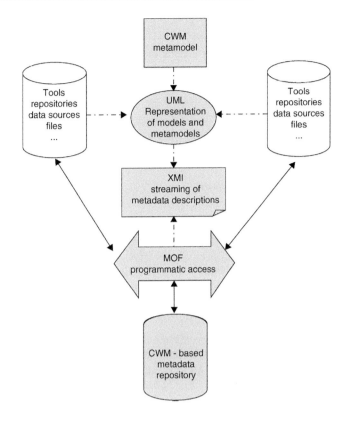

within a CWM information store: generalizations (for parent-child relationships) and associations (for links between objects).

- Instance

    The Instance metamodel contains classes to support the inclusion of data instances with the metadata.

Foundation

The metamodels in the Foundation layer contain general model elements that represent concepts and structures shared by other CWM packages. Metamodels in this layer are not necessarily complete, but serve as a common basis that can be shared with other metamodels.

- Data Types

    The DataTypes metamodel supports definition of metamodel constructs that modelers can use to create the specific data types they need with classes like Enumeration, Union, EnumerationLiteral, UnionMember, etc.

- Expression

    The Expressions metamodel supports the definition of expression trees.

- Keys and Indexes

    This metamodel defines the basic concepts of Index, IndexedFeature, UniqueKey, and KeyRelationship.

- Type Mapping

    This metamodel is used to support the mapping of types between different tools or data sources.

- Business Information

    The Business Information metamodel supports business-oriented information about model elements with classes like ResponsibleParty, Contact, ResourceLocater etc.

- Software Deployment

    The Software Deployment metamodel contains classes like SoftwareSystem, Component, Site to record how the software in a data warehouse is used.

Resource

The metamodels in the resource layer define the type of data sources and formats that are supported.

- Relational
  The Relational metamodel describes relational data this is accessed through an interface like ODBC, JDBC or the native interface of a relational database.
- Record
  The Record metamodel describes the concept of a record and its structure that can be applied to data records stored in files and databases, or to programming language structured data types.
- Multi-Dimensional
  The Multi-Dimensional metamodel describes a generic representation of a multidimensional database using classes like Schema, Dimension, Member etc.
- XML
  The XML metamodel describes the metadata of XML data with classes like ElementType, Attribute etc.

Analysis

The metamodels in the analysis layer define the types of interaction with metadata that are supported.

- Transformation
  The Transformation metamodel contains classes to describe common transformation metadata used in Extract, Transform, Load (ETL) tools and processes.
- OLAP
  The OLAP metamodel contains classes to describe common analysis metadata using in OLAP processing with classes like MemberSelection and CubeDeployment.
- Data Mining
  The Data Mining metamodel provide the necessary abstractions to model generic representations of both data mining tasks and

models (i.e., mathematical models produced or generated by the execution of data mining algorithms).
- Information Visualization
  The Information Visualization metamodel contains classes like Rendering, RenderedObject, to describe metadata associated with the display of data.
- Business Nomenclature
  The Business Nomenclature metamodel supports the definition of terms used in capturing business requirements with classes like Nomenclature, BusinesDomain, Taxonomy, Glossary, Term, Concept, etc.

Management

The metamodels in the management layer define two aspects of warehouse management.

- Warehouse Process
  The Warehouse Process metamodel supports the documentation of process flows used to execute transformations. A process flow can associate a transformation with a set of events, which will be used to trigger the execution of the transformation.
- Warehouse Operation
  The Warehouse Operation metamodel contains classes for the day-to-day operation and maintenance of the warehouse including scheduled activities, measurements, and change requests.

Key Applications

Vendors of data warehouse tools, conform to the CWM specification to ensure that the metadata in their tools is open and accessible to any CWM compliant tool.

Large organizations leverage the specification in order to be able to manage and maintain there data warehouses in a common metadata repository. By using the CWM specification IT administrators and system integrators can extract and link metadata from different vendors tools.

Oracle, IBM, SA, Informatica, Meta Integration Technology Incorporated are among several industry leaders who have data warehouse tools that are CWM compliant to facilitate interoperability.

Cross-References

► Data Warehouse Metadata
► Meta Object Facility
► Metadata
► Metadata Interchange Specification
► Metadata Registry, ISO/IEC 11179
► Metamodel
► Unified Modeling Language
► XML Metadata Interchange

Recommended Reading

1. Common Warehouse Model (CWM) Specification, Version 1.1, Object Management Group. Needham, 2 March 2003. http://www.omg.org/technology/documents/formal/cwm.htm
2. CWM Metadata Interchange Patterns Specification, Version 1.0, Object Management Group. Needham, 25 March 2004.
3. Grossman RL, Hornick MF, Meyer G. Data mining standards initiatives. Commun ACM. 2002;45(8): 59–61.
4. Poole J, Chang D, Tolbert D, Mellor D. Common warehouse metamodel: an introduction to the standard for data warehouse integration. New York: Wiley; 2002.
5. Vaduva A, Dittrich KR. Metadata management for data warehousing: between vision and reality. In: Proceeding of the International Conference on Database Engineering and Applications; 2001. p. 0129.

Comparative Visualization

Hans Hinterberger
Department of Computer Science, ETH Zurich, Zurich, Switzerland

Synonyms

Comparative analysis

Definition

Comparative visualization refers to

1. Methods that support the process of understanding in what way different datasets are similar or different
2. Methods that allow comparing different characteristics of a given dataset
3. Methods that allow a comparison of different types of (linked) data graphics

Key Points

Comparisons of datasets may occur in different ways. Data value to data value (entries of different datasets are compared to one another based on their values), derived quantity to derived quantity (these could be statistical moments of data fields or topological characteristics), methodology to methodology (comparisons of methodologies involve quantifying differences in experiment or simulation parameters), and, if the data are visualized, image to image (such comparisons quantify the differences in the visualizations produced by a given graphical method).

Comparative visualization methods have been developed as an enabling technology for computational and experimental scientists whose ability to collect and generate data far outpaces their ability to analyze and understand such data. The Visualization and Analysis Center for Enabling Technologies (http://www.vacet.org), for example, provides (publicly available) comparative data visualization software [1] for the scientists at the various research labs associated with the US Department of Energy.

Graphical displays that readily allow simultaneous comparisons of several characteristics of multivariate datasets – the parallel coordinate display, for example – are sometimes referred to as "comparative graphs."

There is evidence that visual explorations into a dataset's structure are particularly effective when the data can be compared by simultaneously observing different visualizations of the same data. Today's data visualization packages routinely include several different graphic methods to allow such comparisons. To be truly effective, the different graphics should be operationally linked. See [2, 3] for two examples among others.

Cross-References

▶ Exploratory Data Analysis
▶ Parallel Coordinates

Recommended Reading

1. Bavoil L, Callahan SP, Crossno PJ, Freire J, Scheidegger CE, Silva CT, Vo HT. VisTrails: enabling interactive multiple-view visualizations. In: Proceedings of the IEEE Visualization; 2005.
2. Schmid C, Hinterberger H. Comparative multivariate visualization across conceptually different graphic displays. In: Proceedings of the 7th International Working Conference on Scientific and Statistical Database Management; 1994.
3. Siirtola H. Combining parallel coordinates with the reorderable matrix. In: Proceedings of the International Conference on Coordinated & Multiple Views in Exploratory Visualization; 2003.

Compensating Transactions

Greg Speegle
Department of Computer Science, Baylor University, Waco, TX, USA

Definition

Given a transaction T, and its compensating transaction C, then for any set of transactions H executing concurrently with T, the database state D resulting from executing THC is equivalent to the database state D' resulting from executing H alone. Typically, equivalent means both D and D' satisfy all database consistency constraints, but D and D' do not have to be identical.

A compensating transaction is defined in terms of its corresponding failed transaction, and once started, must be completed. This may involve re-executing the compensating transaction multiple times. The result of compensation is application dependent.

Key Points

A *compensating transaction* is a set of database operations that perform a logical undo of a failed transaction. The goal of the compensating transaction is to restore any database consistency constraints violated by the failed transaction without adversely affecting other concurrent transactions (e.g., *cascading aborts*). However, it does not require the database to be in the exact same state as if the transaction had never executed as with traditional *ACID* properties. A compensating transaction also removes the externalized affects of a failed transaction [2].

Compensating transactions can best be understood by comparing them to traditional atomicity requirements. Under traditional atomicity, either all effects of a transaction are present in the database, or none of them are. Thus, if a transaction T_1 updates a data item and transaction T_2 reads that update, in order to remove all of the effects of T_1, T_2 must also be removed. With compensating transactions, the abort of T_2 is not be required.

Consider an example application of a company manufacturing widgets. The transaction for buying widgets consists of two subtransactions, one to order the widgets and another to pay for them. Since this business is very efficient, as soon as the widgets are ordered, another transaction starts producing the desired widgets. It is possible to compensate for the ordered widgets by simply removing the order from the system. The extra widgets would be produced, but they will be consumed by later orders. Under traditional atomicity requirements, the production transaction would have to be aborted if the buying transaction failed after the order was placed (for example, if the customer could not pay for the widgets).

Compensating transactions are used in long duration transactions called *Sagas* [1], and other applications that require *semantic atomicity*. Unfortunately, compensation is not universally possible – the common example of an externalized event that cannot be undone is the launching of a missile – or may be very complex. Thus, compensating transactions are used when the benefits of avoiding cascading aborts and early externalization of results outweigh the difficulty in determining the compensation.

Cross-References

► ACID Properties
► Sagas
► Semantic Atomicity

Recommended Reading

1. Garcia-Molina H, Salem K. SAGAS. In: Proceedings of the ACM SIGMOD International Conference on Management of Data; 1987. p. 249–59.
2. Korth HF, Levy E, Silberschatz A. A formal approach of recovery by compensating transactions. In: Proceedings of the 16th International Conference on Very Large Data Bases; 1990. p. 95–106.

Complex Event

Opher Etzion
IBM Software Group, IBM Haifa Labs, Haifa University Campus, Haifa, Israel

Synonyms

Composite event; Derived event

Definition

A complex event is an event derived from a collection of events by either aggregation or derivation function [3].

Key Points

A complex event [1, 2] is a derived event; it can be derived by various means:

1. Explicit concatenation of a collection of events,
   • Example: Create an event that contains all the events that are related to the 2008 USA presidential elections.
2. Derivation of an aggregated value from a collection of events from the same type.
   • Example: Create an event that contains the average, maximal and minimal value of a certain stock during a single trade day.
3. Derivation [4] of an event as a function of other events that is a result of event pattern detection.
   • Example: Whenever a sequence of three complain-events from the same customer occurs within a single week, create an event "angry customer" with the customer-id.

Note that this event may or may not contain the raw complain events.

Cross-References

► Complex Event Processing
► Event Pattern Detection

Recommended Reading

1. Ericsson AM, Pettersson P, Berndtsson M, Seiriö M. Seamless formal verification of complex event processing applications. In: Proceedings of the Inaugural International Conference on Distributed Event-Based Systems; 2007. p. 50–61.
2. Luckham D. The power of events. Boston: Addison-Wesley; 2002.
3. Luckham D, Schulte R, editors. EPTS event processing glossary version 1.1. http://complexevents.com/?p=409
4. Zimmer D, Unland R. On the semantics of complex events in active database management systems. In: Proceedings of the 15th International Conference on Data Engineering; 1999. p. 392–99.

Complex Event Processing

Opher Etzion
IBM Software Group, IBM Haifa Labs, Haifa
University Campus, Haifa, Israel

Synonyms

Event processing; Event stream processing

Definition

Complex event processing deals with various
types of processing complex events.

Key Points

Figure 1 shows that the different applications of
the CEP technology are not monolithic, and can
be classified into five different solution segments,
which differ in their motivation, from the user's
perspective, they are:

- *RTE (Real-Time Enterprise)*: The processing
  should affect business processes while they
  are still running. For example, stop an instance
  of a workflow that deals with trading a certain
  stock, if the trade request has been withdrawn.
- *Active Diagnostics*: Finding the root-cause of
  a problem based on events that are symptoms.
- *Information Dissemination*: A personalized
  subscription that enables subscriptions in
  lower granularity, where the subscription
  does not match the published event, but a
  combination of event. For example, notify me
  when IBM stock has gone up 2% within 1 h.
- *BAM (Business Activity Management)*: Monitor for exceptional behavior, by defining KPI
  (Key Performance indicators) and other exceptional behavioral constraints. For example,
  the delivery has not been shipped by the deadline.
- *Prediction*: Mitigate or eliminate future predicted events.

Figure 2 shows the relations among the different terms around complex event processing.
Complex event may be a derived event, but the

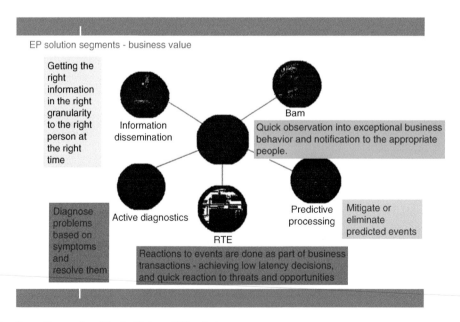

Complex Event Processing, Fig. 1 Various CEP solution segments

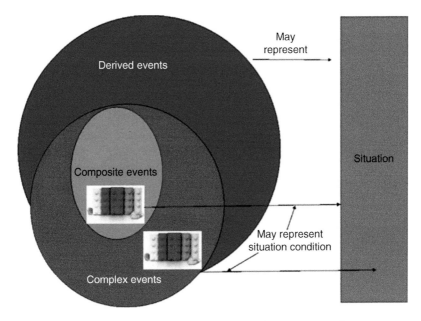

Complex Event Processing, Fig. 2 Relationships among major complex event processing terms

overlapping among them is partial, the complex event processing is materialized by detecting patterns which may correspond to situations (cases that require action). The exact definitions of terms can be found in the EPTS glossary.

Cross-References

▶ Complex Event
▶ Event Pattern Detection

Recommended Reading

1. Etzion O. Event processing, architecture and patterns, tutorial. In: Proceedings of the 2nd International Conference on Distributed Event-Based Systems; 2008.
2. Event processing glossary. Available at: http://www.epts.com
3. Luckham D. The power of events. Reading: Addison-Wesley; 2002.
4. Sharon G, Etzion O. Event processing networks – model and implementation. IBM Syst J. 2008;47(2):321–34.
5. Zimmer D, Unland R. On the semantics of complex events in active database management systems. In: Proceedings of the 15th International Conference on Data Engineering; 1999. p. 392–9.

Composite Event

AnnMarie Ericsson[1], Mikael Berndtsson[2,3], and Jonas Mellin[2,3]
[1]University of Skövde, Skövde, Sweden
[2]University of Skövde, The Informatics Research Centre, Skövde, Sweden
[3]University of Skövde, School of Informatics, Skövde, Sweden

Definition

A composite event is a set of events matching an event specification.

Key Points

Pioneering work on composite events was done in the HiPAC project [1], and the ideas were extended and refined in most proposals for active object-oriented databases during the early 1990s.

A composite event is composed according to an event specification (in an event algebra), where the composition is performed using a set of event operators such as disjunction, conjunction, and sequence. More advanced event operators have been suggested in literature, e.g., [2–4].

The *initiator* of a composite event is the event initiating the composite event occurrence, and the *terminator* is the event terminating the composite event occurrence.

Events contributing to composite events may carry parameters (e.g., temporal) in which the event is said to occur. Events contributing to composite events are also referred to as constituent events.

Composite events need to be composed according to some event context that defines which event can participate in the detection of a composite event. The event context is an interpretation of the streams of contributing events. The seminal work by Chakravarthy et al. [2, 5] defines four different event contexts (or consumption policies): recent, chronicle, continuous, and cumulative.

In the recent event context, only the most recent constituent events will be used to form composite events. The recent event context is, for example, useful if calculations must be performed on combinations of the last measured values of temperature and pressure in a tank [2, 5].

In the chronicle event context, events are consumed in chronicle order. The earliest unused initiator/terminator pair is used to form the composite event. The chronicle event context is, for example, useful if sensors are placed along a conveyor-belt monitoring objects traveling along the belt, and combination of sensor events triggered by the same object is needed. In that case events must be combined in occurrence order since the first event from the first sensor and the first event from the second sensor are likely triggered by the same object [2, 5].

In the continuous event context, each initiator starts the detection of a new composite event, and a terminator may terminate one or more composite event occurrences. The difference between continuous and chronicle event contexts is that in the continuous event context, one terminator can detect more than one occurrence of the composite event.

In the cumulative event context, all events contributing to a composite event are accumulated until the composite event is detected. When the composite event is detected, all contributing events are consumed [2, 5].

Cross-References

▶ Active Database, Active Database (Management) System
▶ Active Database Execution Model
▶ Active Database Knowledge Model
▶ Active Database, Active Database (Management) System
▶ ECA Rules
▶ Event Detection
▶ Event Specification

Recommended Reading

1. Dayal U, Blaustein B, Buchmann A, Chakravarthyand S, et al. HiPAC: a research project in active, time-constrained database management. Technical report CCA-88-02. Xerox Advanced Information Technology, Cambridge. 1988.
2. Chakravarthy S, Mishra D. Snoop: an expressive event specification language for active databases. Data Knowl Eng. 1994;14(1):1–26.
3. Gatziu S. Events in an active object-oriented database system. PhD thesis, University of Zurich, Switzerland. 1994.
4. Gehani N, Jagadish HV, Smueli O. Event specification in an active object-oriented database. In: Proceedings of the ACM SIGMOD International Conference on Management of Data; 1992. p. 81–90.
5. Chakravarthy S, Krishnaprasad V, Anwar E, Kim SK. Composite events for active databases: semantics contexts and detection. In: Proceedings of the 20th International Conference on Very Large Data Bases; 1994. p. 606–17.

Composition

W. M. P. van der Aalst
Eindhoven University of Technology, Eindhoven, The Netherlands

Synonyms

Process composition; Service composition

Definition

In computer science, *composition* is the act or mechanism to combine simple components to build more complicated ones. Composition exists at different levels. For example, one can think of the usual composition of functions in mathematics, i.e., the result of the composed function is passed to the composing one via a parameter. If one has to functions f and g, these can be combined into a new function $h = f.g$, i.e., $h(x) = f(g(x))$. Another level of abstraction is the level of activities. Here all kinds of process modeling languages can be used to compose activities into processes (e.g., Petri nets, BPMN, etc.). Typical composition operators are sequential composition, parallel composition, etc. Process composition is related to business process management, workflow management, etc. Yet another level of abstraction is provided by services, i.e., more complex services can be composed from simpler ones even when they do not reside in the same organization. Service composition is sometimes also referred to as orchestration and a typical language used for this purpose is BPEL.

Key Points

The composition of more complex components from simpler components has been common practice in computer science right from the start. It is clear that composition is needed to allow for "divide and conquer" strategies and reuse. One of the most complex issues is the composition of processes. There are basically two types of composition approaches: graphs-based languages and process algebras. Examples of graph-based languages are Petri nets, state charts, BPMN, EPCs, etc. In these languages activities and subprocesses are connected to impose some ordering relations. For example two transitions in a Petri net can be connected by a place such that the first one triggers the second one [3]. Process algebras enforce a more structured way of modeling processes. Typical operations are sequential composition ($x.y$, i.e., x is followed by y), alternative composition

($x + y$, i.e., there is a choice between x and y), and parallel composition ($x||y$, i.e., x and y are executed in parallel) [1, 2]. Languages like BPEL provide a mixture of both styles, e.g., operators such as sequence, switch, while and pick correspond to the typical process-algebraic operators while the flow construct defines in essence an acyclic graph.

The principle of compositionality states that the meaning of a composite is determined by the meanings of its constituent parts and the rules used to combine them. For example, if a process is composed of parts that have certain properties, then these properties should be preserved by the composition and should not depend on lower-level interactions. Such properties can be obtained by simplifying the language used or restricting the compositions allowed.

Cross-References

▶ Abstraction
▶ Business Process Execution Language
▶ Business Process Management
▶ Business Process Modeling Notation
▶ Orchestration
▶ Petri Nets
▶ Web Services
▶ Workflow Management
▶ Workflow Patterns

Recommended Reading

1. Baeten JCM, Weijland WP. Process algebra, Cambridge tracts in theoretical computer science, vol. 18. Cambridge: Cambridge University Press; 1990.
2. Milner R. Communicating and mobile systems: the Pi-calculus. Cambridge, UK: Cambridge University Press; 1999.
3. van der Aalst WMP. Business process management demystified: a tutorial on models, systems and standards for workflow management. In: Desel J, Reisig W, Rozenberg G, editors. Lectures on concurrency and petri nets, LNCS, vol. 3098. Berlin: Springer; 2004. p. 1–65.

Comprehensions

Peter M. D. Gray
University of Aberdeen, Aberdeen, UK

Synonyms

Calculus expression; List comprehension; Set abstraction; ZF-expression

Definition

The comprehension comes from ideas of mathematical set theory. It originated as a way of defining sets of values so as to avoid the famous paradoxes of early set theory, by starting from other well-defined sets and using some carefully chosen constructors and filters. The values in the sets could be tuples of basic values, which suits the *relational model*, or they could be object identifiers, which fits with *ODMG object data models* [2], or they could be tagged variant records which fit well with *semi-structured data*. They could even be sets, lists or bags defined by other comprehensions.

The abstract structure of a comprehension precisely describes almost all the computations done in functional query languages, despite their very different surface syntax. Better still, it allows many optimizations to be expressed as well defined mathematical transformations.

Key Points

Consider an example, using SQL syntax, to find the set of surnames of persons whose forename is "Jim":

SELECT surname FROM person WHERE forename = "Jim"

Using a list comprehension this can be written as:

[surname(p) | p <- person; f <- forename(p); f = "Jim"]

This denotes the list of values of the expression to the left of the vertical bar. This expression usually includes variables such as *p* which are instantiated by generators to the right of the bar. It can be transliterated as:

The set of values of the surname of p *such that* p *is in the set* person *and* f *is in the set* of forenames of p *and* f is equal to "Jim". Here *forename(p)* could alternatively be written *p.forename* or *(forename p)*.

Thus, the vertical bar can be read as *such that* and the semicolons as conjunctions (*and*). The arrows act as *generators*, supplying alternative possible values, subject to restrictions by predicate terms to the right, acting as *filters*. Thus *p* is generated from the set of persons but is only chosen where the forename of *p* satisfies the test of equaling "Jim".

In the above syntax the arrow operator is overloaded , so that if a function such as *forename* delivers a single value instead of a set then the arrow just assigns that single value to the variable on its left. Strictly, one should make a singleton set containing this value, and then extract it:

[surname(p) | p <- person; f <- [forename(p)]; f = "Jim"]

This wasteful operation would be compiled away to give this equivalent form:

[surname(p) | p <- person; forename(p) = "Jim"]

The term "list comprehension" is commonly used, but one should really distinguish between lists, sets and bags [1]. Thus comprehensions are usually represented internally as lists, but often the order is ignored, as in sets, and sometimes it is necessary to keep duplicates and form a bag, especially when totaling up the contents! Particular classes of operator used in comprehensions give rise to *monad comprehensions* and *monoid comprehensions* with valuable mathematical properties.

Cross-References

▶ OQL

Recommended Reading

1. Buneman P, Libkin L, Suciu D, Tannen V, Wong L. Comprehension syntax. ACM SIGMOD Rec. 1994;23(1):87–96.
2. Fegaras L, Maier D. Towards an effective calculus for Object Query Languages. In: Proceedings of the ACM SIGMOD International Conference on Management of Data; 1995. p. 47–58.

Compression of Mobile Location Data

Goce Trajcevski[1], Ouri Wolfson[2,3], and Peter Scheuermann[1]
[1]Department of ECpE, Iowa State University, Ames, IA, USA
[2]Mobile Information Systems Center (MOBIS), The University of Illinois at Chicago, Chicago, IL, USA
[3]Department of CS, University of Illinois at Chicago, Chicago, IL, USA

Synonyms

Location sensing and compression; Spatiotemporal data reduction

Definition

Miniaturization of computing, sending, and networking devices has provided the technological foundation for applications which generate huge volumes of location-in-time data – order of petabytes (PB) annually from smart phones alone [12]. In moving objects databases (MOD) [9], the data pertaining to the whereabouts of a given mobile object is commonly represented as a sequence of *(location, time)* points, ordered by the temporal dimension. Depending on the application's settings, such points may be obtained by different means, e.g., an onboard GPS-based system, RFID sensors, roadside sensors [18], base stations in a cellular architecture, etc. The main motivation for compressing the location data of a given (collection of) moving object(s) is twofold: (1) Reducing the storage requirements, in addition to smart phones [12], location samples from onboard GPS devices taken once every 5 s, can still generate hundreds of terabytes (TB) of data for one million vehicles. Thus, if a given point, say, (x, y, t) can be eliminated from the representation of the particular trajectory without prohibitively sacrificing the accuracy of its representation, then the space required for that point's storage can be saved; (2) If a particular point along a given trajectory can be eliminated as soon as it is "generated" – i.e., when the *location* value is obtained by the onboard GPS at a given *time* – based on some additional context value – yet another type of savings can be achieved, namely, the *(location, time)* value for that point need not be transmitted then to a given server, thus reducing the bandwidth consumption. The latter is especially important in environments such as wireless sensor networks where communication is one of the major sources of draining the batteries of the sensing devices. This entry explains the basic problems involved in compressing spatiotemporal data corresponding to trajectories of mobile objects and outlines the foundations of the approaches that have addressed some of those problems.

Historical Background

The field of *data compression* originated in the works of Shannon, Fano, and Huffman in the 1940s [14], and its main goal is to represent information in as compact form as possible. Some popular forms of data compression have, historically, been around even earlier, for instance, the Morse code has been used in telegraphy since the mid-nineteenth century. Based on the observation that some letters occur more frequently

Research Supported by the NSF grants CNS0910952 and III 1213038, and the ONR grant N00014-14-10215

Research supported by the National Science Foundation under Grants: DGE-0549489, OII-0611017, 0513736,0326284

than others, the code assigns shorter sequences of (combinations of) "·" and "−" to such letters, thus, for example, "e" → "·", "a" → "· −". On the other hand, the letters which occur less frequently are assigned longer sequences, for example, "q" → "− − · −". In this setting, the frequency of the occurrence of single letters provided statistical structure that was exploited to *reduce the average time* to transmit a particular message since, in practice, the duration of the symbol "−" is (approximately) three times longer than the duration of the · symbol. A natural extension is to use frequency of the *words* over a given alphabet, in order to further compress the encoding of a given text, which is used in the Grad-2 Braille coding. When the probability model of the source is known, the popular approach for encoding a collection of *letters* of a given *alphabet* is the Huffman coding [14]. Contrary to the ASCII/EBDCIC which are *fixed-length* codes, in the sense that every symbol is assigned same number of bits, Huffman code is a *variable-length* one, which assigns shorter codewords to symbols occurring less frequently, in an optimal manner with respect to the entropy of the source. When dealing with texts, some statistical correlations can be detected in terms of the occurrences of *words*. Taking this into consideration, the so-called dictionary techniques for data compression have been obtained, an example of which is the UNIX `compress` command. In computer science, the need for compression techniques was mainly motivated by the reduction of the size of the data for storage and transmission purposes.

Different kinds of data may exhibit different kinds of structure that can be exploited for compression, provided a proper model is developed. For example, given the sequence of numbers $\{9, 11, 11, 11, 14, 13, 15, 17, 16, 17, 20, 21\}$, let x_n denote its n−th element. If we were to transmit the binary representation of each x_i ($i \in \{1, 2, \ldots, 12\}$), 5 bits-per-sample are needed, for a total of 60 bits transmitted. However, if one provides a model represented by the equation $\overline{x}_n = n + 8$, then the difference-sequence (i.e., the residual) of the initial sequence, represented as $e_n = x_n - \overline{x}_n$

becomes: $\{0, 1, 0, -1, 1, -1, 0, 1, -1, -1, 1, 1\}$. This sequence consists of only three different numbers $\{-1, 0, 1\}$ Using the mapping "-1" → "00"; "0" → "− 1"; "1" → "10", each e_i can be encoded with only 2 bits. Hence, the sequence can be transmitted with a total of 24 bits, achieving a compression ratio (equivalently, savings in badnwidth/energy) of 60%. Such intrinsic properties of the underlying domain have been heavily exploited in the areas of speech compression, image compression, etc. [14].

There are several classifications of compression techniques. One example, as mentioned above, is *fixed* vs. *variable* length of the encoding. However, one may also need to distinguish between *static* (the codewords are fixed, say, before the transmission) and *dynamic/adaptive*. The classification that is most relevant for this article is *lossless* vs. *lossy* compression. With lossless compression, the original data can be *exactly* recovered from the compressed one, which it is not the case for the lossy compression.

In addition to the above classifications, there are several different bases for ranking the quality of a given compression method: (1) the complexity of the algorithms, (2) the memory footprint, (3) the amount of compression, and (4) the quality of the data (in lossy compression). The main goal of the methods for compressing *spatiotemporal* data is to strike a good balance between the complexity of the algorithm and the error bound on the compressed data with respect to the original one, with the additional attention to how data error affects the error in the answers to the queries.

There are two research fields that have addressed problems similar in spirit to the ones of compressing mobile location data:

(1) **Cartography**: the goal of the *map generalization* in cartography is to reduce the size-/complexity of a given map for the purpose of simplified representation of the details appropriate to a given scale [20].

(2) **Computational geometry** (CG): in particular, the problem of *polyline* (which is, a sequence of nodes specifying a chain of line segments) simplification [3]. The

problem can be described as follows. Given a polyline PL_1 with vertices $\{v_1, v_2, \ldots, v_n\}$ in a respective k-dimensional Euclidean space, and a tolerance ε, construct another polyline PL_1' with vertices $\{v_1', v_2', \ldots, v_m'\}$ in the same space, such that $m \leq n$ and for every point $P \in PL_1$ its distance from PL_1' is smaller than a given threshold: $dist(P, PL_1') \leq \varepsilon$. In the case that $\{v_1', v_2', \ldots, v_m'\} \subseteq \{v_1, v_2, \ldots, v_n\}$, PL_1' is a *strong* simplification of PL_1; otherwise PL_1' is called a *weak* simplification of PL_1. Complementary to this, there are two other distinct facets of the minimal line simplification problem: (a) Given PL and ε, minimize the number of points m in PL' (known as min-# problem) [5], and (b) given PL and the "budget" m of the vertices in PL', minimize the error ε (known as min-ε problem).

A popular heuristic for polyline simplification in the context of map generalization was proposed by Douglas and Peucker in [6]. Essentially, it recursively approximates a given polyline in a "divide and conquer" manner, where the farthest vertex, according to the distance used, is selected as the "divide" point. Given a *begin_vertex* p_i and an *end_vertex* p_j, if the greatest distance from some vertex p_k to the straight line segment $\overline{p_i p_j}$ is greater than the tolerance ε, then the trajectory is broken into two parts at p_k, and the procedure is recursively called on each of the sub-polylines $\{p_i, \ldots, p_k\}$ and $\{p_k, \ldots, p_j\}$; otherwise, the vertices between p_i and p_j are removed from trajectory, and this segment is simplified as a straight line $\overline{p_i p_j}$. An illustration of the DP heuristic is given in Fig. 1. Although the original version of the algorithm, as presented in [6], has a running time $O(n^2)$, an $O(n \log n)$ algorithm was presented in [10]. However, none of these algorithms can ensure an optimality, in terms of the size of the compression (alternatively, in terms of a minimal ε-error for a fixed reduction factor). An optimal algorithm was presented in [5], with a complexity of $O(n^2)$, subsequently extended for 3D and higher dimensions in [3].

Scientific Fundamentals

Assuming that a given object is moving in a 2D space with respect to a given coordinate system, in the MOD literature, [9, 16, 19] the motion is typically modeled as a *trajectory*. A trajectory is defined as a function $F_t : T \to \mathcal{R}^2$ which maps a given (temporal) interval $[t_b, t_e]$ into a one-dimensional subset of $\mathcal{R}^2$. It is represented as a sequence of 3D points (2D geography + time) (x_1, y_1, t_1), (x_2, y_2, t_2), $\ldots$, (x_n, y_n, t_n), where $t_b = t_1$ and $t_e = t_n$ and $t_1 \leq t_2 \leq \ldots \leq t_n$. Each point (x_i, y_i, t_i) in the sequence represents the 2D location (x_i, y_i) of the object, at the time t_i. For every $t \in (t_i, t_{i+1})$, the *location* of the object is obtained by an *interpolation* between (x_i, y_i) and (x_{i+1}, y_{i+1}) with the ratio $(t - t_i)/(t_{i+1} - t_i)$. Most often, a *linear interpolation* is used, which is, in-between two points, the object is assumed to move along a straight line segment and with a constant speed. However, in certain applications (e.g., airplanes trajectories, hand motion trajectories), Bezier curves or other types of curves may be used to represent the whereabouts in-between two samples [7]. The 2D projection of the trajectory is a polygonal chain with vertices (x_1, y_1), $(x_2, y_2) \ldots (x_n, y_n)$, called a *route*.

Observe that a trajectory may represent both the *past* and the *future* motion, i.e., the *motion plan* of a given object (cf. [9]). Typically, for future trajectories, the user provides the starting location, starting time, and the destination (plus, possibly, a set of to-be-visited) points, and the MOD server uses these, along with the distribution of the speed patterns on the road segments as inputs to a dynamic extension of the Dijkstra's algorithm [16], to generate the shortest travel time trajectory.

One may be tempted to straightforwardly apply the existing results on polyline simplification from the CG literature (e.g., the DP [6, 10] or the optimal algorithm [3, 5]), in order to compress a given trajectory. However, as pointed out in [4], the semantics of the spatial + temporal domain combined raises two major concerns:

(1) What is the *function* used to measure the *distance between points* along trajectories?

**Compression of Mobile
Location Data, Fig. 1**
Douglas-Peucker heuristic

Initial polyline: {P1, P2, P3, P4, ..., P13}

Compressed Polyline: {P1, P6, P10, P13}

(2) How does the choice of that function affect the *error* that the compressed trajectory introduces in the answers of the popular spatiotemporal queries? In the sequel, each of these questions is addressed in a greater detail.

Distance Function: The role of the distance function is to capture how far (or close) are two objects. While there may be some descriptive distances used in certain applications – e.g., "...two city blocks away," typically one expects that the distance function satisfies some mathematical properties (e.g., it is a metric). A popular distance function between two curves, often used in CG applications, is the so-called Hausdorff distance [1]. Given two curves C_1 and C_2, their Hausdorff distance simply looks for the smallest ε such that C_1 is completely contained in the ε-neighborhood of C_2 (i.e., C_1 is completely contained in the Minkowski sum of C_2 and a disk with radius ε) and vice versa. Although it is arguably a very natural distance measure between curves and/or compact sets, the Hausdorff distance is too "static," in the sense that it neither considers any direction nor any dynamics of the motion along the curves. A classical example of the inadequacy of the Hausdorff distance, often used in the CG literature [1, 2], is the *"man walking the dog."* Figure 2 illustrates the corresponding routes of the man *(M-route)* and the dog *(D-route)*, as well as their trajectories *M-trajectory* and *D-trajectory*. Observe that,

ignoring the temporal aspect of their motions, the D-route and the M-route are within Hausdorff distance of e, as exemplified by the points A and B in the XY plane. However, their actual *temporally aware* distance corresponds to the minimal length of the leash that the man needs to hold. The 3D part of Fig. 2 illustrates the discrepancy between the distances among the points along the *routes* and their corresponding counterparts along *trajectories*: when the dog is at the point A, which is at time t, the man is actually at $M(t)$, and their distance is much greater than e (the man is at the geolocation B at the time $t_1 > t$). The Fréchet distance [2] is more general than the Hausdorff one, in the sense that it allows for a variety of possible motion patterns along the given route segments. As an illustration, we point how on the portion of the *D-trajectory* in Fig. 2; the dog may be moving nonuniformly (i.e., accelerating) along a particular route segment.

The discussion above illustrates two extreme points along the spectrum of distance functions for moving objects. Although the Fréchet distance is the most general one, regarding the possible dynamics of motions, it is unnecessarily complex for the common trajectory model in MOD settings. The inadequacy of the L_2 norm as a distance function for spatiotemporal trajectories was pointed out in [4] where, in order to properly capture the semantics of the problem domain, alternative distance functions were introduced. Given a spatiotemporal point $p_m = (x_m, y_m, t_m)$ and a trajectory segment $\overline{p_i, p_j}$

between the vertices $p_i = (x_i, y_i, t_i)$ and $p_j = (x_j, y_j, t_j)$, [4] proposed the E_u and E_t distance functions between the p_m and $\overline{p_i, p_j}$, which are explained next.

- E_u – The three-dimensional time_uniform distance, which is defined when $t_m \in [t_i, t_j]$, as follows:

$$E_u(p_m, \overline{p_i\,p_j}) = \sqrt{(x_m - x_c)^2 + (y_m - y_c)^2}$$

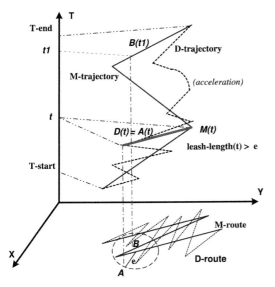

Compression of Mobile Location Data, Fig. 2 Hausdorff vs. Fréchet distance

Compression of Mobile Location Data, Fig. 2 Hausdorff vs. Fréchet distance

Compression of Mobile Location Data, Fig. 3 E_u distance function for trajectory compression

where $p_c = (x_c, y_c, t_c)$ is the unique point on $\overline{p_i\,p_j}$ which has the same *time* value as p_m (i.e., $t_c = t_m$). An illustration of using the E_u distance function for reducing the size of a given trajectory is presented in Fig. 3. Intuitively, the distance is measured at equal *horizontal* planes, for the respective values of the temporal dimension. One can "visually" think of the relationship between the original trajectory and the compressed trajectory as follows: the original trajectory is contained inside the sheared cylinder obtained by sweeping (the center of) a horizontal disk with radius ε along the compressed trajectory.

- E_t – The time distance is defined as $E_t(p_m, \overline{p_i\,p_j}) = |t_m - t_c|$, where t_c is the time of the point on the XY projection $\overline{p'_i\,p'_j}$ of $\overline{p_i\,p_j}$, which is closest (in terms of the 2D Euclidean distance) to the XY projection p'_m of p_m. Intuitively, to find the time distance between p_m and $\overline{p_i\,p_j}$, one needs to:

(i) project each of them on the XY plane;
(ii) find the point $p'_c \in \overline{p'_i, p'_j}$ that is closest to p'_m;
(iii) find the difference between the corresponding times of p_c and p_m.

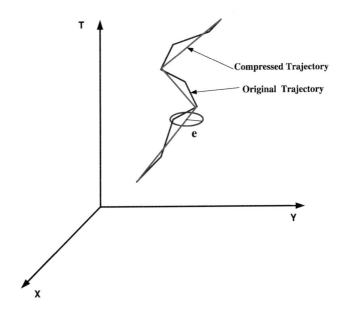

An important observation regarding the *computation* of the compressed version of a given original trajectory as an input is that both the DP [6] and the optimal algorithm [5] can be used, provided they are appropriately modified to reflect the distance function used. Experimental results in [4] demonstrated that the DP heuristics yields a compression factor that is very comparable to the one obtained by the optimal algorithm; however, its execution is much faster.

Spatiotemporal Queries and Trajectory Compression: The most popular categories of spatiotemporal queries, whose efficient processing has been investigated by many MOD researchers [9], are:

- *where_at(T, t)* – returns the expected location at time t.
- *when_at(T, x, y)* – returns the time t at which a moving object on trajectory T is expected to be at location (x, y).
- *intersect(T, P, t_1, t_2)* – is *true* if the trajectory T intersects the polygon P between the times t_1 and t_2. This is an instance of the so-called spatiotemporal range query.
- *nearest_neighbor(T, O, t)* – The operator is defined for an arbitrary set of trajectories O, and it returns a trajectory T' of O. The object moving according to T', at time t, is closest than any other object of O to the object moving according to T.
- *join(O, Θ)* – O is a set of trajectories, and the operator returns the pairs (T_1, T_2) such that their distance, according to the distance function used, is less than a given threshold $Θ$.

An important practical consideration for compressing trajectory data is how the (im)precision generated by the compression affects the answers of the spatiotemporal queries. As it turns out, the distance function used in the compression process plays an important role, and toward this, the concept of *soundness* [4] of a distance function with respect to a particular query was introduced in [4]. A pair *(distance_function, query)* is called *sound* if the error of the *query*-answer, when processed over the compressed trajectory, is bounded. In case the error is *unbounded*, which is, although the compression itself guarantees a distance-error of ε between the points on the compressed trajectory with respect to the original one, the error of the answer to the query can grow arbitrarily large, the pair is called *unsound*. Table 1 below (adapted from [4]) summarizes the soundness properties of three distance functions with respect to five categories of spatiotemporal queries.

As one can see, there is no single distance function that is sound for all the possible spatiotemporal queries.

The compression techniques for spatiotemporal data presented thus far implicitly assumed that the trajectories are available in their entirety, i.e., they are past-motion trajectories. However, in practice, it is often the case that the *(location, time)* data is generated onboard mobile units and is transmitted to the MOD server in real time [21]. *Dead reckoning* is a policy which essentially represents an agreement between a given moving object and the MOD server regarding the updates transmitted by that particular object. The main idea is that the *communication* between them can be reduced (consequently, network bandwidth can be spared) at the expense of the *imprecision* of the data in the MOD representing the object's motion. In order to avoid an unbounded error of the object's location data, the agreement specifies a threshold δ that is a param-

Compression of Mobile Location Data, Table 1 Distance soundness and error bound on spatiotemporal query-answers

	Where_at	When_at	Intersect	Nearest neighbor
E_2 (L_2 over routes)	Unsound	Unsound	Unsound	Unsound
E_u	Sound (ε)	Unsound	Sound (ε)	Sound (2ε)
E_t	Unsound	Sound (ε)	Unsound	Unsound

eter of the policy, which can be explained as follows:

– The object sends its *location* and the *expected velocity* to the MOD server, and, as far as the MOD server is concerned, the future trajectory of that object is an infinite ray originating at the update point and obtained by extrapolation, using the velocity vector.

– The information that the MOD server has is the *expected trajectory* of the moving object. However, each moving object is aware of its *actual location*, by periodically sampling it, e.g., using an onboard GPS.

– For as long as its actual location at a given time t_i does not deviate by more than δ from the location that the MOD estimates at t_i using the information previously transmitted, the object does not transmit any new updates. When the actual distance deviates by more than δ from its location on the expected trajectory, the object will send another *(location, time, velocity)* update.

The policy described above is commonly known as a *distance-based* dead reckoning, and an illustration is given in Fig. 4. At time t_0, the object sent its location and the predicted velocity (arrowed line) to the MOD server. The dashed line extending the vector indicates the expected trajectory of the moving object, and the squares along it indicate the object's positions at six times instances, as estimated by the MOD, while the shaded circles indicate the actual positions of the object. Typically, the actual trajectory is obtained by connecting the GPS points with straight line segments, assuming that in-between two updates, the object was moving with a constant speed. As illustrated, at t_6, the distance between the actual position and the MOD-estimated one exceeds the threshold agreed upon ($d_6 > \delta$), and the object sends a new update, at which point the MOD changes the expected trajectory, based on that update. Thus, at t_6, the MOD server actually performs two tasks:

(1) corrects its own "knowledge" about the recent past and approximates the actual trajectory between t_0 and t_6 with a straight line segment, which defines the *actual simplification* of the near-past trajectory;

(2) generates another infinite ray corresponding to the future-expected trajectory, starting at the last update point, and using the newly received velocity vector for extrapolation.

Various trade-offs between the update costs and the (impacts on the) imprecision of the MOD data for several different variants of dead reckoning are investigated in [21]. The dead reckoning, in a sense, achieves in real time both of the goals of compression: reduces the communication and enables the MOD server to store only a subset of the actual trajectory. Assuming that a dead-reckoning policy with threshold δ was used

Compression of Mobile Location Data, Fig. 4
Distance-based dead-reckoning policy

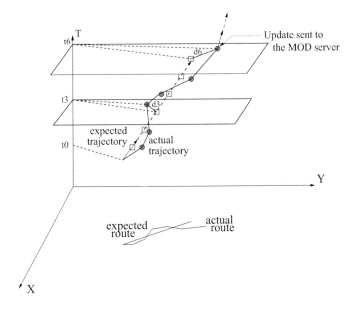

in real time, clearly, the MOD has obtained a compressed past trajectory, say Tr_m^c, of a given mobile object o_m. If o_m was to transmit every single GPS-based update, i.e., no dead reckoning applied, the MOD would have an uncompressed trajectory Tr_m available. The results in [17] have established that Tr_m^c is a strong simplification of Tr_m, with an error-bound 2δ.

We conclude the scientific fundamentals with an observation regarding the impact of incorporating a more *global context* in the process if the mobile data compression. Namely, by incorporating the knowledge about an existing road network – i.e., its graph-based representation with speed values associated with the edges – there is a possibility of a MOD-based data compression [13]. Clearly, this could further increase the overall benefits, in terms of the overall MOD-wide savings.

Key Applications

The compression of moving objects trajectories data is of interest in several scientific and application domains.

Wireless Sensor Networks (WSN)

Wireless sensor networks consist of a large number of *sensors* – devices that are capable of measuring various phenomena, performing elementary calculations, and communicating with each other, organizing themselves in an ad hoc network. A particularly critical aspect of the WSN is the efficient management of the energy reserves, given that the communication between two nodes drains a lot more battery power than the operations of sensing and (local) computing. Consequently, in many tracking applications that can tolerate delays and imprecision in the *(location, time)* data, performing distributed/local compression of the trajectory data before it is sent to the dedicated sink can reduce the size of the transmitted data (i.e., the number of packets), thereby yielding an increase in the overall networks' lifetime. The impact and trade-offs among different policies for in-network trajectory data reduction are presented in [8].

Location-Based Services (LBS)

A variety of applications in LBS depend on the data for mobile objects with different mobility properties (e.g., pedestrians, private vehicles, taxis, public transportation, etc.). Typically, LBS are concerned with a context-aware delivery of the data which matches the preferences of users based on their locations [15]. In order to provide faster response time, and more relevant information, the LBS should be able to predict, based on the motion patterns, what kind of data will be relevant/requested in a near future by given users. This, in turn, implies some accumulated knowledge about the mobility patterns of the users in the (near) past. However, keeping such data in its entirety can impose prohibitively high storage requirements.

Geographic Information Systems (GIS)

Recently, a plethora of services and devices has emerged for providing path planning and navigation for the mobile users: from MapQuest and Google maps through Garmin and iPaq Travel Companion. Each of these services relies on some traffic-based information in order to generate the optimal (in distance or travel time) path for their users. However, as the traffic conditions fluctuate, the future portions of the routes may need to be recalculated. In order to better estimate the impact of the traffic fluctuations, some knowledge from the past is needed which ultimately means storing some past information about trajectories. However, as observed in the literature [12], storing the uncompressed trajectory data corresponding to GPS-equipped mobile entities could generate PBs of data.

Spatiotemporal Data Mining

Clustering is a process of grouping a set of (physical or abstract) objects into classes of similar objects, and its purpose is to facilitate faster data analysis in a given domain of interest. With the recent advances in miniaturization of computing devices and communications technologies, the sheer volume makes it very costly to apply clustering to the original trajectories' data. Compressing such data, especially if one can guarantee a bounded error for the queries of interest, can significantly improve the processing time for many algorithms for trajectories' clustering [11].

Future Directions

Any problem domain that depends on storing large volumes of trajectories' data, in one way and level or another, needs some sort of data compression in order to reduce the storage requirements and to speed up processing of spatiotemporal queries of interest. Clearly, a desirable property of the compression techniques is to ensure a bound on the errors of the answers to the queries.

There are several directions of interest for the future research on mobile data compression. In applications like GIS and LBS, it is a paramount to add some context awareness to the compression techniques. For example, combining the mobile location data compression with the particular tourists attractions and the season/time could provide a speedup in algorithms which are used for generating real-time advertisements, while ensuring that the error (in terms of users that received particular ad) is bounded. An interesting aspect that has been presented in [4] is the so-called aging of the trajectories: a trajectory that is 1 week old could have higher impact on the traffic-impact analysis than a trajectory that was recorded 5 weeks ago. Consequently, one may reduce the older trajectory with a higher error bound, thus further reducing the storage requirements. Automatizing this process in a manner that reflects the specifics of a given problem domain (e.g., context-aware information delivery) is an open question. Despite the large body of works on OLAP and warehousing of traditional data, very little has been done on spatiotemporal OLAP. It is likely that the process of mobile data compression will play an important role in these directions.

Cross-References

▶ Data Compression in Sensor Networks
▶ Data Mining
▶ Moving Objects Databases and Tracking

Recommended Reading

1. Alt H, Guibas L. Discrete geometric shapes: matching, interpolation, and approximation. In: Handbook of computational geometry. Elsevier Science Publishers; 1999.
2. Alt A, Knauer C, Wenk C. Comparison of distance measures for planar curves. Algorithmica. 2004;38.
3. Barequet G, Chen DZ, Deascu O, Goodrich MT, Snoeyink J. Efficiently approximating polygonal path in three and higher dimensions. Algorithmica. 2002;33(2):150–167.
4. Cao H, Wolfson O, Trajcevski G. Spatio-temporal data reduction with deterministic error bounds. VLDB J. 2006;15(3):211–28.
5. Chan W, Chin F. Approximation of polygonal curves with minimum number of line segments or minimal error. Int J Comput Geom Appl. 1996;6(1):59–77.
6. Douglas D, Peucker T. Algorithms for the reduction of the number of points required to represent a digitised line or its caricature. Can Cartogr. 1973;10(2):112–22.
7. Faraway JJ, Reed MP, Wang J. Modelling three-dimensional trajectories by using Bézier curves with application to hand motion. Appl Stat. 2007;56(5):571–85.
8. Ghica O, Trajcevski G, Wolfson O, Buy U, Scheuermann P, Zhou F, Vaccaro D. Trajectory data reduction in wireless sensor networks. IJNGC. 2010;1(1):28–51.
9. Güting RH, Schneider M. Moving objects databases. San Francisco: Morgan Kaufmann; 2005.
10. Hershberger J, Snoeyink J. Speeding up the Douglas-Peuker line-simplification algorithm. In: Proceedings of the 5th International Symposium on Spatial Data Handling; 1992.
11. Jensen CS, Lin D, Ooi BC. Continuous clustering of moving objects. IEEE Trans Knowl Data Eng. 2007;19(9):1161–1174.
12. Mc Kansey Global Institute. Big data: the next frontier for innovation, competition, and productivity; 2011.
13. Popa IS, Zeitouni K, Oria V, Kharrat A. Spatio-temporal compression of trajectories in road networks. GeoInformatica. 2014. https://doi.org.10.1007/s10707-014-0208-4.
14. Sayood K. Introduction to data compression. San Francisco: Morgan Kaufmann; 1996.
15. Schiller J, Voisard A. Location-based services. San Francisco: Morgan Kaufmann; 2004.
16. Trajcevski G, Wolfson O, Hinrichs K, Chamberlain S. Managing uncertainty in moving objects databases. ACM Trans Database Syst. 2004;29(3):463–507.
17. Trajcevski G, Cao H, Wolfson O, Scheuermann P, Vaccaro D. On-line data reduction and the quality of history in moving objects databases. In: Proceedings of the 5th ACM International Workshop on Data Engineering for Wireless and Mobile Access; 2006.

18. Turner-Fairbank Highway Research Center. Traffic detector handbook, vol. I. 3rd ed. McLean: U.S. Department of transportation; 2006.
19. Vlachos M, Hadjielefteriou M, Gunopulos D, Keogh E. Indexing multidimensional time-series. VLDB J. 2006;15(1):1–20.
20. Weibel R. Generalization of spatial data: principles and selected algorithms. In: Algorithmic foundations of geographic information systems. LNCS. Springer; 1998.
21. Wolfson O, Sistla AP, Chamberlain S, Yesha Y. Updating and querying databases that track mobile units. Distrib Parallel Databases. 1999;7(3):257–88.

Computational Media Aesthetics

Chitra Dorai
IBM T. J. Watson Research Center, Hawthorne, NY, USA

Synonyms

CMA; Media semantics; Production-based approach to media analysis

Definition

Computational media aesthetics is defined as the algorithmic study of a variety of image and aural elements in media founded on their patterns of use in film grammar, and the computational analysis of the principles that have emerged underlying their manipulation, individually or jointly, in the creative art of clarifying, intensifying, and interpreting some event for the audience [3]. It is a computational framework to establish semantic relationships between the various elements of sight, sound, and motion in the depicted content of a video and to enable deriving reliable, high-level concept-oriented content annotations as opposed to verbose low-level features computed today in video processing for search and retrieval, and nonlinear browsing of video. This media production knowledge-guided semantic analysis has led to a shift away from a focus on low level features

that cannot answer high level queries for all types of users, to applying the principled approach of computational media aesthetics to analyzing and interpreting diverse video domains such as movies, instructional media, broadcast video, etc.

Historical Background

With the explosive growth of media available on the Web, especially on hugely popular video sharing websites, such as YouTube, managing the digital media collections effectively and leveraging the media content in the archives in new and profitable ways continues to be a challenge to enterprises, big and small. Multimedia content management refers to everything from ingesting, archival, and storage of media to indexing, annotation, and tagging of content for easy access, search, and retrieval, and browsing of images, video, and audio. One of the fundamental research problems in multimedia content management is the semantic gap – that renders all automatic content annotation systems of today brittle and ineffective – between the shallowness of features in their descriptive power that can be currently computed automatically and the richness of meaning and interpretation that users desire search algorithms to associate with their queries for easy searching and browsing of media. Smeulders et al. [8] describe that while "the user seeks semantic similarity, the database can only provide similarity on data processing." This semantic gap is a crucial obstacle that content management systems have to overcome in order to provide reliable media descriptions to drive search, retrieval, and browsing services that can gain widespread user acceptance and adoption. There is a lack of framework to establish semantic relationships between the various elements in the content since current features are frame/shot-representational and far too simple in their expressive power.

Addressing the semantic gap problem in video processing will enable innovative media management, annotation, delivery, and navigational services for enrichment of online shopping, help desk services, and anytime-anywhere training over wireless devices. Creating technologies to annotate content with deep semantics results in an ability to establish semantic relationships between the form and the function in the media, thus for the first time enabling user access to stored media not only in predicted manner but also in unforeseeable ways of navigating and accessing media elements. Semantics-based media annotations will break the traditional linear manner of accessing and browsing media, and support vignette-oriented viewing of audio and video as intended by the content creators. This can lead to new offerings of customized media management utilities for various market segments, such as education and training video archives, advertisement houses, news networks, broadcasting studios, etc.

Foundations

Computational Media Aesthetics advocates an approach that markedly departs from existing methods for deriving video content descriptions by analyzing audio and visual features (for a survey of representative work, see [8]). It proposes that to go beyond describing just what is seen in a video, the visual and emotional impact of how the content is depicted needs to be understood. Both media compositional and aesthetic principles need to guide media analysis for richer, more expressive descriptions of the content depicted and seen.

What are the methodologies for analyzing and interpreting media? Structuralism, in film studies for example, proposes film segmentation followed by an analysis of the parts or sections. Structural elements or portions of a video, when separated from cultural and social connotations can be treated as plain data and therefore, can be studied using statistical and computational tools. Another rich source is production knowledge or film grammar. Directors regularly use accepted rules and techniques to solve problems presented by the task of transforming a story from a written script to a captivating visual and aural narration [2]. These rules encompass a wide spectrum of cinematic aspects ranging from shot arrange-ments, editing patterns and the triangular camera placement principle to norms for camera motion and action scenes. Codes and conventions used in narrating a story with a certain organization of a series of images have become so standardized and pervasive over time that they appear natural to modern day film production and viewing. However, video production mores are found more in history of use, than in an abstract predefined set of regulations, are descriptive rather than prescriptive, and elucidate on ways in which basic visual and aural elements can be synthesized into larger structures and on the relationships that exist between the many cinematic techniques employed worldwide and their intended meaning to a movie audience.

Media aesthetics is both a process of examination of media elements such as lighting, picture composition, and sound by themselves, and a study of their role in manipulating the viewer's perceptual reactions, in communicating messages artistically, and in synthesizing effective media productions [10]. Inspired by it, Dorai and Venkatesh defined Computational media aesthetics [3] as the algorithmic study of a variety of image and aural elements in media guided by the patterns of their use, and the computational analysis of the principles for manipulating these elements to facilitate high-level content annotations.

Computational media aesthetics provides a handle on interpreting and evaluating relative communication effectiveness of media elements in productions through knowledge of film codes that mediate perception of the content shown in the video. It exposes the semantic and semiotic information embedded in the media production by focusing not merely on the representation of perceived content in digital video, but on the semantic connections between the elements and the emotional, visual appeal of the content seen and remembered. It advocates a study of mappings between specific cinematic elements and narrative forms, and their intended visual and emotional import.

In multimedia processing, many research efforts have sought to model and describe specific events occurring in a particular video domain in detail for providing high-level descriptions; computational media aesthetics, on the other hand enables development of video analysis techniques founded upon production knowledge for film/video understanding, for the extraction of high-level semantics associated with the expressive elements and narrative forms synthesized from the cinematic elements, and for the detection of high-level mappings through the use of software models. It highlights the systematic use of film grammar, as motivation and also as foundation in the automated process of analyzing, characterizing, and structuring of produced videos for media search, segment location, and navigational functions.

Computational media aesthetics provides a framework to computationally determine elements of form and narrative structure in videos from the basic units of film grammar namely, the shot, the motion, the recording distances, and from the practices of combination that are commonly followed during the audiovisual narration of a story. At first, primitive computable aspects of cinematographic techniques are extracted. New expressive elements (higher order semantic entities) can then be defined and constructed from these primitive aspects. Both the definition and extraction of these semantic entities are based on film grammar, and these entities are formulated only if directors purposefully design them and manipulate them. The primitive features and the higher order semantic notions thus form the vocabulary of content description language for media.

Key Applications

In seeking to create tools for the automatic understanding of media, computational media aesthetics states the problem as one of faithfully reflecting the forces at play in media production, and interpreting the data with its maker's eye. Several studies have explored the workings of Computational Media Aesthetics when applied to extraction of meaning using many of the aesthetic elements introduced by Zettl [10]: Time, sound and color. Adams et al. [1] took an example of carrying one aspect of film grammar all the way from literature to computable entity, namely tempo and pace for higher level analysis of movies. Adams et al. [1] showed that although descriptive and sometimes fuzzy in scope, film grammar provides rich insights into the perception of subjective time as tempo and pace and its manipulation by the makers of film for drama. Further research [4, 5, 6, 7, 9] has applied this approach pervasively from extracting mood in music, emotion in movies, to adding musical accompaniment to videos and extracting semantic metadata for mobile images at the time of image capture.

Film is not the only domain with a grammar to leverage in analysis. News, sitcoms, educational video, etc., all have more or less complex grammars that may be used to capture their crafted structure and to derive semantic descriptions with automated techniques following the framework of computational media aesthetics.

Cross-References

▶ Video Content Analysis

Recommended Reading

1. Adams B, Dorai C, Venkatesh S. Towards automatic extraction of expressive elements from motion pictures: tempo. In: Proceedings of the IEEE International Conference on Multimedia and Expo; 2000. p. 641–45.
2. Arijon D. Grammar of the film language. Los Angeles: Silman-James Press; 1976.
3. Dorai C, Venkatesh S. Computational media aesthetics: finding meaning beautiful. IEEE Multimedia. 2001;8(4):10–2.
4. Davis M. Editing out video editing. IEEE Multimedia. 2003;10(2):2–12.
5. Mulhem P, Kankanhalli MS, Ji Yi, Hassan H. Pivot vector space approach for audio-video mixing. IEEE Multimedia. 2003;10(2):28–40.
6. Salway A, Graham M. Extracting information about emotions in films. In: Proceedings of the 9th International Conference on Multimedia Modeling; 2003. p. 299–302.
7. Sarvas R, Herrarte E, Wilhelm A, Davis M. Metadata creation system for mobile images. In: Proceedings of the 2nd International Conference Mobile Systems, Applications and Services; 2004. p. 36–48.

8. Smeulders A, Worring M, Santini S, Gupta A. Content based image retrieval at the end of the early years. IEEE Trans Pattern Anal Mach Intell. 2000;22(12):1349–80.
9. Yazhong Feng, Yueting Zhuang, Yunhe Pan. Music information retrieval by detecting mood via computational media aesthetics. In: Proceedings of the IEEE/WIC International Conference on Web Intelligence; 2003. p. 235–41.
10. Zettl H. Sight, sound, motion: applied media aesthetics. Belmont: Wadsworth Publishing; 1999.

Computationally Complete Relational Query Languages

Victor Vianu[1] and Dirk Van Gucht[2]
[1]University of California-San Diego, La Jolla, CA, USA
[2]Indiana University, Bloomington, IN, USA

Synonyms

Complete query languages; Chandra and Harel complete query languages

Definition

A *relational query language* (or query language) is a set of expressions (or programs). The semantics of a query language defines for each of these expressions a corresponding query which is a generic, computable function from finite relation instances to finite relation instances over fixed schemas. A query language is *computationally complete* (or complete) if it defines all queries.

The genericity condition is a consistency criterion requiring that a query commute with isomorphisms of the database domain. Thus, when applied to isomorphic input relation instances, a query returns isomorphic output relation instances. The concept of genericity is based on the well-accepted idea that the result of a query should be independent of the representation of data in a database and should treat the elements of the database as uninterpreted objects [4]. The computability condition requires that the query can be effectively computed; in other words, it must be *implementable* by a program of a Turing-complete programming language under some suitable encoding of relation instances into objects of that language.

Historical Background

The search for an appropriate notion of "complete" query language began soon after the introduction of the relational model by Codd, with its accompanying query languages relational algebra (RA) and relational calculus (RC) [9]. Initially, RA was proposed as a yardstick for query expressiveness. A language was called by Codd "relationally complete" if it was able to simulate RA [10]. Bancilhon and Paredaens independently proposed the notion of BP completeness of a language, using an instance-based approach: a language is BP complete if for every pair of input and output instances satisfying a consistency criterion (The criterion requires that every automorphism of the input be also an automorphism of the output), there exists a query in the language mapping the input instance to the output instance [5, 13]. The notion of genericity was first articulated in the database context by Aho and Ullman [4], although its roots can already be found in the consistency criterion used in the definition of BP completeness and an idea similar to genericity underlies Tarski's concept of "logical notion," introduced in a series of lectures in the mid-1960s [15]. The modern notion of computationally complete query language is due to Chandra and Harel, who also defined the first such language, QL [7].

Foundations

Codd introduced the relational model and its query languages, relational algebra (RA) and relational calculus (RC). These query languages are equivalent, i.e., they define the same set of queries. For example, assuming that **R** and S are relation schemas both of arity 2, then the RA expression $\pi_{1,4}(\sigma_{2=3}(\mathbf{R} \times \mathbf{S})) \cup (\mathbf{R} - \mathbf{S})$ and the RC expression $\{(x, y) \mid \exists z : (\mathbf{R}(x, z) \wedge \mathbf{S}(z, y)) \vee (\mathbf{R}(x, y) \wedge \neg \mathbf{S}(x, y))\}$ define the same computable query Q which maps each relation instance R over $\mathbf{R}$ and each relation instance S over $\mathbf{S}$ to their join unioned with their set difference. Notice that Q is also generic: consider, for example, the input relation instances

$$R_1 = \boxed{\begin{array}{cc} a & b \\ a & c \end{array}} \quad S_1 = \boxed{\begin{array}{cc} a & b \\ b & d \end{array}}$$

and the isomorphic input relation instances

$$R_2 = \boxed{\begin{array}{cc} e & f \\ e & g \end{array}} \quad S_2 = \boxed{\begin{array}{cc} e & f \\ f & h \end{array}}$$

then $Q(R_1, S_1)$ and $Q(R_2, S_2)$ are the isomorphic output relation instances $\boxed{a \ c}$ and $\boxed{e \ g}$, respectively. (As a caveat to genericity, consider the query C defined by the RA expression $\sigma_{1=\mathbf{a}}(\mathbf{R})$, where $\mathbf{a}$ is some *constant* interpreted as a. Then, $C(R_1) = R_1$, but $C(R_2) = \emptyset$. The difficulty is that though R_1 and R_2 are isomorphic, they are not isomorphic by an isomorphism that *fixes* a. If however, the value of e in R_2 is replaced by a, then $C(R_2^{e \leftarrow a}) = R_2^{e \leftarrow a}$. This suggests that when constants are involved, genericity should be modified to isomorphisms that fix these constants. In the literature, this is referred to as $\mathbf{C}$ genericity.)

The development of the relational model led to the introduction of the query language SQL which has, at its logical core, a sub-language pure SQL that is equivalent with RA and RC. For example, in pure SQL, the query Q can be defined by the expression (Here A, B, C, and D are attribute names referring to the first and second columns of $\mathbf{R}$ and the first and second columns of $\mathbf{S}$, respectively.)

```
(SELECT R.A, S.D AS B FROM R, S WHERE R.B = S.C)
UNION
(
(SELECT R.A, R.B FROM R)
EXCEPT
(SELECT S.C AS A, S.D AS B FROM S)
);
```

A natural question is now "Are RA, RC, and pure SQL complete query languages?" The answer is no. To this end, consider the following three queries, which are easily seen to be generic and computable:

1. TC maps each binary relation instance R to its transitive closure R^*.
2. EVEN maps each unary relation instance R to $\{()\}$ (true) if $|R|$ is even and to $\emptyset$ (false), otherwise.
3. EVEN$^<$ maps each pair of a unary relation instance R and a binary relation instance O to EVEN(R) if O defines an ordering on $dom(R)$ and is undefined otherwise. (Here, $dom(R)$ denotes the set of values that occur in the tuples of R.)

It turns out that none of the above queries is expressible in RA (or RC or pure SQL). Consider (1). Fagin showed in [11] that the TC query cannot be defined by any RC expression (the result was later re-proven for RA by Aho and Ullman [4]). Intuitively, the difficulty in computing TC is the following. For each $i \geq 0$, there exists an RA expression $\mathbf{E}^i$ that defines the pairs of elements in R at distance i in the directed graph represented by R. However, there does not exist a *single* RA expression that defines the union of all these pairs, as needed for computing TC. A solution to this problem is to augment RA with an *iteration* construct. This led Chandra and Harel to define the language While (initially introduced as RQ in [8] and LE in [6]). The language uses, in addition to database relations, typed relational variables (of fixed arity) initialized to $\emptyset$, to which RA expressions can be assigned. Iteration is provided by a construct "while change to $\mathbf{R}$ do $\langle program \rangle$ od" whose semantics is to iterate $\langle program \rangle$ as long as the value of the relational variable $\mathbf{R}$ changes. For example, the following While program defines TC:

$\mathbf{TC} := \mathbf{R}$;
while change to $\mathbf{TC}$
    do $\mathbf{TC} := \mathbf{TC} \cup \pi_{1,4}(\sigma_{3=4}(\mathbf{TC} \times \mathbf{R}))$ od.

Here $\mathbf{TC}$ is a binary relation variable (initialized to $\emptyset$). The program first assigns R to **TC**, then loops as long as **TC** changes. Upon termination, this value is R*. One might hope that While is a complete query language. However, this is not the case. Indeed, even though it is easy to write a While program that defines the EVEN< query, Chandra showed that no such program can define the simple linear-time computable EVEN query [6]. Thus, While is not computationally complete. Intuitively, While programs do not have the ability to compute with natural numbers, unless such computations can be simulated by utilizing an ordering on the elements of its input. With such orderings available, While can define precisely the PSPACE-computable queries [17]. The PSPACE upper bound (that holds with or without order) is a consequence of the fact that a program's finite set of variables are of fixed arity and can only hold relation instances built from the elements of its inputs.

To overcome these problems, it appears natural to embed RA into a language that can perform arbitrarily powerful computations. This is in the spirit of "embedded SQL" languages, in which a computationally complete programming language such as C or Java accesses the database using SQL queries. A language called LC (for *Looping+Counters*), abstracting the "embedded SQL" paradigm, was introduced by Chandra [6] (with a variant called WhileN later defined in [1]). The language LC extends While by allowing integer variables (initialized to zero) that can be incremented or decremented. Iteration for computation on integers is provided by an additional While loop of the form "while $i > 0$ do ⟨program⟩ od" which causes ⟨program⟩ to iterate as long as the value of the integer variable i is positive. For example, consider the following program in an LC-like syntax: (The "if-else" statement is a macro that can be e asily written using just the "while-change" construct.)

$$\begin{aligned}
&\mathbf{TC} := \mathbf{R}; \\
&n := 0; \\
&\text{while change to } \mathbf{TC} \\
&\quad \text{do} \\
&\quad\quad n := n + 1; \\
&\quad\quad \mathbf{TC} := \mathbf{TC} \cup \pi_{1,4} (\sigma_{2=3} (\mathbf{TC} \times \mathbf{R})) \\
&\quad \text{od}; \\
&\text{if } n \leq 1 \text{ return } \{()\} \text{ else return}
\end{aligned}$$

At the end of the computation, n contains the number of times the body of the *while* loop was executed. Thus, if R is nonempty, the final value of n is the diameter of the directed graph represented by R (here the diameter means the maximum finite distance between two nodes). The program returns $\{()\}$ (true) if $n \leq 1$ and $\emptyset$ (false) otherwise. Note that since LC is computationally complete on the integers, the condition "$n \leq 1$" could be replaced by *any* computable property of n. Thus, LC can test any computable property of the diameter of R. Clearly, LC can define strictly more queries than While, since all computable functions on natural numbers can be defined and used. This leads to the next question: "Is LC a complete query language?" Again, the answer is no. Indeed, Chandra showed that the EVEN query can still not be defined in this language. This time, the difficulty stems from the fact that even though the values of the relation and natural number variables can depend on each other, LC programs lack the ability to *explicitly coerce* (encode) these values into each other. However, when input relation instances are accompanied by an ordering of the domain, such coercions can be simulated, and Abiteboul and Vianu showed that then LC *is* complete [1].

To obtain a complete language without order, several solutions are possible. A brute force approach to the coercion problem is to augment LC with an encoding function *enc* mapping relations to integers and a decoding function *dec* returning query answers from their encodings and the original input database. Since LC is computationally complete on integers, it can compute the integer encoding of the answer from that of the input. Although this theoretically produces a complete language, manipulating integer encodings of databases is not a satisfying solution, so further discussion of this approach is omitted. Instead, two more appealing alternatives are described, that both go back to While as a starting point and extend it in different ways. Recall that While is limited to PSPACE computations, as a consequence of two facts: (i) only relations of fixed arity are used, and (ii) the relations can be populated by tuples using only elements occurring in the input. The first approach, proposed by Chandra and Harel, breaks the PSPACE space barrier by relaxing (i) it allows untyped relational variables, whose arity can grow arbitrarily. The other approach, introduced by Abiteboul and Vianu, relaxes (ii), and it keeps typed relational variables but allows the introduction of new domain values in the course of the computation. These languages are described next.

The complete language proposed by Chandra and Harel was called QL [6]. Up to minor syntactic differences, QL is very similar to While, only with untyped relation variables. Consider the following QL program (For simplicity, the syntax used here differs slightly from the original QL syntax.) which is strikingly similar to the LC program shown above:

$$\mathbf{TC} := \mathbf{R};$$
$$\mathbf{ONE} := \pi_1(\mathbf{R}) \cup \pi_2(\mathbf{R});$$
$$\mathbf{N} := \{()\};$$
while change to $\mathbf{TC}$
    do
        $\mathbf{N} := \mathbf{N} \times \mathbf{ONE};$
        $\mathbf{TC} := \mathbf{TC} \cup \pi_{1,4}(\sigma_{2=3}(\mathbf{TC} \times \mathbf{R}))$
    od;
if $(\mathbf{N} = \{()\})$ or $(\mathbf{N} = \mathbf{ONE})$ then return $\{()\}$ else return $\emptyset$.

In this program, $\mathbf{R}$ is a binary relation input variable, and $\mathbf{TC}$, $\mathbf{ONE}$, and $\mathbf{N}$ are relation variables. Note that the arities of $\mathbf{TC}$ and $\mathbf{ONE}$ remain fixed throughout the execution of the program, while the arity of $\mathbf{N}$ changes. Integers can be easily simulated using the arity of relations. Thus, starting with relation instance R, the variable $\mathbf{ONE}$ is initialized to dom(R), which plays the role of the natural number 1. The variable $\mathbf{N}$ plays the role of a natural number variable n. The statement $\mathbf{N} := \{()\}$ corresponds the statement $n = 0$, and the statement $\mathbf{N} := \mathbf{N} \times \mathbf{ONE}$ serves to increment n by 1; notice that the $\times$ operator plays the role of the addition operator $+$ over natural numbers, and the decrement operator can be simulated by projection. Similarly to the earlier LC program, the final arity of $\mathbf{N}$ is the diameter of the directed graph represented by $\mathbf{R}$. The final "if" statement compares $\mathbf{N}$ to $\{()\}$ or $\mathbf{ONE}$, and the program returns $\{()\}$ (true) if the diameter of $\mathbf{R}$ is at most 1 and $\emptyset$ (false) otherwise. Observe therefore that this QL program defines the same query as its corresponding LC program.

The above example illustrates how arithmetic on natural numbers can be simulated in QL. So far, this allows simulating LC. Recall that LC is not complete but becomes so if an ordering of the domain is provided. QL is however complete even if an ordering is not provided, because it can construct its own orderings! Indeed, such orderings of the domain can simply be constructed in QL by building one relation whose arity equals the size of the input domain. In such a relation, any tuple that does not contain repeated elements provides a successor relation on the domain, which in turns induces an ordering. The completeness of QL now follows from the completeness of LC on ordered domains. Thus, QL can express all computable queries. But is everything it expresses a query? It is easy to see that all mappings defined by QL programs are computable and generic. The difficulty is to guarantee that a QL program always produces answers of the desired arity. In fact, this property is undecidable for QL programs. Fortunately, there is an effective syntactic restriction guaranteeing that QL programs are "well behaved," i.e., always produce answers of fixed arity. Moreover, all computable queries can be expressed by QL programs satisfying the syntactic restriction.

The language WhileNew, introduced by Abiteboul and Vianu in [3], extends While by allowing the creation of new values throughout the computation. This is achieved by an instruction of the form $\mathbf{S} := \text{new}(\mathbf{R})$, where $\mathbf{R}$ and $\mathbf{S}$ are relational variables and $arity(\mathbf{S}) = arity(\mathbf{R}) + 1$. The semantics is the following. Given a relation instance R over $\mathbf{R}$, the relation instance S over $\mathbf{S}$ is obtained by extending each tuple of R by one distinct new value not occurring in the input, the current state, or in the program. For example, if R is the relation instance in Fig. 1 then S is of the form shown in the same figure. The values α, β, γ are distinct new values. Note that the new construct is, strictly speaking, nondeterministic. Indeed, the new values are arbitrary, so several

$$R = \begin{array}{|cc|} \hline a & b \\ a & c \\ c & a \\ \hline \end{array} \qquad S = \begin{array}{|ccc|} \hline a & b & \alpha \\ a & c & \beta \\ c & a & \gamma \\ \hline \end{array}$$

Computationally Complete Relational Query Languages, Fig. 1 An application of new. Here, $S = \mathrm{new}(R)$

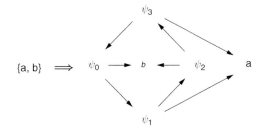

Computationally Complete Relational Query Languages, Fig. 2 A query with new values not expressible in WhileNew

outcomes are possible depending on the choice of values. However, the different outcomes differ *only* in the choice of new values.

The ability to successively introduce new values throughout the computation easily allows simulating integers and arithmetic, yielding the power of LC. Moreover, orderings of the input domain can also be constructed and marked by distinct new values. Since LC is complete on ordered domains, this shows that WhileNew can express all computable queries. As in the case of QL, one must ask whether *all* mappings expressed by WhileNew are in fact queries according to the definition. The difficulty arises from the presence of new values. Indeed, if new values may appear in the outputs of a WhileNew program, the mapping it defines is non-deterministic. Moreover, it is undecidable whether a WhileNew program never contains new values in its output. The solution to this problem is similar to the one for QL: one can impose a syntactic restriction on WhileNew programs guaranteeing that no new value appears in their answers. All generic computable queries can be expressed by WhileNew programs satisfying the syntactic restriction.

As an aside, suppose the definition of query is extended by allowing new values in query answers. This arises naturally in some contexts such as object-oriented databases, where outputs to queries may contain new objects with their own fresh identifiers. One might hope that WhileNew remains complete for this extension. Surprisingly, it was shown by Abiteboul and Kanellakis that the answer is negative [2]. Indeed, WhileNew cannot express the query containing the input/output pair shown in Fig. 2, where $\psi_0, \dots, \psi_3$ are new values. As shown by Abiteboul and Kanellakis, completeness with new values can be achieved by adding to While-New a construct called *duplicate elimination*.

This is however a rather complex construct that encapsulates a test for isomorphism of relations. The search for a language using more natural primitives and complete for queries with new values in the answer remains open.

The relational model, though very simple, is not always the most natural model for databases in certain application domains. In the late 1980s and early 1990s, database researchers considered object-oriented databases as an alternative to the relational model, and a significant amount of theory was developed around the model and its query languages, including the completeness of object-oriented query languages (see [1, 16]). Finally, consistency notions other than genericity can be considered for specialized application domains. This was done, for example, in the context of spatial databases [12, 14].

Key Applications

The theoretical query languages discussed here are closely related to various practical languages. Thus, RA and RC correspond to pure SQL. The language While corresponds to wrapping programming constructs such as loops around SQL, as done in PL/SQL (Oracle); assignment statements can be implemented using SQL insert and delete operations.

The language LC (or WhileN) can again be simulated in PL/SQL augmented with natural number variables (with no coercion allowed). Another approach is to embed SQL in a programming language such as C or Java. In such languages, relational variables must be statically defined and so have fixed arity. One significant feature of the embedded SQL languages that sets them apart from LC is that they allow accessing tuples in relations one at a time, using looping over cursors. In particular, this allows coercing the entire database into a native data structure, and yields computational completeness. However, there is a catch: programs using cursors are generally nondeterministic, in the sense that running the same program on the same database content may yield different results. Unfortunately, it is undecidable whether a given embedded SQL program is deterministic, and no natural syntactic restriction is known that ensures determinism while preserving completeness. Thus, completeness is achieved at the cost of losing the guarantee of determinism.

The computationally complete language QL can be simulated in Dynamic SQL. In this language, one can dynamically create relation variables whose schemas depend on the data in the database. This can be used to support untyped relational variables. The language WhileNew allowing the introduction of new domain values is akin to object-oriented languages that allow the creation of new object identifiers.

Future Directions

The database area is undergoing tremendous expansion and diversification under the impetus of the Web and a host of specialized applications. Consequently, new structures and objects have to be modeled and manipulated. For example, in biological and scientific applications, sequences and matrices occur prominently; in XML databases, text and tree-structured documents are the main objects. This has led to new database models and query languages. Their formal foundations are fast developing but are not yet as mature as for the relational data model. Notions of computationally complete languages for the new models are still emerging and are likely to build upon the theory developed for relational databases.

Cross-References

▶ BP-Completeness
▶ Constraint Query Languages
▶ Ehrenfeucht-Fraïssé Games
▶ Expressive Power of Query Languages
▶ Object Data Models
▶ Query Language
▶ Relational Calculus
▶ Relational Model
▶ Semantic Web Query Languages
▶ Semi-structured Query Languages
▶ SQL
▶ XML
▶ XPath/XQuery

Recommended Reading

1. Abiteboul S, Hull R, Vianu V. Foundations of databases. Reading: Addison-Wesley; 1995.
2. Abiteboul S, Kanellakis PC. Object identity as a query language primitive. J ACM. 1998;45(5): 798–842.
3. Abiteboul S, Vianu V. Procedural languages for database queries and updates. J Comput Syst Sci. 1990;41(2):181–229.
4. Aho AV, Ullman JD. Universality of data retrieval languages. In: Proceedings of the 6th ACM SIGACT-SIGPLAN Symposium on Principles of Programming Languages; 1979. p. 110–20.
5. Bancilhon F. On the completeness of query languages for relational data bases. In: Proceedings of the 7th Symposium on the Mathematical Foundations of Computer Science; 1978. p. 112–23.
6. Chandra A. Programming primitives for database languages. In: Proceedings of the 8th ACM SIGACT-SIGPLAN Symposium on Principles of Programming Languages; 1981. p. 50–62.
7. Chandra A, Harel D. Computable queries for relational data bases. J Comput Syst Sci. 1980;21(2):156–78.
8. Chandra A, Harel D. Structure and complexity of relational queries. J Comput Syst Sci. 1982;25(1): 99–128.
9. Codd E. A relational model for large shared data-banks. Commun ACM. 1970;13(6):377–87.
10. Codd E. Relational completeness of data base sublanguages. In: Rustin R, editor. Data base systems. Englewood Cliffs: Prentice-Hall; 1972. p. 65–98.
11. Fagin R. Monadic generalized spectra. Z Math Logik Grundlagen Math. 1975;21(1):2189–96.
12. Gyssens M, Van den Bussche J, Van Gucht D. Complete geometric query languages. J Comput Syst Sci. 1999;58(3):483–511.

13. Paredaens J. On the expressive power of the relational algebra. Inf Process Lett. 1978;7(2):107–11.
14. Paredaens J. Spatial databases, a new frontier. In: Proceedings of the 5th International Conference on Database Theory; 1995. p. 14–32.
15. Tarski A, Corcoran J. What are logical notions? Hist Philos Logic. 1986;7:154.
16. Van den Bussche J, Van Gucht D, Andries M, Gyssens M. On the completeness of object-creating database transformation languages. J ACM. 1997;44(2):272–319.
17. Vardi MY. The complexity of relational query languages. In: Proceedings of the 14th Annual ACM Symposium on Theory of Computing; 1982. p. 137–46.

Computerized physician order entry (CPOE) is a process by which physicians directly enter medical orders into a computer. CPOE is typically done when the computer is being used to access an electronic health record (EHR), and the physician is creating a treatment plan for a specific patient in a clinical setting.

In many medical institutions, non-physicians such as nurses, dieticians, social workers, pharmacists, therapists, or advanced nurse practitioners can also enter certain types of orders, hence the broader, useful term *"computerized provider order entry."*

Computerized Physician Order Entry

Michael Weiner
Regenstrief Institute, Inc., Indiana University School of Medicine, Indianapolis, IN, USA

Synonyms

Computer-based physician order entry; Computer-based provider order entry; Computerized provider order entry; Computerized order entry (COE); Physician order entry

Definition

In daily medical practice, physicians routinely create plans of diagnosis and treatment for their patients. These plans typically contain specific, formal orders – directives – that are expected to be implemented by other medical professionals, such as nurses or personnel at laboratories or pharmacies. When such personnel are expected to implement part of a physician's diagnosis or treatment plan, corresponding orders must be created and documented in the patient's medical record. Physicians have traditionally used paper-based charting systems to record medical orders.

Historical Background

CPOE was first implemented and described in the latter half of the twentieth century. In the US, early reports came from several institutions, including Harvard Medical School and Brigham and Women's Hospital, the US Veterans Health Administration [16, 20], Vanderbilt University [8], University of Virginia [12], Indiana University, and Regenstrief Institute for Health Care [14].

In 1994, Sittig and Stead published "Computer-based physician order entry: the state of the art" [17], summarizing many of the early results. Many difficulties were reported regarding leadership, delays, cultural resistance, high costs, technical support, workflow, and other operational difficulties for end users.

By the end of 2006, CPOE was on the rise, though adoption rates – often correlated with adoption of EHRs – varied widely throughout the world. Most modern EHR systems, whether developed by public, private, or academic institutions, would be expected to include at least some form of CPOE. In the US, only 10% of hospitals had complete availability of CPOE in 2002 [2]. In the United Kingdom, Australia, and New Zealand [21], the fraction is much higher, since use of EHRs exceeds 80% and is approaching 99% among general practitioners in ambulatory practice.

Foundations

CPOE can be used to order a variety of medical services. In some medical institutions with EHRs, CPOE can be used to order any type of medical service and may be required to generate any order. When not required, providers may have the opportunity to select between writing orders in a paper-based chart or an EHR, or clerks or other professionals may perform CPOE on a provider's behalf or direction.

CPOE is performed via a user interface of some kind, though this may occur on a desktop computer, terminal or thin client, personal digital assistant, other portable computer, or other form of computer. A user would typically authenticate himself or herself, identify a patient, and proceed to enter orders, often by navigating through a series of forms, each of which might facilitate a certain type of order, such as for radiology, laboratory, pharmacy, nursing, or referral (see Fig. 1). The layout or interface seen by the user

is highly variable and may depend on the developer, personal preferences, underlying database structures, or medical or administrative processes generated in response to specific orders.

Many EHRs allow providers to generate orders and non-order documentation in a single computer session. Non-order documentation may include clinical details such as historical information, measurements, other observations, patient's preferences or directives, or narrative notes or reports. The workflow imposed by a CPOE system should be considered carefully in conjunction with the user's baseline workflow. Greatest success with implementation can often be found when the system does not disrupt the user's own pattern of work.

Once a session is completed, the orders generated lead to action. This may occur via simple printing of the orders or through electronic delivery to remote locations, such as a radiology department, consultant's office, laboratory, or pharmacy. Some orders, such as for prescriptions

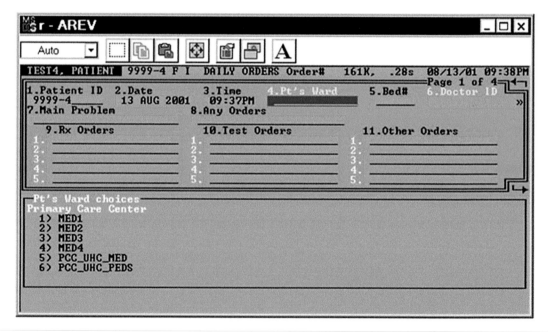

Computerized Physician Order Entry, Fig. 1 A user interface for computerized physician order entry. A user would typically authenticate himself or herself, identify a patient, and proceed to enter orders, often by navigating through a series of forms, each of which might facilitate a certain type of order. This screenshot shows form fields that allow entry of orders for drugs, diagnostic tests, and other types of orders (e.g., nursing). The visual separation of types of orders into categories is done primarily for the user's convenience and organization; it may or may not reflect underlying data models

for drugs or lifestyle changes, may be provided to patients for direct implementation or delivery elsewhere. A session's orders are then typically archived in the EHR. Consistent with traditional medical documentation, orders from CPOE are generated once and cannot be modified or deleted once finalized, though what is being ordered can often later be modified or discontinued with a subsequent order.

CPOE has been developed and implemented for a variety of reasons. Many advantages have been postulated, including rectification of substantial legibility problems with handwritten orders. CPOE systems have the potential to refer to all of a patient's medical history as well as all available medical knowledge, to improve the quality of medical care in real time, at the point of care. One of the most important potentials of CPOE is inclusion of *clinical decision support*, by which the computer can be programmed to suggest tailored orders *de novo* (e.g., for a vaccination recommended by clinical guidelines) or respond to specific orders, such as in the event of

a possible drug reaction [11] or contraindication to a procedure (see Fig. 2). CPOE can reduce the rate of certain medication errors by more than half [4]. Removing the healthcare provider from electronic order entry, or moving CPOE outside the point of care, could negate these large potential benefits.

Some institutions are starting with, focus, or limit their computer-based development to electronic prescribing, or "e-prescribing." This is a form of CPOE. E-prescribing is targeted especially because prescribing is frequent, can be targeted by CPOE algorithms, and represents a most common form of medical error [13, 18]. In the US, the Institute of Medicine has recommended e-prescribing of drugs, in conjunction with clinical decision support [9].

Medical orders symbolize but also allow and direct the operations behind medical care. By encoding and electronically documenting orders, CPOE improves capabilities to assess and improve quality of care and to conduct clinical research related to diagnosis and treatment. Med-

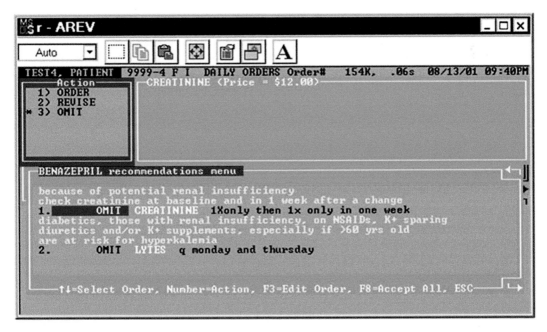

Computerized Physician Order Entry, Fig. 2 Example of clinical decision support. In this instance, the user has prescribed benazepril. The software responds, as shown, by prompting the user to decide about ordering blood tests, to monitor for possible side effects from the drug. To enable this capability, the application has been programmed with clinical rules that combine guidelines or medical knowledge with this patient's medical history, evaluation, or treatment

ical practices, governments, and other authorities can gather and study data from CPOE systems to improve knowledge about how medical care is formulated and delivered. CPOE can also facilitate billing processes that depend on orders, such as for certain procedures or drugs. Query languages or systems must be designed to accommodate data models used for CPOE.

Customization of CPOE systems may allow individual providers or groups of providers to create templates or order sets, which are groups of orders often used or often used together. This could save time, improve standardization, and improve care.

CPOE does have costs, risks, and unintended consequences [5]. A CPOE system must be developed thoughtfully and be maintained frequently and regularly, to ensure that it accommodates the latest tests, treatments, and guidelines, both locally and more broadly. If an institution does not stock a particular drug, the system might not allow that drug to be ordered or might at least alert the provider about the issue. Institutional changes and policies that can affect CPOE are frequent and so must lead to corresponding modifications to the CPOE system.

CPOE can increase the time required to generate a medical order [3, 15]. Studies of this have reported mixed findings, with increases in some and decreases in others. Increased time for initial learning and ongoing use can cause dissatisfaction among providers and even complete failures of systems. Increasing time to generate or implement orders can have adverse clinical effects. Adverse effects might be expected especially in emergencies or acute care, when life-saving drugs may be needed rapidly. Errors in processing electronic orders could also be expected to lead to adverse effects in at least some cases. In 2005, Koppel et al. reported that one CPOE system design often facilitated medication-related errors, such as by providing inadequate views of medications and increasing inappropriate dosing and incompatible orders [10].

If not implemented effectively, increased use of computers in healthcare might distract providers, causing them to spend less time with patients or decrease patient's satisfaction [7, 19]. This could have an adverse effect on patient-provider relationship or patient's health. Provider-to-provider communication might also suffer if appropriate internal communications systems are not used.

Clinical decision support, a key feature of CPOE systems, can backfire by presenting too many or inappropriate alerts. Effective solutions to "alert overload" are not yet well developed or widespread, though some solutions have been discussed [18]. Several recent studies of interventions in decision support have had negative results and require further investigation.

Financial costs of implementing CPOE can be high, especially initially. Developing cost-effectiveness analyses of EHRs and CPOE systems are thus complex, because benefits or harms can occur much later than implementation of a system and later than the time of initial care or clinical presentation.

Moving orders from paper to computers has created situations that require new handling. Computer programs, for example, must know the authority of the authenticated user and whether the user has permission to generate the requested orders. This need also exists with paper systems but is handled in those environments by people, rather than computers. In addition, the use of templates or order sets has not been heavily studied. Templates may in some cases decrease quality of care if they are adopted hastily or used in the wrong setting. In ultimately pooling or sharing data across institutions, it will be important to use standards and customary terminologies to represent orders.

Technical Issues

"Prescription" is another term for medical order, though the term is used conventionally to refer to providing instructions to patients. The structure of a traditional drug prescription provides a useful framework for understanding the primary components of medical orders. Drug orders have a superscription (including timestamp and patient's identifier), inscription (name and amount or strength of ingredient), subscription (formulation or method to prepare), and *signa* ("sig," or directions including route and frequency). Orders of any other type have analogous components, though some may have additional or somewhat different components. CPOE systems should handle components of orders with agility. Below are discussed a few key technical issues that present themselves in the design, study, and implementation of CPOE systems.

Data Models for CPOE. One must consider what data model would best support CPOE. For example, should orders be categorized and, if so, how? Many institutions have found that categories of orders, such as laboratory, consultative, pharmacy, nursing, and radiological, are clinically and informationally logical but may also be necessary from the standpoint of linking disparate data systems. A key goal in the design is the ability to accommodate future expansion of order types and categories even before those types are developed or identified. This can prompt a somewhat "flat" model, in which the nature, type, or category of an order is a value of a database field, rather than a field or variable itself. Other important aspects of orders that may have implications for the data model are the indication for the order, urgency (i.e., when it should be implemented), and who is expected to implement it.

Standardization of text in an order can be helpful for both accuracy of implementation and research. For example, an order that can have multiple forms, such as "take two 40-mg tablets by mouth twice daily" and "take one 80-mg tablet by mouth every 12 hours" can complicate both clinical care and research. Allowing providers to add narratives or free text is essential for tailoring to patient's needs, but this demands effective handling in data storage and clinical decision support. A system that can standardize the order accurately without hindering the user's experience is desirable but challenging to create. Standardization of terminology used in orders should accommodate query systems. One difficulty is that each type of order – such as for a drug, diet, radiological procedure, or laboratory test – may have unique "domains" or components. For example, only drugs have a dose, yet the dose may need to be a discrete, searchable component of a

query. Therefore, an ideal data model can handle all types of orders, as well as new types, but it can also identify and distinguish between various values of key components of orders, regardless of how widespread those components are across various types of orders.

Order sets can often translate directly into a group of individual orders, but users often desire the ability to customize order sets. This may mean maintaining a base of order sets but also the customizations that are unique to each provider or role. Order sets also need to be integrated with any available decision support systems, and many providers seek to share customized order sets with each other. There are thus aspects of order sets that pertain to the system itself and to particular patients, groups of providers, and individual providers. Associated with each order set is also generally information about conditions under which the order set applies, such as a diagnosis, age group, or other criteria found in a clinical guideline – hence the possible need for order sets to be linked to ontologies or terminology systems that in turn link to such guidelines.

Specific types of orders often require further processing or delivery to specific clinical departments. Therefore, the processing needed must be encoded into the system, though it might be a part of the main data engine more than a core part of the data model or record. In any case, if laboratory orders, for example, need to be delivered to the laboratory, then the data system must support this well enough so that all laboratory orders – and only laboratory orders – are processed in this way. Similarly, systems that notify particular professionals or departments should be modular enough that those systems can be updated readily as personnel or even departments change.

An audit trail is important for documentation, accreditation, and quality and safety of care. Included in the data system should be a method to indicate formally not just who created a record and when, but what happened to the record: where it was sent and who accessed it later. Whether this is part of CPOE or the larger records system may depend on the circumstances and design of a system. Many alert systems that stem from orders do not currently provide effective prioritization, so this is an area of important research.

Key Applications

CPOE continues to undergo development and will for the foreseeable future. Due to the difficulty and time required to generate electronic orders, various forms of data entry are being explored. These include transcription with or without voice recognition and input using portable devices, digital pens, or tablets. Due to capability for electronic communication of orders, development is also occurring in remote areas or environments with limited access to healthcare, such as for rural or homebound patients.

CPOE is undergoing significant development especially in the US, the United Kingdom and other parts of Europe, Asia, Australia, and New Zealand. It can be expected to grow throughout the world, even as developing countries create EHR systems.

Future Directions

The largest looming issues for CPOE are how to maximize efficiency of data entry and effectiveness of decision support in the complex environment. There are also unmet needs for CPOE to be linked to access to general medical knowledge. Health policy, attention to quality, patient safety, and reimbursement will likely gain importance in driving uses of CPOE, especially in areas where its use is currently low. The precise roles, usefulness, impact, and specifications of incentives for healthcare providers to adopt health information technologies such as CPOE are not yet clear.

Cross-References

▶ Clinical Decision Support
▶ Electronic Health Record

Recommended Reading

1. Agency for Healthcare Research and Quality. AHRQ National Resource Center for Health Information Technology. 2007. Available online at: http://healthit. ahrq.gov. Accessed 29 Aug 2007.
2. Ash JS, Gorman PN, Seshadri V, Hersh WR. Computerized physician order entry in U.S. hospitals: results of a 2002 survey. J Am Med Inform Assoc. 2004;11(2):95–9.
3. Bates DW, Boyle DL, Teich JM. Impact of computerized physician order entry on physician time. In: Proceedings of the Annual Symposium on Computer Applications in Medical Care; 1994. p. 996.
4. Bates DW, Leape LL, Cullen DJ, et al. Effect of computerized physician order entry and a team intervention on prevention of serious medication errors. JAMA. 1998;280(15):1311–6.
5. Campbell EM, Sittig DF, Ash JS, Guappone KP, Dykstra RH. Types of unintended consequences related to computerized provider order entry. J Am Med Inform Assoc. 2006;13(5):547–56.
6. Certification Commission for Healthcare Information Technology. 2007. Available online at: http://www. cchit.org. Accessed 29 Aug 2007.
7. Frankel R, Altschuler A, George S, et al. Effects of exam-room computing on clinician-patient communication: a longitudinal qualitative study. J Gen Intern Med. 2005;20(8):677–82.
8. Geissbuhler A, Miller RA. A new approach to the implementation of direct care-provider order entry. In: Proceedings of the AMIA Annual Fall Symposium; 1996. p. 689–93.
9. Institute of Medicine. Crossing the quality chasm: a new health system for the 21st century. Washington, DC: The National Academies Press; 2001.
10. Koppel R, Metlay JP, Cohen A, et al. Role of computerized physician order entry systems in facilitating medication errors. JAMA. 2005;293(10):1197–203.
11. Kuperman GJ, Bobb A, Payne TH, et al. Medication-related clinical decision support in computerized provider order entry systems: a review. J Am Med Inform Assoc. 2007;14(1):29–40.
12. Massaro TA. Introducing physician order entry at a major academic medical center: I. Impact on organizational culture and behavior. Acad Med. 1993;68(1):20–5.
13. Miller RA, Gardner RM, Johnson KB, Hripcsak G. Clinical decision support and electronic prescribing systems: a time for responsible thought and action. J Am Med Inform Assoc. 2005;12(4):403–9.
14. Overhage JM, Mamlin B, Warvel J, Warvel J, Tierney W, McDonald CJ. A tool for provider interaction during patient care: G-CARE. In: Proceedings of the Annual Symposium on Computer Applications in Medical Care; 1995. p. 178–82.
15. Overhage JM, Perkins S, Tierney WM, McDonald CJ. Controlled trial of direct physician order entry: effects on physicians' time utilization in ambulatory

primary care internal medicine practices. J Am Med Inform Assoc. 2001;8(4):361–71.

16. Payne TH. The transition to automated practitioner order entry in a teaching hospital: the VA Puget Sound experience. In: Proceedings of the AMIA Annual Symposium; 1999. p. 589–93.

17. Sittig DF, Stead WW. Computer-based physician order entry: the state of the art. J Am Med Inform Assoc. 1994;1(2):108–23.

18. Teich JM, Osheroff JA, Pifer EA, Sittig DF, Jenders RA. Clinical decision support in electronic prescribing: recommendations and an action plan: report of the joint clinical decision support workgroup. J Am Med Inform Assoc. 2005;12(4):365–76.

19. Weiner M, Biondich P. The influence of information technology on patient-physician relationships. J Gen Intern Med. 2006;21(Suppl 1):S35–9.

20. Weir C, Lincoln M, Roscoe D, Turner C, Moreshead G. Dimensions associated with successful implementation of a hospital based integrated order entry system. In: Proceedings of the Annual Symposium on Computer Applications in Medical Care; 1994. p. 653–7.

21. Wells S, Ashton T, Jackson R. Electronic clinical decision support. 2005. Updated Oct 2005. Available via Internet at: http://www.hpm.org/nz/a6/2.pdf. Accessed 7 Aug 2014.

Conceptual Modeling Foundations: The Notion of a Model in Conceptual Modeling

Bernhard Thalheim
Christian-Albrechts University, Kiel, Germany

Synonyms

Model; Modeling

Definition

A model is a well-formed, adequate, and dependable artifact that represents other origin artifacts. Its criteria of well-formedness, adequacy, and dependability must be commonly accepted by its community of practice within some context and correspond to the functions that a model fulfills in utilization scenarios and use spectra. As an artifact, a model is grounded in its community's subdiscipline and is based on elements chosen from the subdiscipline. A conceptual model is based on abstract concepts and their interrelationships [2]. A conceptual database model represents the structure and the integrity constraints of a database within the given utilization scenario.

Main Text

The conceptual modeling community widely uses models for constructing information systems. Conceptual modeling is a widely applied practice, and its application has led to a large body of useful constructs and methods for creating artifacts that describe some abstraction of reality and serve as a prescription for the development of a database system.

Science and technology widely uses models in a variety of in utilization scenarios. Models function as an instrument in those utilization scenarios. The main function of a conceptual database model is description-prescription. In this case, a conceptual model is used as a mediator between a reality and an abstract reality that developers of a database system intend to build. Other functions of a model, besides the description-prescription function, are the explanation, the optimization-variation, the validation-verification-testing, the reflection-optimization, the explorative, the hypothetical, and the documentation-visualization functions. The functions of a model determine the purposes of the deployment of the model.

Models have several *essential properties*:

- A model's artifact is *well formed* if it satisfies a well-formedness criterion. Typically [2], conceptual models have predicate calculus as their formal foundation. Concepts are predicates and relationships among concepts are n-ary predicates ($n \geq 2$). Constraints are well-formed formulas expressing constraints with cardinality constraints and generalization/specialization constraints being the most common. For ease of understanding, the underlying predicate calculus is usually rendered as a conceptual-model diagram. In typical conceptual-model diagrams, named boxes represent sets of objects, lines connecting boxes represent relationships among objects, and embellishments associated with lines such as cardinality-constraint notes, arrowheads, and generalization/specialization triangles denote constraints over the objects and their relationships.

- Well-formedness enables an artifact to be *adequate* for a collection of origin artifacts if (i) it is analogous to the origin artifacts to be represented according to some analogy criterion, (ii) it is more focused (e.g., simpler, truncated, more abstract or reduced) than the origin artifacts being modeled, and (iii) it is sufficient to satisfy its purpose.
- Well-formedness also enables an artifact to be *justified* (i) by an empirical corroboration according to its objectives, supported by some argument calculus, (ii) by rational coherence and conformity explicitly stated through formulas, (iii) by falsifiability that can be given by an abductive or inductive logic, and (iv) by stability and plasticity explicitly given through formulas. An artifact is *sufficient* as measured by a *quality* characterization for internal quality, for external quality, and for quality in use through quality characteristics that support correctness, generality, usefulness, comprehensibility, parsimony, robustness, novelty, etc. Sufficiency is typically combined with some assurance evaluation (tolerance, modality, confidence, and restrictions). A well-formed artifact is *dependable* if it is justified and sufficient.
- A *model* is a well-formed, adequate, and dependable artifact by adherence to its underlying formalisms if it is commonly accepted by its community of practice within some context and if it fulfills the functions in utilization scenario and use spectra. Adequacy and dependability characterize the utility of a model for deployment, reliability, and degree of precision efficiency for satisfying the use necessities.

Not only should a model faithfully represent a collection of origin artifacts by being well formed, adequate, and dependable; it should also provide facilities or methods for its use. A model is *functional* if there are methods for utilization of the artifact to achieve the objectives for which an artifact might serve. Typical task objectives include defining, constructing, evolving, migrating, exploring, communicating, understanding, replacing, substituting, documenting, negotiating, replacing, optimizing, validating, verifying, testing, reporting, and accounting. We call a model *effective* if it can be deployed according to its objectives.

Models satisfy several properties that make them functional and effective [1, 3]:

- *Mapping* property: The model can be defined through a mapping from the origin artifacts that it represents.
- *Analogy* property: The model is analogous to the origin artifact based on some analogy criterion.
- *Truncation* property: The model lacks some of the ascriptions made to the origin artifacts and thus functions by abstraction by disregarding the irrelevant.
- *Pragmatic* property: The model use is only justified for particular model users, the tools of investigation, and the period of time.
- *Amplification* property: Models use specific extensions which are not observed in the origin artifacts.
- *Idealization* property: Modeling abstracts from reality by scoping the model to the ideal state of affairs.
- *Utilization* property: The model functions well within its intended scenarios of usage according to its capacity and potential.
- *Divergence* property: models deliberately diverge from reality in order to simplify salient properties of interest, transforming them into artifacts that are easier to work with.
- *Added value* property: Models provide a value or benefit based on their utility, capability, and quality characteristics.

- *Purpose* property: Models and conceptual models are governed by the purpose. The model preserves the purpose.

Cross-References

Data Model

    (a) Semantic Data Model
    (b) Conceptual Modeling
    (c) Entity-Relationship Model
    (d) Conceptual Data Model

▸ Database Design

    (a) Conceptual Schema Design
    (b) Schema Deployment

Recommended Reading

1. Thalheim B. The conceptual model ≡ an adequate and dependable artifact enhanced by concepts. Ser Front Artif Intell Appl. 2014;260:241–54.
2. Thalheim B. The conceptual framework to user-oriented content management. Ser Front Artif Intell Appl. 2007;154:30–49.
3. Mahr B. Information science and the logic of models. Softw Syst Model. 2009;8(3):365–83.

Conceptual Schema Design

Alexander Borgida[1] and John Mylopoulos[2]
[1]Rutgers University, New Brunswick, NJ, USA
[2]Department of Computer Science, University of Toronto, Toronto, ON, Canada

Definition

Conceptual schema design is the process of generating a description of the contents of a database in high-level terms that are natural and direct for users of the database. The process takes as input information requirements for the applications that will use the database, and produces a schema expressed in a conceptual modeling notation, such as the Extended Entity-Relationship (EER) Data Model or UML class diagrams. The challenges in designing a conceptual schema include: (i) turning informal information requirements into a *cognitive model* that describes unambiguously and completely the contents of the database-to-be; and (ii) using the constructs of a data modeling language appropriately to generate from the cognitive model a conceptual schema that reflects it as accurately as possible.

Historical Background

The history of conceptual schema design is intimately intertwined with that of conceptual data models (aka semantic data models). In fact, for many years researchers focused on the design of suitable *languages* for conceptual schemas, paying little attention to the design process itself. Jean-Raymond Abrial proposed the *binary semantic model* in 1974 [1], shortly followed by Peter Chen's *entity-relationship model* (ER for short) [4]. Both were intended as advances over logical data models proposed only a few years earlier, and both emphasized the need to model more naturally the contents of a database. The ER model and its extensions were relatively easy to map to logical schemas for relational databases, making EER [9] the first conceptual modeling notation to be used widely by practitioners. On the other hand, Abrial's semantic model was more akin to object-oriented data models that became popular more than a decade later.

The advent of object-oriented software analysis techniques in the late 1980s revived interest in object-oriented data modeling and led to a number of proposals. Some of these, notably OMT [7] adopted many ideas from the EER model. These ideas were consolidated into the *Unified Modeling Language* (UML), specifically UML class diagrams.

The process of designing conceptual schemas by using such modeling languages was not studied until the late 1970s, see for instance [8]. In the early 1980s, the DATAID project proposed a state-of-the-art design process for databases, including conceptual schema design [2].

Throughout this history, research on knowledge representation in Artificial Intelligence (AI) has advanced a set of concepts that overlaps with those of conceptual data models. Notably, semantic networks, first proposed in the 1960s, were founded on the notions of *concept*, *link* and *isA* hierarchy (analogously to *entity*, *relationship* and *generalization* for the EER model). The formal treatment of these notations has led to modern modeling languages such as Description Logics, including OWL, for capturing the semantics of web data. In fact, Description Logics have been shown to be able to capture the precise semantics of conceptual modeling notations such as EER diagrams and UML class diagrams.

Another important recent development has been the rise of *ontological analysis*. An ontology is a specification of a conceptualization of a domain. As such, an ontology offers a set of concepts for modeling an application, and foundational ontologies strive to uncover appropriate cognitive primitives with which to describe and critique conceptual modeling notations [5]. Foundational ontologies have been used to analyze the appropriate use of EER constructs [10] and UML [6]. Based on this work, a two-phase perspective is adopted here by distinguishing between the design of a *cognitive model* based on cognitive primitives, and the design of a corresponding *conceptual schema* that is based on the constructs of a conceptual model such as the EER or UML class diagrams.

Foundations

Building a cognitive model. Information requirements for a database-to-be are generally expressed informally, based on multiple sources (e.g., applications/queries that need to be supported, existing paper and computerized systems). Information requirements describe some part of the world, hereafter the *application domain* (or *universe of discourse*). A *cognitive model* is a human conceptualization of this domain, described in terms of cognitive primitives that underlie human cognition. The

following are some of the most important among these primitives.

An *object* is anything one may want to talk about, and often represents an *individual* ("my dog," "math422"). Usually, individual objects persist over many states of the application domain. Moreover, they have *qualities* (such as size, weight), and can be related to other individuals by *relations* (e.g., "friendOf," "part of," "between"). Individuals may be concrete (such as "Janet," or "that tree"), abstract (e.g., "the number 12," "cs422"), or even hypothetical (e.g., "Santa," "the king of the USA"). Individuals have a notion of *identity*, allowing them, for example, to be distinguished and counted. For some individuals such as "Janet," identity is an intrinsic notion that distinguishes her from all other objects. *Values* are special individuals whose identity is determined by their structure and properties (e.g., numbers, sets, lists and tuples). The number "7," for example, is the unique number that comes after "6," while "{a, b, c}" is the unique set with elements "a," "b" and "c."

Individuals can be grouped into *categories* (also called *concepts*, *types*), where each category (e.g., "Book") captures common properties of all its instances (e.g., "my DB textbook," "Ivanhoe"). Categories themselves are usefully structured into taxonomies according to their generality/specificity. For instance, "Book" is a specialization of "LibraryMaterial," along with "Journal" and "DVD." Moreover, "Book" has its own specializations, such as "Hardback," "Paperback."

Many categories (e.g., "Person") are *primitive*, in the sense that they don't have a definition. By implication, there is no algorithm to determine whether a given individual object is an instance of such a category – one must be told explicitly such facts. Other categories are *defined*, in the sense that instances can be recognized based on some rule. For example, an instance of "Teenager" can be recognized given an instance of "Person" and its age.

Relations relate two or more objects, for example "book45 is on loan to Lynn," or "Trento is between Bolzano and Verona" and can represent among others a dependence of some sort between these objects. Each relation is characterized by a positive integer n-its *arity*-representing the number of objects being related. In the example above, "onLoanTo" is binary, while "between" is ternary. Predicate logic notation is used to express specific relations between individuals, e.g., "between (Trento, Bolazano, Verona)." Like their individual counterparts, relations can be grouped into relation categories. In turn, relation categories can be organized into subcategory hierarchies: "brotherOf" and "sisterOf" are subcategories of "siblingOf," which in turn is a subcategory of "relativeOf." A binary relation category has a *domain* and a *range*. A cognitive model can specify arbitrary constraints between relations and categories, which describe the valid states ("semantics") of the application domain. Cardinality constraints, for instance, specify upper/lower bounds on how many instances of the range of a relation can be associated to an instance of the domain, and vice versa.

A subtle complexity arises when one wants to describe information *about* a relation, for example when did it become or ceased to be true. In such cases the modeler can resort to *reification*. For example, the reification of "Trento is between Bolzano and Verona" consists of creating a new individual, "btw73" which is an instance of the category "Between" and is related to its three arguments via three functional relations: "refersTo (btw73,Trento)," "source(btw73,Bolzano)," "destination(btw73,Verona)." Note that this representation allows another instance of "Between," say "btw22," with functions to the same three individuals. To avoid this redundancy, one needs suitable constraints on "Between." One can now model other information about such reified relations, e.g., "believes(yannis,btw73)."

There are several categories of relations that deserve special consideration.

PartOf (with inverse *hasPart*) represents the part-whole relation that allows composite conceptualizations consisting of simpler parts. PartOf actually represents several distinct relations with different formal properties [5]. Most prominent among them are the relations *componentOf* (e.g., cover is a component of book) and *memberOf* (e.g., player member of a team). PartOf is frequently confused with other relations, such as *containment, connectedness, hasLocation*. A useful diagnostic test for this confusion is to check that if A is part of B, and B is damaged, then A is also considered damaged. Note how "love-note placed inside book" fails this test.

The relation between an object and a category is often called *instanceOf*, and the set of all instances of a category are called its *extension*. The *isA* (*subcategory*) relation represents a taxonomic ordering between categories, e.g., "isA(HardcoverBook,Book)." The isA relation is transitive and anti-symmetric and interacts with instanceOf in an important way: if "isA(A, B)" and "instanceOf(x,A)" then "instanceOf(x,B)." Any general statements one associates with a category, apply to all its specializations. For example, "every book has a title" will automatically apply to subcategories ("every HardcoverBook has a title"); this is called *inheritance* of constraints. In many cases, a group of subcategories are mutually *disjoint* (e.g., "Hardcover," "Paperback," are disjoint subcategories of the category "Book"). Sometimes, a set of subcategories *covers* their common parent category in the sense that every instance of the latter is also an instance of at least one subcategory (for example, "Male," "Female" cover "Person"). When a collection of subcategories partitions a parent category, it is often because some relation (e.g., "gender") takes on a single value from an enumerated set (e.g., {"M," "F"}).

When building an isA hierarchy, it is useful to start by building a backbone consisting of primitive, disjoint concepts that describe the basic categories of individuals in the application domain. Some categories can be distinguished from others by their *rigidity* property: an instance of the category remains an instance throughout its lifetime. For example, "Person" is such a category (in the sense of "once a person, always a person"), but "Student" is not. Once a backbone taxonomy has been constructed, other categories, such as role categories, can be added to the hierarchy as specializations of the categories. Welty and Guarino [11] present a principled approach to the construction of taxonomies. A useful rule for building meaningful isA hierarchies is to ensure that the children of any category are all at the same level of granularity. A leaf category in an isA hierarchy should be further refined if it has instances that can be usefully grouped into subcategories, which participate in new relations or which are subject to new constraints.

Categories may also be seen as objects, and can then be grouped into meta-categories that capture common meta-properties of their instances. For example, the meta-category "LibraryMaterialType" has instances such as "Book," "Hardback" and "DVD," and may have an integer-valued meta-property such as "number-on-order."

In most worlds there are not just enduring objects but also occurrences of events, activities, processes, etc., such as borrowing, renewing or returning a book. These phenomena are called *events*. An event can be described from a number of perspectives: First, there are the *participants* in an event: the material borrowed, the patron who did the borrowing, the library from which the material was borrowed, the date when the material is required to be returned, etc. Often these participants are given names describing the *role* they play in the event: "borrower," "lender," "due date." Second, every event takes place in time, so its temporal aspects can be represented: starting and possibly ending time if it is of a long duration, cyclic nature, etc. Third, events may also have parts – the sub-events that need to take place (e.g., taking the book to the counter, proffering the library card, . . .). There may also be special relations, such as causality or temporal precedence that hold among events.

In database modeling, one often ignores events because they are transient, while the database is supposed to capture persistence. However, events are in fact present in the background: relations between objects other than "partOf" are usually established or terminated by events. The "onLoan" relation, for example, is created by a "borrow" event and terminated by a "return" event. And values for many qualities (e.g., the size or weight of an object) are established through events representing acts of observation. As a result, rela-

tions often carry information about the events. Thus information about the "borrow" activity's participants is present in the arguments of the "onLoan" relation. And since "renew" shares the main participants of "borrow," its traces can also be attached to "onLoan," through an additional temporal argument, such as "lastRenewedOn."

Note that it is application requirements that determine the level of detail to be maintained in a cognitive model. For example, whether or not one records the time when the borrowing and renewal occurred, or whether it is sufficient to have a "dueDate" attribute on the "onLoan" relation. In addition, the details of the subparts of "borrow" will very likely be suppressed. On the other hand, semantic relations between events, such as the fact that a book renewal can only occur after the book has been borrowed, do need to be captured.

Because databases often have multiple sets of users, there may be several conflicting interpretations of terms and information needs. The important process of reconciling such conflicts, known in part as "view integration" is not addressed in this entry.

From a cognitive model to a conceptual schema. The above account has focused on the construction of a model that captures information requirements in terms of cognitive primitives, with constraints expressed in a possibly very rich language. Every effort was made to keep the modeling and methodology independent of a particular modeling language. Next, one must tackle the problem of producing a conceptual schema expressed in some particular formal notation, in this case the EER. Comparable discussions apply if the target was UML class diagrams, or even OWL ontologies.

The basic mapping is quite straightforward: *categories of individuals* in the conceptual model are mapped to ER *entity sets*; relation categories in the conceptual model are modeled directly as relationship sets, with the participating entities playing (potentially named) roles in the relationship set. Qualities and values related to individuals by binary relations, or appearing as arguments of relations become *attributes*. Cardinality constraints of the cognitive model are mapped directly to the conceptual schema.

One complex aspect of EER schema development is the definition of *keys* consisting of one or more attributes that uniquely distinguish each individual instance of an entity set. Moreover, the values of these attributes must be stable/unchanging over time. For example, "isbnNr" or "callNumber" would be a natural key attributes for "Book." Globally unique identifiers are actually relatively rare. Instead, entities are often identified with the help of intermediate relationships (and attributes). For example, "BookCopy," has attribute "copyNr," which surely does not identify a particular book copy; but if "BookCopy" is represented as a *weak entity*, related to "Book" via the "copyOf" identifying relationship then "BookCopy" will have a composite identifier. In other situations where one would need a large set of attributes to identify an entity, and especially if these attributes are not under the control of database administrators, the designer may introduce a *new* surrogate entity attribute specifically for identification purposes (e.g., "studentId" for the entity set "Student").

Sub-categories, with possible disjoint and coverage constraints, are represented in a direct manner in EER since it supports these notions. Significantly, in most version of EER key attribute(s) can only be specified for the top-most entity in a hierarchy, and must then be inherited by all subentities. Therefore one cannot have "Employee" with key "ssn," while sub-class "TeachingAssistant" has key "studentId." The reason for this is to avoid multiple ways of referring to what would otherwise be the same individual.

Since the EER model supports n-ary relationships, reified relationships are normally only required in case the model needs to make "meta" statements about relationships, e.g., recording that a particular loan was verified by a clerk. In variants of the EER that allow aggregate relationships, this is modeled by relating entity "Clerk" via relationship "hasVerified" to the aggregate representing the "lentTo" relationship. In impoverished variants of EER that do not support aggregates, this can be encoded using weak entities that reify the relationship.

The EER notation (in constrast to UML, say), does not provide support for distinguishing *partOf* relationships, nor for relationship hierarchies. However, the designer may encode these using reified relationships.

Conceptual Schema Design in Practice

There is a plethora of commercial tools for conceptual schema design based on the EER model or UML class diagrams. These support the drawing, documenting and layout of conceptual schemas, and even the automatic generation of standard logical (relational) schemas to support database design.

More advanced tools, such as icom (http://www.inf.unibz.it/~franconi/icom/), allow not just the drawing of conceptual schemas, but their translation to formal logic. Advantages of such tools include (i) precise, formal semantics for all the constructs of the conceptual model; (ii) the ability to check the conceptual schema for consistency-an issue which is particularly interesting in the case of *finite models*; (iii) the ability to augment a conceptual schema with constraints expressed in the underlying logic.

Key Applications

Conceptual schema design is the first-and for many the most important-step in the design of any database.

Future Directions

One of the major research challenges of the new century is data integration, and the semantics of data captured in conceptual models has been shown to form a useful foundation for merging multiple information sources. In this context, one can imagine using a conceptual schema as the mediating schema to access different database sources.

The advent of the Web has made databases a public resource that can be shared world-wide. In such a setting, where the user may know nothing about a database she is accessing, the issues that dominate are (i) encapsulating the semantics of data along with the data, so that the user can interpret the data; (ii) ensuring data quality, so that the user can determine whether the data are suitable for her purposes; (iii) ensuring compliance with privacy, security and governance regulations. In turn, these new requirements are going to redefine the scope of database design in general and conceptual schema design in particular.

Specifically, extended conceptual modeling languages and conceptual schema design techniques are envisioned where quality, privacy, security and governance policies can be expressed explicitly and can be accommodated during the design process to produce conceptual schemas that are well-suited to address such concerns. We also envision extensions to conceptual modeling languages that introduce primitive concepts for modeling dynamic, intentional and social aspects of an application domain. These aspects constitute an important component of the semantics of any database and a prerequisite for dealing with data quality and regulation compliance.

Cross-References

▶ Description Logics
▶ Semantic Data Model

Recommended Reading

1. Abrial J-R. Data semantics. In: Koffeman K, editor. Data management systems. Amsterdam: North-Holland; 1974.
2. Atzeni P, Ceri S, Paraboschi S, Torlone R. Database systems: concepts, languages & architectures. New York: McGraw Hill; 1999.
3. Batini C, Ceri S, Navathe S. Conceptual database design. Menlo Park: Benjamin/Cummings Publishing Company; 1991.
4. Chen PP. The entity-relationship model – toward a unified view of data. ACM Trans Database Syst. 1976;1(1):9–36.
5. Gangemi A, Guarino A, Masolo C, Oltramari A, Schneider L. Sweetening ontologies with DOLCE. In: Proceedings of the 12th International Conference on Knowledge Engineering and Knowledge Management: Ontologies and the Semantic Web; 2002. p. 166–81.
6. Guizzardi G, Herre H, Wagner G. Towards ontological foundations for UML conceptual models. In: Proceedings of the Confederated International Conference on DOA, CoopIS and ODBASE; 2002. p. 1100–7.
7. Rumbaugh J, Blaha M, Premerlani W, Eddy F, Lorensen W. Object-oriented modeling and design. Englewood Cliffs: Prentice Hall; 1991.
8. Sakai H. Entity-relationship approach to the conceptual schema design. In: Proceedings of the ACM SIGMOD International Conference on Management of Data; 1980. p. 1–8.
9. Teorey T, Yang D, Fry J. A logical design methodology for relational databases using the extended entity-relationship model. ACM Comput Surv. 1986;18(2):197–222.
10. Wand Y, Storey VC, Weber R. An ontological analysis of the relationship construct in conceptual modeling. ACM Trans Database Syst. 1999;24(4):494–528.
11. Welty C, Guarino N. Supporting ontological analysis of taxonomic relationships. Data Knowl Eng. 2001;39(1):51–74.

Concurrency Control for Replicated Databases

Bettina Kemme
School of Computer Science, McGill University, Montreal, QC, Canada

Synonyms

Isolation in Replicated Databases

Definition

Data replication is a core technology to achieve fault tolerance, high availability, and increased performance. Each "logical" data item has one or more physical data copies, also called replicas, that are distributed across the database servers in the system. *Replica control* is the task of translating the read and write operations on logical data items into operations on the physical data copies. When data is accessed with transactional context, replica control has to be combined with concurrency control in order to provide global isolation of concurrent transactions across the entire system. Just as centralized database systems, replicated database systems offer several levels of isolation and replica consistency. One-copy serializability was developed as a first – and very strong – correctness criterion requiring that the concurrent execution of transactions in a replicated system is equivalent to a serial execution of these transactions over a single logical copy of the database. A straightforward extension of strict two-phase locking can achieve one-copy serializability. Many other, weaker correctness criteria have been defined since then, and there exists now a wide range of concurrency control mechanisms that can be categorized as pessimistic or optimistic; many of them rely on each replica having multiple versions.

Historical Background

Replication became a hot topic in the early 1980s. In their book *Concurrency Control and Recovery in Database Systems* [3], Bernstein et al. presented a thorough formalism to reason about the correctness of transaction execution and concurrency control mechanisms in central, distributed, and replicated systems. Their definitions of serializability and one-copy serializability (1SR) are still used to reason about the correctness of transactional systems. Several baseline concurrency control strategies used in non-replicated or distributed databases have been extended and combined with replica control in order to provide one-copy serializability [3, 4]. Furthermore, the correctness of execution despite site or communication failures has been analyzed thoroughly [1, 2]. Work done in this early phase is very visible in textbooks on database systems and distributed systems [7] and builds part of the foundations of academic education in this area. In 1996, Gray et al. [5] indicated that these traditional approaches provide poor performance and do not scale as they commit transactions only if they have executed all their operations on all (available) physical data copies. Since then, several new correctness criteria, such as one-

copy snapshot isolation [10], as well as advanced replica control schemes, have been developed [6, 11, 12]. This more recent work became particularly relevant in the context of cloud environments where data is replicated across cloud locations for better local access and availability and within a cloud for better scalability and fault tolerance.

Scientific Fundamentals

Transactions in a Non-replicated System

The formalism that describes transactions and their execution in a replicated database is derived from the transaction model in a non-replicated system. In a non-replicated system, a database consists of a set of data items $x, y, \ldots$. A transaction T_i is a sequence of read operations $r_i(x)$ and write operations $w_i(x)$ on data items that build a logical unit. In real database systems, clients submit SQL statements that can access and manipulate many records of different tables. However, the abstraction into simple read and write operations on data items allows for a clear and powerful reasoning framework.

The database system should provide the transactional properties atomicity, consistency, isolation, and durability (see the entry ▶ ACID properties). In this entry, we are interested in the isolation property. Isolation requires that even if transactions execute concurrently in the system, each transaction should have the impression it executes isolated on the data. In particular, when two operations conflict, i.e., they are from different transactions and want to access the same data item and at least one is a write, the execution order matters. Given a set of transactions, a history describes the order in which the database server executes the operations of these transactions.

The traditional correctness criterion in a non-replicated system is *serializability* (see entry ▶ Serializability). It requires a history to be equivalent to a serial history where the same transactions are executed serially one after the other. Equivalence typically refers to executing all conflicting operations in the same order.

A slightly weaker correctness criterion is *snapshot isolation*. In here, transactions read a snapshot of the data that was committed at the time the transaction started, i.e., transactions see all updates from transactions that committed before their start but no updates from concurrent transactions. This is called the *snapshot read property*. Furthermore, if there are two concurrent transactions (neither started after the other committed) and they both want to update the same data item, one of them must abort. This is the *snapshot write property*.

Concurrency control is in charge of guaranteeing the desired level of isolation. Concurrency control can be pessimistic, typically by requiring that a transaction acquires locks before accessing data items, or optimistic, in which case a validation at the end of transaction checks whether any conflicts occurred during execution requiring to abort the transaction. As shown in Fig. 1a, the concurrency control component works together with the transaction manager that keeps track of active transaction and the data manager that performs the individual read and write operations submitted by the clients.

Transaction Execution in a Replicated System

In a replicated database, there is a set of database servers $A, B, \ldots$, also referred to as sites, and each logical data item x has a set of physical copies x^A, x^B, ... where the index refers to the server on which the copy resides. In full replication, each data item has a copy on each server, while using partial replication, it has only copies on a subset of the servers.

Execution Model and Architecture

As replication should be transparent to clients, they continue to submit operations on the logical data items. Replica control has to map an operation $o_i(x)$, $o_i \in \{r, w\}$, of transaction T_i into operations on the physical copies of x, e.g., $o_i(x^A), o_i(x^B), \ldots$. Given the mapping for a set of transactions, when executing these transactions in the replicated system, each database server A produces a local history

showing the execution order of all the operations performed on the copies maintained by A.

The most common execution model for transactions in a replicated environment is to perform a read operation on one data copy while write operations update all copies. This is referred to as ROWA (or read one write all). As most database applications typically have more read than write operations, it makes sense to provide fast read access and only penalize write operations.

A problem of ROWA is that if one copy is not accessible, write operations cannot be performed anymore on the data item. In order to be able to continue even if failures occur, ROWAA (read one write all available) needs to be used. It does not require to perform updates on copies that are currently not available. In particular, many current approaches, based on Paxos [9], consider a write operations successful as soon as a majority of replicas has confirmed that the write operation can be executed. This entry does not further look at failures but focuses on providing isolation in a ROWA system. ROWAA and quorums are discussed in the entries *quorum systems* and *replication for availability and fault tolerance*. There is also a special entry on the *two-phase commit protocol*.

While a logical operation might now imply many physical operations, the overall execution still needs to isolate concurrent transactions so that users, ideally, have the impression their transactions are completely isolated from others. Thus, replica control has to work together with a distributed concurrency control component. There are many ways to distribute concurrency and replica control tasks across the system.

One possibility is a centralized architecture (Fig. 1b). All transactions are submitted to a global transaction middleware that keeps track of active transactions and performs concurrency and replica control. The individual data stores might only have a data manager interface (left component) or have their own local transaction and concurrency control management (right). The latter might be the case if non-replicated off-the-shelf database systems are used in a black box format on top of which a replicated architecture is built. Clearly, the tasks of the

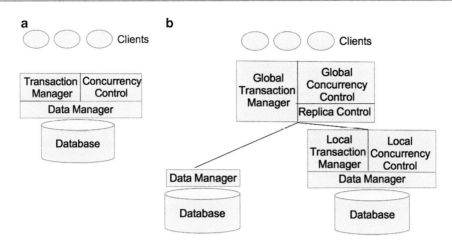

Concurrency Control for Replicated Databases, Fig. 1 Concurrency and replica control architecture. (**a**) Non-replicated. (**b**) Centralized

global component will be different depending on whether the local data stores have their own transaction management or not.

Global transaction management can also be distributed. Again, there might be only one integrated transaction management layer (a global one) that works in tight collaboration with the data store to provide a holistic solution (Fig. 2a). Or, the distributed global transaction middleware is on top of unchanged off-the-shelf database systems so that the global transaction management components have to work together with their local counterparts (Fig. 2b).

Concurrency Control for One-Copy Serializability

One-copy serializability was the first correctness criterion for replicated histories. The execution of a set of transactions in a replicated environment is one-copy serializable if it is equivalent to a serial execution over a single, non-replicated (logical) database. Defining equivalence is not straightforward since the replicated system executes on physical copies while the non-replicated on the logical data items. The handling of failures further complicates the issue. But assuming a ROWA system, it means that whenever any local history at one site, e.g., A, executes operation $o_i(x^A)$ before conflicting operation $o_j(x^B)$ (at least one is a write), then the equivalent serial execution over the logical database executes $o_i(x)$

before $o_j(y)$. A formal definition is given in the definitional entry *one-copy serializibility*.

Pessimistic Concurrency Control: Strict Two-Phase Locking

Strict two-phase locking (S2PL) is probably the best known concurrency control mechanism to achieve serializability (see also entry *2-Phase-Locking*). In this case, the concurrency control module implements a lock manager.

Principles

Using S2PL, a transaction has to acquire a shared lock on data item x before performing a read on x and an exclusive lock on x before writing x. An exclusive lock on x conflicts with other shared and exclusive locks on x. If a transaction requests a lock and another transaction holds a conflicting lock, the requesting transaction has to wait. Only when a transaction terminates it releases all its locks, which then can be granted to waiting transactions. S2PL guarantees serializability because the order in which locks are granted for the first pair of conflicting operations between two transactions determines the serialization order. Deadlocks can occur. A deadlock involving two transactions can happen if the transactions have two pairs of conflicting operations and execute them in different order.

Applying S2PL in a replicated system with a centralized architecture (Fig. 1b), clients submit

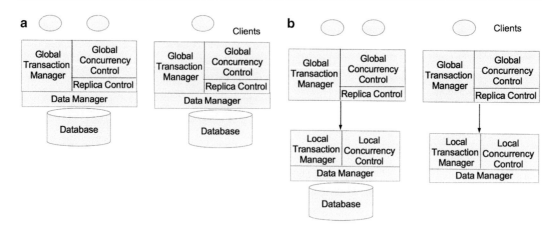

Concurrency Control for Replicated Databases, Fig. 2 Distributed transaction management. (**a**) Integrated. (**b**) Blackbox

their operations to the global transaction manager. The transaction manager gets the appropriate lock via the lock manager, and then the replica control module transfers each read operation to one database site with a copy of the data item and write operations to all sites with copies. In the distributed architecture (Fig. 2a), a client connects to any site. The execution of individual operations is illustrated in Fig. 3a. The figure shows the message exchange between a client and two sites. When the client submits a read operation on logical data item x to the local site A, a local shared lock is acquired on the local physical copy $(sl(x^A))$ and the read operation executes locally. If it is a write operation, an exclusive lock $(xl(x^A))$ is acquired locally and the operation executes locally. At the same time, the operation is forwarded to site B. B, upon receiving the request, acquires a lock on the local copy, performs the operation, and sends a confirmation back to A. When A has received all confirmations, it sends the confirmation back to the client.

Architectural Comparison

Comparing how well S2PL maps to the two architectures reflects well the principle trade-offs between a centralized and a decentralized architecture.

In Favor of a Centralized Architecture

In principle, a central component makes the design of coordination algorithms often simpler. It directs the flow of execution and has the *global knowledge* of where copies are located. One central concurrency control module serializes all operations. In the distributed architecture, the flow of execution is more complex as no single component has the full view of what is happening in the system.

While the centralized architecture acquires one lock per each operation on a logical data item, the distributed architecture acquires locks per data copies. Thus, a write operation involves many exclusive locks, adding to the complexity. The distributed architecture has the additional disadvantage of potential *distributed deadlocks*: there is no deadlock at any site locally, but globally, a deadlock has occurred. Figure 3b depicts an example execution where a distributed deadlock occurs although the transactions both only access a single data item (something not even possible with the centralized architecture). T_1 first acquires the lock on site A which forwards the request to site B. Concurrently, T_2 acquires the lock first on B and then requests it on A. At A, T_2 has to wait for T_1 to release the lock, on B, T_1 waits for T_2. The deadlock is distributed since no single site observes a deadlock. Such deadlocks need to be resolved

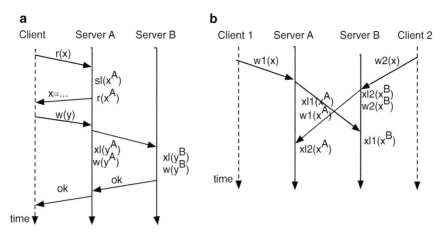

Concurrency Control for Replicated Databases, Fig. 3 Distributed transaction execution. (**a**) Execution example. (**b**) Distributed deadlock

via timeout or a deadlock detection mechanism, which in turn, could be implemented centrally or distributed.

In Favor of a Decentralized Architecture

A central middleware is a potential bottleneck and a single point of failure. In contrast, in the distributed architecture, if ROWAA is used, the system can continue executing despite the failure of individual sites. Furthermore, the middleware approach has an extra level of indirection. In the above example algorithm, this leads to four messages per read operation (a pair of messages between clients and middleware and a pair between middleware and one database site). In contrast, the distributed architecture has two messages (between the client and one database site).

Middleware vs. Tight Integration

If locks are set at the middleware layer (be it centralized or distributed), there is no access to the data manager modules of the database servers. Now assume clients submit SQL statements. The middleware cannot know what records will actually be accessed by simply looking at the SQL statement. Such information is only available during the execution. Thus, a middleware-based lock manager might need to set locks on entire tables instead of individual records. In contrast, when concurrency control is integrated as in Fig. 2a, the local site can execute an SQL statement

first locally, lock only the records that are updated, and then forward the update requests on the specific records to the other sites, allowing for a finer-grained concurrency control. Generally, a tighter coupling often allows for better optimization.

Optimistic Concurrency Control

Optimistic concurrency control (OCC) assumes conflicts are rare. Thus, it executes a transaction first and only detects conflicts at commit time. In the most well-known mechanism [8] developed for a non-replicated system, a write operation on x generates a local copy of x. A read on x either reads the local copy (if the transaction has previously written x) or the last committed version of x. At commit time of a transaction T_i, validation is performed. If validation succeeds, a write phase turns T_i's local copies into committed versions and T_i commits. Otherwise, T_i aborts. In the simplest form of OCC, validation and write phases are executed in a critical section. The validation order determines the serialization order. Therefore, validation of T_i fails if there is a committed transaction T_j that is concurrent to T_i (committed after T_i started), and T_j's writeset (data items written by T_j) overlaps with T_i's readset (data items read by T_i). As T_i validates after T_j it should be serialized after T_j, and thus, read what T_j has written. In the concurrent execution, however, it might have read an earlier version. If this is the case, it needs to be aborted.

An OCC for a fully replicated database is proposed in [11]. A transaction can first execute completely locally at one database server. At commit time, the writeset (all updated data items) and the readset (the identifiers of all read items) of the transaction are sent to all sites via a special multicast that delivers messages at all sites in the same order. That is, if there is a set of concurrent transactions, each executing first locally at one site and then multicasting the commit message, all sites will receive these commit requests in the same order. Each site now performs the validation and write phase in a critical section in the order the commit requests are received. Thus, they will all decide on the same outcome for each transaction. In order to determine concurrent transactions, each transaction can receive a monotonically increasing commit timestamp. When a transaction starts, it receives as start timestamp the current commit timestamp. A transaction is concurrent to all transactions whose commit timestamp is larger than its own start timestamp.

The advantage of this approach compared to locking is that only a single commit message needs to be exchanged among replicas. The correct execution order is guaranteed through the total order multicast. However, sending readsets in order to perform validation can be difficult, in particular in SQL systems where a single read operation could access many different records. Snapshot isolation, described below, makes handling of reads much easier and, thus, is extremely popular in replicated environments.

Concurrency Control for One-Copy Snapshot Isolation

One-copy snapshot isolation [10] extends the idea of the snapshot isolation level for a ROWA system where a transaction executes all its read operation at one of the sites, called its local site. The execution of a set of transactions in a replicated environment respects one-copy snapshot isolation if (i) the execution at each site follows snapshot isolation semantics and (ii) there is an execution over a non-replicated (logical) database

that respects snapshot isolation semantics such that whenever any local execution at one site, e.g., A, executes operation $o_i(x^A)$ before conflicting operation $o_j(x^B)$ (at least one is a write), then the equivalent serial execution over the logical database executes $o_i(x)$ before $o_j(y)$. In this definition, the snapshot of a transaction T_i is defined as the set of updates that were committed at T_i's local site at the time T_i started at this site.

A replicated snapshot isolation protocol similar to the OCC described above can be found in [6]. A transaction executes first locally at one database server reading from a local snapshot as of its start time. Writes are performed on local copies. At commit time, only the write set is sent via a total order multicast to all sites. All sites perform validation and write phase in a critical section. Validation of transaction T_i fails if there is a committed transaction T_j that is concurrent to T_i and the writesets of both transactions overlap.

Other Isolation Levels

Indirect Dependencies
A particular problem in a replicated environment is that two non-conflicting transactions might be indirectly ordered through read operations. Assume for instance transaction T_1 updating x and T_2 updating y. These two transactions do not conflict, and so it appears that it does not matter if site A executes T_1 before T_2, and site B executes T_2 before T_1. But now assume there is a read-only transaction T_3 at site A that reads both x^A and y^A after T_1's update but before T_2's update. Similarly, another read-only transaction T_4 performs the same reads between the execution of T_2 and T_1 at B. In this case, there is no serial or snapshot isolation execution over the logical database that can order all conflicting operations the same way. In order to obey the execution order at A, it would have to execute first T_1 then T_3 and then T_2. But to follow the execution order at B, it would have to execute first T_2, then T_4 and then T_1.

The above OCC and snapshot isolation protocols avoid this problem because they execute the updates of all transactions in the same order at all replicas, whether they conflict or not. Thus, snapshots advance at all sites in the same way. However, a total order multicast among all nodes is expensive and especially unattractive if the data is not fully replicated, but each site has only replicas of some data items.

Therefore, the weaker isolation level of *parallel snapshot isolation* [12] allows such inconsistency in order to not impose the same execution order of non-conflicting update transactions at all replicas. It guarantees the snapshot read property by letting a transaction T_i see the snapshot of committed updates at T_i's local site at the time T_i started at this site, and it guarantees the snapshot write property by avoiding that any two transactions that are concurrent at least one site update the same data item. Furthermore, it provides a sort of causality by guaranteeing that if a transaction T_i started at one site after T_j committed, T_i does not commit before T_j at any site. As concurrent transactions that do not have conflicting updates are allowed to commit in different orders at different sites, read-only transactions at different sites might see different snapshots: one might include the updates of an update transaction T_1 but not the updates of T_2 while a snapshot at another site might contain T_2's but not T_1's updates.

The Walter system [12] implements parallel snapshot isolation. It maintains a vector clock to maintain causal dependencies and snapshots. In order to detect write conflicts, a distributed validation phase similar to a two-phase commit is executed at the end of the transaction. Each data item in the system has a dedicated primary replica. In the validation phase, the transaction coordinator requests all primary replicas to check for conflicts with previously committed transaction. If there is a conflict on a primary replica, the replica votes *no*, otherwise yes. If there is one or more *no* votes, the transaction has to abort.

Session Consistency

While the OCC and snapshot isolation algorithms presented above execute updates of all transactions in the same order at all replicas, there is some asynchrony. For instance, a client could submit a transaction T_1 that updates x to site A. A executes T_1 sends and receives the writeset, validates T_1 and returns the commit to the client. The client now submits T_2 to site B to read x. B, however, might not yet have committed T_1. Thus, T_2 will read an older version of x. In a non-replicated system, this is not possible, but in a replicated system, it is possible to get an older snapshot. For some applications, this might not be a problem. However, others might require what is called *session consistency*: additionally to the proper isolation level, a client should also be able to read the writes the client performed in previous transactions. Session consistency can be achieved by not returning a commit to the user until it is confirmed that the updates have been committed at all replicas. Alternatively, a client transaction can piggyback the commit timestamp of the last update transaction committed by the client. Then, at any replica, a read operation of the new transaction will be delayed until the updates of this latest update transaction of the client are reflected on the data copies. While the first strategy delays the updates of all update transactions, the second only potentially delays follow-up read operations of the same client.

Summary

In summary, most concurrency control mechanism need some form of communication among replicas in order to guarantee isolation. This can happen at the beginning, during, or at the end of the transaction. Order can be enforced through a total order multicast, or in the form of some voting phase similar in concept to two-phase commit. Locking mechanisms are less desirable in distributed and replicated environment because the communication increases lock duration and thus, contention. Instead, most concurrency control mechanism rely on a multi-version system in order to more flexibly serve read requests, and are optimistic, that is, they contain some validation, typically at the end of transaction, to determine conflicts. The concurrency control mechanism has to be adjusted to the architecture in place, be it centralized or distributed, tightly integrated with the underlying data store or built on top as a middleware layer.

Experimental Results

Gray et al. [5] show that traditional approaches do not scale well. They analyzed the distributed locking approach and determined that the potential of deadlock increases quickly with the number of replicas in the system, the message overhead becomes too high, and transaction response times become too long. The goal of more recent research into replica and concurrency control has aimed at reducing the overhead by either providing lower levels of correctness or by developing more efficient ways to control the flow of execution in the system.

Cross-References

▶ ACID Properties
▶ Concurrency Control: Traditional Approaches
▶ Distributed Concurrency Control
▶ One-Copy-Serializability
▶ Replica Control
▶ Replicated Database Concurrency Control
▶ Replication Based on Group Communication
▶ Replication for Availability and Fault Tolerance
▶ Replication for Scalability
▶ Transaction Models: The Read/Write Approach
▶ Two-Phase Locking

Recommended Reading

1. Abbadi AEl, Toueg S. Availability in partitioned replicated databases. In: Proceedings of the 5th ACM SIGACT-SIGMOD Symposium on Principles of Database Systems; 1986. p. 240–51.
2. Bernstein PA, Goodman N. An algorithm for concurrency control and recovery in replicated distributed databases. ACM Trans Database Syst. 1984;9(4):596–615.
3. Bernstein PA, Hadzilacos V, Goodman N. Concurrency control and recovery in database systems. Reading: Addison Wesley; 1987.
4. Carey MJ, Livny M. Conflict detection tradeoffs for replicated data. ACM Trans Database Syst. 1991;16(4):703–46.
5. Gray J, Helland P, O'Neil P, Shasha D. The dangers of replication and a solution. In: Proceedings of the ACM SIGMOD International Conference on Management of Data; 1996. p. 173–82.
6. Kemme B, Alonso G. A suite of database replication protocols based on group communication primitives. In: Proceedings of the 18th IEEE International Conference on Distributed Computing Systems; 1998. p. 156–63.
7. Kindberg T, Coulouris GF, Dollimore J. Distributed systems: concepts and design. 4th ed. Harlow/New York: Addison Wesley; 2005.
8. Kung HT, Robinson JT. On optimistic methods for concurrency control. ACM Trans Database Syst. 1981;6(2):213–26.
9. Lamport L. The part-time parliament. ACM Trans Comput Syst. 1998;16(2):133–69.
10. Lin Y, Kemme B, Jiménez-Peris R, Patiño-Martínez M, Armendáriz-Íñigo JE. Snapshot isolation and integrity constraints in replicated databases. ACM Trans Database Syst. 2009;34(2):Article 11.
11. Pedone F, Guerraoui R, Schiper A. The database state machine approach. Distrib Parallel Databases. 2003;14(1):71–98.
12. Sovran Y, Power R, Aguilera MK, Li J. Transactional storage for geo-replicated systems. In: Proceedings of the 23rd ACM Symposium on Operating System Principles; 2011. p. 385–400.

Concurrency Control Manager

Andreas Reuter
Heidelberg Laureate Forum Foundation,
Schloss-Wolfsbrunnenweg 33, Heidelberg,
Germany

Synonyms

Concurrency control manager; Lock manager; Synchronization component

Definition

The concurrency control manager (CCM) synchronizes the concurrent access of database transactions to shared objects in the database. It is responsible for maintaining the guarantees regarding the effects of concurrent access to the shared database, i.e., it will protect each transaction from anomalies that can result from the fact that other transactions are accessing the same data at the same time. Ideally, it will make sure that the result of transactions running in parallel is identical to the result of some serial execution of the same transactions. In real applications, however, some transactions may opt for lower levels of synchronization, thus trading protection from side effects of other transactions for performance. The CCM is responsible for orchestrating all access requests issued by the transactions such that each transaction receives the level of protection it has asked for. The CCM essentially implements one of a number of different synchronization protocols, each of which ensures the correct execution of parallel transactions, while making different assumptions regarding prevalent access patterns, frequency of conflicts among concurrent transactions, percentage of aborts, etc. The protocol that most CCMs are based on is using locks for protecting database objects against (inconsistent) parallel accesses; for that reason, the CCM is often referred to as the "lock manager." Some CCMs distinguish between multiple versions of a data object (e.g., current version, previous version) in order to increase the level of parallelism [2].

Key Points

The CCM monitors all access requests issued by higher-level components, be it to the primary data (tuples, records), or to access paths, directory data, etc. on behalf of the database transactions. This information (transaction **X** accesses object **O** in order to perform action **A** at time **T**) is employed in different ways, depending on the synchronization protocol used. In case of locking protocols, each access request has to be explicitly granted by the CCM. If no conflict will arise by performing action A on object O, the access request is granted. If, however, a conflict is detected (e.g., A is an update request, and some other transaction Y is already updating object O), the request is not granted, and the CCM will record the fact that X has to wait for the completion of transaction Y. In case of optimistic protocols, the requests are granted right away, but when

X wants to commit, all its accesses are checked for conflicts with accesses performed by other transactions that are either still running or have committed while X was active. In those situations the time T of the access request is relevant. So the basic data structure maintained by the CCM is a table of accesses/access requests. For locking protocols, the CCM will also maintain a list of which transaction waits for the completion of which other transaction(s). That list is tested by the CCM for deadlocks. If a deadlock or, in case of optimistic protocols, a conflict is detected, the CCM decides which transaction will be aborted. It then informs the transaction manager, who will initiate the abort; rollback of the operations is performed by the recovery manager [1].

Recommended Reading

1. Gray J, Reuter A. Transaction processing – concepts and techniques. San Mateo: Morgan Kaufmann; 1993.
2. Weikum G, Vossen G. Transactional information systems: theory, algorithms, and the practice of concurrency control. San Mateo: Morgan Kaufmann; 2001.

Concurrency Control: Traditional Approaches

Gottfried Vossen
Department of Information Systems,
Westfälische Wilhelms-Universität, Münster,
Germany

Synonyms

Concurrency control; Page locking; Scheduling; Transaction execution; Two-phase locking

Definition

The core requirement on a transactional server is to provide the ACID properties for transactions, which requires that the server includes a *concurrency control* component as well as a *recovery component*. Concurrency control essentially guarantees the isolation properties of transactions, by giving each transaction the impression that it operates alone on the underlying database and, more technically, by providing serializable executions. To achieve serializability, a number of algorithms have been proposed. Traditional approaches focus on the read-write model of transactions and devise numerous ways for correctly scheduling read-write transactions. Most practical solutions employ a variant of the two-phase locking protocol.

Key Points

Concurrency control for transactions is done by the *transaction manager* of a database system and within that component by a *scheduler* and has originally been developed for single-server, centralized systems. The scheduler implements an algorithm that takes operations from active transactions and places them in an interleaving or *schedule* that must obey the correctness criterion of serializability. As illustrated in Fig. 1, which indicates the "positioning" of a transaction

manager within the multilayer architecture of a database system, the scheduler receives steps from multiple transactions, one after the other, and tries to attach them to the schedule already output. This is possible if the new schedule is still serializable; otherwise, the scheduler can block or even reject a step (thereby causing the respective transaction to abort) [1, 5, 6]. The algorithm a scheduler follows must be such that serializability can be tested on the fly; moreover, it must be very efficient so that high throughput rates (typically measured in [committed] transactions per minute) can not only be achieved, but even be guaranteed.

Classification of Approaches

Scheduling algorithms for database transactions can be classified as "traditional" if they concentrate on scheduling read-write transactions; nontraditional schedulers take semantic information into account which is not available at the syntactic layer of read and write page operations. Traditional schedulers generally fall into two categories: A scheduler is *optimistic* or aggressive if it mostly lets steps pass and rarely blocks; clearly,

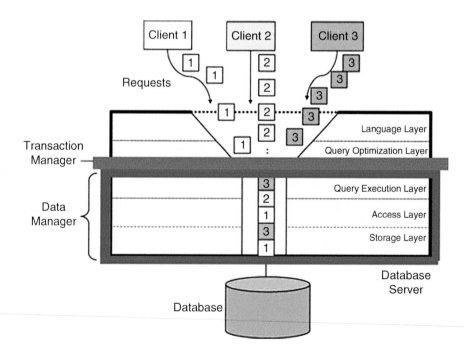

Concurrency Control: Traditional Approaches, Fig. 1 Illustration of a transaction scheduler

this bears the danger of "getting stuck" eventually when the serializability of the output can no longer be guaranteed. An optimistic scheduler is based on the assumption that conflicts between concurrent transactions are rare; it will only test from time to time whether the schedule produced so far is still serializable, and it takes appropriate measures if the schedule is not.

On the other hand, a scheduler is *pessimistic* or conservative if it mostly blocks (upon recognizing conflicts); in the extreme yet unlikely case that all transactions but one have been blocked, the output would become a serial schedule. This type of scheduler is based on the assumption that conflicts between transactions are frequent and therefore need to be constantly observed. Pessimistic schedulers can be *locking* or *non-locking* schedulers, where the idea of the former is to synchronize read or write access to shared data by using locks which can be set on and removed from data items on behalf of transactions. The intuitive meaning is that if a transaction holds a lock on a data object, the object is not available to transactions that execute concurrently. Non-locking schedulers replace locks, for example, by timestamps that are attached to transactions. Among locking schedulers, the most prominent protocol is based on a *two-phase* approach (and hence abbreviated two-phase locking or 2PL) in which, for each transaction, a first phase during which locks are obtained is strictly separated from a second phase where locks can only be released. Non-two-phase schedulers replace the two-phase property, for example, by an order in which transactions may access data objects. Figure 2 summarizes the major classes of concurrency control protocols. Beyond the approaches shown in Fig. 2, the concurrency control problem can even be broken into two subproblems, which could then be solved individually by possibly distinct protocols: (i) read operations are synchronized against write operations or vice versa; (ii) write operations are synchronized against other write operations, but not against reads. If these synchronization tasks are distinguished, a scheduler can be thought of as consisting of two components, one for each of the respective synchronization tasks. Since the

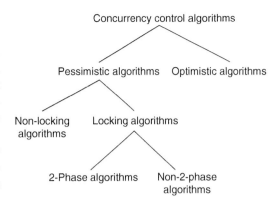

Concurrency Control: Traditional Approaches, Fig. 2
Overview of concurrency control protocol classes

two components need proper integration, such a scheduler is called a hybrid scheduler. From an application point of view, most classes of protocols surveyed above, including the hybrid ones, have not achieved great relevance in practice. Indeed, 2PL is by far the most important concurrency control protocol, since it can be implemented with low overhead, it can be extended to abstraction levels beyond pure page operations, and it has always outperformed any competing approaches [2–4].

Cross-References

▸ B-Tree Locking
▸ Concurrency Control for Replicated Databases
▸ Distributed Concurrency Control
▸ Eventual Consistency
▸ Locking Granularity and Lock Types
▸ Performance Analysis of Transaction Processing Systems
▸ Serializability
▸ Snapshot Isolation
▸ Tuning Concurrency Control
▸ Two-Phase Locking

Recommended Reading

1. Bernstein PA, Hadzilacos V, Goodman N. Concurrency control and recovery in database systems. Reading: Addison-Wesley; 1987.

2. Bernstein PA, Newcomer E. Principles of transaction processing for the systems professional. San Francisco: Morgan Kaufmann; 1997.
3. Claybrook B. OLTP – online transaction processing systems. New York: Wiley; 1992.
4. Gray J, Reuter A. Transaction processing: concepts and techniques. San Francisco: Morgan Kaufmann; 1993.
5. Papadimitriou CH. The theory of database concurrency control. Rockville: Computer Science; 1986.
6. Weikum G, Vossen G. Transactional information systems – theory, algorithms, and the practice of concurrency control and recovery. San Francisco: Morgan Kaufmann; 2002.

attribute in R maps to a variable or a constant, and φ maps to β. In a multirelational database schema, there are multi-tables, meaning in effect that variables can be shared between tables (just as constants are). An example of a 2-multitable is shown below. The conditions φ are all *true*, and it so happens that the two tables do not share any variables.

T_1	A	B	C	φ
	a	b	x	*true*
	e	f	g	*true*
T_2	B	C	D	φ
	y	c	d	*true*

Conditional Tables

Gösta Grahne
Concordia University, Montreal, QC, Canada

Synonyms

C-tables; Extended relations

Definition

A conditional table [4] generalizes relations in two ways. First, in the entries in the columns, variables, representing unknown values, are allowed in addition to the usual constants. The second generalization is that each tuple is associated with a condition, which is a Boolean combination of atoms of the form $x = y$, $x = a$, $a = b$, for x, y null values (variables), and a, b constants. A conditional table essentially represents an existentially quantified function free first order theory.

Formally, let *con* be a countably infinite set of constants, and *var* be a countably infinite set of variables, disjoint from *con*. Let U be a finite set of attributes, and $R \subseteq U$ a relational schema. A tuple in a c-table over R is a mapping from R, and a special attribute, denoted φ, to *con* $\cup$ *var* $\cup$ β, where β is the set of all Boolean combinations of equality atoms, as above. Every

Key Points

It is now possible to extend the complete set of regular relational operators $\{\pi, \sigma, \bowtie, \cup, -, \rho\}$ to work on c-tables. To distinguish the operators that apply to tables from the regular ones, the extended operators are accented by a dot. For instance, c-table join is denoted $\dot\bowtie$. The extended operators work as follows: projection $\dot\pi$ is the same as relational projection, except that the condition column φ can never be projected out. Selection $\dot\sigma_{A=a}(T)$ retains all tuples t in table T, and conjugates the condition $t(A) = a$ to $t(\varphi)$. A join $T \dot\bowtie T'$ is obtained by composing each tuple $t \in T$ with each tuple $t' \in T'$. The new tuple $t \cdot t'$ has condition $t(\varphi) \wedge t'(\varphi) \wedge \delta(t,t')$, where condition $\delta(t,t')$ states that the two tuples agree on the values of the join attributes. The example below serves as an illustration of this definition. The union $\dot\cup$ is the same as relational union, and so is renaming $\dot\rho$, except that the φ-column cannot be renamed. Finally, the set difference, say $T \dot- T'$ is obtained by retaining all tuples $t \in T$ and conjugating to them the condition stating that the tuple t differs from each tuple t' in T'. A tuple t differs from a tuple t', if it differs from t' in at least one column.

Let T_1 and T_2 be as in the figure above. The three c-tables in the figure below, illustrate the result of evaluating $\dot\sigma_{C=g}(T_1)$, $T_1 \dot\bowtie T_2$, and $\dot\pi_{BC}(T_2) \dot- \dot\pi_{BC}(T_1)$, respectively.

A	B	C	φ
a	b	x	x = g
e	f	g	g = g

A	B	C	D	φ
a	y	x	d	b = y ∧ x = c
e	y	g	d	f = y ∧ g = c

Note that the second tuple in the first c-table has a tautological condition. Likewise, since any two constants differ, the condition of the second tuple in the middle c-table is contradictory.

B	C	φ
y	c	$(y \neq b \vee c \neq x) \wedge (y \neq f \vee c \neq g)$

So far, nothing has been said about what the c-tables mean. In the *possible worlds* interpretation, an incomplete database is a usually infinite *set* of ordinary databases, one of which corresponds to the actual (unknown) database. Considering c-tables, they serve as finite representations of sets of possible databases. One (arbitrary) such database is obtained by instantiating the variables in the c-table to constants. Each occurrence of a particular variable is instantiated to the same constant. Formally, the instantiation is a valuation v: $con \cup var \rightarrow con$, that is identity on the constants. Valuations are extended to tuples and conditions tables in the obvious way, with the caveat that given a particular valuation v, only those tuples t for which $v(t(\varphi)) \equiv true$, are retained in $v(T)$. Consider for instance the first table above. For those valuations v, for which $v(x) = g$, there will be two tuples in $v(T)$, namely (a,b,g) and (e,f,g). For valuations v', for which $v(x) \neq g$, there will only be the tuple (e,f,g) in $v'(T)$.

The remarkable property of c-tables is that for all c-tables T and relational expressions E, it holds that $v(\dot{E}(T)) = E(v(T))$ for all valuations v. In other words, the extended algebra is a Codd sound and complete inference mechanism for c-tables. Furthermore, c-tables are closed under relational algebra, meaning that the result of ap-

plying any relational expression on any (schema-wise appropriate) c-table can be represented as another c-table. The extended algebra actually computes this representation, as was seen in the example above.

Needless to say, all of this comes with a price. Testing whether a c-table is satisfiable, that is, whether there exists at least one valuation v, such that $v(T) \neq \varnothing$ is an NP-complete problem [2]. Furthermore, even if one starts with a simple c-table where all variables are distinct, and all conditions are *true*, applying even a monotone relational expression to such a c-table can result in quite a complex table, so here again [2] satisfiability of the resulting table is NP-complete [2]. To make matters even worse, testing containment of c-tables is Π_2^p-complete. A c-table T is contained in a c-table T', if every for every valuation v of T, there exists a valuation v' of T', such that $v(T) = v(T')$. Nonetheless, c-tables possess a natural robustness. For instance, it has been shown [1, 3] that the set of of possible databases defined by a set of materialized views, can be represented as a c-table.

Cross-References

► Certain (and Possible) Answers
► Incomplete Information
► Naive Tables

Recommended Reading

1. Abiteboul S, Duschka OM. Complexity of answering queries using materialized views. In: Proceedings of the 17th ACM SIGACT-SIGMOD-SIGART Symposium on Principles of Database Systems; 1998. p. 254–63.
2. Abiteboul S, Kanellakis PC, Grahne G. On the representation and querying of sets of possible worlds. Theor Comput Sci. 1991;78(1):158–87.
3. Grahne G, Mendelzon AO. Tableau techniques for querying information sources through global schemas. In: Proceedings of the 7th International Conference on Database Theory; 1999. p. 332–47.
4. Imielinski T, Lipski Jr W. Incomplete information in relational databases. J ACM. 1984;31(4):761–91.

Conjunctive Query

Val Tannen
Department of Computer and Information
Science, University of Pennsylvania,
Philadelphia, PA, USA

Synonyms

Horn clause query; SPC query

Definition

Conjunctive queries are first-order queries that both practically expressive and algorithmically relatively tractable. They were studied first in [2] and they have played an important role in database systems since then.

As a subset of the relational calculus, conjunctive queries are defined by formulae that make only use of atoms, conjunction, and existential quantification. As such they are closely related to Horn clauses and hence to logic programming. A single Datalog rule can be seen as a conjunctive query [1].

Optimization and reformulation for various purposes is quite feasible for conjunctive queries, as opposed to general relational calculus/algebra queries. The equivalence (and indeed the *containment*) of conjunctive queries is decidable, albeit NP-complete [1].

Key Points

This entry uses terminology defined in the entry Relational Calculus.

Conjunctive queries are first-order queries of a particular form: $\{\langle e_1, \dots, e_n \rangle \mid \exists x_1 \dots x_m \psi\}$ where ψ is an (equality-free) conjunction of relational atoms, i.e., atoms of the form $R(d_1, \dots, d_k)$ (where $d_1, \dots, d_k$ are variables of constants). In addition, it is required that any variable among $e_1, \dots, e_n$ must also occur in one of the relational atoms of ψ. This last condition, called *range restriction*

[3] is necessary for *domain independence*, e.g., consider $\{\langle x, y\rangle \mid R(x)\}$, and, in fact, it is also sufficient.

The semantics of a conjunctive query in an instance $\mathcal{I}$, as a particular case of first-order queries, involves assignments μ defined on the variables among $e_1, \dots, e_n$ such that $\mathcal{I}, \mu \models \exists x_1 \dots x_m \psi$. Note however that this is the same as extending μ to a *valuation* ν defined in addition on $x_1 \dots x_m$ and such that $\mathcal{I}, \nu \models \psi$. Since ψ is a conjunction of relational atoms, this amounts to ν being a *homomorphism* from ψ seen as a relational instance (the *canonical instance* associated to the query) into $\mathcal{I}$. This simple observation has many useful consequences, including some that lead to the decidability of equivalence (containment). It also leads to an alternative way of looking at conjunctive queries, related to logic programming. Here is an example of a conjunctive query in both relational calculus form and in a Prolog-like, or "rule-based," formalism, also known as a *Datalog rule* [1, 3]:

$$\{\langle x, c, x\rangle \mid \exists y \ R(c, y) \wedge S(c, x, y)\}$$

$$ans(x, c, x) : -R(c, y), S(c, x, y)$$

In the spirit of rule-based/logic programming, the output tuple of a conjunctive query is sometimes called the "head" of the query and the atom conjunction part the "body" of the query. So far, this discussion has considered only the class CQ of conjunctive without equality in the body. The class $CQ^=$ which allows equalities in the body defines essentially the same queries but there are a couple of technical complications. First, the range restriction condition must be strengthened since, for example, $\{\langle x, y\rangle \mid \exists z \ R(x) \wedge y = z\}$ is domain dependent. Therefore, for $CQ^=$ it is required that any variable among $e_1, \dots, e_n$ must equal, as a consequence of the atomic equalities in ψ, some constant, or some variable that occurs in one of the relational atoms of ψ.

$CQ^=$ has the additional pleasant property (shared, in fact, with the full relational calculus) that query heads can be restricted to consist of just distinct variables.

Clearly CQ $\subseteq$ CQ$^=$. The converse is "almost true." It is possible to get rid of equality atoms in a conjunctive query if the query is *satisfiable* i.e., there exists some instance on which the query returns a non-empty answer. All the queries in CQ are satisfiable (take the canonical instance). Queries in CQ$^=$ are satisfiable if the equalities in their body do not imply the equality of distinct constants. Thus, for conjunctive queries (of both kinds) satisfiability is decidable, as opposed to general first-order queries. Now, any satisfiable query in CQ$^=$ can be effectively translated into an equivalent query in CQ.

The conjunctive queries correspond to a specific fragment of the relational algebra, namely the fragment that uses only the selection, projection, and Cartesian product operations. This fragment is called the *SPC algebra*. There is an effective translation that takes every conjunctive query into an equivalent SPC algebra expression. There is also an effective translation that takes every SPC algebra expression into an equivalent conjunctive query.

Via the translation to the SPC algebra it can be seen that conjunctive queries correspond closely to certain SQL programs. For example, the CQ$^=$ query $ans(x, y)$: $-R(x, z)$, $x = c$, $S(x, y, z)$ corresponds to

```
select r.1, s.2
from R r, S s
where r.1=c and s.1=r.1 and r.3=s.2
```

Such SQL programs, in which the "where" clause is a conjunction of equalities arise often in practice. So, although restricted, conjunctive queries are important.

Cross-References

▶ Datalog
▶ Relational Algebra
▶ Relational Calculus

Recommended Reading

1. Abiteboul S, Hull R, Vianu V. Foundations of databases: the logical level. Reading: Addison Wesley; 1994.
2. Chandra AK, Merlin PM. Optimal implementation of conjunctive queries in relational data bases. In Proceedings of the 9th Annual ACM Symposium on Theory of Computing; 1977. p. 77–90.
3. Ullman JD. Principles of database and knowledge-base systems, vol. I. Rockville: Computer Science Press; 1988.

Connection

Sameh Elnikety
Microsoft Research, Redmond, WA, USA

Synonyms

Database socket

Definition

A connection is a mechanism that allows a client to issue SQL commands to a database server. In a typical usage, the client software opens a connection to the database server, and then sends SQL commands and receives responses from the server.

To open a connection, the client specifies the database server, database name, as well as the client's credentials. Establishing a connection includes a handshake between the client software and the database server. For example, the client credentials, such as a simple user name and password, are examined by the database server to authorize the connection. Further information may also be negotiated such as the specific protocol and data encoding.

Key Points

Handling and servicing connections is an important part of database servers because connections are the main source of concurrency.

Database servers limit the number of connections they can accept and may provide differentiated service to connections from high priority clients (e.g., from database administrators).

Connections are implemented using inter-process (e.g. pipes) or remote (e.g., TPC sockets) communication mechanisms. Database vendors and third-party providers supply libraries that client programs use to open connections to database servers. Several standards are used including ODBC (Open Database Connectivity) [2], JDBC (Java Database Connectivity) [3], and ADO.NET (data access classes in Microsoft.NET platform) [1].

When client software uses a database system extensively, it employs a connection pool to reuse a group of open connections, allowing multiple concurrent SQL commands. Using a connection pool avoids closing and reopening connections, as well as opening too many connections that tie up resources at both ends of the connection.

Cross-References

▶ Session

Recommended Reading

1. Adya A, Blakeley J, Melnik S, Muralidhar S. Anatomy of the ADO.NET Entity Framework. In: Proceedings of the ACM SIGMOD International Conference on Management of Data; 2007. p. 877–88.
2. Data Management: SQL Call Level Interface (CLI), Technical Standard C451-15/10/1993. The Open Group.
3. Sun Microsystems, The Java Database Connectivity (JDBC). Available at: http://java.sun.com/javase/technologies/database/.

Consistency Models for Replicated Data

Alan Fekete
University of Sydney, Sydney, NSW, Australia

Synonyms

Memory consistency; Replica consistency

Definition

When a distributed database system keeps several copies or replicas for a data item, at different sites, then the system may ensure that the copies are always consistent (that is, they have the same value), or the system may allow temporary discrepancy between the copies. Even if the copies are not the same, the algorithms that manage the data may be able to hide the discrepancies from clients. A consistency model defines the extent to which discrepancies can exist or be observed, between the copies. If the system offers a strong consistency model, then clients will not be aware of the fact that the system has replicated data, while a weak consistency model requires more careful programming of the clients, so they can cope with the discrepancies they observe.

Historical Background

Most replication research in the 1970s aimed to provide the illusion of an unreplicated database offering serializability. In the early 1980s, Bernstein and colleagues formalized this notion as a strong consistency model [2].

The late 1980s and early 1990s focused on systems that offered weak consistency in various definitions. Eventual consistency was introduced in the work of Demers et al. [3], while consistency models in which reads might see stale values were also explored by several groups [1, 5].

Since 2000, a new strong consistency model, one-copy SI, was introduced [4], and it has attracted much attention.

Foundations

There are many different system architectures that can be used for a distributed database with replicated data. For example, clients may submit operations directly to the different databases, or instead requests may all go through a middleware layer; the local databases may communicate directly with one another, or only with the clients or middleware; the requests that arrive at a local database may be a read or write on the local replica of an item, or they may be a SQL statement (which might involve many reads and/or writes), or indeed a whole transaction may form a single request; a read may be performed on one replica and writes on all replicas (read-one-write-all), or else a complex quorum rule may determine where reads and writes are performed; and each site may perform the operations once only, or else there may be possibilities for operations to be done tentatively, then (after conflicting information is received) the system might be able to roll back some operations and then replay them in a different order. Sometimes several system designs offer clients the same functionality, so the choice would be based only on performance, or the validity of assumptions in the design (such as the ability to know in advance which transactions access which items). If the client cannot learn, by the values returned or the operations which are allowed, which system design is used, then one can say that the different designs offer the same consistency model. However, sometimes a difference in design does change the functionality of the whole system, in ways that clients can detect. If one abstracts away the details of the system design, and instead focuses on what the essential features are that distinguish between the properties, then one is describing the consistency model offered by the system. For example, some system designs allow clients to learn about the existence of several copies. Perhaps one client might read the same item several times and see different values in each read. These values may have been taken from different older transactions. This can't happen in a system where all the data is at one site, with one copy for each item. Thus this system provides a consistency model which reveals the existence of copies to the client.

A system provides a strong consistency model if it provides clients with the illusion that there is a single copy of each piece of data, hiding all evidence of the replication. There are in fact several variants among strong models, because there are several different isolation models used by different DBMS platforms, and because the formal definition of isolation doesn't always capture exactly the properties of an implementation.

For example, one-copy serializability (q.v.) was defined in [2] as a consistency model in which clients have the illusion of working with a single-site unreplicated database which uses a concurrency control algorithm that offer serializability (q.v.) as the isolation level. In contrast, one-copy SI [4] is a different strong consistency model, where clients see the same behaviors as in an unreplicated system where concurrency control is done by Snapshot Isolation (q.v.).

In contrast to strong consistency models which maintain an illusion of a single-site system, in weaker models the clients are able to see that the system has replicas. Different models are characterized by the ways in which the divergence between replicas is revealed. The best-known weak consistency model is eventual consistency (q.v.) which is suitable for replicated databases where an updating transaction can operate at any replica, and the changes are then propagated lazily to other replicas through an epidemic mechanism. Eventually, each replica learns about the updates, and this consistency model ensures that a reconciliation mechanism resolves conflicting information, so that when the system quiesces, the values in all the replicas of a logical item eventually converge to the same value. In a system providing eventual consistency, there is not much that can be said about the value seen by a read, before convergence has been reached.

A different weak consistency model is common in systems where there is a single master copy for each item, and all updates are done first at the master, before being propagated in order, to the replicas. In this model, writes happen in a well-defined order, and each read sees a value from some prefix of this order; however, a read can see a value that is stale, that is, it does not include the most recent updates to the item.

Key Applications

The commercial DBMS vendors all offer replication mechanisms with their products. The performance impact of strong consistency models is usually seen as high, and these are typically provided only within a single cluster. For replication across dispersed machines, most platforms offer some form of weak consistency. There are also a range of research prototypes which give the user a choice between several consistency models; in general the user sees a tradeoff, where improved performance comes from accepting weaker consistency models.

Future Directions

Effective database replication is not yet a solved problem; the existing proposals compromise somehow among many desired properties, such as scalability for read-heavy workloads, scalability for update-heavy workloads, availability in face of failures or partitions, generality of the clients supported, ease of system programming, capacity to use varied local databases as black boxes, and the consistency provided. Thus the design space of possible systems is still being actively explored, and sometimes a new design achieves a consistency model different from those previously seen. One topic for ongoing research is how users can express their requirements for performance and for different levels of consistency, and how a system can then choose the appropriate replica control mechanism to provide the user with what they need. Research is also likely in consistency models that deal, to some extent, with malicious (often called "Byzantine") sites.

Cross-References

▶ Data Replication
▶ Strong Consistency Models for Replicated Data
▶ Weak Consistency Models for Replicated Data

Recommended Reading

1. Alonso R, Barbará D, Garcia-Molina H. Data caching issues in an information retrieval system. ACM Trans Database Syst. 1990;15(3):359–84.
2. Attar R, Bernstein PA, Goodman N. Site initialization, recovery, and backup in a distributed database system. IEEE Trans Softw Eng. 1984;10(6):645–50.
3. Demers AJ, Greene DH, Hauser C, Irish W, Larson J, Shenker S, Sturgis HE, Swinehart DC, Terry DB. Epidemic algorithms for replicated database maintenance. In: Proceedings of the ACM SIGACT-SIGOPS 6th Symposium on the Principles of Distributed Computing; 1987. p. 1–12.
4. Plattner C, Ganymed GA., Scalable replication for transactional web applications. In: Proceedings of the ACM/IFIP/USENIX 5th International Middleware Conference; 2004. p. 155–74.
5. Sheth AP, Rusinkiewicz M. Management of interdependent data: specifying dependency and consistency requirements. In: Proceedings of the Workshop on the Management of Replicated Data; 1990. p. 133–6.

Consistent Query Answering

Leopoldo Bertossi
Carleton University, Ottawa, ON, Canada

Definition

Consistent query answering (CQA) is the problem of querying a database that is inconsistent, i.e., that fails to satisfy certain integrity constraints, in such a way that the answers returned by the database are consistent with those integrity constraints. This problem involves a characterization of the semantically correct or consistent answers to queries in an inconsistent database.

Key Points

Databases may be inconsistent in the sense that certain desirable integrity constraints (ICs) are not satisfied. However, it may be necessary to still use the database, because it contains useful information, and, most likely, most of the data is still consistent, in some sense. CQA, as introduced in [1], deals with two problems. First, with the logical characterization of the portions of data that are consistent in the inconsistent database. Secondly, with developing computational mechanisms for retrieving the consistent data. In particular, when queries are posed to the database, one would expect to obtain as answers only those answers that are semantically correct, i.e., that are consistent with the ICs that are violated by the database as a whole.

The consistent data in the database is characterized [1] as the data that is invariant under all the database instances that can be obtained after making minimal changes in the original instance with the purpose of restoring consistency. These instances are the so-called (minimal) *repairs*. In consequence, what is consistently true in the database is what is *certain*, i.e., true in the collection of possible worlds formed by the repairs. Depending on the queries and ICs, there are different algorithms for computing consistent answers. Usually, the original query is transformed into a new query, possibly written in a different language, to be posed to the database at hand, in such a way that the usual answers to the latter are the consistent answers to the former [1]. For surveys of CQA and specific references, c.f. [2, 3].

Cross-References

▶ Database Repair
▶ Inconsistent Databases

Recommended Reading

1. Arenas M, Bertossi L, Chomicki J. Consistent query answers in inconsistent databases. In: Proceedings of the 18th ACM SIGACT-SIGMOD-SIGART Symposium on Principles of Database Systems; 1999. p. 68–79.
2. Bertossi L. Consistent query answering in databases. ACM SIGMOD Rec. 2006;35(2):68–76.
3. Chomicki J. Consistent query answering: five easy pieces. In: Proceedings of the 11th International Conference on Database Theory; 2007. p. 1–17.

Constraint Databases

Floris Geerts
University of Antwerp, Antwerp, Belgium

Definition

Constraint databases are a generalization of relational databases aimed to store possibly infinite-sized sets of data by means of a finite representation (*constraints*) of that data. In general, constraints are expressed by *quantifier-free first-order formulas* over some fixed vocabulary Ω and are interpreted in some Ω-structure $\mathcal{M} = \langle \mathbb{U}, \Omega \rangle$. By varying Ω and $\mathcal{M}$, constraint databases can model a variety of data models found in practice including traditional relational databases, spatial and spatiotemporal databases, and databases with text fields (strings). More formally, let Ω be a fixed vocabulary consisting of function, predicate, and constant symbols, and let $\mathcal{R} = \{R_1, \dots, R_\ell\}$ be a relational schema, where each relation name R_i is of arity $n_i > 0$. An Ω-*constraint database* $\mathbf{D}$ with schema $\mathcal{R}$ maps each relation $R_i \in \mathcal{R}$ to a quantifier-free formula $\varphi_{R_i}^{\mathbf{D}}(x_1, \dots, x_{n_i})$ (with n_i free variables $x_1, \dots, x_{n_i}$) in *first-order logic* over Ω. When interpreted over an Ω-structure $\mathcal{M} = \langle \mathbb{U}, \Omega \rangle$, an Ω-constraint database $\mathbf{D}$ with schema $\mathcal{R}$ corresponds to the collection of the $\mathcal{M}$-*definable sets* $[R_i]_{\mathcal{M}}^{\mathbf{D}} = \{(a_1, \dots, a_{n_i}) \in \mathbb{U}^{n_i} \mid \mathcal{M} \models \varphi_{R_i}^{\mathbf{D}}(a_1, \dots, a_{n_i})\}$, for $R_i \in \mathcal{R}$. Constraint query languages have been devised to manipulate and query constraint databases.

Key Points

The primary motivation for constraint databases comes from the field of spatial and spatiotemporal databases where one wants to store an infinite set of points in the real Euclidean space and query it as if all (infinitely) points are present [3, 4, 5]. In the spatial context, the constraints used to finitely represent data are *Boolean combinations of polynomial inequalities*. For instance, the infinite set of points in the real plane $\mathbb{R}^2$ depicted in Fig. 1a can be described by means of a disjunction of polynomial inequalities with integer coefficients as follows: $\varphi(x, y) = (x^2/25 + y^2/16 = 1) \lor (x^2 + 4x + y^2 - 2y \leq 4) \lor (x^2 - 4x + y^2 - 2y \leq -4) \lor (x^2 + y^2 - 2y = 8 \land y < -1)$. In the language of constraint databases, $\varphi(x, y)$ is a quantifier-free first-order formula over $\Omega = (+, \cdot, 0, 1, <)$, and Fig. 1a represents the $\mathcal{M}$-definable set in $\mathbb{R}^2$ corresponding to the formula φ for the Ω-structure $\mathcal{M} = \langle \mathbb{R}, \Omega \rangle$. If $\mathcal{R}$ is a

Constraint Databases, Fig. 1 Example of set definable by (**a**) polynomial constraints and (**b**) linear constraints

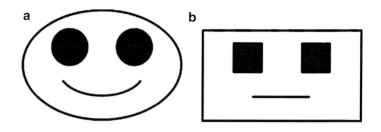

relational schema consisting of a binary relation R, then the Ω-constraint database **D** with schema $\mathcal{R}$ defined by $R \mapsto \varphi(x, y)$ "stores" the set in Fig. 1a. In this case, the $\mathcal{M}$-definable sets are also known as semi-algebraic sets [2].

When *Boolean combination of linear inequalities* suffices, such as in geographical information systems (GIS), one considers constraint databases over $\Omega = (+,0,1,<)$ and $\mathcal{M} = \langle \mathbb{R}, \Omega \rangle$. Figure 1b shows an example of a set defined by means of a first-order formula over $\Omega = (+,0,1,<)$. The advantage of the constraint approach to represent spatial data is the uniform representation of the various spatial entities. Whereas in GIS one normally defines a special data type for each spatial object such as line, poly-line, circle, etc., each of those is now represented by constraints in the same constraint language.

Other common scenarios of constraint databases include *dense order constraints over the rationals*, where $\Omega = (<, (c)_{c \in \mathcal{Q}})$ and $\mathcal{M} = \langle \mathbb{Q}, \Omega \rangle$. That is, rational numbers with order and constants for every $c \in Q$; and *constraints over strings*, where $\Omega = ((f_a)_{a \in \Sigma}, \prec, \mathrm{el})$ and $\mathcal{M} = \langle \sigma^*, \Omega \rangle$ [1]. Here, Σ is a finite alphabet, f_a is a function that adds a at the end of its argument, $\prec$ is the prefix relation, and $\mathrm{el}(x, y)$ is a binary predicate that holds if $|x| = |y|$, where $|\cdot|$ stands for the length of a finite string. In the latter case, the $\mathcal{M}$-definable sets are precisely the regular languages over Σ.

Finally, standard relational databases with schema $\mathcal{R}$ can be considered as constraint databases over *equality constraints over an arbitrary infinite domain* $\mathbb{U}$, where $\Omega = ((c)_{c \in \mathbb{U}})$ and $\mathcal{M} = \langle \mathbb{U}, \Omega \rangle$. Indeed, consider a tuple $t = (a_1,..., a_n)$ consisting of some constants $a_i \in \mathbb{U}$, for $i \in [1,n]$. The tuple t can be expressed by the formula $\varphi_t (x_1,..., x_n) = (x_1 = a_1) \wedge ... \wedge$ $(x_n = a_n)$ over the signature $\Omega = ((c)_{c \in \mathbb{U}})$. More generally, an instance $I = \{t_1,..., t_N\}$ over $R \in \mathcal{R}$ corresponds to $\varphi_I = \mathrm{V}_{i=1}^{N} \varphi_{t_i}$. Therefore, a relational instance $(I_1,...,I_\ell)$ over $\mathcal{R}$ can be represented as the constraint database **D** defined by $R_i \mapsto \varphi_{I_i} (x_1, ..., x_{n_i})$, for $i \in [1,\ell]$. This shows that constraint databases indeed generalize standard relational databases.

Cross-References

▶ Constraint Query Languages
▶ Geographic Information System
▶ Relational Model
▶ Spatial Data Types

Recommended Reading

1. Benedikt M, Libkin L, Schwentick T, Segoufin L. Definable relations and first-order query languages over strings. J ACM. 2003;50(5):694–751.
2. Bochnak J, Coste M, Roy MF. Real algebraic geometry. Berlin: Springer; 1998.
3. Kanellakis PC, Kuper GM, Revesz PZ. Constraint query languages. J Comput Syst Sci. 1995;51(1): 26–52.
4. Kuper GM, Libkin L, Paredaens J, editors. Constraint databases. Berlin: Springer; 2000.
5. Revesz PZ. Introduction to constraint databases. Berlin: Springer; 2002.

Constraint Query Languages

Floris Geerts
University of Antwerp, Antwerp, Belgium

Definition

A constraint query language is a query language for constraint databases.

Historical Background

The field of constraint databases was initiated in 1990 in a paper by Kanellakis, Kuper, and Revesz [1]. The goal was to obtain a database-style, optimizable version of constraint logic programming. It grew out of the research on datalog and constraint logic programming. The key idea was that the notion of tuple in a relational database could be replaced by a conjunction of constraints from an appropriate language and that many of the features of the relational model could then be extended in an appropriate way. In particular, standard query languages such as those based on first-order logic and datalog could be extended to such a model.

It soon became clear, however, that recursive constraint query languages led to noneffective languages. The focus therefore shifted to non-recursive constraint query languages. The standard query language is the constraint relational calculus (or equivalently, the constraint relational algebra). The study of this query language turned out to lead to many interesting research problems. During the period from 1990 to 2000, the constraint setting has been studied in great generality which led to deep connections between constraint databases and embedded finite model theory. Also, the potential application of constraint databases in the spatial context led to numerous theoretical results and concrete implementations such as the DEDALE and the DISCO systems. The connection with so-called o-minimal geometry underlies many of the results in the spatial setting. The success of this research led to the publication of a comprehensive survey of the area in 2000 [2] and a textbook in 2002 [3].

In recent years, constraint query languages have been studied in new application domains such a strings and spatiotemporal and moving objects.

Foundations

In the constraint model, a database is viewed as a collection of constraints specified by quantifier-free first-order logic formulas over some fixed vocabulary Ω. When interpreted over an Ω-structure $\mathcal{M} = \langle \mathbb{U}, \Omega \rangle$, each constraint corresponds to an $\mathcal{M}$-definable set. Consequently, when interpreted over $\mathcal{M}$, an Ω-constraint database corresponds to a collection of $\mathcal{M}$-definable sets. For instance, consider the vocabulary $\Omega = (+, \cdot, 0, 1, <)$ and $\mathcal{M} = \langle \mathbb{U}, \Omega \rangle$. Constraints in first-order logic over Ω, denoted by FO(Ω), correspond to Boolean combinations of polynomial inequalities with integer coefficients. The corresponding $\mathcal{M}$-definable sets are better known as semi-algebraic sets. Let $\mathcal{R} = \{R, S\}$ be a relational schema consisting of two binary relations R and S, and let $\mathbf{D}$ be the constraint database that maps $R \mapsto \varphi_R$ $(x, y) = (x^2 + y^2 \leq 1) \wedge (y - x \geq 0)$ and $S \mapsto \varphi_S$ $(x, y) = (x^2 + y^2 \leq 1) \wedge (-y - x \geq 0)$. The two $\mathcal{M}$-definable sets in $\mathbb{R}^2$ corresponding to φ_R and φ_S are shown in Fig. 1a, b, respectively.

A constraint database can therefore be viewed from two different perspectives: First, one can simply look at the finite representations (constraints) stored in them; second, one can regard them as a set of definable sets. Whereas in traditional relational databases, a query is simply a mapping that associates with each database an answer relation, in the constraint setting, the two different perspectives give rise to two different notions of queries.

Indeed, for a fixed vocabulary Ω, relational schema $\mathcal{R}$ consisting of relation names $R_1, \ldots, R_\ell$, where each relation R_i is of arity $n_i > 0$, and natural number k, a k-ary constraint query with schema $\mathcal{R}$ over Ω, is a (partial) function Q that maps each Ω-constraint databases $\mathbf{D}$ with schema $\mathcal{R}$ to a k-ary Ω-constraint relation $Q(\mathbf{D})$. That is, a constraint query works entirely on the representational (constraint) level. On the other hand, given an additional Ω-structure $\mathcal{M} = \langle \mathbb{U}, \Omega \rangle$, a k-ary unrestricted query with schema $\mathcal{R}$ over $\mathcal{M}$ is a (partial) function Q that maps each collection $\mathbf{D}$ of sets in $\mathbb{U}^{n_i}$, for $i \in [1, \ell]$, to a set $Q(\mathbf{D})$ in $\mathbb{U}^k$. Such a collection of sets $\mathbb{U}^{n_i}$, for $i \in [1, \ell]$, is called an unrestricted database with schema $\mathcal{R}$ over $\mathcal{M}$.

For instance, consider again $\Omega = (+, \cdot, 0, 1, <)$ and $\mathcal{R} = \{R, S\}$. The mapping Q_1 that associates each Ω-constraint database $\mathbf{D}$ over $\mathcal{R}$

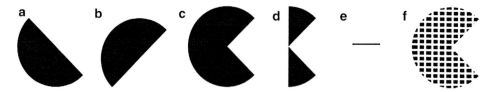

Constraint Query Languages, Fig. 1 : Sets in $\mathbb{R}^2$ defined by $\varphi_R\,(x,\,y)$ **(a)**; by $\varphi_S\,(x,\,y)$ **(b)**; by $\varphi_R\,(x,\,y) \vee \varphi_S\,(x,\,y)$ **(c)**; and by $\varphi_1(x,\,y)$ **(d)**. The set in $\mathbb{R}$ defined by $\varphi_2(x)$ **(e)**. An example of a non-definable set in $\mathbb{R}^2$ **(f)**

with the binary Ω-constraint relation defined by taking the disjunction of the constraints in R and S is an example of a two-ary constraint query over Ω. When applied on the database **D** given above, $Q_1(\mathbf{D})$ is mapped to $\varphi_R\,(x,\,y) \vee \varphi_S\,(x,\,y)$. Similarly, the mapping Q_2 that maps **D** to the constraint in R or S that contains the polynomial with the largest coefficient (if there is no such unique constraint, then Q_2 is undefined) is also a constraint query. It will be undefined on the example database **D** since both R and S consist of a polynomial with coefficient one.

So far, only constraint queries have been considered. To relate constraint and unrestricted queries requires some care. Clearly, a constraint query only makes sense if it corresponds to an unrestricted query. In this case, a constraint query is called *consistent*. More formally, a constraint query Q is called consistent if there exists an unrestricted query Q_0 such that for any constraint database **D** and any unrestricted database $\mathbf{D}_0$, if **D** represents $\mathbf{D}_0$, then $Q(\mathbf{D})$ is defined if and only if $Q_0(\mathbf{D}_0)$ is defined and, furthermore, $Q(\mathbf{D})$ represent $Q_0(\mathbf{D}_0)$. One also says that Q *represents* Q_0.

For instance, consider again $\Omega = (+, \cdot, 0, 1, <)$, $\mathcal{M} = \langle \mathbb{R}, \Omega \rangle$ and $\mathcal{R} = \{R, S\}$. The mapping $\tilde{Q}_1$ that assigns to any two sets $A \subseteq \mathbb{R}^2$ and $B \subseteq \mathbb{R}^2$, corresponding to R and S, respectively, their union $A \cup B \subseteq \mathbb{R}^2$ is an unrestricted query. It is clear that Q_1 and $\tilde{Q}_1$ satisfy the condition of consistency, and therefore Q_1 is consistent. Figure 1c shows $\tilde{Q}_1(\mathbf{D}_0)$ for the unrestricted database $\mathbf{D}_0$ shown in Fig. 1a, b. This set is indeed represented by the constraint relation $Q_1(\mathbf{D}) \mapsto \varphi_R\,(x,\,y) \vee \varphi_S\,(x,\,y)$. On the other hand, it is easily verified that Q_2 is not consistent. Indeed, it suffices to consider the behavior of Q_2 on **D** defined above and $\mathbf{D}'$ defined by $R \mapsto \varphi'_R\,(x, y) = (x^2 + y^2 \le 1) \wedge (6(y - x) \ge 0)$ and $S \mapsto \varphi'_S$

$(x, y) = (x^2 + y^2 \le 1) \wedge (-y - x \ge 0)$. While both **D** and $\mathbf{D}'$ represent the same unrestricted database, note that $Q_2(\mathbf{D})$ is undefined while $Q_2(\mathbf{D}') \mapsto \varphi_R$. Hence, no unrestricted query that is consistent with Q_2 can exist.

Finally, unrestricted queries are defined without any reference to the class of $\mathcal{M}$-definable sets. A desirable property, however, is that when an unrestricted query Q is defined on an unrestricted database **D** that consists of $\mathcal{M}$-definable sets, then also $Q(\mathbf{D})$ is an $\mathcal{M}$-definable set. Such unrestricted queries are called *closed*. Note that an unrestricted query that is represented by a consistent constraint query is uniquely defined and moreover is trivially closed. An example of an unrestricted query for $\Omega = (+, \cdot, 0, 1, <)$ and $\mathcal{M} = \langle \mathbb{R}, \Omega \rangle$ that is not closed is the query Q that maps any $\mathcal{M}$-definable set A in $\mathbb{R}^2$ to its intersection $A \cap \mathbb{Q}^2$. Figure 1f shows (approximately) the result of this query on $\tilde{Q}_1(\mathbf{D}_0)$ (i.e., Fig. 1c). Since this is not a semi-algebraic set in $\mathbb{R}^2$, it cannot be defined by means of a quantifier-free FO(Ω)-formula. As a consequence, Q is not closed.

Now that the notion of query is defined in the setting of constraint databases, the basic *constraint query language* is introduced. This language, in the same spirit as the relational calculus for traditional relational databases, is the *relational calculus* or first-order logic of the given class of constraints. More specifically, given a vocabulary Ω and relational schema $\mathcal{R}$, a *relational calculus formula* over Ω is a first-order logic formula over the expanded vocabulary $(\Omega, \mathcal{R})$ obtained by expanding Ω with the relation names (viewed as predicate symbols) of the schema $\mathcal{R}$. This class of queries is denoted by FO($\Omega, \mathcal{R}$) or simply FO(Ω) when $\mathcal{R}$ is understood from the context.

For instance, for $\Omega = (+, \cdot, 0, 1, <)$ and $\mathcal{R} = \{R, S\}$, the expressions $\varphi_1(x, y) = (R(x, y) \lor S(x, y)) \land x > 0$ and $\varphi_2(x) = \exists y\, \varphi_1(x, y)$ are formulas in FO$(+, \cdot, 0, 1, <, R, S)$.

Given an Ω-structure $\mathcal{M} = \langle \mathbb{U}, \Omega \rangle$, formulas in FO$(\Omega, \mathcal{R})$ express (everywhere defined) unrestricted queries with schema $\mathcal{R}$ over $\mathcal{M}$. Indeed, a formula $\varphi(x_1, \ldots, x_k) \in$ FO$(\Omega, \mathcal{R})$ defines the k-ary unrestricted query Q over $\mathcal{M}$ as follows: consider the expansion of $\mathcal{M}$ to a structure $\langle \mathcal{M}, \mathbf{D} \rangle = \langle \mathbb{U}, \Omega, \mathbf{D} \rangle$ over the expanded vocabulary $(\Omega, \mathcal{R})$ by adding the sets in the unrestricted database $\mathbf{D}$ to $\mathcal{M}$ for each $R_i \in \mathcal{R}$. Then, $Q(\mathbf{D}) = \{(a_1, \ldots, a_k) \in \mathbb{U}^k \mid 0\ \langle \mathcal{M}, \mathbf{D} \rangle \models \varphi(a_1, \ldots, a_k)\}$.

For instance, for $\Omega = (+, \cdot, 0, 1, <)$ and $\mathcal{M} = \langle \mathbb{R}, \Omega \rangle$, the formula $\varphi_1(x, y)$ defined above corresponds to the unrestricted query Q_1 that takes the union of the two sets in $\mathbb{R}^2$ corresponding to R and S, respectively, restricted to those points in $\mathbb{R}^2$ with strictly positive x-coordinate, similarly for φ_2, but with an additional projection on the x-axis. The results of these two unrestricted queries have been shown in Fig. 1d, e, respectively.

The previous example raises the following two questions: (i) are the unrestricted queries expressed by formulas in first-order logic closed, and (ii) if so, can one find a corresponding constraint query that is effectively computable? The fundamental mechanism underlying the use of first-order logic as a constraint query language is the following observation that provides an answer to both questions.

Every relational calculus formula φ expresses a consistent, effectively computable, total constraint query that represents the unrestricted query expressed by φ, if and only if $\mathcal{M}$ admits *effective quantifier elimination*.

Here, an Ω-structure $\mathcal{M}$ admits effective quantifier elimination if there exists an effective algorithm that transforms any first-order formula in FO(Ω) to an equivalent (in the structure $\mathcal{M}$) *quantifier-free* first-order formula in FO(Ω).

Consider the two FO$(+, \cdot, 0, 1, <, R, S)$ formulas φ_1 and φ_2 given above. It is known that the structure $\langle \mathbb{R}, +, \cdot, 0, 1, < \rangle$ admits effective quantifier elimination. In case of φ_1, it is easy to see that the result of corresponding constraint query is obtained by "plugging" in the constraints for R (resp. S) as given by the constraint database into the expression for φ_1. That is, on the example database $\mathbf{D}$, φ_1 corresponds to the constraint query that maps $\mathbf{D}$ to $(\varphi_R(x, y) \lor \varphi_S(x, y)) \land (x > 0)$, which is a two-ary Ω-constraint relation. In case of φ_2, however, first plug in the descriptions of the constraints as before, resulting in $\exists y (\varphi_R(x, y) \lor \varphi_S(x, y)) \land (x > 0)$. In order to obtain an Ω-constraint relation, one needs to perform quantifier elimination. It is easily verified that in this example, a corresponding constraint query is one that maps $\mathbf{D}$ to $(0 < x) \land (x \leq 1)$ which is consistent with Fig. 1e.

For Ω-structure $\mathcal{M}$ that admits effective quantifier elimination, this suggests the following effective evaluation mechanism for constraint relational calculus queries φ on a constraint database $\mathbf{D}$: (i) plug in the contents of $\mathbf{D}$ in the appropriate slots (relations). Denote the resulting formula by plug$(\varphi, \mathbf{D})$, and (ii) eliminate the quantifiers in plug$(\varphi, \mathbf{D})$. Since $\mathbf{D}$ consists of quantifier-free formulas, the number of quantifiers that need to be eliminated is the same as in φ and is therefore independent of $\mathbf{D}$. For many structures $\mathcal{M}$, this implies that the evaluation of constraint queries can be done in polynomial data complexity, which is a desirable property for any query language.

It is important to point out that the classical equivalence between the relational calculus and the relational algebra can be easily extended to the constraint setting. That is, for a fixed Ω and schema $\mathcal{R}$, one can define a *constraint relational algebra* and show that every constraint relational calculus formula can be effectively converted to an equivalent constraint relational algebra expression, and vice versa. This equivalence is useful for concrete implementations of constraint database systems.

The study of expressivity of FO$(\Omega, \mathcal{R})$ for various Ω-structures, $\mathcal{M}$ has led to many interesting results. In particular, the impact of the presence of the "extra" structure on the domain elements in $\mathbb{U}$ has been addressed when $\mathbf{D}$ consists of an ordinary finite relational database that takes values from $\mathbb{U}$ [4]. In particular, the correspondence between natural and active-domain semantics has been revisited. That is, conditions are identified for $\mathcal{M} = \langle \mathbb{U}, \Omega \rangle$ such that the language $\text{FO}(\Omega, \mathcal{R})$ is equal to $\text{FO}_{\text{act}}(\Omega, \mathcal{R})$, the query language obtained by interpreting $\forall x$ and $\exists x$ over the active domain of **D** instead of over $\mathbb{U}$. Such structures are said to admit the *natural-active collapse*. Similarly, ordered structures $\mathcal{M}$ are identified that admit the *active-generic collapse*. That is, $\text{FO}_{\text{act}}(\Omega, \mathcal{R})$ is equal to $\text{FO}_{\text{act}}(<, \mathcal{R})$ with respect to the class of generic queries. In other words, every generic query definable under active-domain semantics with Ω-constraints is already definable with just order constraints. Finally, structures $\mathcal{M}$ are considered that allow the *natural-generic collapse*. This is the same as the active-generic collapse but with natural domain semantics instead of active-domain semantics. The study of these collapse properties for various structures not only sheds light on the interaction of the structure on $\mathbb{U}$ and the query language; it is also helpful to understand the expressiveness of constraint query languages [4,2].

Indeed, let $\Omega = (+, \times, 0, 1, <)$ and $\mathcal{M} = \langle \mathbb{R}, \Omega \rangle$. It can be shown that $\mathcal{M}$ admits all three collapses because it is a so-called o-minimal structure. As a consequence, the query EVEN that returns **yes** if the cardinality of **D** is even and **no** otherwise is not expressible in $\text{FO}(\Omega, \mathcal{R})$. Indeed, if it would be expressible by a query φ in $\text{FO}(\Omega, \mathcal{R})$, it would already have been expressible by a query in $\text{FO}_{\text{act}}(<, \mathcal{R})$, which is known not to be true in the traditional database setting.

The expressivity of $\text{FO}(\Omega, \mathcal{R})$ has been studied extensively as well when **D** corresponds to sets of infinite size. In particular, expressiveness questions have been addressed in the spatial setting where $\Omega = (+, \cdot, 0, 1, <)$ and $\mathcal{M} = \langle \mathbb{R}, \Omega \rangle$ (polynomial constraints), and $\Omega' = (+, 0, 1, <)$ and $\mathcal{M} = \langle \mathbb{R}, \Omega' \rangle$ (linear constraints). In this setting, many reductions are presented in [5] to expressiveness questions in the finite case. Combined with the collapse results mentioned above, these reductions were used to show that, for example, topological connectivity of Ω- (resp. Ω'-) constraint databases is not expressible in first-order logic. Indeed, a proof of this results relies on the fact that the EVEN query is not expressible in $\text{FO}(\Omega, \mathcal{R})$ (resp. $\text{FO}(\Omega', \mathcal{R})$) [5].

An interesting line of work in the spatial context concerns the expressive power of $\text{FO}(\Omega, \mathcal{R})$ with respect to queries that preserve certain geometrical properties. More formally, let $\mathcal{G}$ be a group of transformations of $\mathbb{R}^k$. A query Q is called $\mathcal{G}$-generic if, for every transformation $g \in \mathcal{G}$ and for any two databases **D** and **D'**, $g(\mathbf{D}) = \mathbf{D'}$ implies $g(Q(\mathbf{D})) = Q(\mathbf{D'})$. Transformation groups and properties of the corresponding generic queries have been studied for the group of homeomorphisms, affinities, similarities, isometries, among others [6]. Especially the study of the topologically queries (those that are generic under homeomorphisms) has received considerable attention [7, 8].

To conclude, both for the historical reasons mentioned above and in view of the limited expressive power of $\text{FO}(\Omega, \mathcal{R})$, various recursive extensions of $\text{FO}(\Omega, \mathcal{R})$ have been proposed such as constraint transitive closure logic [9], constraint DATALOG [1], and $\text{FO}(\Omega, \mathcal{R})$ extended with a WHILE-loop [6]. The interaction of recursion with the structure on $\mathbb{U}$ imposed by Ω leads in most cases to computationally complete query languages. Worse still, queries defined in these languages may not be closed or even terminate. To remedy this, special-purpose extensions of $\text{FO}(\Omega, \mathcal{R})$ have been proposed that guarantee both termination and closure. Characteristic examples include $\text{FO}(\Omega, \mathcal{R}) + \text{AVG}$ and $\text{FO}(\Omega, \mathcal{R}) + \text{SUM}$ in the context of aggregation [10]. In the spatial setting, extensions of $\text{FO}(\Omega, \mathcal{R})$ with various connectivity operators have been proposed [11].

Results concerning constraint query languages have been both extended to great generality and applied to concrete settings. Refer to [12] for a gentle introduction and to [2] for a more detailed survey of this research area up to 2000. Some more recent results are included in Chapter 5 of [13] for the general constraint setting and Chapter 12 in [14] for the spatial setting.

Key Applications

Manipulation and querying of constraint databases, querying of spatial data.

Cross-References

Recommended Reading

1. Kanellakis PC, Kuper GM, Revesz PZ. Constraint query languages. J Comput Syst Sci. 1995;51(1): 26–52.
2. Kuper GM, Libkin L, Paredaens J. Constraint databases. Berlin: Springer; 2000.
3. Revesz PZ. Introduction to constraint databases. New York: Springer; 2002.
4. Benedikt M, Libkin L. Relational queries over interpreted structures. J ACM. 2000;47(4):644–80.
5. Grumbach S, Su J. Queries with arithmetical constraints. Theor Comput Sci. 1997;173(1):151–81.
6. Gyssens M, Van den Bussche J, Van Gucht D. Complete geometric query languages. J Comput Syst Sci. 1999;58(3):483–511.
7. Kuijpers B, Paredaens J, Van den Bussche J. Topological elementary equivalence of closed semi-algebraic sets in the real plane. J Symb Log. 2000;65(4): 1530–55.
8. Benedikt M, Kuijpers B, Löding C, Van den Bussche J, Wilke T. A characterization of first-order topological properties of planar spatial data. J. ACM. 2006;53(2):273–305.
9. Geerts F, Kuijpers B, Van den Bussche J. Linearization and completeness results for terminating transitive closure queries on spatial databases. SIAM J Comput. 2006;35(6):1386–439.
10. Benedikt M, Libkin L. Aggregate operators in constraint query languages. J Comput Syst Sci. 2002;64(3):628–54.
11. Benedikt M, Grohe M, Libkin L, Segoufin L. Reachability and connectivity queries in constraint databases. J Comput Syst Sci. 2003;66(1):169–206.
12. Van den Bussche J. Constraint databases. A tutorial introduction ACM SIGMOD Record. 2000;29(3): 44–51.
13. Libkin L. Embedded finite models and constraint databases. In: Grädel E, Kolaitis PG, Libkin L, Marx M, Spencer J, Vardi MY, Venema Y, Weinstein S, editors. Finite Model Theory and Its Applications. Berlin/Heidelberg: Springer; 2007.
14. Geerts F, Kuijpers B. Real algebraic geometry and constraint databases. In: Aiello M, Pratt-Hartmann I, Van Benthem J, editors. Handbook of Spatial Logics. Dordrecht: Springer; 2007.

Constraint-Driven Database Repair

Wenfei Fan
University of Edinburgh, Edinburgh, UK
Beihang University, Beijing, China

Synonyms

Data reconciliation; Data standardization; Minimal-change integrity maintenance

Definition

Given a set Σ of integrity constraints and a database instance D of a schema R, the problem of *constraint-driven database repair* is to find an instance D' of the same schema R such that (i) D' is *consistent*, i.e., D' satisfies Σ, and moreover, (ii) D' *minimally differs* from the original database D, i.e., it takes a minimal number of repair operations or incurs minimal cost to obtain D' by updating D.

Historical Background

Real-life data is often dirty, i.e., inconsistent, inaccurate, stale, or deliberately falsified. While the prevalent use of the Web has made it possible, on an unprecedented scale, to extract and integrate data from diverse sources, it has also increased the risks of creating and propagating dirty data. Dirty data routinely leads to misleading or biased analytical results and decisions and incurs loss of revenue, credibility, and customers. With this comes the need for finding repairs of dirty data and editing the data to make it consistent. This is the data cleaning approach that US national statistical agencies, among others, have been practicing for decades [10].

The notion of constraint-based database repairs is introduced in [1], highlighting the use of integrity constraints for characterizing the consistency of the data. In other words, constraints are used as data quality rules, which detect inconsistencies as violations of the constraints. Prior work on constraint-based database repairs has mostly focused on the following issues. (i) Integrity constraints used for repair. Earlier work considers traditional functional dependencies, inclusion dependencies, and denial constraints [1, 2, 4, 7, 16]. Extensions of functional and inclusion dependencies, referred to as conditional functional and inclusion dependencies, are recently proposed in [3, 11] for data cleaning. (ii) Repair semantics. Tuple deletion is the only repair operation used in [6], for databases in which the information is complete but not necessarily consistent. Tuple deletion and insertion are considered in [1, 4] for databases in which the information may be neither consistent nor complete. Updates, i.e., attribute-value modifications, are proposed as repair operations in [16]. Cost models for repairs are studied in [2, 9]. (iii) Algorithms. The first algorithms for finding repairs are developed in [2], based on traditional functional and inclusion dependencies. Algorithms for repairing and incrementally repairing databases are studied in [9], using conditional functional dependencies. Algorithms for finding certain fixes, i.e., repairs with 100% confidence, are presented in [13], based on master data. There have also been algorithms that repair data by improving both data consistency and data accuracy [5] and by exploring the interaction between data repairing and entity resolution [12]. The repair model adopted by these algorithms supports updates as repair operations. (iv) Fundamental issues associated with constraint-based repairs. One issue concerns the complexity bounds on the database repair problem [2, 9]. Another issue concerns the static analysis of constraint consistency [3, 11] for determining whether a given set of integrity constraints is dirty or not itself. We refer the interested reader to [10] for a recent survey.

Constraint-based database repairs are one of the two topics studied for constraint-based data cleaning. The other topic, also introduced in [1], is *consistent query answers*. Given a query Q posed on an inconsistent database D, it is to find tuples that are in the answer of Q over every repair of D. There has been a host of work on consistent query answers [1,5,6,11,13] (see [4,7] for comprehensive surveys).

Foundations

The complexity of the constraint-based database repair problem is highly dependent upon what integrity constraints and repair model are considered.

Integrity Constraints for Characterizing Data Consistency

A central technical issue for data cleaning concerns how to tell whether the data is dirty or clean. Constraint-based database repair characterizes inconsistencies in terms of violations of integrity constraints. Constraints employed for data cleaning include functional dependencies, inclusion dependencies, denial constraints, conditional functional dependencies, and conditional inclusion dependencies. To illustrate these constraints, consider the following relational schema R, which consists of three relation schemas:

customer(name, country code, area code, phone, city, street, zip)
order(name, country code, area code, phone, item id, title, price, item type)
book(isbn, title, price, format)

Traditional functional dependencies and inclusion dependencies defined on the schema R include:

FD: customer(country code, area code, phone → city, street, zip)
IND: order(name, country code, area code, phone) ⊆ customer (name, country code, area code, phone)

The functional dependency asserts that the phone number (country code, area code, and phone) of a customer uniquely determines her address (state, city, street, zip). That is, for any two customer tuples, if they have the same country code, area code, and phone number, then they must have the same state, city, street, and zip code. The inclusion dependency asserts that for any order tuple t, there must exist a customer tuple t' such that t and t' match on their name, country code, area code, and phone attributes. In other words, an item cannot be ordered by a customer who does not exist.

Consider a set Σ consisting of the FD and IND given above. One may want to use Σ to specify the consistency of database instances of R. An example instance D of R is shown in Fig. 1. This database is inconsistent, because tuples t_3 and t_4 in D violate the FD. Indeed, while t_3 and t_4 have the same country code, area code, and phone number, they differ in their street attributes. In other words, t_3, t_4, or both of them may be dirty.

One may also want to add a denial constraint to the set Σ:

DC: $\forall nm, cc, ac, ph, id, tl, tp, pr \neg (order(nm,$ $cc, ac, ph, id, tl, tp, pr) \wedge pr > 100)$

Here $nm, cc, ac, ph, id, tl, tp$ and pr stand for name, country code, area code, phone number, item id, title, item type, and price, respectively. This constraint says that no items in the order table may have a price higher than 100. In the database D of Fig. 1, tuple t_6 violates the constraint: the price of the CD is too high to be true. In general denial constraints can be expressed as universally quantified first-order logic sentences of the form:

$$\forall \bar{x}_1, \dots, \bar{x}_m \neg (R_1(\bar{x}_1) \wedge \dots \wedge$$
$$R_m(\bar{x}_m) \wedge \varphi(\bar{x}_1, \dots, \bar{x}_m)),$$

where R_i is a relation symbol for $i \in [1,m]$, and φ is a conjunction of built-in predicates.

Now consider an instance D' of D by removing t_3 from D and changing $t_6[price]$ to, e.g., 7.99. Then the database D' satisfies Σ. However, D' is not quite clean: it violates each of the following constraints. In other words, if one further extends Σ by including the following constraints, then D' no longer satisfies Σ.

CFD1: customer(country code = 44, zip → street)

CFD2: customer(country code = 44, area code = 131, phone → city = EDI, street, zip)

CFD3 : customer(country code = 01, area code = 908, phone → city = MH, street, zip)

CIND1: order(id, title, price, item type = book) ⊆ book(isbn, title, price)

Here CFD1, CFD2, and CFD3 are conditional functional dependencies defined on the customer relation. The constraint CFD1 asserts that for each customer in the UK, i.e., when the country code is 44, her zip code uniquely determines

	Name	Country-code	Area-code	Phone	City	Street	Zip
t_1:	M. Smith	44	131	1233444	NYC	Mayfield	EH8 8LE
t_2:	L. Webber	44	131	2344455	NYC	Crichton	EH8 8LE
t_3:	M. Hull	01	908	3456788	NYC	Mountain Ave	07974
t_4:	R. Xiong	01	908	3456788	NYC	Main St	07974

a   Example customer data

	Name	Country-code	Area-code	Phone	Item-id	Title	Price	Item-type
t_5:	M.Smith	44	131	1233444	b23	H. Porter	17.99	book
t_6:	M.Hull	01	908	3456788	c68	John Denver	2000	CD

b   Example order data

	isbn	Title	Price	Format
t_7:	b23	Harry porter	17.99	hard-cover
t_8:	b88	Snow white	7.94	audio

c   Example book data

Constraint-Driven Database Repair, Fig. 1 Example database instance

her street. In contrast to traditional functional dependencies, CFD1 does not hold on the entire customer relation. Indeed, it does not hold on customer tuples with, e.g., country code = 01. Instead, it is applicable only to the set of customer tuples with country code = 44. Constraints CFD2 and CFD3 refine the traditional functional dependency given earlier: CFD2 requires that when the country code is 44 and area code is 131, the city must be Edinburgh (EDI), similarly for CFD3. None of these can be expressed as traditional functional dependencies.

Constraint CIND1 is a conditional inclusion dependency, asserting that when the type of an item t in the order table is book, there must exist a corresponding tuple t' in the book table such that t and t' match on their id, title, and price. Again this is a constraint that only holds conditionally. Indeed, without the condition item type = book, a traditional inclusion dependency order(id, title, price) $\subseteq$ book(isbn, title, price) does not make sense since, among other things, it is unreasonable to require each CD item in the order table to match a tuple in the book table.

These conditional dependencies tell us that the database D' is not clean after all. Indeed, tuples t_1, t_2 in the customer table violate CFD1: while they both represent customers in the UK and have the same zip code, they differ in their street attributes. Furthermore, each of t_1 and t_2 violates CFD2: while its area code is 131, its city is NYC instead of EDI. Similarly, t_4 violates CFD3. From these one can see that while it takes two tuples to violate a traditional functional dependency, a single tuple may violate a conditional functional dependency. The inconsistencies in D' are not limited to the customer table: while tuple t_5 in the order table has item type = book, there exists no tuple t' in the book table such that t_5 and t' match on their id, title, and price attributes. Thus, either t_5 in the order table is not error-free or the book table is incomplete or inconsistent. The inconsistency across different tables is detected by CIND1.

Conditional functional and inclusion dependencies are extensions of traditional functional and inclusion dependencies, respectively. In their general form, each conditional functional (resp.

inclusion) dependency is a pair comprising of (i) a traditional functional (resp. inclusion) dependency and (ii) a pattern tableau consisting of tuples that enforce binding of semantically related data values. Traditional functional (resp. inclusion) dependencies are a special case of conditional functional (resp. inclusion) dependencies, in which the tableaux do not include tuples with patterns of data values. As opposed to traditional functional and inclusion dependencies that were developed mainly for schema design, conditional functional and inclusion dependencies aim to capture inconsistencies of the data, for data cleaning. As shown by the example above, conditional dependencies are capable of detecting more errors and inconsistencies than what their traditional counterparts can find.

In summary, integrity constraints specify a fundamental part of the semantics of the data. Indeed, errors and inconsistencies in real-world data often emerge as violations of integrity constraints. The more expressive the constraints are, the more errors and inconsistencies can be caught. On the other hand, as will be seen shortly, the expressive power of the constraints often comes with the price of extra complexity for finding database repairs.

Repair Models

Consider functional dependencies, inclusion dependencies, denial constraints, conditional functional dependencies, and conditional inclusion dependencies. Given a database instance D of a schema R, if D violates a set Σ consisting of these constraints, one can always edit D to obtain a consistent instance D' of R, such that D' satisfies Σ. An extreme case is to delete all tuples from D and thus get an empty D'. Such a fix is obviously impractical: it removes inconsistencies as well as correct information. Apparently database repairs should not be conducted with the price of losing information of the original data. This motivates the criterion for database repairs to minimally differ from the original data.

Several repair models have been proposed [1, 6]. One model allows tuple deletions only, assuming that the information in the database D is inconsistent but is complete. In this model, a repair D' is a maximal subset of D that satisfies Σ. For example, consider Σ consisting of the FD given above and the database D shown in Fig. 1. Then a repair of D can be obtained by removing either t_3 or t_4 from the customer table.

Another model allows both tuple deletions and insertions. In this model, a repair D' is an instance of R such that (i) D' satisfies Σ, and (ii) the difference between D and D', i.e., $(D \smallsetminus D') \cup (D' \smallsetminus D)$, is minimal when D' ranges over all instances of R that satisfy Σ. As an example, let Σ consist of the FD and IND given above and D be the database of Fig. 1. Then one can obtain a repair of D either by removing both t_3 and t_5 or by removing t_3 but inserting a tuple t' to the book table such that t_5 and t' agree on their id, title, and price attributes.

A more practical model is based on updates, i.e., attribute-value modifications. To illustrate this, let us consider Σ consisting of all the constraints that have been encountered, i.e., FD, IND, DC, CFD1, CFD2, CFD3, and CIND given above, and the database D of Fig. 1. Observe that every tuple in the customer relation violates at least one of the (conditional) functional dependencies in Σ. In the two models mentioned above, the only way to find a repair is by removing all tuples from the customer table. However, it is possible that only some fields in a customer tuple are not correct, and thus it is an overkill to remove the entire tuple. A more reasonable fix is to update the tuples by, e.g., changing t_1[city] and t_2[city] to EDI (for CFD2), t_1[street] to Crichton (for CFD1), t_3[city] and t_4[city] to MH (for CFD3), t_3[street] to Mountain Ave (for FD), t_6[price] to 7.99 (for DC), and t_5[title] to Harry Porter (for CIND). This yields a repair in the update model.

An immediate question about the update model concerns what values should be changed and what values should be chosen to replace the old values. One should make the decisions based on both the accuracy of the attribute values to be modified and the "closeness" of the new value to the original value. Following the practice of US national statistical agencies [10], one can define a cost metric as follows [2, 8]. Assuming that a *weight* in the range [0,1] is associated with each attribute A of each tuple t in D, denoted by $w(t, A)$ (if $w(t, A)$ is not available, a default weight can be used instead). The weight reflects the confidence of the *accuracy* placed by the user in the attribute $t[A]$ and can be propagated via data provenance analysis in data transformations. For two values v, v' in the same domain, assume that a *distance function*(v, v') is in place, with lower values indicating greater similarity. The cost of changing the value of an attribute $t[A]$ from v to v' can be defined to be:

$$\text{cost}(v, v') = w(t, A) \cdot \text{dis}(v, v') / \max(|v|, |v'|),$$

Intuitively, the more accurate the original $t[A]$ value v is and the more distant the new value v' is from v, the higher the cost of this change. The similarity of v and v' is measured by $\text{dis}(v, v')/(|v|, |v'|)$, where $|v|$ is the length of v, such that longer strings with one-character difference are closer than shorter strings with one-character difference. The cost of changing the value of a tuple t to t' is the sum of $\text{cost}(t[A], t'[A])$ when A ranges over all attributes in t for which the value of $t[A]$ is modified. The cost of changing D to D', denoted by $\text{cost}(D', D)$, is the sum of the costs of modifying tuples in D. A repair of D in the update model is an instance D' of R such that (i) D' satisfies Σ, and (ii) $\text{cost}(D', D)$ is minimal when D' ranges over all instances of R that satisfy Σ.

The accuracy of a repair can be measured by precision and recall metrics, which are the ratio of the number of errors correctly fixed to the total number of changes made and the ratio of the number of errors correctly fixed to the total number of errors in the database, respectively.

Methods for Finding Database Repairs

It is prohibitively expensive to find a repair of a dirty database D by manual effort. The objective of constraint-based database repair is to automatically find candidate repairs of D. These candidate repairs are subject to inspection and change by human experts. There have only been preliminary results on methods for finding quality candidate repairs, as outlined below.

Given a set Σ of integrity constraints, either defined on a schema R or discovered from sample instances of R, one first wants to determine whether or not Σ is dirty itself. That is, before Σ is used to find repairs of D, one has to check, at compile time, whether or not Σ is consistent or makes sense, i.e., whether or not there exists a nonempty database instance of R that satisfies Σ. For traditional functional and inclusion dependencies, this is not an issue: one can specify arbitrary functional and inclusion dependencies without worrying about their consistency. While conditional inclusion dependencies alone retain this nice property, it is no longer the case when it comes to conditional functional dependencies. For example, consider the following conditional functional dependencies: $\psi_1 = R(A = \text{true} \rightarrow B = b_1)$, $\psi_2 = R(A = \text{false} \rightarrow B = b_2)$, $\psi_3 = R(B = b_1 \rightarrow A = \text{false})$, and $\psi_4 = R(B = b_2 \rightarrow A = \text{true})$, where the domain of attribute A is Boolean. While each of these constraints can be separately satisfied by a nonempty database instance, there exists no nonempty instance that satisfies all of these constraints. Indeed, for any tuple t in an instance, no matter what Boolean value $t[A]$ has, these constraints force $t[A]$ to take the other value from the Boolean domain.

The consistency problem is already NP-complete for conditional functional dependencies alone [9], and it becomes undecidable for conditional functional and inclusion dependencies taken together [3]. In light of the complexity, the consistency analysis is necessarily conducted by effective heuristic methods, ideally with performance guarantee. There has been approximate algorithms developed for checking the consistency of conditional functional dependencies and heuristic algorithms for conditional functional and inclusion dependencies taken together (see [10] for details).

After Σ is confirmed consistent, one needs to detect the inconsistencies in the database D, i.e., to find all tuples in D that violate one or more constraints in Σ. It has been shown that it is possible to automatically generate a fixed number of SQL queries from Σ, such that these queries can find all violations in D. This strategy works when Σ consists of functional dependencies, in-

clusion dependencies, conditional functional, and inclusion dependencies [10]. Better yet, the size of the queries is dependent upon neither the number of constraints in Σ nor the size of the pattern tableaux in the conditional dependencies in Σ.

After all violations are identified, the next step is to find an accurate repair of D by fixing these violations. This is challenging: in the repair model based on attribute-value updates, the problem of finding a database repair is already NP-complete even when the database schema is fixed, and only a fixed number of traditional functional (or inclusion) dependencies are considered [2].

To cope with the tractability of the problem, several heuristic algorithms have been developed, based on the cost model given above. A central idea of these algorithms is to separate the decision of which attribute values should be equal from the decision of what value should be assigned to these attributes. Delaying value assignment allows a poor local decision to be improved in a later stage of the repairing process and also allows a user to inspect and modify a repair. To this end an equivalence class $eq(t, A)$ can be associated with each tuple t in the dirty database D and each attribute A in t. The repairing is conducted by merging and modifying the equivalence classes of attributes in D. For example, if tuples t_1, t_2 in D violate a functional dependency $R(X \rightarrow A)$, one may want to fix the inconsistency by merging $eq(t_1, A)$ and $eq(t_2, A)$ into one, i.e., by forcing t_1 and t_2 to match on their A attributes. If a tuple t_1 violates an inclusion dependency $R_1[X] \subseteq R_2[Y]$, one may want to resolve the conflict by finding an existing tuple t_2 in the R_2 relation or inserting a new tuple t_2 into the R_2 table, such that for each corresponding attribute pair (A, B) in $[X]$ and $[Y]$, $t_1[A] = t_2[B]$ by merging $eq(t_1, A)$ and $eq(t_2, B)$ into one. A target value is assigned to each equivalence class when no more merging is possible.

Based on this idea, effective heuristic algorithms have been developed for repairing databases using traditional functional and inclusion dependencies (e.g., [9]). The algorithms modify tuple attributes in the right-hand side of a functional or inclusion dependency in the presence of a violation. This strategy, however, no longer works for conditional functional dependencies: the process may not even terminate if only tuple attributes in the right-hand side of a conditional functional dependency can be modified. Heuristic algorithms for repairing conditional functional dependencies have been developed [9], which are also based on the idea of equivalence classes but may modify tuple attributes in either the left-hand side or right-hand side of a conditional functional dependency in the presence of a violation.

Data repairing methods are typically heuristic. They attempt to fix all the errors in the data, but do not guarantee that the generated fixes are correct. Worse still, new errors may be introduced when trying to repair the data. In practice, we often want to find *certain fixes*, i.e., fixes that are guaranteed to be correct, although we might not be able to fix *all* the errors in the data. The need for certain fixes is particularly evident when repairing critical data, e.g., medical data, in which a seemingly minor error may mean life or death. To this end, repairing algorithms with certain fixes have recently been developed [13].

Data repairing and entity resolution (a.k.a. record matching, data deduplication) are typically treated as independent processes. However, the two processes often interact with each other: repairing helps us resolve entities and vice versa. This suggests that we unify repairing and entity resolution by interleaving their operations. Algorithms for both data repairing and entity resolution have recently been studied in [12], based on integrity constraints.

Closely related to data consistency is *data accuracy*. Given a set I_e of tuples pertaining to the same entity e, data accuracy aims to find the most accurate values for e. More specifically, it is to compute a tuple t_e, referred to as the *target tuple for e from* I_e, such that for each attribute A of e, $t_e[A]$ is a value in I_e that is closest to the *true A-value* of e. There have been algorithms for deciding relative accuracy of attributes and for deducing the true values of entities [5], which yield another approach to data repairing.

Key Applications

Constraint-based database repairs have a wide range of applications in, e.g., data standardization, data quality tools, data integration systems, master data management, and credit card fraud detection. The need for data repairing tools is more evident when we deal with big data, for which the scale of data inconsistencies is far more daunting than for traditional databases.

Future Directions

The study of constraint-based database repair is still in its infancy. There is naturally much more to be done. One topic for future research is to identify new integrity constraints that are capable of detecting inconsistencies and errors commonly found in practice, without incurring extra complexity. The second topic is to develop more accurate and practical repair models. The third topic is to find heuristic methods, with performance guarantees, for reasoning about integrity constraints used for data cleaning, such as their consistency and implication analyses. The fourth yet the most challenging topic is to develop scalable distributed algorithms for finding database repairs when the dataset is "big," with performance guarantee such as to guarantee that the precision and recall of the repairs found are above a predefined bound with a high confidence.

Cross-References

▶ Data Cleaning
▶ Data Quality Models
▶ Database Dependencies
▶ Database Repair
▶ Functional Dependency
▶ Inconsistent Databases
▶ Record Linkage

Recommended Reading

1. Arenas M, Bertossi LE, Chomicki J. Consistent query answers in inconsistent databases. In: Proceedings of the 18th ACM SIGACT-SIGMOD-SIGART Sym-

posium on Principles of Database Systems; 1999. p. 68–79.

2. Bohannon P, Fan W, Flaster M, Rastogi R. A cost-based model and effective heuristic for repairing constraints by value modification. In: Proceedings of the ACM SIGMOD International Conference on Management of Data; 2005. p. 143–54.

3. Bravo L, Fan W, Ma S. Extending dependencies with conditions. In: Proceedings of the 33rd International Conference on Very Large Data Bases; 2007. p. 243–54.

4. Calì A, Lembo D, Rosati R. On the decidability and complexity of query answering over inconsistent and incomplete databases. In: Proceedings of the ACM SIGACT-SIGMOD-SIGART Symposium on Principles of Database Systems; 2003. p. 260–71.

5. Cao Y, Fan W, Yu W. Determining the relative accuracy of attributes. In: Proceedings of the ACM SIGMOD International Conference on Management of Data; 2013. p. 565–76.

6. Chomicki J. Consistent query answering: five easy pieces. In: Proceedings of the 11th International Conference on Database Theory; 2007. p. 1–17.

7. Chomicki J, Marcinkowski J. Minimal-change integrity maintenance using tuple deletions. Inf Comput. 2005;197(1–2):90–121.

8. Chomicki J, Marcinkowski J. On the computational complexity of minimal-change integrity maintenance in relational databases. Inconsistency Tolerance. 2005. Lecture Notes in Computer Science 3300:119–150.

9. Cong G, Fan W, Geerts F, Jia X, Ma S. Improving data quality: consistency and accuracy. In: Proceedings of the 33rd International Conference on Very Large Data Bases; 2007. p. 315–26.

10. Fan W, Geerts F. Foundations of data quality management. Synthesis lectures on data management. Morgan & Claypool Publishers; 2012.

11. Fan W, Geerts F, Jia X, Kementsietsidis A. Conditional functional dependencies for capturing data inconsistencies. ACM Trans Database Syst. NY, USA: 2008;33(2):1–48.

12. Fan W, Li J, Ma S, Tang N, Yu W. Interaction between record matching and data repairing. In: Proceedings of the ACM SIGMOD International Conference on Management of Data; 2011. p. 469–80.

13. Fan W, Li J, Ma S, Tang N, Yu W. Towards certain fixes with editing rules and master data. VLDB J. 2012;21(2):213–38.

14. Fellegi I, Holt D. A systematic approach to automatic edit and imputation. J Am Stat Assoc. 1976;71(353):17–35.

15. Lopatenko A, Bertossi LE. Complexity of consistent query answering in databases under cardinality-based and incremental repair semantics. In: Proceedings of the 11th International Conference on Database Theory; 2007. p. 179–93.

16. Wijsen J. Database repairing using updates. ACM Trans Database Syst. 2005;30(3):722–68.

Content-and-Structure Query

Thijs Westerveld
Teezir Search Solutions, Ede, Netherlands

Synonyms

CAS query; CO+S query

Definition

A content-and-structure query is a formulation of an information need in XML retrieval or, more generally, in semi-structured text retrieval that includes explicit information about the structure of the desired result.

Key Points

Content-and-structure query is a term from semi-structured text retrieval, used predominantly for XML retrieval. The term refers to a specific way of querying a structured document collection. In addition to describing the (topical) content of the desired result, content-and-structure queries include explicit hints about the structure of the desired result or the structure of the context it appears in. Content-and-structure queries are useful for users who have knowledge about the collection structure and want to express the precise structure of the information they are after. For example, they can express the granularity of the desired results, e.g., return *sections* about architecture, or they can express the structural context of the information they are looking for, e.g., return *sections* about architecture within documents about *Berlin*. It is up to the retrieval system to decide how to use the structural hints in locating the most relevant information. In INEX, the Initiative for the Evaluation of XML Retrieval [1], content-and-structure queries are known as *CAS* queries or CO+S queries (*Content-Only queries* with structural hints) and expressed in the NEXI language [2]. More information on query languages, including content-only and content-and-structured queries in the field of XML search can be found in [1].

Cross-References

▸ Content-Only Query
▸ Narrowed Extended XPath I
▸ XML Retrieval

Recommended Reading

1. Amer-Yahia S, Lalmas M. XML search: languages, INEX and scoring. ACM SIGMOD Rec. 2006;35(4):16–23.
2. Trotman A, Sigurbjörnsson B. Narrowed extended xpath i (NEXI). In: Fuhr N, Lalmas M, Malik S, Szlavik Z, editors. Advances in XML Information Retrieval: Third International Workshop of the Initiative for the Evaluation of XML Retrieval, INEX 2004, Dagstuhl Castle, 6–8 Dec 2004. Revised Selected Papers, Vol. 3493. Berlin/Heidelberg/New York: Springer; 2005. GmbH. http://www.springeronline.com/3-540-26166-4.

Content-Based Publish/Subscribe

Hans-Arno Jacobsen
Department of Electrical and Computer Engineering, University of Toronto, Toronto, ON, Canada

Definition

Content-based publish/subscribe is a communication abstraction that supports selective message dissemination among many sources and many sinks. The publication message content is used to make notification decisions. Subscribers express interest in receiving messages based on specified filter criteria that are evaluated against publication messages. Content-based publish/subscribe is an instance of the publish/subscribe concept.

Key Points

Content-based publish/subscribe is an instance of the publish/subscribe concept. In the content-based model clients interact by publishing messages and subscribing to messages through the publish/subscribe system. The key difference between the content-based and other publish/subscribe models is that the content of publication messages is used as the basis for disseminating publications to subscribers. Subscriptions consist of filters that specify subscriber interests making reference to publication message content. The publish/subscribe system matches publications against subscriptions by evaluating the publication message content against the filters expressed in subscriptions. The kind of content to subscribe to that exists in content-based publish/subscribe systems is either out-of-band information and must be know to clients, or is dynamically discoverable by clients based on additional support provided by the system.

A publication message published to the content-based publish/subscribe system is delivered to all subscribers with matching subscriptions. A subscription is a Boolean function over predicates. A publication matches a subscription, if the Boolean function representing it evaluates to true, otherwise the publication does not match the subscription.

As in the other publish/subscribe models, the content-based publish/subscribe model decouples the interaction among publishing data sources and subscribing data sinks. The same decoupling characteristics as discussed under the general publish/subscribe concept apply here as well. Specific realizations of this model found in practice vary in the exact decoupling offered. To properly qualify as publish/subscribe, at least the anonymous communication style must exist. That is publishing clients must not be aware of who the subscribing clients are and how many subscribing clients exist, and vice versa. Thus, content-based publish/subscribe enables the decoupled interaction of n sources with m sinks for $n, m \geq 1$.

In content-based publish/subscribe, the publication data model is defined by the data model underlying the definition of publication messages. The publication data model defines the structure and the type of publication messages processed by the system. In many approaches, a publication is a set of attribute-value pairs, where values are explicitly or implicitly typed. In explicit typing, each attribute-value pair has an additional type component specifying the type of the value. In implicit typing, no type is specified, and the type interpretation for matching is conveyed by the operator specified in the subscription referencing the attribute. The type-based publish/subscribe concept is a refinement of this based on programming language type theory.

Some content-based publish/subscribe approaches exist that define multi-valued attributes, where an attribute may have more than one value. Also, publication schemas and patterns have been introduced that specify certain attributes as required, while others are optional. Besides representing publications as attribute-value pairs, many other representations of publications have been introduced in the literature, such as XML, RDF, and strings.

The subscription language model is closely tied to the publication data model and defines the subscriptions the publish/subscribe system processes. The subscription language model that corresponds to the above described attribute-value pair-based publication data model, represents subscriptions as Boolean functions over predicates. Most content-based publish/subscribe systems process conjunctions of predicates only. In these systems, more general subscriptions must be represented as separate conjunctive subscriptions. A predicate is an attribute- operator-value triple that evaluates to true or false. Besides representing subscriptions as Boolean functions over predicates, many other representations of publications have been introduced in the literature, such as XPath, RQL, regular expressions, and keywords.

Subscription language and publication data model define subscriptions and publications processed by the publish/subscribe system. The matching semantic defines when a publication matches a subscription. Commonly the matching semantic is crisp; that is a publication matches

a subscription or it does not. However, other semantics have been explored, such as an approximate match, a similarity-based match, or even a probabilistic match.

The publish/subscribe matching problem is stated as follows: Given a set of subscriptions and a publication, determine the subscriptions that match for the given publication. The publish/subscribe system can be interpreted as a filter that based on the subscriptions it stores, publications that do not match are filtered out, while those that match are forwarded to subscribers that have expressed interest in receiving information by registering subscriptions. The challenge is to efficiently solve this problem without computing a separate match between all subscriptions and the given publication. This is possible since in many applications, subscriptions share predicates, subsumption relationships exist among different subscriptions, and the evaluation of one predicate may allow to determine the result of other predicates without requiring explicit computation.

Content-based publish/subscribe systems differ in the publication data model, the subscription language model, the matching semantic, and the system architecture. In a system based on a centralized architecture, all publishers and subscribers connect to one and the same publish/subscribe system. In a system based on a distributed architecture, publishers and subscribers connect to one of many publish/subscribe systems that are interconnected in a federation. The federated publish/subscribe system offers the same functionality and solves the same matching problem as the centralized one.

Content-based publish/subscribe differs from topic-based publish/subscribe in that the entire message content is used for matching, while in a topic-based approach only the topic associated with a message is used.

Content-based publish/subscribe differs from database stream processing in that the publish/subscribe system processes publications of widely varying schemas. In the extreme case, every publication processed by the system could be based on a different schema. In stream processing, each data stream follows one and the same schema, which is an important assumption in the design of the stream query engine.

Content-based publish/subscribe has been an active area of research, since at least the late 1990s. The early work in the area was influenced from approaches in active databases, network management, and distributed systems. Many academic and industry research projects have developed content-based publish/subscribe systems. Various standards exhibit elements of the above described content-based publish/subscribe model. These standards are the CORBA Notification Service [3], the OMG Data Dissemination Service [4], the OGF's Info-D specification [2], and the Advanced Message Queuing Protocol [1].

Content-based publish/subscribe intends to support applications that need to highly selectively disseminate messages from one or more data sources to several data sinks. The mapping of sources to sinks can change dynamically with every message published and is fully determined by the publication content and the at publication time existing subscriptions. Given no change in the subscription set, one and the same message published twice, is sent to the same recipient set. Applications that require fine-grained filtering capabilities are ideally suited for realization with content-based publish/subscribe. Most existing publish/subscribe systems allow the application to dynamically change subscriptions at run-time. There are many applications that follow these characteristics. Examples include selective information dissemination, information filtering, database trigger processing, application-level firewalls, intrusion detection systems, and notification and altering services. Furthermore, recently it was demonstrated how higher-level applications can be build effectively with content-based publish/subscribe. Scenarios in this category are business activity monitoring, business process execution, monitoring and control of service level agreements, and automatic service discovery.

In the literature, the term content-based publish/subscribe refers to the above-described model and encompasses the centralized as well as the distributed realization of the publish/-subscribe concept. In this context the terms matching and filtering are used interchangeably. The term content-based routing is reserved for the distributed realization of the model, where the publish/subscribe system is also referred to as a router, a broker, or a publish/subscribe message broker. In information retrieval, the publish/subscribe matching problem is referred to as information filtering. Subscriptions are then referred to as profiles or filters.

Cross-References

▶ Publish/Subscribe
▶ Type-Based Publish/Subscribe

Recommended Reading

1. AMQP Consortium. Advanced message queuing protocol specification, version 0–10 edition; 2008.
2. OGF. Information dissemination in the grid environment base specifications; 2007.
3. OMG. Notification service specification, version 1.1, formal/04-10-11 edition; Oct 2004.
4. OMG. Data distribution service for real-time systems, version 1.2, formal/07-01-01 edition; Jan 2007.

Content-Based Video Retrieval

Cathal Gurrin
Dublin City University, Dublin, Ireland

Synonyms

Digital video retrieval; Digital video search

Definition

Content-based Video Retrieval refers to the provision of search facilities over archives of digital video content, where these search facilities are based on the outcome of an analysis of digital video content to extract indexable data for the search process.

Historical Background

As the volume of digital video data in existence constantly increases, the resulting vast archives of professional video content and UCC (User Created Content) are presenting an opportunity for the development of content-based video retrieval systems. Content-based video retrieval system development was initially lead by academic research such as the Informedia Digital Video Library [3] from CMU and the Físchlár Digital Video Suite [6] from DCU (Dublin City University). Both of these systems operated over thousands of hours of content, however digital video search has now become an everyday WWW phenomenon, with millions of items of digital video being indexed by the major WWW search engines and video upload sites. The early research focused content-based digital video systems, such as the offerings from CMU and DCU, exploited aspects of text search and content-based image search in order to provide intelligent indexing, retrieval, summarization and visualization of digital video content. In recent years, the emerging WWW search engines have focused on the textual indexing of large quantities of digital video, at the expense of performing complex and time-consuming visual content analysis.

Foundations

The aim of a content-based video search system is to answer user queries with a (ranked) list of appropriate video content. Many different sources of video content exist and each needs to be treated differently. Firstly there is professional created content such as TV news content, documentaries, TV programmes, sports video and many others. Professional content is directed and polished content with many visual effects, such as a movie, music video or TV programme. Secondly there is the increasing quantities of UCC (User Created Content), such as home movie

content or amateur/semi-professional content and finally there is security video content, which is increasingly being captured as the number of surveillance cameras in use increases. In addition to the type of content, another factor of key importance for content-based video search is the unit of retrieval.

Unit of Retrieval

Different content types will require different units of retrieval. In most WWW video search systems such as YouTube or Google Video, the unit of retrieval is the entire video content, which is sensible because most of the video content is short UCC or UUC (User Uploaded Content) clips. Retrieval of entire video units is not ideal for other types of content, for example TV news video, where the logical unit of retrieval would be a news story. There are a number of units of retrieval that are typically employed in addition to the entire video unit.

A *shot* in digital video is a sequence of continuous images (frames) from a single camera. A shot boundary is crossed when a recording instance ends and a new one begins. In many content-based video search systems the (automatically segmented) shot is the preferred unit of retrieval due to the fact that it is relatively easy to split a video file into its constituent shot in an automatic process called shot boundary detection [1]. Once a video stream has been segmented into shots, it can be browsed or indexed for subsequent search and retrieval, as shown in Fig. 1. A *scene* in digital video is a logical combination of video shots that together comprise some meaningful semantic unit. A *news story* is a special type of scene that is found in the context of news video. News stories can be automatically segmented from news video in a process called story-segmentation. This, like scene segmentation, is not a simple process, though reasonable accuracy can be achieved by replying on a number of individual cues from the video content and can be improved by exploiting the unique video production techniques of a particular broadcaster. For some video content, generating a *summary* or a *video skim* is a logical unit of retrieval. These summaries can be independent of any user need

Content-Based Video Retrieval, Fig. 1 StoryBoard Interface from the Físchlár Video Retrieval System [2]

(context or query) or can be generated in response to a user need. Summaries have been successfully employed in the domain of field sports [4] or news summaries of reoccurring news topics [2]. Finally, as mentioned earlier, the entire video content may be returned in response to a user query, as is the case on many WWW video search engines in 2008.

Representing Video Content Visually on Screen

The quality of the interface to a content-based video retrieval system has a great effect on the usefulness of the system. In order to represent digital video visually, one or more keyframes (usually JPEG images) are usually extracted from the video to represent the content. These keyframes can then be employed for visual analysis of the video content by representing a video clip (typically a shot) by its keyframe and applying visual analysis techniques to the keyframe. In addition, these keyframes may be employed for visual presentation of video contents to support a degree of random access into the content. By processing video into a sequence of shots/scenes, and representing each shot/scene with one or more keyframes allows for the display of an entire video as a sequence of keyframes in what is called a StoryBoard interface, as shown in Fig. 1. Clicking on any keyframe would typically begin video playback from that point.

However, relying on simply presenting keyframes in screen can still require browsing through a very large information entity for long videos, maybe having over a hundred keyframes (shots) per hour. Therefore the ability to search within video content to locate a desired section of the video is desirable.

Searching Archives of Digital Video

The goals of supporting search through digital video archives are to (i) understand video content and (ii) understand how relevant content is likely to be to a user's query and to (iii) present the most highest ranked content for user consideration. We try to achieve these goals by indexing video content utilizing a number of sources (textual, audio and visual), either alone or in any combination. The unit of retrieval can be shots, scenes, stories, entire video units or any other unit of retrieval required.

Content-Based Retrieval using Text Sources

The most common searching technology used for video retrieval in the WWW is content searching using proven text search techniques. This implies that it is possible to generate textual content (often referred to as a text surrogate) for the video. There are a number of sources of text content that can be employed to generate these textual surrogates, for example sources based on analyzing the digital video or the broadcast video stream:

- *Spoken words*, generated by utilizing a speech-to-text tool. The spoken words will provide an indication of the content of the video.
- *Written words*, extracted using a process of OCR (OCR – Optical Character Recognition) from the actual visual content of the video frames.
- *Professional closed caption annotation*, which are the closed caption (teletext) transcripts of video that accompanies much broadcast video content.

In addition there are many sources of textual evidence that can be employed that do not directly rely on the content of the digital video stream, and typically, these would be available for publicly available WWW digital video content:

- *Professionally annotated metadata* from the content provider which, if available, provides a valuable source of content for the textual surrogate.
- *Community annotated metadata* from general users of the content. On WWW video sharing sites users are encouraged to annotate comments about the video content and these annotations can be a valuable source of indexable content.

All of these sources of textual data can be employed alone, or in any combination to generate textual surrogates for video content (shots, scenes, stories or entire videos). Users can query such systems with conventional text queries and this is the way that most WWW video search engines operate. Text search through video archives is a very effective way to support search and retrieval and relies on well-known and proven text search techniques.

Content-Based Retrieval using Visual Sources

Digital Video, being a visual medium, can also be analyzed using visual analysis tools, which typically operate over individual keyframes to visually index each clip (typically a shot or scene). The visual content analysis tools are often borrowed from the domain of visual image analysis. The first generation of video analysis systems relied on modeling video with easily extractable low-level visual features such as color, texture and edge detection. However a significant "semantic gap" exists between these low-level visual features and the semantic meaning of the video content, which is how a typical user would like to query a video search system. To help bridge this semantic gap, video content in the current generation of video search systems is processed to seek more complex semantic (higher-level or derived) visual concepts, such as people (faces, newsreaders), location (indoor/outdoor, cityscape/landscape), objects (buildings, cars, airplanes), events (explosions, violence) and production techniques such as camera motion. The output of these higher-level concept detectors can, with sufficient development and training, be successfully integrated (mainly as filters) into content-based video retrieval systems. However, the development of these concept detectors can be a difficult process and it is not reasonable to assume the development of tens of thousands of concept detectors to cover all concepts for the video archive. Research carried out by the Informedia team at CMU suggest that "concept-based" video retrieval with fewer than 5,000 concepts, detected with minimal accuracy of 10% mean average precision is likely to provide high accuracy results, comparable to text retrieval on the web, for a typical broadcast news video archive. Extending into other domains besides broadcast news may require some additional concepts. A review of image analysis techniques will provide more details of these semantic visual concept detectors and how they can be developed.

The output of easily extracted low-level feature analysis can be also employed in a content-based video retrieval system to support linking between visually similar content, though it is unlikely to be used to support direct user querying. Semantic features can form part of a user query, whereby a user, knowing the semantic factures extracted from a video archive, can specify semantic features that are required/not required in the result of a video search. For example, a user may request video content concerning forest fires, that also contains the feature "fire."

Content-Based Retrieval using Audio Sources

Apart from the speech-to-text there are other uses of audio sources for content-based video retrieval. For example, security video to identify non-standard audio events, such as a window breaking to provide a special access point to security video at this point. Key events in sports video can be identified using visual analysis (e.g., goalmouth detection, or onscreen scoreboard changing) but also using audio analysis, for example crowd noise level or commentator excitement level.

Effective Retrieval

As can be seen from the current generation of WWW video search engines, most content-based video retrieval relies on user text queries to operate over text surrogates of video content. In typical situations the use of visual sources does not achieve any noticeable improvement in performance over using textual sources (such as CC text or ASR text). However, combining both sources of evidence can lead to higher performance than using either source alone. Figure 2 summarizes a typical shot-level content-based indexing process for digital video and illustrates some of the indexing options available.

Often successful academic video search systems allow the user to search to find the location in a piece of video which is likely to be of interest, with the user being encouraged to browse this area of the video by presenting keyframes from shots in the general video area (before and after).

Key Applications

The key application areas can be broadly divided into two categories; domain dependent video retrieval and generic (non-domain) video retrieval. In *domain dependent video retrieval*, the domain of the search system is limited, thereby allowing the search tool to exploit any domain dependent knowledge to develop a tailored and more effective content-based video retrieval system. Typical domains include news video where the unit of retrieval would a news story. Domain dependent additions for news video retrieval include anchor person detection to aid in the identification of news story bounds, inter-story linkage, story trails and timeline story progression. An example of a typical news story retrieval system is the Físchlár-News Digital Video Library that was operational from 2001 to 2004 at Dublin City University, and shown in Fig. 3.

Another example domain dependent application area is sports video, where research has been progressing on generating automatic summaries of field sports events and a third example is security video where research is ongoing into the automatic analysis of security footage to identify events of interest or even to identify and track individuals and objects in the video streams from many cameras in a given location.

Domain independent video retrieval attempts to index all types of content, such as general TV programmes or generic UCC. Given that there are no domain specific clues to exploit, retrieval is usually on textual indexing of a textual surrogate or extracted visual concepts, such as objects [5], locations, people, etc. The unit of retrieval would typically be a shot or an entire video clip, but could also be a non-shot unit that matches the user request. Domain independent video retrieval is most commonly seen in WWW video search

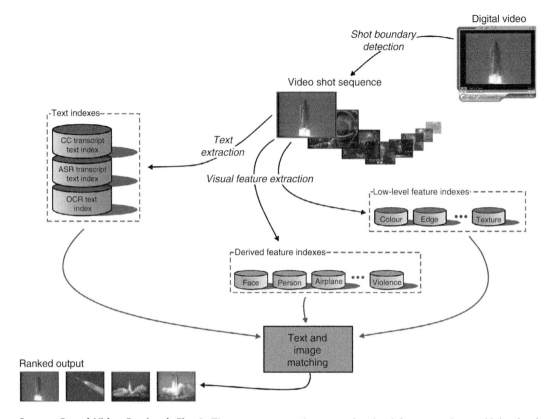

Content-Based Video Retrieval, Fig. 2 The content-based indexing process for digital video (from Físchlár system at TRECVid in 2004) showing some text extraction, some low-level features and some higher level (derived) features (concepts)

engines, which index content based on text surrogates and returns entire videos in the result set.

Future Directions

Future applications of, and research into content-based video search will likely focus on developing techniques for providing access to large archives of digital video content as broadcasters continue the process of digitizing their huge archives of programmes and the raw video content (rushes) that is used in the making of TV programmes. For rushes content especially, one will not be able to rely on text transcripts for indexing purposes. In addition, the ever increasing volume of UCC requires content-based retrieval techniques to be developed that will provide an improved semantic search facility over this content. Finally, the third point of research into the future will likely be in migrating content-based retrieval tools onto consumer devices (PVRs for example), which themselves are becoming capable of storing hundreds of hours of recorded video and UCC.

Experimental Results

In the field of content-based video retrieval there exists an annual, worldwide forum for the evaluation of techniques for video search, called TRECVid [7] which began in 2001. The TRECVid workshop is (2007) part of the TREC [8] conference series is sponsored by the National Institute of Standards and Technology (NIST). In 2007, 54 teams from Europe, the Americas, Asia, and Australia participated in TRECVid. Over the course of the TRECVid evaluations, data employed has been either

Content-Based Video Retrieval, Fig. 3 Físchlár-News, a domain dependent content-based video retrieval system

TV news, documentaries, educational video and rushes (Rushes content, is the unproduced content that is used to prepare TV programming.) content. TRECVid has organized a number of tasks for the annual evaluations which may change each year. The tasks evaluated, 2001, have included shot boundary determination, interactive and automatic (no query modification or browsing) video search, high-level concept detection, story boundary determination for TV news and camera motion analysis, among others.

In addition to TRECVid, other evaluation forums also exist such as Video Analysis and Content Extraction (VACE) which is a US program that addresses the lack of tools to assist human analysts monitor and annotate video for indexing. The video data used in VACE is broadcast TV news, surveillance, Unmanned Aerial Vehicle, meetings, and ground reconnaissance video. Other evaluation forums such as the French ETISEO and EU PETS evaluations have evaluated content-based retrieval (event detection and object detection) from surveillance video. ARGOS, sponsored by the French government, evaluated tasks similar to TRECVid and employed video data from TV news, scientific documentaries and surveillance video archives.

Some summary findings from content-based video retrieval research are that employing visual analysis of the video content does not provide a significant increase in search performance over using text transcripts, that text transcripts provide the single most important clue for searching content, that employing as many text sources as possible aids text search quality, and finally that, incorporating visual content search can improve retrieval over that of text indexing alone. The performance of visual indexing tools suggests that this is an unsolved problem with much research needed. As an example, the highest accuracy attained (in terms of Inferred Average Precision) for the twenty visual concepts evaluated at TRECVid in 2007 are shown in Table 1.

Finally, it should be noted that the interface to an interactive video search system (for example [3, 6]) can make a huge difference for effective

Content-Based Video Retrieval, Table 1 Inferred Average Precision (infAP (In terms of infAP, a value of 1.0 infers that the technique locates only correct examples of the concept, whereas a value of 0.0 infers that the technique only locates incorrect examples)) measurement for the top performing techniques for visual concept detection at the TRECVid workshop in 2007

Concept	infAP	Concept	infAP
Sports	0.144	Computer/TV screen	0.209
Weather	0.062	US flag	0.041
Office	0.222	Airplane	0.226
Meeting	0.279	Car	0.265
Desert	0.155	Truck	0.108
Mountain	0.12	Boat/ship	0.212
Waterscape/ waterfront	0.374	People marching	0.104
Police/security	0.046	Explosion/fire	0.069
Military personnel	0.081	Maps	0.236
Animal	0.249	Charts	0.225

content-based video search and retrieval. Content searching through text transcripts can locate the area of the video, but a good storyboard interface to find the exact video shot of interest is a valuable addition.

Data Sets

The TRECVid evaluation framework provides a number of datasets to support the comparative and repeatable evaluation of TREC. Since 2001, these datasets, along with the associated queries and relevance judgements are available. The video data employed in these datasets comes from various sources, such as the video from the Movie Archive of the Internet Archive, news video data in a number of languages (English, Arabic and Chinese) and rushes content. Datasets used in other evaluation forums are also available

Cross-References

▶ Video Content Analysis
▶ Video Content Modeling
▶ Video Metadata

- ▶ Video Representation
- ▶ Video Scene and Event Detection
- ▶ Video Segmentation
- ▶ Video Shot Detection
- ▶ Video Summarization

Recommended Reading

1. Browne P, Smeaton AF, Murphy N, O'Connor N, Marlow S, Berrut C. Evaluating and combining digital video shot boundary detection algorithms. In: Proceedings of the IMVIP 2000 – Irish Machine Vision and Image Processing Conference; 2000. p. 93–100.
2. Christel MG, Hauptmann AG, Wactlar HD, Ng TD. Collages as dynamic summaries for news video. In: Proceedings of the 10th ACM International Conference on Multimedia; 2002. p. 561–9.
3. Hauptmann A. Lessons for the future from a decade of informedia video analysis research, image and video retrieval. In: Proceedings of the 4th International Conference Image and Video Retrieval; 2005. p. 1–10.
4. Sadlier D, O'Connor N. Event detection in field sports video using audio-visual features and a support vector machine. IEEE Trans Circuits Syst Video Technol. 2005;15(10):1225–33.
5. Sivic J, Zisserman A. Video Google: a text retrieval approach to object matching in videos. In: Proceedings of the 9th IEEE Conference Computer Vision; 2003. p. 1470–7.
6. Smeaton AF, Lee H, Mc Donald K. Experiences of creating four video library collections with the Físchlár system. Int J Digit Libr. 2004;4(1):42–4.
7. Smeaton AF, Over P, Kraaij W. Evaluation campaigns and TRECVid. In: Proceedings of the 8th ACM SIGMM International Workshop on Multimedia Information Retrieval; 2006. p. 321–30.
8. http://trec.nist.gov Last visited June '08.

Content-Only Query

Thijs Westerveld
Teezir Search Solutions, Ede, Netherlands

Synonyms

CO query; Content-only query

Definition

A content-only query is a formulation of an information need in XML retrieval or, more generally, in semi-structured text retrieval that does not contain information regarding the structure of the desired result.

Key Points

Content-only query or CO query is a term from semi-structured text retrieval, used predominantly for XML retrieval. The term refers to a specific way of querying a semi-structured document collection. Content-only queries ignore the structure of the collection and only refer to the (topical) content of the desired result. In that sense, they are similar to the keyword queries typically used in traditional information retrieval systems or in web search engines. The fact that structural information is lacking from the query formulation does not mean structure does not play a role. When a content-only query is posed, it is up to the retrieval system to decide the appropriate level of granularity to satisfy the information need. This contrasts so-called *content-and-structure queries* where the user specifies structural clues regarding the desired result. More information on query languages, including content-only and content-and-structured queries in the field of XML search can be found in [1].

Cross-References

- ▶ Content-and-Structure Query
- ▶ Narrowed Extended XPath I
- ▶ Xml Retrieval

References

1. Amer-Yahia S, Lalmas M. XML search: languages, INEX and scoring. ACM SIGMOD Rec. 2006;35(4):16–23.

Context

Opher Etzion
IBM Software Group, IBM Haifa Labs, Haifa
University Campus, Haifa, Israel

Synonyms

Life-span (in part); Space-span (in part)

Definition

A context is a collection of semantic dimensions
within which the event occurs. These dimensions
may include: temporal context, spatial context,
state-related context and reference-related context.

Key Points

Event processing is being done within context,
which means that an event is interpreted differently in different contexts, and may trigger different actions in different contexts, or be irrelevant
in certain context. In the event processing network, each agents operates within a single context. While the term context has been associated
with the spatial dimension, in event processing
it is most strongly associated with the temporal
dimension.

Each context-dimension may be specified either explicitly, or by using higher level abstractions.

Examples are:

- Temporal context:
  - Explicit: Everyday between 8 AM and
    5 PM EST.
  - Implicit: From sunrise to sunset.
  - Mixed: Within 2 h from admission to the
    hospital.
- Spatial context:
  - Explicit: Within 1 KM from coordinate
    $+ 51° 3' 45.71'', -1° 18' 25.56''$.

- Implicit: Within the borders of the city of
    Winchester.
  - Mixed: Within 1 KM north of the border
    between Thailand and Laos.
- State-oriented context:
  - Explicit: When "red alert" is present.
  - Implicit: During traffic jam in the area.
- Reference-oriented context:
  - Explicit: Context-instance for each
    platinum-customer with credit-limit
    > $1 M.
  - Implicit: Context-instance for each "angry
    customer."

Note that the state-oriented dimension is different, since it does not relate to the event itself, and is global in nature. A context may
consist of one dimension only or combination
of dimensions. The reference-oriented context is
mainly used to partition the event space. Context
instances may or may not cover the entire space
of possibilities, a context can also be created from
binary operations on contexts (union, intersection, difference).

Cross-References

▶ Complex Event Processing
▶ Event Processing Network
▶ Retrospective Event Processing

Recommended Reading

1. Adi A, Biger A, Botzer D, Etzion O, Sommer Z.
   Context awareness in Amit. In: Proceedings of the
   5th Annual Workshop on Active Middleware Services;
   2003. p. 160–7.
2. Barghouti NS, Krishnamurthy B. Using event contexts and matching constraints to monitor software
   processes. In: Proceedings of the 17th International
   Conference on Software Engineering; 1995. p. 83–92.
3. Buvac S. Quantificational logic of context. In: Proceedings of the 10th National Conference on AI; 1996.
   p. 600–6.
4. Hong C, Lee K, Suh Y, Kim H, Kim H, Lee H.
   Developing context-aware system using the conceptual context model. In: Proceedings of the 6th IEEE
   International Conference on Information Technology;
   2006. p. 238.

5. Rakotonirainy A, Indulska J, Loke SW, Zaslavsky A. Middleware for reactive components: an integrated use of context, roles, and event based coordination. In: Proceedings of the IFIP/ACM International Conference on Distributed Systems Platforms; 2001. p. 77–98.

Contextualization in Structured Text Retrieval

Jaana Kekäläinen, Paavo Arvola, and
Marko Junkkari
University of Tampere, Tampere, Finland

Definition

In information retrieval, contextualization refers to a (re)scoring method, where the relevance of a retrievable unit (e.g. document, image, sentence or passage) is estimated by taking its context into account. The context of the unit may consist of surrounding text or external texts associated with the unit by links. In contextualization the unit's retrieval status value (RSV) is not calculated in isolation, but depending on its explicitly defined context.

Historical Background

In hyperlinked semi-structured documents, context is considered *external* in the form of citations and hyperlinks and *internal* in the form of the document's structure and these sources of information are exploited as contextual evidence. It is hypothesized that units in a good context (having strong contextual evidence) should be better candidates to be relevant to the posed query, than those in a poor context. The term contextualization in this domain was introduced by Arvola and others [1].

Focused retrieval addresses the possibility to return the most specific (and exhaustive) passages, instead of full documents [8]. Thus, internally, the text parts (sentences, paragraphs)

of a document form a context of each other within a document's internal structure. In structured documents, such as XML documents, the hierarchy is explicitly marked such that ancestor elements include the content of their descendants. As a consequence, small but relevant elements down in the hierarchy have a lower probability to match the query [1]. In other words, they may have too little textual evidence, if any in the event of non-text elements [6]. This problem, known as the vocabulary mismatch, is typical for text retrieval, and is caused by natural language allowing several ways to refer to objects. However, elements are often dependent on each other because of textual cohesion. Thus, one solution is to use the context of the element to give more evidence about the subject of the element.

A commonly used, and obvious context for elements is their containing document, i.e. the root element [12]. Contextualization based on the document is the initial contextualization model in semi-structured retrieval [1, 12]. Contextualization by ancestors utilizes a hierarchy vertically. Later on other contextualization models have also been introduced [3, 9, 11].

Apart from documents internal structure, the external link and bibliographical structure of a document collection can be a massive source of implicit contextual evidence. Thus, in external contextualization, the in-links and out-links of a unit in the citation graph are used as context [11]. There is a long tradition of link analysis in informetrics and web search, and a well-established set of related retrieval methods. Retrieval approaches in hypertext structure utilize link structures as contextual source also in the retrieval of non-textual items [6].

Foundations

Inside a (structured) document, two main structural dimensions can be distinguished: vertical and horizontal [3]. The vertical dimension is based on the hierarchy of a document. For example a newspaper article has a basic hierarchical division of sections subsections and paragraphs.

This structure gives natural contexts of different sizes. A paragraph can be viewed in the context of an article, a section, and a possible subsection and so on. The horizontal dimension is based on the sequential document order, where text parts follow each other consecutively. In addition, marked structure can be utilized in ad hoc contextualization, and graph theory based methods which are not necessarily based on vertical or horizontal dimensions.

In *vertical (hierarchical) contextualization* [1, 3], a rough classification of different context levels can be formed according to the vertical distance of ancestor elements. The nearest context of the element is the parent element. Likewise, the furthermost context is the whole document, i.e. the root element, which possesses no explicit internal context. In tower contextualization all ancestors of an elements are taken into account. The vertical contextualization is applicable on marked structured documents, and root contextualization in non-structured documents as well.

Horizontal contextualization [3]. In the document order, text passages form a sequence from the first to the last one, where each one of them is preceding or following another as neighbors as the nearest context. This approach considers term proximity both in structured or unstructured documents, thus it has been applied to passage and sentence retrieval [4, 5].

Kinship contextualization [9] considers the preceding and following elements in a branch as context and thus covers both horizontal and vertical dimensions within a structured document.

Ad hoc contextualization [3] contains a number of contextualization possibilities, where the contextualizing elements are selected arbitrarily from a known structure. A typical usage of this kind of contextualization is query specific. For example queries containing source element constraints can actually be considered as a form of contextualization. This can be illustrated with a query asking for paragraphs about "XML" from articles including a title with a word "contextualization". The interpretation of Content-and-structure queries is a representative example of ad hoc contextualization [2].

In addition, in hierarchical and other network structures, suitable graph based methods are viable in determining the context. Norozi and others [11] introduced a method using random walks for internal and external contexts.

Independent of the abovementioned contextualization method, a straightforward way to implement contextualization is calculating the context score as the linear combination of scores of the contextualizing elements and adding this score to the initial score. This can be done using a contextualization vector, where the role of each of the contextualizing elements in element scoring can be tuned based on e.g. their proximity from the contextualized element [3]. Other contextualization approaches consider language models [7] and probabilistic models [5].

Key Applications

Semi-structured retrieval, web retrieval, retrieval of non-textual objects, sentence retrieval, passage retrieval.

Experimental Results

The effectiveness of contextualization has been experimented in several test settings, and all the contextualization types mentioned above have been examined [1, 2, 3, 5, 7, 9, 10, 11, 12]. Typically, the results show notable improvement over a non-contextualized baseline in terms of precision and recall, and many of the contextualization methods seem to work in general. However, it is collection or document dependent, which contextualization approach perform best. Lengthy documents with one or few subjects are more amenable for the method than short documents, or documents with diverse subjects; contextualization might not work in an encyclopedia with short entries, but can improve effectiveness, say, in the retrieval of the elements of scientific articles.

Cross-References

▶ Content-and-Structure Query
▶ INitiative for the Evaluation of XML Retrieval
▶ Relationships in Structured Text Retrieval
▶ Structured Document Retrieval
▶ Term Statistics for Structured Text Retrieval
▶ XML Retrieval

Recommended Reading

1. Arvola P, Junkkari M, Kekäläinen J. Generalized contextualization method for XML information retrieval. In: Proceedings of the International Conference on Information and Knowledge Management; 2005. p. 20–7.
2. Arvola P, Junkkari M, Kekäläinen J. Query evaluation with structural indices. In: Fuhr N, Lalmas M, Malik S, Kazai G, editors. Advances in XML information retrieval and evaluation, LNCS vol. 3977. 2006. p. 134–45.
3. Arvola P, Kekäläinen J, Junkkari M. Contextualization models for XML retrieval. Inform Process Manag. 2011;47(5):762–76.
4. Callan JP. Passage-level evidence in document retrieval. In: Proceedings of the 30th Annual International ACM SIGIR Conference on Research and Development in Information Retrieval; 2004. p. 302–10.
5. Carmel D, Shtok A, Kurland O. Position-based contextualization for passage retrieval. In: Proceedings of the International Conference on Information and Knowledge Management; 2013. p. 1241–4.
6. Dunlop MD, Van Rijsbergen CJ. Hypermedia and free text retrieval. Inform Process Manag. 1993;29:287–98.
7. Fernandez RT, Losada DE, Azzopardi LA. Extending the language modeling framework for sentence retrieval to include local context. Inform Retrieval. 2011;14:355–89.
8. Kazai G, Lalmas M, Reid J. Construction of a test collection for the focused retrieval of structured documents. In: Sebastiani F editor. Advances in information retrieval. LNCS vol. 2633. Heidelberg: Springer; 2003. p. 88–103.
9. Norozi MA, Arvola P. Kinship contextualization: utilizing the preceding and following structural elements. In: Proceedings of the Annual International Conference on Research and Development in Information Retrieval; 2013a. p. 837–40.
10. Norozi MA, Arvola P. When is the structural context effective? In: Proceedings of the 13th Dutch-Belgian Workshop on Information Retrieval; 2013b. p. 72–5.
11. Norozi MA, de Vries AP, Arvola P. Contextualization using hyperlinks and internal hierarchical structure of Wikipedia documents. In: Proceedings of the International Conference on Information and Knowledge Management; 2012. p. 1291–300.
12. Sigurbjörnsson B, Kamps J, De Rijke M. An element-based approach to XML retrieval. In: Proceedings of the 2nd International Workshop of the Initiative for the Evaluation of XML Retrieval; 2003. p. 19–26. https://inex.mmci.uni-saarland.de/static/proceedings/INEX2003-preproceedings.pdf. Accessed 27 June 2014.

Continuous Data Protection

Kazuhisa Fujimoto
Hitachi Ltd., Tokyo, Japan

Synonyms

CDP; Continuous backup

Definition

CDP is a data protection service capturing data changes to storage, often providing the capability of restoring any point in time copies.

Key Points

CDP differs from usual backups in that users do not need to specify the point in time until they recover data from backups. From an application point of view, every time when it updates data in an original volume, CDP keeps updates. In case of recovery, when users specify the point in time, CDP creates the point in time copy from an original volume and updates.

In several CDP implementations, users can specify the granularities of restorable objects which help them to specify the point in time easily. For example, restorable objects range from crash-consistent images to logical objects such as files, mail boxes, messages, database files, or logs.

Cross-References

▶ Backup and Restore

Recommended Reading

1. Laden G, et al. Architectures for controller based CDP. In: Proceedings of the 5th USENIX Conference on File and Storage Technologies; 2007. p. 107–21.

Continuous Monitoring of Spatial Queries

Kyriakos Mouratidis
Singapore Management University, Singapore, Singapore

Synonyms

Spatiotemporal stream processing

Definition

A continuous spatial query runs over long periods of time and requests constant reporting of its result as the data dynamically change. Typically, the query type is range or nearest neighbor (NN), and the assumed distance metric is the Euclidean one. In general, there are multiple queries being processed simultaneously. The query points and the data objects move frequently and arbitrarily, i.e., their velocity vectors and motion patterns are unknown. They issue location updates to a central server, which processes them and continuously reports the current (i.e., updated) query results. Consider, for example, that the queries correspond to vacant cabs and that the data objects are pedestrians that ask for a taxi. As cabs and pedestrians move, each free taxi driver wishes to know his/her closest client. This is an instance of continuous NN monitoring. Spatial monitoring systems aim at minimizing the processing time at the server and/or the communication cost incurred by location updates. Due to the time-critical nature of the problem, the data are usually stored in main memory to allow fast processing.

Historical Background

The first algorithms in the spatial database literature process snapshot (i.e., one-time) queries over static objects. They assume disk-resident data and utilize an index (e.g., an *R-tree*) to restrict the search space and reduce the I/O cost. Subsequent research considered spatial queries in client-server architectures. The general idea is to provide the user with extra information (along with the result at query time) in order to reduce the number of subsequent queries as he/she moves (see entry "▶ Nearest Neighbor Query"). These methods assume that the data objects are either static or moving linearly with known velocities. Due to the wide availability of positioning devices and the need for improved location-based services, the research focus has recently shifted to continuous spatial queries. In contrast with earlier assumed contexts, in this setting (i) there are multiple queries being evaluated simultaneously, (ii) the query results are continuously updated, and (iii) both the query points and the data objects move unpredictably.

Foundations

The first spatial monitoring method is called *Q-index* [13] and processes static range queries. Based on the observation that maintaining an index over frequently moving objects is very costly, *Q-index* indexes the queries instead of the objects. In particular, the monitored ranges are organized by an *R-tree*, and moving objects probe this tree to find the queries that they influence. Additionally, *Q-index* introduces the concept of *safe regions* to reduce the number of location updates. Specifically, each object p is assigned a circular or rectangular region, such that p needs to issue an update only if it exits this area (because, otherwise, it does not influence the result of any

Continuous Monitoring of Spatial Queries, Fig. 1 Circular and rectangular safe regions

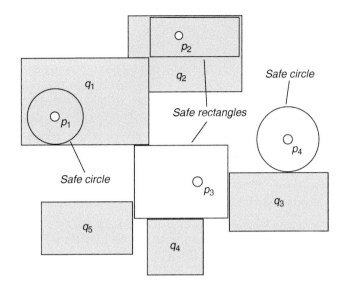

query). Figure 1 shows an example, where the current result of query q_1 contains object p_1, that of q_2 contains p_2, and the results of q_3, q_4, and q_5 are empty. The safe regions for p_1 and p_4 are circular, while for p_2 and p_3, they are rectangular. Note that no query result can change unless some objects fall outside their assigned safe regions. Kalashnikov et al. [4] show that a grid implementation of *Q-index* is more efficient (than *R-trees*) for main memory evaluation.

Monitoring Query Management (MQM) [1] and *MobiEyes* [2] also monitor range queries. They further exploit the computational capabilities of the objects to reduce the number of updates and the processing load of the server. In both systems, the objects store locally the queries in their vicinity and issue updates to the server only when they cross the boundary of any of these queries. To save their limited computational capabilities, the objects store and monitor only the queries they may affect when they move. MQM and *MobiEyes* employ different strategies to identify these queries. The former applies only to static queries. The latter can also handle moving ones, making however the assumption that they move linearly with fixed velocity.

Mokbel et al. [7] present *Scalable INcremental hash-based Algorithm* (SINA), a system that monitors both static and moving ranges. In contrast with the aforementioned methods, in SINA the objects do not perform any local processing.

Instead, they simply report their locations whenever they move, and the objective is to minimize the processing cost at the server. SINA is based on *shared execution* and *incremental evaluation*. Shared execution is achieved by implementing query evaluation as a spatial join between the objects and the queries. Incremental evaluation implies that the server computes only updates (i.e., object inclusions/exclusions) over the previously reported answers, as opposed to reevaluating the queries from scratch.

The above algorithms focus on ranges, and their extension to NN queries is either impossible or nontrivial. The systems described in the following target NN monitoring. Hu et al. [3] extend the safe region technique to NN queries; they describe a method that computes and maintains rectangular safe regions subject to the current query locations and kNN results. Mouratidis et al. [11] propose *Threshold-Based algorithm* (TB), also aiming at communication cost reduction. To suppress unnecessary location updates, in TB the objects monitor their distance from the queries (instead of safe regions). Consider the example in Fig. 2, and assume that q is a continuous 3-NN query (i.e., $k = 3$). The initial result contains p_1, p_2, and p_3. TB computes three thresholds (t_1, t_2, t_3) which define a range for each object. If every object's distance from q lies within its respective range, the result of the query is guaranteed to remain unchanged. Each threshold is set

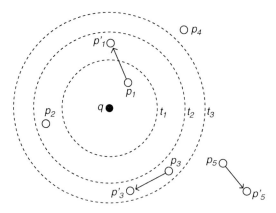

Continuous Monitoring of Spatial Queries, Fig. 2 TB example ($k = 3$)

in the middle of the distances of two consecutive objects from the query. The distance range for p_1 is $[0, t_1)$, for p_2 is $[t_1, t_2)$, for p_3 is $[t_2, t_3)$, and for p_4 and p_5 is $[t_3, \infty)$. Every object is aware of its distance range, and when there is a boundary violation, it informs the server about this event. For instance, assume that p_1, p_3, and p_5 move to positions p'_1, p'_3 and p'_5, respectively. Objects p_3 and p_5 compute their new distances from q and avoid sending an update since they still lie in their permissible ranges. Object p_1, on the other hand, violates its threshold and updates its position to the server. Since the order between the first two NNs may have changed, the server requests for the current location of p_2 and updates accordingly the result and threshold t_1. In general, TB processes all updates issued since the last result maintenance, and (if necessary) it decides which additional object positions to request for, updates the kNNs of q, and sends new thresholds to the involved objects.

All the following methods aim at minimizing the processing time. Koudas et al. [6] describe *aDaptive Indexing on Streams by space-filling Curves* (DISC), a technique for e-approximate kNN queries over streams of multidimensional points. The returned (e-approximate) kth NN lies at most e distance units farther from q than the actual kth NN of q. DISC partitions the space with a regular grid of granularity such that the maximum distance between any pair of points in a cell is at most e. To avoid keeping all

arriving data in the system, for each cell c, it maintains only K points and discards the rest. It is proven that an exact kNN search in the retained points corresponds to a valid ekNN answer over the original dataset provided that $k \leq K$. DISC indexes the data points with a *B-tree* that uses a *space-filling curve* mechanism to facilitate fast updates and query processing. The authors show how to adjust the index to (i) use the minimum amount of memory in order to guarantee a given error bound e or (ii) achieve the best possible accuracy, given a fixed amount of memory. DISC can process both snapshot and continuous ekNN queries.

Yu et al. [17] propose a method, hereafter referred to as YPK-CNN, for continuous monitoring of exact kNN queries. Objects are stored in main memory and indexed with a regular grid of cells with size $\delta \times \delta$. YPK-CNN does not process updates as they arrive, but directly applies them to the grid. Each NN query installed in the system is reevaluated every T time units. When a query q is evaluated for the first time, a two-step NN search technique retrieves its result. The first step visits the cells in an iteratively enlarged square R around the cell c_q of q until k objects are found. Figure 3a shows an example of a single NN query where the first candidate NN is p_1 with distance d from q; p_1 is not necessarily the actual NN since there may be objects (e.g., p_2) in cells outside R with distance smaller than d. To retrieve such objects, the second step searches in the cells intersecting the square SR centered at c_q with side length $2 \cdot d + \delta$ and determines the actual kNN set of q therein. In Fig. 3a, YPK-CNN processes p_1 up to p_5 and returns p_2 as the actual NN. The accessed cells appear shaded.

When reevaluating an existing query q, YPK-CNN makes use of its previous result in order to restrict the search space. In particular, it computes the maximum distance d_{max} among the current locations of the previous NNs (i.e., d_{max} is the distance of the previous neighbor that currently lies furthest from q). The new SR is a square centered at c_q with side length $2 \cdot d_{max} + \delta$. In Fig. 3b, assume that the current NN p_2 of q moves to location p'_2. Then, the rectangle defined by $d_{max} = \text{dist}\left(p'_2, q\right)$ is guaranteed to contain at

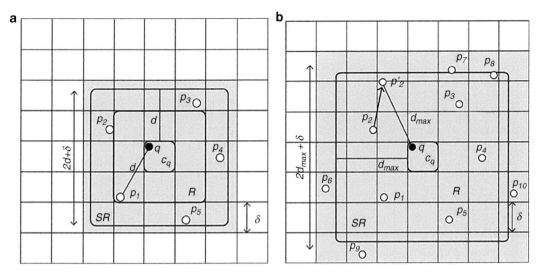

Continuous Monitoring of Spatial Queries, Fig. 3 YPK-CNN examples

least one object (i.e., p_2). YPK-CNN collects all objects (p_1 up to p_{10}) in the cells intersecting SR and identifies p_1 as the new NN. Finally, when a query q changes location, it is handled as a new one (i.e., its NN set is computed from scratch).

Xiong et al. [16] propose *Shared Execution Algorithm for Continuous NN queries* (SEA-CNN). SEA-CNN focuses exclusively on monitoring the NN changes, without including a module for the first-time evaluation of an arriving query q (i.e., it assumes that the initial result is available). Objects are stored in secondary memory, indexed with a regular grid. The *answer region* of a query q is defined as the circle with center q and radius NN_dist (where NN_dist is the distance of the current kth NN). Bookkeeping information is stored in the cells that intersect the answer region of q to indicate this fact. When updates arrive at the system, depending on which cells they affect and whether these cells intersect the answer region of the query, SEA-CNN determines a circular search region SR around q, and computes the new kNN set of q therein. To determine the radius r of SR, the algorithm distinguishes the following cases:

(i) If some of the current NNs move within the answer region or some outer objects enter the answer region, SEA-CNN sets $r =$ NN_dist and processes all objects falling in the answer region in order to retrieve the new NN set.

(ii) If any of the current NNs moves out of the answer region, processing is similar to YPK-CNN, i.e., $r = d_{max}$ (where d_{max} is the distance of the previous NN that currently lies furthest from q), and the NN set is computed among the objects inside SR. Assume that in Fig. 4a, the current NN p_2 issues an update reporting its new location p_2'. SEA-CNN sets $r = d_{max} = \mathrm{dist}\left(p_2', q\right)$, determines the cells intersecting SR (these cells appear shaded), collects the corresponding objects (p_1 up to p_7), and retrieves p_1 as the new NN.

(iii) Finally, if the query q moves to a new location q', then SEA-CNN sets $r = NN\_dist + \mathrm{dist}(q, q')$ and computes the new kNN set of q by processing all the objects that lie in the circle centered at q' with radius r. For instance, in Fig. 4b, the algorithm considers the objects falling in the shaded cells (i.e., objects from p_1 up to p_{10} except for p_6 and p_9) in order to retrieve the new NN (p_4).

Mouratidis et al. [9] propose another NN monitoring method, termed *conceptual partitioning monitoring* (CPM). CPM assumes the same system architecture and uses similar indexing and

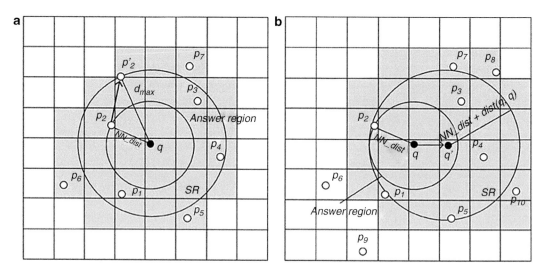

Continuous Monitoring of Spatial Queries, Fig. 4 SEA-CNN update handling examples

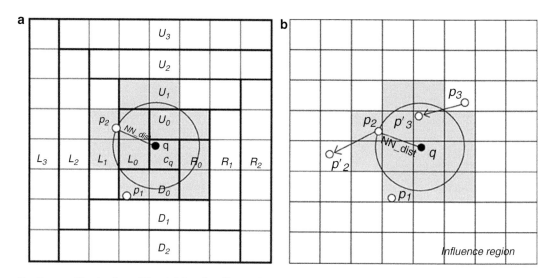

Continuous Monitoring of Spatial Queries, Fig. 5 CPM examples

bookkeeping structures as YPK-CNN and SEA-CNN. When a query q arrives at the system, the server computes its initial result by organizing the cells into conceptual rectangles based on their proximity to q. Each rectangle *rect* is defined by a *direction* and a *level number*. The direction is U, D, L, or R (for up, down, left, and right), and the level number indicates how many rectangles are between *rect* and q. Figure 5a illustrates the conceptual space partitioning around the cell c_q of q. If *mindist*(c,q) is the minimum possible distance between any object in cell c and q,

the NN search considers the cells in ascending *mindist*(c, q) order. In particular, CPM initializes an empty heap H and inserts (i) the cell of q with key equal to 0 and (ii) the level zero rectangles for each direction *DIR* with key *mindist*(DIR_0, q). Then, it starts de-heaping entries iteratively. If the de-heaped entry is a cell, it examines the objects inside and updates accordingly the NN set (i.e., the list of the k closest objects found so far). If the de-heaped entry is a rectangle DIR_{lvl}, it inserts into H (i) each cell $c \in DIR_{lvl}$ with key *mindist*(c, q) and (ii) the next level rectangle

DIR_{lvl+1} with key $mindist(DIR_{lvl+1}, q)$. The algorithm terminates when the next entry in H (corresponding either to a cell or a rectangle) has key greater than the distance NN_dist of the kth NN found. It can be easily verified that the server processes only the cells that intersect the circle with center at q and radius equal to NN_dist. This is the minimal set of cells to visit in order to guarantee correctness. In Fig. 5a, the search processes the shaded cells and returns p_2 as the result.

The encountered cells constitute the *influence region* of q, and only updates therein can affect the current result. When updates arrive for these cells, CPM monitors how many objects enter or leave the circle centered at q with radius NN_dist. If the *outgoing* objects are more than the *incoming* ones, the result is computed from scratch. Otherwise, the new NN set of q can be inferred by the previous result and the update information, without accessing the grid at all. Consider the example of Fig. 5b, where p_2 and p_3 move to positions p_2' and p_3', respectively. Object p_3 moves closer to q than the previous NN_dist, and, therefore, CPM replaces the outgoing NN p_2 with the incoming p_3. The experimental evaluation in [11] shows that CPM is significantly faster than YPK-CNN and SEA-CNN.

Key Applications

Location-Based Services

The increasing trend of embedding positioning systems (e.g., GPS) in mobile phones and PDAs has given rise to a growing number of location-based services. Many of these services involve monitoring spatial relationships among mobile objects, facilities, landmarks, etc. Examples include location-aware advertising, enhanced 911 services, and mixed-reality games.

Traffic Monitoring

Continuous spatial queries find application in traffic monitoring and control systems, such as on-the-fly driver navigation, efficient congestion detection and avoidance, as well as dynamic traffic light scheduling and toll fee adjustment.

Security Systems

Intrusion detection and other security systems rely on monitoring moving objects (pedestrians, vehicles, etc.) around particular areas of interest or important people.

Future Directions

Future research directions include other types of spatial queries (e.g., *reverse nearest neighbor* monitoring [5, 15]), different settings (e.g., NN monitoring over sliding windows [10]), and alternative distance metrics (e.g., NN monitoring in *road networks* [12]). Similar techniques and geometric concepts to the ones presented above also apply to problems of a nonspatial nature, such as continuous skyline [14] and top-k queries [8, 18].

Experimental Results

The methods described above are experimentally evaluated and compared with alternative algorithms in the corresponding reference.

Cross-References

- ▶ B+-Tree
- ▶ Nearest Neighbor Query
- ▶ Nonrelational Streams
- ▶ Reverse Nearest Neighbor Query
- ▶ Road Networks
- ▶ R-Tree (and Family)
- ▶ Space-Filling Curves for Query Processing
- ▶ Streaming Analytics
- ▶ Top-k Queries

Recommended Reading

1. Cai Y, Hua K, Cao G. Processing range-monitoring queries on heterogeneous mobile objects. In: Proceedings of the 5th IEEE International Conference on Mobile Data Management; 2004, p. 27–38.
2. Gedik B, Liu L. MobiEyes: distributed processing of continuously moving queries on moving objects in a

mobile system. In: Advances in Database Technology, Proceedings of 9th International Conference on Extending Database Technology; 2004, p. 67–87.

3. Hu H, Xu J, Lee D. A generic framework for monitoring continuous spatial queries over moving objects. In: Proceedings of the ACM SIGMOD International Conference on Management of Data; 2005, p. 479–90.

4. Kalashnikov D, Prabhakar S, Hambrusch S. Main memory evaluation of monitoring queries over moving objects. Distrib Parallel Databases. 2004;15(2):117–35.

5. Kang J, Mokbel M, Shekhar S, Xia T, Zhang D. Continuous evaluation of monochromatic and bichromatic reverse nearest neighbors. In: Proceedings of the 23rd International Conference on Data Engineering. 2007. p. 806–15.

6. Koudas N, Ooi B, Tan K, Zhang R. Approximate NN queries on streams with guaranteed error/performance bounds. In: Proceedings of the 30th International Conference on Very Large Data Bases. 2004. p. 804–15.

7. Mokbel M, Xiong X, Aref W. SINA: scalable incremental processing of continuous queries in spatio-temporal databases. In: Proceedings of the ACM SIGMOD International Conference on Management of Data. 2004. p. 623–34.

8. Mouratidis K, Bakiras S, Papadias D. Continuous monitoring of top-k queries over sliding windows. In: Proceedings of the ACM SIGMOD International Conference on Management of Data. 2006. p. 635–46.

9. Mouratidis K, Hadjieleftheriou M, Papadias D. Conceptual partitioning: an efficient method for continuous nearest neighbor monitoring. In: Proceedings of the ACM SIGMOD International Conference on Management of Data. 2005. p. 634–45.

10. Mouratidis K, Papadias D. Continuous nearest neighbor queries over sliding windows. IEEE Trans Knowledge Data Eng. 2007;19(6):789–803.

11. Mouratidis K, Papadias D, Bakiras S, Tao Y. A threshold-based algorithm for continuous monitoring of k nearest neighbors. IEEE Trans Knowl Data Eng. 2005;17(11):1451–64.

12. Mouratidis K, Yiu M, Papadias D, Mamoulis N. Continuous nearest neighbor monitoring in road networks. In: Proceedings of the 32nd International Conference on Very Large Data Bases. 2006. p. 43–54.

13. Prabhakar S, Xia Y, Kalashnikov D, Aref W, Hambrusch S. Query indexing and velocity constrained indexing: scalable techniques for continuous queries on moving objects. IEEE Trans Comput. 2002;51(10):1124–40.

14. Tao Y, Papadias D. Maintaining sliding window skylines on data streams. IEEE Trans Knowl Data Eng. 2006;18(3):377–91.

15. Xia T, Zhang D. Continuous reverse nearest neighbor monitoring. In: Proceedings of the 22nd International Conference on Data Engineering; 2006.

16. Xiong X, Mokbel M, Aref W. SEA-CNN: scalable processing of continuous k-nearest neighbor queries in spatio-temporal databases. In: Proceedings of the 21st International Conference on Data Engineering; 2005. p. 643–54.

17. Yu X, Pu K, Koudas N. Monitoring k-nearest neighbor queries over moving objects. In: Proceedings of the 21st International Conference on Data Engineering. 2005. p. 631–42.

18. Zhang D, Du Y, Hu L. On monitoring the top-k unsafe places. In: Proceedings of the 24th International Conference on Data Engineering; 2008. p. 337–45.

Continuous Multimedia Data Retrieval

Jeffrey Xu Yu
Department of Systems Engineering and Engineering Management, The Chinese University of Hong Kong, Hong Kong, China

Definition

Continuous multimedia is widely used in many applications nowadays. Continuous multimedia objects, such as audio and video streams, being stored on disks with different requirements of bandwidths, are required to be retrieved continuously without interruption. The response time is an important measurement in supporting continuous multimedia streams. Several strategies are proposed in order to satisfy the requirements of all users in a multi-user environment where multiple users are trying to retrieve different continuous multimedia streams together.

Historical Background

Several multimedia data retrieval techniques are proposed to support the real-time display of continuous multimedia objects. There are three categories [6]. The first category is to sacrifice the quality of the data in order to guarantee the required bandwidth of multimedia objects. The existing techniques either use lossy compression techniques (such as predictive [15], frequency oriented [11], and importance oriented [10]), or use a low resolution device. The second category is to use the placement techniques to satisfy the continuous requirement by arranging the data to appropriate disk locations. In other words, it is to organize multimedia data across the surface of a disk drive to maximize its bandwidth when it is retrieved [4, 5, 16, 22, 20]. The third category is to increase the bandwidth of storage device by using parallelism. The basic idea is to employ the aggregate bandwidth of several disk drives by putting an object across multiple disks, for example, a Redundant Arrays of Inexpensive Disk (RAID) [17]. The existing works [9, 19] focus on this direction.

Foundations

This section focuses on the second and third categories, and discusses multimedia data retrieval regarding single/multiple stream(s) and single/multiple disk(s).

Retrieval of a Single Stream on a Single Disk

For the retrieval of a single multimedia stream on a single disk, the stream data is read into a first-in-first-out queue (FIFO) continuously first, and then is sent to the display devices, possibly via a network, at the appropriate rate. In order to satisfy the real-time requirements - to display multimedia data continuously on a display, it is required to keep the FIFO non empty. In other words, there is some multimedia data to be displayed in the FIFO in the duration of the playback. As pointed in [6], pre-fetching all the data into the FIFO before playback is not a feasible solution because the size of the stream can be very large.

Suppose that a read request of a large multimedia data is issued. The starting time and the minimum buffer space, to display the retrieved multimedia data continuously, are determined as follows, under the following conditions: (i) the timing of data retrieval is known in advance, (ii) both the transfer rate and consumption rate are constant, and (iii) the transfer rate of the storage device is at least as great as the consumption rate. Consider Fig. 1. First, the amount of data, that needs to be consumed by a display, is illustrated as the dotted line marked *data read*. The vertical line segments show the amount of data that needs to be consumed in order to continuously display, and the horizontal line shows the time periods such amount of data is consumed on a display. Second, the solid zigzag line, marked *data buffered*, shows the data to be accessed in the data buffers. The vertical line segments show the data to be read into the buffers followed by the line segments that show data is consumed in the buffer during a certain time interval. Here, in the solid zigzag line, there is a minimum point (marked *minimum-shift up to zero*), which is a possible negative value and is denoted as $z(< 0)$. Third, the dotted zigzag line (marked *shifted buffer plot*) is the line by shifting the entire solid zigzag line up by the amount of $|z|$ where $z < 0$. Finally, the starting time to display is determined as the point at which the shifted-up dotted zigzag line (*shifted buffer plot*) and the dotted line (*data read*) intersect, which is indicated as *intersection - start time* in Fig. 1. Also, the minimum buffer size is the maximum value of in the line of *shifted buffer plot*, which is indicated as *required buffer space* in Fig. 1. Details are discussed in [7].

Retrieval of a Single Stream on Multiple Disks

The multimediadata retrieval using multiple disks is a technique to retrieve a data stream continuously at the required bandwidth. The main idea behind is to de-cluster the data stream into several fragments [2, 14], and distribute these fragments across multiple processors (and disks). By combining the I/O bandwidths of several disks, a system can provide the required retrieval rate to display a continuous multimedia stream in real-time. Assume that the required retrieval rate is B and the bandwidth of each disk is B_D. The degree of de-clustering can be calculated as $M = \left\lceil \frac{B}{B_D} \right\rceil$, which implies the number of disks that is needed to satisfy the required retrieval rate.

When the degree of de-clustering is determined, the fragments can be formed using a

Continuous Multimedia Data Retrieval, Fig. 1
Finding minimum buffer space and start time (Fig. 2 in [6])

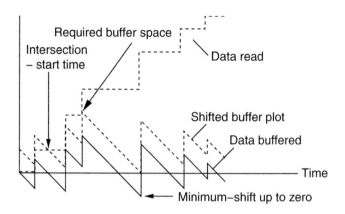

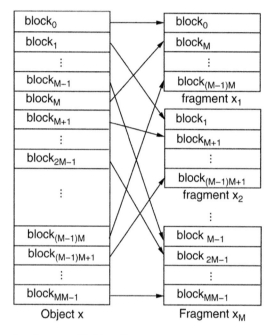

Object x   Fragment x_M

Continuous Multimedia Data Retrieval, Fig. 2
Round-robin partitioning of object x (Fig. 5 in [9])

round-robin partitioning strategy as illustrated in Fig. 2, where an object x is partitioned into M fragments stored on M disks. The round-robin partitioning is conducted as follows. First, the object x is divided in N blocks (disk pages) depending on the disk-page size allowed on disks. In Fig. 2, the number of blocks is $N = M \cdot M$. The first $block_0$ is assigned to first fragment indicated as x_1 in Fig. 2, and the second $block_1$ is assigned to the second fragment indicated as x_2. The first M blocks from $block_0$ to $block_{M-1}$ are assigned

to the M fragments one by one. In next run, the next set of blocks, from $block_M$ to $block_{2M-1}$ will be assigned to the M fragments in the similar fashion. The process repeats until all data blocks are assigned to the fragments in a round-robin fashion.

Retrieval of Multiple Streams on a Single Disk

In a multi-user environment, several users may retrieve data streams simultaneously. Therefore, there are multiple data streams requested on a single disk. The data streams are retrieved in rounds, and each stream is allowed a disk access or a fixed number of disk accesses at one time. All data retrieval requests need to be served in turn. Existing solutions include SCAN, round-robin, EDF, and Sorting-Set algorithms. The round-robin algorithm retrieves data for each data retrieval request, in turn, in a predetermined order. The SCAN algorithm moves the disk head back and forth, and retrieves the requested blocks when the disk head passes over the requested blocks [18]. The EDF (earliest-deadline-first) algorithm serves the request with the earliest deadline first, where a deadline is given to a data stream [13]. The sorting-set algorithm is designed to exploit the trade-off between the number of rounds between successive reads for a data stream and the length of the round [8, 21], by assigning each data stream to a sorting set. Fixed time slots are allocated to a sorting set in a round during which its requests are possibly processed.

Continuous Multimedia Data Retrieval, Fig. 3 Retrieval of multiple streams on multiple disks [6]

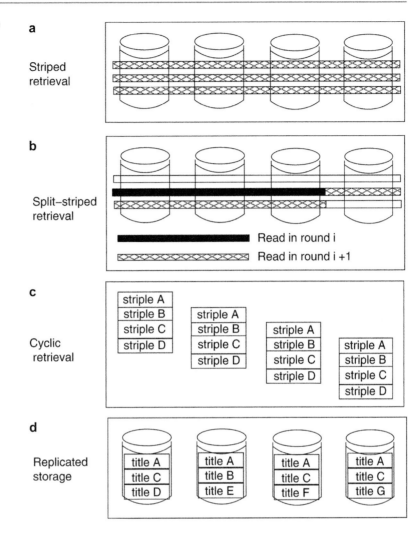

a — Striped retrieval

b — Split–striped retrieval

Read in round i
Read in round i +1

c — Cyclic retrieval

d — Replicated storage

C

Retrieval of Multiple Streams on Multiple Disks

In order to support multiple stream retrieval, making use of parallel disks is an effective method, where a data stream is striped across the parallel disks. There are several approaches to retrieve data streams when they are stored on parallel disks (Fig. 3). It is important to note that the main issue here is to increase the number of data streams to be retrieved simultaneously. It is not to speed up retrieval for an individual data stream using multiple disks. Consider the striped retrieval as shown in Fig. 3a, where a data stream is striped across m parallel disks. Suppose that each disk has r_c bandwidth, m parallel disks can be together used to increase the bandwidth up to $m \cdot r_c$. However, the issue is the system capacity in terms of the number of data streams it can serve, for example, from n data streams to $m \cdot n$ data streams using m parallel disks. Suppose that each data stream will be served in turn. When it increases the number of data streams from n to $m \cdot n$, in the striped retrieval, the round length (or in other words consecutive reads for a single data stream) increases proportionally from n to $m \cdot n$. It implies that, in order to satisfy the required retrieval rate, it needs to use a larger buffer, which also implies a larger startup delay. An improvement over striped retrieval is to use split-stripe retrieval which allows partial stripes to be used (Fig. 3b), in order to reduce the buffer size required in the striped retrieval. But, it has its limit to significantly reduce startup delay and buffer space.

Observe the data transfer patterns in the striped retrieval and the split-striped retrieval, which show busty patterns for data to be read into the buffer. For instance, consider Fig. 3a, an entire strip for a single data stream will be read in and be consumed in a period which is related to the round length. It requests larger buffer sizes, because it needs to keep the data to be displayed continuously until the next read, in particular when the number of streams increases from n to $m \cdot n$. Instead, an approach is proposed to read small portion of data frequently, in order to reduce the buffer space required. The approach is called cyclic retrieval. As shown in Fig. 3c, the cyclic retrieval tries to read multiple streams rather than one stream at one time. Rather than retrieving an entire stripe at once, the cyclic retrieval retrieves each striping unit of a stripe consecutively [1, 3]. Using this approach, the buffer space is significantly reduced. But the reduction comes with cost. The buffer space reduction is achieved at the expense of cuing (*a stream is said to be cued if it is paused and playback may be initiated instantaneously*) and clock skew tolerance.

As an alternative to striped (split-striped) or cyclic retrieval, it can deal with each disk independently rather than treating them as parallel disks. Here, each disk stores a number of titles (data streams). When there is a multimedia data retrieval request, a disk that contains the data stream will respond. The data streams that are frequently requested may be kept in multiple disks using replication. The number of replications can be determined based on the retrieval frequency of data streams [12], as shown in Fig. 3d. Based on the replicated retrieval, both the startup delay time and buffer space can be reduced significantly. It is shown that it is easy to scale when the number of data streams increase at the expense of more disk space required. [9] discusses data replication techniques.

A comparison among striped-retrieval, cyclic retrieval, and replicated retrieval in supporting n streams is shown in Table 1.

Continuous Multimedia Data Retrieval, Table 1 A comparison of multi-disk retrieval strategies supporting n streams (Table 1 in [6])

	Striped	Cyclic	Replicated
Instant restart	yes	no	yes
Clock skew tolerance	yes	no	yes
Easy scaling	no	no	yes
Capacity	per-system	per-system	per-title
Startup delay	$O(n)$	$O(n)$	$O(1)$
Buffer space	$O(n^2)$	$O(n)$	$O(n)$

Key Applications

Continuous multimedia data retrieval is used in many real-time continuous multimedia streams such as audio and video through the network. Especially in a multi-user environment, the continuous multimedia data retrieval techniques are used to support simultaneous display of several multimedia objects in real-time.

Cross-References

▶ Buffer Management
▶ Buffer Manager
▶ Multimedia Data Buffering
▶ Multimedia Data Storage
▶ Multimedia Resource Scheduling
▶ Scheduling Strategies for Data Stream Processing
▶ Storage Management
▶ Storage Manager

Recommended Reading

1. Berson S, Ghandeharizadeh S, Muntz R, Ju X. Staggered striping in multimedia information systems. In: Proceedings of the ACM SIGMOD International Conference on Management of Data; 1994. p. 79–90.
2. Carey MJ, Livny M. Parallelism and concurrency control performance in distributed database machines. ACM SIGMOD Rec. 1989;18(2):122–33.
3. Chen MS, Kandlur DD, Yu PS. Storage and retrieval methods to support fully interactive playout in a disk-array-based video server. Multimedia Systems. 1995;3(3):126–35.

4. Christodoulakis S, Ford DA. Performance analysis and fundamental performance tradeoffs for CLV optical disks. ACM SIGMOD Rec. 1988;17(3):286–94.

5. Ford DA, Christodoulakis S. Optimizing random retrievals from CLV format optical disks. In: Proceedings of the 17th International Conference on Very Large Data Bases; 1991. p. 413–22.

6. Gemmell DJ. Multimedia information storage and management, chap. 1. In: Disk scheduling for continuous media. Norwell: Kluwer; 1996.

7. Gemmell J, Christodoulakis S. Principles of delay-sensitive multimedia data storage retrieval. ACM Trans Inf Syst. 1992;10(1):51–90.

8. Gemmell DJ, Han J. Multimedia network file servers: multichannel delay-sensitive data retrieval. Multimedia Systems. 1994;1(6):240–52.

9. Ghandeharizadeh S, Ramos L. Continuous retrieval of multimedia data using parallelism. IEEE Trans Knowl Data Eng. 1993;5(4):658–69.

10. Green JL. The evolution of DVI system software. Commun ACM. 1992;35(1):52–67.

11. Lippman A, Butera W. Coding image sequences for interactive retrieval. Commun ACM. 1989;32(7):852–60.

12. Little TDC, Venkatesh D. Popularity-based assignment of movies to storage devices in a video-on-demand system. Multimedia Systems. 1995;2(6):280–7.

13. Liu CL, Layland JW. Scheduling algorithms for multiprogramming in a hard real-time environment. In: Tutorial: hard real-time systems. Los Alamitos: IEEE Computer Society; 1989. p. 174–89.

14. Livny M, Khoshafian S, Boral H. Multi-disk management algorithms. SIGMETRICS Perform Eval Rev. 1987;15(1):69–77.

15. Luther AC. Digital video in the PC environment. 2nd ed. New York: McGraw-Hill; 1991.

16. McKusick MK, Joy WN, Leffler SJ, Fabry RS. A fast file system for UNIX. Comput Syst. 1984;2(3):181–97.

17. Patterson DA, Gibson GA, Katz RH. A case for redundant arrays of inexpensive disks (RAID). In: Proceedings of the ACM SIGMOD International Conference on Management of Data; 1988. p. 109–16.

18. Teorey TJ, Pinkerton TB. A comparative analysis of disk scheduling policies. In: Proceedings of the 3rd ACM Symposium on Operating System Principles; 1971. p. 114.

19. Tsai WJ, Lee SY. Storage design and retrieval of continuous multimedia data using multi-disks. In: Proceedings of the 1994 International Conference on Parallel and Distributed Systems; 1994. p. 148–53.

20. Wong CK. Minimizing expected head movement in one-dimensional and two-dimensional mass storage systems. ACM Comput Surv. 1980;12(2):167–78.

21. Yu PS, Chen MS, Kandlur DD. Grouped sweeping scheduling for DASD-based multimedia storage management. Multimedia Systems. 1993;1(3):99–109.

22. Yue PC, Wong CK. On the optimality of the probability ranking scheme in storage applications. J ACM. 1973;20(4):624–33.

Continuous Queries in Sensor Networks

Yong Yao and Johannes Gehrke
Cornell University, Ithaca, NY, USA

Synonyms

Long running queries

Definition

A powerful programming paradigm for data acquisition and dissemination in sensor networks is a declarative query interface. With a declarative query interface, the sensor network is programmed for long term monitoring and event detection applications through continuous queries, which specify what data to retrieve at what time or under what conditions. Unlike snapshot queries which execute only once, continuous queries are evaluated periodically until the queries expire. Continuous queries are expressed in a high-level language, and are compiled and installed on target sensor nodes, controlling when, where, and what data is sampled, possibly filtering out unqualified data through local predicates. Continuous queries can have a variety of optimization goals, from improving result quality and response time to reducing energy consumption and prolonging network lifetime.

Historical Background

In recent years sensor networks have been deployed successfully for a wide range of applications from environmental sensing to

Continuous Queries in Sensor Networks, Fig. 1 Database view of sensor networks

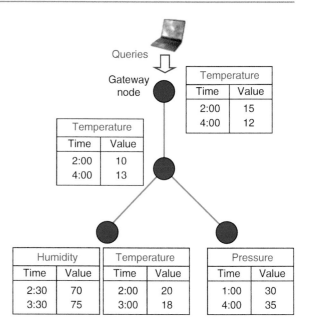

process monitoring. A database approach to programming sensor networks has gained much importance: Clients program the network through queries without knowing how the results are generated, processed, and returned to the client. Sophisticated catalog management, query optimization, and query processing techniques abstract the client from the physical details of contacting the relevant sensors, processing the sensor data, and sending the results to the client. The concept of a sensor network as a database was first introduced in [3]. A number of research projects, including TinyDB [9] and Cougar [14] have implemented continuous queries as part of their database languages for sensor networks. In these systems time is divided into epochs of equal size, and continuous queries are evaluated once per epoch during their lifetime. Fig. 1 shows this database view of sensor networks.

Two properties are significant to continuous query processing in sensor networks: energy conservation and fault-tolerance in case of failures of sensors, both topics that are not of importance in traditional database systems or data stream systems. Advanced query processing techniques have been proposed to enable energy-efficient query processing in the presence of frequent node and communication failures. For example, a lot of research has been dedicated to in-network query

processing [6, 9, 14] to reduce the amount of data to be transmitted inside the network. Another approach is to permit approximate query processing [4, 5], which produces approximate query answers within a pre-defined accuracy range, but consumes much less energy. Sensor data is correlated in time and space. Data compression in sensor networks and probabilistic data models [1, 7, 8] exploit data correlation and remove redundant data from intermediate results.

Next generation sensor network may consist of media-rich and mobile sensor nodes, which result in new challenges arise for continuous query processing such as mobility and high data rates. ICEDB [15] describes a new framework for continuous query processing in sensor networks with intermittent network connectivity and large amount of data to transfer.

Foundations

Continuous queries are a natural approach for data fusion in sensor networks for long running applications as they provide a high-level interface that abstracts the user from the physical details of the network. The design and implementation of continuous queries needs to satisfy several requirements. First, it has to preserve the scarce resources such as energy and bandwidth in battery-powered sensor networks. Thus the simple approach of transmitting all relevant data back to a central node for query evaluation is prohibitive for sensor networks of non-trivial size, as communication using the wireless medium consumes a lot of energy. Since sensor nodes have the ability to perform local computation, communication can be traded for computation by moving computation from the clients into the sensor network, aggregating partial results or eliminating irrelevant data. Second, sensor network applications usually have different QoS requirements, from accuracy, energy consumption to delay. Therefore the continuous query model needs to be flexible enough to adopt various processing techniques in different scenarios.

Sensor Data Model

In the view of a sensor network as a database, each sensor node is modeled as a separate data source that generates records with several fields such as the sensor type, location of the sensor node, a time stamp, and the value of the reading. Records of the same sensor type from different nodes have the same schema, and these records collectively form a distributed table of sensor readings. Thus the sensor network can be considered as a large distributed database system consisting of several tables of different types of sensors.

Sensor readings are samples of physical signals whose values change continuously over time. For example, in environmental monitoring applications, sensor readings are generated every few seconds (or even faster). For some sensor types (such as PIR sensors that sense the presence of objects) their readings might change rapidly and thus may be outdated rather quickly, whereas for other sensors, their value changes only slowly over time as for temperature sensors that usually have a small derivative. Continuous queries recompute query results periodically and keep query results up-to-date. For applications that require only approximate results, the system can cache previous results and lower the query update rate to save energy.

Instead of querying raw sensor data, most applications are more interested in composite data which captures high-level events monitored by sensor networks. Such composite data is produced by complex signal processing algorithms given raw sensor measurements as inputs. Composite data usually has a compact structure and is easier to query.

Continuous Query Models

In TinyDB and Cougar, continuous queries are represented as a variant of SQL with a few extensions. A simple query template in Cougar is shown in the figure below. (TinyDB uses a very similar query structure.)

SELECT{attribute, aggregate}
FROM{Sensordata S}
WHERE {predicate}
GROUP BY {attribute}
HAVING {predicate}
DURATION time interval
EVERY time span *e*

The template can be extended to support nested queries, where the basic query block shown below can appear within the *WHERE* or *HAVING* clause of another query block. The query template has an obvious semantics: the *SELECT* clause specifies attributes and aggregates from sensor records, the *FROM* clause specifies the distributed relation describing the sensor type, the *WHERE* clause filters sensor records by a predicate, the *GROUP BY* clause classifies sensor records into different partitions according to some attributes, and the *HAVING* clause eliminates groups by a predicate. Join queries between external tables and sensor readings are constructed by including the external tables and sensor readings in the *FROM* clause and join predicates in the *WHERE* clause.

Two new clauses introduced for continuous queries are *DURATION* and *EVERY*; The *DURATION* clause specifies the lifetime of the continuous query, and the *EVERY* or clause determines the rate of query answers. TinyDB has two related clauses: *LIFETIME* and *SAMPLE INTERVAL*, specifying the lifetime of the query and the sample interval, respectively. The *LIFETIME* clause will be discussed in more detail a few paragraphs later.

In event detection applications, sensor data is collected only when particular events happen. The above query template can be extended with a condition clause as a prerequisite to determine when to start or stop the main query. Event-based queries have the following structure in TinyDB:

> *ON EVENT* {event(arguments)}:
> {*query body*}

Another extension to the basic query template is lifetime-based queries, which have no explicit *EVERY* or *SAMPLE INTERVAL* clause; only the query lifetime is specified through a *LIFETIME* clause [9]. The system automatically adjusts the sensor sampling rate to the highest rate possible with the guarantee that the sensor network can process the query for the specified lifetime. Lifetime-based queries are more intuitive in some mission critical applications where user queries have to run for a given period of time, but it is hard to predict the optimal sampling rate in advance. Since the sampling rate is adjusted continuously according to the available power and the energy consumption rate in the sensor network, lifetime-based queries are more adaptive to unpredictable changes in sensor networks deployed in a harsh environment.

Common Types of Continuous Queries in Sensor Networks Select-All Queries

Recent sensor network deployments indicate that a very common type of continuous queries is a select-all query, which extracts all relevant data from the sensor network and stores the data in a central place for further processing and analysis. Although select-all queries are simple to express, efficient processing of select-all queries is a big challenge. Without optimization, the size of the transmitted data explodes quickly, and thus the power of the network would be drained in a short time, especially for those nodes acting as bridge to the outside world; this significantly decreases the lifetime of the sensor network.

One possible approach is to apply model-based data compression at intermediate sensor nodes [7]. For many types of signals, e.g., tem-

perature and light, sensor readings are highly correlated in both time and space. Data compression in sensor networks can significantly reduce the communication overhead and increase the network lifetime. Data compression can also improve the signal quality by removing unwanted noise from the original signal. One possible form of compression is to construct and maintain a model of the sensor data in the network; the model is stored both on the server and on sensor nodes in the network. The model on the server can be used to predicate future values within a pre-defined accuracy range. Data communication happens to synchronize the data model on the server with real sensor measurements [7].

Aggregate Queries

Aggregate queries return aggregate values for each group of sensor nodes specified by the *GROUP BY* clause. Below is is an example query that computes the average concentration in a region every 10 s for the next hour:

> *SELECT*   AVG(R.concentration)
> *FROM*   ChemicalSensor R
> *WHERE*   R.loc *IN* region
> *HAVING*   AVG(R.concentration) > T
> *DURATION* (now,now+3600)
> *EVERY*   10

Data aggregation in sensor networks is well-studied because it scales to sensor networks with even thousands of nodes. Query processing proceeds along a spanning tree of sensor nodes towards a gateway node. During query processing, partial aggregate results are transmitted from a node to its parent in the spanning tree. Once an intermediate node in the tree has received all data from nodes below it in a round, the node compute a partial aggregate of all received data and sends that output to the next node. This solution works for aggregate operators that are incrementally computable, such as avg, max, and moments of the data. The only caveat is that this in-network computation requires synchronization between sensor nodes along the communication path, since a node has to "wait" to receive results to be aggregated. In networks with high loss rates, broken links are hard to differentiate from long delays due to high loss rates, making synchronization a non-trivial problem [13].

Join Queries

In a wide range of event detection applications, sensor readings are compared to a large number of time and location varying predicates to determine whether a user-interesting event is detected [1]. The values of these predicates are stored in a table. Continuous queries with a *join* operator between sensor readings and the predicate table are suitable for such applications. Similar join queries can be used to detect defective sensor nodes whose readings are inaccurate by checking their readings against readings from neighboring sensors (again assuming spatial correlation between sensor readings). Suitable placement of the join operator in a sensor network has also been examined [2].

Key Applications

Habitat Monitoring

In the Great Duck Island experiment, a network of sensors was deployed to monitor the microclimate in and around nesting burrows used by birds, with the goal of developing a habitat monitoring kit that would enable researchers worldwide to engage in non-intrusive and non-disruptive monitoring of sensitive wildlife and habitats [10]. In a more recent experiment, a sensor network was deployed to densely record the complex spatial variations and the temporal dynamics of the microclimate around a 70-m tall redwood tree [12].

The Intelligent Building

Sensor networks can be deployed in intelligent buildings for the collection and analysis of structural responses to ambient or forced excitation of the building's structure, for control of light and temperature to conserve energy, and for monitoring of the flow of people in critical areas. Continuous queries are used both for data collection and for event-based monitoring of sensitive areas and to enforce security policies.

Industrial Process Control

Idustrial manufacturing processes often have strict requirements on temperature, humidity, and other environmental parameters. Sensor networks can be deployed to monitor the production environment without expensive wires to be installed. Continuous join queries compare the state of the environment to a range of values specified in advance and send an alert when an exception is detected [1].

Cross-References

▶ Approximate Query Processing
▶ Data Acquisition and Dissemination in Sensor Networks
▶ Data Aggregation in Sensor Networks
▶ Data Compression in Sensor Networks
▶ Data Fusion in Sensor Networks
▶ Database Languages for Sensor Networks
▶ Distributed Database Systems
▶ In-Network Query Processing
▶ Sensor Networks

Recommended Reading

1. Abadi D, Madden S, Lindner W. REED: robust, efficient filtering and event detection in sensor networks. In: Proceedings of the 31st International Conference on Very Large Data Bases; 2005. p. 768–80.
2. Bonfils B, Bonnet P. Adaptive and decentralized operator placement for in-network query processing. In: Proceedings of the 2nd International Workshop on International Processing in Sensor Networks; 2003. p. 47–62.
3. Bonnet P, Gehrke J, Seshadri P. Towards sensor database systems. In: Proceedings of the 2nd International Conference on Mobile Data Management; 2001. p. 3–14.
4. Chu D, Deshpande A, Hellerstein J, Hong W. Approximate data collection in sensor networks using probabilistic models. In: Proceedings of the 22nd International Conference on Data Engineering; 2006.
5. Considine J, Li F, Kollios G, Byers J. Approximate aggregation techniques for sensor databases. In: Proceedings of the 20th International Conference on Data Engineering; 2004. p. 449–60.
6. Deligiannakis A, Kotidis Y, Roussopoulos N. Hierarchical in-network data aggregation with quality guarantees. In: Advances in Database Technology,

Proceedings of the 9th International Conference on Extending Database Technology; 2004. p. 658–75.

7. Deshpande A, Guestrin C, Madden S, Hellerstein J, Hong W. Model-driven data acquisition in sensor networks. In: Proceedings of the 30th International Conference on Very Large Data Bases; 2004. p. 588–99.

8. Kanagal B, Deshpande A. Online filtering, smoothing and probabilistic modeling of streaming data. In: Proceedings of the 24th International Conference on Data Engineering; 2008. p. 1160–9.

9. Madden S, Franklin M, Hellerstein J, Hong W. The design of an acquisitional query processor for sensor networks. In: Proceedings of the ACM SIGMOD International Conference on Management of Data; 2003. p. 491–502.

10. Mainwaring A, Polastre J, Szewczyk R, Culler D, Anderson J. Wireless sensor networks for habitat monitoring. In: Proceedings of the 1st ACM International Workshop on Wireless Sensor Networks and Applications; 2002. p. 88–97.

11. Stoianov I, Nachman L, Madden S, Tokmouline T. PIPENET: a wireless sensor network for pipeline monitoring. In: Proceedings of the 6th International Symposium on Information Processing in Sensor Networks; 2007. p. 264–73.

12. Tolle G, Polastre J, Szewczyk R, Culler D, Turner N, Tu K, Burgess S, Dawson T, Buonadonna P, Gay D, Hong W. A macroscope in the redwoods. In: Proceedings of the 3rd International Conference on Embedded Networked Sensor Systems; 2005.

13. Trigoni N, Yao Y, Demers AJ, Gehrke J, Rajaraman R. Wave scheduling and routing in sensor networks. ACM Trans Sensor Netw. 2007;3(1):2.

14. Yao Y, Gehrke J. Query processing in sensor networks. In: Proceedings of the 1st Biennial Conference on Innovative Data Systems Research; 2003.

15. Zhang Y, Hull B, Balakrishnan H, Madden S. ICEDB: intermittently connected continuous query processing. In: Proceedings of the 23rd International Conference on Data Engineering; 2007. p. 166–75.

Continuous Query

Shivnath Babu
Duke University, Durham, NC, USA

Synonyms

Standing query

Definition

A continuous query Q is a query that is issued once over a database D, and then logically runs continuously over the data in D until Q is terminated. Q lets users get new results from D without having to issue the same query repeatedly. Continuous queries are best understood in contrast to traditional SQL queries over D that run once to completion over the current data in D.

Key Points

Traditional database systems expect all data to be managed within some form of persistent data sets. For many recent applications, where the data is changing constantly (often exclusively through insertions of new elements), the concept of a continuous data stream is more appropriate than a data set. Several applications generate data streams naturally as opposed to data sets, e.g., financial tickers, performance measurements in network monitoring, and call detail records in telecommunications. Continuous queries are a natural interface for monitoring data streams. In network monitoring, e.g., continuous queries may be used to monitor whether all routers and links are functioning efficiently.

The Tapestry system [3] for filtering streams of email and bulletin-board messages was the first to make continuous queries a core component of a database system. Continuous queries in Tapestry were expressed using a subset of SQL. Barbara [2] later formalized continuous queries for a wide spectrum of environments. With the recent emergence of general-purpose systems for processing data streams, continuous queries have become the main interface that users and applications use to query data streams [1].

Materialized views and triggers in traditional database systems can be viewed as continuous queries. A materialized view V is a query that needs to be reevaluated or incrementally updated whenever the base data over which V is defined changes. Triggers implement event-condition-action rules that enable database systems to take appropriate actions when certain events occur.

Cross-References

▶ Database Trigger
▶ ECA Rules

Recommended Reading

1. Babu S, Widom J. Continuous queries over data streams. ACM SIGMOD Rec. 2001;30(3): 109–20.
2. Barbara D. The characterization of continuous queries. Int J Coop Inform Syst. 1999;8(4):295–323.
3. Terry D, Goldberg D, Nichols D, Oki B. Continuous queries over append-only databases. In: Proceedings of the ACM SIGMOD International Conference on Management of Data; 1992. p. 321–30.

ConTract

Andreas Reuter
Heidelberg Laureate Forum Foundation,
Schloss-Wolfsbrunnenweg 33, Heidelberg,
Germany

Definition

A ConTract is an extended transaction model that employs transactional mechanisms in order to provide a run-time environment for the reliable execution of long-lived, workflow-like computations. The focus is on durable execution and on correctness guarantees with respect to the effects of such computations on shared data.

Key Points

The notion of a ConTract (concatenated transactions) combines the principles of workflow programing with the ideas related to long-lived transactions. The ConTract model is based on a two-tier programing approach. At the top level, each ConTract is a script describing a (long-lived) computation. The script describes the order of execution of so-called steps. A step is a predefined unit of execution (e.g., a service invocation) with no visible internal structure. A step can access shared data in a database, send messages, etc.

A ConTract, once it is started, will never be lost by the system, no matter which technical problems (short of a real disaster) will occur during execution. If completion is not possible, all computations performed by a ConTract will be revoked, so in a sense ConTracts have transactional behaviour in that they will either be run to completion, or the impossibility of completion will be reflected in the invocation of appropriate recovery measures.

The ConTract model draws on the idea of Sagas, where the notion of compensation is employed as a means for revoking the results of computations beyond the boundaries of ACID transactions. In a ConTract, by default each step is an ACID transaction. But it is possible to group multiple steps (not just linear sequences) into a transaction. Compensation steps must be supplied by the application explicitly.

The ideas of the ConTract model have selectively been implemented in some academic prototypes, but a full implementation has never been attempted. It has influenced many later versions of "long-lived transaction" schemes, and a number of its aspects can be found in commercial systems such as BizTalk.

Cross-References

▶ Extended Transaction Models and the ACTA Framework
▶ Sagas

Recommended Reading

1. Reuter A, Waechter H. The ConTract model. In: Stonebraker M, Hellerstein J, editors. Readings in database in database systems. 2nd ed. Los Altos: Morgan Kaufmann; 1992. p. 219–63.

Control Data

Nathaniel Palmer
Workflow Management Coalition, Hingham,
MA, USA

Synonyms

Workflow control data; Workflow enactment service state data; Workflow engine state data

Definition

Data that is managed by the Workflow Management System and/or a Workflow Engine. Such data is internal to the workflow management system and is not normally accessible to applications.

Key Points

Workflow control data represents the dynamic state of the workflow system and its process instances.

Workflow control data examples include:

- State information about each workflow instance.
- State information about each activity instance (active or inactive).
- Information on recovery and restart points within each process, etc.

The workflow control data may be written to persistent storage periodically to facilitate restart and recovery of the system after failure. It may also be used to derive audit data.

Cross-References

- ► Activity
- ► Process Life Cycle
- ► Workflow Management and Workflow Management System
- ► Workflow Model

Convertible Constraints

Carson Kai-Sang Leung
Department of Computer Science, University of
Manitoba, Winnipeg, MB, Canada

Synonyms

Convertible antimonotone constraints; Convertible antimonotonic constraints; Convertible monotone constraints; Convertible monotonic constraints

Definition

A constraint C is *convertible* if and only if C is convertible antimonotone or convertible monotone. A constraint C is *convertible antimonotone* provided there is an order R on items such that when an ordered itemset S satisfies constraint C, so does any prefix of S. A constraint C is *convertible monotone* provided there is an order R' on items such that when an ordered itemset S' violates constraint C, so does any prefix of S'.

Key Points

Although some constraints are neither antimonotone nor monotone in general, several of them can be converted into antimonotone or monotone ones by properly ordering the items. These *convertible constraints* [1–3] possess the following nice properties. By arranging items according to some proper order R, if an itemset S satisfies a *convertible antimonotone constraint* C_{cam}, then all prefixes of S also satisfy C_{cam}. Similarly, by arranging items according to some proper order R' if an itemset S violates a *convertible monotone constraint* C_{com}, then any prefix of S also violates C_{com}. Examples of convertible constraints include $avg(S.Price) \geq \$50$, which expresses that the average price of all items in an itemset S is at least $50. By arranging items in nonascending order R of price, if the average price of items in an itemset S is at least $50, then the average price of items in any prefix of S would not be lower than that of S (i.e., all prefixes of S satisfying a convertible antimonotone constraint C_{cam} also satisfy C_{cam}). Similarly, by arranging items in nondescending order R^{-1} of price, if the average price of items in an itemset S falls below \$50, then the average price of items in any prefix of S would not be higher than that of S (i.e., any prefix of S violating a convertible monotone constraint C_{com} also violates C_{com}). Note that (i) any *antimonotone constraint* is also convertible antimonotone (for any order R and (ii) any *monotone constraint* is also convertible monotone (for any order R$'$).

Cross-References

► Frequent Itemset Mining with Constraints

Recommended Reading

1. Pei J, Han J. Can we push more constraints into frequent pattern mining? In: Proceedings of the 6th ACM SIGKDD International Conference on Knowledge Discovery and Data Mining; 2000. p. 350–4.
2. Pei J, Han J, Lakshmanan LVS. Mining frequent item sets with convertible constraints. In: Proceedings of the 17th International Conference on Data Engineering; 2001. p. 433–42.
3. Pei J, Han J, Lakshmanan LVS. Pushing convertible constraints in frequent itemset mining. Data Mining Knowl Discov. 2004;8(3):227–52. https://doi.org/10.1023/B:DAMI.0000023674.74932.4c.

model puts much more emphasis on communication and cooperation than computation.

Key Points

Turing machines are a nice illustration of the classical "computation-oriented" view of systems. However, this view is too limited for many applications (e.g., web services). Many systems can be viewed as a collection of interacting entities (e.g., communicating Turing machines). For example, in the context of a service-oriented architecture (SOA), coordination is more important than computation. There exist many approaches to model and support coordination. Linda is an example of a language to model coordination and communication among several parallel processes operating upon objects stored in and retrieved from a shared, virtual, associative memory [1]. Linda attempts to separate coordination from computation by only allowing interaction through tuplespaces. However, one could argue that this is also possible in classical approaches such as Petri nets (e.g., connect processes through shared places), synchronized transition systems/automata, process algebra, etc. Coordination also plays an important role in agent technology [2].

Some authors emphasize the interdisciplinary nature of coordination [3]. Coordination is indeed not a pure computer science issue and other disciplines like organizational theory, economics, psychology, etc., are also relevant.

Coordination

W. M. P. van der Aalst
Eindhoven University of Technology, Eindhoven, The Netherlands

Definition

Coordination is about managing dependencies activities, processes, and components. Unlike the classical computation models, a coordination

Cross-References

► Business Process Management
► Choreography
► Web Services
► Workflow Management

Recommended Reading

1. Gelernter D, Carriero N. Coordination languages and their significance. Commun ACM. 1992;35(2): 97–107.

2. Jennings NR. Commitments and conventions: the foundation of coordination in multi-agent systems. Knowl Eng Rev. 1993;8(3):223–50.
3. Malone TW, Crowston K. The interdisciplinary study of coordination. ACM Comput Surv. 1994;26(1): 87–119.

Copyright Issues in Databases

Michael W. Carroll
Villanova University School of Law, Villanova, PA, USA

Synonyms

Intellectual property; License

Definition

Copyright is a set of exclusive rights granted by law to authors of original works of authorship. It applies automatically as soon as an original work is created and fixed in a tangible medium of expression, such as when it is stored on a hard disk. Originality requires independent creation by the author and a modicum of creativity. Copyright covers only an author's original expression. Facts and ideas are not copyrightable. Copyright usually applies only partially to databases. Copyrightable expression usually is found in database structures, such as the selection and arrangement of field names, unless these do not reflect any creativity or are standard within an area of research. Copyright will also apply to creative data, such as photographs or expressive and sufficiently long text entries. By and large, the rule on facts and ideas means that most numerical data, scientific results, other factual data, and short text entries are not covered by copyright.

Historical Background

Copyright has evolved from a limited right to control the unauthorized distribution of a limited class of works, primarily books, to a more expansive set of rights that attach automatically to any original work of authorship. Copyright law has always been national in scope, but through international treaties most nations now extend copyright to non-resident copyright owners. To comply with these treaties, copyright is now also automatic in the USA, which has abandoned requirements that a copyright owner register the work with the Copyright Office or publish the work with the copyright symbol – © – in order to retain copyright.

Foundations

Copyright

Copyright attaches to an original work of authorship that has been embodied in a fixed form. The "work" to which copyright attaches can be the structure of the database or a relatively small part of a database, including an individual data element, such as a photograph. It is therefore possible for a database to contain multiple overlapping copyrighted works or elements. To the extent that a database owner has a copyright, or multiple copyrights, in elements of a database, the rights apply only to those copyrighted elements. The rights are to reproduce, publicly distribute or communicate, publicly display, publicly perform, and prepare adaptations or derivative works.

Standards for Obtaining Copyright

Originality
Copyright protects only an author's "original" expression, which means expression independently created by the author that reflects a minimal spark of creativity. A database owner may have a copyright in the database structure or in the user interface with the database, whether that be a report form or an electronic display of field names associated with data. The key is whether the judgments made by the person(s) selecting and arranging the data require the exercise of sufficient discretion to make the selection or arrangement "original." In *Feist Publications, Inc. v. Rural Telephone Service Company*, the US Supreme Court held that a white pages telephone

directory could not be copyrighted. The data – the telephone numbers and addresses – were "facts" which were not original because they had no "author." Also, the selection and arrangement of the facts did not meet the originality requirement because the decision to order the entries alphabetically by name did not reflect the "minimal spark" of creativity needed.

As a practical matter, this originality standard prevents copyright from applying to complete databases – i.e., those that list all instances of a particular phenomenon – that are arranged in an unoriginal manner, such as alphabetically or by numeric value. However, courts have held that incomplete databases that reflect original selection and arrangement of data, such as a guide to the "best" restaurants in a city, are copyrightable in their selection and arrangement. Such a copyright would prohibit another from copying and posting such a guide on the Internet without permission. However, because the copyright would be limited to that particular selection and arrangement of restaurants, a user could use such a database as a reference for creating a different selection and arrangement of restaurants without violating the copyright owner's copyright.

Copyright is also limited by the merger doctrine, which appears in many database disputes. If there are only a small set of practical choices for expressing an idea, the law holds that the idea and expression merge, and the result is that there is no legal liability for using the expression.

Under these principles, metadata is copyrightable only if it reflects an author's original expression. For example, a collection of simple bibliographic metadata with fields named "author," "title," "date of publication," would not be sufficiently original to be copyrightable. More complex selections and arrangements may cross the line of originality. Finally, to the extent that software is used in a database, software is protectable as a "literary work." A discussion of copyright in executable code is beyond the scope of this entry.

Fixation

A work must also be "fixed" in any medium, permitting the work to be perceived, reproduced, or otherwise communicated for a period of more than a transitory duration. The structure and arrangement of a database may be fixed any time that it is written down or implemented. For works created after January 1, 1978 in the USA, exclusive rights under copyright shower down upon the creator at the moment of fixation.

The Duration of Copyright

Under international treaties, copyright must last for at least the life of the author plus 50 years. Some countries, including the USA, have extended the length to the life of the author plus 70 years. Under US law, if a work was made as a "work made for hire," such as a work created by an employee within the scope of employment, the copyright lasts for 120 years from creation if the work is unpublished or 95 years from the date of publication.

Ownership and Transfer of Copyright

Copyright is owned initially by the author of the work. If the work is jointly produced by two or more authors, such as a copyrightable database compiled by two or more scholars, each has a legal interest in the copyright. When a work is produced by an employee, ownership differs by country. In the USA, the employer is treated as the author under the "work made for hire" doctrine, and the employee has no rights in the resulting work. Elsewhere, the employee is treated as the author and retains certain moral rights in the work while the employer receives the economic rights in the work. Copyrights may be licensed or transferred. A non-exclusive license, or permission, may be granted orally or even by implication. A transfer or an exclusive license must be done in writing and signed by the copyright owner. Outside of the USA, some or all of the author's moral rights cannot be transferred or terminated by agreement. The law on this issue varies by jurisdiction.

The Copyright Owner's Rights

The rights of a copyright owner are similar throughout the world although the terminology differs as do the limitations and exceptions to these rights.

Reproduction

As the word "copyright" implies, the owner controls the right to reproduce the work in copies. The reproduction right covers both exact duplicates of a work and works that are "substantially similar" to the copyrighted work when it can be shown that the alleged copyist had access to the copyrighted work. In the USA, some courts have extended this right to cover even a temporary copy of a copyrighted work stored in a computer's random access memory ("RAM").

Public Distribution, Performance, Display or Communication

The USA divides the rights to express the work to the public into rights to distribute copies, display a copy, or publicly perform the work. In other parts of the world, these are subsumed within a right to communicate the work to the public.

Within the USA, courts have given the distribution right a broad reading. Some courts, including the appeals court in the Napster case, have held that a download of a file from a server connected to the internet is both a reproduction by the person requesting the file and a distribution by the owner of the machine that sends the file. The right of public performance applies whenever the copyrighted work can be listened to or watched by members of the public at large or a subset of the public larger than a family unit or circle of friends. Similarly, the display right covers works that can be viewed at home over a computer network as long as the work is accessible to the public at large or a subset of the public.

Right of Adaptation, Modification or Right to Prepare Derivative Works

A separate copyright arises with respect to modifications or adaptations of a copyrighted work so long as these modifications or adaptations themselves are original. This separate copyright applies only to these changes. The copyright owner has the right to control such adaptations unless a statutory provision, such as fair use, applies.

Theories of Secondary Liability

Those who build or operate databases also have to be aware that copyright law holds liable certain parties that enable or assist others in infringing copyright. In the USA, these theories are known as contributory infringement or vicarious infringement.

Contributory Infringement

Contributory copyright infringement requires proof that a third party intended to assist a copyright infringer in that activity. This intent can be shown when one supplies a means of infringement with the intent to induce another to infringe or with knowledge that the recipient will infringe. This principle is limited by the so-called *Sony* doctrine, by which one who supplies a service or technology that enables infringement, such as a VCR or photocopier, will be deemed not to have knowledge of infringement or intent to induce infringement so long as the service or technology is capable of substantial non-infringing uses.

Two examples illustrate the operation of this rule. In *A&M Records, Inc. v. Napster, Inc.*, the court of appeals held that peer-to-peer file sharing is infringing but that Napster's database system for connecting users for peer-to-peer file transfers was capable of substantial non-infringing uses and so it was entitled to rely on the *Sony* doctrine. (Napster was held liable on other grounds.) In contrast, in *MGM Studios, Inc. v. Grokster, Ltd.*, the Supreme Court held that Grokster was liable for inducing users to infringe by specifically advertising its database service as a substitute for Napster's.

Vicarious Liability for Copyright Infringement

Vicarious liability in the USA will apply whenever (i) one has control or supervisory power over the direct infringer's infringing conduct and (ii) one receives a direct financial benefit from the infringing conduct. In the Napster case, the court held that Napster had control over its users, because it could refuse them access to the Napster server and, pursuant to the Terms of Service Agreements entered into with users, could terminate access if infringing conduct was discovered.

Other courts have required a greater showing of actual control over the infringing conduct.

Similarly, a direct financial benefit is not limited to a share of the infringer's profits. The Napster court held that Napster received a direct financial benefit from infringing file trading because users' ability to obtain infringing audio files drew them to use Napster's database. Additionally, Napster could potentially receive a financial benefit from having attracted a larger user base to the service.

Limitations and Exceptions

Copyrights' limitations and exceptions vary by jurisdiction. In the USA, the broad "fair use" provision is a fact-specific balancing test that permits certain uses of copyrighted works without permission. Fair use is accompanied by some specific statutory limitations that cover, for example, certain uses in the classroom use and certain uses by libraries. The factors to consider for fair use are: (i) the purpose and character of the use, including whether such use is of a commercial nature or is for nonprofit educational purposes; (ii) the nature of the copyrighted work; (iii) the amount and substantiality of the portion used in relation to the copyrighted work as a whole; and (iv) the effect of the use upon the potential market for or value of the copyrighted work. The fact that a work is unpublished shall not itself bar a finding of fair use if such finding is made upon consideration of all the above factors.

Countries whose copyright law follows that of the United Kingdom, a more limited "fair dealing" provision enumerates specific exceptions to copyright. In Europe, Japan, and elsewhere, the limitations and exceptions are specified legislatively and cover some private copying and some research or educational uses.

Remedies and Penalties

In general, a copyright owner can seek an injunction against one who is either a direct or secondary infringer of copyright. The monetary consequences of infringement differ by jurisdiction. In the USA, the copyright owner may choose between actual or statutory damages. Ac-

tual damages cover the copyright owner's lost profits as well as a right to the infringer's profits derived from infringement. The range for statutory damages is $750–$30,000 per copyrighted work infringed. If infringement is found to have been willful, the range increases to $150,000. The amount of statutory damages in a specific case is determined by the jury. There is a safe harbor from statutory damages for non-profit educational institutions if an employee reproduces a copyrighted work with a good faith belief that such reproduction is a fair use.

A separate safe harbor scheme applies to on-line service providers when their database is comprised of information stored at the direction of their users. An example of such a database would be YouTube's video sharing database. The service provider is immune from monetary liability unless the provider has knowledge of infringement or has control over the infringer and receives a direct financial benefit from infringement. The safe harbor is contingent on a number of requirements, including that the provider have a copyright policy that terminates repeat infringers, that the provider comply with a notice-and-takedown procedure, and that the provider have an agent designated to receive notices of copyright infringement.

Key Applications

In cases arising after the *Feist* decision, the courts have faithfully applied the core holding that facts are in the public domain and free from copyright even when substantial investments are made to gather such facts. There has been more variation in the characterization of some kinds of data as facts and in application of the modicum-of-creativity standard to the selections and arrangements in database structures.

On the question of when data is copyrightable, a court of appeals found copyrightable expression in the "Red Book" listing of used car valuations. The defendant had copied these valuations into its database, asserting that it was merely copying unprotected factual information. The court disagreed, likening the valuations to expressive opinions and finding a modicum of originality in these. In addition, the selection and arrangement of the data, which included a division of the market into geographic regions, mileage adjustments in 5,000-mile increments, a selection of optional features for inclusion, entitled the plaintiff to a thin copyright in the database structure.

Subsequently, the same court found that the prices for futures contracts traded on the New York Mercantile Exchange (NYMEX) probably were not expressive data even though a committee makes some judgments in the setting of these prices. The court concluded that even if such price data were expressive, the merger doctrine applied because there was no other practicable way of expressing the idea other than through a numerical value, and a rival was free to copy price data from NYMEX's database without copyright liability.

Finally, where data are comprised of arbitrary numbers used as codes, the courts have split. One court of appeals has held that an automobile parts manufacturer owns no copyright in its parts numbers, which are generated by application of a numbering system that the company created. In contrast, another court of appeals has held that the American Dental Association owns a copyright in its codes for dental procedures.

On the question of copyright in database structures, a court of appeals found that the structure of a yellow pages directory including listing of Chinese restaurants was entitled to a "thin" copyright, but that copyright was not infringed by a rival database that included 1,500 of the listings, because the rival had not copied the plaintiff's data structure. Similarly, a different court of appeals acknowledged that although a yellow pages directory was copyrightable as a compilation, a rival did not violate that copyright by copying the name, address, telephone number, business type, and unit of advertisement purchased for each listing in the original publisher's directory. Finally, a database of real estate tax assessments that arranged the data collected by the assessor into 456 fields grouped into 34 categories was sufficiently original to be copyrightable.

Cross-References

▶ European Law in Databases
▶ Licensing and Contracting Issues in Databases

Recommended Reading

1. American Dental Association v. Delta Dental Plans Ass'n, 126 F.3d 977 (7th Cir.1997).
2. Assessment Technologies of WI, LLC v. WIRE data, Inc., 350 F.3d 640 (7th Cir. 2003).
3. Bellsouth Advertising & Publishing Corp. v. Donnelly Information Publishing, Inc., 999 F.2d 1436 (11th Cir. 1993) (en banc).
4. CCC Information Services, Inc. v. MacLean Hunter Market Reports, Inc., 44 F.3d 61 (2d Cir. 1994).
5. Feist Publications, Inc. v. Rural Telephone Service Co., 499 U.S. 340 (1991).
6. Ginsburg JC. Copyright, common law, and sui generis protection of databases in the United States and abroad. Univ Cinci Law Rev. 1997;66:151–76.
7. Key Publications, Inc. v. Chinatown Today Publishing Enterprises, Inc., 945 F.2d 509 (2d Cir. 1991).
8. New York Mercantile Exchange, Inc. v. Intercontinental Exchange, Inc., 497 F.3d 109, (2d Cir. 2007).
9. Southco, Inc. v. Kanebridge Corp., 390 F.3d 276 (3d Cir. 2004) (en banc).

CORBA

Aniruddha Gokhale
Vanderbilt University, Nashville, TN, USA

Synonyms

Common object request broker architecture; Object request broker

Definition

The Common Object Request Broker Architecture (CORBA) [1, 2] is standardized by the Object Management Group (OMG) for distributed object computing.

Key Points

The CORBA standard specifies a platform-independent and programming language-independent architecture and a set of APIs to simplify distributed application development. The central idea in CORBA is to decouple the interface from the implementation. Applications that provide services declare their interfaces and operations in the Interface Description Language (IDL). IDL compilers read these definitions and synthesize client-side stubs and server-side skeletons, which provide data marshaling and proxy capabilities.

CORBA provides both a type-safe RPC-style object communication paradigm called the Static Invocation Interface (SII) and a more dynamic form of communication called the Dynamic Invocation Interface (DII), which allows creation and population of requests dynamically via reflection capabilities. The DII is often used to bridge different object models. CORBA defines a binary format for on-the-wire representation of data called the Common Data Representation (CDR). CDR has been defined to enable programming language-neutrality.

The CORBA 1.0 specification (October 1991) and subsequent revisions through version 1.2 (December 1993) defined these basic capabilities; however, they lacked any support for interoperability across different CORBA implementations.

The CORBA 2.0 specification (August 1996) defined an interoperability protocol called the General Inter-ORB Protocol (GIOP), which defines the packet formats for data exchange between communicating CORBA entities. GIOP is an abstract specification and must be mapped to the underlying transport protocol. The most widely used concrete mapping of GIOP is called the Internet Inter-ORB Protocol (IIOP) used for data exchange over TCP/IP networks.

Despite these improvements, the earlier versions of CORBA focused only on the client-side portability and lacked any support for server-side portability. This limitation was addressed in the CORBA 2.2 specification (August 1996) through the Portable Object Adapter (POA) concept. The POA enables server-side transparency to applications and server-side portability. The POA provides a number of policies that can be used to manage the server-side objects.

The CORBA specification defines compliance points for implementations to ensure interoperability. The CORBA specification has also been enhanced with additional capabilities that are available beyond the basic features, such as the real-time CORBA specification [3]. Implementations of these specifications must provide these additional capabilities.

In general, CORBA enhances conventional procedural RPC middleware by supporting object oriented language features (such as encapsulation, interface inheritance, parameterized types, and exception handling) and advanced design patterns for distributed communication. The most recent version of CORBA specification at the time of this writing is 3.3 (January 2008), which also includes support for a component architecture.

Cross-References

▶ Client-Server Architecture
▶ DCE
▶ DCOM
▶ .NET Remoting
▶ Request Broker
▶ SOAP

Recommended Reading

1. Object Management Group, Common Object Request Broker Architecture (CORBA), Version 3.1, OMG Document No. formal/2008-01-08, January 2008.
2. Soley RM, Stone CM. Object management architecture guide. 3rd ed. Object Management Group; 1995.
3. Object Management Group, Real-Time CORBA Specification, Version 1.2, OMG Document No. formal/2005-01-04, January 2005.

Correctness Criteria Beyond Serializability

Mourad Ouzzani[1], Brahim Medjahed[2], and
Ahmed K. Elmagarmid[1,3]
[1]Qatar Computing Research Institute, HBKU,
Doha, Qatar
[2]The University of Michigan – Dearborn,
Dearborn, MI, USA
[3]Purdue University, West Lafayette, IN, USA

Synonyms

Concurrency control; Preserving database consistency

Definition

A *transaction* is a logical unit of work that includes one or more database access operations such as insertion, deletion, modification, and retrieval [7]. A *schedule* (or history) S of n transactions $T_1, \ldots, T_n$ is an ordering of the transactions that satisfies the following two conditions: (i) the operations of T_i (i = 1,...,n) in S must occur in the same order in which they appear in T_i, and (ii) the operations of T_j (j ≠ i) may be interleaved with T_i's operations in S. A schedule S is *serial* if for every two transactions T_i and T_j that appear in S, either all the operations of T_i appear before all the operations of T_j or vice versa. Otherwise, the schedule is called *nonserial* or *concurrent*. Nonserial schedules of transactions may lead to issues with the correctness of the schedule due to concurrency such as lost update, dirty read, and unrepeatable read. For instance, the lost update problem occurs whenever two transactions, while attempting to modify a data item, both read the item's old value before either one writes the item's new value [2].

The simplest way for controlling concurrency is to allow only serial schedules. However, with no concurrency, database systems make poor use of their resources and hence will be inefficient, resulting in smaller transaction execution rates,

for example. To broaden the class of allowable transaction schedules, *serializability* has been proposed as the major correctness criterion for concurrency control [8, 11]. Serializability ensures that a concurrent schedule of transactions is equivalent to some serial schedule of the same transactions [12]. While serializability has been successfully used in traditional database applications, e.g., airline reservations and banking, it has been proven to be restrictive and hardly applicable in advanced applications such as computer-aided design (CAD), computer-aided manufacturing (CAM), office automation, and multidatabases. These applications introduce new requirements that either prevent the direct use of serializability, e.g., violation of local autonomy in multidatabases, or make the use of serializability inefficient, e.g., long-running transactions in CAD/CAM applications. These limitations have motivated the introduction of more flexible correctness criteria that go beyond the traditional concept of serializability.

Historical Background

Concurrency control began appearing in database systems in the early to mid-1970s. It emerged as an active database research thrust starting from 1976 as witnessed by the early influential papers published by Eswaren et al. [5] and Gray et al. [8]. A comprehensive coverage of serializability theory has been presented in 1986 by Papadimitriou in [12]. Simply put, serializability theory is a mathematical model for proving whether a concurrent execution of transactions is correct. It gives precise definitions and properties that nonserial schedules of transactions must satisfy to be serializable. Equivalence between a concurrent and serial schedule of transactions is at the core of the serializability theory. Two major types of equivalence have then been defined: *conflict* equivalence and *view* equivalence. We should know that if two schedules are conflict equivalent then they are view equivalent. The converse is generally not true.

Conflict equivalence has initially been introduced by Gray et al. in 1975 [8]. A concurrent schedule of transactions is *conflict equivalent* to a serial schedule of the same transactions (and hence *conflict serializable*) if both schedules order conflicting operations in the same way, i.e., they have the same precedence relations of conflicting operations. Two operations are said to be *conflicting* if they are from different transactions upon the same data item and at least one of them is a `write` operation. If two operations conflict, their execution order matters. For instance, the value returned by a `read` operation depends on whether or not that operation precedes or follows a particular `write` operation on the same data item. Conflict serializability is usually tested by analyzing the acyclicity of the graph derived from the execution of the different transactions in a schedule. This graph, called *serializability graph*, is a directed graph that models the precedence of conflicting operations in the transactions.

View equivalence has been proposed by Yannakakis in 1984 [15]. A concurrent schedule of transactions is *view equivalent* to a serial schedule of the same transactions (and hence *view serializable*) if the respective transactions in the two schedules read and write the same data values. View equivalence is based on the following two observations: (i) if each transaction reads each of its data items produced by the same `write` operations, then all `write` operations write the same value in both schedules; and (ii) if the final `write` operation on each data item is the same in both schedules, then the final value of all data items will be the same in both schedules. View serializability is usually expensive to check. One approach is to check the acyclicity of a special graph called *polygraph*. A polygraph is a generalization of the precedence graph that takes into account all precedence constraints required by view serializability.

Foundations

The limitations of the traditional serializability concept combined with the requirement of advanced database applications triggered a wave of new correctness criteria that go beyond serializability. These criteria aim at achieving one or several of the following goals: (i) accept non-serializable but correct executions by exploiting the semantics, structure, and integrity constraints of the transactions, (ii) allow inconsistencies to appear in a controlled manner which may be acceptable for some applications, (iii) limit conflicts by creating a new version of the data for each update, and (iv) treat transactions accessing more than one database, in the case of multi-databases, differently from those accessing one single database and maintain overall correctness. While a large number of correctness criteria have been presented in the literature, this entry will focus on those that had a considerable impact on the field. These criteria will be presented as described in their original versions since several of these criteria have been either extended, improved, or applied to specific contexts. Table 1 summarizes the correctness criteria outlined in this section.

Multiversion Serializability

Multiversion databases aim at increasing the degree of concurrency and providing a better system recovery. In such databases, whenever a transaction writes a data item, it creates a new version of this item instead of overwriting it. The basic idea of *multiversion serializability* [1] is that some schedules can be still seen as serializable if a read is performed on some older version of a data item instead of the newer modified version. Concurrency is increased by having transactions read older versions, while other concurrent transactions are creating newer versions. There is only one type of conflict that is possible, when a transaction reads a version of a data item that was written by another transaction. The two other conflicts (write, write) and (read, write) are not possible since each write produces a new version, and a data item cannot be read until it has been produced, respectively. Based on the assumption that users expect their transactions to behave as if there were just one copy of each data item, the notion of a *one-copy serial* schedule is defined. A schedule is one-copy serial if for all i, j, and x, if a transaction T_j reads x from a transaction T_i, then either $i = j$ or T_i is the last transaction preceding t_j that writes into any version of x. Hence, a schedule

Correctness Criteria Beyond Serializability, Table 1 Representative correctness criteria for concurrency control

Correctness criterion	Basic idea	Examples of application domains	Reference
Multiversion serializability	Allows some schedules as serializable if a read is performed on some older version of a data item instead of the newer modified version	Multiversion database systems	[1]
Semantic consistency	Uses semantic information about the transactions to accept some non-serializable but correct schedules	Applications that can provide some semantic knowledge	[6]
Predicatewise serializability	Focuses on data integrity constraints	CAD database and office information systems	[9]
Epsilon serializability	Allows inconsistencies to appear in a controlled manner by attaching a specification of the amount of permitted inconsistency to each transaction	Applications that tolerate some inconsistencies	[13]
Eventual consistency	Requires that duplicate copies are consistent at certain times but may be inconsistent in the interim intervals	Distributed databases with replicated or interdependent data	[14]
Quasi-serializability	Executes global transactions in a serializable way while taking into account the effect of local transactions	Multidatabase systems	[4]
Two-level serializability	Ensures consistency by exploiting the nature of the integrity constraints and the nature of the transactions in multidatabase environments	Multidatabase systems	[10]

is defined as *one-copy serializable* (1-SR) if it is equivalent to a one-serial schedule. 1-SR is shown to maintain correctness by proving that a multiversion schedule behaves like a serial non-multiversion schedule (there is only one version for each data item) if the multiversion schedule is one-serializable. The one-copy serializability of a schedule can be verified by checking the acyclicity of the multiversion serialization graph of that schedule.

Semantic Consistency

Semantic consistency uses semantic information about the transactions to accept some non-serializable but correct schedules [6]. To ensure that users see consistent data, the concept of *sensitive transactions* has been introduced. Sensitive transactions output only consistent data and thus must see a consistent database state. A semantically consistent schedule is one

that transforms the database from a consistent state to another consistent state and where all sensitive transactions obtain a consistent view of the database with respect to the data accessed by these transactions, i.e., all data consistency constraints of the accessed data evaluate to True. Enforcing semantic consistency requires knowledge about the application which must be provided by the user. In particular, users will need to group actions of the transactions into steps and specify which steps of a transaction of a given type can be interleaved with the steps of another type of transactions without violating consistency. Four types of semantic knowledge are defined: (i) transaction semantic types, (ii) compatibility sets associated with each type, (iii) division of transactions into steps, and (iv) countersteps to (semantically) compensate the effect from some of the steps executed within the transaction.

Predicatewise Serializability

Predicatewise serializability (PWSR) has been introduced as a correctness criterion for CAD database and office information systems [9]. PWSR focuses solely on data integrity constraints. In a nutshell, if database consistency constraints can be expressed in a conjunctive normal form, a schedule is said to be PWSR if all projections of that schedule on each group of data items that share a disjunctive clause (of the conjunctive form representing the integrity constraints) are serializable. There are three different types of restrictions that must be enforced on PWSR schedules to preserve database consistency: (i) force the transactions to be of *fixed structure*, i.e., they are independent of the database state from which they execute, (ii) force the schedules to be *delayed read*, i.e., a transaction T_i cannot read a data item written by a transaction T_j until after T_j has completed all of its operations, or (iii) the conjuncts of the integrity constraints can be ordered in a way that no transaction reads a data item belonging to a higher numbered conjunct and writes a data item belonging to a lower numbered conjunct.

Epsilon Serializability

Epsilon serializability (ESR) [13] has been introduced as a generalization to serializability where a limited amount of inconsistency is permitted. The goal is to enhance concurrency by allowing some non-serializable schedules. ESR introduces the notion of *epsilon transactions* (ETs) by attaching a specification of the amount of permitted inconsistency to each (standard) transaction. ESR distinguishes between transactions that contain only read operation, called query epsilon transaction or query ET, and transactions with at least one update operation, called update epsilon transaction or update ET. Query ETs may view uncommitted, possibly inconsistent, data being updated by update ETs. Thus, update ETs are seen as exporting some inconsistencies, while query ETs are importing these inconsistencies. ESR aims at bounding the amount of imported and exported inconsistency for each ET. An *epsilon-serial* schedule is defined as a schedule where (i) the update ETs form a serial schedule if con-sidered alone without the query ET and (ii) the entire schedule consisting of both query ETs and update ETs is such that the non-serializable conflicts between query ETs and update ETs are less than the permitted limits specified by each ET. An epsilon-serializable schedule is one that is equivalent to an epsilon-serial schedule. If the permitted limits are set to zero, ESR corresponds to the classical notion of serializability.

Eventual Consistency

Eventual consistency has been proposed as an alternative correctness criterion for distributed databases with replicated or interdependent data [14]. This criterion is useful in several applications like mobile databases, distributed databases, and large-scale distributed systems in general. Eventual consistency requires that duplicate copies are consistent at certain times but may be inconsistent in the interim intervals. The basic idea is that duplicates are allowed to diverge as long as the copies are made consistent periodically. The times where these copies are made consistent can be specified in several ways which could depend on the application, for example, at specified time intervals, when some events occur, or at some specific times. A correctness criterion that ensures eventual consistency is the *current copy serializability*. Each update occurs on a current copy and is asynchronously propagated to other replicas.

Quasi-Serializability

Quasi-serializability (QSR) is a correctness criterion that has been introduced for multidatabase systems [4]. A multidatabase system allows users to access data located in multiple autonomous databases. It generally involves two kinds of transactions: (i) Local transactions that access only one database; they are usually outside the control of the multidatabase system, and (ii) global transactions that can access more than one database and are subject to control by both the multidatabase and the local databases. The basic premise is that to preserve global database consistency, global transactions should be executed in a serializable way while taking into account the effect of local transactions. The effect of local transactions appears in the form of indirect conflicts that these local transactions introduce between global transactions which may not necessarily access (conflict) the same data items. A *quasi-serial* schedule is a schedule where global transactions are required to execute serially and local schedules are required to be serializable. This is in contrast to global serializability where all transactions, both local and global, need to execute in a (globally) serializable way. A global schedule is said to be quasi-serializable if it is (conflict) equivalent to a quasi-serial schedule. Based on this definition, a quasi-serializable schedule maintains the consistency of multidatabase systems since: (i) a quasi-serial schedule preserves the mutual consistency of globally replicated data items, based on the assumptions that these replicated data items are updated only by global transactions, and (ii) a quasi-serial schedule preserves the global transaction consistency constraints as local schedules are serializable and global transactions are executed following a schedule that is equivalent to a serial one.

Two-Level Serializability

Two-level serializability (2LSR) has been introduced to relax serializability requirements in multidatabases and allow a higher degree of concurrency while ensuring consistency [10]. Consistency is ensured by exploiting the nature of integrity constraints and the nature of transactions in multidatabase environments. A global schedule, consisting of both local and global transactions, is 2LSR if all local schedules are serializable, and the projection of that schedule on global transactions is serializable. Local schedules consist of all operations, from global and local transactions, that access the same local database. Ensuring that each local schedule is serializable is already taken care of by the local database. Furthermore, ensuring that the global transactions are executed in a serializable way can be done by the global concurrency controller using any existing technique from centralized databases like the two-phase-locking (2PL) protocol. This is possible since the global

transactions are under the full control of the global transaction manager. [10] shows that under different scenarios, 2LSR preserves a strong notion of correctness where the multidatabase consistency is preserved and all transactions see consistent data. These different scenarios differ depending on: (i) which kind of data items, local or global, global and local transactions are reading or writing, (ii) the existence of integrity constraints between local and global data items, and (iii) whether all transaction are preserving the consistency of local databases when considered alone.

Key Applications

The major database applications behind the need for new correctness criteria include distributed databases, mobile databases, multidatabases, CAD/CAM applications, office automation, cooperative applications, and software development environments. All of these applications introduced requirements and limitations that either prevent the use of serializability like the violation of local autonomy in multidatabases or make the use of serializability inefficient like blocking long-running transactions.

Future Directions

A recent trend in transaction management focuses on adding transactional properties (e.g., isolation, atomicity) to business processes [3]. A business process (BP) is a set of tasks that are performed collaboratively to realize a business objective. Since BPs contain activities that access shared and persistent data resources, they have to be subject to transactional semantics. However, it is not adequate to treat an entire BP as a single "traditional" transaction mainly because BPs: (i) are of long duration and treating an entire process as a transaction would require locking resources for long periods of time, (ii) involve many independent database and application systems and enforcing transactional properties across the entire process would require expensive coordination among these systems, and (iii) have external effects and using conventional transactional rollback mechanisms is not feasible. These characteristics open new research issues to take the concept of correctness criterion and how it should be enforced beyond even the correctness criteria discussed here.

Cross-References

▶ ACID Properties
▶ Transaction Management
▶ Two-Phase Commit
▶ Two-Phase Locking

Recommended Reading

1. Bernstein PA, Goodman N. Multiversion concurrency control – theory and algorithms. ACM Trans Database Syst. 1983;8(4):465–83.
2. Bernstein PA, Hadzilacos V, Goodman N. Concurrency control and recovery in database systems. Reading: Addison-Wesley; 1987.
3. Dayal U, Hsu M, Ladin R. Business process coordination: state of the art, trends, and open issues. In: Proceedings of the 27th International Conference on Very Large Data Bases; 2001. p. 3–13.
4. Du W, Elmagarmid AK. Quasi serializability: a correctness criterion for global concurrency control in Interbase. In: Proceedings of the 15th International Conference on Very Large Data Bases; 1989. p. 347–55.
5. Eswaran KP, Gray J, Lorie RA, Traiger IL. The notions of consistency and predicate locks in a database system. Commun ACM. 1976;19(11):624–33.
6. Garcia-Molina H. Using semantic knowledge for transaction processing in a distributed database. ACM Trans Database Syst. 1983;8(2):186–213.
7. Gray J, Reuter A. Transaction processing: concepts and techniques. Los Altos: Morgan Kaufmann; 1993.
8. Gray J, Lorie RA, Putzolu GR, Traiger IL. Granularity of locks in a large shared data base. In: Proceedings of the 1st International Conference on Very Data Bases; 1975. p. 428–51.
9. Korth HF, Speegle GD. Formal model of correctness without serializability. In: Proceedings of the ACM SIGMOD International Conference on Management of Data; 1988. p. 379–86.
10. Mehrotra S, Rastogi R, Korth HF, Silberschatz A. Ensuring consistency in multidatabases by preserving two-level serializability. ACM Trans Database Syst. 1998;23(2):199–230.
11. Papadimitriou CH. The serializability of concurrent database updates. J ACM. 1979;26(4):631–53.
12. Papadimitriou CH. The theory of database concurrency control. Rockville: Computer Science; 1986.
13. Ramamritham K, Pu C. A formal characterization of epsilon serializability. IEEE Trans Knowl Data Eng. 1995;7(6):997–1007.
14. Sheth A, Leu Y, Elmagarmid A. Maintaining consistency of interdependent data in multidatabase systems. Technical Report CSD-TR-91-016, Purdue University. 1991. http://www.cs.toronto.edu/georgem/ws/ws.ps
15. Yannakakis M. Serializability by locking. J ACM. 1984;31(2):227–44.

Cost and Quality Trade-Offs in Crowdsourcing

Lei Chen
Hong Kong University of Science and Technology, Hong Kong, China

Synonyms

Incentive and performance trade-offs; Payment and quality trade-offs

Definition

In crowdsourcing, some tasks are conducted by the crowd due to enjoyment [8] or social reward [6]. However, arbitrary tasks are seldom enjoyable, and social award is often associated to some specific tasks, such as Wikipedia (https://en.wikipedia.org/wiki/Main_Page) and Stack Overflow (http://stackoverflow.com/). Thus, given an arbitrary task, a requester often needs to offer incentive (i.e., the cost of the task) to motivate workers to conduct the task. The cost per task is often paid in the form of financial compensation, a few cents per task. The quality of a crowdsourcing task is often referred as accuracy. Since the workers are humans, which may make errors when they perform tasks, the results returned by the crowd will have errors as a consequence. The trade-offs between cost and quality refer to the relationships between the financial incentive and the performance.

Historical Background

Wikipedia, the well-known online crowdsourcing example was launched in 2001. Since then, many other social media platforms also emerged in the form of crowdsourcing, such as Flickr and del.icio.us. One common feature of these crowdsourcing platforms is that workers perform tasks for free. They either do the work for enjoyment (fun) or for gaining some reputation (social rewards). However, the free tasks are quite few, and the requester often needs to offer incentive to motivate workers to work on arbitrary tasks. Amazon Mechanical Turk (AMT), launched on November 2, 2005, is a crowdsourcing Internet marketplace enabling requesters to pay incentive to the crowd for performing tasks that computers are currently unable to do effectively, such as image labeling and language translation. In a survey conducted in February 2009, 91% of turkers preferred monetary reward over any other incentive. Out of everyone, only 42% of turkers were doing tasks just for fun. In another survey, it was reported that 25% of Indian turkers and 13% of US turkers were relying on different tasks in AMT as their primary source of income. These numbers are despite the fact that the requester has the entire right to reject completed work [10]. It seems straightforward that higher payment will lead workers to contribute high-quality work. However, the recent experimental studies have shown that financial incentives may undermine "intrinsic motivation" (e.g., enjoyment, desire to help out) or even leading to poor outcomes [3] and undermining actual performance through "choking effect" [1]. Based on the results on two real experimental studies on AMT, Mason and Watts [5] show that increasing incentive (paying more) to the crowd will not affect the quality (accuracy) of the results; it only increases the quantity of the completed tasks.

Scientific Fundamentals

In the field of crowdsourcing, where a specific set of functions of a company is outsourced to an undefined set of people over the Internet, cost spent and the output quality of a task is highly correlated. In traditional economic theory, most of the times, higher pay ensures higher quality output. But the same principle does not work in crowdsourcing all the time. The workers used in crowdsourcing are recruited over the Internet without any interview like traditional office culture. Hence, there is no way to ensure the authenticity of any worker. Also, no one can identify the expertise of a worker before assigning them for a task. This creates a huge problem as the requestor, one who pays, might end up spending a lot if the workers recruited does not have enough expertise. If the workers are not good enough, the requestor will have to wait until an expert in the field gives a correct answer, which could happen at any point in time.

Also, the workers pool will have different types of workers including spammers and genuine users. Many workers will try to finish up the job fast to earn money and maximize profit. All of these will result in spending more without any guarantee on the output [1]. Also, this will lead to the generation of a large amount of useless content [7]. So in crowdsourcing, the higher cost might increase the participation but not necessarily the quality.

At the same time, paying less could also lead to so-called task starvation problem, where tasks are not accepted by any workers [2]. A trend of rushing to finish the job to get money can be observed as well, which is demonstrated in the study conducted by Dr. Gabriella Kazai [4]. The study showed an improvement in quality by 126% when the right amount is paid. It also showed that the spam count was higher when the amount offered was smaller. In the experiment, 47% of the worker were unsatisfied with the initial payment, while 72% were happy with the improved payment.

Future Directions

Since the cost is not directly related to the quality of the crowdsourcing results, one interesting future research direction is to consider the budget allocation (incentive allocation) according to difficulty of tasks and workers' background [9]. Novice workers cannot contribute more than a limit and as their solutions could be of lower quality compared to that of experts. Thus, with the right incentive assigned to the proper workers, the quality of the crowdsourcing results can be improved.

Recommended Reading

1. Ariely D, Gneezy U, Loewenstein G, Mazar N. Large stakes and big mistakes. Rev Econ Stud. 2009;76:451–69.
2. Faradani S, Hartmann B, Ipeirotis PG. What's the right price? Pricing tasks for finishing on time. In: Proceedings of the 2011 AAAI Conference on Artificial Intelligence. 2011. p. 26–31.
3. Gneezy U, Rustichini A. Pay enough or don't pay at all. Q J Econ. 2000;115(3):791–810.
4. Kazai G. An exploration of the influence that task parameters have on the performance of crowds. In: Proceedings of the First International Conference on Crowdsourcing. 2010.
5. Mason W, Watts DJ. Financial incentives and the performance of crowds. In: ACM SIGKDD human computation. 2009. p. 100–08.
6. Nov O, Naaman M, Ye C. What drives content tagging: the case of photos on Flickr. In: Proceedings of the ACM Conference on Human Factors in Computing Systems; 2008. p. 1097–1110.
7. Snow R, O'Connor B, Jurafsky D, Ng AY. Cheap and fast – but is it good?: evaluating non-expert annotations for natural language tasks. In: Proceedings of the Conference on Empirical Methods in Natural Language Processing. 2008. p. 254–63.
8. von Ahn L. Games with a purpose. Computer. 2006;39(6):92–4.
9. Xie H, Lui JCS, Jiang JW, Chen W. Incentive mechanism and protocol design for crowdsourcing systems. In: Proceedings of the 2014 Annual Allerton Conference on Communication, Control, and Computing. 2014. p. 140–47.
10. Xintong G, Hongzhi W, Song Y, Hong G. Brief survey of crowdsourcing for data mining. Expert Syst Appl. 2014;41(17):7987–94.

Cost Estimation

Stefan Manegold
CWI, Amsterdam, The Netherlands

Definition

Execution costs, or simply *costs*, is a generic term to collectively refer to the various goals or objectives of database query optimization. Optimization aims at finding the "cheapest" ("best" or at least a "reasonably good") query execution plan (QEP) among semantically equivalent alternative plans for the given query. Cost is used as a metric to compare plans. Depending on the application, different types of costs are considered. Traditional optimization goals include minimizing response time (for the first answer or the complete result), minimizing resource consumption (like CPU time, I/O, network bandwidth, or amount of memory required), or maximizing throughput, i.e., the number of queries that the system can answer per time. Other, less obvious objectives – e.g., in a mobile environment – may be to minimize the power consumption needed to answer the query or the on-line time being connected to a remote database server.

Obviously, evaluating a QEP to measure its execution cost does not make sense. *Cost estimation* refers to the task of predicting the (approximate) costs of a given QEP a priori, i.e., without actually evaluating it. For this purpose, mathematical algorithms or parametric equations, commonly referred to as *cost models*, provide a simplified "idealized" abstract description of the system, focusing on the most relevant components. In general, the following three cost components are distinguished.

1. *Logical costs* consider only the data distributions and the semantics of relational algebra operations to estimate intermediate result sizes of a given (logical) query plan.

2. *Algorithmic costs* extend logical costs by taking also the computational complexity (expressed in terms of *O*-classes) of the algorithms into account.

3. *Physical costs* finally combine algorithmic costs with system/hardware specific parameters to predict the total costs, usually in terms of execution time.

Next to query optimization, cost models can serve another purpose. Especially algorithmic and physical cost models can help database developers to understand and/or predict the performance of existing algorithms on new hardware systems. Thus, they can improve the algorithms or even design new ones without having to run time and resource consuming experiments to evaluate their performance.

Since the quality of query optimization strongly depends on the quality of cost estimation, details of cost estimation in commercial database products are usually well kept secrets of their vendors.

Historical Background

Not all aspects of database cost estimation are treated as independent research topic of their own. Mainly selectivity estimation and intermediate result size estimation have received intensive attention yielding a plethora of techniques proposed in database literature. Discussion of algorithmic costs usually occurs with the proposal of new or modified database algorithms. Given its tight coupling with query optimization, physical cost estimation has never been an independent research topic of its own. Apart from very few exceptions, new physical cost models and estimation techniques are usually published as "by-products" in publications that mainly deal with novel optimization techniques.

The first use of (implicit) cost estimation were complexity analyses that led to heuristic optimization rules. For instance, a join is always considered cheaper than calculating first the Cartesian product, followed by a selection. Likewise, linear operations that tend to reduce the data stream (selections, projections) should be evaluated as early as data dependencies allow, followed by (potentially) quadratic operations that do not "blow-up" the intermediate results (semijoins, foreign-key joins). More complex, and hence expensive, operations (general joins, Cartesian products) should be executed as late as possible.

Since a simple complexity metric does not necessarily reflect the same ranking of plans as the actual execution costs, first explicit cost estimation in database query optimization aimed at estimating intermediate result sizes. Initial works started with simplifications such as assuming uniform data distributions and independence of attribute values. Over time, the techniques have been improved to model non-uniform data distributions. Till date, effective treatment of (hidden) correlations is still an open research topic.

The following refinement was the introduction of physical costs. With I/O being the dominating cost factor in the early days of database management systems, the first systems assessed query plans by merely estimating the number of I/O operations required. However, I/O systems exhibit quite different performance for sequential and randomly placed I/O operations. Hence, the models were soon refined to distinguish between sequential and random accesses, weighing them with their respective costs, i.e., time to execute one operation.

With main memory sizes growing, more and more query processing work is done within main memory, minimizing disk accesses. Consequently, CPU and memory access costs can no longer be ignored. Assuming uniform memory access costs, memory access has initially been covered by CPU costs. CPU costs are estimated in terms of CPU cycles. Scoring them with the CPU's clock speed yields time, the common unit to combine CPU and I/O costs to get the overall physical costs.

Only recently with the advent of CPU caches and extended memory hierarchies, the impact of memory access costs has become so significant that it needs to be modeled separately [15, 16]. Similarly to I/O costs, memory access costs are estimated in terms of number of memory accesses (or cache misses) and scored by their penalty to achieve time as common unit.

In parallel and distributed database systems, also network communication, costs are considered as contributing factors to the overall execution costs.

Foundations

Different query execution plans require different amounts of effort to be evaluated. The objective function for the query optimization problems assigns every execution plan a single non-negative value. This value is commonly referred to as *costs* in the query optimization business.

Cost Components

Logical Costs/Data Volume

The most important cost component is the amount of data that is to be processed. Per operator, three data volumes are distinguished: input (per operand), output, and temporary data. Data volumes are usually measured as cardinality, i.e., number of tuples. Often, other units such as number of I/O blocks, number of memory pages, or total size in bytes are required. Provided that the respective tuple sizes, page sizes, and block sizes are known, the cardinality can easily be transformed into the other units.

The amount of input data is given as follows: For the leaf nodes of the query graph, i.e., those operations that directly access base tables stored in the database, the input cardinality is given by the cardinality of the base table(s) accessed. For the remaining (inner) nodes of the query graph, the input cardinality is given by the output cardinality of the predecessor(s) in the query graph.

Estimating the output size of database operations – or more generally, their *selectivity* – is anything else but trivial. For this purpose, DBMSs usually maintain statistic about the data stored in the database. Typical statistics are

1. Cardinality of each table,
2. Number of distinct values per column,

3. Highest/lowest value per column (where applicable).

Logical cost functions use these statistics to estimate output sizes (respectively selectivities) of database operations. The simplest approach is to assume that attribute values are uniformly distributed over the attribute's domain. Obviously, this assumption virtually never holds for "real-life" data, and hence, estimations based on these assumption will never be accurate. This is especially severe, as the estimation errors compound exponentially throughout the query plan [9]. This shows, that more accurate (but compact) statistics on data distributions (of base tables as well as intermediate results) are required to estimate intermediate results sizes.

The importance of statistics management has led to a plethora of approximation techniques, for which [6] have coined the general term "*data synopses*". Such techniques range from advanced forms of *histograms* (most notably, *V-optimal histograms* including multidimensional variants) [7, 10] over *spline synopses* [11, 12], *sampling* [3, 8], and *parametric curve-fitting techniques* [4, 20] all the way to highly sophisticated methods based on *kernel estimators* [1] or *Wavelets* and other transforms [2, 17].

A logical cost model is a prerequisite for the following two cost components.

Algorithmic Costs/Complexity

Logical costs only depend on the data and the query (i.e., the operators' semantics), but they do not consider the algorithms used to implement the operators' functionality. Algorithmic costs extend logical costs by taking the properties of the algorithms into account.

A first criterion is the algorithm's complexity in the classical sense of complexity theory. Most unary operator are in $O(n)$, like selections, or $O(n \log n)$, like sorting; n being the input cardinality. With proper support by access structures like indices or hash tables, the complexity of selection may drop to $O(\log n)$ or $O(1)$, respectively. Binary operators can be in $O(n)$, like a union of sets that does not eliminate duplicates, or, more often, in $O(n^2)$, as for instance join operators.

More detailed algorithmic cost functions are used to estimate, e.g., the number of I/O operations or the amount of main memory required. Though these functions require some so-called "physical" information like I/O block sizes or memory pages sizes, they are still considered algorithmic costs and not physical cost, as these informations are system specific, but not hardware specific. The standard database literature provides a large variety of cost formulas for the most frequently used operators and their algorithms. Usually, these formulas calculate the costs in terms of I/O operations as this still is the most common objective function for query optimization in database systems [5, 13].

Physical Costs/Execution Time

Logical and algorithmic costs alone are not sufficient to do query optimization. For example, consider two algorithms for the same operation, where the first algorithm requires slightly more I/O operations than the second, while the second requires significantly more CPU operations than the first one. Looking only at algorithmic costs, both algorithms are not comparable. Even assuming that I/O operations are more expensive than CPU operations cannot in general answer the question which algorithm is faster. The actual execution time of both algorithms depends on the speed of the underlying hardware. The physical cost model combines the algorithmic cost model with an abstract hardware description to derive the different cost factors in terms of time, and hence the total execution time. A hardware description usually consists of information, such as CPU speed, I/O latency, I/O bandwidth, and network bandwidth. The next section discusses physical cost factors on more detail.

Cost Factors

In principle, physical costs are considered to occur in two flavors, *temporal* and *spatial*. Temporal costs cover all cost factors that can easily be related to execution time, e.g., by multiplying the number of certain events with there respective cost in terms of some time unit. Spatial costs contain resource consumptions that cannot directly (or not at all) be related to time. The following

briefly describes the most prominent cost factors of both categories.

Temporal Cost Factors

Disk-I/O This is the cost of searching for, reading, and writing data blocks that reside on secondary storage, mainly on disk. In addition to accessing the database files themselves, temporary intermediate files that are too large to fit in main memory buffers and hence are stored on disk also need to be accessed. The cost of searching for records in a database file or a temporary file depends on the type of access structures on that file, such as ordering, hashing, and primary or secondary indexes. I/O costs are either simply measured in terms of the number of block-I/O operations, or in terms of the time required to perform these operations. In the latter case, the number of block-I/O operations is multiplied by the time it takes to perform a single block-I/O operation. The time to perform a single block-I/O operation is made up by an initial seek time (*I/O latency*) and the time to actually transfer the data block (i.e., block size divided by *I/O bandwidth*). Factors such as whether the file blocks are allocated contiguously on the same disk cylinder or scattered across the disk affect the access cost. In the first case (also called *sequential I/O*), I/O latency has to be counted only for the first of a sequence of subsequent I/O operations. In the second case (*random I/O*), seek time has to be counted for each I/O operation, as the disk heads have to be repositioned each time.

Main-Memory Access These are the costs for reading data from or writing data to main memory. Such data may be intermediate results or any other temporary data produced/used while performing database operations.

Similar to I/O costs, memory access costs can be modeled be estimating the number of memory accesses (i.e., cache misses) and scoring them with their respective penalty (latency) [16].

Network Communication In centralized DBMSs, communication costs cover the costs of shipping the query from the client to the server and the query's result back to the client. In distributed, federated, and parallel DBMSs, communication costs additionally contain all costs for shipping (sub-) queries and/or (intermediate) results between the different hosts that are involved in evaluating the query.

Also with communication costs, there is a latency component, i.e., a delay to initiate a network connection and package transfer, and a bandwidth component, i.e., the amount of data that can be transfer through the network infrastructure per time.

CPU Processing This is the cost of performing operations such as computations on attribute values, evaluating predicates, searching and sorting tuples, and merging tuples for join. CPU costs are measured in either CPU cycles or time. When using CPU cycles, the time may be calculated by simply dividing the number of cycles by the CPU's clock speed. While allowing limited portability between CPUs of the same kind, but with different clock speeds, portability to different types of CPUs is usually not given. The reason is, that the same basic operations like adding two integers might require different amounts of CPU cycles on different types of CPUs.

Spatial Cost Factors

Usually, there is only one spatial cost factor considered in database literature: *memory size*. This cost is the amount of main memory required to store intermediate results or any other temporary data produced/used while performing database operations.

Next to not (directly) being related to execution time, there is another difference between temporal and spatial costs that stems from the way they share the respective resources. A simple example shall demonstrate the differences. Consider to operations or processes each of which consumes 50% of the available resources (i.e., CPU power, I/O-, memory-, and network bandwidth). Further, assume that when run one at a time, both tasks have equal execution time. Running both tasks concurrently on the same

system (ideally) results in the same execution time, now consuming all the available resources. In case each individual process consumes 100% of the available resources, the concurrent execution time will be twice the individual execution time. In other words, if the combined resource consumption of concurrent tasks exceed 100%, the execution time extends to accommodate the excess resource requirements. With spatial cost factors, however, such "stretching" is not possible. In case two tasks together would require more than 100% of the available memory, they simply cannot be executed at the same time, but only after another.

Types of (Cost) Models

According to their degree of abstraction, (cost) models can be classified into two classes: *analytical models* and *simulation models*.

Analytical Models In some cases, the assumptions made about the real system can be translated into mathematical descriptions of the system under study. Hence, the result is a set of mathematical formulas that is called an analytical model. The advantage of an analytical model is that evaluation is rather easy and hence fast. However, analytical models are usually not very detailed (and hence not very accurate). In order to translate them into a mathematical description, the assumptions made have to be rather general, yielding a rather high degree of abstraction.

Simulation Models Simulation models provide a very detailed and hence rather accurate description of the system. They describe the system in terms of (a) simulation experiment(s) (e.g., using event simulation). The high degree of accuracy is charged at the expense of evaluation performance. It usually takes relatively long to evaluate a simulation base model, i.e., to actually perform the simulation experiment(s). It is not uncommon, that the simulation actually takes longer than the execution in the real system would take.

In database query optimization, though it would appreciate the accuracy, simulation models are not feasible, as the evaluation effort is far too high. Query optimization requires that costs of numerous alternatives are evaluated and compared as fast as possible. Hence, only analytical cost models are applicable in this scenario.

Architecture and Evaluation of Database Cost Models

The architecture and evaluation mechanism of database cost models is tightly coupled to the structure of query execution plans. Due to the strong encapsulation offered by relational algebra operators, the cost of each operator, respectively each algorithm, can be described individually. For this purpose, each algorithm is assigned a set of *cost functions* that calculate the three cost components as described above. Obviously, the physical cost functions depend on the algorithmic cost functions, which in turn depend on the logical cost functions. Algebraic cost functions use the data volume estimations of the logical cost functions as input parameters. Physical cost functions are usually specializations of algorithmic cost functions that are parameterized by the hardware characteristics.

The cost model also defines how the single operator costs within a query have to be combined to calculate the total costs of the query. In traditional sequential DBMSs, the single operators are assumed to have no performance side-effects on each other. Thus, the cost of a QEP is the cumulative cost of the operators in the QEP [18]. Since every operator in the QEP is the root of a sub-plan, its cost includes the cost of its input operators. Hence, the cost of a QEP is the cost of the topmost operator in the QEP. Likewise, the cardinality of an operator is derived from the cardinalities of its inputs, and the cardinality of the topmost operator represents the cardinality of the query result.

In non-sequential (e.g., distributed or parallel) DBMSs, this subject is much more complicated, as more issues such as scheduling, concurrency, resource contention, and data dependencies have to considered. For instance, in such environments, more than one operator may be executed at a time, either on disjoint (hardware) resources, or (partly) sharing resources. In the first case, the total cost (in terms of time) is calculated as the maximum of the costs (execution times) of all operators running concurrently. In the second case, the operators compete for the same resources, and hence mutually influence their performance and costs. More sophisticated cost function and cost models are required here to adequately model this resource contention [14, 19].

Cross-References

► Distributed Query Optimization
► Multi-query Optimization
► Optimization and Tuning in Data Warehouses
► Parallel Query Optimization
► Process Optimization
► Query Optimization
► Query Optimization (in Relational Databases)
► Query Optimization in Sensor Networks
► Query Plan
► Selectivity Estimation
► Spatiotemporal Selectivity Estimation
► XML Selectivity Estimation

Recommended Reading

1. Blohsfeld B, Korus D, Seeger B. A comparison of selectivity estimators for range queries on metric attributes. In: Proceedings of the ACM SIGMOD International Conference on Management of Data; 1999. p. 239–50.
2. Chakrabarti K, Garofalakis MN, Rastogi R, Shim K. Approximate query processing using wavelets. In: Proceedings of the 26th International Conference on Very Large Data Bases; 2000. p. 111–22.
3. Chaudhuri S, Motwani R, Narasayya VR. On random sampling over joins. In: Proceedings of the ACM SIGMOD International Conference on Management of Data, Philadephia; 1999. p. 263–74.
4. Chen CM, Roussopoulos N. Adaptive selectivity estimation using query feedback. In: Proceedings of the ACM SIGMOD International Conference on Management of Data; 1994. p. 161–72.
5. Garcia-Molina H, Ullman JD, Widom J. Database systems: the complete book. Englewood Cliffs: Prentice Hall; 2002.
6. Gibbons PB, Matias Y. Synopsis data structures for massive data sets. In: Proceedings of the 10th Annual ACM-SIAM Symposium on Discrete Algorithms; 1999. p. 909–10.

7. Gibbons PB, Matias PB, Poosala V. Fast incremental maintenance of approximate histograms. In: Proceedings of the 23th International Conference on Very Large Data Bases; 1997. p. 466–75.
8. Haas PJ, Naughton JF, Seshadri S, Swami AN. Selectivity and cost estimation for joins based on random sampling. J Comput Syst Sci. 1996;52(3): 550–69.
9. Ioannidis YE, Christodoulakis S. On the propagation of errors in the size of join results. In: Proceedings of the ACM SIGMOD International Conference on Management of Data; 1991. p. 268–77.
10. Ioannidis YE, Poosala V. Histogram-based approximation of set-valued query-answers. In: Proceedings of the 25th International Conference on Very Large Data Bases; 1999. p. 174–85.
11. König AC, Weikum G. Combining histograms and parametric curve fitting for feedback-driven query result-size estimation. In: Proceedings of the 25th International Conference on Very Large Data Bases; 1999. p. 423–34.
12. König AC, Weikum G. Auto-tuned spline synopses for database statistics management. In: Proceedings of the International Conference on Management of Data; 2000.
13. Korth H, Silberschatz A. Database systems concepts. New York/San Francisco/Washington, DC: McGraw-Hill; 1991.
14. Lu H, Tan KL, Shan MC. Hash-based join algorithms for multiprocessor computers. In: Proceedings of the 16th International Conference on Very Large Data Bases; 1990. p. 198–209.
15. Manegold S. Understanding, modeling, and improving main-memory database performance. PhD thesis, Universiteit van Amsterdam, Amsterdam; 2002.
16. Manegold S., Boncz PA, Kersten ML. Generic database cost models for hierarchical memory systems. In: Proceedings of the 28th International Conference on Very Large Data Bases; 2002. p. 191–202.
17. Matias Y, Vitter JS, Wang M. Wavelet-based histograms for selectivity estimation. In: Proceedings of the ACM SIGMOD International Conference on Management of Data; 1998. p. 448–59.
18. Selinger PG, Astrahan MM, Chamberlin DD, Lorie RA, Price TG. Access path selection in a relational database management system. In: Proceedings of the ACM SIGMOD International Conference on Management of Data; 1979. p. 23–34.
19. Spiliopoulou M, Freytag J-C. Modelling resource utilization in pipelined query execution. In: Proceedings of the European Conference on Parallel Processing; 1996. p. 872–80.
20. Sun W, Ling Y, Rishe N, Deng Y. An instant and accurate size estimation method for joins and selection in a retrieval-intensive environment. In: Proceedings of the ACM SIGMOD International Conference on Management of Data; 1993. p. 79–88.

Count-Min Sketch

Graham Cormode
Computer Science, University of Warwick, Warwick, UK

Synonyms

CM Sketch

Definition

The Count-Min (CM) Sketch is a compact summary data structure capable of representing a high-dimensional vector and answering queries on this vector, in particular point queries and dot product queries, with strong accuracy guarantees. Such queries are at the core of many computations, so the structure can be used in order to answer a variety of other queries, such as frequent items (heavy hitters), quantile finding, join size estimation, and more. Since the data structure can easily process updates in the form of additions or subtractions to dimensions of the vector (which may correspond to insertions or deletions, or other transactions), it is capable of working over streams of updates, at high rates.

The data structure maintains the linear projection of the vector with a number of other random vectors. These vectors are defined implicitly by simple hash functions. Increasing the range of the hash functions increases the accuracy of the summary, and increasing the number of hash functions decreases the probability of a bad estimate. These tradeoffs are quantified precisely below. Because of this linearity, CM sketches can be scaled, added, and subtracted, to produce summaries of the corresponding scaled and combined vectors.

Historical Background

The Count-Min sketch was first proposed in 2003 [1] as an alternative to several other sketch techniques, such as the Count sketch [2] and the AMS sketch [3]. The goal was to provide a simple sketch data structure with a precise characterization of the dependence on the input parameters. The sketch has also been viewed as a realization of a counting *Bloom filter* or Multistage-Filter [4], which requires only limited independence randomness to show strong, provable guarantees. The simplicity of creating and probing the sketch has led to its wide use in disparate areas since its initial description.

Foundations

The CM sketch is simply an array of counters of width w and depth d, $CM\,[1,\,1]$ ... $CM\,[d, w]$. Each entry of the array is initially zero. Additionally, d hash functions

$$h_1 \ldots h_d : \{1 \ldots n\} \to \{1 \ldots w\}$$

are chosen uniformly at random from a pairwise-independent family. Once w and d are chosen, the space required is fixed: the data structure is represented by wd counters and d hash functions (which can each be represented in $O(1)$ machine words [5]).

Update Procedure

Consider a vector a, which is presented in an implicit, incremental fashion (this abstract model captures a wide variety of data stream settings, see entries on "▶ Data Stream" for more details). This vector has dimension n, and its current state at time t is $\alpha(t) = [a\alpha_1(t), \ldots \alpha_1(t), \ldots \alpha_n(t)]$. Initially, a is the zero vector, 0, so $\alpha_i(0)$ is 0 for all i. Updates to individual entries of the vector are presented as a stream of pairs. The t-th update is (i_t, c_t), meaning that

$$\alpha_{i_t}(t) = \alpha_{i_t}(t-1) + c_t$$

$$\alpha_i{'}(t) = \alpha_i{'}(t-1)$$

$$i' \neq i_t$$

This procedure is illustrated in Fig. 1. In the remainder of this article, t is dropped, and the current state of the vector is referred to as just a for convenience. It is assumed throughout that although values of $a\,i$ increase and decrease with updates, each $a\,i \geq 0$. The Count-Min sketch also applies to the case where $a\,i$ s can be less than zero, with small factor increases in space. Here, details of these extensions are omitted for simplicity of exposition (full details are in [1]).

When an update (i_t, c_t) arrives, $c\,t$ is added to one count in each row of the Count-Min sketch; the counter is determined by (h_j). Formally, given (i_t, c_t), the following modifications are performed:

Because computing each hash function takes $O(1)$ (constant) time, the total time to perform an update is $O(d)$, independent of w. Since d is typically small in practice (often less than 10), updates can be processed at high speed.

Point Queries

A *point query* is to estimate the value of an entry in the vector a_i. The point query procedure is similar to updates: given a query point i, an estimate is found as $a_i = \min_{1 \leq j \leq d} CM[j, h_j(i)]$. Since the space used by the sketch is typically much smaller than that required to represent the vector exactly, there is necessarily some approximation in the estimate, which is quantified as follows:

Count-Min Sketch, Fig. 1
Each item i is mapped to one cell in each row of the array of counts: when an update of $c\,t$ to item $i\,t$ arrives, $c\,t$ is added to each of these cells

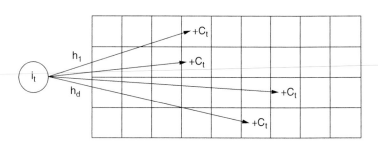

Theorem 1 (Theorem 1 from [1]) *If* $w = \lceil \frac{e}{\varepsilon} \rceil$, *and* $d = \lceil ln\frac{1}{\delta} \rceil$, *the estimate* $\hat{a}_i$ *the following guarantees:* $a_i \leq \hat{a}_i$; *and, with probability at least* $1 - \delta$,

$$\hat{a}_i \leq a_i + \varepsilon |\alpha|_1$$

The proof follows by considering the estimate in each row and observing that the expected error in using $CM[j, h_j(i)]$ as an estimate has expected (nonnegative) error $|\alpha|_1/w$. By the Markov inequality [5], the probability that this error exceeds $\varepsilon|\alpha|_1$ is at most $\frac{1}{e}$ (where e is the base of the natural logarithm, i.e., $2.71828\ldots$, a constant chosen to optimize the space for fixed accuracy requirements). Taking the smallest estimate gives the best estimator, and the probability that this estimate has error exceeding $\varepsilon|\alpha|_1$ is the probability that *all* estimates exceed this error, i.e., $e^{-d} \leq \delta$.

This analysis makes no assumption about the distribution of values in α. However, in many applications, there are Zipfian, or power law, distributions of item frequencies. Here, the (relative) frequency of the ith most frequent item is proportional to i^{-z}, for some parameter z, where z is typically in the range 1–3 ($z = 0$ gives a perfectly uniform distribution). In such cases, the skew in the distribution can be used to show a stronger space/accuracy tradeoff:

Theorem 2 (Theorem 5.1 from [6])
For a Zipf distribution with parameter z, the space required to answer point queries with error $\varepsilon|\alpha|_1$ with probability at least $1 - \delta$ is given by $O(\varepsilon^{-\min\{1,1/z\}} \ln 1/\delta)$.

Moreover, the dependency of the space on z is optimal:

Theorem 3 (Theorem 5.2 from [6])
The space required to answer point queries correctly with any constant probability and error at most $\varepsilon|\alpha|_1$ is $\Omega(\varepsilon^{-1})$ over general distributions, and $\Omega(\varepsilon^{-1/z})$ for Zipf distributions with parameter z, assuming the dimension of a, n is $\Omega(\varepsilon^{-\min\{1,1/z\}})$.

Range, Heavy Hitter, and Quantile Queries

A *range query* is to estimate $\sum_{i=l}^{r} a_i$ for a range $[l...r]$. For small ranges, the range sum can be estimated as a sum of point queries; however, as the range grows, the error in this approach also grows linearly. Instead, $\log n$ sketches can be kept, each of which summarizes a derived vector a^k where

$$a^k[j] = \sum_{i=j2^k}^{(j+1)2^k - 1} a_i$$

for $k = 1...\log n$. A range of the form $j2^k \ldots (j+1)2^k - 1$ is called a *dyadic range*, and any arbitrary range $[l...r]$ can be partitioned into at most $2\log n$ dyadic ranges. With appropriate rescaling of accuracy bounds, it follows that:

Theorem 4 (Theorem 4 from [1])
Count-Min sketches can be used to find an estimate $\hat{r}$ for a range query on $l...r$ such that

$$\hat{r} - \varepsilon|a|_1 \leq \sum_{i=l}^{r} a_i \leq \hat{r}$$

The right inequality holds with certainty, and the left inequality holds with probability at least $1 - \delta$. The total space required is $O\left(\frac{\log^2 n}{\varepsilon} \log \frac{1}{\delta}\right)$.

Closely related to the range query is the ϕ-*quantile query*, which is to find a point j such that

$$\sum_{i=1}^{j} a_i \leq \phi|a|_1 \leq \sum_{i=1}^{j+1} a_i$$

A natural approach is to use range queries to binary search for a j which satisfies this requirement approximately (i.e., tolerates up to error in the above expression) given ϕ. In order to give the desired guarantees, the error bounds need to be adjusted to account for the number of queries that will be made.

Theorem 5 (Theorem 5 from [1])
ε-approximate ϕ-quantiles can be found with probability at least $1 - \delta$ by keeping a data

structure with space $O\left(\frac{1}{\varepsilon}\log^2(n)\log\left(\frac{\log n}{\delta}\right)\right)$. The tiem for each insert or delete operation is $O\left(\log(n)\log\left(\frac{\log n}{\delta}\right)\right)$, and the time to find each quantile on demand is $O\left(\log(n)\log\left(\frac{\log n}{\delta}\right)\right)$.

Heavy Hitters are those points i such that $a_i \geq \phi|a|_1$ for some specified ϕ. The range query primitive based on Count-Min sketches can again be used to find heavy hitters, by recursively splitting dyadic ranges into two and querying each half to see if the range is still heavy, until a range of a single, heavy, item is found. Formally,

Theorem 6 (Theorem 6 from [1]) *Using space* $O\left(\frac{1}{\varepsilon}\log(n)\log\left(\frac{2\log(n)}{\delta\phi}\right)\right)$, *and time* $O\left(\log(n)\log\left(\frac{2\log(n)}{\delta\phi}\right)\right)$ *per update, a set of approximate heavy hitters can be output so that every item with frequency at least $(\phi+\varepsilon)|a|_1$ is output, and with probability e $1-\delta$ no item whose frequency is less than $\phi|a|_1$ is output.*

For skewed Zipfian distributions, as described above, with parameter $z >1$, it is shown more strongly that the top-k most frequent items can be found with relative error ε using space only $\widetilde{O}\left(\frac{k}{\varepsilon}\right)$[6].

Inner Product Queries

The Count-Min sketch can also be used to estimate the inner product between two vectors; in database terms, this captures the (equi)join size between relations. The inner product $a \cdot b$ can be estimated by treating the Count-Min sketch as a collection of d vectors of length w and finding the minimum inner product between corresponding rows of sketches of the two vectors. With probability $1-\delta$, this estimate is at most an additive quantity $\epsilon|a|_1|b|_1$ above the true value of $a \cdot b$. This is to be compared with AMS sketches which guarantee additive $\epsilon|a|_2|b|_2$ error, but require space proportional to $\frac{1}{\epsilon^2}$ to make this guarantee.

Interpretation as Random Linear Projection

The sketch can also be interpreted as a collection of inner-products between a vector representing the input and a collection of random vectors defined by the hash functions. Let a denote the vector representing the input, so that $a[i]$is the sum of the updates to the i-th location in the input. Let $r_{j,k}$ be the binary vector such that $r_{j,k}[1] = 1$ if and only if $h_j(i) = k$. Then it follows that $CM[j,k] = a \cdot r_{j,k}$. Because of this linearity, it follows immediately that if sketches of two vectors, a and b, are built then (i) the sketch of $a + b$ (using the same $w,d,h\,j$) is the (component-wise) sum of the sketches and (ii) the sketch of λa for any scalar λ is λ times the sketch of α. In other words, the sketch of any linear combination of vectors can be found. This property is useful in many applications which use sketches. For example, it allows distributed measurements to be taken, sketched, and combined by only sending sketches instead of the whole data.

Conservative Update

If only positive updates arrive, then an alternate update methodology may be applied, known as conservative update (due to Estan and Varghese [4]). For an update (i,c), $\widehat{a}_i$ is computed, and the counts are modified according to $\forall 1 \leq j \leq d : CM\left[j, h_j(i)\right] \leftarrow \max\left(CM\left[j, h_j(i)\right], \widehat{a}_i + c\right)$. It can be verified that procedure still ensures for point queries that $a_i \leq \widehat{a}_i$ and that the error is no worse than in the normal update procedure; it is remarked that this can improve accuracy "up to an order of magnitude" [4]. Note however that deletions or negative updates can no longer be processed, and the additional processing that must be performed for each update could effectively halve the throughput.

Key Applications

Since its description and initial analysis, the Count-Min Sketch has been applied in a wide variety of situations. Here is a list of some of the ways in which it has been used or modified.

- Lee et al. [7] propose using least-squares optimization to produce estimates from Count-Min Sketches for point queries (instead of returning the minimum of locations where the item was mapped). It was shown that this approach can give significantly improved estimates, although at the cost of solving a convex optimization problem over n variables (where n is the size of the domain from which items are drawn, typically 2^{32} or higher).

- The "skipping" technique, proposed by Bhattacharrya et al. [8], entails avoiding adding items to the sketch (and saving the cost of the hash function computations) when this will not affect the accuracy too much, in order to further increase throughout in high-demand settings.

- Indyk [9] uses the Count-Min Sketch to estimate the residual mass after removing a subset of items. That is, given a (small) set of indices I, to estimate $\sum_{i \notin I} a_i$. This is needed in order to find clusterings of streaming data.

- The *entropy* of a data stream is a function of the relative frequencies of each item or character within the stream. Using Count-Min Sketches within a larger data structure based on additional hashing techniques, Lakshminath and Ganguly [10] showed how to estimate this entropy to within relative error.

- Sarlós et al. [11] gave approximate algorithms for personalized page rank computations which make use of Count-Min Sketches to compactly represent web-size graphs.

- In describing a system for building selectivity estimates for complex queries, Spiegel and Polyzotis [12] use Count-Min Sketches in order to allow clustering over a high-dimensional space.

- Rusu and Dobra [13] study a variety of sketches for the problem of inner-product estimation and conclude that Count-Min sketch has a tendency to outperform its theoretical worst-case bounds by a considerable margin and gives better results than some other sketches for this problem.

- Many applications call for tracking *distinct* counts: that is, a_i should represent the number of distinct updates to position i. This can be achieved by replacing the counters in the Count-Min sketch with approximate Count-Distinct summaries, such as the *Flajolet-*

Martin sketch. This is described and evaluated in [14, 15].

- Privacy preserving computations ensure that multiple parties can cooperate to compute a function of their data while only learning the answer and not anything about the inputs of the other participants. Roughan and Zhang demonstrate that the Count-Min Sketch can be used within such computations, by applying standard techniques for computing privacy preserving sums on each counter independently [16].

Related ideas to the Count-Min Sketch have also been combined with group testing to solve problems in the realm of Compressed Sensing and finding significant changes in dynamic streams.

Future Directions

As is clear from the range of variety of applications described above, Count-Min sketch is a versatile data structure which is finding applications within Data Stream systems, but also in Sensor Networks, Matrix Algorithms, Computational Geometry, and Privacy-Preserving Computations. It is helpful to think of the structure as a basic primitive which can be applied wherever approximate entries from high dimensional vectors or multisets are required, and one-sided error proportional to a small fraction of the total mass can be tolerated (just as a Bloom filter should be considered in order to represent a set wherever a list or set is used and space is at a premium). With this in mind, further applications of this synopsis can be expected to be seen in more settings.

As noted below, sample implementations are freely available in a variety of languages, and integration into standard libraries will further widen the availability of the structure. Further, since many of the applications are within high-speed data stream monitoring, it is natural to look to hardware implementations of the sketch. In particular, it will be of interest to understand how modern multicore architectures can take advantage of the natural parallelism inherent in the Count-Min Sketch (since each of the d rows are essentially independent) and to explore the implementation choices that follow.

Experimental Results

Experiments performed in [6] analyzed the error for point queries and F_2 (self-join size) estimation, in comparison to other sketches. High accuracy was observed for both queries, for sketches ranging from a few kilobytes to a megabyte in size. The typical parameters of the sketch were a depth d of 5 and a width w of a few hundred to thousands. Implementations on desktop machines achieved between and two and three million updates per second. Other implementation has incorporated Count-Min Sketch into high speed streaming systems such as Gigascope [17] and tuned it to process packet streams of multigigabit speeds.

Lai and Byrd report on an implementation of Count-Min sketches on a low-power stream processor [12], capable of processing 40 byte packets at a throughput rate of up to 13 Gbps. This is equivalent to about 44 million updates per second.

URL to Code

Several example implementations of the Count-Min sketch are available. C code is given by the MassDal code bank: http://www.cs.rutgers.edu/~muthu/massdal-code-index.html. C++ code due to Marios Hadjieleftheriou is available from http://hadjieleftheriou.com/sketches/index.html. Further implementations are listed at https://sites.google.com/site/countminsketch/code.

Cross-References

▸ AMS Sketch
▸ Data Sketch/Synopsis
▸ Data Stream
▸ FM Synopsis
▸ Frequent Items on Streams
▸ Quantiles on Streams

Recommended Reading

1. Cormode G, Muthukrishnan S. An improved data stream summary: the count-min sketch and its applications. J Algorith. 2005;55(1):58–75.
2. Charikar M, Chen K, Farach-Colton M. Finding frequent items in data streams. In: Proceedings of the 29th International Colloquium on Automata, Languages, and Programming; 2002. p. 693–703.
3. Alon N, Matias Y, Szegedy M. The space complexity of approximating the frequency moments. In: Proceedings of the 28th Annual ACM Symposium on Theory of Computing; 1996. p. 20–9. Journal version in J Comput Syst Sci. 1999;58(1):137–47.
4. Estan C, Varghese G. New directions in traffic measurement and accounting. In: Proceedings of the ACM International Conference of the on Data Communication; 2002. p. 323–38.
5. Motwani R, Raghavan P. Randomized algorithms. Cambridge: Cambridge University Press; 1995.
6. Cormode G, Muthukrishnan S. Summarizing and mining skewed data streams. In: Proceedings of the 2005 SIAM International Conference on Data Mining; 2005.
7. Lee GM, Liu H, Yoon Y, Zhang Y. Improving sketch reconstruction accuracy using linear least squares method. In: Proceedings of the 5th ACM SIGCOMM Conference on Internet Measurement; 2005. p. 273–8.
8. Bhattacharrya S, Madeira A, Muthukrishnan S, Ye T. How to scalably skip past streams. In: Proceedings of the 1st International Workshop on Scalable Stream Processing Systems; 2007. p. 654–63.
9. Indyk P. Better algorithms for high-dimensional proximity problems via asymmetric embeddings. In: Proceedings of the ACM-SIAM Symposium on Discrete Algorithms; 2003.
10. Lakshminath B, Ganguly S. Estimating entropy over data streams. In: Proceedings of the 14th European Symposium on Algorithms; 2006. p. 148–59.
11. Sarlós T, Benzúr A, Csalogány K, Fogaras D, Rácz B. To randomize or not to randomize: space optimal summaries for hyperlink analysis. In: Proceedings of the 15th International World Wide Web Conference; 2006. p. 297–306.
12. Spiegel J, Polyzotis N. Graph-based synopses for relational selectivity estimation. In: Proceedings of the ACM SIGMOD International Conference on Management of Data; 2006. p. 205–16.
13. Rusu F, Dobra A. Statistical analysis of sketch estimators. In: Proceedings of the ACM SIGMOD International Conference on Management of Data; 2007. p. 187–98.
14. Cormode G, Muthukrishnan S. Space efficient mining of multigraph streams. In: Proceedings of the 24th ACM SIGACT-SIGMOD-SIGART Symposium on Principles of Database Systems; 2005. p. 271–82.
15. Kollios G, Byers J, Considine J, Hadjieleftheriou M, Li F. Robust aggregation in sensor networks. Q Bull IEEE TC Data Eng. 2005;28(1):26–32.

16. Roughan M, Zhang Y. Secure distributed data mining and its application in large-scale network measurements. Computer Communication Review. 2006;36(1):7–14.
17. Cormode G, Korn F, Muthukrishnan S, Johnson T, Spatscheck O, Srivastava D. Holistic UDAFs at streaming speeds. In: Proceedings of the ACM SIGMOD International Conference on Management of Data; 2004. p. 35–46.
18. Lai Y-K, Byrd GT. High-throughput sketch update on a low-power stream processor. In: Proceedings of the ACM/IEEE Symposium on Architecture for Networking and Communications Systems; 2006. p. 123–32.

Coupling and Decoupling

Serge Mankovski
CA Labs, CA Inc., Thornhill, ON, Canada

Definition

Coupling is a measure of dependence between components of software system.

De-coupling is a design or re-engineering activity aiming to reduce coupling between system elements.

Key Points

Coupling of system components refers to a measure of dependency among them. Coupled components might depend on each other in different ways. Some examples of the dependencies are:

- One component might depend on syntax, format, or encoding of data produced by another component.
- One component might depend on the execution time within another component.
- One component might depend on state of another component.

Notion of coupling is connected to notion of cohesion. Cohesion is a measure of how related and focused are responsibilities of a software component. For example a highly cohesive component might group responsibilities

- Using the same syntax, format or encoding of data.
- Performed at the same time.
- Executed in the same state.

Highly cohesive components lead to fewer dependencies between components and voice versa.

Notions of coupling and cohesion were studied in structured and object oriented programming. The research developed software tools to calculate coupling and cohesion metrics.

Low coupling is often desirable because it leads to reliability, easy of modification, low maintenance costs, understandability, and reusability. Low coupling can be achieved by deliberately designing system with low values of coupling metric. It can also be achieved by re-engineering of existing software system through re-structuring of system into a set of more cohesive components. These activates are called de-coupling.

Cross-References

▶ Temporal Object-Oriented Databases

Covering Index

Donghui Zhang
Paradigm4, Inc., Waltham, MA, USA

Definition

Given an SQL query, a covering index is a composite index that includes all of the columns referenced in SELECT, JOIN, and WHERE clauses of this query. Because the index contains all the data needed by the query, to execute the query the actual data in the table does not need to be accessed.

Key Points

Covering indexes [1] support index-only execution plans. In general, having everything indexed tends to increase the query performance (in number of I/Os). However, using a covering index with too many columns can actually degrade performance. Typically, multi-dimensional index structures, e.g., the R-tree, perform poorer than linear scan with high dimensions. Some guidelines of creating a covering index are: (i) Create a covering index on frequently used queries. There are overheads in creating a covering index, which is often more significant than creating a regular index with fewer columns. Hence, if a query is seldom used, the overhead to create a covering index on it is not substantiated. This corresponds to Amdahl's law: improve the "interesting" part to receive maximum overall benefit of a system. (ii) Try to build a covering index by expanding an existing index. For instance, if there already exists an index on "age" and "salary," and one needs a covering index on "age," "salary," and "income," it is often better to expand the existing index rather than building a new index, which would share two columns with the existing index.

The term "covering index" is sometimes used to mean the collection of single-column, non-clustered indexes on all the columns in a table. This is due to the "index intersection" technique incorporated into the Microsoft SQL Server's query optimizer [1]. In particular, the query optimizer can build, at run time, a hash-based "covering index" to speedup queries on a frequently used table. This covering index is really a hash table, which is built based on multiple existing indexes. Creating single-column indexes on all columns encourages the query optimizer to perform index intersection, i.e., to build dynamic covering indexes.

Recommended Reading

1. McGehee B. Tips on optimizing covering indexes. 2007. http://www.sql-server-performance.com/tips/covering_indexes_p1.aspx.

Crash Recovery

Theo Härder
University of Kaiserslautern, Kaiserslautern, Germany

Synonyms

Backward recovery; Failure handling; Media recovery; Online recovery; Restart processing; System recovery

Definition

In contrast to transaction aborts, a crash is typically a major failure by which the state of the current database is lost or parts of storage media are unrecoverable (destroyed). Based on log data from a stable log, also called temporary log file, and the inconsistent and/or outdated state of the permanent database, system recovery has to reconstruct the most recent transaction-consistent database state. To limit the amount of redo steps after a crash, some form of periodic checkpointing is mandatory. Nevertheless, DBMS restart may take too long to be masked for the user; hence, a denial of service may be observed. Recovery from media failures relies on the availability of (several) backup or archive copies of earlier DB states – organized according to the generation principle – and archive logs (often duplexed) covering the processing intervals from the points of time the backup copies were created. Archive recovery usually causes much longer outages than system recovery.

Historical Background

Log data delivering the needed redundancy to recover from failures was initially stored on non-volatile core memory to be reclaimed at restart by a so-called log salvager [4] in the "pre-transaction area". Advances in VLSI technology enabled the use of cheaper and larger but volatile semiconductor memory as the computers' main memory. This technology change triggered by 1971 in industry – driven by database product adjustments – the development of new and refined concepts of logging such as log sequence numbers (LSNs), write-ahead log protocol (WAL), log duplexing, and more. Typically, these concepts were not published; nevertheless, they paved the way toward the use of ACID transactions. As late as 1978, Jim Gray documented the design of such a logging system implemented in IMS in a widely referenced publication [6].

Many situations and dependencies related to failures and recovery from those in databases have been thoroughly explored by Lawrence Bjork and Charles Davies in their studies concerning DB/DC systems back in 1973 leading to the so-called spheres of control [2]. The first published implementation of the transaction concept by a full-fledged DBMS recovery manager was that of System R, started in 1976 [5]. It refined the Do-Undo-Redo protocol and enabled automatic recovery for new recoverable types and operations. In 1981, Andreas Reuter presented in his Ph.D. dissertation further investigations and refinements of concepts related to failure handling in database systems [12]. Delivering a first version of the principles of transaction-oriented database recovery [Härder and Reuter 1979], including the *Ten Commandments* [7], this classification framework, defining the paradigm of transaction-oriented recovery and coining the acronym ACID for it [8], was finally published in 1983. The most famous and most complete description of recovery methods and their implementation was presented by C. Mohan et al. in the ARIES paper [10] in 1992, while thorough treatment of all questions related to this topic appeared in many textbooks, especially those of Bernstein et al. [1], Gray and Reuter [4], and Weikum and Vossen [15]. Most solutions implemented for crash recovery in industrial-strength DBMSs are primarily disk based. A different approach is based on main-memory logging on multiple computing nodes with independent failure modes to avoid disk access delays during commit processing [9]. For a long time, proposals to use "safe RAM", for example,

were not widely accepted. But with the advent of disruptive technologies such as storage-class memory or NVRAM [11], this reluctance may quickly disappear.

Foundations

The most difficult failure type to be recovered from is the system failure or system crash (see Logging and Recovery). Due to some (expected, but) unplanned failure event (a bug in the DBMS code, an operating system fault, a power or hardware failure, etc.), the *current database* – comprising all objects accessible to the DBMS during normal processing – is not available anymore. In particular, the in-memory state of the DBMS (lock tables, cursors and scan indicators, status of all active transactions, etc.) and the contents of the database buffer and the log buffer are lost. Furthermore, the state lost may include information about LSNs, ongoing commit processing with participating coordinators and participants, as well as commit requests and votes. Therefore, restart cannot rely on such information and has to refer to the temporary log file (stable log) and the *permanent* (*materialized*) *database*, that is, the state the DBMS finds after a crash at the nonvolatile storage devices (magnetic disks or solid-state disks (SSDs)) without having applied any log information.

Consistency Concerns

According to the *ACID principle*, a database is consistent if and only if it contains the results of successful transactions – called transaction-consistent database. Because a DBMS application must not lose changes of committed transactions and all of them have contributed to the DB state, the goal of crash recovery is to establish the most recent transaction-consistent DB state. For this purpose, redo and undo recovery is needed, in general. Results of committed transactions may not yet be reflected in the database, because execution has been terminated in an uncontrolled manner, and the corresponding pages containing such results were not propagated to the permanent DB at the time of the crash. Therefore, they must be repeated, if necessary – typically by means of log information. On the other hand, changes of incomplete transactions may have reached the permanent DB state on persistent storage. Hence, undo recovery has to completely roll back such uncommitted changes.

Because usually many interactive users rely in their daily business on DBMS services, crash recovery is very time-critical. Therefore, crash-related interruption of DBMS processing should be masked for them as far as possible. Although today DBMS crashes are rather rare events and may occur several times a month or a year – depending on the stability of both the DBMS and its operational environment – their recovery should take no more than a number of seconds or at most a few minutes (as opposed to archive recovery), even if TByte or PByte databases with thousands of users are involved.

Forward Recovery

Having these constraints and requirements in mind, which kind of recovery strategies can be applied? Despite the presence of so-called nonstop systems (giving the impression that they can cope with failures by forward recovery), rollforward is very difficult, if not impossible in any stateful system. To guarantee atomicity in case of a crash, rollforward recovery had to enable all transactions to resume execution so that they can either complete successfully or require to be aborted by the DBMS. Assume the DB state containing the most recent successful DB operations could be made available, that is, all updates prior to the crash have completely reached the permanent DB state. Even then rollforward would be not possible, because a transaction cannot resume in "forward direction" unless its local state is restored. Moreover in a DBMS environment, the in-memory state lost makes it entirely impossible to resume from the point at the time the crash occurred. For these reasons, a rollback strategy for active transactions is the only choice in case of crash recovery to ensure atomicity (wiping out all traces of such transactions); later on these transactions are started anew either by the user or the DBMS environment. The only opportunities for forward

actions are given by redundant structures where it is immaterial for the logical DB content whether or not modifying operations are undone or completed. A typical example is the splitting operation of a B-tree node.

Logging Methods and Rules

Crash recovery – as any recovery from a failure – needs some kind of redundancy to detect invalid or missing data in the permanent database and to "repair" its state as required, i.e., removing modifications effected by uncommitted transactions from it and supplementing it with updates of complete transactions. For this task, the recovery algorithms typically rely on log data collected during normal processing. Different forms of logging are conceivable. *Logical logging* is a kind of operator logging; it collects operators and their arguments at a higher level of abstraction (e.g., for internal operations (actions) or operations of the data management language (DML)). While this method of logging may save I/O to and space in the log file during normal processing, it requires at restart time a DB state that is level consistent with respect to the level of abstraction used for logging, because the logged operations have to be executed using data of the permanent database. For example, action logging and DML-operation logging require action consistency and consistency at the application programming interface (API consistency), respectively [7]. Hence, the use of this kind of methods implies the atomic propagation (see below) of all pages modified by the corresponding operation which can be implemented by shadow pages or differential files. *Physical logging* – in the simplest form collecting the before- and after-images of pages – does not expect any form of consistency at higher DB abstraction levels and, in turn, can be used in any situation, in particular, when nonatomic propagation of modified pages (*update-in-place*) is performed. However, writing before- and after-images of all modified pages to the log file is very time-consuming (I/O) and not space economical at all. Therefore, a combination of both kinds leads to the so-called physiological logging, which can be roughly characterized as "physical to a page and logical within a page".

It enables compact representation of log data (logging of elementary actions confined to single pages, entry logging) and leads to the practically most important logging/recovery method; nonatomic propagation of pages to disk is sufficient for the application of the log data. Together with the use of *log sequence numbers* in the log entries and in the headers of the data pages (for combined use of LSNs and PageLSNs, see ARIES Protocol), simple and efficient checks at restart detect whether or not the modifications of elementary actions have reached the permanent database, that is, whether or not undo or redo operations have to be applied.

While, in principle, crash recovery methods do not have specific requirements for forcing pages to the permanent DB, sufficient log information, however, must have reached the stable log. The following rules (for forcing of the log buffer to disk) have to be observed to guarantee recovery to the most recent transaction-consistent DB state:

- Redo log information must be written at the latest in phase 1 of commit.
- WAL (*write-ahead logging*) has to be applied to enable undo operations, before uncommitted (dirty) data is propagated to the permanent database.
- Log information must not be discarded from the temporary log file, unless it is guaranteed that it will no longer be needed for recovery; that is, the corresponding data page has reached the permanent DB. Typically, sufficient log information has been written to the archive log, in addition.

Taxonomy of Crash Recovery Algorithms

Forcing log data as captured by these rules yields the necessary and sufficient condition to successfully cope with system crashes. Specific assumptions concerning page propagation to the permanent database only influence performance issues of the recovery process. When dirty data can reach the permanent DB (steal property), recovery must be prepared to execute undo steps and, in turn, redo steps when data modified by a transaction is not forced at commit or before (no-force property). In contrast, if propagation of dirty data is prevented (no-steal property), the permanent DB only contains clean (but potentially missing or old) data, thus making undo steps unnecessary. Finally, if all transaction modifications are forced at commit (force property), redo is never needed at restart.

Hence, these properties concerning buffer replacement and update propagation are maintained by the buffer manager/transaction manager during normal processing and lead to four cases of crash recovery algorithms which cover all approaches so far proposed:

1. *Undo/Redo*: This class contains the *steal/no-force* algorithms which have to observe no other requirements than the logging rules. However, potentially undo and redo steps have to be performed during restart after a crash.
2. *Undo/NoRedo*: The so-called *steal/force* algorithms guarantee at any time that all actions of committed transactions are in the permanent DB. However, because of the steal property, dirty updates may be present, which may require undo steps, but never redo steps during restart.
3. *NoUndo/Redo*: The corresponding class members are known as *no-steal/no-force* algorithms which guarantee that dirty data never reaches the permanent DB. Dirty data pages are either never replaced from the DB buffer or, in case buffer space is in short supply, they are displaced to other storage areas outside the permanent DB. Restart after a crash may require redo steps, but never undo steps.
4. *NoUndo/NoRedo*: This "magic" class of the so-called *no-steal/force* algorithms always guarantees a state of the permanent DB that corresponds to the most recent transaction-consistent DB state. It requires that no modified data of a transaction reaches the permanent DB before commit and that all transaction updates are atomically propagated (forced) at commit. Hence, neither undo nor redo steps are ever needed during restart.

The discussion of these four cases is summarized in Fig. 1 which represents a taxonomy of crash recovery algorithms.

Crash Recovery, Fig. 1
Taxonomy of crash
recovery algorithms

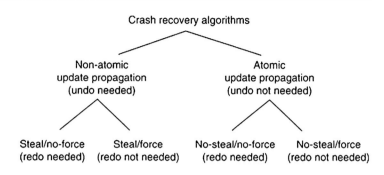

Implementation Implications

The latter two classes of algorithms (NoUndo) require a mechanism which can propagate a set of pages in an atomic way (with regard to the remaining DBMS processing). Such a mechanism needs to defer updates to the permanent DB until or after these updates become committed and can be implemented by various forms of shadowing concepts or differential file approaches.

Algorithms relying on redo steps, i.e., without the need to force committed updates to the permanent DB, have no control about the point of time when committed updates reach the permanent DB. While the buffer manager will propagate back most of the modified pages soon after the related update operations, a few hot-spot pages are modified again and again, and, since they are referenced so frequently, have not been written from the buffer. These pages potentially have accumulated the updates of many committed transactions, and redo recovery will therefore have to go back very far on the temporary log. As a consequence, restart becomes expensive and the DBMS's out-of-service time unacceptably long. For this reason, some form of *checkpointing* is needed to make restart costs independent of mean time between failures. Generating a checkpoint means collecting information related to the DB state in a safe place, which is used to define and limit the amount of redo steps required after a crash. The restart logic can then return to this checkpoint state and attempt to recover the most recent transaction-consistent state.

From a conceptual point of view, the algorithms of class 4 seem to be particularly attractive, because they always preserve a transaction-consistent permanent DB. However, in addition to the substantial cost of providing atomic update propagation, the need of forcing all updates at commit, necessarily in a synchronous way which may require a large amount of physical I/Os and, in turn, extend the lock duration for all affected objects, makes this approach rather expensive. Furthermore, with the typical disk-based DB architectures, pages are units of update propagation, which has the consequence that a transaction updating a record in a page cannot share this page with other updaters, because dirty updates must not leave the buffer and updates of complete transactions must be propagated to the permanent DB at commit. Hence, no-steal/force algorithms imply at least *page locking* as the smallest lock granule.

One of these cost factors – either synchronously forced updates at commit or atomic updates for NoUndo – applies to the algorithms of class 2 and 3 each. Therefore, they were not a primary choice for the DBMS vendors competing in the today's market.

Hence, the laissez-faire solution "steal, no-force" with nonatomic update propagation (update-in-place) is today's favorite solution, although it always leaves the permanent DB in a "chaotic state" containing dirty and outdated data pages and keeping the latest version of frequently used pages only in the DB buffer. Hence, with the optimistic expectation that crashes become rather rare events, it minimizes recovery provisions during normal processing. Checkpointing is necessary, but the application of direct checkpoints flushing the entire buffer at a time is not advisable anymore, when buffers of several GByte are used. To affect normal processing as little as possible, so-called fuzzy checkpoints are written; only a few pages with metadata concerning the DB buffer state have to be synchronously propagated, while data pages are "gently" moved to the permanent DB in an asynchronous way.

Archive Recovery

So far, data of the permanent DB was assumed to be usable or at least recoverable using the redundant data collected in the temporary log. This is illustrated by the upper path in Fig. 2. If any of the participating components is corrupted or lost because of other hardware or software failure, archive recovery – characterized by the lower path – must be tried. Successful recovery also implies independent failure modes of the components involved.

The creation of an archive copy, that is, copying the online version of the DB, is a very expensive process; for example, creating a transaction-consistent DB copy would interrupt update operation for a long time which is unacceptable for most DB applications. Therefore, two base methods – fuzzy dumping and incremental dumping – were developed to reduce the burden of normal DB operation while an archive copy is created. A fuzzy dump copies the DB on the fly in parallel with normal processing. The other method writes only the changed pages to the incremental dump. Of course, both methods usually deliver inconsistent DB copies such that log-based post-processing is needed to apply incremental modifications. In a similar way, either type of dump can be used to create a new, more up-to-date copy from the previous one, using a separate offline process such that DB operation is not affected.

Archive copies are "hopefully" never or very infrequently used. Therefore, they may be susceptible to magnetic decay. For this reason, redundancy is needed again, which is usually solved by keeping several generations of the archive copy.

So far, all log information was assumed to be written only to the temporary log file during normal processing. To create the (often duplexed) archive log, usually an independent and asynchronously running process copies the redo data from the temporary log. To guarantee successful recovery, failures when using the archive copies must be anticipated. Therefore, archive recovery must be prepared to start from the oldest generation and hence the archive log must span the whole distance back to this point in time.

Key Applications

Recovery algorithms, and in particular for crash recovery, are a core part of each commercial-strength DBMS and require a substantial fraction of design/implementation effort and of the code base: "A recoverable action is 30% harder and requires 20% more code than a non-recoverable action" (J. Gray). Because the occurrence of failures cannot be excluded and all data driving the daily business are managed in databases, mission-critical businesses depend on the recoverability of their data. In this sense, provisions for crash recovery are indispensable in such DBMS-based applications. Another important application area of crash recovery techniques are file systems, in particular their metadata about file existence, space allocation, etc.

Future Directions

So far, crash recovery provisions were primarily disk oriented. Future storage management based on nonvolatile RAM [11] will provide many optimization opportunities for logging and

Crash Recovery, Fig. 2
Two ways of DB crash recovery and the components involved

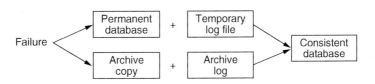

recovery tasks, e.g., latency will be dramatically reduced for log writes, and, in turn, commit protocols may be completed much faster. However, workload characteristics and hardware improvements – in particular, larger memories and multi-core CPUs – have a strong impact on the time required for recovery from a crash. Furthermore, the steadily increased throughput in high-performance transaction systems may lead to more recovery steps to be done during restart. To guarantee fast recovery times, logging and recovery algorithms must be designed to enable recovery provisions to happen concurrently with normal transaction processing and to incur minimal overhead [13]. Ideally, such an algorithm would make the restart time to be ready for new transactions independent of the amount of undo and redo steps necessary to restore the database to the least-recent transaction-consistent state.

For this reason, by using the mechanisms proposed in [3], we have developed novel techniques and algorithms for transaction commit, logging, recovery, and page flushing, which maintain the persistent state of the database (both log and data pages) always in a committed state [14]. Hence, undo steps are not needed anymore during restart. Crash and archive recovery require only redo steps on pages not up-to-date when the failure occurred. This redo recovery at the page level can happen concurrently with processing of new transactions. Therefore, normal transaction processing can resume as soon as the conventional analysis phase of the restart protocol is completed. This design focusing on high availability by providing minimal restart times works in traditional disk environments, but it is greatly accelerated when specific hardware configurations such as nonvolatile RAM or flash are used.

Cross-References

▶ ACID Properties
▶ Application Recovery
▶ B-Tree Locking
▶ Buffer Management
▶ Logging and Recovery
▶ Multilevel Recovery and the ARIES Algorithm

Recommended Reading

1. Bernstein PA, Hadzilacos V, Goodman N. Concurrency control and recovery in database systems. Reading: Addison-Wesley; 1987.
2. Davies CT. Data processing spheres of control. IBM Syst J. 1978;17(2):179–98.
3. Graefe G, Kuno HA. Definition, detection, and recovery of single-page failures, a fourth class of database failures. Proc VLDB Endowment; 2012;5(7): 646–55.
4. Gray J, Reuter A. Transaction processing: concepts and techniques. San Francisco: Morgan Kaufmann; 1993.
5. Gray J, McJones P, Blasgen M, Lindsay B, Lorie R, Price T, Putzolu F, Traiger IL. The recovery manager of the System R database manager. ACM Comput Surv. 1981;13(2):223–42.
6. Gray J. Notes on data base operating systems. Advanced course: operating systems. Berlin: Springer; 1978. p. 393–481. LNCS 60.
7. Härder T. DBMS architecture – still an open problem. In: Proceedings of the German National Database Conference; 2005. p. 2–28.
8. Härder T, Reuter A. Principles of transaction-oriented database recovery. ACM Comput Surv. 1983;15(4):287–317.
9. Hvasshovd S-O. Recovery in parallel database systems. 2nd ed. Burlington: Morgan Kaufmann; 1999.
10. Mohan C, Haderle DJ, Lindsay BG, Pirahesh H, Schwarz PM. ARIES: a transaction recovery method supporting fine-granularity locking and partial rollbacks using write-ahead logging. ACM Trans Database Syst. 1992;17(1):94–162.
11. Pelley S, Wenisch TF, Gold BT, Bridge B. Storage management in the NVRAM era. Proc VLDB Endowment; 2013;7(2):121–32.
12. Reuter A. Fehlerbehandlung in Datenbanksystemen. Munich: Carl Hanser; 1981. p. 456.
13. Sauer C, Graefe G, Härder T. An empirical analysis of database recovery costs. In: Proceedings of the Sigmod Workshops: RDSS; 2014.
14. Sauer C, Härder T. A simple recovery mechanism enabling fine-granular locking and fast, REDO-only recovery. CoRR abs/1409.3682, 2014.
15. Weikum G, Vossen G. Transactional information systems: theory, algorithms, and the practice of concurrency control and recovery. San Francisco: Morgan Kaufmann; 2002.

Cross-Language Mining and Retrieval

Wei Gao[1] and Cheng Niu[2]
[1]Qatar Computing Research Institute, Doha, Qatar
[2]Microsoft Research Asia, Beijing, China

Synonyms

Cross-language informational retrieval; Cross-language text mining; Cross-language web mining; Translingual information retrieval

Definition

Cross-language mining is a task of text mining dealing with the extraction of entities and their counterparts expressed in different languages. The interested entities may be of various granularities from acronyms, synonyms, cognates, proper names to comparable or parallel corpora. Cross-Language Information Retrieval (CLIR) is a sub-field of information retrieval dealing with the retrieval of documents across language boundaries, i.e., the language of the retrieved documents is not the same as the language of the queries. Cross-language mining usually acts as an effective means to improve the performance of CLIR by complementing the translation resources exploited by CLIR systems.

Historical Background

CLIR addresses the growing demand to access large volumes of documents across language barriers. Unlike monolingual information retrieval, CLIR requires query terms in one language to be matched with the indexed keywords in the documents of another language. Usually, the cross-language matching can be done by making use of bilingual dictionary, machine translation software, or statistical model for bilingual words association. CLIR generally takes into account but not limited to the issues like how to translate query terms, how to deal with the query terms nonexistent in a translation resource, and how to disambiguate or weight alternative translations (e.g., to decide that "traitement" in a French query means "treatment" but not "salary" in English, or how to order the French terms "aventure," "business," "affaire," and "liaison" as relevant translations of English query "affair"), etc. The performance of CLIR can be measured by the general evaluation metrics of information retrieval, such as recall precision, average precision, and mean reciprocal rank, etc.

The first workshop on CLIR was held in Zürich during the SIGIR-96 conference. Workshops have been held yearly since 2000 at the meetings of CLEF (Cross Language Evaluation Forum), following its predecessor workshops of TREC (Text Retrieval Conference) cross-language track. The NTCIR (NII Test Collection for IR Systems) workshop is also held each year in Japan for CLIR community focusing on English and Asian languages.

The study of cross-language mining appears relatively more lately than CLIR, partly due to the increasing demands on the quality of CLIR and machine translation, as well as the recent advancement of text/Web mining techniques. A typical early work on cross-lingual mining is believed to be PTMiner [14] that mines parallel text from the Web used for query translation. Other than parallel data mining, people also tried to mine the translations of Out-of-Vocabulary (OOV) terms from search results returned from search engine [5, 18] or from web anchor texts and link structures [12]. Based on phonetic similarity, transliteration (the phonetic counterpart of a name in another language, e.g., "Schwarzenegger" is pronounced as "shi wa xin ge" in Chinese pinyin) of foreign names also could be extracted properly from the Web [10]. These methods are proposed to alleviate the OOV problem of CLIR since there is usually lack of appropriate translation resources for new terminologies and proper names, particularly in the scenario of cross-language web search.

Foundations

Most approaches to CLIR perform query translation followed by monolingual retrieval. So the retrieval performance is largely determined by the quality of query translation. Queries are typically translated either using a bilingual dictionary [15], a machine translation (MT) software [7], bilingual word association model learned from parallel corpus [6, 14], or recently a query log of a search engine [9]. Despite the types of the resources being used, OOV translation and translation disambiguation are the two major bottlenecks for CLIR. On one hand, translation resources can never be comprehensive. Correctly translating queries, especially Web queries, is difficult since they often contain new words (e.g., new movies, brands, celebrities, etc.) occurring timely and frequently, yet being OOV to the system; On the other hand, many words are polysemous, or they do not have a unique translation, and sometimes the alternative translations have very different meanings. This is known as translation ambiguity. Selecting the correct translation is not trivial due to the shortage of context provided in a query, and effective techniques for translation disambiguation are necessary.

It should be mentioned that document translation with MT in the opposite direction is an alternative approach to CLIR. However, it is less commonly used than query translation in the literature mainly because MT is computationally expensive and costly to develop, and the document sets in IR are generally very large. For cross-language web search, it is almost impractical to translate all the web pages before indexing. Some large scale attempts to compare query translation and document translation have suggested no clear advantage for either of the approaches to CLIR [12]. But they found that compared with extremely high quality human query translations, it is advantageous to incorporate both document and query translation into a CLIR system.

Cross-Language Web Mining

Mining Parallel Data

The approaches of mining parallel text make extensive use of bilingual websites where parallel web pages corresponding to the specified language pair can be identified and downloaded. Then the bilingual texts are automatically aligned in terms of sentences and words by statistical aligning tools, such as GIZA++ (http://www.fjoch.com/GIZA++.html). The word translation probabilities can be derived with the statistics of word pairs occurring in the alignments, after which one can resort to statistical machine translation models, e.g., IBM model-1 [4], for translating given queries into the target language. The typical parallel data mining tools include PTMiner [14], STRAND [16] and the DOM-tree-alignment-based system [17].

Mining OOV Term Translation

Web pages also contain translations of terms in either the body texts or the anchor texts of hyper-links pointing to other pages. For example, in some language pairs, such as Chinese-English or Japanese-English, the Web contains rich body texts in a mixture of multiple languages. Many of them contain bilingual translations of proper nouns, such as company names and person names. The work of [5, 16] exploits this nice characteristic to automatically extract translations from search result for a large number of unknown query terms. Using the extracted bilingual translations, the performance of CLIR between English and Chinese is effectively improved. Both methods select translations based on some variants of co-occurrence statistics.

The anchor text of web pages' hyperlinks is another source for translational knowledge acquisition. This is based on the observation that the anchor texts of hyperlinks pointing to the same URL may contain similar descriptive texts. Lu et al. [11] uses anchor text of different languages to extract the regional aliases of query terms for constructing a translation lexicon. A probabilistic inference model is exploited to estimate the similarity between query term and extracted translation candidates.

Query Translation Disambiguation

Translation disambiguation or ambiguity resolution is crucial to the query translation accuracy. Compared to the simple dictionary-based translation approach without addressing translation disambiguation, the effectiveness of CLIR can be 60% lower than that of monolingual retrieval [3]. Different disambiguation techniques have been developed using statistics obtained from document collections, all resulting in significant performance improvement. Zhang et al. [19] give concise review on three main translation disambiguation techniques. These methods include using term similarity [1], word co-occurrence statistics of the target language documents, and language modeling based approaches [20]. In this subsection, we introduce these approaches following the review of Zhang et al. [19].

Disambiguation by Term Similarity

Adriani [1] proposed a disambiguation technique based on the concept of statistical term similarity. The term similarity is measured by the Dice coefficient, which uses the term-distribution statistics obtained from the corpus. The similarity between term x and y, $SIM(x, y)$, is calculated as:

$$SIM(x, y)$$
$$= 2 \sum_{i=1}^{n} \left(w_{xi} w_{yi} \right) / \left(\sum_{i=1}^{n} w_{xi}^2 + \sum_{i=1}^{n} w_{yi}^2 \right)$$

where w_{xi} and w_{yi} is the weights of term x and y in document i. This method computes the sum of maximum similarity values between each candidate translation of a term and the translations of all other terms in the query. For each query term, the translation with the highest sum is selected as its translation. The results of Indonesian-English CLIR experiments demonstrated the effectiveness of this approach. There are many variant term association measures like Jaccard, Cosine,

Overlap, etc. that can be applied similarly for calculating their similarity.

Disambiguation by Term Co-occurrence

Ballesteros and Croft [3] used co-occurrence statistics obtained from the target corpus for resolving disambiguation. They assume the correct translations of query terms should co-occur in target language documents and incorrect translations tend not to co-occur. Similar approach is studied by Gao et al. [8]. They observed that the correlation between two terms is stronger when the distance between them is shorter. They extended the previous co-occurrence model by incorporating a distance factor $D(x, y) = e^{-\alpha(Dis(x,y)-1)}$. The mutual information between term x and y, $MI(x, y)$, is calculated as:

$$MI(x, y) = \log \left(\frac{f_w(x, y)}{f_x f_y} + 1 \right) \times D(x, y)$$

where $f_w(x, y)$ is the co-occurrence frequency of x and y that occur simultaneously within a window size of w in the collection, f_x is the collection frequency of x, and f_y is the collection frequency of y. $D(x, y)$ decreases exponentially when the distance between the two terms increases, where α is the decay rate, and $D(x, y)$ is the average distance between x and y in the collection. The experiments on the TREC9 Chinese collection showed that the distance factor leads to substantial improvements over the basic co-occurrence model.

Disambiguation by Language Modeling

In the work of [20], a probability model based on hidden Markov model (HMM) is used to estimate the maximum likelihood of each sequence of possible translations of the original query. The highest probable translation set is selected among all the possible translation sets. HMM is a widely used for probabilistic modeling of sequence data. In their work, a smoothing technique based on absolute discounting and interpolation method is adopted to deal with the zero-frequency problem during probability estimation. See [20] for details.

Pre-/Post-Translation Expansion

Techniques of OOV term translation and translation disambiguation both aim to translate query correctly. However, it is arguable that precise translation may not be necessary for CLIR. Indeed, in many cases, it is helpful to introduce words even if they are not direct translations of any query word, but are closely related to the meaning of the query. This observation has led to the development of cross-lingual query expansion (CLQE) techniques [2, 13]. [2] reported the enhancement on CLIR by post-translation expansion. [13] made performance comparison on various CLQE techniques, including pre-translation expansion, post-translation expansion and their combinations. Relevance feedback, the commonly used expansion technique in monolingual retrieval, is also widely adopted in CLQE. The basic idea is to expand original query by additional terms that are extracted from the relevant retrieval result initially returned. Amongst different relevance feedback methods, explicit feedback requires documents whose relevancy is explicitly marked by human; implicit feedback is inferred from users' behaviors that imply the relevancy of the selected documents, such as which returned documents are viewed or how long they view some of the documents; blind or "pseudo" relevance feedback is obtained by assuming that top n documents in the initial result are relevant.

Cross-Lingual Query Suggestion

Traditional query translation approaches rely on static knowledge and data resources, which cannot effectively reflect the quickly shifting interests of Web users. Moreover, the translated terms can be reasonable translations, but are not popularly used in the target language. For example, the French query "aliment biologique" is translated into "biologic food," yet the correct formulation nowadays should be "organic food." This mismatch makes the query translation in the target language ineffective. To address this problem, Gao et al. [9] proposed a principled framework called Cross-Lingual Query Suggestion (CLQS), which leverages cross-lingual mining and translation disambiguation techniques to suggest related queries found in the query log of a search engine.

CLQS aims to suggest related queries in a language different from the original query. CLQS is closely related to CLQE, but is distinct in that it suggests full queries that have been formulated by users so that the query integrity and coherence are preserved in the suggested queries. It is used as a new means of query "translation" in CLIR tasks. The use of query log for CLQS stems from the observation that in the same period of time, many search users share the same or similar interests, which can be expressed in different manners in different languages. As a result, a query written in a source language is possible to have an equivalent in the query log of the target language. Especially, if the user intends to perform CLIR, then original query is even more likely to have its correspondent included in the target language log. Therefore, if a candidate for CLQS appears often in the query log, it is more likely to be the appropriate one to suggest. CLQS is testified being able to cover more relevant documents for the CLIR task.

The key problem with CLQS is how to learn a similarity measure between two queries in different languages. They define cross-lingual query similarity based on both translation relation and monolingual similarity. The principle for learning is, for a pair of queries, their cross-lingual similarity should fit the monolingual similarity between one query and the other query's translation. There are many ways to obtain a monolingual similarity between queries, e.g., co-occurrence based mutual information and χ^2. Any of them can be used as the target for the cross-lingual similarity function to fit. In this way, cross-lingual query similarity estimation is formulated as a regression task:

$$sim_{CL}\left(q_f, q_e\right) = w \cdot \varphi\left(f\left(q_f, q_e\right)\right)$$
$$= sim_{ML}\left(T_{q_f}, q_e\right)$$

where given a source language query q_f, a target language query q_e, and a monolingual query similarity between them sim_{ML}, the cross-lingual query similarity sim_{CL} can be calculated as an inner product between a weight vector and the feature vector in the kernel space, and φ is the mapping from the input feature space onto the kernel space, and w is the weight vector which can be learned by support vector regression training. The monolingual similarity is measured by combining both query content-based similarity and click-through commonality in the query log.

This discriminative modeling framework can integrate arbitrary information sources to achieve an optimal performance. Multiple feature functions can be incorporated easily into the framework based on different translation resources, such as bilingual dictionaries, parallel data, web data, and query logs. They work uses co-occurrence-based dictionary translation disambiguation, IBM translation model-1 based on parallel corpus, and Web-based query translation mining as means to discover related candidate queries in the query log. Experiments on TREC6 French-English CLIR task demonstrate that CLQS-based CLIR is significantly better than the traditional dictionary-based query translation with disambiguation and machine translation approaches.

Latent Semantic Index (LSI) for CLIR

Different from most of the alternative approaches discussed above, LSI for CLIR [6] provides a method for matching text segments in one language with the segments of similar meaning in another language without having to translate either. Using a parallel corpus, LSI can create a language-independent representation of words. The representation matrix reflects the patterns of term correspondences in the documents of two languages. The matrix is factorized by Singular Value Decomposition (SVD) for deriving a latent semantic space with a reduced dimension, where similar terms are represented by similar vectors. In latent semantic space, therefore, the monolingual similarity between synonymous terms from one language and the cross-lingual similarity between translation pairs from different languages tend to be higher than the similarity with irrelevant terms. This characteristic allows relevant documents to be retrieved even if they do not share any terms in common with the query, which makes LSI suitable for CLIR.

Key Applications

Cross-language mining and retrieval is the foundation technology for searching web information across multiple languages. It can also provide the cross-lingual functionality for the retrieval of structured, semi-structured and un-structured document databases of specific domains or in large multinational enterprises.

Experimental Results

In general, for every presented work, there is an accompanying experimental evaluation in the corresponding reference. Especially, the three influential international workshops held annually, i.e., CLEF, NTCIR and TREC, defines many evaluation tasks for CLIR, and there are a large number of experimental results being published based on these benchmark specifications.

Data Sets

Data sets for benchmarking CLIR are released to the participants of TREC, CLEF and NTCIR workshops annually with license agreements.

Cross-References

▶ Anchor Text
▶ Average Precision
▶ Document databases
▶ Document Links and Hyperlinks
▶ Evaluation Metrics for Structured Text Retrieval
▶ Information Extraction
▶ Information Retrieval
▶ MAP
▶ Mean Reciprocal Rank
▶ Query Expansion for Information Retrieval
▶ Query Translation
▶ Singular Value Decomposition
▶ Snippet
▶ Stemming
▶ Stoplists
▶ Term Statistics for Structured Text Retrieval

▸ Term Weighting
▸ Text Indexing and Retrieval
▸ Text Mining
▸ Web Information Extraction
▸ Web Search Relevance Feedback

Recommended Reading

1. Adriani M. Using statistical term similarity for sense disambiguation in cross-language information retrieval. Inf Retr. 2000;2(1):71–82.
2. Ballestors LA, Croft WB. Phrasal translation and query expansion techniques for cross-language information retrieval. In: Proceedings of the 20th Annual International ACM SIGIR Conference on Research and Development in Information Retrieval; 1997. p. 84–91.
3. Ballestors LA, Croft WB. Resolving and ambiguity for cross-language information retrieval. In: Proceedings of the 21st Annual International ACM SIGIR Conference on Research and Development in Information Retrieval; 1998. p. 64–71.
4. Brown PF, Pietra SAD, Pietra VDJ, Mercer RL. The mathematics of machine translation: parameter estimation. Comput Linguist. 1992;19(2):263–312.
5. Cheng PJ, Teng JW, Chen RC, Wang JH, Lu WH, Chien LF. Translating unknown queries with Web corpora for cross-language information retrieval. In: Proceedings of the 30th Annual International ACM SIGIR Conference on Research and Development in Information Retrieval; 2004. p. 146–53.
6. Dumais ST, Landauer TK, and Littman ML. Automatic cross-linguistic information retrieval using latent semantic indexing. In: Proceedings of the ACM SIGIR Workshop on Cross-Linguistic Information Retrieval; 1996. p. 16–23.
7. Fujii A, Ishikawa T. Applying machine translation to two-stage cross-language information retrieval. In: Proceedings of the 4th Conference on Association for Machine Translation in the Americas; 2000. p. 13–24.
8. Gao J, Zhou M, Nie, JY, He H, Chen W. Resolving query translation ambiguity using a decaying co-occurrence model and syntactic dependence relations. In: Proceedings of the 25th Annual International ACM SIGIR Conference on Research and Development in Information Retrieval; 2002. p. 183–90.
9. Gao W, Niu C, Nie JY, Zhou M, Hu J, Wong KF, Hon HW. Cross-lingual query suggestion using query logs of different languages. In: Proceedings of the 33rd Annual International ACM SIGIR Conference on Research and Development in Information Retrieval; 2007. p. 463–70.
10. Jiang L, Zhou M, Chien LF, Niu C. Named entity translation with Web mining and transliteration. In: Proceedings of the 20th International Joint Conference on AI; 2007. p. 1629–34.
11. Lu WH, Chien LF, Lee HJ. Translation of web queries using anchor text mining. ACM Trans Asian Lang Information Proc. 2002;1(2):159–72.
12. McCarley JS. Should we translate the documents or the queries in cross-language information retrieval? In: Proceedings of the 27th Annual Meeting of the Association for Computational Linguistics; 1999. p. 208–14.
13. McNamee P, Mayfield J. Comparing cross-language query expansion techniques by degrading translation resources. In: Proceedings of the 25th Annual International ACM SIGIR Conference on Research and Development in Information Retrieval; 2002. p. 159–66.
14. Nie JY, Smard M, Isabelle P, Durand R. Cross-language information retrieval based on parallel text and automatic mining of parallel text from the Web. In: Proceedings of the 22nd Annual International ACM SIGIR Conference on Research and Development in Information Retrieval; 1999. p. 74–81.
15. Pirkola A, Hedlund T, Keshusalo H, Järvelin K. Dictionary-based cross-language information retrieval: problems, methods, and research findings. Inf Retr. 2001;3(3–4):209–30.
16. Resnik P, Smith NA. The Web as a parallel corpus. Comput Linguist. 2003;29(3):349–80.
17. Shi L, Niu C, Zhou M, Gao J. A DOM Tree alignment model for mining parallel data from the Web. In: Proceedings of the 44th Annual Meeting of the Association for Computational Linguistics; 2006. p. 489–96.
18. Zhang Y, Vines P. Using the Web for automated translation extraction in cross-language information retrieval. In: Proceedings of the 30th Annual International ACM SIGIR Conference on Research and Development in Information Retrieval; 2004. p. 162–9.
19. Zhang Y, Vines P, Zobel J. An empirical comparison of translation disambiguation techniques for Chinese-English Cross-Language Information Retrieval. In: Proceedings of the 3rd Asia Information Retrieval Symposium; 2006. p. 666–72.
20. Zhang Y, Vines P, Zobel J. Chinese OOV translation and post-translation query expansion in Chinese-English cross-lingual information retrieval. ACM Trans Asian Lang Information Proc. 2005;4(2):57–77.

Cross-Modal Multimedia Information Retrieval

Qing Li and Yu Yang
City University of Hong Kong, Hong Kong, China

Synonyms

Cross-media information retrieval; Multi-modal information retrieval

Definition

Multimedia information retrieval tries to find the distinctive multimedia documents that satisfy people's needs within a huge dataset. Due to the vagueness on the representation of multimedia data, usually the user may only have some clues (e.g., a vague idea, a rough query object of the same or even different modality as that of the intended result) rather than concrete and indicative query objects. In such cases, traditional multimedia information retrieval techniques as Query-By-Example (QBE) fails to retrieve what users really want since their performance depends on a set of specifically defined features and carefully chosen query objects. The cross-modal multimedia information retrieval (CMIR) framework consists of a novel multifaceted knowledge base (which is embodied by a layered graph model) to discover the query results on multiple modalities. Such cross-modality paradigm leads to better query understanding and returns the retrieval result which meets user need better.

Historical Background

Previous works addressing multimedia information retrieval can be classified into two groups: approaches on single-modality, and those on multi-modality integration.

Retrieval Approaches on Single-Modality
The retrieval approach in this group only deals with a single type of media, so that most content-based retrieval (CBR) approaches [1–5] fall into this group. These approaches differ from each other in either the low-level features extracted from the data, or the distance functions used for similarity calculation. Despite the differences, all of them are similar in two fundamental aspects: (i) they all rely on low-level features; (ii) they all use the query-by-example paradigm.

Retrieval Approaches on Multi-Modality Integration
More recently there are some works that investigate the integration of multi-modality data, usually between text and image, for better retrieval performance. For example, iFind [6] proposes a unified framework under which the semantic feature (text) and low-level features are combined for image retrieval, whereas the 2M2Net [7] system extends this framework to the retrieval of video and audio. WebSEEK [5] extracts keywords from the surrounding text of image and videos, which is used as their indexes in the retrieval process. Although these systems involve more than one media, different types of media are not actually integrated but are on different levels. Usually, text is only used as the annotation (index) of other medias. In this regard, cross-modal multimedia information retrieval (CMIR) enables an extremely high degree of multi-modality integration, since it allows the interaction among objects of any modality in any possible ways (via different types of links).

MediaNet [8] and multimedia thesaurus (MMT) [9] seek to provide a multimedia representation of semantic concept - a concept described by various media objects including text, image, video, etc - and establish the relationships among these concepts. MediaNet extends the notion of relationships to include even perceptual relationships among media objects. Both approaches can be regarded as "concept-centric" approaches since they realize an organization of multi-modality objects around semantic concepts. In contrast, CMIR is "concept-less" since it makes no attempt to identify explicitly the semantics of each object.

Foundations

The cross-modality multimedia information retrieval (CMIR) mechanism shapes a novel scenario for multimedia retrieval: The user starts the search by supplying a set of seed objects as the hints of his intention, which can be of any modality (even different with the intended objects), and are not necessarily the eligible results by themselves. From the seeds, the system figures out the user's intention and returns a set of cross-modality objects that potentially satisfy this intention. The user can give further hints by identifying the results approximating his need, based on which the system improve its estimation about the user intention and refines the results

Cross-Modal Multimedia Information Retrieval, Table 1 CBR paradigms, drawbacks, and suggested remedies in CMIR

	CBR paradigms	Drawbacks	Suggested remedies in CMIR
Interaction	Highly representative sample object	Vague idea, or clear idea without appropriate samples	Cross-modality seed objects, only as hints
Data index	Low-level features	Inadequate to capture semantics	Multifaceted knowledge (user behaviors, structure, content)
Results	Single-modality, perceptually similar objects	Looks like or sounds like, but not what user actually needs	Cross-modality, semantically related objects

towards it. This scenario can be also interpreted as a cooperative process: the user tries to focus the attention of the system to the objects by giving hints on the intended results, while the system tries to return more reasonable results that allows user to give better hints. A comparison between CMIR and the current CBR approaches is shown in Table 1.

To support all the necessary functionalities for such an ideal scenario, a suite of unique models, algorithms and strategies are developed in CMIR. As shown in Fig. 1, the foundation of the whole mechanism is a multifaceted knowledge base describing the relationships among cross-modality objects. The kernel of the knowledge base is a layered graph model, which characterizes the knowledge on (i) history of user behaviors, (ii) structural relationships among media objects, and (iii) content of media objects, at each of its layers. Link structure analysis-an established technique for web-oriented applications-is tailored to the retrieval of cross-modality data based on the layered graph model. A unique relevant feedback technique that gears with the underlying graph model is proposed, which can enrich the knowledge base by updating the links of the graph model according to user behaviors. The loop in Fig. 1 reveals the hill-climbing nature of the CMIR mechanism, i.e., it enhances its performance by learning from the previously conducted queries and feedbacks.

Layered Graph Model

As the foundation of the retrieval capability, the multifaceted knowledge base accommodates a broad range of knowledge indicative of data semantics, mainly in three aspects: (i) user be-

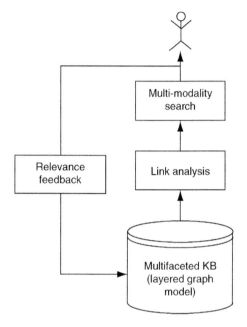

Cross-Modal Multimedia Information Retrieval, Fig. 1 Overview of the CMIR mechanism

haviors in the user-system interaction, (ii) structural relationships among media objects, and (iii) content of each media object. The kernel of the knowledge base is a three-layer graph model, with each layer describing the knowledge in one aspect, called knowledge layer. Its formal definition is given as follows.

Definition 1

A knowledge layer is a undirected graph G = (V, E), where V is a finite set of vertices and E is a finite set of edges. Each element in V corresponds to a media object $O_i \in O$, where O is the collection of media objects in the database. E is a ternary relation defined on $V \times V \times$

R, where R represents real numbers. Each edge in E has the form of <O_i, O_j, r>, denoting a semantic link between O_i and O_j with r as the weight of the link. The graph corresponds to a $|V| \times |V|$ **adjacency matrix** (The adjacency matrix defined here is slightly different from the conventional definition in mathematics, in which each component is a binary value indicating the existence of the corresponding edge.) $M = [m_{ij}]$, where $m_{ij} = m_{ji}$ always holds. Each element m_{ij} = r if there is an edge <O_i, O_j, r>, and $m_{ij} = 0$ if there is no edge between O_i and O_j. The elements on the diagonal are set to zero, i.e., $m_{ii} = 0$.

Each semantic link between two media objects may have various interpretations, which corresponds to one of the three cases: (i) a user has implied the relevance between the two objects during the interaction, e.g., designating them as the positive example in the same query session; (ii) there is a structural relationships between them, e.g., they come from the same or linked web page(s); or (iii) they resemble each other in terms of their content. The multifaceted knowledge base seamlessly integrates all these links into the same model while preserving their mutual independence.

Definition 2

*The multifaceted knowledge base is a **layered graph model** consisting of three superimposed knowledge layers, which from top to bottom are **user layer**, **structure layer**, and **content layer**. The vertices of the three layers correspond to the same set of media objects, but their edges are different either in occurrences or in interpretations.*

Figure 2 illustrates the layered graph model. Note that the ordering of the three layers is immutable, which reflects their priorities in terms of knowledge reliability. The user layer is placed uppermost since user judgment is assumed most reliable (not necessarily always reliable). Structure links is a strong indicator of relevance, but not as reliable as user links. The lowest layer is the content layer. As a generally accepted fact in CBR area, content similarity does not entail any well-defined mapping with semantics.

A unique property of the layered graph model is that it stores the knowledge on the links (or re-lationships) among media objects, rather than on the nodes (media objects) upon which most existing retrieval systems store the data index. All the algorithms based of this model can be interpreted as manipulation of links: to serve the user query, relevant knowledge is extracted from this graph model by analyzing the link structure; meanwhile, user behaviors are studied to enrich the knowledge by updating the links. An advantage of such link-based approach is that the retrieval can be performed in a relatively small locality connected via links instead of the whole database, and therefore it can afford more sophisticated retrieval algorithms.

Link Analysis Based Retrieval

As illustrated in Fig. 3, the retrieval process can be described as a circle: the intended objects are retrieved through the upper semicircle, and the user evaluations are studied and incorporated into the knowledge base though the lower half-circle, which initiates a new circle to refine the previously retrieved results based on the updated knowledge. Consequently, it is a hill-climbing approach in that the performance is enhanced incrementally as the loop is repeated.

The retrieval process consists of five steps (as shown in Fig. 3): (i) generate the seed objects as the hints of the user's intention; (ii) span the seeds to a collection of candidate objects via the links in the layered graph model; (iii) distill the results by ranking the candidates based on link structure analysis, (iv) update the knowledge base to incorporate the user evaluation of the current results, and (v) refine the results based on user evaluations.

Key Applications

Multimedia Information Retrieval System

For multimedia data, the modalities supported can be texts (surrounding or tagged), images, videos and audios. An ongoing prototype [10] utilizes the primitive features and similarity functions for these media shown in Table 2. The experimental results prove the usefulness of the approach for better query understanding.

Cross-Modal Multimedia Information Retrieval, Fig. 2 The Layered graph model as multifaceted knowledge base

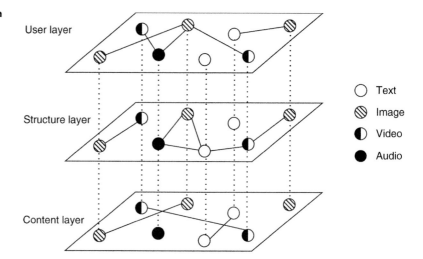

Cross-Modal Multimedia Information Retrieval, Fig. 3 Overview of the link analysis based retrieval algorithm.

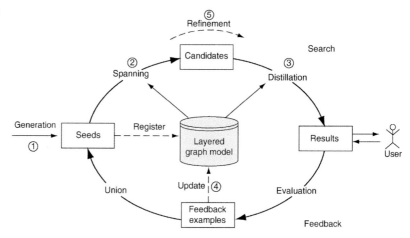

Cross-Modal Multimedia Information Retrieval, Table 2 Primitive features and similarity function used in prototype

	Text	Image	Video
Primitive features	Keywords, weighted by TF*IDF	256-d HSV color histogram, 64-d LAB color coherence, 32-d Tamura directionality	First frame of each shot as key-frame, indexing key-frame as an image
Similarity function	Cosine distance	Euclidean distance for each feature, linear combination of different similarities	Key-frame (image) similarity as shot similarity, average pair-wise shot similarity as video similarity

Future Directions

Due to the generality and extensibility of the CMIR, there are many potential directions that can be implemented on it:

Navigation The graph model provides abundant links through which the user can traverse from an object to its related objects. An intuitive scenario for navigation is when the user is looking at a certain object, he is recommended with the objects that are linked to it in the graph model, ranked by their link weights and link types, from which he may select one as the next object he will navigate to.

Clustering Clustering cross-modality objects into semantically meaningful groups is also an important and challenging issue, which requires an underlying *similarity function* among objects, along with a method that produces clusters based on the similarity function. The layered graph model provides knowledgeable and rich links, based on which different similarity functions can be easily formulated. Meanwhile, many existing approaches can be employed as the clustering method, such as simulated and deterministic annealing algorithm [11]. Moreover, CMIR inherently allows the clustering of cross-modality objects, rather than single-modality objects that most previous classification approaches can deal with.

Personalized Retrieval The user layer of the graph model characterizes the knowledge obtained from the behaviors of the whole population of users, and allows a query from a single user to benefit from such common knowledge. However, each user may have his/her personal interests, which may not agree with each other. The "multi-leveled user profile" mechanism [12] leads a good direction for future study.

Cross-References

▶ Multimedia Data
▶ Multimedia Information Retrieval Model

Recommended Reading

1. Chang SF, Chen W, Meng HJ, Sundaram H, Zhong D. VideoQ: an automated content based video search system using visual cues. In: Proceedings of the 5th ACM International Conference on Multimedia; 1997.
2. Flickner M, Sawhney H, Niblack W, Ashley J. Query by image and video content: the QBIC system. IEEE Comput. 1995;28(9):23–32.
3. Huang TS, Mehrotra S, Ramchandran K. Multimedia analysis and retrieval system (MARS) project. In: Proceedings of the 33rd Annual Clinic on Library Application of Data Processing-Digital Image Access and Retrieval; 1996.
4. Smith JR, Chang SF. VisualSEEk: a fully automated content-based image query system. In: Proceedings of the 4th ACM International Conference on Multimedia; 1996.
5. Smith JR, Chang SF. Visually searching the web for content. IEEE Multimedia Mag. 1997;4(3):12–20.
6. Lu Y, Hu CH, Zhu XQ, Zhang HJ, Yang Q. A unified framework for semantics and feature based relevance feedback in image retrieval systems. In: Proceedings of the 8th ACM International Conference on Multimedia; 2000. p. 31–8.
7. Yang J, Zhuang YT, Li Q. Search for multi-modality data in digital libraries. In: Proceedings of the Second IEEE Pacific-Rim Conference on Multimedia; 2001.
8. Benitez AB, Smith JR, Chang SF. MediaNet: a multimedia information network for knowledge representation. In: Proceedings of the SPIE Conference on Internet Multimedia Management Systems; 2000. p. 1–12.
9. Tansley R. The multimedia thesaurus: an aid for multimedia information retrieval and navigation. Master thesis, Computer Science, University of Southampton; 1998.
10. Yang J, Li Q, Zhuang Y. Octopus: aggressive search of multi-modality data using multifaceted knowledge base. In: Proceedings of the 11th International World Wide Web Conference; 2002. p. 54–64.
11. Hofmann T, Buhmann JM. Pairwise data clustering by deterministic annealing. IEEE Trans Pattern Anal Mach Intell. 1997;19(1):1–14.
12. Li Q, Yang J, Zhuang YT. Web-based multimedia retrieval: balancing out between common knowledge and personalized views. In: Proceedings of the 2nd International Conference on Web Information Systems Eng; 2001.

Cross-Validation

Payam Refaeilzadeh[1], Lei Tang[2], and Huan Liu[3]
[1]Google Inc., Los Angeles, CA, USA
[2]Chief Data Scientist, Clari Inc., Sunnyvale, CA, USA
[3]Data Mining and Machine Learning Lab, School of Computing, Informatics, and Decision Systems Engineering, Arizona State University, Tempe, AZ, USA

Synonyms

Rotation estimation

Definition

Cross-validation is a statistical method of evaluating and comparing learning algorithms by dividing data into two segments: one used to learn or train a model and the other used to validate the model. In typical cross-validation, the training and validation sets must cross over in successive rounds such that each data point has a chance of being validated against. The basic form of cross-validation is k-fold cross-validation. Other forms of cross-validation are special cases of k-fold cross-validation or involve repeated rounds of k-fold cross-validation.

In k-fold cross-validation, the data is first partitioned into k equally (or nearly equally) sized segments or folds. Subsequently k iterations of training and validation are performed such that within each iteration a different fold of the data is held out for validation, while the remaining $k-1$ folds are used for learning. Figure 1 demonstrates an example with $k = 3$. The darker section of the data is used for training, while the lighter sections are used for validation. In data mining and machine learning, tenfold cross-validation ($k = 10$) is the most common.

Cross-validation is used to evaluate or compare learning algorithms as follows: in each iteration, one or more learning algorithms use $k-1$ folds of data to learn one or more models, and subsequently the learned models are asked to make predictions about the data in the validation fold. The performance of each learning algorithm on each fold can be tracked using some predetermined performance metric-like accuracy. Upon completion, k samples of the performance metric will be available for each algorithm. Different methodologies such as averaging can be used to obtain an aggregate measure from these samples, or these samples can be used in a statistical hypothesis test to show that one algorithm is superior to another.

Historical Background

In statistics or data mining, a typical task is to learn a model from available data. Such a model may be a regression model or a classifier. The problem with evaluating such a model is that it may demonstrate adequate prediction capability on the training data but might fail to predict future unseen data. Cross-validation is a procedure for estimating the generalization performance in this context. The idea for cross-validation originated in the 1930s [6]. In the paper, one sample is used for regression and a second for prediction. Mosteller and Tukey [9] and various other people further developed the idea. A clear statement of cross-validation, which is similar to current version of k-fold cross-validation, first appeared in [8]. In the 1970s, both Stone [12] and Geisser [4] employed cross-validation as a means for choosing proper model parameters, as opposed to using cross-validation purely for estimating model performance. Currently, cross-validation is widely accepted in data mining and machine

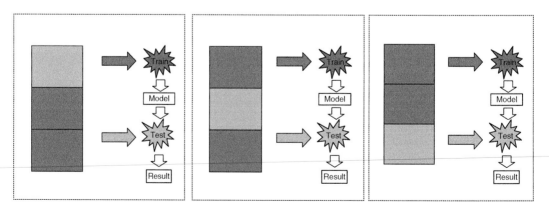

Cross-Validation, Fig. 1 Procedure of threefold cross-validation

learning community and serves as a standard procedure for performance estimation and model selection.

Foundations

There are two possible goals in cross-validation:

- To estimate performance of the learned model from available data using one algorithm. In other words, to gauge the generalizability of an algorithm
- To compare the performance of two or more different algorithms and find out the best algorithm for the available data or alternatively to compare the performance of two or more variants of a parameterized model

The above two goals are highly related, since the second goal is automatically achieved if one knows the accurate estimates of performance. Given a sample of N data instances and a learning algorithm A, the average cross-validated accuracy of A on these N instances may be taken as an estimate for the accuracy of A on unseen data when A is trained on all N instances. Alternatively if the end goal is to compare two learning algorithms, the performance samples obtained through cross-validation can be used to perform two-sample statistical hypothesis tests, comparing a pair of learning algorithms.

Concerning these two goals, various procedures are proposed:

Resubstitution Validation

In resubstitution validation, the model is learned from all the available data and then tested on the same set of data. This validation process uses all the available data but suffers seriously from overfitting. That is, the algorithm might perform well on the available data yet poorly on future unseen test data.

Hold-Out Validation

To avoid over-fitting, an independent test set is preferred. A natural approach is to split the available data into two non-overlapped parts: one for training and the other for testing. The test data is held out and not looked at during training. Hold-out validation avoids the overlap between training data and test data, yielding a more accurate estimate for the generalization performance of the algorithm. The downside is that this procedure does not use all the available data and the results are highly dependent on the choice for the training/test split. The instances chosen for inclusion in the test set may be too easy or too difficult to classify and this can skew the results. Furthermore, the data in the test set may be valuable for training, and if it is held-out prediction, performance may suffer, again leading to skewed results. These problems can be partially addressed by repeating hold-out validation multiple times and averaging the results, but unless this repetition is performed in a systematic manner, some data may be included in the test set multiple times while others are not included at all, or conversely some data may always fall in the test set and never get a chance to contribute to the learning phase. To deal with these challenges and utilize the available data to the max, k-fold cross-validation is used.

K-Fold Cross-Validation

In k-fold cross-validation, the data is first partitioned into k equally (or nearly equally) sized segments or folds. Subsequently k iterations of training and validation are performed such that within each iteration a different fold of the data is held out for validation, while the remaining $k-1$ folds are used for learning. Data is commonly stratified prior to being split into k folds. Stratification is the process of rearranging the data as to ensure each fold is a good representative of the whole. For example in a binary classification problem where each class comprises 50% of the data, it is best to arrange the data such that in every fold, each class comprises around half the instances.

Leave-One-Out Cross-Validation

Leave-one-out cross-validation (LOOCV) is a special case of k-fold cross-validation where k equals the number of instances in the data. In other words, in each iteration nearly all the data except for a single observation are used for training, and the model is tested on that single observation. An accuracy estimate obtained using LOOCV is known to be almost unbiased, but it has high variance, leading to unreliable estimates [3]. It is still widely used when the available data are very rare, especially in bioinformatics where only dozens of data samples are available.

Repeated K-Fold Cross-Validation
To obtain reliable performance estimation or comparison, a large number of estimates are always preferred. In k-fold cross-validation, only k estimates are obtained. A commonly used method to increase the number of estimates is to run k-fold cross-validation multiple times. The data is reshuffled and re-stratified before each round.

Pros and Cons
Kohavi [5] compared several approaches to estimate accuracy, cross-validation (including regular cross-validation, leave-one-out cross-validation, stratified cross-validation) and bootstrap (sample with replacement), and recommended *stratified tenfold cross-validation* as the best model selection method, as it tends to provide less biased estimation of the accuracy.

Salzberg [11] studies the issue of comparing two or more learning algorithms based on a performance metric and proposes using k-fold cross-validation followed by appropriate hypothesis test rather than directly comparing the average accuracy. Paired t-test is one test which takes into consideration the variance of training and test data and is widely used in machine learning. Dietterich [2] studied the properties of tenfold cross-validation followed by a paired t-test in detail and found that such a test suffers from higher than expected type I error. In this study, this high type I error was attributed to high variance. To correct for this, Dietterich proposed a new test: fivefold × twofold cross-validation. In this test twofold cross-validation is run five times resulting in ten accuracy values. The data is reshuffled and re-stratified after each round. All ten values are used for average accuracy estima-

tion in the t-test, but only values from one of the five twofold cross-validation rounds are used to estimate variance. In this study fivefold × twofold cross-validation is shown to have acceptable type I error but not to be as powerful as tenfold cross-validation and has not been widely accepted in data mining community.

Bouckaert [1] also studies the problem of inflated type I error with tenfold cross-validation and argues that since the samples are dependent (because the training sets overlap), the actual degrees of freedom are much lower than theoretically expected. This study compared a large number of hypothesis schemes and recommend tenfold × tenfold cross-validation to obtain 100 samples, followed with t-test with degree of freedom equal to 10 (instead of 99). However, this method has not been widely adopted in data mining field either, and tenfold cross-validation remains the most widely used validation procedure.

A brief summary of the above results is presented in Table 1.

Why Tenfold Cross-Validation: From Ideal to Reality
Whether estimating the performance of a learning algorithm or comparing two or more algorithms in terms of their ability to learn, an ideal or statistically sound experimental design must provide a *sufficiently large* number of *independent* measurements of the algorithm(s) performance.

To make independent measurements of an algorithm's performance, one must ensure that the factors affecting the measurement are independent from one run to the next. These factors are (i) the training data the algorithm learns from and (ii) the test data one uses to measure the algorithm's performance. If some data is used for testing in more than one round, the obtained results, for example, the accuracy measurements from these two rounds, will be dependent, and a statistical comparison may not be valid. In fact, it has been shown that a paired t-test based on taking several random train/test splits tends to have an extremely high probability of type I error and should never be used [2].

Cross-Validation, Table 1 Pros and cons of different validation methods

Validation method	Pros	Cons
Resubstitution validation	Simple	Over-fitting
Hold-out validation	Independent training and test	Reduced data for training and testing, large variance
k-fold cross-validation	Accurate performance estimation	Small samples of performance estimation, overlapped training data, elevated type I error for comparison, underestimated performance variance or overestimated degree of freedom for comparison
Leave-one-out cross-validation	Unbiased performance estimation	Very large variance
Repeated k-fold cross-validation	Large number of performance estimates	Overlapped training and test data between each round, underestimated performance variance or overestimated degree of freedom for comparison

Not only must the datasets be independently controlled across different runs, there must not be any overlap between the data used for learning and the data used for validation in the same run. Typically, a learning algorithm can make more accurate predictions on a data that it has seen during the learning phase than those it has not. For this reason, an overlap between the training and validation set can lead to an overestimation of the performance metric and is forbidden. To satisfy the other requirement, namely, a sufficiently large sample, most statisticians call for 30+ samples.

For a truly sound experimental design, one would have to split the available data into $30 \times 2 = 60$ partitions to perform 30 truly independent train/test runs. However, this is not practical because the performance of learning algorithms and their ranking is generally not invariant with respect to the number of samples available for learning. In other words, an estimate of accuracy in such a case would correspond to the accuracy of the learning algorithm when it learns from just 1/60 of the available data (assuming training and validation sets are of the same size). However, the accuracy of the learning algorithm on unseen data when the algorithm is trained on all the currently available data is likely much higher since learning algorithms generally improve in accuracy as more data becomes available for learning. Similarly, when comparing two algorithms A and B, even if A is discovered to be the superior algorithm when using 1/60 of the available data, there is no guarantee that it will also be the superior algorithm when using all the available data for learning. Many high-performing learning algorithms use complex models with many parameters, and they simply will not perform well with a very small amount of data. But they may be exceptional when sufficient data is available to learn from.

Recall that two factors affect the performance measure: the training set and the test set. The training set affects the measurement indirectly through the learning algorithm, whereas the composition of the test set has a direct impact on the performance measure. A reasonable experimental compromise may be to *allow for overlapping training sets*, *while keeping the test sets independent*. K-fold cross-validation does just that.

Now the issue becomes selecting an appropriate value for k. A large k is seemingly desirable, since with a larger k (i) there are more performance estimates and (ii) the training set size is closer to the full data size, thus increasing the possibility that any conclusion made about the learning algorithm(s) under test will generalize to the case where all the data is used to train the learning model. As k increases, however, the overlap between training sets also increases. For example, with fivefold cross-validation, each training set shares only 3/4 of its instances with each of the other four training sets, whereas with tenfold cross-validation, each training set shares 8/9 of its instances with each of the other nine training sets. Furthermore, increasing k shrinks the size of the test set, leading to less precise, less fine-grained measurements of the performance metric. For example, with a test set size of 10 instances, one can only measure accuracy to the nearest 10%, whereas with 20 instances the accuracy can be measured to the nearest 5%. These competing factors have all been considered, and the general consensus in the data mining community seems to be that $k = 10$ is a good compromise. This value of k is particularity attractive because it makes predictions using 90% of the data, making it more likely to be generalizable to the full data.

Key Applications

Cross-validation can be applied in three contexts: performance estimation, model selection, and tuning learning model parameters.

Performance Estimation

As previously mentioned, cross-validation can be used to estimate the performance of a learning algorithm. One may be interested in obtaining an estimate for any of the many performance indicators such as accuracy, precision, recall, or F-score. Cross-validation allows for all the data to be used in obtaining an estimate. Most commonly one wishes to estimate the accuracy of a classifier in a supervised-learning environment. In such a setting, a certain amount of labeled data is available, and one wishes to predict how well a certain classifier would perform if the available data is used to train the classifier and subsequently ask it to label unseen data. Using tenfold cross-validation one repeatedly uses 90% of the data to build a model and test its accuracy on the remaining 10%. The resulting average accuracy is likely somewhat of an underestimate for the true accuracy when the model is trained on all data and tested on unseen data, but in most cases this estimate is reliable, particularly if the amount of labeled data is sufficiently large and if the unseen data follows the same distribution as the labeled examples.

Model Selection

Alternatively cross-validation may be used to compare a pair of learning algorithms. This may be done in the case of newly developed learning algorithms, in which case the designer may wish to compare the performance of the classifier with some existing baseline classifier on some benchmark dataset, or it may be done in a generalized model selection setting. In generalized model selection, one has a large library of learning algorithms or classifiers to choose from and wish to select the model that will perform best for a particular dataset. In either case the basic unit of work is pair-wise comparison of learning algorithms. For generalized model selection, combining the results of many pair-wise comparisons to obtain a single *best* algorithm may be difficult, but this is beyond the scope of this entry. Researchers have shown that when comparing a pair of algorithms using cross-validation, it is best to employ proper two-sample hypothesis testing instead of directly comparing the average accuracies. Cross-validation yields k pairs of accuracy values for the two algorithms under test. It is possible to make a null hypothesis assumption that the two algorithms perform equally well and set out to gather evidence against this null hypothesis using a two-sample test. The most widely used test is the paired t-test. Alternatively the nonparametric sign test can be used.

A special case of model selection comes into play when dealing with non-classification model selection, for example, when trying to pick a feature selection [7] algorithm that will maximize a classifier's performance on a particular dataset. Refaeilzadeh et al. [10] explore this issue in detail and explain that there are in fact two variants of cross-validation in this case: performing feature selection before splitting data into folds (OUT) or performing feature selection k times inside the cross-validation loop (IN). The paper explains that there is potential for bias in both cases: With OUT, the feature selection algorithm has looked at the test set, so the accuracy estimate is likely inflated. On the other hand, with IN the feature selection algorithm is looking at less data than would be available in a real experimental setting, leading to underestimated accuracy. Experimental results confirm these hypotheses and further show that:

- In cases where the two feature selection algorithms are not statistically differentiable, IN tends to be more truthful.
- In cases where one algorithm is better than another, IN often favors one algorithm and OUT the other.

OUT can in fact be the better choice even if it demonstrates a larger bias than IN in estimating accuracy. In other words, *estimation bias is not necessarily an indication of poor pair-wise comparison*. These subtleties about the potential for bias and validity of conclusions obtained through cross-validation should always be kept in mind, particularly when the model selection task is a complicated one involving preprocessing as well as learning steps.

Tuning

Many classifiers are parameterized, and their parameters can be *tuned* to achieve the best result with a particular dataset. In most cases it is easy to learn the proper value for a parameter from the available data. Suppose a naïve Bayes classifier is being trained on a dataset with two classes: $\{+, -\}$. One of the parameters for this classifier is the prior probability $p(+)$. The best value for this parameter according to the available data can be obtained by simply counting the number of instances that are labeled positive and dividing this number by the total number of instances. However, in some cases parameters do not have such intrinsic meaning, and there is no good way to pick a best value other than trying out many values and picking the one that yields the highest performance. For example, support vector machines (SVMs) use soft margins to deal with noisy data. There is no easy way of learning the best value for the soft margin parameter for a particular dataset other than trying it out and seeing how it works. In such cases, cross-validation can be performed on the training data as to measure the performance with each value being tested. Alternatively a portion of the training set can

be reserved for this purpose and not used in the rest of the learning process. But if the amount of labeled data is limited, this can significantly degrade the performance of the learned model, and cross-validation may be the best option.

Cross-References

- ► Classification
- ► Evaluation Metrics for Structured Text Retrieval
- ► Feature Selection for Clustering

Recommended Reading

1. Bouckaert RR. Choosing between two learning algorithms based on calibrated tests. In: Proceedings of the 20th International Conference on Machine Learning; 2003. p. 51–8.
2. Dietterich TG. Approximate statistical tests for comparing supervised classification learning algorithms. Neural Comput. 1998;10(7):1895–923.
3. Efron B. Estimating the error rate of a prediction rule: improvement on cross-validation. J Am Stat Assoc. 1983;78(382):316–31.
4. Geisser S. The predictive sample reuse method with applications. J Am Stat Assoc. 1975;70(350):320–8.
5. Kohavi R. A study of cross-validation and bootstrap for accuracy estimation and model selection. In: Proceedings of the 14th International Joint Conference on AI; 1995. p. 1137–45.
6. Larson S. The shrinkage of the coefficient of multiple correlation. J Educat Psychol. 1931;22(1):45–55.
7. Liu H, Yu L. Toward integrating feature selection algorithms for classification and clustering. IEEE Trans Knowl Data Eng. 2005;17(4):491–502.
8. Mosteller F, Tukey JW. Data analysis, including statistics. In: Handbook of social psychology. Reading: Addison-Wesley; 1968.
9. Mosteller F, Wallace DL. Inference in an authorship problem. J Am Stat Assoc. 1963;58(302):275–309.
10. Refaeilzadeh P, Tang L, Liu H. On comparison of feature selection algorithms. In: Proceedings of AAAI-07 Workshop on Evaluation Methods in Machine Learning II; 2007. p. 34–9.
11. Salzberg S. On comparing classifiers: pitfalls to avoid and a recommended approach. Data Min Knowl Disc. 1997;1(3):317–28.
12. Stone M. Cross-validatory choice and assessment of statistical predictions. J Royal Stat Soc. 1974;36(2):111–47.
13. Tang L, Liu H. Community detection and mining in social media. Morgan & Claypool Publishers, San Rafael; 2010.

14. Zafarani R, Abbasi MA, Liu H. Social media mining: an introduction. Cambridge University Press, New York; 2014.

Crowd Database Operators

Beth Trushkowsky
Department of Computer Science, Harvey Mudd College, Claremont, CA, USA

Synonyms

Crowd-based operators; Crowd-powered operators

Definition

Crowd database operators are query plan operators in which all or part of the computation is done by humans, via crowdsourcing. They are alternate implementations of traditional relational operators, like sort or select, for use in hybrid human/machine query processing systems like crowd database systems. The use of crowdsourcing enables these systems to perform query operations that are well suited for people to compute, such as subjective comparisons, fuzzy matching for predicates and joins, entity resolution, etc., that leverage human perception, knowledge, and experience. The implementation of these operators typically includes user interfaces for collecting input from crowd workers, strategies to combine data received from multiple workers, as well as techniques to balance the cost of paying workers and the quality of the operator's output.

Historical Background

Crowdsourcing has emerged as a paradigm for leveraging human intelligence and activity at large scale. Platforms like Amazon's Mechanical Turk (AMT) crowdsourcing marketplace offer access to human workers via programmatic interfaces (APIs). The marketplace allows requesters to post small units of work, called micro-tasks, which workers complete for a small reward, typically a few cents each. The APIs provide an opportunity to create hybrid human/computer systems that can tackle problems the machine alone cannot. Members of the database systems community have explored the potential of such hybrid systems for database query processing [1–3]. In an operator-based relational query engine, crowd-based processing can be encapsulated into operators that can be used along with traditional machine-based operators in query plans.

Scientific Fundamentals

Crowd-based operators differ from traditional operators in several key ways. With people involved, new and revised algorithms are needed that consider human behavior and errors, as well as balance the cost versus quality trade-off that is inherent in paid crowdsourcing.

Filtering Operators

A crowd-based filter uses human computation to determine which of a set of items satisfy a given constraint, or predicate. It is a modification of the SELECT operator in a traditional query plan.

In the Qurk [2] system, users write user-defined functions (UDFs) to specify custom filters to apply on a set of items. These UDFs are automatically translated into a task that can be posted on Amazon's Mechanical Turk (AMT) crowdsourcing marketplace. Crowd workers are asked to indicate with a yes/no response if each item satisfies the predicate. By default, Qurk sends each item to be evaluated by five workers. These responses can be reconciled via custom "combiners" like the majority vote, which chooses the final answer as the response given by the majority of the workers.

Parameswaran et al. [4] look at incorporating a filter's selectivity, i.e., how likely a given item is to satisfy the filter, as well as workers' false positive and false negative error rates in a strategy for crowd-based filtering. Given these three statistics, the objective is to determine a strategy that minimizes the number of questions asked to the crowd while keeping the total error below a threshold. A *strategy* for filtering an item is defined as a sequence of yes/no responses from workers. At each point in this sequence, e.g., after one yes and two no responses, a strategy dictates whether it will continue gathering responses or return with a "pass" or "fail" indicating whether the item meets the given constraint. They propose algorithms for determining the approximate optimal deterministic strategy and the optimal probabilistic strategy.

The above definition of a strategy assumes that items are equally difficult to filter, that there is no prior knowledge about whether items will pass the filter, and that workers have homogeneous skills and are of equal cost. Parameswaran et al. [5] augment the representation of a strategy's state beyond a count of yes/no responses to include worker identity and allow each item to have its own selectivity. With these changes, a strategy becomes a Markov decision process (MDP), which can be solved via linear programming (LP). However, the LP solution, while optimal, can be intractable due to the amount of state maintenance required for workers and their abilities. To ameliorate this issue, they propose a "posterior-based approach," which describes a strategy's state as the probability of a "pass" given the cost incurred so far; this representation does not grow with the number of workers, providing tractability.

Das Sarma et al. [6] investigate the cost versus latency trade-off for filtering with the crowd. Given a large set of items, the goal is to find k items that satisfy a predicate. The lowest cost algorithm is sequential: send one item at a time to be evaluated by the crowd, waiting for the response between phases, and stop after receiving k correct items. However, the lowest latency algorithm has the crowd process the entire dataset in one parallel phase. Ideally, a hybrid algorithm would find the balance between these low-cost/high-latency and the high-cost/low-latency options. They devise several approaches for slightly "parallelizing" the sequential version to gain some of the benefit of the extreme parallel version; the user can customize the aggressiveness of the parallelization strategy.

They explore both a deterministic setting, in which humans do not make mistakes, as well as an uncertain setting in which each item may need multiple evaluations to deduce correctness.

Sorting, Top-k, and MAX Operators

Crowd-based sorting operators use human computation to order a given dataset by the value of one or more attributes, for example, sorting a set of images for those that best depict the Golden Gate Bridge. Related operators include finding the top-k items and the MAX (i.e., the first of the top items), from a set of ordered items sorted by the particular attribute(s). Work in this space looks into task design as well as modeling of the quality versus budget trade-off for these operators.

One strategy to achieve a crowd-based sort is to replace the comparator method used in a traditional sorting algorithm with a method that asks people to evaluate the comparison, in a *comparison task*, to see which is "better" with respect to some description. For example, CrowdDB's CrowdCompare operator [1] generates a comparison task in which workers compare two items. In general, a comparison task asks a worker to sort some number s of items.

For the sort operator for Qurk, Marcus et al. [7] investigate task interfaces for sorting a dataset with reasonable crowdsourcing cost. A full sort of N items based on comparison tasks, in which each task a [small] number of items are compared, would necessitate $O\left(\begin{array}{c} N \\ 2 \end{array}\right)$ comparisons because the transitive property may not hold in crowdsourced comparisons. In other words, there would be a cycle in the graph that has edges from item i to j when $i > j$. An alternate task interface in Qurk asks workers to rate individual items, rather than comparing them, using a 7-point Likert scale. Workers are also shown a small random sample of other items in the dataset, to give a sense of the total distribution. This approach requires only $O(N)$ crowdsourcing tasks; however, experimental results show that the ratings approach is less accurate than the comparison approach for items that are similar. They compare the comparison-only and ratings-only approach with an additional hybrid strategy that first uses the crowd for ratings and then iteratively performs crowd-based comparisons on select "windows" of items to improve quality while keeping to a specified budget. The hybrid strategy outperforms the other strategies because it fixes local inconsistencies in addition to having the power to shift items that are far away in the sort order from where they shouldbe.

To execute a crowd-powered top-k or MAX query, an algorithm could simply sort the whole dataset and return the top value(s). However, this approach may waste money performing crowd-based comparisons between items that may not be necessary. Naturally, there is a trade-off between the cost of performing more comparison tasks and achieving higher quality, i.e., approaching the true top-k that would result from executing all comparisons.

Algorithms based on tournament sort have been investigated in the literature for implementing crowd-based top-k [8–11]. In general, the sort algorithm operates as follows: the dataset is partitioned into disjoint subsets of size s, with $s > k$. The crowd completes comparison tasks that each contains s items; multiple worker responses per task are typically gathered for quality control purposes. After worker responses are combined (see below) for each of the subsets, the top-k items from each subset advance to the next round of the tournament, in which new subsets of size s are formed and compared by the crowd, and so on. Two questions arise in the implementation of this algorithm: (1) how many workers should be asked to complete a comparison task for a given subset of s items and (2) how should the results of multiple comparison tasks be combined?

Polychronopoulos et al. [8] address the second question using a technique called *median rank aggregation*, which assigns the rank for an item as the median of its position in the different rankings produced by the workers for the s items. Then, by determining if the median rank metric's uncertainty is beyond a threshold, their algorithm adaptively decides that more workers should evaluate a particular subset s of items.

Experimental results show the adaptive strategy outperforms a basic strategy that uses a fixed number of comparison tasks per item subset, because fewer (more) comparison tasks are used for dissimilar (similar) items. When compared to the comparison-only strategy from Marcus et al. [7] described above, their algorithm costs an order of magnitude less to achieve the same accuracy.

To address the first question above, Davidson et al. [9] represent worker error, the likelihood a worker will answer a comparison task incorrectly, with a variable error model in which the probability of error decreases the further apart items are in the true ordering. For example, a query might ask for the top-k most recent pictures of the same person; it is easier to detect that a picture of the person as an adult was taken more recently than a picture of that person as child, versus comparing pictures taken only a week apart. The error model can be learned with training examples with known true comparison values. The sorting algorithm is based on tournament sort, with the tournament tree divided into upper and lower levels.

Comparisons in the earlier, lower level are evaluated using one crowd comparison. Matchups in the upper level get N_L votes, with N_L based on the characteristics of the variable error model (i.e., exponential, linear, or logarithmic). Incorporating the variable error model, they provide formal bounds on the number of value questions needed to determine the correct top-k or MAX item(s) with high probability.

Guo et al. [10] look at assessing the state of a tournament sorting algorithm at a particular stage in its execution, specifically for determining the maximum item (evaluating MAX). The *judgment problem* is to determine which item has the greatest likelihood of being the maximum at the current stage. The *next votes problem* decides how to allocate more rounds of comparisons. In a particular round of a tournament-like sorting algorithm, workers may be extremely delayed in casting their votes, or in the worst case may not vote at all. However, if more votes are needed, the structure of the tournament (i.e., the scheduled matchups) can be rearranged – the challenge is constructing this rearrangement.

They propose maximum likelihood approaches to both problems and show that each is

NP-hard. They then develop and compare several heuristics to address them. A heuristic based on the PageRank algorithm proves effective for the judgment problem. For the next votes problem, two heuristics worked well.

The first is a "greedy" algorithm, which weighs all eligible item pairs by the product of the scores derived from the PageRank algorithm. A heuristic called "complete tournament" first executes a single tournament among the K highest scoring items. Then the $(K+1)$st item is paired with each of the first K and the highest scoring pairs according to the greedy algorithm are compared. The parameter K is chosen so both steps of the heuristic have sufficient crowdsourcing budget to complete.

Venetis et al. [11] present a generalization of the approaches to computing MAX with the crowd by describing a framework of families of algorithms; a family contains all possible parameterizations for a given algorithm. They describe a tournament-based algorithm using comparison tasks, as well as *bubble max*. The bubble max algorithm compares a pivot item with a set of random items not yet processed. The winner of this round becomes the new pivot, and the process continues until all items have been processed.

Parameters include r_i and s_i, the number of human responses and number of items in a comparison task, respectively, for the ith iteration of an algorithm. The goal is to determine the optimal parameterization for an algorithm that maximizes quality, subject to cost and time (i.e., rounds of crowdsourcing). However, the optimal settings of r_i and s_i are heavily influenced by the worker error model and quality/budget constraints of a particular scenario. They describe a number of strategies for determining the optimal parameter settings. Simulation results show that the best strategy first finds the optimal r and s assuming each remains constant across iterations of the algorithm and then performs a hill-climbing search technique to determine if "stealing" units of r and/or s from certain rounds and giving them to others yield quality improvements. This strategy outperforms alternate strategies for both tournament-based and bubble max algorithms.

Tournament is also superior to bubble max given the same budget.

Group-by Operators

The group-by operator effectively partitions, or clusters, a relation based on the value of one or more attributes. Davidson et al. [9], in addition to top-k, investigate using the crowd to cluster a given set of items into distinct types, where the number of types is assumed to be fixed but is unknown. They used "type" questions: a worker is shown two items and asked if they match for a particular type, e.g., do two photographs depict the same person. Given that each question asked incurs crowdsourcing costs, the goal is to deduce how many questions will be needed for a particular dataset. Workers are assumed to have a *constant* error model, i.e., they will answer questions correctly with probability $> 1/2 + \varepsilon$. They provide bounds on the number of type questions needed to achieve correct clustering with high probability.

Gomes et al. [12] look at a slightly different goal: can the crowd be used to discover categories, or clusters, in a large dataset? Workers are shown a small subset of the data and asked to label, via color coding, which items belong in the same category. The objective is to form a global categorization based on these individual clusterings. While the subsets given to workers overlap to facilitate aggregation, workers may disagree on a clustering or number of clusters even given the same subset. They propose a Bayesian model for aggregating the clusterings in which workers are pairwise binary classifiers in the space of items (pairwise item assessments can be derived from the workers' clusterings). Experimental results show superior performance over two existing clustering aggregation methods.

Join Operators

There are several ways described in the literature in which the crowd can be involved in a crowd-based join. In CrowdDB [1] the CrowdJoin operator performs an index nested-loop join between two relations, at least one of which must be a CROWD table (i.e., its tuples are provided by the crowd). For each tuple of the outer relation, new tuples for the inner relation are gathered from crowd workers; the task UI is prepopulated with the join column values from the outer relation.

The crowd can also help evaluate the join predicate(s) between relations. In Qurk [7], workers are tasked with deciding if tuples from two relations match with respect to one or more attributes. A naive implementation of this join would attempt to ask the crowd about every possible pairing of tuples between the relations, i.e., the Cartesian product. To potentially reduce the number of tuples participating in the join, Qurk allows users to specify additional predicates that must be true for each relation if the join predicate is true; evaluating these predicates first has a cost only linear in the size of the relations. The additional predicates are also UDFs and are specified in the query using the keyword POSSIBLY. Depending on how costly each predicate is to evaluate, the optimizer may decide to forego one or more of them (hence, the predicates are each possibly evaluated). The particular join implementation is a block nested-loop join: partitions of one relation are iteratively compared for matches with tuples from the other relation.

Three task UIs are compared and evaluated for the crowd-based join. The first is the simple single page task that presents the worker with one item from each relation and asks if they have the same value for the given predicate. The second, "naive batching" places multiple of these pairs in the same task.

The "smart batching" interface presents two columns of items and asks the worker to indicate all pairs that match. Experimental results show the accuracy among the different UIs was roughly equivalent, highlighting the potential of cost savings due to batching. The highest accuracy was attained with the smart batching UI using two items from each relation, coupled with the quality control mechanisms that model worker error and bias described in [13].

Scan Operators

The operators discussed thus far assume there is a set of data items that crowd workers are processing. However, the crowd can also be helpful for assembling the dataset itself. In a relational query plan, this data collection would be akin to the physical SCAN operator that retrieves content from a relational table. Rather than reading tuples from disk storage, a crowd-based SCAN gathers tuples using the crowd. The results of the SCAN can be persisted to disk for future use. Existing hybrid database systems support the SCAN operator [1–3]. It is triggered at query time, typically by a query that asks for results from a table that does not yet have any, or enough, tuples to satisfy the request.

Trushkowsky et al. [14] investigate reasoning about the quality of results for queries that involve a crowd-based SCAN, called crowdsourced enumeration queries. In enumeration queries, the query result can keep growing with more responses from the crowd; this raises a fundamental question: when is the result set complete? Adapting algorithms from literature on *species estimation*, they provide a technique to estimate the expected size of the result set, and thus query progress, that uses the frequency counts of each item in the set of workers' responses received thus far. The estimation algorithm incorporates knowledge about human behaviors, particularly compensating for workers who provide a disproportionately large number of responses, called "streakers." Experimental results show that the algorithm provides reasonable estimates of the result set size for crowdsourced enumeration queries.

Key Applications

Crowd Database Systems

Crowd database operators can be used in place of traditional relational operators as part of a query processor in existing or new hybrid human/machine database systems.

Users and Developers of Crowdsourcing Applications

Many of the techniques and algorithms used to implement crowd database operators can also be used on their own by users who want to process a dataset in a particular manner, or developers can incorporate them into other software systems.

Cross-References

▶ Cost and Quality Trade-Offs in Crowdsourcing
▶ Crowd Database Operators
▶ Crowd Database Systems
▶ Human Factors Modeling in Crowdsourcing

Recommended Reading

1. Franklin MJ, Kossmann D, Kraska T, Ramesh S, Xin R. CrowdDB: answering queries with crowdsourcing. In: Proceedings of the ACM SIGMOD International Conference on Management of Data; 2011.
2. Marcus A, Wu E, Madden S, Miller R. Crowdsourced databases: query processing with people. In: Proceedings of the 5th Biennial Conference on Innovative Data Systems Research; 2011.
3. Parameswaran AG, Park H, Garcia-Molina H, Polyzotis N, Widom J. Deco: declarative crowdsourcing. In: Proceedings of the 21st ACM International Conference on Information and Knowledge Management; 2012.
4. Parameswaran AG, Garcia-Molina H, Park H, Polyzotis N, Ramesh A, Widom J. Crowdscreen: algorithms for filtering data with humans. In: Proceedings of the ACM SIGMOD International Conference on Management of Data; 2012.
5. Parameswaran A, Boyd S, Garcia-Molina H, Gupta A, Polyzotis N, Widom J. Optimal crowdpowered rating and filtering algorithms. Proc VLDB Endowment. 2014;7(9):685–696.
6. Das Sarma A, Parameswaran A, Garcia-Molina H, Halevy A. Crowd-powered find algorithms. In: Proceeings of the IEEE International Conference on Data Engineering.
7. Marcus A, Wu E, Karger D, Madden S, Miller R. Human-powered sorts and joins. Proc VLDB Endowment 2011;5(1):13–24.
8. Polychronopoulos V, de Alfaro L, Davis J, Garcia-Molina H, Polyzotis N. Human – powered top-k lists. In: Proceedings of the 11th International Workshop on the World Wide Web and Databases; 2013.
9. Davidson SB, Khanna S, Milo T, Roy S. Using the crowd for top-k and group-by queries. In: Proceedings of the 15th International Conference on Database Theory; 2013.
10. Guo S, Parameswaran A, Garcia-Molina H. So who won? Dynamic max discovery with the crowd. In: Proceedings of the ACM SIGMOD International Conference on Management of Data; 2012.
11. Venetis P, Garcia-Molina H, Huang K, Polyzotis N. Max algorithms in crowdsourcing environments. In: Proceedings of the 21st international conference on World Wide Web; 2012.
12. Gomes R, Welinder P, Krause A, Perona P. Crowdclustering. In: Advances in Neural Information Proceedings of the Systems 24, Proceedings of the 25th Annual Conference on Neural Information Proceedings of the Systems; 2011.
13. Ipeirotis PG, Provost F, Wang J. Quality management on Amazon mechanical turk. In: Proceedings of the ACM SIGKDD Workshop on Human Computation, 2010.
14. Trushkowsky B, Kraska T, Franklin MJ, Sarkar Purnamrita. Crowdsourced enumeration queries. In: Proceedings of the 29th International Conference on Data Engineering; 2013.

Crowd Database Systems

Ju Fan[1], Meihui Zhang[2], and Beng Chin Ooi[3]
[1]DEKE Lab and School of Information, Renmin University of China, Beijing, China
[2]Information Systems Technology and Design, Singapore University of Technology and Design, Singapore, Singapore
[3]School of Computing, National University of Singapore, Singapore, Singapore

Synonyms

Crowd-powered database systems; Crowdsourcing data analytics systems; Declarative crowdsourcing systems; Human-powered database systems

Definition

Crowdsourcing database systems are designed to add crowd functionality into traditional database management systems (DBMSs) for processing queries that cannot be answered by machines only. The systems typically take declarative queries written in SQL-like query language as input and process over stored relational data together with the collective knowledge gathered on-demand from the crowd. A typical crowdsourcing database system includes a query parser, which compiles the input query; a query optimizer, which generates the optimized query plan; an executor, which manages the query execution; and an HIT manager, which interacts with the public crowd.

Historical Background

While relational database system offers a powerful tool for data management, it imposes limitations in some situations. One situation is when there is missing information in the stored data. Querying the missing data will certainly return empty or incomplete answer. People, on the other hand, are good at seeking information especially when they are provided with tools such as search engines. People are also far more professional at making subjective comparisons or performing computationally difficult tasks comparing to computers. For instance, it is relatively easy for a person to provide information like "the head-quarter of Google" or find images of the same person. Crowdsourcing has been widely used since microtask crowdsourcing platforms such as Amazon Mechanical Turk (AMT) were launched publicly. It is proved to be an effective tool for harnessing human intelligence to solve problems that computers cannot perform well. It is then natural to explore the possibility of integrating the power of the crowd to extend the capabilities of relational databases to answer queries that need human input. To this end, crowdsourcing database systems have been designed and built. The primary systems developed by different research groups include CrowdDB [1, 2] from UC Berkeley and ETH Zurich, Qurk [3–5] from MIT, Deco [6, 7] from Stanford and UCSC, and CDAS [8–11, 15, 16] from NUS. By integrating the crowdsourcing with databases, these systems can systematically create/optimize the crowdsourcing tasks and manage the data sourced from the crowd.

Scientific Fundamentals

Architecture of Crowdsourcing Database Systems

A typical crowdsourcing system specifies how data are stored and how queries are processed. The system usually includes the following components, as shown in Fig. 1:

- *Crowdsourced data* is represented and stored in the databases based on a well-defined data model. In particular, the model specifies which part of the data needs to be crowdsourced.
- *Crowdsourcing queries* are formulated in a specified query language to capture users' intent. The language allows users to indicate which data to crowdsource and provide crowdsourcing instructions.
- The *query processing* module specifies how queries are processed. The basic workflow of query processing consists of query plan generation and execution, at large following a traditional approach. In particular, a parser is first applied to parse a query into an initial logical plan, which is then optimized by an optimizer. Next, a query executor runs the optimized plan by generating human intelligence tasks (HITs). The HIT manager instantiates user interfaces of HITs by utilizing UI templates and some standard HIT parameters (e.g., number of assignments, price) and publishes the HITs on crowdsourcing platforms to collect results from the crowd. Finally, the query processing module finishes execution and outputs the results to answer the query.

Next, four crowdsourcing systems in line with the above architecture, namely Qurk, CrowdDB, Deco, and CDAS, will be introduced in chronological order of their publication.

Data Model and Query Language

The basic data model in crowdsourcing database systems is *relational model*. This enables developers or users to build their applications and

Crowd Database Systems, Fig. 1 Architecture of crowdsourcing database systems

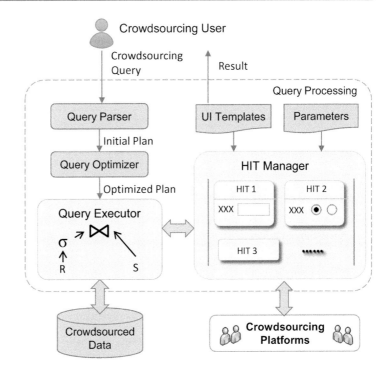

query the data in the traditional way. To encapsulate the complexities of dealing with the crowd and provide an interface that is familiar to most database users, crowdsourcing database systems adopt an SQL-like declarative query interface. However, to support the need of interacting with the crowd, existing systems have designed their own mechanisms.

Qurk [3–5] uses an SQL-based query language with user-defined functions (UDFs) to encapsulate the input from the crowd. Instead of requiring the users to implement the UDFs in raw HTML format, Qurk has predefined several task templates that can generate the UIs for posting different kinds of tasks to the crowd. Typical crowdsourcing tasks include:

1. Filter: produces tuples that satisfy the conditions specified in the UDF. The UI template is yes/no question.
2. Sort: ranks the input tuples according to the UDFs specified in the ORDER BY clause.
3. Join: compares input tuples and performs join according to the UDF. Qurk extends the traditional join syntax with a POSSIBLY keyword that indicates the features that may help filter the join.
4. Generative: allows workers to generate data for multiple fields.

An example of "finding black cars" is shown below. The Combiner field specifies how to resolve the inconsistencies in the crowd responses.

```
SELECT i.image
FROM car_image i
WHERE isBlack(i)
TASK isBlack(field) TYPE Filter:
Prompt: "<table><tr> \
<td><img src='%s'></td> \
<td>Is the car in the image in
      black color?</td> \
</tr></table>", tuple[field]
YesText: "Yes"
NoText: "No"
Combiner: MajorityVote
```

CrowdDB [1, 2] has designed CrowdSQL, another form of SQL extension, to support crowdsourcing missing data and performing subjective comparisons. There are two possible cases of missing data: (1) entities are known, but certain properties of the entities need to be crowdsourced; (2) the entities are unknown, and hence the entire tuples need to be crowdsourced. CrowdDB captures both cases by introducing a new keyword CROWD.

1. Crowdsourced column:

   ```
   CREATE TABLE car_review (
   review STRING,
   make CROWD STRING,
   model CROWD STRING,
   sentiment CROWD STRING);
   ```

2. Crowdsourced table:

   ```
   CREATE CROWD TABLE car (
   make STRING,
   model STRING,
   color STRING,
   style STRING,
   PRIMARY KEY (make, model));
   ```

To support the functionality of performing subjective comparison, CrowdDB designs two new built-in functions, CROWDEQUAL and CROWDORDER.

1. CROWDEQUAL: $\sim=$ symbol

   ```
   SELECT review FROM car_review
   WHERE sentiment ~= "pos";
   ```

2. CROWDORDER:

   ```
   SELECT image i FROM car_image
   WHERE subject = "Volvo S60"
   ORDER BY CROWDORDER(i, "Which image
       visualizes better %subject");
   ```

Deco [6, 7] has clear separation between the user and system view. *Conceptual relations* are the logical relations specified by the schema designer and queried by the end user. *Raw schema* is the schema for data table actually stored in the underlying RDBMS and is invisible to the schema designer and users. Deco designs *fetch rules* that allow the schema designers to specify how data are collected from the crowd. A fetch rule takes the following form $A_1 \Rightarrow A_2 : P$, where A_1 is a set of attributes used as fetch conditions, A_2 are a set of attributes to be fetched, and P is a fetch procedure that implements the generation of crowdsourcing tasks. Examples:

```
image ⟹ color: Ask for the color
    given an image of a car
∅ ⇒ model: Ask for a new car model
```

The schema designers can also specify *resolution rules* such as de-duplication and majority voting to resolve the inconsistencies in the collected data. Examples:

```
image ⟹ color: majority-of-3
∅ ⇒ model: dupElim
```

To avoid the empty-answer syndrome, Deco extends SQL query language with an AtLeast n clause that requires a minimum number of tuples in the query result.

CDAS [8–11, 15, 16] employs relational model. Different from traditional databases, some attributes of tuples in CDAS databases are unknown before executing crowdsourcing. These attributes are specified by the schema designer a priori and typically are those that can be easily recognized by humans, e.g., review.sentiment and image.quality. A CDAS query is an SQL query over the designated relations. Currently, three types of queries are supported in CDAS.

1. Selection query: applies one or more human-recognized selection conditions over the tuples in a single relation. An example of "finding high-quality images of black Volvo cars" is shown below.

   ```
   SELECT i.image
   FROM car_image i
   WHERE make = "Volvo" AND color
       = "black" AND quality = "high"
   ```

2. Join query: leverages human intelligence to combine tuples from two or more relations according to certain join conditions. An example of "linking cars with the corresponding images" is shown below. (CDAS allows users to specify additional criteria by extending the standard SQL with a JoinFilter keyword, which may help filter possible join candidates. For instance, c.style = i.style as a join-filter in the example indicates that any image-car pairs with different values on style will not be in the join result.)

```
SELECT c.*, i.image
FROM car c, car_image i
WHERE c.make = i.make AND
      c.model = i.model
JoinFilter c.style = i.style
```

3. Complex query: supports more general queries that contain both selections and joins and expresses more complex crowdsourcing intent. An example of "finding cars with black color and positive reviews" is shown below.

```
SELECT c.*, i.image, r.review
FROM car c,car_image i,car_review r
WHERE r.sentiment = "pos" AND
      i.color = "black"
AND c.make = i.make AND
    c.model = i.model
AND c.make = r.make AND
    c.model = r.model
```

Query Processing and Optimization

The performance of query processing in crowdsourcing database systems is typically measured by the following three metrics:

- *Monetary cost*: A better query processing scheme is expected to invoke lower monetary cost. Typically, the monetary cost depends on two factors: price of each HIT and number of HITs.
- *Quality*: Crowdsourcing may yield relatively low-quality results or even noise, if there are spammers or malicious workers. To improve the quality, existing crowdsourcing database systems exploit a duplicate-based strategy that assigns a HIT to multiple workers and combines their answers via majority voting or other sophisticated algorithms [9, 10].
- *Latency*: As the crowd takes time to complete HITs, latency is used to quantify the speed of query execution. Query processing with lower latency is expected in crowdsourcing database systems.

Crowdsourcing database systems implement all operators of relational algebra, such as join, index scans, etc. Moreover, to facilitate the interaction with the crowd, the systems also implement *crowd operators* that encapsulate operations from the crowd.

The basic functionality of all crowd operators can be described as follows. These operators are initialized with predefined UI templates and some standard HIT parameters used for crowdsourcing. At runtime, a crowd operator consumes a set of tuples and instantiates crowdsourcing tasks by filling values of the tuples into the corresponding UI templates. Depending on the type of the crowd operator, the generated tasks can be used to crowdsource new data or conduct subjective comparison. Besides, the crowd operator can batch several tuples into a single HIT. In addition to the creation of HITs, the crowd operator also processes the results from the crowd and carries on quality control. To implement the functionality of crowd operators, existing systems have introduced their own schemes.

Qurk [3–5] focuses on implementing join and sort in its query model.

- Implementation of join: Qurk implements a block nested loop join and crowdsources the tuples from two tables for evaluating if they satisfy join conditions. In particular, Qurk studies the techniques for batching multiple comparisons into one HIT. Figure 2 shows three batching strategies proposed in Qurk. Simple join only puts one comparison into an HIT, while naïve batching displays multiple comparisons. In smart batching, two columns of tuples (images) are displayed, and the workers are asked to link pairs of them that satisfy join conditions.
- Implementation of sort: Qurk implements two basic approaches to execute sort. The comparison-based approach solicits the crowd to directly specify the ordering of items. This approach may be expensive for large datasets due to the quadratic comparison. Another cheaper approach is rating based, which asks the crowd to rate each item along a numerical scale. The drawback of this approach is that ratings may not be fully consistent with the ordering that would result if workers directly compare the pair.

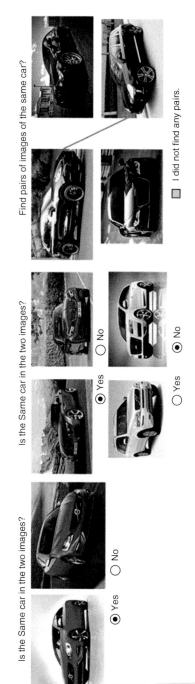

Crowd Database Systems, Fig. 2 Implementation of join operator in Qurk

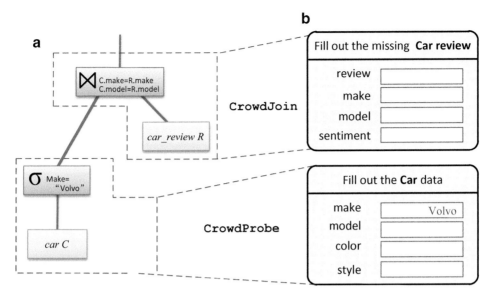

Crowd Database Systems, Fig. 3 An example of query processing in CrowdDB. (**a**) Logical plan. (**b**) Physical plan

CrowdDB [1, 2] has introduced the following three crowd operators:

- CrowdProbe: This operator is designed to collect from the crowd missing information of CROWD columns or new tuples. The typical user interface of CrowdProbe is a form with several fields for collecting information from the crowd.
- CrowdJoin: This operator implements an index nested loop join over two tables, at least one of which is crowdsourced. In particular, the inner relation in the join must be a CROWD table, and the user interface is used to crowd-source new tuples of inner relation which can be joined with the tuples in outer relation.
- CrowdCompare: This operator is designed to implement the two functions, CROWDEQUAL and CROWDORDER defined in CrowdDB's query model. The interface of the operator crowdsources two tuples and leverages the crowd to compare these tuples.

Figure 3 illustrates how CrowdDB answers query "finding information and reviews of 'Volvo' cars." To execute the left logical plan, CrowdDB utilizes a CrowdProbe operator to crowdsource missing attributes of Volvo cars, such as color and style. Next, it employs a CrowdJoin to crowdsource the "car_review" tuple that can be joined with the obtained "car" tuple.

Deco [6, 7] focuses on crowdsourcing missing values of tuples or new tuples based on the defined fetch rules. Given a fetch rule with the form $A_1 \Rightarrow A_2 : P$, Deco presents the values of attributes in A_1 and asks the crowd to fill out the values of attributes in A_2. For instance, to execute "make, model $\Rightarrow$ style," Deco presents cars and collects their styles from the crowd. In particular, if A_1 is empty, the system fetches new values of attributes in A_2. For instance, "$\varnothing \Rightarrow$ make, model" asks for a new car.

As inconsistency may exist in the collected data, Deco applies certain resolve rules on top of the fetched values. Besides, operators in relational algebra, such as Dependent Left Outerjoin (DLOJoin), Filter, and Scan, are also supported in Deco. Figure 4 provides an example in Deco for processing the query "collecting style and color of at least eight cars with 'Volvo' as their manufacturer." This plan fetches cars (operator 7), colors of cars (operator 10), and styles of cars (operator 13) separately. The fetched data

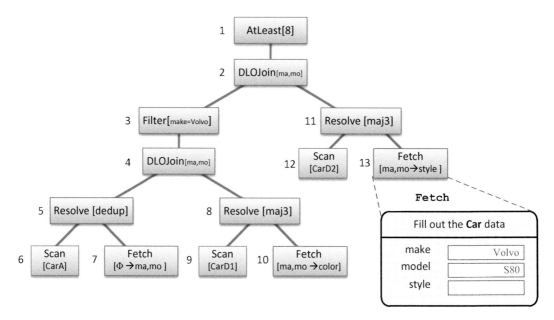

Crowd Database Systems, Fig. 4 An example query plan in deco system

may be resolved with different resolve rules, such as de-duplication in operator 5 and majority voting in operator 11. The collected data are then integrated by DLOJoins and filtered by condition "make='Volvo'."

CDAS [8–11, 15, 16] has introduced three crowd operators:

– CrowdSelect (CSelect): This operator abstracts the human operation of filtering tuples satisfying certain conditions. A typical user interface of CrowdSelect presents a tuple (e.g., an image) and several selection conditions and collects yes/no answers from the crowd.
– CrowdJoin (CJoin): This operator solicits crowdsourcing to directly compare tuples to be joined. The user interface of CrowdJoin presents two tuples, each from a relation, and Yes/No options for the crowd to choose.
– CrowdFill (CFill): This operator crowdsources missing attributes in tuples. To this end, the crowd can either choose a value from attribute's value domain or fill a new value if no value in the domain fits the tuple.

Figure 5 illustrates how CDAS processes a complex query of "finding black cars with high-quality images and positive reviews" and shows the interfaces of the three operators. CDAS generates an execution plan consisting of crowd operators described as above. In this plan, CrowdSelect operators are applied to crowdsource selections over images and reviews. Then, CrowdJoin operators, in collaboration with CrowdFill operators, are used to implement the joins among images: CrowdFill operators are used to fill tuple values and filter tuples with different values, while CrowdJoin operators directly compare tuples.

Moreover, CDAS develops effective quality control techniques [9, 10] for each crowd operator to produce reliable results even if the crowdsourcing workers have errors.

Key Applications

Crowdsourcing database systems have broad applications in acquiring and managing crowd intelligence, including knowledge acquisition [12], data integration [8, 13], and solving a wide variety of problems over Web data [14].

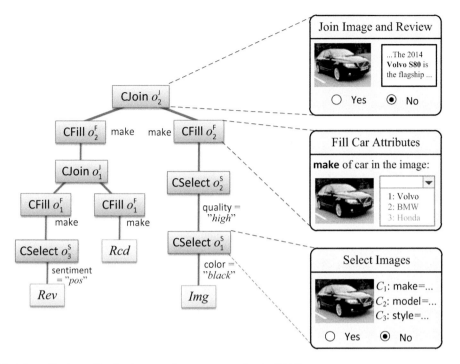

Crowd Database Systems, Fig. 5 An example query processing in CDAS

Cross-References

▸ Cost and Quality Trade-Offs in Crowdsourcing
▸ Crowd Database Operators

Recommended Reading

1. Feng A, Franklin MJ, Kossmann D, Kraska T, Madden S, Ramesh S, Wang A, Xin R. CrowdDB: query processing with the VLDB crowd. Proc VLDB Endowment. 2011;4(12):1387–90.
2. Franklin MJ, Kossmann D, Kraska T, Ramesh S, Xin R. CrowdDB: answering queries with crowdsourcing. In: Proceedings of the SIGMOD Conference; 2011. p. 61–72.
3. Marcus A, Wu E, Karger DR, Madden S, Miller RC. Demonstration of Qurk: a query processor for humanoperators. In: Proceedings of the SIGMOD Conference; 2011. p. 1315–8.
4. Marcus A, Wu E, Madden S, Miller RC. Crowdsourced databases: query processing with people. In: Proceedings of the 5th Biennial Conference on Innovative Data Systems Research; 2011. p. 211–4.
5. Marcus A, Wu E, Karger DR, Madden S, Miller RC. Human-powered sorts and joins. Proc VLDB Endowment. 2011;5(1):13–24.
6. Parameswaran AG, Park H, Garcia-Molina H, Polyzotis N, Widom J. Deco: declarative crowdsourcing.

In: Proceedings of the 21st ACM International Conference on Information and Knowledge Management; 2012. p. 1203–12.
7. Park H, Widom J. Query optimization over crowdsourced data. Proc VLDB Endowment. 2013;6(10):781–92.
8. Fan J, Lu M, Ooi BC, Tan W-C, Zhang M. A hybrid machine-crowdsourcing system for matching web tables. In: Proceedings of the 30th International Conference on Data Engineering; 2014. p. 976–87.
9. Gao J, Liu X, Ooi BC, Wang H, Chen G. An online cost sensitive decision-making method in crowdsourcing systems. In: Proceedings of the ACM SIGMOD International Conference on Management of Data; 2013. p. 217–28.
10. Liu X, Lu M, Ooi BC, Shen Y, Wu S, Zhang M. CDAS: a crowdsourcing data analytics system. Proc VLDB Endowment. 2012;5(10):1040–51.
11. CDAS: https://www.comp.nus.edu.sg/~cdas
12. Kumar KS, Triantafillou P, Weikum G. Combining information extraction and human computing for crowdsourced knowledge acquisition. In: Proceedings of the 30th International Conference on Data Engineering; 2014. p. 988–99.
13. Wang J, Kraska T, Franklin MJ, Feng J. CrowdER: crowdsourcing entity resolution. Proc VLDB Endowment. 2012;5(11):1483–94.
14. Doan AH, Ramakrishnan R, Halevy AY. Crowdsourcing systems on the World-Wide Web. Commun ACM. 2011;54(4):86–96.

15. Fan J, Li G, Ooi BC, Tan KL, Feng J. iCrowd: an adaptive crowdsourcing framework. In: Proceedings of the ACM SIGMOD International Conference on Management of Data; 2015. p. 1015–30.
16. Fan J, Zhang M, Kok S, Lu M, Ooi BC. CrowdOp: query optimization for declarative crowdsourcing systems. IEEE Trans Knowl Data Eng. 2015;27(8):2078–92.

Crowd Mining and Analysis

Yael Amsterdamer[1] and Tova Milo[2]
[1]Department of Computer Science, Bar Ilan University, Ramat Gan, Israel
[2]School of Computer Science, Tel Aviv University, Tel Aviv, Israel

Synonyms

Crowd data mining, Crowd mining

Definition

Crowd mining is the process of identifying data patterns in human knowledge, with the assistance of a crowd of web users. The focus is on domain areas in which data is partially or entirely undocumented and where humans are the main source of knowledge, such as data that involves people's habits, experiences, and opinions. A key challenge in mining such data is that the human knowledge forms an open world and it is thus difficult to know what kind of information one should be looking for.

In classic databases, a similar problem is addressed by data mining techniques that identify interesting patterns in recorded data such as relational databases or textual documents. These techniques, however, are not suitable for the crowd. This is mainly due to properties of the human memory, such as the tendency to remember simple trends and summaries rather than exact details, which should be taken into consideration when gathering and analyzing human knowledge.

In order to harvest knowledge from humans, crowd mining processes generate questions, pose them to the crowd via crowdsourcing platforms, and aggregate and analyze the obtained answers. Question generation is guided by considerations such as effective exploitation of the crowd capabilities, human effort minimization and computational efficiency, data dependencies, and uncertainty minimization. Related issues that arise due to the crowd involvement include the development of natural language interfaces, dealing with user sessions and multiple workers in parallel, spam detection and quality assurance, and the management of contributor recruitment and rewarding.

Historical Background

Crowdsourcing platforms and frameworks that involve crowdsourcing have become a widespread phenomenon in the last decade. These include some well-known examples such as Wikipedia, the Free Encyclopedia, recommendation and reviewing websites, and general-purpose crowdsourcing platforms such as Amazon Mechanical Turk and CrowdFlower. The main reason for this emergence was the sharp increase in the number of internet users and the time they spend online. It has been observed that this crowd of web users outperforms computer processors in certain types of tasks, e.g., natural language processing, and that engaging the crowd is often more cost-efficient than hiring experts.

In recent years, the success and abundance of crowdsourcing frameworks have gained the attention of the academic community. Many research studies have sought sound scientific foundations and a principled approach toward crowdsourcing, in contrast to the ad hoc solutions that were implemented by different crowdsourcing frameworks. In particular, one line of work focused on *crowd data sourcing*, namely, the involvement of the crowd in performing data management tasks such as data procurement, data processing, and data cleaning (see, for example, [1]). This body of work has set the foundations of automatically producing tasks for the crowd as a part of a complex computational process. For example, in [2–7], the crowd is engaged to evaluate operators in a database query. Crowd mining may be viewed as a branch of these approaches, which on the one hand deals with similar crowd-related aspects such as cost and uncertainty and on the other hand brings data procurement to a new level, by allowing to identify important or interesting data patterns in unrecorded knowledge domains.

The first research paper about crowd mining [8] presents a formal model and a generic framework for mining the crowd. Then, it suggests one possible implementation for this framework, *CrowdMiner*, which focuses on minimizing the uncertainty about the significance of the discovered patterns, as well as the exploration of the data domain. The work of [9] has further studied the impact of concept taxonomies and semantic data dependencies on the complexity of crowd mining algorithms. In [10], a new query language, *OASSIS-QL*, was developed, for declaratively specifying data patterns of interest, to be mined from the crowd. This paper also dealt with practical aspects of implementing an evaluation engine for *OASSIS-QL*, such as the order in which questions are posed to a crowd worker, and handling multiple workers in parallel.

Scientific Fundamentals

Every process of crowd mining includes a few key components, which are outlined below. The components are generic and leave room for particular modeling and implementation choices, depending on the mining goals, the desired type of data patterns, prior knowledge about the data, etc.

A Formal Model for the Crowd

Crowd mining is a computational process that involves the crowd as a principal data source. In order to successfully interact with the crowd, the process requires a formal model to capture the capabilities and limitations of human workers.

The model should on the one hand be generic enough to allow exploring unknown domains, and on the other hand, it should reasonably reflect the behavior of the crowd, to allow accurate performance predictions and the overall success of the process. The model can typically be divided into (i) a model of the crowd knowledge and (ii) of the access methods to this knowledge, which are targeted toward the identification of particular types of data patterns.

The crowd knowledge can be viewed as a *virtual distributed data repository*: each person is associated a data set which reflects his or her personal knowledge, but often there is no full, direct access to this data. Consider the *personal history* of a crowd member. This history of a crowd member u is modeled in crowd mining as a bag (multiset) D_u of *transactions*, where each transaction $t \in D_u$ contains data about a particular occasion in this crowd member's past. As a simple example, if we wish to analyze the culinary habits of a person, each transaction $t \in D_u$ in our model may correspond to a particular meal and may contain data items $i_1, \ldots, i_{n_t}$ corresponding to the food items consumed by u in this meal. Alternatively, each transaction may contain *facts* about the occurrences in a particular occasion, using some standard representation, e.g., OWL, the W3C standard.

People typically cannot recall all the details about their knowledge or history, and even if they could, it would not be reasonable to collect this amount of data from a person. However, social studies show that people can often provide simple summaries for this data, even more complicated ones when asked concrete questions [11]. These summaries may be viewed as *data patterns that apply to individuals* and, as explained below, can be used to identify overall significant patterns. Crowd mining processes thus generate questions to the crowd about their personal data patterns, often along with some measure that indicates the significance of the pattern in their data. Returning to the culinary habit example, assume that we are interested in identifying sets of food items that are often consumed together. This data may be obtained from the crowd by asking a *concrete question* of the form "How often do you consume $i_1, i_2, \ldots$ and i_n together?" The answer of a crowd worker u will include (roughly) the frequency in which $i_1, \ldots, i_n$ are consumed by u, i.e., the % of transactions $t \in D_u$ such that $i_1, \ldots, i_n \in t$.

In addition, the data items themselves can be obtained from the crowd, by asking them *open questions*. For example, consider questions of the form "What do you consume together with i_1 and how often do you do that?" Such questions are useful for discovering new data items and prominent data patterns. However, since people recall data more easily when asked concrete questions, it is preferable to interleave open and concrete questions, as also demonstrated by the experimental study in [8, 10].

Overall Significant Data Patterns

The significance measure of a particular data pattern for different crowd members should be aggregated in order to identify data patterns that are *overall significant* in some population. Different aggregation functions can be used for this purpose, e.g., minimum, average, majority, weighted average (according to expertise or trust), etc. In [8], average was used and proven to have useful properties for crowd mining. Since it is usually impossible to pose every question to every member of a particular population, crowd mining techniques compute an estimate for the aggregated significance of a data pattern. This is done by posing questions only to a small sample of the population – the crowd contributors – and by using a statistical model to estimate both whether a pattern is significant and the uncertainty in this answer [8].

Exploring the Data Domain

Let $\mathbf{I} = \{i_1, i_2, i_3, \ldots\}$ be a domain of data items relevant to the mining task, out of which data patterns can be constructed. For example, to learn about culinary habits, $\mathbf{I}$ may contain all of the possible food dishes, food ingredients, etc. Since $\mathbf{I}$ represents items in human knowledge, one can distinguish between three cases: (A) $\mathbf{I}$ is given as input to the crowd mining process, e.g., via a vocabulary (B) $\mathbf{I}$ is unknown, and must be discovered as a part of the mining process (C)

only a subset of $\mathbf{I}$ is given as input. In cases (B) and (C), the mining process may discover unknown data items via open questions to the crowd, as described above. See [6] for more details about the collection of answers to open questions from the crowd.

Question Generation

The generation of the next questions to the crowd given the collected knowledge is at the core of the crowd mining process. The selection of questions is done incrementally, after processing previous answers obtained from the crowd, and is guided by a target function. In [8], the next question is chosen to minimize the overall error probability, namely, the number of patterns that are misclassified as significant or insignificant. This technique uses a formulation for the expected error probability for each possible data pattern after asking one more question and generates the questions that yield the largest expected decrease in error. This technique can also be extended for selecting the K-best next questions.

In [9, 10], it is assumed that the mining process is given, as an input, a taxonomy of semantic subsumption relation between data items, which implies a predetermined semantic subsumption relation between data patterns. As a simple example from the culinary domain, if macaroni is a type of pasta, then every occasion of eating macaroni with tomato sauce is also a case of eating pasta with tomato sauce. The subsumption relation between data patterns can be exploited in the process of searching for significant patterns. For instance, if the pattern which represents eating macaroni with sauce is significant, it can be inferred that less specific patterns, like eating macaroni or eating pasta with sauce, are also significant. This inference scheme was used for classic data mining [12] and further developed and adapted to crowd mining in [9]. The latter technique generated the next crowd question which would provably lead to the eventual classification of many dependent patterns as (in)significant.

Incorporating Existing Knowledge

In some cases, in order to successfully mine relevant patterns, the mining process must take into consideration both general, ontological knowledge (e.g., nutritional values of certain food dishes) as well as personal knowledge (e.g., the culinary habits of crowd members). To account for this, the crowd mining query language OASSIS-QL [10] allows specifying the desired data patterns using both types of knowledge. The query evaluation process described in [10] first extracts the relevant data patterns based on an ontology of general facts and then generates crowd questions to identify the significant ones among these patterns.

Key Applications

Social scientists, microeconomists, journalists, marketers, public health specialists, and politicians alike routinely analyze people's behaviors to identify new trends, understand, and document behaviors. Crowd mining is a generic tool which can be used for this broad spectrum of potential applications. One concrete example, namely, learning about the culinary habits of some population, which may be of use, e.g., for a dietician, was illustrated above. Additional examples include learning about folk medicine practices in some population, which may be of use for a health researcher or a drug company; learning about the positive or negative patterns in the day-to-day work life of employees in some large organization, which may be of use for the management or for an ergonomics consultant; obtaining travel recommendations from the crowd (popular combinations of activities and places) about some travel destinations, which may be of use for planning a trip; and many more.

Cross-References

▶ Cost and Quality Trade-offs in Crowdsourcing
▶ Crowd Database Operators
▶ Crowd Database Systems
▶ Data Mining

▶ Frequent Itemsets and Association Rules
▶ Human Factors Modeling in Crowdsourcing
▶ OWL: Web Ontology Language

Recommended Reading

1. Doan A, Franklin M, Kossmann D, Kraska T. Crowdsourcing applications and platforms: a data management perspective. Proc VLDB Endowment. 2011;4(12):1508–9.
2. Davidson SB, Khanna S, Milo T, Roy S. Using the crowd for top-k and group-by queries. In: Proceedings of the 16th International Conference on Database Theory; 2013. p. 225–36.
3. Franklin MJ, Kossmann D, Kraska T, Ramesh S, Xin R. CrowdDB: answering queries with crowdsourcing. In: Proceedings of the ACM SIGMOD International Conference on Management of Data; 2011. p. 61–72.
4. Marcus A, Wu E, Karger DR, Madden S, Miller RC. Human-powered Sorts and Joins. Proc VLDB Endowment. 2011;5(1):13–24.
5. Parameswaran AG, Park H, Garcia-Molina H, Polyzotis N, Widom J. Deco: declarative crowdsourcing. In: Proceedings of the 21st ACM International Conference on Information and Knowledge Management; 2012. p. 1203–12.
6. Trushkowsky B, Kraska T, Franklin MJ, Sarkar P. Crowdsourced enumeration queries. In: Proceedings of the 29th International Conference on Data Engineering; 2013. p. 673–84.
7. Venetis P, Garcia-Molina H, Huang K, Polyzotis N. Max algorithms in crowdsourcing environments. In: Proceedings of the 21st International World Wide Web Conference; 2012. p. 989–98.
8. Amsterdamer Y, Grossman Y, Milo T, Senellart P. Crowd mining. In: Proceedings of the ACM SIGMOD International Conference on Management of Data; 2013. p. 241–52.
9. Amarilli A, Amsterdamer Y, Milo T. On the complexity of mining itemsets from the crowd using taxonomies. In: Proceedings of the 17th International Conference on Database Theory; 2014. p. 15–25.
10. Amsterdamer Y, Davidson SB, Milo T, Novgorodov S, Somech A. OASSIS: query driven crowd mining yael. In: Proceedings of the ACM SIGMOD International Conference on Management of Data; 2014. p. 1–12.
11. Bradburn NM, Rips LJ, Shevell SK. Answering autobiographical questions: the impact of memory and inference on surveys. Science. 1987;236(4798): 158–61.
12. Srikant R, Agrawal R. Mining generalized association rules. In: Proceedings of the 21st International Conference on Very Large Data Bases; 1995. p. 407–19.

Crowdsourcing Geographic Information Systems

Dieter Pfoser
Department of Geography and Geoinformation
Science, George Mason University, Fairfax, VA,
USA

Synonyms

User-generated geospatial content; Volunteered
geographic information (VGI)

Definition

The crowdsourcing of geographic information
addresses the collection of geospatial data contributed by non-expert users and the aggregation
of these data into meaningful geospatial datasets.
While crowdsourcing generally implies a coordinated bottom-up grassroots effort to contribute
information, in the context of geospatial data, the
term volunteered geographic information (VGI)
specifically refers to a dedicated collection effort inviting non-expert users to contribute. A
prominent example here is the OpenStreetMap
effort focusing on map datasets. Crowdsourcing
geospatial data is an evolving research area that
covers efforts ranging from mining GPS tracking
data to using social media content to profile
population dynamics.

Historical Background

With the proliferation of the Internet as the
primary medium for data publishing and
information exchange, we have seen an explosion
in the amount of online content available on the
Web. Thus, in addition to professionally produced
content being offered free on the Internet, the
public has also been encouraged to make content
available online to everyone. The volumes
of such user-generated content (UGC) are
already staggering and constantly growing. With

geospatial content playing an essential UGC role
[13], research in this area takes advantage of this
explosion in volunteered geographic information
(VGI) [6] to produce datasets that complement
authoritative datasets rather than replace them. A
term frequently mentioned when discussing user-
generated content is *crowdsourcing*, which implies a coordinated, bottom-up grassroots effort
to contribute information. VGI refers specifically
to geographic content contributed by non-expert
users and does not require some level of coordination among the individuals who are making
these contributions [6]. User-generated geospatial
content has been categorized in several ways.
Web-based services and tools can provide means
for users through *attentional* (e.g., geo-wikis,
geocoding photos) or *unattentional*, passive
efforts (e.g., GPS traces from their daily com-
mutes) to create vast amounts of data concerning
the real world that contain significant amounts
of information (crowdsourcing). Another way
of looking at user-generated geospatial content
is by way of differentiating between explicit
and implicit content (cf. [5]). *Explicit content* is
generated purposefully in the desired form. An
example here is the OpenStreetMap (http://www.
openstreetmap.org) road network. In contrast,
implicit content reflects derived information as
the original, user-generated content was produced
with a different purpose in mind. However, the
desired content may be extracted from it nevertheless. An example here would be extracting a
road network from GPS tracking data. Implicit
content is also often embedded in social media
contributions (e.g., blogs, microblogs, social multimedia). Here it is referred to as *ambient* (AGI –
[15]) geographical information. Overall, while
initially VGI research focused on explicit datasets
and dedicated applications, the availability of
implicit data by means of the smart phone revolution has lead to a surge in data mining research
focusing on ambient user-generated (geo)content.
It is interesting to see that while volunteered
geographic information (VGI) can be large in
volume, it is comparatively poor in quality. The
full exploitation of crowdsourced information as
a means to complement authoritative datasets is
contingent on the thorough understanding of its

accuracy. This fact has led to a recent surge in research addressing the quality aspect of VGI.

Scientific Fundamentals

Crowdsourcing geospatial data is a broad research area that spans many different efforts. In the discussion that follows, we will possibly not cover the topic in its entirety, but try to discuss representative contributions. The most well-known crowdsourcing examples of geospatial data include platforms like OSM, Wikimapia (http://wikimapia.org), and Google MapMaker (http://www.google.com/mapmaker), which allow non-experts to perform basic cartographic tasks, digitizing and editing road networks, building outlines, and point-of-interest (POI) information. OpenStreetMap (OSM) is a collaborative project to create a free editable *map of the world*. It was inspired by the success of Wikipedia and the lack of availability of affordable map data. The project has grown to over 1.6 million registered users, who can collect data using GPS devices, aerial photography, and other free sources. This crowdsourced data is then made available under the Open Database License. Figure 1 shows how simple it is to contribute edits

to OSM using the browser-based iD editor. The figure shows a map excerpt of George Mason University with features such as buildings, roads, trees, and amenities clearly visible. Wikimapia is also a collaborative mapping project that aims to mark and describe all *geographical objects* in the world. It combines an interactive Web map with a geographically referenced wiki system. As of June 2014, over 23,000,000 objects have been marked by registered users and guests. Google Mapmaker is a commercial effort comparable to OSM. Besides map data, approaches to the collection of geospatial objects include the use of Google (now Trimble) SketchUp (http://www.sketchup.com) and the related 3D Warehouse (https://3dwarehouse.sketchup.com), which creates a geographically tagged database of three-dimensional objects.

Maps and Traffic Data. Crowdsourcing specifically plays an important role for data collection with respect to traffic information and navigation services. Street maps and transportation networks are of fundamental importance in a wealth of applications. In the past, map production was costly and proprietary data vendors such as NAVTEQ (now Nokia) and Tele Atlas (now TomTom) dominated the market. Over the last years, VGI efforts such as OSM have complemented commercial map datasets. However, these efforts still require

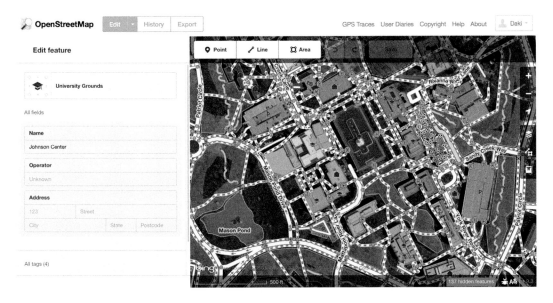

Crowdsourcing Geographic Information Systems, Fig. 1 OpenStreetMap – browser-based iD editor

dedicated users to author maps using specialized software tools. Lately, on the other hand, the commoditization of GPS technology and integration in mobile phones coupled with the advent of low-cost fleet management and positioning software have triggered the generation of vast amounts of tracking data, an example of *unattentional crowdsourced geospatial map data*. Such tracking information can be collected explicitly through location-based social networks such as Waze (https://www.waze.com) or implicitly from GPS enabled devices. In conjunction with the proliferation of GPS technology as an end-user consumer product, GPS devices are being integrated in large-scale enterprise settings for fleet management and asset tracking. One example of this is the emergence of floating car data (FCD) repositories, which refer to using data generated by one vehicle as a sample to assess overall traffic conditions (e.g., "a cork swimming in the river"). Having large amounts of vehicles collecting such data for a given spatial area such as a city (e.g., taxis, public transport, utility vehicles, delivery fleets) can render an accurate picture of the traffic condition in time and space (e.g., [12]). By *map-matching* the tracking data, it is possible to derive travel time related to specific portions of the road network, lending itself to a live assessment of the traffic conditions and future trends. Besides the use of such data in traffic assessment and forecasting, there has been a recent surge of actual *map construction algorithms* that derive not only travel time attributes but also actual road network geometries from tracking data such as shown in Fig. 2 (cf. [3]).

Social Media. The Web has created a number of services that facilitate the collection of *location-relevant content*, i.e., data for which location is just but one and most often even not the most important attribute. Prominent examples include (i) photo-sharing sites such as Flickr, Panoramio, and Instagram; (ii) social media and specifically microblogging sites with geotagging features, e.g., Facebook, Google+, Twitter, as well as related photo-sharing sites (twitpic); and (iii) geospatial check-in services such as Foursquare. *Social networking and microblogging services* such as Twitter provide a continuous source of data from which useful information can be extracted. Aggregating geo-related blog posts allows one to derive *topic information in relation to space*. They describe, for example, social activities occurring within a city. Such topics, as captured by social media, are highly dynamic at all scales, as spaces can change with the latest trends, news headlines, or social movements [15]. In [10] a probabilistic topic model is used to decompose the stream of digital traces contained in textual and event-based data from services such as Twitter and Foursquare into a set of urban topics related to various activities of people during the course of the week. Due to the combined use of implicit textual and movement data, one can obtain semantically rich modalities of the urban dynamics and overcome the drawbacks of traditional methods such as on-site surveys. The results of this work can be used to enrich location-based services with real-time context.

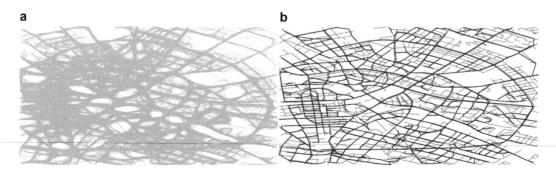

Crowdsourcing Geographic Information Systems, Fig. 2 Map construction example – GPS trajectories of vehicles collected in Berlin, Germany - (**a**) GPS trajectories, (**b**) constructed road network

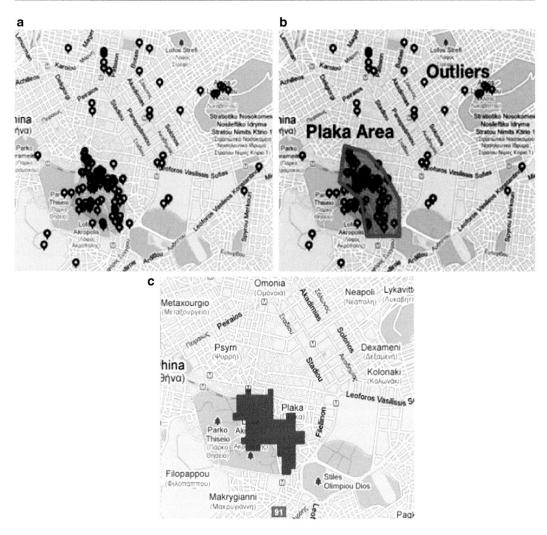

Crowdsourcing Geographic Information Systems, Fig. 3 Crowdsourcing colloquial geospatial objects from user-contributed content - (**a**) Flickr image locations, (**b**) Plaka common knowledge extent, (**c**) clustering result

Point Clouds. Users generate data by means of many different applications in which geospatial data is simply used to index and access the collected data. This data may not always be in the format, quality, and amount we expect it, but by employing intelligent data collection and *mining algorithms*, we are able to discover valuable insights into *urban function*. As mentioned, prominent dataset examples here are photo-sharing sites and microblogging services using geotagging features. To illustrate the potential for geospatial data generation, consider the use of Flickr data for the computation of feature shapes of various spatial objects including city scale, countries, and other colloquial areas [4]. The approach is based on computing primary shapes for point clouds that are grouped together by Yahoo! GeoPlanet WOEIDs (Yahoo! Where On Earth IDs), which are part of the Flickr metadata. Flickr data is used to identify *places* and also *relationships* among them (e.g., containment) based on tag analysis and spatial clustering in [9]. A system that combines browser-based computing with crowdsourcing [11] derives colloquial geospatial objects or POIs from point-cloud data such as Flickr image locations. The user provides the search terms (e.g., the name of the tourist area "Plaka, Athens"), and respective point databases are queried based on their tag information and using Web service APIs to retrieve a point cloud that characterizes this location. This point cloud represents the collective notion – or the colloquial footprint of the area of interest, which then needs to be aggregated to derive the actual location of the sought geospatial object. The specific approach uses a hierarchical grid-based filter-and-refinement approach to retrieve a minimal dataset from the Web data sources and to still produce adequate spatial object geometries (Fig. 3).

Qualitative Geospatial Data Sources. In the geospatial domain, authoring content typically involves quantitative, coordinate-based data. While technology has helped a lot to facilitate geospatial data collection, e.g., all smartphones are equipped with GPS positioning sensors, yet authoring quantitative data requires specialized applications (often part of social media platforms) and/or specialized knowledge, e.g., OSM. This fact hinders the widespread adoption of VGI as an even bigger, large-scale geospatial data source. The broad mass of users contributing content on the Internet are much more comfortable with using *qualitative information*. People do not use coordinates to describe their spatial experiences, but rely on qualitative concepts in the form of toponyms (landmarks) and spatial relationships (near, next, etc.). With spatial reasoning being a basic form of human cognition, narratives expressing such geospatial experiences, e.g., travels blogs, would provide an even bigger source of geospatial data. Typically, spatial information extraction from texts is associated with georeferencing of texts, which typically involves the identification and geocoding of toponyms. Several commercial software packages and services, e.g., Google Places API (https://developers.google.com/places/) and Yahoo! BOSS Geo Services (https://developer.yahoo.com/boss/geo/), do exist. Using this approach, travel blogs (e.g., travelpod.com, travelblog.org) have been mined to give

researchers a person's conceptualization of place based on geo-referenced text to construct regions of thematic saliency [1]. However, the *extraction of qualitative spatial data* in the form of relationships from texts requires the utilization of efficient natural language processing (NLP) tools to automatically extract and map phrases to spatial relations. This has been addressed to some extent in the literature, e.g., [16], but efforts have always been limited by the unclear mapping of spatial language expressions to spatial relationships such as metric, directional, and topological relations. In addition, while computing using a qualitative representation of spatial data has been an active research topic for quite some time, the quantitative, i.e., coordinate-based, representation is still dominant. Hence, a research topic has been the quantitative representation of qualitative relationship data. The authors of [14] introduce a basic expectation-maximization/Gaussian mixture model approach to quantify qualitative spatial data extracted from crowdsourced travel blog narrative. The method automatically extracts qualitative spatial data from texts, quantifies the relations using probability distribution functions, and introduces a location estimation method based on spatial relation fusion. This approach bridges the gap between qualitative and quantitative representation of spatial relations using efficient machine learning techniques and introduces an actual text-to-map application. Figure 4 gives three quantified relationship examples crowdsourced from textual narrative.

Image-Based Scene Generation. Algorithmic advancements have further fostered the collection of 3D data in urban environments based on crowdsourcing. For example, large collections of geotagged imagery can be stitched together to reconstruct 3D scenes using extensions of photogrammetric and computer vision principles. Geotagged images from Flickr are used in [2] to build 3D models of Rome. User-contributed imagery has also been overlaid and embedded in Google Street View. This feature has been replaced by Photo Sphere (https://www.google.com/maps/about/contribute/photosphere/), a mobile phone app that allows for the creation

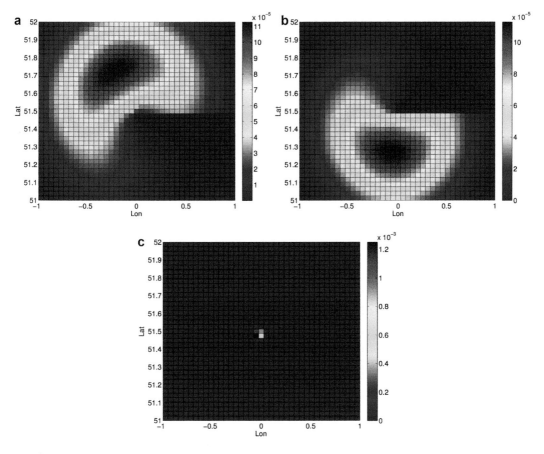

Crowdsourcing Geographic Information Systems, Fig. 4 Probabilistic heat maps for three basic spatial relationships obtained by crowdsourcing spatial relationships from textual narratives - (**a**) North, (**b**) South, (**c**) Near

of 360-degree immersive panoramas. Microsoft Photosynth (http://photosynth.net) provides similar functionality, in which a set of overlapping images showing the same scene are analyzed based on matching features to create a 3D point cloud and a model of the captured scene. 3D panoramas and models have been popularized with the emergence of smartphones that not only provide image capturing capabilities but also have the computing power to generate the 3D scenes as well as the connectivity to communicate the results to respective repositories in real time.

Data Quality. Volunteered data is usually provided with little to no information on mapping standards, quality control procedures, and metadata in general. Understanding and measuring the data quality of information provided by volunteers who may have unreported agendas and/or biases is a significant problem in geography today. Recent studies have already started to assess the *quality of user-generated geospatial content* and here specifically OSM content by comparing it to established authoritative mapping organizations such as the UK's Ordnance Survey road datasets [7]. These studies conclude that such data is comparable to more authoritative sources. Road centerlines in OSM were shown to be within few meters of their Ordnance Survey equivalents. The interesting issue of the localness of the GI contributed data is addressed in [8]. By examining two major websites, Flickr and Wikipedia, the authors find that more than half of Flickr users contribute local information on average, while in Wikipedia the authors' participation is less local.

Key Applications

Mapping
User-generated geospatial data directly contributes to the enrichment of existing map datasets. Data includes automatically generated road networks, point-of-interest data, and colloquial geospatial objects.

Understanding Urban Form and Function
Crowdsourcing geospatial data from social media applications provides for a better understanding of urban form and function, i.e., the existing infrastructure and how it is utilized. Such data is directly relevant to applications such as geomarketing, which try to understand socioeconomic processes.

Traffic Assessment
Using mobile phones of users as traffic sensors allows us to compute actual traffic conditions in terms of travel time. Such data is then used to optimize navigation solutions and to improve routing services.

3D Scene Reconstruction
Crowdsourced images can be used to develop 3D models of places. This effort is complementary to dedicated collection campaigns of, e.g., Google.

Cross-References

▶ Crowd Mining and Analysis
▶ Geographic Information System
▶ Geography Markup Language
▶ Geo-targeted Web Search
▶ Human Factors Modeling in Crowdsourcing
▶ Linked Open Data
▶ Spatial Data Mining
▶ Spatiotemporal Data Mining
▶ Text Mining

Recommended Reading

1. Adams B, McKenzie G. Inferring thematic places from spatially referenced natural language descriptions. In: Sui D, Elwood S, Goodchild M, Crowdsourcing Geographic Knowledge: Volunteered Geographic Information, editors, Crowdsourcing geographic knowledge: Volunteered Geographic Information (VGI) in theory and practice. Netherlands: Springer; 2013, p. 201–21.
2. Agarwal S, Snavely N, Simon I, Seitz SM, Szeliski R. Building rome in a day. In: Proceedings of the 12th IEEE Computer Vision Conference; 2009. p. 72–9.
3. Ahmed M, Karagiorgou S, Pfoser D, Wenk C. Map construction algorithms, Springer International Publishing, Switzerland, 2015.
4. Cope A. The shape of alpha. Available at http://code.flickr.net/2008/10/30/the-shape-of-alpha/, 2008.
5. Crooks AT, Pfoser D, Jenkins A, Croitoru A, Karagiorgou S, Efentakis A, Lamprianidis G, Smith D, Stefanidis A. Crowdsourcing urban form and function. Int J Geogr Inf Sci. 2015;29(5):720–41.
6. Goodchild MF. Citizens as sensors: the world of volunteered geography. GeoJournal. 2007;69(4):211–21.
7. Haklay M. How good is volunteered geographical information? A comparative study of OpenStreetMap and ordnance survey datasets. Environ Plan B. Plan & Design. 2010;37(4):682–703.
8. Hecht B, Gergle D. On the localness of user-generated content. In: Proceedings of the ACM Computer Supported Cooperative Work Conference; 2010, p. 229–32.
9. Intagorn S, Plangprasopchok A, Lerman K. Harvesting geospatial knowledge from social metadata. In: Proceedings of the 7th ISCRAM Conference; 2010.
10. Kling F, Pozdnoukhov A. When a city tells a story: urban topic analysis. In: Proceedings of the ACM SIGSPATIAL GIS Conference; 2012. p. 482–85.
11. Lamprianidis G, Pfoser D. Collaborative geospatial feature search. In: Proceedings of the ACM SIGSPATIAL GIS Conference; 2012. p. 169–78.
12. Pfoser D, Brakatsoulas S, Brosch P, Umlauft M, Tsironis G, Tryfona N. Dynamic travel time provision for road networks. In: Proceedings of the ACM SIGSPATIAL GIS confernce; 2008. p. 475–78.
13. Pfoser D. On user-generated geocontent. Proceeding of 12th SSTD Symposium In: Proceedings of the 12th SSTD Symposium; 2011. p. 458–61.
14. Skoumas G, Pfoser D, Kyrillidis A, Sellis T. Location estimation using crowdsourced spatial relations. ACM Trans Spat Algorithms Syst. 2016;2(2):1–23.
15. Stefanidis T, Crooks AT, Radzikowski J. Harvesting ambient geospatial information from social media feeds. GeoJournal. 2013;78(2):319–38.
16. Zhang Z, Zhang C, Du C, Zhu S. SVM-based extraction of spatial relations in text. In: Proceedings of the IEEE ICSDM Confernncence; 2011. p. 529–533.

Cube

Torben Bach Pedersen
Department of Computer Science, Aalborg University, Aalborg, Denmark

Synonyms

Hypercube

Definition

A cube is a data structure for storing and and analyzing large amounts of multidimensional data, often referred to as *On-Line Analytical Processing (OLAP)*. Data in a cube lives in a space spanned by a number of hierarchical *dimensions*. A single point in this space is called a *cell*. A (non-empty) cell contains the values of one or more *measures*.

Key Points

As an example, a three-dimensional cube for capturing sales may have a Product *dimension* P, a Time dimension T, and a Store dimension S, capturing the product sold, the time of sale, and the store it was sold in, for each sale, respectively. The cube has two *measures*: Dollar Sales and ItemSales, capturing the sales price and the number of items sold, respectively. In a cube, the combinations of a dimension value from each dimension define a *cell* of the cube. The measure value(s), e.g., DollarSales and ItemSales, corresponding to the particular combination of dimension values are then stored stored in the corresponding cells.

Data cubes provide true multidimensionality. They generalize spreadsheets to any number of dimensions, indeed cubes are popularly referred to as "spreadsheets on stereoids." In addition, *hierarchies* in dimensions and formulas are first-class, built-in concepts, meaning that these are supported without duplicating their definitions. A collection of related cubes is commonly referred to as a *multidimensional database* or a *multidimensional data warehouse*.

In a cube, dimensions are first-class concepts with associated domains, meaning that the addition of new dimension values is easily handled. Although the term "cube" implies three dimensions, a cube can have any number of dimensions. It turns out that most real-world cubes have 4–12 dimensions [2]. Although there is no theoretical limit to the number of dimensions, current tools often experience performance problems when the number of dimensions is more than 10–15. To better suggest the high number of dimensions, the term "hypercube" is often used instead of "cube."

Depending on the specific application, a highly varying percentage of the cells in a cube are non-empty, meaning that cubes range from *sparse* to *dense*. Cubes tend to become increasingly sparse with increasing dimensionality and with increasingly finer granularities of the dimension values. A non-empty cell is called a *fact*. The example has a fact for each combination of time, product, and store where at least one sale was made.

Generally, only two or three dimensions may be viewed at the same time, although for low-cardinality dimensions, up to four dimensions can be shown by nesting one dimension within another on the axes. Thus, the dimensionality of a cube is reduced at query time by *projecting* it down to two or three dimensions via *aggregation* of the measure values across the projected-out dimensions. For example, to view sales by Store and Time, data is aggregates over the entire Product dimension, i.e., for all products, for each combination of Store and Time.

OLAP SQL extensions for cubes were pioneered by the proposal of the data cube operators CUBE and ROLLUP [1]. The CUBE operator generalizes GROUP BY, crosstabs, and subtotals using the special "ALL" value that denotes that an aggregation has been performed over all values for one or more attributes, thus generating a subtotal, or a grand total.

Data cubes are implemented either using relational databases – so-called Relational OLAP (ROLAP) systems, dedicated multidimensional data structures – so-called Multidimensional OLAP (MOLAP) systems, or a combination of these – so-called Hybrid OLAP (HOLAP) systems [2, 4].

While data cubes remain the cornerstone technology for OLAP, they are also increasingly used as the preferred data store when doing data mining, so-called On-Line Analytical Mining (OLAM). Data cubes are also increasingly used for complex and non-conventional types of data, such as spatial, temporal, text, multimedia, (Semantic) Web, networks, and graphs [3, 4].

Cross-References

▶ Cube Implementations
▶ Data Warehousing on Nonconventional Data
▶ Dimension
▶ Hierarchy
▶ Measure
▶ Multidimensional Modeling
▶ OLAM
▶ Online Analytical Processing

Recommended Reading

1. Gray J, Chaudhuri S, Bosworth A, Layman A, Venkatrao M, Reichart D, Pellow F, Pirahesh H. Data cube: a relational aggregation operator generalizing group-by, cross-tab and sub-totals. Data Mining Knowl Discov. 1997;1(1):29–54.
2. Jensen CS, Pedersen TB, Thomsen C. Multidimensional databases and data warehousing. Synthesis lectures on data management. San Rafael: Morgan Claypool; 2010.
3. Pedersen TB. Managing complex multidimensional data. In: Aufaure M-A, Zimányi E, editors. Business intelligence – second European summer school, eBISS 2012. Brussels: Springer LNBIB; 2013, 15–21 July 2012, Tutorial Lectures.
4. Vaisman A, Zimányi E. Data warehouse systems – design and smplementation. Springer; 2014.

Cube Implementations

Konstantinos Morfonios[1] and Yannis Ioannidis[2]
[1]Oracle, Redwood City, CA, USA
[2]University of Athens, Athens, Greece

Synonyms

Cube materialization; Cube precomputation

Definition

Cube implementation involves the procedures of computation, storage, and manipulation of a data cube, which is a disk structure that stores the results of the aggregate queries that group the tuples of a fact table on all possible combinations of its dimension attributes. For example in Fig. 1a, assuming that R is a fact table that consists of three dimensions (A, B, C) and one measure M (see definitional entry for *Measure*), the corresponding cube of R appears in Fig. 1b. Each cube node (i.e., view that belongs to the data cube) stores the results of a particular aggregate query as shown in Fig. 1b. Clearly, if D denotes the number of dimensions of a fact table, the number of all possible aggregate queries is 2^D; hence, in the worst case, the size of the data cube is exponentially larger with respect to D than the size of the original fact table. In typical applications, this may be in the order of gigabytes or even terabytes, implying that the development of efficient algorithms for the implementation of cubes is extremely important.

Let *grouping attributes* be the attributes of the fact table that participate in the group-by clause of an aggregate query expressed in SQL. A common representation of the data cube that captures the computational dependencies among all the aggregate queries that are necessary for its materialization is the cube lattice [6]. This is a directed acyclic graph (DAG) where each node represents an aggregate query q on the fact table and is connected via a directed edge with every other node whose corresponding group-by part is missing one of the grouping attributes of q.

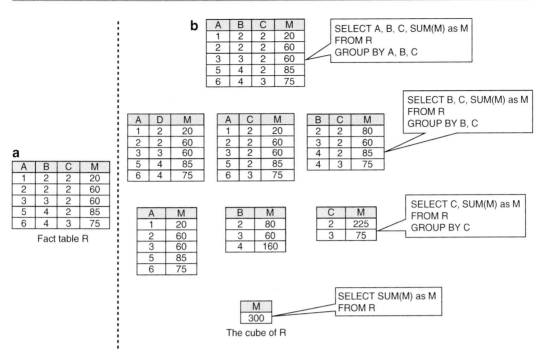

Cube Implementations, Fig. 1 Fact table R and the corresponding data cube

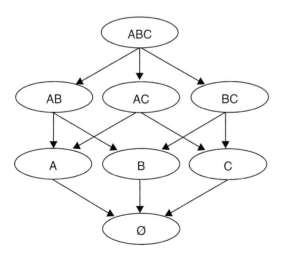

Cube Implementations, Fig. 2 Example of a cube lattice

For example, Fig. 2 shows the cube lattice that corresponds to the fact table R (Fig. 1a).

Note that precomputing and materializing parts of the cube is crucial for the improvement of query-response times as well as for accelerating operators that are common in On-Line Analytical Processing (OLAP), such as drill-down, roll-up, pivot, and slice-and-dice, which make an extensive use of aggregation [3]. Materialization of the entire cube seems ideal for efficiently accessing aggregated data; nevertheless, in real-world applications, which typically involve large volumes of data, it may be considerably expensive in terms of storage space, as well as computation and maintenance time. In the existing literature, several efficient methods have been proposed that attempt to balance the aforementioned tradeoff between query-response times and other resource requirements. Their brief presentation is the main topic of this entry.

Historical Background

Most data analysis efforts, whether manual by analysts or automatic by specialized algorithms, manipulate the contents of database systems in order to discover trends and correlations. They typically involve complex queries that make an extensive use of aggregation in order to group together tuples that "behave in a similar fashion." The response time of such queries over extremely large data warehouses can be prohibitive. This problem inspired Gray et al. [3] to introduce the data-cube operator and propose its off-line computation and storage for efficiency at query time. The corresponding seminal publication has been the seed for a plethora of papers thereafter, which have dealt with several different aspects of the lifecycle of a data cube, from cube construction and storage to indexing, query answering, and incremental maintenance.

Taking into account the format used for the computation and storage of a data cube, the cube-implementation algorithms that have appeared in the literature can be partitioned into four main categories: Relational-OLAP (ROLAP) algorithms exploit traditional materialized views in RDBMSes; Multidimensional-OLAP (MOLAP) algorithms take advantage of multidimensional arrays; Graph-Based methods use specialized graph structures; finally, approximation algorithms use various in-memory representations, e.g., histograms.

The literature also deals with the rest of the cubes lifecycle [11]. Providing fast answers to OLAP aggregate queries is the main purpose of implementing data cubes to begin with, and various algorithms have been proposed to handle different types of queries on the formats above. Moreover, as data stored in the original fact table changes, data cubes must follow suit; otherwise, analysis of obsolete data may result into invalid conclusions. Periodical reconstruction of the entire cube is impractical, hence, incremental-maintenance techniques have been proposed.

The ideal implementation of a data cube must address efficiently all aspects of cube functionality in order to be viable. In the following section, each one of these aspects is further examined separately.

Foundations

In the following subsections, the main stages of the cube lifecycle are analyzed in some detail, including subcube selection, computation, query processing, and incremental maintenance. Note that the references given in this section are only indicative, since the number of related publications is actually very large. A more comprehensive survey may be found elsewhere [12].

Subcube Selection

In real-world applications, materialization of the entire cube is often extremely expensive in terms of computation, storage, and maintenance requirements, mainly because of the typically large fact-table size and the exponential number of cube nodes with respect to the number of dimensions. To overcome this drawback, several existing algorithms select an appropriate subset of the data cube for precomputation and storage [4–6]. Such selection algorithms try to balance the tradeoff between response times of queries (sometimes of a particular, expected workload) and resource requirements for cube construction, storage, and maintenance. It has been shown [6] that selection of the optimum subset of a cube is an NP-complete problem. Hence, the existing algorithms use heuristics in order to find near-optimal solutions.

Common constraints used during the selection process involve constraints on the time available for cube construction and maintenance, and/or on the space available for cube storage. As for the criteria that are (approximately) optimized during selection, they typically involve some form of the benefit gained from the materialization of a particular cube subset.

A particularly beneficial criterion for the selection problem that needs some more attention, since it has been integrated in some of the most efficient cube-implementation algorithms (including Dwarf [17] and CURE [10], which will be briefly presented below) is the so-called redundancy reduction. Several groups of researchers have observed that a big part of the cube data is usually redundant [7, 8, 10, 11, 17, 20]. Formally, a value stored in a cube is *redundant* if it is repeated multiple times in the same attribute in the cube. For example, in Fig. 1b, tuples $\langle 1, 20 \rangle$ of node A, $\langle 1, 2, 20 \rangle$ of AB, and $\langle 1, 2, 20 \rangle$ of AC are redundant, since they can be produced by properly projecting tuple $\langle 1, 2, 2, 20 \rangle$ of node ABC. By appropriately avoiding the storage of such redundant data, several existing cube-implementation algorithms achieve the construction of compressed cubes that can still be considered as fully materialized. Typically, the decrease in the final cube size is impressive, a fact that benefits the performance of computation as well, since output costs are considerably reduced and sometimes, because early identification of redundancy allows pruning of parts of the computation. Furthermore, during query answering, aggregation and decompression are not necessary; instead, some simple operations, e.g., projections, are enough.

Finally, for some applications (e.g., for mining multidimensional association rules), accessing the tuples of the entire cube is not necessary, because they only need those group-by tuples with an aggregate value (e.g. count) above some prespecified minimum support threshold (minsup). For such cases, the concept of *iceberg cubes* has been introduced [2]. Iceberg-cube construction algorithms [2, 16] take into consideration only sets of tuples that aggregate together giving a value greater than minsup. Hence, they perform some kind of subcube selection, by storing only the tuples that satisfy the aforementioned condition.

Cube Computation

Cube computation includes scanning the data of the fact table, aggregating on all grouping attributes, and generating the contents of the data cube. The main goal of this procedure is to place tuples that aggregate together (i.e., tuples with identical grouping-attribute values) in contiguous positions in main memory, in order to compute the required aggregations with as few data passes as possible. The most widely used algorithms that accomplish such clustering of tuples are sorting and hashing. Moreover, nodes connected in the cube lattice (Fig. 2) exhibit strong computational dependencies, whose exploitation is particularly beneficial for the performance of the corresponding computation algorithms. For instance, assuming that the data in the fact table R (Fig. 1a) is sorted according to the attribute combination ABC, one can infer that it is also sorted according to both AB and A as well. Hence, the overhead of sorting can be shared by the computation of multiple aggregations, since nodes $ABC \rightarrow AB \rightarrow A \rightarrow \varnothing$ can be computed with the use of pipelining without reclustering the data. Five methods that take advantage of such node computational dependencies have been presented in the existing literature [1] in order to improve the performance of computation algorithms: smallest-parent, cache-results, amortize-scans, share-shorts, and share-partitions.

Expectedly, both sort-based and hash-based aggregation methods perform more efficiently when the data they process fits in main memory; otherwise, they are forced to use external-memory algorithms, which generally increase the I/O overhead by a factor of two or three. In order to overcome such problems, most computation methods initially apply a step that partitions data into segments that fit in main memory, called *partitions* [2, 10, 15]. Partitioning algorithms distribute the tuples of the fact table in accordance with the principle that tuples that aggregate together must be placed in the same partition. Consequently, they can later process each partition independently of the others, since by construction, tuples that belong to different partitions do not share the same grouping-attribute values.

In addition to the above, general characteristics of cube-computation algorithms, there are some further details that are specific to each of four main categories mentioned above (i.e., RO-LAP, MOLAP, Graph-Based, and Approximate), which are touched upon below.

ROLAP algorithms store a data cube as a set of materialized relational views, most commonly using either a *star* or a *snowflake schema*. Among these algorithms, algorithm CURE [10] seems to be the most promising, since it is the only solution with the following features: It is purely compatible with the ROLAP framework, hence its integration into any existing relational engine is rather straightforward. Also, it is suitable not only for "flat" datasets but also for processing datasets whose dimension values are hierarchically organized. Furthermore, it introduces an efficient algorithm for external partitioning that allows the construction of cubes over extremely large volumes of data whose size may far exceed the size of main memory. Finally, it stores cubes in a compressed form, removing all types of redundancy from the final result.

MOLAP algorithms store a data cube as a multidimensional array, thereby avoiding to store the dimension values in each array cell, since the position of the cell itself determines these values. The main drawback of this approach comes from the fact that, in practice, cubes have a large number of empty cells (i.e., cubes are sparse), rendering MOLAP algorithms inefficient with respect to their storage-space requirements. To overcome this problem, the so-called chunk-based algorithms have been introduced [21], which avoid the physical storage of most of the empty cells, storing only *chunks*, which are nonempty subarrays. ArrayCube [21] is the most widely accepted algorithm in this category. It has also served as an inspiration to algorithm MM-Cubing [16], which applies similar techniques just to the dense areas of the cube, taking into account the distribution of data in a way that avoids chunking.

Graph-Based algorithms represent a data cube as some specialized graph structure. They use such structures both in memory, for organizing data in a fashion that accelerates computation of the corresponding cube, and on disk, for compressing the final result and reducing storage-space requirements. Among the algorithms in this category, Dwarf [17] seems to be the strongest overall, since it is the only one that guarantees a polynomial time and space complexity with respect to dimensionality [18]. It is based on a highly compressed data structure that eliminates prefix and suffix redundancies efficiently. Prefix redundancy occurs when two or more tuples in the cube share the same prefix, i.e., the same values in the left dimensions; suffix redundancy, which is in some sense complementary to prefix redundancy, occurs when two or more cube tuples share the same suffix, i.e., the same values in the right dimensions and the aggregate measures. An advantage of Dwarf, as well as of the other graph-based methods, is that not only does its data structure store a data cube compactly, but it also serves as an index that can accelerate selective queries.

Approximate algorithms assume that data mining and OLAP applications do not require fine grained or absolutely precise results in order to capture trends and correlations in the data; hence, they store an approximate representation of the cube, trading accuracy for level of compression. Such algorithms exploit various techniques, inspired mainly from statistics, including histograms [14], wavelet transformations [19], and others.

Finally, note that some of the most popular industrial cube implementations include Microsoft SQL Server Analysis Services (http://www.microsoft.com/sql/technologies/analysis/default.mspx) and Hyperion Essbase, which has been bought by ORACLE in 2007 (http://www.oracle.com/hyperion).

Query Processing

The most important motivation for cube materialization is to provide low response times for OLAP queries. Clearly, construction of a highly-compressed cube is useless if the cube format inhibits good query answering performance. Therefore, efficiency during query processing should be taken into consideration as well when selecting a specific cube-construction algorithm and its corresponding storage format. Note that the latter determines to a great extent the access methods that can be used for retrieving data stored in the corresponding cube; hence, it strongly affects performance of query processing algorithms over cube data.

Intuitively, it seems that brute-force storage of an entire cube in an uncompressed format behaves best during query processing: in this case, every possible aggregation for every combination of dimensions is precomputed and the only cost required is that of retrieving the data stored in the lattice nodes participating in the query. On the other hand, query processing over compressed cubes seems to induce additional overhead for on-line computation or restoration of (possibly redundant) tuples that have not been materialized in the cube.

Nevertheless, the literature has shown that the above arguments are not always valid in practice. This is mostly due to the fact that indexing an uncompressed cube is nontrivial in real-world applications, whereas applying custom indexing techniques for some sophisticated, more compact representations has been found efficient [2]. Furthermore, storing data in specialized formats usually offers great opportunities for unique optimizations that allow a wide variety of query types to run faster over compressed cubes [2]. Finally, recall that several graph-based algorithms, e.g., Dwarf [17], store the cube in a way that is efficient with respect to both storage space and query processing time.

Incremental Maintenance

As mentioned earlier, in general, fact tables are dynamic in nature and change over time, mostly as new records are inserted in them. Aggregated data stored in a cube must follow the modifications in the corresponding fact table; otherwise, query answers over the cube will be inaccurate.

According to the most common scenario used in practice, data in a warehouse is periodically updated in a batch fashion. Clearly, the window of time that is required for the update process must be kept as narrow as possible. Hence, reconstruction of the entire cube from scratch is practically not a viable solution; techniques for incremental maintenance must be used instead.

Given a fact table, its corresponding cube, and a set of updates to the fact table that have occurred since the last cube update, let *delta cube* be the cube formed by the data corresponding to these updates. Most incremental-maintenance algorithms proposed in the literature for the cube follow a common strategy [20]: they separate the update process into the *propagation* phase, during which they construct the delta cube, and the *refresh* phase, during which they merge the delta cube and the original cube, in order to generate the new cube. Most of them identify the refresh phase as the most challenging one and use specialized techniques to accelerate it, taking into account the storage format of the underlying cube (some examples can be found in the literature [11, 17]). There is at least one general algorithm, however, that tries to optimize the propagation phase [9]. It selects particular nodes of the delta

cube for construction and properly uses them in order to update all nodes of the original cube.

Key Applications

Efficient implementation of the data cube is essential for OLAP applications in terms of performance, since they usually make an extensive use of aggregate queries.

Cross-References

▶ Data Warehouse
▶ Dimension
▶ Hierarchy
▶ Measure
▶ Snowflake Schema
▶ Star Schema
▶ Visual Online Analytical Processing (OLAP)

Recommended Reading

1. Agarwal S, Agrawal R, Deshpande P, Gupta A, Naughton JF, Ramakrishnan R, Sarawagi S. On the computation of multidimensional aggregates. In: Proceedings of the 22th International Conference on Very Large Data Bases; 1996. p. 506–21.
2. Beyer KS, Ramakrishnan R. Bottom-up computation of sparse and iceberg CUBEs. In: Proceedings of the ACM SIGMOD International Conference on Management of Data; 1999. p. 359–70.
3. Gray J, Bosworth A, Layman A, Pirahesh H. Data cube: a relational aggregation operator generalizing group-by, cross-tab, and sub-total. In: Proceedings of the 12th International Conference on Data Engineering; 1996. p. 152–9.
4. Gupta H. Selection of views to materialize in a data warehouse. In: Proceedings of the 6th International Conference on Database Theory; 1997. p. 98–112.
5. Gupta H, Mumick IS. Selection of views to materialize under a maintenance cost constraint. In: Proceedings of the 7th International Conference on Database Theory; 1999. p. 453–70.
6. Harinarayan V, Rajaraman A, Ullman JD. Implementing data cubes efficiently. In: Proceedings of the ACM SIGMOD International Conference on Management of Data; 1996. p. 205–16.
7. Kotsis N, McGregor DR. Elimination of redundant views in multidimensional aggregates. In: Pro-

ceedings of the 2nd International Conference on Data Warehousing and Knowledge Discovery; 2000. p. 146–61.

8. Lakshmanan LVS, Pei J, Zhao Y. QC-Trees: an efficient summary structure for semantic OLAP. In: Proceedings of the ACM SIGMOD International Conference on Management of Data; 2003. p. 64–75.

9. Lee KY, Kim MH. Efficient incremental maintenance of data cubes. In: Proceedings of the 32nd International Conference on Very Large Data Bases; 2006. p. 823–33.

10. Morfonios K, Ioannidis Y. CURE for cubes: cubing using a ROLAP engine. In: Proceedings of the 32nd International Conference on Very Large Data Bases; 2006. p. 379–90.

11. Morfonios K, Ioannidis Y. Supporting the data cube lifecycle: the power of ROLAP. VLDB J. 2008;17(4):729–64.

12. Morfonios K, Konakas S, Ioannidis Y, Kotsis N. ROLAP implementations of the data cube. ACM Comput Surv. 2007;39(4):12.

13. Mumick IS, Quass D, Mumick BS. Maintenance of data cubes and summary tables in a warehouse. In: Proceedings of the ACM SIGMOD International Conference on Management of Data; 1997. p. 100–11.

14. Poosala V, Ganti V. Fast approximate answers to aggregate queries on a data cube. In: Proceedings of the 11th International Conference on Scientific and Statistical Database Management; 1999. p. 24–33.

15. Ross KA, Srivastava D. Fast computation of sparse datacubes. In: Proceedings of the 23th International Conference on Very Large Data Bases; 1997. p. 116–25.

16. Shao Z, Han J, Xin D. MM-Cubing: computing iceberg cubes by factorizing the lattice Space. In: Proceedings of the 16th International Conference on Scientific and Statistical Database Management; 2004. p. 213–22.

17. Sismanis Y, Deligiannakis A, Roussopoulos N, Kotidis Y. Dwarf: shrinking the PetaCube. In: Proceedings of the ACM SIGMOD International Conference on Management of Data; 2002. p. 464–75.

18. Sismanis Y, Roussopoulos N. The complexity of fully materialized coalesced cubes. In: Proceedings of the 30th International Conference on Very Large Data Bases; 2004. p. 540–51.

19. Vitter JS, Wang M. Approximate computation of multidimensional aggregates of sparse data using wavelets. In: Proceedings of the ACM SIGMOD International Conference on Management of Data; 1999. p. 193–204.

20. Wang W, Feng J, Lu H, Yu JX. Condensed cube: an efficient approach to reducing data cube size. In: Proceedings of the 18th International Conference on Data Engineering; 2002. p. 155–65.

21. Zhao Y, Deshpande P, Naughton JF. An array-based algorithm for simultaneous multidimensional aggregates. In: Proceedings of the ACM SIGMOD International Conference on Management of Data; 1997. p. 159–70.

Current Semantics

Michael H. Böhlen[1,2], Christian S. Jensen[3], and Richard T. Snodgrass[4,5]
[1]Free University of Bozen-Bolzano, Bozen-Bolzano, Italy
[2]University of Zurich, Zürich, Switzerland
[3]Department of Computer Science, Aalborg University, Aalborg, Denmark
[4]Department of Computer Science, University of Arizona, Tucson, AZ, USA
[5]Dataware Ventures, Tucson, AZ, USA

Synonyms

Temporal upward compatibility

Definition

Current semantics constrains the semantics of non-temporal statements applied to temporal databases. Specifically, current semantics requires that non-temporal statements behave as if applied to the non-temporal database that is the result of taking the timeslice of the temporal database at the current time.

Main Text

Current semantics [5] requires that queries and views on a temporal database consider the current information only and work exactly as if applied to a non-temporal database. For example, a query to determine who manages the high-salaried employees should consider the current database state only. Constraints and assertions also work exactly as before: they are applied to the current state and checked on database modification.

Database modifications are subject to the same constraint as queries: they should work exactly as if applied to a non-temporal database. Database modifications, however, also have to take into consideration that the current time is constantly moving forward. Therefore, the effect of modifications must persist into the future (until over-written by a subsequent modification).

To define current semantics, we assume a timeslice operator $\tau[t](D^t)$ that takes the snapshot of a temporal database D^t at time t. The timeslice operator takes the snapshot of all temporal relations in D^t and returns the set of resulting non-temporal relations.

Let now be the current time [2] and let t be a time point that does not exceed now. Let D^t be a temporal database instance at time t. Let $M_1, \ldots, M_n$, $n \geq 0$ be a sequence of non-temporal database modifications. Let Q be a non-temporal query. Current semantics requires that for all Q, t, D^t, and $M_1, \ldots, M_n$ the following equivalence holds:

$$Q(M_n(M_{n-1}(\ldots(M_1(D^t)\ldots))))$$
$$= Q(M_n(M_{n-1}(\ldots(M_1(\tau[now](D^t)))\ldots)))$$

Note that for $n = 0$ we do not have any modifications and we get $Q(D^t) = Q(\tau[now](D^t))$, i.e., a non-temporal query applied to a temporal database must consider the current database state only.

An unfortunate ramification of the above equivalence is that temporal query languages that introduce new reserved keywords not used in the non-temporal languages they extend will violate current semantics. The reason is that the user may have previously used such a keyword as an identifier in the database. To avoid being overly restrictive, it is reasonable to consider current semantics satisfied even when reserved words are added, as long as the semantics of all statements that do not use the new reserved words is retained by the temporal query language.

Temporal upward compatibility [1] is a synonym that focuses on settings where the original temporal database is the result of rendering a non-temporal database temporal.

The SQL:2011 standard [3,4] supports current semantics for transaction-time tables, as do the IBM DB2, Oracle 12c, and SQL Server DBMSes. The Teradata DBMS goes further, consistently supporting current semantics for valid-time, transaction-time, and bitemporal tables.

Cross-References

▶ Nonsequenced Semantics
▶ Now in Temporal Databases
▶ Sequenced Semantics
▶ Snapshot equivalence
▶ Temporal Database
▶ Temporal Data Models
▶ Temporal Query Languages
▶ Timeslice Operator

Recommended Reading

1. Bair J, Böhlen MH, Jensen CS, Snodgrass RT. Notions of upward compatibility of temporal query languages. Wirtschaftsinformatik. 1997;39(1):25–34.
2. Clifford J, Dyreson C, Isakowitz T, Jensen CS, Snodgrass RT. On the semantics of "NOW" in databases. ACM Trans Database Syst. 1997;22(2):171–214.
3. Kulkari K, Michels JE. Temporal features in SQL:2011. ACM SIGMOD Rec. 2012;41(3):34–43.
4. Melton J, editor. ISO/IEC 9075, database language SQL:2011 Part 2: SQL/foundation, Dec 2011.
5. Snodgrass RT. Developing time-oriented database applications in SQL. San Francisco: Morgan Kaufmann; 1999.

Curse of Dimensionality

Lei Chen
Hong Kong University of Science and Technology, Hong Kong, China

Synonyms

Dimensionality curse

Definition

The *curse of dimensionality*, first introduced by Bellman [1], indicates that the number of samples needed to estimate an arbitrary function with a given level of accuracy grows exponentially with respect to the number of input variables (i.e., dimensionality) of the function.

For similarity search (e.g., nearest neighbor query or range query), the *curse of dimensionality* means that the number of objects in the data set that need to be accessed grows exponentially with the underlying dimensionality.

Key Points

The *curse of dimensionality* is an obstacle for solving dynamic optimization problems by backwards induction. Moreover, it renders machine learning problems complicated, when it is necessary to learn a state-of-nature from finite number data samples in a high dimensional feature space. Finally, the *curse of dimensionality* seriously affects the query performance for similarity search over multidimensional indexes because, in high dimensions, the distances from a query to its nearest and to its farthest neighbor are similar. This indicates that data objects tend to be close to the boundaries of the data space with the increasing dimensionality. Thus, in order to retrieve even a few answers to a nearest neighbor query, a large part of the data space should be searched, making the multidimensional indexes less efficient than a sequential scan of the data set, typically with dimensionality greater than 12 [2]. In order to break the *curse of dimensionality*, data objects are usually reduced to vectors in a lower dimensional space via some dimensionality reduction technique before they are indexed.

Cross-References

▶ Dimensionality Reduction

Recommended Reading

1. Bellman RE. Adaptive control processes. Princeton: Princeton University Press; 1961.
2. Beyer KS, Goldstein J, Ramakrishnan R, Shaft U. When is "Nearest Neighbor" meaningful? In: Proceedings of the 7th International Conference on Database Theory; 1999. p. 217–35.

Printed by Printforce, the Netherlands